Frontiers in Earth Sciences

Series editors

J.P. Brun, Clermont-Ferrand, France
Onno Oncken, Potsdam, Germany
Helmut Weissert, Zürich, Switzerland
Wolf-Christian Dullo, Kiel, Germany

More information about this series at http://www.springer.com/series/7066

Central Macarena seen from the east. Based upon early field studies within the Macarena range and the Roraima tepuis, Augusto Gansser documented the puzzling relationships between the Guiana Shield and the Andean belt, in the process formulating the "Roraima Problem" from a geomorphologic-stratigraphic standpoint. Subsequent studies served to highlight the "problem" within the Macarena uplift, where Cambrian sediments are exposed at approx. 1000 m above sea level whilst Silurian sediments within the Orinoco low are buried at approx. 2300 m depth. The coexistence of vertical tectonics (over 2000 m displacement) and extensional strain systems (Phanerozoic graben-rift fills), with thrust-and-fold belts, may point to concealed wrench structures in subsurface, undetected by geophysical methods. An updated synthesis of the Roraima Problem, using new surface and sub-surface cartographic and structural data, is presented within Chap. 1 of this volume. *Redrawn from Gansser A., 1941, Central Macarena, Geological Report Shell No. 100, Appendix 12-16 with enclosures.*

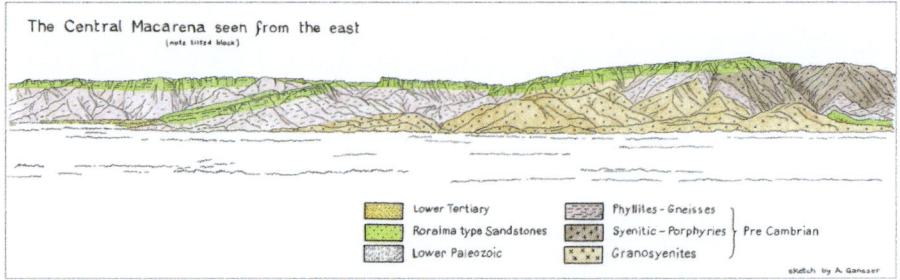

Geological landscape of the Rio Nevado canyon. The juncture between the Santander Massif and the Eastern Cordillera presents some of the most complex structural and stratigraphic relationships to be found anywhere in the Colombian Andes. This scaled geologic sketch of the Rio Nevado canyon was generated during detailed field-based mapping and structural and stratigraphic study, which produced over 20 km of 1:25.000-scale structural cross sections, as revealed in Chap. 9 of this volume. The composition highlights intensely disharmonic folds outcropping along the southern and northern walls of the Rio Nevado canyon, whilst maintaining the relative orientation and geometry of the structures and permitting comparison of the fold styles in both canyon walls. Completed in the style of pioneering Swiss geologist, Arnold Heim, this sketch provides a reliable graphical representation of the complex architecture underlying the Eastern Cordillera. Its production harkens a return to classical methodologies in the understanding and interpretation of natural landscape evolution vs. the indiscriminate use of purely algorithmic methods in the reconstruction of "balanced cross sections". *Adapted from original illustration in ink and water colour on parchment by Laura Román García.*

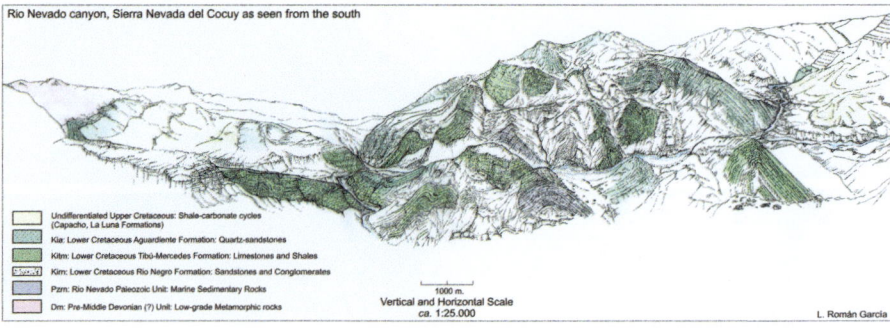

Fabio Cediel • Robert Peter Shaw
Editors

Geology and Tectonics of Northwestern South America

The Pacific-Caribbean-Andean Junction

Editors
Fabio Cediel
Consulting Geologist
Department of Geology
University EAFIT
Medellín, Colombia

Robert Peter Shaw
Consulting Geologist
Kelowna, BC, Canada

ISSN 1863-4621 ISSN 1863-463X (electronic)
Frontiers in Earth Sciences
ISBN 978-3-319-76131-2 ISBN 978-3-319-76132-9 (eBook)
https://doi.org/10.1007/978-3-319-76132-9

Library of Congress Control Number: 2018940188

© Springer Nature Switzerland AG 2019
This work is subject to copyright. All rights are reserved by the Publisher, whether the whole or part of the material is concerned, specifically the rights of translation, reprinting, reuse of illustrations, recitation, broadcasting, reproduction on microfilms or in any other physical way, and transmission or information storage and retrieval, electronic adaptation, computer software, or by similar or dissimilar methodology now known or hereafter developed.
The use of general descriptive names, registered names, trademarks, service marks, etc. in this publication does not imply, even in the absence of a specific statement, that such names are exempt from the relevant protective laws and regulations and therefore free for general use.
The publisher, the authors and the editors are safe to assume that the advice and information in this book are believed to be true and accurate at the date of publication. Neither the publisher nor the authors or the editors give a warranty, express or implied, with respect to the material contained herein or for any errors or omissions that may have been made. The publisher remains neutral with regard to jurisdictional claims in published maps and institutional affiliations.

Printed on acid-free paper

This Springer imprint is published by the registered company Springer Nature Switzerland AG.
The registered company address is: Gewerbestrasse 11, 6330 Cham, Switzerland

Augusto Gansser (1910–2012)
Field Geologist, Worldwide Scientific Explorer
Geologic Analysis: Regional Geology and Tectonics
Colombia 1938–1946 "Unique – a dream"

Highlights
- *Key geological realms in Colombia were the focus of his first enthusiastic geological explorations, including Gorgona Island, Chocó Arc, Sierra Nevada de Santa Marta, La Macarena, and the Guiana Shield: remote regions studied along walking trails, on horseback, in river boats, and light planes, under conditions spanning extreme tropical to glacial.*

- *Gansser was an exceptional draftsman, known by his stylish sketches of landscape, structures, and stratigraphic sections (in the style of Swiss geologist, coworker, and mentor Arnold Heim), giving priority to facts and gathering field data to incorporate in his geologic analysis and regional geological models.*
- *He was a successful explorer of natural resources, particularly oil and gas.*
- *He was a pioneer and leader of the modern geological interpretation of the northern Andes.*
- *Gansser's short exposé on the "Roraima Problem" (1974) demonstrates his capacity for data synthesis, and his continental-scale vision regarding, what even today, remain unanswered questions related to the tectono-sedimentary evolution of the Guiana Shield and its present-day morpho-structural expression.*
- *In 1979, Gansser published several key papers on the origin of the Andean-type "Trans-Himalayan magmatic belt," that was a precursor to the India-Asia collision and uplift of the Himalayas.*

- *In the words of Rasoul Sorkhabi (2012): "Gansser was one of the first geologists to apply plate tectonic theory to the evolution of mountain belts, thanks to his upbringing in Alpine geology where 'mobilist' tectonics characterized by thrust sheets, fold nappes, and compressional forces had been worked out by Swiss geologists, decades before the theory of plate tectonics was universally accepted!"*

Upon his death, Gansser was cremated with his hammer placed along with his body—not merely as a beautiful gesture for the fulfilling life of a great field geologist but also to fulfill a last wish of his:

"Instead of flowers, I would like my geologist's hammer."

Gansser's Principle Publications Related to Northwestern South America

1938 – Der Nevado del Cocuy: Columbianisches Bergerlebnis (self-published). (Note: Gansser immortalized his wife (Toti, his field companion) by naming a

5000 meter peak "Pico Toti," as, while climbing it together, she fell down a slope, but was saved by her rope).
1941 – Geological Report, Shell No. 100. Central Macarena. (Contributors: Renz, O., Hubach, E.) 16 pp. 25 photos 2 Tables 9 Annex. (in-house files).
1945 – Geological Report, Shell Pacific Chocó, (contributors: Poborski, S., Bäclin, R., Swolfs, H., Haanstra, U.) 75 pp. (in-house files).
1950 – Geological and petrographical notes on Gorgona island in relation to northwestern S. America. Schweizerische Mineralogische und Petrographische Mitteilungen. Bulletin Suisse de Mineralogie et Petrographie, v. 30 p. 219–237.
1954 – Observations on the Guiana Shield (S. America). Eclogae Geol. Helvetiae, v. 47 p. 77–112.
1955 – Ein Beitrag zur Geologie und Petrographie der Sierra Nevada de Santa Marta (Kolumbien, Südamarika). Schweiz, Mineralogische und Petrographische Mitteilungen, v. 35 no. 2 p. 209–279.

*1960 – Über Schlammvulkane und Salzdome. Mitteilungen aus dem Geologischen Institut der Eidg. Techn. Hochschule und der Universität Zürich, Serie B, Nr. 15. Vierteljahrsschrift der Naturforschenden Gesellschaft in Zürich, v. 105 no. 1, p. 1–46.
1962 – Lateinamerika - Land der Sorge und der Zukunft. Sozialwissenschaftliche Studien für das Schweizerische Institut für Auslandforschung 9. Erlenbach-Zürich/ Stuttgart: Rentsch. p. 315.
1963 – Quarzkristalle aus den kolumbianischen Anden (Südamerika). Schweizerische Mineralogische und Petrographische Mitteilungen. Bulletin Suisse de Mineralogie et Petrographie, v. 43 p. 91–103.
1969 – The Alps and the Himalayas, in Himalayan and Alpine Orogeny: New Delhi, Report of the Twenty-Second International Geological Congress, 1964, Part XI, Proceedings of Section 11 p. 387–399.
1973a – Facts and theories on the Andes. Twenty-sixth William Smith Lecture (with generalized geological map of the Andes*

1:20,000,000). Journal of the Geological Society of London, v. 129 p. 93–131.

1973b – Orogene Entwicklung in den Anden, im Himalaya und den Alpen, ein Vergleich (Orogenic evolution in the Andes, Himalayas and the Alps; a review) Eclogae Geologicae Helvetiae, v. 66 p. 23–40.

1974a – The ophiolitic melange. A worldwide problem on Tethyan examples. Eclogae Geologicae Helvetiae, v. 63 p. 479–507.

1974b – The Roraima problem (South America). Mitteilungen aus dem Geologischen Institut der Eidg. Technischen Hochschule und der Universität Zürich, Zürich, 177: 80–100.

1981 – Palaeogene komatiites from Gorgona Island, East Pacific: A primary magma for ocean floor basalts? (co-authors: Dietrich, V.J., Sommerauer, J., and Cameron, W.E.) Geochemical Journal v. 15 p. 141–161.

2000 – La Macarena, Massagno, Switzerland (self-published), 111 p.

Biographies

Augusto Gansser (2000) La moglie di un geólogo, 2nd edn Massagno, Switzerland, 236 p.

Ursula Eichenberger (Text), Ursula Markus (Hrsg.) (2008) Augusto Gansser. Aus dem Leben eines Welt-Erkunders (From the life of a world explorer) Zürich: AS Verlag, ISBN 978–3–909,111-58-9.

Rasoul Sorkhabi (2012) Memorial to Augusto Gansser (1910–2012) Geological Society of America Memorials, v. 41 p. 15–21.

Note: The Ganssers had four daughters (Ursula, born in Colombia, 1941; Manuela, 1949; Francesca, 1956; and Rosanna, 1959) and two sons (Mario, born in Colombia, 1943; Luca, also born in Colombia, 1945). In 2000, Gansser's dear wife and companion, Linda Biaggi-Gansser (Toti), died. She had kept diaries and notes of their lifelong journey, which formed the basis for the self-published biographical work "La moglie di un geologo: Augusto Gansser."

Acknowledgments

Beginning in the late 1800s, individual scientific curiosity and a persistent search for answers to questions born of field observation led a select group of geoscientists to study the mountainous regions of northern South America. Names like Sievers, Stübel, Karsten, Hettner, Sheibe, Grosse, Gansser, T. Ospina, Renz, Hubach, G. Botero, and Bürgl, considered by us *the Pioneering Generation*, are synonymous with geological reconnaissance, investigation, comparative geology, and geological analysis in *terra incognita*, as in were, in the new world uncovered by A. von Humboldt and J.B. Boussingault.

Early globalization of the world resource economy, in our case for precious metals and fossil fuels, acted as a catalyst to the appearance of, and important contributions from, the *Escuela de Minas* (Medellín, 1887) and the Department of Geology and Geophysics at the *Unversidad Nacional*, Bogotá (1958). Prominent Northern Andean explorers and geoscientists, as well as their North American and European counterparts (*the Industrial Generation*), supported by new technologies and refined methodologies, have permitted a state-of-the-art and up-to-the-moment understanding of the region, as presented herein in *Geology and Tectonics of Northwestern South America*.

The above-noted geological generations, comprised of individuals and considered as a whole, have contributed to the scientific advances reported herein and to the educational and socioeconomic well-being of the Northern Andean region, and we salute them with our profound and enthusiastic academic gratitude.

The Universidad EAFIT (Department of Geology, Medellín) provided fertile ground and an essential place of gathering, investigation, information exchange, discussion, and debate, during incubation, growth, production, and editing of this volume. The open-mindedness, professionalism, dedication, and all-around support provided by this institution are gratefully acknowledged.

Contents

Part I Regional Overview

1 Phanerozoic Orogens of Northwestern South America:
   Cordilleran-Type Orogens. Taphrogenic Tectonics.
   The Maracaibo Orogenic Float. The Chocó-Panamá Indenter   3
   Fabio Cediel

2 Proterozoic Basement, Paleozoic Tectonics
   of NW South America, and Implications for Paleocontinental
   Reconstruction of the Americas. 97
   Pedro A. Restrepo-Pace and Fabio Cediel

Part II The Guiana Shield and the Andean Belt

3 The Proterozoic Basement of the Western Guiana Shield
   and the Northern Andes 115
   Salomon B. Kroonenberg

Part III Early Paleozoic Tectono-Sedimentary History

4 Ordovician Orogeny and Jurassic Low-Lying Orogen
   in the Santander Massif, Northern Andes (Colombia) 195
   Carlos A. Zuluaga and Julian A. Lopez

Part IV Major Tectono-Magmatic Events

5 Spatial-Temporal Migration of Granitoid Magmatism
   and the Phanerozoic Tectono-Magmatic Evolution of the Colombian
   Andes 253
   Hildebrando Leal-Mejía, Robert P. Shaw,
   and Joan Carles Melgarejo i Draper

6 Phanerozoic Metallogeny in the Colombian Andes:
   A Tectono-magmatic Analysis in Space and Time 411
   Robert P. Shaw, Hildebrando Leal-Mejía,
   and Joan Carles Melgarejo i Draper

7 Paleogene Magmatism of the Maracaibo Block and Its Tectonic
   Significance 551
   José F. Duque-Trujillo, Teresa Orozco-Esquivel, Carlos Javier
   Sánchez, and Andrés L. Cárdenas-Rozo

8 Late Cenozoic to Modern-Day Volcanism
   in the Northern Andes: A Geochronological, Petrographical,
   and Geochemical Review 603
   M. I. Marín-Cerón, H. Leal-Mejía, M. Bernet, and J. Mesa-García

Part V The Northern Andean Orogen

9 Diagnostic Structural Features of NW South America:
   Structural Cross Sections Based Upon Detailed Field Transects 651
   Fabio Colmenares, Laura Román García, Johan M. Sánchez, and
   Juan C. Ramirez

10 Cretaceous Stratigraphy and Paleo-Facies Maps
   of Northwestern South America 673
   Luis Fernando Sarmiento-Rojas

11 Morphotectonic and Orogenic Development
   of the Northern Andes of Colombia: A Low-Temperature
   Thermochronology Perspective 749
   Sergio A. Restrepo-Moreno, David A. Foster, Matthias Bernet,
   Kyoungwon Min, and Santiago Noriega

12 The Romeral Shear Zone 833
   César Vinasco

Part VI Continental Uplift-Drift

13 Exhumation-Denudation History of the Maracaibo Block,
   Northwestern South America: Insights from Thermochronology ... 879
   Mauricio A. Bermúdez, Matthias Bernet, Barry P. Kohn, and
   Stephanie Brichau

Part VII Active Oceanic-Continental Collision

14 The Geology of the Panama-Chocó Arc 901
   Stewart D. Redwood

Part VIII Holocene-Anthropocene

15 Sediment Transfers from the Andes of Colombia during the Anthropocene 935
Juan D. Restrepo

16 The Historical, Geomorphological Evolution of the Colombian Littoral Zones (Eighteenth Century to Present) 957
Iván D. Correa and Cristina I. Pereira

Index... 983

Contributors

Mauricio A. Bermúdez Escuela de Ingeniería Geológica, Universidad Pedagógica y Tecnológica de Colombia, Sogamoso, Colombia

Matthias Bernet ISTerre, Université Grenoble Alps, Grenoble, France

Institut des Sciences de la Terre, Université Grenoble Alpes, Grenoble, France

Stephanie Brichau Geosciences Environment Toulouse, Université Paul Sabatier, Toulouse, France

Andrés L. Cárdenas-Rozo Earth Sciences Department, EAFIT University, Medellín, Colombia

Fabio Cediel Consulting Geologist, Department of Geology University EAFIT, Medellín, Colombia

Fabio Colmenares Geosearch Ltd., Bogotá, Colombia

Iván D. Correa Area de Ciencias del Mar, Universidad EAFIT, Medellín, Colombia

José F. Duque-Trujillo Earth Sciences Department, EAFIT University, Medellín, Colombia

David A. Foster Department of Geological Sciences, University of Florida, Gainesville, FL, USA

Barry P. Kohn School of Earth Sciences, University of Melbourne, Melbourne, VIC, Australia

Salomon B. Kroonenberg Delft University of Technology, Delft, Netherlands

Hildebrando Leal-Mejía Mineral Deposit Research Unit (MDRU), The University of British Columbia (UBC), Vancouver, BC, Canada

Departament de Mineralogia, Petrologia i Geologia Aplicada, Facultat de Ciències de la Terra, Universitat de Barcelona, Barcelona, Catalonia, Spain

Julian A. Lopez Departamento de Geociencias, Universidad Nacional de Colombia, Bogotá, Colombia

M. I. Marín-Cerón Departamento de Ciencias de la Tierra, Universidad EAFIT, Medellín, Colombia

Joan Carles Melgarejo i Draper Departament de Mineralogia, Petrologia i Geologia Aplicada, Facultat de Ciències de la Terra, Universitat de Barcelona, Barcelona, Catalonia, Spain

J. Mesa-García Departamento de Ciencias de la Tierra, Universidad EAFIT, Medellín, Colombia

Geology Department, University of Michigan, Ann Arbor, MI, USA

Kyoungwon Min Department of Geological Sciences, University of Florida, Gainesville, FL, USA

Santiago Noriega Universidad Nacional de Colombia, Facultad de Minas, Departamento de Geociencias y Medio Ambiente, Medellín, Colombia

Teresa Orozco-Esquivel Centro de Geociencias, Universidad Nacional Autónoma de México, Querétaro, Qro., Mexico

Cristina I. Pereira Area de Ciencias del Mar, Universidad EAFIT, Medellín, Colombia

Juan C. Ramirez Geosearch Ltd., Bogotá, Colombia

Stewart D. Redwood Consulting Economic Geologist, Panama City, Panama

Juan D. Restrepo Departamento de Ciencias de la Tierra, Universidad EAFIT, Medellín, Colombia

Sergio A. Restrepo-Moreno Universidad Nacional de Colombia, Facultad de Minas, Departamento de Geociencias y Medio Ambiente, Medellín, Colombia

Department of Geological Sciences, University of Florida, Gainesville, FL, USA

Pedro A. Restrepo-Pace Oilsearch Limited, Sydney, NSW, Australia

Laura Román-García Geosearch Ltd., Bogotá, Colombia

Carlos Javier Sánchez Earth Sciences Department, EAFIT University, Medellín, Colombia

Johan M. Sánchez Geosearch Ltd., Bogotá, Colombia

Luis Fernando Sarmiento-Rojas Independent Consultant, Bogotá, Colombia

Robert P. Shaw Departament de Mineralogia, Petrologia i Geologia Aplicada, Facultat de Ciències de la Terra, Universitat de Barcelona, Barcelona, Catalonia, Spain

César Vinasco Departamento De Geociencias Y Medio Ambiente, Universidad Nacional De Colombia, Facultad De Minas, Medellin, Colombia

Carlos A. Zuluaga Departamento de Geociencias, Universidad Nacional de Colombia, Bogotá, Colombia

Abbreviations

BABB	Back-Arc Basin Basalt
BA-Suarez	Buenos Aires-Suarez
CA-VA	Cajamarca-Valdivia Terrane
CAT	Caribbean Terrane Assemblage
CCOP	Cretaceous Caribbean-Colombian Oceanic Plateau
CCSP	Central Continental Sub-Plate
CHO	Chocó Arc
CLIP	Caribbean Large Igneous Province
CTR	Central Tectonic Realm
ca.	circa, approximately
cm	centimeter
DEM	Digital Elevation Model
dm	decimeter
E	East
EC	Eastern Cordillera (of Colombia)
e.g.	exempli gratia, for example
etc.	et cetera
Fig., Figs.	Figure, Figures
Fm., Fms.	Formation, Formations
GER	General Element Ratio
GS(R)	Guiana Shield (Realm)
Ga.	Giga-annum, billion years
Gp.	Group
g/t	grams per tonne
HFSE	High Field Strength Elements
ICP(ES)	Inductively Coupled Plasma (Emission Spectroscopy)
INGEOMINAS	Instituto de Investigaciones en Geociencias, Minería y Química
i.e.	id est., that is
kbar	kilobar(s)
kg/yr.	kilograms per year
LILE	Large Ion Lithophile Elements

LOI	Loss On Ignition
K-(feld)spar	potassium feldspar
Ka	kilo-annum, thousand years
km	kilometer
m	meter
mm	millimeter
mg/kg	milligrams per kilogram
Ma	mega-annum, million years
m.y.	million years
M	million
MALI	Modified Alkali Lime Index
MMT	Million Metric Tonnes
MORB, N-, E-	Mid Ocean Ridge Basalt, Normal, Enriched
MSP	Maracaibo Sub-Plate Realm
MVT	Mississippi Valley-Type
N	North
NAB	Northern Andean Block
PAT	Pacific Terrane Assemblage
PER	Pearce Element Ratio
PGEs	Platinum Group Elements
PLOCO	Provincia Litosférica Oceánica Cretácica del Occidente de Colombia
ppm	part per million
REE(L, H)	Rare Earth Element(s) (Light, Heavy)
RSZ	Romeral Shear Zone
RTG	Ridge Tholeiitic Granitoid
S	South
SEDEX	Sedimentary Exhalative
T	tonnes
UPME	Unidad de Planeacón Minero Energético
UTM	Universal Transverse Mercator
VAG	Volcanic Arc Granites
VMS	Volcanogenic Massive Sulphide
vs.	versus
W	West
WTR	Western Tectonic Realm
wt%	weight percent

Prologue

....back to basics, back to the source!
....it has been said:

The scientists who study the earth [universe] as a whole, are often in error but never in doubt. Nowadays they're less often in error, but their doubts have grown as big as all outdoors (Ferris 2005)

The northwest corner of continental South America, as well as its Pacific and Caribbean companions, has travelled a long and varied route over the last 540 million years (Phanerozoic), in order to reach its present (temporary) resting point. The scientific curiosity of most researchers today, however, concentrates upon the more recent stages of the journey (let's say, the last 100 m.y. or so, from Late Cretaceous to present). The debate over tectonic models for emplacement of the Caribbean plate remains open. Notwithstanding, an understanding of the geological history of continental South America, and its Phanerozoic interplay with distinct oceanic plates and continental masses, is a critical factor in the formulation of any geological model for today's Caribbean.

By the time of the scientific revolution, the NW corner of South America and its bordering Pacific Ocean and Caribbean Sea were covered by incipient geological maps, although the region remained an as yet untested geophysical laboratory. The search for "ores" (not yet considered geological exploration) and wildcat drilling for oil had revealed promising economic prospects and raised interesting academic questions about the geological history of the region. In Europe and North America, as the debate regarding the validity of plate tectonic theory reached its zenith, the scarce hard geological data available for NW South America were forced to fit the nascent paradigms of the new tectonic era.

The proverbial "complexity" of NW South America has become the repeated, almost clichéd, introduction to any paper written on the geology and tectonic history of the region. Curiously, the main geological features (presumably) are often expressed in terms of the "Andean cordilleras," their intramontane valleys and flatlands (*llanos*), that is, in terms of the most basic physiographic elements of physical geography. Contrary to these physiographic underpinnings, we herein propose the use of stratigraphic-controlled morphostructural features, as a first approach to understanding the large-scale geotectonic framework of NW South America.

The NW corner of South America, including the western Guiana Shield, Northern Andean Block (or North Andes), and neighboring Pacific and Caribbean plates, provides an excellent natural laboratory, of "manageable" scale, suited to the testing of modern tectonic, magmatic, geophysical, and metallogenic models, currently used by many as unquestionable paradigms. To undertake such an exercise we apply and interpret factual data, provided by generations of field-based geological and geophysical surveys, combined with biostratigraphic, geochronological, lithochemical, isotopic, and petrographic research, and not least of all, detailed basin analysis. All of this information is publically available—albeit, not (yet) as a fully comprehensive, integrated, computerized data bank—as critical historic information has yet to be passed from paper to digital format. Notwithstanding, carefully and patiently compiled empirical observations permit the dedicated scientist to present well-founded interpretations, syntheses, and conclusions pertaining to the detailed geological history of the region.

The geological laboratory of NW South America is endowed with the following:

- Proterozoic metamorphic, igneous, and sedimentary rock units, remnants of intracratonic orogens, not yet fully documented let alone understood.
- Paleontological and geochronological data which attest to the presence of lithostratigraphic units belonging to each recorded erathem of the Phanerozoic.
- Morphostructural patterns which reveal the interference and/or superposition of peneconteporaneous or successive tectonic events.
- A history of tectonic interaction between oceanic and continental plates, recorded in diverse marine and terrestrial environments from throughout the Phanerozoic.
- Sedimentary basins filled with tectonostratigraphic sequences that register uplift, basin development, and regional orogenic evolution.
- Widespread magmatic suites, exposed at essentially all levels, generated along extensional, collisional, transcurrent, and consuming plate margins, which record tectonic, petrogenetic, and isotopic interactions within/between oceanic and continental crust and the upper mantle throughout much of the Phanerozoic.
- A highly varied metallogenic record, temporally and spatially reflective of the tectonomagmatic development of the region.
- A complete range of climatic zones and thermal layers, originating at sea-level and extending up to almost 6000 m elevation.

Via this volume, northern Andean tectonostratigraphic and magmatic history may be contrasted with classical cordilleran-type orogenic and magmatic models, such as those used in the Central Andes and elsewhere. Numerous differences are illustrated that render the application of typical "Cordilleran-type" or "Andean-type" models for northern Andean development unacceptable. Throughout this volume, the importance and contribution of underlying Proterozoic through mid-Mesozoic geological and structural elements, in the evolution of Mesozoic through Cenozoic northern Andean orogenic phase tectonics (structural style, uplift mechanisms, basin development, magmatism, etc.), are revealed. These features are exemplified by highly oblique subduction-collision systematics associated with accretion of allochthonous oceanic terranes along the Pacific margin; the detachment, migration, and "plis de fond" style of deformation developed in the Maracaibo tectonic float; and unique inversion systematics culminating in the transpressive pop-up of the Eastern Cordillera, all of which have no clear geological analog in classical Cordilleran-type orogens. A critical revision of subduction models and the generation of related (or unrelated) granitoid arcs in the Northern Andes lead the reader to comparative regional geological analysis beyond the Andes.

Almost 30 years ego, Peter Molnar presented a crucial paper entitled "Continental Tectonics in the Aftermath of Plate Tectonics." He observed: "The success of plate tectonics required an acceptance of continental drift, and thus a reinterpretation of the large-scale geological history of most of the earth. But the basic tenet of plate tectonics, rigid-body movements of large plates of lithosphere, fails to apply to continental interiors, where buoyant continental crust can detach from the underlying mantle to form mountain ranges and broad zones of diffuse tectonic activity."

Today, in the light of Molnar's prognosis, we attempt to understand key geological features, observed in the Guiana Shield and within the basement beneath Cenozoic northern Andean basins, documented via surface geological mapping and detailed geophysical studies. Beyond the multiple deformation processes engraved upon the Guiana Shield, as it partially underlies cratonized Paleozoic and Mesozoic basins, we recognize large scale grabens and rifts, as well as Proterozoic basement involved in basin deformations and the tectonic migration of continental "subplates," or tectonic rafts incorporated in exotic terranes.

The Pacific–Caribbean–Andean Junction presents a multidisciplinary approach to understanding the geological history and tectonic assembly of NW South America, with a focus upon onshore and circum-continental Colombia as the geological keystone of the region. The individual thematic contributions integrated herein are presented by an experienced team of independent, predominantly autochthonous geoscientists from academia and industry, supported by international experts, all with a long-standing, hands-on relationship to the northern Andean geological mosaic. Although the individual works and resulting volume are not free of controversial conclusions, the combined thesis permits geological and tectonic synthesis at a detailed scale, which will in turn permit re-evaluation of historic impasses and the reformulation of critical questions that may lead to higher levels of understanding through new avenues of research.

F. Cediel and R. P. Shaw

Part I
Regional Overview

Chapter 1
Phanerozoic Orogens of Northwestern South America: Cordilleran-Type Orogens. Taphrogenic Tectonics. The Maracaibo Orogenic Float. The Chocó-Panamá Indenter

Fabio Cediel

Abbreviations

CAT	Caribbean terranes
CA-VA	Cajamarca-Valdivia terrane
CCOP	Caribbean-Colombian oceanic plateau
CCR	Colombian Caribbean Realm
CHO	Chocó-Panamá Arc
CTR	Central Tectonic Realm
DAP	Dagua-Piñón
GDFS	Garrapatas-Dabeiba Fault System
GSR	Guiana Shield Realm
GU-FA	Guajira-Falcón terranes
IRLPM	Igneous-related low-pressure metamorphism
ME	Sierra de Mérida ("Venezuelan Andes")
MOF	Maracaibo orogenic float
NAB	North Andean belt
NW SA	Northwestern South America
OPTFS	Oca-El Pilar Transform Fault System
PAT	Pacific Terranes
RO	Romeral Mélange
SNStM	Sierra Nevada de Santa Marta
WETSA	W-E Tectono-Sedimentary Anomaly
WTR	Western Tectonic Realm

F. Cediel (✉)
Consulting Geologist, Department of Geology University EAFIT, Medellín, Colombia

1.1 Introduction

Regional geology of northwestern South America and the link between local and global continental-oceanic geology.

Regional geological syntheses contain, by nature, a certain degree of geological "guess estimation." Such works cannot conceal or ignore the significant lack of fundamental geological information for extensive areas of northwestern South America. In fact, they tend to emphasize this aspect. Notwithstanding, a certain degree of geological speculation, based upon empirical observation at the regional level, is necessary from time to time. Such inference can provide a kind of inventory with which to qualify the state of geological knowledge for specific areas and for a region as a whole. It is hoped that such presentations would encourage new concepts, debates, and syntheses applicable to a better understanding of the geological history of the region as a whole.

No attempt has been made in this synthesis to reconcile differing or contradictory geological, geophysical, or geochemical interpretations, and the alert reader will detect apparent inconsistencies in differing interpretations of the geological record. These discrepancies are derived from the fact that, in some cases, the quality or density of the available data, or differing data sources, affects the nature and validity of the resulting conclusions.

1.2 Tectonic Realms (Figs. 1.1 and 1.2)

Via definition of the various litho-tectonic and morpho-structural domains, I derive a synthesis in terms of tectonic plates, subplates, terranes, composite terranes, and sedimentary basins. This analysis contrasts with the general custom observed in Northern Andean literature, of using major physiographic features such as cordilleras, serranías, valleys, or depressions as geologic reference points (e.g., "Western Cordillera," "Central Cordillera," etc.), thereby incurring the false notion that, for example, a certain cordillera or depression today corresponds to a single litho-tectonic unit or represents a single geological or tectonic time period or event.

1.2.1 Guiana Shield Realm (GSR)

This litho-tectonic realm is comprised of the autochthonous mass of the Precambrian Guiana Shield. The western edge of the GSR extends throughout the subsurface of the Llanos, Guarico, and Barinas-Apure basins of northeastern Colombia and northwestern Venezuela. To the south, the GSR extends beneath the eastern foreland front of Colombia's Eastern Cordillera, through to the Garzón Massif, and under the Putumayo basin. In Ecuador, the GSR underlies the Putumayo-Napo basin, the

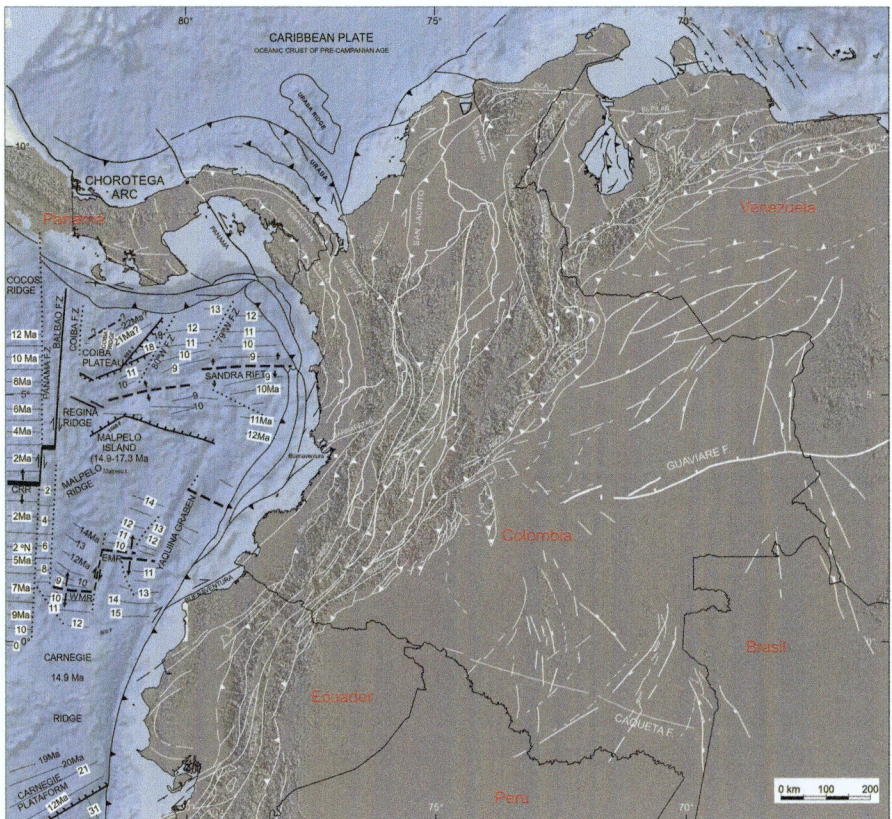

Fig. 1.1 Structural sketch map of NW South America

eastern margin of the Cordillera Real, and extends eastwards into the Amazon basin of both Colombia and Ecuador.

The Guiana Shield formed the backstop for the progressive accretionary continental growth of northwestern South America from the Middle to Upper Proterozoic through to the Holocene. Outcrops of 1300–900 Ma granulite document continental collision, penetrative deformation, and high-grade metamorphism during a broadly Grenvillian-aged orogenic event.

1.2.2 The Central Tectonic Realm (CTR)

The CTR is a composite and temporally and compositionally heterogeneous lithotectonic realm. The Precambrian and Paleozoic constituents of the CTR are allochthonous to parautochthonous with respect to the Guiana Shield autochthon, while

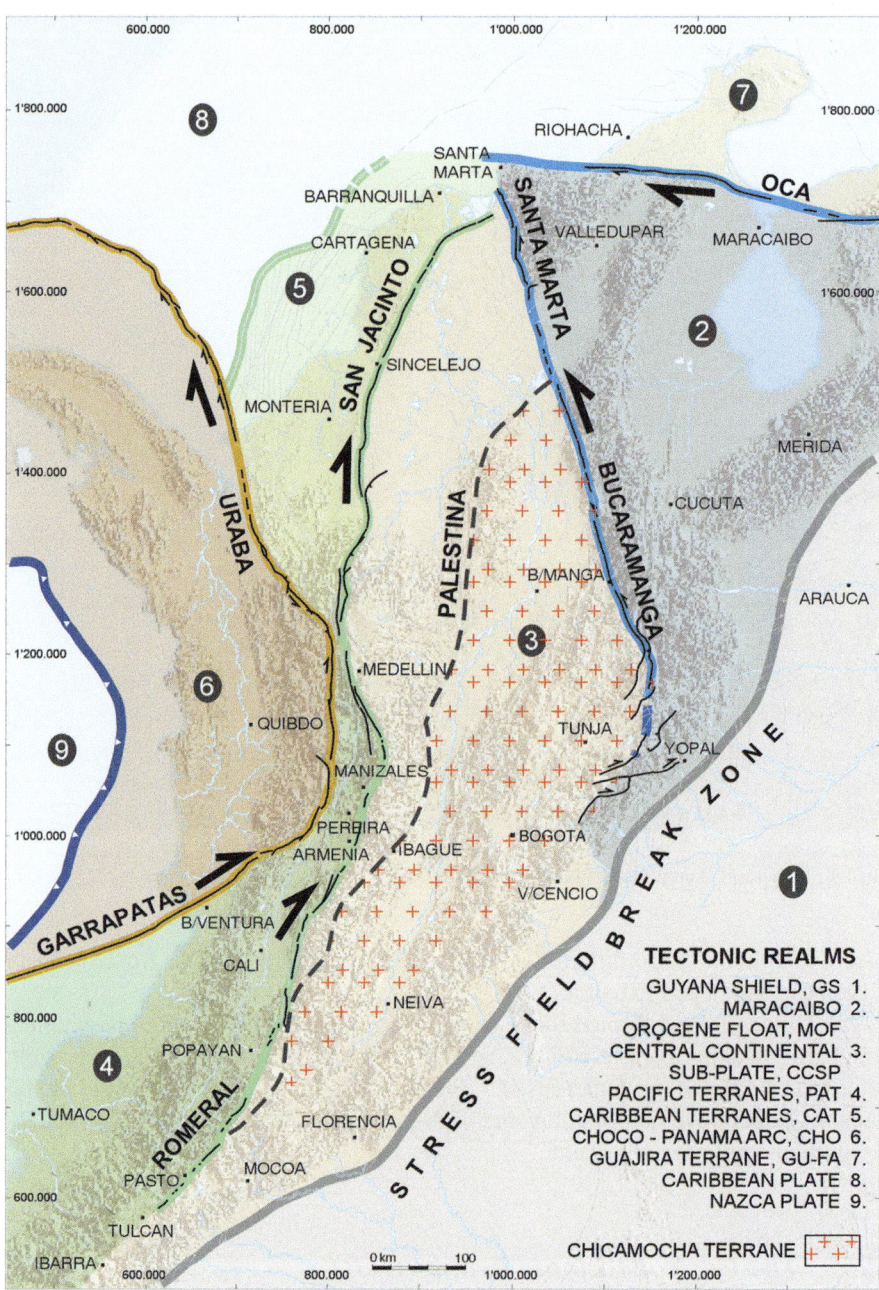

Fig. 1.2 Main tectonic realms of NW South America

the Mesozoic to recent components are considered to be parautochthonous to autochthonous with respect to the CTR.

The CTR has played host to numerous complex geological events from the Early Paleozoic up to the present. These events include a Middle Ordovician-Silurian Cordilleran-type orogeny followed by a period of prolonged regional extension (taphrogenesis), which began in the Mississippian (?) and continued into the Middle Mesozoic. The Mesozoic-Cenozoic transition to transpressional regimes, collisions, and magmatism during the Northern Andean orogeny defines the present structural and morphological character of the CTR.

The oldest constituent of the CTR is the exotic Chicamocha terrane. This Precambrian allochthon, a possible relict of Oaxaquia, was welded directly to the Guiana Shield during a Grenvillian orogenic event. In Colombia, Chicamocha is represented by fragmented granulite-grade bodies of migmatite and quartz-feldspar gneiss. To the west of Chicamocha, the Cajamarca-Valdivia terrane represents the remnant of an oceanic island arc, accreted during the Early Paleozoic, which presently forms the litho-tectonic basement to much of the physiographic Central Cordillera.

1.2.3 Maracaibo Orogenic Float (MOF)

The MOF hosts numerous composite litho-tectonic provinces and morpho-structural features, including the Sierra Nevada de Santa Marta (SM), the Sierra de Mérida (ME, the "Venezuelan Andes"), the Serranía de Perijá and Santander Massif (SP), and the César-Ranchería and Maracaibo basins. The MOF is characterized as a disrupted segment of the northwesternmost Guiana Shield, overlain in this region by extensive Phanerozoic supracrustal sequences. In the Late Cretaceous, the MOF began to migrate northwestward, along the Santa Marta-Bucaramanga and Oca-El Pilar fault systems, in the process uplifting the Sierra de Mérida, the Santander-Perijá belt, and the Sierra Nevada de Santa Marta. Although technically a part of the Guiana Shield, the MOF is distinguished from the GSR by a unique and regionally constrained style of deformation brought about by the evolving Mesozoic-Cenozoic through recent interaction between the Pacific (Nazca) and Caribbean plates and continental South American. The possible causes, timing, and mechanisms behind this migration remain a matter of debate.

1.2.4 Western Tectonic Realm (WTR)

Despite important local data, complete characterization of individual terranes in the WTR, including the definition of their limits and time(s) of collision/accretion with the continent, remains deficient. Regardless, it has been established that all litho-tectonic units comprising the WTR include fragments of Pacific oceanic plateaus, aseismic ridges, intra-oceanic island arcs, and/or ophiolite, and all developed within and/or upon oceanic basement, as demonstrated by paleomagnetic data and paleogeographic reconstructions.

The WTR consists of two composite terrane assemblages: (1) the Pacific (PAT) assemblage, including the Romeral (RO), Dagua (DAP), and Gorgona (GOR) terranes, and (2) to the north, the Caribbean terranes (CAT), including San Jacinto (SJ) and Sinú (SN).

1.3 Pacific Terranes (Romeral, Dagua-Piñón, Gorgona)

The Romeral terrane contains mafic-ultramafic complexes, ophiolite sequences, and oceanic sediments of probable Late Jurassic(?) and Early Cretaceous age. Although an allochthonous origin for some of the constituents of the Romeral assemblage may argue (see discussion in Cediel et al. 2003), much of the Romeral terrane appears to represent the reworked remnants of pericratonic, marginal basin mafic magmatism and continental to marine sedimentation, deposited along the rifted proto-Caribbean margin. Multiple phases of tectonic reworking and translation during the Meso-Cenozoic led to burial, high-pressure metamorphism, dismemberment and obduction/accretion of the marginal basin assemblages along the paleocontinent (represented by metamorphic rocks of the CTR), and the present-day configuration of the Romeral tectonic mélange and shear zone.

To the west of the Romeral assemblage, the Dagua terrane is dominated by basalt and diabase with important thicknesses of flyschoid siliciclastic sediments, including siltstone and graywacke with chert and minor limestone. The chemical characteristics of the Dagua basalt/diabase are unlike those of island arc or marginal basin basalts, and appear to represent accreted fragments of aseismic ridges, and/or oceanic plateaus which numerous authors associate with the Caribbean-Colombian oceanic plateau (or CCOP; e.g., Kerr et al. 1997; Sinton et al. 1998). A Middle to Late Cretaceous age for basalts of the Dagua terrane correlates well with numerous Middle and Upper Cretaceous paleontological dates for this unit from both Colombia and Ecuador.

Farther west, the Gorgona terrane is located on the westernmost margin of northwestern South America; however, it is located mostly offshore. It also appears to represent an accreted oceanic plateau; however, paleomagnetic data and tectonic reconstructions (e.g., Kerr and Tarney 2005) suggest it is far traveled and unrelated to the Dagua terrane and the CCOP in general. It contains massive basaltic flows, pillow lavas, komatiitic lava flows, and a peridotite-gabbro complex; it has been assigned a Late Cretaceous age.

With respect to the Gorgona terrane, studies by McGeary and Ben-Abraham (1989) suggest it also represents an aseismic oceanic ridge or fragment of an oceanic plateau. Notwithstanding, paleomagnetic data presented by Estrada (1995) indicates Gorgona doesn't have any clear correlation with the CCOP. For example, the El Horno basalt (86 ± 3 Ma) was located at about 25° south relative to South America in Late Cretaceous time. Estrada (1995) presented reconstructions and possible trajectories that suggest a longitude of origin near 135° west. Similar conclusions were drawn by Kerr and Tarney (2005), these authors citing a location of origin between 26° and 30° south.

The accretion of the Gorgona terrane is considered to be mid Eocene by Kerr and Tarney (2005) and is certainly pre-Miocene (McGeary and Ben-Abraham 1989). Accretion was possibly followed by strike-slip faulting along the Buenaventura fault zone, resulting in fragmentation of the original terrane.

1.4 Caribbean Terranes (San Jacinto, Sinú)

The San Jacinto terrane includes a thick pile of sedimentary deposits, both marine and terrestrial, ranging in age from Upper Paleocene to Miocene, which unconformably overlie oceanic crust containing ultramafic and mafic volcanic rocks and a fragment of a Coniacian-Campanian island arc (Cansona and Finca Vieja Fms.). Paleomagnetic data for the Coniacian Finca Vieja Fm. indicates a Pacific provenance to the south and west. Petrochemical analyses suggest that the volcanic sequences of the southern Caribbean basalts in general and the Cretaceous Pacific Realm belong to a similar volcanic province.

The Sinú terrane is located outboard of San Jacinto and, as with San Jacinto, contains a thick sequence of marine and terrestrial sedimentary rocks deposited upon oceanic basement. The oldest recognized sedimentary rocks in the Sinú basin are Oligocene in age (as opposed to Paleocene in San Jacinto).

1.4.1 Guajira-Falcón Terranes (GU-FA)

Based on similarities in age and composition, the Guajira-Falcón terranes may be interpreted as tectonically translated segments of the Western Tectonic Realm. The composite Guajira-Falcón terrane is comprised of a collection of fragments of Proterozoic and Paleozoic continental crust, Jurassic sedimentary sequences, and Cretaceous oceanic crust. Detailed studies of the Margarita Complex portion of the GU-FA terrane (e.g., Maresch et al. 2000) demonstrate an accretionary-metamorphic history and migratory path beginning in the Albian. Paleomagnetic studies indicate that volcanic outcrops of the GU-FA, from their Guajira and Greater Antilles sites, occupied latitudes about 108 south of their present positions and possibly off northwestern South America in the Cretaceous.

1.4.2 Chocó-Panamá Arc (CHO)

The composite Chocó Arc assemblage represents the southeastern segment of the Panamá double arc (the western segment of which is the Central American Chorotega Arc). The Chocó Arc maintains a radius and vergence-oriented east-northeast, and the Chorotega Arc maintains an approximately north-directed vergence. In Colombia, the basement of the Chocó Arc is comprised of at least two distinct lithotectonic assemblages: the Cañas Gordas terrane and the El Paso-Baudó terrane which includes the Baudó Range (Cediel et al. 2010). Cañas Gordas consists of mixed volcanic rocks of the Barroso Fm. overlain by fine-grained sedimentary rocks of the Penderisco Fm. Both assemblages contain Barremian through Middle Albian fossil assemblages and are interpreted to represent accreted slivers of Farallon Plate oceanic basement.

The El Paso-Baudó terrane is comprised of Late Cretaceous to Paleogene tholeiitic basalt of N- and E-MORB affinity overlain by minor pyroclastic rocks, chert, and turbidite. The terrane represents a Late Cretaceous sliver of the Caribbean-Colombia oceanic plateau (CCOP) assemblage, interpreted to have formed along the trailing edge of the Caribbean Plate. Development of the San Juan and Atrato basins began in the Paleogene. Collision of the El Paso-Baudó assemblage along the western Cañas Gordas margin and uplift of the Baudó Range are recorded in the Miocene. Faults related to the assembly and accretion of the Chocó Arc, including the Garrapatas-Dabeiba system, reactivate, deform, and/or truncate earlier structures associated with the Romeral and San Jacinto fault systems.

1.5 Phanerozoic Basins in Colombia

Basins represent the end product of polyphase sedimentary evolution, as determined by a variety of tectonic processes, and the nature and composition of basin fill are key components of an orogen. This understanding is critical in deciphering the evolution of Phanerozoic basins throughout the Colombo-Caribbean region. In Colombia, poly-deformed basins are characterized by complex facies architecture, the result of varied, large-scale, discrete to at times overlapping events which demarcate the Phanerozoic tectonic evolution of the northern Andean region.

The geological knowledge pertaining to basin development summarized below is the result of decades of hydrocarbon exploration in Colombia, supported by geological mapping, subsurface drilling, geophysical (seismic, gravity, magnetics) analysis, and detailed paleontological studies. This historic information, when placed within the context of new and ongoing exploration and research, constitutes a significant database for the application of basin analysis to the understanding of the tectonic evolution of the region.

The recognized petroleum occurrences of the NAB are commonly associated with Cretaceous to Cenozoic basins, which are particularly well studied. Regardless, hydrocarbon seeps and manifestations linked to Paleozoic and Early Mesozoic stratigraphy have led to an understanding of basin dynamics throughout the Phanerozoic, and the acquisition of new geological data through continued geological field mapping and remote sensing studies should improve our understanding of basin development in the northern Andean Block.

1.5.1 Pre-cretaceous Basins

Our understanding of pre-Cretaceous sedimentary basins in the northern Andes is still at an adolescent stage, as the need for basic geological mapping of diagnostic pre-Cretaceous outcrops, and stratigraphic, geochemical, and petrophysical studies which will permit the integrated analysis of pre-Cretaceous basins are overshadowed by the apparent urgency to offer new hydrocarbon plays within younger basins. Nevertheless, the limited available data, compiled and viewed at a regional scale, allows preliminary facies analysis and tectonic understanding of sedimentary deposits dating from the pre-Cretaceous. The data highlight important stratigraphy and basin development reflected within:

- Ordovician epicontinental marine deposits
- Pennsylvanian(?) or Middle to Late Permian marine deposits
- Late Triassic marine deposits
- Middle to Late Jurassic marine deposits

1.5.2 Cretaceous Basins

In Colombia, more than 90% of established oil reserves are located within Sub-Andean basins. In this context, the geological history sensu lato of Colombian sedimentary basins is in many respects much better studied and understood than that of many of the outcropping cordilleran sectors.

Cretaceous basins of the northern Andean Block developed under two distinct tectonic regimes: extensional, from the Berriasian to the Aptian-Albian, followed by transpressional from the Cenomanian to the Maastrichtian.

The Lower Cretaceous regime represents the final stages of the extensional environment which dominated the Bolivar Aulacogen. This final phase culminates in deep rifting and epicontinental marine transgression. The crustal architecture inherited from Jurassic and earlier times is clearly reflected in Lower Cretaceous basin geometry.

The Upper Cretaceous transpressional regime is recorded simultaneously in two distinct tectonic realms:

1. In the WTR, linked to the interaction of the Pacific continental margin with Farallon Plate oceanic lithosphere, including subduction, collision, and accretion of allochthonous terranes, and the initial stages of Cordilleran-type orogeny
2. In the Maracaibo subplate, linked to the tectonic migration of the Maracaibo orogenic float

1.5.3 Cenozoic Basins

Flexural subsidence and basin development related to tectonic inversion of the Andean Cordilleras began in the Maastrichtian and continues today. Inversion of the Eastern Cordilleran basin is particularly well recorded in Cenozoic continental sedimentary deposits and in development of two economically important and well-studied intracratonic basins of Cenozoic age which flank the Eastern Cordillera, including the Llanos and Middle Magdalena basins.

Within the oceanic realm, along the Caribbean margin, the Sinú and San Jacinto (SIB and SJAB) basins consist of two distinct Cenozoic continental margin depo-centers, floored by allochthonous Cretaceous oceanic basement and bound by wrench fault systems. Abundant oil and gas exploration has been undertaken in both these basins.

Along the Pacific margin of the Chocó Arc, the San Juan Basin and Atrato basins are floored by Caribbean plateau oceanic crust. These basins were fed by Paleogene to Holocene deltaic sequences containing some of the richest petroleum source rocks in the northern Andean region.

1.6 Geological History of the Colombian Andes

1.6.1 An Annotated Graphical Essay

This is the first attempt to summarize, in near-purely graphical format, the most relevant tectonic, magmatic, and sedimentary events related to the Phanerozoic of Colombia and surrounding areas of Venezuela and Ecuador. The selected format includes unrestored paleogeographic maps which depict the regional structural and sedimentological framework.

The preparation, analysis, and interpretation of geological maps (both surface and subsurface) have seemingly become outdated; even worse, some large-scale regional interpretations seem to ignore available field mapping (*) – the prime and essential tool of geological interpretation – to accommodate conjectural tectonic models (e.g., Kennan and Pindell 2009; Taboada et al. 2000; Gutscher 2002; Syracuse et al. 2016).

(*)"Facts do not cease to exist because they are ignored" (anonymous)
Notwithstanding, I observe that:

- The historic interpretation of the bulk of geochemical analyses for Phanerozoic plutonic and volcanic rocks from the northern Andean region inevitably fall into the "subduction-related" model. Most of these data lack accompanying isotopic (Pb, Sr, Nd, Hf, etc.) analyses which may be used in the interpretation of granitoids outside of a strictly subduction-related setting (see examples in Leal-Mejía et al. 2018).
- During time intervals where no volcanism along an orogenic belt is apparent, the default solution to the problem is often relegated to "flat-slab" subduction,

regardless of the abundance of nonsupportive data derived from tectonic and rheological analyses, or ignoring contrary arguments provided by subsurface and surface structural mapping.
- The presence or absence of magmatism (in particular calc-alkaline volcanism and its assignment to a predefined lithochemical model) has been the argument for the interpretation of "rift-related vs. subduction-related tectonic settings." Precaution with respect to the blanket application of this paradigm is advised, as highlighted, for example, by tectono-magmatic analysis of granitoid magmatism in the northern Andean Block during the latest Early Paleozoic, Carboniferous, Permian, and Triassic-Jurassic and by inconsistencies found within the Mérida Range (Venezuelan "Andes"), Colombia's Eastern Cordillera, and the Western Caribbean orogen.

In the following pages, I dissect geological and geophysical maps and stratigraphic and structural cross sections – all supported by hard data – in order to (1) identify major tectonic events and (2) understand their relationship with accompanying basin development. Understandably, this risky exercise involves questioning previous studies, the examination of diverse geological interpretations, and the application of various kinematic models and/or tectonic paradigms.

It should be borne in mind that the present composite geological framework of the northern Andean Block is the result of the superimposition of paleogeologic maps, representing distinct slices in time, and that the most recent tectonic deformation(s) inherit key structural attributes and patterns related to previous geodynamic events, reactivated and eventually exposed in today's outcrops. Documented examples of regional superimposition in NW South America are observed in Mid-Proterozoic and Early and Late Paleozoic sutures reactivated during Jurassic, Cretaceous, and Cenozoic tectonic events.

In assembling diverse sources of information, otherwise distant in time and space, two goals are accomplished:

1. To compile and present a coherent regional overview which preserves the original source(s) of information
2. To facilitate the rapid visualization of synthesized data, interpretations, and proposed geological models.

Over 79 selected illustrations presented herein transmit geological data and conceptualizations which may be read and interpreted without textual interruption or the implantation of preconceived mental images. Illustrations in which no additional sources of information are cited are of my own authorship.

1.6.1.1 Geological Setting and Morpho-Structural Expression (Figs. 1.3, 1.4, 1.5, 1.6, and 1.7)

General Statement
An updated understanding of the geologic evolution of the region and the definition of key Phanerozoic structural elements should provide substantial contribution

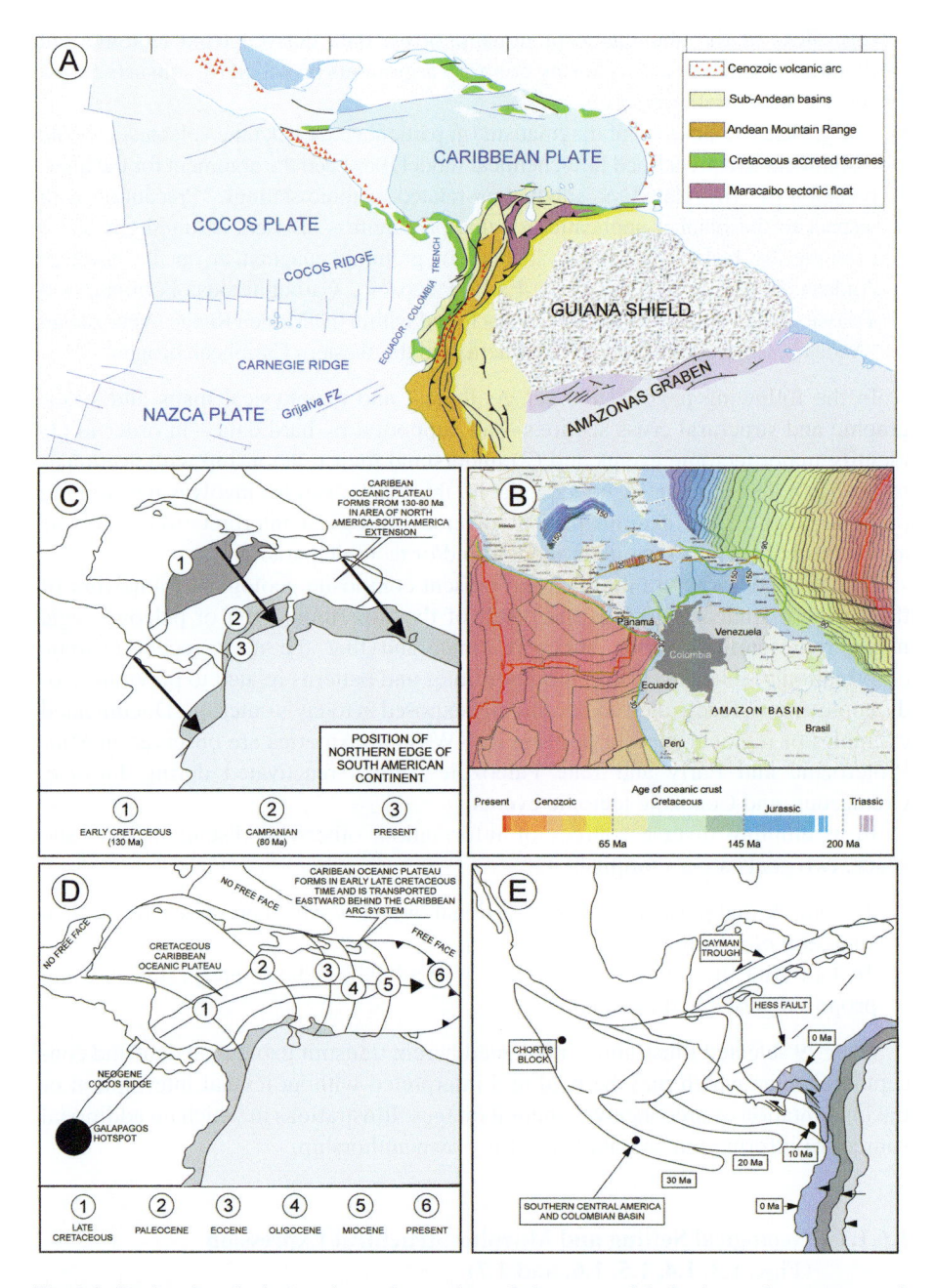

Fig. 1.3 Regional geological setting and tectonic evolutionary models for the northwest corner of South America, including Colombia, Ecuador, Venezuela, Panamá, and the Caribbean Basin. **c–e** after Mann (1995)

1 Phanerozoic Orogens of Northwestern South America: Cordilleran-Type Orogens...   15

Choco-Panama Arc
BAU = Baudo Mountain
PA = Panama Arc
CG = Cañas Gordas Litho-Unit

Pacific Colombian Trench
fab = fore arc basin
ac = accretionary prism
tf = trench fill

Guiana Shield, GS
Amazonia Realm
TB = Table Mountain, Tepui
Orinoquia Realm

Western Tectonic Realm
(Western Cordillera, s.l.)
GOR = Gorgona
DAP = Dagua-Piñon
RM = Romeral Melange
RO = Romeral

Central Tectonic Realm
(Central Cordillera, s.l.)
CA-CV = Cajamarca-Valdivia Litho-Unit
SL = San Lucas Block
IB = Ibague Block
GA = Garzon Massif
CR = Cordillera Real

Eastern Tectonic Realm, EC
(Eastern Cordillera)
pd = piedmonte

Guajira–Falcon Composite Terrane, GU-FA
GU = Guajira Amalgamated Structure
PA = Paraguana Amalgamated Structure

Caribbean Tectonic Realm
Western Caribbean:
SJ = San Jacinto Fault Belt
SN = Sinu Fold Belt
Eastern Caribbean:
CCo = Cordillera de la Costa
IR = Interior Range

Maracaibo Orogenic Float, MOF
ME = Sierra de Merida
BA = Baragua and San Luis Range
SP = Santander Massif-Serrania de Perija
SM = Sierra Nevada de Santa Marta

Cenozoic Basins

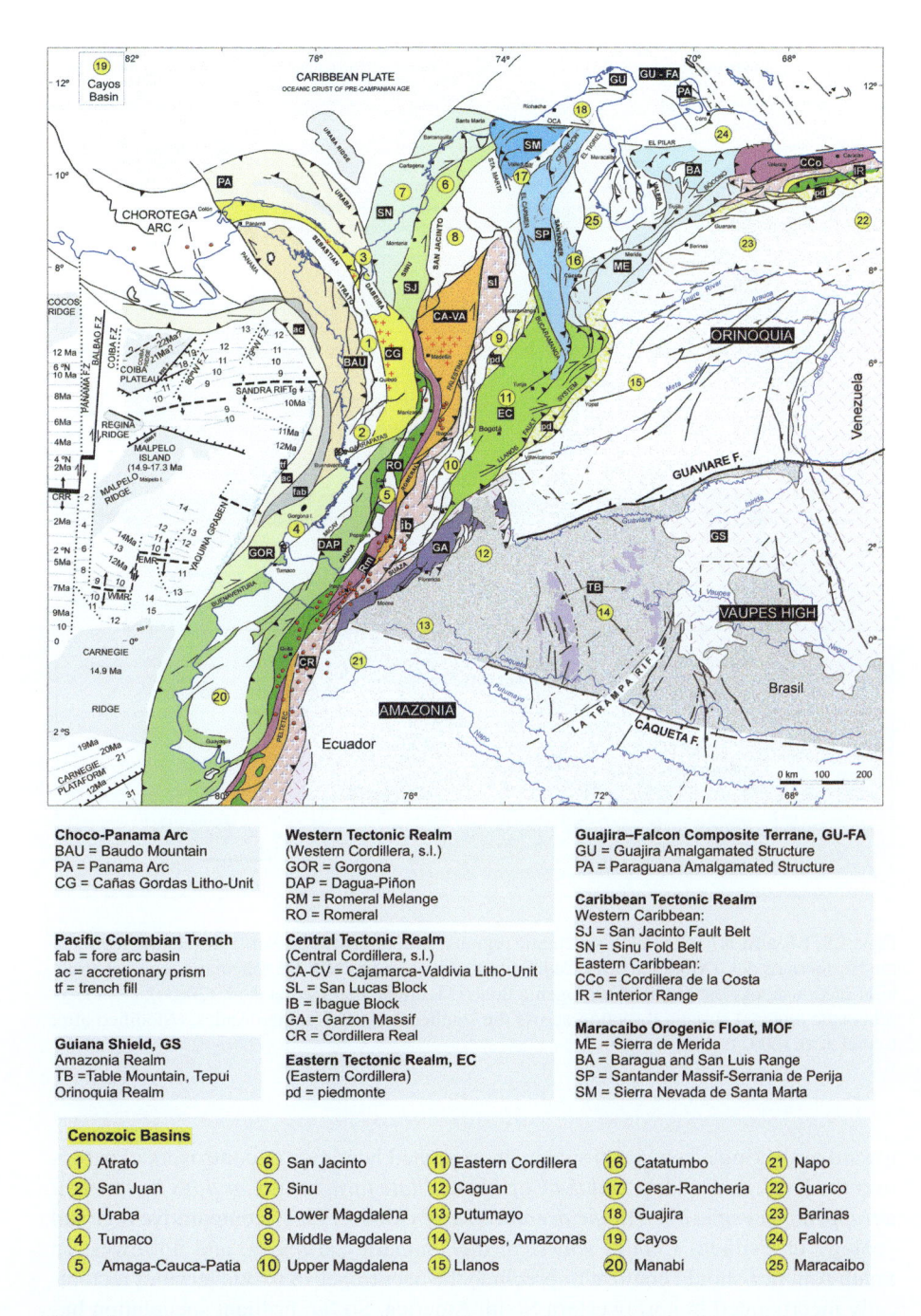

1. Atrato
2. San Juan
3. Uraba
4. Tumaco
5. Amaga-Cauca-Patia
6. San Jacinto
7. Sinu
8. Lower Magdalena
9. Middle Magdalena
10. Upper Magdalena
11. Eastern Cordillera
12. Caguan
13. Putumayo
14. Vaupes, Amazonas
15. Llanos
16. Catatumbo
17. Cesar-Ranchería
18. Guajira
19. Cayos
20. Manabi
21. Napo
22. Guarico
23. Barinas
24. Falcon
25. Maracaibo

Fig. 1.4 Major morpho-structural units and Cenozoic basins of Colombia and the Northern Andean Block. (Modified after Cediel et al. (2003))

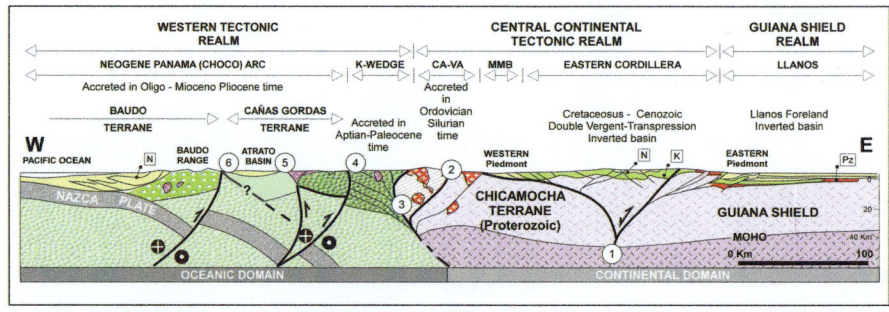

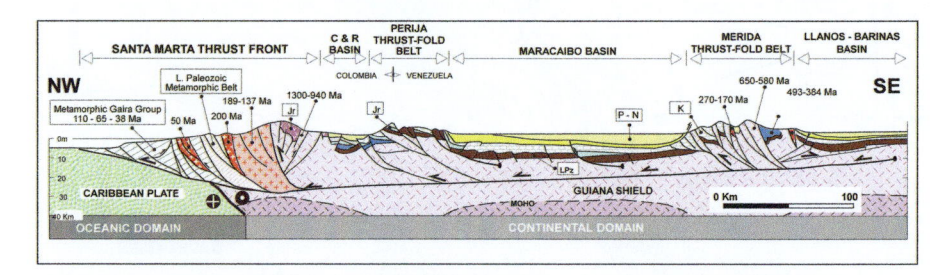

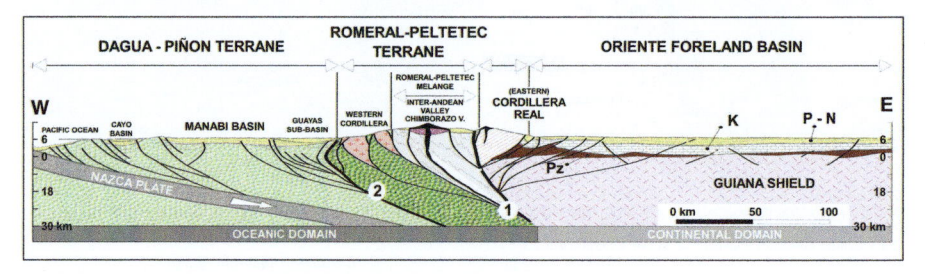

Figs. 1.5, 1.6, and 1.7 (1.5) E-W schematic regional structural section across the central region of the Northern Andes. (Modified after Cediel et al. (2003)). (1.6) NW-SE schematic regional structural section across the Maracaibo orogenic float. (Modified after Cediel et al. (2003)). (1.7) E-W schematic regional structural section across the southernmost Colombian Andes. (Modified after Cediel et al. (2003))

toward answering a key question which originated hundreds of controversial papers over the last four decades: *Did the Caribbean Plate form* in situ, *or does it represent a trapped piece of exotic Pacific oceanic crust?* (Mann 1995). Comparative regional geology of southern Central America, the southern Caribbean, and northwestern South America should confirm or preclude the occurrence of exotic terranes tectonically incorporated in northwestern South America. So far, brilliant speculation has shown untested possibilities.

1.6.1.2 Continent-Continent Collision and Intracontinental Orogens. Meso-Neoproterozoic (Figs. 1.8, 1.9, 1.10, 1.11, and 1.12)

General Statement

Three distinct but tectonically correlative events mark the Proterozoic history of NW SA: (1) the collision of a continental terrane (Oaxaquia?) with the western margin of the Guiana Shield and the formation of a granulite belt, (2) a rift-drift process that leaves behind the Chicamocha terrane and numerous tectonic rafts that

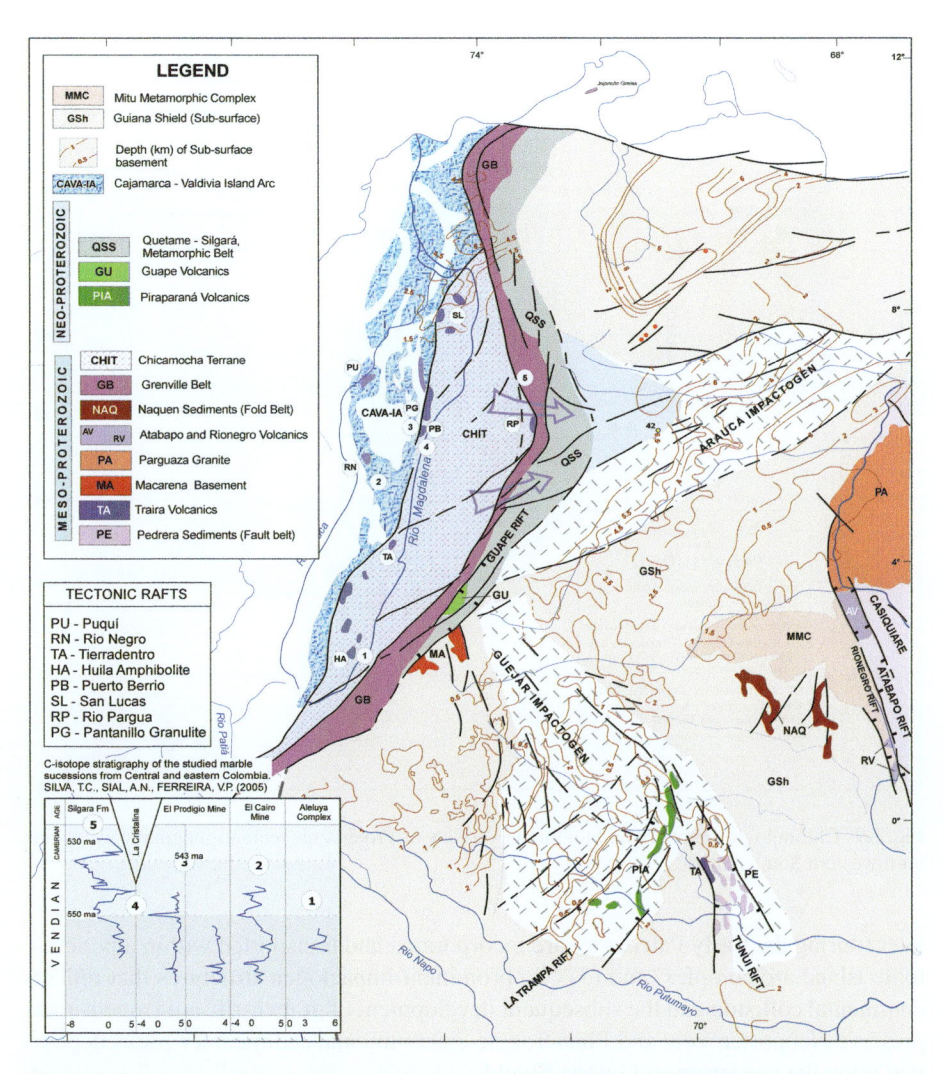

Fig. 1.8 Paleogeographic sketch map depicting relevant Meso- and Neoproterozoic tectonostratigraphic units

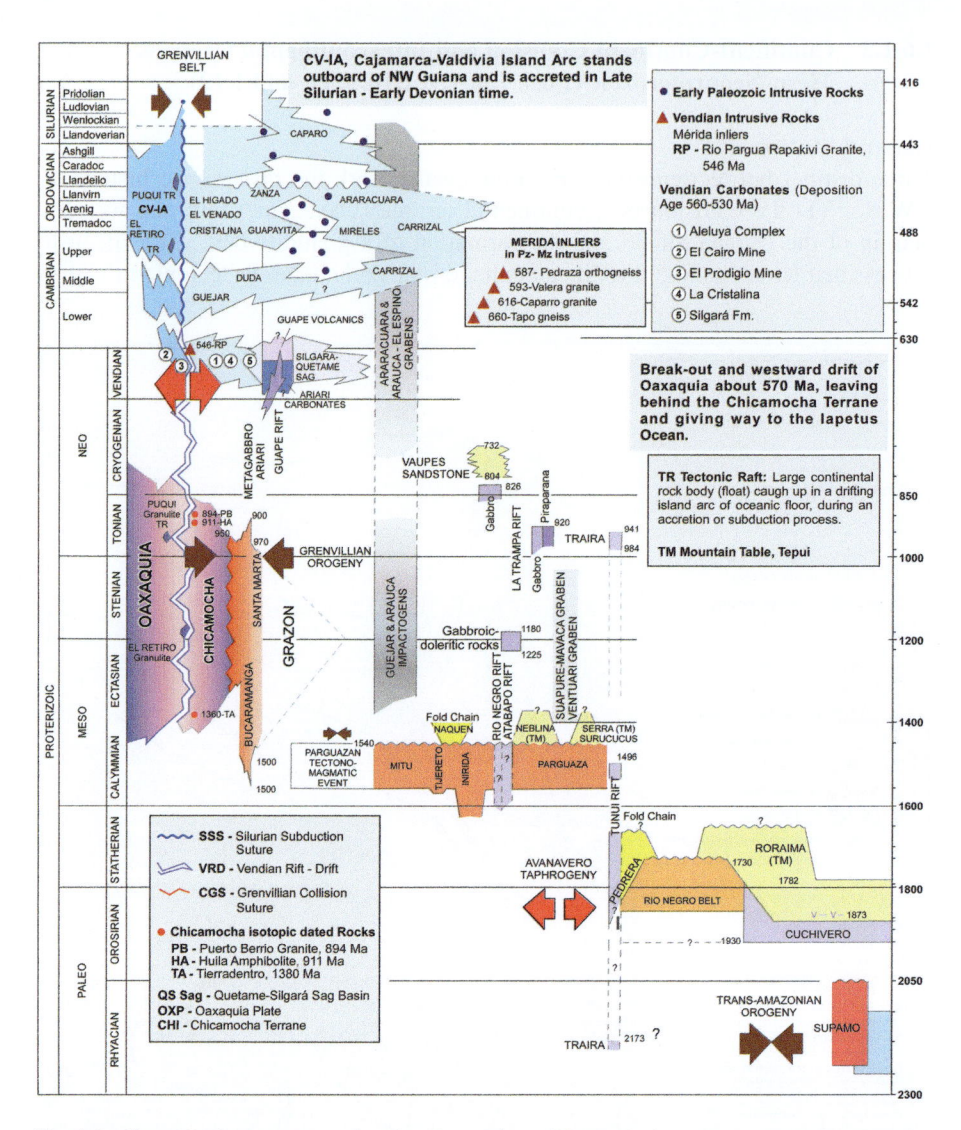

Fig. 1.9 Chronologic summary and regional overview of Proterozoic tectono-magmatic events in northwestern South America

later (during the Early Paleozoic) are incorporated and transported within a pericratonic island arc complex, and (3) two prominent impactogen structures that reflect continental collision and the subsequent development of graben-rift-aulacogens that preserve Neoproterozoic and Phanerozoic epicontinental sedimentary rocks deposited upon the westernmost Guiana Shield.

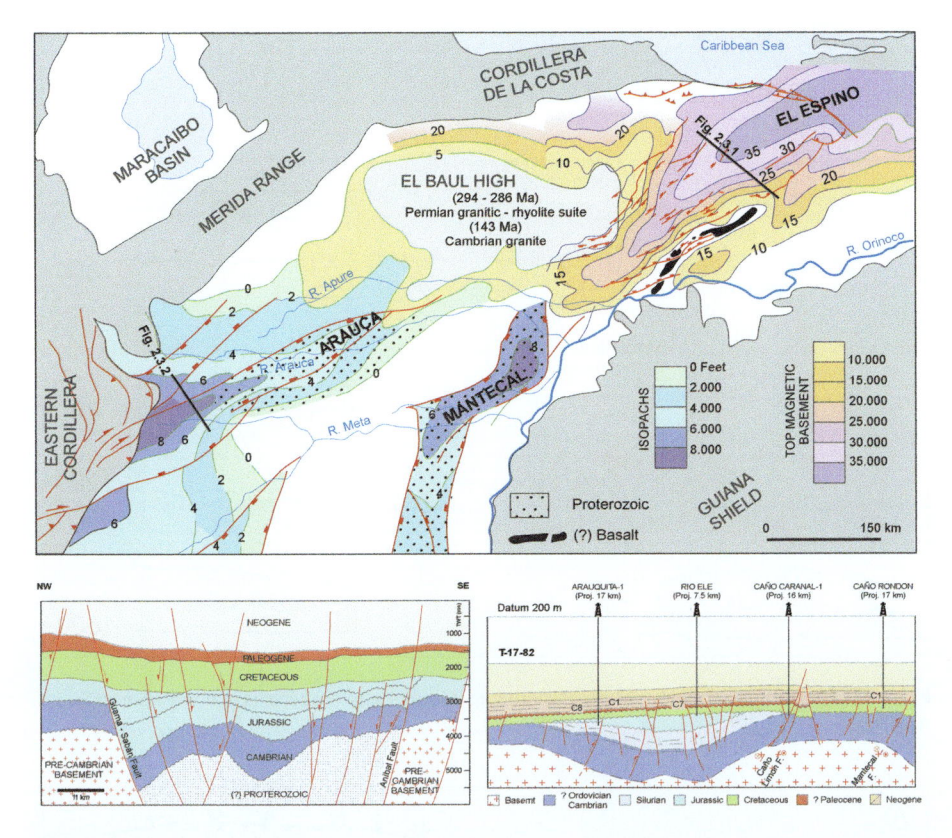

Fig. 1.10 Arauca-El Espino collision-related impactogen and graben-rift-aulacogen. (Compiled after Arminio et al. (2013), Barrios et al. (2011), Viscarret et al. (2009), Cediel and Cáceres (2000), and Geotec (1996))

Differential uplift and denudation resulting from superposed orogenic events and Andean (Meso-Cenozoic) tectonics have left only sparse, relict basement exposures of Neoproterozoic rocks in the northern Andes. The term "basement" used herein refers to the assemblage of rocks which make up the craton and metamorphic units underlying Paleozoic sediments in the Andean realm. In the field, this designation is supported by well-constrained field relationships together with limited but good quality geochronological data.

Despite the numerous more recently acquired geochronological data, however, Late Proterozoic-Early Paleozoic paleogeographic reconstructions of northern South America remain elusive. That is to say, a better understanding of the geological evolution of the distinct mappable units is still to be accomplished. Notwithstanding, the litho-tectonic units depicted in Fig. 1.8 reveal a coherent geological history and permit construction of this first unrestored paleogeographic scheme.

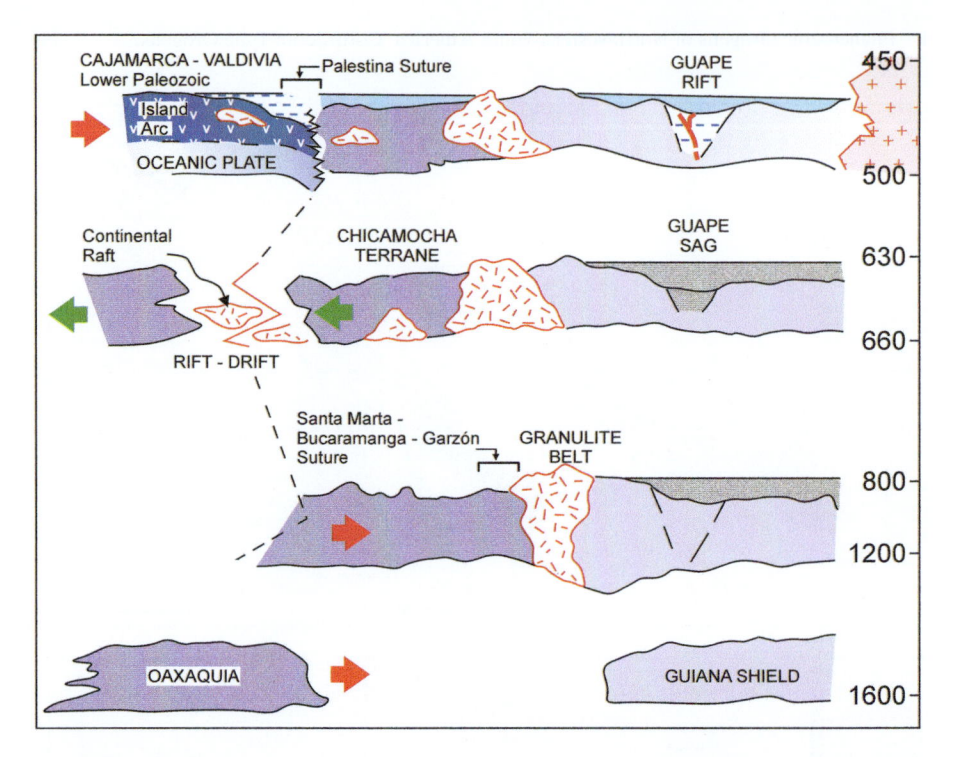

Fig. 1.11 Conceptual model for Meso-Proterozoic (Grenvillian) orogenesis in northwestern South America

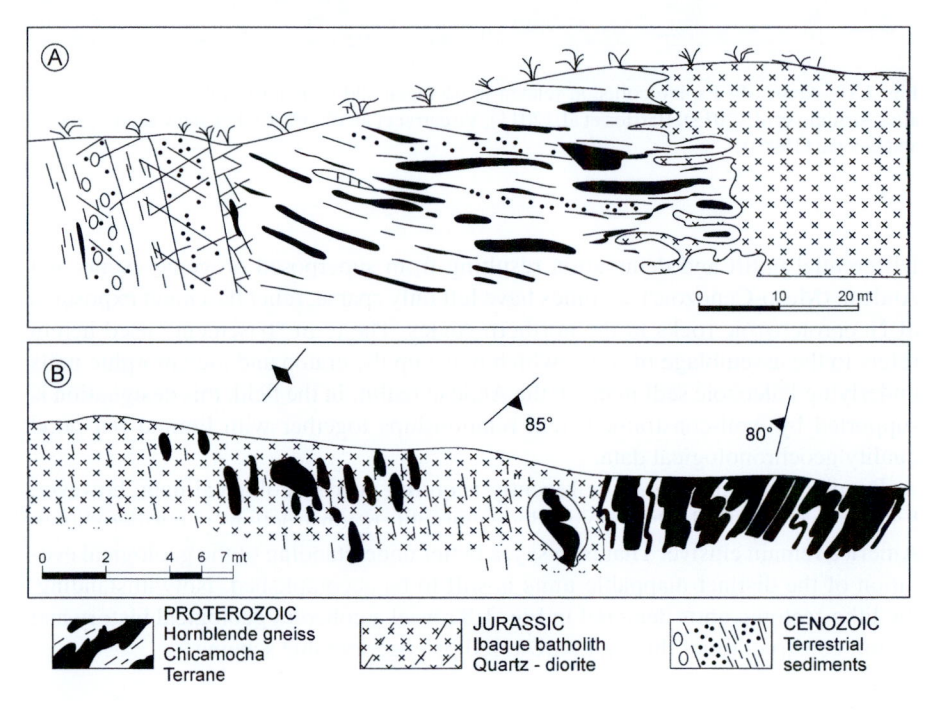

Fig. 1.12 Proterozoic xenoliths incorporated in Jurassic intrusive bodies (Ibague Batholith). (Modified after Muñoz and Vargas (1981))

1 Phanerozoic Orogens of Northwestern South America: Cordilleran-Type Orogens... 21

1.6.1.3 Phanerozoic Orogenic Systems (Figs. 1.13, 1.14, 1.15, 1.16, and 1.17)

General Statement

The integration of surface geology, gravity anomalies, and stratigraphic-sedimentological unconformities in the major sedimentary basins of northwestern South America outlines clear structural boundaries which may be interpreted as sutures. Today, this geological mosaic records at least 11 (Fig. 1.16) orogens or orogenic systems. The differentiation of these events as plate boundary (continental

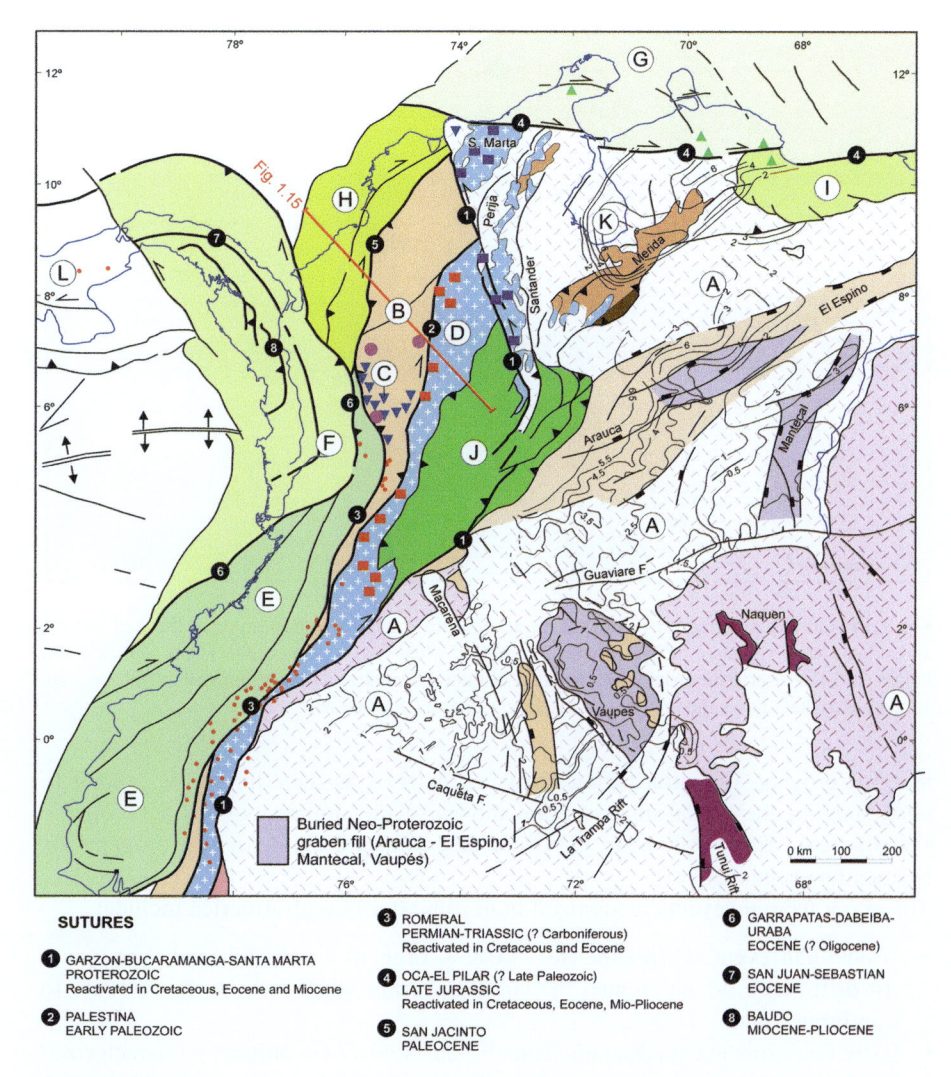

Fig. 1.13 Sutures and associated litho-tectonic units (orogens) that partially outcrop in northwestern South America

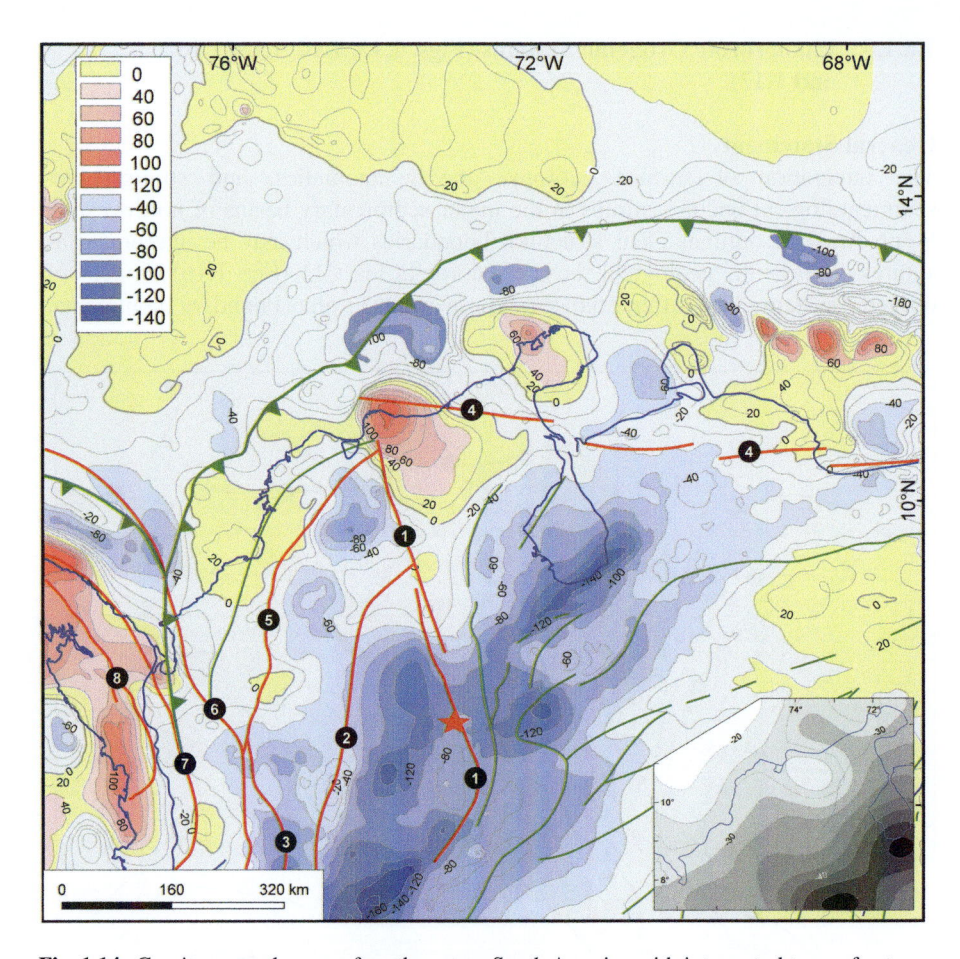

Fig. 1.14 Gravity anomaly map of northwestern South America with interpreted trace of sutures and other major regional fault systems. (Compiled after Cerón et al. (2007) and Sanchez-and Palma (2014)). Inset: Mohorovic discontinuity obtained from gravity inversion with refraction data control. Red star shows the location of the Bucaramanga seismic nest

margin) orogens vs. intracratonic orogens, as well as their timing, is crucial to understanding the spatial superposition and coeval development of distinct orogenic systems (e.g., Early Paleozoic vs. Cretaceous transgression; Andean orogeny vs. Maracaibo orogenic float).

The major tectonic events recognized in northwestern South America include:

- Grenvillian event (Orinoquiense ~1.0 Ga, orogeny): collision and subsequent rift-drift phase which created the structural framework for Late Proterozoic to Cambrian basins
- Early Paleozoic event (Quetame/Caparonensis ~0.47 Ga orogeny): characterized by arc accretion and extensive deposition of Ordovician transgressive marine sequences

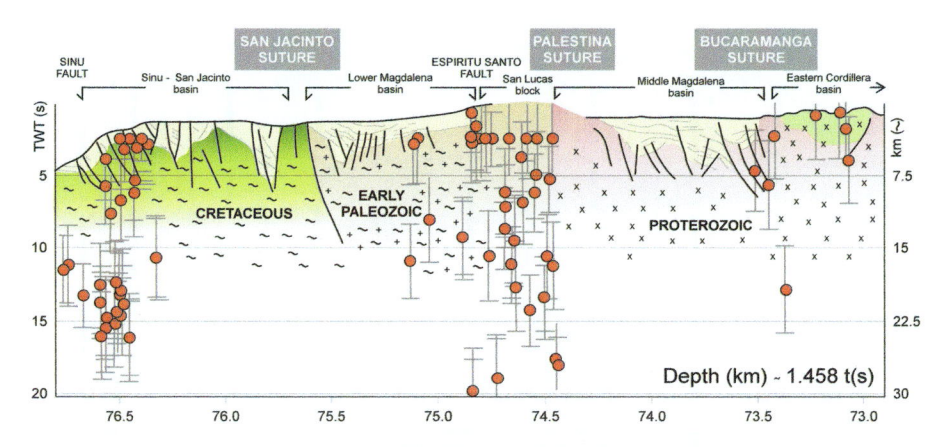

Fig. 1.15 Interpretation of the Trans-Andean seismic-reflection line (ANH). Red dots with vertical bars represent vertical errors hypocenter solutions in a 60 km wide corridor. Depth-time relation has been estimated by using several oil wells. Note: aseismic character of the San Jacinto suture, and vertical distribution of hypocenters. (Modified after Vargas and Mann (2013))

- Late Paleozoic to Early Cretaceous phase: characterized by aulacogen development (Bolivar Aulacogen), including aborted rifting with associated grabens, punctuated by collisional orogeny in the Permo-Triassic
- Late Cretaceous to recent events, generating two spatially independent orogenic domains:
    - A cordilleran, subduction, and accretion-related domain, in the west, (Northern Andean Orogeny)
    - An orogenic float-type domain related to the NW-directed tectonic migration of a disrupted block of the Guiana Shield (Maracaibo orogenic float)
- Cenozoic, Western, and Eastern Caribbean orogens
- Cenozoic Chocó-Panamá indenter

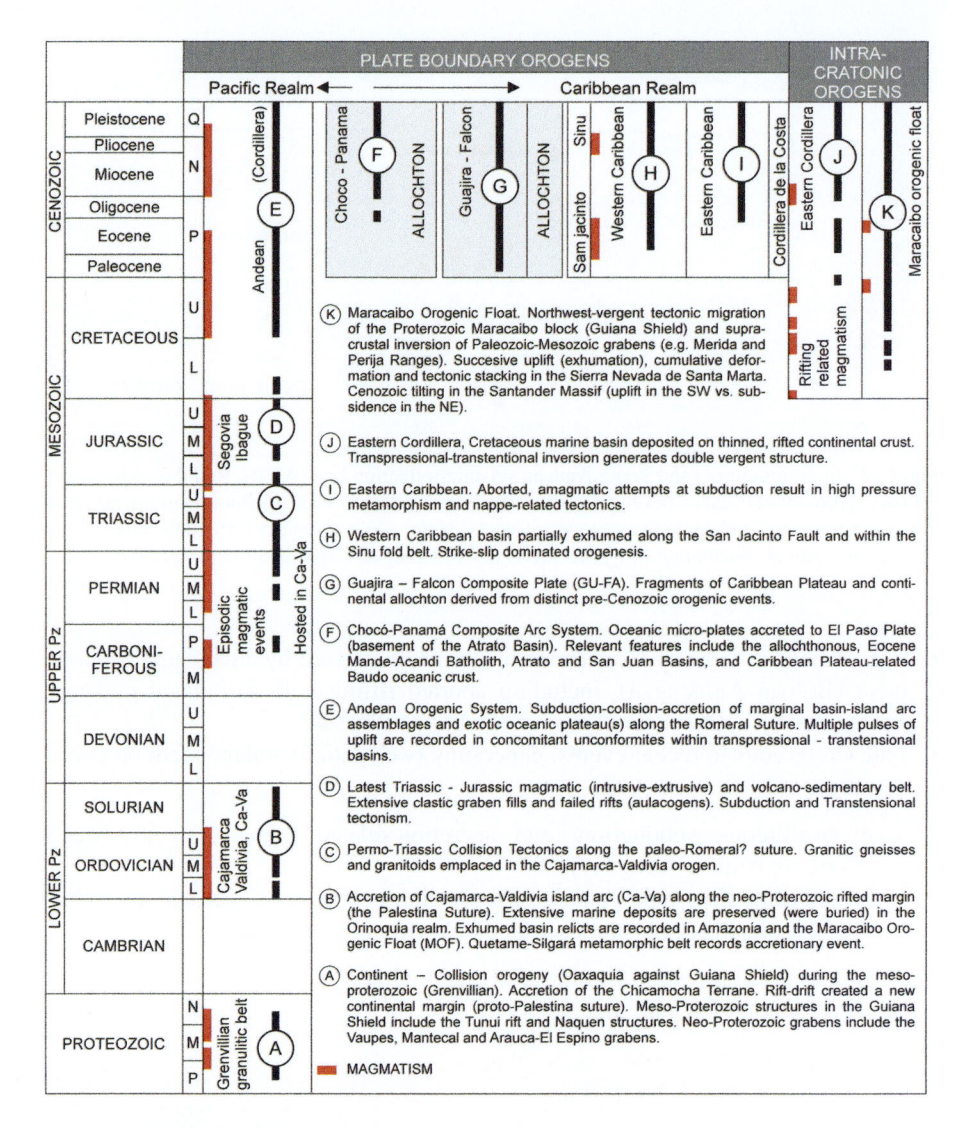

Fig. 1.16 Chronology of plate boundary orogens vs. intracratonic orogens in the Northern Andes. See text for details

(K) Maracaibo Orogenic Float. Northwest-vergent tectonic migration of the Proterozoic Maracaibo block (Guiana Shield) and supra-crustal inversion of Paleozoic-Mesozoic grabens (e.g. Merida and Perija Ranges). Succesive uplift (exhumation), cumulative deformation and tectonic stacking in the Sierra Nevada de Santa Marta. Cenozoic tilting in the Santander Massif (uplift in the SW vs. subsidence in the NE).

(J) Eastern Cordillera, Cretaceous marine basin deposited on thinned, rifted continental crust. Transpressional-transtentional inversion generates double vergent structure.

(I) Eastern Caribbean. Aborted, amagmatic attempts at subduction result in high pressure metamorphism and nappe-related tectonics.

(H) Western Caribbean basin partially exhumed along the San Jacinto Fault and within the Sinu fold belt. Strike-slip dominated orogenesis.

(G) Guajira – Falcon Composite Plate (GU-FA). Fragments of Caribbean Plateau and continental allochton derived from distinct pre-Cenozoic orogenic events.

(F) Chocó-Panamá Composite Arc System. Oceanic micro-plates accreted to El Paso Plate (basement of the Atrato Basin). Relevant features include the allochthonous, Eocene Mande-Acandi Batholith, Atrato and San Juan Basins, and Caribbean Plateau-related Baudo oceanic crust.

(E) Andean Orogenic System. Subduction-collision-accretion of marginal basin-island arc assemblages and exotic oceanic plateau(s) along the Romeral Suture. Multiple pulses of uplift are recorded in concomitant unconformities within transpressional - transtensional basins.

(D) Latest Triassic - Jurassic magmatic (intrusive-extrusive) and volcano-sedimentary belt. Extensive clastic graben fills and failed rifts (aulacogens). Subduction and Transtensional tectonism.

(C) Permo-Triassic Collision Tectonics along the paleo-Romeral? suture. Granitic gneisses and granitoids emplaced in the Cajamarca-Valdivia orogen.

(B) Accretion of Cajamarca-Valdivia island arc (Ca-Va) along the neo-Proterozoic rifted margin (the Palestina Suture). Extensive marine deposits are preserved (were buried) in the Orinoquia realm. Exhumed basin relicts are recorded in Amazonia and the Maracaibo Orogenic Float (MOF). Quetame-Silgará metamorphic belt records accretionary event.

(A) Continent – Collision orogeny (Oaxaquia against Guiana Shield) during the mesoproterozoic (Grenvillian). Accretion of the Chicamocha Terrane. Rift-drift created a new continental margin (proto-Palestina suture). Meso-Proterozoic structures in the Guiana Shield include the Tunui rift and Naquen structures. Neo-Proterozoic grabens include the Vaupes, Mantecal and Arauca-El Espino grabens.

■ MAGMATISM

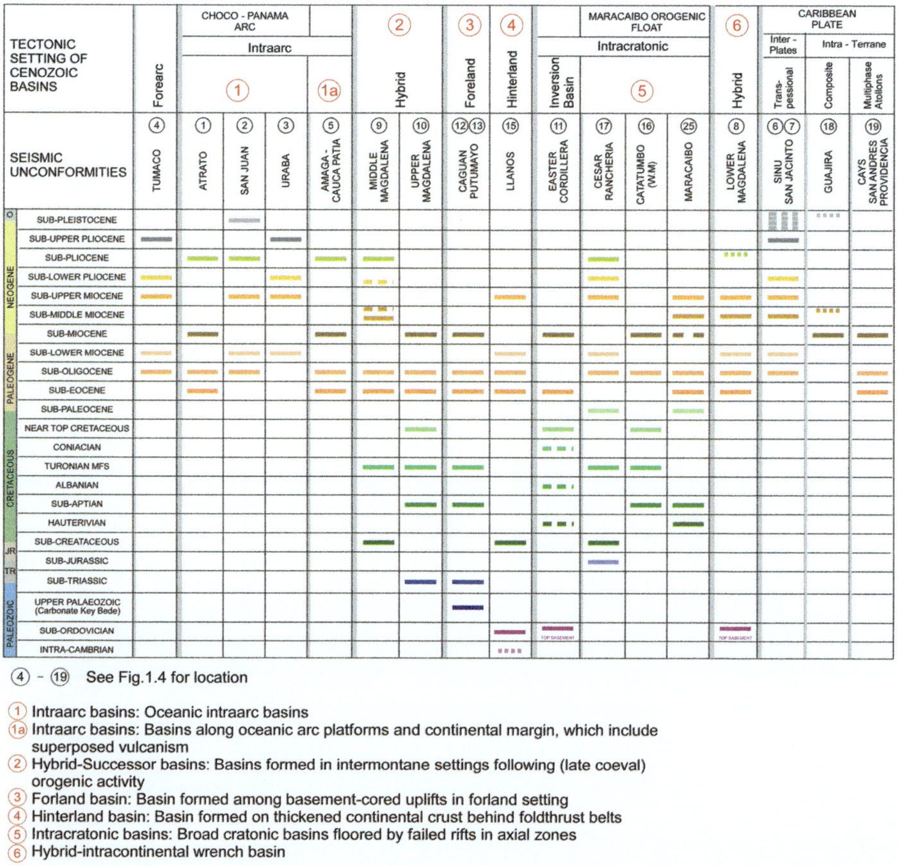

Fig. 1.17 Chronology of seismic-recorded unconformities in the Cenozoic basins of Colombia, with classification of tectonic basins. (Modified after Cediel et al. (1998))

① Intraarc basins: Oceanic intraarc basins
①a Intraarc basins: Basins along oceanic arc platforms and continental margin, which include superposed vulcanism
② Hybrid-Successor basins: Basins formed in intermontane settings following (late coeval) orogenic activity
③ Forland basin: Basin formed among basement-cored uplifts in forland setting
④ Hinterland basin: Basin formed on thickened continental crust behind foldthrust belts
⑤ Intracratonic basins: Broad cratonic basins floored by failed rifts in axial zones
⑥ Hybrid-intracontinental wrench basin

1.6.1.4 Island Arc Accretion and Marine Transgression. Early Paleozoic (Figs. 1.18, 1.19, and 1.20)

General Statement

Surface and subsurface stratigraphic data, seismic-structural maps, and cross sections compiled from data from numerous petroleum exploration campaigns, support compilation, and interpretation of the paleomaps offered herewith. Despite the fact that our knowledge of the Cajamarca-Valdivia Island Arc "complex" is still poor, its regional identity is well established. Deciphering the polyphase metamorphic history of the Cajamarca-Valdivia rock unit and its age equivalent, the Silgará-Quetame metamorphic belt, is a matter of systematic, multidisciplinary work, which should

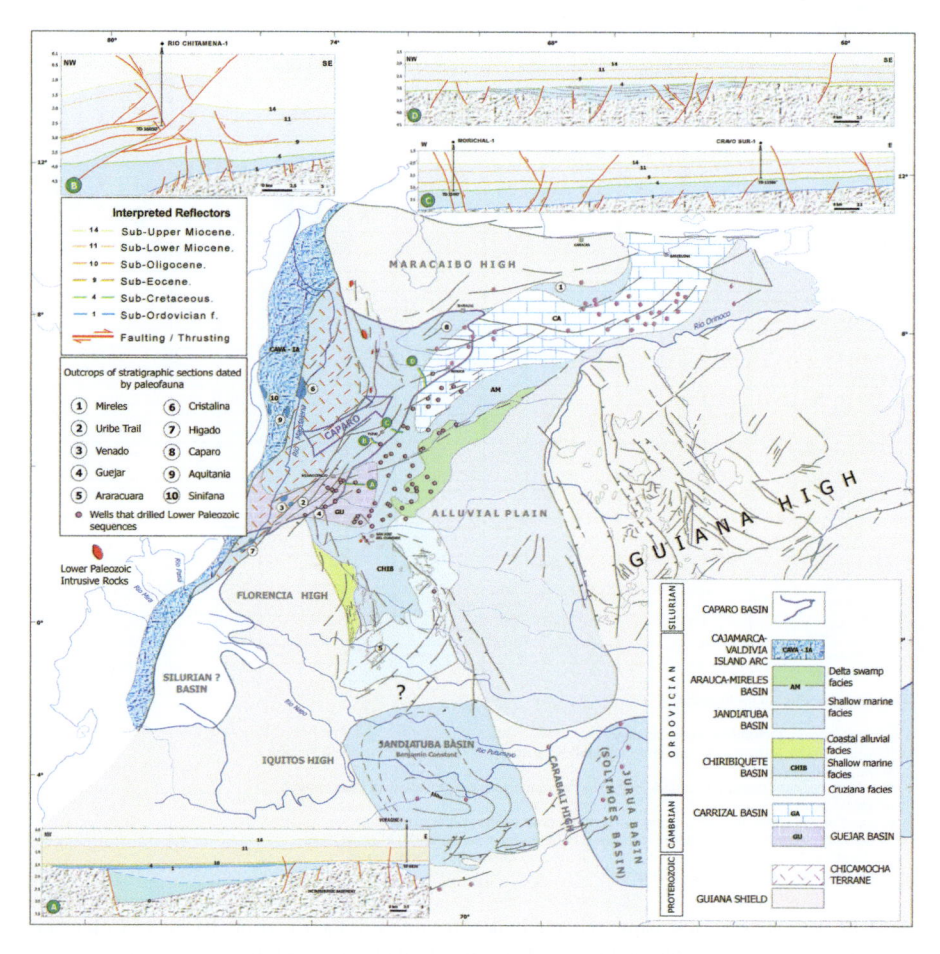

Fig. 1.18 Paleogeographic sketch map depicting relevant Lower Paleozoic (Cambrian, Ordovician, Silurian) tectono-stratigraphic units in Northwestern South America

include possible correlations with localized(?) Lower Paleozoic metamorphic rocks drilled in the Llanos Basin, which remain unexplained. Inherent to the above mentioned rock units are occurrences of Cambrian, Ordovician, and Silurian marine fauna. The plan geometry and character of these supracontinental marine basins reveal Proterozoic structural inheritance. Outcrops and subsurface occurrences of meta-gabbro (and related mafic intrusives) in the Guape Rift (Fig. 1.19) and gravity anomalies interpreted to the SW along the Garzón Massif are of particular significance.

- The Cajamarca-Valdivia Island Arc

The Cajamarca-Valdivia terrane (CA-VA) is a Lower Paleozoic meta-volcanoclastic and metapelite belt that extends from the central Andes of Colombia

1 Phanerozoic Orogens of Northwestern South America: Cordilleran-Type Orogens...

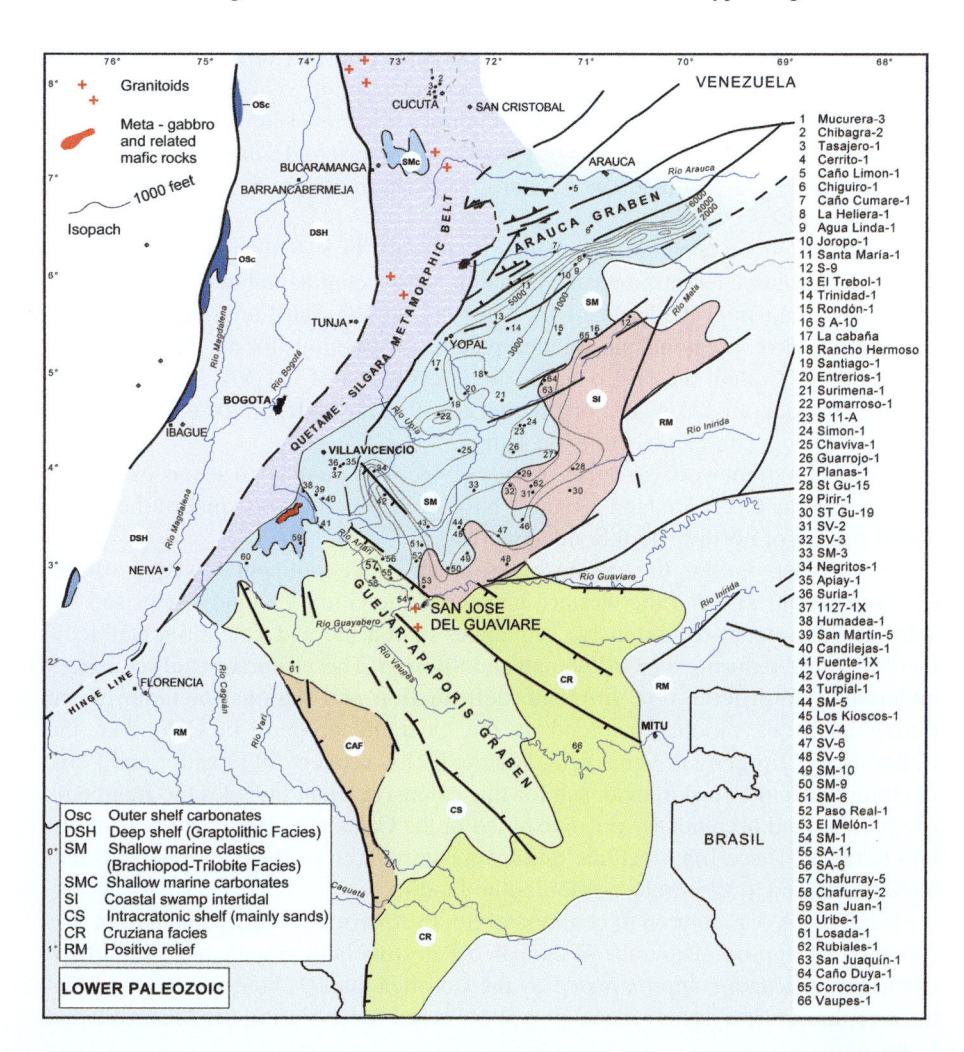

Fig. 1.19 Sedimentary facies distribution in the eastern Lower Paleozoic basins of Colombia

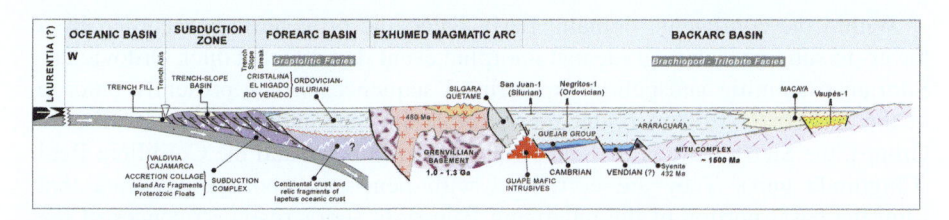

Fig. 1.20 Structural reconstruction of Lower Paleozoic litho-stratigraphic units; a conceptual W-E cross section after Cediel and Caceres (2000)

to northern Perú. A traverse across the physiographic Central Cordillera of Colombia reveals the unit consists mainly of amphibolite and graphitic schists metamorphosed to lower-middle greenschist to epidote-amphibolite facies. Limited studies to date indicate mineral assemblages that suggest a single prograde metamorphic event. Schists are isoclinally folded, and, toward the western margin of the terrane, foliation is transposed.

Major and trace element geochemistry reveals two distinct sources for the terrane's protoliths: an intra-oceanic island arc and a continental margin. Accretion and metamorphism of the terrane took place in Late Silurian-Early Devonian time. Based upon current regional paleogeographic reconstructions for the Paleozoic, two models could explain the origin and tectonic evolution of CA-VA, both involving closure of a back-arc basin. The first model is within an Andean-type margin setting and the second a continental collision setting.

Within this framework (in either of the two tectonic models), it seems likely that the pericratonic CA-VA island arc must have been standing proximal to the continental margin of northwestern South America.

To the east, within the continental domain, the continental wedge of the Chicamocha terrane and the western margin of the Guiana Shield comprised the subsiding basement for extensive sequences of marine and epicontinental sediments deposited during the Ordovician and Silurian. These supracrustal sequences underwent Cordilleran-type orogenic deformation and regional metamorphism during an event variably recorded as the Quetame orogeny in Colombia, the Caparonensis orogeny in Venezuela, and the Ocloy orogeny in Ecuador and Perú. In Colombia and Ecuador, evidence for this extensive event includes the fragments of ophiolite and accretionary prism exposed in the Cajamarca-Valdivia, Loja, and El Oro terranes. These litho-tectonic units were intruded by orogenic granitoids and metamorphosed to greenschist-amphibolite facies.

The CA-VA was sutured to continental South America along a paleomargin that followed the approximate trace of the paleo-Palestina fault system and its southern extension in Ecuador, approximated by the Cosanga fault (?; note that the modified trace of the suture system reflects polyphase reactivation during the Mesozoic and Cenozoic).

- The Quetame and Silgará Groups

Within the continental domain, Early Paleozoic orogenesis is recorded by a lower- to subgreenschist-grade metamorphic event that affected thick Ordovician-Silurian psammitic and pelitic supracrustal sequences which presently outcrop in the Eastern Cordillera (Quetame Group), the Santander-Perijá belt (Silgará Group), the Sierra Nevada de Santa Marta (Sevilla belt), and the Cordillera Real (Chiguinda unit). They are correlated with penecontemporaneous strata that form the basal portion of the onlapping Paleozoic supracrustal sequences of the Putumayo-Napo basins.

The low-grade, subgreenschist nature of metamorphism within the Lower Paleozoic supracrustal sequences has led to challenges in the interpretation of this regional tectono-metamorphic event and, in some instances, the documentation of multiple, more localized events (see discussion and references in Restrepo-Pace 1995). This apparent provinciality with respect to regional Ordovician-Silurian metamorphism in northwestern South America is unfounded and is more an artifact of the mechanisms behind regional metamorphism in general, rather than a reflection of the existence of multiple events. For example, in Colombia's Eastern Cordillera, weakly to non-metamorphosed windows of Ordovician-Silurian strata are observed. These rocks preserve diagnostic marine fauna for identification and dating, and they can be correlated with lower greenschist rocks of the same age that exhibit the imprint of regional metamorphism without having to evoke any major difference in overall tectonic history. A similar, although contrary, form of protolith preservation is observed in the amphibolite-grade Cajamarca-Valdivia terrane to the west, where regional metamorphism of the accretionary prism assemblage has left relicts of Grenville-aged granulite basement lodged and preserved as tectonic rafts in the amphibolite-grade metamorphic assemblages of the Cajamarca and Valdivia terrane.

The concept of "igneous-related low pressure metamorphism," IRLPM, recognized by Restrepo-Pace (1995, p. 27–28) in the Santander Massif during the Late Triassic-Early Jurassic may be applied with equal validity to help explain the provincial nature of Paleozoic regional metamorphism.

- Paleozoic "basement" in the Llanos Basin

Paleozoic sedimentary sequences located beneath Cretaceous-Cenozoic cover in the Llanos Basin of Colombia appear to be predominantly of Early Ordovician age (Llanvirnian and Arenigian stages). Most, if not all, of the deposits are marine as demonstrated by drill core recovered to date, all of which contains marine palynomorphs and graptolite fragments. Paleo-facies interpretations suggest a trend toward deeper water, offshore, or more open marine circulation to the southwest and west.

Ordovician stratigraphy overlies remnants of the Cambrian basins preserved in graben structures (e.g., Güejar and Carrizal, Fig. 1.18). A few wells have drilled flat-lying Silurian marine sediments overlying Ordovician deposits (e.g., San Juan-1).

1.6.1.5 Taphrogenic Tectonics and Plate Collision. Upper Paleozoic (Figs. 1.21, 1.22, 1.23, and 1.24)

General Statement
Abundant marine paleo-fauna is observed in outcrops scattered all over NW South America. These occurrences have been the subject of numerous biostratigraphic studies which document the presence of marine deposits dating from the Middle-Upper Devonian, Pennsylvanian, and Middle-Upper Permian (Stibane 1968; Forero 1968; Chacón et al. 2013; Rabe 1997). Likewise, exploration drilling in the Llanos Basin has retrieved marine paleo-fauna diagnostic of Upper Paleozoic deposits. Facies analysis and paleoenvironmental interpretations, along with geological

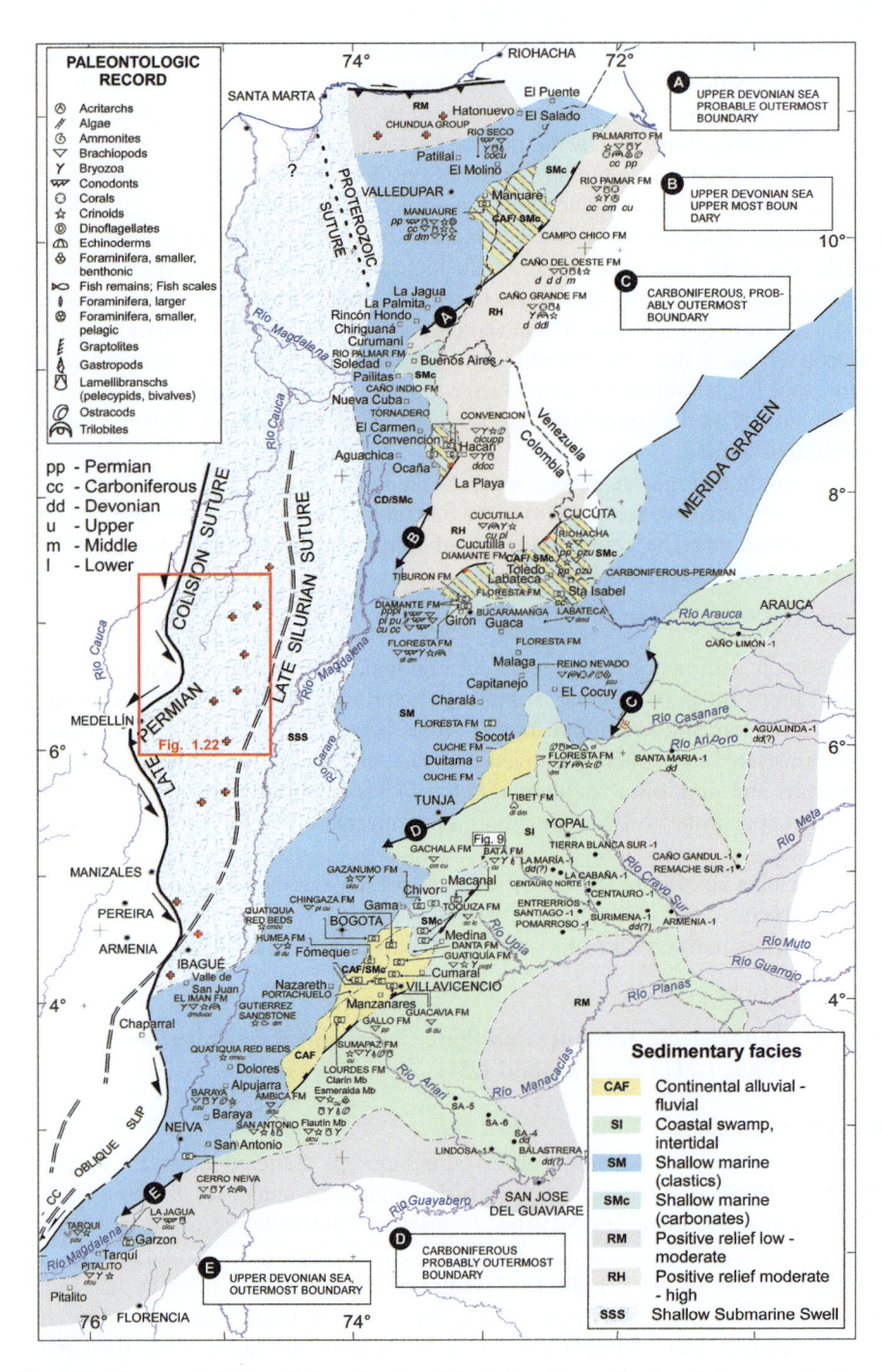

Fig. 1.21 Documented Upper Paleozoic paleontological and general paleo-facies reconstruction. W-E Permian (? Triassic) collision

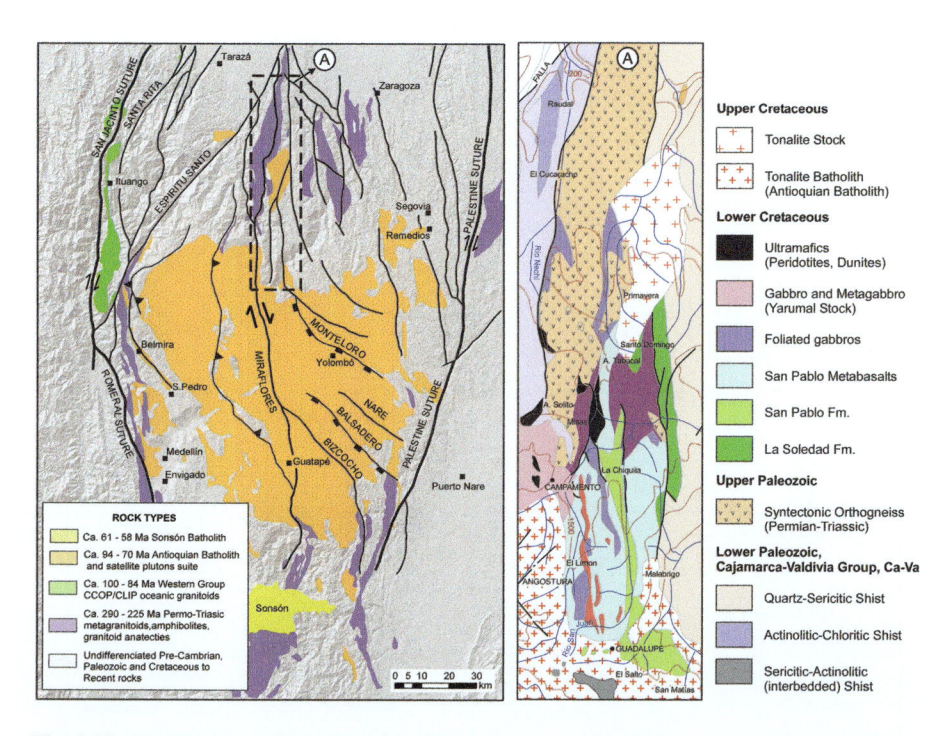

Fig. 1.22 Distribution of Permo-Triassic (290–225 Ma) meta-granitoids, amphibolites, and granitoid anatectites and interpreted emplacement of the Antioquian Batholith (94–70 Ma) through and along the Nechi rift. (see Leal-Mejía et al. 2018). ((A) Modified after Estrada (1967) and Gonzalez (2001))

mapping of a few – but relevant – areas, provide clear evidence of several, fault-restricted basins where interbedded marine and fluvial-deltaic sediments attain hundreds of meters thickness. The lack of geological maps at a suitable scale hinders the construction of paleogeographic maps for any given time slice. Permian volcanic magmatism is known to occur in Paleozoic basins outcrops of today Eastern Cordillera.

Upper Paleozoic taphrogenic development (incipient rifting and graben formation) was a prelude to wholesale extension during Triassic-Jurassic time. In contrast to these extensional kinematics, the northern(?) and western margins of continental NW South America record Permian magmatism and related metamorphism attributed to a poorly contextualized tectono-magmatic event possibly associated with the amalgamation and breakup of Pangea.

For the purposes of the reconstructions presented herein, I define "Late" or "Upper" Paleozoic to include sedimentary sequences biostratigraphically dated from Middle Devonian to Late Permian.

- Devonian fossiliferous, marine sediments are rather abundant along Colombia's Eastern Cordillera but are absent in the Sierra Nevada de Santa Marta and Mérida Andes of Venezuela. The best exposures are found in the Perijá Range, Santander,

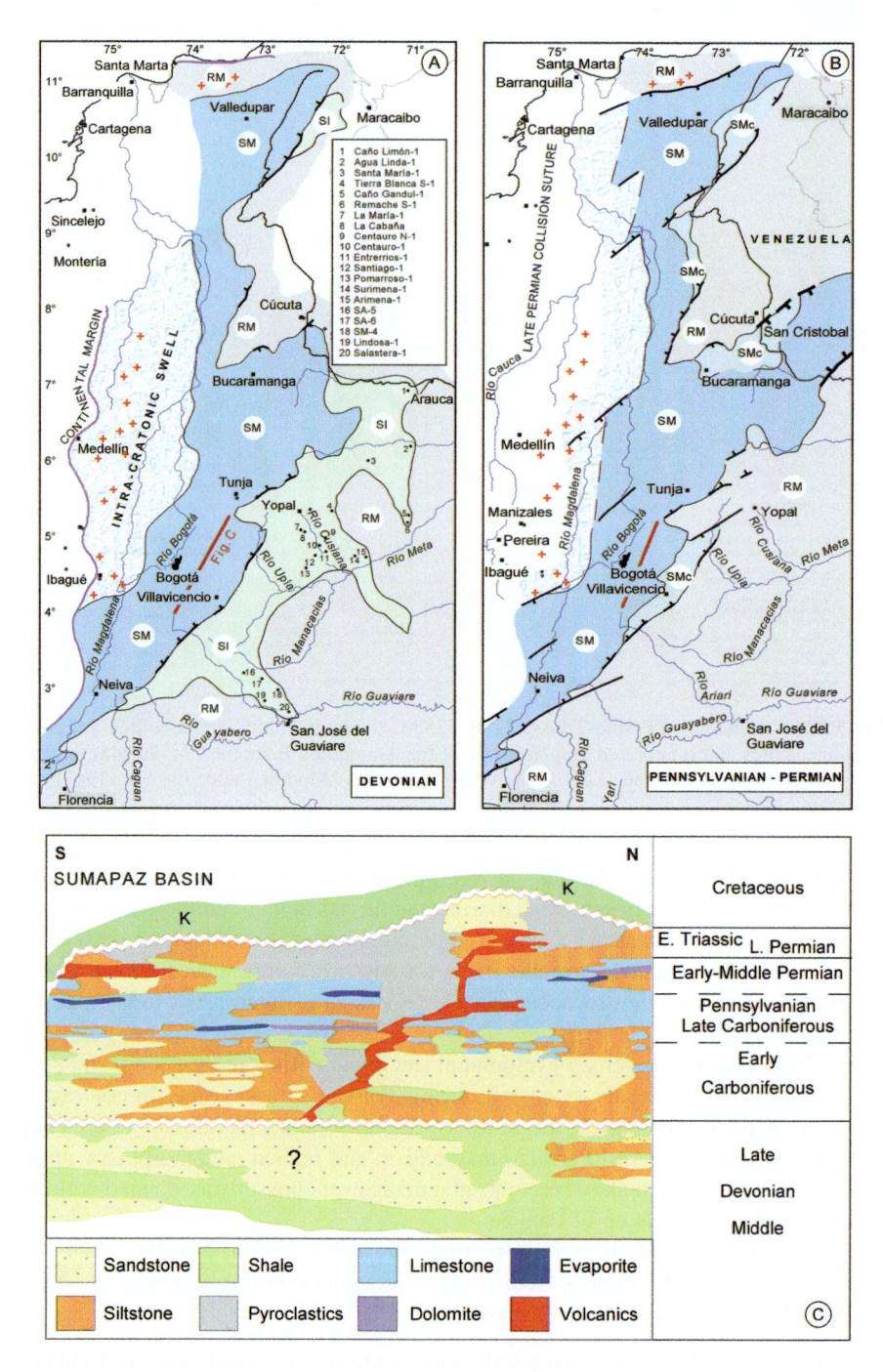

Fig. 1.23 Devonian and Pennsylvanian-Permian litho-stratigraphic composition and distribution of paleo-geographic basins

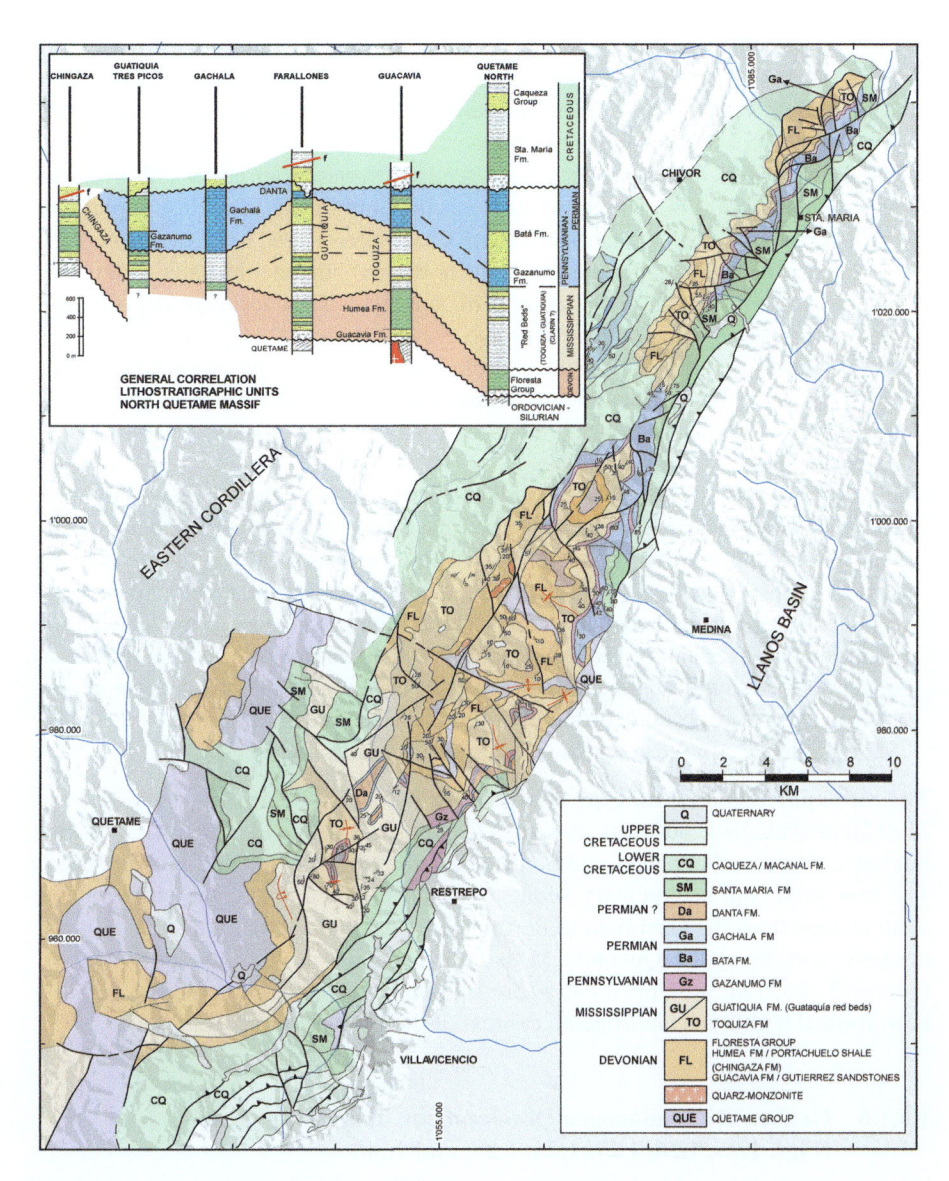

Fig. 1.24 Quetame Massif. Upper Paleozoic basin developed on top of the Quetame metamorphic belt

Floresta and Quetame Massifs, and in the southeastern foothills of the Central Cordillera. Most of the sections begin with a basal conglomerate overlying metamorphic rocks and consist of interlayered marine sandstones, siltstones, and shales with occasional minor limestones.
- The thickness of individual Devonian sections varies from about 200 m up to approximately 800 m. The upper contact, in some of the localities, depicts an apparently conformable relationship with Carboniferous red beds; in others (e.g., at Manaure in the Perijá Range and in the northern Santander Massif), an angular and/or discordant relationship is well documented. The Devonian deposits are regarded as a single, broad transgressive event.
- Mississippian sedimentary rocks – including red beds – are poorly recorded, since very few faunal localities representing this period have been identified. For the most part, it is considered to be a stratigraphic hiatus in the northern Andean region.
- Pennsylvanian to Middle Permian sedimentary rocks are generally exposed in the same localities where Devonian sections have been measured and documented, with the exception of the Floresta Massif. Red beds are observed at the base of known Pennsylvanian sections, which gradually grade into marine deposits consisting of interbedded sandstones, marls, shales, and limestones. These deposits are conformably overlain by Early-Middle Permian marine limestones that may locally attain significant thickness. The Carboniferous to Permian deposits are interpreted to have been deposited within localized, structurally restricted basins. Stratigraphic sections can contain numerous unconformities.

The Sumapaz Basin (Prototype of an Upper Paleozoic Basin)
Most Upper Paleozoic outcrops are linked to exposures within the Garzón, Quetame, Sumapaz, Floresta, Santander and Santa Marta Massifs, and the Serranía de Perijá. The best preserved sedimentary sequences, and the area where the stratigraphic relationships from the Devonian to Permian are best understood, are located in the northern portion of the Quetame Massif (also known as the Sumapaz Massif). Figures 5 illustrate an Upper Paleozoic transtensional basin, interpreted to have been formed within the taphrogenic context of the Bolivar Aulacogen.

1.6.1.6 Graben-Rift-Aulacogen Development. Triassic-Jurassic (Figs. 1.25, 1.26, 1.27, 1.28, 1.29, and 1.30)

General Statement
An intraplate graben system associated with the breakup of Pangea, the separation of the Mexican terranes, and the opening of the proto-Caribbean basin developed in the northern half of Colombia and western Venezuela during the latest Triassic and Jurassic. To the south, an extensional arc – back-arc regime – dominated since Late Triassic time. A conspicuous northwest-west trending discontinuity separates the northern and southern tectonic regimes (Fig. 1.25).

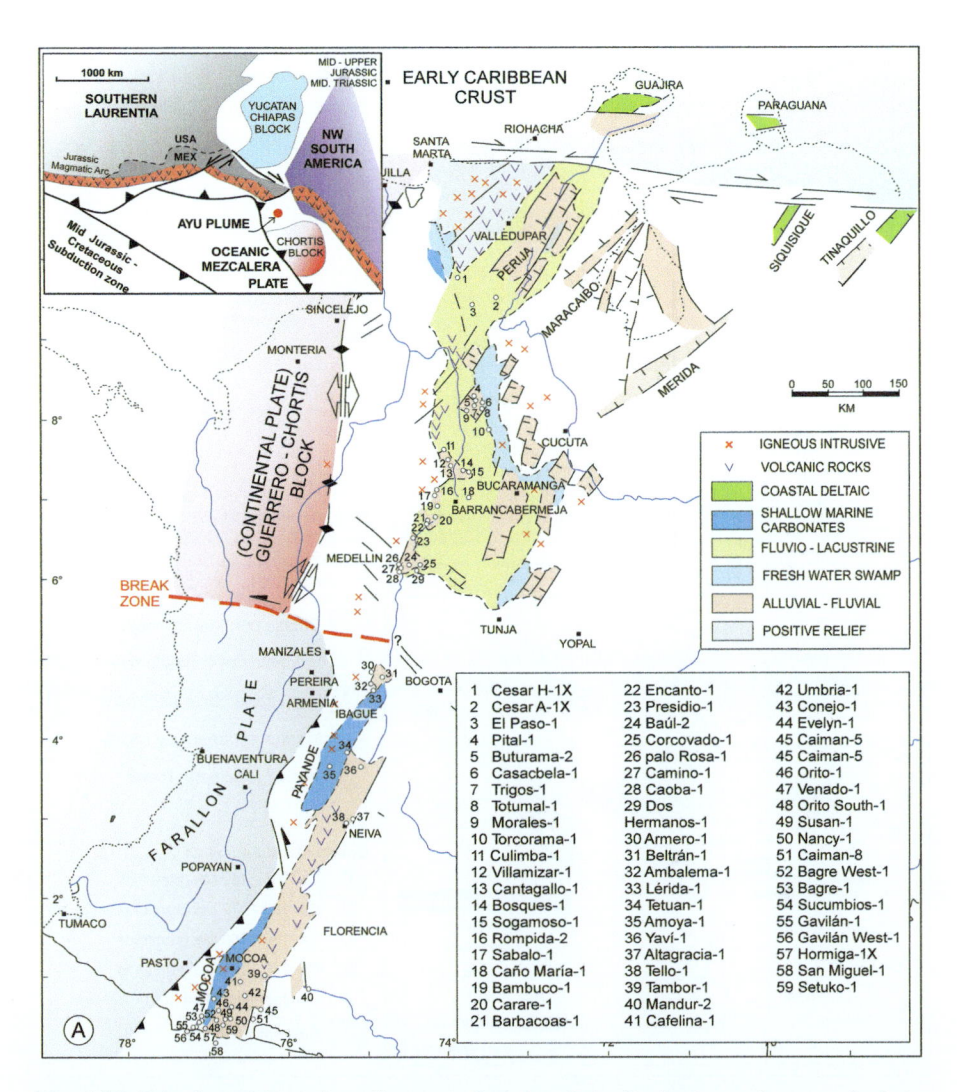

Fig. 1.25 Triassic and Jurassic sedimentary deposits; their distribution, paleo-tectonic facies interpretation and basin types. (Source upper left box: Keppie et al. (2004))

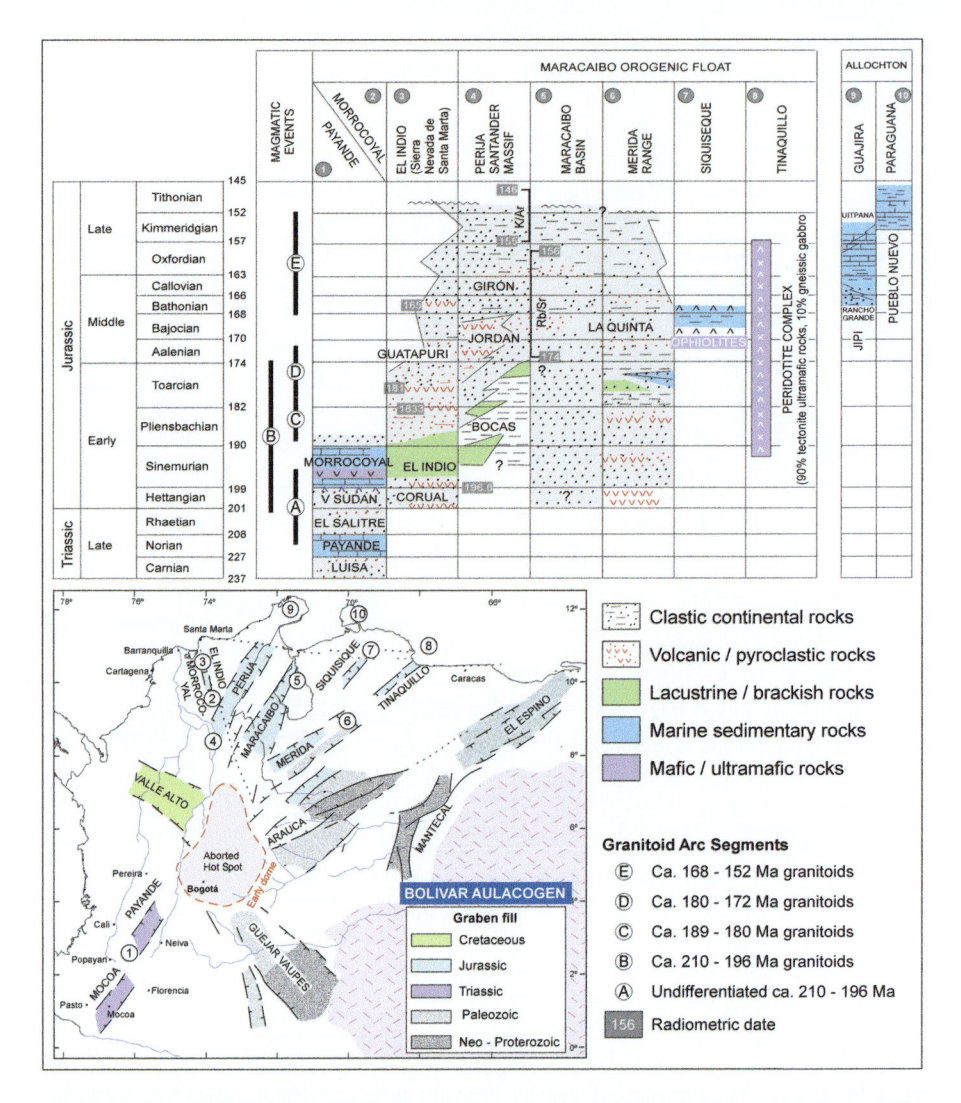

Fig. 1.26 Graben-rift-aulacogen-type deposits in the context of the Bolivar Aulacogen. (Compiled after Bartok et al. (1985), Sung Hi Choi et al. (2017), Geyer (1973), Cediel (1969), Maze (1984), Mendi et al. (2013), Schubert (1986), Leal-Mejía et al. (2018))

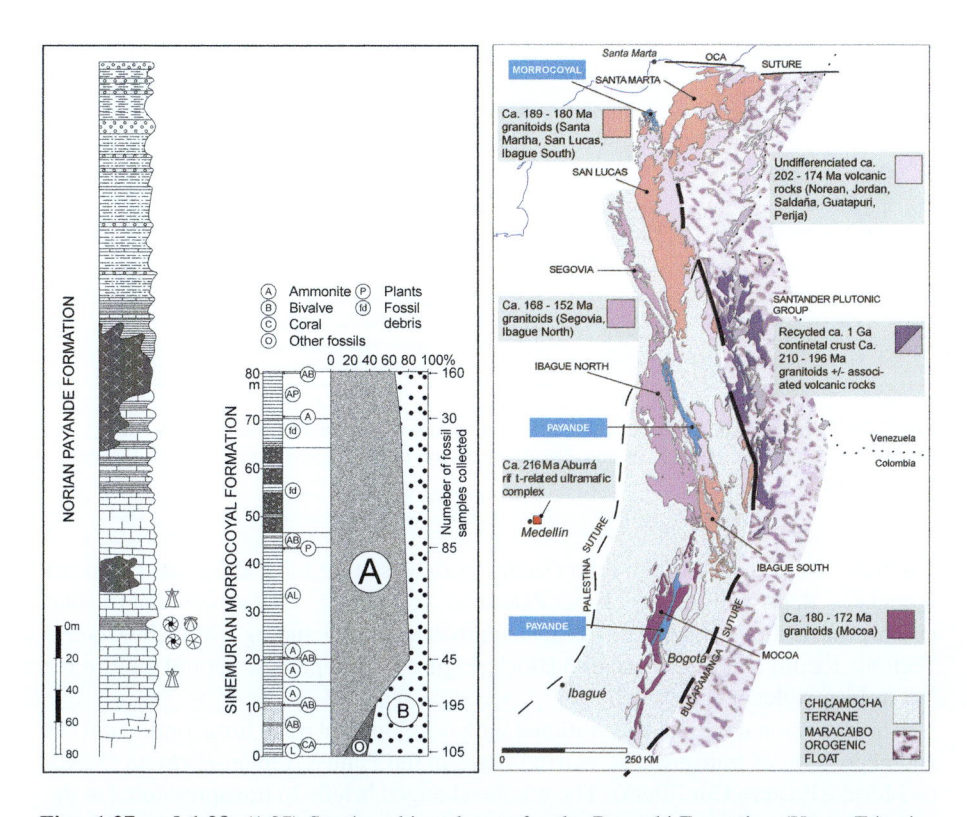

Figs. 1.27 and 1.28 (1.27) Stratigraphic columns for the Payandé Formation (Upper Triassic-Lower Jurassic) and Morrocoyal Formation (Lower Jurassic). (Modified after Geyer (1973) and Cediel et al. (1980)). (1.28) Strike slip-restored reconstruction of major magmatic blocks (Leal-Mejía et al. 2018) and paleo-structural setting of the marine Payandé and Morrocoyal deposits

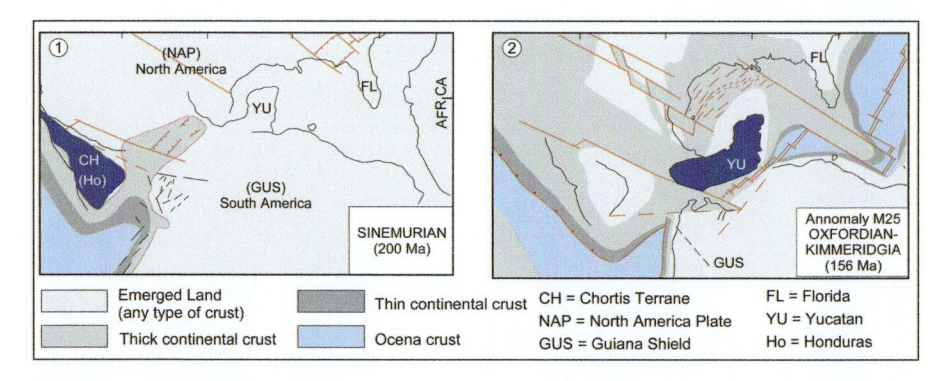

Fig. 1.29 Paleo-tectonic reconstruction of the Sinemurian deposits (in the Chortis Block) and Kimmeridgian deposits (in the Yucatan Block) in relation to continental Northwest South America after Stephan et al. (1990)

The name Bolivar Aulacogen refers to the tectonic setting for widespread failed rift sequences deposited during Late Paleozoic to Middle Cretaceous continental taphrogenesis throughout northwestern South America, including within the Central Tectonic Realm and the Maracaibo Block. Figure 1.26 depicts structural pattern for the Bolivar Aulacogen.

The extensional regime was initiated with development of an intracontinental rift and deposition of transgressive marine strata in the Pennsylvanian-Permian (Sierra de Mérida, Eastern Cordillera). The regime changed briefly to transpressional at the end of the Permian, as recorded by tight folds associated with strike-slip faulting observed in the Sierra de Mérida. I interpret this transpressional regime to reflect the hinterland effects in NW South America of the final assembly of Pangea, tangential to the principle sutures (e.g., Oachita, Marathon) which record Laurentia-Gondwana interaction during the Late Permian (Keppie 2008; Van der Lelij 2013).

Rifting resumed during the Triassic (Payandé Formation) and continued into the Early Jurassic (Morrocoyal rift) and the Middle Jurassic (Siquisique rift). In the Late Jurassic, extensive rifting and extensional arc development is marked by deposition of the continental and volcaniclastic deposits of the Girón, La Quinta, Jordán, and Noreán Formations.

The Mexican terranes (Keppie 2008), dominated by the Guerrero-Chortis and Maya blocks, loosely accumulated along the northwesternmost South American margin during the assembly of Pangea, rifted away from the Colombia margin during latest Triassic time onward.

Important metaluminous (I-type) magmatism of latest Triassic-Jurassic age, associated with extensional arc development during rollback of the Farallon Plate, was also emplaced in the taphrogenic context of the Bolivar Aulacogen (see the detailed analysis of Leal-Mejía et al. 2018). Overall, I envision a complex distribution of temporally and geographically limited extensional (forearc and back-arc?) basins with localized, modified, continental margin magmatic arcs coexisting in a broadly

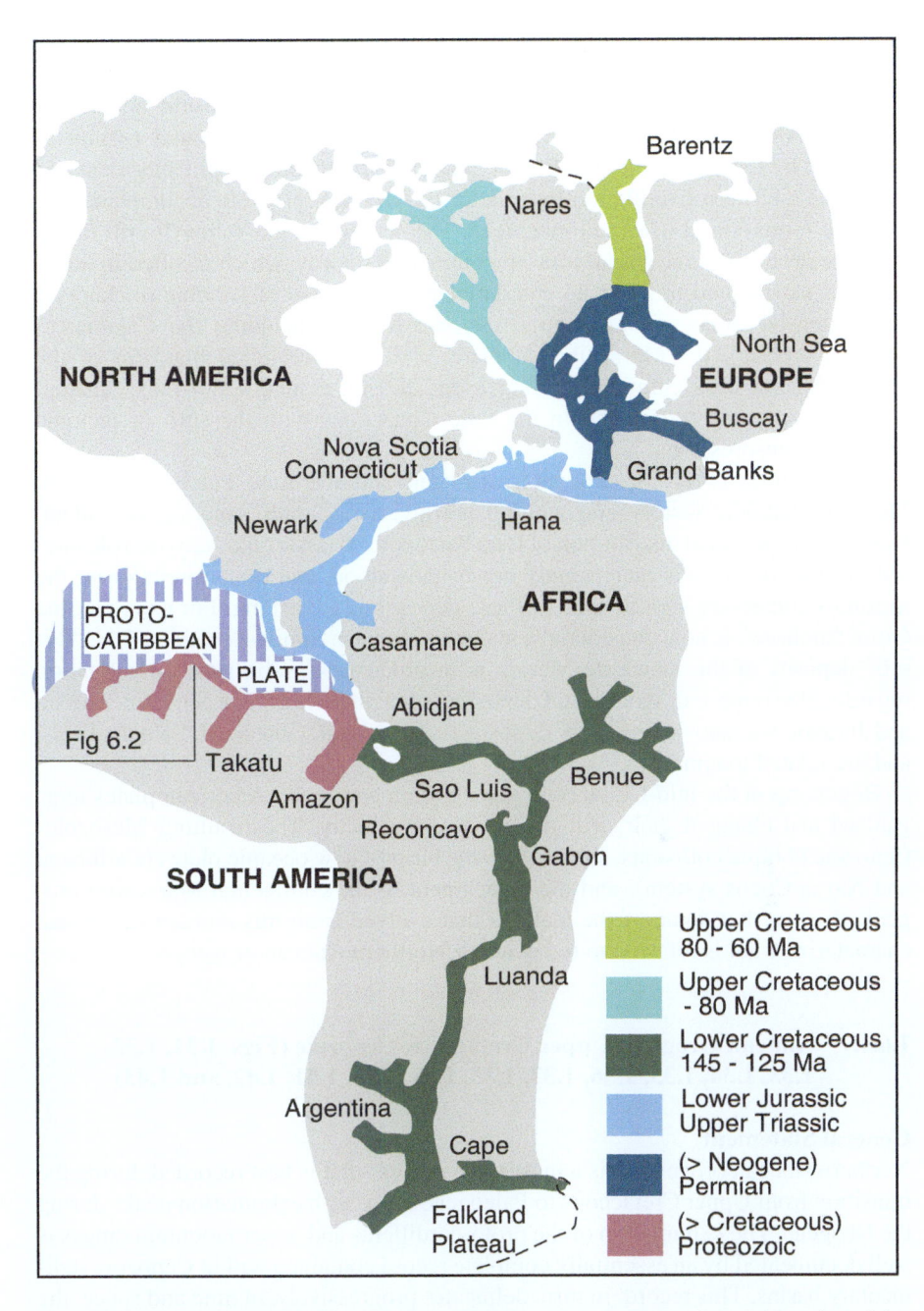

Fig. 1.30 Schematic representation of rifts and aulacogens in Pangean time and location of the proto-Caribbean Plate. (Modified after Burke (1977))

(and ultimately) taphrogenic environment and forming on a markedly thinned, heterogeneous, Proterozoic-Paleozoic metamorphic basement.

The Bolivar Aulacogen culminated in the Early Cretaceous with the opening of the Valle Alto rift. This last event was marked by deep continental rifting, as evidenced by the emplacement of mafic magmatism related to rapid subsidence in Colombia's Eastern Cordilleran basin after ca. 135 Ma, possibly accompanied by the local formation of oceanic lithosphere. The opening of the Valle Alto rift facilitated the invasion of the Cretaceous epicontinental seaway, which resulted in deposition of marine and transitional epicontinental sequences of variable thicknesses over extensive areas of the Central Tectonic Realm (including the Cajamarca-Valdivia terrane), the Maracaibo subplate, and the continental platform of the Guiana Shield. This culminant rifting event did not extend south into Ecuador. Regional extension terminated in the Late Cretaceous with the shift of tectonic regime to transpressional.

The complexity of the Bolivar Aulacogen and the Late Paleozoic through Mesozoic tectonic history surrounding northwestern South America is evident. However, the regional distribution of Late Paleozoic, Triassic, and Jurassic volcanic-sedimentary deposits is increasingly better understood, and reinterpretation of the tectono-sedimentary significance of these deposits is advancing. For example, the Girón "molasse" is now considered a syn-rift sequence. Similar revision of the "flysch" deposits of the Sierra de Mérida is in order, as is substantial investigation regarding the temporal, spatial, and depositional relationships between Late Triassic and Jurassic volcano-sedimentary deposits (e.g., Noreán, Guatapurí, Saldaña Fms.) and arc-related magmatism.

Beginning in the mid-Cretaceous, the Farallon and South American plates reorganized and changed their drift direction and velocity. The resulting Mesozoic-Cenozoic oblique collisions, subduction, the birth of new oceanic plates (Caribbean and Nazca-Cocos system), and the detachment of the continental Maracaibo orogenic float are but some of the features that evolved from this reorganization and characterize what is referred to today as the Northern Andean orogeny.

1.6.1.7 Andean Orogeny. Upper Cretaceous-Cenozoic (Figs. 1.31, 1.32, 1.33, 1.34, 1.35, 1.36, 1.37, 1.38, 1.39, 1.40, 1.41, 1.42, and 1.43)

General Statement

Northern Andean orogenesis is a multistage process that is best recorded during the transition from Upper Cretaceous to Paleogene time, with exhumation peaks during the Neogene. The exhumation of the proto-cordilleras and lesser mountain ranges is well documented by an essentially complete record contained within Cenozoic sedimentary basins. This record, in turn, delineates progressively, in time and space, the location, migration, and evolution of intracratonic as well as pericratonic uplift and sedimentation.

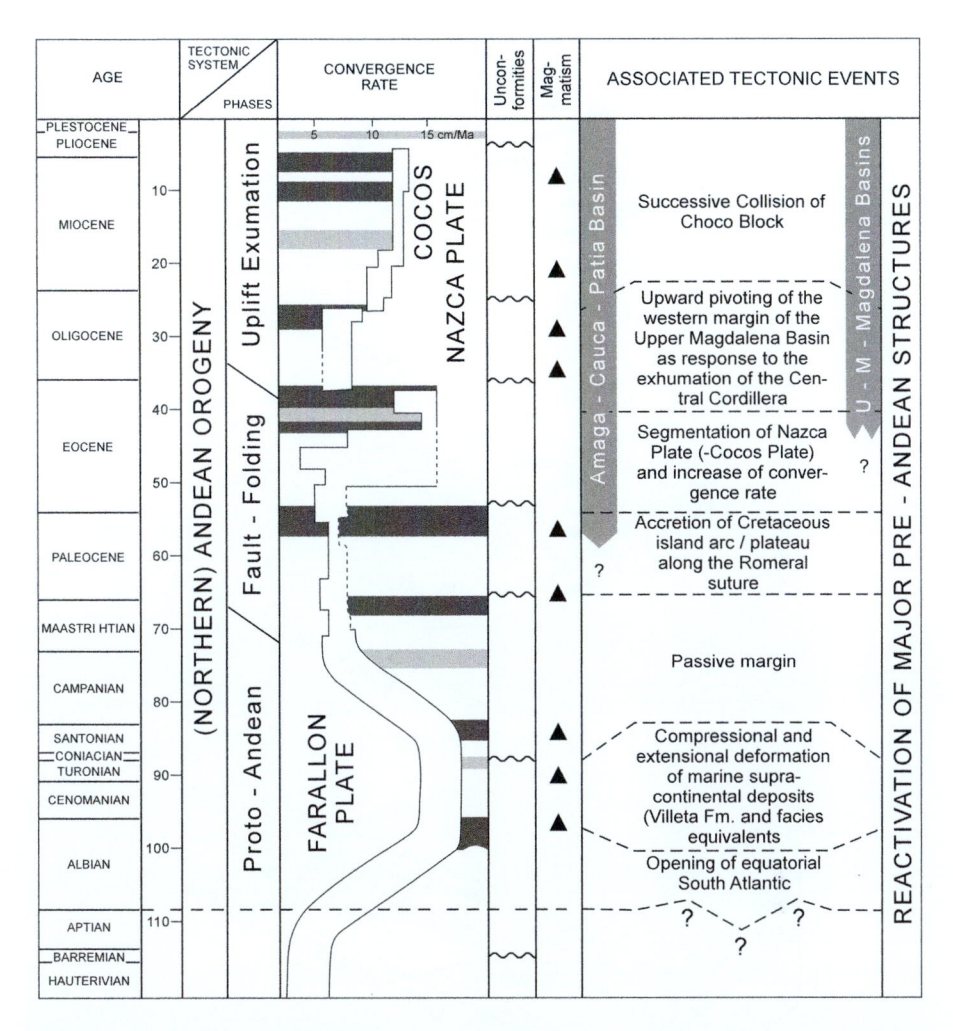

Fig. 1.31 Chronology of the Northern Andean Orogeny and associated tectonic events. (Modified after Jaimes and de Freitas (2006))

The age and facies distribution documented within northern Andean sedimentary basins reflect not only particular tectonic regimes but also record the nonsynchronous, transtemporal character of structural deformation throughout the region. The reactivation of ancient deep-seated fault systems and sutures, and the inversion of normal faults within transpressional-transtensional regimes, became the dominant kinematic style. This tectonic style remains important, even today.

The tectonic assembly of the northern Andean region is characterized by a prolonged, heterogeneous, regionally versus temporally punctuated series of orogenic events. These events record the interaction of no fewer than four distinct plate systems: the South American, the proto-Caribbean, the Pacific (Farallon-Nazca), and

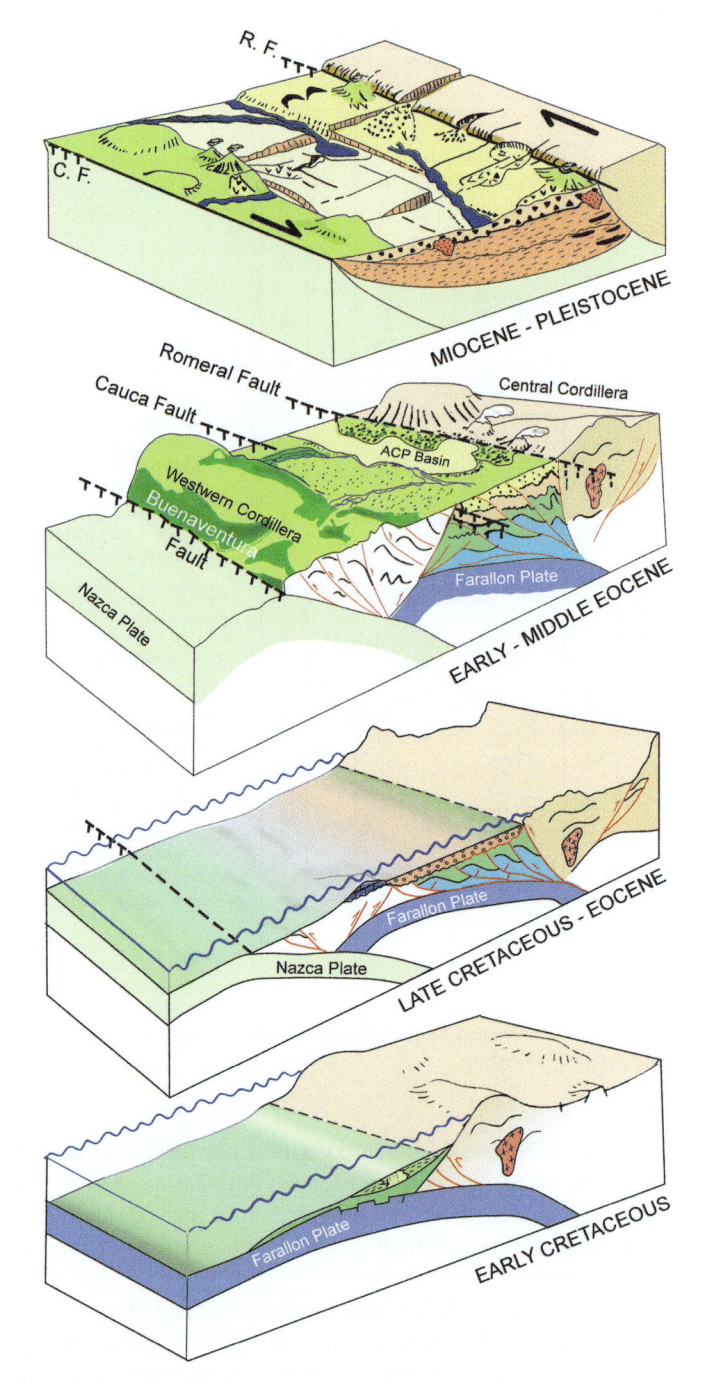

Fig. 1.32 Evolution model for the Farallones-Nazca Plates and tectonic development of the continental plate margin. (Sierra 2011)

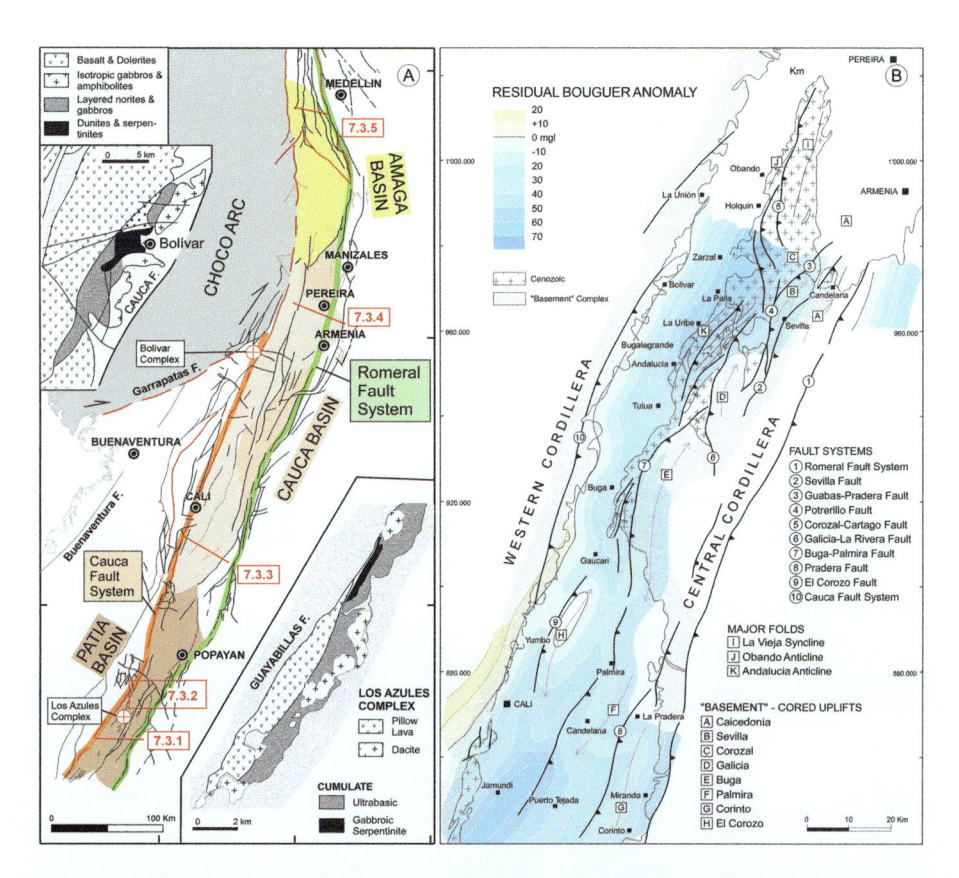

Fig. 1.33 (**a**) Structure of the Amagá-Cauca-Patia Basin. Data from the Bolívar and Los Azules complexes after Kerr et al. (2004) and Espinosa (1980). (**b**) Residual Bouguer anomaly map of the Cauca-Patia basin

the Caribbean oceanic plateau. The oceanic constituents have acted to a large degree independently over time on the corresponding South American continental margin.

Northern Andean orogenic events since the transition from a generally extensional regime during the Bolivar Aulacogen to a compressive (transpressive) regime beginning in the Aptian-Albian and up to the Holocene have been formulative in the present-day litho-tectonic and morpho-structural configuration of the northern Andean region.

During construction of the present geotectonic framework, I have favored the use of existing biostratigraphic and radiometric information and the application of field observations and geochemical investigations regarding the various litho-tectonic components of the region. Available data (although incomplete) provide a firm basis for interpretation of the interaction between allochthonous litho-tectonic components (PAT, CAT, and CHO, etc.) as defined in the tectonic realms of Fig. 1.2

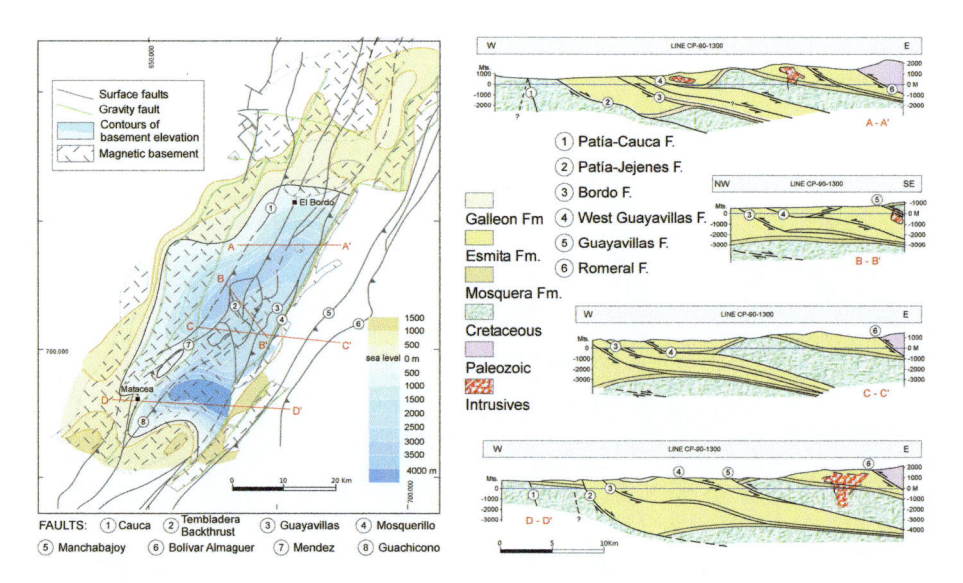

Fig. 1.34 Structure of the Patia Basin

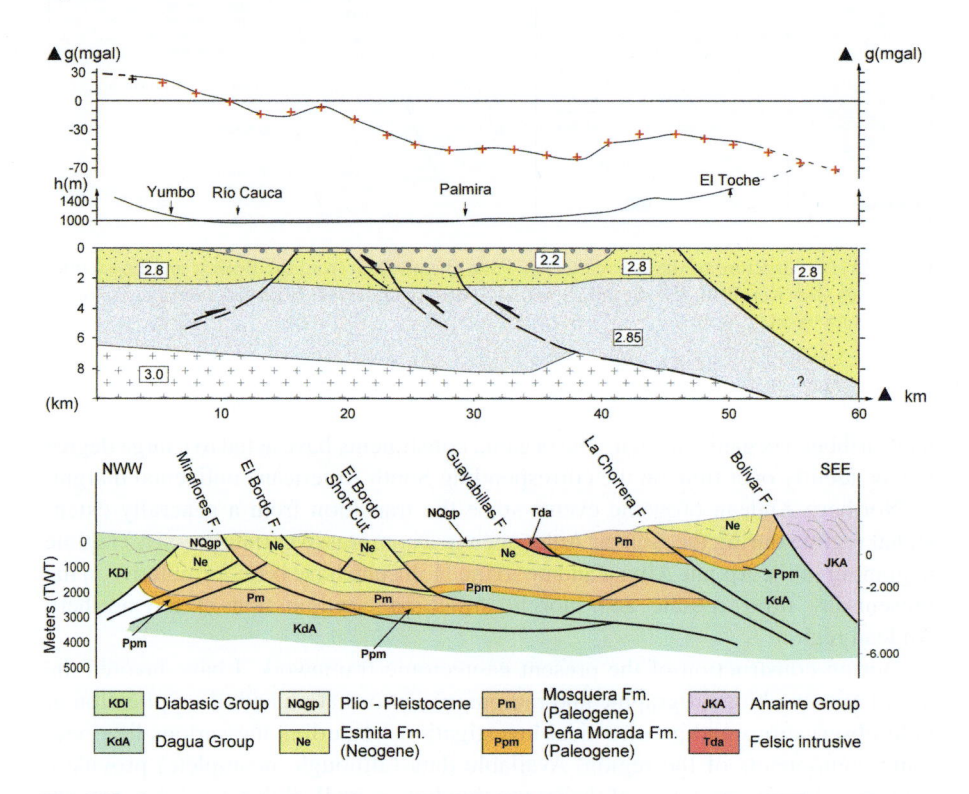

Fig. 1.35 Structure of the Cauca Basin. (Modified after Barrero and Laverde (1998) and Barrero et al. (2006))

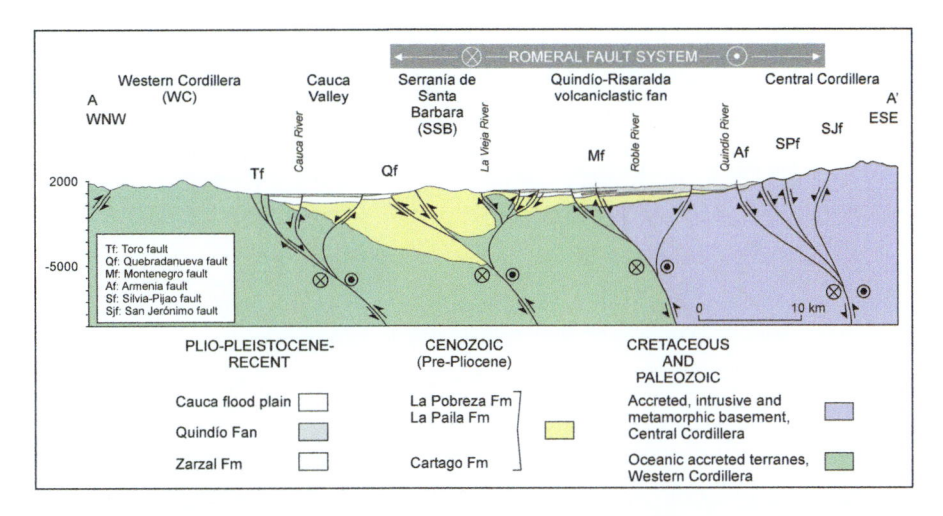

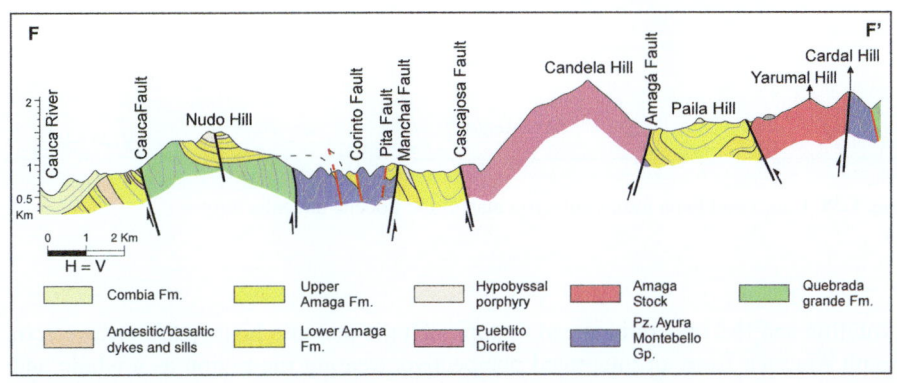

Figs. 1.36 and 1.37 (1.36) Structural section across the Romeral Fault System (Cauca Basin). (Modified after Suter et al. (2008)). (1.37) Structural section across the Amaga Basin. (Modified after Sierra (2011))

and the South American continental autochthon (including the Central Tectonic Realm and the morpho-structural components of the Maracaibo subplate) during the Meso-Cenozoic. A schematic synthesis of the time-space evolution of the northern Andean region during northern Andean orogenesis is presented in four time slices in Fig. 1.43.

In this context, I interpret the progressive, although temporally and geographically isolated, tectonic events spanning the Late Mesozoic and Cenozoic, as the Northern Andean Orogeny (e.g., Cediel et al. 2003). In doing so, I emphasize the complex, prolonged, and regionally punctuated nature of northern Andean tectonic

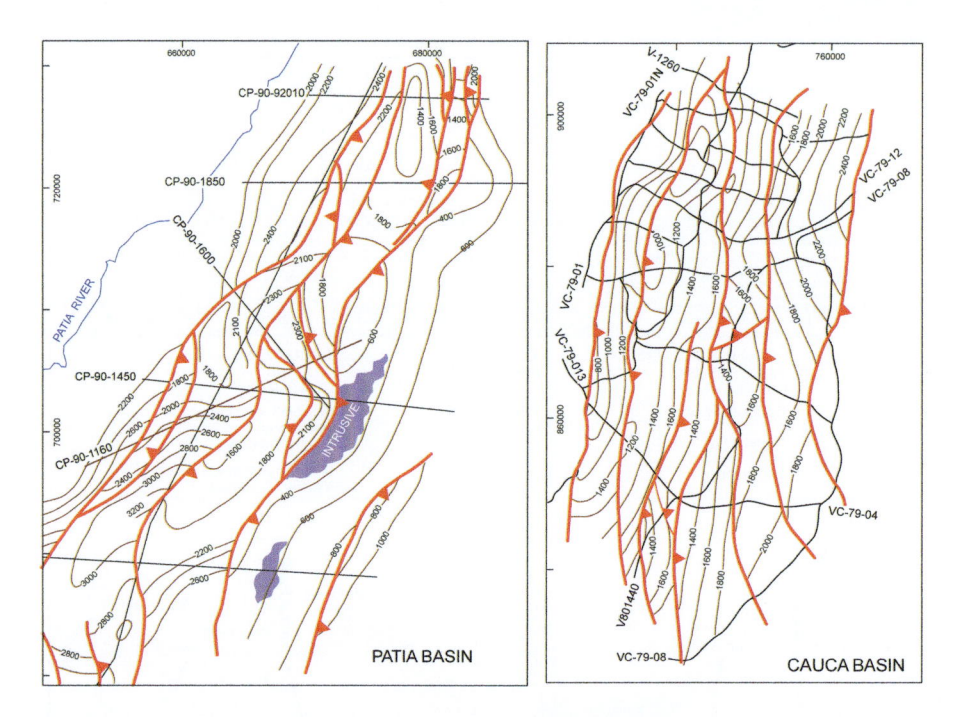

Fig. 1.38 Cauca and Patia basins sub-crop map of the Eocene unconformity

evolution and the imperative need to approach the tectonic history of northwestern South America from an integrated perspective, treating the region as a whole and integrating the components of the Northern Andean Block, from Perú to Venezuela and Panamá, into an internally coherent framework.

Figure 1.43 demonstrates how the Western Tectonic Realm and Maracaibo orogenic float act simultaneously and by distinct tectonic mechanisms, to generate their own individual deformational styles. Enormous transpression was exerted upon the Central Tectonic Realm resulting in exhumation and uplift of the Cajamarca-Valdivia terrane, which forms the core of the physiographic Central Cordillera, and tectonic inversion, exhumation and uplift, of much the rift-related Cretaceous sedimentary basin, which is today exposed within the physiographic Eastern Cordillera. In this context the CTR has developed its own distinct morpho-structural expression. This scheme for Meso-Cenozoic tectonic assembly and evolution in the Colombia Andes proposes a critical reevaluation of the application of typical "Andean-type" orogenesis to the geotectonic evolution of the northern Andean region.

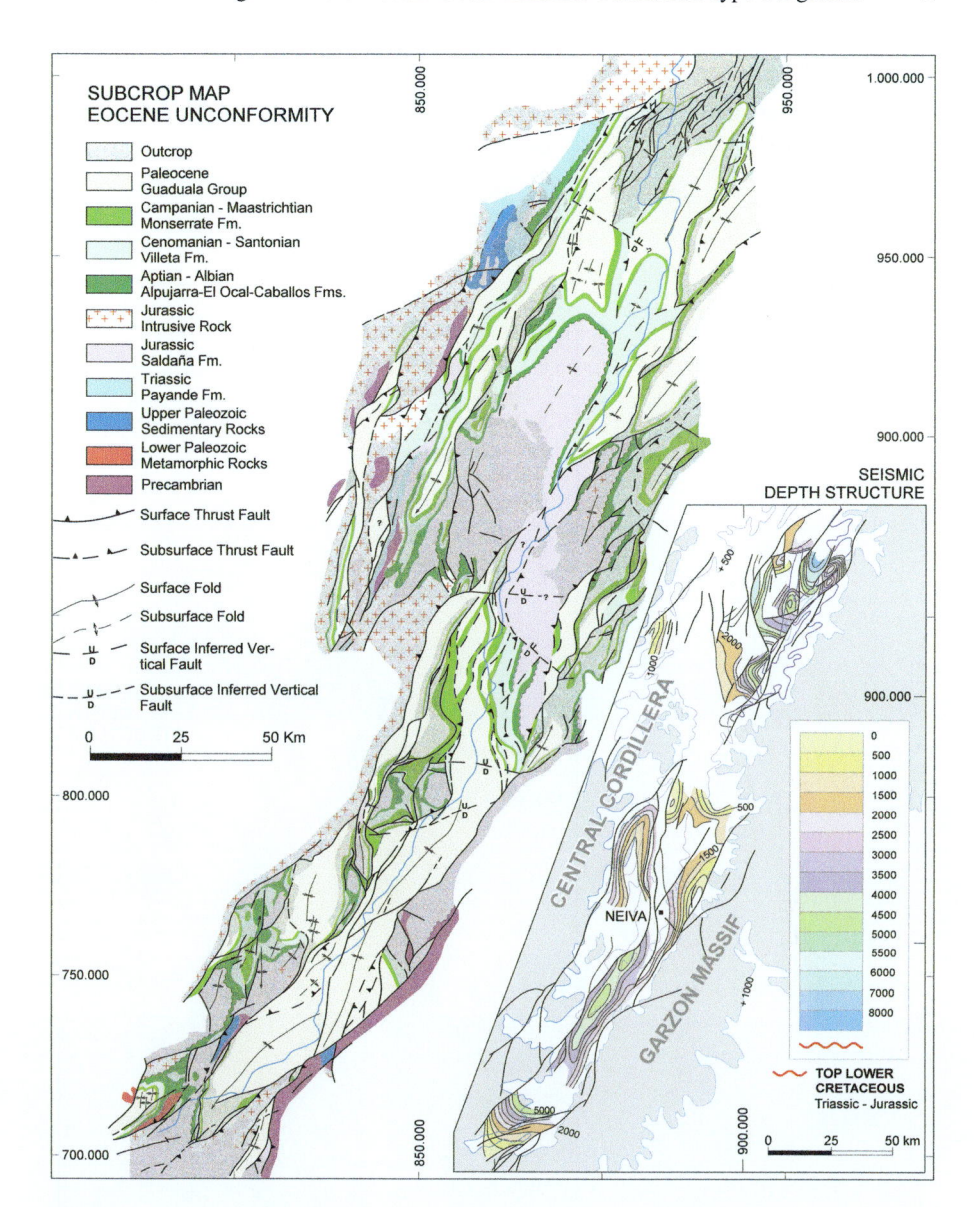

Fig. 1.39 Upper Magdalena Basin sub-crop map, Eocene unconformity and seismic depth to structure. (Source: Geotec (1998))

Fig. 1.40 Middle Magdalena Basin sub-crop map, Eocene unconformity and seismic depth to structure. (Source: Geotec (1998))

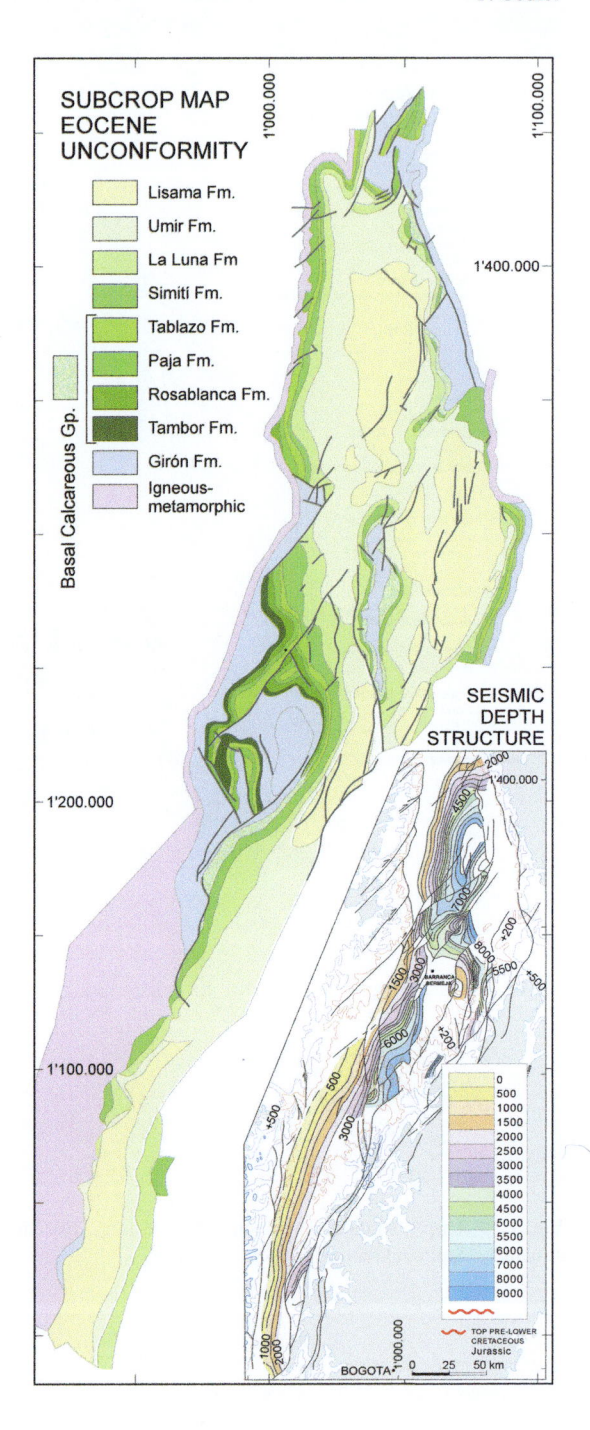

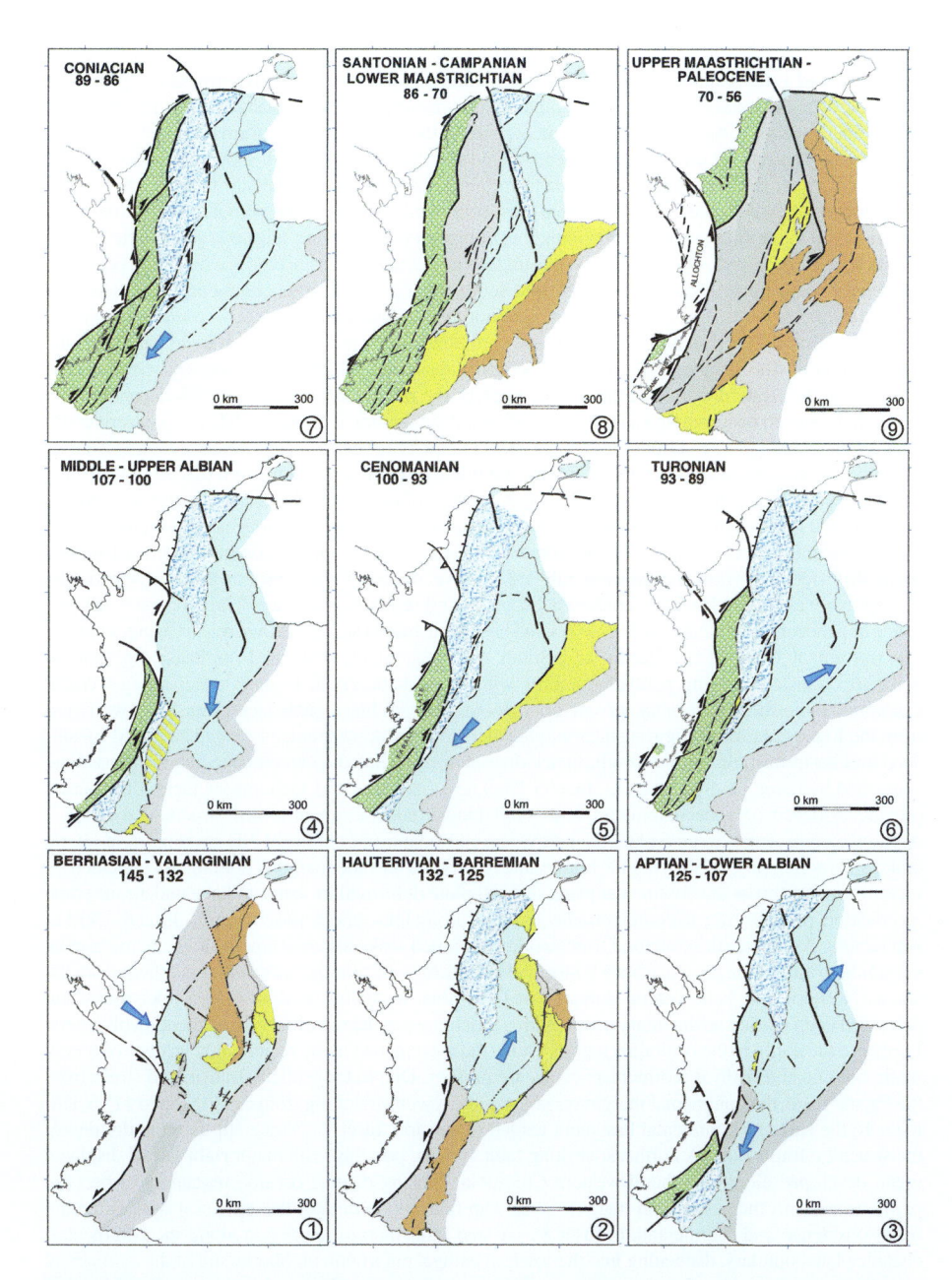

Figs. 1.41 and 1.42 Unrestored paleo-geographic litho-facies sketch maps of Cretaceous-Cenozoic marine, transitional and continental deposits in Colombia with interpreted plate tectonic setting. (Modified after Etayo et al. (1994)). (A) Transtensional regime. Inset 1 – Facies boundaries coincide with both old tectonic lineaments (N trend) and new fracture zones (NW trend), which represent the continent-ward extension of oceanic transform faults. "Bimodal" development: 1. From ca. 5° N southward: Oblique-slip (transform) continental margin. 2. To the N: Divergent (passive) continental margin. Related to late stages of breakup of the Chortis (Yaquí) Block (i.e., the separation of North and South America). A stage of rapid subsidence and submergence is recorded by the basal Cretaceous (Berriasian) sedimentary infilling, confined to the NW-trending "Cundinamarca Trough," a rift-related, aulacogen-like depression in central Colombia. An orthogonal, normal fault system which controlled the shape of the trough is revealed by marine conglomerates and coarser clastic sediments, as well as by mafic igneous activity restricted to the trough, and connected to ancient zones of crustal weakness

Figs. 1.41 and 1.42 (continued) within the basement. Inset 2 – Two distinctive structural regimes are indicated by sedimentary patterns: 1. Dominantly downward, vertical displacements in the N half of Colombia. 2. Upward, high-angle displacements behind the continental transform margin. From ca. 5° N southward, and along the modern-day eastern side of the Cauca valley, a subtle change in relative motion occurred, from an oblique-slip transform fault to a left lateral strike-slip fault. To the N, as a consequence of persistent extension, oceanic crust continued forming on the outboard side of mainland Colombia. In the northern continental block, the arrangement of discrete, adjacent, sedimentary facies produces a mosaic pattern that reflects the underlying structural framework, which has persisted, unchanged from Early Mesozoic time. To the south, a strip of subsided continental basement was flanked by block faults. Inset 3 – Dynamic metamorphism affecting marginal composite crust in SW Colombia. The largest structural features in the continental crust are welts which trend parallel to the passive and transform margins. K-Ar and Ar-Ar uplift ages from exhumed high-pressure metamorphic rocks along the western margin of the southern half of the present-day Central Cordillera provide evidence that a switch occurred, from an older SE-directed subduction regime, to a NE-oriented strike-slip fault regime, in the process imparting a mega shear character to the so-called Romeral fault system. A contrary regime, containing the passive continental margin, dominated the western margin of the (present-day) Central Cordillera to the north. The configuration of preexisting intraplate structures continues to control the location of two main depositional domains: 1. A narrow and elongated basin, trending to the southwest (the present-day Upper Magdalena Valley). 2. To the north, a complex, diamond-shaped basin that stretches from the western inboard passive margin to the eastern Guiana Shield. (B) Transpressional regime.Inset 4 – Thickness (isopach) and sedimentary facies distribution provide evidence of penecontemporaneous fault blocks and basement-rooted structural dynamics. Changes in relative movement between the Pacific and South American plates generated northeast, right-lateral strike-slip translation along parallel fractures within ensimatic crust, in deep water along western Colombia. In northern Colombia, a contemporaneous sandy to black-shale facies transition, westward from the Llanos platform, suggests internal plate deformation accompanied by rapid eastward subsidence and basin infilling. To the south, facies distribution is related to an embayed eastern coastline, controlled by rising basement blocks. Inset 5 Thickness (isopach) and sedimentary facies distribution provide evidence of penecontemporaneous fault blocks and basement-rooted structural dynamics. Along the continental-oceanic plate boundary, accommodation of continual NE-striking, right-lateral strike-slip motion produced ensimatic borderland basins and ensialic marginal doming. As indicated by facies variation across the continental plate, internal plate deformation seems to have undergone some reactivation, probably due to contemporaneous strain along lithospheric plate margins. Inset 6 – Strike-slip faulting dominates the structural framework. Continued strike-slip movement along the margins of the southern half of the present Cauca Valley created a furrow of "composite crust" that differs from the simatic western and sialic central domains of Colombia, in having wedges of sea floor rocks and deformed deep marine sedimentary cover, with a much greater degree of felsic basement involvement. Further to the north, in the Antioquia region, and to the west in the Pacific region, only wedges of purely mafic oceanic and bathyal sedimentary rocks are present. Due to tangentially transmitted stress from the Pacific Plate, the continental margin was flexed into a welt stretching from ca. 3° N up to 11° N latitude. To the east, the continental basement tended to subside. Inset 7 – Strike-slip faulting dominates the structural framework. A northeast-striking fault system, possibly with major right-lateral displacement, developed along peripheral western Colombia, causing diverse oceanic sequences to become partly accreted to the continental margin. Strike-slip faulting dominates the structural style, affecting both subsidence and sedimentation. Inset 8 – A step fault margin developed along the continental-oceanic plate boundary, deepening into the western bathyal environment. Shortening of the continental plate due to continued northeast movement of the western oceanic plate, resulted in slicing of the sliding plates along major strike-slip features, such as the dextral paleo-Palestina fault. The observed overall tilting of the sedimentary domain, dipping from southwest to northeast, is the result of progressive SW-NE plate movement. Inset 9 – Conjugated fault zones delineate a braided pattern of uplifted and depressed blocks, tilted toward the east on the continental plate. Convergent collision between a slab of the Farallon Plate and the Colombian continental plate from ca. 5° N northward led to off-scraping of seafloor deposits and subduction of oceanic crust. The development of horizontal compressional stresses resulted in depression of the NE deep marine "basin" and elevation of its continental margin. Inset 10 – Factual record of a wrench-thrusting mechanism. Sliding of opposite margins (western allochton and continental margin). With the fragmentation of the Farallon Plate during the Eocene, rapid northwestward transport of the remaining Farallon Plate took place. Subduction of the proto-Nazca (trailing-edge Farallon) Plate is accompanied by the development of the Mandé volcanic arc.

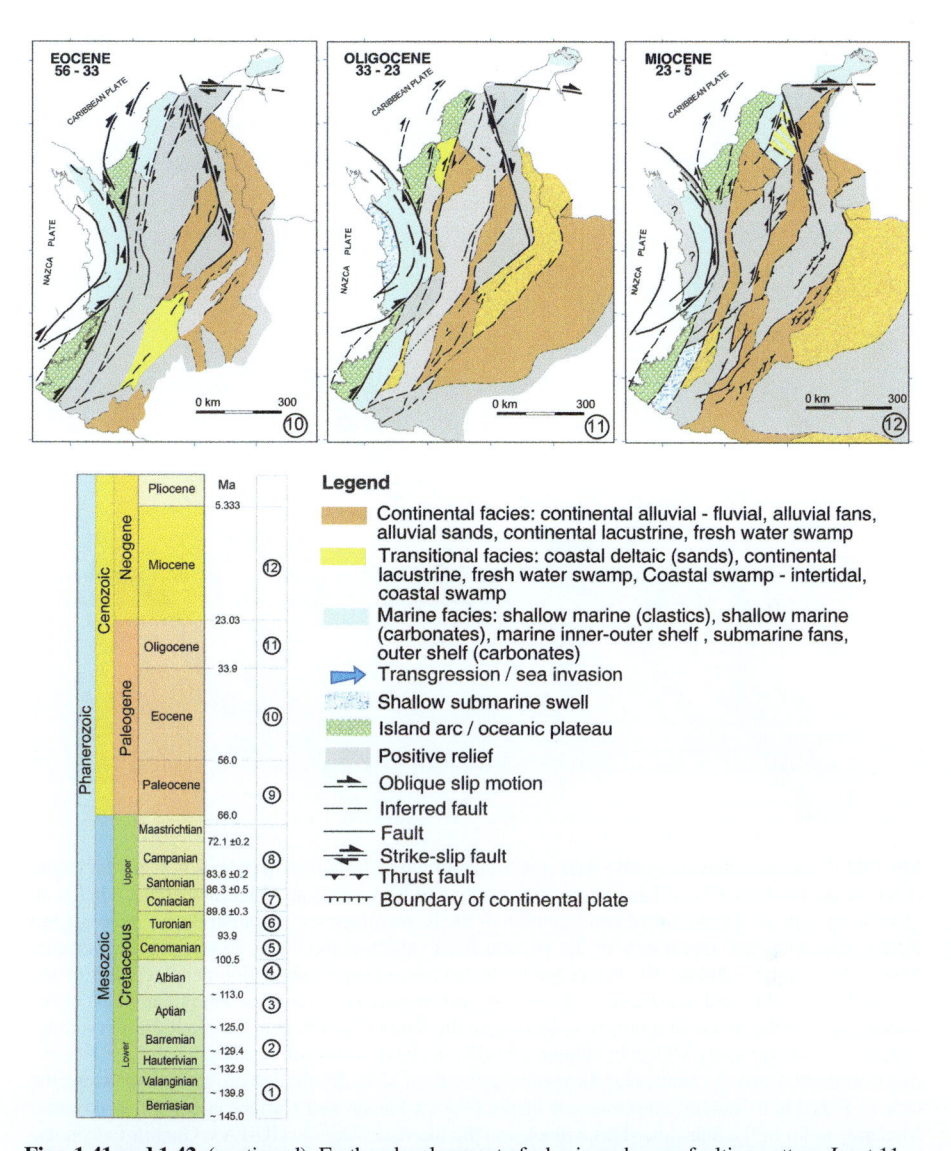

Figs. 1.41 and 1.42 (continued) Further development of a basin and range faulting pattern. Inset 11 – Tectonic activity, particularly thrusting, is less accentuated than in the preceding Eocene time. Along the western margin of Colombia, a two-fold differentiation is observed: 1. From Urabá to Buenaventura, the Atrató-San Juan Basin was being filled with slump-type deposits which conceal the stacking of slices of plutonic rocks, submarine basalts, and hemipelagic sediments. 2. Farther south, along the Patía Basin, terrigenous turbidites covered oceanic basement. Intraplate tectonics: the preceding transpressional tectonic regime appears to undergo a quiescent phase. Notwithstanding, the deformational pattern persists. Inset 12 – Onset of a transpressional regime that embraces and shapes the embryonic Andean belt. Movement along strike-slip faults with large thrust or reverse-slip components are understood as convergent wrenching. The suture zone depicted on the Pacific northwestern edge of Colombia marks the end of the collision of the oceanic-continental plates in the Late Miocene. A new subduction zone develops within the Nazca plate. The Caribbean-South America plate boundary acts as a sinistral oblique-slip transform margin. During the Miocene, the Guajira allochton migrated from west to east, to its present position, along the strike-slip Oca fault. Tectonism north of the Ibagué fault is manifest by two geomorphologic features: (1) the ancestral Cauca and Magdalena valleys that are bordered either by large thrusts or strike-slip faults and (2) south of the Ibagué fault, a volcanic arc (central range) separates a western marginal marine domain from an eastern continental domain

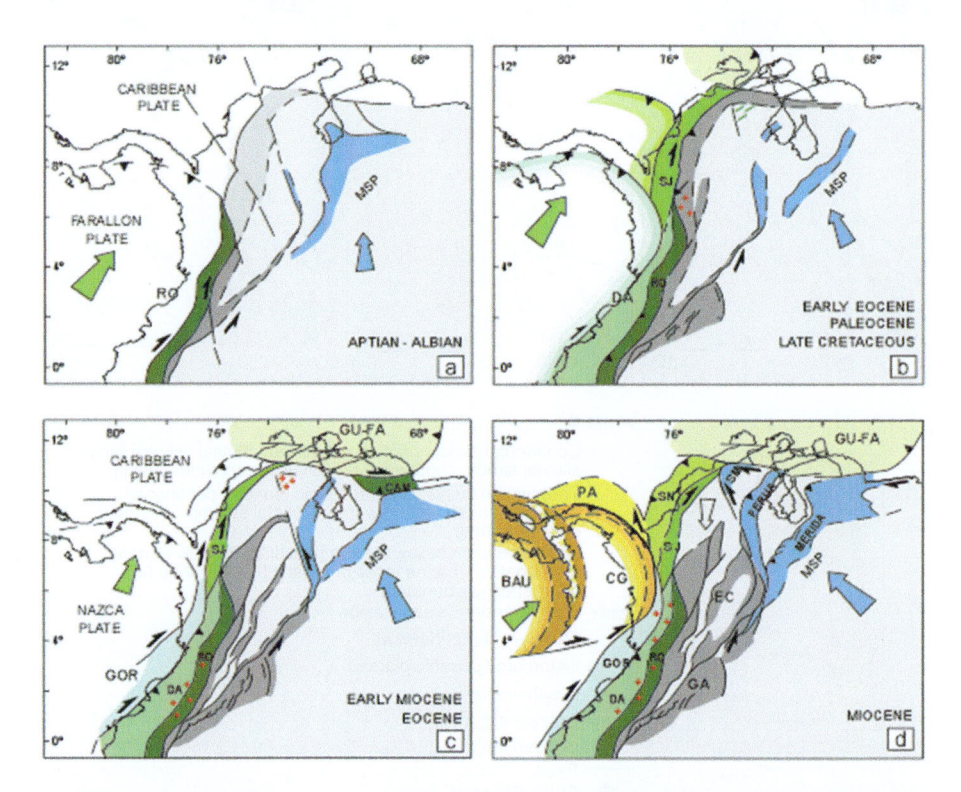

Fig. 1.43 Contemporaneous tectono-stratigraphic evolution of the northern Andean orogen, Maracaibo orogenic float and the Chocó-Panamá Arc from Aptian to Miocene time. (Modified after Cediel et al. (2003)). (**a**) Aptian-Albian; initial configuration of the Romeral terrane (Farallon Plate), mélange and fault system and first appearance of the Mérida Arch (blue) in the MOF. RO = Romeral terrane; MSP = Maracaibo subplate. (**b**) Paleocene-Early Eocene; oblique subduction and accretion of the Dagua-Piñón (DAP) and San Jacinto (SJ) terranes and metamorphic deformation (green lines) of the leading edge of the Maracaibo orogenic float along the Santa Marta thrust front. Red crosses = magmatism. (**c**) Eocene-Early Miocene; oblique subduction and accretion of the Gorgona terrane. Eocene magmatism (red crosses) punctuates the metamorphic front of the Maracaibo orogenic float. Along the Oca-El Pilar Fault System, emplacement of the Guajira-Falcón and Caribbean Mountain terranes. Moderate uplift of the Santander-Perijá block and the Sierra de Mérida. GU-FA = Guajira-Falcón terrane; CAM = Caribbean Mountain terrane; Other abbreviations as for 11a and 11b. (**d**) Miocene oblique collision of the Sinú block and tangential accretion of the Cañas Gordas and later Baudó terranes. Subduction of the Nazca Plate south of the Panamá- Chocó Arc (CG-BAU). Further uplift of the Sierra de Mérida, Serranía de Perijá, and Sierra Nevada de Santa Marta (SM). Late Miocene-Pliocene pop-up of the Eastern Cordillera (EC). Dextral-oblique thrusting in the Garzón Massif (GA). Continued northwest migration of the Maracaibo orogenic float. Near complete modern configuration. BAU = Baudó terrane; CG = Cañas Gordas terrane; PA = Panamá terrane; SN = Sinú block; other abbreviations as for 11a–c. Grey shaded areas in all time slices represent paleo-topographic swells, elevated and/or emergent areas. Red crosses represent magmatism

1.6.1.8 Southern Caribbean. Western Caribbean and Eastern Caribbean (Figs. 1.44, 1.45, 1.46, 1.47, 1.48, 1.49, 1.50, 1.51, 1.52, and 1.53)

General Statement

The western Caribbean, extending from the Chocó segment of the Panamá double arc to the leading apex of the Maracaibo Block, is the unavoidable cornerstone in any attempt to explain the geologic evolution (i.e., *allochthonous* vs. in situ *origin*) of the Caribbean Plate. Hence, it is not surprising that the Colombian Caribbean Realm (CCR) has become a literary battlefield and the subject of contradictory interpretations, some of which ignore factual surface geological, subsurface borehole, and detailed geophysical data.

A comprehensive analysis of all available data including stratigraphic sections controlled by field mapping and borehole logging, stratigraphy (including igneous and metamorphic rocks), surface structural mapping combined with detailed subsurface structural maps derived from diagnostic seismic lines, paleo-geographic reconstructions, gravimetric and paleomagnetic data, and local field records leads us to question:

- Geotectonic models "adjusted" exclusively to the *subduction* paradigm. The "big picture" of the Caribbean Colombian Realm (including the oceanic and continental components) illustrates a distinct geotectonic history, from that recorded along the southern Caribbean margin north of the Oca-El Pilar Fault System (Colombia-Venezuela).
- An autochthonous origin for the Guajira and Falcón-Paraguaná peninsulas and Margarita Island composites fragments of continental and oceanic crust, including Proterozoic to Cenozoic, igneous, metamorphic, and sedimentary rocks, that many authors correlate with age equivalent litho-stratigraphic units in continental northern South America.

Discussions regarding the origin of the Caribbean Plate are considered beyond the scope of this limited presentation. Notwithstanding, a summary of important observations which may pertain to the topic, as supported by the documented geology observed along the continental margin of northwestern South America, includes the following points:

- The presence of (at least seven) rift-related grabens, close to the continental edge of northwestern South America, including aulacogen arms cut by the Oca-Pilar transform fault (e.g., Fig. 1.26), attests to the Jurassic age of the extensional regime that gave birth to the proto-Caribbean Plate (Fig. 1.30).

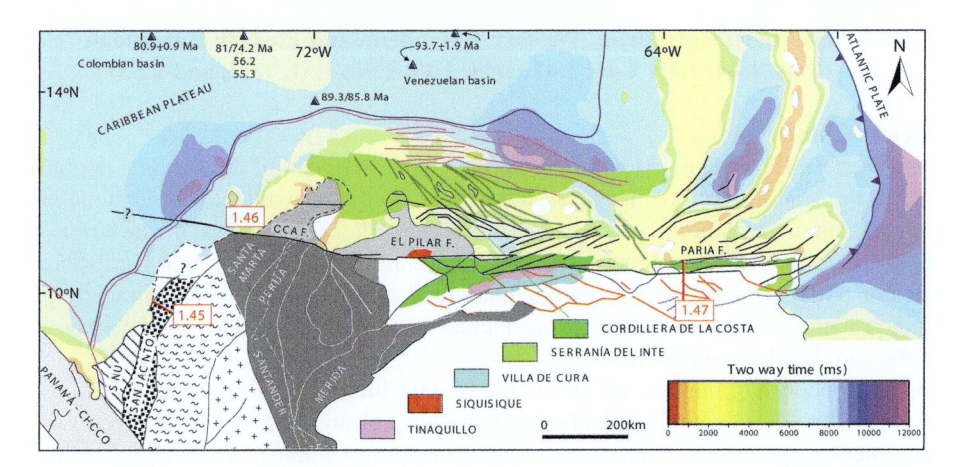

Fig. 1.44 Diagnostic features of the regional tectonic contact between the southern Caribbean Plate and the SW South American Plate. (Compiled after Escalona and Mann (2011))

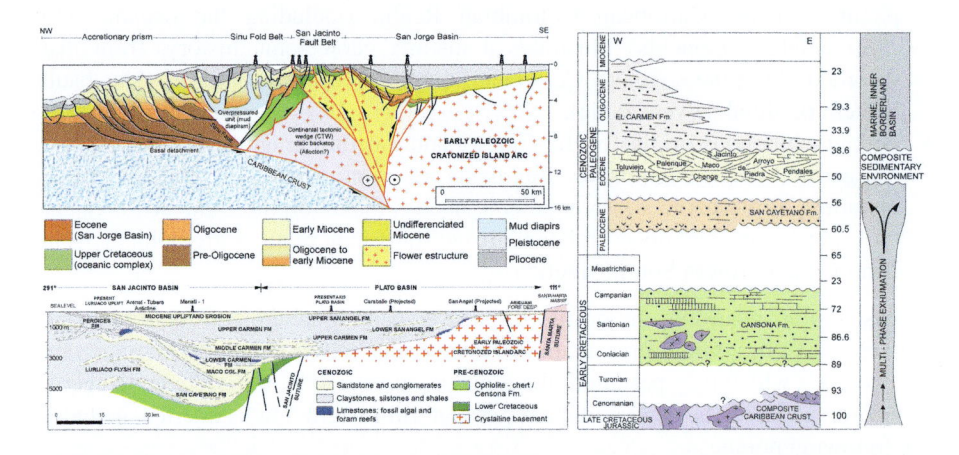

Fig. 1.45 Key tectonic features of the Western Caribbean Orogen and its concomitant stratigraphic development. The San Jacinto Fault Belt is a right-lateral, strike-slip orogen, and as such, non-magmatic. The compressional component (transpressive faulting) lead to development of a positive flower structure. The Sinú deep water fold-thrust belt is a thin-skinned, amagmatic and aseismic, compressional belt, and is not associated with subduction. This type of compressional belt does not represent any mountain-building event (orogen); its seismic architecture and deformational pattern reveals gravity-driven dynamics. (Compiled after Mantilla (2007) and Moreno et al. (2009))

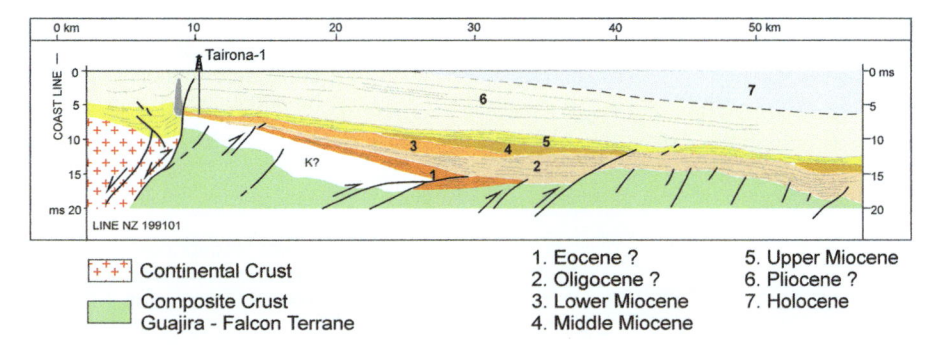

1 Phanerozoic Orogens of Northwestern South America: Cordilleran-Type Orogens... 55

Continental Crust
Composite Crust Guajira - Falcon Terrane

1. Eocene ?
2. Oligocene ?
3. Lower Miocene
4. Middle Miocene
5. Upper Miocene
6. Pliocene ?
7. Holocene

Fig. 1.46 Seismic structure along the tectonic contact between the Maracaibo Block (Sierra Nevada de Santa Marta) and the southern Caribbean Plate

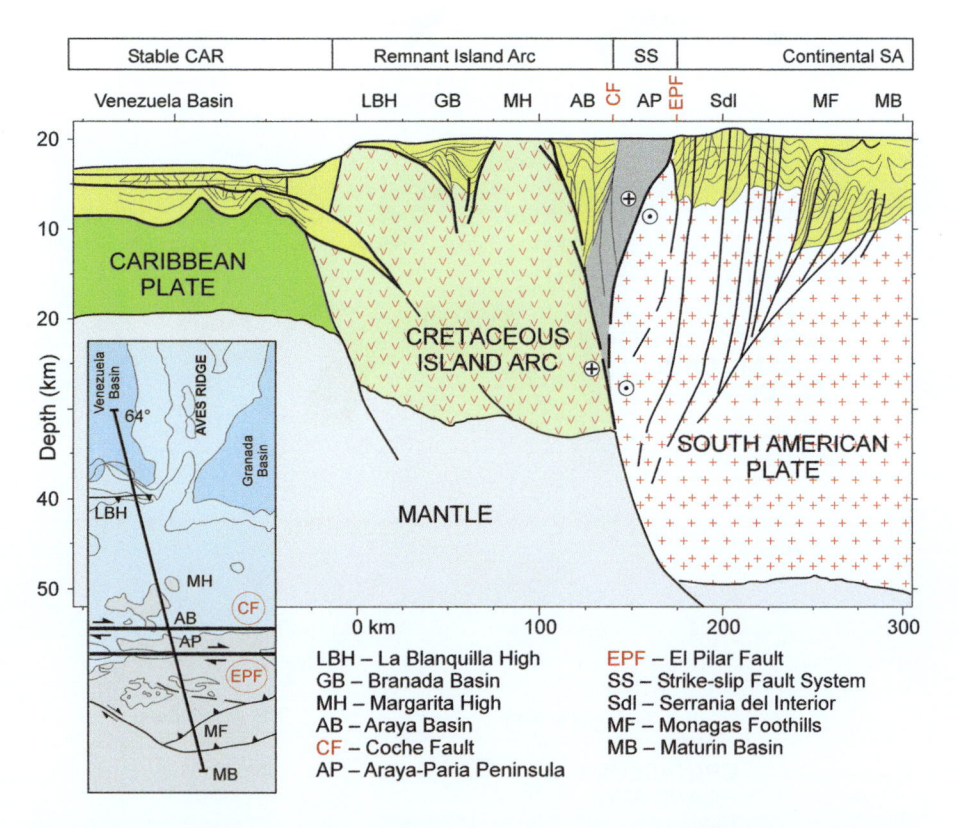

LBH – La Blanquilla High
GB – Branada Basin
MH – Margarita High
AB – Araya Basin
CF – Coche Fault
AP – Araya-Paria Peninsula
EPF – El Pilar Fault
SS – Strike-slip Fault System
Sdl – Serrania del Interior
MF – Monagas Foothills
MB – Maturin Basin

Fig. 1.47 Structural interpretation of wide-angle seismic velocity data along Line 64°W, a 460-km long, approximately north-south, onshore-offshore reflection/ refraction transect. The profile extends across the transform plate boundary between the southeastern Caribbean (CAR) and South American (SA) plates. (Compiled after Clark (2007) and Clark et al. (2008))

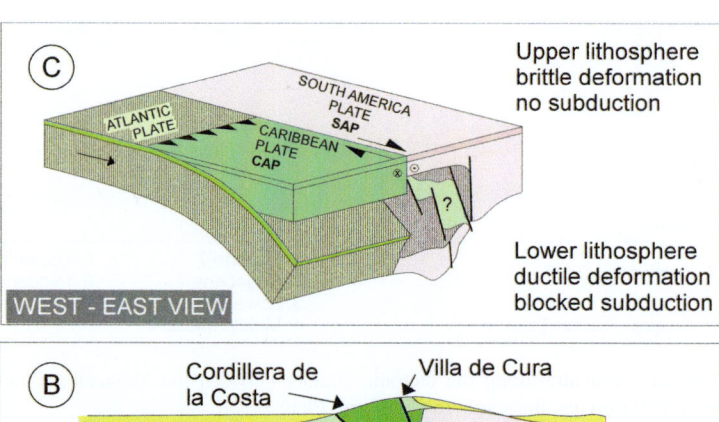

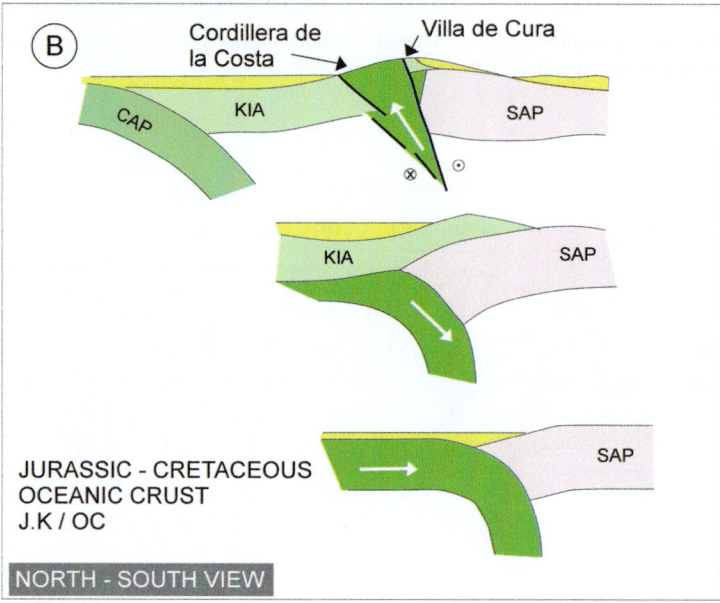

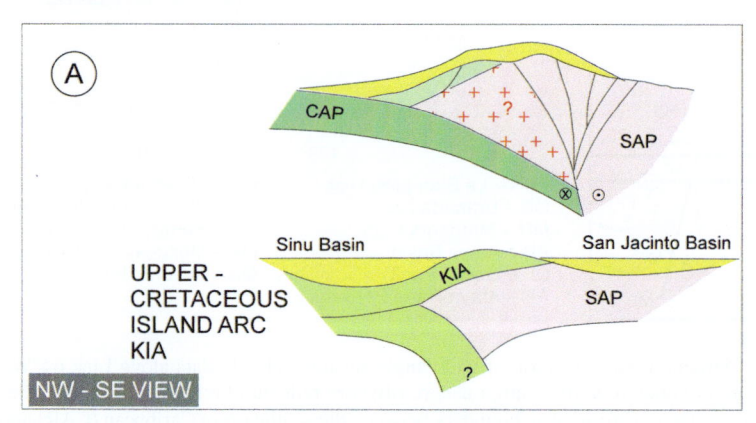

Fig. 1.48 Schematic illustration of the blocked subduction process along the southern Caribbean – South America plate margin: (**a**) Colombian-Western Caribbean. (**b**) Eastern Caribbean, (**c**) Atlantic-Caribbean subduction and Caribbean-South America wrench structure. (Modified after Clark et al. (2008))

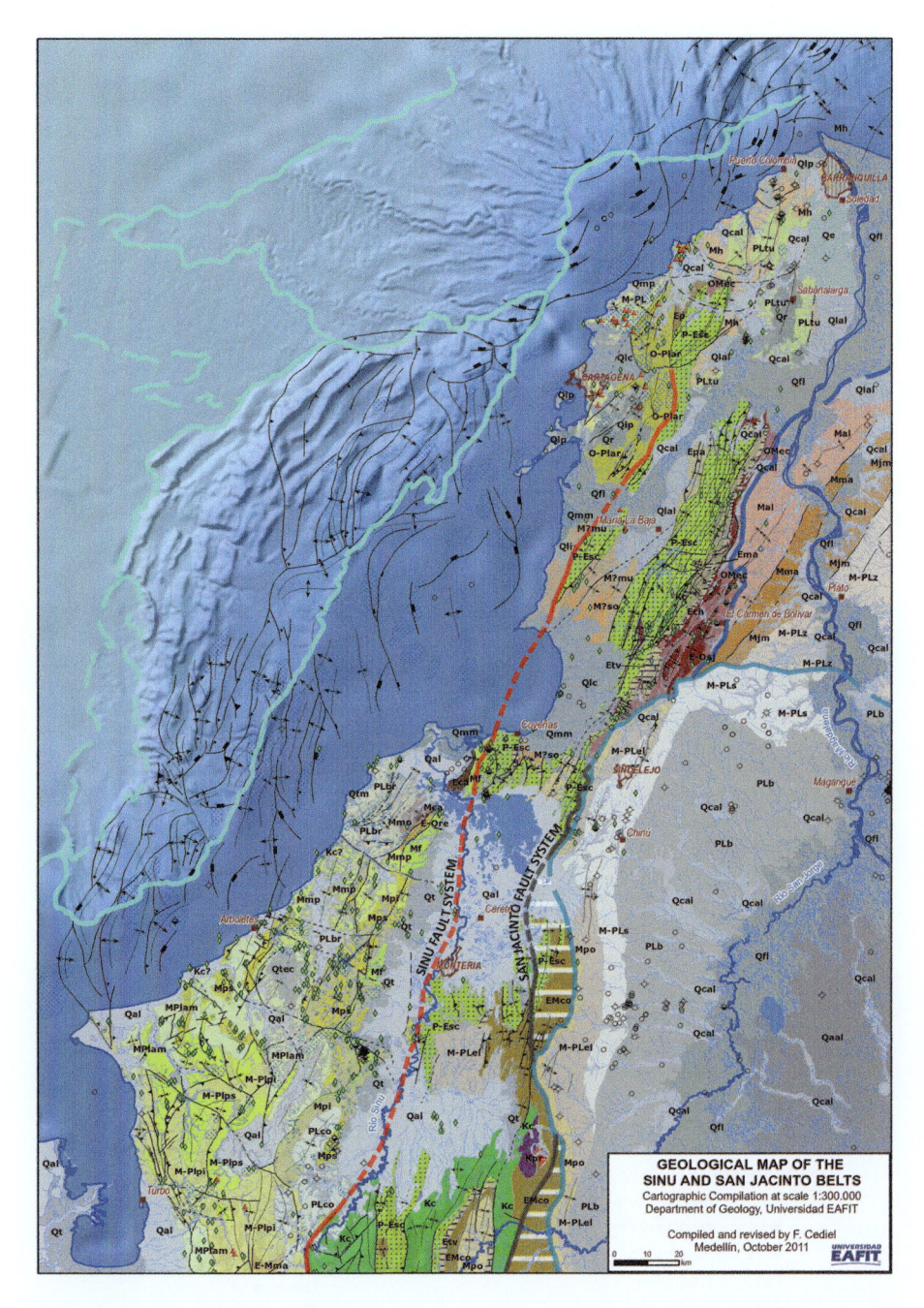

Fig. 1.49 Geological map of the western Caribbean Orogen (Sinú and San Jacinto belts). (Cediel 2010)

SAN JACINTO FAULT BELT

	K c	Cansona Fm.
	K pr	Planeta Rica periodites
	K np	Nuevo Paraíso basalts
	P-E sc	San Cayetano Fm.
	E ma	Maco Fm.
	E pa	Palenque Fm.
	E p	Pendales Fm.
	E mo	El Morro Fm.
	E tv	Toluviejo Fm.
	E ap	Arroyo de Piedra Fm.
	E ch	Chengue Fm.
	E-O sj	San Jacinto Fm.
	O-M ec	El Carmen Fm.
	Q t	Tenerife Fm.
	Q lp	La Popa Fm.
	Q r	Rotinet Fm.
	Q ag	Arroyo Grande Fm.

SINÚ FOLD BELT

	K c?	Cansona Fm (?)
	E ca	Candelaria Fm.
	E lr	La Risa Fm.
	E-O re	Resbalosa Fm.
	E-M ma	Maralú Fm.
	O-Pl ar	Arjona Fm.
	M-Pl pi	Lower Pavo Fm.
	M-Pl ps	Upper Pavo Fm.
	M ca	Campano Fm.
	M f	Floresanto Fm.
	M pi	Lower Pajuil Fm.
	M ps	Upper Pajuil Fm.
	M mo	Moñitos Fm.
	M mp	Morrocoy Fm.
	M-Pl am	Arenas Monas Fm.
	Pl b	Broqueles Fm.
	Pl coi	Lower Corpa Fm.
	Pl cos	Upper Corpa Fm.
	PLE lp	La Popa Fm.
	Q tec	El Carmelo Terrace

LURUACO BLOCK

	P-E sc	San Cayetano Fm.
	M h	Hibácharo Fm.
	M-PL	Bayunca Fm.
	PL tu	Tubará Fm.
	PLE lp	La Popa Fm.

QUATERNARY DEPOSITS

	Q aal	Alluvial fan
	Q al	Alluvium
	Q cal	Colluvial allivium
	Q e	Eolic deposits
	Q fl	Fluvial lacustrine deposits
	Q lal	Alluvial plain
	Q lc	Coastal plain
	Q li	Flood plain
	Q mm	Mangrove substrata
	Q mp	Beach deposits
	Q t	Terraces
	Q tm	Marine fluvial lacustrine deposits

CIÉNAGA DE ORO FAN DELTA

	Em co	Ciénaga de Oro Fm.

RANCHO FAN DELTA

	M al	Alférez Fm.
	M alb	Barcelona Member
	M ma	Mandatú Fm.
	M jm	Jesús del Monte Fm.
	M-Pl z	Zambrano Fm.

PORQUERA FAN DELTA

	M po	Porquera Fm.
	M-Pl el	El Cerrito Fm.
	M-Pl s	Sincelejo Fm.
	Pl b	Betulia Fm.
	M ?so	San Onofre Fm.
	M ?mu	Mucacal Fm.

K	Cretaceous
P	Paleocene
E	Eocene
E-O	Eocene-Oligocene
O	Oligocene
O-M	Oligocene-Miocene
M	Miocene
M-Pl	Miocene-Pliocene
Pl	Pliocene
PL	Pleistocene
Q	Quaternary

◆	Gas seeps
▲	Oil seeps
○	Exploratory well

Fig. 1.50 Map legend, Fig. 1.49

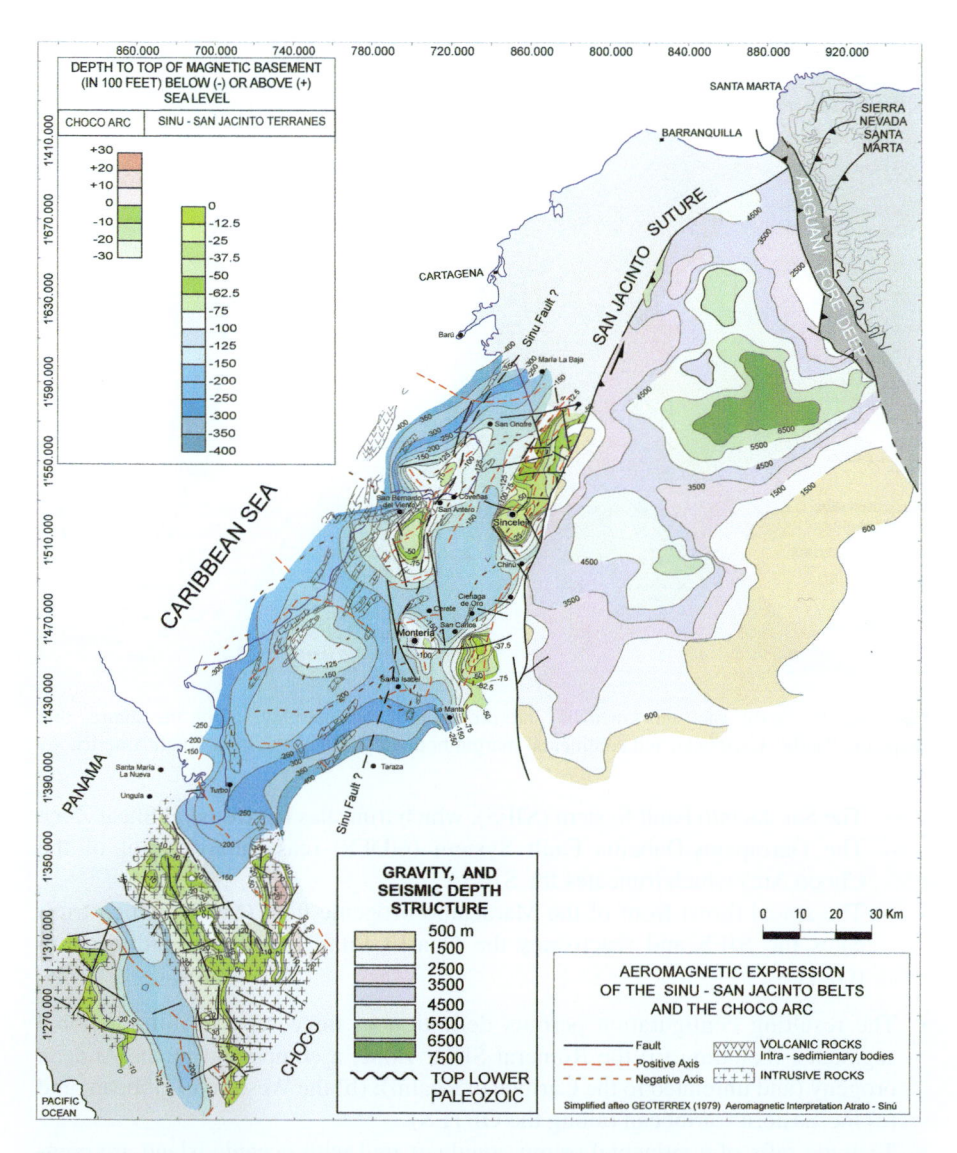

Fig. 1.51 Aeromagnetic expression of the San Jacinto-Sinú belts and the Chocó Arc, together with the gravity and seismic depth structure of the Lower Magdalena basin. (Simplified after GEOTERREX (1979))

- The documented oceanic rock units accreted to continental South America are either Cretaceous oceanic plateaus or Meso-Cenozoic oceanic island arcs, which outcrop, at least in part, along the present-day Pacific and Caribbean borderlands.
- Three successive tectonic events, as recorded by relative movement along major fault systems, intervene, modify, or contribute to shape the litho-tectonic and morpho-structural characteristics of this broad ocean-continent contact zone. The fault systems include (see Fig. 1.52):

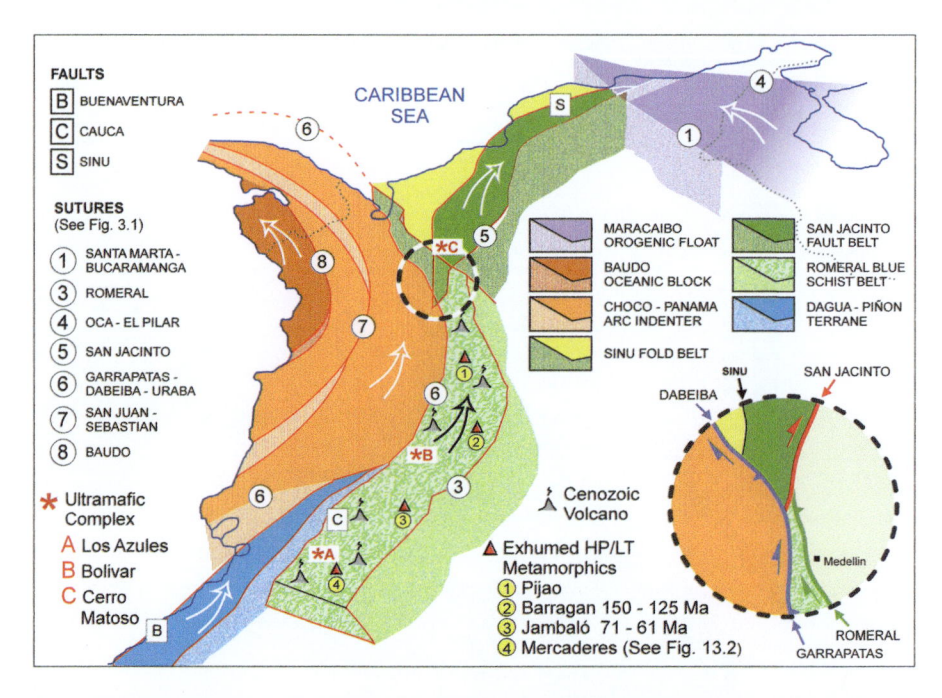

Fig. 1.52 Block diagram and kinematic interpretation of the interplay among the sutures that shape the Pacific, Caribbean and continental margin boundary of northwestern South America

- The San Jacinto Fault System (SJFS), which truncates the Romeral Shear Zone
- The Garrapatas-Dabeiba Fault System (GDFS) (easternmost limit of the Chocó Arc) which truncates the SJFS
- The apical thrust front of the Maracaibo orogenic float (MOF), which truncates the SJFS and reactivates the Oca-El Pilar Transform Fault System (OPTFS).

- The resulting configuration permits definition of three separate litho-tectonic segments, including (a) the Romeral Shear Zone, a component of the Andean orogeny (and unrelated to the Caribbean Realm); (b) the Western Caribbean; and (c) the Eastern Caribbean (along the OPTFS).
- Tectonic rafts of continental origin, amalgamated with oceanic island arc complexes, are not always fully detected by geophysical methods.
- Mafic-ultramafic assemblages such as the Los Azules and Bolivar complexes and accreted island arcs (e.g., Sabanalarga, Buga) are of Pacific provenance, as is the Cerro Matoso ultramafic complex which forms part of the Cansona island arc, hosted within San Jacinto terrane basement.
- The nature and age of Meso-Cenozoic marine-continental basins documented within the San Jacinto and Sinú terranes or the Western Caribbean segment, and within and along the Maracaibo orogenic float to the east, do not provide evidence of subduction-related deposition or deformation.

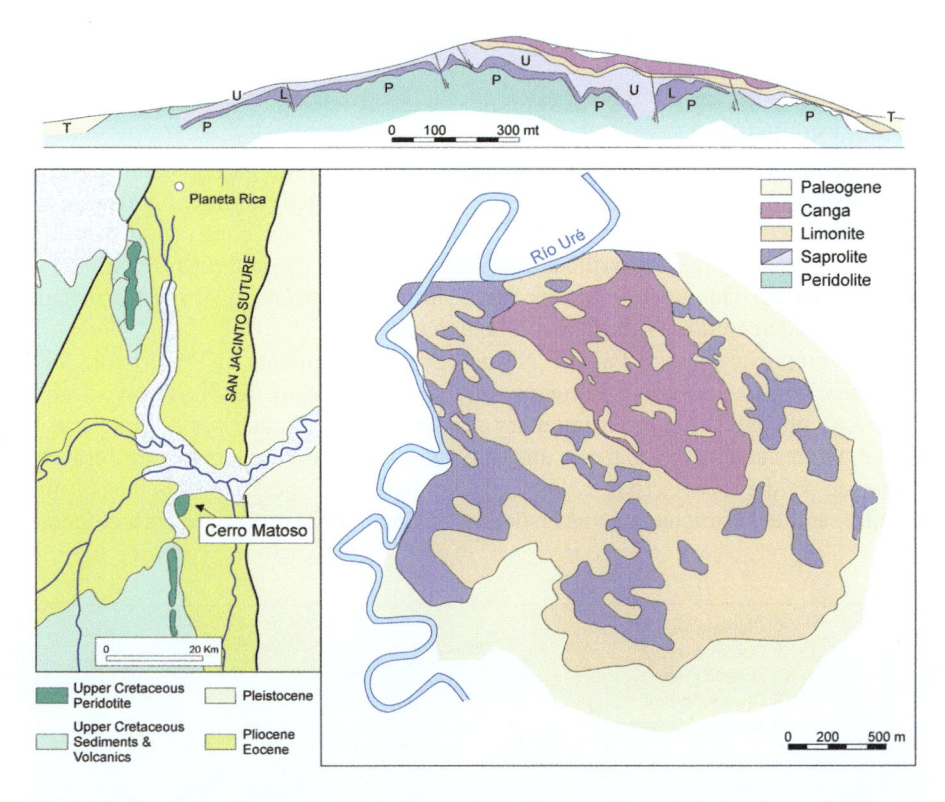

Fig. 1.53 The Upper Cretaceous peridotites at Cerro Matoso. (Compiled after Lopez (1986) and Gleeson et al. (2004))

- The nature of the crust required to feed a potential subduction zone and produce a continental volcanic arc in the West and East Caribbean segments is unknown, but no volcanic arc, active, eroded, or otherwise, has been shown to exist. The occurrence of minor, localized Miocene basalt along the Western Caribbean margin does not constitute a continental volcanic arc. Thickened, buoyant Farallon-Caribbean plateau oceanic crust of Pacific provenance, which forms basement to the Western and Eastern Caribbean segments, is not conducive to wholesale subduction.
- Recent paleo-geographic reconstructions (e.g., Nerlich et al. 2014) suggest that Farallon-Caribbean plateau crust was docked (i.e., static) with respect to the Western and Eastern Caribbean continental margin during the Early Eocene (ca. 54.5 Ma). Right-lateral strike-slip along the southern (both Western and Eastern) Caribbean boundary is documented during Oligocene-Neogene time. This movement vector is again not conducive to the wholesale subduction of Farallon-Caribbean plateau crust.

1.6.1.9 Guajira-Falcón (GU-FA), Composite Terranes (Figs. 1.54, 1.55, 1.56, and 1.57)

General Statement

The Guajira Peninsula, the Falcón-Paraguaná Block, and Margarita Island terranes are considered to represent the amalgamation of disrupted, northeast and west to east translated fragments and rafts of mixed oceanic and continental affinity, presently stranded in the Caribbean and docked along the margin of continental South America. This operation took place during emplacement of the Caribbean plate, prior to accretion of the Chocó-Panamá indenter.

The composite Guajira-Falcón terrane contains fragments of Proterozoic and Paleozoic continental crust (remnants of the separation of the North and South American plates), Jurassic sedimentary sequences, and Cretaceous oceanic crust. Based upon facies associations and the contained fossil record, the Jurassic sequences of the GU-FA (particularly, of Kimmeridgian age, in the Cocinas basin, Guajira, and the Paraguaná Jurassic deposits; Geyer 1973) correlate with contem-

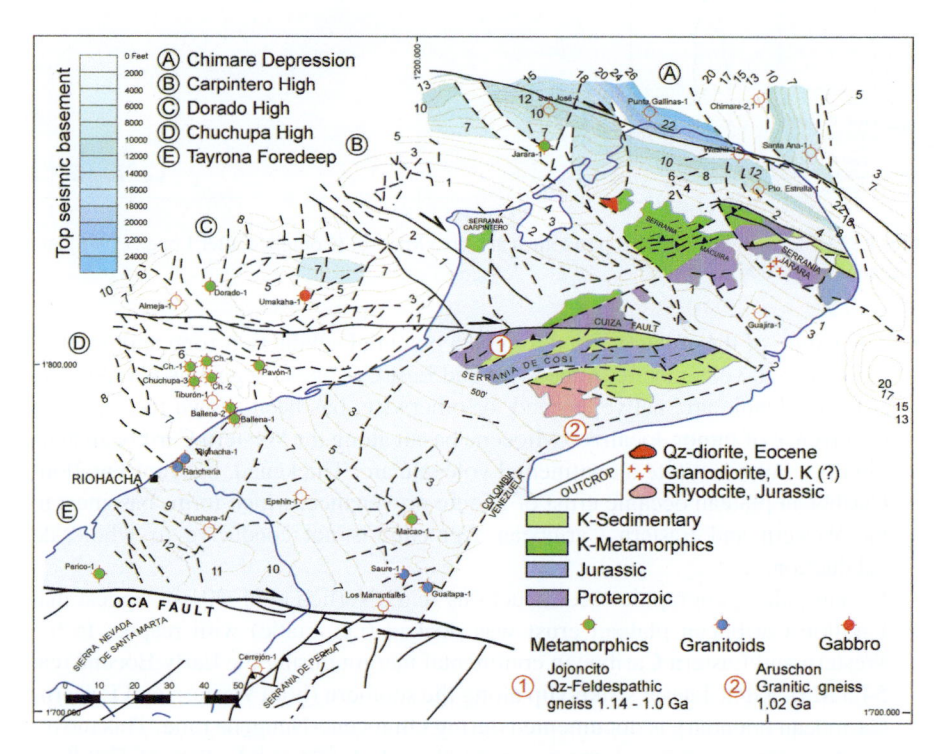

Fig. 1.54 Seismic depth structure and pre-Cenozoic outcrops of the Guajira peninsula. (Compiled after Londoño et al. (2015), Zuluaga et al. (2015), Geotec (1986), and Baquero et al. (2015))

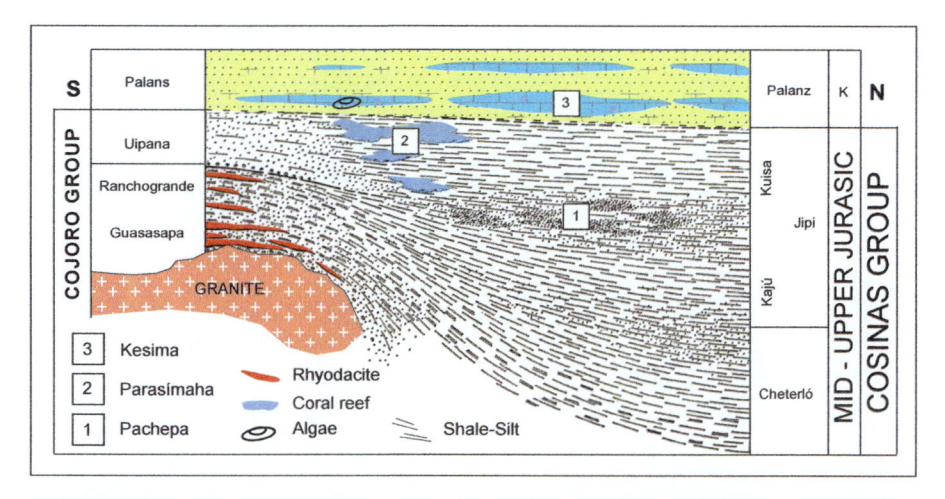

Fig. 1.55 Middle and Upper Jurassic stratigraphic relations between igneous rocks and sedimentary deposits in the Upper Guajira peninsula. (Modified after Geyer (1973))

poraneous deposits presently exposed in the Yucatan Peninsula and NE Mexico (e.g., Gonzales and Holguin 1991; Villaseñor et al. 2012). Kimmeridgian-aged deposits of marine affinity were never deposited in continental South America. Paleomagnetic studies presented by MacDonald and Opdyke (1972) indicate volcanic outcrops of the GU-FA from their Guajira, and Greater Antilles sites occupied latitudes about 10S south of their present positions and possibly off northwestern South America in the Cretaceous. Detailed petrographic studies of the Margarita Complex portion of the GU-FA terrane by Stoeckhert et al. (1995), in addition to studies by Maresch et al. (2009), demonstrate an accretionary-metamorphic history and migratory path beginning in the Albian for this heterogeneous association of rocks.

The present position of the GU-FA is an important testimony to the post-Jurassic (post-Albian?) emplacement of the Caribbean plate, an emplacement history in which the San Jacinto and Oca-El Pilar fault systems have played a critical role.

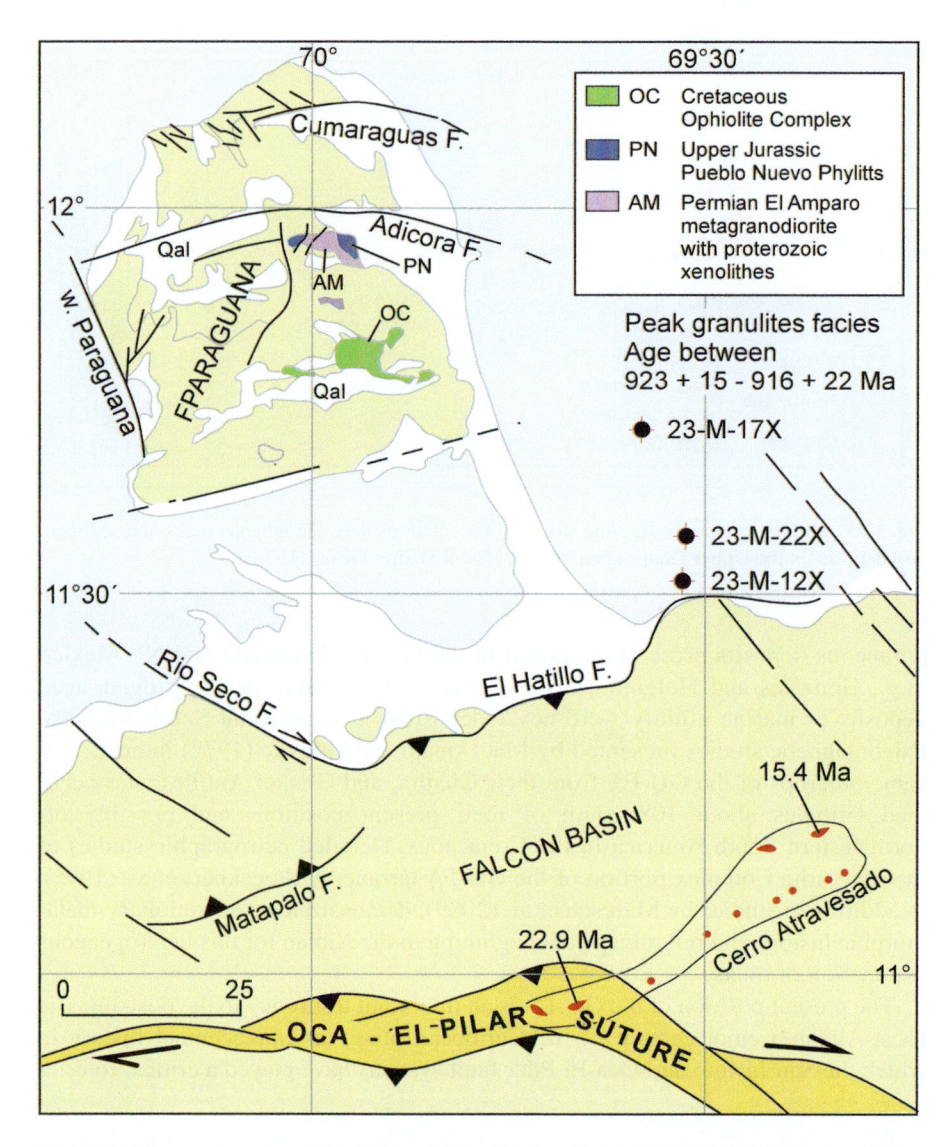

Fig. 1.56 Structural setting and geochronology of rock units in the Paraguaná peninsula and the Falcón Basin (Falconia terrane; see Fig. 1.57). (Compiled after Mendi et al. (2013), Benkovics and Asensio (2015) and Baquero et al. (2015))

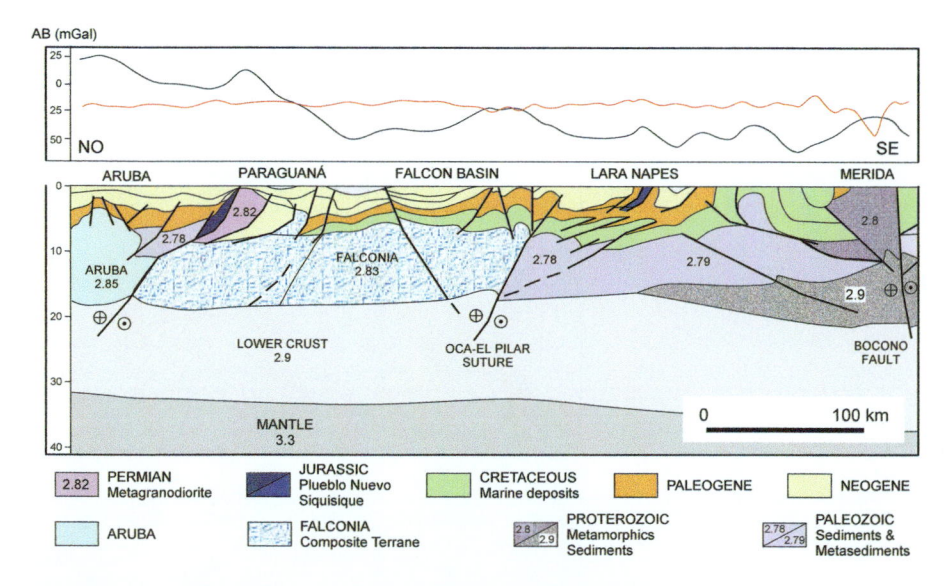

Fig. 1.57 Structural setting and litho-stratigraphic interpretation of the Falconia terrane as depicted along a N-S gravimetric profile. (Modified after Linares et al. (2014))

1.6.1.10 Chocó-Panamá Indenter. Composite Arc System (Figs. 1.58, 1.59, 1.60, 1.61, 1.62, 1.63, and 1.64)

General Statement

The formation and development of the Chocó Arc took place within a sequence of events, which are schematically outlined in Fig. 1.60. The geological characterization of these events remains incomplete due to a scarcity of geological and geophysical data, especially in Colombia. Regardless, evaluation of the available information within a regional context permits recognition of the principle litho-tectonic elements and tectono-stratigraphic events.

The Farallon Plate, containing the Caribbean-Colombian oceanic plateau (CCOP), is considered a composite, diachronous litho-tectonic unit. Farallon forms basement to the Caribbean Plate and varies from ca. 144 Ma (latest Jurassic?) in the east, younging westwards to ca. 75 Ma, in the westernmost Caribbean (e.g., Nerlich et al. 2014). Radiometric age dates for accreted CCOP rocks in northern South America suggest that plateau-related mafic-ultramafic magmatism (superimposed upon the Farallon Plate) was extruded in three stages including (1) a limited phase at ca. 100 Ma, (2) widespread eruptions from ca. 92 and 87 Ma (Kerr et al. 1997, 2003; Sinton et al. 1998; Hastie and Kerr 2010; Nerlich et al. 2014), and (3) lesser magmatism from ca. 77 and 72 Ma (Kerr et al. 1997; Sinton et al. 1998).

Fig. 1.58 Geological map of the Chocó Arc

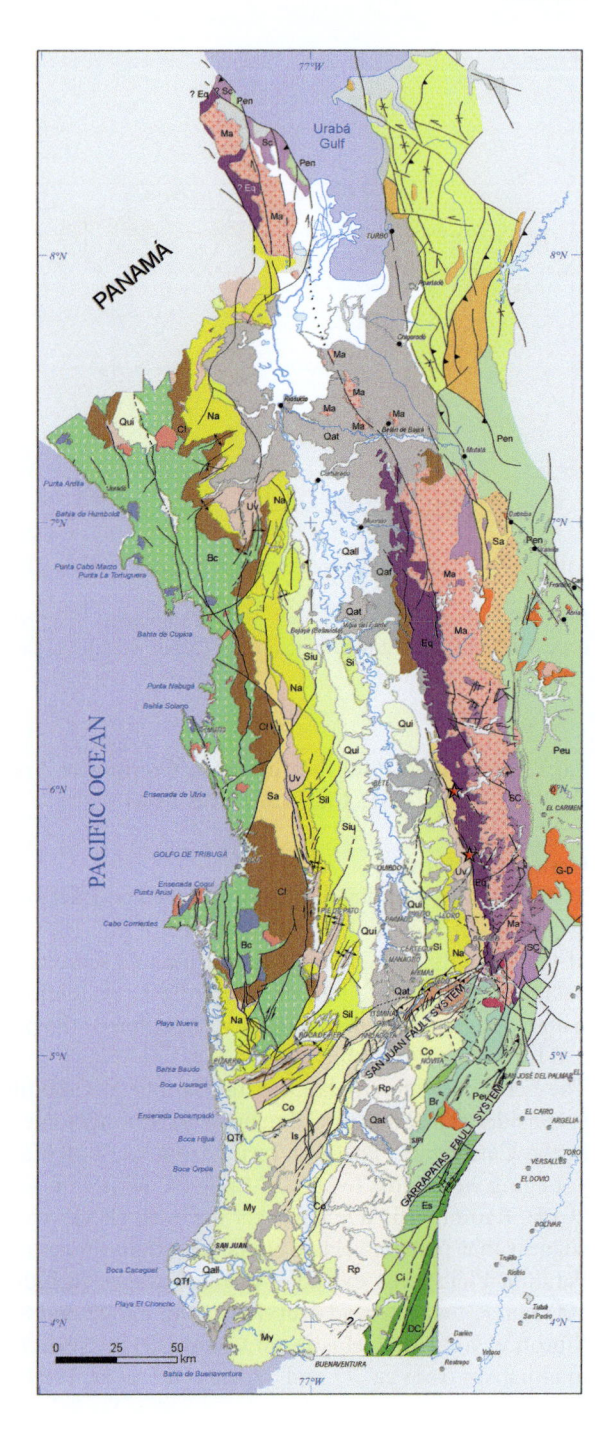

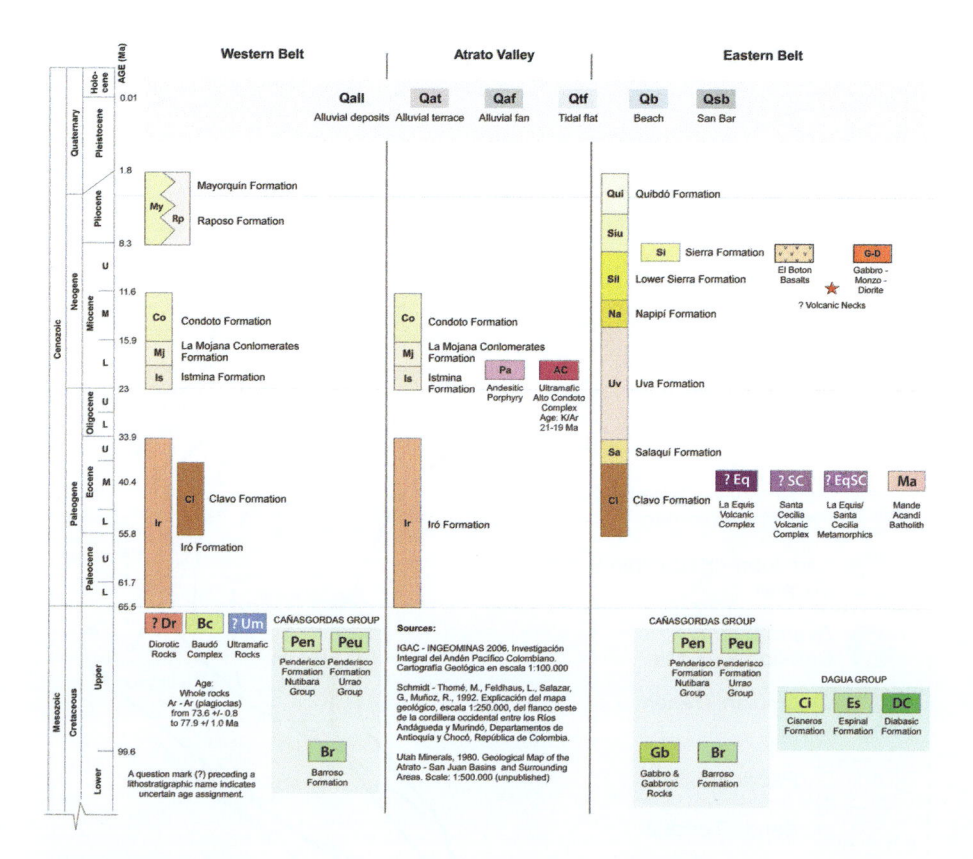

Fig. 1.59 Map legend

In Colombia accreted rocks of the eastern Chocó Arc are preserved in the Cañas Gordas terrane which consists of mixed tholeiitic to calc-alkaline volcanic rocks of the Barroso Fm. overlain by fine-grained sedimentary rocks of the Penderisco Fm. Sedimentary interbeds in the Barroso Fm., and the overlying Penderisco Fm., contain Barremian, Middle Albian, and Upper Cretaceous fossil assemblages (González 2001 and references cited therein), suggesting they represent intercalated, structurally complex slivers of accreted Farallon Plate.

To the west, the El Paso-Baudó assemblage contains Late Cretaceous to Paleogene sections of tholeiitic basalt of N- and E-MORB affinity (Goossens et al. 1977; Kerr et al. 1997), interbedded and overlain by minor pyroclastic rocks, chert, and turbidite of Late Mesozoic-Early Cenozoic age. El Paso-Baudó may represent a Late Cretaceous sliver of the CCOP assemblage, which formed along the trailing edge of the Caribbean Plate. The Mandé (Acandí) was emplaced within El Paso oceanic basement between ca. 60 and 42 Ma (Leal-Mejía 2011; Montes et al. 2012, 2015), broadly coincident with the accretion of the western Chocó Arc to in the Eocene.

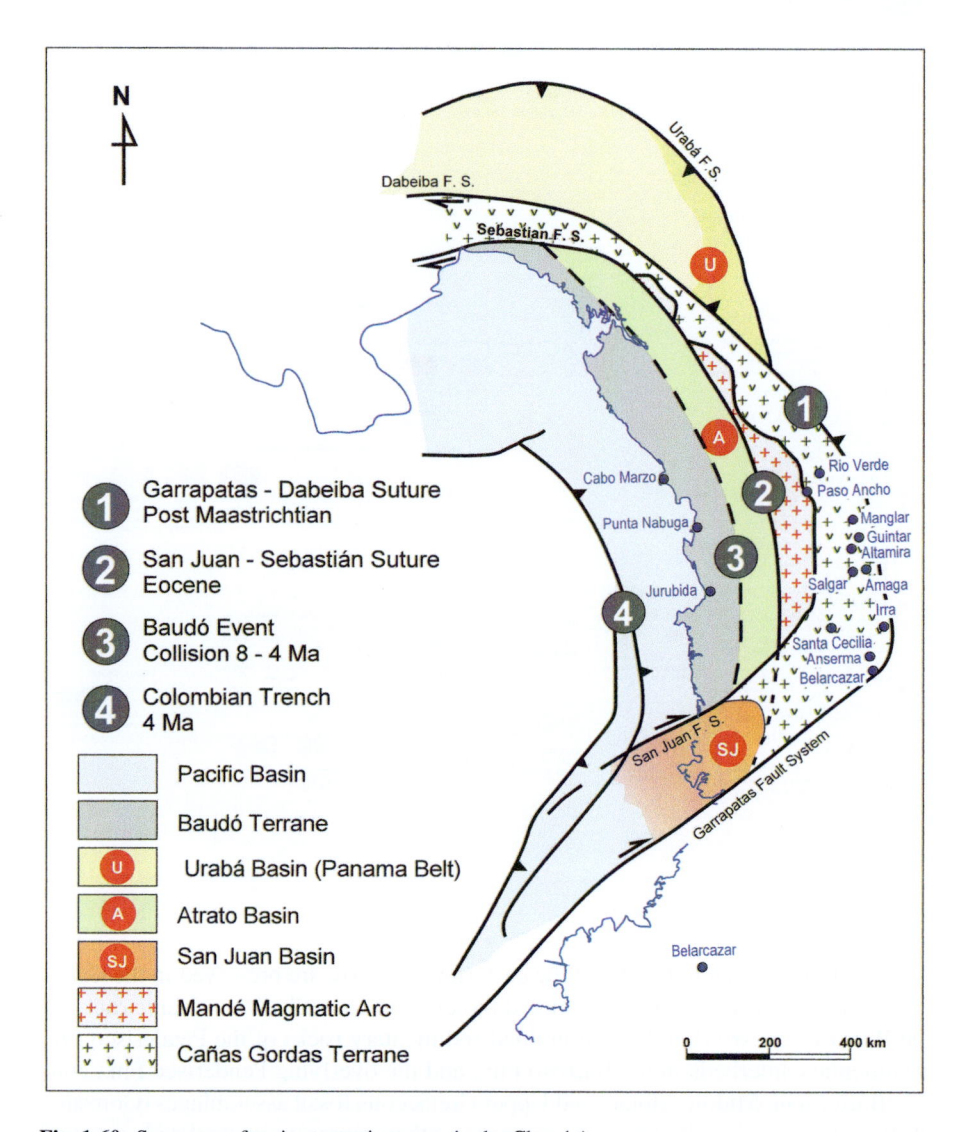

Fig. 1.60 Sequence of major tectonic events in the Chocó Arc

The structural architecture of the Chocó Arc includes the following features:

1. The Garrapatas-Dabeiba Suture (Late Cretaceous)

The youngest paleontological age recorded in the Cañasgordas Group is the Late Cretaceous (Maastrichtian). Accretion of the Cañasgordas terrane to the continental margin may have taken place in incremental slivers, along the Garrapatas-Dabeiba Suture, during the Maastrichtian and possibly continued into the Paleocene.

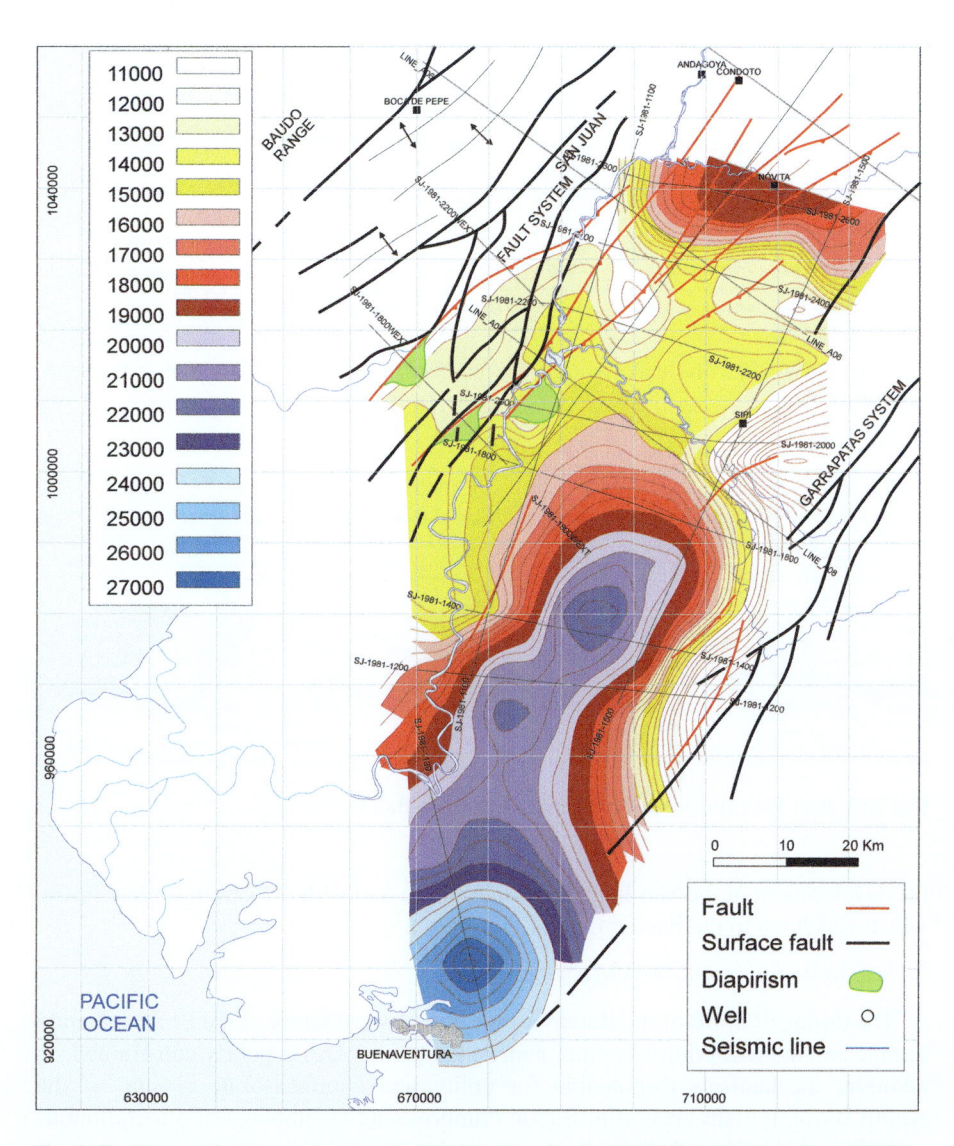

Fig. 1.61 Structural map, top basement in the San Juan basin. (Modified after Petrobras (1990))

2. The San Juan-Sebastián Suture (Eocene)

This second regional-scale suture was generated during emplacement of the El Paso-Baudó terrane. The feature has not been mapped in detail at surface. The Paleocene-Eocene Mandé (Acandí) magmatic arc, hosted within El Paso-Baudó, was generated via Chilean-type subduction processes, penecontemporaneous with

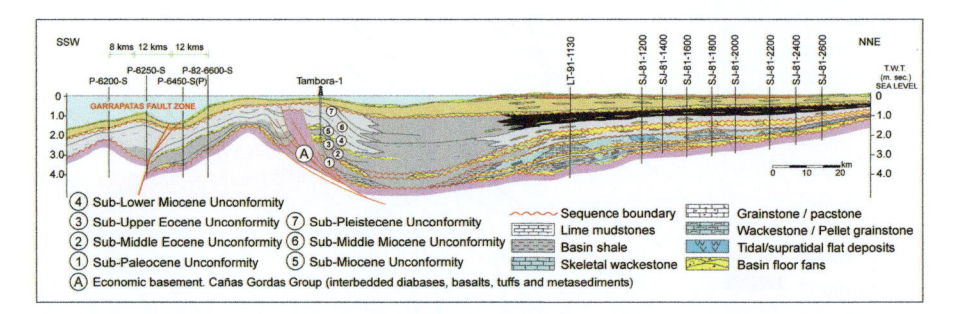

Fig. 1.62 SSW-NNE structural section along the San Juan Basin. (Modified after Petrobras, Ecopetrol (2002))

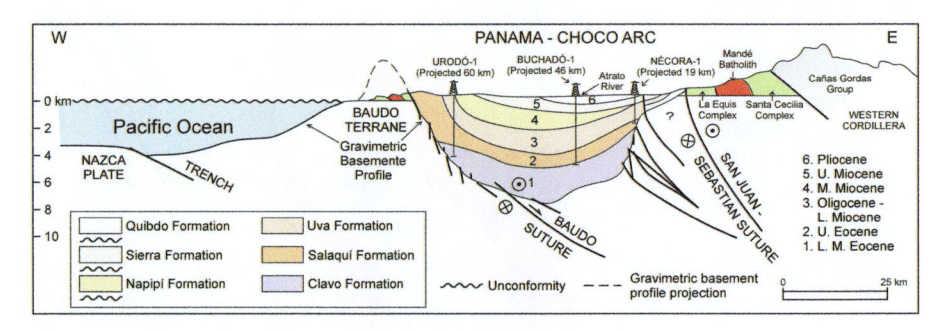

Fig. 1.63 East-West structural section across the Atrato Basin

the development of the San Juan-Sebastián suture and with the Atrato forearc basin, which was open to the Pacific Ocean to the west.

3. Baudó Range Uplift (8–4 Ma?)

The timing of uplift of the Baudó Range, the western sector of the El Paso-Baudó terrane and the westernmost component of the Chocó Arc, is poorly constrained. In addition, mechanisms responsible for uplift, and commensurate closure of the Atrato Basin, remain uncertain. Baudó comprises an assemblage of allochthonous oceanic rocks emplaced along the continental margin by the continuous interaction of the Farallon/CCOP assemblage, with northwestern South America. The following synopsis is offered: subduction of the oceanic plate and Mandé Arc magmatism developed until the relationship between density and buoyancy of the various plates curbed the process, substantially diminishing the rate of subduction. Continued compression produced a positive flexure in the oceanic plate, leading to the uplift of today's Baudó Range and Atrato Basin closure. This mechanism was enhanced by a rapid increase in sedimentary/lithostatic load in the forearc basin.

The following additional data and observations support the postulated tectonic architecture and sequence of events outlined herein and in Fig. 1.60:

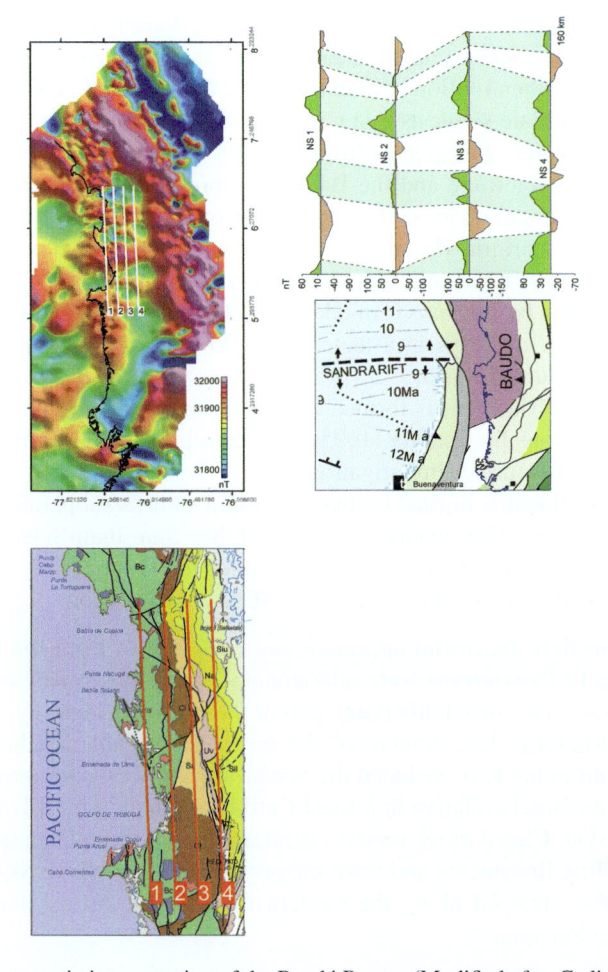

Fig. 1.64 Aeromagnetic interpretation of the Baudó Range. (Modified after Cediel et al. (2010))

- Paleomagnetic data have significantly improved knowledge of the paleogeography and paleo-tectonics of the Chocó Arc. Using paleomagnetic evidence, Estrada (1995) demonstrated the allochthonous character and distinct latitudinal provinces represented by the Cañasgordas terrane and the Baudó Range (referred to as the Western Cordillera terrane and Chocó terrane). The paleolatitudinal origins of these assemblages are directly associated with the tectonic evolution and migration of the eastern Pacific (Farallon, Caribbean) plates. In this sense, it has long been suggested that, since the Late Cretaceous, plate interactions along NW South America have been dominated by interactions with the Farallon Plate (e.g., Pardo-Casas and Molnar 1987, among many others).

- In these and numerous other reconstructions, the Farallon Plate, containing the Caribbean-Colombian oceanic plateau, moved along a north-directed vector into the Paleogene, when motion shifted to mainly NE-directed. The relative motion of the Farallon Plate suggests that terranes accreted against the western edge of South America were transported from latitudes to the south and west.
- The Cañasgordas terrane and the Baudó Arc present two groups of paleomagnetic data with the main group having a mean of about 10, suggesting equatorial paleolatitudes of origin. The nature of the paleomagnetic data is not conclusive, but the geological framework favors a southern provenance.
- The Atrato Basin basement is formed by the El Paso terrane, including the Baudó Complex, which outcrops in a tectonic window in the Istmina-Condoto High (along the San Juan Suture).
- San Juan Basin basement is formed by the Cañasgordas terrane (or in the case that interpretations by Estrada (1995) are correct, by the Gorgona terrane; this ascertain requires further investigation).
- The San Juan Basin is limited by two important sutures/subparallel transcurrent fault systems (the Garrapatas-Dabeiba and the San Juan-Sebastián sutures). These structures controlled sedimentation since the Oligocene (?) and gave rise to a deltaic system which prograded in a NE to SW direction.

It is evident that the initial approach and collision of both the Cañasgordas terrane and the El Paso terrane were orthogonal. During subsequent tectonic migration, a NW rotation occurred, liberating part of the collisional energy and leading to the morpho-structural development of the present-day Panamá-Chocó Arc. This tectonic migrationmay have included the northward and westward movement of the South American Block, relative to a fixed Caribbean Plate (e.g., Silver et al. 1990; Farris et al. 2011). Chocó block rotation is inferred from the existence of tear faults and E-W trending lineaments and from the progressive SW-NE to SE-NW orientation of fold axes mapped along the western flank of the Atrato Basin and in its extensions into Panamá.

1.6.1.11 Maracaibo Orogenic Float (Figs. 1.65, 1.66, 1.67, 1.68, 1.69, 1.70, and 1.71)

General Statement
The large-scale Neogene features of the Maracaibo Block can be assembled in a quantitative kinematic block-mosaic that reveals internally consistent relationships between strike-slip faulting, compression, and uplift (elevation). This fundamental observation led Laubscher (1957) to postulate "the kinematic puzzle of the Neogene Northern Andes." The "puzzle" is manifest via the simple comparison of the structural grain and tectonic style of the Maracaibo Block versus that of the Andean domain located to the southwest.
The structural style and tectonic evolution of the Maracaibo Block is best understood through an integrated geological and geophysical analysis of all its litho-tectonic

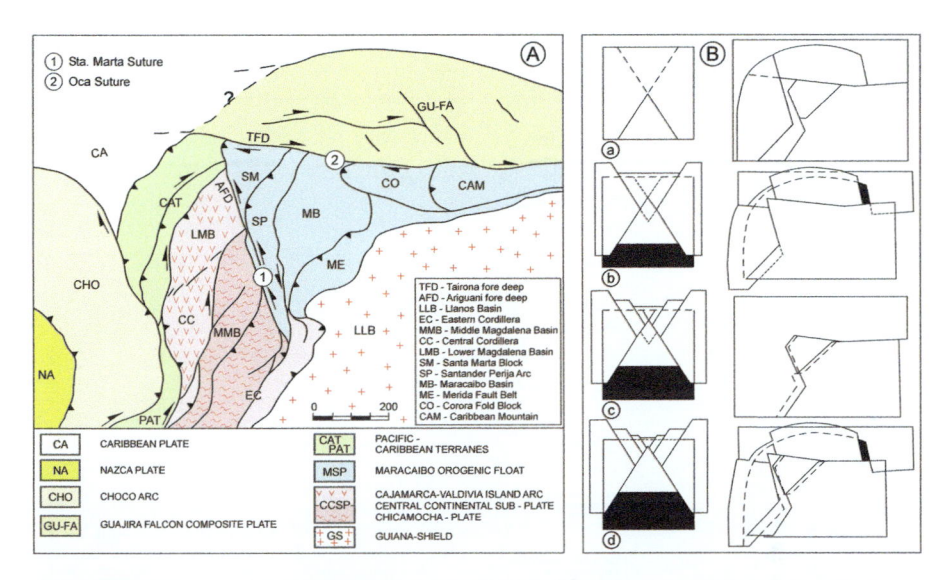

1 Phanerozoic Orogens of Northwestern South America: Cordilleran-Type Orogens... 73

Fig. 1.65 (a) Geological setting and kinematic model of the Maracaibo orogenic float (b) Paper cut-out model for the simplified Maracaibo Block mosaic (White = compression; black = extension). (Modified after Laubscher (1987))

and morpho-structural components, including not only the topographically elevated Mérida Range, Santander Massif, Perijá Range, and Sierra Nevada de Santa Marta (SNStM) but also the intervening and surrounding basins (e.g., César-Rancheriá, Guajira, Maracaibo; Fig. 1.4).

The SNStM, which forms the apex of the Maracaibo orogenic float, is a pyramid-like range which rises to an elevation of 5800 m above sea level, over a horizontal distance of just 45 km from the Caribbean coastline. The range then drops at equal distance from the coastline to depths beyond 2500 m bathymetry. A strong, negative Bouguer gravity anomaly is associated with the SNStM, implying that the range is "rootless," that is, isostatically unbalanced.

Most of the sterile debate and confusion in the formulation of a tectonic model for the Maracaibo orogenic float arises as a consequence of the obstinate application of models involving "subduction" of the Caribbean Plate, irreverent of existing rheological analyses, and notwithstanding the absence of deep geophysical data in and around the Santa Marta Massif. Revised geological mapping (Geosearch 2008), elastic geomechanic modeling of the Bucaramanga, and Oca Faults (Florez and Mavko 2001) and channel flow extrusion of Eocene granitoids (Godin 2006; Piraquive 2016) are key new data pieces to apply in the resolution of Laubscher's puzzle.

A conscious seismic-structural evaluation of the Mérida Range completed by Monod et al. (2010) negates the application of Andean-type deformational models

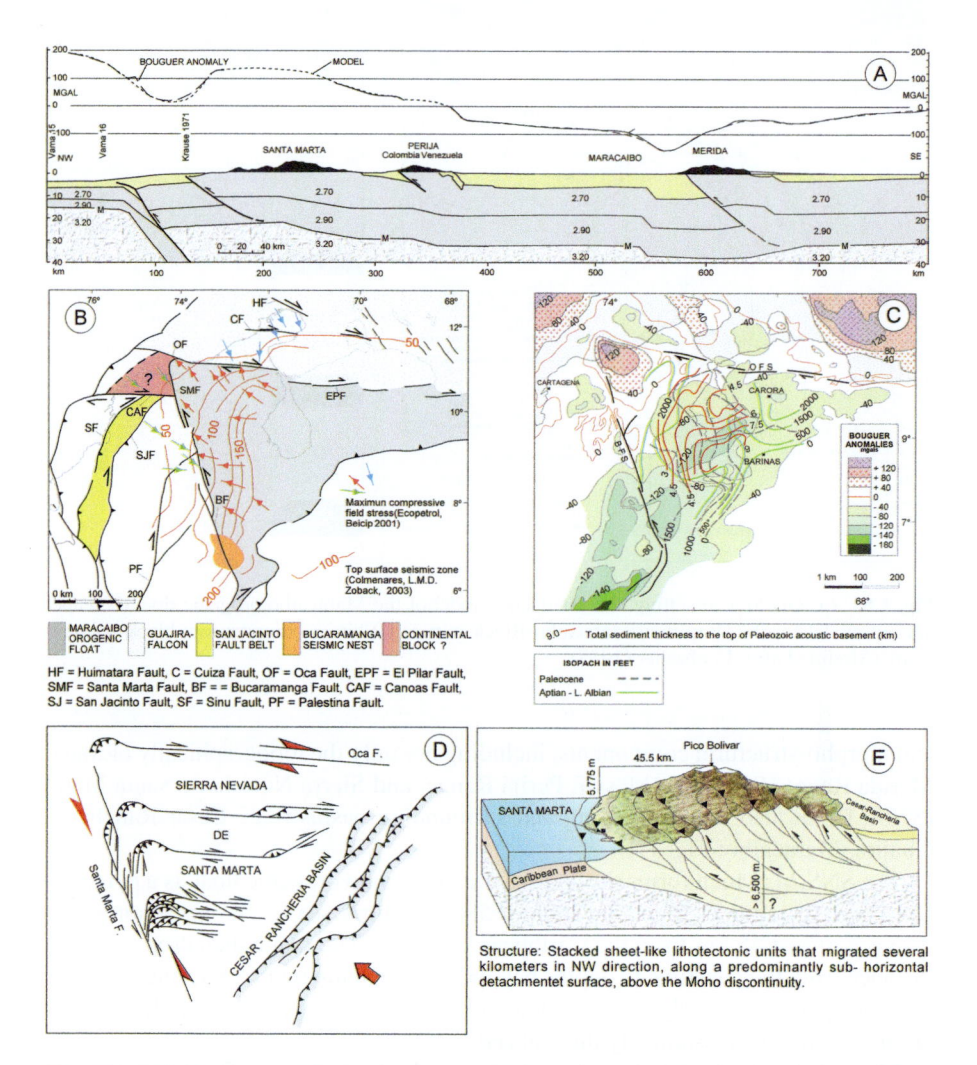

Fig. 1.66 Diagnostic geophysical and structural features of the Maracaibo Orogenic. (**a** Modified after Kellogg and Bonini (1982); **b** Compiled after Colmenares and Zoback (2003); **c** Modified after Cediel et al. (2003); **d** Modified after Geosearch (2008))

and has opened the way to updated regional tectonic interpretation. Notwithstanding, answers to ongoing questions, like the type and age of the metamorphism affecting Upper Paleozoic sedimentary sequences (Marechal 1983), are needed to complete the Pre-Cretaceous geological history.

An updated understanding of the Serranía de Perijá and Santander Massif (Chap. 4), integrated with the detailed geological history of productive oil and gas basins in and around the Maracaibo Block (Mann et al. 2006; Cediel 2011),

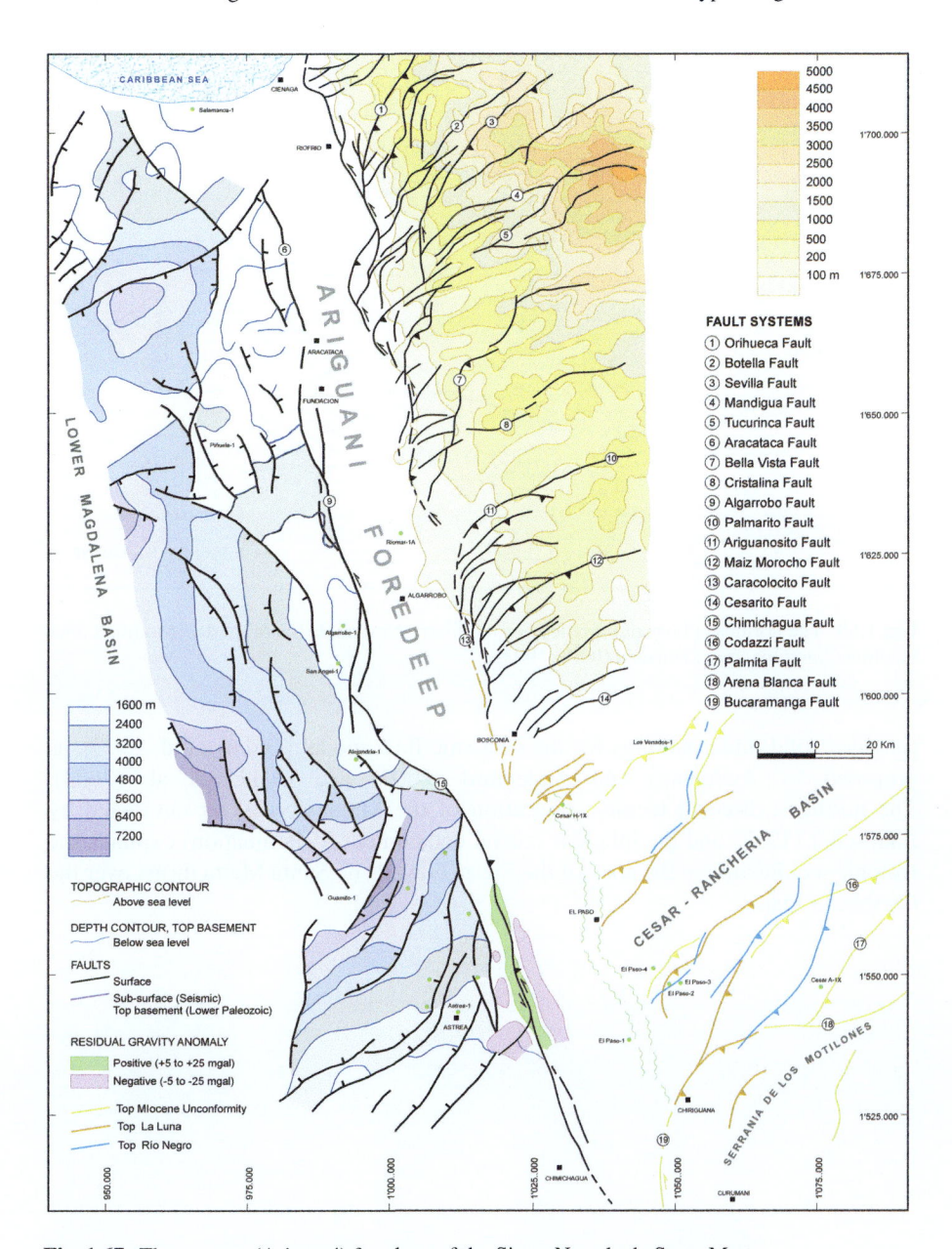

1 Phanerozoic Orogens of Northwestern South America: Cordilleran-Type Orogens...

Fig. 1.67 The western (Ariguani) foredeep of the Sierra Nevada de Santa Marta

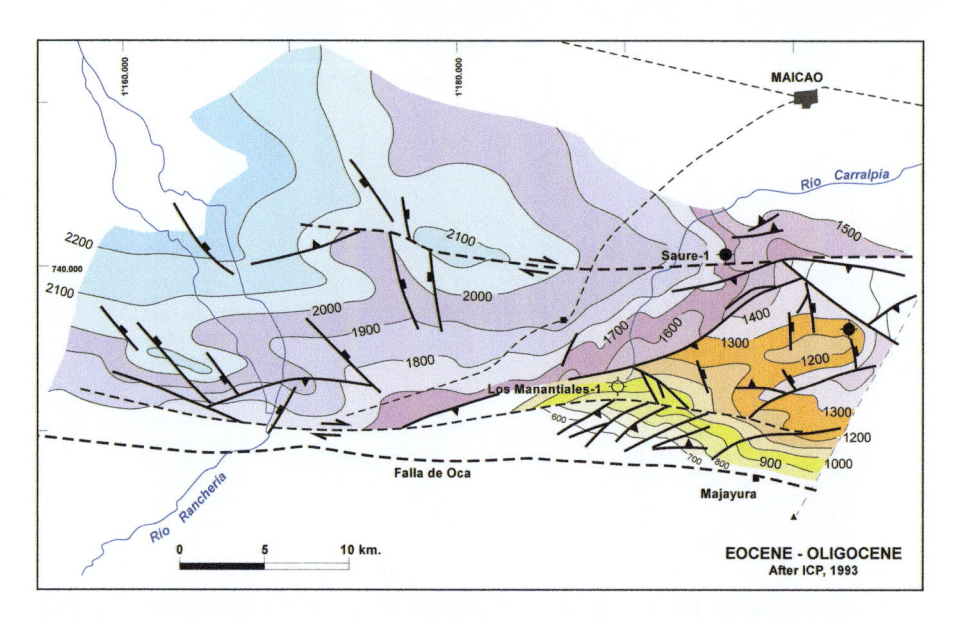

Fig. 1.68 The northern (Tayrona) foredeep of the Sierra Nevada de Santa Marta. (Modified after Instituto Colombiano del Petroleo (ICP) (1993))

provides solid underpinnings for the orogenic float model (Oldow et al. 1990), as proposed by Audemard and Audemard (2002) and Cediel et al. (2003). The northeast-directed tectonic migration of the Maracaibo Block envisaged by Laubscher (1957) and the inherent crustal detachment (delamination) explains the tectonic architecture at the apex of the Sierra Nevada de Santa Marta thrust over the Caribbean Plate.

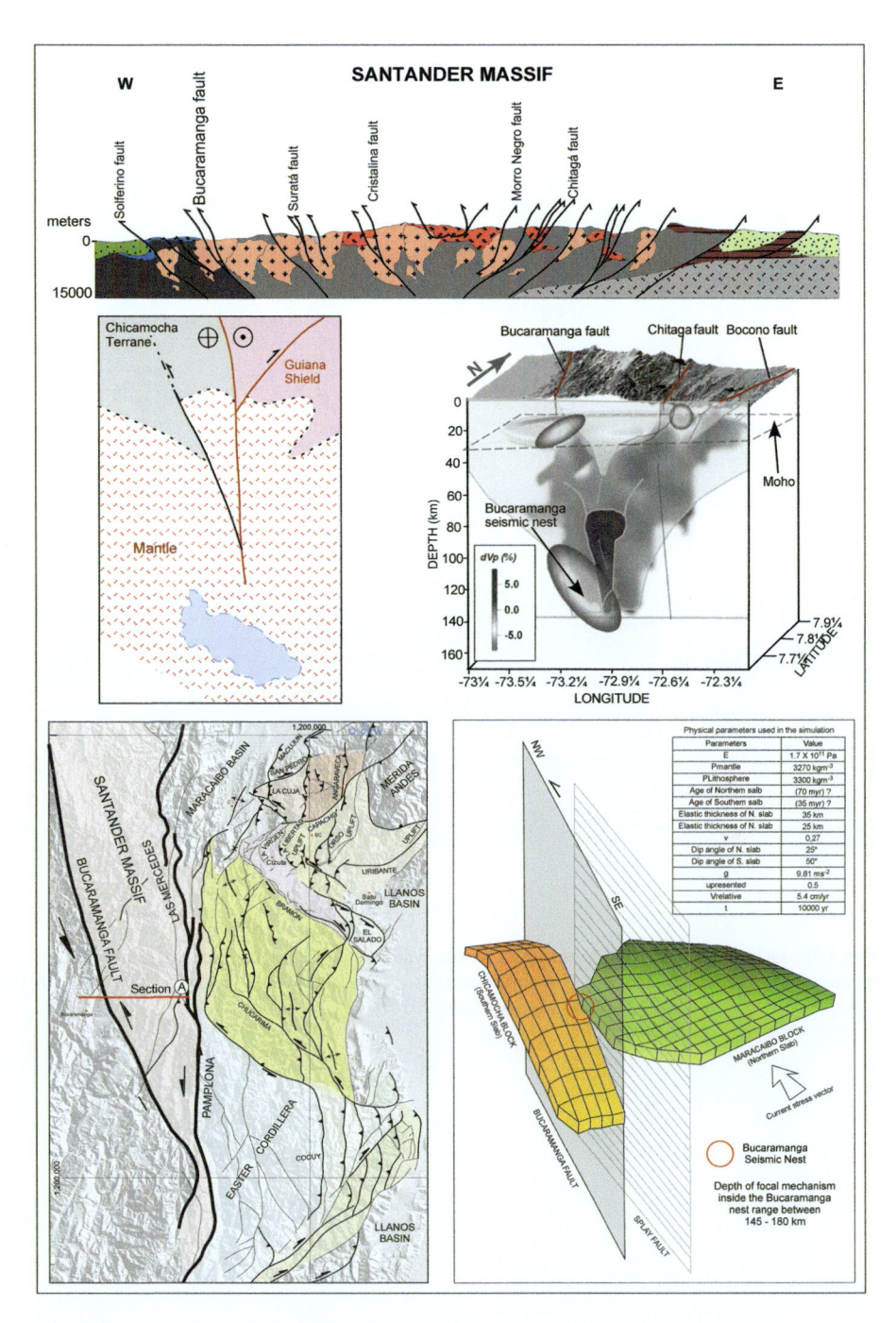

Fig. 1.69 Geological and geophysical setting of the Bucaramanga Fault (suture) and Bucaramanga seismic nest. Note use of rhombochasm as slip-marker. (Compiled after Zarifi et al. (2007), Londoño et al. (2010) and Restrepo-Pace (1995))

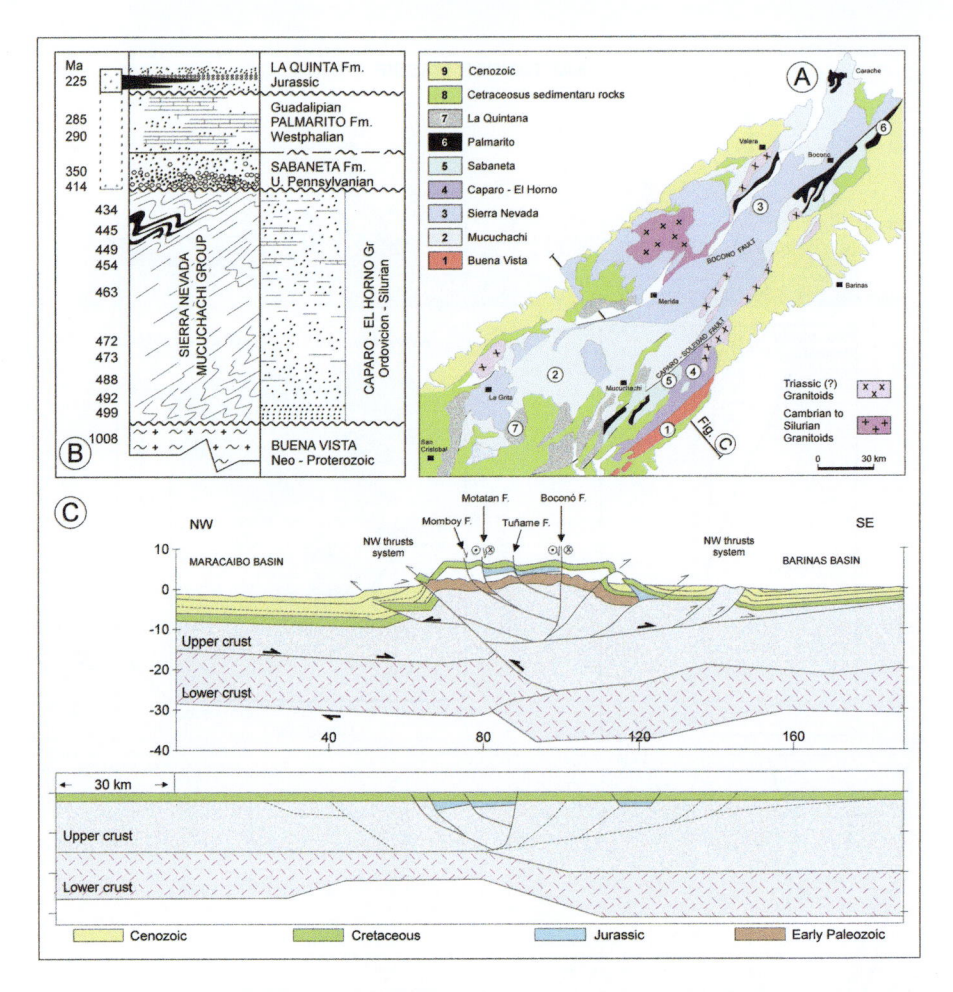

Fig. 1.70 Geological sketch map and stratigraphic synthesis underpinning the structural interpretation of the Mérida Range. This interpretation precludes subduction beneath the range. (**b** Compiled after Marechal (1983) and Monod et al. (2010))

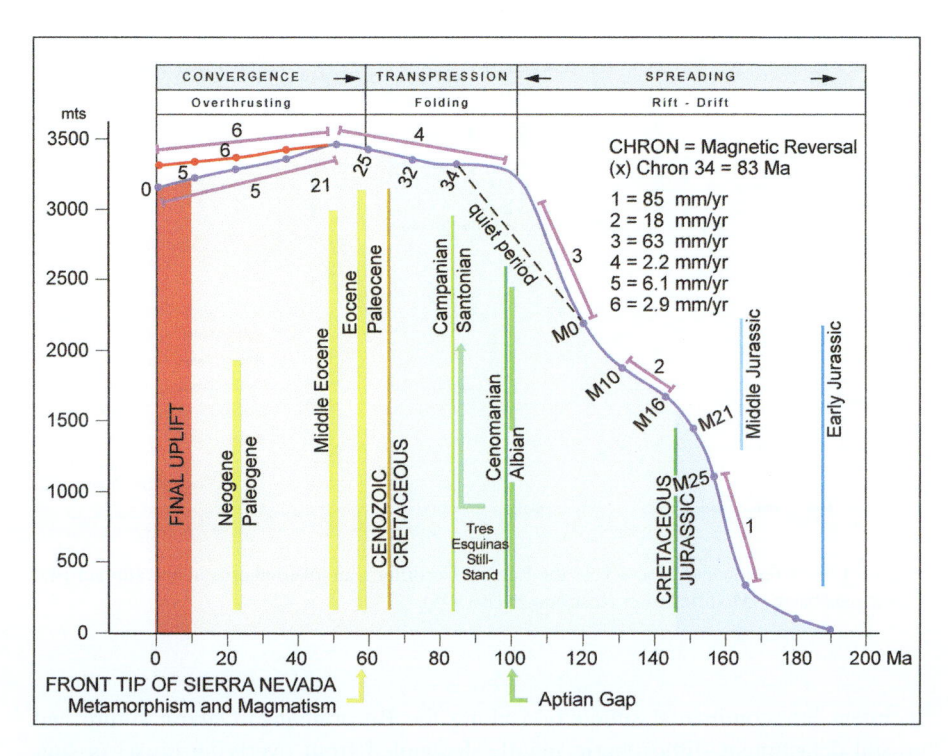

Fig. 1.71 Relative motion and displacement curve (spreading-convergence and chronology vs. elevation) of the Maracaibo Block. (Compiled after Klitgord and Schouten (1986) and Nürnberg and Müller (1991))

1.6.1.12 Eastern Cordillera (Figs. 1.72, 1.73, 1.74, and 1.75)

General Statement

The Eastern Cordillera is a Late Jurassic-Cretaceous rift-related basin caught between and inverted during development of two coeval orogenic systems: the Maracaibo orogenic float to the NNE and the Andean orogeny to the south and west. The axis of basin inversion is the buried Bucaramanga-Garzón fault system and collateral Jurassic-aged rift-related faults and grabens. Composite geological mapping and detailed geophysical analysis permit the interpretation of a thick-skinned, divergent intracontinental orogen.

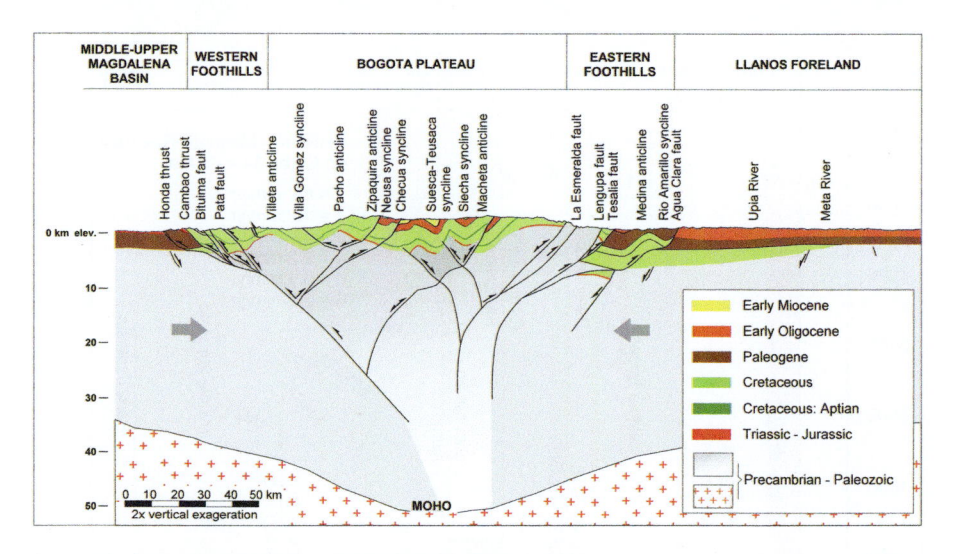

Fig. 1.72 E-W regional section across the Eastern Cordillera, an inverted crustal and supracrustal Cretaceous basin. (Modified after Restrepo-Pace 1995)

Incipient intracontinental subduction along the Bucaramanga-Garzón suture and crustal detachment (lithospheric mantle decoupled from overlying crust) is suggested by local gravity models.

The Eastern Cordillera, considered a litho-tectonic unit, is overridden by the Maracaibo orogenic float to the north and by the Garzón Massif to the south (see Fig. 1.13).

Fig. 1.73 Valanginian to Miocene structural and magmatic evolution of the Eastern Cordillera. (Compiled after Vásquez et al. (2010) and Geosearch (2008))

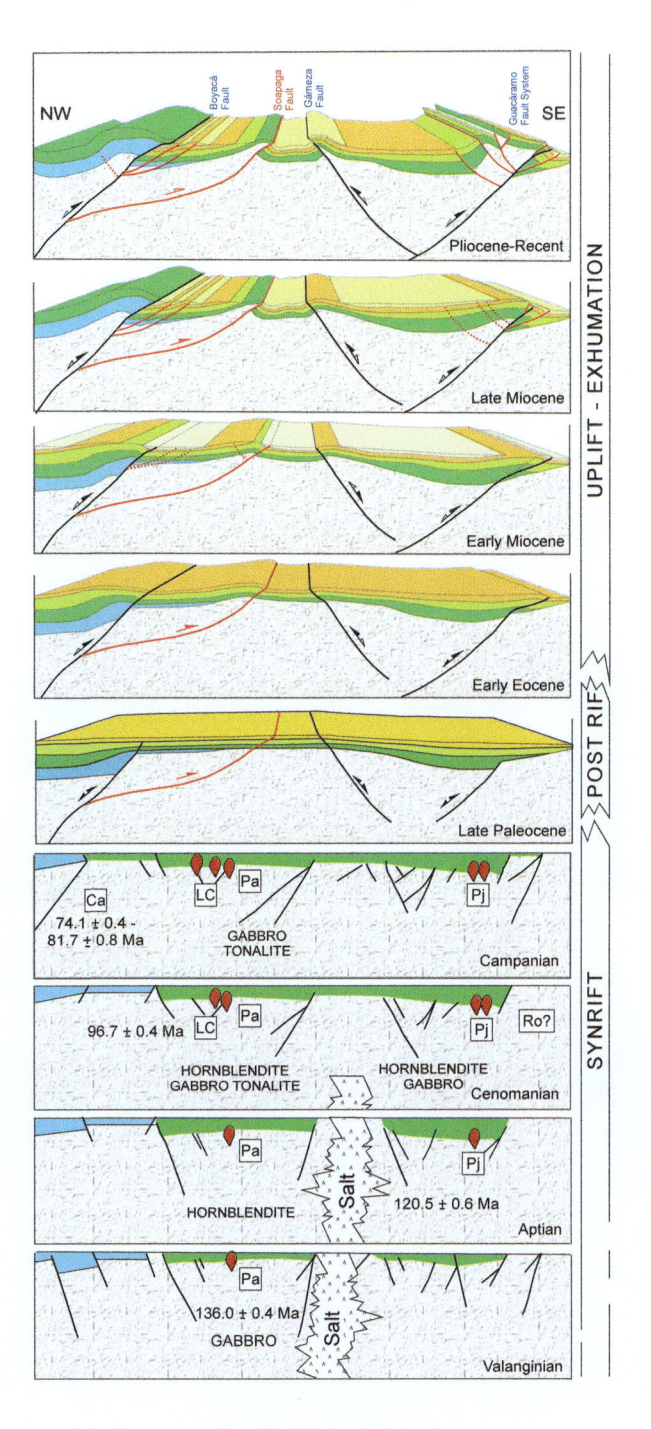

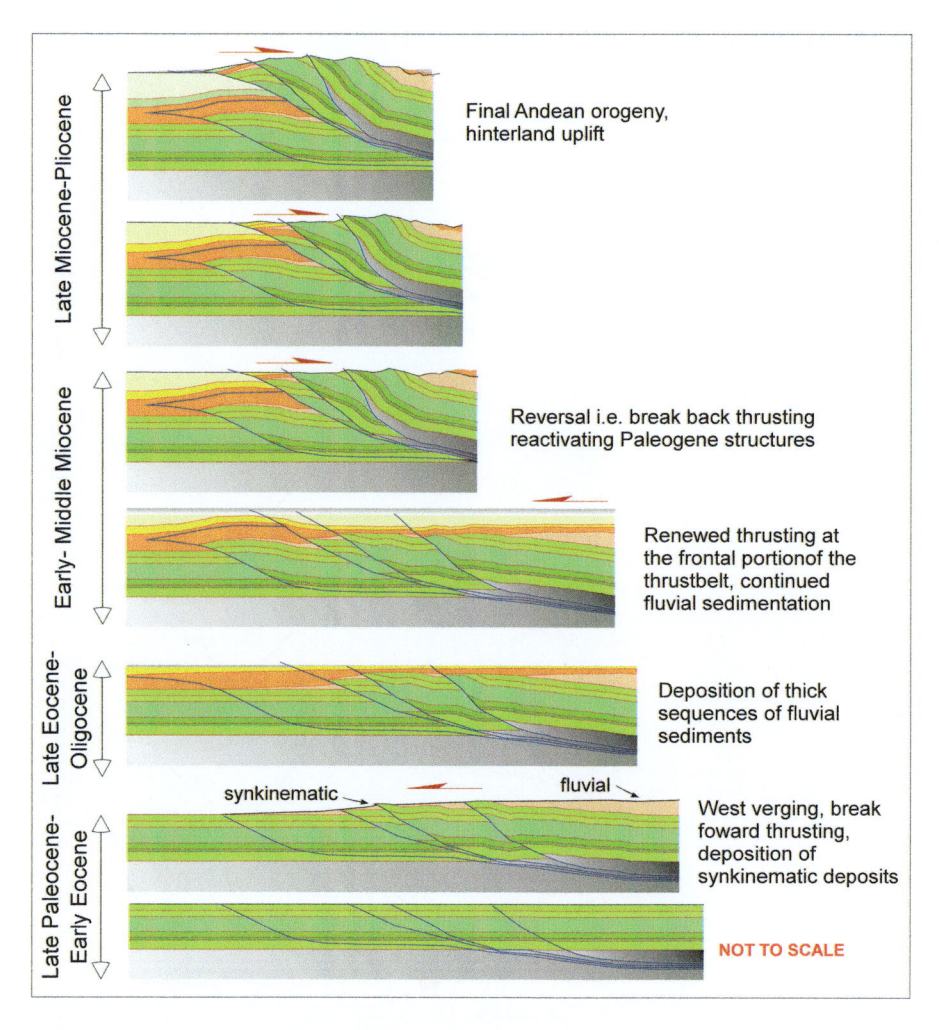

Fig. 1.74 Cenozoic west-vergent tectonic evolution of the Eastern Cordillera's western foothills. (Modified after Restrepo-Pace et al. (2004))

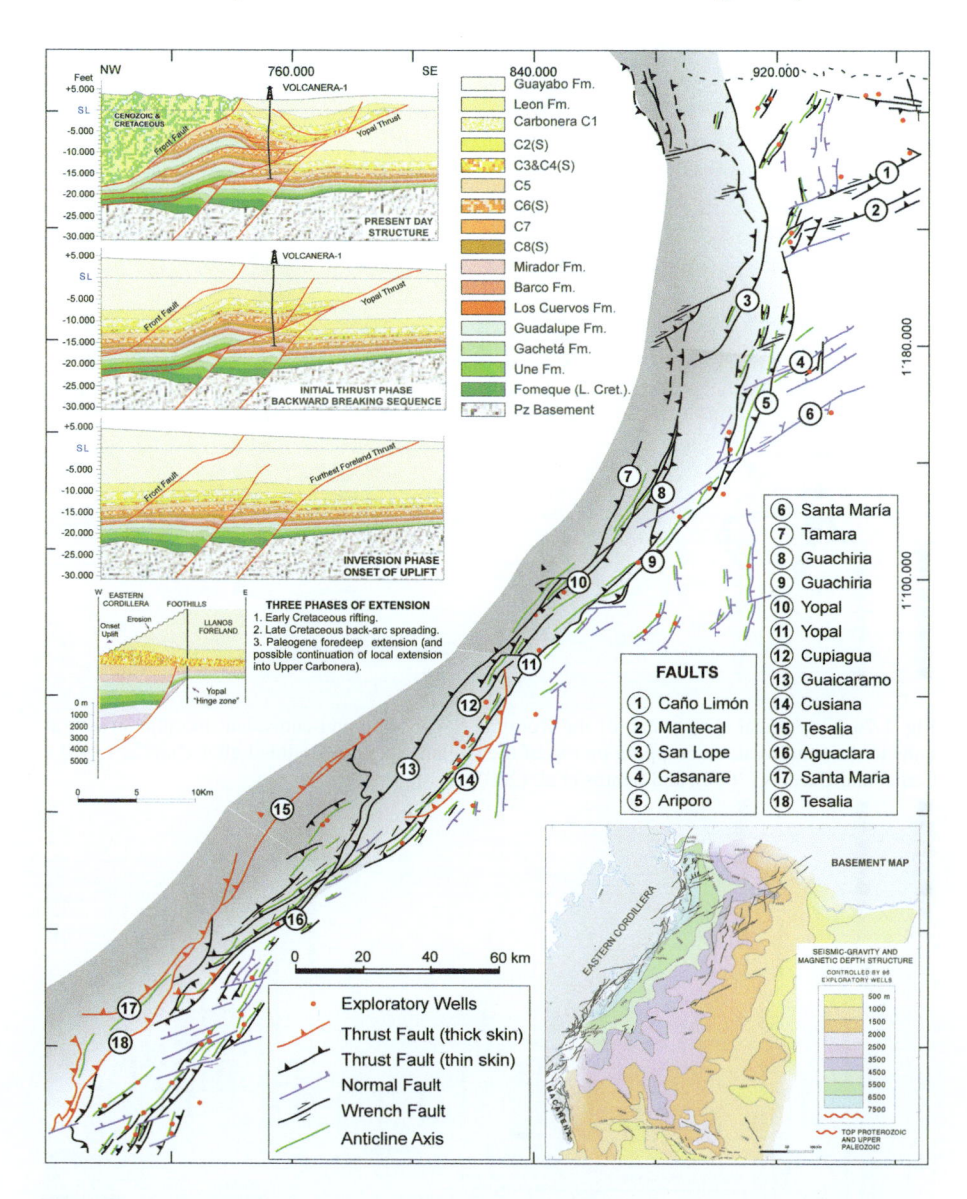

Fig. 1.75 East-vergent fault patterns in the eastern foothills of the Eastern Cordillera (Geotec (1996))

1.6.1.13 Kinematics of the Guiana Shield and Its Western Mobile Belt. The Roraima Tectono-Sedimentary Problem (Figs. 1.76, 1.77, 1.78, 1.79, and 1.80)

General Statement

In 1974 Gansser, departing from his field studies in the Macarena and the Roraima tepuis (Tafelbergs, table mountains, etc.), formulated the "Roraima Problem" from a geomorphologic- stratigraphic point of view.

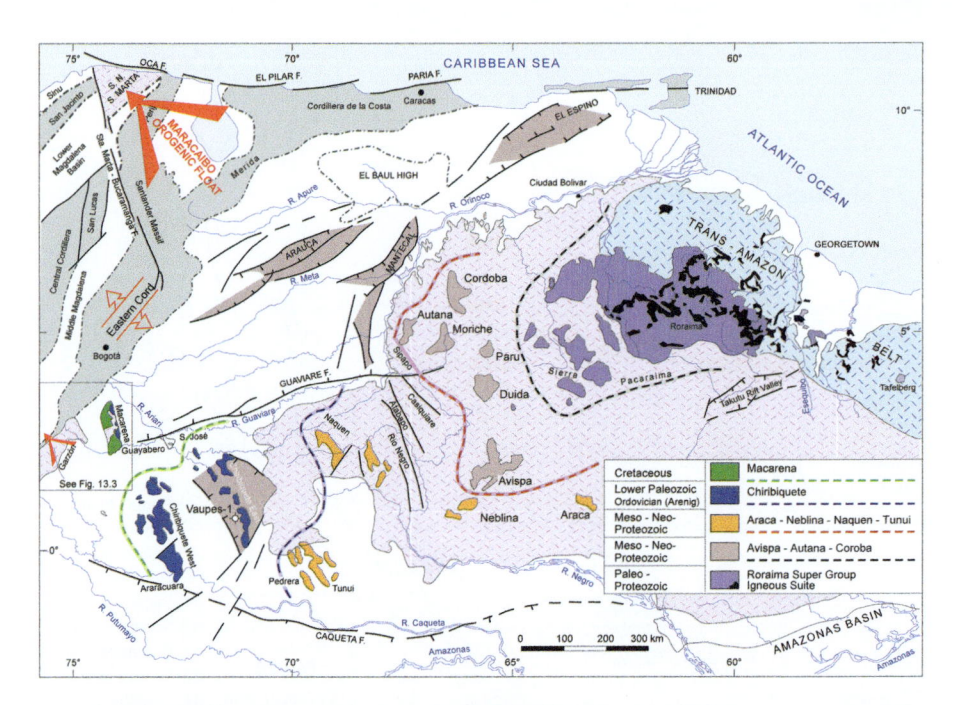

Fig. 1.76 Geological sketch map of the Roraima Supergroup and equivalent morpho-structural units to the southwest, outcropping on top of the Guiana Shield. (Modified after Gansser (1974), Cediel and Cáceres (2000), and Santos et al. (2003))

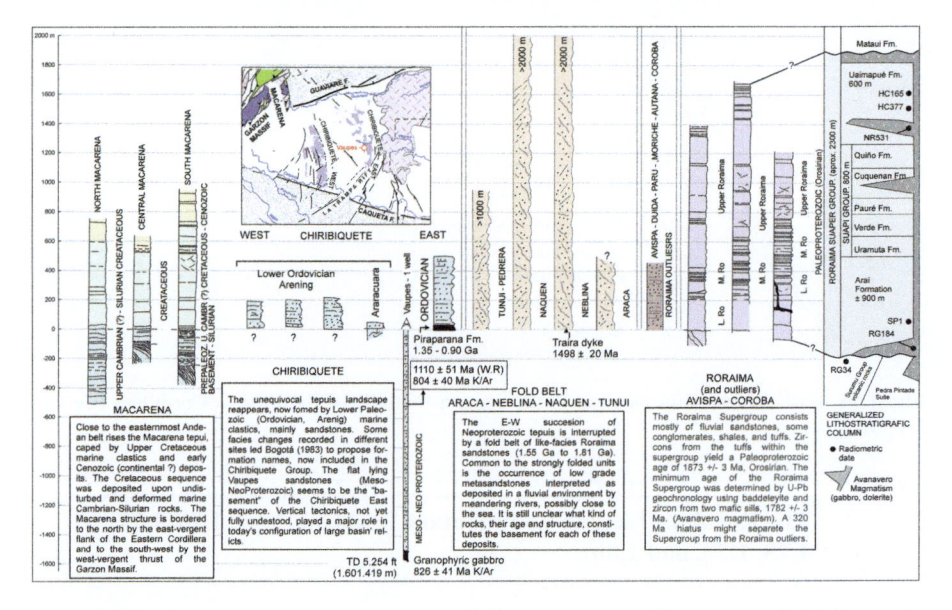

Fig. 1.77 Lithostratigraphy of Proterozoic to Cretaceous tepuis. (Modified after Bogotá (1983), Gansser (1974), and Santos et al. (2003))

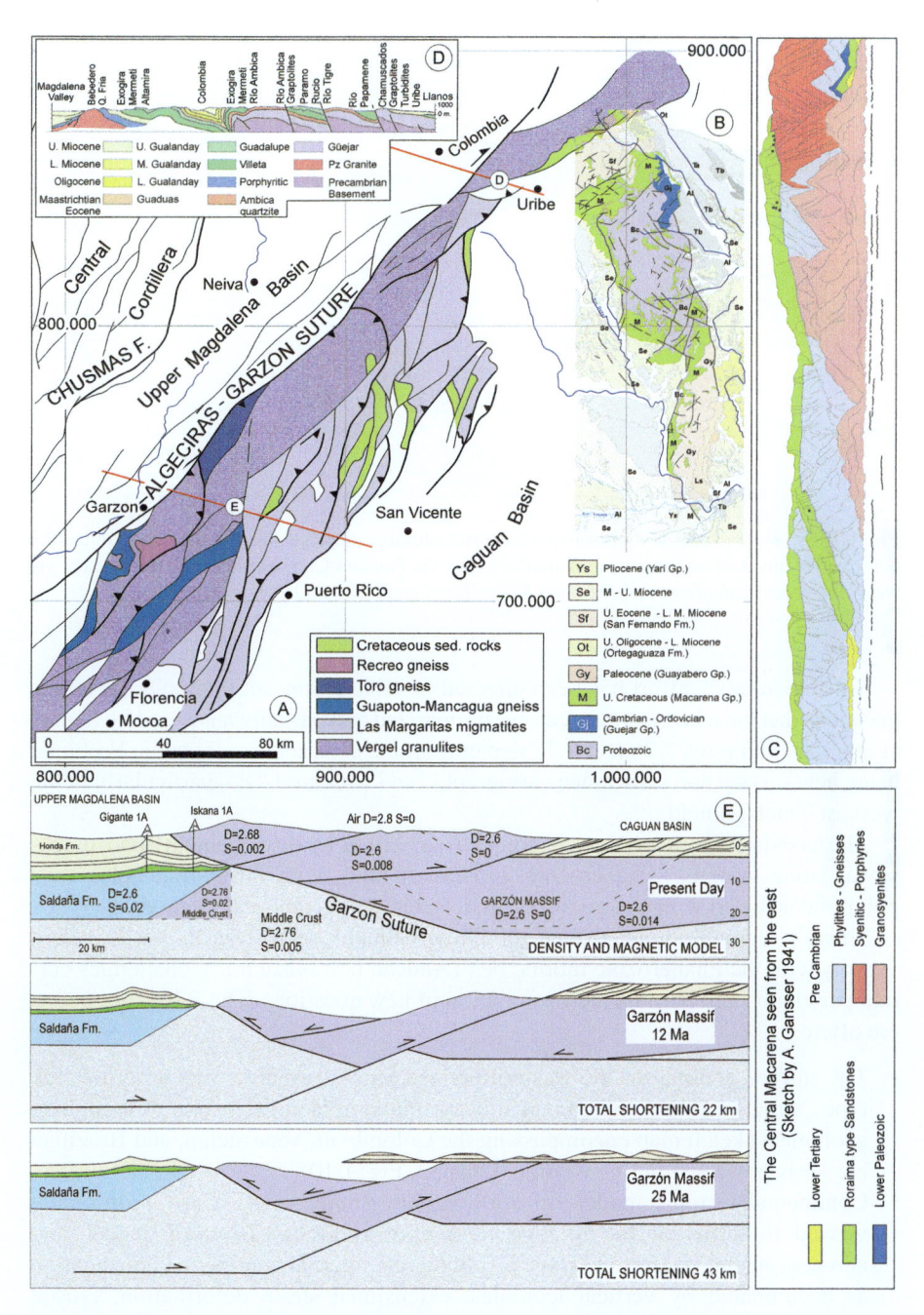

Fig. 1.78 Geological map of the Garzón-Macarena ranges. Gravimetric structural restoration of the Garzón Massif. (Compiled after Bakioglu (2014), Ibanez-Mejia (2010), Jimenez-Mejia et al. (2006), Gansser (1941), Cediel and Caceres (2000))

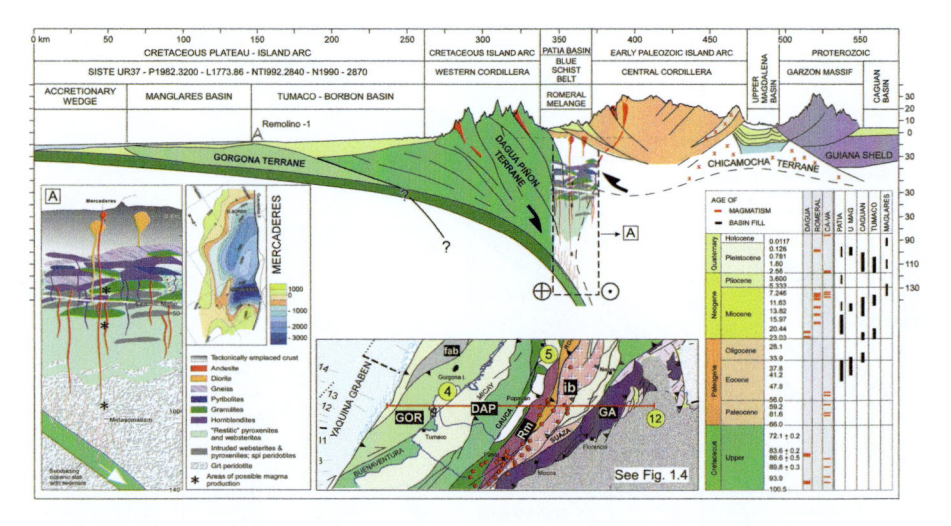

Fig. 1.79 E-W structural section across southern Colombia, depicting the principle tectonic units deformed by the Andean orogeny and unaffected by the Panamá-Chocó indenter. Note the east- vs. west-convergent compressional regime. (A) Modified after Weber (1998)

Since then much effort has been directed to solve chronostratigraphic questions, questions that have been more recently summarized and partially answered by Santos et al. (2003). Figures 1.76 and 1.77 depict a synthesis of today's Roraima Problem, using new surface and subsurface cartographic and structural data, particularly in the western Guiana Shield.

The coexistence of vertical tectonics (over 2000 m displacement) and extensional strain systems (Phanerozoic graben-rift fills), with thrust-fold belts, may point to concealed wrench structures, undetected by geophysical methods.

Today's regional knowledge of the Guiana Shield, seen from its northwestern vicinity with the Phanerozoic mobile belt (Andean belt sensu *lato*), challenges current paradigms and permits the formulation of new questions. The following points are offered:

- The surface geology of the western Guiana Shield reveals a vast array of rock types, structurally assembled in a hitherto untold geological history. A combined geological sketch map encompassing the Colombian, Venezuelan, and Brazilian border is presented in this volume (Chap. 3, Fig. 1.10).
- Continental tectonic models in northwestern South America are traditionally viewed in terms of the relative motion of rigid blocks along paleo- and neo-continental borders but must also integrate internal continental deformation as documented by vertical tectonics, extensional strain deformation, crustal delamination, and supracrustal thrust faulting and folding.

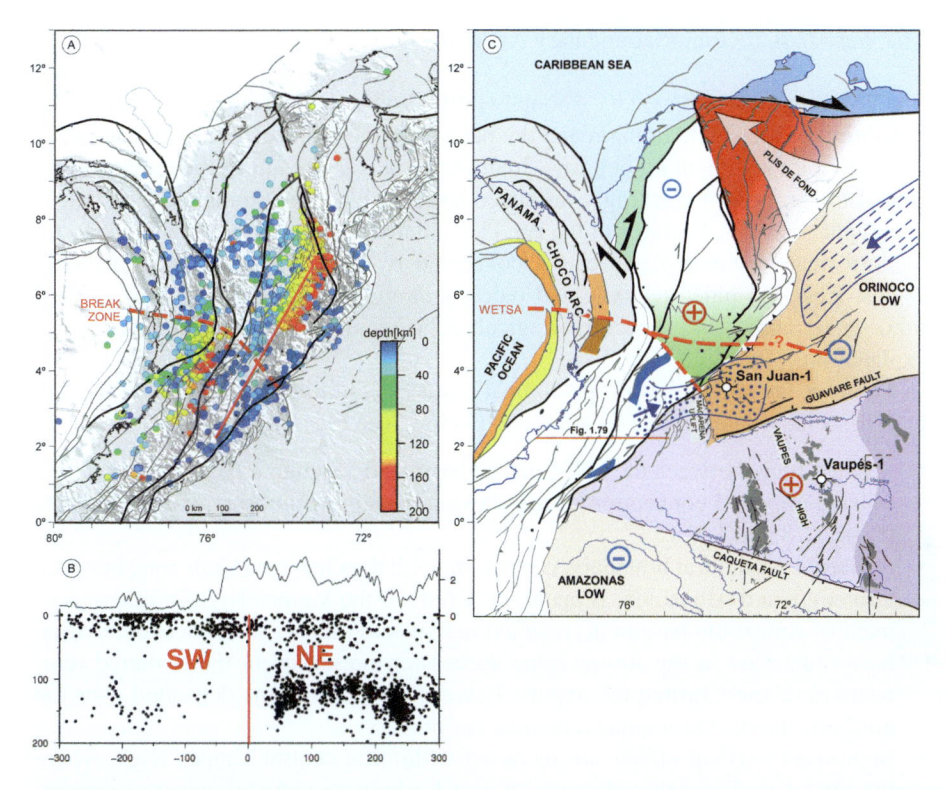

Fig. 1.80 (a) Earthquake locations and vertical cross sections along strike of the Eastern Cordillera. Red vertical line in (b) indicates the inferred "slab tear" separating the SW and NE sectors. (Modified after Seccia (2012)). (c) The W-E Tectono-Sedimentary Anomaly, WETSA, and major regional geotectonic element

- Within the context of available plate tectonic models, Proterozoic intracratonic orogenic events (e.g., the 1.3 Ga Sunsas Orogeny, e.g., Santos et al. 2003) are far from being fully understood.
- Significant discrete as well as penetrative deformation within the western Guiana Shield is recorded during continental collision (with Oaxaquia), manifested as Neoproterozoic impactogen structures and a Grevillian granulite-grade metamorphic belt.
- The Phanerozoic mobile belt seems to exert no direct, visible deformation on the Guiana Shield. On the contrary, a segment of the shield overrides the Caribbean Plate (Maracaibo orogenic float), and the Garzón Massif is thrusted over the Upper Magdalena basin.

The West-East Tectono-Sedimentary Anomaly

The W-E Tectono-Sedimentary Anomaly (WETSA) is a broad corridor affecting the continuity of the shield domain and the paleo-geographic distribution of Cambrian to Quaternary sedimentary deposits. Interpretation is the result of the temporal and spatial integration of seemingly unrelated structural, sedimentological, and paleo-geographic features. Features of the WETSA aid in deciphering the kinematic puzzle of northwestern South America, as concealed by successively younger tectonic events. Restored paleo-tectonic and paleo-geological maps derived from verified litho-stratigraphic and geochronological data seem to point to a satisfactory explanation.

Some of the most relevant geological features resulting from the WETSA and their associated interpretations are related below (see Fig. 1.80):

- A paleo-high, between the NW Carrizal basin and SE Güejar basin, is interpreted to have impeded connection of these two advancing Cambrian marine deposits (Figs. 1.18 and 1.80).
- A morpho-structural boundary, coinciding with the Guaviare Fault zone between Orinoquía and Amazonía, is apparent. Uplift of the Vaupes High (northern segment of Amazonía basement) resulted in exposure of a Lower Paleozoic marine basin, recorded in the stratigraphic succession preserved in the scattered remnants in of the Chiribiquete tepuis. Lower Paleozoic basins deposited over the northern Block (Orinoquía) remained buried.
- Significant vertical offsets are recorded within the central Vaupes High, where the top of Ordovician sediments (e.g., Chiribiquete tepui), located at approx. 1000 m elevation, outcrops close to flat-lying buried Proterozic sandstones (well Vaupes-1)*.
- Similarly, within the Macarena uplift Cambrian sediments are exposed at approx. 1000 m above sea level vs. Silurian sediments within the Orinoco low, buried at approx. 2300 m depth (well San Juan-1)** (Table 1.1).
- E-W morpho-structural control of Triassic depo-centers is observed to the south of the WETSA. To the north, similar control is observed in Jurassic basins (Fig. 1.25).
- E-W morpho-structural control of the development of Lower Cretaceous basins is observed to the north of the WETSA. To the south, similar control is observed in Upper Cretaceous basins (e.g., Middle vs. Upper Magdalena basins; see Figs. 1.41 and 1.42).

The western extrapolation of the WETSA may be reflected in:

A documented seismic high or boundary in the Chocó Arc, between the Atrato basin to the north and the San Juan basin to the south (Fig. 1.80)

An E-W crustal discontinuity or "tear" through the central Colombian Andes, as interpreted by Vargas and Mann (2013), among others.

Table 1.1 Summarized logs for wells Vaupes-1 and San Juan-1 (Fig. 1.80)

Name	Location North	Location West	Total depth (feet)	Logged depth (feet)	Geological description
Vaupes-1	01 09′ 45″ N	71 00′ 30″ W	5254	0–110	Quaternary sediments
				110–4892	Quartz arenites, feldspathic sublitharenites, illitic subarkose sample 400 ft., 804 ± 40 K/Ar, Ma.
				4892–5053	Contact, meta-sandstone
				5053–5099	Granophyric diabase
				5099–5254	Two-pyroxene (tholeiitic) olivine gabbro, 826 ± 41 K/Ar, Ma
(*)San Juan-1			7004	5100–5750	Lower Oligocene, delta plain, brackish
				6017–6258	Upper Eocene, delta plain, brackish
				6390	Upper cretaceous, reworked(?)
				6495	Coniacian-lower Campanian, continental
				6860	Paleozoic(?)
				6905	Upper Silurian, marine

(*)Black shale with acritarchs as dominant elements. Domasia bispinosa, baltisphaeridium gordonense, B. molium, B. ramusculosum, and Multiplicisphaeridium ramusculosum indicate an Upper Silurian age (Robertson Research 1982)

This discontinuity is reflected by the manifestation of Pliocene to Recent arc-related volcanism to the south and the absence of volcanism to the north (see detailed analysis by Leal-Mejía et al. 2018).

E-W displacement of the forearc basin, accretionary prism, and trench fill along the Pacific Colombian trench, at the eastern end of the axis of the Sandra rift, may be a recent manifestation of the effects of the WETSA.

References

Arminio JF, Yoris F, Quijada C, Lugo JM, Shaw D, Keegan JB, Marshall JE (2013) Evidence for Precambrian stratigraphy in Graben basins below the Eastern Llanos Foreland, Colombia. Search and Discovery Article #50874

Audemard EF, Audemard F (2002) Structure of the Mérida Andes, Venezuela: relations with the South America-Caribbean geodynamic interaction. Tectonophysics 345:299–327. https://doi.org/10.1016/S0040-1951(01)00218-9

Bakioglu KB (2014) Garzón massif basement tectonics: a geopyhysical study, Upper Magdalena valley, Colombia. Master's Thesis University of South Carolina, Columbia. Retrieved from http://scholarcommons.sc.edu/etd/2782

Baquero M, Grande S, Urbani F, Cordani U, Hall C, Armstrong R (2015) New evidence for Putumayo crust in the basement of the Falcon Basin and Guajira Peninsula, Northwestern Venezuela. In: Bartolini C, Mann P (eds) Petroleum geology and potential of the Colombian

Caribbean margin, AAPG Memoir, vol 108. The American Association of Petroleum Geologists, Tulsa, pp 103–136

Barrero D, Laverde F (1998) Estudio Integral de evaluación geológica y potencial de hidrocarburos de la cuenca "intramontana" Cauca- Patia, ILEX- Ecopetrol report, Inf. No. 4977

Barrero D, Laverde F, Ruiz CA, Alfonso C (2006) Oblique collision and basin formation in Western Colombia: the origin, evolution and petroleum potencial of Cauca-Patia basin. IX Simposio Bolivariano de Cuencas Sub-Andinas, Cartagena de Indias, CD

Barrios YA, Baptista N, Gonzales G (2011) New exploration traps in the Espino Graben, Eastern Venezuela Basin. Search and Discovery Article #10333

Bartok PE, Renz O, Westermann GEG (1985) The Siquisique ophiolites, Northern Lara State: a discussion on their Middle Jurassic ammonites and tectonic implications. Geol Soc Am Bull 96(8):1050–1055

Benkovics L, Asensio A (2015) New evidences of an active strike-slip fault system in northern Venezuela, near offshore Perla field. In: Bartolini C, Mann P (eds) Petroleum geology and potential of the Colombian Caribbean margin, AAPG Memoir, vol 108. The American Association of Petroleum Geologists, Tulsa, pp 749–764

Bogotá J (1983) Estratigrafía del Paleozóico Inferior en el area amazónica de Colombia. Geología Norandina, Marzo, pp 29–38

Burke K (1977) Aulacogens and continental breakup. Annu Rev Earth Planet Sci 5:371–396

Cediel F (1969) Die Giron-Gruppe: Eine frueh-mesozoische Molasse der Ostkordillere Kolumbiens. Neues Jahrbuch Geologie und Palaeontologie, Abhandlungen, Munchen 133(2):111–162

Cediel F (2010) Geologic map of the Sinu and San Jacinto Belts, North West Colombia. Cartographic compilation at scale 1:300.000. Department of Geology, EAFIT University

Cediel F (ed) (2011) Petroleum geology of Colombia. (35 authors and co-authors) vol 15 (basin by basin) 1262 p 1348 figs. EAFIT University, Medellin

Cediel F, Cáceres C (2000) Geological map of Colombia, 3rd edn. Geotec Ltd, Bogotá

Cediel F, Mojica J, Macia C (1980) Definicion Estratigrafica del Triasico en Colombia, Suramerica, Formaciones Luisa, Payande y Saldana. Newsl Stratigr 9(2):73–104, Berlin, Stuttgart

Cediel F, Barrero D, Caceres C (1998) Seismic Atlas of Colombia: seismic expression of structural styles in the basins of Colombia, Robertson Research International, UK, ed, vol 1 to 6. Geotec Ltd, Bogota

Cediel F, Shaw R, Cáceres C (2003) Tectonic assembly of the northern Andean block. In: Bartolini C, Buffer RT, Blickwede J (eds) The circum—Gulf of Mexico and the Caribbean: hydrocarbon habitats, basin formation and plate tectonics, AAPG Memoir, vol 79. American Association of Petroleum Geologists, Tulsa, pp 815–848.1

Cediel F, Restrepo I, Marín-Cerón MI, Duque-Caro H, Cuartas C, Mora C, Montenegro G, García E, Tovar D, Muñoz G (2010) Geology and hydrocarbon potential, Atrato and San Juan Basins, Chocó (Panamá) Arc, Colombia, Tumaco Basin (Pacific Realm), Colombia, Agencia Nacional de Hidrocarburos (ANH)-EAFIT, Medellín

Cerón JF, Kellogg JN, Ojeda GY (2007) Basement configuration of the northwestern South America – Caribbean margin from recent geophysical data. C T F Cienc Tecnol Futuro, Bucaramanga, 3(3):25–50

Chacón PA, Reyes J, Cáceres C, Sarmiento G, Cramer T (2013) Análisis estratigráfico de la sucesión del Devónico-Pérmico al oriente de Manaure y San José de Oriente (Serranía del Perijá, Colombia). Geología Colombiana vol 38. Bogotá Colombia, pp 5–24

Clark SA (2007) Characterizing the southeast Caribbean-South American plate boundary at 64°W. PhD Thesis, 79 pp, Rice University, Houston

Clark SA, Zelt CA, Magnani MB, Levander A (2008) Characterizing the Caribbean–South American plate boundary at 64°W using wide-angle seismic data. J Geophys Res 113:B07401. https://doi.org/10.1029/2007JB005329

Colmenares L, Zoback MD (2003) Stress field and seismotectonics of northern South America. Geology 31(8):721–724

Escalona A, Mann P (2011) Tectonics, basin subsidence mechanisms and paleogeography of the Caribbean-South American plate boundary zone. Mar Pet Geol 28:8–30

Espinosa A (1980) Sur les roches basiques et ultrabasiques du bassin du Patia (Cordillere Occidentale des Andes colombiennes): Etude Geologique et Petrographique. These de Docteur Sciences de la Terre, Departement de Mineralogie, Universite de Geneve

Estrada A (1967) Asociacion magmatica basica del Nechi: Tesis de grado, Facultad Nacional de Minas, Medellin, 88 pp

Estrada JJ (1995) Paleomagnetism and accretion events in the Northern Andes. PhD Thesis, State University of New York

Etayo F, Cáceres C, Cediel F (1994) Facies distribution and tectonic setting through the Phanerozoic of Colombia: INGEOMINAS, ed., Geotec Ltd., Bogotá (17time-slices/maps in scale 1:2,000.000)

Farris DW, Jaramillo C, Bayona G, Restrepo-Moreno SA, Montes C, Cardona A, Mora A, Speakman RJ, Glascock MD, Valencia V (2011) Fracturing of the Panamanian Isthmus during initial collision with South America. Geology 39(11):1007–1010

Florez JM, Mavko G (2001) Elastic geomechanic model of the Bucaramanga and Oca Faults, and the origin of the Sierra Nevada de Santa Marta; Northern Andes. Stanford Rock Physics Laboratory, Department of Geophysics, Stanford University

Forero A (1968) Estratigrafía del Pre-Cretáceo en el borde occidental de la Serranía de Perijá. Geología Colombiana 7:7–78

Gansser A (1941) (Contributors: Renz, O., Hubach, E.) Geological report, Shell No. 100. Central Macarena. 16 pp, 25 photos, 2 tabs., 9 Annex. (in-house files)

Gansser A (1974) The Roraima problem (South America). Mitteilungen aus dem Geologischen Institut der Eidg. Technischen Hochschule und der Universität Zürich, Zürich 177:80–100

Geosearch (2008) Evolucion geohistorica de la Sierra Nevada de Santa Marta. Bogota

Geotec Ltda (1986) In-house files

Geotec Ltda (1998) The foothills of the Eastern Cordillera, Colombia. Geological Map, Scale 1:100.000

GEOTERREX (1979) Aeromagnetic Interpretation Atrato – Sinú

Geyer OF (1973) Das Praekretazische Mesozoikum von Kolumbien. Geologisches Jarbuch 5:1–56

Gleeson SA, Herrington RJ, Durango J, Velasquez CA, Koll G (2004) The mineralogy and geochemistry of Cerro Matoso, S.A. Ni-laterite deposit, Montelibano, Colombia. Econ Geol 99:1197–1121

Godin, L., Grujic, D., Law, R. D. & Searle, M. P. (2006) Channel flow, ductile extrusion and exhumation in continental collision zones: an introduction. In: Law RD, Searle MP, Godin L (eds) Channel flow, ductile extrusion and exhumation in continental collision zones, Special publications, 268. Geological Society, London, The Geological Society of London, pp 1–23

González H (2001) Mapa geológico del Departamento de Antioquia: Geología, recursos minerales y amenazas potenciales, Memoria Explicativa. INGEOMINAS, Bogotá

González GR, Holguín QN (1991) Geology of the source rocks of Mexico. Source rock geology. XIII World Petroleum Congress, Buenos Aires, Argentina, Topic 2, Forum with Posters, pp 1–10

Goossens PJ, Rose WI, Flores D (1977) Geochemistry of Tholeiites of the basic igneous complex of north-western South America. Bull Geol Soc Am 88:1711–1720

Gutscher M (2002) Andean subduction styles and their effect on thermal structure and interplate coupling. J S Am Earth Sci 15:3–10

Hastie AR, Kerr AC (2010) Mantle plume or slab window?: physical and geochemical constraints on the origin of the Caribbean oceanic plateau. Earth Sci Rev 8(3):283–293

Ibanez-Mejia M (2010) New U-Pb geochronological insights into the Proterozoic tectonic evolution of Northwestern South America: the Meso- neoproterozoic Putumayo orogen of Amazonia and implications for Rodinia reconstructions. Master of Science in the Graduate College the University of Arizona

Jaimes E, de Freitas M (2006) An Albian–Cenomanian unconformity in the northern Andes: evidence and tectonic significance. J S Am Earth Sci 21:466–492

Jimenez-Mejia DM, Caetano J, Cordani UG (2006) P–T–t conditions of high-grade metamorphic rocks of the Garzon Massif, Andean basement, SE Colombia. J S Am Earth Sci 21:322–336

Kellogg JN, Bonini WE (1982) Subduction of the Caribbean Plate and basement uplifts in the overriding South American Plate. Tectonics 1(3):251–276

Kennan L, Pindell J (2009) Dextral shear, terrane accretion and basin formation in the Northern Andes: best explained by interaction with a Pacific-derived Caribbean Plate? In: James KH, Lorente MA, Pindell JL (eds) The origin and evolution of the Caribbean Plate, Special publications 328. Geological Society, London, pp 487–531

Keppie JD (2008) Terranes of Mexico revisited: a 1.3 billion year odyssey. In: Keppie JD, Murphy JB, Ortega-Gutierrez F, Ernst WG (eds) Middle American terranes, potential correlatives, and orogenic processes. CRC Press – Taylor & Francis Group, Boca Raton, pp 7–36

Keppie J, Duncan R, Damian NJ, Dostal A, Ortega-Rivera BV, Miller D, Fox J, Muise JT, Powell SA, Mumma SA, JWK L (2004) Mid-Jurassic tectonothermal event superposed on a Paleozoic geological record in the Acatlán Complex of Southern Mexico: hotspot activity during the breakup of Pangea. Gondwana Res 7(1):239–260. © 2004 International Association for Gondwana Research, Japan. ISSN: 1342-937X

Kerr AC, Tarney J (2005) Tectonic evolution of the Caribbeanand northwestern South America: the case for accretion of two late cretaceous oceanic plateaus. Geology 33(4):269–272

Kerr AC, Tarney J, Marriner GF, Nivia A, Saunders AD (1997) The Caribbean–Colombian cretaceous igneous province: the internal anatomy of an oceanic plateau. In: Mahoney JJ, Coffin MF (eds) Large igneous provinces: continental, oceanic, and planetary flood volcanism, Geophysical monograph 100. American Geophysical Union, Washinghton, pp 123–144

Kerr AC, White RV, Thompson PME, Tarney J, Saunders AD (2003) No oceanic plateau—no Caribbean plate? The seminal role of an oceanic plateau in Caribbean plate evolution. In: Bartolin C, Buffler RT, Blickwede J (eds) The circum-Gulf of Mexico and the Caribbean: hydrocarbon habitats, basin formation, and plate tectonics, AAPG Memoir 79, pp 126–168

Kerr AC, Tarney J, Kempton PD, Pringle M, Nivia A (2004) Mafic pegmatites intruding oceanic plateau gabbros and ultramafic cumulates from Bolıvar, Colombia: evidence for a 'wet' mantle plume? J Petrol 45(9):1877–1906. https://doi.org/10.1093/petrology/egh037

Klitgord KD, Schouten H (1986) Plate kinematics of Central Atlantic. In: Vogt PR, Tucholke BE (eds) The geology of North America. The Western North Atlantic Region, Geol Soc Am 22, pp 351–378

Laubscher HP (1957) The kinematic puzzle of the Neogene Northern Andes. In: Schaer JP, Rodgers J (eds) The anatomy of mountain ranges, chapter 11. Princeton Legacy Library. Princeton, New Jersey

Leal-Mejía H (2011) Phanerozoic Gold Metallogeny in the Colombian Andes: A tectono-magmatic approach. PhD Thesis, Universitat de Barcelona

Leal-Mejía H, Shaw RP, Melgarejo JC (2018) Spatial-temporal migration of granitoid magmatism and the Phanerozoic tectono-magmatic evolution of the Colombian Andes. In: Cediel F, Shaw RP (eds) Geology and tectonics of Northwestern South America. Springer, Cham, pp 253–397

Linares F, Orihuela N, García AY, Audemard F (2014) Generación del mapa de basamento de la Cuenca de Falcón a partir de datos gravimétricos de modelos combinados. Geociencias Aplicadas Latinoamericanas 1:9–19. https://doi.org/10.3997/2352-8281.20140002

Londoño JM, Bohorquez OP, Ospina LF (2010) Tomografía sísmica 3D del sector de Cúcuta, Colombia. Boletín de Geología, Bucaramanga, 32(1):107–124

Londoño J, Schiek C, Biegert E (2015) Basement architecture of the Southern Caribbean Basin, Guajira Offshore, Colombia. In: Bartolini C, Mann P (eds) Petroleum geology and potential of the Colombian Caribbean margin, AAPG Memoir 108. The American Association of Petroleum Geologists, Tulsa, pp 85–102

Lopez J (1986) Geology, mineralogy and geochemistry of the Cerro Matoso nickeliferous laterite, Cordoba, Colombia. Department of Earth Resources, Master of Science, Colorado State University

MacDonald WD, Opdyke ND (1972) Tectonic rotations suggested by paleomagnetic results from northern Colombia, South America: VI Conferencia de Geología del Caribe, Margarita, Venezuela, Memorias, 301 p

Mann P (ed) (1995) Geologic and tectonic development of the Caribbean plate boundary in Southern Central America, Geological Society of America special paper 295, pp Vii–XXii

Mann P, Escalona A, Castillo V (2006) Regional geologic and tectonic setting of the Maracaibo supergiant basin, western Venezuela. AAPG Bull 90(4):445e477

Mantilla A (2007) Crustal structure of the southwestern Colombian Caribbean margin. Geological interpretation of geophysical data. Dissertation Dr rer nat. Chemisch-Geowissenschaftlichen Fakultät der Friedrich-Schiller-Universität Jena

Marechal P (1983) Les Temoins de Chaine Hercynienne dans le Noyau Ancien des Andes de Merida (Venezuela): Structure et evolution tectometamorphique. Doctoral Thesis, Universite de Bretagne Occidentale France 176: 11.6.C

Maresch WV, Stoeckhert B, Baumann A, Kaiser C, Kluge R, Krueckhans-Lueder G, Brix MR, Thomson M (2000) Crustal history and plate tectonic development in the southern Caribbean, Zeitschrift fuer Angewandte Geologie. Sonderheft Zeitschrift Angewandete Geologie 1:283–289

Maresch WV, Kluge R, Baumann A, Pindell JL, Krückhans-Lueder G, Stanek K (2009) The occurrence and timing of high-pressure metamorphism on Margarita Island, Venezuela: a constraint on Caribbean-South America interaction. Geol Soc London Spec Publ 328(1):705–741

Maze WB (1984) Jurassic La Quinta Formation in the Sierra de Perija, northwestern Venezuela: geology and tectonic environment of red beds and volcanic rocks. GSA Memoirs 162:263–282

McGeary S, Ben-Abraham S (1989) The accretion of Gorgona Island, Colombia: multichannel seismic evidence. In: Howel DG (ed) Tectonostratigraphic terranes of the circum-Pacific region, Earth sciences series 1. Circum-Pacific Council for Energy and Mineral Resources, Houston, pp 543–554

Mendi D, Baquero ML, Oliveira EP, Urbani F, Pinto J, Grande S, Valencia V (2013) Petrography and U-Pb Zircon Geochronology of Geological Units of the Mesa de Cocodite, Península de Paraguaná, Venezuela. American Geophysical Union, Spring Meeting 2013, abstract #V53A-02

Monod B, Damien D, Hervouët Y (2010) Orogenic float of the Venezuelan Andes. Tectonophysics 490:123–135

Montes C, Cardona A, McFadden R, Morón SE, Silva CA, Restrepo-Moreno S, Ramírez DA, Hoyos N, Wilson J, Farris D, Bayona G, Jaramillo CA, Valencia V, Bryan J, Flores JA (2012) Evidence for middle Eocene and younger land emergence in central Panama: implications for Isthmus closure. Geol Soc Am Bull 124(5–6):780–799

Montes C, Cardona A, Jaramillo C, Pardo A, Silva JC, Valencia V, Ayala C, Pérez-Angel LC, Rodríguez-Parra LA, Ramirez V, Niño H (2015) Middle Miocene closure of the American seaway. Science 348(6231):226–229

Moreno O, Guerrero C, Rey A, Gómez P, Audemard F, Fiume G (2009) Modelo Alternativo para el desarrollo del Frente Deformado Costa fuera del Caribe Colombiano. ACGGP, X Simposio Bolivariano Exploración Petrolera en Cuencas Subandinas, Cartagena, pp 1–6

Munoz J and Vargas H (1981) Petrologia de las anfibolitas y neises Precambricos presentes entre los ríos Mendarco y Ambeima, Tolima, Colombia. Tesis, Universidad Nacional de Colombia

Nerlich R, Clark SR, Bunge HP (2014) Reconstructing the link between the Galapagos hotspot and the Caribbean Plateau. GeoResJ 1–2:1–7

Nürnberg D, Müller DR (1991) The tectonic evolution of the South Atlantic from late Jurassic to present. Tectonophysics 191:27–53

Oldow JS, Albert W, Bally HG, Lallemant A (1990) Transpression, orogenic float, and lithospheric balance. Geology 18:991–994

Pardo-Casas F, Molnar P (1987) Relative motion of the Nazca (Farallon) and South American plates since late Cretaceous time. Tectonics 6(3):233–248

Piraquive A (2016) Marco estructural deformaciones y exhumación de los Esquistos de Santa Marta: la acreción e historia de deformación de un terreno Caribeño al norte de la Sierra Nevada de Santa Marta. Tesis doctoral. Universidad Nacional de Colombia, 262 p

Rabe EH (1997) Contributions to the stratigraphy of the East-Andean Area of Colombia. PhD Dissertation, Universität Giessen

Restrepo-Pace P (1989) Restauración de la sección geológica Cáqueza Puente Quetame: Moderna interpretación estructural de la deformación del flanco este de la Cordillera Oriental. Universidad Nacional de Colombia, Bogotá

Restrepo-Pace P (1995) Late Cambrian to early Mesozoic tectonic evolution of the Colombian Andes, based on new geochronological, geochemical and isotopic data. PhD Dissertation, Departmente of Geosciences, University of Arizona

Restrepo-Pace P, Colmenares F, Higuera C, Mayorga M (2004) A foldand thrust belt along the western flank of the Eastern Cordillera of Colombia: style, kinematics and timing constraints derived from seismic data and detailed surface mapping. In: KR MC Clay (ed) Thrust tectonics and hydrocarbon systems, AAPG Memoir, 82:598–613

Robertson Research RR (1982) Paleozoic sections in the Llanos of Colombia. Unpublish report

Sanchez J, Palma M (2014) Crustal density structure in northwestern South America derived from analysis and 3-D modeling of gravity and seismicity data. Tectonophysics 634:97–115

Santos JO, Potter PE, Reis NJ, Hartmann LA, Fletcher IR, McNaughton NJ (2003) Age, source and regional stratigraphy of the Roraima Supergroup and Roraima-like outliers in northern Sout America base don U-Pb geochronology. GSA Bull 115(3):331–348. 15 figs., 3 tab

Schubert C (1986) Stratigraphy of the Jurassic La Quinta Formation, Merida Andes, Venezuela: Type Section. Zeitschrift der Deutschen Geologischen Gesellschaft Band 137:391–411

Seccia D (2012) Deep geometry of subduction below the Andean belt of Colombia as revealed by seismic tomography. Dottorato di Ricerca in Geofisica Ciclo XXIV: Geofisica della terra solida. Alma Mater Studiorum – Universita' di Bologna

Sierra G (2011) Amaga, Cauca and Patia basins. Geology and hydrocarbon potential. In: Cediel F (ed) Geology and hydrocarbon potential, Regional Geology of Colombia. Department of Geology, University EAFIT, Medellín

Silver EA, Reed DL, Tagudin JE, Heil DJ (1990) Implications of the north and south Panama thrust belts for the origin of the Panama orocline. Tectonics 9:261–281

Sinton CW, Duncan RA, Storey M, Lewis J, Estrada JJ (1998) An oceanic flood basalt province within the Caribbean plate. Earth Planet Sci Lett 155(3):221–235

Stephan JF, Mercier de Lepinay B, Calais E, Tardy M, Beck C, Carfantan JC, Olivet JL, Vila JM, Bouysse P, Mauffret A, Bourgois J, Thery M, Tournon J, Blanchet R, Dercourt J (1990) Paleogeodynamic maps of the Caribbean; 14 steps from Lias to present. Bulletin de la Société Géologique de France, (8), t VI(6):915–919

Stibane FR (1968) Zur Geologie von Kolumbien, Südamerika. Das Quetame- und Garzón Massiv. Geotek Forsch, 30 I-II + 1–85, 26 Abb 3 Taf Stuttgart

Stoeckhert B, Maresch WV, Brix M, Kaiser C, Toetz A, Kluge R, Krueckhans-Lueder G (1995) Crustal history of Margarita Island (Venezuela) in detail: Constraint on the Caribbean Plate-Tectonic scenario. Geology 5(9):787–790

Sung Hi Choi, Mukasa SB, Andronikov AV, Marcano MC (2017) Extreme Sr–Nd–Pb–Hf isotopic compositions exhibited by the Tinaquillo peridotite massif, northern Venezuela: implications for geodynamic setting. Contrib Mineral Petrol 153:443–463. https://doi.org/10.1007/s00410-006-0159-3

Suter F, Neuwerth R, Gorin G, Guzmán C (2008) (Plio-) Pleistocene alluvial lacustrine basin infill evolution in a strike-slip active zone (Northern Andes, Western-Central Cordilleras, Colombia). Geol Acta 6(3):231–249

Syracuse EM, Maceira M, Prieto GA, Zhang H, Ammon CJ (2016) Multiple plates subducting beneath Colombia, as illuminated by seismicity and velocity from the joint inversion of seismic and gravity data. Earth Planet Sci Lett 444:139–149

Taboada A, Rivera L, Fuenzalida M, Cisterna A, Philip H, Bijwaard H, Olaya J, Rivera C (2000) Geodynamics of the northern Andes: Subductions and intracontinental deformation (Colombia). Tectonics 19(5):787–813

Van der Lelij R (2013) Reconstructing north-western Gondwana with implications for the evolution of the Iapetus and Rheic Oceans: a geochronological, thermochronological and geochemical study. PhD Thesis, Université de Genève

Vargas CA, Mann P (2013) Tearing and breaking off of subducted slabs as the result of collision of the Panama arc-indenter with Northwestern South America. Bull Seismol Soc Am 103(3):2025–2046

Vásquez M, Uwe R, Sudo M, Moreno JM (2010) Magmatic evolution of the Andean Eastern Cordillera of Colombia during the cretaceous: influence of previous tectonic processes. J S Am Earth Sci 29:171–186

Villaseñor AB, Olóriz F, López Palomino I, López-Caballero I (2012) Updated ammonite biostratigraphy from Upper Jurassic deposits in Mexico. Revue de Paléobiologie, Genève Vol. spéc 11:249–267

Viscarret P, Wright J, Urbani F (2009) New U-Pb zircon ages of El Baúl Massif, Cojedes State, Venezuela. Rev Téc Ing Univ Zulia 32(3):210–221

Weber MBI (1998) The Mercaderes-Rio Mayo xenoliths, Colombia: Their bearing on mantle and crustal processes in the Northern Andes. PhD Thesis, Faculty of Science of the University of Leicester

Zarifi Z, Havskov J, Hanyga A (2007) An insight into the Bucaramanga Nest. Tectonophysics 443:93–105

Zuluaga C, Pinilla A, Mann P (2015) Jurassic silicic volcanism and associated Continental-arc Basin in northwestern Colombia (southern boundary of the Caribbean plate). In: Bartolini C, Mann P (eds) Petroleum geology and potential of the Colombian Caribbean margin, AAPG Memoir, vol 108. The American Association of Petroleum Geologists, Tulsa, pp 137–160

Chapter 2
Proterozoic Basement, Paleozoic Tectonics of NW South America, and Implications for Paleocontinental Reconstruction of the Americas

Pedro A. Restrepo-Pace and Fabio Cediel

Abbreviations

CCB	Cauarane-Coeroeni belt
CGMW	Commission for the Geological Map of the World
COGEMA	Compagnie Générale des Matières nucléaires
CPRM	Companhia de Pesquisa de Recursos Minerais, (Serviço Geológico do Brasil)
Ga	Giga-annum, billion (10^9) years
ITD	Isothermal decompression
LA-(MC)-ICP-MS	Laser ablation (multicCollector) inductively coupled plasma mass spectrometry
Ma	Mega-annum, million (10^6) years
PRORADAM	Proyecto Radargramétrico del Amazonas
REE	Rare earth elements
RNJ	Rio Negro-Juruena (geological province)
SHRIMP	Sensitive high-resolution ion micro probe
TDM	Depleted mantle age
TTG	Tonalite-trondhjemite-granodiorite
UHT	Ultrahigh temperature

P. A. Restrepo-Pace (✉)
Oilsearch Limited, Bligh Street- level 23, Sydney, NSW, Australia
e-mail: pedro.restrepo@oilsearch.com

F. Cediel
Consulting Geologist, Department of Geology University EAFIT, Medellín, Colombia

2.1 Introduction

Pre-Jurassic paleocontinental reconstructions are largely built from circumstantial geological evidence, given the absence of contiguous oceanic crust between intervening continental fragments. Gathering such evidence is fraught with greater difficulty from rocks that have undergone strong and recent orogenic overprints as in the case of the Andes. The Rondinia paleocontinental reconstruction of Hoffman (1991) provided a framework to investigate the interplay of the major continental fragments since Late Proterozoic time. Of particular interest here, it has been the long-standing geological debate this reconstruction generated regarding the interactions of the margins of Amazonia and Laurentia in Late Proterozoic to Early Paleozoic time (Bond et al. 1984; Kent and Van der Voo 1990; Hoffman 1991; Keppie et al. 1991; Keppie 1993; Dalla Salda et al. (1992a, b); Park 1992; Dalziel et al. 1994 and others).

Geological, geochronological data summarized here constrains the consolidation of the proto-Andean orogen in the northern Andes (Colombia-Venezuela) during the Grenvillian-Orinoquiense (~1.0 Ga) and Caparonensis-Quetame (~0.47–0.43 Ga) orogenic events. Data also seem to support that these discrete events extend along the Andes in Ecuador, Peru, and Argentina. A fragment of the northern South American basement may have become attached to Mexico in Late Paleozoic time as suggested by Yañez et al. (1991) and Restrepo-Pace (1995). Provincial fauna from Paleozoic sediments further constrains paleocontinental positions with a major shift in affinity by mid-Paleozoic time. The Rodinia model of Hoffman and its suggestion that the proto-Andes consists of remobilized pericratonic sequences seems to honor geological data from northern South America here presented.

2.2 Andean Basement of Colombia-Venezuela

Differential uplift and denudation resulting from Andean (Meso-Cenozoic) tectonics have left but sparse basement exposures in the northern Andes. Bordering the Andean realm, the basement crops out as isolated massifs in El Baúl in Venezuela and the Macarena in Colombia (Fig. 2.1). In the Andean domain proper, the basement occurs in the cores of regional inversion anticlinoria in the Mérida Andes (Venezuela), Santander, and Garzón (also referred to as massifs in local geological literature). The Borde Llanero Fault System is a present structural boundary between the Andean domain to the west and the cratonic domain to the east. The Borde Llanero Fault System is a deep-seated inversion system—with varying degrees of along strike oblique slip—that accounts for the present relief along the Eastern Andean chain. Paleozoic metamorphic units exist to the west of this structural boundary and are absent in the eastern cratonic domain. The cratonic domain characterized by the presence of Amazonian basement rocks dated 2.5–1.5 Ga (Priem et al. 1982) which are unconformably overlain by Upper Cambrian

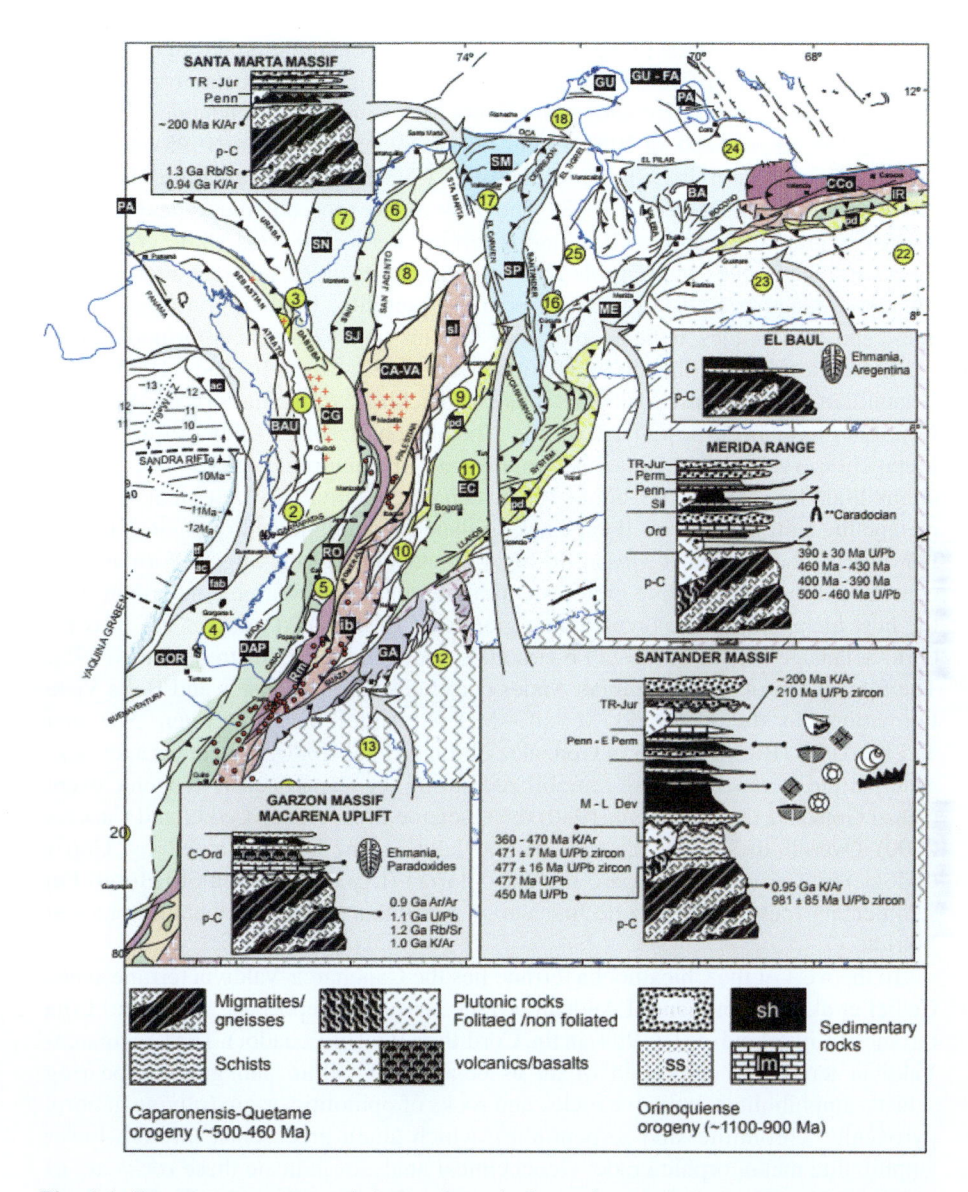

Fig. 2.1 Basement exposures of the northern Andes and present day structural boundaries. Summary of stratigraphic relationships for the basement exposures of the northern Andes. Age constraints derived from field relationships

to Upper Ordovician marine sedimentary rocks. The Cambro-Ordovician sequence is exposed in the Macarena-El Baúl localities and has been detected by numerous wells and seismically mapped in the subsurface of the Andean foreland basin. In subsurface it is estimated to consist of over 2 km folded marine Vendian and Cambro-Ordovician sediments subcropping the Mesozoic strata (Dueñas 2001).

In the west of the Borde Llanero Fault System, the core of the Andes of Colombia-Venezuela consists of Grenville-age (~1.2–1.0 Ga) high-grade metamorphic rocks exposed at the Garzón, Santander, and Santa Marta massifs (Tschantz et al. 1974; Alvarez 1981; Kroonenberg 1982; Priem et al. 1989; Restrepo-Pace 1995) as well as in the Colorado Massif in eastern Mérida Andes (Sierra Nevada Formation, González de Juana et al. 1980; Marechal 1983). The older units exposed in the Andean domain consist of granulitic charnockites, garnetiferous charnockitic-enderbitic granulites, metacalcsilicate rocks and hornblende-biotite augen gneisses, and rare anorthosites. The assemblage overall is of pelitic-psamitic protolith. U-Pb zircon ages, Rb-Sr ages, and Ar-Ar ages indicate that they belong to a Grenvillian-Orinoquienese (~1.0 Ga) tectonothermal event (Kroonenberg 1982; Restrepo-Pace 1995). The Grenvillian-Orinoquiense rocks that make the backbone of the Eastern Cordillera of Colombia have been designated as the Chicamocha terrane (Cediel et al. 2003). Metapelitic rocks of greenschist-amphibolite metamorphic grade overlie the Grenvillian basement. The contact between the Grenvillian basement and the metapelitic suite is never well exposed, so their exact relationship remains cryptic. However, at the Santander Massif the age of metapelites of the Silgará Formation is constrained by calk-alkaline granites exhibiting a strong foliation concordant with the host metapelitic suite (syntectonic granites). The foliated granites are of early Ordovician age (477 ± 16 Ma U/Pb zircon crystallization age, Restrepo-Pace 1995). The Silgará Fm of the Colombian Andes correlates with the Tostós and Bella Vista Formations in the Colorado Massif—Mérida Andes dated between 500 and 475 Ma U/Pb (Burkley 1976 in González de Juana et al. 1980). The latter ages suggest that the metapelites were remobilized during the Caparonensis orogenic event (sensu González de Juana et al. 1980) the Quetame event (sensu Cediel and Caceres 2000) Overall these. The closure of the latter event is constrained by Upper Ordovician (Caradocian) Caparo Fm and Silurian (Llandovery-Wenlok) Horno Fm sedimentary rocks (González de Juana et al. 1980) that overlie the metamorphics in Mérida.

To the west of the Chicamocha terrane lies the Cajamarca-Valdivia terrane sensu Cediel et al. 2003 or Central Andean terrane sensu Restrepo-Pace 1992, and Loja terrane of Litherland et al. (1994) in the Cordillera Real of Ecuador). The Cajamarca-Valdivia terrane is composed of an association of pelitic and graphite-bearing schists, amphibolites, intrusive rocks, and rocks of ophiolitic origin (olivine gabbro, pyroxenite, chromitite, and serpentinite), which attain greenschist through lower amphibolite metamorphic grade. Geochemical analyses indicate these rocks are of intraoceanic arc and continental margin affinity (Restrepo-Pace, 1992). They form a parautochthonous accretionary prism of Ordovician-Silurian (?) age, sutured to the Guiana Shield in the south, along the Palestina and Cosanga fault systems.

Silurian rocks have been reported in few localities in the Eastern Cordillera (Grösser and Prössl 1991). In the Mérida Andes of Venezuela and in Ecuador, the Silurian is well developed, along the Caparo and Pumbuiza basins, respectively, while the Devonian is largely absent. This contrasting exposure or preservation may be the result of differential uplift/denudation and/or subsidence following the

Quetame-Caparonensis Orogenic episode. It should be noted however, that the Silurian period is relatively short (~20 Ma), and the meager Silurian fauna thus far recovered in the Colombian localities may not been particularly diagnostic to constrain it. Post-tectonic granites along the Eastern Andes of Colombia also include a suite of non-foliated granites with ages between 470 and 360 Ma (Goldsmith et al. 1971; Etayo-Serna and Barrero 1983; Boinet et al. 1985; Maya 1992; Restrepo-Pace 1995). In the Venezuelan Andes U-Pb ages indicate two distinct magmatic events from 460 to 430 Ma and from 400 to 390 Ma (Burkley 1976; Shagam 1977; Benedetto 1982; Benedetto and Ramírez 1985). Middle to Late Devonian sediments containing critical diagnostic paleogeographic tracer fauna and Pennsylvanian to Permian marine sequences overlie unconformably the metamorphic basement.

2.3 Late Precambrian-Paleozoic Forensics of the Northern Andes of South America

The most important tectonic events that lead to the consolidation of the Andean basement of Northwestern South America followed the docking of the Oaxaquia and the Cajamarca-Valdivia terranes (Fig. 2.2). As Oaxaquia accreted to the South American margin, the Guejar and Arauca impactogens were generated. Subsequently trench rollback allowed for arc development—proto Cajamarca-Valdivia terrane-, the subsidence of the remobilized Chicamocha basement, rifting of the Guape and Arauca and for epicontinental sequences to be deposited in Cambro-Ordovician time. It is during the Early Paleozoic that the continental wedge of the Chicamocha terrane and the western margin of the Guiana Shield developed at its subsiding margin extensive sequences of marine and epicontinental sediments. These supracrustal sequences underwent Cordilleran-type orogenic deformation and regional metamorphism during an event variably recorded as the Quetame orogeny in Colombia, the Caparonensis orogeny in Venezuela, and the Ocloy orogeny in Ecuador and Peru. In Colombia and Ecuador, evidence for this extensive event includes the fragments of ophiolite and accretionary prism exposed in the Cajamarca-Valdivia, Loja, and El Oro terranes.

The Cajamarca-Valdivia (Loja) terrane was sutured to continental South America along a paleomargin that followed the approximate trace of the paleo-Palestina fault system and its southern extension in Ecuador, approximated by the Cosanga fault (note that the modified trace of the Palestina system reflects reactivation during the Mesozoic). The continuation of this suture into southern Ecuador can be inferred based on occurrence of the pre-Jurassic Zumba ophiolite (Litherland et al. 1994). Farther east (inland), this orogeny is recorded by a lower- to subgreenschist-grade metamorphic event that affected the thick psammitic and pelitic Ordovician-Silurian supracrustal sequences. These metamorphosed sequences outcrop in the Eastern Cordillera (Quetame group), the Santander-Perija´ belt (Silgara´ group), the Sierra Nevada de Santa Marta, the Sierra de Mérida, and the Cordillera Real (Chiguinda

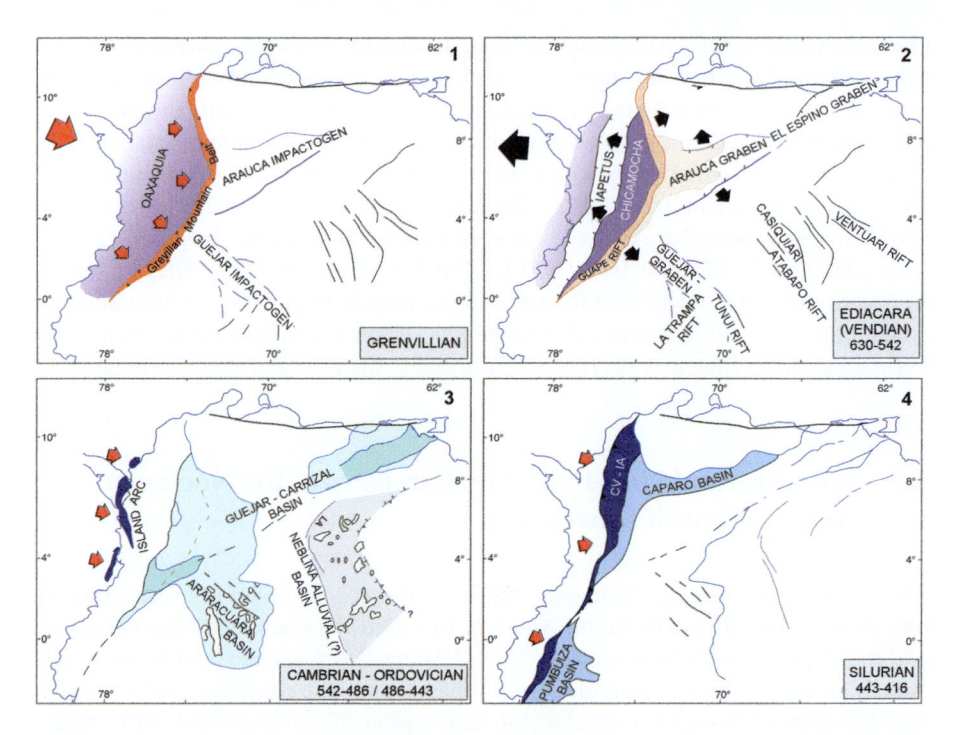

Fig. 2.2 Paleogeography and tectonic evolution of NW South America. For explanation refer to text

unit). They are correlated with penecontemporaneous strata that form the basal portion of the onlapping Paleozoic supracrustal sequences of the Maracaibo, Llanos, Barinas-Apure, and Putumayo-Napo basins.

The low-grade, subgreenschist nature of the metamorphism outlined above has led to problems in correlating this regional event and, in some instances, the interpretation of multiple, more localized events (see discussion and references in Restrepo-Pace 1995). We feel that this apparent provinciality with respect to Ordovician-Silurian regional metamorphism in northwestern South America is unfounded and is more an artefact of the mechanisms behind regional metamorphism in general than a reflection of the existence of multiple events. For example, in the Eastern Cordillera, weakly to nonmetamorphosed windows of Ordovician-Silurian strata are observed. These rocks preserve diagnostic marine fauna for identification and dating, and they can be correlated with lower greenschist rocks of the same age that exhibit the imprint of regional metamorphism without having to evoke any major difference in overall tectonic history. The concept of "igneous-related low-pressure metamorphism" recognized by Restrepo-Pace (1995, pp. 27–28) in the Santander massif during the Late Triassic to Early Jurassic may be applied with equal validity to help explain the provincial nature of Paleozoic regional metamorphism. A similar, although contrary, form of protolith preservation is observed in the amphibolite-grade Cajamarca-Valdivia terrane to the west. Here, regional metamorphism of the accretionary prism assemblage has left relicts of Orinoco (Grenville)-aged granulite basement lodged and preserved in the amphibolite-grade metamorphic assemblages of the Cajamarca and Valdivia groups (Cediel and Caceres 2000). The collision and amalgamation of the Cajamarca-Valdivia arc mark the closure and consolidation of the basement in this part of the Andes.

2.4 The Bigger Picture

The Grenvillian age (~1.0 Ga) basement of the Andes of northern South America, could be traced further south into Ecuador, Peru, Bolivia, and northern Argentina (Ramos 1988; Wasteneys 1994, Litherland et al. 1989; Restrepo-Pace 1995; Restrepo-Pace et al. 1997; Chew et al. 2007). Lower Ordovician syntectonic granites with ages ranging from 500 to 475 Ma date the climax of the Caparonesis orogenic episode in northern South America. Rocks involved in this tectonothermal event can be traced in the central Andes and the southern Andes as well. The Caparonensis event correlates with the early stages of the Famatinian Orogenic cycle (Guandacol phase Rapela et al. 1990) of the Puna of northern Argentina and southern Bolivian Andes. In the latter, it is marked by numerous syntectonic intrusions with ages ranging from 480 to 460 Ma and low to medium pressure high-temperature metamorphism (Aceñolaza 1982; Rapela et al. 1990 and others). Closure of the Caparonensis-Quetame event in northern South America is constrained by the presence of (unmetamorphosed) Upper Ordovician (Caradocian), Caparo Fm and Silurian (Llandovery-Wenlok), and Horno Fm sedimentary rocks (González de Juana et al. 1980) overlying the metamorphic basement. A regional unconformity at the base of the Late Ordovician marine clastic sequences is observed in the San Juan region Argentina which marks the closure of a similar event in the southern Andes (Baldis et al. 1992, p. 348).

A Late Carboniferous-Early Permian deformational event is reported to have involved basement rocks in the Mérida Andes. This event is characterized by the local development of low-grade metamorphism accompanied by plutonism (Marechal 1983). The Upper Mississippian (?) - Lower Carboniferous clastics comprising a "molassic-facies" consisting of conglomeratic and tectonic breccia deposits (Mérida facies - Sabaneta Fm of Shagam et al., 1970) represent the closure of the Late Paleozoic orogenic event. Such an episode is not clear from the rock record in the Colombian Andes. Intracontinental back-arc extension within north to northwest trending structures occurred during Pennsylvanian to Permian time (Cediel et al. 2003). A Permian magmatic arc developed along the present day position of the Central Cordillera of Colombia (Vinasco 2004). Gently folded Carboniferous strata underlie the basal Cretaceous sediments (Trumpy 1949). It is difficult to reconcile the lack of evidence in the rock record for a strong and widespread Late

Paleozoic deformation in Colombia and Venezuela, with plate tectonic reconstructions depicting northwestern South America impinging on the Ouachita embayment.

By Devonian time the northern Andes faced the Appalachian region as evidenced by their faunal affinities. The Permian magmatic arc developed along the central Andes of Colombia (Vinasco 2004) continues into central-western Mexico signaling the onset of amalgamation of Pangea (Dickinson and Lawton 2001, Vega-Carrillo et al. 2007, Restrepo-Pace et al. 1997). The present-day basement Southern Mexico was then attached from northwestern South America during Caparonensis-Quetame orogenic event at the time of closure of Pangea.

2.5 Constraints on the Relative Position of NW South America from Paleozoic Faunal Assemblages

The controls exerted by the paleoenvironment on faunal provinciality or cosmopolitanism of a given species or assemblage is still a matter of debate. Nonetheless, the relative paleo-positions of continental fragments derived primarily from paleomagnetic studies can be refined by comparing time correlative provincial fossil assemblages. In the case of northern South America in Paleozoic time, the first order conclusion is that early Cambro-Ordovician fauna is dominantly Gondwanan with minor Acado-Baltic affinity, whereas Siluro-Devonian fauna is distinctively Appalachian (Fig. 2.3). The Middle Cambrian limestones from the Macarena uplift contain trilobites of the genus *Ehmania* (Harrington and Kay 1951) and *Paradoxides* (Rushton 1962). The former is represented by two species of the *Amecephalina* (Harrington and Kay 1951; Rushton 1962; Borello 1971; Forero-Suárez 1990) or *Bathyuriscus-Elrathina* Zones (Borello 1971) in the Precordillera of northwestern Argentina. The latter, an Acado-Baltic trilobite, can be found within the Carolina Slate belt (Secor et al. 1993), in the Paradoxides zone of eastern New England (Devine 1985), eastern Newfoundland, New Brunswick, and Avalon Peninsula (North 1971; Palmer 1983).

Ordovician marine sedimentary rocks from El Baúl are marked by the presence of Parabolina Argentina, a zonal index for the Lower Tremadoc in northwestern Argentina (Frederikson 1948; Aceñolaza 1982). The Clarenville Fm. in Random Island, Eastern Newfoundland also yields Parabolina Argentina together with Angelina (Dean 1985). In the Ordovician sequence at the Macarena uplift, Colombia, fauna also relates to northern Argentina and southern Bolivia: *Geragnostus tilcuyensis*, *Kainella colombiana*, *Pseudokaianella maracanae*, and *Parabolinopsis* sp. together with *Lingulella desiderata*, *Acrotreta aequatorialis*, *Nanortis* sp., and *Obolus* sp. recall the Kaianella fauna of Argentina-Bolivia (Harrington and Kay 1951).

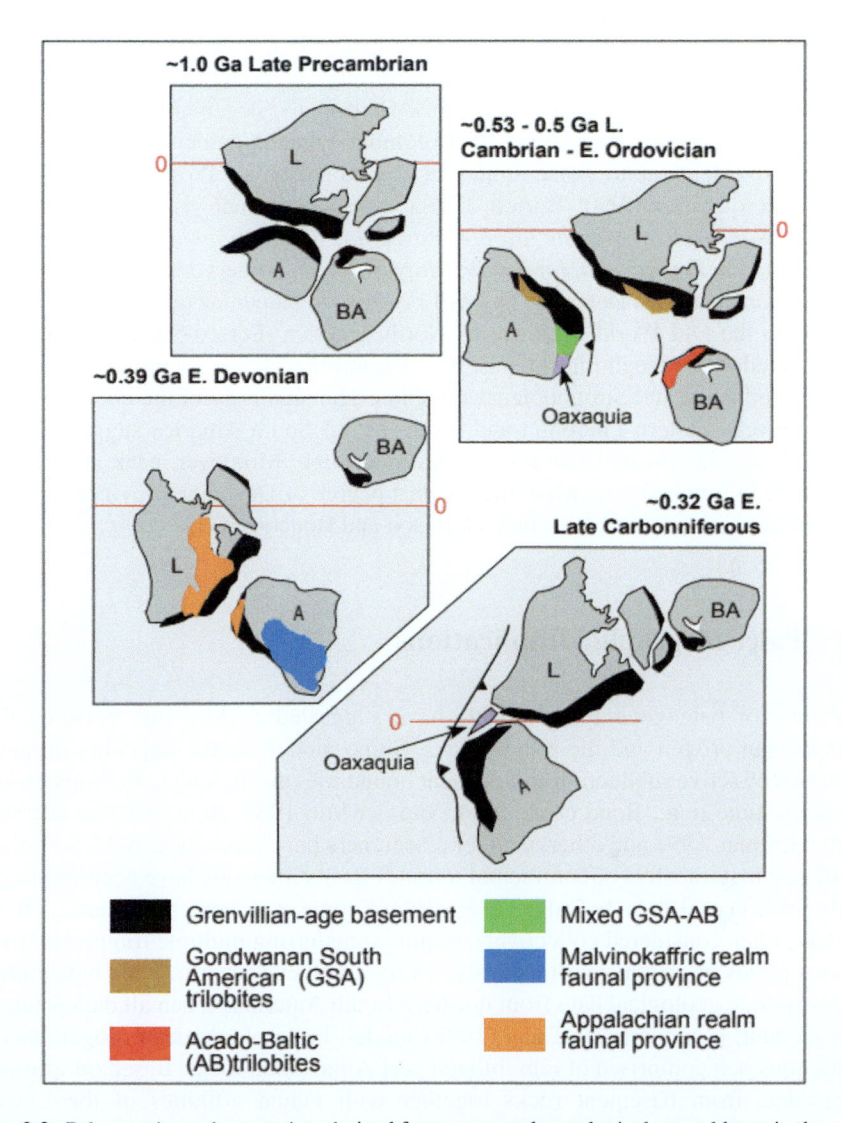

Fig. 2.3 Paleocontinental constraints derived from tracer paleontological assemblages in the context of Hoffman 1991 reconstruction. (Modified from Restrepo-Pace et al. 1994)

The dominantly Gondwanan character of Lower Paleozoic fauna in northwestern South America shifts to Appalachian-affinity by early Devonian time (i.e., not related to the Malvinokaffric Realm). Emsian to Siegenian (~394–374 Ma) sedimentary rocks in the Andes of Colombia contain associations of benthic brachiopods identical to those found in the Appalachian Province (Harrington 1967; Forero-Suárez 1990; Barrett 1988a, b). Genera such as *Cyrtina, Elytha, Atrypa, Nucleospira, Meristella, Megastrophia, Cymostrophia, Stropheodonta, Chonostrophia, Leptocoelia, Iphigenia, Platyorthys*, and others are closely related to Appalachian taxa. Late Devonian fauna of Frasnian to Famenian age (~374–360 Ma) belongs to the Old World Province of North America (Forero-Suárez 1990). The latter includes Schizophoria amanaensis, Carinifella alleni, Laminatia laminata, Devonoproductus, and Strophopleura notabilis. The similarity of the above benthic fauna between eastern Laurentia and northwestern South America suggests proximity of these continental margins in Devonian time. Moreover, peak similarities occur here in Emsiam time when the greatest degree of Devonian provinciality was reached for marine fauna as a whole (Johnson and Boucot 1973).

2.6 Paleogeographic Implications

A variety of paleogeographic models have suggested a close link between the Appalachian orogen and the proto-Andes. Some models depict opposing orogens separated by active subduction and/or shear boundaries throughout Late Proterozoic-Paleozoic time (e.g., Bond et al. 1984; Van der Voo 1988; Kent and Van der Voo 1990; Hoffman 1991 and others). Other researchers have taken these models further to suggest that transfers of continental terranes from either side have occurred (e.g., Dalla Salda et al. 1992a, b; Dalziel et al. 1994; Keppie et al. 1991 and others). These models, when considered collectively, require transferring multiple fragments from various points of origin simultaneously, a very complex scenario. Most have failed to incorporate geological data from northern South America. When all data is taken into account, it supports Hoffman (1991) model: the proto-Andean orogen was a contiguous belt comprised of remobilized peri-Amazonian rocks. Based on isotopic tracer data from basement rocks together with faunal affinities of the Lower Paleozoic sequences, this orogenic system is extended into southern Mexico. Identical Pb isotopic compositions of the Grenville-age basement of Colombia and southern Mexico imply continuity of these widely spaced basement terranes (Ruiz et al. 1999). Nd model (TDM) ages for the Colombian basement rocks range from 1.9 to 1.45 point to an Amazonian provenance for the basement here (Restrepo-Pace et al. 1997). The assemblage of Late Cambrian-Tremadocian trilobites of southern Mexico is akin to northwestern South American and Argentinean faunal assemblages (Frederikson 1948; Robison and Pantoja-Alor 1968; Aceñolaza 1982; Moya et al. 1993; Landing et al. 2007) (Fig. 2.4). These are tied together by the presence of Parabolina Argentina, a zonal index for the Lower Tremadoc in northwestern Argentina. The Tremadoc Parabolina Argentina is present in the Tiñú Formation of

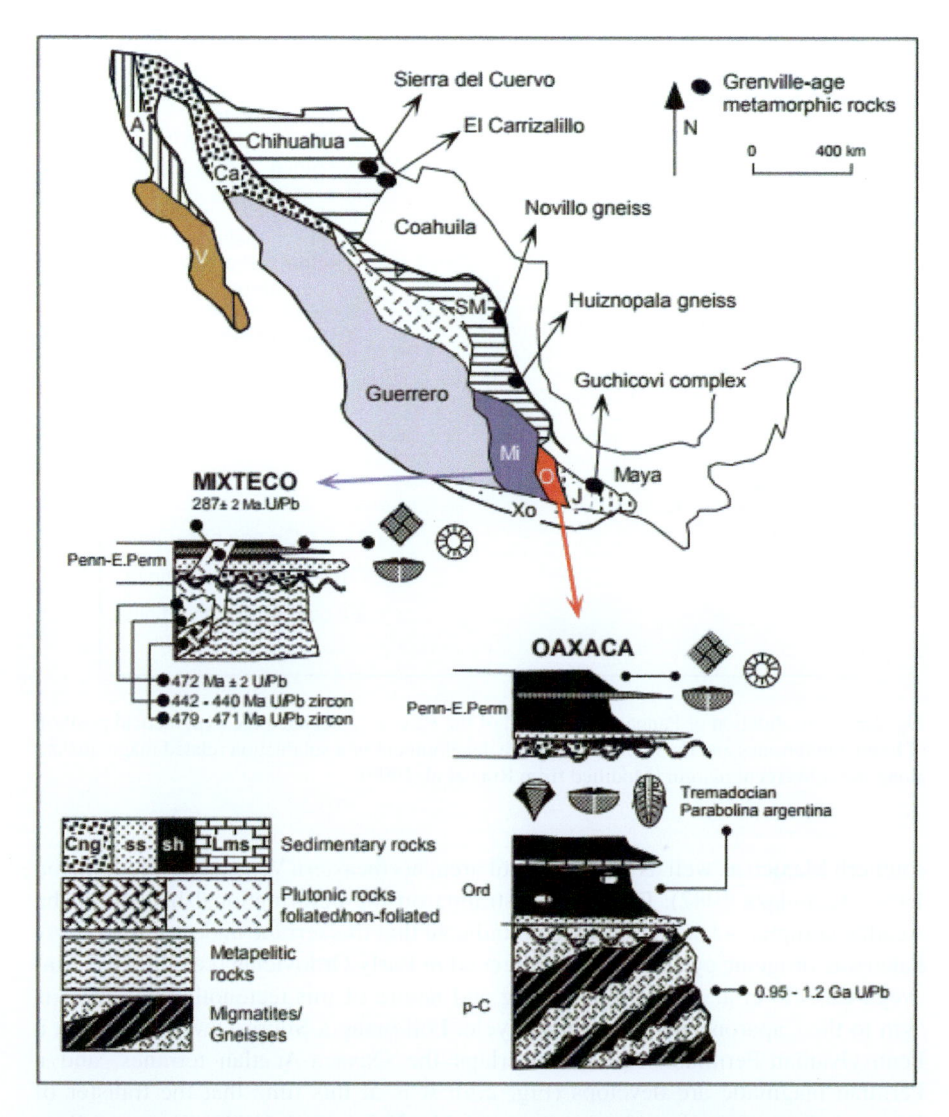

Fig. 2.4 Terrane map of Mexico depicting the basement remnants of probable South American provenance, attached to the south eastern Mexico in Late Paleozoic time (Modified from Restrepo-Pace 1995)

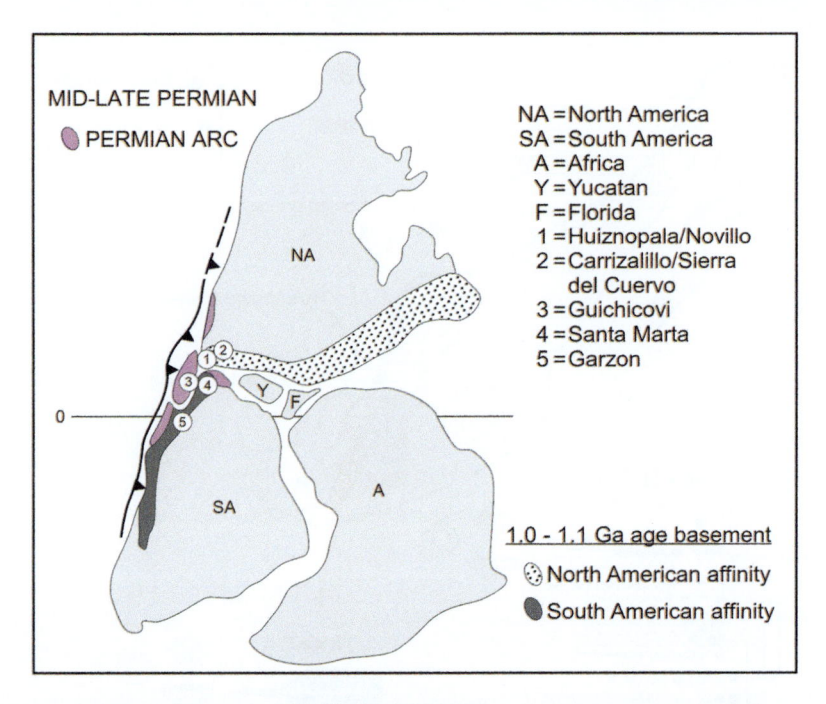

Fig. 2.5 Consolidation of Pangaea by the end of the Paleozoic depicted the hypothetical position of basement terranes and the implication of the development of a subduction related magmatic arc along the convergent margin (Modified from Ruiz et al. 1999)

southern Mexico as well as in the El Baúl area, northeastern Venezuela (Frederikson 1948; Aceñolaza 1982). Detailed constraints on the deformational history of the Acatlán complex – southern Mexico – indicate that this terrane underwent an Early Paleozoic orogenic cycle which commenced in Early Ordovician (ca. 490–477 Ma) (Vega-Carrillo et al. 2007). The timing and nature of this tectonothermal event is akin to the Caparonensis-Famatinian cycle. Following a Siluro-Devonian hiatus, a Pennsylvanian-Permian sequence overlaps the Oaxaca-Acatlán terranes, and a Permian magmatic arc develops (Fig. 2.5). It is at this time that the transfer of Oaxaquia basement takes place as suggested by Yañez et al. (1991), Restrepo-Pace et al. (1997), and Ruiz et al. (1999).

A summary of events for the northern Andes is presented in Fig. 2.6. Data support the differentiation of two tectonic events of regional significance that consolidated the basement of the northern Andes: Orinoquiense and Quetame (~1.0 Ga and ~0.47 Ga, respectively). These discrete tectonothermal events appear to be traceable along the Eastern Andes of South America: a Grenvillian-Orinoquiense event (~1.0 Ga) and a Caparonensis-Famatinian event (~0.47–0.43 Ga). Both intimately associated with the assemblage of the Rondinia and Pangaea as suggested by Hoffman (1991).

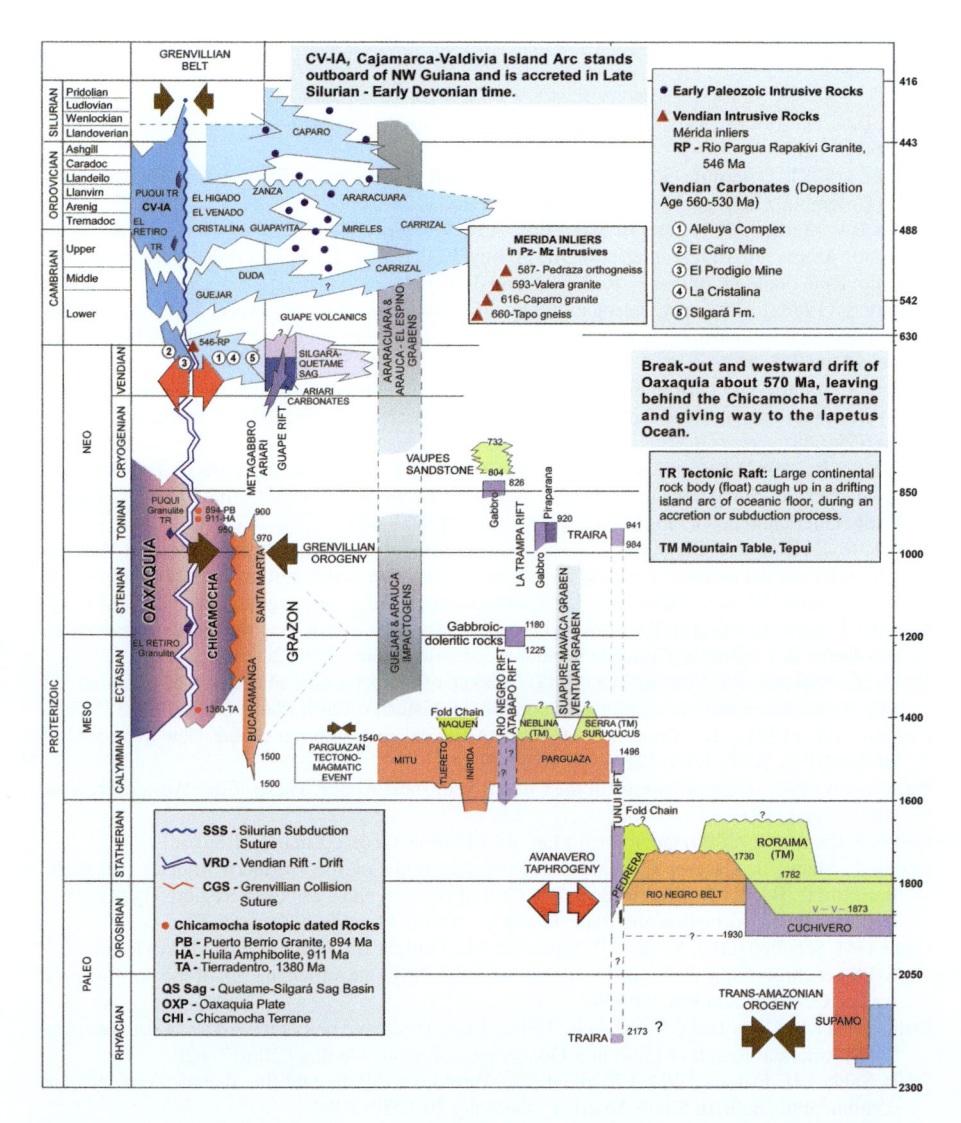

Fig. 2.6 Summary tectonic events for Late Precambrian-Paleozoic in northern South America

References

Aceñolaza FG (1982) The Ordovician system of South America. Zbl Geol Paläent Teil I 5/6:627–645

Alvarez J (1981) Determinación de edad Rb/Sr en rocas del Macizo de Garzón, Cordillera Oriental de Colombia. Geología Norandina (Colombia) 4:31–38

Baldis BA, Martínez RD, Pereyra ME, Pérez AM, Villegas CR (1992) Ordovician events in the South American Andean platform. In: Webby BD, Laurie JD (eds) Global perspectives on ordovician geology. Balkema, Rotterdam, pp 345–353

Barrett S (1988a) Devonian Paleogeography of South America. In: Proceedings on the 2nd International Symposium on the Devonian System. Canadian Society of Petroleum Geologists, vol 14, pp 705–717

Barrett S (1988b) The Devonian system in Colombia. In: Proceedings on the 2nd International Symposium on the Devonian System. Canadian Society of Petroleum Geologists, vol 14, pp 655–668

Benedetto JL (1982) Las unidades Tecto-estratigráficas Paleozoicas del norte de Sur América, Apalaches del Sur y noreste de Africa: comparación y discusión. Actas del Quinto Congreso Latinoamericano de Geología, Argentina, vol 1, pp 469–488

Benedetto JL, Ramírez PE (1985) La secuencia sedimentaria Precambrico-Paleozoico Inferior pericratónica del extremo norte de Sudamerica y sus relaciones con las cuencas del norte de Africa. Actas del Quinto Congreso Latinoamericano de Geología, Argentina, vol 2, pp 411–425

Boinet T, Burgois J, Bellon H, Toussaint JF (1985) Age et repartition du magmatisme Premesozoique des Andes de Colombia. Comptes Rendus Academie Science Paris 2(10):445–450

Bond GC, Nickeson PA, Kominz MA (1984) Breakup of a supercontinent between 625 Ma and 555 Ma: Ne evidence and implications for continental histories. Earth Planet Sci Lett 70:325–345

Borello, A.V. (1971). The Cambrian of South America, in Cambrian of the New World. C.H. Holland. Ed., Wiley-Interscience, pp 405–408

Burkley LA (1976) Geochronology of the Central Venezuela Andes: Thesis. Case Western Reserve University, p 150

Cediel F, Caceres C (2000) Geological Map of Colombia. Geotec Ltda. 3a. Edición

Cediel F, Shawn J, Caceres C (2003) Tectonic Assembly of the Northern Andean Block, In: Bartolini C, Buffler R, Blickwede J, The Gulf of Mexico and Caribbean Region: Hydrocarbon Habitats, Basin Formation and Plate Tectonics, AAPG Memoir 79 Chapter 37

Chew DM, Schaltegger U, Kosler J, Whitehouse MJ, Gutjahr M, Spikings RA, Miskovíc A (2007) U-Pb geochronologic evidence for the evolution of the Gondwanan margin of the north-central Andes. Geol Soc Am Bull 119:697–711

Dalla Salda LH, Cingolani CA, Varela R (1992a) Early Paleozoic belt of the Andes in southwestern South America: Result of Laurentia-Gondwana collision? Geology 20:617–620

Dalla Salda LH, Dalziel IWD, Cingolani CA, Varela R (1992b) Did the Taconic Appalachians continue into southern South America ? Geology 20:1059–1062

Dalziel IWD, Dalla Salda LH, Gahagan LM (1994) Paleozoic Laurentia-Gondwana interaction and the origin of the Appalachian-Andean mountain system. Geol Soc Am Bull 106:243–252

Dean WT (1985) Relationships of Cambrian – Ordovician fauna in the Caledonide-Appalachian Region, with particular reference to trilobites. In: Geyer RA (ed) The Tectonic Evolution of the Caledonide-Appalachian Orogen. Wiesbaden Vieweg, Braunschweig, pp 27–29

Devine CM (1985) Ancient Avalonia and the trilobite fauna of Jamestown RI: fossils. Quaterly Summer:27–28

Dickinson WR, Lawton TF (2001) Carboniferous to Cretaceous assembly and fragmentation of Mexico. GSA Bull 113(9):1142–1160

Dueñas H (2001) Asociaciones Palinologicas y Posibilidades de Hidrocarburos del Paleozoico de la Cuenca de los Llanos Orientales. Colombia

Etayo-Serna F, Barrero D (1983) Mapa de terrenos geológicos de Colombia: Ingeominas special publication, Colombia, vol 14, pp 48–56

Forero-Suárez A (1990) The basement of the eastern cordillera, Colombia: an allochtonous terrane in northwestern South America. J South Am Earth Sci 3(2/3):144

Frederikson EA (1948) Lower Tremadocian trilobites from Venezuela. J Paleontol 32:541–543

Goldsmith F, Marvin RF, Menhert HH (1971) Radiometric ages in the Santander Massif, Eastern Cordillera, Colombia. U.S. Geol Surv Prof Pap 750D:44–49

González de Juana C, Iturralde J, Picard X (1980) Geología de Venezuela y de sus cuencas petrolíferas: Ediciones Fonives. Caracas 1:407

Grösser JR, Prössl KF (1991) First evidence of the Silurian in Colombia: Palynostratigraphic data from the Quetame Massif, Cordillera Oriental. J S Am Earth Sci 4(3):231–238

Harrington H (1967) Devonian of South America: international symposium on the Devonian system. Alberta Society of Petroleum Geologists, Calgary, vol 1, pp 651–671

Harrington JH, Kay M (1951) Cambrian and Ordovician Fauna of eastern Colombia. J Paleontol 25:655–668

Hoffman PA (1991) Did the birth of North America turn Gondwana inside out? Science 252:1409–1411

Johnson JG, Boucot AL (1973) Devonian brachiopods. In: Hallam A (ed) Atlas of paleobiogeography. Elsevier Scientific Publishing Company, pp 89–96

Kent DV, Van der Voo R (1990) Paleozoic paleogeography from paleomagnetism of the Atlantic-bordering continents. In: Mckerrow WS, Scotese CR (eds), Paleozoic Paleogeography and Biogeography. Geological Society of America Memoir, 12:49–56

Keppie JD (1993) Transfer of the northeastern Appalachians (Meguma, Avalon, Gander, and Exploits terranes) from Gondwana to Laurentia during Middle Paleozoic continental collision. In: Proceedings of the first Circum-Pacific and Circum-Atlantic terrane conference, Guanajuato, México, Instituto de Geología Universidad Nacional Autónoma de México, pp 71–73

Keppie JD, Nance RD, Murphy JB, Dostal J (1991) Northern Appalachians: Avalon and Meguma Terranes. In: Dallmeyer RD, Lécorché JP (eds) The West African Orogens and Circum-Atlantic correlatives. Springer, Berlin/Heidelberg, pp 316–333

Kroonenberg SB (1982) A Grenville granulite belt in the Colombian Andes and its relation to the Guiana Shield. Geologie Mijnbouw 61:325–333

Landing E, Westrop SR, Keppie JD (2007) Terminal Cambrian and lowest Ordovician succession of Mexican West Gondwana: biotas and sequence stratigraphy of the Tiñu Formation. Geol Mag 144:909–936

Litherland MJA, Annels RN, Darbyshire DPF, Fletcher CJN, Hawkins MP, Klink BA, Mitchell WI, O'Connor EA, Pitfield PEJ, Power G, Webb BC (1989) The Proterozoic of Eastern Bolivia and its relationship to the Andean mobile belt. Precambrian Res 43:157–174

Litherland M, Aspden JA, Jemielita RA (1994) The metamorphic belts of Ecuador: Overseas Memoir of the British Geological Survey No. 11, p 147

Marechal P (1983) Les temoins de chaine Hercynienne dans le noyau ancien des Andes de Merida (Venezuela): structure et evolution tectonometamorphique. Ph.D. Disseratation. Universite de Bretagne Occidentale, p 176

Maya M (1992) Catálogo de dataciones isotópicas en Colombia. Boletín Geológico Ingeominas (Colombia) 32:135–187

Moya MC, Malanca S, Hongn FD, Bahlburg H (1993) El Tremadoc Temprano en la Puna occidental Argentina: Actas del XII Congreso Geológico Argentino y II Congreso de exploración de Hidrocarburos, vol 2, pp 20–30

North FK (1971) In: Holland CH (ed) The Cambrian of Canada and Alaska in Cambrian of the New World. Wiley-Interscience, pp 231–242. Acatlán, Estado de Puebla: Boletín de la Sociedad Geológica Mexicana, vol 39, pp 27–28

Palmer, A. R., (1983). The decade of North American geology (DNAG) geologic time scale. Geology, 11:503–504

Park RG (1992) Plate kinematic history of Baltica during the middle to late Proterozoic: a model. Geology 20:725–728

Priem HNA, Andriessen P, Boelrijk A, De Boorder H, Hebeda E, Huguett E, Verdumen E, Verschure R (1982) Precambrian Amazonas región of southeastern Colombia (western Guiana Shield). Geol Mijnb 61:229–242

Priem HNA, Kroonenberg SB, Boelrijk NAIM, Hebeda EH (1989) Rb-Sr and K-Ar evidence for the presence of a 1.6 Ga basement underlying the 1.2 Ga Garzón-Santa Marta Granulite belt in the Colombian Andes. Precambrian Res 42:315–324

Ramos VA (1988) Late Proterozoic-early Paleozoic of South America – a collisional history. Episodes 11(3):168–173

Rapela CW, Tosselli A, Heaman L, Saavedra J (1990) Granite plutonisms in the Sierras Pampeanas; an inner cordilleran Paleozoic arc in the southern Andes. In: Kay SM, Rapela CW (eds) Plutonism from Antartica to Alaska. Geological Society of America Special Paper, 241

Restrepo-Pace PA (1992) Petrotectonic characterization of the Central Andean Terrane, Colombia. J S Am Earth Sci 5(1):97–116

Restrepo-Pace PA (1995) Late Precambrian to early Mesozoic tectonic evolution of the Colombian Andes, based on new geochronological, geochemical and isotopic data. Ph. D. Thesis. University of Arizona, p 195

Restrepo-Pace PA, Ruiz J, Cosca M (1994) The transfer of terranes from South to North America based on the Proterozoic evolution of Colombia and southern México: Eighth International Conference on Geochronology, Cosmochronology and Isotope Geology, U.S. Geological Survey Circular 1107, p 266

Restrepo-Pace PA, Ruiz J, Gehrels G, Cosca M (1997) Geochronology and Nd isotopic data of Grenville-age rocks in the Colombian Andes: new constraints for late Proterozoic-early Paleozoic paleocontinental reconstructions of the Americas. Earth Planet Sci Lett 150:427–441

Robison RA, Pantoja-Alor J (1968) Tremadocian trilobites from the Nochixtlán region, Oaxaca, México. J Paleontol 42:767–800

Ruiz J, Tosdal R, Restrepo PA, Murillo-Muñetón G (1999) Pb isotopic evidence for Colombia-southern Mexico connections before Pangea. In Laurentia-Gondwana connections before Pangea Geological Society of America 336, pp 183–197

Rushton AWA (1962) Paradoxides from Colombia. Geological Magazine 100:255–257

Secor DT Jr, Samson SL, Snoke AW, Palmer AR (1993) Confirmation of the Carolina Slate Belt as an exotic terrane. Science 221:649–650

Shagam R (1977) Evolución tectónica de los Andes Venezolanos: Memorias de V Congreso Geológico de Venezuela, vol 11, pp 855–877

Trumpy D (1949) Geology of Colombia. N.V. de Bataafsche Petroleum Maatschappij, The Hague, pp 3–6

Tschantz CM, Marvin RF, Cruz J, Menhert H, Cebulla G (1974) Geologic evolution of the Sierra Nevada de Santa Marta area, Colombia. Geol Soc Am Bull 85:273–284

Van der Voo R (1988) Paleozoic paleogeography of North America, Gondwana and intervening displaced terranes: comparisons of paleomagnetism with paleoclimatology and biogeographical patterns. Geol SocAm Bull 100:311–324

Vega-Carrillo R, Talavera-Mendoza O, Meza-Figueroa D, Ruiz J, Gehrels GE, López-Martínez M, de la Cruz-Vargas JE (2007) Pressure-temperature-time evolution of Paleozoic high-pressure rocks of the Acatlán Complex (southern Mexico): implications for the evolution of the Iapetus and Rheic Oceans. Geol Soc Am Bull 119(9–10):1249–1264

Vinasco CJ (2004) Evolucao crustal e historia tectonica dos granitoides Permo-Triassicos Dos Andes do norte. Universidade de São Paulo, Brazil, Ph.D. dissertation

Wasteneys HA (1994) Geochronology of the Arequipa Massif, Perú: correlation with Laurentia. Abstracts of the eight International Conference on Geochronology, Cosmochronology and Isotope geology. USGS circular, 1107, p 350

Yañez P, Ruiz J, Patchett JP, Ortega-Gutiérrez F, Gehrels G (1991) Isotopic studies of the Acatlán Complex, southern México: implications for Paleozoic North American tectonics. Geol Soc Am Bull 103:817–828

Part II
The Guiana Shield and the Andean Belt

Part II
The Guiana Shield and the Andean Belt

Chapter 3
The Proterozoic Basement of the Western Guiana Shield and the Northern Andes

Salomon B. Kroonenberg

3.1 The Amazonian and Orinoquian Basement

3.1.1 General Geology of the Guiana Shield

The Colombian Precambrian basement forms the westernmost extension of the Guiana Shield, the northern half of the Amazonian Craton (Fig. 3.1). Apart from two Archean nuclei, the Imataca high-grade belt in Venezuela (2.74–2.63 Ga; Tassinari et al. 2004a, b) and the Amapá high-grade belt in northern Brazil (2.65–2.60 Ga: Rosa-Costa et al. 2003), the largest part of the shield was formed in the Paleoproterozoic during the Trans-Amazonian Orogeny between 2.26 and 1.98 Ga. This orogeny resulted from the collision of the Archean parts of Amazonia with those of the West African Craton (Bispo-Santos et al. 2014). Two younger orogenic events are recorded along its western extremity, the Querarí Orogeny (1.86–1.72 Ga) in Colombia, western Venezuela and northwestern Brazil and the Grenvillian Orogeny in Neoproterozoic slivers in the Colombian Andes and the Andean foredeep (1.3–1.0 Ga, called Putumayo by Ibáñez-Mejía et al. 2011). Several phases of anorogenic magmatism have been distinguished as well, one around 1.89–1.81 Ga along the southern border of the shield and one Mesoproterozoic around 1.59–1.51 Ga in the western part (Fig. 3.2). All ages cited in this paper are U-Pb or Pb-Pb zircon ages unless otherwise stated.

The Trans-Amazonian Orogeny has developed in three phases, each of them producing distinguishing geological units (Kroonenberg et al. 2016). During the first phase between 2.26 and 2.09 Ga, a 2000 km long greenstone belt developed along the whole northern border of the Guiana Shield, from Venezuela (Pastora-Carichapo Group) through Guyana (Barama-Mazaruni Group), Suriname (Marowijne Greenstone Belt),

---

S. B. Kroonenberg (✉)
Delft University of Technology, Delft, Netherlands
e-mail: S.B.Kroonenberg@tudelft.nl

Fig. 3.1 Guiana Shield and Brazilian Shield together form the Amazonian Craton. In black outcrops of Andean Precambrian. (Modified after Cordani and Sato 1999)

French Guiana and Amapá in Brazil (Vila Nova Group). It consists of a series of ocean-floor (ultra)mafic metavolcanics, island-arc intermediate and felsic metavolcanics, followed by turbiditic metagreywackes and epicontinental meta-arenites. The whole sequence is folded into broad synclinoria and intruded by tonalite, trondhjemite and granodiorite (TTG) plutons (Gibbs and Barron 1993; Sidder and Mendoza 1991; Delor et al. 2003; Cordani and Sato 1999; Cordani et al. 2000; Cordani and Teixeira 2007; Kroonenberg and De Roever 2010; Kroonenberg et al. 2016).

A second phase of the Trans-Amazonian Orogeny is evidenced by a discontinuous 2.08–1.98 Ga belt of high-grade rocks, consisting of the sinuous Cauarane-Coeroeni belt roughly parallel to the greenstone belt and the Bakhuis granulite belt intersecting it. It represents a rifting phase followed by sedimentation, volcanism and ultimately high-grade metamorphism with an anti-clockwise cooling path. The Cauarane-Coeroeni belt, defined by Fraga et al. (2008, 2009a) (formerly also called Central Guiana Granulite Belt), can be followed from southwestern Suriname

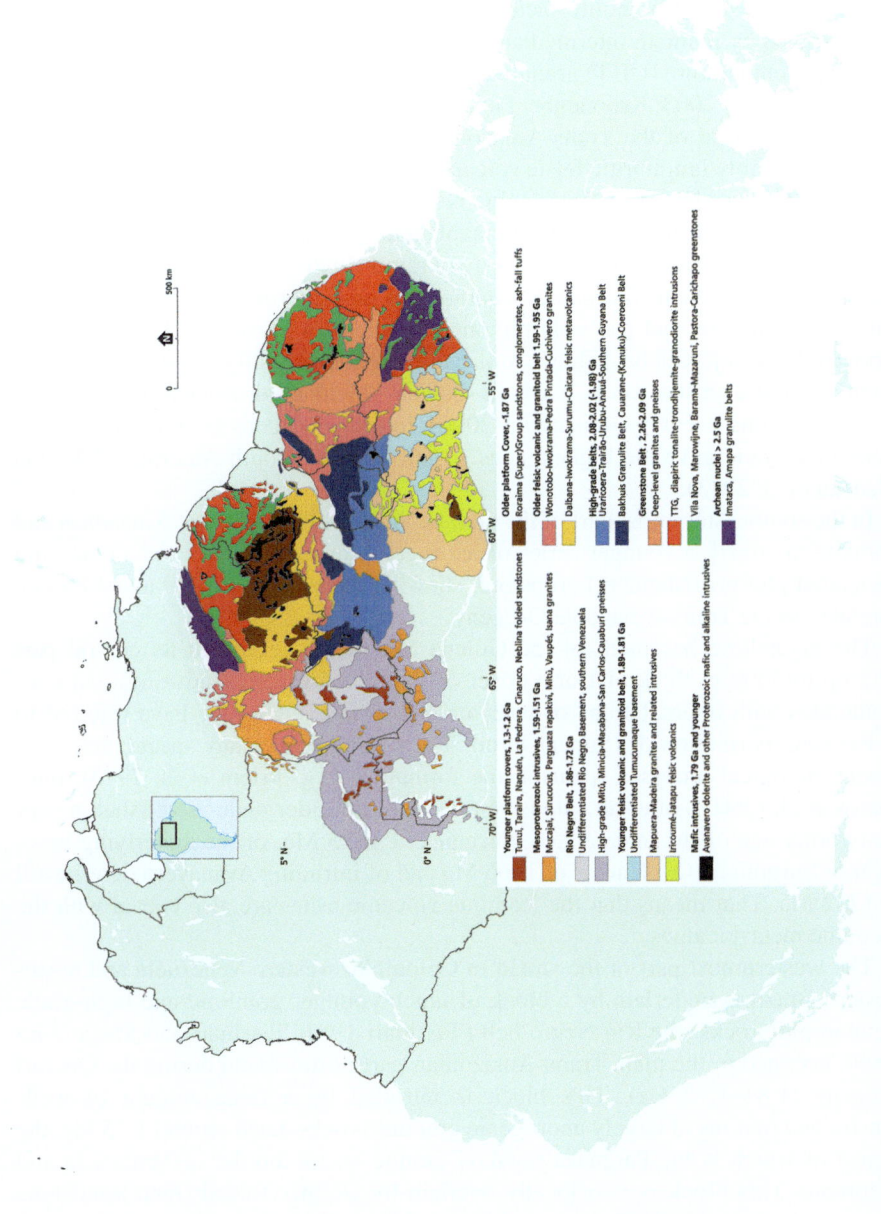

3 The Proterozoic Basement of the Western Guiana Shield and the Northern Andes   117

Fig. 3.2 Lithological-chronological-geological map of the Guiana Shield (Kroonenberg et al. 2016). © Cambridge University Press. Reprinted with permission

(Kroonenberg 1976; Priem et al. 1977), through Guyana (Berrangé 1977; Gibbs and Barron 1993), into the state of Roraima in Brazil (Fraga et al. 2008, 2009a, 2011). It consists of an essentially supracrustal sequence of pelitic and quartzofeldspathic metasediments, amphibolites, quartzites and calcsilicate rocks, metamorphosed to amphibolite facies to granulite facies. The Bakhuis granulite belt in Suriname is characterized by mafic to intermediate granulites and metapelitic gneisses showing ultra-high temperature (UHT) granulite-facies metamorphism around 2.08–2.03 Ga (De Roever et al. 2003; Kroonenberg et al. 2016).

The third phase of the Trans-Amazonian Orogeny is characterized by a huge outpour of mainly ignimbritic felsic volcanics and associated granitoid rocks around 1.99–1.95 Ga, in a broad W-E stretching belt, equally about 2000 km long, roughly parallel to the greenstone belt. The metavolcanics and associated plutons go by the name Caicara/Cuchivero in Venezuela, Iwokrama/Kuyuwini in Guyana, Surumú in Brazil, and Dalbana in Suriname. Charnockite and anorthosite intrusions in the Bakhuis Mountains and gabbroic plutons elsewhere in Suriname (Lucie Gabbro, formerly De Goeje Gabbro) show similar ages, together testifying of an important magmatic pulse in the whole northern Guiana Shield in an Andean-type setting, called Orocaima event by Reis et al. (2000). Inherited zircons from the Iwokrama rocks in Guyana gave the highest ages so far found in South America of 4.2 Ga (Nadeau et al. 2013).

In the southeasternmost part of the Guiana Shield, in the states of Amazonas and Roraima in Brazil, a younger series of anorogenic felsic volcanics (Iricoumé) and associated plutons (Mapuera) crops out, showing ages between 1.89 and 1.81 Ga, unrelated to the Trans-Amazonia Orogeny.

The crystalline basement of the Guiana Shield is overlain in its central part by a up to 3000 m thick platform cover of Paleoproterozoic sandstones and conglomerates with intercalations of volcanic ash, which since long have referred to as Roraima Formation or (Super)Group. There have been many speculations and geochronological analyses spent on the formation (e.g. Priem et al. 1973), until Santos et al. (2003), after an extensive review of all older data, established a very trustworthy age of the intercalated volcanics of 1873 Ma, of the underlying basement of Surumu metavolcanics of 1966 Ma and of intruding Avanavero dolerite sill of 1782 Ma. That means that the Roraima volcanic ashes are also coeval with the Iricoumé metavolcanics.

The westernmost part of the shield in Colombia, western Venezuela and northwestern Brazil is underlain by a block of much younger granitoid and high-grade metamorphic rocks, the Río Negro belt (Tassinari 1981; Tassinari and Macambira 1999), accreted to the main Trans-Amazonian part of the shield during the Querarí Orogeny (1.84–1.72 Ga). This block is intruded by a large amount of well-constrained plutons of largely anorogenic granitoid rocks dated around 1.55 Ma, the largest of which is the Parguaza rapakivi granite on the border of Venezuela and Colombia. This block is also locally overlain by slightly folded (meta)sandstone covers as in the Naquén, Pedrera and Tunuí ridges. This area will be discussed in more detail below.

Many rocks in the western part of the Guiana Shield suffered intense shearing and low-grade thermal metamorphism around 1.3–1.1 Ga (Priem et al. 1968, 1971; Gibbs and Barron 1993) probably caused by the continental collision of Amazonia and Laurentia during the Grenvillian Orogeny as evidenced by the Grenvillian granulites in the Colombian Andes and the basement in the adjacent Putumayo foredeep (Kroonenberg 1982; Cordani et al. 2010; Ibáñez-Mejía et al. 2011; see also par. 3.2.).

Several generations of mafic and alkaline intrusions have been recognized in the Guiana Shield, including the Avanavero one referred to above, but there are also younger generations such as the Käyser dolerite (1500 Ma) in Suriname and at last the ~200 Ma Jurassic dykes that mark the separation of South America and Africa (Deckart et al. 2005).

3.1.2 The Colombian Part of the Guiana Shield

In Colombia the basement crops out in large areas of eastern Amazonia and the eastern Llanos Orientales and is also exposed in many cataracts in major and minor rivers. Further westwards, towards the Andes, and southwards, towards the Amazon River, the basement is progressively covered by younger sediments of Ordovician to Cenozoic age. Nevertheless, drilling by oil companies into the Subandean foreland basins frequently struck basement (Ibáñez-Mejía et al. 2011), confirming its continuity beneath the sedimentary cover. Within the Andean cordilleras, large slices of Proterozoic rocks have been incorporated during later orogenies (Fig. 3.1).

The Colombian Precambrian constitutes the westernmost part of the Guiana Shield and comprises a small fragment of a mid-Paleoproterozoic (Late Trans-Amazonian) basement and large tracts of late Paleoproterozoic metamorphic basement, intruded by late Proterozoic syntectonic granites and Mesoproterozoic anorogenic granites. It is covered by low-grade metamorphosed and non-metamorphic sandstone plateaus and intruded by small Neoproterozoic basic and alkaline intrusions.

The first systematic description of the rocks of the Colombian part of the Guiana Shield has been published by Galvis et al. (1979) and Huguett et al. (1979) in the framework of the mapping project PRORADAM. The crystalline rocks of the Guiana Shield in Colombia south of the Guaviare River were designated by them as *Complejo Migmatítico de Mitú*. They describe it as having formed by 'sedimentation, volcanism and probably plutonism; later, the whole complex was metamorphosed and at last suffered mainly potassic metasomatism that affected the metamorphic rocks, imparting a granitoid aspect to the major part of the complex'. In their, now outdated, view, migmatization is a solid-state metasomatic process, not an anatectic process as nowadays considered. Unfortunately, their metasomatic conception coloured many descriptions, making it difficult to understand them in a modern way. In an excellent review of the Colombian Amazonian geology, Celada et al. (2006) reject the name as such, because of the inappropriate use of the term migmatitic, as many rocks in the area are clearly intrusive and not migmatitic in either sense. They propose to call the complex simply 'Complejo Mitú', a position later supported by López et al. (2007) and López (2012). We will retain the latter designation, inasmuch as we restrict the use of it to the high-grade metamorphic part of the basement.

Galvis et al. (1979) distinguish the following rock units in the Mitú complex: (1) Atabapo-Río Negro gneisses (including gneisses s.s., amphibolites, amphibolic gneisses, quartzites and quartz gneisses, quartzofeldspathic gneisses, aluminous gneisses and blastomylonites), (2) migmatitic granites and (3) Araracuara gneisses. However, these units have not been mapped separately.

Additional data are given in unpublished reports by De Boorder (1976, 1978). Kroonenberg (1985) revised the petrography of the PRORADAM samples. After PRORADAM, several mapping projects have been carried out in the basement. Bogotá (1981) and Bruneton et al. (1983) give a detailed description of the geology of the Guainía and Vichada departments in the framework of a mineral exploration project by COGEMA.

In the framework of the production of 1:100,000 geological map sheets of the country, the Servicio Geológico Colombiano has published a limited number of sheets in the Guiana Shield, in the area around Mitú (Rodríguez et al. 2010, 2011b) and near Puerto Inírida (López et al. 2010) and Puerto Carreño (Ochoa et al. 2012). On the Venezuelan side, the UGSG map of Hackley et al. (2005) is a major source of information and on the Brazilian side the 1:1 M map sheets NA.19 and SA.19 (CPRM 2004a, b). Preliminary 1:1 M geological maps of the same sheets showing the combined geology of the three countries have been prepared by the Commission for the Geological Map of the World (2009a, b). In the framework of this book, a combined map of the western Guiana Shield has been prepared, as well as a description of the sequence of events (Figs. 3.3, 3.4, and 3.5).

The following descriptions of major rock types are synthesis of observations by the authors mentioned above and own field and petrographic observations in 1979–1981 and 1985–1991.

3.1.2.1 Mid-Paleoproterozoic Caicara Metavolcanics

Along the Atabapo River and parts of the Río Negro river, fine-grained banded acid to intermediate metavolcanic rocks occur, which by their macroscopic aspect (fiamme, agglomeratic sections, banding) appear to be largely of ignimbritic origin (Figs. 3.6, 3.7, and 3.8; Kroonenberg 1985). Microscopically the very fine-grained granoblastic matrix testifies to the metavolcanic origin as well and shows that their metamorphic grade is much lower than in the other parts of the metamorphic basement. They are characterized by euhedral, normally zoned plagioclase phenocrysts and locally also bipyramidal quartz phenocrysts with deep embayments, in a typical fine-grained granoblastic groundmass (Figs. 3.9 and 3.10). Alkali feldspar phenocrysts engulfed finer-grained matrix grains. Metamorphism is evident from the preferred orientation of biotite crystals. Similar rocks also occur much further west along the rivers Yari, Mesay and Caquetá near the Araracuara Plateau.

Not all previous authors have recognized these rocks as metavolcanic. Galvis et al. 1979 and Huguett et al. 1979 call them Neises del Atabapo-Río Negro and consider them as blastomylonitic gneiss; Barrios (1985) and Barrios et al. (1985) describe them as Atabapo migmatites. López et al. (2010), referring to them as diatexites, show beautiful microscopic examples of outgrown phenocrysts and

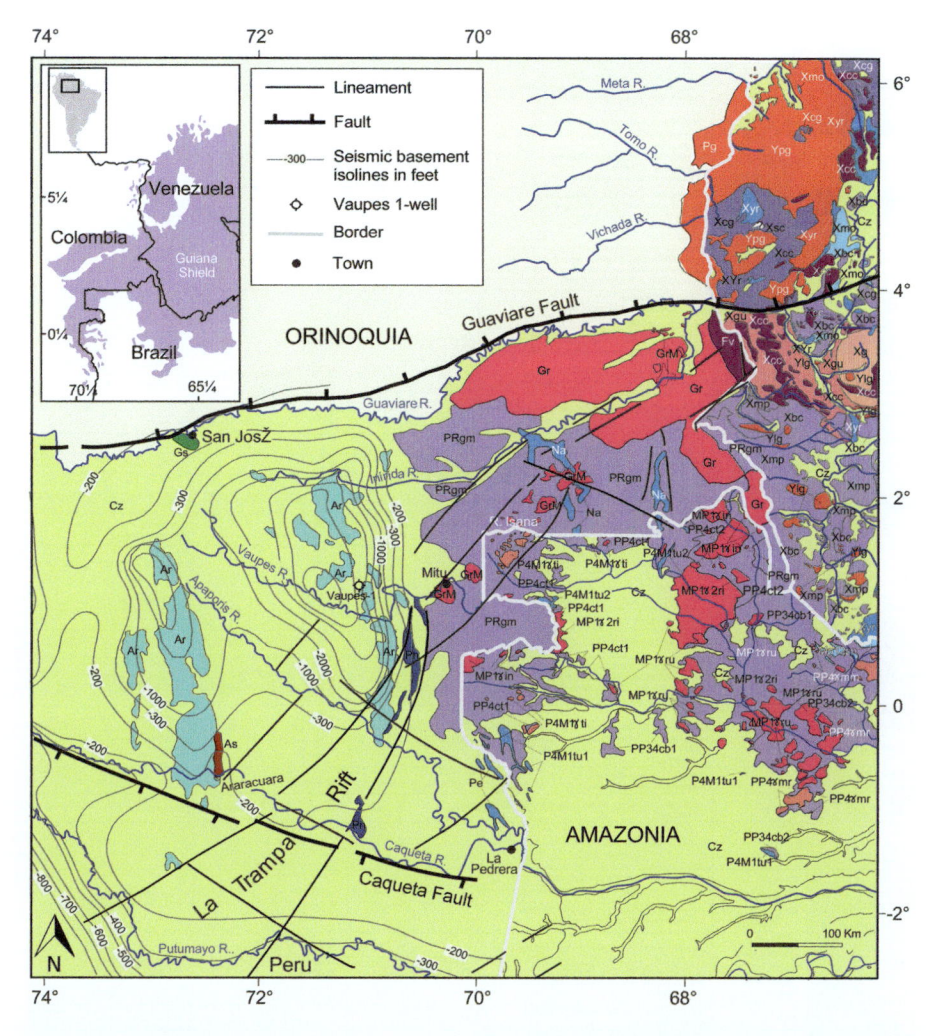

Fig. 3.3 Combined geological sketch map of Colombian, Venezuelan and Brazilian border. (Based on Bruneton et al. 1983; Gómez et al. 2007; Hackley et al. 2005; Almeida 2014 and unpublished own data)

recrystallized groundmasses from the Atabapo River outcrops without recognizing the metavolcanic character of the rocks. However, Bogotá (1981) and Bruneton et al. (1983) had already confirmed their metavolcanic origin and mapped them separately as such ('Atabapo Gneiss').

On the Venezuelan map of Hackley et al. (2005), the same rocks along the upper Atabapo River are mapped as the metavolcanic Caicara Formation (legend unit Cox et al. 1993; Wynn 1993), a name coined already by Ríos (1972), cited by Sidder and Mendoza (1991). We follow their nomenclature. In Venezuela only Rb-Sr isochrons for these rocks have been obtained: 1782 ± 72 Ma (Barrios et al. 1985) and 1793 ± 98 Ma (Gaudette and Olszewski 1985). However, in Brazil, Schobbenhaus

UNIT		Lithology	Age (Ma)	Method	Author
Cenozoic	Cz				
Araracuara Formation	Ar	Mudstones, shales, siliceous siltstones, metasiltsnones, feldspathic metasandstones and metasandstones with marble lenses.	500 - 435	Based on Acritarchs	Théry, J. M., Peniguel, T., & Haye, G. (1988)
Guaviare Syenite	Gs	Nepheline Syenite - Monzosyenite.	494 ± 5 577.8 ± 6.3 - 9	Ar-Ar U-Pb	Arango, M. I (2012)
Vaupés Basin	Subsurface	Sandstones Contact metamorphosed sandstone Granophyric diabase Intrusive granophyric gabbro	1) 804 ± 40 2) 1110 ± 51 3) 826 ± 41	1) K-Ar (Sandstones) 2) W R (Sandstones) 3) K-Ar (Gabbro)	Franks (1988)
Piraparaná Formation	Pn	Rhyodacitic volcano-sedimentary rocks quartzarenites and feldspathic sandstones.	920 ± 90	Rb-Sr	Priem et al. (1982)
Siliceous Intrusive Rocks	Ylg				
Parguaza Granite	Pg Ypg	Pink, rather dark-coloured very coarse grained rock with the typical rapakivi texture with large pink ovoid potassium feldspar megacrysts. Biotite and hornblende are the main mafic minerals.	1) 1545 ± 20 2) 1531 ± 39 3) 1392 ± 5 4) 1401 ± 2	1) U-Pb 2) Rb-Sr 3) U-Pb 4) U-Pb	1) Gaudette et al. (1978) 2) Gaudette et al. (1978) 3) Bonilla et al. (2013a, b) 4) Bonilla et al. (2013a, b)
Naquén Formation	Na	Metaquartzarenites. Metaconglomerates and metasandstones.			
Pedrera Formation	Pe	Metaquartzarenites. Metaconglomerates and metasandstones.	Deposition interval: 1580 - 1350	1) U-Pb (underlying basement)	
Tunuí Group	P4M1tu1 (Tunuí facies)	Quartzite and sericite-quartzite, with subordinated sericite-andalusite Quartzite, ferruginous quartzite, metapelite, graphite bearing pelite, phyllite and quartzarenite.	1) 1593 ± 6 2) 1880 3) 1225 4) 1334 ± 2	2) U-Pb (youngest detrital zircon) 3) Rb-Sr (intruding mafic dyke) 4) Ar-Ar (mica age-metamorphic event)	1) Ibáñez-Mejía (2011) 2) Santos et al. (2003) 3) Priem et al. (1982) 4) Santos et al. (2003)
	P4M1tu2 (Taiuacu-Cauera facies)	Banded, polideformed and migmatitic paragneisses.			
Cinaruco	Xyr	Micaceous quartzite, phyllite and conglomerate			McCandless(1962) & Cox et al (1991)
Neblina Formation	Mp12sn	Quartzarenite, quartzite and metaconglomerate.	1400 - 543	?	CPRM (2004b)
Moriche, Esmeralda Formations (undifferentiated)	Xmo	Ferruginous quartzites.			
Inhamoin Intrusive Suite	MP1½in	Porphyritic biotite monzogranite with titanite.	1) 1483 ± 2 2) 1536 ± 4	1) Pb-Pb 2) Pb-Pb	1) Almeida et al. (2013) 2) CPRM (2004b)
Rio Uaupés Intrusive Suite	MP1½ru	Porphyritic biotite monzogranite with titanite.	1518 ± 25	U-Pb	Santos et al. (2000)
Rio Icana Intrusive Suite	MP1½ri	Muscovite-biotite granite, generally sheared and with magmatic flow structures, associated with a series of para-derived migmatitic sequence with cordierite, biotite and sillimanite.	1) 1521 ± 32 to 1536 ± 4 2) 1530 ± 21 to 1578 ± 27	1) Pb-Pb 2) U-Pb	1) Almeida et al. (1997) 2) Ibáñez-Mejía (2011)
Porphyroblastic Granite	Gr	Calc-alkaline basement.	1500 - 1550	?	?
Mitú Granite	GrM	Coarse-grained homogeneous unmetamorphosed biotite granite with pink alkali feldspar megacrysts and with very large titanite crystals as a typical microscopic characteristic.	1) 1552 2) 1574 ± 10	1) U-Pb 2) U-Pb	1) Priem et al. (1982) 2) Ibáñez-Mejía (2011)
Araracuara Syenogranite	As	Syenogranite.	1732 ± 17 1756 ± 08	U-Pb	Ibañez - Mejía, et al (2011)
Tiquié Granite	P4M1½ti	Biotite-monzogranite, syenogranite and rarely grey-pink alkali feldspar granite, locally porphyritic.	1) 1746 ± 6 and 1756 ± 12 2) 1749 ± 5	1) Pb-Pb 2) Pb-Pb	1) CPRM (2004a) 2) CPRM (2004b)
Marie - Mirim Intrusive Suite	PP4½mm	Biotites syenogranite, monzogranite to orthogranite with riebeckite.	1756 ± 12	Pb-Pb	CPRM (2004b)
Marauiá Intrusive Suite	PP4½mr	Biotite (leuco) monzogranite, leucosyenogranite with riebeckite.	1746 ± 6	Pb-Pb	CPRM (2004b)
Calc - alkaline Granite	Xg				
Intrusive rocks (undifferentiated)	Xgu				
Cumati Complex	PP4ct1 (Querari facie)	Hornblende-biotite (meta) granitoids and monzogranitic to dioritic orthogneisses.	1) 1777 ± 4 2) 1785 ± 2	1) U-Pb 2) Pb-Pb	1) Almeida et al. (2013) 2) Almeida et al. (2013)
	PP4ct2 (Tonú facie)	Tonalitic to granodioritic biotite orthogneiss, polideformed and locally migmatitic.			
Cauaburi Complex	PP34cb1 (Tarsira facie)	Granitoids and monzogranitic augengneisses.	1) 1807 ± 6 2) 1795 ± 2	1) U-Pb 2) Pb-Pb	1) Santos et al. (2003) 2) Santos et al. (2003)
	PP34cb2 (Santa Isabel facie)	Monzogranitic to tonalitic (mega) granitoids and orthogneisses, with subordinate amphibolites and migmatites.			
San Carlos Complex	Xmp	Granite, granitic gneiss, augen gneiss and pegmatites.			
Mitú Complex	PRgm	Quartzofeldspathic gneisses, metapelitic gneisses, amphibolites, granulites.	1) 1850 2) 1740 ± 5	1) U-Pb 2) U-Pb	1) Gaudette and Olszewski (1985) 2) Cordani 2011, pers com
Basement Complex	Xbc	Granitics gneisses to foliated granodioritic and migmatitic gneisses.			
Felsic volcanics	Fv	Felsic volcanics, tuffs, agglomerates, red beds.			
Caicara Formation	Xcc	Ignimbritic acid metavolcanics.	1) 1782 ± 72 2) 1793 ± 98 3) 1966 ± 9 4) 1984 ± 9	1) Rb-Sr 2) Rb-Sr 3) U-Pb (obtained from the correlated Surumú Group 4) U-Pb (obtained from the correlated Surumú Group)	1) Barrios et al. (1985) 2) Gaudette and Olszewski (1985) 3) Schobbenhaus et al. (1994) 4) Santos et al. (2003)
Cuchivero Group	Xcg	Biotite granites, hypabyssals and granodiorites.	1) 1956 - 1932	1) Rb-Sr (Santa Rosalia and San Pedro granites)	1) Gaudette et al. (1978)

Fig. 3.4 Legend of geological map of 3.3

et al. (1994) published a first conventional U-Pb age of the equally correlated acid metavolcanic Surumú Group at 1966 ± 9 Ma (conventional U-Pb), and Santos et al. (2003) published a SHRIMP U-Pb age 1984 ± 9 Ma for a Surumu rhyodacite from the Roraima Province, immediately south of the Venezuelan Amazonas territory. Therefore I consider these rocks as not belonging to the Mitú complex but to an older Late Trans-Amazonian basement.

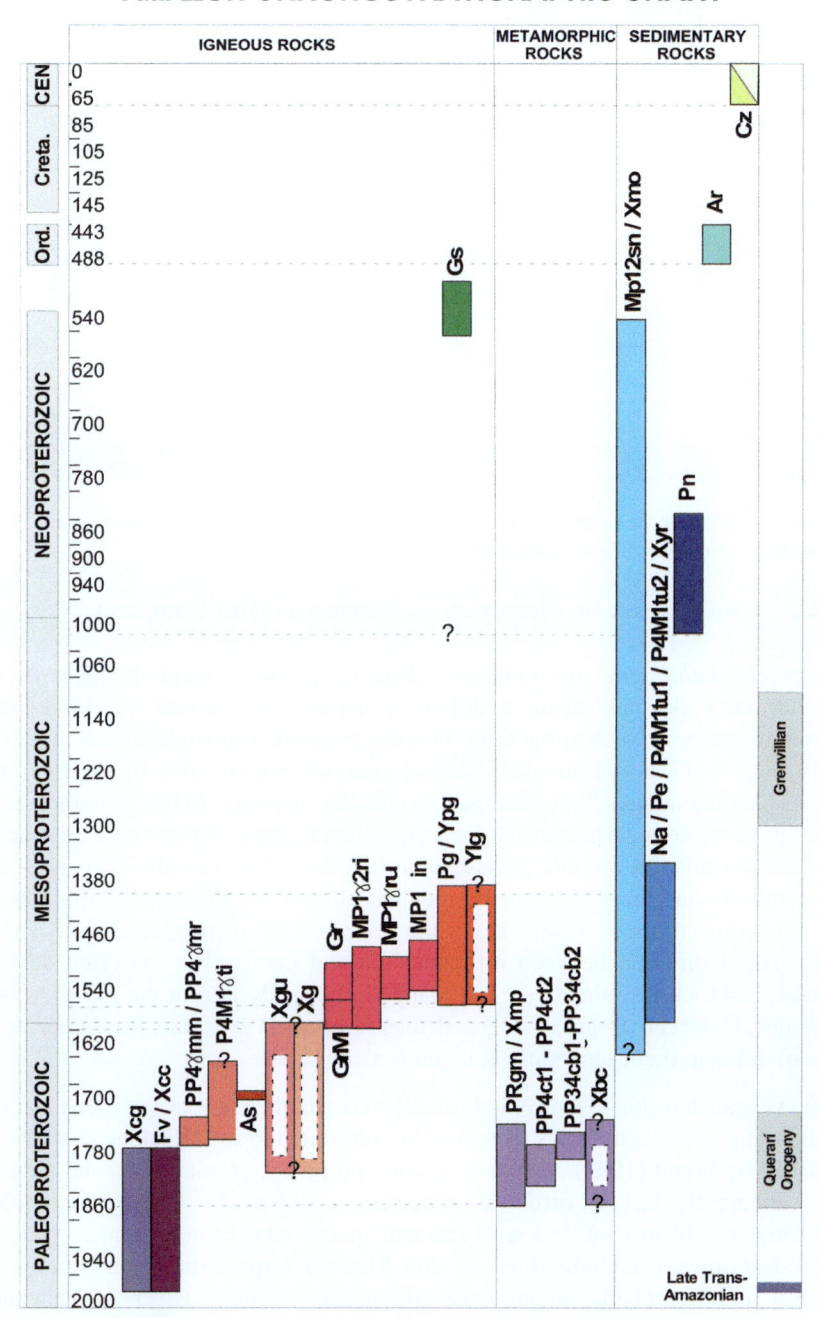

Fig. 3.5 Sequence of events in the Colombian Amazonian Precambrian

Fig. 3.6 Metaignimbrite with fiamme, Caicara Formation, río Orinoco near mouth Caño Guachapana, Venezuela. (Photo: Kroonenberg)

3.1.2.2 Late Proterozoic Metamorphic Basement (Mitú Complex)

Quartzofeldspathic gneisses Quartzofeldspathic gneisses form the bulk of the metamorphic rocks, comprising both homogeneous orthogneisses with large alkali feldspar megacrysts, such as the Caño Yí gneisses defined by Rodríguez et al. (2010, 2011b, Fig. 3.11), and migmatitic banded gneisses, which often by their compositional banding suggest a supracrustal origin (De Boorder 1978). Bruneton et al. (1983) present chemical arguments for a supracrustal origin of these rocks. Common types are (hornblende)-biotite gneisses, biotite-plagioclase (tonalitic) gneisses and biotite-muscovite gneisses, usually metamorphosed in the amphibolite facies. The latter crop out extensively along the Vaupés, Cuduyarí, Querarí and Papurí rivers. The distinction between orthogneisses and paragneisses is often difficult to make, and therefore they were not mapped separately during the PRORADAM campaign. However, Bogotá (1981), Bruneton et al. (1983) and Rodríguez et al. (2011b) did map them separately at larger scales.

On the Venezuelan side of the Guainía and Río Negro, these rocks have been mapped as belonging to the San Carlos metamorphic-plutonic terrane (Hackley et al. 2005), described by Wynn (1993) as granite, granite porphyry, granitegneiss and augengneiss, apparently largely ortho- in appearance, and to the basement complex: well-foliated, chloritized and well-foliated quartz-rich biotite-granite gneisses. Older descriptions include those of the Minicia migmatitic gneiss along the Orinoco and Macabana augengneiss along the Ventuari River in Venezuela (Figs. 3.12, 3.13, and 3.14; Rivas 1985).

On the Brazilian side of the Vaupés and Traira areas, they correspond with the facies Querarí of the Cumati series (hornblende-biotite (meta) granitoids and

Fig. 3.7 Primary layering in Caicara acid metavolcanic sequence, Guarinuma, Raudal Chamuchina, Río Atabapo. (Photo: Kroonenberg)

monzogranitic to dioritic orthogneisses: CPRM 2004a, b; Commission for the Geological Map of the World 2009a, b). In the Brazilian part of the Río Negro border area, they correspond with the Tonú facies of the Cumati series (tonalitic to granodioritic biotite orthogneisses, polydeformed, locally migmatitic) and the Cauaburí series, facies Santa Izabel (monzogranitic to tonalitic (meta)granitoids and orthogneisses, with subordinate amphibolites and migmatites).

Table 3.1 shows the U-Pb radiometric ages for the quartzofeldspathic gneisses in Colombia, Brazil and Venezuela. Priem et al. (1982) gave a conventional U-Pb age of 1846 Ma from a biotite gneiss along the Guainía River but discarded this age because of presumed older radiogenic lead. However, in view of the similar ages obtained by Gaudette and Olszewski (1985) and others, this date may indeed be a reliable age. The quartzofeldspathic gneisses in the basement therefore show a range in ages between 1.86 and 1.72 Ga. Recently ϵ_{Nd} values between +0,78 and −2,24 and T_{DM} ages between 2,40 Ga and 1,99 Ga have been obtained for these rocks, suggesting a largely juvenile character for them (Almeida et al. 2013).

Fig. 3.8 Recrystallized metavolcanic rock with grey plagioclase phenocrysts and dendritic biotite; Guarinuma, Raudal Chamuchina, Río Atabapo. (Photo: Kroonenberg)

Fig. 3.9 Embayed quartz phenocryst in recrystallized groundmass in Atabapo metavolcanite (López et al. 2010)

Metapelitic gneisses Migmatitic biotite-(muscovite) gneisses of metapelitic composition, evidenced by the presence of aluminous minerals as sillimanite, andalusite, cordierite and locally also garnet, occur in isolated outcrops near Puerto Colombia in the Guainía River, in the upper Cuduyarí, in the Río Paca/Rio Papurí and in the Vaupés River just upstream from Mitú, but they have nowhere been mapped separately (Fig. 3.15; Galvis et al. 1979; Huguett et al. 1979; Kroonenberg 1980;

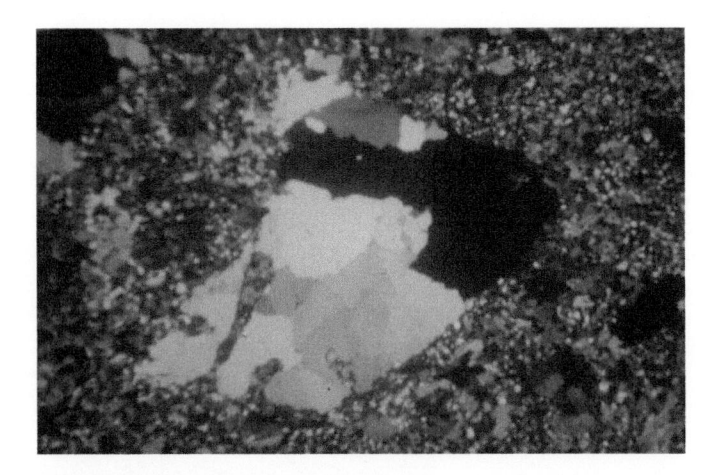

Fig. 3.10 Deformed bipyramidal quartz phenocryst with embayments in acid metavolcanic gneiss, IGM 130464 Araracuara. (Photo: Kroonenberg)

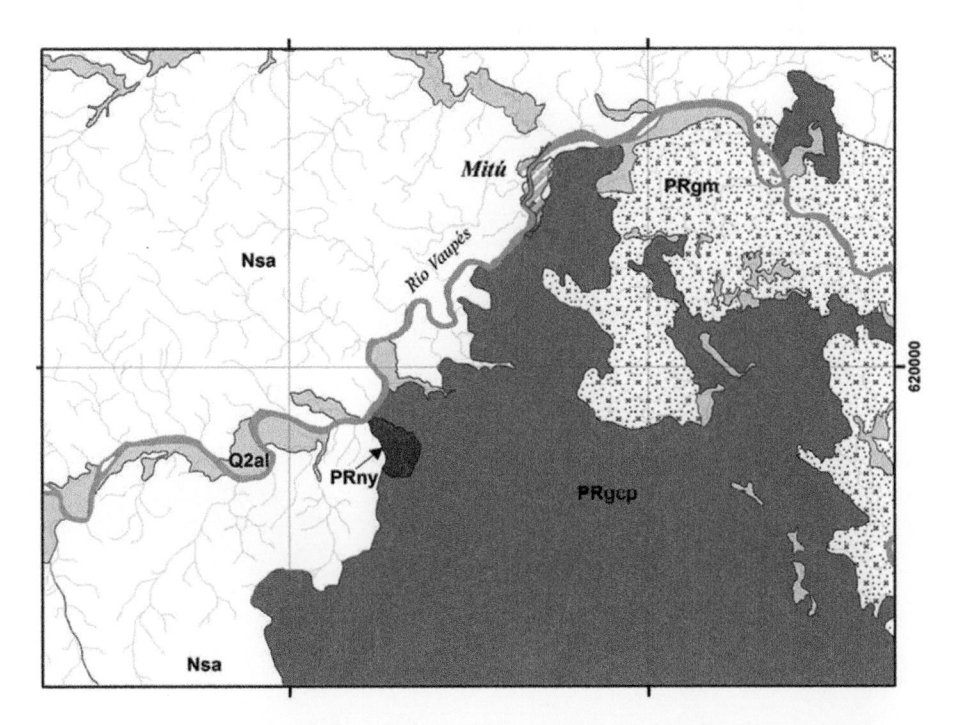

Fig. 3.11 Geological map of sheet 443, Mitú, showing PRgm, monzogranito de Mitú, PRny gneiss del caño Yi, PR gcp, granofels del Cerro Pringamosa. (After Rodríguez et al. 2011b)

Bruneton et al. 1983). Locally there is green spinel as an accessory. Metamorphic grade is in the amphibolite facies. Replacement of cordierite by higher-pressure minerals might indicate a later static phase of metamorphism (Kroonenberg 1980). No geochronological data of these rocks have been published.

Fig. 3.12 Augengneiss with aplite vein traversed by *en echelon* quartz veins: at least three phases of deformation. Guyanese geologist Chris Barron, Río Atabapo near Boca Caño Caname 1981. (Photo: Kroonenberg)

Fig. 3.13 Minicia supracrustal migmatitic quartzofeldspathic gneiss with crosscutting pegmatite vein, río Orinoco, Venezuela. (Photo: Kroonenberg)

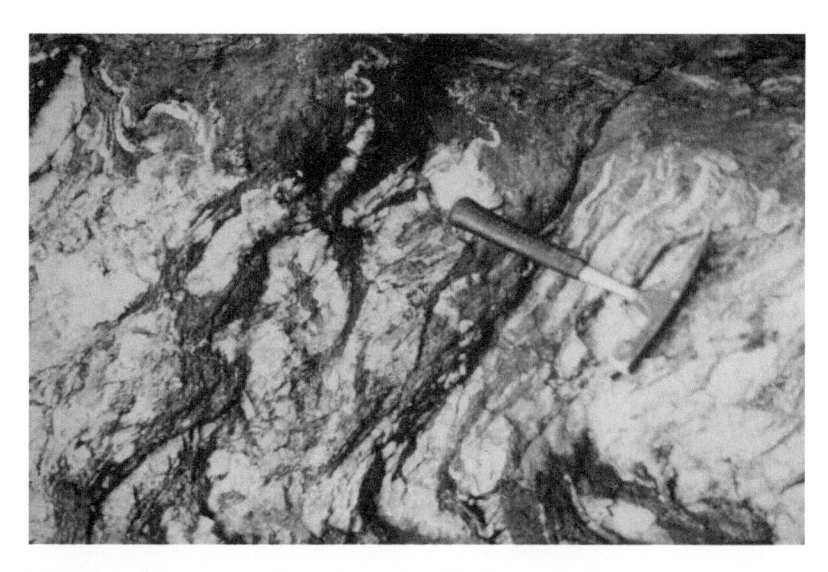

Fig. 3.14 Migmatitic quartzofeldspathic gneiss, El Chorro, río Caquetá, Araracuara. (Photo: Kroonenberg)

Table 3.1 Radiometric ages (U-Pb) for quartzofeldspathic gneisses in Colombia, Brazil and Venezuela

Sample nr.	Location, rock type	Method	Age (Ma)	Author
8697	Minicia gneiss (bi-gar)	Conventional U-Pb	1859	Gaudette and Olszewski (1985)
8699B	Macabana Gn. (bi-hbl)	Conventional U-Pb	1823	Gaudette and Olszewski (1985)
PRA 21	Guainía R, bi-gneiss	Conventional U-Pb	1846 ± 95	Priem et al. (1982)
6850/6085	Casiquiare R., tonalite	Pb-Pb SHRIMP	1834 ± 24	Tassinari et al. (1996)
MS63	Cauaburi gneisses	U-Pb SHRIMP	1807 ± 6	Santos et al. (2003)
CG8	Cauaburi gneisses	Pb-Pb evaporation	1795 ± 2	Santos et al. (2003)
	Cumati gneisses	Pb-Pb evaporation	1785 ± 2	Almeida et al. (2013)
	Cumati gneisses	U-Pb SHRIMP	1777 ± 4	Almeida et al. (2013)
J-263	Caquetá, bi-granite	La-mc-ICP-MS	1732 ± 17	Ibáñez-Mejía et al. (2011)
PR-3215	Mesay, bi gneiss	La-mc-ICP-MS	1756 ± 8	Ibáñez-Mejía et al. (2011)
EP2Mi	Caquetá bi ms gneiss	ICP-MS	1721 ± 9.6	Cordani et al. (2016)
HB-667	Vaupés bi hbl gneiss	ICP-MS	1779 ± 3.7	Cordani et al. (2016)
J-36	Cuduyarí bi-ms granite	ICP-MS	1739 ± 38	Cordani et al. (2016)
J-127	CañoNaquén bi hbl gn	ICP-MS	1775 ± 3.7	Cordani et al. (2016)
J-199	Guainia bi-hbl gneiss	ICP-MS	1796 ± 3.7	Cordani et al. (2016)
PR-3001	Cuduyarí mig bi plag gn	ICP-MS	1740 ± 5	Cordani et al. (2016)

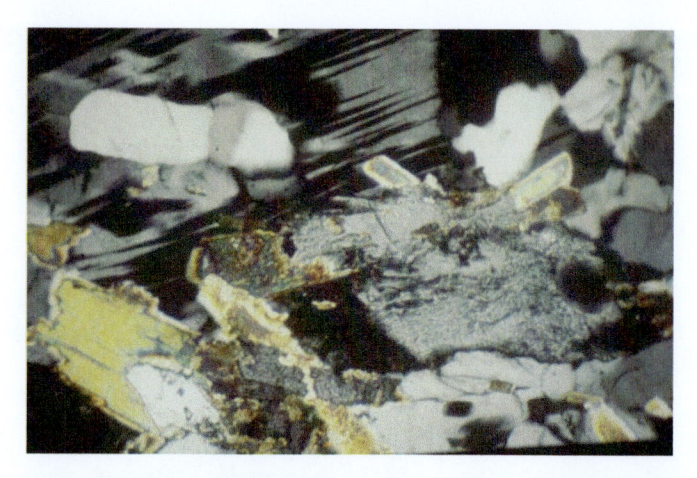

Fig. 3.15 Cordierite-sillimanite-andalusite biotite gneiss, IGM 130356, Puerto Colombia, Río Guainía. (Photo: Kroonenberg)

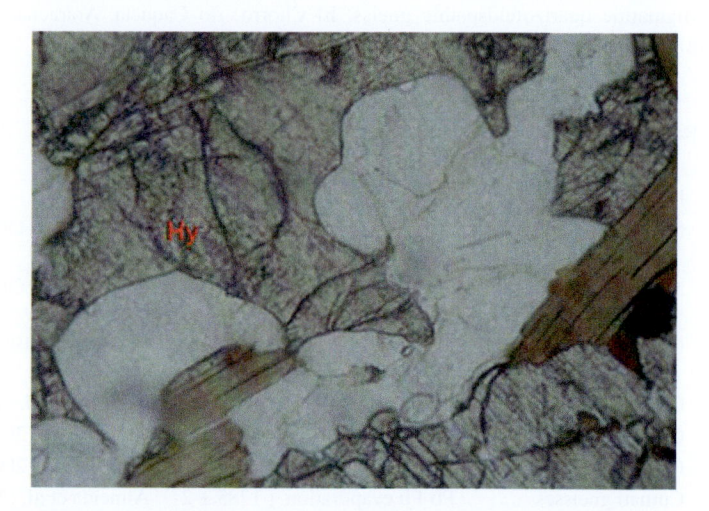

Fig. 3.16 Orthopyroxene in granulite, IGM 5000372. (Rodríguez et al. 2011b)

Amphibolites Amphibolites, consisting of hornblende, plagioclase +/− quartz and sometimes clinopyroxene or biotite, occur in thin bands and boudins intercalated in gneissic rocks, e.g. at the confluence of Querarí and Vaupés rivers.

Granulites A single granulite sample with orthopyroxene, biotite, plagioclase, quartz and subordinate alkali feldspar (Figs. 3.11 and 3.16) was described from the Sierra de Pringamosa south of Mitú by Rodríguez et al. (2010, 2011b). It is the only indication of granulite-facies metamorphism in the Colombian Amazones. No age data are available.

3.1.2.3 Late Paleoproterozoic Older Granites (Tiquié Granite)

Tiquié granite Along the Isana River in the Brazilian-Colombian border area, CPRM maps the Tiquié granite (biotite-monzogranite, syenogranite and rarely grey-pink alkali feldspar granite, locally porphyritic). Granite plutons have been mapped in this area on morphological grounds by Botero (1999), which coincide with outcrops of coarse-grained biotite granites along the Guainía River upstream from Manacacías. These A-type granites have been dated in Brazil in a range of 1746 ± 6 Ma and 1756 ± 12 Ma with some inherited ages from the Cumati-Cauaburi basement of 1784 ± 7 Ma e 1805 ± 8 Ma (Pb-Pb evaporation, CPRM 2004a). Sm-Nd data show an ϵ_{Nd} value of +4.05 and a T_{DM} model age of 1.82 Ga.

3.1.2.4 Mesoproterozoic Younger Granites (Mitú, Içana, Atabapo and Other Granites)

Mitú granite (or monzogranite; Rodríguez et al. 2011a, b) This is a coarse-grained homogeneous unmetamorphosed biotite granite with pink alkali feldspar megacrysts (up to 15 cm according to De Boorder 1976) and with very large titanite crystals as a typical microscopic characteristic (Figs. 3.17 and 3.18). The description resembles those of the El Remanso granite of the Inírida river and the San Felipe granite from the Río Negro of Bruneton et al. (1983) and on the Venezuelan side the San Carlos granite of Martínez (1985). Chemical analyses by Rodríguez et al. (2011a, b) show the metaluminous and anorogenic (A-type) character of these intrusions. On the Brazilian side of the border in the Vaupés-Papurí and Río Negro areas, these granites are mapped as Inhamoin granite and Uaupés granite, porphyritic biotite monzogranite with titanite (Dall'Agnol and Macambira 1992; CPRM (2004a, b); Reis et al. 2006; CGMW 2009a). Priem et al. (1982) obtained a conventional U-Pb zircon age of 1552 Ma for the Mitú granite. The Uaupés and Inhamoin granites have been dated at 1518 ± 25 Ma (Santos et al. 2000; CPRM 2004a) and 1483 ± 2 Ma (Pb-Pb evaporation), respectively (Almeida et al. 2013). A recent U-Pb LA-MC-ICPMS age for the Mitú granite of 1574 ± 10 Ma was obtained by Ibáñez-Mejía et al. 2011. Sm-Nd data show ϵ_{Nd} values between −1.85 and −2.37 and T_{DM} model ages between 2.05 and 1.97 Ga (Almeida et al. 2013).

Tijereto granophyre Another undeformed intrusive exposed along the Caquetá River, individualized by the PRORADAM authors as Granófiro de Tijereto, of intermediate and slightly alkaline composition (with magnesioriebeckite), shows a model Rb-Sr age of 1495 Ma according to Priem et al. (1982) and therefore fits in the same category of younger, Mesoproterozoic intrusives.

Içana medium-grained bi-mica granites Bruneton et al. (1983) mapped two distinct areas along the Río Negro, as consisting of medium-grained two-mica granites, without alkali feldspar megacrysts but locally with large muscovite flakes and

Fig. 3.17 Mitú granite, río Vaupés near Mitú hospital. (Photo: De Boorder 1976)

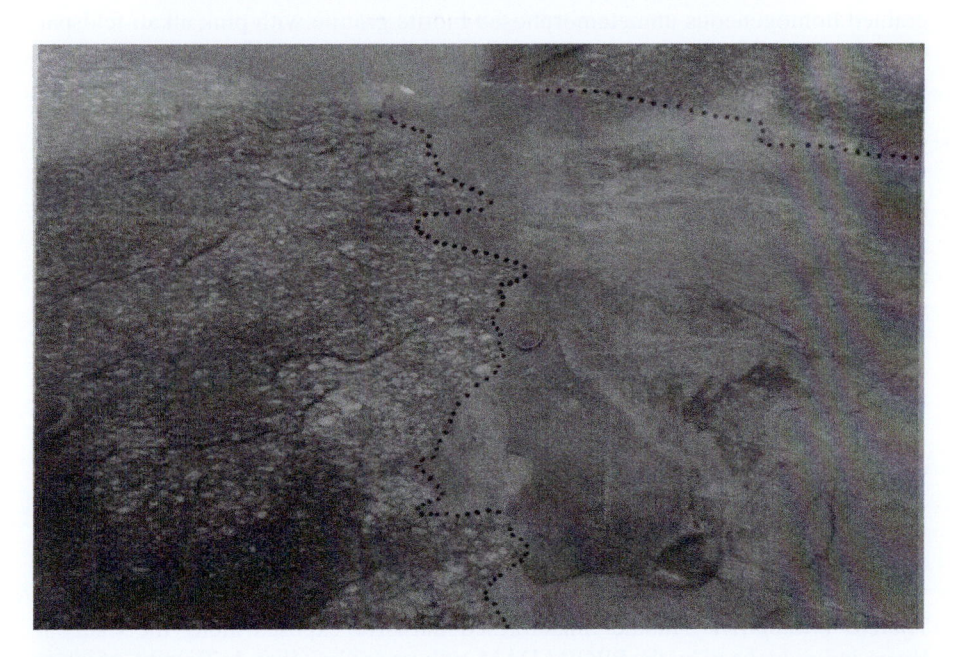

Fig. 3.18 Intrusive contact of megacryst granite into fine-grained gneisses, Río Papurí. (Photo: De Boorder 1976)

sometimes sillimanite. This would obviously be an S-type granite. This rock seems comparable with the Brazilian Río Içana Intrusive Suite, a muscovite-biotite granite, generally sheared, and with magmatic flow structures, associated with a series of para-derived migmatitic sequences with cordierite, biotite and sillimanite. The Brazilians map it along the Río Içana close to the Colombian border in the Río Negro area and along the Papurí border river near Yavaraté (CPRM 2004a; Reis et al. 2006). The Içana granites show ages ranging from 1521 ± 32 Ma (Almeida et al. 2007) to 1536 ± 4 Ma (Pb-Pb evaporation), besides some inherited ages from different basement types (1745 ± 13 Ma and 1803 ± 9 Ma; Almeida et al. 2013). Recently, Ibáñez-Mejía et al. (2011) obtained U-Pb LA-MC-ICPMS crystallization ages on zircons from two bi-mica monzogranites from the middle Apaporis River of 1530 ± 21 Ma and 1578 ± 27, apart from a considerable quantity of inherited zircons. Sm-Nd data show ϵ_{Nd} values of −3.05 and a T_{DM} model age of 2,04 Ga (Almeida et al. 2013).

Atabapo granite At San Fernando de Atabapo, a greyish-pink coarse-grained inequigranular leucocratic calcalkaline granite with characteristic blue quartz crops out over ~120 km^2 (Bruneton et al. 1983; Rivas 1985). Rb-Sr data indicate an age of 1617 ± 90 Ma (Gaudette and Olszewski 1985) or 1669 Ma (Barrios et al. 1985).

La Campana fine-grained (subvolcanic) granites Bruneton et al. (1983) distinguish various types of non-mappable fine-grained granites to aplites, supposedly late crystallization phases of the main magmatic pulses. Along the Yarí River near Araracuara, fine-grained granitic intrusions occur which appear to be related to the acid metavolcanics in this area (Fig. 3.19).

3.1.2.5 Mesoproterozoic Parguaza Rapakivi Granite

The Parguaza rapakivi granite forms a huge batholith of over 30,000 km^2, straddling the border of Venezuela and Colombia north of the Guaviare River. Most of the batholith is situated in Venezuela; in Colombia it only occupies isolated inselbergs in the Vichada department and cataracts in the Orinoco River (Gaudette et al. 1978; Bangerter 1985; Rivas 1985; Herrera-Bangerter 1989; Bonilla et al. 2013).

As Galvis et al. (1979) and Huguett et al. (1979) limit the Mitú complex to the crystalline basement *south of the Guaviare River*, the Parguaza rapakivi granite would strictly speaking not belong to the Mitú complex. In Venezuela the Parguaza rapakivi granite is known to intrude into a Paleoproterozoic basement older than the Mitú complex, the Caicara metavolcanics and the Santa Rosalia and San Pedro granites of the Cuchivero Group (Mendoza 1974; Sidder and Mendoza 1991; see above). Contact metamorphic aureoles are absent; most of the contacts are tectonic, though some apophyses of the Parguaza granite into Cuchivero rocks have been found along the Suapure River in Venezuela (Mendoza 1974; Herrera-Bangerter 1989).

The main granite is a pink, rather dark-coloured very coarse-grained rock with the typical rapakivi texture with large pink ovoid potassium feldspar megacrysts up to 8 cm, surrounded by a thin greenish plagioclase mantle (Figs. 3.20 and 3.21). Biotite and hornblende are the main mafic minerals. Apart from this main, wyborgite type, there are smaller bodies of less coarse pyterlite rapakivi granite, clinopyroxene- or sodic amphibole-bearing alkali granite and syenite (Bruneton et al. 1983;

Fig. 3.19 Fine-grained subvolcanic granite, La Campana cataract, Yari river near Araracuara. (Photo: Kroonenberg)

Fig. 3.20 Parguaza rapakivi granite, Caño Cupavén, Venezuela. (Photo: Kroonenberg)

Bangerter 1985; Herrera-Bangerter 1989; González and Pinto 1990; Bonilla et al. 2013, Bonilla-Pérez et al. 2013). In the main Venezuelan body, there are numerous xenoliths, abundant pink and green aplite veins, several, partly columbite-/tantalite-bearing pegmatites and late thin olivine basalt dykes with complex relationships to each other (Herrera-Bangerter 1989). Chemically it is a typical anorogenic peralkaline granite, with high FeO/MgO as many other rapakivi granites in the world. Gaudette et al. (1978) report a conventional U-Pb zircon age of 1545 ± 20 Ma and a

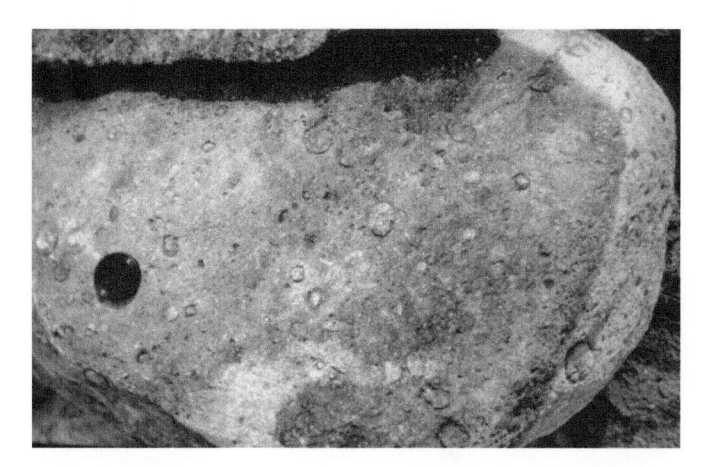

Fig. 3.21 Weathered surface of Parguaza rapakivi granite, showing differential weathering of unstable plagioclase rims around more stable alkali feldspars. Caño Cupavén, Venezuela. (Photo: Kroonenberg)

Rb-Sr isochron age of 1531 ± 39 Ma. Younger Rb-Sr isochron ages of about 1380 Ma were reported from around Puerto Ayacucho and San Pedro by Barrios et al. (1985); Bonilla-Pérez et al. 2013) present new LA-ICPMS data from the Colombian part of the batholith between 1392 ± 5 Ma and 1401 ± 2 Ma, i.e. considerably younger than the ages obtained by earlier authors. These ages not necessarily invalidate older data, as Mirón-Valdespino and Álvarez (1997) deduce from magnetic data and the distribution of Barrios (1985) Rb-Sr radiometric ages that the intrusion and cooling history of the batholith encompasses a prolonged period between 1480 and 1240 Ma, starting from an older core and a younger rim (Fig. 3.22).

3.1.2.6 Mesoproterozoic Tunuí Folded Metasandstone Formations

In the eastern part of the Colombian Guiana Shield, prominent N-S to NW-SE oriented ridges of folded low-grade metasandstones arise above the lowlands, the Naquén (Caparro in Brazil) and Caracanoa (or Raudal Alto) ridges in the Guainía Department and the Libertad (La Pedrera) and Machado (Taraíra) ridges in the Vaupés Department. The metasediments are strongly tilted, faulted and folded and form impressive escarpments up to 800 m (Fig. 3.23). Such ridges were first identified in Brazil as Tunuí Formation (Pinheiro et al. 1976; Renzoni 1989a), the name of a ridge in the continuation of the Naquén ridge into Brazil, and we will continue to use this name for the ensemble of the metasandstone formations (with the exclusion of the Piraparaná Formation, which will be discussed later). Unlike Almeida et al. (2002), we do not include in the Tunuí Formation the higher-grade migmatitic gneisses and amphibolites described by that author downstream from the Tunuí type locality in Brazil. Galvis et al. (1979) and Huguett et al. (1979) call the northern metasandstone occurrences Roraima Formation, which is unfortunate because the

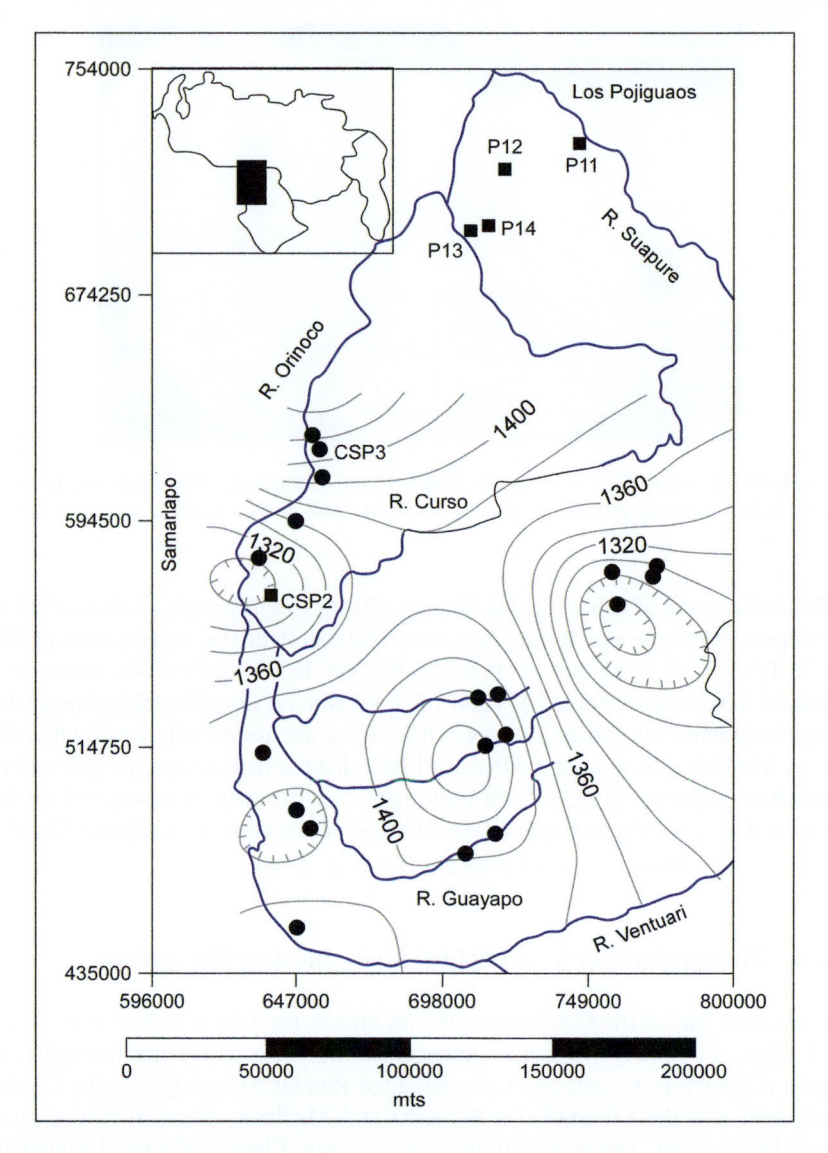

Fig. 3.22 Chrontour's Parguaza granite shows a crystallization history of >100 Ma (Mirón-Valdespino and Álvarez 1997). Reproduced with permission

Roraima Formation in Brazil, Venezuela, Guyana and Suriname is unmetamorphosed, though older (Priem et al. 1973; Santos et al. 2003).

The southern metasandstone occurrences in Colombia have received the name La Pedrera Formation from Galvis et al. (1979) because of slightly different lithologies, though the same authors admit that they offer great similarity with the northern occurrences. We include this formation into the Tunuí Formation. All metasandstone

Fig. 3.23 Northern extremity of Sierra de Naquén. (Source: Google Earth)

formations rest with unconformable and sheared contacts on top of the Complejo Mitú. Detailed stratigraphical and sedimentological studies have been made since then because of the discovery of gold in the conglomeratic sections of these formations.

Sedimentology and stratigraphy The *Naquén section* (Renzoni 1989a, b; Fig. 3.24; Galvis 1993) has a cumulative thickness of about 2000 m and consists of a non-fossiliferous series of ten fining-upwards sequences of quartz-rich metaconglomerates, metaquartzarenites and metamudstones, the latter often black and locally containing pyrite.

These sequences have been interpreted by Renzoni (1989b) as having been deposited in a fluvial environment by meandering rivers, possibly close to the sea, as some lenticular flaser-like sandstone laminae may point to tidal influence. Some of the coarse conglomerates may have been deposited in braided patterns in an alluvial fan environment. Based on the prograding character of the series, provenance of the sediments is probably from the north or northeast, though no paleocurrent data are available. The combination of fluvial with tidal characteristics leads Renzoni to infer a deltaic environment, though from the description of his sections,

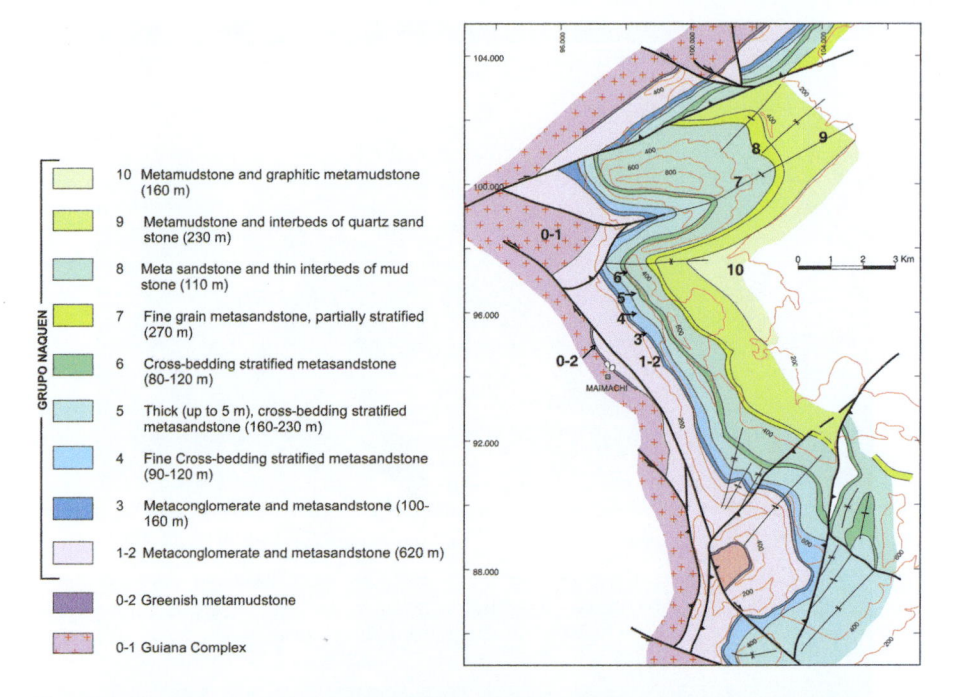

Fig. 3.24 Stratigraphy of the Tunuí Formation in the Naquén ridge, after Renzoni (1989a)

the fluvial character is largely predominant. Gold is usually concentrated in the conglomerates and conglomeratic sandstones, but also locally occurs in organic-rich mudstones close to unconformities, and is not only detrital but also remobilized by hydrothermal and supergene processes. Low-grade metamorphism is expressed in the lower parts of the sequence by complete welding of detrital grains in the sandstones and the development of coarse muscovite, though a preferred orientation is not evident. In the higher parts, metamorphism is less well expressed or not at all, and the grains are not welded (Galvis et al. 1979).

The *Caracanoa* or *Raudal Alto ridge* equally consists of at least 1000 metres of whitish quartz conglomerates and cross-bedded quartzites with phyllites at the base which rest unconformably upon the Complejo Mitú. The series is intruded by undated 'Campoalegre' diabase dykes (Galvis 1993; Carrillo 1995).

The *La Libertad range* north of the Apaporis River and close to La Pedrera has been studied in detail by Coronado and Tibocha (2000), also because of its gold potential (Fig. 3.25). The ridge is a southeast-plunging anticlinal-synclinal fold structure. They studied an 88 m sequence in which two major units are distinguished, a lower one consisting of monotonous metaquartzarenites (Fig. 3.26) and an upper one consisting of metaquartzarenites with phyllite intercalations. The quartzarenites often show trough cross-bedding and are transected by quartz veins and locally sheared. Phyllites consist mainly of muscovite. The sediments are thought to have been deposited in a fluvial to tidal environment (Fig. 3.27).

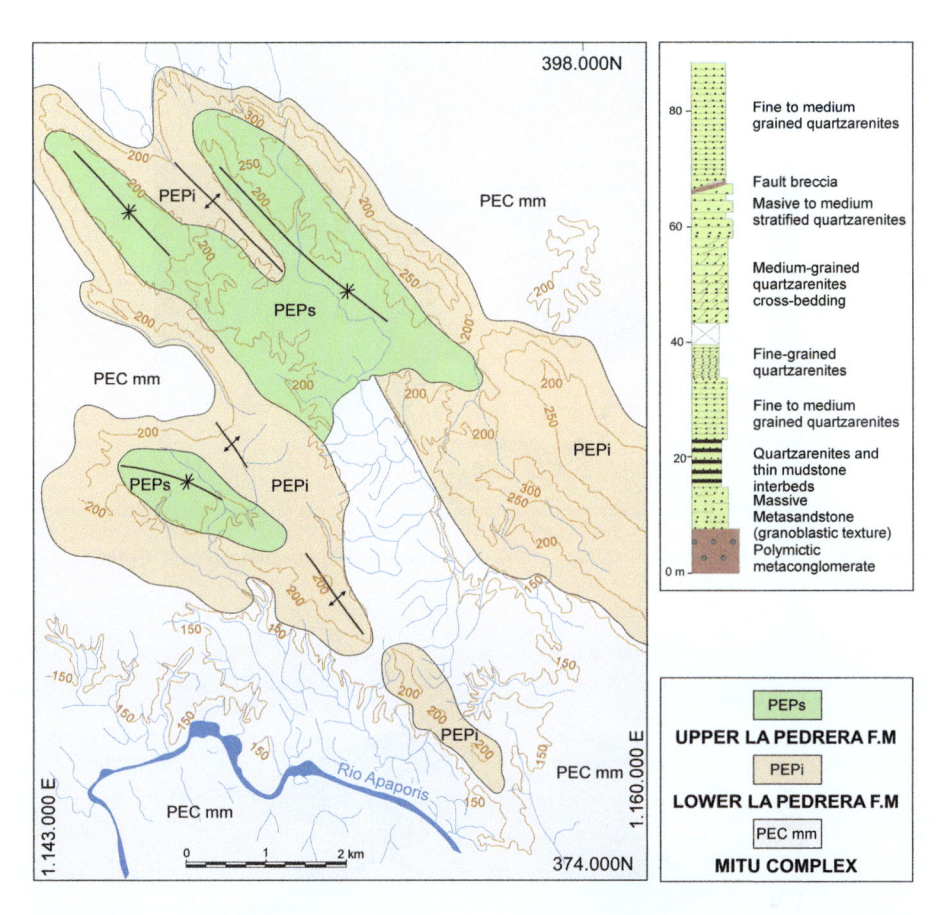

Fig. 3.25 Geological map La Libertad ridge (La Pedrera Formation), after Coronado and Tibocha (2000)

Low-grade metamorphism is evidenced by muscovite growth and locally andalusite blastesis in the finer sediments, especially in the lower parts of the sequence. Gold is mainly present in disseminated form and in narrow quartz veins in the lower part of the sequence. The contact with the underlying Complejo Mitú was observed by Galvis et al. (1979) as containing detachment folds due to shearing.

The *Machado ridge* in the Taraíra area forms a ca. 1000 m thick moderately SW-dipping monoclinal sequence (Figs. 3.28 and 3.29). It differs in several aspects from the three ridges described before. It has been explored for gold extensively by several companies, including Mineralco, Minercol, Cosigo and HorseShoe (Leal 2003; Ashley 2011), and small-scale mining is active. The sequence starts with up to 250 m of rhyolitic tuff (Mirador member of Carrillo 1995, Complejo Volcánico de Taraira de Cuéllar et al. 2003), and only on top of them the sequence of quartzconglomerates and quartz arenites starts. Two major members have been distinguished, a lower Peladero member with volcanic intercalations and horizons

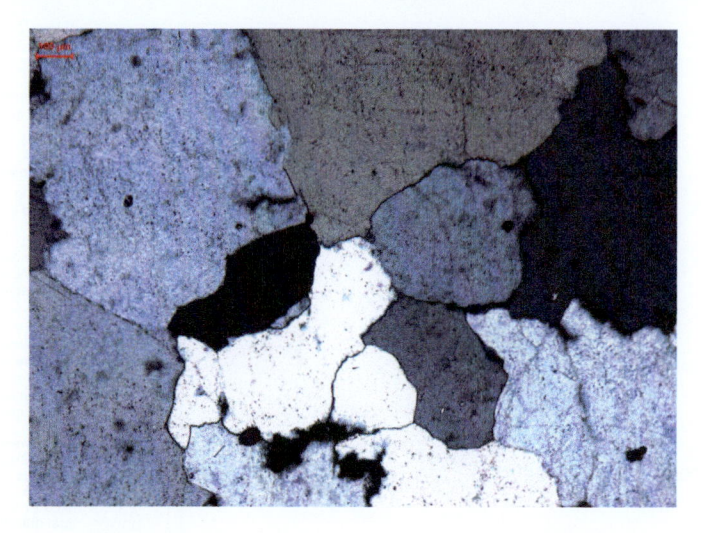

Fig. 3.26 Interlocking detrital quartz grains in La Pedrera metaquartzarenite, Quinché, río Caquetá. (Photo: Kroonenberg)

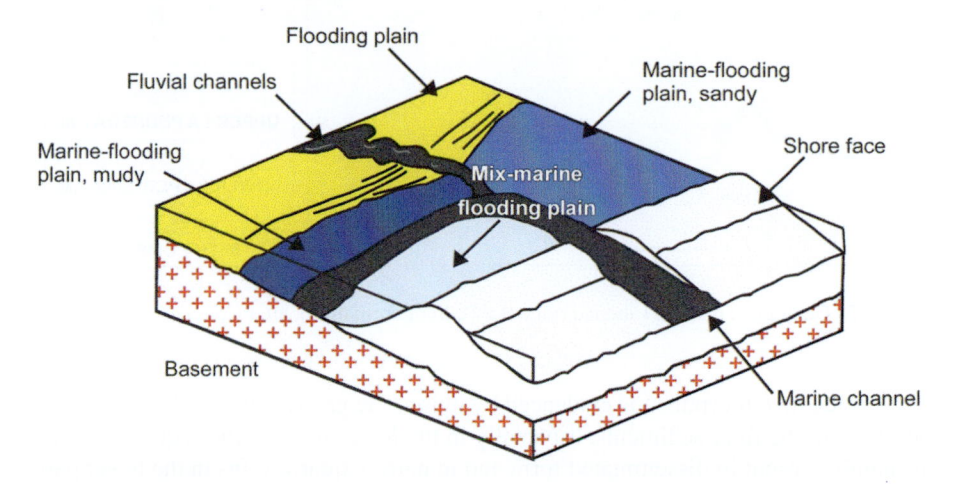

Fig. 3.27 Sedimentary environment of the La Pedrera Formation as interpreted by Coronado and Tibocha (2000)

with silica enrichment and an upper Cerro Rojo member; the latter called this way because of strong red coloration with hematite and other iron oxides, a feature not observed in the other metasandstone ridges. The base of the Cerro Rojo member is a polymict alluvial fan conglomerate, with apart from quartz also volcanic clasts. This member shows a fining-upwards sequence, terminating with finely laminated sandstone and mudstone beds, interpreted as subtidal to intertidal deposits. On top of the sequence, another sandstone formation has been distinguished, the Machado Formation, equally with strong concentrations of specular hematite

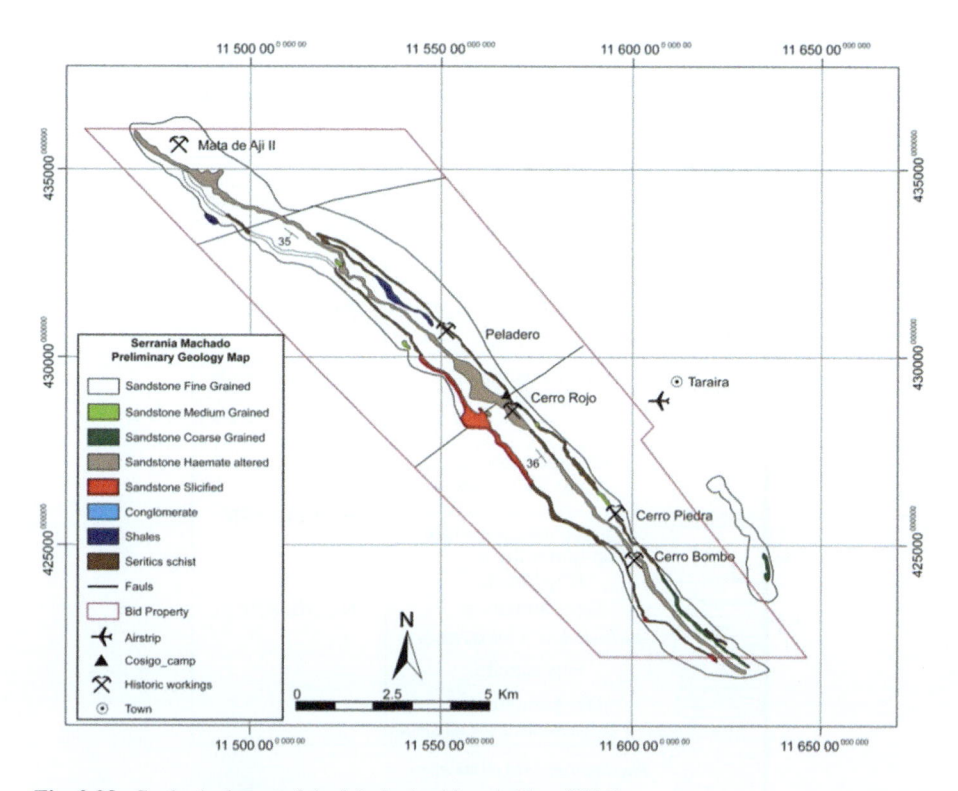

3 The Proterozoic Basement of the Western Guiana Shield and the Northern Andes 141

Fig. 3.28 Geological map of the Machado ridge, Ashley (2011)

(Cuéllar et al. 2003; Ashley 2011), though the designation as banded iron formation by Galvis and Gómez (1998) seems not justified (Cuéllar et al. 2003). Diabase dykes up to 10 m thick intrude into the series.

At the *Cerro El Carajo* in the Llanos Orientales of the Vichada department (Fig. 3.30), fine to coarse quartz meta-arenites with parallel and cross-bedding define a NW-striking monoclinal structure. Just like in the Tunuí sandstones, they show andalusite as a typical metamorphic mineral (González and Pinto 1990; De la Espriella et al. 1990; Ochoa et al. 2012). The contact with the crystalline basement is not exposed, but Ghosh (1985) observed andalusite in similar Cinaruco meta-arenites (Venezuela) in contact with the Parguaza rapakivi granite without stating what kind of contact.

The well Vaupés-1 drilled by Amoco in the 1980s to investigate the hydrocarbon potential of the Vaupés-Amazonas basin struck mainly Mesoproterozoic (contact) metamorphosed sandstones, intruded by a Neoproterozoic gabbro (Fig. 3.31; Franks 1988; see par. 3.1.3.9 below).

Geochronology. The age of the Tunuí metasediments has long been a controversial issue, mainly due its incorrect association with the Roraima sandstones by Galvis et al. (1979) and Huguett et al. (1979). Age data come from four different sources:

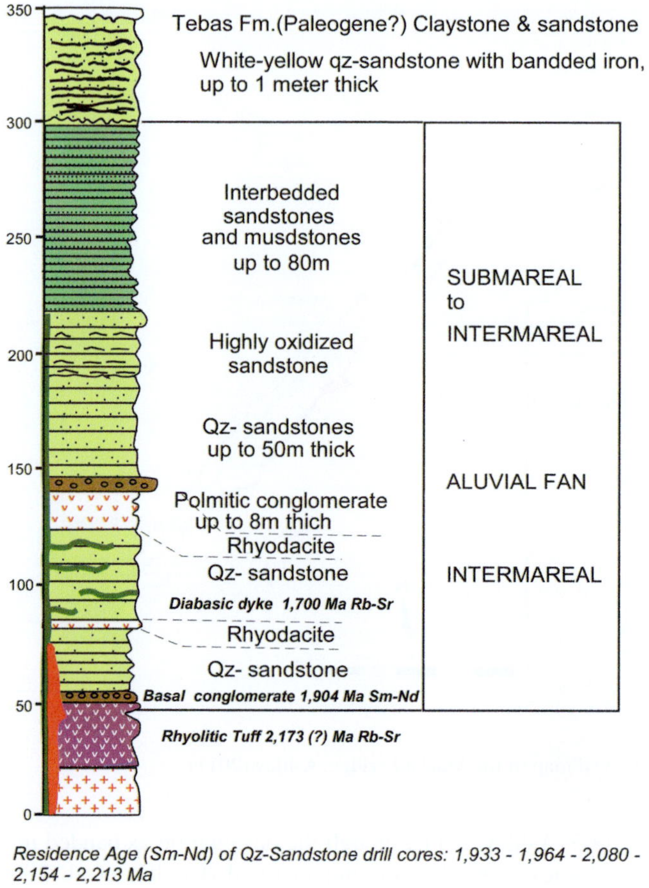
350
300
250
200
150
100
50
0
Tebas Fm.(Paleogene?) Claystone & sandstone
White-yellow qz-sandstone with bandded iron, up to 1 meter thick
Interbedded sandstones and musdstones up to 80m
Highly oxidized sandstone
Qz- sandstones up to 50m thick
Polmitic conglomerate up to 8m thich
Rhyodacite
Qz- sandstone
Diabasic dyke 1,700 Ma Rb-Sr
Rhyodacite
Qz- sandstone
Basal conglomerate 1,904 Ma Sm-Nd
Rhyolitic Tuff 2,173 (?) Ma Rb-Sr
SUBMAREAL to INTERMAREAL
ALUVIAL FAN
INTERMAREAL
Residence Age (Sm-Nd) of Qz-Sandstone drill cores: 1,933 - 1,964 - 2,080 - 2,154 - 2,213 Ma

Fig. 3.29 Geological column of the Machado ridge, after Cuéllar et al. (2003)

the age of the basement underlying the sandstones, the age of detrital grains within the sandstones, the age of younger dykes intruding the sandstones and the age of metamorphism, as analysed by Santos et al. (2003).

Recent data show that the granitic basement on which the Taraira metasediments have been deposited have a U-Pb zircon crystallization age of 1593 ± 6 Ma (Ibáñez-Mejía et al. 2011), showing that the metasediments are at least 300 Ma younger than the Roraima, now dated at 1873 ± 3 Ma (Santos et al. 2003).

The youngest detrital zircon grains found in the Tunuí-like Aracá sandstone further to the east in Brazil show ages around 1.88 Ga, also younger than the age of the Roraima sandstones (Santos et al. 2003). In Brazil recently three detrital zircons populations from the Brazilian part of the Naquén (Caparro) have been dated at 1720 ± 11, 1780 ± 8 and 1916 ± 57 Ma, suggesting that the metasand-

Fig. 3.30 Large-scale cross-bedding in Cerro El Carajo metasandstone, Vichada (Ochoa et al. 2012)

stones are at least younger than the youngest of these ages (Almeida et al. 2013). No geochronological data are available from the Caracanoa and La Libertad metasandstone ridges. Fernandes et al. (1977) established the age of unmetamorphosed felsic subvolcanic rocks with quartz pebble xenoliths from the Traira (Taraira) River, associated with the Tunuí Group at 1427 ± 29 Ma (whole-rock Rb-Sr isochron). This is apparently the same age as 1498 ± 20 Ma cited by Santos et al. (2003) using modern decay constants. Fernandes et al. (1977) consider the volcanites to be younger than the metasediments because of the quartzite xenoliths, an observation confirmed by Bogotá (1981), but as discussed above there are also acid volcanics at the base of the metasediments. A mafic dyke intruding into the metasandstones in the Raudal Tente in the Taraira River fits in a 1225 Ma Rb-Sr isochron (Priem et al. 1982), whereas whole-rock K-Ar ages of 941 ± 14 and 984 ± 12 Ma (Cujubim diabase) have been obtained by Fernandes et al. (1977).

Muscovites from the Tunuí sediments have been K-Ar dated at 1293 ± 18 and 1045 ± 19 Ma by Fernandes et al. (1977), and modern Ar-Ar datings on muscovites from the Aracá sandstones in Brazil by Santos et al. (2003) give 1334 ± 2 Ma. These ages, including the mica ages from other rocks by Pinson et al. (1962) and Priem et al. (1982), are now all attributed to later metamorphism related with the K'Mudku-Nickerie Metamorphic Episode (Priem et al. 1982; Kroonenberg 1982; Santos et al. 2003; Cordani et al. 2005; Kroonenberg and De Roever 2010).

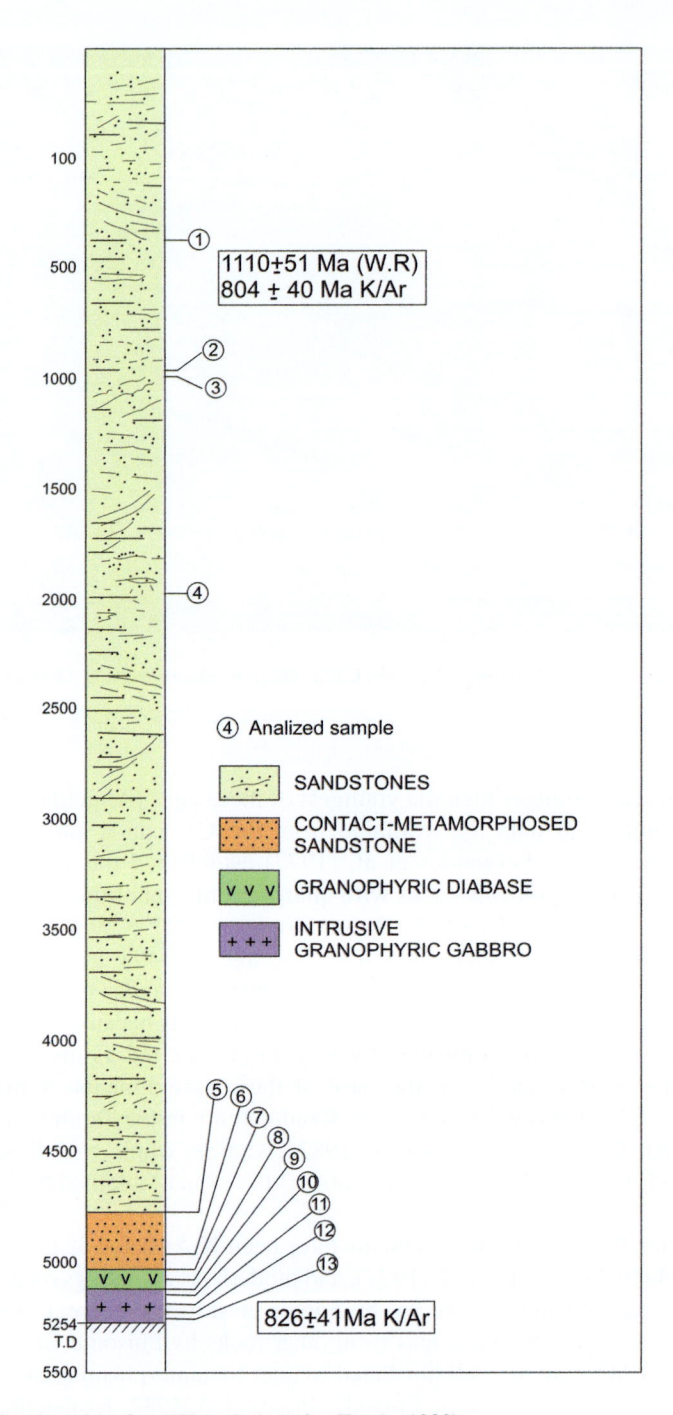

Fig. 3.31 Stratigraphy of well Vaupés-1. (After Franks 1988)

So while the field relations are not entirely clear in all cases, it is evident that the Tunuí metasediments have been deposited in the Mesoproterozoic somewhere in the interval between 1580 and 1350 Ma and if the old Brazilian field data of Fernandes et al. (1977) on crosscutting unmetamorphosed volcanics are correct, even between 1580 and 1480 Ma. In spite of its different, partly volcanic and volcaniclastic facies, the Machado-Taraira metasandstones seem to be coeval with the other metasandstones. The fact that all these metasandstone occurrences show gold mineralization also pleads for a common origin as molassic deposits in a Mesoproterozoic basin following the intrusion of the younger granites and deformed and metamorphosed during the Grenvillian Orogeny (see below).

3.1.2.7 Mesoproterozoic Mylonitization

Large areas in the Colombian Amazones and elsewhere in the Guiana Shield are traversed by important mylonite zones, often with WSW-ENE orientation (see review by Cordani et al. 2010). Although this deformation event did not result in specific mappable rock units, it is recorded geochronologically in many preexisting older rocks through a rejuvenation of mica ages. Already in the first K-Ar and Rb-Sr radiometric age, determinations on micas in rocks from Colombian Amazones gave ages around 1205 ± 60 Ma (Pinson et al. 1962); Priem et al. (1982) recorded mica ages between 1150 and 1350 for over 50 rock samples from the whole Colombian Guiana Shield and correlated this with the Nickerie Metamorphic Episode, coined by him on the base of similar mica age resetting associated with widespread shearing and mylonitization in the Precambrian of Suriname (Priem et al. 1971). Santos et al. (2003) show mica age resetting in the Aracá sandstone plateau in Brazil around 1334 Ma.

3.1.2.8 Neoproterozoic (?) Piraparaná Formation

The Piraparaná Formation has been defined by Galvis et al. (1979) and Huguett et al. (1979) in the course of the PRORADAM project as a folded series of westwards-dipping reddish volcanosedimentary rocks, cropping out in a wide arc from the Yaca-Yacá cataract in the Vaupés River along the Piraparaná river to the south, including a few outcrops along the Caquetá River. Along the Apaporis River, the formation has been seen to unconformably overlie the Complejo Mitú, and at the Raudal Jirijirimo in the same river, it is unconformably overlain by the Paleozoic Araracuara Formation. At the type locality, a thickness of 80 m has been established.

In contrast to the Tunuí rocks, the Piraparaná sediments are unmetamorphosed. They consist of polymict conglomerates (Fig. 3.32) and arkosic sands (Figs. 3.33, 3.34, and 3.35), mixed with pyroclastic material. In some levels the clasts consist

Fig. 3.32 Piraparaná conglomerate, Raudal Carurú, Río Piraparaná. (Photo De Boorder 1978)

largely of granite; elsewhere they also contain volcanites, quartzites and sandstones. At one site Galvis et al. (1979) claim to have observed carbonate cement and carbonate clasts, though the author of the present report only has seen secondary replacement by calcite in thin section. The sandstones contain feldspars, diminishing in abundance towards the top. No detailed sedimentological nor stratigraphical studies have been made, but the PRORADAM authors suppose a continental depositional environment on the base of the red coloration.

At the Raudal Yacá-Yacá in the Vaupés River, a reddish rhyodacitic lava crops out that has been included by Galvis et al. (1979) and Huguett et al. (1979), in the Piraparaná Formation on the basis of its similar colour, though no contact relations with the sediments themselves have been observed in the field. From this rock a crude six-point Rb-Sr isochron of 920 ± 90 Ma has been obtained by Priem et al. (1982). Whether this age indeed refers to the formation as a whole therefore remains uncertain. Also the relations of the Yaca-yacá lavas and Piraparaná sediments with rhyodacitic volcanics in the Machado ridge remain to be established.

Fig. 3.33 Piraparaná sandstone, quartz-cemented, río Caquetá, 1 N .(Photo: Kroonenberg)

Fig. 3.34 Piraparaná sandstone, note quartz outgrowth around detrital grain, río Caquetá. (Photo: Kroonenberg)

Ibáñez-Mejía (2010) proposes 'that the Piraparana formation could represent either (1) foreland basin deposits related to Putumayo [~ Grenvillian, see below] orogenic development inboard in Amazonia, or (2) Neoproterozoic syn-rift sedimentation and volcanism associated with early extensional events of the Neoproterozoic Güejar-Apaporis graben preceding the collapse of the Putumayo orogen and related Grenville-age belts. Only detailed sedimentary provenance studies in the Piraparana formation will allow us to test these hypotheses'.

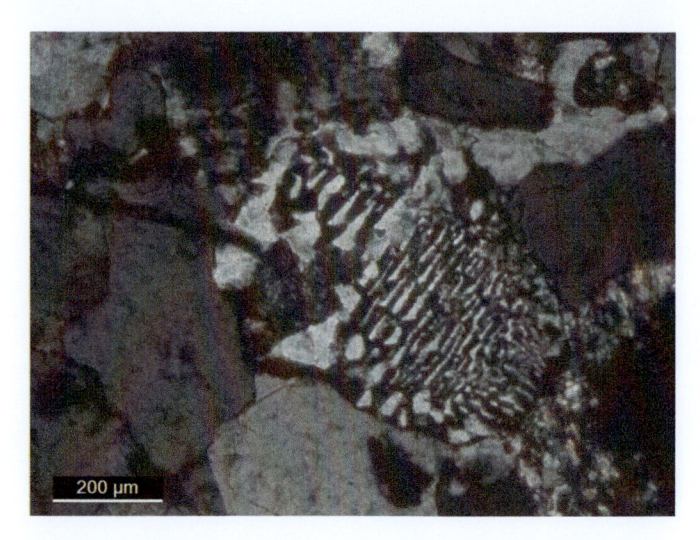

Fig. 3.35 Piraparaná sandstone, granophyre clast, río Caquetá. (Photo: Kroonenberg)

3.1.2.9 Meso-Neoproterozoic Mafic Intrusives

During the PRORADAM reconnaissance, at least 15 unmetamorphosed diabase (dolerite) dykes have been found to intrude the Complejo Mitú and the Tunuí metasediments. Petrographically they are usually pigeonite dolerites without orthopyroxene, while locally (Caño Tí) coarser, olivine-bearing granophyric gabbros occur. Priem et al. (1982) obtained a crude Rb-Sr isochron from five of them between 1225 and 1180 Ma. At the bottom of boring Vaupés-1 in Mesoproterozoic (meta)sandstone, a two-pyroxene olivine-bearing granophyric gabbro was encountered which was K-Ar dated at 826 ± 41 Ma (Franks 1988). The significance of this isolated age cannot be evaluated without additional data using other analytical methods but could fit in the same age group as the ~900 Ma diabase dykes found in the Taraira area and the 920 Ma Yacá-Yacá lavas.

3.1.2.10 Ediacaran San José del Guaviare Nepheline Syenite

In low hills near San José de Guaviare, a conspicuous body of nepheline syenite is exposed, partly unconformably overlain by a semihorizontal Paleozoic (?) sandstone sequence. This body was long considered to be of Paleozoic age as well on the base of K-Ar biotite ages between 485 ± 25 Ma and 445 ± 22 obtained by Pinson et al. (1962). However, recent U-Pb dating of zircon and ^{40}Ar-^{39}Ar dating of biotite by Arango et al. (2012) indicate an age of 577.8 ± 6.3–9 Ma (Ediacaran) crystallization and of 494 ± 5 Ma (late Cambrian) cooling.

3.1.3 Structure

3.1.3.1 Folding of the Basement Rocks

Unfortunately very little attention has been paid to the structural analysis of syntectonic deformation of the basement. The Caicara metavolcanics along the Atabapo River present generally NW strikes (N50°W, Galvis et al. 1979). The metamorphic basement of the Mitú complex shows very variable foliations. Bruneton et al. (1983) note that the foliation in the Atabapo-Río Negro is generally N110°–120°E. Along the Vaupés River, N10°E–N40°E strikes predominate elsewhere, also N70°E and N80°E, in the Papurí River however between N110°E and N170°E (De Boorder 1976). Fold axes of the Tunuí metasandstones are oriented N30°W–N50°W, and in the Piraparaná monoclinal, they are N-S to N20°E. Data are insufficient to present a deformational history of the basement. More attention has been paid to lineaments.

3.1.3.2 Lineaments

As a part of the PRORADAM project, an extensive study of lineaments from 1:200,000 radar imagery was undertaken by De Boorder (1980, 1981). At that time no geophysical information was available, and even up to now, it is the only structural information that appears on national geological maps. De Boorder distinguishes larger regional lineaments 100–300 km long, such as the WNW Carurú lineament more or less parallel to the Vaupés River, which is based on the parallel lineation of scarps of the Paleozoic sandstone plateaus. It runs more or less parallel to the grain of the Vaupés swell. Similar lineaments occur parallel to the Apaporis and Caquetá rivers. Furthermore, there are six major lineaments with orientations between NNE-SSW and ENE-WSW (Figs. 3.3 and 3.36). Surprisingly the prominent NNW-SSE alignment of the elongate Ordovician sandstone plateaus of Araracuara and Chiribiquete has not been indicated as a lineament.

A major feature is the La Trampa (The Trap) wedge, a curved segment between prominent NE-SW lineaments running from the Vaupes southwards to the Putumayo River, and identified on the basis of lineaments a.o. along the Pirá river, the large southwards bends in the Putumayo River and the occurrence of several deep earthquake foci in this area (Fig. 3.36). According to De Boorder, this could represent a possible rift structure which might be prospective for hydrocarbons. However, the aeromagnetic data do not support the presence of a rift structure (see below). The distribution of smaller lineaments in the area could give clues to important tectonic or lithological discontinuities below the cover of younger sediments (De Boorder 1980, 1981). A study of lineaments on the basis of a more detailed aeromagnetic survey in the Vichada and Guainía provinces (Obando 2006 en Celada et al. 2006) confirms the importance of the lineaments inferred by De Boorder. Older magnetic surveys are of insufficient quality to deduce structural detail (Kroonenberg and Reeves 2012).

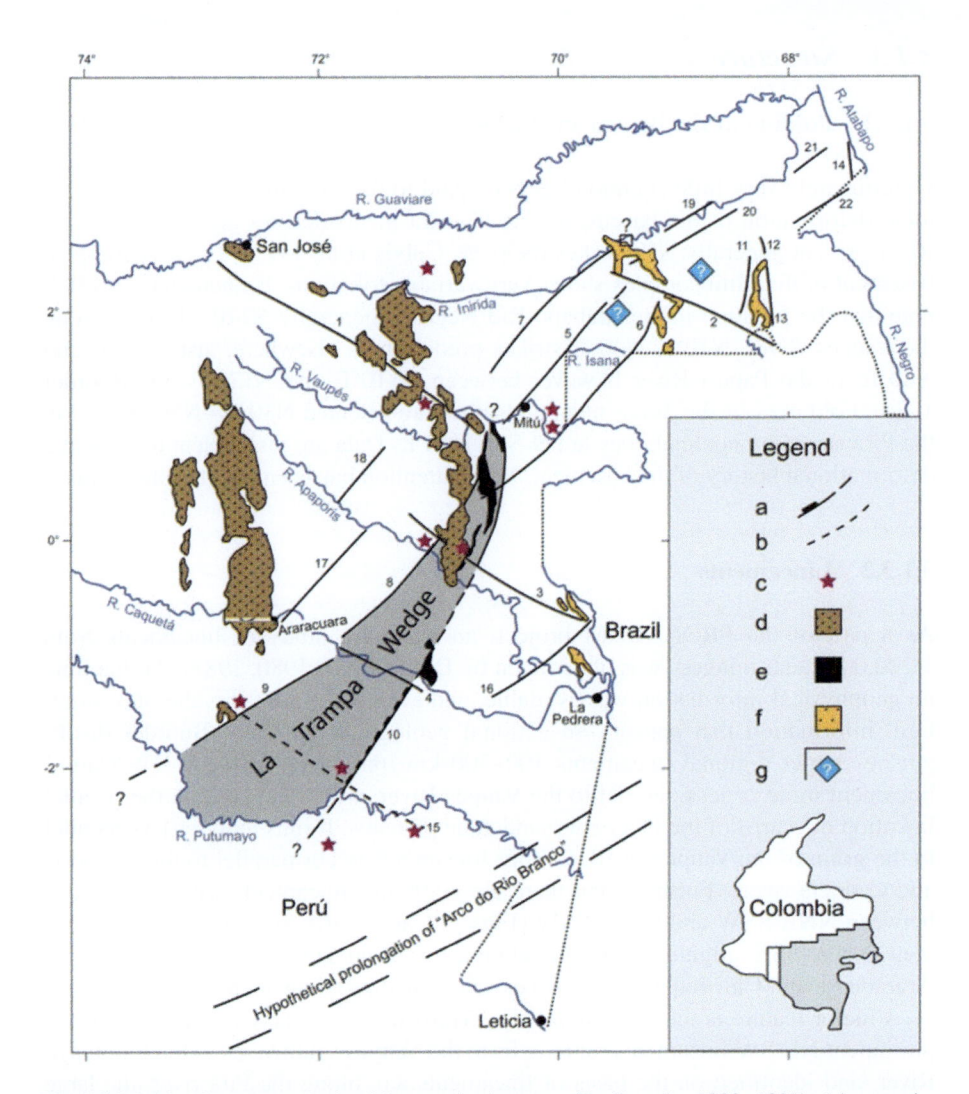

Fig. 3.36 Major lineaments derived from radar imagery (De Boorder 1980, 1981). (**a**) tectonic lineament, mainly from radar imagery; (**b**) lineament deduced from alignment epicentres of deep earthquakes; (**c**) epicentres of deep earthquakes; (**d**) major outcrop of Araracuara Formation; (**e**) major outcrop of Piraparaná Formation; (**f**) major outcrops of Tunuí Formation; (**g**) outline of area for microlineament studies

3.1.4 Geochronological Provinces in the Amazonian Craton: A Discussion

There are at least three different views on the role of the Colombian basement in the evolution of the Guiana Shield, as stated above, at least in part due to the scarcity of available data.

1. Tassinari (1981) defines it as a separate unit, the Río Negro-Juruena (RNJ) geochronological province, based on the fact that rocks from both the Río Negro area north of the Amazon basin and the Juruena area south of it all plot together in a Rb-Sr reference isochron between 1750 and 1500 Ma. Low initial Sr ratios suggested that all this material is juvenile. The RNJ province would have been the result of a volcanic arc accreted onto an older core, the supposedly Archean Central Amazonia Province (CAP) on the east, which also includes the Parguaza rapakivi granite and the basement in which it intrudes. The suture between the two provinces would roughly follow the upper course of the Orinoco River in Venezuela. In later papers (Tassinari et al. 1996; Tassinari and Macambira 1999; Tassinari et al. 2000), he maintains this vision, on the basis of additional material. Also recent ϵ_{Nd} values and T_{DM} ages suggest a largely juvenile character for the rocks of this province (Almeida et al. 2013).
2. Tassinari's vision was challenged repeatedly by Santos et al. (2000, 2006) and most eloquently in Santos (2003). In the first place, he dislodges the Juruena part from Tassinari's RNJ province, on the base of differences in lithology, structure and age, giving the Río Negro Province an identity of its own. He enlarges it considerably, encompassing almost the whole Amazonas Province of Venezuela as well as the Parguaza rapakivi granite. The Río Negro Province is now no longer bordered in the east by the Central Amazonian Province, but a new province has been squeezed between them, the Tapajós-Párima Province, characterized by the presence of the gold-bearing Parima greenstone belt in the northern part and the equally gold-bearing ~2.0 Ga Jacareacanga greenstone belt south of the Amazon basin (Santos et al. 2004). Furthermore, there appears to be no evidence at all of any Archean crust either in Santos's new Tapajós-Párima Province or in Tassinari's old CAP (Santos et al. 2004; Reis et al. 2006; Kroonenberg 2014). Therefore, whether the Río Negro indeed has accreted on the west side of an Archean nucleus has become highly questionable. Even though Santos (2003) supports the juvenile character of the rocks of the Río Negro Province, its geodynamic origin remains uncertain.
3. An entirely different view is possible if we take the Fraga et al. (2008, 2009a, b) interpretation of the structure of the Guiana Shield in consideration. As stated above, the 2.04–1.99 Ga Cauarane-Coeroeni belt (Fig. 3.2) is a major high-grade belt stretching E-W through the shield, cross-cutting all major geochronological provinces defined by Tassinari et al. (1996), Tassinari and Macambira (1999) and Santos et al. (2006). It divides the shield in a northern part with ages 2.2–1.98 Ga and a southern part, with ages generally between 1.89 and

1.74 Ga. The westernmost known extremity of the CCB is in the Complexo Urariquera in the northernmost Brazilian state of Roraima (Reis et al. 2003; Fraga et al. 2008, 2009a, b). Whether and how it continues in southern Venezuela and Colombia is unknown. On the Venezuelan map, Hackley et al. (2005) show the continuation as San Carlos metamorphic-plutonic terrane, the same unit that crops out along the Río Negro and corresponds with the gneisses of the Mitú complex at the other side of that river. No modern age data are available for the San Carlos terrane, and from the Mitú, Minicia gneisses no ages >1.85 Ga have been found, i.e. at least 100 Ma younger than the youngest CCB ages. However, the cordierite-bearing metapelitic gneisses along the Guainía River have a similar metamorphic history as those in the Cauarane-Coeroeni belt (Kroonenberg 1980), so it becomes important to date those rocks: they might correspond with the westernmost extension of the CCB. Also the recently discovered presence of granulites near Mitú (Rodríguez et al. 2011a, b) deserves further investigation.

In spite of these alternatives, the geochronological evidence available at present supports Tassinari's (1981) original concept of a younger unit accreted at the western side of a pre-existing basement (Fig. 3.2). This older basement, however, is not Archean but Paleoproterozoic in age, and some elements such as the metavolcanics along the Atabapo River and the metapelites along the Guainía River may still belong to that older basement.

3.1.5 Geological Evolution of the Colombian Part of the Guiana Shield and Adjacent Areas

The sequence of events in the Colombian part of the Guiana Shield, as appears from the descriptions above, is summarized in Table 3.2.

3.1.5.1 Late Trans-Amazonian Orogeny

The Trans-Amazonian Orogeny, as defined originally by Hurley et al. (1967), is represented in the Colombian part of the Guiana Shield only by the Caicara metavolcanics of the Cuchivero Group exposed along the upper Atabapo River. The Caicara metavolcanics are considered to be older than the Mitú complex on the base of its comagmatic association with the Santa Rosalia and San Pedro granites in Venezuela (1956–1732 Ma, Rb-Sr, Gaudette et al. 1978) and the 1.98–1.97 Ga U-Pb ages from the Surumú metavolcanics in Brazil (Schobbenhaus et al. 1994; Santos et al. 2003). The Cuchivero Group might have constituted the basement onto which the younger basement of the Mitú complex accreted, and it forms also the basement in which the Parguaza rapakivi granite intruded. Geochemically these rocks straddle the boundary between volcanic arc granites and within plate granite in the trace element discrimination diagram of Pearce et al. (1984; Fig. 3.37).

3 The Proterozoic Basement of the Western Guiana Shield and the Northern Andes   153

Table 3.2 Sequence of events in the Colombia Amazonian Precambrian

Age(Ma)	Formation	Events	Context
600–800	Nepheline Syenite S José, dolerite dykes, gabbro	Alkaline and mafic magmatism	Anorogenic
900?	Piraparaná	Fluvial sedimentation	Grenvillian Molasse?
1300–1100	Putumayo orogen (Subandean foreland)	Deformation, medium-high-grade metamorphism	Grenvillian collision Laurentia-Amazonia
1300–1100	K'Mudku-Nickerie metamorphic episode	WSW-ENE mylonite, thermal resetting mineral ages, deformation low-grade metam. Tunuí	Grenvillian collision Laurentia-Amazonia
1500–1400	Tunuí, etc.; metasandstones	Fluviodeltaic sedimentation	Molasse?
1550–1400	Parguaza rapakivi granite	Anorogenic magmatism	Rifting?
1550–1500	Mitú, Içana, Tijereto, Inhamoin porphyritic titanite granites	Anorogenic magmatism	Rifting?
1850–1740	Mitú complex, Minicia, Macabana gneisses; Atabapo metavolcanics? Tiquié granites	Deformation, medium-high-grade metamorphism, anatexis, syntect. Intrusives	Querarí orogeny
>1850?	Mitú complex, Minicia etc. protoliths	Deposition of graywackes(?), pelitic rocks, acid volcanics?	Continental margin? Back-arc basin? Rift?
1980	Cuchivero Gp, Caicara metavolc	Acid volcanism and shallow intrusions	Late trans-Amazonian magmatism

Fig. 3.37 Discrimination diagram of Venezuelan Caicara volcanics (triangles) according to the Pearce et al. (1984) classification (Sidder and Mendoza 1991)

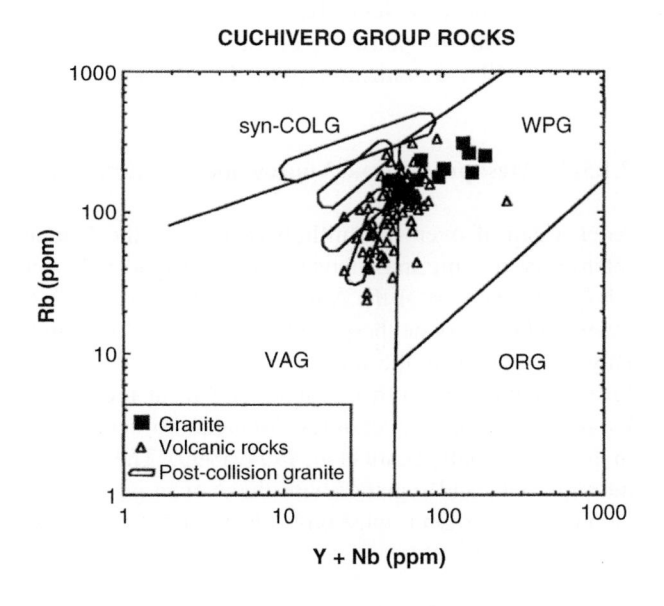

3.1.5.2 Querarí Orogeny, 1.86–1.72 Ga: Deposition, Deformation and Metamorphism of the Mitú Complex Supracrustals

The quartzofeldspathic nature of most of the supracrustal rocks in the Mitú complex suggests that they were originally immature sediments, possibly of greywacke and/or acid to intermediate volcanogenic composition. This may point to an origin in either a passive continental margin setting or an island-arc environment. The scarcity of mafic rocks precludes an origin in a back-arc basin. Orthogneisses may represent early syntectonic intrusions. Deformation and metamorphism took place during an orogenic event between 1.86 and 1.72 Ga.

Priem et al. (1982) state that it seems obvious to correlate the 'pre-Parguazan' history of the Mitú complex with the Trans-Amazonian Orogeny. Now that many more modern U-Pb ages have been obtained from the metamorphics (see Table 3.1 above), it becomes clear that if we accept Priem's view, the Trans-Amazonian Orogeny would span almost half a billion years, from 2.2 Ga to 1.7 Ga, more than a full-fledged Wilson cycle. Moreover, nowhere else in the Guiana Shield high-grade metamorphic supracrustals with ages between 1.86 and 1.72 Ga have been found. Therefore we prefer to consider the deposition, deformation and metamorphism of the Mitú metamorphics as a separate event.

Almeida et al. (2013) recognize even two orogenic events in the adjacent part of Brazil, the Cauaburí Orogeny of 1.81–1.75 Ga and the Querarí Orogeny (1.74–1.70 Ga). In Colombia there are no compelling field or geochronological reasons to distinguish *two* orogenic events in this interval; we see rather a continuum of these ages, and therefore I propose to retain the name *Querarí Orogeny* for the whole series of deformation and metamorphic events between 1.86 and 1.72 Ga. Moreover, the Querarí river is largely situated in Colombian territory. This orogeny led to accretion of the Río Negro belt to the older Paleoproterozoic basement and was accompanied by the intrusion of the late-syntectonic S-type Tiquié granites. This marked the final cratonization of the Guiana Shield.

3.1.5.3 Mesoproterozoic Anorogenic Granitoid Magmatism: 1.55–1.4 Ga

After a gap of over 100 million years after the Querarí Orogeny, an episode of intense anorogenic magmatism started around 1.55 Ga that is widespread in the whole western part of the Amazonian Craton (Figs. 3.2 and 3.3; Dall'Agnol et al. 1999, 2006; Kroonenberg and de Roever 2010 and references therein). The Parguaza granite is the most conspicuous representative example, but the Mitú, Içana, Atabapo and other granites are from the same time interval, and typical Parguaza-like rapakivi granites (Mucajaí, Surucucus) also occur much further east in the shield (Dall'Agnol et al. 1994, 1999). There is no link with any coeval metamorphic belt, and together with the A-type geochemical characteristics of these granites, their origin is most probably related to an extensional phase in the evolution of the Guiana Shield.

3.1.5.4 Mesoproterozoic Sedimentation of Tunuí Sandstone: 1.58–1.35 Ga?

The widespread occurrence of epicontinental, partly coarse-clastic sedimentary sequences up to 2000 m in thickness over the whole western half of the shield, not only in Colombia, Venezuela (Cinaruco Formation) and adjacent Brazil but also much further eastwards in the Aracá plateau in Brazil (Fig. 3.38), suggests an episode of post-orogenic erosion and sedimentation in a huge molasse-like basin after the Querarí Orogeny, at least between 1580 and 1350 Ma.

Some occurrences may be older, as the youngest detrital zircons in the Naquén-Caparra plateau were only 1720 Ma. While sandstone plateaus rest unconformably on the crystalline basement, others may have been intruded by the anorogenic granites, as occasionally contact-metamorphic andalusite was reported at the contact with the Parguaza granite (De la Espriella et al. 1990; Ochoa et al. 2012).

3.1.5.5 K'Mudku-Nickerie Tectonometamorphic Episode: 1.3–1.1 Ga

The mylonitization and mica age rejuvenation event that affected all previously mentioned rock units was first described in Guyana by Barron (1969) as K'Mudku event and since then recognized in many areas of the shield (Fig. 3.39; Gibbs and Barron 1993; Cordani et al. 2010). Priem et al. (1971) showed that only the easternmost part of the shield was not affected by this event, called Nickerie Metamorphic Episode by him. Kroonenberg (1982) interpreted this as a result of the Grenvillian Amazonia-Laurentia collision along the western border of the Guiana Shield around 1200–1000 Ma (see below). A further correlation is possible with the 1350–1300 Rondonian-San Ignacio belt and the 1250–1000 Sunsás belt in the southwestern part of the Amazonia Craton, close to the border with Bolivia, which equally testify to the Laurentia-Amazonia collision in Elsevirian and Grenvillian times, respectively (Cordani et al. 2010). Recently similar ages around 1000 Ma were obtained from basement rocks from drill cores into the Subandean basin (Ibáñez-Mejía et al. 2011; see below). Assigning a specific geochronological province across the Guiana Shield to the K'Mudku event, as Santos et al. (2006) suggest, however, goes against existing field and geochronological data.

3.1.5.6 Late Proterozoic-Phanerozoic Events

In the Neoproterozoic the Piraparaná epicontinental rocks were deposited, possibly a far effect of the Grenvillian Orogeny along the western border of the shield. Furthermore, several isolated mafic and alkaline intrusions and extrusions took place, obviously in an intraplate setting but without clear geotectonic context. In the Phanerozoic, Ordovician sandstone plateaus (Araracuara Formation) and Neogene sediments covered large parts of the basement.

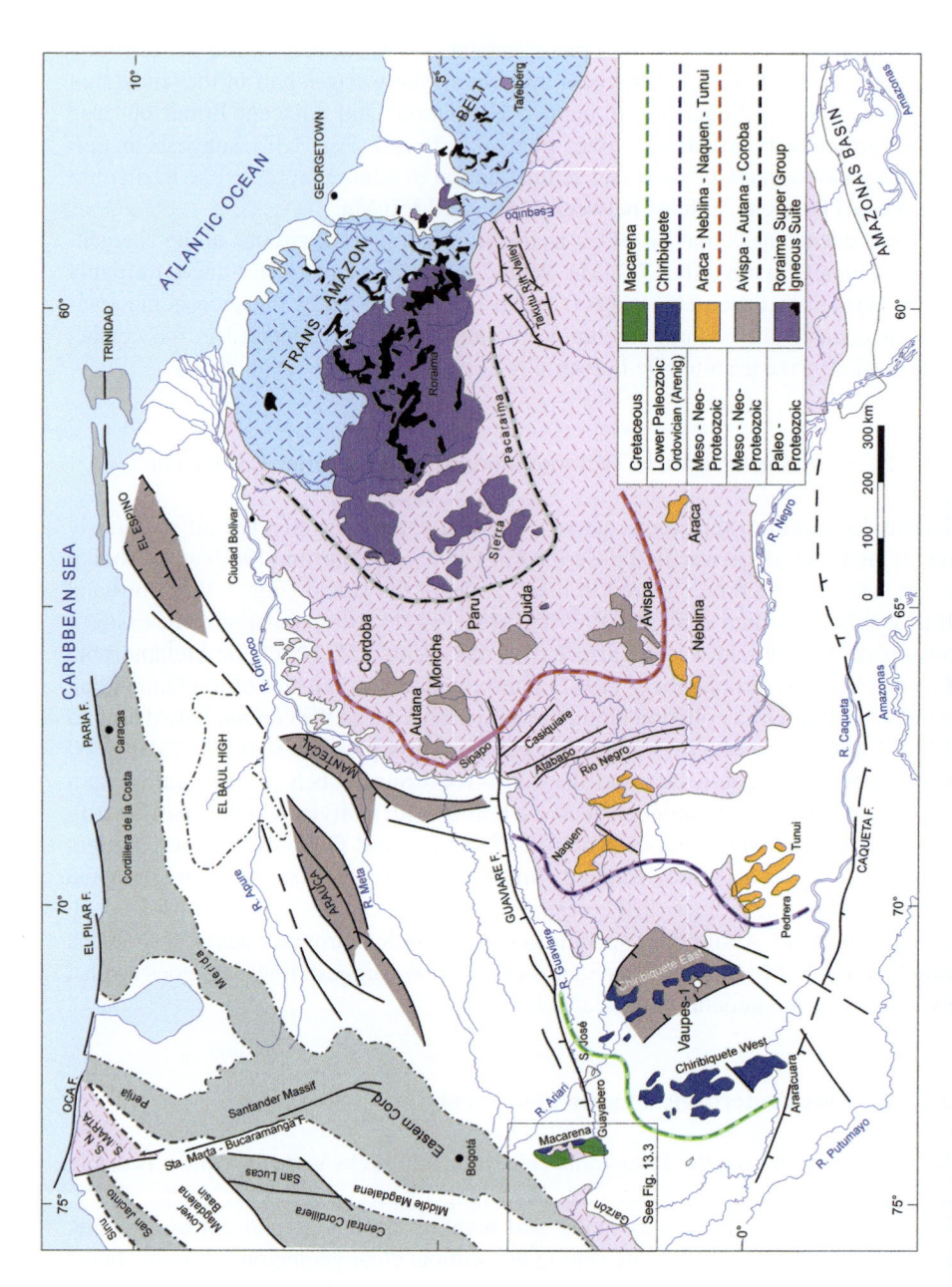

Fig. 3.38 Distribution of (meta)sandstone plateaus (tepuis) in the Guiana Shield. (see Cediel 2018)

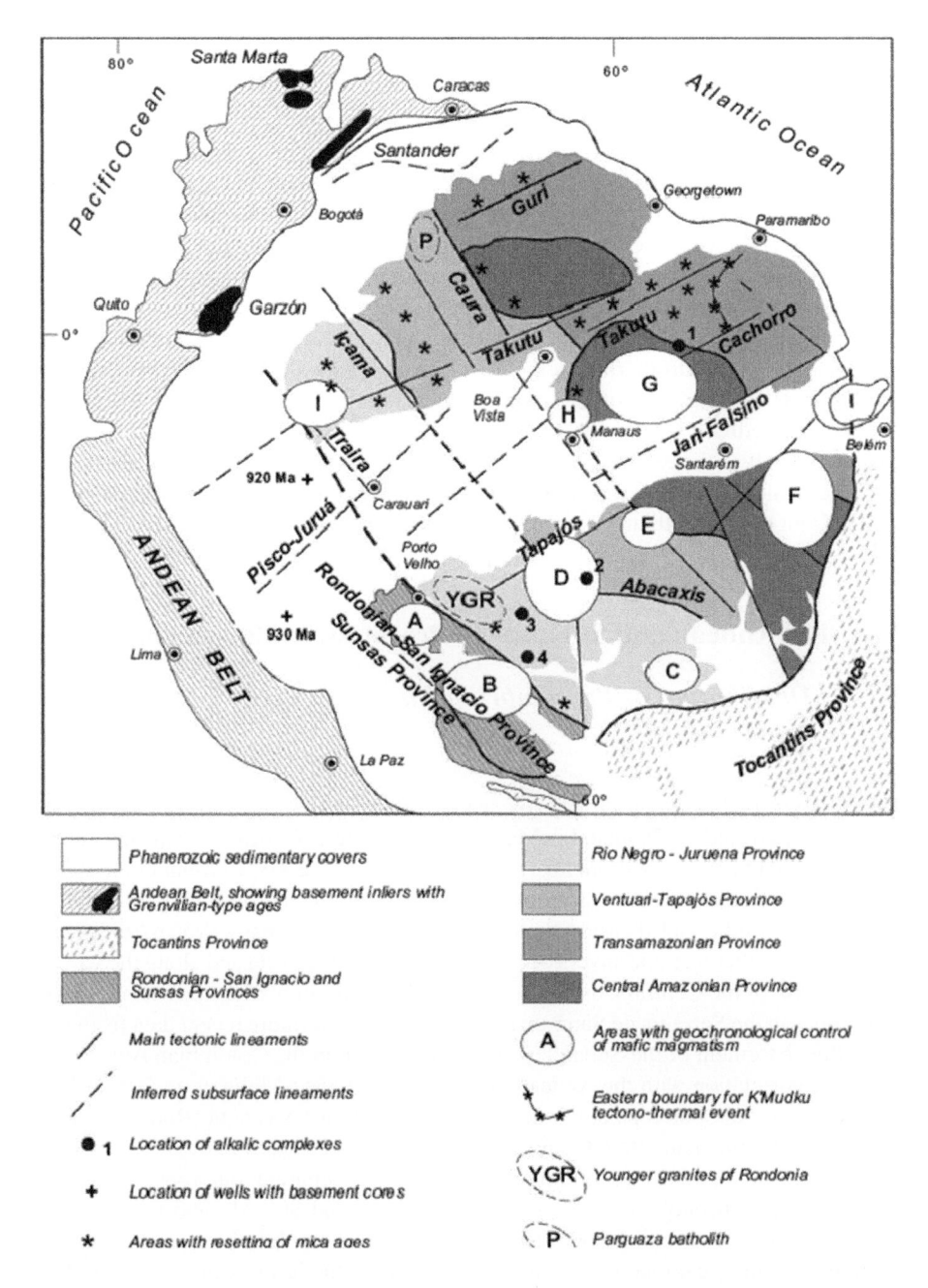

Fig. 3.39 Main lineaments and areas with mica age resetting in the Amazonian Craton (Cordani et al. 2010). Reproduced with permission

3.1.6 Geoeconomic Potential

The most important mineralizations in the area are columbite-tantalite in the Parguaza rapakivi granite and gold in the Tunuí sandstone plateaus. Columbite-tantalite occurs in coarse crystals in 'quartz-pegmatites' which never have been seen in outcrop but only as float on top of the presumed veins in the Venezuelan part of the batholith. Heavy mineral concentrates from neighbouring creeks contain up to 73% of cassiterite, further 15% of partly Ta-rich rutile and 8% of columbite-tantalite (Pérez et al. 1985; Herrera-Bangerter 1989; Bonilla et al. 2013).

Part of the gold in the Tunuí metasandstone plateaus is derived from Proterozoic paleoplacers, but hydrothermal remobilization also plays a role. Also wolframite occurrences have been reported from the metasandstone plateaus (Ashley 2011). The nearest bedrock source for the gold placers in these plateaus is in the Parima greenstone belt in the extreme NW of Roraima state in Brazil (cf. Reis et al. 2003).

Proterozoic diamondiferous kimberlites occur in the Guaniamo area, Venezuela, not far from the Colombian border (Fig. 3.40).

3.2 The Andean and Subandean Precambrian Basement

3.2.1 Distribution of Precambrian Basement in the Colombian Andes

Three major upthrusts of Proterozoic rocks exist in the Colombian Andes: the Garzón Massif and the Santander Massif in the Eastern Cordillera and the Sierra Nevada de Santa Marta (Kroonenberg 1982; Cediel et al. 2003; Cordani et al. 2005; Ordóñez-Carmona et al. 2006; Ramos 2010). The Serranía de Macarena, an isolated NW-trending outlier uplift east of the Eastern Cordillera, also has a Proterozoic basement core. Smaller tectonic slivers occur in the Guajira Peninsula and along the whole eastern flank of the Central Cordillera from the Ecuadorian border up to its northern extremities in the Serranía de San Lucas (Fig. 3.41). Furthermore recent data from the crystalline basement of the Subandean Putumayo basin in the Colombian Amazones suggest a correlation with the Andean Precambrian (Ibáñez-Mejía et al. 2011). The belt of Proterozoic outcrops continues into northwestern Venezuela (Rodríguez and Áñez 1978; Priem et al. 1989; Grande 2012; Grande and Urbani 2009).

There is no physical continuity between those separate outcrops, but their Grenvillian geochronological history between 1100 and 900 Ma and their generally high grade of metamorphism (granulite-facies or amphibolite facies) suggest a common geological history. Granulite-facies xenoliths have been erupted by the Nevado Del Ruiz volcano (Jaramillo 1978, 1980), suggesting that the Proterozoic basement of the Andes is at least present below the Central Cordillera. High-grade metamorphic rocks, partly granulites, have also been reported from the western flank of the Central Cordillera, such as Puquí, Caldas-La Miel, Nechí, San Isidro

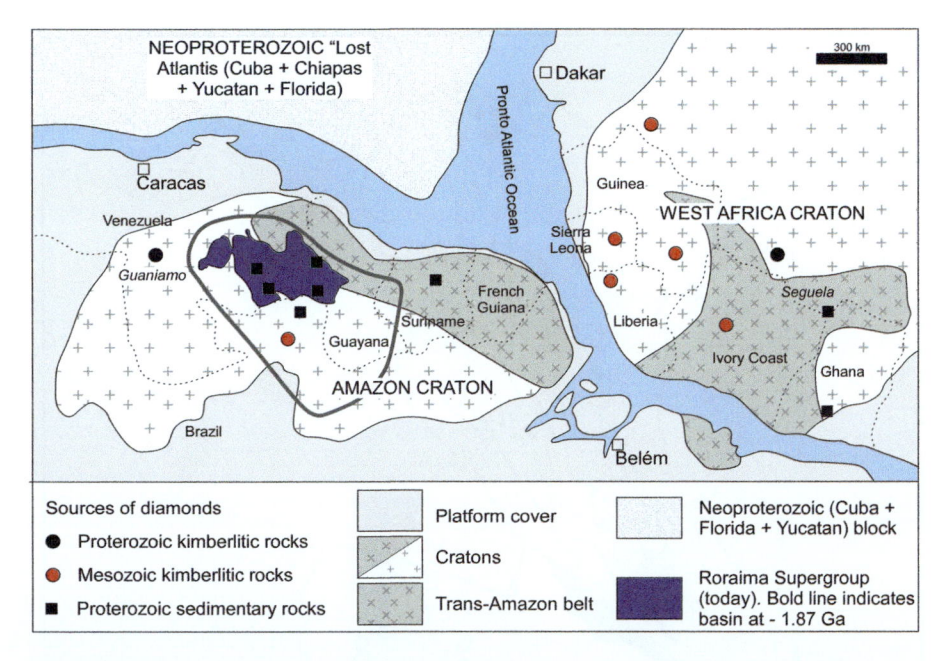

Fig. 3.40 Diamond occurrences in the Guiana Shield (Santos et al. 2003). Reproduced with permission

and Las Palmas, but so far they have been dated as Triassic, not Precambrian (Ordóñez-Carmona et al. 2001; Restrepo et al. 2009, 2011; Rodríguez et al. 2012).

Two models have been proposed for the geotectonic significance of the Proterozoic outcrops, an autochthonous and an allochthonous one. The autochthonous model considers the Garzón-Santa Marta Granulite Belt as a juvenile accretion to the Guiana Shield during the collision of Laurentia and Amazonia during the Grenvillian Orogeny (Kroonenberg 1982; Restrepo-Pace et al. 1997; Cediel et al. 2003). The argument is based mainly on the lithological similarity of the two belts and on shearing, mylonitization and thermal mineral resetting at the same time in the adjacent Guiana Shield, interpreted as indentation tectonics. The autochthonous model is also supported by the Paleozoic history of the Santander Massif (Van der Lelij 2013; Van der Lelij et al. 2016).

On the other hand, scientists especially used to accretionary tectonics in the Western Cordillera and the Serranía de Baudó prefer to subdivide the Colombian Andes in terms of fault-bounded accreted *terranes* (Etayo et al. 1983; Toussaint 1993; Ordóñez-Carmona et al. 2006). The 'terrane' concept was originally developed along the Pacific coasts of California, British Columbia and Alaska, where allochthonous, totally unrelated tectonic blocks have been displaced parallel to the mainland for hundreds to a thousand kilometres along transform faults until they became yuxtaposed into their present position. Toussaint (1993) considers only the Garzón Massif as part of the (almost) autochthonous 'Andaquí terrane' and the other massifs as part of the allochthonous 'Chibcha terrane'. Moreover, Forero (1990) considers

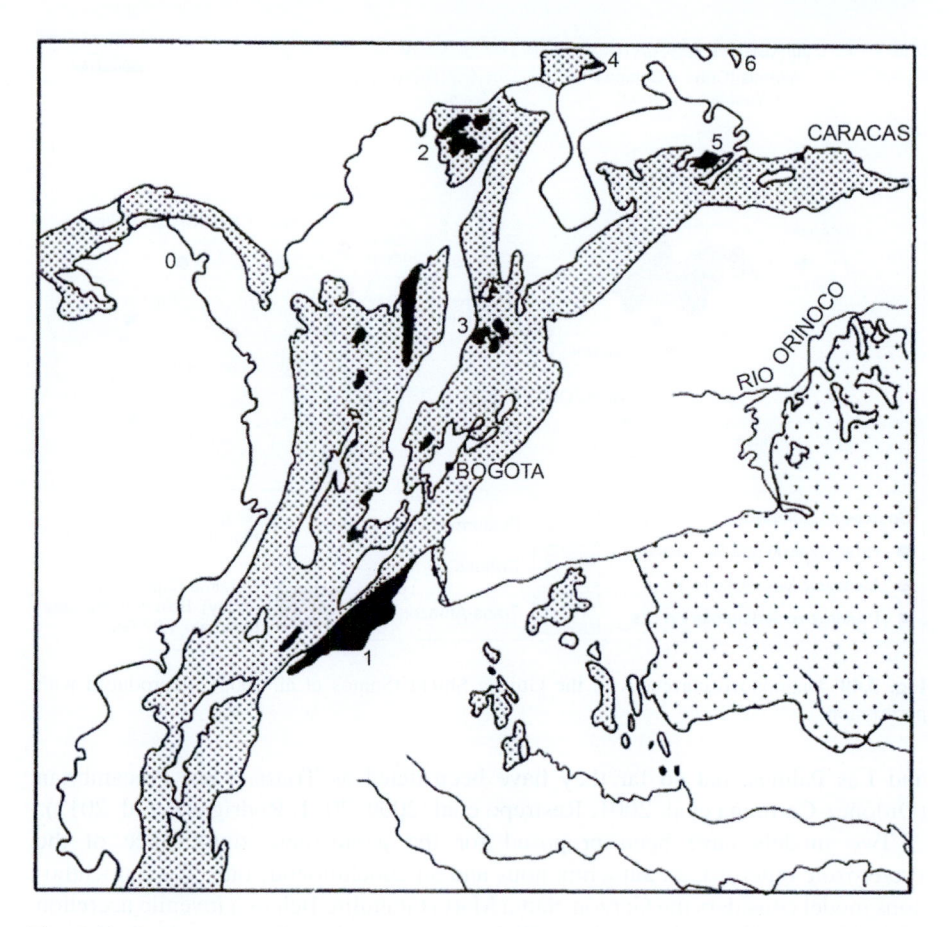

Fig. 3.41 Outcrops of Andean Grenvillian in Colombia and adjacent Venezuela: (1) Garzón Massif, (2) Sierra Nevada de Santa Marta, (3) Santander Massif, (4) Guajira, (5) Venezuelan occurrences, (6) San Lucas (Kroonenberg 1982)

on the base of paleontological evidence that the Paleozoic of the Eastern Cordillera belongs to Laurentia, and not to South America, and accreted to the Guiana Shield in Silurian-Devonian times. This is also the line followed by Cordani et al. (2005). Furthermore, Bayona et al. (2010) present palaeomagnetic evidence from the Sierra Nevada de Santa Marta for large-scale northwards along-margin displacements of basement-cored tectonic blocks in Jurassic-Cretaceous times.

However, in our view the common protoliths, metamorphism and age history plead against an allochthonous character. The lateral displacements along still active major faults do not invalidate the fact that all Grenvillian segments along the whole length of the Colombian Eastern and Central Cordilleras originally formed a continuous belt along the western margin of the Guiana Shield. Nor is there any sign of unrelated microcontinents which were docked against the mainland. The strongest argument for the integrity of the Andean Precambrian is the fact that the Grenvillian basement continues eastwards at the base of the Subandean Putumayo foredeep, beyond the eastern boundary thrust fault of the Eastern Cordillera (Ibáñez-Mejía et al. 2011), and hence forms an integral part of the Guiana Shield since the Grenvillian Orogeny. It is not illogical to suppose that a continuous Grenvillian basement is present in the deeper continental crust below the Eastern and eastern Central Cordillera. Below we discuss the Precambrian outcrops, first in the Eastern Cordillera and the Subandean basement, then in northern Colombia and at last in the eastern flank of the Central Cordillera. At the end we will discuss their wider geodynamic significance.

3.2.2 The Garzón Massif

The Garzón Massif forms the backbone of the southern part of the Eastern Cordillera over a distance of over 250 km, covering about 10,000 km^2 and reaching elevations up to about 3000 m (Fig. 3.42). Both its eastern and western boundaries are thrust faults, in which the Proterozoic basement is thrust over Mesozoic and Tertiary rocks. Towards the north and south, the massif pinches out between other thrust faults. Small slivers reappear further north, such as the El Barro Gneiss near the village of Alpujarra (Fuquén and Osorno 2002). Final uplift of the Garzón Massif took place between 12 and 3.3 Ma (Van der Wiel 1991).

Lithology The Garzón Massif consists mainly of Proterozoic banded granulites of charnockitic-enderbitic composition, mafic and ultramafic granulites, metapelitic granulites, marbles and quartzites (Fig. 3.43a–d). Compositional banding testifies to a supracrustal origin of the rocks. Moreover, their migmatitic aspects testify of incipient melting, and in some areas advanced anatexis has proceeded to a certain homogenization of the rocks. Metamorphic grade is in the granulite facies, but along the peripheries of the massif, also amphibolite-facies rocks are common. Two bodies of syntectonic megacryst granites have been described, the Guapotón-Mancagua granites. Discordant pegmatite and aplite veins are common (Kroonenberg 1982; Restrepo-Pace et al. 1997; Murcia 2002; Jiménez et al. 2006; Ibáñez-Mejía et al. 2011). The Proterozoic sequence is locally overlain by Upper Paleozoic unmetamorphosed sediments (Stibane and Forero 1969; Mojica et al. 1987) and is intruded by various large Triassic-Jurassic granitic batholiths and small lamprophyric dykes.

Subdivision The first comprehensive description of the Garzón rocks was by Luigi Radelli (1962a, 1967), who concentrated on the migmatitic aspect, but does not mention the presence of granulites, although in one sample he describes orthopyroxene. He interprets the rocks as being of metasomatic origin ('granitization'). Kroonenberg (1982, 1983) subdivided the Proterozoic rocks in the Garzón Group (the banded granulites and associated rocks) and the syntectonic Guapotón-Mancagua granites (later called augengneisses; Priem et al. 1989).

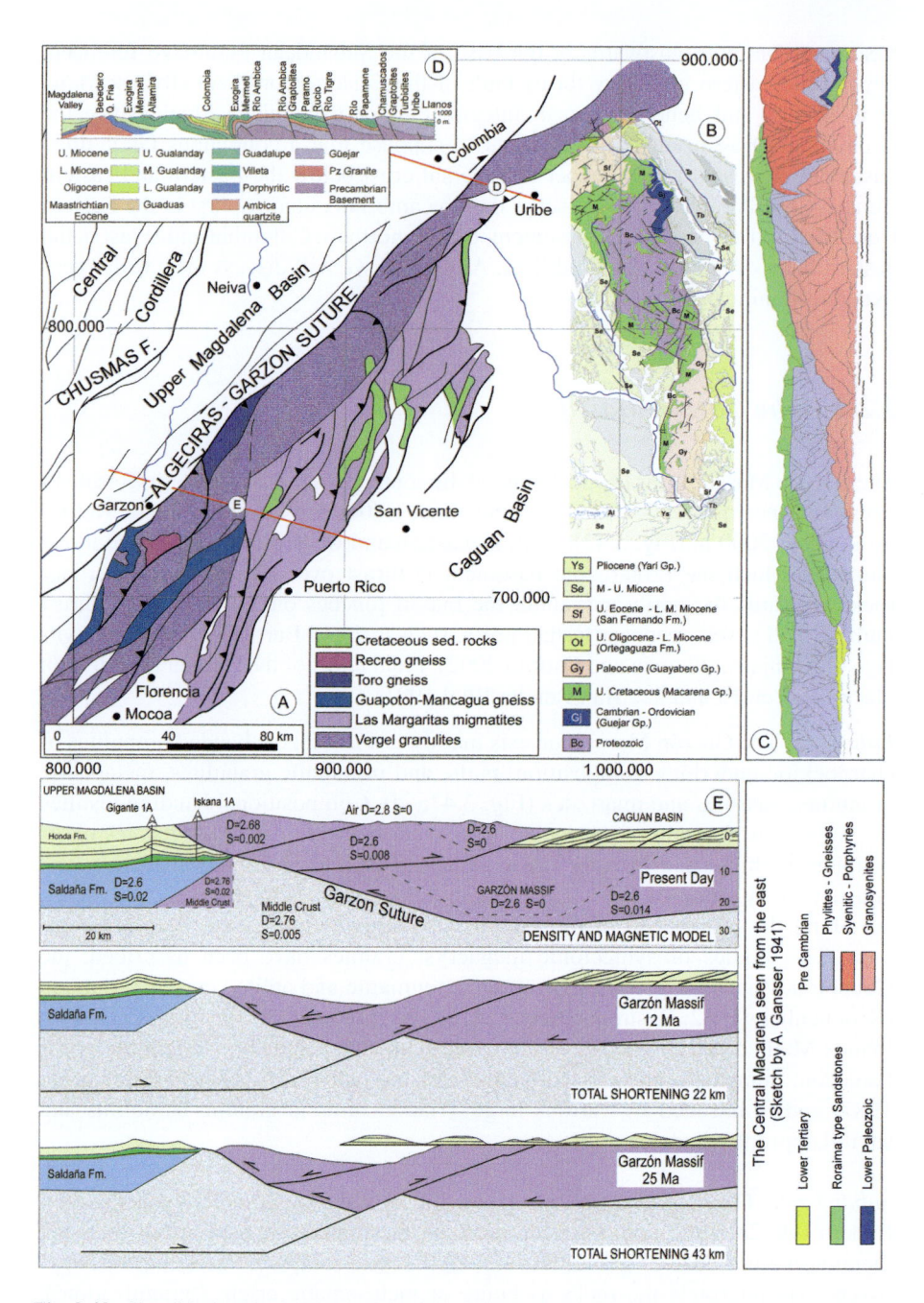

Fig. 3.42 Simplified geological map of the Garzón Massif and Sierra de la Macarena. (see Cediel 2018)

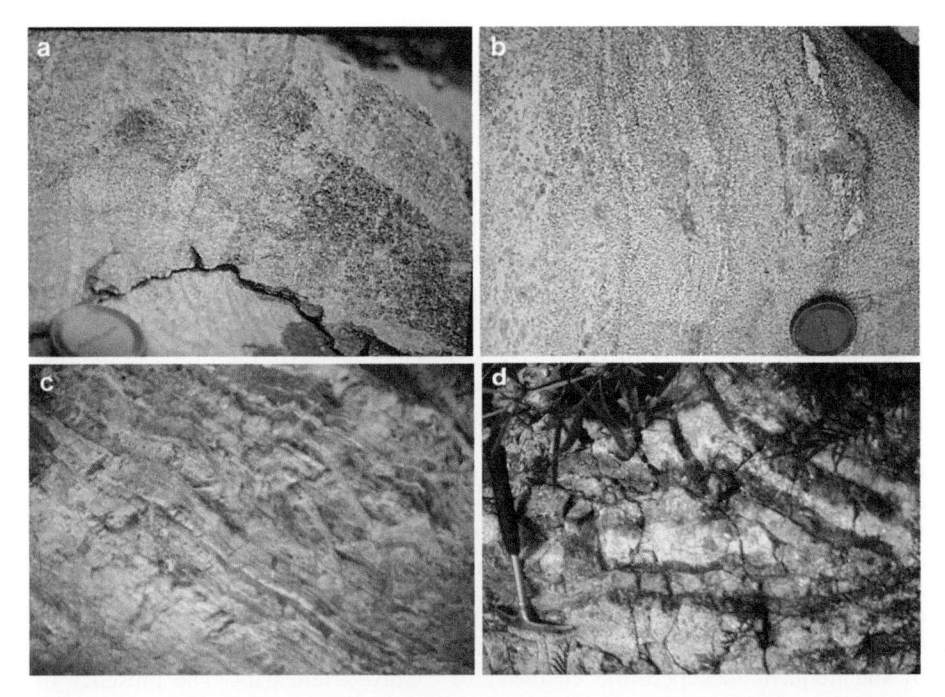

Fig. 3.43 Typical outcrops of (**a**) grey charnockitic granulites, (**b**) migmatitic mafic granatiferous granulites, (**c**) migmatitic and compositionally banded granulites, and (**d**) folded forsterite marble and calcsilicate rocks (Photo: Kroonenberg)

The Geological Survey of Colombia Ingeominas (now Servicio Geológico Colombiano) started a mapping campaign in the 1990s, resulting in the publication of several 1:100,000 map sheets of the area. In that framework Rodríguez (1995a) distinguished an additional unit in the map sheet Garzón, the El Recreo Anatectic Granite, for the more homogenized granulites in the highest part of the massif, but invoking, as Radelli, a metasomatic origin, unfortunately based on incorrect and outdated petrogenetic concepts. Transitions between the Garzón Group and the El Recreo Anatectic Granite are gradual. In a later mapping campaign of the Garzón map sheet, Velandia et al. (2001) reformulate the name as El Recreo Granite. Ingeominas and Geoestudios (1998–2001) map adjacent areas using only macroscopic descriptions of the rocks; change the name into El Recreo Gneiss, because of its more metamorphic than igneous character; and introduce new units, Toro Gneiss, Las Margaritas Gneiss and El Vergel Granulites. Fuquén and Osorno (2002) distinguish the El Barro Gneiss near the town of Alpujarra. Jiménez (2003) drew a detailed map of the whole Garzón Massif based on the subdivisions of Ingeominas and Geoestudios (Fig. 3.42). Rodríguez et al. (2003) change the name Garzón Group into Garzón complex and divide it into El Recreo granite-granofels and Florencia migmatites, discarding the names introduced of Ingeominas and Geoestudios (1998–2001) on the basis of their new petrographic data. Amidst this confusion and in the absence of clear-cut distinguishing criteria between the proposed subunits, we prefer to retain the old twofold subdivision in modern in Garzón complex and Guapotón-Mancagua Gneiss.

Geochemistry No whole-rock geochemical data have been published so far from the Garzón Massif, except for a few graphs in Kroonenberg (1990) and Restrepo-Pace (1995) (Figs. 3.44, 3.45, and 3.46). In principle the common discrimination diagrams are meant for igneous rocks, so interpretation of the data for the Garzón Massif granulites should be taken with caution because of the superimposed effect of metamorphism and concomitant mobility of several elements. Nevertheless, the bulk of the granulites plot in the calc-alkaline field in the K_2O-SiO_2 diagram of Peccerillo and Taylor (1975; Fig. 3.44), suggesting a possible volcano-sedimentary origin in an active continental-margin setting of the protoliths. This is in harmony with mafic granulites plotting in the calc-alkaline field of the Ti-Zr-Sr triangular plot and intermediate samples plotting in or near the orogenic granite field in the discrimination diagrams by Pearce et al. (1984; Fig. 3.45).

REE spider diagrams of charnoenderbitic and mafic granulites and of the Guapotón orthogneiss show weak Eu anomalies, suggesting an origin by fractional differentiation from a plagioclase-rich magma source; a single ultramafic granulite (opx-cpx-hbl-spinel) SK 132 shows an almost flat REE profile (Fig. 3.46a–d). The same wide range in profiles is seen in Restrepo-Pace (1995).

Metamorphism Granulite-facies metamorphism is evident from the ubiquitous development of granoblastic orthopyroxene in both felsic and mafic granulites and by the frequent mesoperthitic character of exsolved feldspars. The presence of orthopyroxene in the leucosomes indicates that anatexis also took place in the granulite facies (Kroonenberg 1982, 1983). According to Jiménez et al. (2006), the geothermobarometric data define a clockwise, nearly isothermal decompression path (ITD) for rocks from Las Margaritas migmatites, ranging from 780–826 °C and 6.3–8.0 kbar down to 630 °C and 4 kbar (cf. Fig. 3.47). For a garnet-bearing charnockitic gneiss from the Vergel Granulites, the path is counterclockwise, from 5.3–6.2 kbar and 700–780 °C to 6.2–7.2 kbar and 685–740 °C. Altenberger et al. (2012) argue for much higher values in the Vergel Granulites, reaching UHT (ultra-high temperature) conditions, up to 900–1000 °C, on the basis of ternary feldspar diagrams, titanium in quartz and mineral chemistry of exsolved pyroxenes.

Geochronology The first radiometric ages on charnockitic granulites of the Garzón Massif were obtained by Álvarez and Cordani (1980) and Álvarez (1981) and show a Rb-Sr isochron of 1180 Ma, while a hornblende K-Ar age of 925 ± 50 Ma was obtained from a basic granulite by Álvarez and Linares (1984). Priem et al. (1989) show a 1172 ± 90 Ma Rb-Sr errorchron but do not exclude the presence of an older basement on the basis of a six-point best-fit line of 1596 ± 300 Ma for the Guapotón gneisses. Rb-Sr mineral ages of 918 ± 27 for phlogopite and 895 ± 16 for K-feldspar were obtained, next to Phanerozoic biotite ages. The first U-Pb zircon ages were published by Restrepo-Pace et al. (1997), showing an age of 1088 ± 6 Ma for El Vergel Granulites. Cordani et al. (2005) obtained SHRIMP U-Pb zircon ages of 1158 ± 23 Ma for igneous cores of zircons from the Guapotón Gneisses and 1000 ± 25 Ma for their metamorphic rims, 1015 ± 8 Ma for the leucosome of Las Margaritas gneisses and for the Vergel Granulites a protolith age of ~1100 Ma and

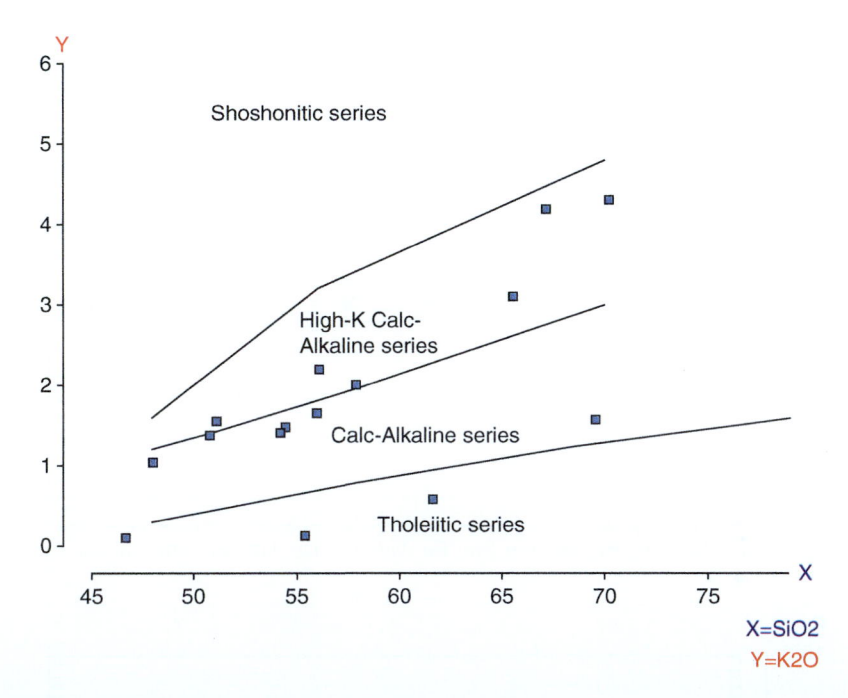

Fig. 3.44 K2O-SiO2 diagram after Peccerillo and Taylor (1975) showing calc-alkaline nature of charnockitic, enderbitic and mafic granulites. Unpublished XRF data and Kroonenberg (1990). Analyst F, Stephan, Utrecht (1982)

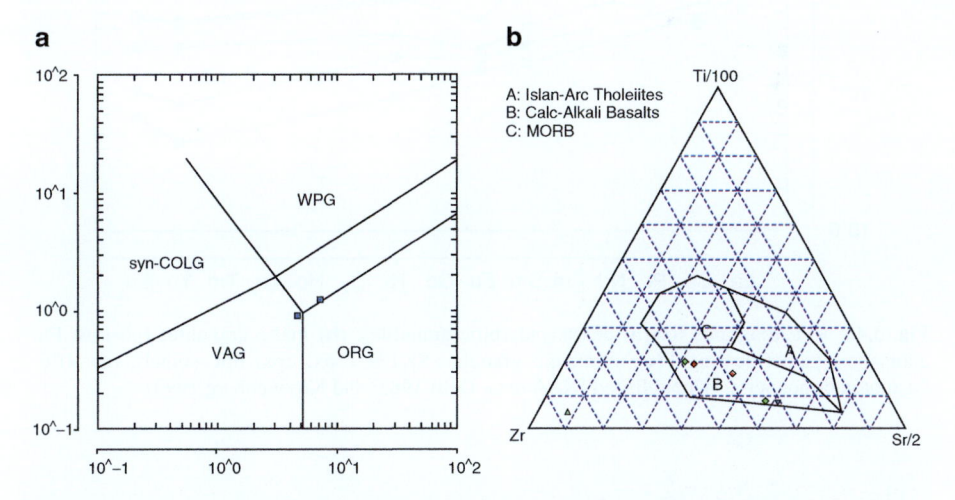

Fig. 3.45 Discrimination diagrams for intermediate and mafic granulites according to Pearce et al. (1984). Unpublished XRF data and Kroonenberg (1990). Analyst F. Stephan, Utrecht (1982)

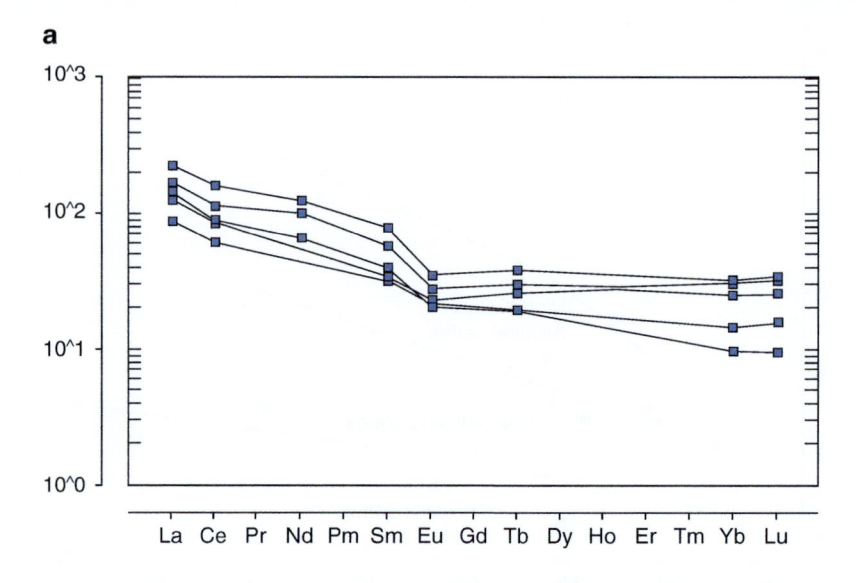

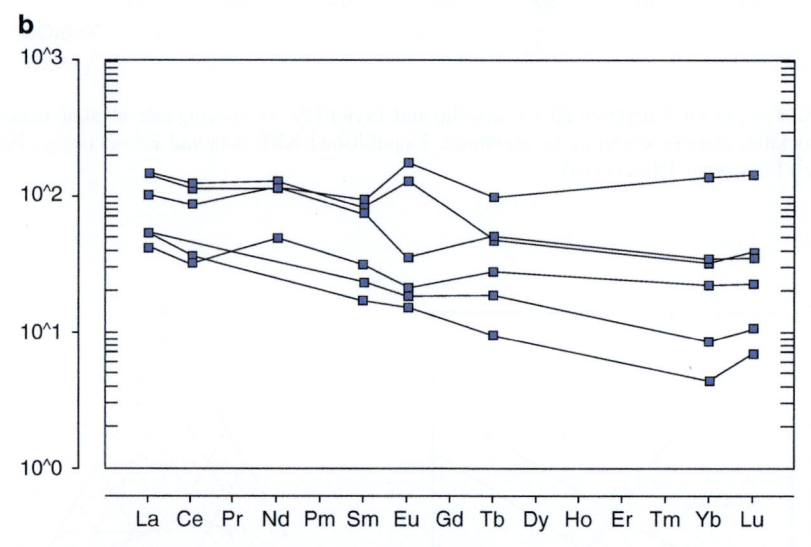

Fig. 3.46 REE diagrams for (**a**) charnoenderbitic granulites, (**b**) mafic granulites (positive Eu anomalies: garnet-bearing), (**c**) ultramafic granulite SK132 (opx, cpx, hbl, spinel), (**d**) REE Guapotón orthogneiss. (Unpublished INAA data, Delft 1983; and Kroonenberg 1990)

c

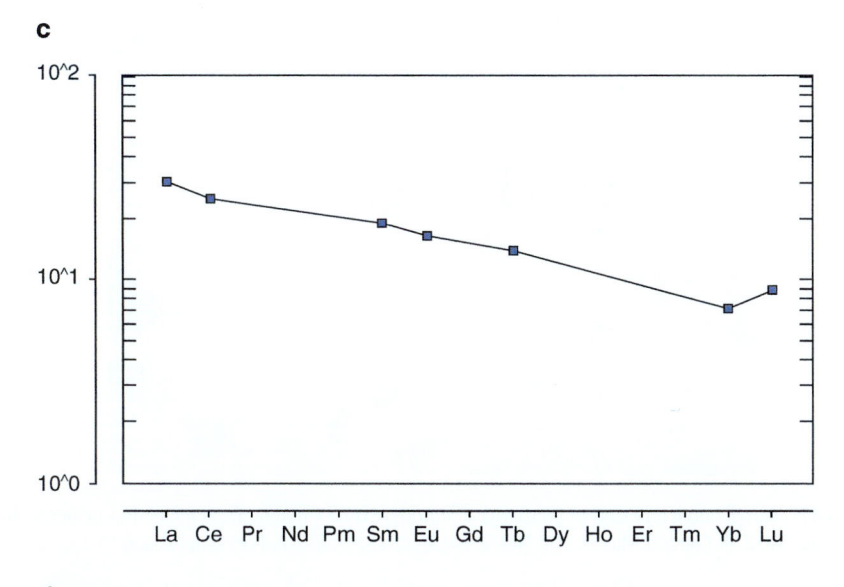

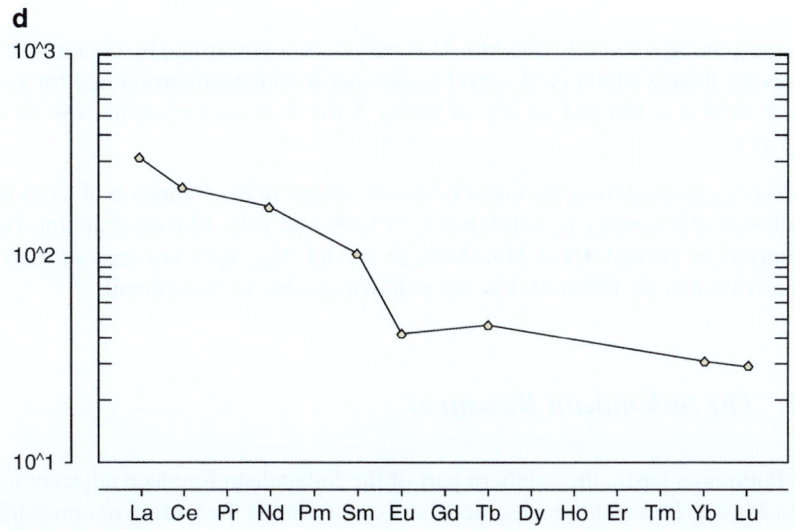

d

Fig. 3.46 (continued)

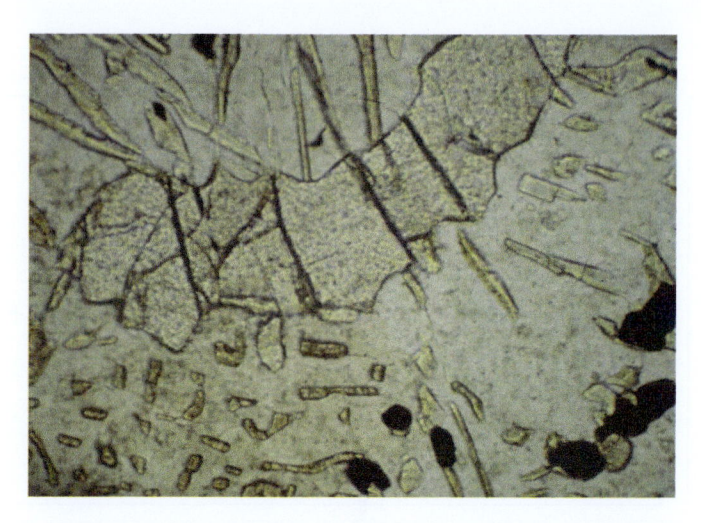

Fig. 3.47 Garnet being replaced by cordierite + orthopyroxene + magnetite symplectites in metapelitic granulite, SK 274, Garzón Massif, Río Neiva: evidence for isothermal decompression?

a metamorphic age around 1000 Ma. This pattern was confirmed by the most recent analyses by Ibáñez-Mejía et al. (2011), showing a youngest detrital age for zircon cores of 1135 ± 4 Ma and an age of 990 ± 5 for their metamorphic overgrowths (Fig. 3.48).

The ages obtained from the Garzón Massif concur in the formation of a calcalkaline volcano-sedimentary protolith between 1200 and 1100 Ma and granulite-facies metamorphism around 1000 Ma. Average model T_{DM} ages are around 1.55 Ga (Restrepo-Pace et al. 1997). Below we will discuss this in more detail.

3.2.3 The Subandean Basement

In the Putumayo basin, the southern part of the Subandean foredeep adjacent to the Garzón Massif, Precambrian basement has been found at the bottom of cores drilled by oil companies at depths between 940 and 2350 m (Ibáñez-Mejía et al. 2011, 2015). The basement in the Payara-1 well consists of granulite-facies metapelitic gneisses, from which igneous cores of zircons have been dated at 1606 ± 6 Ma and the metamorphic overgrowths at 986 ± 17 Ma (Ibáñez-Mejía et al. 2011). This author considers the protolith as igneous because of the zoned character of the zircons, but in view of the mineral paragenesis of the rock with orthopyroxene, garnet and sillimanite, it is rather a metapelitic gneiss with detrital zircons from a common igneous source rock. The Solita-1 well shows amphibolite-facies migmatitic amphibole gneisses, with zircons showing a metamorphic event at 1046 ± 43 Ma and with xenocrystic cores up to 1.85 Ga. Migmatitic gneisses from the Mandur-2 well show

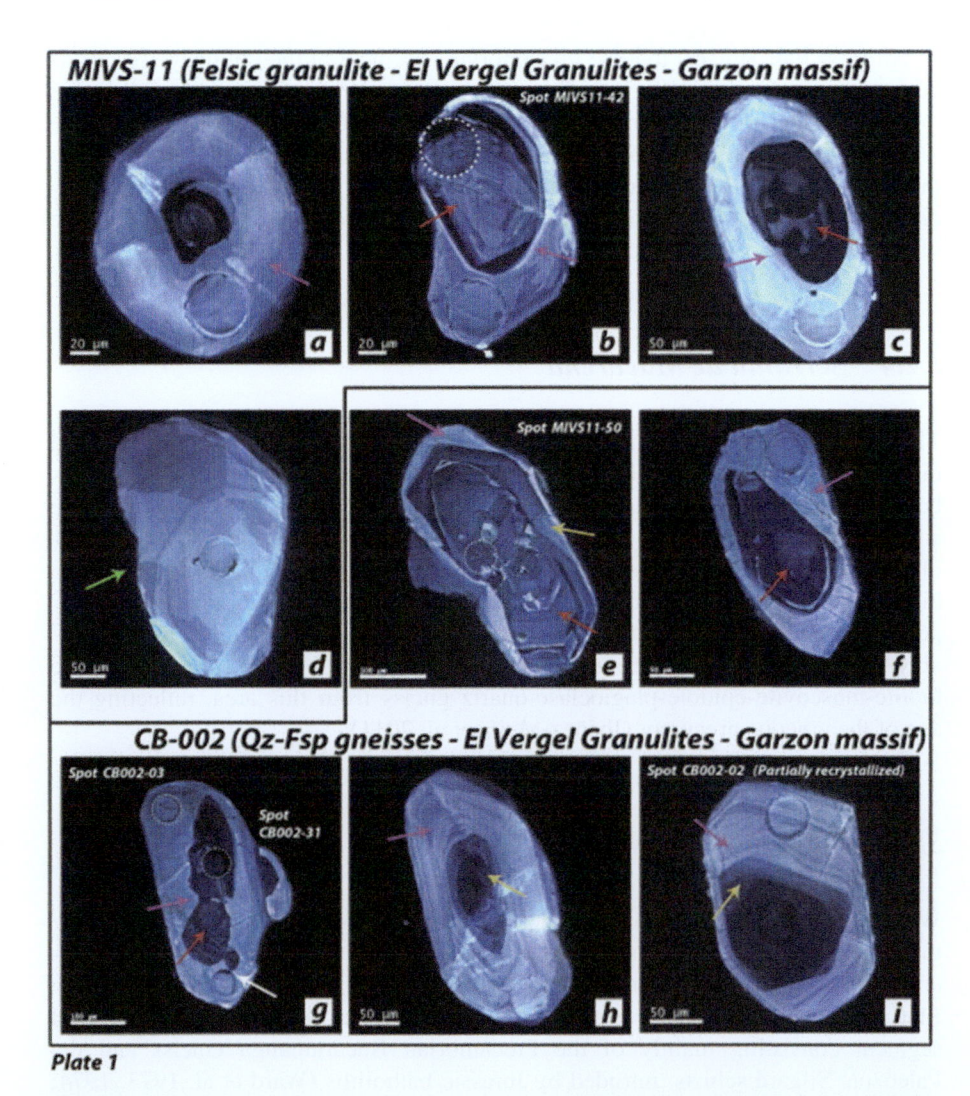

Fig. 3.48 Zircons from granulites of the Garzón Massif: igneous and/or detrital core, metamorphic overgrowths (Ibáñez-Mejía et al. 2011). Reproduced with permission (Elsevier)

amphibolite-facies metamorphism of 1019 ± 8 Ma in overgrowth rims in zircons from the melanosome and 1592 ± 8 Ma from their protolith cores. Leucosome zircons show ages of 1017 ± 4 Ma. The Caiman well consists of migmatitic biotite gneisses cut by leucogranite. The metamorphic overgrowths on zircons from the migmatites gave an age 989 ± 14 Ma; xenocrystic cores gave ages between 1470 and 1680 Ma. The crystallization age of the leucogranite was 952 ± 21 Ma, while xenocrystic zircon cores range between 1440 and 1700 Ma (Ibáñez-Mejía et al. 2011, 2015).

All these data suggest that the Precambrian basement of the Putumayo basin has been metamorphosed by the same Grenvillian event as in the Garzón Massif but that the ages of the protoliths are in the same order of those of the adjacent Guiana Shield. As Ibáñez-Mejía et al. (2011, 2015) concluded, this supports the idea that the Mesoproterozoic Amazonian basement stretched all the way to the Andean cordilleras. It also lends more confidence to the hypothesis of the autochthonous nature of the Garzón Massif.

3.2.4 Serranía de Macarena

The Serranía de Macarena forms an NNW-SSE oriented fault-bounded uplifted outlier of the Eastern Cordillera, projecting into the Llanos Orientales. Little has been published on the geology of this area after Trümpy (1943). The basement here consists of 'mica schists and alkali feldspar gneisses, hornblende gneisses, amphibolites, and injection gneisses with all intermediate types from sericitic schist to highly injected granosyenitic gneiss' (Trümpy 1943). A Precambrian age was suspected because Cambrian-Ordovician sediments cover the basement unconformably. Recently a zircon U-Pb age of 1461 ± 10 Ma was obtained from a mylonitic biotite-muscovite-epidote-plagioclase-quartz gneiss from this area, reflecting the age of the igneous precursor (Ibáñez-Mejía et al. 2011).

3.2.5 Santander Massif

While the Eastern Cordillera in southern Colombia strikes approximately NE, near the town of Bucaramanga, it suddenly turns NW. The NNW striking western boundary fault of the Eastern Cordillera in this area, the sinistral Bucaramanga-Santa Marta fault, also forms the western limit of the Santander Massif, an uplifted crustal segment consisting mainly of the Precambrian Bucaramanga Gneiss and the Paleozoic Silgará schists, intruded by Jurassic batholiths (Ward et al. 1973, 1974; Restrepo-Pace et al. 1997) and covered by younger rocks (Fig. 3.49). Three main fault-bounded blocks have been mapped, one east and north of Bucaramanga, a second one near the town of Berlín and a small one near Chitagá.

The main rock types distinguished by Ward et al. (1973) in the Bucaramanga Gneiss are metapelitic gneisses with biotite, locally muscovite, and often cordierite and sillimanite, semipelitic gneisses, sillimanite-biotite quartzites, meta-arenitic (quartzofeldspathic) biotite gneisses, calcsilicate rocks, marbles and locally hornblende gneisses and amphibolites. Migmatitic character is common. Garnet is rare except in the garnetiferous amphibolites from the second zone, which may also contain pyroxene (Urueña and Zuluaga 2011). These authors also present a detailed geochemical study of leucosomes, mesosomes and melanosomes of the migmatites from the second block. They reconstruct a metamorphic history under amphibolite-facies conditions between 660 and 750 °C and from 5.5 to 7.2 kbar. Amaya (2012) reports the presence of orthopyroxene-bearing garnetiferous mafic granulites and reconstructs a clockwise metamorphic history – still essentially within the amphibolite facies – with a prograde part ranging from 580 to 670 °C, and 6.7 to 8.6 Kbar, caused by injection of leucosome liquids.

The first Precambrian radiometric datings from the Bucaramanga Gneiss give a Rb-Sr whole-rock age of 680 ± 140 Ma for a biotite gneiss and a K-Ar hornblende age of 945 ± 40 Ma (Goldschmidt et al. 1971). Two hornblendes from an amphibolitic gneiss dated by Restrepo-Pace et al. (1997) gave integrated Ar-Ar ages of 574 ± 8 Ma and 668 ± 9 Ma. Restrepo-Pace and Cediel (2010) show a 981 ± 85 Ma U-Pb concordia age for a migmatite, apparently already obtained in 1995. U-Pb SHRIMP data by Cordani et al. (2005) show a great range of zircon ages, between 1550 and 900 Ma, of which perhaps the most tell-tale are a cluster of three zircons around 1057 ± 28 Ma and a single one of 1112 ± 24 Ma. A younger group shows ages around 864 ± 66 Ma, possibly related to a later metamorphic episode.

3.2.6 Sierra Nevada de Santa Marta

The Sierra Nevada de Santa Marta is a triangular massif, reaching from the Caribbean coast up to 5775 m, the highest coastal relief in the world. It is bounded by the left-lateral Bucaramanga-Santa Marta fault in the west, the right-lateral Oca fault along the coast in the north and the right-lateral Cerrejón fault in the southeast: a Colombian promontory that has projected itself already for over 100 km in a NW direction into the Caribbean Sea since the Tertiary (Tschanz et al. 1974; Montes et al. 2010). The Neogene uplift history has been reconstructed thermogeochronologically by Cardona et al. (2011), Villagómez (2010), and Villagómez et al. (2011).

It has a complex geological structure, in which three geological provinces separated by thrust faults have been distinguished: from NW to SE the Santa Marta Province (the NW promontory of the massif), the Sevilla Province and the Sierra Nevada Province which forms the core of the complex (Fig. 3.50). The Cesar-Ranchería depression along the SE border is still underlain by Sierra Nevada rock units (Villagómez et al. 2011). Precambrian basement crops out on five widely spaced sites within the Sierra Nevada Province, separated by huge Jurassic batholiths, as well as on the western and northern side of the Sevilla Province (Tschanz et al. 1974; Ordóñez et al. 2002; Cardona et al. 2006; Colmenares et al. 2007).

The basement rocks have been denominated Los Mangos Granulites by Tschanz et al. (1974), a name retained by Ordóñez et al. (2002) and by the recent extensive mapping project in the Sierra Nevada de Santa Marta of the Servicio Geológico

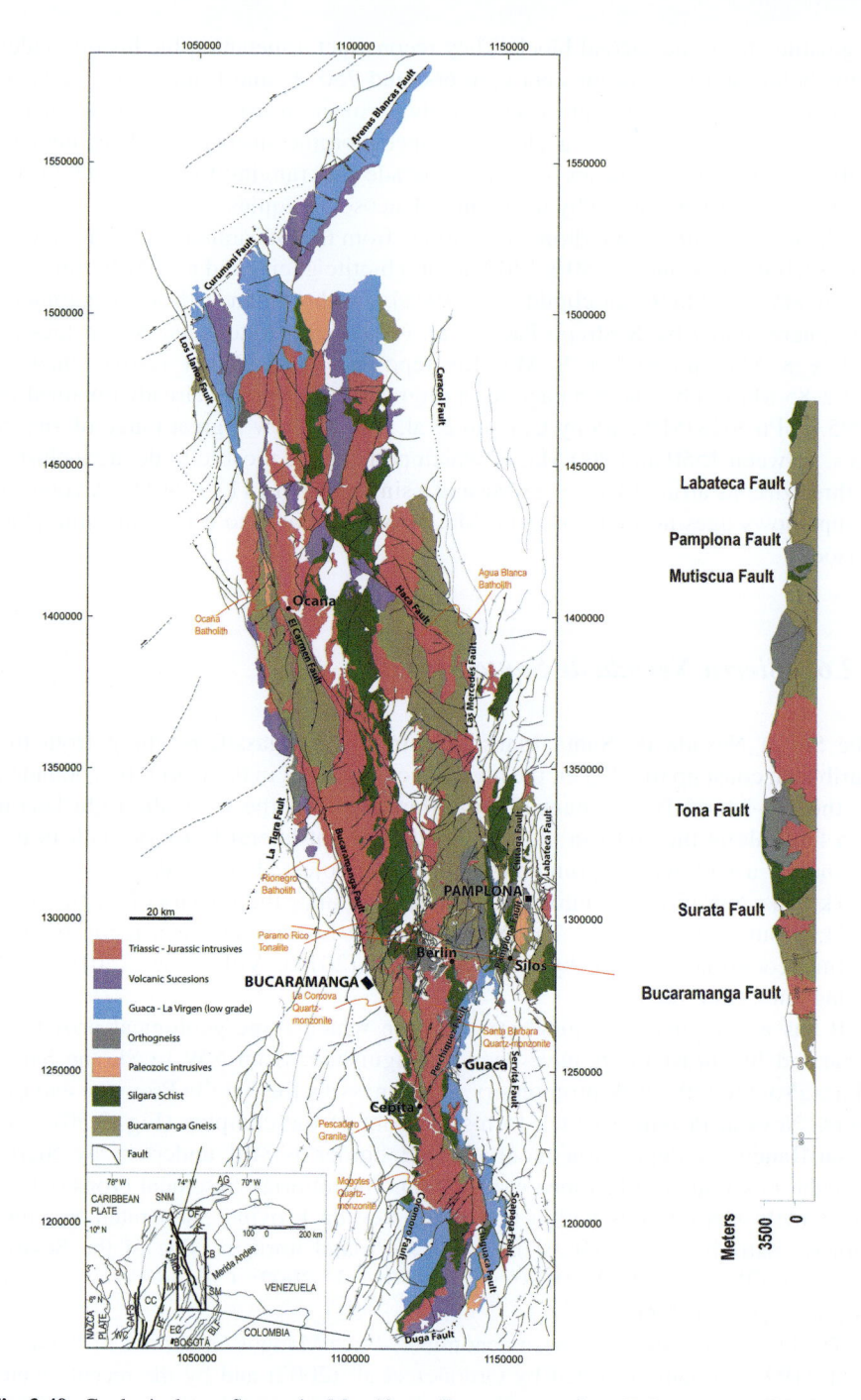

Fig. 3.49 Geological map Santander Massif near Bucaramanga, from Zuluaga et al., (2017)

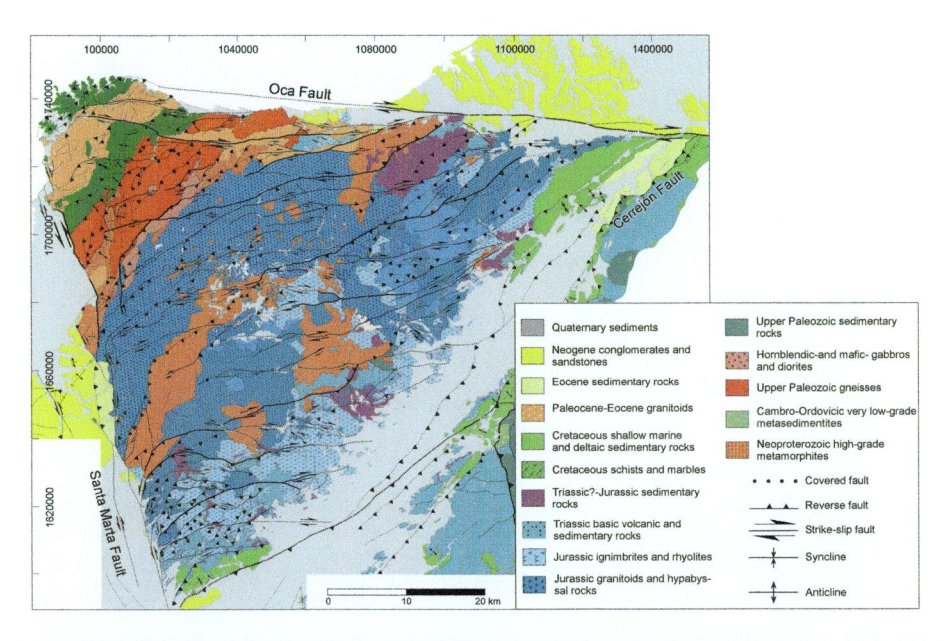

Fig. 3.50 Simplified geological map of the Sierra Nevada de Santa Marta. (see Colmenares et al. 2018)

Colombiano (Colmenares et al. 2007). The Los Mangos Granulites consist of banded and often migmatitic, granoblastic rocks including quartz-perthite granulites; intermediate granulites; mafic, calcareous and ultramafic granulites; garnet-rich granulites; and anorthosites. The migmatitic character of these units was already described by Radelli (1962b, 1967). Colmenares et al. (2007) describe also hornblende gneisses, garnetiferous biotite-muscovite gneisses, amphibolites and granulites. No orthopyroxene is mentioned, and the criteria used by Colmenares et al. (2007) to postulate granulite-facies metamorphism are insufficient. Tschanz et al. (1974) and Ordóñez et al. (2002) state that many granulites contain orthopyroxene. Also the apparent absence of metapelitic rocks is unusual. Amphibole-plagioclase thermobarometry on amphibolites indicates minimum metamorphic conditions of 6.0–7.6 kbar and 760–810° within the amphibolite-granulite-facies transition (Cordani et al. 2005). Anorthosites and anorthositic gneisses consisting almost exclusively of calcic plagioclase with accessory amphiboles and uralitized pyroxenes occur as separate concordant bands up to 1 metre in thickness within banded hornblende gneisses and garnet-biotite gneisses of the Los Mangos granulites in the Sevilla Province on the W side of the massif (Fig. 3.51; Cortes, 2013).

MacDonald and Hurley (1969) obtained a Rb-Sr isochron 1300–1400 Ma for a biotite-plagioclase gneiss and a hornblende gneiss (Dibulla Gneiss) near the northern shore of the Sierra Nevada Province. Tschanz et al. (1974) give Rb-Sr whole-rock ages of 752 and 1300 for two widely separated but similar quartz-perthite

Fig. 3.51 Río Sevilla anorthositic gneiss with amphibole lenses, Road to El Palmor. (From Colmenares et al. 2007).

granulites (Los Mangos Granulite) in the Sierra Nevada Province and a K-Ar age of 940 ± 30 Ma for a hornblende from hornblende-pyroxene-garnet-plagioclase gneiss from the western side of the Sevilla Province. Restrepo-Pace et al. (1997) give an integrated Ar-Ar age for biotite from quartz-pyroxene-garnet-biotite gneisses or granulites of 561 ± 6 Ma and a total fusion age of 845 Ma for another biotite. The upper and lower intercepts on the discordia line in the U-Pb concordia diagram of nine abraded zircons from a garnet-pyroxene-biotite-quartz-plagioclase granulite from the Guatapurí River are 1513 ± 35 Ma and 456 ± 60 Ma, respectively, but their significance is not clear because of the large error margins (Restrepo-Pace et al. 1997). Sm-Nd systematics show T_{DM} ages of 1.72–1.77 Ga. Ordóñez et al. (2002) show a Sm-Nd isochron for garnet and whole rock of 971 ± 8 Ma, and T_{DM} model ages between 1.47 and 1.92 Ga, so in the same order of magnitude as Restrepo-Pace et al. (1997). U-Pb SHRIMP analyses by Cordani et al. (2005) on rounded zircons from a biotite gneiss show apparent ages between 1400 Ma and 980 Ma. Five typical magmatic zircons yielded an age of 1374 ± 13 Ma, two other nearly concordant zircons yielded an age of 1145 ± 14 Ma and two more concordant grains presented 1081 + 14 Ma and 991 + 12 Ma. According to Cordani et al. (2005), the c. 1370 Ma age can be attributed to the magmatic crystallization of the zircons within a magmatic protolith. The zircon ages around 1140 ± 14 Ma might be related to a strong metamorphic event and the 991 ± 12 Ma age to a younger metamorphic event.

3.2.7 Guajira Peninsula

In the northernmost Guajira Peninsula, two pre-Mesozoic rock units have been recognized as possibly Precambrian, the Uray Member and the Jojoncito leucogranite (Fig. 3.52; MacDonald 1964; Lockwood 1965; Álvarez 1967; see review by Rodríguez and Londoño 2002 and López and Zuluaga 2012). The Uray Gneiss in the Macuira Mountains is a (often garnetiferous) hornblende-plagioclase gneiss body with incipient migmatitic character (cf Radelli 1961, 1967), calcsilicate rocks and diopside marbles, mostly metamorphosed under amphibolite-facies conditions, with some retrograde features. The Uray Member forms part of the Macuira Formation and is intruded by a Triassic (?) Siapana granodiorite body, but further contact relations are unclear (MacDonald 1964). A Precambrian age is suspected by Radelli (1961), but so far only Phanerozoic ages have been obtained.

A second unit, the Jojoncito leucogranite in the Simarua range (Álvarez 1967), is a leucocratic quartzofeldspathic gneiss with mesoperthite as a striking petrographic feature, suggesting granulite-facies metamorphism, but without its diagnostic minerals (Rodríguez and Londoño 2002). A 1250 Ma zircon age from this unit was mentioned by Irving (1971) and Case and MacDonald (1973), without further detail. Cordani et al. (2005) analysed zircons from the Jojoncito leucogranite and found three main groupings with apparent U-Pb SHRIMP ages of 1529 ± 43 Ma, 1342 ± 25 Ma and 1236 ± 16 Ma. These groups might reflect detrital ages from a sedimentary parent rock. High-grade metamorphic overgrowth rims gave ages of c. 1165 ± Ma and 916 ± 19 Ma. Sm-Nd systematics show a T_{DM} model age of 1.85 Ga. (Cordani et al. 2005). In the eastern, Venezuelan part of the Guajira Peninsula, also Grenvillian rocks have been reported, as well as offshore in the adjacent Venezuelan Falcon basin (Grande and Urbani 2009; Baquero et al. 2015).

3.2.8 Eastern Flank of Central Cordillera

Along the whole eastern flank of the Central Cordillera, numerous small outcrops of Precambrian rocks occur, often isolated fault-bounded uplifted blocks, often intruded by younger plutons or covered with younger deposits. From south to north, the following units have been distinguished.

Río Téllez-La Cocha Migmatitic Complex South of the Garzón Massif, in the western flank of the Central Cordillera, extending to the frontier with Ecuador, several small, elongate, fault-bounded outcrops of partly migmatitic biotite-hornblende gneisses, muscovite gneisses and garnet-sillimanite-biotite schists showing amphibolite-facies or greenschist-facies metamorphism have been reported by Ponce (1979), París and Marín (1979) and Núñez (2003). Ponce (1979) considers these rocks as Precambrian, and though Jiménez (2003) suggests that these rocks

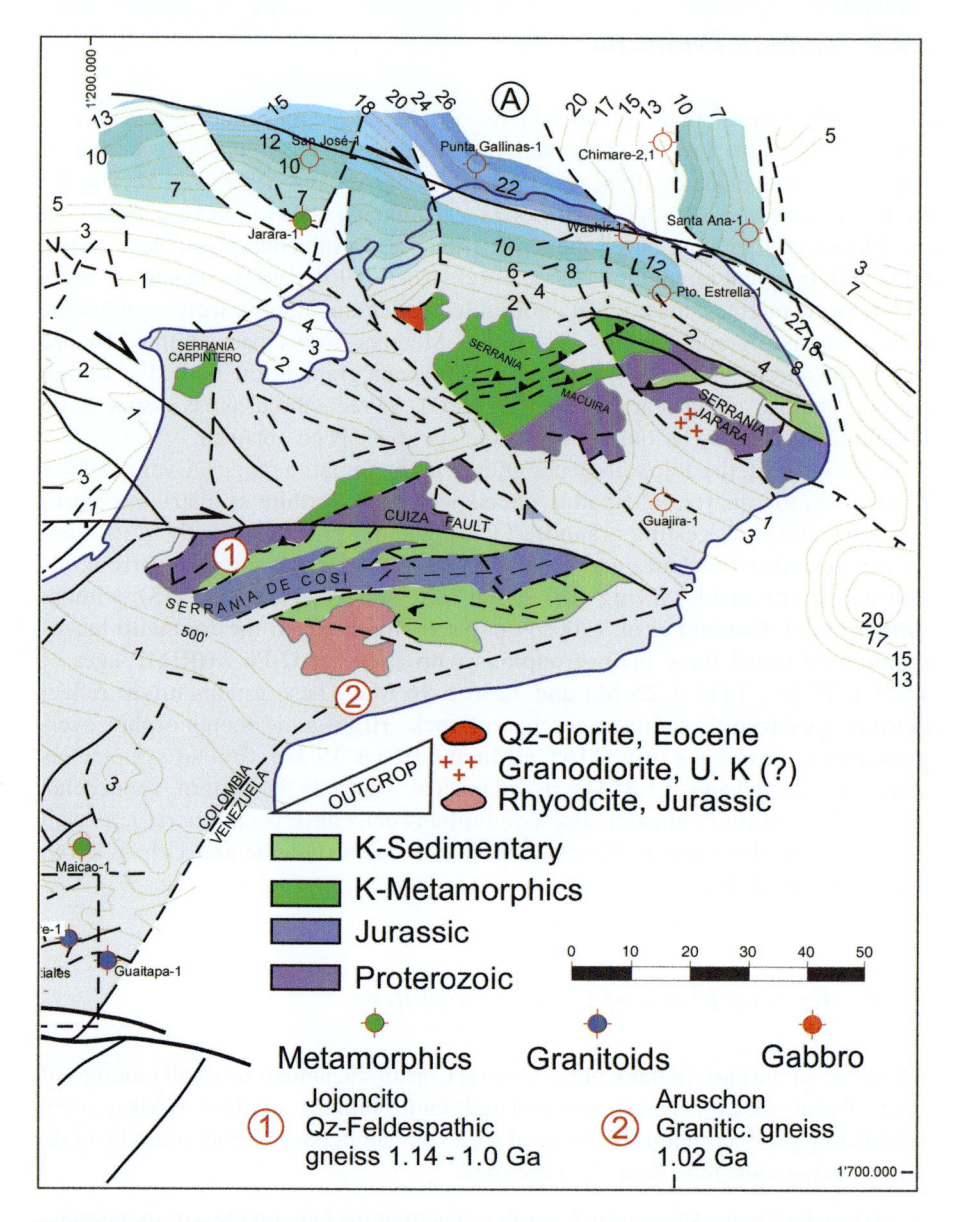

Fig. 3.52 Simplified geological map of the Guajira Peninsula Macuira, Uray rocks: dense vertical hatching, 4 is Jojoncito gneissic leucogranite. (Case and MacDonald 1973; Cediel 2018)

are much younger on the base of a U-Pb age of 166 ± 3.8 Ma from a granodiorite, we suspect that this age refers to a Jurassic intrusive body and prefer to maintain the Precambrian age of this unit, in harmony with the opinion of Ordóñez-Cardona et al. (2006).

Las Minas Massif and La Plata Massif Along the eastern flank of the Central Cordillera, just west of the Garzón Massif in the Eastern Cordillera, two smaller fault-bounded Precambrian Massif have been mapped, the Las Minas Massif and the La Plata Massif. The Las Minas Massif consists of migmatitic biotite gneisses; hornblende gneisses and amphibolites, partly garnet-bearing; and calcsilicate rocks. Slightly further north the La Plata Massif shows hornblende-biotite gneisses, orthopyroxene- and clinopyroxene-bearing quartzofeldspathic granulites as well as anatectic monzogranites (Kroonenberg 1982, 1985; Priem et al. 1989; Velandia et al. 2001; Marquínez et al. 2002a, b; Rodríguez 1995b; Ibáñez-Mejía et al. 2011). Restrepo-Pace et al. 1997 obtained an Ar-Ar hornblende cooling age obtained from a Las Minas amphibolite of 911 ± 2 Ma. Ibáñez-Mejía et al. (2011) obtained a U-Pb zircon detrital age of 1005 ± 23 Ma for a felsic gneiss near Pital and a detrital age of 1088 ± 24 Ma and a metamorphic age of 972 ± 12 for a mafic gneiss of the Las Minas Massif.

Icarcó Complex (Muñoz and Vargas 1981a, b; Murillo et al. 1982; Esquivel et al. 1987). In the southern part of the Tolima Department between the rivers Mendarco and Ambeima, three different outcrops of Precambrian have been mapped, designated Icarcó Complex by Esquivel et al. (1987). They consist of amphibolites, migmatitic hornblende gneisses, quartzofeldspathic gneisses and biotite-sillimanite gneisses and furthermore garnet-bearing quartzites, granulites and virtually pure marble lenses. On the basis of major elements of chemistry, the amphibolites are thought to be of igneous origin; the other rocks are metavolcanic-metasedimentary deposited in a continental shelf environment (Muñoz and Vargas 1981a, b). The mineral parageneses indicate mainly amphibolite-facies and locally granulite-facies metamorphism. The main foliation strikes between N-S and N10°E, and there is a pervasive cataclastic foliation striking 70–90°. Contacts with surrounding rock units are partly tectonic, but locally the migmatites are intruded by the Jurassic Ibagué batholith (Muñoz and Vargas 1981a, b). Roof pendants of similar rocks within the Ibagué batholith are mapped as Davis Biotite Gneisses (Esquivel et al. 1987). No radiometric data are available.

Tierradentro gneisses and amphibolites Migmatitic biotite gneisses (locally with muscovite and sillimanite), quartzofeldspathic gneisses, hornblende gneisses and amphibolites and occasionally quartzites and marbles have been described from the Río Coello near Ibagué (Tolima) by Barrero (1969), Barrero and Vesga (1976) and Mosquera et al. (1982) (Fig. 3.53). This unit is intruded by the Jurassic Ibagué batholith. In the absence of radiometric data, these rocks have been correlated with the granulites of the Sierra Nevada de Santa Marta (Barrero 1969; Kroonenberg 1985). West of Lérida and Armero another fault-bounded sliver of Precambrian rocks has been mapped by Barrero and Vesga (1976, 2010), continuing northwards along the hanging wall of the eastern boundary fault of the Eastern Cordillera at least as far north as Honda. They consist of schists, quartzofeldspathic biotite gneisses and amphibolites. The only available radiometric age is a K-Ar age of 1365 ± 270 Ma on hornblende from an amphibolite (Barrero and Vesga 1976; Vesga and Barrero 1978).

Fig. 3.53 Tierradentro amphibolites and migmatitic gneisses, Río Coello, Tolima. (Photo: Kroonenberg)

San Lucas Metamorphic Complex West of Puerto Berrio, the strip of Precambrian rocks in the eastern foothills of the Central Cordillera reappears, but it continues far northwards, with some interruptions, to form the western flank of the San Lucas Mountains, the northernmost extremity of the Central Cordillera (Fig. 3.54; Bogotá and Aluja 1981; Toussaint 1993; Ordóñez-Carmona et al. 1999, 2006; Figueroa et al. 2006; Cuadros et al. 2014; Clavijo et al. 2008). The Otú fault on the western side of the Serranía de San Lucas is generally considered as the westernmost limit of the Precambrian basement in this part of the Colombian Andes (Feininger et al. 1972; Ordóñez-Carmona et al. 1999, 2006; Clavijo et al. 2008; Cuadros et al. 2014). The basement is unconformably overlain by graptolite-bearing Ordovician shales (Feininger et al. 1972 and references cited therein). However, also west of this fault, occasionally high-grade metamorphic rocks occur, such as the Puquí gneiss and the Pantanillo granulite, from which so far only Phanerozoic Ar-Ar whole rock and K-Ar hornblende ages have been obtained (Rodríguez et al. 2012; Rodríguez and Albarracin 2012).

Lithologically the San Lucas rocks are migmatitic quartzofeldspathic gneisses, amphibolites, marbles, mafic granulites, leucogranite gneiss and metaquartzmonzonite apparently intruding the other rocks (Feininger et al. 1972; Ordóñez et al. 1999; Clavijo et al. 2008; Zapata et al. 2014; Cuadros et al. 2014). Ordóñez-Carmona et al. (1999) obtained a Rb-Sr isochron for the El Vapor mylonite of 894 ± 36 Ma, a single zircon Pb-Pb (Kober method) age of 1100 Ma and Sm-Nd T_{DM} model ages of 1829 and 1757 Ma. Similar values were obtained by Figueroa et al. (2006): they obtained a zircon U-Pb age of 1124 ± 22 Ma age, a whole-rock Sm-Nd age of 1312.5 ± 3.2 Ma and a T_{DM} model age of 1.6 Ga on a granulite near the Poporopo Pb-Zn mine.

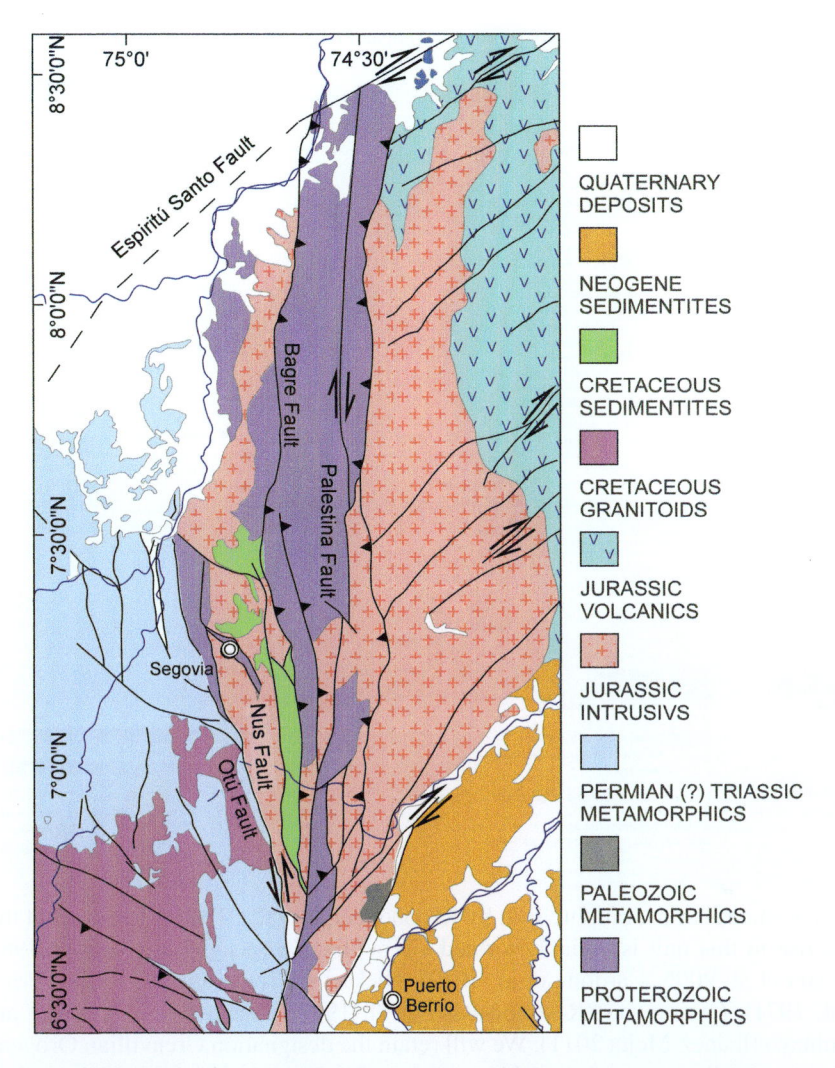

Fig. 3.54 Precambrian of the Serranía de San Lucas modified after Cuadros et al. 2014

3.2.9 Geological Evolution of the Andean Precambrian: The Grenvillian Orogeny

From the data presented above, it is clear that the Andean Precambrian in Colombia differs strongly from the Amazonian Precambrian in age, lithology and metamorphism. The great majority of the rocks show zircon U-Pb ages between 1150 and 950 Ma, granulites and gneisses of widely different compositions predominate and granulite-facies metamorphism is widespread (Table 3.3). This warrants the distinction of the Andean Precambrian as a separate geological province, termed the

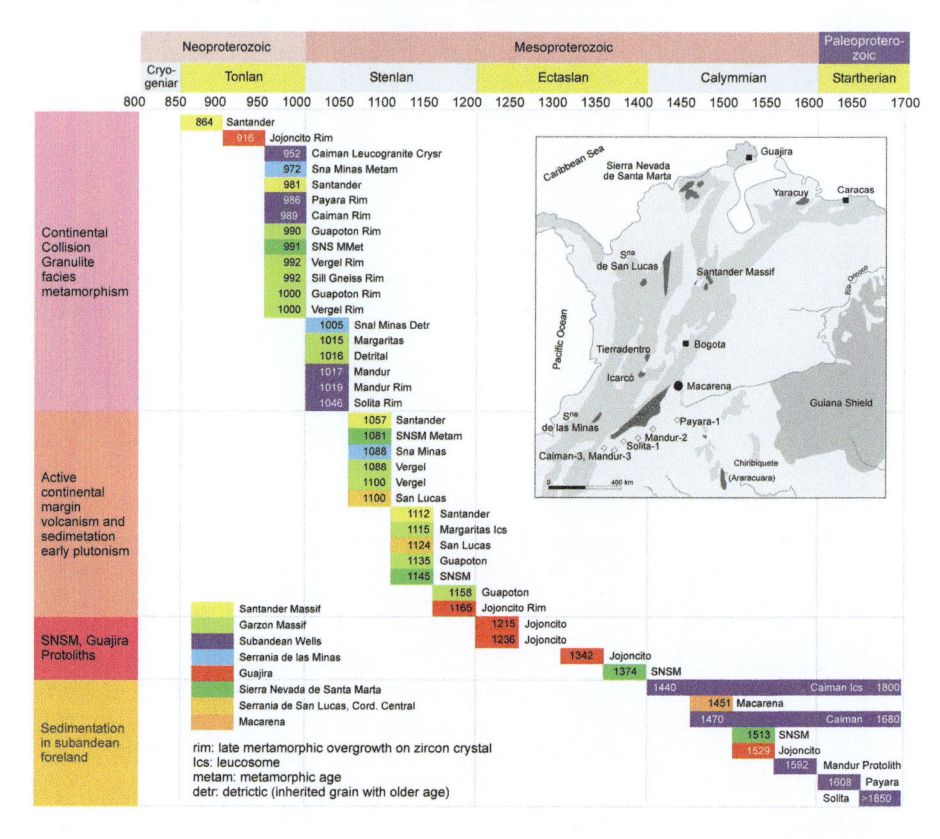

Table 3.3 U-Pb chronogram of the Colombian Andean Precambrian

Garzón-Santa Marta Granulite Belt by Kroonenberg (1982). The orogenic event that gave rise to this unit is variably termed Grenvillian Orogeny (Kroonenberg 1982; Cordani et al. 2005; Cardona et al. 2010), Nickerian (Toussaint 1993, after Priem et al. 1971), Orinoquian (Restrepo-Pace et al. 1997; Martín-Bellizzia 1972) and Putumayo (Ibánez-Mejía 2011). We will retain the designation Grenvillian Orogeny, as it is generally accepted that this orogeny was the result of the collision of the Amazonian Craton with Laurentia and one of the key events in the assembly of Rodinia (Kroonenberg 1982; Cordani et al. 2005; Cardona et al. 2010; Ramos 2010).

Eastern boundary of Grenvillian Orogeny The boundary between the Amazonian and Andean Precambrian is hidden below the sediment cover of the Subandean foreland basins. The Precambrian basement rocks retrieved from boreholes in the basin by Ibáñez-Mejía et al. (2011) are largely metasedimentary gneisses and granulites subjected to Grenvillian metamorphism. They contain detrital zircons apparently derived from the adjacent Amazonian basement, but the age of sedimentation is unknown so far. The maximum age of sedimentation is given by the 1444 ± 15 Ma age of detrital zircons from the Caiman-3 well (Ibáñez-Mejía et al. 2011). There is no firm evidence that Amazonian basement rocks themselves have been subjected to Grenvillian high-grade metamorphism, and so how far the Amazonian basement extends westwards and the Grenvillian high-grade metamorphism eastwards is still unknown. However, far-field effects of the Grenvillian Orogeny are well discernible in almost the whole Guiana Shield through shearing, mylonitization and thermal resetting of mineral ages: the K'Mudku, Nickerie or Orinoquian event (see above).

1530–1230 Ma: Early stages of the Grenvillian Orogeny In spite of the common characteristics of all Andean Precambrian tectonic blocks, there are also interesting differences between them. Detrital zircons between 1530 and 1230 Ma are known from the Guajira Peninsula, the Sierra Nevada de Santa Marta and the Serranía de Macarena, but not from the other Andean outcrops. Furthermore, Cordani et al. (2005) suggest a magmatic protolith around 1370 in the Sierra Nevada de Santa Marta. The significance of those isolated early dates cannot be evaluated but suggests that some tectonic activity already started at that time, as is the case in the Grenville Province in Laurentia (Rivers 1997).

1150–1050 Ma: Active continental margin sedimentation and early magmatic activity There is a great similarity in the lithology of all Andean outcrops; quartzofeldspathic gneisses and granulites predominate, while metapelitic, metabasic, calcsilicate and quartzitic lithologies are also common. They point to a largely supracrustal, metasedimentary origin of the precursor rocks. Compositional banding on centimetre to metre scale, apart from migmatitic effects, is also evidence of a supracrustal origin. The bulk of the sediments is feldspar-rich, suggesting an immature character of the sediments. In view of the calc-alkaline affinities of the Garzón quartzofeldspathic granulites, it is logical to suppose that there is an important volcanogenic contribution, probably deposited as greywackes in an active continental margin (Kroonenberg 1982; Jiménez et al. 2006; Cordani et al. 2005). Some mafic rocks may represent metamorphosed basaltic sills or dykes, or synsedimentary lava flows into the basins, but their general scarcity does not favour an important back-arc spreading stage as envisaged by Ibáñez-Mejía et al. (2011). Only the anorthosites in the Sierra Nevada de Santa Marta are unknown from the other areas: their significance as individual bands within gneiss-granulite complexes has still to be evaluated. They also occur in Precambrian outliers in western Venezuela (Grande and Urbani 2009). Metapelitic rocks have not been recorded from the Sierra Nevada.

Orthogneisses like the Guapotón-Mancagua augengneisses intruded between 1158 and 1135 Ma may represent the deeper substructures of acid volcanic edifices. Also the early Jojoncito leucogranites (~1215–1236) may belong to this category. Early metamorphism and anatexis around Ma 1115 are evident from the Margaritas leucosomes in the Garzón Massif and in the Sierra Nevada de Santa Marta.

1050–950 Ma: Continental collision, granulite-facies metamorphism and migmatization Peak metamorphism in the granulites and gneisses is recorded in the metamorphic rims of zircons between 1050 and 950 Ma within all blocks of Andean Colombia as well as in the Subandean basement. Granulite-facies metamorphism is often concomitant with migmatization, as is evident from the presence of orthopyroxene in leucosome and from leucosome zircon dates, but anatexis did not result in large-scale plutonism. The clockwise metamorphic history of the Vergel

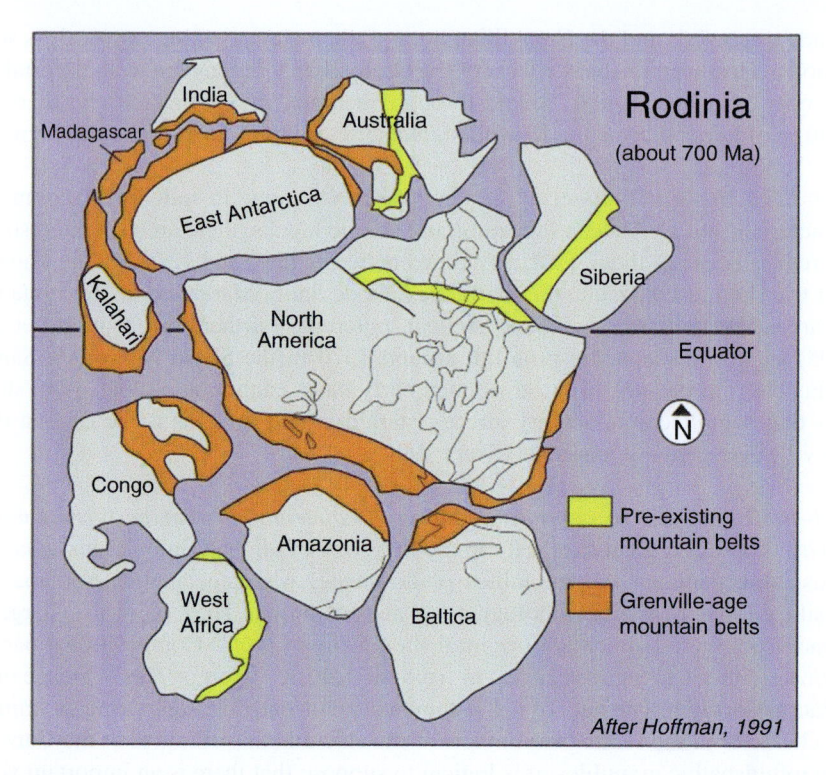

Fig. 3.55 Position of Amazonia and Laurentia in Rodinia supercontinent. (After Hoffman 1991)

Granulites suggests an isobaric cooling path, caused by thickening of the crust as a result of the collision (Jiménez et al. 2006). Younger Ar-Ar and Rb-Sr mineral ages, not included in Table 3.3, reflect different stages of cooling.

The continental collision between Amazonia and Laurentia plays a key role in the assembly of the Rodinia supercontinent around 1 Ga (Fig. 3.55; Hoffman 1991). Other continental fragments involved are the Oaxaquia and Baltica (Ruiz et al. 1999; Cordani et al. 2005, 2010; Ibáñez-Mejía et al. 2011; Geraldes et al. 2015), and there is discussion to which part of Laurentia Amazonia collided, but this discussion remains outside the scope of this chapter.

References

Almeida ME, Pinheiro SS, Luzardo R (2002) Reconhecimento geológico ao longo dos rios Negro, Xié e Içana Missão Tunuí, noroeste do Estado do Amazonas. Projeto Gis Brasil Província Rio Negro Folhas 1:1.000.000 NA.19 Pico Da Neblina e SA.19 Içá. CPRM Manaus, 2002

Almeida ME, Macambira MJB, Oliveira EC (2007) Geochemistry and zircon geochronology of the I-type high-K calc-alkaline and S-type granitoid rocks from southeastern Roraima, Brazil: Orosirian collisional magmatism evidence 1.97–1.96 Ga in central portion of Guyana shield. Precambrian Res 155(2007):69–97

Almeida ME, Macambira MJB, Santos JOS, do Nascimento RSC, Paquette JL (2013) Evolução crustal do noroeste do Cráton Amazônico Amazonas, Brasil baseada em dados de campo, geoquímicos e geocronológicos. Anais do 13° Simpósio de Geologia da Amazônia, 201–204

Almeida (2014) GIS SOUTH AMERICA 1:1 M: NA.19 Pico da Neblina and SA.19 Içá sheets. Memorias Geological Map of South America Workshop, Villa de Leyva, Colombia: 32 and 473–480

Altenberger UD, Mejia Jimenez M, Günter C, Sierra Rodríguez GI, Scheffler F, Oberhänsli R (2012) The Garzón Massif, Colombia-a new ultrahigh-temperature metamorphic complex in the Early Neoproterozoic of northern South America. Miner Petrol 105:171–185

Álvarez W (1967) Geology of the Simarua and Carpintero areas, Guajira Peninsula, Colombia. Tesis Ph.D., Princeton Univ., 168 pp

Álvarez J (1981) Determinación de la edad Rb-Sr en rocas del Macizo de Garzón. Geología norandina Bogotá 4:31–38

Álvarez J, Cordani UG (1980) Precambrian basement within the septentrional Andes: age and geological evolution; 26th Int.Geol. Congr. Paris, Abstract 1:10

Álvarez J, Linares E (1984) Una edad K/Ar del macizo de Garzón, Departamento del Huila, Colombia. Geología norandina 9:31–33

Amaya S (2012) Caracterización Petrográfica y Petrológica de los Neises, Migmatitas y Granulitas del Neis de Bucaramanga, en el Macizo de Santander, Departamento de Santander. Tesis de maestría, Universidad nacional Bogotá, Colombia, 130 pp

Arango MI, Zapata G, Martens U (2012) Caracterización petrográfica, geoquímica y edad de la sienita nefelínica de San José del Guaviare. Boletín de Geología 34:15–26

Ashley RM (2011) Technical report Machado Project, Vaupés Department, Colombia, Horseshoe Gold Mining Inc., 91 pp

Bangerter G (1985) Estudio sobre la petrogénesis de las mineralizaciones de niobio, tántalo y estaño en el granito Rapakivi de Parguaza y sus diferenciaciones. Memoria I simposium Amazonico, Puerto Ayacucho, Venezuela; Boletín de Geología, Publicación Especial no. 10, 175–185

Baquero M, Grande S, Urbani F, Cordani UG, Hall C, Armstrong R (2015) New Evidence for Putumayo Crust in the Basement of the Falcon Basin and Guajira Peninsula, Northwestern Venezuela. in C. Bartolini and P. Mann, eds., Petroleum geology and potential of the Colombian Caribbean Margin: AAPG Memoir 108, p. 103–136

Barrero D (1969) Petrografía del stock de Payandé y metamorfitas asociadas. Boletín geológico 17:113–144

Barrero D, Vesga CJ (1976) Mapa geológico del cuadrángulo K-9 Armero y parte sur del J-9 La Dorada. Escala 1:100.000. INGEOMINAS. Bogotá

Barrios F, (1985) Geología de la Subregión Atabapo-Guarinuma, Territorio Federal Amazonas. Memoria I Simposium Amazónico, Puerto Ayacucho, Venezuela; Boletín de Geología, Publicación Especial no. 10, 9–21

Barrios F, Rivas D, Cordani U, Kawashita K (1985) Geocronología del Territorio Federal Amazonas. Memoria I Simposium Amazónico, Puerto Ayacucho, Venezuela; Boletín de Geología, Publicación Especial no 10, 22–31

Barron CN (1969) Notes on the stratigraphy of Guyana. Proceedings Seventh Guiana Geological Conference, Paramaribo, 1966. Records Geological Survey Guyana, 6, II, 1–28

Bayona G, Jiménez G, Silva C, Cardona A, Montes C, Roncancio J, Cordani U (2010) Paleomagnetic data and K–Ar ages from Mesozoic units of the Santa Marta massif: a preliminary interpretation for block rotation and translations. J S Am Earth Sci 29:817–831

Berrangé JP (1977) The geology of southern Guyana, South America. Inst. Geol. Sciences, London, Overseas Division, Memoir 4, 112 pp

Bispo-Santos F, D'Agrella-Filho MS, Janikian L, Reis NJ, Trindade RIF, Reis MAAA (2014) Towards Columbia: Paleomagnetism of 1980–1960 Ma Surumu volcanic rocks, Northern Amazonian Craton. Precambrian Research 244: 123–138

Bogotá J (1981) Sintesis geología regional de las zonas limitrofes Colombia-Brasil-Venezuela: COGEMA Compañía General de Materias Nucleares

Bogotá J, Aluja J (1981) Geología de la Serranía de San Lucas. Geología norandina 4:49–55

Bonilla-Pérez A, Frantz JC, Charão-Marques J, Cramer T, Franco-Victoria JA, Mulocher E, Amaya-Perea Z (2013) Petrografía, geoquímica y geocronología del Granito de Parguaza en Colombia. Boletín de Geología 35:83–104

Bonilla A, Frantz JC, Charão J, Cramer T, Mulocher E, Franco JA, Amaya Z (2013) Magmatismo rapakivi en el NW del Craton Amazonico. Anais do 13° Simpósio de Geologia da Amazônia, 246–249

Botero P (Ed.) (1999) Paisajes fisiográficos de la Orinoquia-Amazonia ORAM Colombia. Análisis Geográficos IGAC, Bogotá 27–28, 1–361 and maps

Bruneton P, Pallard B, Duselier D, Varney E, Bogotá J, Rodríguez C, Martín E (1983) Contribución a la geología del oriente de las Comisarías del Vichada y del Guainía Colombia. Geología norandina 6:3–12

Cardona A, Cordani UG, MacDonald WD (2006) Tectonic correlations of pre-Mesozoic crust from the northern termination of the Colombian Andes, Caribbean region. J S Am Earth Sci 21:337–354

Cardona A, Chew D, Valencia VA, Bayona G, Mišković A, Ibañez-Mejía M (2010) Grenvillian remnants in the Northern Andes: Rodinian and Phanerozoic paleogeographic perspectives. J S Am Earth Sci 29:92–104

Cardona A, Valencia V, Weber M, Duque J, Montes C, Ojeda G, Reiners P, Domanik K, Nicolescu S, Villagómez D (2011) Transient Cenozoic tectonic stages in the southern margin of the Caribbean plate: U-Th/He thermochronological constraints from Eocene plutonic rocks in the Santa Marta massif and Serranía de Jarara, northern Colombia. Geol Acta 9:445–466

Carrillo VM (1995) Sobre la edad de la secuencia metasedimentaria que encaja las mineralizaciones auríferas vetiformes en la región del Taraira Vaupés. Geología colombiana 19:73–81

Case JE, MacDonald WD (1973) Regional gravity anomalies and crustal structure in Northern Colombia. Geol Soc Am Bull 84:2905–2916

Cediel F (2018) Phanerozoic orogens of Northwestern South America: cordilleran-type orogens, taphrogenic tectonics and orogenic float. Springer, Cham, pp. 3–89

Cediel F, Shaw RP, Cáceres C (2003) Tectonic assembly of the Northern Andean Block. In : C. Bartolini, R.T. Buffler and J. Blickwede eds., The Circum-Gulf of Mexico and the Caribbean: Hydrocarbon habitats, basin formation and plate tectonics: AAPG Memoir 79, 815–848

Celada CM, Garzón M, Gómez E, Khurama S, López JA, Mora M, Navas O, Pérez R, Vargas O, Westerhof AB (2006) Potencial de recursos minerales en el Oriente colombiano: compilación y análisis de la información geológica disponible fase 0 versión 1.0. Ingeominas, 233 pp

Clavijo J, Mantilla L, Pinto J, Bernal L, Pérez A (2008) Evolución geológica de la Serranía de San Lucas, norte del Valle Medio del Magdalena y noroeste de la Cordillera Oriental Boletín de Geología, 30, 45–62

Colmenares FH, Mesa AM, Roncancio JH, Arciniegas EG, Pedraza PE, Cardona A, Romero AJ, Silva CA, Alvarado SI, Romero OA, Vargas AF (2007) Geología de la planchas 11, 12, 13, 14, 18, 19, 20, 21, 25, 26, 27, 33 y 34. Proyecto: "Evolución geohistórica de la Sierra Nevada de Santa Marta". 401 pp

Colmenares F, Román-García L, Sánchez JM, Ramirez JC (2018) Diagnostic structural features of NW South America: Structural crosssections based upon detail field transects. In: Cediel F and Shaw RP (eds) Geology and Tectonics of Northwestern South America: The Pacific-Caribbean-Andean Junction, Springer, Cham, pp. 651–670

Commission for the Geological Map of the World (2009a) Geological and mineral resources map of South America Sheet NA.19

Commission for the Geological Map of the World (2009b) Geological and mineral resources map of South America Sheet SA.19

Cordani UG, Sato K (1999) Crustal evolution of the South American Platform, based on Nd isotopic systematics on granitoid rocks. Episodes, 167–173

Cordani UG, Sato K, Teixeira W, Tassinari CCG, Basei MAS (2000) Crustal evolution of the South American Platform. In: Cordani U.G. et al. eds Tectonic evolution of South America, p. 19–40

Cordani UG, Cardona A, Jiménez JM, Liu D, Nutman AP (2005) Geochronology of Proterozoic basement inliers in the Colombian Andes: tectonic history of remnants of a fragmented Grenville belt. In: Vaughan et al. eds. Terrane processes at the margins of Gondwana. Geol Soc. London Spec Pub 246, 329–346

Cordani UG, Teixeira W (2007) Proterozoic accretionary belts in the Amazonian Craton.in: R.D. Hatcher et al. 4D framework of continental crust.. Geol. Soc Amer Memoir 200, 297–320

Cordani UG, Fraga LM, Reis N, Tassinari CCG, Brito-Neves BB (2010) On the origin and tectonic significance of the intra-plate events of Grenvillian-type age in South America: a discussion. J S Am Earth Sci 29:143–159

Cordani UG, Sato K, Sproessner W, Santos Fernandes F (2016) U-Pb zircon ages of rocks from the Amazonas Territory of Colombia and their bearing on the tectonic history of the NW sector of the Amazonian Craton. Brazilian J Geol 46(Suppl 1):5–35. June 2016

Coronado JA, Tibocha EV (2000) Reconocimiento geológico y caracterización de areas potenciales para oro en un sector de la Serrania de la Libertad, distrito minero de Taraira, Vaupés-Colombia. Tesis de grado, Universidad Nacional de Colombia, 85 pp

Cortes E (2013) Análisis petrogenético de las denominadas "Anortositas" aflorantes en la vertiente occidental de La Sierra Nevada de Santa Marta – Sector Rio Sevilla – El Palmor – (Colombia). Tesis de Maestría Universidad nacional, Bogotá, 189 pp

Cox DC, Wynn JC, Skidder GB, Page NJ (1993) Geology of the Venezuelan Guayana Shield. In: Geology and Mineral Resource Assessment of the Venezuelan Guayana Shield, by USGS and CVG Técnica Minera CA. USGS Bulletin 2062. 9–15

CPRM 2004a Carta Geológica do Brasil ao milionésimo. Folha NA.19 Pico da Neblina

CPRM 2004b Carta Geológica do Brasil ao milionésimo. Folha SA.19 Içá

Cuadros FA, Botelho NF, Ordóñez O, Matteini M (2014) Mesoproterozoic crust in the San Lucas range (Colombia): an insight into the crustal evolution of the northern Andes. Precambrian Res 245:186–206

Cuéllar JV, Orozco VM, Castro F (2003) La geología de la Serranía de Machado en el distrito aurífero de Taraira Vaupés. Unpublished manuscript

Dall'Agnol R, Macambira MJB (1992) Titanita-biotita granitos do baixo Rio Uaupés, Província Rio Negro, Amazonas. Parte I: geologia petrografia e geocronologia. Revista Brasileira de Geociências 22, 3–14

Dall'Agnol R, Lafon JM, Macambira MJB (1994) Proterozoic anorogenic magmatism in the central Amazonian province, Amazonian craton: geochronological, petrological and geochemical aspects. Mineral Petrol 50:113–138

Dall'Agnol R, Costi HT, da S. Leite AA, de Magalhães MS, Teixeira NP (1999) Rapakivi granites from Brazil and adjacent areas. Precambrian Res 95:9–39

Dall'Agnol R, Costi HT, Lamarão CN, Teixeira NP, Bettencourt JS, Fraga LM (2006) Granitóides proterozóicos e suas implicações na evolução crustal do Cráton Amazônico

De Boorder H (1976) Informe sobre el reconocimiento de la geología del trayecto Mitú-Yavaraté-Montfort-Acaricuara-Mitú, a lo largo de los ríos Vaupés, Papurí, y Paca y el caño Yi en la Comisaría especial del Vaupés. Unpublished report PRORADAM, Bogotá, 46 pp

De Boorder H (1978) Contribución a la geología de las cuencas de los ríos Vaupés, Piraparaná y Taraira. Unpublished report PRORADAM, Bogotá, 50 pp

De Boorder H (1980) Contribución preliminar al estudio de la estructura geológica de la Amazonia colombiana. Revista CIAF 5(1):49–96

De Boorder H (1981) Structural-geological interpretation of SLAR imagery of the Colombian Amazones. Trans Inst Min Metall 90:B145–B152

De la Espriella R, Florez C, Galvis J, González CF, Marino J, Pinto H (1990) Geologia Regional del Norte de la Comisariadel Vichada. Geología Colombiana 17:93–106

Deckart K, Bertrand H, Liégeois JP (2005) Geochemistry and Sr, Nd, Pb isotopic composition of the Central Atlantic Magmatic Province CAMP in Guyana and Guinea. Lithos 82:289–314

Delor C, de Roever EWF, Lafon JM, Lahonère D, Rossi P, Cocherie A, Guerrot C, Potrel A (2003) The Bakhuis ultra-high temperature granulite belt Suriname : II implications for late trans-Amazonian crustal stretching in a revised Guiana shield framework. Geologie de la France 2,3 4:207–230

De Roever EWF, Lafon JM, Delor C, Rossi P, Cocherie A, Guerrot C, Potrel A (2003) The Bakhuis Ultra-high temperature granulite belt : I Petrological and geochronological evidence for a counterclockwise P-T path at 2.07–2.05 Ga. Géologie de la France 2003, 2,3,4: 175–205

Esquivel J, Núñez A, Flores G (1987) Geología y Prospección Geoquímica de la Plancha 281 – Rioblanco (Tolima), escala 1:100,000. Memoria explicative. Ingeominas, Ibagué, 217 pp

Etayo F, Barrero D, Lozano H, Espinosa A, González H, Orrego A, Ballesteros I, Forero H, Ramírez C (1983) Mapa de terrenos geológicos de Colombia. Publ Esp Ingeominas 14. 235 pp

Feininger T, Barrero D, Castro N (1972) Geología de parte de los departamentos de Antioquia Caldas, sub-zona II-B. Boletín Geológico 20(2):1–173

Fernandes PECA, Pinheiro SdS, de Montalvão RMG, Issler RSA, Abreu AS, Tassinari CCG (1977) Geologia, in: Projeto Radambrasil. Levantamento de Recursos Naturais Vol 14, Folha SA.19, Içá, 17–123

Figueroa et al (2006) Cartografía geológica de 9.600 km2 de la Serranía de San Lucas: Planchas 55 (El Banco), 64 (Barranco De Loba), 85 (Simití) Y 96 (Bocas Del Rosario): Aporte al conocimiento de su evolución geológica. Ingeominas, Bogotá, 192 pp

Forero A (1990) The basement of the Eastern Cordillera, Colombia: an allochthonous terrane in northwestern South America. J S Am Earth Sci 3:141–151

Fraga LM, Reis NJ, Dall'Agnol R, Armstrong R (2008) The Cauarane-Coeroene belt, the tectonic southern limit of the preserved Rhyacian crustal domain in the Guyana Shield, northern Amazonian Craton. Abstract 33th IGC Oslo, symposium AMS-07, paper 1344505

Fraga LM, Reis NJ, Dall'Agnol R (2009a) The Cauarane-Coeroene belt, the main tectonic feature of the central Guyana Shield, northern Amazonian Craton. SBG Núcleo Norte, Simpósio de Geologia da Amazônia 11, Manaus, Expanded Abstract, 3 pp

Fraga LM, Macambira MJB, Dall'Agnol R, Costa JBS (2009b) 1.94–1.93 Ga charnockitic magmatism from the central part of the Guiana Shield, Roraima, Brazil: single zircon evaporation data and tectonic implications. J S Am Earth Sci 27:247–257

Fraga LM, Dreher AM, Grazziottin H, Reis NJ (2011) Suíte Trairão – Arco Magmático de 2,03–2,04 Ga, na parte norte do Craton Amazônico. 12° Simpósio de Geologia da Amazônia, Boa Vista, Roraima: 1–4

Franks PC (1988) Radiometric dating, petrography, and thermal history of core samples from Vaupes NO. 1 Well, Leticia license, Colombia Amoco Production Company

Fuquén JA, Osorno JF (2002) Geología de la plancha 303 Colombia Departamentos de Huila, Tolima y Meta escala 1:100.000. Memoria explicativa. Ingeominas, Bogotá, 90 pp

Galvis J (1993) Los sedimentos precámbricos del Guainía y el origen de las ocurrencias auríferas en el Borde Occidental el Escudo de Guayanas. Geología Colombiana 18:119–136

Galvis J, Gómez LM (1998) Hierro bandeado en Colombia. Rev Acad Colomb Cienc 22(85):485–496

Galvis J, Huguett A, Ruge P (1979) Geología de la Amazonia Colombiana. Boletín Geológico INGEOMINAS 22(3):3–86

Gaudette H, Olszewski WJ, Hurley PM, Fairbairn HW (1978) Geology and age of the Parguaza Rapakivi granite, Venezuela. Geol Soc Am Bull 89:1335–1340

Gaudette HE, Olszewski WJ Jr (1985) Geochronology of the basement rocks, Amazonas Territory, Venezuela, and the tectonic evolution of the western Guiana shield. Geol Mijnb 64:131–143

Geraldes MC, Tavares A, Dos Santos A (2015) An overview of the Amazonian craton evolution: insights for Paleocontinental reconstruction. Int J Geosci 2015(6):1060–1076

Ghosh SK (1985) Geology of the Roraima Group and its implications, in Memoria Simposium Amazonico, 1st, Venezuela, 1981: Caracas, Venezuela, Dirección General Sectorial de Minas y Geologia, Publicación Especial 10, p. 22–30

Gibbs AK, Barron CN (1993) Geology of the Guiana shield. Oxford University Press, New York. 246 pp

Goldschmidt R, Marvin RF, Mehnert HH (1971) Radiometric ages in the Santander Massif, Eastern Cordillera, Colombian Andes. US Geol Surv Prof Pap 750-D:D44–D49

Gómez J, Nivia A, Montes NE, Jiménez DM, Sepúlveda MJ, Gaona T, Osorio J, Diederix H, Mora M, Velásquez ME (2007) Atlas Geológico de Colombia 1:500,000, Planchas 5–20 and 5–23, Ingeominas

González CF, Pinto H (1990) Petrografía del Granito de Parguaza y otras rocas precámbricas en el Oriente de Colombia. Geología Colombiana 17:107–121

Grande S (2012) Petrología y petrogénesis de las rocas neoproterozóicas del terreno Falconia. Geos 42:60–63

Grande S, Urbani F (2009) Presence of high-grade rocks in NW Venezuela of possible Grenvillian affinity. In: James, K. H., Lorente, M. A. & Pindell, J. L. (eds) The Origin and Evolution of the Caribbean Plate. Geological Society, London, Special Publications, 328, 533–548

Hackley PC, Urbani F, Karlsen AW, Garrity CP (2005) Geologic shaded relief map of Venezuela. USGS Open File Rep:2005–1038

Herrera-Bangerter (1989) Die proterozoischen rapakivigranite von El Parguaza, südliches Venezuela. Inaugural-Dissertation Universität Zürich, 213 pp
Hoffman PF (1991) Did the breakout of Laurentia turn Gondwanaland inside-out? Science 252:1409–1412
Huguett A, Galvis J, Ruge P (1979) Geología. In: La Amazonia colombiana y sus recursos. Proyecto Radargramétrico del Amazonas, Bogotá, 29–92
Hurley PM, de Almeida FFM, Melcher GC, Cordani UG, Rand JR, Kawashita K, Vandoros P, Pinson WH, Fairbairn HW (1967). Test of continental drift by comparison of radiometric ages. Science 157:495–500.
Ibáñez-Mejía M (2010) New U-Pb Geochronological insights into the Proterozoic tectonic evolution of northwestern South America: The Mesoneoproterozoic Putumayo orogen of Amazonia and implications for Rodinia reconstructions. Unpublished MSc thesis, University of Arizona, 67 pp
Ibáñez-Mejía M, Ruiz J, Valencia VA, Cardona A, Gehrels GE, Mora AR (2011) The Putumayo Orogen of Amazonia and its implications for Rodinia reconstructions: new U–Pb geochronological insights into the Proterozoic tectonic evolution of northwestern South America. Precambrian Res 191:58–77
Ibañez-Mejia M, Pullen A, Arenstein J, Gehrels GE, Valley J, Ducea Mihai N, Mora Andres R, Mark P, Joaquin R (2015) Unraveling crustal growth and reworking processes in complex zircons from orogenic lower-crust: the Proterozoic Putumayo Orogen of Amazonia. Precambrian Res 267:285–310
Ingeominas Geoestudios (1998–2001) Mapa geológico de Colombia: Geología de las planchas 367, 368, 389, 390, 411, 412, 414, 430, 431, 448, 449, 465, Escala 1:100,000, Informes inéditos, Ingeominas, Bogotá
Irving EM (1971) La evolución estructural de los Andes mas septentrionales de Colombia. Boletín de Geología (Bogota) 19(2):1–90
Jaramillo JM (1978) Rocas metamórficas de alto grado – granulites en algunas lavas del Nevado del Ruíz, Colombia. Resúmenes 2ndo Congreso Colombiano de Geología Bogotá, 16
Jaramillo JM (1980) Petrology and geochemistry of the Nevado del Ruiz volcano, northern Andes, Colombia. PhD thesis, Houston University, 167 pp
Jiménez DM (2003) Caracterização metamórfica e geocronológica das rochas proterozóicas do Maciço de Garzón – Sudeste dos Andes da Colômbia. Dissertação de mestrado, Universidade de São Paulo, Brasil, 167 pp
Jiménez DM, Juliani C, Cordani UG (2006) P-T-t conditions of high-grade metamorphic rocks of the Garzón Massif, Andean basement, SE Colombia. J S Am Earth Sci 21:322–336
Kroonenberg SB (1976) Amphibolite-facies and granulite-facies metamorphism in the Coeroeni-Lucie área, SW Surinam. Thesis Amsterdam, Mededelingen Geologisch Mijnbouwkundige Dienst Suriname, 25, 109–289
Kroonenberg SB (1980) Petrografía y edad de algunos gneises cordieríticos del Guainía, Amazonia colombiana. Revista CIAF Bogotá 5(1):213–218
Kroonenberg SB (1982) A Grenvillian granulite belt in the Colombian Andes and its relation to the Guiana shield. Geologie & Mijnbouw 61:325–333
Kroonenberg SB (1983) Litología, metamorfismo y origen de las granulitas del Macizo de Garzón. Cordillera Oriental Colombia Geología norandina Bogotá 6(39):46
Kroonenberg SB (1985) El borde occidental del Escudo de Guayana en Colombia. Memoria I Simposium Amazonico, Puerto Ayacucho. Venezuela; Boletin de Geología Caracas, Publicación Especial no. 10:51–63
Kroonenberg SB (1990) Geochemistry of the Garzón granulites in the Colombian Andes: Evidence for Proterozoic calc alkaline magmatism. Symposium International. 'Géodynamique Andine' 15 17 Mai 1990, Grenoble, France. pp. 383–385
Kroonenberg SB (2014) Geological evolution of the Amazonian Craton: Forget about geochronological provinces. Memorias Geological Map of South America Workshop, Villa de Leyva, Colombia: 22 and 109–131
Kroonenberg SB, de Roever EWF (2010) Geological Evolution of the Amazonian Craton, in: Amazonia, Landscape and Species Evolution. Edited by C. Hoorn and F.P. Wesselingh. Wiley, p. 9–28

Kroonenberg SB, Reeves CV (2012) Geology and petroleum potential, Vaupés-Amazonas Basin, Colombia. In: F. Cediel (Ed.,) Petroleum Geology of Colombia, 15. Universidad EAFIT, Medellín, 92 pp

Kroonenberg SB, de Roever EWF, Fraga LM, Reis NJ, Faraco MT, Lafon JM, Cordani UG, Wong TE (2016) Paleoproterozoic evolution of the Guiana shield in Suriname: a revised model. Netherlands J Geosci – Geologie en Mijnbouw 95:491–522

Leal H (2003) Estudio metalogenético de las mineralizaciones auríferas de Cerro Rojo, distrito minero de Taraira, departamento de Vaupés, Colombia. Universidad Nacional de Colombia, Bogotá

Lockwood JP (1965) Geology of the Serranía de Jarara Area. Guajira Peninsula, Colombia. Tesis Ph.D., Princeton Univ. New Jersey, 167p

López JA (2012) Unidades, petrografía y composición química del Complejo Migmatítico de Mitú en los alrededores de Mitú: Réplica Boletín de Geología, 34, 101–103

López JA, Khurama S, Bernal LE, Cuéllar M (2007) El Complejo Mitú: Una nueva perspectiva. Memorias XI Congreso Colombiano de Geología, CD Room. Bucaramanga, Santander

López J, Mora BM, Jiménez DM, Khurama S, Marín E, Obando G, Páez TI, Carrillo LE, Bernal VLE, Celada CM (2010) Cartografía geológica y muestreo geoquímico de las Planchas 297 – Puerto Inírida, 297 Bis – Merey Y 277 Bis – Amanaven, Departamento del Guainia. Ingeominas, Bogotá. 158 pp

López JA, Zuluaga C (2012) Neis de Macuirá: evolución tectónica de las rocas metamórficas paleozoicas de la Alta Guajira, Colombia. Boletín de Geología 34:15–36

MacDonald WD (1964) Geology of the Serranía de Macuira area, Guajira Peninsula, Colombia. Ph.D. dissertation, Princeton Univ., 167 pp

MacDonald WD, Hurley PM (1969) Precambrian gneisses from northern Colombia, South America. Geol Soc Am Bull 80:1867–1872

Marquínez G, Morales CJ, Caicedo JC (2002a) Mapa Geológico de Colombia Plancha 344 Tesalia Escala 1:100.000 Memoria Explicativa. Ingeominas, Bogotá, 154 pp

Marquínez G, Rodríguez YJ, Fuquen JA (2002b) Mapa Geológico de Colombia Plancha 365 Coconuco, Escala 1:100.000, Memoria explicative. Ingeominas, Bogotá, 116 pp

Martín-Bellizzia C (1972) Paleotectónica de1 Escudo de Guayana. Conf. Geol. Inter-Guayanas IX. Ciudad Guayana, Venezuela, Memoria. Bol Geol Publ Espec 6:251–305

Martínez JM (1985) Geología de la Subregión de San Carlos de Río Negro, Territorio Federal Amazonas. . Memoria I Simposium Amazonico, Puerto Ayacucho, Venezuela; Boletin de Geología Caracas, Publicación Especial no. 10, 65–71

Mendoza V (1974) Evolución tectónica del Escudo de Guayana. Segundo Congreso Latinoamericano de Geologia, Bol. Geol Caracas Publ Esp. no. 7, III, 2237–2270

Mirón-Valdespino OE, Álvarez VC (1997) Paleomagnetic and rock magnetic evidence for inverse zoning in the Parguaza batholith (southwestern Venezuela) and its implications about tectonics of the Guyana. Precambrian Res 85:1–25

Mojica J, Villaroel C, Macia C (1987) Nuevos afloramientos fosilíferos del Ordovícico Medio (Fm. El Higado) al oeste de Tarqui, Valle Superior del Magdalena (Huila, Colombia). Geología Colombiana 16:95–97

Montes C, Guzmán G, Bayona G, Cardona A, Valencia V, Jaramillo C (2010) Clockwise rotation of the Santa Marta massif and simultaneous Paleogene to Neogene deformation of the Plato-San Jorge and Cesar-Ranchería basins. J S Am Earth Sci 29:832–848

Mosquera D, Núñez A, Vesga CJ (1982) Reseña explicativa del mapa geológico preliminar Plancha 244 Ibagué Escala 1:100.000. Ingeominas, Bogotá. 27 pp

Muñoz J, Vargas H (1981a) Petrología de las anfibolitas y neises precámbricos presentes entre los ríos Mendarco y Ambeima, Tolima, Colombia. Universidad Nacional de Colombia, Bogotá

Muñoz J, Vargas H (1981b) Petrología de las anfibolitas y neises precámbricos presentes entre los ríos Mendarco-Ambeima. Tercer Congreso Colombiano de Geología, (p. 43). Medellín

Murillo A, Esquivel J, Flores D, Arboleda C (1982) Geología de la Plancha 281 Río Blanco. Ingeominas, Bogotá

Murcia A (2002) Reconocimiento geológico en el Macizo de Garzón. Publicaciones geológicas especiales de Ingeominas 24, 58 pp

Nadeau S, Chen W, Reece J, Lachhman D, Ault R, Faraco MTL, Fraga LM, Reis NJ, Betiollo LM (2013) Guyana: the lost Hadean crust of South America? Brazilian J Geol 43:601–606

Núñez A (2003) Cartografía geológica de las zonas Andina Sur y Garzón – Quetame (Colombia). Reconocimiento geológico regional de las planchas 411 La Cruz, 412 San Juan De Villalobos, 430 Mocoa, 431 Piamonte, 448 Monopamba, 449 Orito Y 465 Churuyaco Departamentos de Caquetá, Cauca, Huila, Nariño y Putumayo. Ingeominas, Bogotá. 298 pp

Obando GJ (2006) Procesamiento e interpretación de magnetometría aérea, proyecto Oriente Colombiano. In: Celada, C.M. et al. (eds) 2006, Potencial de recursos minerales en el Oriente colombiano: compilación y análisis de la información geológica disponible, 183–215

Ochoa A, Ríos P, Cardozo AM, Cubides JV, Giraldo DF, Rincón HD, Mendivelso D (2012) Cartografia geológica y muestreo geoquímico de las planchas 159, 160, 161, 179, 180 y 181 Puerto Carreño, Vichada. Memoria Explicativa. Ingeominas, Bogotá 2012, 127 pp

Ordóñez O, Pimentel MM, de Moraes R, Restrepo JJ (1999) Rocas grenvillianas en la región de Puerto Berrío, Antioquia. Revista de la Academia Colombiana de. Ciencias 23(87):225–232

Ordóñez-Carmona O, Pimentel MM, de Moraes R, Restrepo JJ (1999) Rocas grenvillianas en la región de Puerto Berrío, Antioquia. Revista de la Academia Colombiana de. Ciencias 23(87):225–232

Ordóñez-Carmona O, Pimentel MM, Correa AM, Martens UC, Restrepo JJ (2001) Edad Sm/Nd del metamorfismo de alto grado de El Retiro (Antioquia). In: VIII Congreso Colombiano de Geología, Manizales, Colombia, CD-ROM

Ordóñez O, Pimentel MM, de Moraes R (2002) Granulitas de Los Mangos, un fragmento grenvilliano en la parte oriental de la Sierra Nevada de Santa Marta. Revista de la Academia Colombiana de Ciencias Exactas, Físicas y Naturales 26(99):169–179

Ordóñez-Carmona O, Restrepo-Alvarez JJ, Pimentel PM (2006) Geochronological and isotopical review of pre-Devonian crustal basement of the Colombian Andes. J S Am Earth Sci 21:372–382

París G, Marín PA (1979) Generalidades acerca de la geología del Departamento del Cauca. Ingeominas, Bogotá. 38 pp

Pearce JA, Harris NBW, Tindle AG (1984) Trace element discrimination diagrams for the tectonic interpretation of granitic rocks. J Petrol 25:956–983

Peccerillo A, Taylor SR (1975) Geochemistry of Upper Cretaceous volcanic rocks from the Pontic chain, northern Turkey. Bull Volcanol 39:557–569

Pérez H, Salazar R, Peñaloza A, Rodríguez S (1985) Evaluación preliminar geo-económica de los aluviones presentando minerals de Ti, Sn, Nb y Ta del territorio de Boquerones y Aguamena, Distrito cedeño, Estado Bolívar y Territorio Federal Amazonas. Memoria I Simposium Amazonico, Puerto Ayacucho, Venezuela; Boletin de Geología Caracas, Publicación Especial no 10, 587–602

Pinheiro SdaS, Fernandes PECA, Pereira ER, Vasconcelos EG, Pinto AdoC, de Montalvão RMG, Issler RS, Dall'Agnol R, Teixeira W, Fernandes CAC (1976) Geologia, in: Projeto Radambrasil. Levantamento de Recursos Naturais Vol 11, Folha NA.19, Pico da Neblina, 17–137

Pinson WH, Hurley PM, Mencher E, Fairbairn HW (1962) K-Ar and Rb-Sr ages of biotites from Colombia, South America. Geol Soc Am Bull 73:807–910

Ponce A (1979) Anotaciones sobre la geología del Suroriente del Departamento de Nariño. Informe inédito 1769, Ingeominas, 44 pp

Priem HNA, Hebeda EH, Boelrijk NAIM, Verschure RH, Verdurmen EAT (1968) Isotopic age determination on Surinam rocks, 4 ages of basement rocks in north-western Surinam and of the Roraima tuff at Tafelberg. Geol Mijnb 47:191–196

Priem HNA, Boelrijk NAIM, Hebeda EH, Verdurmen EAT, Verschure RH (1971) Isotopic ages of the trans-Amazonian acidic magmatism and the Nickerie episode in the Precambrian basement of Surinam, South America. Geol Soc Am Bull 82:1667–1680

Priem HNA, Boelrijk NAIM, Hebeda EH, Verdumen EAT, Verschure RH (1973) Age of the Precambrian Roraima formation in northeastern South America: evidence from isotopic dating of Roraima pyroclastic volcanic rocks in Suriname. Geol Soc Am Bull 84:1677–1684

Priem HNA, Boelrijk NAIM, Hebeda EH, Kroonenberg SB, Verdurmen EAT, Verschure RH (1977) Isotopic ages of the high-grade metamorphic Coeroeni group, SW Suriname. Geol Mijnb 56:155–160

Priem HNA, Andriessen PAM, Boelrijk NAIM, De Boorder H, Hebeda EH, Huguett A et al (1982) Geochronology of the precambrian in the Amazonas region of southwestern Colombia western Guiana shield. Geologie & Mijnbouw:229–242

Priem HNA, Kroonenberg SB, Boelrijk NAIM, Hebeda EH (1989) Rb Sr evidence for the presence of 1.6 Ga basement underlying the 1.2 Ga Garzón Santa Marta Granulite Belt in the Colombian Andes. Precambrian Res 42:315–324

Radelli L (1961) El basamento cristalino de la Península de La Guajira. Boletín geológico 8:5–32

Radelli L (1962a) Introducción al estudio de la petrografía del Macizo de Garzón (Huila – Colombia). Geología Colombiana 3:17–46

Radelli L (1962b) Introducción al estudio de la geología y petrografía del Macizo de Santa Marta. Geología Colombiana 2:41–115

Radelli L (1967) Géologie des Andes Colombiennes. Travaux du Laboratoire Sciences de la Terre, Grenoble, no. 6, 483 pp

Ramos V (2010) The Grenville-age basement of the Andes. J S Am Earth Sci 29:77–91

Reis NJ, Almeida ME, Riker SL, Ferreira AL (2006) Geologia e Recursos minerais do Estado do Amazonas. (Convênio CPRM/CIAMA). 125 p., il. Escala 1:1.000.000. Manaus, CPRM

Reis NJ, de MSG F, Fraga LM, Haddad RC (2000) Orosirian calc-alkaline volcanism and the Orocaima event in the Northern Amazônian Craton, Eastern Roraima State. Brazil Rev Bras Geociênc 30(3):38–383

Reis NJ, Fraga LM, de Faria MSG, Almeida ME (2003) Geologia do Estado de Roraima, Brasil. Géologie de la France 2003 2-3-4:121–134

Renzoni G (1989a) Comparación entre las secuencias metasedimentarias de la Serranía de Naquén y de la Serra Da Jacobina. Boletín Geológico 302:25–42

Renzoni G (1989b) La secuencia aurífera de la Serranía de Naquén. Boletín Geológico 302:43–103

Restrepo JJ, Ordóñez-Carmona O, Armstrong R, Pimentel MM (2011) Triassic metamorphism in the northern part of the Tahamí Terrane of the central cordillera of Colombia. J S Am Earth Sci 32(4):497–507

Restrepo JJ, Ordóñez-Carmona O, Martens U, Correa AM (2009) Terrenos, complejos y provincias en la Cordillera Central de Colombia. I+D, 9, 2, 49–56

Restrepo-Pace PA (1995) Late Precambrian to Early Mesozoic tectonic evolution of the Colombian Andes, based on new geochronological geochemical and isotopic data. Dissertation, Univ. Arizona, 198 pp

Restrepo-Pace PA, Ruiz J, Gehrels G, Costa M (1997) Geochronology and Nd isotopic data of Grenville-age rocks in the Colombian Andes: new constraints for Late Proterozoic-Early Paleozoic paleocontinental reconstructions of the Americas. Earth Planet Sci Lett 150:427–441

Restrepo-Pace PA, Cediel F (2010) Northern South America basement tectonics and implications for paleocontinental reconstructions of the Americas. J S Am Earth Sci 29:764–771

Ríos JH (1972) Geología de la región de Caicara, Estado Bolívar: IV Congreso Geológico Venezolano, Caracas 1971, Memoria, Publicación Especial 5(3):1759–1782

Rivas D (1985) Geología de la subregion Atabapo, Territorio Federal Amazonas, Venezuela. Memoria I Simposium Amazonico, Puerto Ayacucho, Venezuela; Boletin de Geología, Publicación Especial no. 10, 122–139

Rivers T (1997) Lithotectonic elements of the Grenville Province: review and tectonic implications. Precambrian Res 86:117–154

Rodríguez G (1995a) Petrografía y microtexturas del Grupo Garzón y el granito de anatexis de El Recreo, Macizo de Garzon, Cordillera Oriental – Colombia. Revista Ingeominas 5:17–36

Rodríguez G (1995b) Petrografía del Macizo de La Plata. Revista Ingeominas 5:5–16

Rodríguez G, Albarracin HA (2012) Cartografía, petrografía y geoquímica de granulitas básicas en el segmento norte de la Cordillera Central de Colombia, denominada Granulitas de San Isidro. Geología Colombiana 37:95–110

Rodríguez G, González H, Restrepo JJ, Martens U, Cardona JD (2012) Occurrence of granulites in the northern part of the Western Cordillera of Colombia. Boletín de Geología 34:37–53

Rodríguez G, Londoño AC (2002) Mapa geológico del Departamento de La Guajira. Geología, recursos minerales y amenazas potenciales, Escala 1:250.000. Medellín, 259 pp

Rodríguez G, Sepúlveda J, Ortiz FH, Ramírez C, Ramos K, Bermúdez JG, Sierra MI (2010) Mapa Geológico Plancha 443 Mitú – Vaupés. Ingeominas

Rodríguez G, Zapata M, Velásquez E, Cossio U, Londoño AC (2003) Geología de las planchas 367 Gigante, 368 San Vicente del Caguán, 389 Timaná, 390 Puerto Rico, 391 Lusitania (Parte Noroccidental)y 414 El Doncello, Departamentos de Caquetá y Huila. Ingeominas, Medellín. 168 pp

Rodríguez G, Sepúlveda J, Ramírez C, Ortiz FH, Ramos K, Bermúdez JG, Sierra MI (2011a) Cartografía geológica y exploración geoquímica de la plancha 443 Mitú. INGEOMINAS, Bogotá. 169p

Rodríguez G, Sepúlveda J, Ramírez C, Ortiz FH, Ramos K, Bermúdez JG, Sierra MI (2011b) Unidades, petrografía y composición química del Complejo Migmatítico de Mitú en los alrededores de Mitú. Boletín de Geología 33:27–42

Rodríguez SE, Áñez G (1978) Los depósitos de mena titanífera de San Quintín Central, Estado Yaracuy; génesis, caracteres geológicos y estimación de reservas. Min. Energía y Minas, Dir. General Sect. Minas y Geol. Venezuela 13:83–181

Rosa-Costa LT, Ricci PSF, Lafon J-M, Vasquez ML, Carvalho JMA, Klein EL, Macambira EMB (2003) Geology and geochronology of Archean and Paleoproterozoic domains of southwestern Amapá and north-western Pará, Brazil, southeastern Guiana shield. Géologie de la France 2-3-4: 101–120

Ruiz J, Tosdal RM, Restrepo PA, Murillo-Muñetón G (1999) Pb isotope evidence for Colombia-southern México connections in the Proterozoic. Geol Soc Am Spec Pap 336:183–197

Santos JOS (2003) Geotectônica dos Escudos das Guianas e Brasil-Central. In: Bizzi LA, Schobbenhaus C, Vidotti RM, Gonçalves JH (eds) Geologia, Tectônica e Recursos Minerais do Brasil. CPRM, Brasília, pp 169–195

Santos JOS, Hartmann LA, Gaudette HE, Groves DI, McNaughton NJ, Fletcher IR (2000) A new understanding of the provinces of the Amazon craton based on integration of field mapping and U-Pb and Sm-Nd geochronology. Gondwana Res 3:453–488

Santos JOS, Potter PE, Reis NJ, Hartmann LA, Fletcher IR, McNaughton NJ (2003) Age, source and regional stratigraphy of the Roraima Supergroup and Roraima-like outliers in northern South America based on U-Pb geochronology. Geol Soc Am Bull 115:331–348

Santos JOS, Reis NJ, Chemale F, Hartmann LA, Pinheiro SS, McNaughton NJ (2004) Paleoproterozoic evolution of northwestern Roraima state – absence of Archean crust, based on U-Pb and Sm-Nd isotopic evidence. Short papers IV South American symposium on isotope. Geology:278–281

Santos JOS, Hartmann LA, Faria MS, Riker SR, Souza MM, Almeida ME, McNaughton NJ (2006) Compartimentação do Cráton Amazonas em províncias: avanços ocorridos no período 2000–2006. Simpósio de Geologia da Amazônia, vol. 9, Sociedade Brasileira de Geologia, Bel'em, Brazil, Resumos Expandidos, CD ROM

Schobbenhaus C, Hoppe A, Lork A, Baumann A (1994) Idade U/Pb do magmatismo Uatumã no norte do Cráton Amazônico, Escudo das Guianas, Brasil: primeiros resultados. In: Congr. Bras. Geol. 38. Anais. Balneário Camboriu, 2, 395–397

Sidder GB, Mendoza V (1991) United States Department of the interior U.S. geological survey geology of the Venezuelan Guayana Shield and its relation to the Entire Guayana Shield Open-File Report 91–141, 62 pp

Stibane F, Forero A (1969) Los afloramientos del Paleozóico en La Jagua (Huila) y Río Nevado (Santander del Sur). Geología Colombiana 6:31–66

Tassinari CCG (1981) Evolução geotectônica da Província rio Negro-Juruena na região amazônica. Dissertaçao de mestrado, Instituto de Geociências, Universidade de São Paulo, 99 pp

Tassinari CCG, Cordani UG, Nutman AP, van Schmus WR, Bettencourt JS, Taylor PN (1996) Geochronological systematics on basement rocks from the Rio Negro-Juruena Province Amazonian Craton, and tectonic implications. Int Geol Rev 38:161–175

Tassinari CCG, Macambira MJB (1999, 1999) Geochronological provinces of the Amazonian Craton. Episodes:174–182

Tassinari CCG, Bettencourt JS, Geraldes MC, Macambira MJB, Lafon JM (2000) The Amazonian Craton. In: Cordani, U.G. et al. (Eds.), Tectonic evolution of South America. 31st International Geological Congress, Rio de Janeiro, Brazil: 41–95

Tassinari CCG, Munhá JMU, Teixeira W, Nutman A, Palacios T, Sosa SC, Calado BO (2004a) Thermochronological history of the Imataca complex, NW Amazonian Craton. Short papers IV South American Symposium on Isotope Geology :121–123

Tassinari CCG, Munhá JMU, Teixeira W, Palacios T, Nutman A, Sosa SC, Santos AP, Calado BO (2004b) The Imataca complex, NW Amazonian Craton, Venezuela: crustal evolution and integration of geochronological and petrological cooling histories. Episodes 27:3–12

Toussaint JF (1993) Evolución geológica de Colombia. Universidad Nacional de Colombia, Medellín. 129 pp

Trümpy D (1943) PreCretaceous of Colombia. Geol Soc Am Bull 54:1281–1304

Tschanz CM, Marvin RR, Cruz J, Mehnert H, Cebula GT (1974) The geologic evolution of the sierra Nevada de Santa Marta, northeastern Colombia. Geol Soc America Bull 85:273–284

Urueña CL, Zuluaga CA (2011) Petrografía del Neis de Bucaramanga en cercanías a Cepitá, Berlín y Vetas – Santander. Geología Colombiana 36:37–56

Van der Lelij R (2013) Reconstructing north-western Gondwana with implications for the evolution of the Iapetus and Rheic Oceans: a geochronological, thermochronological and geochemical study. Thèse de doctorat : Univ. Genève, 2013, no. Sc. 4581

Van der Lelij R, Spikings R, Ulianov A, Chiaradia M, Mora A (2016) Palaeozoic to Early Jurassic history of the northwestern corner of Gondwana, and implications for the evolution of the Iapetus, Rheic and Pacific oceans. Gondwana Res 31:271–294

Van der Wiel AM (1991) Uplift and volcanism of the SE Colombian Andes in relation to Neogene sedimentation in the Upper Magdalena Valley. Thesis, Wageningen, 208 pp

Velandia F, Ferreira P, Rodríguez G, Núñez A (2001) Levantamiento geológico de la plancha 366 Garzón, escala 1:100.000. Memoria explicativa, Ingeominas, Bogotá, 82 pp

Vesga CJ, Barrero D (1978) Edades K/Ar en rocas igneas y metamórficas de la Cordillera Central de Colombia y su implicación geológica. Resúmenes II Congreso colombiano de Geología, Bogotá, p 19

Villagómez D (2010) Thermochronology, geochronology and geochemistry of the Western and Central cordilleras and Sierra Nevada de Santa Marta, Colombia: the tectonic evolution of NW South America, Thèse de doctorat : Univ. Genève, no. Sc.4277, 166 pp

Villagómez D, Spikings R, Mora A, Guzmán G, Ojeda G, Cortés E, van der Lelij R (2011) Vertical tectonics at a continental crust-oceanic plateau plate boundary zone: Fission track thermochronology of the Sierra Nevada de Santa Marta, Colombia. Tectonics, 30, TC4004. https://doi.org/10.1029/2010TC002835

Ward DE, Goldschmidt R, Cruz J, Restrepo H (1973) Geología de los cuadranulos H-12 Bucaramanga y H-13 Pamplona Departamento Santander, Colombia. Boletín Geológico 1973, 1–3, 1–132

Ward DE, Goldsmith R, Cruz J, Restrepo A (1974) Geology of Quadrangles H-12, H-13, and parts of 1-12 and 1-13,(zone III) in northeastern Santander department, Colombia. US Geol Surv Open File Rep 74-258:466

Wynn JC (1993) Geophysics of the Venezuelan Guayana Shield, In: Geology and Mineral Resource Assessment of the Venezuelan Guayana Shield, by USGS and CVG Técnica Minera CA. USGS Bulletin 2062: 17–27

Zapata S, Cardona A, Montes C, Valencia V, Vervoort J, Reiners P (2014) Provenance of the Eocene Soebi Blanco formation, Bonaire, Leeward Antilles: correlations with post-Eocene tectonic evolution of northern South America. J S Am Earth Sci 52:179–193

Zuluaga CA, Amaya S, Urueña C, Bernet M (2017) Migmatization and low-pressure overprinting metamorphism as record of two pre-Cretaceous tectonic episodes in the Santander Massif of the Andean basement in northern Colombia (NW South America). Lithos 274–275:123–146

Part III
Early Paleozoic Tectono-Sedimentary History

Part III

Chapter 4
Ordovician Orogeny and Jurassic Low-Lying Orogen in the Santander Massif, Northern Andes (Colombia)

Carlos A. Zuluaga and Julian A. Lopez

Abbreviations

bt	Biotite
CAP	Continental Arcs Potassic
grt	Garnet
hbl	Hornblende
IK	Kübler crystallinity index
IOP	Initial Oceanic Arcs Potassic
kfs	Potassium feldspar
LOP	Late Oceanic Arcs Potassic
MGV	"Metasedimentitas de Guaca, La Virgen"
ms	Muscovite
PAP	Post-collisional Arcs Potassic
pl	Plagioclase
PT	Pressure and temperature
qz	Quartz
sil	Sillimanite
S_{n+1}	Oldest recognized metamorphic foliation, can be followed by progressively younger foliations (S_{n+2}, S_{n+3}, etc.)
syn-COLG	Syn-collisional granites
ttn	Titanite
VAG	Volcanic arc granites
WR	Whole rock

C. A. Zuluaga (✉) · J. A. Lopez
Departamento de Geociencias, Universidad Nacional de Colombia, Bogotá, Colombia
e-mail: cazuluagacas@unal.edu.co

4.1 Introduction

This chapter focuses on the pre-Cretaceous tectonic history of the crystalline basement of the Santander Massif. Particularly, we discuss the presence of an Ordovician orogenic event in Colombia and the implications of the presence of the western Pangea subduction zone and a magmatic arc setting and its possible relation to Triassic-Jurassic magmatic arc in north-central Colombia. We provide arguments for a Triassic-Jurassic low-lying magmatic arc product of oblique convergence. Evidence is drawn from observations in igneous and metamorphic rocks from a crystalline core in the Santander Massif. This study has implications for the characteristics of the development of Mesozoic sedimentary basins in Colombia that has been explained by extensional tectonics and a major marine transgression coetaneous with the growth of magmatic arcs toward the western border of the basins (Sarmiento-Rojas et al. 2006). The understanding of the crystalline basement is relevant to address the spatial and temporal relationship between plate tectonics, a major extensional tectonic event, and the beginning of basin development during Early to Middle Jurassic.

4.2 Geologic Background

Several large isolated basement blocks with presumably Proterozoic rocks are observed in the Colombian Andean system along the eastern and central ranges. The Santander Massif (Fig. 4.1) is one of these major blocks, the massif is located close to where the Eastern Cordillera branches to the Merida Andes of Venezuela, and it is characterized as an uplifted block located between an east vergent thrust system and a NNW sinistral strike slip fault with an inverse west vergent component (Bucaramanga fault). Migmatitic gneisses with reported <1.71 Ga protolith ages (Cordani et al. 2005; Ordóñez-Carmona et al. 2006) are the oldest lithologies in the massif. Two major tectonic pulses (early Paleozoic and Late Triassic-Early Jurassic) have been identified in the rocks of the massif. The early Paleozoic pulse occurred during the Ordovician; this event has been correlated with the Caparonensis orogeny in Venezuela and with the Famatinian orogenic event in southern South America (Mantilla et al. 2016; Restrepo-Pace 1995); this pulse is named as Quetame-Caparonensis by Restrepo-Pace and Cediel (2010). The Ordovician orogenic pulse in the massif is characterized by the occurrence of amphibolite to granulite facies metamorphism and migmatization (Zuluaga et al. 2017). The Late Triassic-Early Jurassic pulse is related to a low-lying orogen with numerous shallow granitoids (van der Lelij et al. 2016; Zuluaga et al. 2015). These granitoids were generated in a major magmatic episode that probably lasted from 200 to 129 Ma (Bustamante et al. 2016) in a stationary arc active for ~ 40 m.y. (Cochrane et al. 2014). The magmatic episode also produced low-grade metamorphism of Early Devonian sedimentary rocks and overprinted the early Paleozoic amphibolite-granulite facies basement

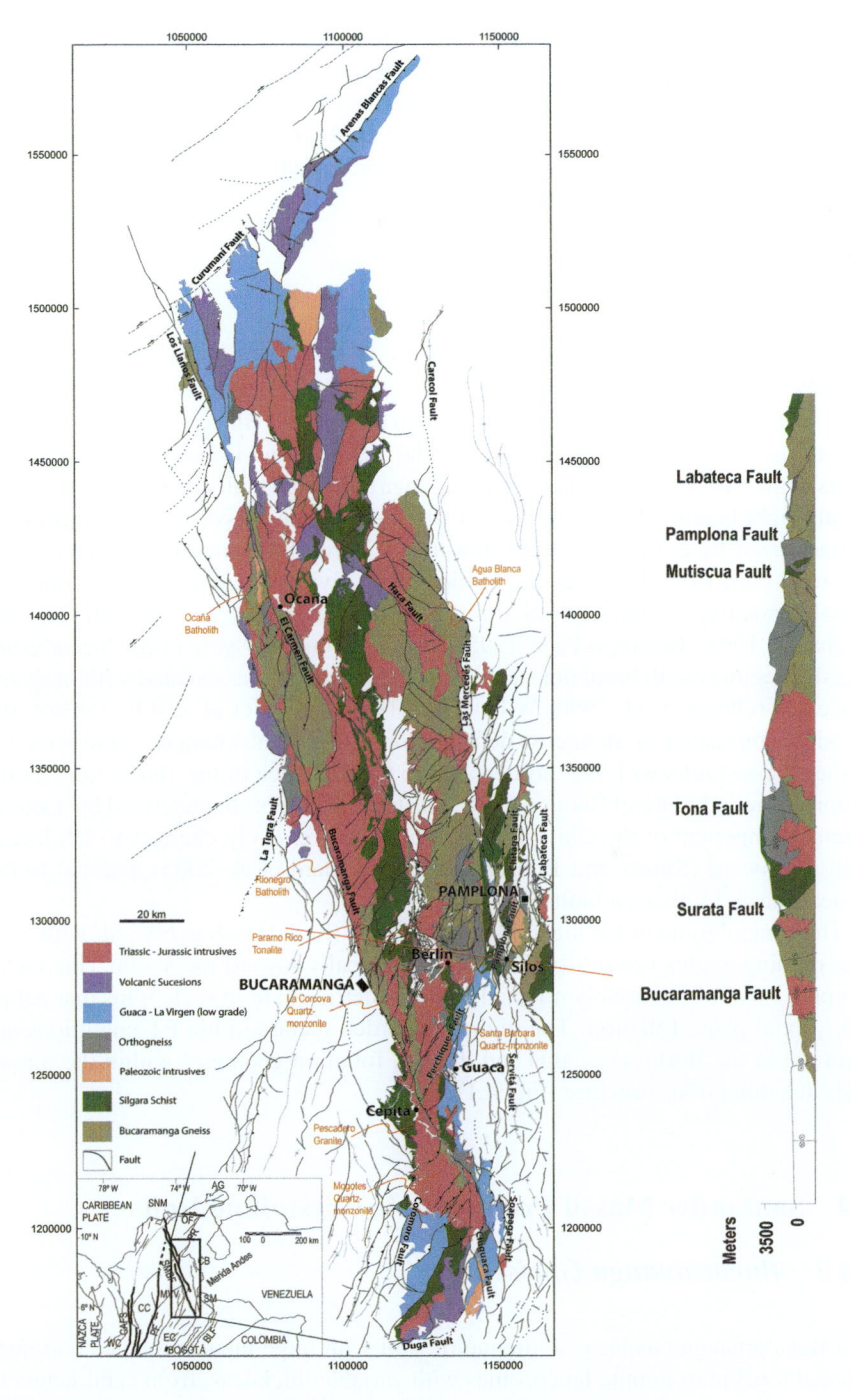

Fig. 4.1 Basement geology map of the Santander Massif with crystalline units and main structures. Compiled and modified from Ward et al. (1977a, b, c, d), Daconte and Salinas (1980a, b), Vargas and Arias (1981a, b), Vargas et al. (1984), Pulido (1985), and Fúquen et al. (2010)

rocks (Zuluaga et al. 2017). Several other tectonic events affected the northwestern South America margin after Jurassic arc magmatism; data and interpretations for these events are presented by several authors and will not be addressed here (see, e.g., Mantilla et al. 2009, 2011, 2013; Bissig et al. 2014; Leal et al. 2011; Amaya et al. 2017; Van der Lelij et al. 2016 and references therein).

4.3 Santander Massif Structure

The main structural features that define the boundaries of the Santander Massif are the NW Bucaramanga-El Carmen-Los Llanos fault system to the west (left lateral with a reverse component), the NE Curumaní and Arenas Blancas faults to the north (right lateral and reverse, respectively), the NS Soapaga-Labateca-Las Mercedes faults to the east (reverse), and the NE Soapaga and Duga faults to the south (reverse, locally right lateral) (Fig. 4.1). Present structures in the massif imprint a geometry characterized by converging faults toward the core of the massif (Kammer 1993; Restrepo-Pace 1995; Kammer and Mojica 1996) giving the appearance of compressional horsts like the known back thrust and shortcut thrust geometry (McClay and Buchanan 1992; Restrepo-Pace 1995). These geometries generated during compressive regimes with basal detachments and are probably associated with orogenic processes (Oldow et al. 1990; Willet et al. 1993; Butler et al. 2011). Inward, the massif is dominated by an intense deformation, locally with changes in the vergence of the reverse faults with NS attitude, due to the influence of the "horse tail" termination of the right lateral Boconó fault. This termination is characterized by a strong reverse component of the Chitaga fault, which progressively changes to left lateral fault southward (Singer and Beltrán 1996; Audemard et al. 2005), parallel to the Chucarima and Labateca faults.

The general trend of the metamorphic foliation (S_{n+1}) is N-S to NW-SE with variable dipping angles toward the E and W and locally toward the south. This metamorphic foliation is variably affected by asymmetric cm to m scale folds that define a S_{n+2} axial plane foliation. The highest recorded metamorphic PT conditions are recorded in the Berlín area, this and the S_{n+1} foliation geometry within the massif hints at a dome-like structure (Fig. 4.1).

4.4 Santander Massif Pre-mesozoic Crystalline Units

4.4.1 Bucaramanga Gneiss

The Bucaramanga Gneiss is a migmatitic unit with large tabular foliated and non-foliated tonalite to granite leucosomes with variable thickness, from centimeters to meters. Foliated leucosomes are commonly concordant with the general foliation trend, while non-foliated leucosomes show discordant relationships with mesosome foliation (Fig. 4.2a). Small leucosome lenses and patches are also observed mostly in pelites and quartz-feldspathic lithologies, and, in some cases, they are observed as thin layers in stromatic migmatites (Fig. 4.2b, c); these leucosomes are abundant in the Berlín area in the core of the dome-like structure. Mesosomes are represented by hornblende-bearing gneiss/schist, amphibolite, mica-rich gneiss/schist, and quartz-feldspar-rich gneiss with variable amounts of cordierite, sillimanite, and garnet (Fig. 4.2d–i); these lithologies have regional continuity in NW-SE to NE-SW trending elongated packets that extends N-S within the massif. Bucaramanga Gneiss outcrops in the Berlín area contain up to 50–60% of leucosomes; the leucosome proportion reduces to less than 10% at the edges of the massif (i.e., in the Bucaramanga area and in the Cepitá area). Temperature-pressure estimations

Fig. 4.2 Outcrop photographs (top), hand-sample photographs (middle), and thin-section micrographs (bottom) showing some characteristics of the Bucaramanga Gneiss. (**a**) Several injected concordant and discordant leucosomes within an amphibolite, (**b**) a concordant leucosome in a quartz-feldspar gneiss with abundant leucosome lenses, (**c**) anatexite with abundant leucosomes, (**d, g**) hbl-bt-kfs-pl-qz gneiss, (**e, h**) ttn-pl-hbl amphibolite, (**f, i**) sil-grt-pl-bt-ms-qz schist; these images exemplify the most common mineral assemblages of the Bucaramanga Gneiss. Mineral abbreviations after Whitney and Evans (2010)

indicate that this unit reached metamorphic peak conditions with temperatures up to ca. 800 °C above the wet pelite solidus (Fig. 4.8) consistent with the abundance of leucosomes in the metamorphic core (Zuluaga et al. 2017). The age of metamorphism is interpreted by van der Lelij et al. (2016) in the range 490–450 Ma according to U-Pb zircon ages from gneisses and leucosomes in migmatites from the Bucaramanga Gneiss.

4.4.2 "Orthogneiss" Unit

This unit groups several felsic to intermediate orthogneiss bodies observed in the Santander Massif (Ward et al. 1973). Foliation in these bodies is concordant with the foliation of the other two medium- to high-grade metamorphic units present in the massif (Silgará Schist and Bucaramanga Gneiss). One of the largest bodies is found at the core of the massif in the Berlín area (informally called here Berlín Orthogneiss; Fig. 4.3a–c); there, the orthogneiss shows a strong compositional similarity with leucosomes from the migmatitic Bucaramanga Gneiss. The Berlín Orthogneiss is mainly a mesoscopically layered coarse- to fine-grained quartz+plagioclase+K-feldspar gneiss with variable amounts of garnet, biotite, muscovite, sillimanite, and magnetite; it contains prominent quartz+feldspar lenses within an anastomosing mica-rich matrix (Fig. 4.3a).

The protolith of the Berlín Orthogneiss is an S-type granitoid (granite to tonalite) with a calc-alkaline peraluminous trend (Jiménez 2016). In the Berlín area, the contact between the orthogneiss and the Bucaramanga Gneiss is transitional with lenticular and tabular foliated leucosomes compositionally similar to the foliated leucosomes within the Bucaramanga Gneiss and blocks of pelitic gneisses within the orthogneiss. The gneiss has commonly the mineral association plagioclase + quartz + K-feldspar + biotite with variable content of amphibole, garnet, sillimanite, cordierite, magnetite, titanite, clinopyroxene (diopside), and chrysoberyl (Fig. 4.3d–i). Reported ages in the Berlín Orthogneiss and other related bodies include (i) 450 ± 80 Ma (WR Rb/Sr) in a gneissic granite from the Caraba River (Goldsmith et al. 1971), (ii) 413 ± 30 Ma (Hb K/Ar) in a metadiorite near the Ocaña city (Goldsmith et al. 1971), (iii) 472 ± 3.4 Ma (zircon U/Pb laser ablation; Van der Lelij 2013), and (iv) 465 to 421 Ma (WR K/Ar; Forero 1990).

4.4.3 Silgara Schist

The Silgará Schist is largely a pelitic to semipelitic metasedimentary sequence with mineral associations that indicate epidote to amphibolite facies metamorphism (Fig. 4.4). It consists of five broad lithological groups: marble and calc-silicate schist, amphibolite, quartzite, and pelitic to semipelitic schist.

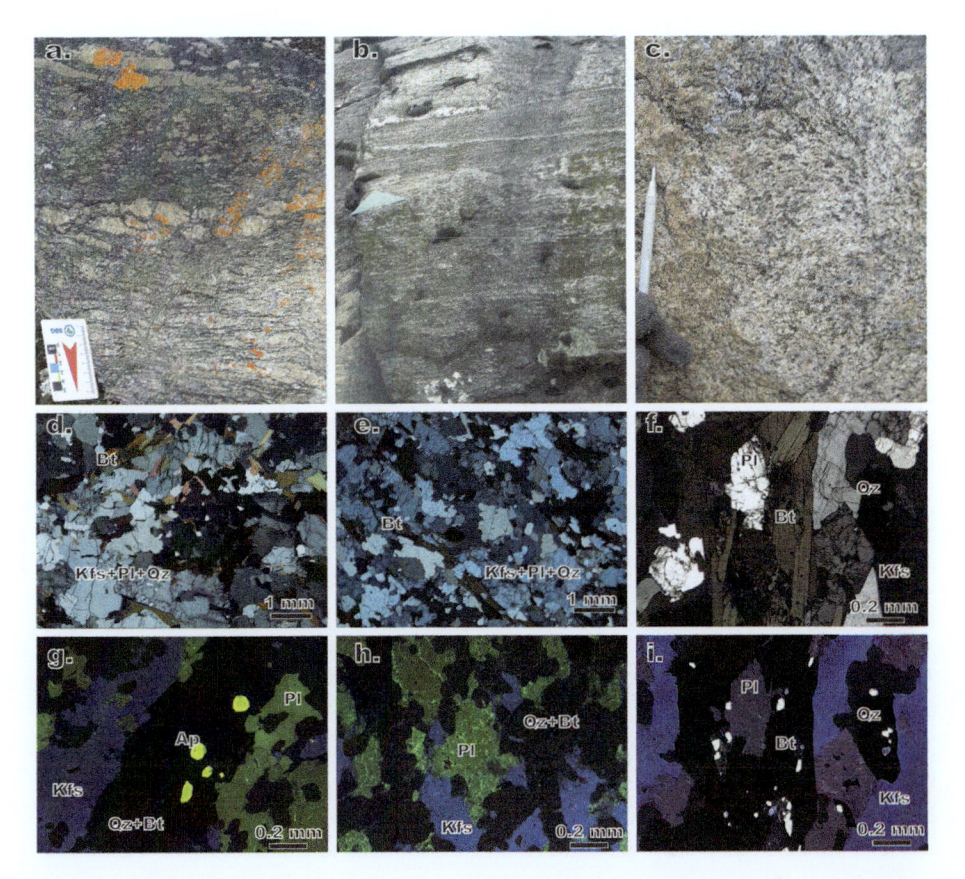

Fig. 4.3 Outcrop photographs (top), thin-section micrographs (middle), and cathodoluminescence images (bottom) showing some characteristics of the Berlín Orthogneiss. (**a, d**, and **e**) Layered coarse-grained bt-kfs-pl-qz gneiss. (**b, f**, and **g**) Layered fine-grained bt-kfs-pl-qz gneiss. (**c, h**, and **i**) Fine-grained pl-bt-kfs-qz gneiss; this lithology shows no layering and has more abundant biotite than layered gneisses

In the Mutiscua area, the sequence consists of pelitic schist (Fig. 4.4a), quartzite, amphibolite, and calc-silicate rocks with increasing metamorphic grade in the W-SW direction, passing through the garnet, staurolite, and kyanite zones in pelitic rocks and through Ca-amphibole, zoisite, and diopside zones in calc-silicate rocks. In the Cepitá area, the base of the sequence is rich in quartzite, the top shows more predominance of semipelites and pelites, and amphibolite lenses locally present in the middle to the base of the sequence (Fig. 4.4b–i); metamorphic grade increases toward the east from the garnet zone in the westernmost outcrops to the staurolite zone near the Chicamocha River. Reported temperature-pressure estimates range between ~500 and ~650 °C with pressures above 4 kbars probably reaching pressures as high as 7.2 kbars in the high-amphibolite facies rocks (Fig. 4.8; Castellanos

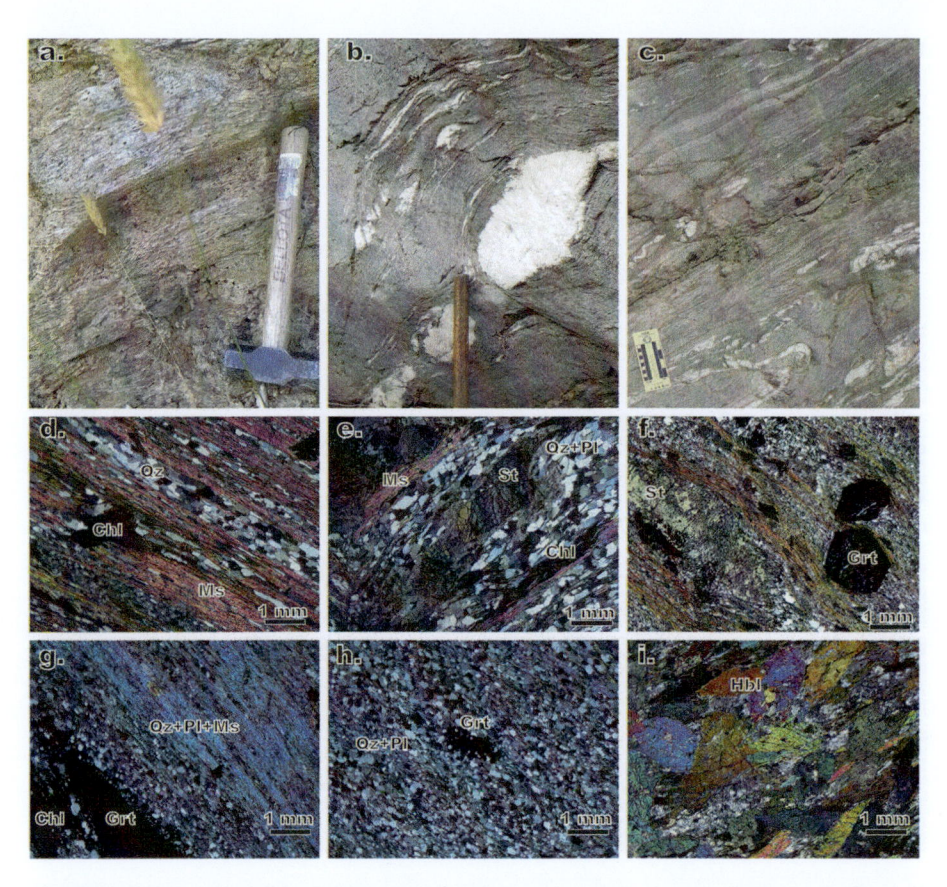

Fig. 4.4 Outcrop photographs (top) and thin-section micrographs (middle and bottom) showing some common lithologic types and microstructures observed in the Silgará Schist. (**a**) White to gray ms-qz-pl schist with 0.5–1 cm plagioclase porphyroblasts. (**b, c**) Quartzite and grt-bt-ms-qz schist with abundant quartz lenses (5 mm–5 cm); note also cm-scale folding. (**d**) Quartz microlithons separated by muscovite (P) domains and large chlorite porphyroblasts crossing foliation. (**e**) Chlorite closely associated with muscovite domains and large staurolite porphyroblasts. (**f**) Garnet and staurolite porphyroblasts in a muscovite, plagioclase, and quartz matrix. (**g**) Stretched garnet porphyroblast in a muscovite, plagioclase, and quartz mylonitic matrix; locally secondary chlorite is present replacing garnet. (**h**) Anhedral garnet with corroded edges in a fine-grained quartz-rich muscovite matrix. (**i**) Randomly oriented hornblende grains in a massive amphibolite

et al. 2008; García et al. 2005). The C- and O-isotope composition of calcareous lithologies from the unit suggests that protolith sedimentation occurred from Neoproterozoic to early Cambrian times (Arenas 2004; Silva et al. 2004). These results are consistent with reported U/Pb laser ablation results from detrital zircons that yielded ages from 906.5 ± 10.5 to 1610.3 ± 9.8 Ma in the Cepitá area and from 506.7 ± 9.3 to 2586.9 ± 10.2 Ma in the Matanzas area (Mantilla et al. 2016).

4.4.4 "Metasedimentitas de Guaca, La Virgen" (MGV)

This unit overlies unconformably the Silgará Schists and is represented by slate and phyllite that commonly contain the assemblage quartz-muscovite-chlorite with minor content of kaolinite, illite, plagioclase, and potassium feldspar (Fig. 4.5). Lithologies commonly show evidence of a superimposed dynamic metamorphism (e.g., quartz veins, iron oxides and mica veins, undulose extinction, and mechanical twinning). Kübler crystallinity indexes (IK, Fig. 4.8) indicate that metamorphism occurred in the early and late anchizone (transition zone between diagenesis and metamorphism). Three main ages had been proposed for the photolith: a Middle Devonian age (Emsian; Ward et al. 1973), an early Silurian age (Ludlow; Forero 1990), and a Carboniferous-Permian age (Moreno-Sánchez et al. 2005).

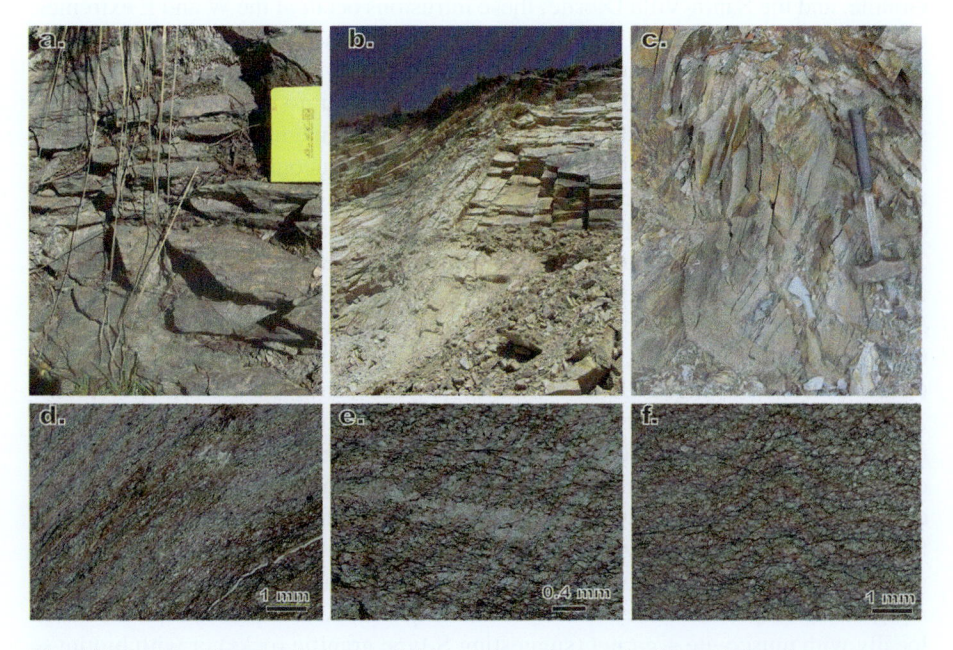

Fig. 4.5 Outcrop photographs (top) and thin-section micrographs (bottom) showing some common characteristics of the low-grade metasedimentary Guaca – La Virgen unit. Medium- (**a**) to fine-grained gray phyllite (**b, c**). In (**a**) foliation surfaces are slightly undulate due to grain size. In (**c**) the development of pencil-like shards formed due to the intersection of cleavage with a fracture set. It is primarily composed of quartz, sericite, and chlorite with minor graphite and chlorite (**d–f**)

4.5 Magmatism

4.5.1 Magmatic Belts

A N-S Triassic-Jurassic magmatic belt cut the Santander Massif metamorphic crystalline core parallel to the elongated direction of the massif (Fig. 4.1). There are also some Paleozoic and Cenozoic intrusives (Goldsmith et al. 1971; Ward et al. 1973; Boinet et al. 1985; Dörr et al. 1995; Restrepo-Pace 1995; Ordóñez-Calderón 2003; Ordóñez et al. 2006; Mantilla et al. 2009, 2011, 2012, 2013; Leal et al. 2011). Intrusives have a wide range of composition between gabbro and granite (Ordóñez et al. 2006; Mantilla et al. 2013; van der Lelij 2013; van der Lelij et al. 2016), but most commonly they are granites, tonalites, granodiorites, and quartz-monzonites (Plutonic Santander Group; Ward et al. 1973). The emplacement structural level, limits, and shape of the different igneous facies of the batholiths are not well known; furthermore, their crystallization history is unknown.

The main Paleozoic non-foliated plutons are the Durania Granite, the Pamplona Granite, and the Sanín-Villa Diorite; these intrusions occur at the W and E extremes of the massif and have small sizes compared to the Triassic-Jurassic plutons. Note that irregularly shaped foliated Paleozoic leucogranites were explained in the preceding section within the "Orthogneiss" unit. Triassic-Jurassic intrusions are represented by a set of batholiths and stocks defining a magmatic belt with a N-S direction. This magmatic belt includes granitic (some leucogranites), granodioritic, tonalitic, and locally charnockitic plutons and represents the oldest calc-alkaline arc of Triassic-Jurassic age of the Colombian Andes (Jaillard et al. 1990; Leal et al. 2011). The main Triassic-Jurassic named intrusives are the Onzaga Granodiorite, the Guaca River Diorite, the Suratá River Diorite and Granodiorite, the Pescadero Monzogranite (Granodiorite), the Santa Bárbara Quartz-monzonite, the Mogotes Quartz-monzonite, the La Corcova Quartz-monzonite, the Páramo Rico Tonalite (Granodiorite), the Agua Blanca Granite (Agua Blanca Batholith), the Ocaña Alkaline Granite (Ocaña Batholith), and the Rionegro Batholith. In general, granitic rocks are phaneritic, mainly inequigranular, with a medium to fine grain size (locally coarse), and locally porphyritic, and intruded by pegmatitic, granitic (aplitic), and dark tonalitic dikes (Figs. 4.6 and 4.7). The textural features of these rocks include the presence of perthites, zoned K-feldspar, myrmekitic and granophyric intergrowths mainly in the aplitic dikes, and local solid-state deformation toward the margins of the plutons. They usually contain quartz, K-feldspar, and plagioclase, locally with muscovite + garnet (suggesting S-type granitic rocks) or with biotite + hornblende + epidote (suggesting I-type granitoids). Some of this granitoids contain two micas (muscovite and biotite), which along with garnet suggest a peraluminous character; others have the association hornblende + epidote suggesting a metaluminous character (Fig. 4.8).

Published radiometric ages are compiled here in Table 4.1. The Paleozoic intrusions have U-Pb zircon crystallization ages between ~483 Ma and ~430 Ma; cooling ages (or ages of superposed thermal events) obtained from micas, amphiboles, and

Fig. 4.6 Outcrop (top and middle) and hand-sample photographs (bottom) showing some characteristics of the Jurassic granitoids. (**a**) Santa Barbara Quartz-monzonite with microgranular tonality dikes. (**b**) Contact between a fine-grained rhyolite (left) and a granitoid (right). (**c**) Coarse-grained gray quartz-monzonite of the Rionegro Tonalite, Granodiorite, and Quartz-monzonite. (**d–g**) Coarse to fine-grained granitoids of the Santa Bárbara Quartz-monzonite

whole rock Ar-Ar/K-Ar range between ~461 Ma and ~350 Ma (Ward et al. 1973; Boinet et al. 1985). U-Pb zircon crystallization ages of the Triassic-Jurassic belt range between ~212 Ma and ~196 Ma (Dörr et al. 1995; Mantilla et al. 2013; van der Lelij 2013; van der Lelij et al. 2016); micas K-Ar cooling ages range between ~196 Ma and ~172 Ma (Ward et al. 1973).

The mineralogical content in both magmatic belts evidences three groups: (i) granitoids with garnet + muscovite +/− biotite and without epidote + titanite, (ii) granitoids with muscovite +/− biotite and without garnet, and (iii) granitoids with epidote (allanite) + biotite +/− hornblende (Fig. 4.7). We present here a summarized description of mapped granitic rocks that are a sample of their variability.

Fig. 4.7 Thin-section micrographs (left) and cathodoluminescence images (right) showing some microstructures of the Jurassic granitoids. (**a, b**) Páramo Rico Tonalite (Granodiorite); note that K-feldspar occurs interstitial between quartz and plagioclase crystals. (**c, d**) Santa Bárbara Quartz-monzonite, large K-feldspar crystals grew around quartz and plagioclase. (**e, f**) Santa Bárbara Quartz-monzonite; note the presence of biotite and abundant small apatite crystals. (**g, h**) Santa Bárbara Quartz-monzonite, graphic granite

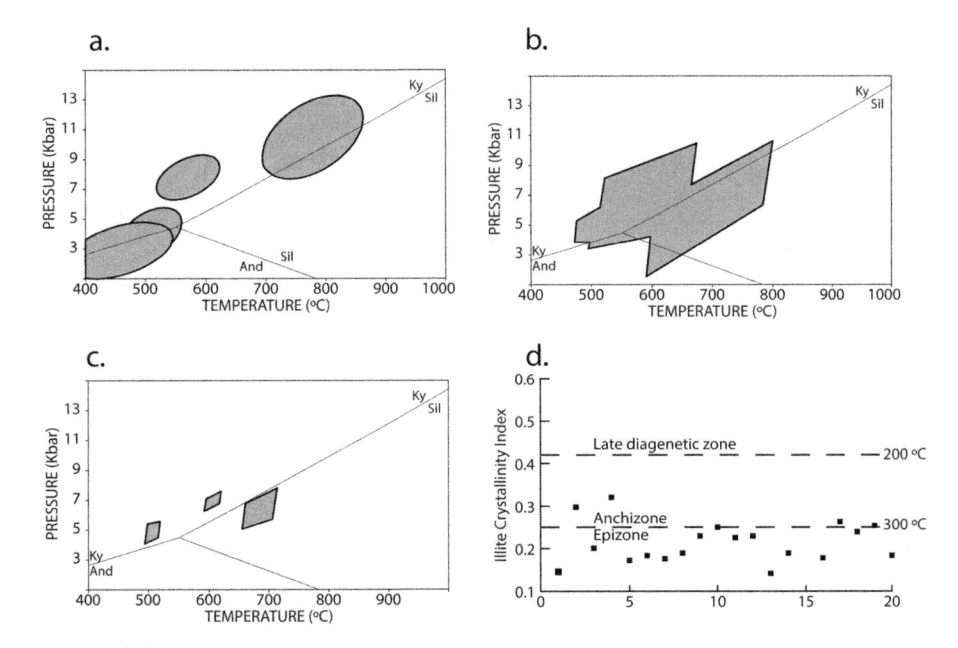

Fig. 4.8 Summary of PT estimations from metamorphic units in the Santander Massif. (**a**) PT estimations for the Bucaramanga Gneiss from Zuluaga et al. (2017). (**b**) PT estimations for pelitic schist and gneiss from the Berlín and Mutiscua areas from Castellanos et al. (2008); note that the highest PT estimations correspond to samples located to what is inferred here to be the Bucaramanga Gneiss. (**c**) PT estimations for Silgará Schist from the Cepitá area from García et al. (2005); note that the highest temperature estimation corresponds to a sample located near the contact with the Bucaramanga Gneiss, and the lowest is located near the top of the sequence. (**d**) Temperature estimations from the illite crystallinity index for the MGV unit, approximate temperature values taken from Merriman and Frey (1999)

Durania Granite This is an irregular-shaped elongated N-S white muscovite granite pluton with the presence of garnet and tourmaline; this pluton has locally a gneissic fabric. It intrudes the Silgará Schist with a border zone defined by the presence of pegmatitic dikes with muscovite, tourmaline, and locally garnet.

Sanín-Villa Diorite This is an elongated N-S hornblende diorite pluton with minor biotite and titanite that intrudes the Bucaramanga Gneiss and the Silgará Schist to the west of Rio de Oro.

Guaca River Diorite This is a medium-grained hornblende-epidote diorite stock emplaced in the Bucaramanga Gneiss and intruded by the Mogotes Quartz-monzonite.

Table 4.1 Published radiometric ages of crystalline units in the Santander Massif

Sample	Coordinates	Age	Error	Method	Material	Lithology	Unit	Reference
IMN-33199	7°15′59.99″ −72°54′11.14″	680	140	Rb-Sr	WR	Bt Gneiss	Bucaramanga Gneiss	Goldsmith et al. (1971) Ward et al. (1973)
IMN-33199	7°15′59.99″ −72°54′11.14″	198	7	K-Ar	Bt	Bt Gneiss	Bucaramanga Gneiss	Goldsmith et al. (1971) Ward et al. (1973)
IMN-33199	7°15′59.99″ −72°54′11.14″	189	4	K-Ar	Bt	Bt Gneiss	Bucaramanga Gneiss	Goldsmith et al. (1971) Ward et al. (1973)
IMN-12263	8°16′54.77″ −73°25′32.48″	945	40	K-Ar	Hbl	Hbl Gneiss	Bucaramanga Gneiss	Goldsmith et al. (1971) Ward et al. (1973)
IMN-12256	7°14′28.09″ −72°47′35.99″	450	80	Rb-Sr	WR	Orthogneiss	Berlín Orthogneiss	Goldsmith et al. (1971) Ward et al. (1973)
IMN-12262	8°16′54.01″ −73°24′21.08″	413	30	K-Ar	Hbl	Metadiorite	Bucaramanga Gneiss	Goldsmith et al. (1971) Ward et al. (1973)
OT-1 Integrated	8°18′03.83″ −73°26′09.45″	574	8	Ar-Ar	Hbl	Qz-Hbl-Pl Gneiss	Bucaramanga Gneiss	Restrepo-Pace (1995) Restrepo-Pace et al. (1997)
OT-2 Integrated	8°18′03.83″ −73°26′09.45″	668	9	Ar-Ar	Hbl	Qz-Hbl-Pl Gneiss	Bucaramanga Gneiss	Restrepo-Pace (1995) Restrepo-Pace et al. (1997)
CSB-2a	6°57′06.82″ −72°58′30.44″	177	4	Ar-Ar	Bt integrated	Qz-Pl-Bt Gneiss (Paragneiss)	Bucaramanga Gneiss	Restrepo-Pace (1995)
CSB-3a	6°57′00.83″ −72°55′43.43″	175.4	1.7	Ar-Ar	Bt integrated	Qz-Pl-Bt Gneiss (Paragneiss)	Bucaramanga Gneiss	Restrepo-Pace (1995)
BP-2	7°24′12.82″ −72°49′11.44″	198.3	0.5	Ar-Ar	Bt plateau	Qz-Bt-Hbl-Pl Gneiss (Orthogneiss)	Berlín Orthogneiss	Restrepo-Pace (1995)
BP-2	7°24′12.82″ −72°49′11.44″	477	16	U-Pb	Zr	Qz-Bt-Hbl-Pl Gneiss (Orthogneiss)	Berlín Orthogneiss	Restrepo-Pace (1995)
BP-2	7°24′12.82″ −72°49′11.44″	196.6		Ar-Ar	Bt plateau	Qz-Bt-Hbl-Pl Gneiss (Orthogneiss)	Berlín Orthogneiss	Restrepo-Pace (1995)

Sample	Coordinates	Age	Error	Method	Material	Lithology	Unit	Reference
BV-2	7°14′19.81″ −72°54′08.43″	190.6	2	Ar-Ar	Hbl plateau	Orthogneiss	Berlín Orthogneiss	Restrepo-Pace (1995)
BV-2	7°14′19.81″ −72°54′08.43″	194	2	Ar-Ar	Hbl plateau	Hbl-Pl-Qz gneiss (Orthogneiss)	Berlín Orthogneiss	Restrepo-Pace (1995)
BV-3	7°14′47.83″ −72°54′45.45″	184	3	Ar-Ar	Bt plateau	Qz-Bt-Pl Gneiss (Orthogneiss)	Berlín Orthogneiss	Restrepo-Pace (1995)
BV-3	7°14′47.83″ −72°54′45.45″	184	4	Ar-Ar	Bt plateau	Qz-Bt-Pl Gneiss (Orthogneiss)	Berlín Orthogneiss	Restrepo-Pace (1995)
BV-4	7°06′01.83″ −72°53′54.43	187	4	Ar-Ar	Bt integrated	Qz-Bt-Pl Gneiss (Orthogneiss)	Berlín Orthogneiss	Restrepo-Pace (1995)
BV-5	7°14′28.82″ −72°54′04.43″	175	3	Ar-Ar	Bt integrated	Qz-Bt-Pl Gneiss (Orthogneiss)	Berlín Orthogneiss	Restrepo-Pace (1995)
CB-7	7°06′19.80″ −72°41′47.42″	204	0.6	Ar-Ar	Hbl integrated	Hbl-Pl-Qz Gneiss (Orthogneiss)	Berlín Orthogneiss	Restrepo-Pace (1995)

Sample	Coordinates	Age	Error	Method	Material	Lithology	Unit	Reference
CB-7	7°06′19.80″ −72°41′47.42″	203.4		Ar-Ar	Hbl integrated	Hbl-Pl-Qz Gneiss (Orthogneiss)	Berlín Orthogneiss	Restrepo-Pace (1995)
–		900–985		U-Pb	Zr superior intercept		Bucaramanga Gneiss	Restrepo-Pace in Keppie et al. (2001 – Personal com.)
NB-1	8°16′54.77″ −73°25′32.48″	1760		Sm-Nd	WR model age	Qz-Fsp-Bt Gneiss	Bucaramanga Gneiss	Ordóñez et al. (2006)
NB-3	8°16′51.88″ −73°24′21.09″	1710		Sm-Nd	WR model age	Hbl Gneiss	Bucaramanga Gneiss	Ordóñez et al. (2006)
PCM-1105	7°17′57.83″ −72°53′17.86″	1540	12	U.Pb	Zr SHRIMP	Bt Gneiss	Bucaramanga Gneiss	Cardona (2003) Cordani et al. (2005)
PCM-1106	7°17′57.83″ −72°53′17.86″	1558	18	U.Pb	Zr SHRIMP	Bt Gneiss	Bucaramanga Gneiss	Cardona (2003) Cordani et al. (2005)

(continued)

Table 4.1 (continued)

Sample	Coordinates	Age	Error	Method	Material	Lithology	Unit	Reference
PCM-1107	7°17'57.83" −72°53'17.86"	1465	15	U-Pb	Zr SHRIMP	Bt Gneiss	Bucaramanga Gneiss	Cardona (2003) Cordani et al. (2005)
PCM-1108	7°17'57.83" −72°53'17.86"	1186	35	U-Pb	Zr SHRIMP	Bt Gneiss	Bucaramanga Gneiss	Cardona (2003) Cordani et al. (2005)
PCM-1109	7°17'57.83" −72°53'17.86"	1138	13	U-Pb	Zr SHRIMP	Bt Gneiss	Bucaramanga Gneiss	Cardona (2003) Cordani et al. (2005)
PCM-1110	7°17'57.83" −72°53'17.86"	1071	20	U-Pb	Zr SHRIMP	Bt Gneiss	Bucaramanga Gneiss	Cardona (2003) Cordani et al. (2005)
PCM-1111	7°17'57.83" −72°53'17.86"	1035	26	U-Pb	Zr SHRIMP	Bt Gneiss	Bucaramanga Gneiss	Cardona (2003) Cordani et al. (2005)
PCM-1112	7°17'57.83" −72°53'17.86"	930	17	U-Pb	Zr SHRIMP	Bt Gneiss	Bucaramanga Gneiss	Cardona (2003) Cordani et al. (2005)
PCM-1113	7°17'57.83" −72°53'17.86"	864	66	U-Pb	Zr SHRIMP	Bt Gneiss	Bucaramanga Gneiss	Cardona (2003) Cordani et al. (2005)
PCM-815	7°18'13.81" −72°53'18.44"	208.6	0.5	Ar-Ar	Bt plateau	Bt-Hbl Gneiss	Bucaramanga Gneiss	Cardona (2003) Cordani et al. (2005)
PCM-816	7°18'13.81" −72°53'18.44"	203.8	0.7	Ar-Ar	Bt plateau	Bt-Hbl Gneiss	Bucaramanga Gneiss	Cardona (2003) Cordani et al. (2005)
PCM-817	7°18'13.81" −72°53'18.44"	202.0	0.6	Ar-Ar	Bt plateau	Bt-Hbl Gneiss	Bucaramanga Gneiss	Cardona (2003) Cordani et al. (2005)
PCM-818	7°18'13.81" −72°53'18.44"	201.9	0.5	Ar-Ar	Hbl plateau	Bt-Hbl Gneiss	Bucaramanga Gneiss	Cardona (2003) Cordani et al. (2005)
PCM-819	7°18'13.81" −72°53'18.44"	198.2	0.6	Ar-Ar	Hbl plateau	Bt-Hbl Gneiss	Bucaramanga Gneiss	Cardona (2003) Cordani et al. (2005)
PCM-820	7°18'13.81" −72°53'18.44"	200.1	0.6	Ar-Ar	Hbl plateau	Bt-Hbl Gneiss	Bucaramanga Gneiss	Cardona (2003) Cordani et al. (2005)
PCM-1102	7°18'13.81" −72°53'18.44"	193.6	0.5	Ar-Ar	Bt plateau	Bt-Hbl Gneiss	Bucaramanga Gneiss	Cardona (2003) Cordani et al. (2005)

Sample	Coordinates	Age	Error	Method	Material	Lithology	Unit	Reference
PCM-1103	7°18′13.81″ −72°53′18.44″	160.7	1.0	Ar-Ar	Bt plateau	Bt-Hbl Gneiss	Bucaramanga Gneiss	Cardona (2003) Cordani et al. (2005)
PCM-1104	7°18′13.81″ −72°53′18.44″	154.5	1.6	Ar-Ar	Bt plateau	Bt-Hbl Gneiss	Bucaramanga Gneiss	Cardona (2003) Cordani et al. (2005)
PCM-1105	7°17′57.83″ −72°53′17.86″	212.7	0.5	Ar-Ar	Hbl plateau	Bt-Hbl Gneiss	Bucaramanga Gneiss	Cardona (2003) Cordani et al. (2005)
PCM-1106	7°17′57.83″ −72°53′17.86″	193.8	0.4	Ar-Ar	Hbl integrated	Bt-Hbl Gneiss	Bucaramanga Gneiss	Cardona (2003) Cordani et al. (2005)
PCM-1107	7°17′57.83″ −72°53′17.86″	181.8	1.2	Ar-Ar	Hbl integrated	Bt-Hbl Gneiss	Bucaramanga Gneiss	Cardona (2003) Cordani et al. (2005)
PCM-1105	7°17′57.83″ −72°53′17.86″	200.2	0.4	Ar-Ar	Bt plateau	Bt Gneiss	Bucaramanga Gneiss	Cardona (2003) Cordani et al. (2005)
PCM-1106	7°17′57.83″ −72°53′17.86″	199.1	0.3	Ar-Ar	Bt plateau	Bt Gneiss	Bucaramanga Gneiss	Cardona (2003) Cordani et al. (2005)
PCM-1107	7°17′57.83″ −72°53′17.86″	199.2	0.4	Ar-Ar	Bt plateau	Bt Gneiss	Bucaramanga Gneiss	Cardona (2003) Cordani et al. (2005)
PCM-1105	7°17′57.83″ −72°53′17.86″	1780		Sm-Nd	WR model age	Bt Gneiss	Bucaramanga Gneiss	Cardona (2003) Cordani et al. (2005)
PCM-1106	7°17′57.83″ −72°53′17.86″	1160		Sm-Nd	WR	Bt Gneiss	Bucaramanga Gneiss	Cardona (2003)
PCM-1107	7°17′57.83″ −72°53′17.86″	1619		Sm-Nd	WR model age	Bt Gneiss	Bucaramanga Gneiss	Cardona (2003)
PCM-1108	7°17′57.83″ −72°53′17.86″	1080		Sm-Nd	WR max. age sed.	Bt Gneiss	Bucaramanga Gneiss	Cardona (2003)
PCM-1178		1890		Sm-Nd	WR model age	Gabbro	Bucaramanga Gneiss	Cardona (2003) Cordani et al. (2005)
PCM-1179		1160		Sm-Nd	WR	Gabbro	Bucaramanga Gneiss	Cardona (2003)
AL-08-12 drill hole	7°23′11.17″ −72°53′29.22″	462.5	13.1	U-Pb	Zr LA-MC-ICP-MS	Qz-Fsp Gneiss	Bucaramanga Gneiss	Leal et al. (2011)

(continued)

Table 4.1 (continued)

Sample	Coordinates	Age	Error	Method	Material	Lithology	Unit	Reference
AL-07-09 drill hole	7°23′18.87″ −72°53′29.21″	460	8.9	U-Pb	Zr LA-MC-ICP-MS	Qz-Fsp Gneiss	Bucaramanga Gneiss	Leal et al. (2011)
AL-07-09 drill hole	7°23′18.87″ −72°53′29.21″	462–474		U-Pb	Zr	Granitoid Gneiss	Bucaramanga Gneiss	Leal et al. (2011)
GE-58-M1	7°23′27.57″ −72°53′28.66″	~477.0	4.1	U-Pb	Zr LA-MC-ICP-MS	Metadiorite		Mantilla et al. (2012)
GE-58-M2	7°23′27.57″ −72°53′28.66″	480.5	7.7	U-Pb	Zr LA-MC-ICP-MS	Metadiorite		Mantilla et al. (2012)
GE-58-M3	7°23′27.57″ −72°53′28.66″	480.5	14.1	U-Pb	Zr LA-MC-ICP-MS	Metadiorite		Mantilla et al. (2012)

Sample	Coordinates	Age	Error	Method	Material	Lithology	Unit	Reference
GE-58-M4	7°23′27.57″ −72°53′28.66″	481.5	11.8	U-Pb	Zr LA-MC-ICP-MS	Metadiorite		Mantilla et al. (2012)
GE-58-M5	7°23′27.57″ −72°53′28.66″	477.1	10.7	U-Pb	Zr LA-MC-ICP-MS	Metadiorite		Mantilla et al. (2012)
GE-58-M6	7°23′27.57″ −72°53′28.66″	469.1	8.9	U-Pb	Zr LA-MC-ICP-MS	Metadiorite		Mantilla et al. (2012)
GE-58-M7	7°23′27.57″ −72°53′28.66″	474.9	10.2	U-Pb	Zr LA-MC-ICP-MS	Metadiorite		Mantilla et al. (2012)
GE-58-M8	7°23′27.57″ −72°53′28.66″	458.0	11.4	U-Pb	Zr LA-MC-ICP-MS	Metadiorite		Mantilla et al. (2012)
GE-58-M9	7°23′27.57″ −72°53′28.66″	462.9	11.3	U-Pb	Zr LA-MC-ICP-MS	Metadiorite		Mantilla et al. (2012)
GE-58-M10	7°23′27.57″ −72°53′28.66″	207.3	4.6	U-Pb	Zr LA-MC-ICP-MS	Metadiorite		Mantilla et al. (2012)

GE-58-M11	7°23'27.57" −72°53'28.66"	223.9	6.2	U-Pb	Zr LA-MC-ICP-MS	Metadiorite	Mantilla et al. (2012)
GE-58-M12	7°23'27.57" −72°53'28.66"	239.0	4.6	U-Pb	Zr LA-MC-ICP-MS	Metadiorite	Mantilla et al. (2012)
GE-58-M13	7°23'27.57" −72°53'28.66"	756.9	27.1	U-Pb	Zr LA-MC-ICP-MS	Metadiorite	Mantilla et al. (2012)
GE-58-M14	7°23'27.57" −72°53'28.66"	765.5		U-Pb	Zr LA-MC-ICP-MS	Metadiorite	Mantilla et al. (2012)
GE-58-M15	7°23'27.57" −72°53'28.66"	1371.5	27.3	U-Pb	Zr LA-MC-ICP-MS	Metadiorite	Mantilla et al. (2012)
GE-58-M16	7°23'27.57" −72°53'28.66"	1381.2		U-Pb	Zr LA-MC-ICP-MS	Metadiorite	Mantilla et al. (2012)
GH-72-M2	7°23'27.57" −72°53'28.66"	~477.2		U-Pb	Zr LA-MC-ICP-MS	Metadiorite	Mantilla et al. (2012)
GH-72-M3	7°23'27.57" −72°53'28.66"	478.6	15.8	U-Pb	Zr LA-MC-ICP-MS	Metadiorite	Mantilla et al. (2012)
GH-72-M4	7°23'27.57" −72°53'28.66"	480.0	11.7	U-Pb	Zr LA-MC-ICP-MS	Metadiorite	Mantilla et al. (2012)
GH-72-M5	7°23'27.57" −72°53'28.66"	481.0	12.2	U-Pb	Zr LA-MC-ICP-MS	Metadiorite	Mantilla et al. (2012)
GH-72-M6	7°23'27.57" −72°53'28.66"	485.8	11.1	U-Pb	Zr LA-MC-ICP-MS	Metadiorite	Mantilla et al. (2012)
GH-72-M7	7°23'27.57" −72°53'28.66"	484.9	15.2	U-Pb	Zr LA-MC-ICP-MS	Metadiorite	Mantilla et al. (2012)
GH-72-M8	7°23'27.57" −72°53'28.66"	208.8	6.0	U-Pb	Zr LA-MC-ICP-MS	Metadiorite	Mantilla et al. (2012)

(continued)

Table 4.1 (continued)

Sample	Coordinates	Age	Error	Method	Material	Lithology	Unit	Reference
GH-72-M9	7°23′27.57″ −72°53′28.66″	211.4	4.9	U-Pb	Zr LA-MC-ICP-MS	Metadiorite		Mantilla et al. (2012)
GH-72-M10	7°23′27.57″ −72°53′28.66″	213.9	5.4	U-Pb	Zr LA-MC-ICP-MS	Metadiorite		Mantilla et al. (2012)
GH-72-M11	7°23′27.57″ −72°53′28.66″	223.2	9.8	U-Pb	Zr LA-MC-ICP-MS	Metadiorite		Mantilla et al. (2012)
GH-72-M12	7°23′27.57″ −72°53′28.66″	502.0	15.1	U-Pb	Zr LA-MC-ICP-MS	Metadiorite		Mantilla et al. (2012)
GH-72-M13	7°23′27.57″ −72°53′28.66″	506.6	12.7	U-Pb	Zr LA-MC-ICP-MS	Metadiorite		Mantilla et al. (2012)
GI-60-M2	7°23′27.57″ −72°53′28.66″	~481.9	6.1	U-Pb	Zr LA-MC-ICP-MS	Metadiorite – Calc-alkaline foliated diorite		Mantilla et al. (2012)
GI-60-M3	7°23′27.57″ −72°53′28.66″	482.6	9.3	U-Pb	Zr LA-MC-ICP-MS	Metadiorite – Calc-alkaline foliated diorite		Mantilla et al. (2012)
GI-60-M4	7°23′27.57″ −72°53′28.66″	473.8	12.1	U-Pb	Zr LA-MC-ICP-MS	Metadiorite – Calc-alkaline foliated diorite		Mantilla et al. (2012)
GI-60-M5	7°23′27.57″ −72°53′28.66″	468.6	10.2	U-Pb	Zr LA-MC-ICP-MS	Metadiorite – Calc-alkaline foliated diorite		Mantilla et al. (2012)
GI-60-M6	7°23′27.57″ −72°53′28.66″	467.4	8.6	U-Pb	Zr LA-MC-ICP-MS	Metadiorite – Calc-alkaline foliated diorite		Mantilla et al. (2012)
GI-60-M7	7°23′27.57″ −72°53′28.66″	477.6	9.2	U-Pb	Zr LA-MC-ICP-MS	Metadiorite – Calc-alkaline foliated diorite		Mantilla et al. (2012)
GI-60-M8	7°23′27.57″ −72°53′28.66″	481.8	7.9	U-Pb	Zr LA-MC-ICP-MS	Metadiorite – Calc-alkaline foliated diorite		Mantilla et al. (2012)
GI-60-M9	7°23′27.57″ −72°53′28.66″	478.3	9.0	U-Pb	Zr LA-MC-ICP-MS	Metadiorite – Calc-alkaline foliated diorite		Mantilla et al. (2012)
GI-60-M10	7°23′27.57″ −72°53′28.66″	477.8	10.0	U-Pb	Zr LA-MC-ICP-MS	Metadiorite – Calc-alkaline foliated diorite		Mantilla et al. (2012)

Sample	Coordinates	Age	Error	Method	Material	Lithology	Unit	Reference
GI-60-M11	7°23′27.57″ −72°53′28.66″	480.5	8.5	U-Pb	Zr LA-MC-ICP-MS	Metadiorite – Calc-alkaline foliated diorite		Mantilla et al. (2012)
GI-60-M12	7°23′27.57″ −72°53′28.66″	211.3	3.1	U-Pb	Zr LA-MC-ICP-MS	Metadiorite – Calc-alkaline foliated diorite		Mantilla et al. (2012)
GI-60-M13	7°23′27.57″ −72°53′28.66″	1485		U-Pb	Zr LA-MC-ICP-MS	Metadiorite – Calc-alkaline foliated diorite		Mantilla et al. (2012)
RSC-1	7°19′17.82″ −73°07′40.45″	133	3	Ar-Ar	Bt	Paragneiss	Bucaramanga Gneiss	Restrepo-Pace (1995)
10VDL23	6°56′34.99″ −72°57′59.99″	461.0	2.1	U-Pb	Zr LA-ICP-MS	Bt-Hbl Gneiss	Bucaramanga Gneiss	Van der Lelij (2013)
10VDL37	7°11′23.99″ −72°58′40.99″	451.5	1.3	U-Pb	Zr LA-ICP-MS	Bt-Hbl Gneiss	Berlín Orthogneiss	Van der Lelij (2013)

Sample	Coordinates	Age	Error	Method	Material	Lithology	Unit	Reference
10VDL37	7°11′23.99″ −72°58′40.99″	167.1	15	AFT	Zr LA-ICP-MS	Bt-Hbl Gneiss	Berlín Orthogneiss	Van der Lelij (2013)
10VDL39	7°10′46.01″ −72°59′48.02″	208.8	1.2	U-Pb	Zr LA-ICP-MS	Pegmatite	Berlín Orthogneiss	Van der Lelij (2013)
10VDL43	7°14′17.99″ −72°49′30.02″	209.2	3.4	U-Pb	Zr LA-ICP-MS	Migmatite	Berlín Orthogneiss	Van der Lelij (2013)
10VDL43	7°14′17.99″ −72°49′30.02	209.2	3.4	U-Pb	Zr LA-ICP-MS	Orthogneiss?	Berlín Orthogneiss	Van der Lelij (2013)
10VDL44	7°20′40.00″ −72°42′58.99″	473.5	2.5	U-Pb	Zr LA-ICP-MS	Orthogneiss?	Berlín Orthogneiss	Van der Lelij (2013)
10VDL50	7°15′22.01″ −72°53′44.99″	477	5.3	U-Pb	Zr LA-ICP-MS	Paragneiss migmatite	Bucaramanga Gneiss	Van der Lelij (2013)
10VDL51	7°15′17.01″ −72°53′48.01″B	472.5	3.4	U-Pb	Zr LA-ICP-MS	Gneiss	Berlín Orthogneiss	Van der Lelij (2013)

(continued)

Table 4.1 (continued)

Sample	Coordinates	Age	Error	Method	Material	Lithology	Unit	Reference
WR194	7°23′13.81″ −72°53′28.99″	462.5	13.1	U-Pb	Zr LA-MC-ICP-MS	Granitic Gneiss	Berlín Orthogneiss	Leal et al. (2011)
WR-195	7°23′12.69″ −72°53′26.90″	460	18.9	U-Pb	Zr LA-MC-ICP-MS	Granitic Gneiss	Berlín Orthogneiss	Leal et al. (2011)
IMN-12255	7°06′49.83″ −72°51′47.43″	198	8	K-Ar	WR	Phyllite	Silgará schist	Goldsmith et al. (1971) Ward et al. (1973)
IMN-12257	8°03′41.23″ −72°58′01.56″	221	8	K-Ar	WR	Phyllite	Silgará Schist	Goldsmith et al. (1971) Ward et al. (1973)
AB-2a	6°49′46.99″ −73°00′39.99″	208	4	Ar-Ar	Ms integrated	Qz-Chl-Ms-Grt Schist	Silgará Schist	Restrepo-Pace (1995)
AB-6	6°49′46.99″ −73°00′39.99″	213	3	Ar-Ar	Ms plateau	Qz-Chl-Ms-Grt-St Schist	Silgará Schist	Restrepo-Pace (1995)
AB-7	6°49′49.00″ −73°00′40.02″	242	5	Ar-Ar	Hbl integrated	Hbl-Qz Foliated Dyke	Silgará Schist	Restrepo-Pace (1995)
AB-7	6°49′49.00″ −73°00′40.02″	183	4	Ar-Ar	Hbl min. age	Hbl-Qz Foliated Dyke	Silgará Schist	Restrepo-Pace (1995)
AB-7	6°49′49.00″ −73°00′40.02″	340	7	Ar-Ar	Hbl max. age	Hbl-Qz Foliated Dyke	Silgará Schist	Restrepo-Pace (1995)
BM-1	7°06′19.84″ −73°03′20.43″	185	3	Ar-Ar	Ms plateau	Qz-Pl-Grt-Ms Gneissic Schist	Silgará Schist	Restrepo-Pace (1995)
BM-1	7°06′19.84″ −73°03′20.43″	185	2	Ar-Ar	Ms plateau	Qz-Pl-Grt-Ms Gneissic Schist	Silgará Schist	Restrepo-Pace (1995)
BM-4	7°06′34.84″ −73°03′20.43″	184	2	Ar-Ar	Ms integrated	Qz-Bt-Ms Schist	Silgará Schist	Restrepo-Pace (1995)
BP-3	7°13′55.83″ −72°48′29.43″	197.6		Ar-Ar	Ms plateau	Qz-Bt-Sill-Grt Schist	Silgará Schist	Restrepo-Pace (1995)

Sample	Coordinates	Age	Error	Method	Detail	Lithology	Unit	Reference
BP-3	7°13′55.83″ −72°48′29.43″	199.3	0.5	Ar-Ar	Ms plateau	Ms-Qz Schist	Silgará Schist	Restrepo-Pace (1995)
PC-1	7°28′03.83″ −72°41′47.40″	383.4		Ar-Ar	Bt Tot. fussion	Qz-Bt-Ms Schist	Silgará Schist	Restrepo-Pace (1995)
RSC-1	7°19′17.82″ −73°07′40.45″	133	3	Ar-Ar	Ms integrated	Qz-Bt-Sill-Grt Gneiss	Silgará Schist	Restrepo-Pace (1995)
SBb-1	7°11′25.83″ −72°45′37.43″	195	0.5	Ar-Ar	Ms plateau	Qz-Ms Schist	Silgará Schist	Restrepo-Pace (1995)
SBb-1	7°11′25.83″ −72°45′37.43″	193.3		Ar-Ar	Ms plateau	Qz-Ms Schist	Silgará Schist	Restrepo-Pace (1995)
SBb-2	7°11′25.83″ −72°45′37.43″	197	6	Ar-Ar	Ms integrated	Qz-Ms Schist	Silgará Schist	Restrepo-Pace (1995)
10VDL49	7°29′20.00″ −72°42₁15.99″	479.8	3.1	U-Pb	Zr LA-ICP-MS	Bt-Hbl Gneiss	Silgará Schist	Van der Lelij (2013)
10VDL49	7°29′20.00″ −72°42₁15.99″	197.7	24.1	ZFT		Bt-Hbl Gneiss	Silgará Schist	Van der Lelij (2013)
AMVO25		304.9	7.1	U-Pb	Zr LA-ICP-MS	Grt.St Schist	Silgará Schist	Cardona et al. (2016)
AMVO25		310.4	6.4	U-Pb	Zr LA-ICP-MS	Grt.St Schist	Silgará Schist	Cardona et al. (2016)
AMVO25		414		U-Pb	Zr LA-ICP-MS	Grt.St Schist	Silgará Schist	Cardona et al. (2016)
AMVO26		333		U-Pb	Zr LA-ICP-MS	Ab-Ms Schist	Silgará Schist	Cardona et al. (2016)
AMVO26		368		U-Pb	Zr LA-ICP-MS	Ab-Ms Schist	Silgará Schist	Cardona et al. (2016)
AMVO26		400		U-Pb	Zr LA-ICP-MS	Ab-Ms Schist	Silgará Schist	Cardona et al. (2016)
AMVO26		481		U-Pb	Zr LA-ICP-MS	Ab-Ms Schist	Silgará Schist	Cardona et al. (2016)
AMVO27		362		U-Pb	Zr LA-ICP-MS	Ms-Ab Schist	Silgará Schist	Cardona et al. (2016)
AMVO27		409		U-Pb	Zr LA-ICP-MS	Ms-Ab Schist	Silgará Schist	Cardona et al. (2016)
AMVO27		478		U-Pb	Zr LA-ICP-MS	Ms-Ab Schist	Silgará Schist	Cardona et al. (2016)
AMVO33		932–1543		U-Pb	Zr LA-ICP-MS	Ser Schist	Silgará Schist	Cardona et al. (2016)
AMVO37		1358–1755		U-Pb	Zr LA-ICP-MS		Silgará Schist	Cardona et al. (2016)

(continued)

Table 4.1 (continued)

Sample	Coordinates	Age	Error	Method	Material	Lithology	Unit	Reference
AMVO40		936–1564		U-Pb	Zr LA-ICP-MS	Bt-St Schist	Silgará Schist	Cardona et al. (2016)
AMVO41		457	9.5	U-Pb	Zr LA-ICP-MS	Chl Schist	Silgará Schist	Cardona et al. (2016)
AMVO41		489–634		U-Pb	Zr LA-ICP-MS	Chl Schist	Silgará Schist	Cardona et al. (2016)
AMVO41		1815		U-Pb	Zr LA-ICP-MS	Chl Schist	Silgará Schist	Cardona et al. (2016)
AMVO43		512–550		U-Pb	Zr LA-ICP-MS	Ms Schist	Silgará Schist	Cardona et al. (2016)
AMVO43		1348–1755		U-Pb	Zr LA-ICP-MS	Ms Schist	Silgará Schist	Cardona et al. (2016)
AMVO45		981–1637		U-Pb	Zr LA-ICP-MS	Ms Schist	Silgará Schist	Cardona et al. (2016)
PS-7-1	7°18′06.41″ −72°03′12.07″	696.8	10.7	U-Pb	Zr LA-ICP-MS	Quartzite	Silgará Schist	Mantilla et al. (2016)
PS-7-1	7°18′06.41″ −72°03′12.07	1720.6	24.5	U-Pb	Zr LA-ICP-MS	Quartzite	Silgará Schist	Mantilla et al. (2016)
PS-4-1	6°47′48.01″ −73°00′47.09″	455.1	6.1	U-Pb	Zr LA-ICP-MS	Quartzite	Silgará Schist	Mantilla et al. (2016)
PS-4-1	6°47′48.01″ −73°00′47.09″	506.7	9.3	U-Pb	Zr LA-ICP-MS	Quartzite	Silgará Schist	Mantilla et al. (2016)
PS-4-1	6°47′48.01″ −73°00′47.09″	2586.9	10.2	U-Pb	Zr LA-ICP-MS	Quartzite	Silgará Schist	Mantilla et al. (2016)
PS-1-1	6°43′20.09″ −73°00′07.01″	451.6	7.7	U-Pb	Zr LA-ICP-MS	Meta-sandstone	Silgará Schist	Mantilla et al. (2016)
PS-1-1	6°43′20.09″ −73°00′07.01″	1611.5	13.6	U-Pb	Zr LA-ICP-MS	Meta-sandstone	Silgará Schist	Mantilla et al. (2016)
IMN-12262	8°16′54.81″ −73°24′21.08″	413	30	K-Ar	Hbl	Metadiorite	Sanín Villa Diorite	Goldsmith et al. (1971) Ward et al. (1973)
IMN-14362	7°09′49.82″ −72°36′47.43″	457	13	K-Ar	Ms	Pegmatite into Bucaramanga Gneiss	Durania Granite	Goldsmith et al. (1971) Ward et al. (1973)

Sample	Coordinates	Age	Error	Method	Material	Lithology	Unit	Reference
IMN-14362	7°09'49.82" -72°36'47.43"	432	8	K-Ar	Ms	Pegmatite into Bucaramanga Gneiss	Durania Granite	Goldsmith et al. (1971) Ward et al. (1973)
IMN-14362	7°09'49.82" -72°36'47.43"	439	12	K-Ar	Ms	Pegmatite into Bucaramanga Gneiss	Durania Granite	Goldsmith et al. (1971) Ward et al. (1973)
		461	10	K-Ar	Ms isochron?		Durania Granite	Ward et al. (1973)
–		456	23	K-Ar		Gabbro		Boinet et al. (1985)

Sample	Coordinates	Age	Error	Method	Material	Lithology	Unit	Reference
–		350	18	K-Ar		Quartz-monzonite		Boinet et al. (1985)
–		394	23	K-Ar		Monzonite	Onzaga Stock	Etayo et al. (1983, U. Cordani written com.)
BP-2	7°24'12.82" -72°49'11.43"	477	16	U-Pb	Zr Sup. Intercept	Syn-deformational granitoid		Restrepo-Pace (1995)
BP-2	7°24'12.82" -72°49'11.43"	254	60	U-Pb	Zr Inf. intercept	Syn-deformational granitoid		Restrepo-Pace (1995)
BP-2	7°24'12.82" -72°49'11.43"	428	9	Pb-Pb	Zr	Syn-deformational granitoid		Restrepo-Pace (1995)
BP-2	7°24'12.82" -72°49'11.43"	462	11	Pb-Pb	Zr	Syn-deformational granitoid		Restrepo-Pace (1995)
BP-2	7°24'12.82" -72°49'11.43"	465	6	Pb-Pb	Zr	Syn-deformational granitoid		Restrepo-Pace (1995)
10VDL46	7°16'28.99" -72°39'55.00"	439.2	4.7	U-Pb	Zr LA-ICP-MS	Granite?	Pamplona Granite	Restrepo-Pace (1995)
10VDL47	7°28'22.99" -72°41'43.99"	483.7	5.9	U-Pb	Zr LA-ICP-MS	Gabbro-diorite		Van der Lelij (2013)
10VDL61	7°09'59.00" -73°05'16.99"	1018.3	8.9	U-Pb	Zr LA-ICP-MS	Enclave (Xenolith?)	Suratá River Granodiorite	Van der Lelij (2013)

(continued)

Table 4.1 (continued)

Sample	Coordinates	Age	Error	Method	Material	Lithology	Unit	Reference
BU1101	7°29′35.75″ −72°44′52.85″	456	22.8	K-Ar	WR	Gabbro		Boinet et al. (1985)
BU1132	7°20′57.76″ −72°35′25.86″	349.4	17.5	K-Ar	WR	Quartz-monzonite		Boinet et al. (1985)
APD-47-1	7°42′44.89″ −72°39′20.39″	442.6	+7.2/−6.3	U-Pb	Zr LA-ICP-MS	Granite	Durania Granite	Botello et al. (2014)
YGI-6-2	7°16′39.11″ −72°39′45.71″	442.6	+7.4/−6.0	U-Pb	Zr LA-ICP-MS	Granite	Durania Granite	Mantilla et al. (2016)
IMN-10953	8°17′44.06″ −73°21′50.98″	127	3	K-Ar	Sa	Porphyritic Rhyolite		Goldsmith et al. (1971) Ward et al. (1973)
IMN-10894	7°22′11.94″ −73°07′19.48″	177	6	K-Ar	Bt	Porphyritic Granodiorite	Rionegro Batholith	Goldsmith et al. (1971) Ward et al. (1973)
IMN-10894	7°22′11.94″ −73°07′19.48	172	6	K-Ar	Bt	Porphyritic Granodiorite	Rionegro Batholith	Goldsmith et al. (1971) Ward et al. (1973)
IMN-11547	6°47′05.13″ −72°59′12.86″	193	6	K-Ar	Bt	Granite	Mogotes Batholith	Goldsmith et al. (1971) Ward et al. (1973)
IMN-13201	8°08′26.08″ −72°54′05.76″	196	7	K-Ar	Bt	Quartz-monzonite	Agua Blanca Batholith	Goldsmith et al. (1971) Ward et al. (1973)
IMN-10924	6°54′29.47″ −72°56′17.35″	192	7	K-Ar	Bt	Quartz-monzonite	Santa Bárbara Quartz-monzonite	Goldsmith et al. (1971) Ward et al. (1973)

Sample	Coordinates	Age	Error	Method	Material	Lithology	Unit	Reference
IMN-11045	6°52′43.83″ −72°54′30.52″	194	7	K-Ar	Bt	Quartz-monzonite	Santa Bárbara Quartz-monzonite	Goldsmith et al. (1971) Ward et al. (1973)
IMN-13197	7°07′32.62″ −73°01′59.46″	111	4	K-Ar	Bt	Quartz-monzonite	La Corcova Quartz-monzonite	Goldsmith et al. (1971) Ward et al. (1973)

Sample	Coordinates	Age	Error	Method	Material	Rock type	Unit	Reference
IMN-13197	7°07′32.62″ −73°01′59.46″	195		K-Ar	Ms	Quartz-monzonite	La Corcova Quartz-monzonite	Goldsmith et al. (1971) Ward et al. (1973)
IMN-12264	7°38′49.83″ −73°17′05.89″	160	7	Rb-Sr	WR	Riebeckite Granite	Rionegro Batholith	Goldsmith et al. (1971) Ward et al. (1973)
7-1-1-89	7°12′46.79″ −72°53′31.03″	208.9	30	U-Pb	Zr	Tonalite	Páramo Rico Tonalite – Granodiorite	Dörr et al. (1995)
7-1-1-89	7°12′46.79″ −72°53′31.03″	211.1		U-Pb	Zr	Tonalite	Páramo Rico Tonalite – Granodiorite	Dörr et al. (1995)
10-1-4-89	7°20′32.77″ −72°56′49.05″	205	+5/−9	U-Pb	Zr Inf. intercept	Granodiorite	Páramo Rico Tonalite – Granodiorite	Dörr et al. (1995)
LD13-1	7°20′30.38″ −72°57′09.98″	208.8	4.1	U-Pb	Zr LA-MC-ICP-MS	Rhyodacite Zr recycled	Páramo Rico Tonalite – Granodiorite	Mantilla et al. (2009)
–		205–210		U-Pb	Zr LA-MC-ICP-MS	Granitoids		Leal et al. (2011)
ALR035	7°22′38.69″ −72°54′21.11″	210.6	3.5	U-Pb	Zr TIMS	Quartz-monzonite (Alaskite I)	Páramo Rico Tonalite – Granodiorite	Mantilla et al. (2012)
ALR035	7°22′38.69″ −72°54′21.11″	~201		U-Pb	Zr TIMS Min. age inf. intercept	Quartz-monzonite (Alaskite I)	Páramo Rico Tonalite – Granodiorite	Mantilla et al. (2012)
GE-20-M1	7°19′00.29″ −72°53′56.39″	204.3	+2.7/−3.3	U-Pb	Zr LA-MC-ICP-MS	Quartz-monzonite dike (Alaskite I)	Páramo Rico Tonalite – Granodiorite	Mantilla et al. (2012)
GI-47-M1	7°21′21.09″ −72°54′25.81″	202.2	+5.3/−3.3	U-Pb	Zr LA-MC-ICP-MS	Quartz-monzonite (Alaskite I)	Páramo Rico Tonalite – Granodiorite	Mantilla et al. (2012)

(continued)

Table 4.1 (continued)

Sample	Coordinates	Age	Error	Method	Material	Lithology	Unit	Reference
TBQ-002	7°20′55.60″ −72°56′04.59″	199.1	+2.5/−2.6	U-Pb	Zr LA-MC-ICP-MS	Quartz-monzonite (Alaskite I)	Páramo Rico Tonalite – Granodiorite	Mantilla et al. (2012)
TPD-71	7°19′18.58″ −72°54′10.91″	199.2	+2.8/−2.7	U-Pb	Zr LA-MC-ICP-MS	Diorite – Granodiorite	Páramo Rico Tonalite – Granodiorite	Mantilla et al. (2012)
TBQ-005	7°20′57.11″ −72°56′49.21″	199.0	+2.5/−2.6	U-Pb	Zr LA-MC-ICP-MS	Diorite – Granodiorite	Páramo Rico Tonalite – Granodiorite	Mantilla et al. (2012)
TBQ-003	7°20′55.60″ −72°56′04.59″	198.4	+2.5/−2.6	U-Pb	Zr LA-MC-ICP-MS	Quartz-monzonite (Alaskite II)	Páramo Rico Tonalite – Granodiorite	Mantilla et al. (2012)
TBQ-004	7°21′01.68″ −72°56′15.98″	198.7	+2.6/−2.9	U-Pb	Zr LA-MC-ICP-MS	Quartz-monzonite (Alaskite II)	Páramo Rico Tonalite – Granodiorite	Mantilla et al. (2012)
TBQ-001	7°19′55.09″ −72°54′56.81″	196.7	+2.9/−2.8	U-Pb	Zr LA-MC-ICP-MS	Quartz-monzonite (Alaskite II)	Páramo Rico Tonalite – Granodiorite	Mantilla et al. (2012)
BOC-1	8°11′45.41″ −73°19′27.29	268	26	Sm-Nd	WR isochron	Quartz-monzonite – Granite	Ocaña Batholith	Ordóñez (2001)

Sample	Coordinates	Age	Error	Method	Material	Lithology	Unit	Reference
BOC-2, BOC-3	8°10′59.59″ −73°19′16.21″	268	26	Sm-Nd	WR isochron	Quartz-monzonite – Granite	Ocaña Batholith	Ordóñez (2001)
BOC-4	8°10′53.78″ −73°18′51.36″	268	26	Sm-Nd	WR isochron	Quartz-monzonite – Granite	Ocaña Batholith	Ordóñez (2001)

–		168	7	Rb-Sr	WR isochron	Granodiorite	Pescadero Granodiorite	Ordóñez-Calderón (2003)
–		214	12	Rb-Sr	WR isochron	Quartz-monzonite	Santa Bárbara Quartz-monzonite	Ordóñez-Calderón (2003)
–		212	15	Rb-Sr	WR isochron	Quartz-monzonite	La Corcova Quartz-monzonite	Ordóñez-Calderón (2003)
10VDL05	7°06′02.99″ −73°00′35.99″	198.3	1.8	U-Pb	Zr LA-ICP-MS	Granodiorite dike	La Corcova Granodiorite	Van der Lelij (2013)
10VDL22	6°49′47.99″ −72°59′27.01″	199.1	1.3	U-Pb	Zr LA-ICP-MS	Granodiorite	Pescadero Granodiorite	Van der Lelij (2013)
10VDL22	6°49′47.99″ −72°59′27.01″	172.0	16.4	ZFT	Zr LA-ICP-MS	Granodiorite	Pescadero Granodiorite	Van der Lelij (2013)
10VDL28	6°22′30.99″ −72°49′05.99″	200.4	0.7	U-Pb	Zr LA-ICP-MS	Granodiorite	Onzaga Granodiorite	Van der Lelij (2013)
10VDL31	6°24′27.99″ −72°49′08.01″	201.0	0.9	U-Pb	Zr LA-ICP-MS	Granodiorite	Onzaga Granodiorite	Van der Lelij (2013)
10VDL32	6°25′22.04″ −72°49′28.99″	198.0	0.8	U-Pb	Zr LA-ICP-MS	Granodiorite	Mogotes Granodiorite	Van der Lelij (2013)
10VDL32	6°25′22.04″ −72°49′28.99″	132.2	12.9	ZFT	Zr LA-ICP-MS	Granodiorite	Mogotes Granodiorite	Van der Lelij (2013)
10VDL35	7°10′21.99″ −73°05′07.98″	201.1	1.4	U-Pb	Zr LA-ICP-MS	Diorite	Suratá River Diorite	Van der Lelij (2013)
10VDL52	7°13′53.99″ −72°53′53.99″	227.2	22.1	ZFT	Zr LA-ICP-MS	Tonalite	Páramo Rico Tonalite – Granodiorite	Van der Lelij (2013)
10VDL52	7°13′53.99″ −72°53′53.99″	199.8	1.2	U-Pb	Zr LA-ICP-MS	Tonalite	Páramo Rico Tonalite – Granodiorite	Van der Lelij (2013)

(continued)

Table 4.1 (continued)

10VDL54	8°09′45.01″ −73°17′58.99″	195.8	1.5	U-Pb	Zr LA-ICP-MS	Granite	Ocaña Batholith	Van der Lelij (2013)
10VDL56	8°07′07.99″ −72°56′51.01″	202.2	1	U-Pb	Zr LA-ICP-MS	Granite	Agua Blanca Batholith	Van der Lelij (2013)
10VDL58	7°26′27.01″ −73°13′24.01″	250.7	4.3	U-Pb	Zr LA-ICP-MS	Rhyolite dike	Bocas? Formation	Van der Lelij (2013)
10VDL59	7°17′13.01″ −73°08′46.01″	196.0	1.1	U-Pb	Zr LA-ICP-MS	Tonalite	Rionegro Batholith	Van der Lelij (2013)
10VDL61	7°23′59.00″ −73°05′16.99″	200.0	1.1	U-Pb	Zr LA-ICP-MS	Granodiorite	Suratá River Granodiorite	Van der Lelij (2013)

Santa Rosita Quartz-monzonite This is an N-S elongated medium-grained equigranular to locally porphyritic biotite-epidote quartz-monzonite pluton, with variations to granite and granodiorite. This body is presumably intruding the MGV unit.

Santa Bárbara Quartz-monzonite This is an N-S elongated coarse-grained pink biotite-hornblende quartz-monzonite to monzogranite pluton west of the Bucaramanga fault and intrudes mainly the Bucaramanga Gneiss and the Berlín Orthogneiss. It has local compositional variations to medium-grained syenogranite and granodiorite. It is also characterized by the presence of pegmatitic and aplitic granitic dikes and locally dark fine-grained tonalitic dikes.

Mogotes Quartz-monzonite This is a N-S elongated pluton with several igneous facies. The principal igneous facies are an orange to pink biotite quartz-monzonite and a medium-grained equigranular granite. Other characteristic facies are a coarse-grained quartz-monzonite, a fine-grained monzogranite, and a medium- to coarse-grained granodiorite. This pluton intrudes the Bucaramanga Gneiss, the Silgará Schist, and the MGV.

La Corcova Quartz-monzonite This is an elongated N-S pluton and intrudes the Bucaramanga Gneiss. It has four igneous facies: a coarse-grained granodiorite, a porphyritic biotite monzogranite-granodiorite, a medium-grained biotite monzogranite, and a medium-grained hornblende-biotite granodiorite.

Páramo Rico Tonalite (Granodiorite) This is an irregular granodiorite, tonalite, and monzogranite pluton with a N-S elongation. It intrudes the Bucaramanga Gneiss, the Silgará Schist, and the Santa Bárbara Quartz-monzonite.

Agua Blanca Granite (Agua Blanca Batholith) This is a muscovite-biotite quartz-monzonite pluton elongated in the N-S direction. It intrudes and it is faulted against the Silgará Schist. It also has locally granitic and microgranitic dikes.

Onzaga Granodiorite This is an N-S elongated biotite granodiorite and quartz-monzonite pluton.

Suratá River Pluton This is a small medium-grained biotite-hornblende diorite, tonalite, and granodiorite pluton. It is characterized by the presence of microgranular enclaves. It is emplaced in the Bucaramanga Gneiss and locally in the Silgará Schist.

Ocaña Alkaline Granite (Ocaña Batholith) It is a fine- to coarse-grained pink to gray biotite-muscovite quartz-monzonite and monzogranite. It intrudes the Bucaramanga Gneiss. It locally has variations to porphyritic rhyolite. It also has locally rhyolite, diabase (dolerite), and basalt dikes.

Rionegro Batholith This is a medium- to coarse-grained gray to pink quartz-monzonite to monzogranite pluton with elongated N-S direction that intrudes the Silgará Schist and locally the Bucaramanga Gneiss. It has compositional variations to granodiorite, tonalite, and locally to charnockite. It also has locally granite and tonalite dikes.

Pescadero Monzogranite This is an elongated leucocratic pluton of irregular shape with five igneous facies: a coarse-grained biotite monzogranite, a porphyritic to granophyric muscovite monzogranite, a medium-grained biotite monzogranite, and locally granodiorite.

4.5.2 Geochemistry

We present here an interpretation from a geochemical dataset of most of the Triassic-Jurassic plutons compiled from the literature (Dörr et al. 1995; van der Lelij 2013; Mantilla et al. 2013; Bissig et al. 2014; van der Lelij et al. 2016) and complemented with data from new samples from some of the less well-studied plutons (Table 4.2).

4.5.2.1 Discriminant Tectonic Setting and Magma Affinity

Paleozoic plutons display major element compositional variation with SiO_2 content from 55 to 79 wt%, A/CNK Shand's index from 0.7 to 2.7, and agpaitic index (non-peralkaline rocks) from 0.2 to 0.7. Similarly, Triassic-Jurassic plutons have SiO_2 content ranging from 51 to 78 wt%, A/CNK Shand's index from 0.7 to 3.4, and agpaitic index from 0.2 to 0.9. Most plutons can be classified as silica-rich granitoids (SiO_2 > 70 wt. %; Frost et al. 2016); however, there are some normal granitoids (SiO_2 < 70 wt. %). The compositional variation is reflected in the observed geochemical classification ranging from monzogabbro to granite (Fig. 4.9a); however, note that most lithologies fall in the fields of granite, granodiorite, and adamellite (Fig. 4.9a). All plutons fall within the subalkaline series with metaluminous to peraluminous character (Fig. 4.9b). Most Paleozoic rocks are calc-alkaline; this contrasts to the high-K calc-alkaline to shoshonitic character of Triassic-Jurassic granitoids (Fig. 4.9c, d). Paleozoic plutons (Durania and Pamplona) are also characterized by a highly felsic peraluminous character with major proportion of muscovite than biotite (Fig. 4.9e, f). Triassic-Jurassic plutons have a wide range in aluminum balance from highly felsic peraluminous to highly peraluminous with major proportion of muscovite than biotite (felsic portions of Páramo Rico, Rionegro, and Pescadero), highly peraluminous to moderately peraluminous with major proportion of biotite than muscovite (La Corcova, Santa Bárbara, Pescadero, Rionegro), low peraluminous to medium peraluminous with hornblende (La Corcova, Santa Bárbara, Pescadero, Mogotes, Páramo Rico) to metaluminous (Santa Bárbara, Páramo Rico) (Fig. 4.9e, f). Paleozoic plutons are mainly

Table 4.2 Major (wt%) and trace element (ppm) compositions of Jurassic igneous rocks reported in this study and discussed in the text (Páramo Rico Pluton and Rionegro Batholith)

Sample	J989643	J989644	J989645	J989646	J989647	J989648
Pluton	Páramo Rico					
Coordinates	7°16′25.76″ −72°56′15.22″	7°16′15.68″ −72°55′13.12″	7°16′15.68″ −72°55′13.12″	7°16′20.49″ −72°56′54.59″	7°16′26.41″ −72°56′54.59″	7°16′40.09″ −72°57′00.88″
SiO_2	59.9	59.7	73.9	60.5	71.6	71.2
Al_2O_3	16.55	17.35	12.70	17.25	12.85	11.20
Fe_2O_3	7.08	6.89	1.30	7.75	5.27	5.63
MgO	2.80	2.57	0.30	1.74	1.26	1.60
CaO	4.10	3.61	0.68	0.70	0.29	0.70
Na_2O	2.46	3.52	2.61	1.19	1.13	3.17
K_2O	1.91	4.33	0.1	3.07	1.90	1.85
TiO_2	0.99	0.99	0.14	0.97	0.73	0.70
P_2O_5	0.39	0.33	0.04	0.07	0.10	0.18
MnO	0.13	0.14	0.03	0.18	0.10	0.12
Cr_2O_3	0.01	B.D.	B.D.	0.01	0.01	B.D.
SrO	0.05	0.06	0.03	0.02	0.01	0.01
BaO	0.11	0.09	0.12	0.20	0.07	0.05
C	0.03	0.10	0.07	0.05	0.01	0.03
S	0.02	0.10	0.02	0.03	0.01	0.10
LOI	3.18	2.99	1.00	3.59	2.59	2.89
Ba	1000	715	1050	1735	578	405
Ce	42.3	139.5	42.4	135.5	85.6	30.2
Co	17	12	2	15	11	11
Cs	3.91	4.67	3.23	8.79	4.74	1.04
Cu	26	24	15	24	22	2310
Dy	2.96	4.45	1.72	7.32	5.41	2.78

(continued)

Table 4.2 (continued)

Sample	J989643	J989644	J989645	J989646	J989647	J989648
Er	1.59	2.53	0.98	5.71	3.26	1.56
Eu	1.41	1.90	0.73	1.53	1.31	0.76
Ga	22.3	22.7	12.7	23.1	18.6	15.3
Gd	3.77	6.03	2.02	6.91	5.75	2.97
Hf	9.3	9.4	2.9	6.7	9.5	3.7
Ho	0.60	0.91	0.35	1.83	1.17	0.56
La	22.8	76.6	23.0	69.2	43.7	14.5
Lu	0.24	0.38	0.19	1.01	0.53	0.22
Nb	18.1	12.4	7.2	20.2	15.0	10.3
Nd	22.3	55.7	15.7	52.2	37.1	15.3
Ni	24	8	B.D.	32	20	12
Pb	21	17	85	15	18	444
Pr	5.31	14.95	4.31	14.15	9.53	3.60
Rb	151.0	115.5	2.7	194.0	158.5	80.2
Sm	4.43	8.48	2.83	8.90	7.11	3.20
Sr	420.0	540.0	236.0	195.0	124.0	110.5
Ta	1.2	0.9	0.7	1.8	1.2	1.0
Tb	0.52	0.77	0.27	1.01	0.86	0.44
Th	2.35	19.05	16.95	31.6	19.25	2.97
Tm	0.22	0.38	0.17	0.96	0.52	0.23
U	1.96	2.82	5.30	4.87	2.28	2.56
V	129	139	14	165	87	99
Y	16.6	24.0	9.8	47.7	31.3	14.9
Yb	1.41	2.29	1.12	6.60	3.31	1.42
Zn	129	97	15	91	84	304
Zr	390	389	95	240	353	150

Pluton	Páramo Rico					
Coordinates	7°16′33.73″ −72°55′47.17″	7°16′26.55″ −72°54′48.58″	7°16′31.25″ −72°55′44.53″	7°16′28.28″ −72°57′13.26″	7°16′14.25″ −72°55′16.25″	7°16′32.88″ −72°55′34.23″
SiO_2	51.9	63.6	58.5	75.5	73.6	65.3
Al_2O_3	18.30	16.75	17.65	11.05	10.75	14.10
Fe_2O_3	9.70	6.65	7.39	3.32	5.84	6.32
MgO	4.10	1.20	2.72	0.45	1.32	2.45
CaO	6.69	0.79	4.08	0.09	3.01	4.44
Na_2O	3.08	1.40	2.89	0.03	1.48	2.24
K_2O	2.39	3.71	2.18	2.84	1.91	2.39
TiO_2	1.29	0.86	1.18	0.38	0.66	0.82
P_2O_5	0.42	0.10	0.39	0.07	0.13	0.27
MnO	0.17	0.13	0.13	0.01	0.09	0.13
Cr_2O_3	0.01	0.01	0.01	B.D.	0.01	B.D.
SrO	0.08	0.02	0.06	0.01	0.05	0.05
BaO	0.10	0.11	0.07	0.02	0.06	0.07
C	0.09	0.08	0.10	0.05	0.05	0.01
S	0.07	0.01	0.07	0.01	0.01	0.13
LOI	2.18	4.20	4.07	4.37	1.49	1.60
Ba	867	932	585	178.5	518	651
Ce	88.7	112.0	86.5	99.6	52.8	61.8
Co	18	16	17	8	10	11
Cs	3.66	6.88	5.04	24.0	2.22	2.07
Cu	22	39	39	1	7	23
Dy	8.01	7.15	4.01	4.64	2.23	5.81
Er	4.41	4.48	2.27	2.47	1.24	3.20

(continued)

Table 4.2 (continued)

Sample	J989700	J989701	J989702	J989703	J989710	J989711
Eu	2.18	1.56	1.63	1.31	1.05	1.39
Ga	25.6	22.8	21.7	12.9	17.4	19.9
Gd	9.26	7.83	4.65	5.97	2.80	5.98
Hf	9.1	6.2	4.5	7.2	7.9	4.7
Ho	1.66	1.54	0.81	0.92	0.44	1.20
La	39.3	57.3	26.8	53.0	26.8	26.5
Lu	0.59	0.66	0.31	0.37	0.21	0.43
Nb	13.0	16.4	12.7	8.8	8.5	10.2
Nd	49.5	48.1	26.7	40.8	22.2	33.8
Ni	8	27	12	11	10	5
Pb	10	27	11	2	11	8
Pr	11.20	12.40	6.46	11.15	6.03	8.07
Rb	113.0	140.5	101.0	169.5	94.8	97.1
Sm	10.30	9.07	5.11	7.80	4.01	7.50
Sr	655	202	4676	69.3	430	398
Ta	0.8	1.4	1.1	0.9	0.7	0.9
Tb	1.31	0.77	0.66	0.80	0.41	0.98
Th	4.68	23.2	6.99	20.5	10.75	4.00
Tm	0.62	0.67	0.33	0.39	0.19	0.47
U	1.63	5.73	3.50	2.91	3.25	2.96
V	233	120	157	53	114	174
Y	41.9	42.0	21.2	24.2	11.6	31.3
Yb	3.90	4.23	2.11	2.44	1.28	2.95
Zn	112	81	90	17	82	80
Zr	385	221	167	257	266	168

Pluton	Páramo Rico				Rionegro Batholith	
Coordinates	7°16′55.86″ −72°56′01.88″	7°15′24.12″ −72°55′00.21″	7°15′01.32″ −72°55′01.32″	7°15′38.32″ −72°55′01.32″	7°29′46.89″ −73°13′57.51″	7°30′52.35″ −73°15′24.75″
SiO_2	52.2	64.1	71.9	49.6	77.7	76.8
Al_2O_3	17.20	14.00	12.20	17.05	11.6	10.45
Fe_2O_3	10.10	7.62	4.38	12.30	1.08	3.06
MgO	4.33	2.92	1.04	3.20	0.07	0.19
CaO	7.94	4.25	0.74	2.87	0.24	0.05
Na_2O	2.90	2.44	1.27	1.11	3.81	2.38
K_2O	2.41	2.11	3.02	3.35	4.38	4.78
TiO_2	1.52	0.99	0.58	1.42	0.1	0.13
P_2O_5	0.19	0.33	0.09	0.17	0.01	0.01
MnO	0.13	0.13	0.05	0.12	0.01	0.04
Cr_2O_3	0.01	B.D.	0.01	0.03	B.D.	B.D.
SrO	0.06	0.07	0.01	0.02	B.D.	0.01
BaO	0.14	0.10	0.06	0.09	0.01	0.02
C	0.01	0.01	0.02	0.66	0.01	0.02
S	0.12	0.08	0.26	0.90	0.08	0.14
LOI	2.40	2.59	2.40	7.07	0.39	1.1
Ba	1285	856	529	870	75.3	212
Ce	61.6	89.8	64.5	27.7	63.2	111
Co	30	13	9	31	1	B.D.
Cs	2.0	5.39	5.54	5.79	0.2	0.15
Cu	93	996	25	167	B.D.	4
Dy	2.70	5.17	4.95	3.81	4.76	21.4

(continued)

Table 4.2 (continued)

Sample	J989712	J989714	J989716	J989717	I388935	I388944
Er	1.44	2.72	3.10	2.08	3.34	14.85
Eu	1.20	1.56	1.10	1.20	0.23	0.61
Ga	23.1	21.2	17.5	25.1	17.7	28.6
Gd	3.36	5.88	4.78	3.25	3.49	14.05
Hf	4.5	5.9	8.8	4.8	4.2	29.9
Ho	0.56	1.06	1.10	0.82	1.05	4.62
La	33.4	43.0	30.4	13.4	33.6	47.9
Lu	0.23	0.37	0.53	0.27	0.57	2.36
Nb	8.4	13.9	14.1	22.7	23.4	50.8
Nd	25.4	42.2	29.9	13.7	22.4	52.2
Ni	56	5	16	81	B.D.	B.D.
Pb	10	23	21	32	6	B.D.
Pr	6.86	10.85	7.82	3.43	6.75	13.55
Rb	96.2	108.5	154.5	191.0	138	98.4
Sm	4.53	8.12	6.00	3.38	4.17	13.45
Sr	510	514	127.5	160.0	32.5	65.8
Ta	0.6	0.8	1.0	1.5	2.2	3.4
Tb	0.50	0.88	0.78	0.61	0.72	2.92
Th	6.83	5.83	13.65	3.52	16.05	14.05
Tm	0.21	0.40	0.49	0.31	0.52	2.32
U	0.98	2.07	3.84	2.27	3.16	2.6
V	356	189	70	291	B.D.	B.D.
Y	14.3	26.3	30.2	21.3	32.1	130.5
Yb	1.38	2.39	3.29	1.86	3.92	15.5
Zn	87	138	60	123	6	28

4 Ordovician Orogeny and Jurassic Low-Lying Orogen in the Santander Massif...

Zr	169	217	304	166	107	1170
Sample	I388923	I388924	I388927	I388928	I388929	I388984
Pluton	Rionegro Batholith					
Coordinates	7°41′17.09″ −73°04′00.33″	7°41′42.07″ −73°04′51.66″	7°43′29.29″ −73°02′53.52″	7°41′57.25″ −73°04′56.19″	7°42′05.57″ −73°04′40.41″	7°41′04.89″ −73°03′37.59″
SiO_2	74.9	73.5	75.6	66.8	70.6	70.5
Al_2O_3	13.9	12.2	13.5	15.35	14.45	14.65
Fe_2O_3	1.67	3.07	1.07	4.48	2.28	3.56
MgO	0.37	0.69	0.25	0.93	0.31	0.99
CaO	0.67	0.63	0.59	1.96	0.52	2.26
Na_2O	3.07	1.52	3.44	3.42	3.49	2.89
K_2O	4.48	3.84	4.31	3.45	4.47	3.49
TiO_2	0.17	0.45	0.1	0.6	0.23	0.43
P_2O_5	0.17	0.14	0.12	0.45	0.11	0.15
MnO	0.05	0.06	0.05	0.13	0.04	0.06
Cr_2O_3	B.D.	B.D.	B.D.	B.D.	B.D	B.D.
SrO	0.01	0.01	0.01	0.02	0.01	0.03
BaO	0.05	0.08	0.04	0.09	0.09	0.1
C	0.03	0.05	0.02	0.03	0.04	0.03
S	B.D.	0.01	0.01	B.D.	0.04	B.D.
LOI	1.67	1.17	0.99	2.49	1.49	0.99
Ba	456	688	355	731	743	828
Ce	40.9	54.4	23.4	134	73.9	89.4
Co	1	6	8	7	2	9
Cs	4.66	3.94	5.35	5.09	3.16	5.27

(continued)

Table 4.2 (continued)

Sample	I388923	I388924	I388927	I388928	I388929	I388984
Cu	8	2	9	19	15	24
Dy	3.34	5.15	1.93	5.51	4.4	3.25
Er	1.72	3.04	1.1	2.95	2.39	1.66
Eu	0.67	1.17	0.5	1.58	0.94	1.31
Ga	17.5	16.2	14.8	18.9	16.9	19.6
Gd	3.27	4.94	1.89	7.31	5.03	4.9
Hf	3.1	6.8	1.9	8.5	4.4	5.3
Ho	0.61	1.03	0.37	1	0.83	0.58
La	21.4	26.4	12.9	66.9	39.7	48.2
Lu	0.21	0.39	0.16	0.35	0.33	0.22
Nb	15.8	14.7	9.5	21.7	15.7	14.5
Nd	16.5	24.1	9.1	53.1	28.5	32.9
Ni	B.D.	10	1	1	B.D.	12
Pb	18	20	20	19	16	16
Pr	4.4	6.01	2.47	14.1	7.71	9.65
Rb	181.5	142	166.5	151.5	176	127.5
Sm	3.38	5.01	1.93	8.37	5.26	5.67
Sr	93.6	102	120	172	112.5	233
Ta	2	1.1	1.9	1.8	1.6	1.1
Tb	0.55	0.81	0.32	1.01	0.77	0.68
Th	6.43	10.2	3.79	11.95	13.3	12.45
Tm	0.23	0.42	0.16	0.38	0.34	0.23
U	2.97	2.99	2.16	6.49	2.85	2.06
V	13	33	10	49	18	41
Y	17.2	27.7	10.8	26.2	23.7	15.9
Yb	1.52	2.84	1.11	2.61	2.34	1.45

Sample	30	55	19	75	35	65
Zn	30	55	19	75	35	65
Zr	111	266	59	412	168	207
Sample	I388985	I388986	I388991	I388992	I388994	I388995
Pluton	Rionegro Batholith					
Coordinates	7°41′04.89″ −73°03′37.59″	7°41′26.79″ −73°04′06.74″	7°41′58.69″ −73°04′50.31″	7°42′12.62″ −73°03′46.00″	7°42′49.02″ −73°03′40.99″	7°42′51.81″ −73°03′40.59″
SiO_2	77.0	77.4	67.2	56.7	68.7	66.9
Al_2O_3	13.65	13.4	15	19.55	14.25	14.3
Fe_2O_3	0.79	0.76	3.65	7.92	2.98	3.41
MgO	0.11	0.14	0.85	2.24	0.6	1.01
CaO	0.4	0.35	1.68	0.54	1.37	1.2
Na_2O	3.37	3.26	3.25	1.28	2.94	3.09
K_2O	4.35	5.03	3.96	3.96	3.52	3.51
TiO_2	0.03	0.06	0.46	0.96	0.37	0.42
P_2O_5	0.11	0.07	0.27	0.16	0.016	0.017
MnO	0.07	0.02	0.07	0.56	0.1	0.06
Cr_2O_3	B.D.	B.D.	B.D.	0.01	B.D	B.D.
SrO	B.D.	0.01	0.02	0.01	0.02	0.02
BaO	B.D.	0.04	0.12	0.08	0.1	0.11
C	0.03	0.02	0.02	0.75	0.04	0.08
S	0.02	B.D.	0.01	0.14	0.01	0.01
LOI	0.68	0.88	1.98	4.47	3.21	3.43
Ba	26	369	1015	673	815	913
Ce	5.4	17.1	118	111.5	78.5	89.8
Co	B.D.	B.D.	6	13	6	6
Cs	7.57	2.85	3.3	7.76	3.79	1.38

(continued)

Table 4.2 (continued)

Sample	I388985	I388986	I388991	I388992	I388994	I388995
Cu	44	2	22	37	23	27
Dy	1.32	2.57	4.46	7.49	2.51	2.9
Er	0.75	1.6	2.35	4.24	1.25	1.45
Eu	0.08	0.58	1.65	1.79	1.13	1.14
Ga	18.9	13.3	18.7	26.8	18.9	17.9
Gd	0.93	1.8	6.98	8.93	4.44	5.17
Hf	1	1.3	7.3	4.5	5.2	5.8
Ho	0.25	0.51	0.8	1.45	0.44	0.51
La	2.8	9.3	59.9	53.9	41.7	48.3
Lu	0.14	0.23	0.33	0.57	0.15	0.19
Nb	15.5	5.6	28.2	19	24	13
Nd	2.5	6.8	45	45.8	29.3	33.6
Ni	B.D.	B.D.	2	24	14	16
Pb	15	17	22	27	16	28
Pr	0.69	1.94	13.25	12.7	8.64	9.96
Rb	274	185.5	142.5	186	184.5	117.5
Sm	0.92	1.61	7.28	8.69	4.71	5.43
Sr	16.9	67.2	203	117	184	174
Ta	3.4	0.8	1.3	1.2	10.7	1.1
Tb	0.21	0.39	0.9	1.34	0.53	0.62
Th	1.84	2.82	11.6	16.45	10.3	12.2
Tm	0.13	0.24	0.31	0.59	0.16	0.18
U	6.75	1.59	2.04	4.38	1.62	2.03
V	B.D.	5	39	142	38	45
Y	7.6	15.5	22.1	37.9	11.7	13.6
Yb	1.01	1.55	2.14	3.82	0.99	1.19
Zn	11	7	66	108	60	76

Sample	I388985	I388986	I388991	I388992	I388994	I388995	
Zr	16	37	323	169	186	235	

Sample	St SY-4	St OREAS-13P	St MA-1b	St UTS-1	St STSD-4	St OREAS-45c	Detection limit
SiO_2	51.5	46.8					0.01%
Al_2O_3	20.1	18.55					0.01%
Fe_2O_3	6.16	10.5					0.01%
MgO	0.49	5.26					0.01%
CaO	7.83	9.28					0.01%
Na_2O	6.94	2.5					0.01%
K_2O	1.57	0.55					0.01%
TiO_2	0.28	0.53					0.01%
P_2O_5	0.12	0.16					0.01%
MnO	0.11	0.14					0.01%
Cr_2O_3	B.D.	0.02					0.01%
SrO	B.D.	0.04					0.01%
BaO	B.D.	0.03					0.01%
C			2.46				0.01%
S				0.99			0.01%
LOI					11.5		0.01%
Ba	354	266					0.5 ppm
Ce	129.5	27.4					0.5 ppm
Co						100	1 ppm
Cs	1.56	0.28					0.01 ppm
Cu						712	1 ppm
Dy	19.9	3.57					0.05 ppm
Er	15.4	2.01					0.03 ppm

(continued)

Table 4.2 (continued)

Sample	St SY-4	St OREAS-13P	St MA-1b	St UTS-1	St STSD-4	St OREAS-45c	Detection limit
Eu	2.12	1.45					0.03 ppm
Ga	36.7	18.1					0.1 ppm
Gd	14.1	3.73					0.05 ppm
Hf	11.2	2.1					0.2 ppm
Ho	4.56	0.71					0.01 ppm
La	62.2	12.2					0.5 ppm
Lu	2.16	0.26					0.01 ppm
Nb	14.3	3.1					0.2 ppm
Nd	61.2	15.1					0.1 ppm
Ni						332	1 ppm
Pb						23	2 ppm
Pr	15.85	3.37					0.03 ppm
Rb	54.7	15.3					0.2 ppm
Sm	13.25	3.66					0.03 ppm
Sr	1250	344					0.1 ppm
Ta	0.8	0.2					0.1 ppm
Tb	2.84	0.58					0.01 ppm
Th	1.29	1.75					0.05 ppm
Tm	2.26	0.28					0.01 ppm
U	0.89	0.36					0.05 ppm
V	B.D.	125					5 ppm
Y	123.5	18.8					0.5 ppm
Yb	15.65	1.67					0.03 ppm
Zn						90	2 ppm
Zr	589	91					2 ppm

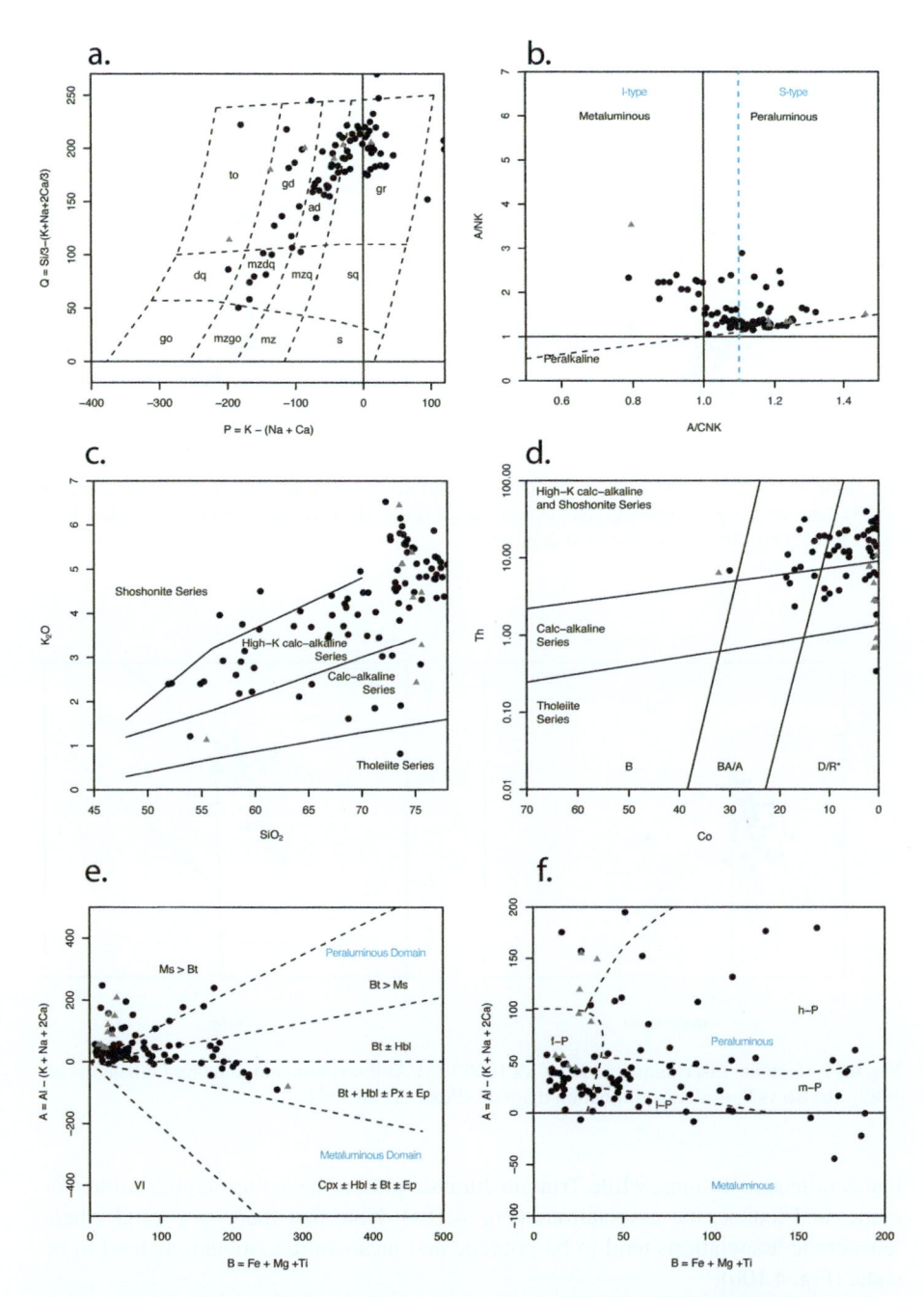

Fig. 4.9 Chemical composition of Jurassic granitoids. (**a**) P-Q geochemical classification multicationic plot (Debon and Le Fort 1983). (**b**) A/NK vs A/CNK discrimination diagram (Shand 1943). (**c**) SiO_2 vs K_2O discriminant diagram (Peccerillo and Taylor 1976). (**d**) Co vs Th discriminant diagram (Hastie et al. 2007). (**e, f**) B-A multicationic plots (Debon and Le Fort 1983; Villaseca et al. 1998)

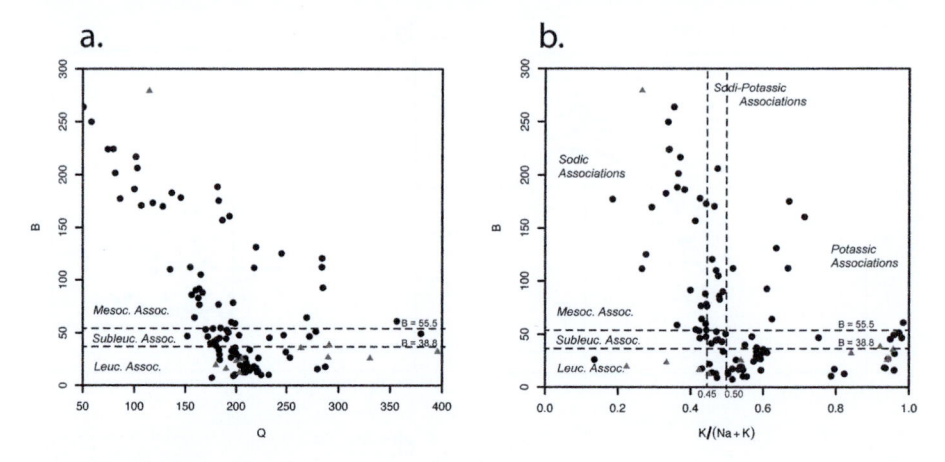

Fig. 4.10 Classification of igneous associations following the scheme of Debon and Le Fort (1983) that uses the parameters Q (quartz content), B (Fe + Mg + Ti), and alkali ratio (K/K + Na). (**a**) Q-B diagram. (**b**) K/(Na + K) vs B diagram

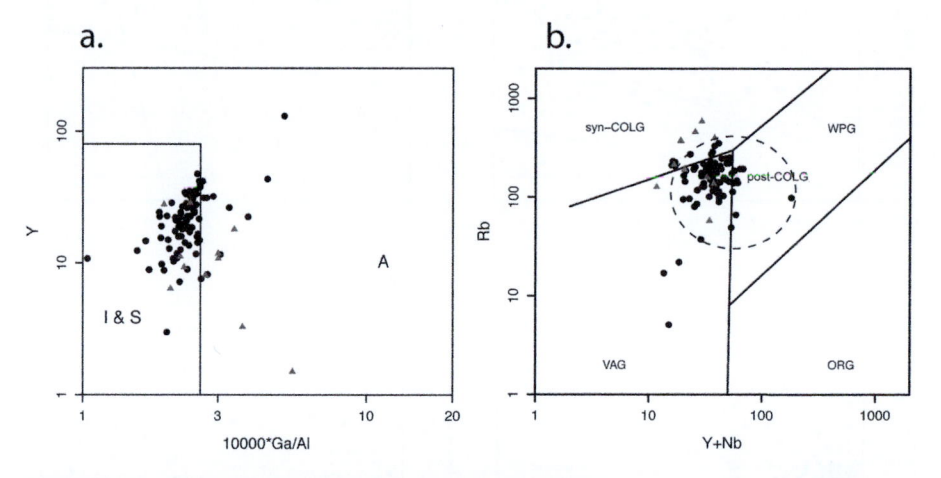

Fig. 4.11 Tectonic discrimination. (**a**) Y vs 10000*Ga/Al discrimination diagram (Whalen et al. 1987). (**b**) Rb vs Y + Nb discrimination diagram (Pearce et al. 1984)

leucocratic associations, while Triassic-Jurassic plutons have mesocratic, subleucocratic, and leucocratic associations (Fig. 4.10a). Note that there is a trend where leucocratic associations tend to be potassic and mesocratic associations tend to be sodic (Fig. 4.10b).

Most plutons have an I&S signature (Fig. 4.11a) with chemistry pointing to a volcanic arc granites (VAG) tectonic settings (Fig. 4.11b). However, the diagram of Fig. 4.11b also indicates the presence of lithologies with geochemical signatures consistent with syn-collisional granites (syn-COLG) and within plate granites (WPG).

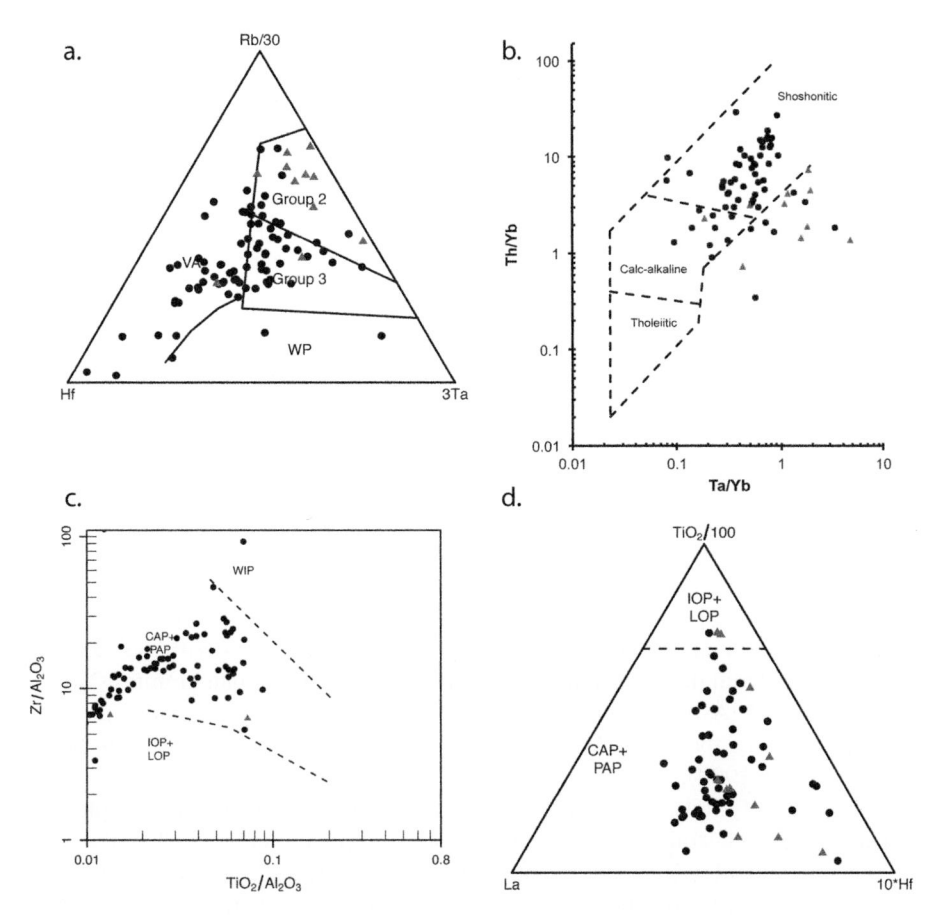

Fig. 4.12 Discrimination between I- and S-type granite signatures. (**a**) Hf-Rb/30-3Ta diagram (Harris et al. 1986). (**b**) Th/Yb vs Ta/Yb discrimination diagram (Pearce 1982). (**c, d**) Zr/Al_2O_3 vs TiO_2/Al_2O_3 and La-$TiO_2/100$-$10Hf$ tectonic discrimination diagrams (Müller et al. 1992)

The pluton compositional diversity is also evidenced in the Hf-Rb-Ta discrimination diagram of Harris et al. (1986) where samples are distributed in all tectonic setting fields of the diagram (Fig. 4.12a). For example, samples from the Páramo Rico and Rionegro plutons fall in three different fields: calc-alkaline volcanic arcs precollision, calc-alkaline syn-collisional with a crustal source, and calc-alkaline late or post-collisional with both crustal and mantle sources. However, sample geochemistry from other plutons indicates a more consistent tectonic setting in the diagram, i.e., the Durania pluton has a calc-alkaline syn-collisional with a crustal source signature and the Santa Bárbara, La Corcova, and Pescadero plutons fall in the calc-alkaline late or post-collisional with both crustal and mantle source fields. The majority of the Paleozoic and Triassic-Jurassic plutons have a high-K calc-alkaline to shoshonitic affinity (Fig. 4.12b) which might be problematic for discrimination of tectonic setting. Müller et al. (1992) highlight the importance of reviewing the different tectonic settings of rocks with $K_2O > 0.4$ wt% and proposed a tectonic discrimination following a hierarchical scheme in which the distinctive settings are successively differentiated. Following this approach, samples are first differentiated between two groups, one that includes initial and late oceanic arcs potassic (IOP + LOP) and the other that includes continental and post-collisional arcs potassic (CAP + PAP); here, most samples fall in the continental and post-collisional arcs potassic fields (CAP + PAP; Fig. 4.12c, d). The next step involves differentiation between CAP and PAP using discrimination diagrams with Zr, Nb, Ce, P, and Ti, which indicate continental arcs potassic (CAP) affinity.

From the Paleozoic group of plutons, only a gabbro-diorite sample has a Shand's index indicating I-type granitoids; the extremely felsic portions of the Durania pluton fall within the A-type granite field of Fig. 4.11a; however, these are further classified as A2-type (Eby 1992) which are related to post-collisional extensional settings. The rest of the samples have an S-type signature (e.g., Durania and Pamplona); this signature is consistent with the garnet, tourmaline, and muscovite mineralogical content of those lithologies. Although highly fractionated I- and S-type granites can present an A-type signature, the character of the extremely felsic portions of the Durania pluton is consistent with the fact that some samples from this pluton fall within the field of within plate granites (WPG; Fig. 4.11b) since A-type granites have been related to anorogenic magmatism in rifted portions of the crust (Whalen et al. 1987; Eby 1990, 1992; Bonin 1990; Barbarin 1990, 1999).

The extremely felsic portions of the Rionegro and Santa Bárbara plutons (Triassic-Jurassic), similarly to the Paleozoic group of plutons, fall within the A-type granite field of Fig. 4.11a. Note that there is no clear differentiation between S- and I-type granites for the Triassic-Jurassic plutons as visualized in the A/CNK vs Fe_2O_3 + FeO and the Na_2O vs K_2O diagrams (Pearce et al. 1984; Chappell and White 1974; Fig. 4.13a, b). The distribution of geochemical signatures for the Triassic-Jurassic plutons between S- and I-type granites might suggest that some bodies (e.g., Páramo Rico Tonalite-Granodiorite and Tonalite-Granodiorite-Quartz-monzonite) were emplaced as nested plutons. However, the P_2O_5 vs Rb, P_2O_5 vs SiO_2, Y vs Rb, and Th vs Rb diagrams (Chappell 1999) show a possible I-type trend for the Triassic-Jurassic magmatic belt and a possible S-type trend for the Paleozoic magmatic belt (Fig. 4.13c, d).

4.5.2.2 Trace Elements and Isotopic Relations

Paleozoic and Triassic-Jurassic plutons show similar trace element patterns in a spider plot normalized to chondrites in that they have a flat MREE to HREE pattern, enrichment in LREE, and negative anomalies for Ti and Nb (Fig. 4.14). However, Paleozoic plutons have a distinctive negative Th anomaly and show less REE

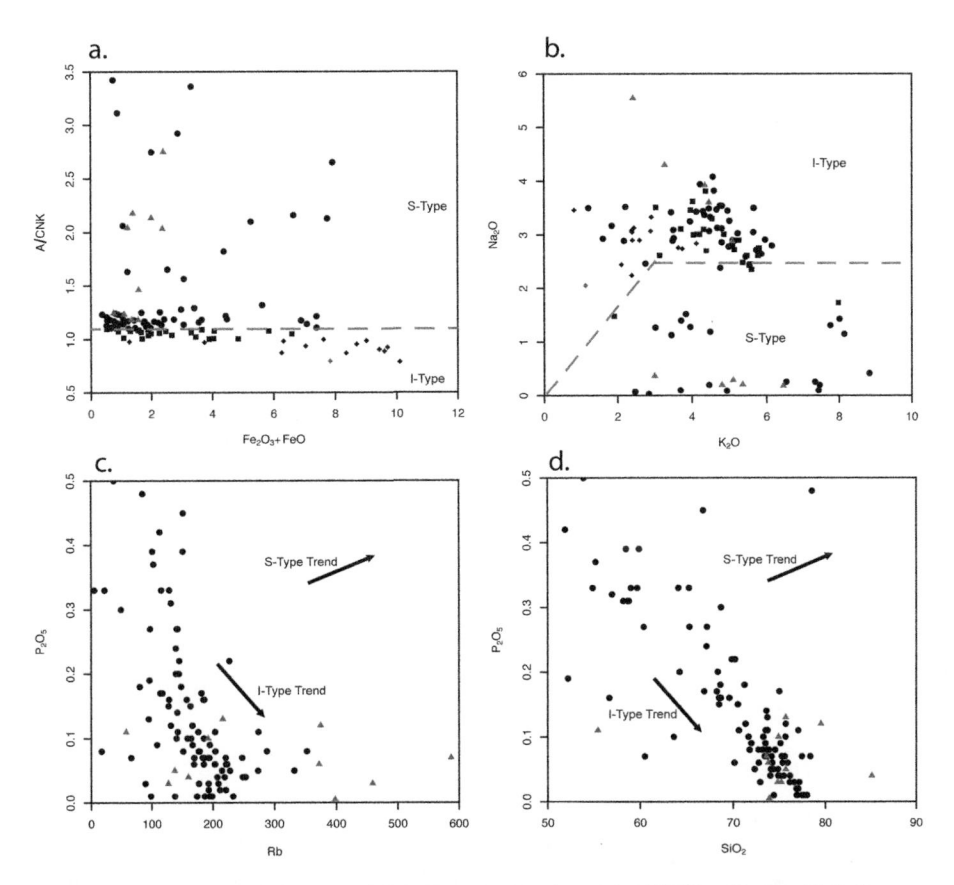

Fig. 4.13 Tectonic discrimination of high-K granitoid suites. (**a**) A/CNK vs Fe_2O_3 + FeO diagram (Pearce et al. 1984). (**b**) Na_2O vs K_2O diagram (Chappell and White 1974). (**c**) Rb vs P_2O_5 diagram (Chappell 1999). (**d**) SiO_2 vs P_2O_5 diagram (Chappell 1999)

enrichment, significant Ba depletion associated with a positive Rb anomaly, and no clear Ta negative anomaly. The Triassic-Jurassic plutons have negative to positive Th anomaly and the associated Ta negative anomaly. The Nb-Ta negative anomaly is consistent with the arc setting interpretation for the Triassic-Jurassic plutons. However, note also that negative Ti anomalies can be compatible with contamination by crustal melts (Chappell and White 1992; Thuy et al. 2004) or subduction settings (Pearce 1996). Triassic-Jurassic plutons also have $^{86}Sr/^{87}Sr$ and ε_{Nd} values suggesting an important crustal component in the parental magma (Ordóñez-Calderón 2003; Ordóñez et al. 2006; van der Lelij 2013). Paleozoic plutons have $^{86}Sr/^{87}Sr$ values suggesting typical compositions of S-type granites with an important crustal component, although some plutons have much less evolved sources, including igneous and sedimentary protoliths and minor depleted mantle-derived material (Ordóñez-Calderón 2003; van der Lelij 2013).

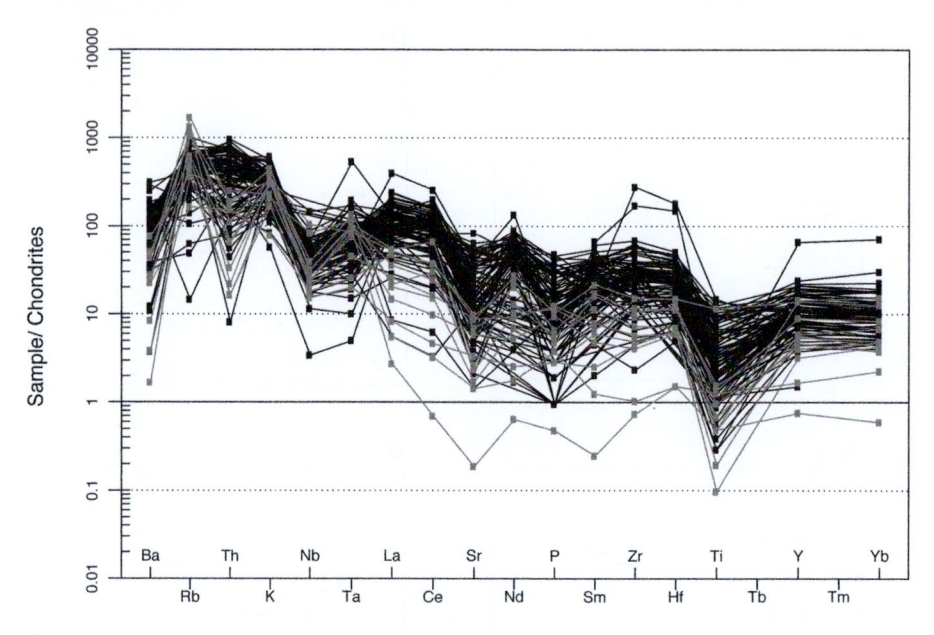

Fig. 4.14 Trace element spider plot normalized to chondrites (Thompson 1982)

4.6 Discussion

The metamorphic core of the Santander Massif (Silgará Schist, Bucaramanga Gneiss, and Berlín Orthogneiss) records an early Paleozoic tectonic pulse coetaneous with the Famatinian orogenic event in southern South America; this pulse is characterized in the massif by the occurrence of greenschist, amphibolite, and granulite facies metamorphism and migmatization (Zuluaga et al. 2017). In general, metamorphic field gradients show increasing grade from the eastern and western boundaries of the massif toward an NNE axis extending from the north of Cepitá to the Berlín area and likely northward of there; this observation and the abundance of leucosomes likely crystallizing from modified partial melts in the Berlín area hint at a dome-like structure with one of the deepest exhumed parts of the massif in the Berlín area. This observation is also consistent with the presence of large orthogneiss bodies in the proposed axis (e.g., Berlín Orthogneiss). Note that this interpretation does not preclude a previous metamorphism in the Bucaramanga Gneiss since protolith ages (<1.71 Ga; Cordani et al. 2005; Ordóñez-Carmona et al. 2006) allow time for known tectonic episodes to have affected the Bucaramanga Gneiss; however, the metamorphism during the Ordovician orogeny erased most evidence from previous episodes. The pelitic to semipelitic metasedimentary protolith sequence of the Silgará Schist has maximum ages of deposition of ca. 507 Ma (Mantilla et al. 2016) which indicates that this unit could have not been affected by metamorphic episodes occurring before the Ordovician orogeny.

The low abundance of late Paleozoic plutons does not allow us to have a strong conclusion about a possible late Paleozoic magmatic arc. However, note that most Paleozoic rocks are calc-alkaline contrasting with the high-K calc-alkaline to shoshonitic character of Triassic-Jurassic granitoids. For example, the Durania pluton has a calc-alkaline syn-collisional signature, and geochemistry indicates a strong crustal contamination.

The Jurassic granitoids in the Santander Massif belong to an identified regional Jurassic magmatic arc in Colombia (Altenberger and Concha-Perdomo 2005; Aspden et al. 1987; Dörr et al. 1995; Sillitoe et al. 1982; Tschanz et al. 1974; van der Lelij et al. 2016; Zuluaga et al. 2015) and are characterized by high-K calc-alkaline to shoshonitic affinity and low Nb-Ta typical of a subduction zone magmatism; this is consistent with $^{86}Sr/^{87}Sr$ and εNd values giving a continental signature (Ordóñez-Calderón 2003; Ordóñez et al. 2006; van der Lelij 2013). These characteristics are interpreted here as magma product of destruction of oceanic crust in a subduction setting (e.g., Müller et al. 1992).

The inconsistency between geochemical character (pointing to medium to high peraluminous character) and mineralogical composition (suggesting metaluminous and peraluminous magmatic series) and the inconsistency between ASI values (pointing to S-type granitoids) and Na_2O-K_2O contents (typical of I-type granitoids) can be related to contamination in relation to various sources of melts as suggested by van der Lelij (2013). For example, peraluminous leucogranites (portions of Páramo Rico and Rionegro plutons) could be the product of partial melting of pelitic rocks (Frost et al. 2016), ferroan calc-alkalic granitoids (portions of Rionegro, Ocaña, Páramo Rico, Pescadero, and La Corcova) could be the product of partial melting of tonalite or granodiorite (Frost et al. 2016), and ferroan alkali-calcic and magnesian granitoids could be the product of differentiation of mafic to intermediate magmas (Frost et al. 2016).

The N-S elongated shapes of the plutons suggest that the generation of space for the accommodation of the magmatic pulses was related to a trans-tensional structural setting, possibly associated to oblique convergence (oblique subduction). This interpretation is consistent with the extensional setting in the northwest corner of South America during Triassic-Jurassic times as reported by different authors (Kennan and Pindell 2009; Mojica et al. 1996; Cediel et al. 2003; van der Lelij 2013; van der Lelij et al. 2016). The extensional settings promoted crustal thinning and favored a high-temperature thermal regimen in the upper crust; this was probably linked to decompression melting. The interpretation of this mechanism together with the oblique subduction tectonic setting is supported by the high-K calc-alkaline to shoshonitic character, generating high-temperature S-type granitic rocks with a peraluminous character (Barbarin 1990), and geochemical signatures that suggest crustal contamination (van der Lelij et al. 2016). Additionally, the characteristics of the arc in the northern part of Colombia were recently interpreted as a low-lying magmatic arc with its axis along the intrusive belt (Zuluaga et al. 2015).

The low elevation of the arc and the high-temperature thermal regimen caused a regional low-pressure metamorphic event which partially overprinted the greenschist to granulite facies rocks of the metamorphic core as indicated by mineral microstructures (e.g., growth of cordierite), pressure-temperature estimations, Pb loss in zircons (van der Lelij et al. 2016), and early Jurassic K/Ar (biotite) dates from a biotite gneiss (Goldsmith et al. 1971). This event also produced low-grade metamorphism of Early Devonian sedimentary rocks (Zuluaga et al. 2017).

Acknowledgments This work received financial support from Colciencias (grant number 036-2013) and Universidad Nacional de Colombia (grant numbers 17296 and 28170). We thank Fabio Cediel for his critical reading of a first version of the manuscript.

References

Altenberger U, Concha-Perdomo AE (2005) Late lower to early middle Jurassic arc magmatism in the northern Ibagué-Batholith/Colombia. Geología Colombiana 30:87–97

Amaya S, Zuluaga C, Bernet M (2017) New fission-track age constraints on the exhumation of the central Santander Massif: implications for the tectonic evolution of the Northern Andes, Colombia. Lithos 282:388–402

Arenas C (2004) Litología y petrología de las metamorfitas carbonatadas y metasedimentitas asociadas de la Formación Silgará, faja noroeste de Mutiscua (Macizo de Santander), norte de Santander. Undergraduate Thesis, Bogotá, 156 p. Universidad Nacional de Colombia, Colombia

Aspden JA, McCourt WJ, Brook M (1987) Geometrical control of subduction-related magmatism: the Mesozoic and Cenozoic plutonic history of Western Colombia. J Geol Soc Lond 144:893–905

Audemard FA, Romero G, Rendón H, Cano V (2005) Quaternary fault kinematics and stress tensors along the southern Caribbean from fault-slip data and focal mechanism solutions. Earth Sci Rev 69:181–233

Barbarin B (1990) Granitoids: main petrogenetic classifications in relation to origin and tectonic setting. Geol J 25:227–238

Barbarin B (1999) A review of the relationships between granitoid types, their origins and their geodynamic environments. Lithos 46:605–626

Bissig T, Mantilla LC, Hart CJR (2014) Petrochemistry of igneous rocks of the California-Vetas mining district, Santander, Colombia: implications for northern Andean tectonics and porphyry Cu (–Mo, Au) metallogeny. Lithos 200-201:355–367

Boinet T, Bourgois J, Bellon H, Toussaint JF (1985) Age et repartition du magmatisme Premesozoique des Andes de Colombia. Comptes Rendus de l'Academie des Sciences Paris 300(II):445–450

Bonin B (1990) From orogenic to anorogenic settings: evolution of granitoid suites after a major orogenesis. Geol J 25:261–270

Botello F, Mantilla FLC, Colegial JD (2014) Edad U-Pb en zircones y contexto tectónico de formación delGranito de Durania (Macizo de Santander, Colombia). Memorias XI Semana Técnica de Geología y IGeosciences anual meeting. UIS. Bucaramanga

Bustamante C, Archanjo CJ, Cardona A, Vervoort JD (2016) Late Jurassic to early cretaceous plutonism in the Colombian Andes: a record of long-term arc maturity. Geol Soc Am Bull 128:B31307–B31301

Butler JP, Beaumont C, Jamieson RA (2011) Crustal emplacement of exhuming (ultra)high-pressure rocks: will that be pro- or retro-side? Geology 39:635–638

Cardona A (2003) Correlações entre fragmentos do embasamento pre-Mesozoíco da terminação setentrionaldos Andes Colombianos, com base em dados isotópicos e geocronológicos. Dissertação de Mestrado, Universidade de São Paulo, Brazil, p 119

Cardona A, Valencia VA, Lotero A, Villafañez Y, Bayona G (2016) Provenance of middle to late-Palaeozoic sediments in the northeastern Colombian Andes: implications for Pangea reconstruction. Int Geol Rev 58(15):1914–1939

Castellanos O, Ríos C, Akira T (2008) A new approach on the tectonometamorphic mechanisms associated with P–T paths of the Barrovian-type Silgará formation at the Central Santander Massif, Colombian Andes. Earth Sci Res J 12:125–155

Cediel F, Shaw RP, Cáceres C (2003) Tectonic assembly of the northern andean block. In: Bartolini C, Buffler RT, Blickwede J (eds) The Circum-Gulf of Mexico and the Caribbean: hydrocarbon habitats, basin formation, and plate tectonics, American Association of Petroleum Geologists memoir, vol 79. American Association of Petroleum Geologists, Tulsa, pp 815–848

Chappell BW (1999) Aluminium saturation in I- and S-type granites and the characterization of fractionated haplogranites. Lithos 46:535–551

Chappell BW, White AJR (1974) Two contrasting granite types. Pac Geol 8:173–174

Chappell BW, White AJR (1992) I- and S-type granites in the Lachlan Fold Belt. Trans R Soc Edinb Earth Sci 83:1–26

Cochrane R, Spikings R, Gerdes A, Ulianov A, Mora A, Villagómez D, Putlitz B, Chiaradia M (2014) Permo-Triassic anatexis, continental rifting and the disassembly of western Pangaea. Lithos 190–191:383–402

Cordani UG, Cardona A, Jimenez DM, Liu D, Nutman AP (2005) Geochronology of Proterozoic basement inliers in the Colombian Andes: tectonic history of remnants of a fragmented Grenville belt. Geol Soc Spec Pub 246:329–346

Daconte R, Salinas R (1980a) Geología de la Plancha 76 Ocaña Escala 1:100000. Instituto Colombiano de Geología y Minería

Daconte R, Salinas R (1980b) Geología de la Plancha 66 Miraflores Escala 1:100000. Instituto Colombiano de Geología y Minería

Debon F, Le Fort P (1983) A chemical-mineralogical classification of common plutonic rocks and associations. Trans R Soc Edinb Earth Sci 73:135–149

Dörr W, Grösser JR, Rodríguez GI, Kramm U (1995) Zircon U-Pb age of the Paramo Rico tonalite-granodiorite, Santander Massif (ordillera riental, Colombia) and its geotectonic significance. J S Am Earth Sci 8:187–194

Eby GN (1990) The A-type granitoids: a review of their occurrence and chemical characteristics and speculations on their petrogénesis. Lithos 26:115–134

Eby GN (1992) Chemical subdivision of the A-type granitoids: petrogenetic and tectonic implications. Geology 20:641–644

Etayo F, Barrero D et al (1983) Mapa de Terrenos de Colombia, Publicaciones Geológicas Especialesdel Ingeominas 14(1):235

Forero A (1990) The basement of the Eastern Cordillera Colombia: an allochthonous terrane in northwestern South America. J S Am Earth Sci 3:141–151

Frost CD, Frost BR, Beard JS (2016) On silica-rich granitoids and their eruptive equivalents. Am Mineral 101:1268–1284

Fúquen J, Ceballos L, Pedraza A, Marín E (2010) Geología de la Plancha 98 Durania Escala 1:100000. Instituto Colombiano de Geología y Minería

García C, Ríos C, Castellanos O (2005) Medium-pressure metamorphism in the central Santander Massif, Eastern Cordillera, Colombian Andes: constraints for a collision model. Boletín de Geología 27:43–68

Goldsmith R, Marvin RF, Mehnert HH (1971) Radiometric ages in the Santander Massif, Eastern Cordillera, Colombian Andes. US Geol Surv Prof Paper 750D:D44–D49

Harris NBW, Pearce JA, Tindle AG (1986) Geochemical characteristics of collision-zone magmatism. In: Coward MP, Ries AC (eds) Collision tectonics, Geological Society special publication, vol 19. Geological Society, London, pp 67–81

Hastie AR, Kerr AC, Pearce JA, Mitchell SF (2007) Classification of altered volcanic Island arc rocks usingImmobile trace elements: development of the Th-Co discrimination diagram. J Petrol 48:2341–2357

Jaillard E, Soler P, Carlier G, Mourier T (1990) Geodynamic evolution of the northern and central Andes during early to middle Mesozoic times: a Tethyan model. J Geol Soc 147:1009–1022

Jiménez C (2016) Caracterización petrológica y geoquímica de la unidad Ortoneis, Macizo de Santander, Colombia. MSc Thesis, Bogotá, 103 p. Universidad Nacional de Colombia, Colombia

Kammer A (1993) Steeply dipping basement faults and associated structures of the Santander Massif, Eastern Cordillera, Colombian Andes. Geologia Colombiana 18:47–64

Kammer A, Mojica J (1996) Una comparación de la tectónica de basamento de las cordilleras central y oriental. Geología Colombiana 20:93–106

Kennan L, Pindell JL (2009) Dextral shear, terrane accretion and basin formation in the Northern Andes: best explained by interaction with a Pacific-derived Caribbean plate? In: James KH, Lorente MA, Pindell JL (eds) The origin and evolution of the Caribbean plate, Geological Society special publications, vol 328. Geological Society, London, pp 487–531

Leal H, Melgarejo I, Draper JC, Shaw R (2011) Phanerozoic gold metallogeny in the Colombian Andes. In: Proceedings Let's talk ore deposits, Society for Geology Applied to mineral deposits, SGA biannual meeting, Antofagasta, Chile. Extended abstracts, pp 209–211

Mantilla LC, Valencia VA, Barra F, Pinto J, Colegial J (2009) Geocronología U-Pb del Distrito Aurífero de Vetas-California (Depto de Santander, Colombia). Boletín de Geología 31:31–43

Mantilla LC, Mendoza H, Bissig T, Hart CJR (2011) Nuevas evidencias sobre el magmatismo Miocénico en el Distrito Minero de Vetas-California (Macizo de Santander, Cordillera Oriental, Colombia). Boletín de Geología 33:43–58

Mantilla LC, Bissig T, Cottle JM, Hart CJR (2012) Remains of early Ordovician mantle-derived magmatism in the Santander Massif (Colombian Eastern Cordillera). J S Am Earth Sci 38:1–12

Mantilla LC, Bissig T, Valencia V, Hart CJR (2013) The magmatic history of the California-Vetas Mining District, Santander Massif, Eastern Cordillera, Colombia. J S Am Earth Sci 45:235–249

Mantilla LC, García CA, Valencia V (2016) Propuesta de escisión de la denominada 'Formación Silgará' (Macizo de Santander, Colombia), a partir de edades U-Pb en zircones detríticos. Boletín de Geología 38:33–47

McClay KR, Buchanan PG (1992) Thrust faults in inverted extensional basins. In: McClay KR (ed) Thrust tectonics. Chapman & Hall, London, pp 93–121

Merriman RJ, Frey M (1999) Patterns of very low-grade metamorphism in metapelitic rocks. In: Robinson D, Frey D (eds) Low-grade metamorphism. Blackwell, London, pp 61–107

Mojica J, Kammer A, Ujueta G (1996) El Jurásico del Sector Noroccidental de Suramérica y Guía de laExcursión al Valle Superior del Magdalena (Nov.1- 4/95), Regiones de Payandé y Prado, Departamento delTolima. Geología Colombiana 21:3–40

Moreno-Sánchez M, Gómez-Cruz AJ, Castillo-González H (2005) La "Formación Floresta Metamorfoseada" (sensu Ward et al, 1973) no es la Formación Floresta sin metamorfosear. X Congreso Colombiano de Geología, Bogotá, Memorias CD, pp 1–7

Müller D, Rock NMS, Groves DI (1992) Geochemical discrimination between shoshonitic and potassic volcanic rocks in different tectonic settings: a pilot study. Mineral Petrol 46:259–289

Oldow JS, Channell JET, Catalano R, D'Argenio B (1990) Contemporaneous thrusting and large-scale rotations in the western Sicilian fold and thrust belt. Tectonics 9:661–681

Ordóñez O (2001) Caracterização isotópica Rb-Sr e Sm-Nd dos principais eventos magmáticos nos AndesColombianos. Tesis de Doctorado (inédita). Universidad de Brasilia, p 176

Ordóñez O, Restrepo JJ, Pimentel MM (2006) Geochronological and isotopical review of pre-Devonian crustal basement of the Colombian Andes. J S Am Earth Sci 21:372–382

Ordóñez-Calderón JC (2003) Petrology of the granitoid rocks in the Santander Massif, Northeast Colombia. MSc thesis, Shimane University, 122 p

Ordóñez-Carmona O, Restrepo-Alvarez JJ, Pimentel MM (2006) Geochronological and isotopical review of pre-Devonian crustal basement of the Colombian Andes. J S Am Earth Sci 21:372–382

Pearce JA (1982) Trace element characteristics of lavas from destructive plate boundaries. In: Thorpe RS (ed) Andesites. Wiley, New York, pp 525–548

Pearce JA (1996) Sources and settings of granitic rocks. Episodes 19:120–125

Pearce JA, Harris NBW, Tindle AG (1984) Trace element discrimination diagrams for the tectonic interpretation of granitic rocks. J Petrol 25:956–983

Peccerillo A, Taylor SR (1976) Geochemistry of Eocene calc-alkaline volcanic rocks from the Kastamonu area, Northern Turkey. Contrib Mineral Petrol 58:63–81

Pulido O (1985) Geología de la Plancha 135 San Gil Escala 1:100000. Instituto Colombiano de Geología y Minería

Restrepo-Pace PA (1995) Late Precambrian to Early Mesozoic tectonic evolution of the Colombian Andes, based on new geochronological, geochemical and isotopic data. PhD thesis, University of Arizona, 195 p

Restrepo-Pace PA, Ruiz J, Gehrels G, Cosca M (1997) Geochronology and Nd isotopic data of Grenville-agerocks in the Colombian Andes: new constraints for late Proterozoic–early Paleozoic paleocontinentalreconstructions of the Americas. Earth and Planetary Science Letters 150:427–441

Restrepo-Pace PA, Cediel F (2010) Northern South America basement tectonics and implications for paleocontinental reconstructions of the Americas. J S Am Earth Sci 29:764–771

Sarmiento-Rojas LF, Van Wess JD, Cloetingh S (2006) Mesozoic transtensional basin history of the Eastern Cordillera, Colombian Andes: inferences from tectonic models. J S Am Earth Sci 21:383–411

Shand SJ (1943) Eruptive rocks their genesis, composition, classification, and their relation to ore-deposits with a chapter on meteorite. Wiley, New York

Sillitoe RH, Jaramillo L, Damon IE, Shtiqullah M, Escovar R (1982) Setting, characteristics, and age of the Andean porphyry copper belt in Colombia. Econ Geol 77:1837–1850

Silva TJC, Sial A, Ferreira V, Estrada JJ (2004) C isotope stratigraphy of a Vendian carbonate succession in northwestern Andes: implications for the NW Andes-Mexico connection. Geos 24:SE02–SE02

Singer A, Beltrán C (1996) Active faulting in the southern Venezuelan Andes and Colombian borderland. In: 3rd international symposium on Andean Geodynamics, St Malo, pp 243–246

Thompson RN (1982) Magmatism of the British Tertiary volcanic province. Scott J Geol 18:49–107

Thuy NTB, Satir M, Siebel W, Vennemann T, Long TV (2004) Geochemical and isotopic constraints on the petrogenesis of granitoids from the Dalat zone, southern Vietnam. J Asian Earth Sci 23:467–482

Tschanz CM, Marvin RF, Cruz J, Mehnert HH, Cebula GT (1974) Geologic evolution of the Sierra Nevada de Santa Marta, Northeastern Colombia. Geol Soc Am Bull 85:273–284

Van der Lelij R (2013) Reconstructing north–western Gondwana with implications for the evolution of the Iapetus and Rheic Oceans: a geochronological, thermochronological and geochemical study. PhD thesis, Université de Genève, 221 p

Van der Lelij R, Spikings R, Ulianov A, Chiaradia M, Mora A (2016) Palaeozoic to early Jurassic history of the northwestern corner of Gondwana, and implications for the evolution of the Iapetus, Rheic and Pacific oceans. Gondwana Res 31:271–294

Vargas R, Arias A (1981a) Geología de la Plancha 97 Cáchira Escala 1:100000. Instituto Colombiano de Geología y Minería

Vargas R, Arias A (1981b) Geología de la Plancha 86 Abrego Escala 1:100000 . Instituto Colombiano de Geología y Minería

Vargas R, Arias A, Jaramillo L, Tellez N (1984) Geología de la Plancha 136 Málaga Escala 1:100000. Instituto Colombiano de Geología y Minería

Villaseca C, Barbero L, Herreros V (1998) A re-examination of the typology of peraluminous granite types in intracontinental orogenic belts. Trans R Soc Edinb Earth Sci 89:113–119

Ward ED, Goldsmith R, Cruz BJ, Restrepo H (1973) Geología de los cuadrángulos H-12 Bucaramanga y H-13 Pamplona. Boletín Geológico de INGEOMINAS 21:1–132

Ward D, Goldsmith R, Cruz J, Jaramillo L, Vargas R (1977a) Geología de la Plancha 121 Cerrito Escala 1:100000. Instituto Colombiano de Geología y Minería

Ward D, Goldsmith R, Jimeno A, Cruz J, Restrepo H, Gómez E (1977b) Geología de la Plancha 120 Bucaramanga Escala 1:100000. Instituto Colombiano de Geología y Minería

Ward D, Goldsmith R, Cruz J, Jaramillo L, Vargas R (1977c) Geología de la Plancha 110 Pamplona Escala 1:100000. Instituto Colombiano de Geología y Minería

Ward D, Goldsmith R, Jimeno A, Cruz J, Restrepo H, Gómez E (1977d) Geología de la Plancha 109 Rio Negro Escala 1:100000. Instituto Colombiano de Geología y Minería

Whalen JB, Currie KL, Chappell BW (1987) A-type granites: geochemical characteristics, discrimination and petrogénesis. Contrib Mineral Petrol 95:407–419

Whitney D, Evans B (2010) Abbreviations for names of rock-forming minerals. Am Mineral 95:185–187

Willett S, Beaumont C, Fullsack P (1993) Mechanical model for the tectonics of doubly vergent compressionalorogens. Geology 21:371–374

Zuluaga C, Pinilla A, Mann P (2015) Jurassic silicic volcanism and associated continental- arc basin in Northwestern Colombia (southern boundary of the Caribbean plate). In: Bartolini C, Mann P (eds) Petroleum geology and potential of the Colombian Caribbean margin, AAPG memoir, vol 108. American Association of Petroleum Geologists, Tulsa, pp 137–160

Zuluaga CA, Amaya S, Urueña C, Bernet M (2017) Migmatization and low-pressure overprinting metamorphism as record of two pre cretaceous tectonic episodes in the Santander Massif of the Andean basement in northern Colombia (NW South America). Lithos 274-275:123–146

Part IV
Major Tectono-Magmatic Events

Part IV
Major Tectono-Magmatic Events

Chapter 5
Spatial-Temporal Migration of Granitoid Magmatism and the Phanerozoic Tectono-Magmatic Evolution of the Colombian Andes

Hildebrando Leal-Mejía, Robert P. Shaw,
and Joan Carles Melgarejo i Draper

5.1 Introduction

Granitoid magmatic rocks form important constituents of the Phanerozoic record of the Colombian Andes. From the viewpoint of modern-day geological exposure, granitoids manifest as volumetrically abundant plutonic, hypabyssal and volcanic rocks, occurring as major batholiths, stocks, dike swarms and extensive volcanic sequences. During both proto-Andean and Andean times, these granitoid expressions form integral components of the temporal-tectonic evolution of Colombia and of Northwestern South America as a whole.

Granitoid magmatism sensu lato and its relationships to plate tectonics and global tectonic setting have become increasingly understood (e.g. Wilson 1989; Barbarin 1999; Hamilton 1994; Stern 2002). In this context, an integral analysis of the distribution, age, lithogeochemical and isotopic composition of granitoid rocks in the Colombian Andes can in due process shed light upon the nature, timing and framework of Colombian tectonic development.

Granitoid magmatism in the Colombian Andes includes a lithogeochemically and texturally diverse suite of holocrystalline phaneritic plutonic, hypabyssal

---

Electronic supplementary material: The online version of this chapter (https://doi.org/10.1007/978-3-319-76132-9_5) contains supplementary material, which is available to authorized users.

H. Leal-Mejía (✉)
Mineral Deposit Research Unit (MDRU), The University of British Columbia (UBC), Vancouver, BC, Canada

Departament de Mineralogia, Petrologia i Geologia Aplicada, Facultat de Ciències de la Terra, Universitat de Barcelona, Barcelona, Catalonia, Spain
e-mail: hlealmej@eoas.ubc.ca

R. P. Shaw · J. C. Melgarejo i Draper
Departament de Mineralogia, Petrologia i Geologia Aplicada, Facultat de Ciències de la Terra, Universitat de Barcelona, Barcelona, Catalonia, Spain

porphyritic and volcanic/volcano-sedimentary rocks and occasionally their metamorphosed equivalents, which, with respect to age, span the entire Phanerozoic. Although the cartographic limits of most plutons and volcanic units have been established and many have been described or studied on an individual basis, there are few works which assess the regional distribution vs. temporal development of Colombian Phanerozoic-aged granitoid magmatism as a whole. This is certainly due to the fact that there are limited studies in Colombia which combine modern-day (post-1995) lithogeochemical, geochronological, petrographic and isotopic studies to multiple intrusives of various ages on a regional level, in an attempt to specifically trace the tectono-magmatic evolution of the entire region over the breadth of the entire Phanerozoic.

The present synthesis of granitoid magmatism throughout the Colombian Andes is based primarily upon geological analysis combined with zircon U-Pb age dates, major-minor-trace element lithogeochemical analyses and isotope geochemistry studies for granitoid intrusive and volcanic rocks, generated by various authors since ca. 1995. The majority of the new data has been published since ca. 2006. The composite data set provides a more precise framework upon which to reconstruct the temporal-spatial development and migration of Phanerozoic granitoid magmatism in Colombia. When this information is integrated with updated kinematic models for the tectonic evolution of the Northern Andean region, refined conclusions can be drawn about the distribution, nature, timing, migration and controls upon Phanerozoic granitoid magmatism in the Colombian Andes.

The cornerstone of this presentation is derived from our own published (e.g. Cediel et al. 2003; Leal-Mejía et al. 2011) and unpublished investigations (e.g. Cediel et al. 1994; Cediel and Cáceres 2000; Leal-Mejía 2011), coupled with important newer studies from numerous third-party authors (cited below), the combined data for which permit a deeper understanding of Colombian and Northern Andean tectono-magmatic development during the period spanning the early Paleozoic through to the conformation of the active volcanic arcs of the present-day Northern Andes.

The widespread distribution of granitoid rocks in the geologic record of the Colombian Andes was recognized in numerous historic works (e.g. Trumpy 1949; Singewald 1950; Gansser 1955, 1973; Nelson 1957; Campbell 1974), and the approximate age of many of the major batholiths and extensive volcanic sequences was broadly inferred based upon field relationships. With the advent of more precise lithogeochemical analyses and radiometric age dating and the initiation of integrated regional geological studies (e.g. Goldsmith et al. 1971; Feininger et al. 1972; Tschanz et al. 1974; Feininger and Botero 1982; Sillitoe et al. 1982; Alvarez 1983; Aspden et al. 1987), a more detailed picture of Phanerozoic granitoid magmatism in the Colombian Andes emerged.

Early lithogeochemical studies, however, did not include important batholiths in the northern and eastern portion of the Colombian cordilleran system (e.g. the Sierra Nevada de Santa Marta, the Santander Massif and Mocoa), in the Serranía de San Lucas, Segovia or Chocó Arc. Additionally, in the existing database, most trace elements, including the rare-earth elements, were often not analysed, and results were generally not placed into a tectonic framework. With respect to radiometric age dating, the early database from the 1970s and 1980s consisted almost exclusively of K-Ar and Rb-Sr (isochron) ages, which had large margins of error and were subsequently proven in many cases, in Colombia and elsewhere, to be imprecise, erratic and of poor repeatability.

In a first integrated attempt to place Meso-Cenozoic granitoids in Colombia into a modern-day tectonic framework using available distribution, lithogeochemical and radiometric age data, Aspden et al. (1987) presented a well-conceived synthesis of subduction-related magmatism in Colombia. Based upon radiometric age data compiled from numerous sources (see Maya 1992), they identified five magmatic episodes (Triassic, Jurassic, Cretaceous, Paleogene and Neogene) and outlined a regional tectonic framework for the evolution of subduction-related magmatism in the Northern Andes. Additionally, they identified some of the key factors and controls influencing granitoid arc development during the Meso-Cenozoic.

Studies presenting high-quality U-Pb (zircon) dating techniques for magmatic rocks permit a more accurate assessment of the crystallization and inheritance age(s) of granitoids. Such studies in the Colombian Andes, at times combined with incipient Sr, Nd and Pb isotope data and more complete lithogeochemical analyses, began to trickle in ca. 1995. Initial studies addressed specific plutons or sub-regions and included a limited number of samples per intrusive body. Examples of such works include Dörr et al. (1995) for the Paramo Rico and Santa Barbara batholiths of the Santander Massif; Ordoñez et al. (2001) for the Sonsón Batholith; Altenberger and Concha (2005) for the northern Ibagué Batholith (K-Ar ages only); Vinasco (2004) and Vinasco et al. (2006) for the Permo-Triassic granitoids found throughout the northern Central Cordillera and elsewhere; Ibañez-Mejía et al. (2007) and Restrepo-Moreno et al. (2007) for the Antioquia Batholith; Ordoñez-Carmona et al. (2007a, b) for the Antioquia, Segovia and Sabanalarga batholiths; Weber et al. (2015) for the Santa Fé Batholith; Mejía et al. (2008), Duque (2009), and Cardona et al. (2011) for the late Cretaceous-Paleocene intrusives of the Sierra Nevada de Santa Marta (i.e. Santa Marta Batholith); Correa et al. (2006) for the Altavista and San Diego Stocks (satellites to the Antioquia Batholith); Villagómez (2010) and Villagómez et al. (2011) for the Ibagué, Antioquia and Buga batholiths; Bustamante et al. (2010) and Zapata et al. (2016) for plutons of the southern Ibagué, Mocoa and Garzón suites; Mantilla et al. (2012) and Bissig et al. (2014) for Phanerozoic intrusive rocks of the Santander Massif; Bayona et al. (2012) and Bustamante et al. (2017) for Paleocene plutons of the Central Cordillera; and Montes et al. 2012, 2015) for the Mandé and Acandí batholiths and associated plutons. Some of the earlier works represented incomplete or in-process studies and were presented in conference-related abstracts containing limited background information which are difficult to access and which do not permit the full geological evaluation of the numerical data or of the derived conclusions. Others are published in well-circulated international journals and are readily accessible for detailed review.

Leal-Mejía (2011) presented an integrated investigation of Phanerozoic granitoid magmatism related to gold metallogeny in the Colombian Andes. This temporal, lithogeochemical and tectono-magmatic study was based upon a review and compilation of the historic to recent data cited above, in addition to the presentation of 107 new high-precision U-Pb (zircon) dates for intrusive and volcanic rocks, supported by new K-Ar, Ar-Ar and Re-Os ages, as well as Sr, Nd and Pb isotope data and 282 research-quality whole-rock major-minor-trace-REE lithogeochemical analyses. The study included new data from many previously un- or understudied Phanerozoic granitoids in the Colombian Andes such as the Pueblo Bello, Norosí-San Martín de Loba, Segovia, Ibagué (north and south), Mariquita and Antioquia, Buga, Sonsón-Nariño, Mandé, Piedrancha-La Llanada and Farallones batholiths. It identified various Permo-Triassic granitoids which had previously been mapped as Precambrian, Jurassic or Paleocene in age. Additionally, many smaller holocrystalline stocks such as El Carmen, Mocoa, Irra, Jejenes, Frontino and Támesis, and numerous clusters of Neogene hypabyssal porphyry stocks observed along the margins of the Central and Western Cordilleras and in the Santander Massif, were analysed. The present chapter draws heavily on the data and conclusions presented by Leal-Mejía (2011).

Important contributions to the U-Pb age date, isotopic and lithogeochemical database for Northern Andean Paleozoic through Jurassic granitoid intrusions have recently been supplied by researchers from the University of Geneva. Such works include Villagómez (2010) and Villagómez et al. (2011), applicable to portions of the Western and Central Cordilleras and Sierra Nevada de Santa Marta; Van der Lelij (2013), Van der Lelij et al. (2016) and Spikings et al. (2015), applicable to the Santander Massif and the Sierra de Mérida (Venezuela); and Cochrane (2013), Cochrane et al. (2014a, b) and Spikings et al. (2015), applicable to the Garzón Massif and various Permo-Triassic and Jurassic granitoids outcropping along the margins of Colombia's Central Cordillera. Pertinent data from these works have been reviewed, and conclusions from these authors have been integrated into the ensuing text and graphics of this chapter.

5.2 Phanerozoic Tectonic Framework of the Colombian Andes

The Colombian Andes is contained within the North Andes (Bird 2003) or Northern Andean Block (Cediel et al. 2003; Cediel 2011; Fig. 5.1). From a geographic standpoint, the Northern Andean Block includes the northernmost Peruvian Andes (north of the Huancabamba Deflection), in addition to the cordilleran systems of Ecuador, Colombia and Venezuela and the eastern Chocó Arc segment of the Panamá double arc. The evolution of the region and its modern-day geologic, tectonic and physiographic expression is the result of complex interactions between oceanic and continental tectonic plates, beginning in the mid-Proterozoic (e.g. Cediel et al. 1994; Ramos 1999; Restrepo-Pace and Cediel 2010; Cediel 2011). Since the Meso-Cenozoic, no less than four plates, including the South American continental block (western Guiana Shield) and the Farallon, Caribbean and Nazca-Cocos oceanic plates, have been involved.

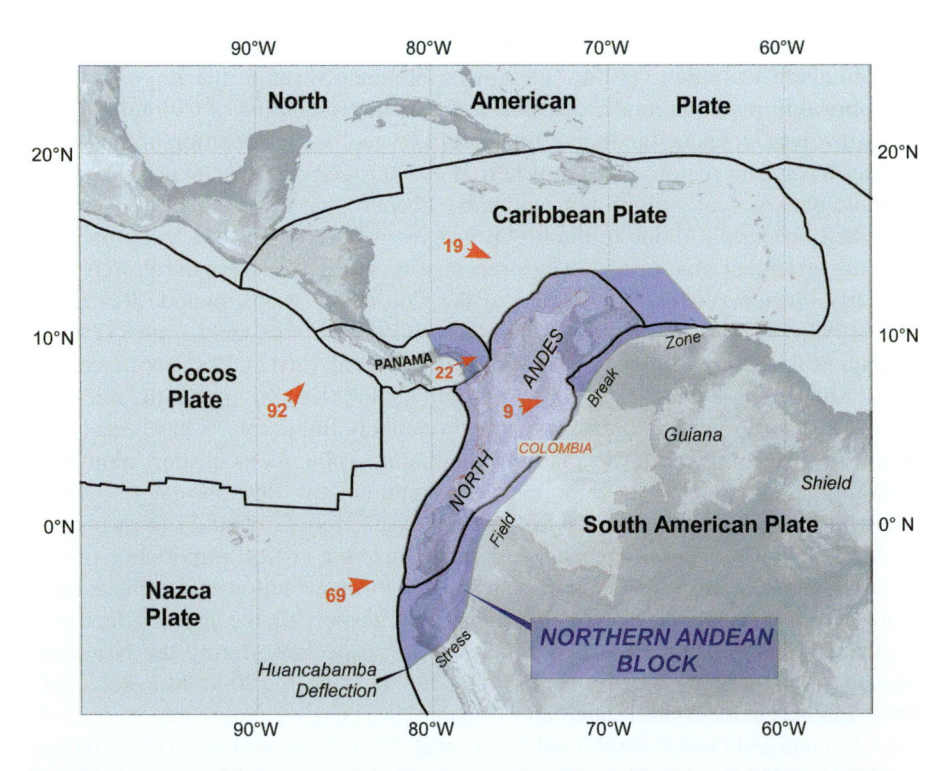

5 Spatial-Temporal Migration of Granitoid Magmatism and the Phanerozoic...

Fig. 5.1 Location of Colombia and the Northern Andean Block (Cediel et al. 2003) in relation to microplates of northwestern South America and surrounding region as defined by Bird (2003). Present-day microplate relative movement vectors and velocities in mm/a after Bird (2003)

Internally, the Northern Andean Block itself comprises a microplate hosting numerous allochthonous and parautochthonous tectonic slivers of mixed oceanic, peri-cratonic and continental affinity. The allochthonous and parautochthonous crustal fragments have accumulated tectonically along the northwest margin of South America during successive accretionary events and are presently broadly bound by the Guiana Shield segment of the South American Plate to the east, the Pacific (Nazca) Plate to the west and the Caribbean Plate to the north (Fig. 5.1).

Numerous tectonic models for the development of the Colombian Andes and adjacent Pacific and Caribbean regions have been proposed over to past 40 years (e.g. Etayo-Serna et al. 1983; Burke et al. 1984; Kellogg et al. 1985; Restrepo and Toussaint 1988; Pindell et al. 1988; Cediel et al. 1994; Taboada et al. 2000). It is now recognized that the present-day Northern Andean configuration is the result of a history involving collision and accretion of allochthonous terranes and the development of subduction-related arcs along the northwestern South American margin (Restrepo and Toussaint 1988). Although emphasizing locally important features regarding Colombian tectonic development, early tectonic models suffered from a variety of geographic and/or temporal limitations, and in some cases the conclusions are now known to be invalid.

More recent regional tectonic models recognize the fact that an accurate account of Colombian tectonic evolution cannot be obtained through the imposition of geopolitical limits upon model construction. The understanding of Colombian tectonics involves an understanding of the integrated geological evolution of the entire Northern Andean region, from northern Perú through Venezuela and Panamá. Tectonic models dealing with pre- and early Phanerozoic time remain elusive and necessarily generalized due to the highly fragmented geological record remaining from this extended time period. More recent works which address the Proterozoic and early Phanerozoic tectonic record of the Colombian Andes include Restrepo-Pace (1992, 1995), Cediel et al. (1994), Ramos (1999), Cediel and Cáceres (2000), Cordani et al. (2005), Keppie (2008), Ibañez-Mejía et al. (2011), Restrepo-Pace and Cediel (2010), Cediel (2011), and Van der Lelij et al. (2016). Similarly, the mid-late Paleozoic record remains controversial, and relatively limited work has been completed for this time period, especially in the metamorphic rocks of the Colombian Central Cordillera, beyond important contributions by Restrepo-Pace (1992), Vinasco (2004), Vinasco et al. (2006), and Cochrane et al. (2014a). With respect to the Meso-Cenozoic, various recent works recognize the critical importance of the evolution and demise of the Farallon Plate and the birth, evolution and emplacement of the Caribbean and Nazca-Cocos plates, with respect to the inboard tectono-magmatic development of the Colombian Andes, especially during the Northern Andean Orogeny (e.g. Cediel et al. 1994; Cediel and Cáceres 2000; Maresch et al. 2000; Taboada et al. 2000; Pindell and Kennan 2001; Cediel et al. 2003; Kerr et al. 2003; Kennan and Pindell 2009; Leal-Mejía et al. 2011; Montes et al. 2012; Spikings et al. 2015).

In a Northern Andean analysis spanning the Proterozoic to the present, Cediel et al. (2003) describe more than 30 litho-tectonic and morpho-structural units (terranes, terrane assemblages, physiographic and morpho-structural domains, etc.), contained within four major tectonic realms (Fig. 5.2). Each realm records distinct and in some cases unique internal deformation styles as a response to progressive westward accretionary continental growth along the northwestern Guiana Shield (Amazon Craton) margin. Tectonic realms and terrane assemblages are delimited by important regional-scale sutures and fault systems. Within this chapter, we will adhere primarily to the tectonic nomenclature presented by and updated from Cediel et al. (2003; Fig. 5.2), much of which is derived from historic detailed analyses of Colombian tectonics, as presented by Etayo-Serna et al. (1983) and Cediel et al. (1994).

The kinematic evolution of the Caribbean Plate and resulting large-scale Northern Andean-Caribbean Plate interactions have been depicted in Meso-Cenozoic tectonic reconstructions presented by various authors including Cediel and Cáceres (2000), Pindell and Kennan (2001), Kerr et al. (2003), Cediel et al. (2003), Kennan and Pindell (2009), Wright and Wyld (2011), Montes et al. (2012), and Nerlich et al. (2014), amongst others.

Cediel and Cáceres (2000), Cediel et al. (2003), and Cediel (2011) observed that the geotectonic evolution of Colombia can be separated into pre-Northern Andean Orogeny events (i.e. events *prior* to approximately the Aptian) and

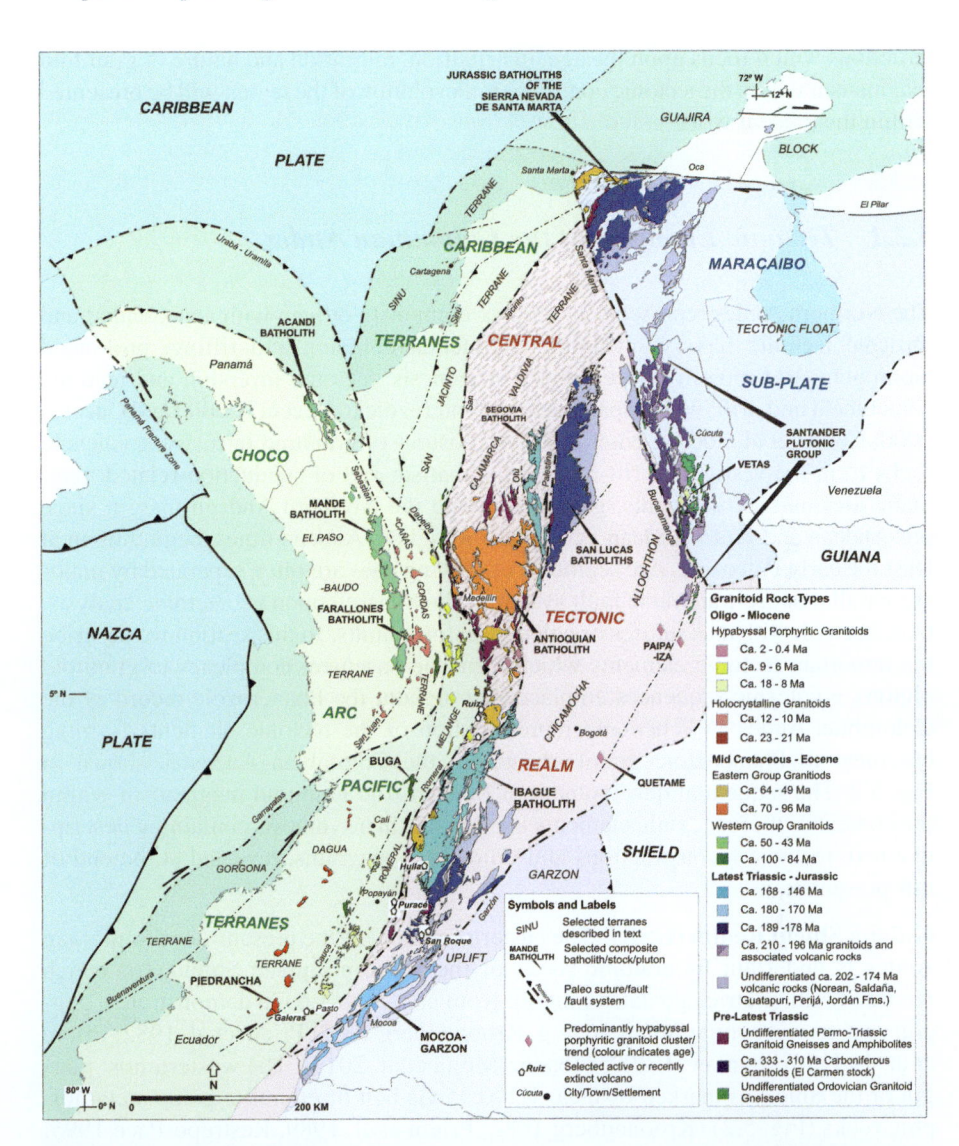

5 Spatial-Temporal Migration of Granitoid Magmatism and the Phanerozoic... 259

Fig. 5.2 Distribution of Phanerozoic granitoids in relation to the major litho-tectonic and morpho-structural elements of the Colombian Andes. (Granitoid shapes modified after Cediel and Cáceres 2000; Gómez et al. 2007; Gómez et al. 2015a. Litho-tectonic base map adapted from Cediel et al. 2003). Age ranges based upon U-Pb (zircon) age dates compiled herein

Northern Andean Orogeny-related events beginning in the Aptian onwards. In this context, these authors present tectono-stratigraphic reconstructions for the mid-late Proterozoic (Orinoquiense event), the early to middle Paleozoic (Quetame/Caparonesis event), the mid- and late Paleozoic to early and mid-Mesozoic (Bolívar Aulacogen) and the Aptian to recent (Northern Andean Orogeny). Similar reconstructions which focus upon the age, distribution, migration and nature of granitoid magmatism within the tectonic configuration/evolution of the region will be presented within the present work.

5.2.1 Tectonic Elements of the Colombian Andes

The Northern Andes represents a complex composite orogen with an extended and intricate tectonic history involving continental collision and rifting, prolonged taphrogeny, transpressive accertionary orogenesis, tectonic inversion and tectonic detachment and drift, which span the mid-Proterozoic to Recent (Cediel and Cáceres 2000; Cediel et al. 2003; Cediel 2011). Exposures of granitoid intrusive ± volcanic rocks form the vestiges of rift-related magmatism and of subduction-related magmatic arcs and arc segments, generated along the Northern Andean margin since pre-Andean and proto-Andean as well as throughout Andean times, beginning in at least the early Paleozoic. Arc segments of varying ages are often separated by major arc-parallel or arc-transverse fault systems which, based upon geotectonic analysis, have been identified as sutures or paleo-transform faults. In this section we describe the important tectonic elements which form the basement complexes to granitoid plutons ± volcanic sequences emplaced throughout the Phanerozoic record of the Colombian Andes. A schematic representation of the tectonic elements forming basement to Phanerozoic granitoids within the Colombian Andes is shown in Fig. 5.2. The sequential development of Phanerozoic granitoid magmatism within the context of these tectonic elements is proposed in a synthesis containing descriptive text, tectonic reconstructions and time-space diagrams, revealed at the end of this presentation.

Guiana Shield Realm (GSR) The eastern foreland of the present-day Colombian Andes is underlain by cratonic rocks of the western Guiana Shield, for which recorded radiometric age dates ranging from ca. 2.5 to 1.5 Ga demonstrate a general east-to-west younging trend (e.g. Kroonenberg 1982; Priem et al. 1982; Priem et al. 1989; Cordani et al. 2005; Ibañez-Mejía et al. 2011). The westernmost margin of the Shield is marked by a ca. 1.2–0.95 Ga belt of granulite grade metamorphic rocks (Fig. 5.2) (Kroonenberg 1982; Priem et al. 1989; Restrepo-Pace 1995; Cordani et al. 2005; Cardona et al. 2010a; Restrepo-Pace and Cediel 2010; Ibañez-Mejía et al. 2011), recorded in outcrops in the Sierra Nevada de Santa Marta and Santander and Garzón Massifs and considered to represent Grenvillian-age continent-continent interaction along the Bucaramanga–Santa Marta–Garzón fault and suture system (Cediel et al. 2003; Cediel 2011) during the final assembly of Rhodinia (Cordani et al. 2005; Ibañez-Mejía et al. 2011). The resulting tectonothermal metamorphic event has been referred to in Colombia as the Orinoquiense Orogen (Kroonenberg 1982; Restrepo-Pace 1995; Cediel and Cáceres 2000; Restrepo-Pace and Cediel 2010; Cediel 2011) or Putumayo Orogen (Ibañez-Mejía et al. 2011).

Maracaibo Sub-plate Realm (MSP) The MSP is a composite tectonic realm also underlain by the Guiana Shield, but much of its uplift history is linked to the Meso-Cenozoic tectonic evolution of the region. Its northern limit, in contact with the Caribbean Plate, is defined by the dextral Oca-El Pilar fault system and the Santa Marta thrust front (Fig. 5.2), whilst its west margin, in contact with the Central Tectonic Realm, is defined by the reactivated Bucaramanga-Santa-Marta fault. Topographic relief is provided by the Santander and Quetame Massifs, the Sierra de Mérida, the Serrania de Perijá and the Sierra Nevada de Santa Marta, the uplift history of which is linked to detachment and NW-vergent tectonic float during the Meso-Cenozoic (Cediel and Cáceres 2000; Cediel et al. 2003). The MSP contains numerous litho-tectonic and morpho-structural components, including exhumed Proterozoic and early Paleozoic basement massifs (Santander, Quetame, Floresta). Late Triassic-Jurassic ensialic extensional volcano-sedimentary basins are exposed along the Santander Massif, Sierra Nevada de Santa Marta and Serranía de Perijá. Uplift of the Santander Massif and Sierra Nevada de Santa Marta has unroofed important holocrystalline granitoid batholiths of early Paleozoic and latest Triassic-Jurassic age.

Central Tectonic Realm (CTR) The CTR (originally termed Central Continental Sub-plate (CCSP) by Cediel et al. 2003) is a composite, temporally and compositionally heterogeneous realm which occupies a wedge located between the Guiana Shield Realm, the Maracaibo Sub-plate Realm, and the Western Tectonic Realm (Fig. 5.2). It forms the basement complex which underlies the entire central portion of the Colombian Andes. The CTR is comprised of a variety of litho-tectonic and morpho-structural entities. Its composite metamorphic basement consists of the Proterozoic Chicamocha Terrane and the Paleozoic to early Mesozoic Cajamarca-Valdivia Terrane (CA-VA). Superimposed upon these core components are Jurassic magmatic arc segments including the San Lucas, Ibagué and Segovia blocks, the late Cretaceous Antioquian Batholith and additional Paleocene plutons to the south, and the Pleistocene to Recent Northern Andean volcanic arc, all of which dominate Colombia's physiographic Central Cordillera. The Lower, Middle and Upper Magdalena basins and Colombia's geologic Eastern Cordillera (EC) were also developed upon CTR metamorphic basement. The Chicamocha and Cajamarca-Valdivia constituents of the CTR are allochthonous to parautochthonous with respect to the Guiana Shield autochthon, having been sutured to the shield in pre-Andean times. The Mesozoic to Recent components of the CTR are considered to be autochthonous with respect to Chicamocha-CA-CV metamorphic basement.

The oldest constituent of the CTR is the Precambrian Chicamocha Terrane, interpreted as an embedded fragment of composite parautochthonous to allochthonous continental crust containing relict fragments of Oaxaquia basement (Keppie and Ortega-Gutierrez 2010) and Oaxaquian-Colombian fringing volcano-magmatic arcs (Ibañez-Mejía et al. 2011; Cediel 2011), amalgamated with tectonic rafts of the westernmost Guiana Shield and welded to the autochtonous Guiana Shield margin during the ca. 1.0–0.95 Ga event (Cediel 2011). It is represented by Mesoproterozoic inliers exposed along portions of Colombia's Central Cordillera (Cediel and Cáceres 2000; Cordani et al. 2005; Cediel 2011; Leal-Mejía 2011), including in the Serranía de San Lucas where Cuadros et al. (2014) documented early Mesoproterozoic basement containing a bimodal assemblage of within-plate granitoids and oceanic island and possibly underplate metamafic rocks of juvenile, mantle-derived character. They published zircon U-Pb crystallization ages ranging from ca. 1.54 to 1.50 Ga for high-grade metagranitoids which exhibit ~1 Ga (Grenvillian) overprinting.

Chicamocha is bound to the west by the composite Cajamarca-Valdivia Terrane (Cediel and Cáceres 2000; Cediel et al. 2003; Cediel 2011) which broadly coincides with the Central Andean Terrane as described by Restrepo-Pace (1992). Cajamarca-Valdivia contains amphibolitic, graphitic and semi-pelitic schists and marbles, metamorphosed to greenschist through epidote amphibolite grade and generally assigned a Neoproterozoic to early Paleozoic age (Feininger et al. 1972; Restrepo-Pace 1992; Cediel and Cáceres 2000; Ordoñez-Carmona et al. 2006; Cediel 2011; Spikings et al. 2015), in agreement with Ediacaran to Cambrian C-isotope stratigraphy ages for contained carbonates, as published by Silva et al. (2005). Based upon geochemical and geological characterization studies presented by Restrepo-Pace (1992), Cajamarca-Valdivia represents a pericratonic island arc and continental margin accretionary prism assemblage, accreted along the western Chicamocha Terrane. Cajamarca-Valdivia forms the basement to Carboniferous and Permian through mid-late Triassic gneissic granitoids, meta-amphibolites and peraluminous granites associated with the assembly and break-up of Pangaea (Vinasco et al. 2006; Cardona et al. 2010b; Cochrane 2013; Cochrane et al. 2014a; Spikings et al. 2015). González (2001) and Cediel (2011) suggest the terrane also contains tectonic floats of Mesoproterozoic metamorphic continental basement rocks, including the Puqui and El Retiro-Rio Negro gneisses, although Ordoñez-Carmona et al. (2006) note that these units produce broadly Permo-Triassic radiometric age dates and hence are better considered members of the Permo-Triassic suite of Vinasco et al. (2006). The Palestina fault system denotes the suture between the Chicamocha and Cajamarca-Valdivia Terrane assemblages. The trace of the modern-day Palestina fault system is the result of various reactivations during the Meso-Cenozoic Northern Andean orogeny (Feininger 1970; Cediel and Cáceres 2000). Associated structures include the Chapeton and Pericos faults.

It is important to note that present-day geological mapping does not permit the precise distribution of Proterozoic, early Paleozoic and Permo-Triassic constituents of CA-VA (Cediel and Cáceres 2000; Gómez et al. 2015a), and zircon-based U-Pb age dating is only beginning to reveal the complexity of the CA-VA assemblage. For example, various units which were formerly thought to be of Proterozoic or early Paleozoic age are now known to belong to the Permo-Triassic assemblage (Restrepo-Pace 1992; Ordoñez-Carmona et al. 2006; Vinasco et al. 2006; Leal-Mejía 2011; Spikings et al. 2015).

The timing of CA-VA assemblage and accretion along the western Chicamocha margin and the origins of the Palestina fault and suture system have been examined by various authors (Feininger 1970; Cediel et al. 1994; Restrepo-Pace 1995; Cediel and Cáceres 2000; Cediel et al. 2003; Restrepo-Pace and Cediel 2010; Cediel 2011).

Regional tectono-sedimentary analysis (Cediel et al. 1994), arc-related magmatic patterns and available metamorphic age dates (Restrepo-Pace 1995; Van der Lelij et al. 2016) suggest early Paleozoic accretion of CA-VA during the Quetame Orogeny (Cediel and Cáceres 2000; Cediel 2011). Peak Barrovian conditions, including emplacement of syn-kinematic granitoids, were attained at ca. 477–472 Ma (Restrepo-Pace 1995; Van der Lelij et al. 2016).

Following amalgamation, important Meso-Cenozoic constituents were superimposed upon of the composite metamorphic basement of the Central Continental Sub-plate. From the late Triassic-Jurassic, these include the San Lucas, Ibagué and Segovia blocks. These composite lithotectons represent temporally/geographically constrained, ensialic extensional basin-continental margin magmatic arc couplets. Volcano-sedimentary basin development and subsequent emplacement of the calc-alkaline San Lucas and Ibagué batholiths appear to have been localized by extensional reactivation of the Palestina suture. A general east-to-west younging trend of major Jurassic arc-related batholiths, from the Santander Plutonic Group through the San Lucas batholiths and into the Segovia Batholith, has been interpreted by various authors (Leal-Mejía 2011; Cochrane 2013; Cochrane et al. 2014b; Spikings et al. 2015) to reflect regional extension due to slab rollback. Cediel and Cáceres (2000) and Cediel et al. (2003) denominated the taphrogenic framework associated with the break-up of Pangaea and the opening of the Proto-Caribbean basin during the Mesozoic and early Cenozoic, the *Bolivar Aulacogen* (see below).

The late Mesozoic history of the CTR is dominated by continued extension and opening of the Valle Alto rift (Cediel et al. 1994; Cediel and Cáceres 2000; Cediel et al. 2003). This period involves deep crustal and continental margin rifting, the emplacement of alkalic-tholeiitic dike suites and mafic magmatism (Fabre and Delaloye 1983; Vásquez et al. 2010), the invasion of the Cretaceous seaway and the deposition of deep to shallow marine sequences over extensive areas of the CTR, the Maracaibo Sub-plate and the continental platform of the Guiana Shield (Cediel et al. 1994; Cediel et al. 2003; Cediel 2011; Cediel 2018; Sarmiento 2018). The latest Jurassic through Cretaceous record has been exhumed and is exposed throughout Colombia's Eastern Cordillera and within erosional relicts such as the San Pablo and Segovia Formations preserved within the physiographic Central Cordillera. The early to mid-Cretaceous, from ca. 145 to 95 Ma, is characterized by an apparent hiatus in significant granitoid magmatism throughout continental Colombia, as suggested by the absence of calc-alkaline arc segments or important volumes of granitoid rocks of any kind dating from this time period.

The late Mesozoic and Cenozoic to Recent components of the Central Tectonic Realm are dominated by subduction-related arc granitoids, associated with assembly and accretion of the Western Tectonic Realm (detailed below) and the conformation of the present-day Northern Andes volcanic arc (Stern 2004). Granitoid assemblages, including the Cretaceous Antioquian Batholith (Feininger and Botero 1982; González 2001; Leal-Mejía 2011) and Paleocene to early Eocene intrusions observed to the south, reflect the subduction-driven and accretionary regime dominant during the late Cretaceous. Oligo-Miocene-Pliocene and Pleistocene to Recent Andean-type magmatism form north–northeast trending belts and clusters of intrusions, partially eroded volcanic edifices and active stratovolcanic cones stretching along the western margin of the Central Tectonic Realm.

Western Tectonic Realm (WTR) The approach, assembly and accretion of the allochthonous Western Tectonic Realm (Fig. 5.2) provided the driving mechanism for granitoid magmatism during the late Meso-Cenozoic Northern Andean orogeny (Cediel and Cáceres 2000; Leal-Mejía 2011). Within the WTR three composite terrane assemblages are recognized, including the Pacific (PAT) and Caribbean (CAT) and the Chocó Arc (CHO). The Romeral and Dagua terranes of the PAT assemblage, and the San Jacinto and Sinú terranes of the CAT assemblage, roughly correspond to litho-tectonic units recognized by Etayo-Serna et al. (1983). The combined PAT and CHO assemblages approximate the *Provincia Litosférica Oceánica Cretácica del Occidente de Colombia* (PLOCO) of Nivia et al. (1996) and form the geographic Western Cordillera of Colombia. All of these tectonic assemblages contain fragments of oceanic crust, oceanic plateaus, aseismic ridges and/or ophiolite with associated marine sedimentary rocks. All developed within/upon oceanic basement and, based upon faunal assemblages (e.g. Etayo-Serna and Rodríguez 1985), paleomagnetic data (e.g. Estrada 1995) and recent paleogeographic reconstructions (e.g. Cediel et al. 1994; Cediel et al. 2003; Kennan and Pindell 2009; Montes et al. 2012), and all, with the exception of components of the Romeral assemblage, are allochthonous with respect to continental South America. The composite terrane assemblages of the Western Tectonic Realm are characterized as follows:

Pacific Terrane Assemblage (PAT) The PAT consists of the Romeral assemblage and Dagua and Gorgona terranes. Romeral may be interpreted as a regional-scale tectonic melange (Cediel and Cáceres 2000), developed within an early Cretaceous, rift-related transtensional basin along the Colombian Pacific margin during the early Cretaceous (Nivia et al. 2006; Kennan and Pindell 2009). The Romeral assemblage includes intensely deformed and fragmented blocks (tectonic floats?) of amphibolite and carbonaceous schist, high-pressure metamorphic rocks (eclogite, blueschist), layered mafic and ultramafic complexes, marine and peri-cratonic arc-related volcanic rocks, ophiolite and meta-sediments, dating from the Paleozoic, Jurassic and early Cretaceous. The suite was assembled along the Pacific margin in tectonic contact with the CTR to the east, along the Romeral fault system (Ego et al. 1995; Cediel et al. 2003). The Romeral assemblage underlies much of the Cauca-Patía intermontane valley (Fig. 5.2), including the northern inter-Andean depression to the north and south of the city of Pasto. Cediel et al. (2003) note that the allochthonous vs. in situ nature of the Romeral mélange remains unclear, but current information suggests the presence of both components of a peri-cratonic nature deposited in a continental margin basin (Nivia et al. 2006; Kennan and Pindell 2009) and allochthonous components formed within an intra-oceanic setting.

To the west of the Romeral melange, the Dagua terrane is comprised of an assemblage of oceanic mafic and ultramafic rocks (Diabasico Group) which forms the basement for important thicknesses of flyschoid silici-clastic sedimentary rocks including chert, siltstone, marlstone and greywacke (Dagua Group).

Lithogeochemical studies (e.g. Kerr et al. 1997; Sinton et al. 1998) indicate the mafic and ultramafic volcanic and intrusive rocks are of oceanic tholeiitic N- and E-MORB affinity, interpreted to represent accreted fragments of oceanic crust, ophiolite, aseismic ridges and/or oceanic plateaus belonging to the Farallon Plate and Cretaceous Caribbean-Colombian Oceanic Plateau (CCOP) or Caribbean Large Igneous Province (CLIP), as described by Kerr et al. (1997) and Sinton et al. (1998), respectively. A mantle plume-hotspot-oceanic flood basalt origin for the CCOP/CLIP assemblage, developed within/upon the Farallon Plate, has been proposed by these authors. Data provided by Nerlich et al. (2014, and references cited therein) indicates that the section of the Farallon Plate which forms basement to CCOP/CLIP plateau rocks varies from ca. 144 Ma in the east, younging westwards to ca. 75 Ma, presently located in the westernmost Caribbean. A summary of radiometric (mostly Ar/Ar) age dates for accreted CCOP/CLIP rocks in northern South America and the Caribbean suggests that plateau-related mafic-ultramafic magmatism superimposed upon the Farallon Plate may be considered in three stages; a volumetrically restricted phase initiated at ca. 100 Ma, followed by the widespread eruption of oceanic plateau rocks dating from ca. 92 to 87 Ma (Kerr et al. 1997; Sinton et al. 1998; Kerr et al. 2003; Hastie and Kerr 2010; Nerlich et al. 2014). Lesser but still significant, plateau-related basaltic magmatism was subsequently recorded between ca. 77 and 72 Ma (Kerr et al. 1997; Sinton et al. 1998), although these dates are considered to represent the wanning stages of CCOP-/CLIP-related magmatism.

Within the Dagua terrane, mid-Cretaceous Ar-Ar ages for oceanic plateau rocks (Kerr et al. 1997; Sinton et al. 1998) are in broad agreement with mid- to late Cretaceous biostratigraphic ages for contained oceanic sedimentary rocks contained within the Dagua Group (Etayo-Serna and Rodríguez 1985). Radiometric age dating of the Bolivar ultramafic complex along the eastern margin of the Dagua terrane returned U-Pb (zircon) ages ranging from ca. 97 to 95 Ma (Villagómez et al. 2011). Additional subduction-related granitoids (our Western Group granitoids; see Sect. 5.3.4.1) appear to form part of the Greater Arc of the Caribbean assemblage (e.g. Pindell and Kennan 2001; Hastie and Kerr 2010; Wright and Wyld 2011; Weber et al. 2015) and were emplaced into the Farallon Plate and Dagua terrane prior to accretion along the Colombian Pacific margin. Approach/collision of the Farallon Plate-CCOP/CLIP assemblage, and accretion of the Dagua terrane, began in the late Cretaceous (see Sect. 5.4.3.2), along the Cauca fault and suture system.

Further west, the Gorgona Terrane is located mostly offshore, on the southwestern margin of the Colombian Pacific. Gorgona also represents an accreted oceanic plateau of mantle plume affinity, containing massive basaltic and spinifex-textured komatiitic lava flows, pillow lavas and a peridotite-gabbro complex (McGeary and Ben-Avraham 1989). Radiometric ages provided by Sinton et al. (1998) range from ca. 87 to 83 Ma; however, paleomagnetic and lithogeochemical data, and paleogeographic reconstructions presented by Estrada (1995), Kerr and Tarney (2005) and Kennan and Pindell (2009), suggest Gorgona has no clear correlation with the CCOP/CLIP. Gorgona is limited to the east by the Buenaventura fault and to the west by the modern-day Colombian trench. Accretion of Gorgona along the western margin of the Dagua terrane took place during the Eocene (Cediel et al. 2003; Kerr and Tarney 2005).

Caribbean Terrane Assemblage (CAT) Two principle terranes are contained within this assemblage, the San Jacinto and Sinú (Fig. 5.2). San Jacinto includes a MORB-type tholeiitic basement considered a fragment of the CCOP/CLIP assemblage, containing upper Cretaceous deep marine carbonaceous cherts, mudstones and marlstones locally intercalated with coarser-grained siliciclastic and felsic pyroclastic material and intruded by minor holocrystalline diorite to quartz diorite plutons of poorly constrained age (e.g. El Alacrán, San Juan de Asís). San Jacinto was accreted to the northern CTR along the San Jacinto fault during the Eocene (Cediel and Cáceres 2000). The Sinú terrane is comprised of similar basement to San Jacinto, overlain by turbidite sequences of Oligocene age. It was juxtaposed along the San Jacinto margin in the Miocene.

The Chocó (Eastern Panamá) Arc (CHO) The Chocó Arc assemblage in Colombia (Duque-Caro 1990; Schmidt-Thomé et al. 1992; Cediel et al. 2010), together with Campanian to Eocene mafic oceanic and intermediate arc-related plutonic rocks of the Darién (San Blas) Range and Azuero Peninsula in Panamá (Wegner et al. 2011; Montes et al. 2012), represents the eastern segments of the Panamá double arc. In Colombia, the basement of the composite Chocó Arc (Fig. 5.2) is comprised of two distinct litho-tectonic assemblages: the Cañas Gordas terrane and the El Paso Terrane which includes the Baudó Range (Cediel et al. 2010; Redwood 2018).

Cañas Gordas consists of mixed volcanic rocks of the Barroso Fm. overlain by fine-grained sedimentary rocks of the Penderisco Fm. The Barroso Fm. is dominated by tholeiitic to calc-alkaline, massive, porphyritic and amygdaloidal basalt, with andesitic flows, tuffs and agglomerates (Rodriguez and Arango 2013). Sedimentary interbeds within the Barroso Fm. mapped near the town of Buriticá contain Barremian through middle Albian fossil assemblages (González 2001 and references cited therein). Weber et al. (2015) consider gabbros belonging to the Barroso Fm. to belong to the CCOP/CLIP plateau assemblage, although biostatigraphic data suggests that Barroso is pre-CCOP and may better represent accreted slivers of the older, Farallon Plate oceanic basement upon which the CCOP rests (Nerlich et al. 2014 and references cited therein). The Penderisco Fm. includes thinly bedded, mudstone, siltstone, marlstone, greywacke and chert. Two members, including Urrao and Nutibara, contain marine fossil assemblages dating from the Aptian-Albian to the upper Cretaceous (González 2001 and references cited therein), again demonstrating the diachronous nature of the Farallon-CCOP/CLIP assemblage. The eastern margin of the Cañas Gordas terrane was intruded by the Buriticá tonalite and the Santa Fé Batholith at ca. 100 Ma and 90 Ma, respectively (Weber et al. 2015). The terrane assemblage was accreted to the continental margin during the late Cretaceous to Paleocene (see Sect. 5.4.3.2).

The El Paso-Baudó components of the Chocó Arc are comprised of late Cretaceous to Paleogene sections of tholeiitic basalt of N- and E-MORB affinity (Goossens et al. 1977; Kerr et al. 1997), overlain by minor pyroclastic rocks, chert and turbidite. El Paso-Baudó represents a late Cretaceous silver of the CCOP/CLIP assemblage, considered to have formed along the trailing edge of the Caribbean

Plate. The Mandé-Acandí arc with associated plutonic, hypabyssal porphyritic stocks and pyroclastic volcanic sequences (La Equis-Santa Cecilia Fms; Sillitoe et al. 1982; Schmidt-Thomé et al. 1992; Leal-Mejía 2011; Montes et al. 2012) was emplaced within El Paso oceanic basement between ca. 60 and 42 Ma (Leal-Mejía 2011; Montes et al. 2012, 2015). Ural-Alaskan-type zoned ultramafic complexes at Alto Condoto and Mumbú were intruded into El Paso Terrane basement in the early Miocene (Tistl 1994; Tistl et al. 1994; Cediel et al. 2010). Development of the San Juan and Atrato basins began in the Paleogene. Final collision of the El Paso-Baudó-Mande assemblage along the western Cañas Gordas margin and uplift of the Baudo Range is recorded in the Miocene (Cediel et al. 2010; Montes et al. 2012, 2015). Faults related to the assembly and accretion of the Chocó Arc, including the Garrapatas-Dabeiba and San Juan-Sebastian systems (Cediel and Cáceres 2000; Cediel et al. 2003, 2010) and Uramita system (Duque-Caro 1991; Montes et al. 2012), reactivate, deform and/or truncate earlier structures associated with the Romeral, Cauca and San Jacinto fault systems.

5.2.2 Structural Framework

The original emplacement of many of the granitoid arc segments in Colombia demonstrates strong structural control, especially with respect to precursor basement architecture during Proterozoic through Paleozoic and early Mesozoic times. Past authors have noted the coaxial nature of both pre-Triassic and late Triassic – early Jurassic structure and the distribution of the major late Triassic – through Pliocene arc segments (Aspden et al. 1987; Cediel et al. 2003; Leal-Mejía 2011). Cediel et al. (1994), and Cediel and Cáceres (2000) demonstrate the reactivation of pre-Triassic structures, including the Bucaramanga–Santa Marta–Suaza (Garzón) and Palestina fault and suture systems, in the development of late Triassic volcano-sedimentary grabens (e.g. the Perijá, Maracaibo and Payande rifts) and the subsequent emplacement of major holocrystalline granitoid batholiths.

Many of the earlier phases of late Triassic to Pliocene granitoid magmatism, especially those from the Jurassic and Cretaceous, have suffered some degree of post-emplacement structural modification during the late Mesozoic-Cenozoic Northern Andean Orogeny (Cediel et al. 2003). A simplified structural framework for the modern-day Colombian Andes is shown in Fig. 5.2. The present-day structural architecture of the region is dominated by a complex array of large-scale strike-slip fault systems with complex transpressive movement vectors, for which, in some cases, origins and reactivations spanning the Proterozoic to the present can be demonstrated (Cediel and Cáceres 2000; Cáceres et al. 2003; Cediel et al. 2003; Cediel 2011). The structures of greatest importance with respect to the post-emplacement history of late Triassic-Pliocene granitoids are the Bucaramanga–Santa Marta–Garzón, Palestina and Romeral fault and suture systems. These fault/suture systems presently record tens of kilometres or more of lateral offset brought on by the sequential accretion of oceanic terranes along the Pacific margin of NW South

America and initiation of tectonic float in the Maracaibo Sub-plate, during stages of the Northern Andean Orogeny (Cediel et al. 2003; Kennan and Pindell 2009). Consequent modification/deformation of late Triassic-Pliocene granitoids ranges from localized to regional shearing and/or thrusting – focussed along intrusive margins – to tectonic segmentation and km-scale sinistral or dextral translation. Beyond these structural modifications, however, and unlike many of the Colombian granitoids of Paleozoic age, the latest Triassic through Plio-Pleistocene granitoids, including their coeval volcano-sedimentary sequences (where preserved), are not regionally metamorphosed nor have they been subjected to regional-scale ductile or penetrative deformation, prograde metamorphism or recrystallization during post-emplacement tectonic events.

5.3 Distribution, Age, Nature and Temporal-Spatial Evolution of Colombian Phanerozoic Granitoid Magmatism

5.3.1 *Introduction*

Figure 5.3 depicts the distribution of all major occurrences of Phanerozoic granitoid plutonic, hypabyssal and volcanic rocks throughout the Colombian Andes, based upon available regional cartographic data as presented by Cediel and Cáceres (2000), Gómez et al. (2007) and Gómez et al. (2015a). The distribution, nature and temporal-spatial evolution of Colombian granitoids will herein be discussed under four age-based headings: Paleozoic-mid-Triassic, latest Triassic-Jurassic, Cretaceous-Eocene and latest Oligocene-Mio-Pliocene. Each of these headings contains detailed information with respect to the age, classification, lithogeochemistry, isotope geochemistry and tectono-magmatic evolution of the respective granitoids. In the interest of space and focus, we have opted to forgo detailed petrographic-minerographic descriptions of the Colombian granitoid suites, and the reader will be referred to specific bibliographic references pertaining the various ages of Colombian granitoids in order to assess this valuable information. The final section of this chapter presents time-space analyses and tectono-magmatic reconstructions which place the granitoids into the integrated, evolving geotectonic framework of the Colombian Andes.

5.3.1.1 U-Pb (Zircon) Age Database: Phanerozoic Magmatic Episodes

Appendix A1 reveals bibliographic reference and geographic location data of granitoids dated by the U-Pb (zircon) method and compiled for use in this study. The quality and provenance of all information were reviewed, and importantly, one of the prerequisites for inclusion herein was that accurate sample location data (normally

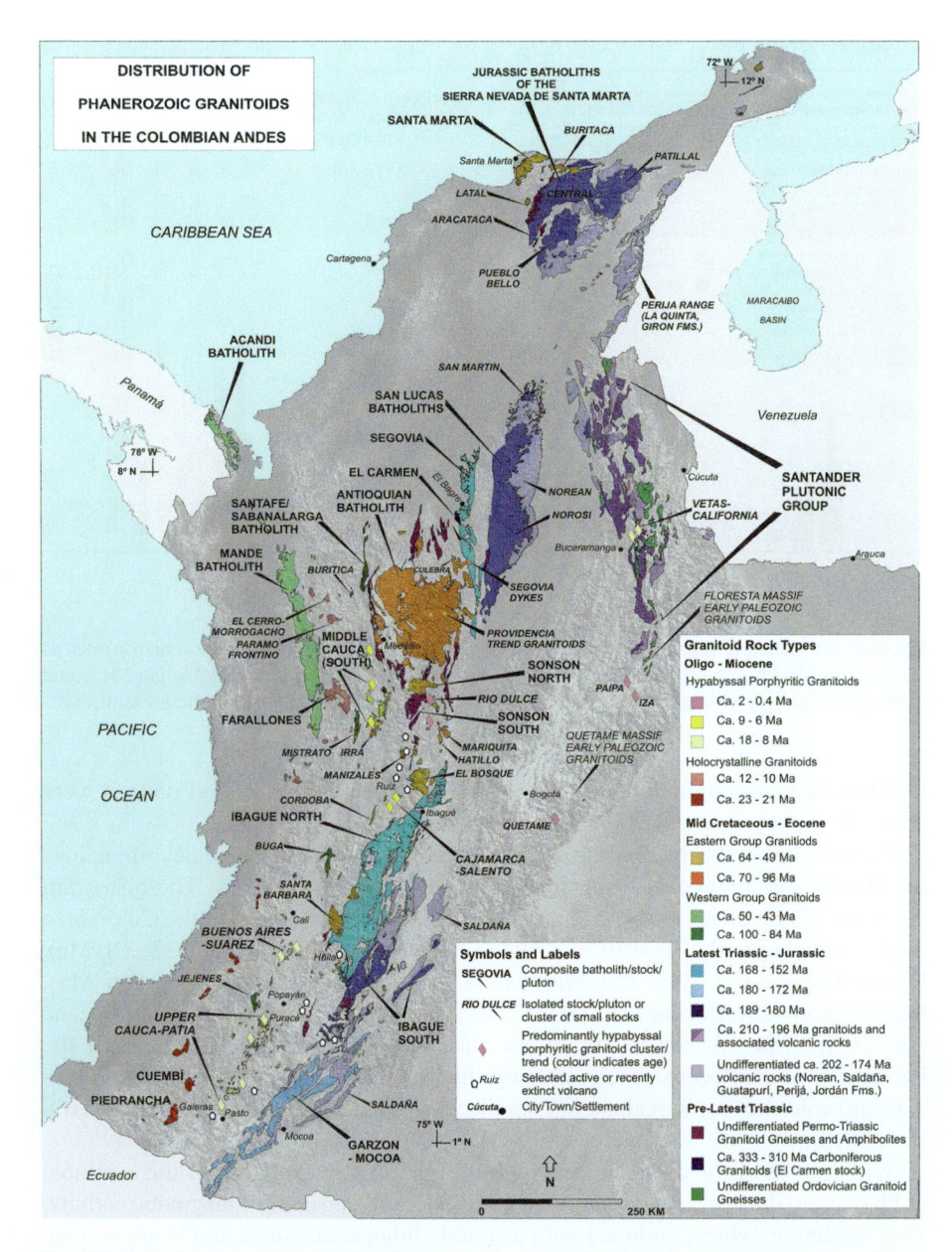

Fig. 5.3 Distribution of the principal Phanerozoic granitoid plutonic, hypabyssal and volcanic rocks in the Colombian Andes in relation to physiographic relief as derived from SRTM 90 m digital elevation model (DEM). (Granitoid shapes adapted from Cediel and Cáceres 2000; Gómez et al. 2007; Gómez et al. 2015a). Age ranges based upon U-Pb (zircon) data compiled herein

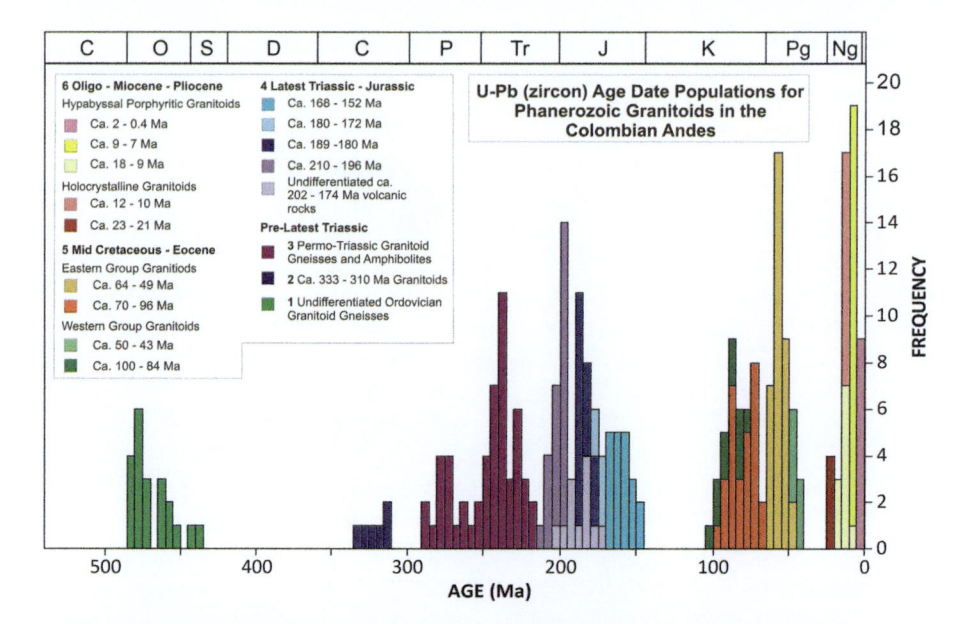

Fig. 5.4 Six principle periods of Phanerozoic granitoid magmatism in the Colombian Andes as derived from U-Pb (zircon) age dates. (Data is compiled primarily from Leal-Mejía (2011; and works cited therein), Gómez et al. 2015b and additional references cited within the present text)

UTM or latitude-longitude coordinates tied to a specified registered datum) were available. Approximately 287 individual data points are represented.

Figure 5.4 plots the distribution of U-Pb (zircon) crystallization age determinations for Phanerozoic granitoids from throughout the Colombian Andes. The histogram highlights six broad episodes of Phanerozoic granitoid magmatism in the Colombian Andes, including early Paleozoic (ca. 485–439 Ma), Carboniferous (ca. 333–310 Ma), Permo-Triassic (ca. 288–223 Ma), latest Triassic-Jurassic (ca. 210–146 Ma), late Cretaceous to Eocene (ca. 100–42 Ma) and latest Oligocene to Mio-Pliocene (ca. 23–1.2 Ma). A continuum of this last episode, from the Pliocene to the Pleistocene and Recent, encompassing the modern-day Colombian (Northern Andean) volcanic arc, is evident, but not discussed in detail herein (see Marín-Cerón et al. 2018).

Within most of the episodes depicted in Fig. 5.4, the resolution of the available U-Pb (zircon) data permits definition of additional sub-episodes of magmatic activity. This resolution, when combined with available lithogeochemical and isotopic data, permits a detailed analysis of the evolution and migration of Phanerozoic granitoid arc segments over time. Individual sub-episodes or pulses can be interpreted to coincide with, or denote, magmatic responses to tectonic developments throughout the region. These sub-episodes and their tectonic significance will be discussed in detail in the Sect. 5.3 of this chapter.

5.3.1.2 Lithogeochemical Database and Data Filtering

Appendix A2 also reveals bibliographic references and sample location data for some 561 Phanerozoic granitoids, for which whole-rock lithogeochemical analyses are presently available. As with the U-Pb (zircon) geochronology samples, accurate sample location data was prerequisite for inclusion within the lithogeochemical sample database utilized herein. Most lithogeochemical samples were collected from surface outcrops, although many samples from the Leal-Mejía (2011) study were taken from underground exposures and diamond drill core. Approximately 85 percent of the entire lithogeochemical sample set database includes a full suite of major, minor, trace and rare-earth elements.

The Colombian Cordilleras evolved at geographically low latitudes and have in many areas been subjected to high rainfall sub-tropical to tropical conditions since at least the late Cretaceous. In this context, samples collected from surface exposures have potentially been subjected to surface weathering and oxidation. Additionally, in the case of samples collected from underground exposures or diamond drill core, the potential exists for syn- or post-crystallization deuteric and/or hydrothermal alteration. In either case, the effects of the above related processes, being visual or cryptic, can significantly alter sample lithogeochemistry.

Based upon the above, the entire 545 sample suites utilized herein were subjected to element mobility analysis, permitting the identification of granitoids which have undergone subsolidus alteration processes such as those which can markedly affect critical alkali and aluminium indices and potentially generate misleading conclusions regarding the petrogenetic trends of the sample set (Davies and Whitehead 2010). The net result of subsolidus alteration is generally increased alkalinity, aluminity and/or silica content. In our particular case, the following general considerations are applicable: (1) high loss on ignition values (LOI > 2.0) may be related to high volatile content (reflecting possible clay, carbonate or sulphide alteration), (2) high SiO_2 content may be related to hydrothermal silicification, and (3) specific alteration and element ratio diagrams such as the alkali-alumina molar ratio plot (Davies and Whitehead 2006), the K-Ca-Na alteration evaluation plot (Warren et al. 2007) and Pearce Element Ratio (PER) and General Element Ratio (GER) diagrams (Stanley and Madeisky 1994, 1996) aid in the detection of altered samples. The result of data filtering was the identification of 212 samples which clearly exhibit the effects of hydrothermal alteration and/or weathering. Although the remaining 349 sample data set may be considered limited with respect to the extensive volume of magma it represents, the data are of high quality and permit the lithogeochemical and petrogenetic characterization of the great majority of the granitoids discussed herein.

For presentation purposes, we illustrate the major, trace and REE lithogeochemical data for samples from each major magmatic episode, using the following diagrams: the AFM diagram (Irvine and Baragar 1971), the K_2O vs. silica plot (Peccerillo and Taylor 1976), the aluminium saturation index diagram (Barton and Young 2002), the modified alkali-lime index (MALI) ($Na_2O + K_2O$-CaO) vs. silica, the $FeO_{tot}/(FeO_{tot} + MgO)$ vs. silica diagrams (Frost et al. 2001; Frost and Frost 2008), the R1-R2 diagram (De la Roche et al. 1980), the C1 chondrite normalized REE plot (McDonough and Sun 1995), the primordial mantle normalized trace element spider diagram (Wood et al. 1979), and the granitoid tectonic discrimination diagram (Ta vs. Yb) (Pearce et al. 1984).

5.3.2 Paleozoic to Mid-Triassic Granitoid Magmatism: Distribution, Age and Nature

Granitoid rocks from this extended time period comprise a texturally, compositionally and petrogenetically diverse suite, which has been subjected to a prolonged and intense tectonic history, both during and post-dating their emplacement and cooling. Given this observation, the geological context of what are commonly referred to as gneissic granitoids, meta-granitoids or foliated granitoids within the Colombian geological literature is complex, and the nature, distribution and genesis of these rocks in the Colombian Andes are relatively poorly understood. These observations may be attributed to various causes. Firstly, although (presumed) Paleozoic through mid-Triassic granitoids are of relatively widespread distribution, especially within the Santander Massif and the physiographic Central Cordillera (Aspden et al. 1987; Ward et al. 1973; Cediel and Caceres 2000; Vinasco et al. 2006; Gómez et al. 2015a), intrusions are limited to relatively small stocks and elongate or irregular-shaped bodies, with complex outcrop patterns, commonly intercalated with other granitoids of older or younger age. Exposure is often inhibited by thick vegetation cover, deep surficial oxidation and soil development or hydrothermal alteration, and the cartographic limits of many of the intrusions have yet to be clearly established. Historically, this situation was exasperated by the fact that few reliable radiometric age dates were available for the gneissic granitoids, and until the more recent application of U-Pb (zircon) dating techniques, many occurrences of these rocks were presumed to be of Precambrian, early Paleozoic or Mesozoic age, based primarily upon field relationships and considerations regarding texture and/or metamorphic grade.

Based upon information provided by more recent geological, age-dating, lithogeochemical and isotopic studies (e.g. Restrepo-Pace 1995; Vinasco et al. 2006; Ibañez-mejía et al. 2008; Cardona et al. 2010b; Horton et al. 2010; Montes et al. 2010; Leal-Mejía 2011; Leal-Mejía et al. 2011; Restrepo et al. 2011; Villagómez et al. 2011; Mantilla et al. 2012; Van der Lelij 2013; Cochrane 2013; Cochrane et al. 2014a; Van der Lelij et al. 2016), three broad populations of granitoids will be highlighted within this section: (1) early Paleozoic, (2) Carboniferous and (3) Permo-Triassic.

5.3.2.1 Distribution of Early Paleozoic to Mid-Late Triassic Granitoids

Figure 5.5 highlights the distribution of early Paleozoic through mid-Triassic granitoids throughout the Colombian Andes, based upon available regional geologic mapping and compilation. For reference, the principle physiographic provinces of the region are also shown. Early Paleozoic rocks described as granitoids, metamorphosed and foliated granitoids and granitic gneisses (orthogneisses) are mostly contained within the Santander, Floresta and Quetame massifs of the eastern Colombian Andes (Horton et al. 2010; Leal-Mejía 2011; Leal-Mejía et al. 2011; Mantilla et al. 2012; Van der Lelij 2013; Van der Lelij et al. 2016). In addition, punctual occurrences of early Paleozoic granitoids have been reported on the northern and western flanks of the Central Cordillera, along the Otú Fault near El Bagre (Leal-Mejía 2011, Leal-Mejía et al. 2011) and along the Cauca River valley (La Miel Orthogneiss; Vinasco et al. 2006; Villagómez et al. 2011; Martens et al. 2014) (Fig. 5.5).

Carboniferous granitoids have been reported at only one locality; the El Carmen-El Cordero Stock, near El Bagre (Leal-Mejía 2011) (Fig. 5.5). These intrusive rocks are of two main types: (1) early, fine-grained melanocratic, phaneritic holocrystalline to weakly porphyritic gabbro-diorites and (2) volumetrically dominant coarse-grained, phaneritic and holocrystalline leucocratic tonalities containing quartz, plagioclase and minor K-feldspar (microcline), with biotite, abundant zircon and ilmenite as accessory minerals. No additional granitoids of similar age have been reported in the Colombian Andes. The El Carmen-El Cordero pluton(s) were historically undifferentiated from Jurassic intrusives of the Segovia Batholith (González 2001) and references contained therein and remain so in all but the most recent geological compilation (e.g. Cediel and Cáceres 2000; Gómez et al. 2015a). Notwithstanding, the geological limits of the El Carmen-El Cordero plutons have yet to be established, and the contacts shown in Figs. 5.3 and 5.5 represent interpretations based upon very preliminary field reconnaissance and the examination of DEM images.

Permian to Triassic granitoids are widely distributed in the Colombian Andes from the border with Ecuador in the south to the Sierra Nevada de Santa Marta and the Guajira peninsula on the Colombian Caribbean coast. The majority of these bodies however are exposed within the northern Central Cordillera (Fig. 5.5). The Permo-Triassic granitoid suite is exposed as relatively small bodies outcropping on the eastern and western flanks of the Central Cordillera. Many of these bodies have been documented under local names, but, due to their small size, do not resolve well within regional-scale geologic maps. From S to N, confirmed granitoid gneisses of Permo-Triassic age include the La Plata orthogranite, La Linea intrusive gneiss, Manizales gneiss, Chinchina gneiss, the southern Sonsón Batholith (i.e. the Nariño Batholith), the Quebrada Pácora stock, the Pantanillo intrusive gneiss, the Cambumbía stock, the Rio Verde intrusive gneiss, the Alto de Minas intrusive gneiss, the Abejorral intrusive gneiss, the El Buey stock, the La Honda stock, the Amagá stock, the Pueblito diorite, the Palmitas granitic gneiss, the Horizontes tonalite gneiss, the Montegrande granitic gneiss, the Naranjales granitic gneiss, the Samaná granitic gneiss, the Santa Isabel gneiss, the Puquí meta-tonalite, the Nechí Gneiss, the Los Muchachitos gneiss and the Uray Gneiss (Vinasco et al. 2006; Ibañez-mejía et al. 2008; Cardona et al. 2010b; Montes et al. 2010; Leal-Mejía 2011; Leal-Mejía et al. 2011; Restrepo et al. 2011; Villagómez et al. 2011; Cochrane et al. 2014a). Some of these bodies are located in Fig. 5.5. The geological limits and age of numerous additional, small, unnamed, undated bodies of granitic orthogneiss have yet to be clearly defined.

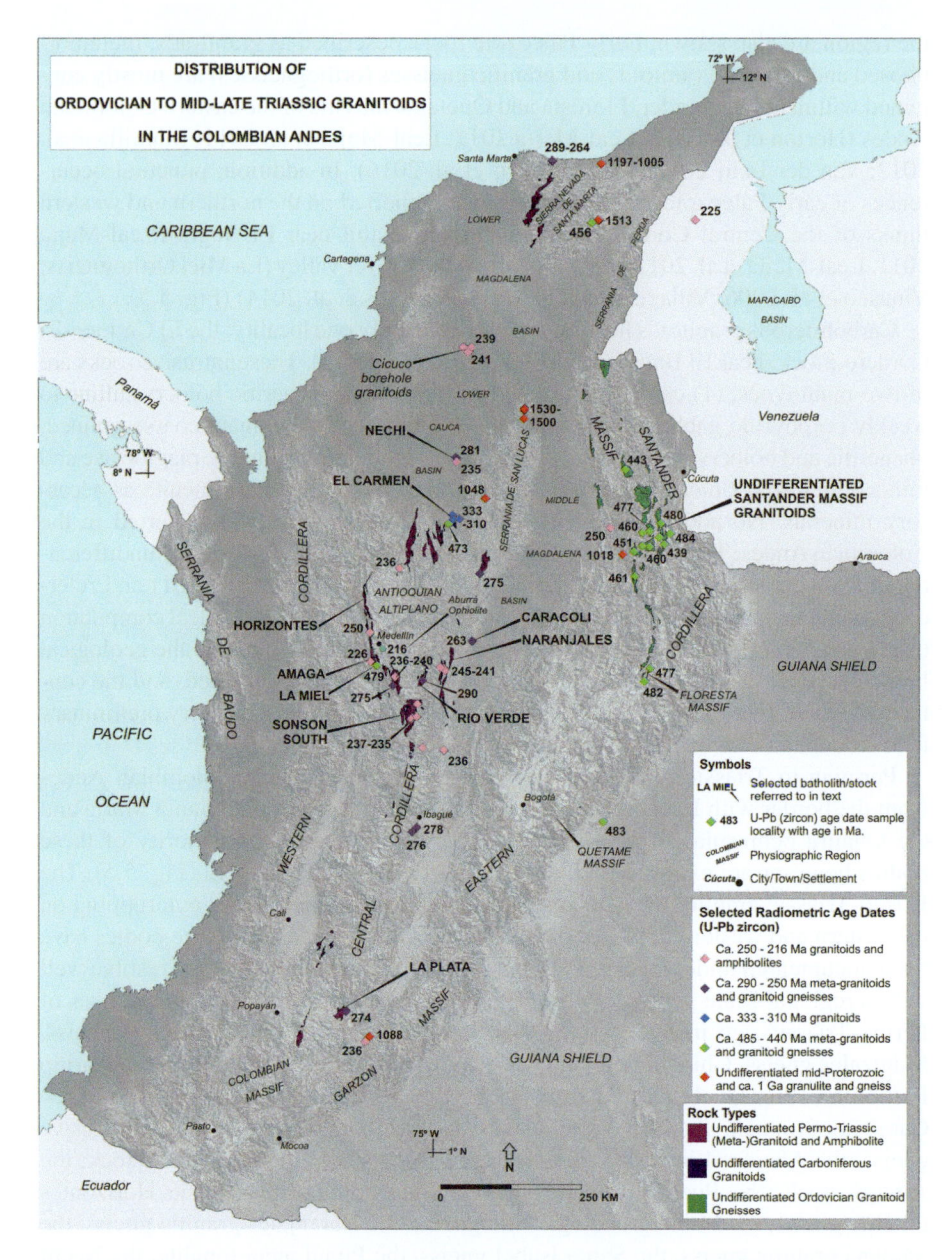

Fig. 5.5 Distribution of early Paleozoic through mid-late Triassic granitoids in the Colombian Andes. Principle modern-day physiographic provinces of the region are shown for reference. (Granitoid shapes modified after Cediel and Cáceres 2000; Gómez et al. 2007; Gómez et al. 2015a)

The gneissic texture observed in many early Paleozoic to mid-Triassic granitoids, especially those from the Permo-Triassic, has led to confusion with respect to their overall abundance and distribution. In various instances, such rocks are mapped as representatives of Precambrian basement, based upon the erroneous assumption that the "gneissic textured" granitoids are necessarily older than their surrounding host rocks. This assumption has been recently disproven using high-precision U-Pb (zircon) dating techniques, which illustrate that in various instances, granitoids previously recorded as "Precambrian" in age in fact belong to the Permo-Triassic suite, hosted within a Paleozoic or Permo-Triassic-aged basement (Vinasco et al. 2006; Ibañez-Mejía et al. 2008; Leal-Mejía 2011; Restrepo et al. 2011; Villagómez et al. 2011, Cochrane et al. 2014a). It is suspected that the abundance of Permo-Triassic granitoids will increase in future studies, at the expense of the "Precambrian" suite. It would appear that modern U-Pb dating techniques will form the best means for the differentiation of the Precambrian and early Paleozoic vs. Permo-Triassic suites.

5.3.2.2 Age Constraints on Paleozoic to Mid-Triassic Granitoid Magmatism

The U-Pb (zircon) age distribution of Phanerozoic granitoids of pre-Jurassic age in the Colombian Andes is presented in histogram format in Fig. 5.6. In terms of age, pre-Jurassic magmatism in the region has been previously well recognized, in two

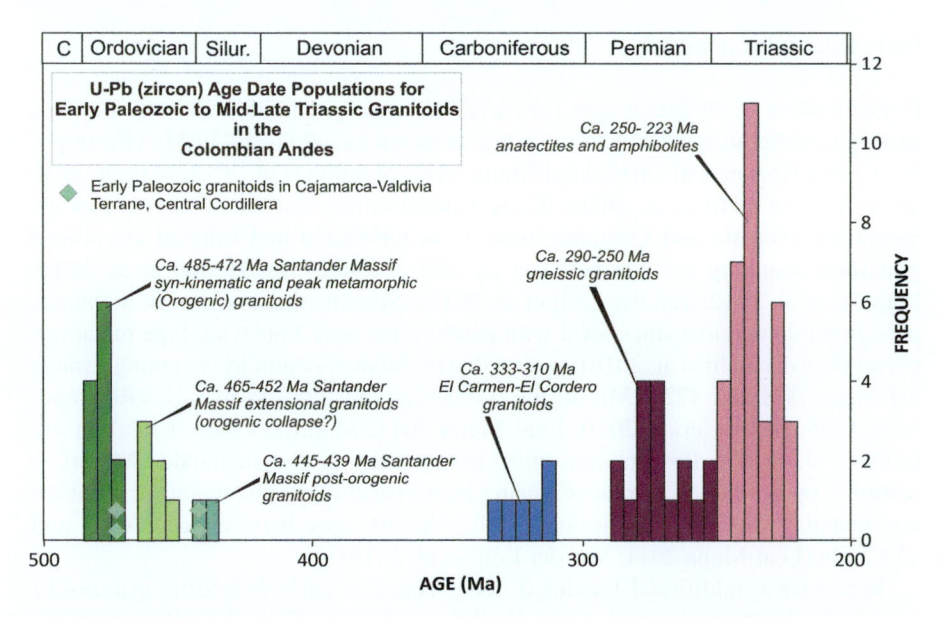

Fig. 5.6 Three principle periods of pre-Jurassic granitoid magmatism in the Colombian Andes, as derived from the distribution of U-Pb (zircon) age dates. Although each period, including respective sub-periods, is well-represented by multiple age dates, the overall distribution of these granitoids is sparse and erratic when compared to granitoids of post latest Triassic age

specific age ranges, including the early Paleozoic (i.e. early to middle Ordovician) and the Permian to mid-late Triassic (e.g. Goldsmith et al. 1971; Boinet et al. 1985; Restrepo-Pace 1995; Ordoñez 2001; Cediel et al. 2003; Vinasco et al. 2006; Ordoñez-Carmona et al. 2006; Ibañez-Mejía et al. 2008; Cardona et al. 2010b; Horton et al. 2010; Montes et al. 2010; Weber et al. 2010). Leal-Mejía (2011) documented previously unrecognized granitoid magmatism of Carboniferous age, in the El Carmen-El Cordero Stock near El Bagre. Additional pre-Jurassic U-Pb zircon ages have more recently been published, for early Paleozoic foliated granitoids in the Angosturas district and other localities in the Santander Massif (Mantilla et al. 2012; Van der Lelij et al. 2016), for the La Miel orthogneiss to the west (Villagómez et al. 2011; Martens et al. 2014) and for Permo-Triassic granitoids and amphibolites of the Central Cordillera (Restrepo et al. 2011; Villagómez et al. 2011; Cochrane et al. 2014a).

The resulting composite U-Pb (zircon) age date database permits definition of three distinct episodes of granitoid magmatism within the Colombian Andes: (1) ca. 485–439 Ma (early Paleozoic; early to mid-Ordovician), (2) ca. 333–310 Ma (Carboniferous) and (3) ca. 289–223 Ma (Permian to mid-late Triassic). Sub-episodes of granitoid magmatism are implicit within the age distribution recorded by each of the major episodes and have been interpreted by the various authors to represent granitoid magmatism within the evolving tectonic framework of the region during the early Phanerozoic, as will be reviewed in Sect. 5.4.

Early Paleozoic Granitoids

Crystallization U-Pb zircon ages for early Paleozoic granitoids within Colombia's eastern cordilleran system span the range between ca. 485 and 439 Ma (Restrepo-Pace 1995; Horton et al. 2010; Leal-Mejía 2011; Mantilla et al. 2012; Martens et al. 2014; Van der Lelij et al. 2016). Early Paleozoic magmatism is recorded in the Santander, Floresta and Quetame massifs, in unfoliated and foliated arc-related granitoids spanning a range between ca. 485 and 482 Ma (Horton et al. 2010; Mantilla et al. 2012; Van der Lelij et al. 2016). Syn-kinematic and peak metamorphic granitoid magmatism, coeval with medium-pressure Barrovian-type metamorphism (Van der Lelij et al. 2016), is recorded by foliated granitoids spanning a range between 479.8 and 472.5 Ma, in the Santander and Floresta massifs (Restrepo-Pace 1995; Horton et al. 2010; Leal-Mejía 2011; Mantilla et al. 2012; Van der Lelij et al. 2016). Post-metamorphic magmatism in the Santander Massif is recorded by granitoids emplaced during post-orogenic extension and/or resumed arc-related magmatism, returning U-Pb (zircon) ages between ca. 462.5 and 439.2 Ma (Leal-Mejía 2011; Van der Lelij et al. 2016).

To the west, additional localized occurrences of early Paleozoic granitoids/ granitic gneisses are exposed at two localities within the Central Cordillera. These include (1) the ca. 479–443 Ma (Villagómez et al. 2011; Martens et al. 2014) La Miel leuco-orthogneiss, composed primarily of k-feldspar, plagioclase, quartz, muscovite and minor biotite, with a clear relict igneous texture, and (2) a

473.4 ± 6.9 Ma (Leal-Mejía 2011) unnamed granodiorite intrusion outcropping along the Otú Fault near El Bagre (Fig. 5.5). The zircon separate for the sample was extracted from saprolitized bedrock, and the cartographic limits and geological context of this granodiorite body have yet to be completely defined.

A notable feature of the early Paleozoic zircon populations in Colombia is the complex internal structure and growth zonation of individual zircon crystals (e.g. Mantilla et al. 2012; Van de Lelij et al. 2016). In this context, beyond the interpreted crystallization ages presented above, many of the early Paleozoic zircon populations present multiple inheritance ages, dating from the early-mid-Proterozoic (Leal-Mejía 2011; Mantilla et al. 2012) and ranging into the late Proterozoic, and suggest a prolonged history of magmatism, metamorphism and crustal recycling during Proterozoic and early Paleozoic times (e.g. Van der Lelij et al. 2016; see below).

Carboniferous Granitoids

The occurrence of the Carboniferous El Carmen-El Cordero granitoids of the northern Central Cordillera was initially recorded by Leal-Mejía (2011). No additional occurrences have appeared in recent literature, so, based upon available geochronological data, this magmatism is presently restricted to the El Carmen-El Cordero occurrences. The El Carmen-El Cordero suite consists of early holocrystalline to weakly porphyritic melanodiorite, which returned a U-Pb (zircon) age of 333.1 ± 4.7 Ma, whilst four samples of Na-rich, quartz, plagioclase ± biotite and K-feldspar leucotonalite, comprising the main El Carmen Stock and associated dikes, returned U-Pb ages ranging from ca. 326 ± 5.6 Ma to 310.6 ± 5.6 Ma (Leal-Mejía 2011). The precise paragenetic relationship between the melanodiorite and the various phases of leucotonalite has yet to be deciphered as the contact between these units is not exposed. The ca. 333–310 Ma U-Pb (zircon) ages recorded by Leal-Mejía (2011) were considered to represent magmatic crystallization ages. Interestingly, unlike the early Paleozoic and Permo-Triassic (see below) granitoids from throughout the Colombian Andes, the Carboniferous granitoids demonstrate no indication of inheritance ages within their U-Pb (zircon) age date profiles.

Permo-Triassic Granitoids

Based upon published U-Pb (zircon) age dates, Permo-Triassic granitoids in the Colombian Andes, including granitoid gneisses and amphibolites, span the range from ca. 290 to 222 Ma (Fig. 5.6). Various authors (e.g. Vinasco 2004; Leal-Mejía 2011; Cochrane 2013) record complex zoning and inheritance patterns for zircons returning Permian through early-mid-Triassic ages. Recognition of the importance of the Permo-Triassic suite was initially revealed in the works of Vinasco (2004) and Vinasco et al. (2006), based upon dating of the La Honda, El Buey, Abejorral and other meta-granitoid intrusions along the Central Cordillera. Subsequent publications expanded the database for Permo-Triassic granitoids in the Central Cordillera to include the Samaná granitic orthogneiss (244.9 ± 4.7 Ma, Ibañez-Mejía et al. 2008), the Santa Isabel gneiss (267.8 ± 3.6 Ma, Restrepo et al. 2011), the Nechí Gneiss (ca. 282–277 Ma, Leal-Mejía 2011; Restrepo et al. 2011), the Las Palmas migmatite (222.0 ± 5.0 Ma, Restrepo et al. 2011), and other localized granitoid, granitic gneiss and amphibolites bodies which have returned ages ranging from ca. 278 to 236 Ma (Leal-Mejía 2011; Villagómez et al. 2011; Cochrane et al. 2014a). Cardona et al. (2010b) reported granitoids ranging from ca. 264 to 288 Ma in the Sierra Nevada de Santa Marta, whilst Montes et al. (2010) reported ca. 240 Ma ages from granitoid samples collected from subsurface drill core from the Lower Magdalena Basin.

Finally, based upon new U-Pb (zircon) age dates and petrochemical and petrographic data, Leal-Mejía (2011) recognized that the Sonsón Batholith (González 2001) consists of at least two composite plutonic bodies, the southern segment of which returns Permo-Triassic U-Pb (zircon) age dates. Based upon early K-Ar dating, González (2001) originally assigned the Sonsón Batholith to the Jurassic. Radiometric age dating by Leal-Mejía (2011), however, returned U-Pb (zircon) ages of 245.4 ± 4.8 Ma and 237.2 ± 4.1 Ma, for samples collected to the south and to the west of the town of Nariño. In this context Leal-Mejía (2011) referred to the granitoids returning early Triassic ages as the Nariño Batholith, however the geologic limits of the Triassic pluton(s) have yet to be formally mapped. The ca. 237 Ma pluton outcropping around the town of Nariño is homogeneous and holocrystalline to weakly foliated. Regional transects across the eastern margin of the pluton to the south of the town of Nariño reveal it is in contact with hornfelsed paragneiss containing early Permian-aged zircons (Leal-Mejía 2011), similar to rocks intercalated in the early Permian Río Verde gneiss complex to the NE of Sonsón (Vinasco et al. 2006). The SW flank of the Nariño Batholith, near San Félix, is unconformably overlain by Aptian-Albian siliciclastic rocks of the Abejorral Formation.

In general, the age of the Nariño Batholith as described herein coincides well with the Permo-Triassic suite documented elsewhere in the Central Cordillera by Vinasco et al. (2006). Based upon presently available mapping, field reconnaissance, regional geological trends and U-Pb age (in addition to available lithogeochemical data), we suggest that the paragneiss along the eastern margin of the Nariño Batholith represents the southern continuation of the Río Verde granitic gneiss, whilst the western sector of the Nariño Batholith appears to represent the southern continuation of Permo-Triassic orthogneiss, granitic gneiss and "post-tectonic granite", along the western margin of the Central Cordillera, as illustrated by Vinasco et al. (2006).

5.3.2.3 Lithogeochemical and Isotopic Characteristics of Early Paleozoic to Mid-Triassic Granitoids

Lithogeochemistry

In terms of whole-rock lithogeochemistry, the Paleozoic to mid-Triassic granitoids exhibit significant differences in composition between the early Paleozoic, Carboniferous and Permo-Triassic age groupings highlighted above (Fig. 5.6).

Available data for each of the three age groups will be summarized separately briefly herein.

Major, trace and rare-earth element lithogeochemical data for early Paleozoic granitoids, as drawn from the data sets of Leal-Mejía (2011) and Van de Lelij et al. (2016), are presented in Figs. 5.7 and 5.8. Whole-rock analyses reveal variable SiO_2 contents ranging from 57.6 to 74.7 wt%, with compositions ranging from gabbrodiorite to tonalite through granite. The sample set defines a calc-alkaline trend, plotting in the medium-K to high-K calc-alkaline fields. Most of the early Paleozoic granitoids are weakly peraluminous, with the most altered samples exhibiting extremely high A/NKC values (>2.0), likely due to post-crystallization hydrothermal alteration. Notwithstanding, two samples (10VDL23 and 10VDL47, Van der Lelij et al. 2016) plot in the metaluminous field. With respect to the classification scheme of Frost et al. (2001), most of early Paleozoic samples plot in the calcic and calc-alkalic fields and are magnesian (oxidized) in composition. Trace element and rare-earth element spider diagrams indicate fractionated arc-related magmatism (volcanic-arc granites (VAG)) with variable negative Eu anomalies.

With respect to the Carboniferous granitoids, petrographic analysis of both the melanodiorite and leucotonalite suites (Leal-Mejía 2011) suggests minor effects brought on by hydrothermal alteration and/or very low-grade regional metamorphism. Hornblende within the melanodiorite has been replaced by pumpellyite, prehnite, chlorite and epidote, whilst the cores of plagioclase have been altered to sericite. Accessory biotite in the leucotonalite has been partially altered to an assemblage containing chlorite, epidote, magnetite and titanite, whilst the cores of plagioclase crystals are strongly sericitized. Alteration plots for the El Carmen suite suggest some degree of major element mobility associated with these effects.

The ca. 333–310 Ma granitoid suite plots in two separate clusters on the lithogeochemistry plots (Fig. 5.7). SiO_2 contents are lower in melano-gabbro/diorites (48.4–48.8 wt%) with respect to leucotonalites (68.8–72.5 wt%). Both groups exhibit notably low K_2O contents (0.04–1.22 wt%) and plot in the compositional ranges of gabbro-norite and tonalite-granodiorite, respectively. The Na (vs. K)-rich, trondhjemitic nature of the leucotonalites becomes particularly evident when samples are plotted on the feldspar triangle of O'Connor (1965; see Leal-Mejía 2011). The melano-gabbro/melanodiorite members of the suite plot clearly tholeiitic, whilst the leucotonalite series presents more evolved calc-alkaline compositions on the AFM diagram. A composite calc-alkaline trend, however, can only be inferred by the data, as there is a clear compositional gap in the differentiation series (Fig. 5.7). This may be a reflection of the limited number of analyses available for the suite ($n = 6$) or alternatively may be a result of the apparent bimodal nature of the suite. Melanodiorite samples plot in the metaluminous field, whereas leucotonalites are weakly peraluminous perhaps due to alteration effects. The entire suite plots in the magnesian and calcic fields of Frost et al. (2001). Trace element diagrams depict large-ion lithophile element enrichment (e.g. Ba, K) and depletion of high-field strength elements (e.g. Nb, Ta, Ti). Chondrite-normalized REE plots reveal flat patterns around 10x chondrite concentrations for the melanodiorites ($\sum$REE = 24.7–34.7 ppm) vs. somewhat more enriched and fractionated patterns for the leucotonalite samples ($\sum$REE = 49–94.1 ppm). Neither rock type produces significant Eu anomalies.

PERMO-TRIASSIC
☐ Undifferentiated Permo-Triassic granitoids
― Amphibolites (ca. 240 - 216 Ma)
☐ Granitoid anatectites (ca. 250 - 223 Ma)
■ Granitoid gneisses, meta-granitoids (ca. 290 - 250 Ma)
CARBONIFEROUS
☐ El Carmen-El Cordero suite (ca. 333-310 Ma)

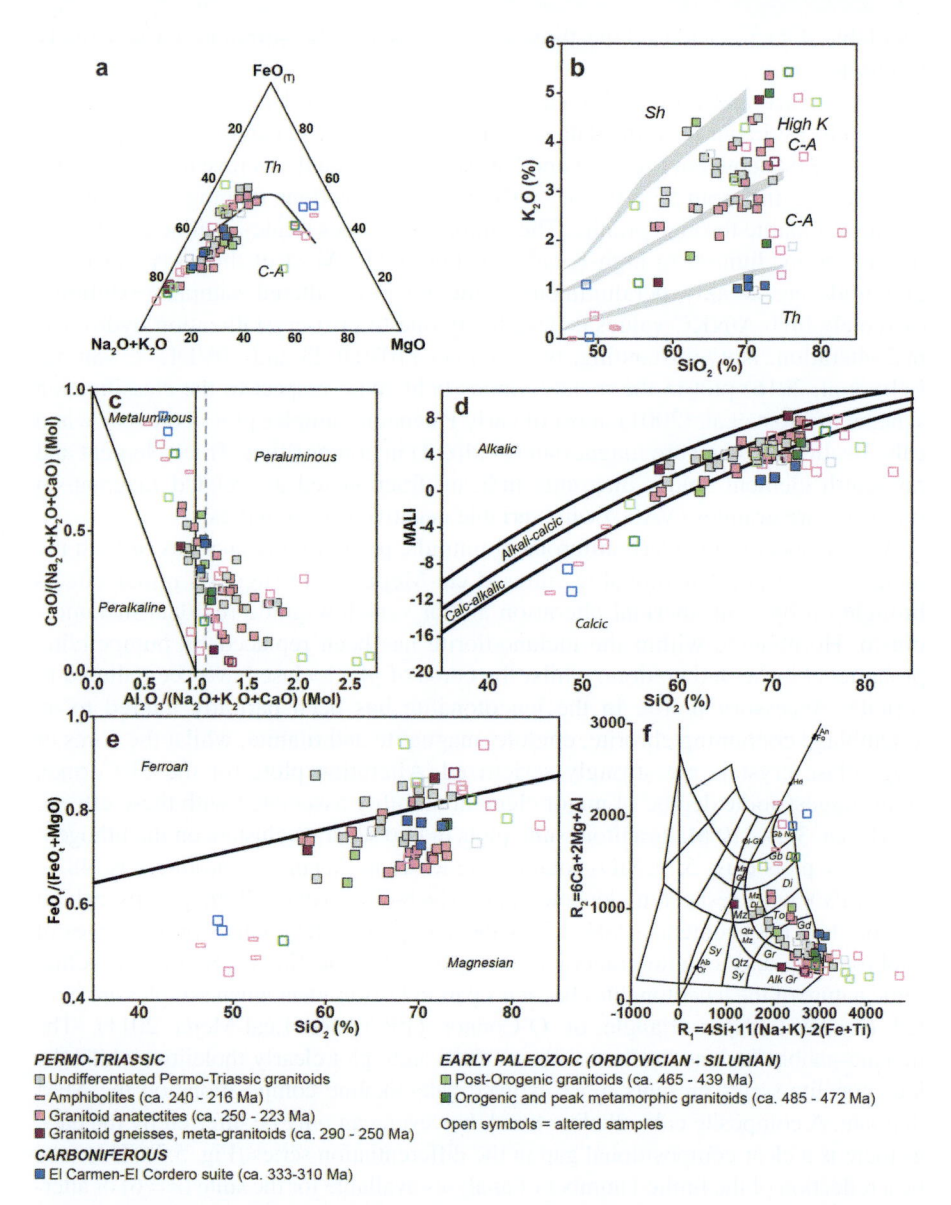

EARLY PALEOZOIC (ORDOVICIAN - SILURIAN)
☐ Post-Orogenic granitoids (ca. 465 - 439 Ma)
☐ Orogenic and peak metamorphic granitoids (ca. 485 - 472 Ma)
Open symbols = altered samples

Fig. 5.7 Major element lithogeochemical plots for pre-Jurassic (i.e. early Paleozoic, Carboniferous and Permo-Triassic) granitoids in the Colombian Andes. (**a**) AFM plot, curve after Irvine and Baragar (1971); (**b**) K_2O vs. SiO_2 plot, boundary fields in grey as summarized by Rickwood (1989); (**c**) alumina saturation plot after Barton and Young (2002); (**d** and **e**) MALI and Fe-index vs. SiO_2 plots, respectively, after Frost et al. (2001); (**f**) R1-R2 classification plot after De La Roche et al. (1980). *Th* tholeiite, *C-A* calc-alkaline, *Sh* shoshonite, *Gb No* gabbro-norite, *Gb Di* gabbro-diorite *Di* Diorite *Mz Di* monzodiorite *Mz* Monzonite *To* Tonalite *Gd* Granodiorite *Gr* Granite *Alk Gr* Alkali Granite

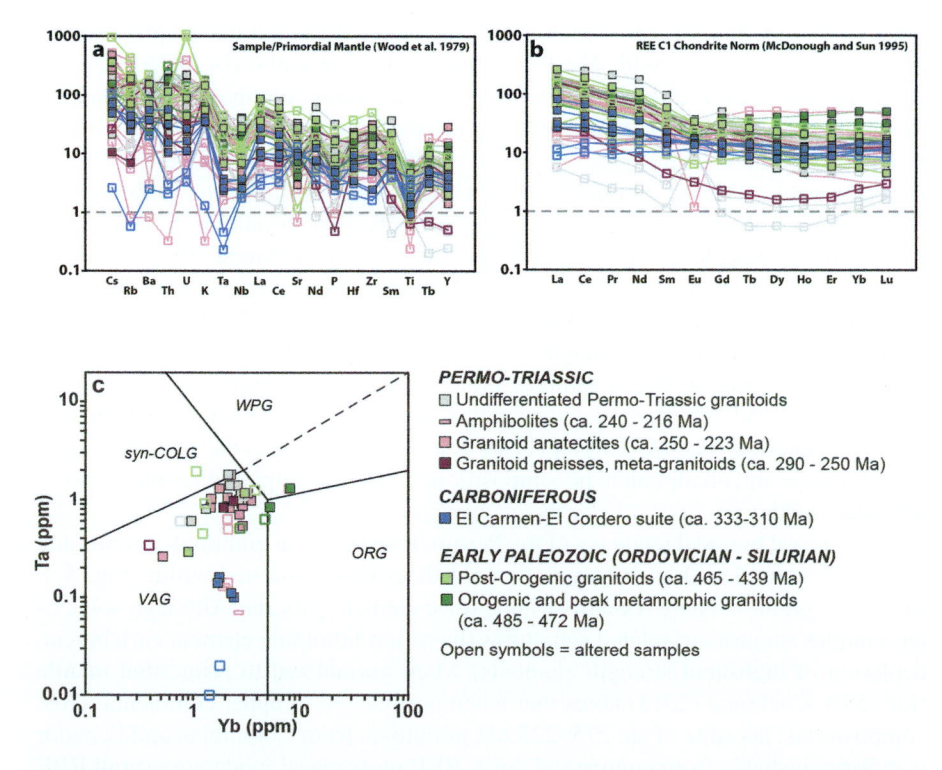

Fig. 5.8 Trace element and REE lithogeochemical plots for pre-Jurassic (i.e. early Paleozoic, Carboniferous and Permo-Triassic) granitoids in the Colombian Andes. (**a** and **b**) Trace Element and REE normalized spider-diagram plots; (**c**) granite discrimination Ta vs. Yb diagram after Pearce et al. (1984). *VAG* volcanic-arc granites, *syn-COLG* syn-collisional granites, *WPG* within-plate granites, *ORG* ocean ridge granites

Lithogeochemical data for Permian to mid-Triassic granitoids, displayed in Figs. 5.7 and 5.8, was sourced from various authors including Saenz (2003), Vinasco et al. (2006), Cardona et al. (2010b), Leal-Mejía (2011), Cochrane et al. (2014a), Rodríguez et al. (2014) and Van der Lelij et al. (2016). The data set represents sample collection within diverse geographic and geologic environments throughout the Colombian Andes and includes some samples from Ecuador and Venezuela. Lithogeochemical interpretation of the Permo-Triassic granitoids is in some respect difficult and complex, given the metamorphic conditions, post-emplacement tectonic history, post-crystallization alteration and, in many cases, the poorly defined geological context of this suite.

The composite data set appears bimodal with respect to SiO_2 contents. The more felsic members yield 57.6–73.5 wt% SiO_2, with anomalous, higher SiO_2 contents (73.7–83.0 wt%) considered to be associated with low temperature alteration (silicification). Coeval Permo-Triassic amphibolites reveal lower SiO_2 contents (46.5–52.5 wt%). The felsic series shows a well-defined medium-K to high-K calc-alkaline trend on the AFM and K_2O vs. SiO_2 diagrams, whilst the amphibolites plot apart in the respective tholeiitic field. The R1-R2 discriminational plot (De la Roche et al. 1980) depicts a clear diorite-tonalite-granodiorite-granite compositional trend for the felsic suite and highlights the general increase in alkalinity and silica content brought on by increasing degrees of post-crystallization alteration. On the same plot the (apparently altered) amphibolites plot in the gabbro-diorite to gabbro-norite field. A remarkable feature of most of the Permo-Triassic granitoids is their generally peraluminous character (e.g. Vinasco et al. 2006; Leal-Mejía 2011; Cochrane et al. 2014a), for both altered and unaltered samples. Again, the amphibolites plot apart in the metaluminous field. With respect to the Frost et al. (2001) classification, the felsic subset straddles the magnesian-ferroan granitoid boundary line, although the majority of the unaltered samples plot clearly on the magnesian side, whilst some altered and undifferentiated samples plot ferroan. The amphibolites plot clearly magnesian. The MALI plot (Frost et al. 2001) indicates that the felsic suite is calc-alkalic and alkali-calcic in composition, whilst the amphibolites plot apart in the calcic field.

The general bimodal tendency of the Permo-Triassic felsic granitoid-amphibolite suite is sustained within the trace element diagrams contained within Fig. 5.8. The felsic granitoid suite reveals variable trace element patterns. Although some of the samples suggest arc-related signatures (large-ion lithophile element enrichment, depletion of high-field strength elements) when normalized to primordial mantle (Fig. 5.8), Cochrane (2013) notes that when normalized to upper continental crust compositions, his suite of ca. 275–225 Ma granitoids from Colombian and Ecuador is indistinguishable from continental crust. REE plots reveal moderate overall REE enrichment and moderately sloping, fractionated trends for the felsic subset. Slightly negative or no Eu anomalies are observed. The amphibolites record essentially flat REE patterns with approximately 10x chondrite concentrations.

Sr-Nd-Pb Isotope Geochemistry

Available Sr, Nd and Pb isotope geochemical data for early Paleozoic, Carboniferous and Permian to mid-Triassic granitoids and amphibolites from the Colombian Andes are shown plotted in Fig. 5.9. For comparative purposes Sr, Nd, and Pb isotope data for early Paleozoic granitoids from the Venezuelan (Merida) Andes (Van der Lelij et al. 2016) and for Permian-mid-Triassic granitoids (Van der Lelij et al. 2016) and amphibolites (Chiaradia et al. 2004; Cochrane et al. 2014a), from Venezuela and Ecuador, respectively, are also plotted.

The early Paleozoic granitoids of the Santander Massif in the eastern Colombian Andes show notably negative ε_{Nd} values ($\varepsilon Nd_{(t)}$ = −6.0 to −1.3) and high initial $^{87}Sr/^{86}Sr$ ratios ($^{87}Sr/^{86}Sr_{(i)}$ = 0.70148–0.71292) (Van der Lelij et al. 2016), suggesting important mixing, assimilation and/or interaction with the upper continental crust. Lead isotope data for the early Paleozoic Santander granitoids show relatively high (radiogenic) values ($^{206}Pb/^{204}Pb$ = 19.01–20.17, $^{207}Pb/^{204}Pb$ = 15.68–17.79, $^{206}Pb/^{204}Pb$ = 38.88–40.67, Van der Lelij et al. 2016) and plot over the upper crust

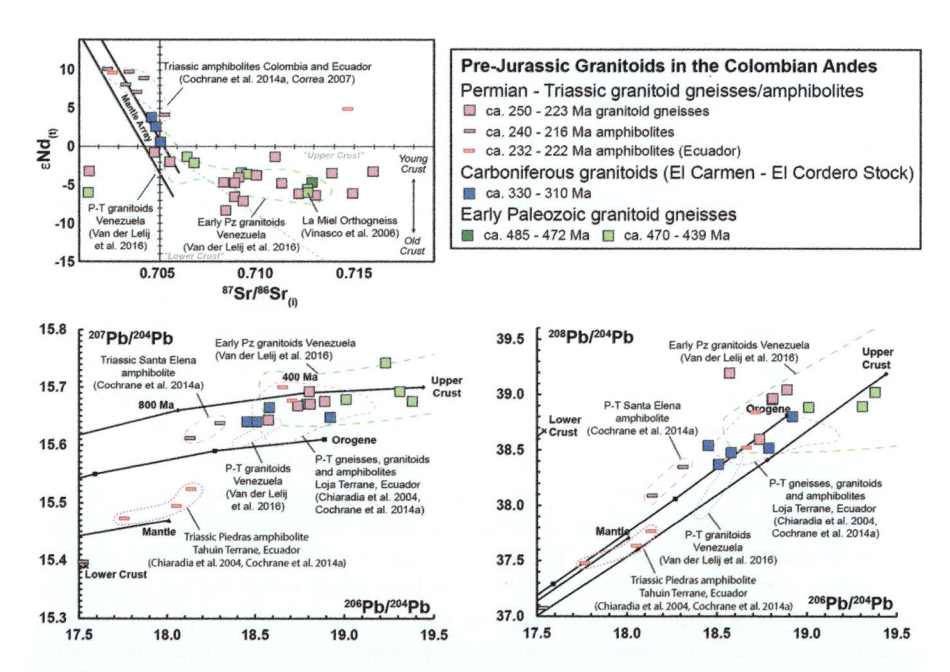

Fig. 5.9 Sr-Nd and Pb isotope plots for pre-Jurassic granitoids in the Colombian Andes. Additional data for igneous and metamorphic suites from the surrounding region are included for reference

lead isotope evolution curve of the plumbotectonics model of Zartman and Doe (1981) (Fig. 5.9). The Santander Massif Pb isotopic compositions compare well with the Pb isotope composition of early Paleozoic granitoids from the Merida Andes in Venezuela.

With respect to the composite Permian to mid-Triassic suite, no Sr-Nd data are available for granitoids within the ca. 290–260 Ma age range. Considering the ca. 250–216 Ma ages, however, a subset of granitic gneisses and coeval amphibolites (including the ca. 240 Ma Santa Elena amphibolite of Cochrane et al. (2014a) and the ca. 216 Ma Aburrá ophiolite of Correa (2007)), from the Colombia's Central Cordillera, is represented. Granitoids from this subset reveal similar, evolved (upper crustal) Sr and Nd isotope compositions ($^{87}Sr/^{86}Sr_{(i)}$ = 0.70150–0.73106, $\varepsilon Nd_{(t)}$ = −8.91 to −0.76, Leal-Mejía 2011), when compared with the early Paleozoic suite (Fig. 5.9). These isotopic compositions contrast markedly with the signatures for the Permo-Triassic amphibolites of Colombia and Ecuador ($^{87}Sr/^{86}Sr_{(i)}$ = 0.70243–0.70535, $\varepsilon_{Nd(t)}$ = +3.37 to +10.18; Cochrane et al. 2014a), reflecting the bimodal nature of the Permo-Triassic suite. A crustal provenance for the Central Cordilleran granitoids, without significant contribution from enriched mantle sources was suggested by Vinasco et al. (2006) and Cochrane (2013), whilst a primarily mantle-derived source for the amphibolites was proposed by Cochrane et al. (2014a; see below).

Pb isotope data for ca. 250–216 Ma Central Cordillera granitoids presented by Leal-Mejía (2011) also plot over the upper crust lead isotope evolution curve of the plumbotectonics model of Zartman and Doe (1981), (^{206}Pb/^{204}Pb = 18.57–18.89, ^{207}Pb/^{204}Pb = 15.64–15.69, ^{206}Pb/^{204}Pb = 38.6–39.2), although the data exhibit less radiogenic compositions than those observed for the early Paleozoic granitoid suite (Fig. 5.9). The Colombian Permo-Triassic granitoids compare well with Pb isotope compositions for (meta-)granitoids and granitoid gneisses of similar age, hosted within the Loja Terrane, Ecuador (Chiaradia et al. 2004; Cochrane et al. 2014a), and also with compositions for Permo-Triassic granitoids from the Merida Andes, Venezuela (Van der Lelij et al. 2016) (Fig. 5.9). Ecuador's Loja Terrane (Aspden et al. 1992) is considered the geological equivalent and southern extension of the Cajamarca-Valdivia Terrane, which forms the basement to Colombia's Central Cordillera (Cediel et al. 2003). With respect to the Permo-Triassic amphibolites, the Pb isotope composition of the ca. 240 Ma Santa Elena amphibolite (Cochrane et al. 2014a) is notably less radiogenic than the bulk of the Permo-Triassic meta-granitoid suite (Fig. 5.9), suggesting a mantle-derived component in the source region.

In contrast to both the early Paleozoic and Permo-Triassic granitoid suites, the ca. 333–310 Ma El Carmen-El Cordero granitoids exhibit ^{87}Sr/^{86}Sr$_{(i)}$ and ε_{Nd} values plotting up the mantle array (Fig. 5.9). The observed low initial ^{87}Sr/^{86}Sr ratios and positive $\varepsilon_{Nd(t)}$ values (^{87}Sr/^{86}Sr$_{(i)}$ = 0.70441–0.70516, $\varepsilon Nd_{(t)}$ = +0.58 to +3.79) led Leal-Mejía (2011) to suggest a primitive, mantle-derived source for the Carmen-El Cordero suite, without the presence of a significant crustal component. The same suite exhibits somewhat less radiogenic Pb isotope values than early Paleozoic and Permo-Triassic granitoids (^{206}Pb/^{204}Pb = 18.45–18.92, ^{207}Pb/^{204}Pb = 15.64–15.67, ^{206}Pb/^{204}Pb = 38.37–38.79, Leal-Mejía 2011), supporting this conclusion. The El Carmen-El Cordero suite intrudes Cajamarca-Valdivia Terrane basement. Lead isotope values for El Carmen-El Cordero can be interpreted to plot on a mixing curve between more mantelic Pb isotope values, as represented by the Santa Elena and Rio Piedras (Ecuador) amphibolites (Fig. 5.9), and crustally derived Pb sources as reflected in the Pb isotope composition of the early Paleozoic and Permo-Triassic meta-granitoid suites.

Additional Isotopic Studies for Early Paleozoic and Permo-Triassic Granitoids

In recent years, an important set of Lu-Hf isotope data has become available for some of the early Phanerozoic granitoid suites of the Colombian Andes. Lu-Hf isotope data, when combined with other isotope analyses (e.g. Rb-Sr, Sm-Nd), have been used to shed additional light upon the potential source regions for granitoid magmas subject to diverse and prolonged geological histories (e.g. Stevenson and Patchett 1990; Deckart et al. 2010; Kurhila et al. 2010). Mantilla et al. (2012) provided Lu-Hf data for early Ordovician granitoids from the Vetas-California district of the Santander Massif. Van der Lelij (2013) and Van der Lelij et al. (2016) combined Lu-Hf data with additional Sr, Nd and Pb isotope analyses for early Paleozoic granitoids from throughout the Santander Massif and Mérida Andes (Venezuela), whilst Cochrane (2013), Cochrane et al. (2014a) and Spikings et al. (2015) supplied Hf isotope data for various Permo-Triassic granitoids and amphibolites in Colombia's Central Cordillera and Ecuador.

Mantilla et al. (2012) interpret radiogenic, initial epsilon Hf (εHf_i) values of >0 for a ca. 477 Ma, calc-alkaline meta-diorite collected near Vetas-California, to be indicative of a depleted mantle source. In this context they interpret the Vetas-California granitoid to represent mantle-derived magmas formed within a supra-subduction zone setting.

Notwithstanding, Van der Lelij (2013) and Van de Lelij et al. (2016) evaluated early Paleozoic granitoid magmatism in the Santander Massif and Mérida Andes (Venezuela) based upon more extensive Lu-Hf (zircon) data, combined with whole-rock Rb-Sr and Sm-Nd isotope analyses. These authors indicate that Lu-Hf model ages of >1.3 Ga are restricted to syn-orogenic (arc- and collision-related) granitoids which formed during Barrovian metamorphism and crustal thickening between ca. 499 and 472 Ma. They note that these same granitoids yield high initial $^{87}Sr/^{86}Sr$ ratios, suggesting a melt derived from evolved, Rb-rich middle to upper crust. A possible crustal end member source for this crust includes Precambrian basement units which are exposed in the Garzón Massif and adjacent regions and sedimentary rocks that host detritus derived from these units (Van der Lelij 2013).

Furthermore, Van der Lelij (2013) and Van de Lelij et al. (2016) indicate that subsequent early Paleozoic granitoids, which crystallized between ca. 472 and 452 Ma, yield younger Lu-Hf model ages, with low initial $^{87}Sr/^{86}Sr$ ratios, suggesting that they were derived at least in part from more juvenile, Rb-poor sources. They conclude that the overall isotopic composition of post-472 Ma granitoids suggests melt derived from recycling of variable, lower to upper crustal end members with unquantified contributions from enriched mantle sources (Van der Lelij 2013).

With respect to the bimodal suite of Permian through Triassic meta-granitoids and amphibolites, Cochrane (2013) and Cochrane et al. (2014a) provided Hf isotope data for zircon separates from 14 granitoids and 4 amphibolites collected in Ecuador and in Colombia's Central Cordillera. Based upon composite lithogeochemical data, Cochrane (2013) considers the ca. 275–225 Ma meta-granitoids to be S-type and to have been derived from an upper crustal source. He notes that coeval zircons in most of the granitoids yield extremely large intra-sample εHfi variations (e.g. +3.2 to −11). He considers these variations to be too large to be representative of magmatic zircons that crystallized from a single, well-mixed source. He suggests the εHf_i variations for the meta-granitoid zircons could be accounted for by source mixing, although he acknowledges that disequilibration reactions which fractionate Hf within zircon could also be responsible for some of the variation. In terms of source mixing, Cochrane (2013) indicates that xenocrystic zircon cores within the meta-granitoid zircon population return ages ranging from ca. 275 Ma to 1.2 Ga. He considers these ages to be representative of the range of meta-sedimentary proto-liths involved in crustal anatexis during petrogenesis of the ca. 275–225 Ma meta-granitoids.

Ca. 240–223 Ma amphibolites studied by Cochrane (2013) were found to yield εHf_i values that negatively correlate with their $^{206}Pb/^{238}U$ zircon ages; that is, the older, ca. 240–232 Ma, amphibolites produced overall less positive εHfi values. He notes that the ca. 240–232 Ma amphibolites contain complex, oscillatory zoned zircons which produce both positive and negative εHf_i values. Cochrane (2013) interprets the data to reflect crustal contamination during older amphibolite emplacement. Conversely, he notes that younger (ca. 225–223 Ma) amphibolites contain only unzoned zircons which exhibit no intra-sample zircon εHf_i variation and return the most juvenile (i.e. positive) εHf_i values (+13 to +15), which approach the depleted mantle array. He further observes that the least radiogenic volumes of zircons extracted from the amphibolites overlap with the Hf isotopic signatures of the meta-granitoids ("crustal anatectites"). He concludes that crustal contamination during emplacement was an important process in the petrogenesis of the older (ca. 240–232 Ma) amphibolites but became progressively less important over time, as reflected in the isotopic composition of the younger amphibolites.

5.3.3 Latest Triassic-Jurassic Granitoid Magmatism: Distribution, Age and Nature

Late Triassic-Jurassic granitoids represent the most extensive period of magmatic activity recorded within the present-day geological exposure of the Colombian Andes. The belt is comprised of a SSW- to NNE-oriented array of volcano-plutonic arc segments extending from the Ecuador border to the Sierra Nevada de Santa Marta on the Caribbean coast (Aspden et al. 1987; Cediel and Cáceres 2000; Gómez et al. 2007; Gómez et al. 2015a). It forms the northern extension of a more extensive system of late Triassic-Jurassic volcano-plutonic arc segments which continue into southernmost Ecuador and northern Perú (Litherland et al. 1994; Cediel et al. 2003; Cochrane 2013; Cochrane et al. 2014b; Spikings et al. 2015).

5.3.3.1 Distribution of Late Triassic to Jurassic Granitoids

The distribution of late Triassic-Jurassic granitoid batholiths, stocks and associated volcanic sequences in the Colombian Andes is shown in Fig. 5.10, whilst Table 5.1 summarizes the nomenclature, ages and morpho-tectonic position of the major batholiths and coeval volcanic/volcano-clastic sequences.

Within the Colombian Andes, volcano-plutonic rocks of late Triassic-Jurassic age cropout in the Garzón and Santander Massifs, the Sierra Nevada de Santa Marta and within the Central Cordillera and the Serranía de San Lucas. Major batholiths and associated stocks within the Jurassic belt include, from south and east to north and west, the Mocoa Batholith in the Garzón Massif (Alfonso 2000); the Santa Bárbara-Rionegro-Mogotes batholiths and the Pescadero, La Corcova and Páramo Rico plutons in the Santander Massif (the "Santander Plutonic Group" of Ward et al. 1973, also described by Royero and Clavijo 2001); the Ibagué Batholith (Nelson 1957; Núñez 1998; Altenberger and Concha 2005); the Norosí (Guamocó)-San Martín batholiths of the Serranía de San Lucas (the "San Lucas granitoids" described by Clavijo et al. 2008); the Segovia Batholith (Feininger et al. 1972); and

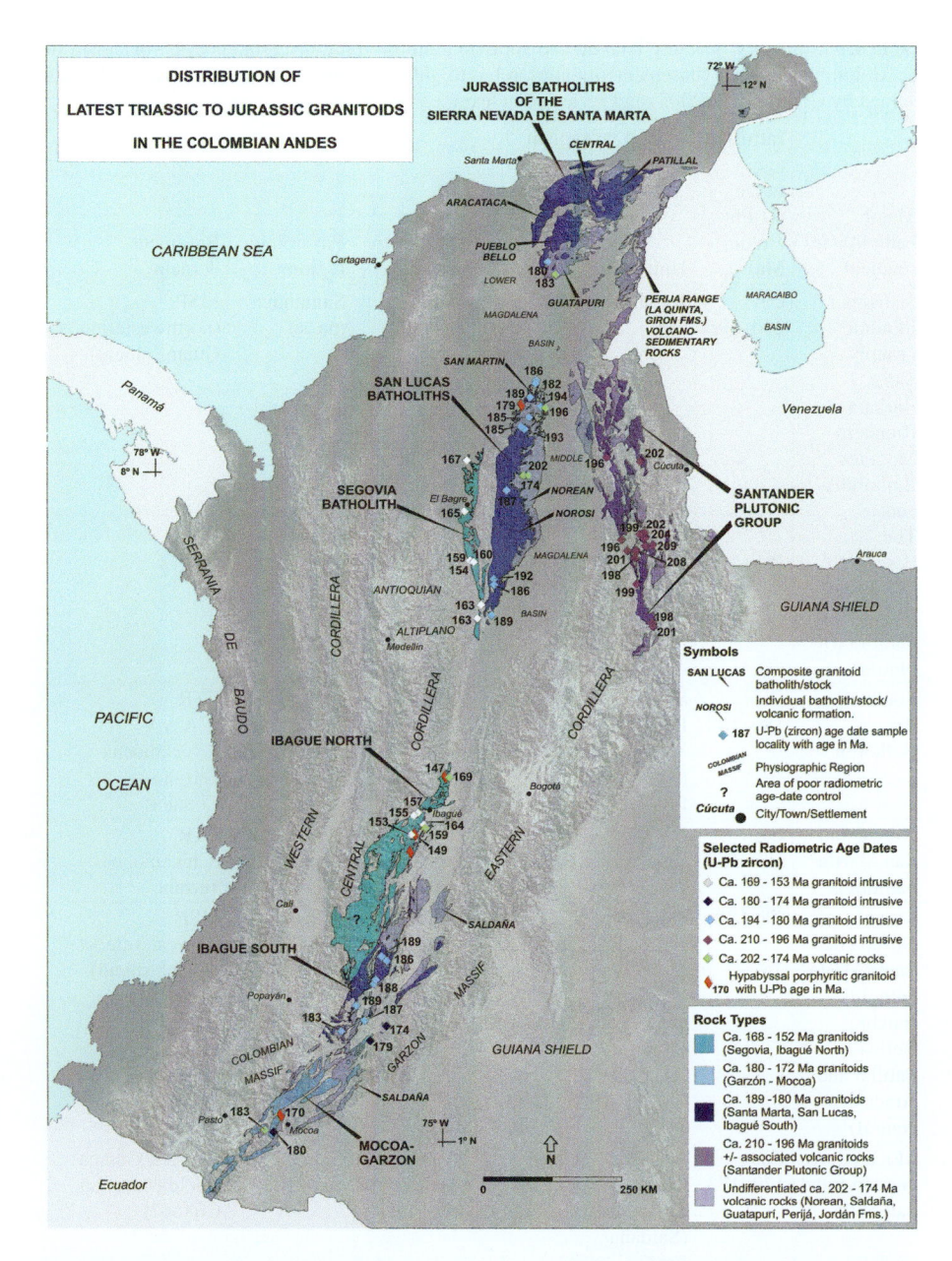

Fig. 5.10 Distribution of latest Triassic through Jurassic granitoids in the Colombian Andes. Principle modern-day physiographic provinces of the region are shown for reference. (Granitoid shapes modified after Cediel and Cáceres 2000; Gómez et al. 2007; Gómez et al. 2015a)

Table 5.1 Summary of latest Triassic and Jurassic granitoid plutonism and coeval volcanism in the Colombian Andes (basement domains refer to litho-tectonic and morpho-structural units defined by Cediel et al. (2003) and reviewed in text (see Fig. 5.2))

Major batholith/arc segment	Intrusive age range (U-Pb zircon, Ma)	Coeval volcanism Unit	Age range (U-Pb zircon, Ma)	Related porphyritic intrusions	Physiologic region	Basement domain
Santander Plutonic group: Santa Bárbara - Rionegro - Mogotes Batholiths and The Pescadero, La Corcova and Páramo Rico plutons	Ca. 210–196	Jordán Fm.	No available age dates	None known	Santander massif	MSP (northwesternmost Guiana shield)
Southern Ibagué Batholith	Ca. 189–182	Southern Ibagué Volcanics (Saldaña Fm.[a])	No available age dates	None known	Central Cordillera	CTR (CA-VA -Chicamocha Terrane Contact)
Norosí and San Martín batholiths	Ca. 189–182	Noreán Fm.	Ca. 201– 174 Ma	Santa Cruz (ca. 178 Ma)	Serranía de san Lucas	CTR (Chicamocha Terrane
Sierra Nevada de Santa Marta batholiths (pueblo Bello-Patillal and Aracataca-central)	Ca. 180	Guatapurí Fm.	Ca. 183 Ma	None known	Sierra Nevada de Santa Marta	MSP (N westernmost Guiana shield)
Mocoa-Garzón trend batholiths	Ca. 179–173	Mocoa trend Volcanics (Saldaña Fm.[a])	Ca. 185 Ma	Mocoa (ca. 170 Ma)	Garzón massif	Western Guiana shield (Amazon Craton?)

(continued)

Table 5.1 (continued)

Major batholith/arc segment	Intrusive age range (U-Pb zircon, Ma)	Coeval volcanism Unit	Age range (U-Pb zircon, Ma)	Related porphyritic intrusions	Physiologic region	Basement domain
Northern Ibagué Batholith	Ca. 169–152	Northern Ibagué Volcanics (Saldaña Fm.[a])	Ca. 158 Ma	Infierno-Chilí, Rovira, Chaparral (ca. 149–146 Ma)	Central Cordillera	CTR (CA-VA -Chicamocha Terrane Contact)
Segovia Batholith	Ca. 167–158	None Identified	No associated volcanic rocks	None known	Northern Central Cordillera	(CTR) Ca-VA

MSP Maracaibo Sub-Plate, *CTR* Central Continental Realm, *CA-VA* Cajamarca-Valdivia Terrane
[a]The Saldaña Fm. as presently understood appears to be a regionally extensive but diachronous unit which requires further subdivision

the Aracataca, Central, Pueblo Bello and Patillal batholiths of the Sierra Nevada de Santa Marta (Tschanz et al. 1974).

In addition to the major suites of plutonic rocks, important deposits of associated volcanic and volcano-sedimentary strata of late Triassic-Jurassic age are preserved. These deposits include those related to the Mocoa and Ibagué batholiths (e.g. Saldaña Fm.), those bordering the Santander Massif (i.e. Jordán Fm.), those associated with the Norosí Batholith of the Serranía de San Lucas (i.e. Noreán Fm.) and those observed along the south-eastern flank of the Sierra Nevada de Santa Marta (i.e. Guatapurí Fm.). These sequences are considered to be generally penecontemporaneous in age with their neighbouring batholiths.

5.3.3.2 Age Constrains on Late Triassic to Jurassic Granitoid Magmatism

Geologic field relationships and historic radiometric whole-rock or mineral separate age dates (i.e. Maya 1992; Gómez et al. 2015b) have, as a whole, provided temporal constraint upon the emplacement of late Triassic-Jurassic granitoids in the Colombian Andes. Regardless, lack of analytical resolution, large margins of error and overlap in the historic data have led to the assignment of the entire suite to a broad interval spanning the late Triassic to early Cretaceous, as reflected in the undifferentiated intrusives displayed upon regional geologic maps (e.g. Cediel and Cáceres 2000; Gómez et al. 2007; Gómez et al. 2015a). The present work has assessed Colombian late Triassic-Jurassic granitoid magmatism from a regional perspective. Sixty-eight high-precision zircon U-Pb age dates have been compiled from Dörr et al. (1995), Bustamante et al. (2010), Leal-Mejía (2011), Villagómez et al. (2011), Cochrane (2013), Mantilla et al. (2013), Van der Lelij (2013), Bissig et al. (2014), Cochrane et al. (2014b), Van der Lelij et al. (2016) and Zapata et al. (2016), providing data for the Norosí and San Martín batholiths, the southern and northern segments of the Ibagué Batholith, the Segovia Batholith, the Pueblo Bello-Patillal Batholith and holocrystalline and porphyritic intrusive in the Garzón Massif and Mocoa Batholith. In addition, 12 zircon U-Pb ages for late Triassic-Jurassic volcano-sedimentary sequences, including the Noreán (eastern flank of Norosí Batholith), Guatapurí (southeastern flank of Pueblo Bello Batholith) and Saldaña (southern Ibagué Batholith) Fms., were compiled.

The temporal distribution of zircon U-Pb ages for late Triassic-Jurassic granitoids is displayed in Fig. 5.11. This histogram permits the definition of four magmatic sub-episodes spanning the ca. 210 and 146 Ma time period, represented by granitoid batholith emplacement in six spatially separate arc segments (Fig. 5.10). The oldest, ca. 210–196 Ma sub-episode, is confined to the batholiths and stocks of the Santander Plutonic Group (Dörr et al. 1995; Mantilla et al. 2013; Bissig et al. 2014; Van der Lelij 2013; Van der Lelij et al. 2016). A second ca. 189–180 Ma sub-episode is recorded by the Norosí and San Martín batholiths, the Pueblo Bello-Patillal Batholith and the southern Ibagué Batholith (Leal-Mejía 2011). Zircon

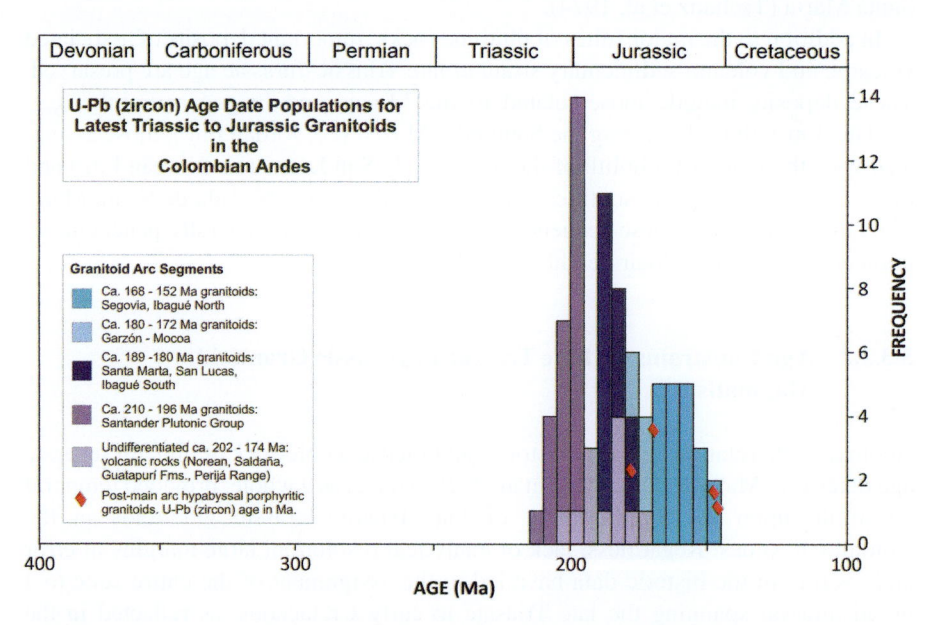

Fig. 5.11 U-Pb (zircon) age date populations for latest Triassic through Jurassic granitoids in the Colombian Andes. Note the clustering of age dates for individual arc segments and how the age populations support the overall east-to-west migration of the granitoid arc axis over time. See text for further discussion (also see Figs. 5.10, 5.31, and 5.32)

U-Pb analyses for the Guatapurí and southern Ibagué volcanic formations return ages penecontemporaneous with the age of spatially related holocrystalline plutons. Zircon separates from volcanic rocks of the Noreán Fm. return a wider range of ages, spanning ca. 202–172 Ma. The base of the Noreán Fm., to the east of the Norosí Batholith, is comprised of andesite flows and felsic pyroclastic rocks with associated diorite dikes and felsic plugs. This bimodal assembly dates from ca. 201 to 193 Ma.

A third sub-episode, emplaced at ca. 180–172 Ma, is revealed in the Mocoa Batholith and intrusive and volcanics exposed along the margins of the Garzón Massif (Bustamante et al. 2010; Leal-Mejía 2011; Cochrane et al. 2014b; Zapata et al. 2016). Rhyolite tuff of spatially related volcanic rocks returned a U-Pb (zircon) age of 181.5 ± 1.6 Ma (Cochrane et al. 2014b). Previous work by Sillitoe et al. (1982) provided K-Ar (magmatic biotite) ages of 210 ± 4 Ma and 198 ± 4 Ma for the Mocoa Batholith. U-Pb (zircon) data do not support the Sillitoe et al. (1982) K-Ar ages, although we note that the U-Pb samples were collected significantly (>10 km) to the west of the Sillitoe et al. (1982) locations. Regardless, no evidence for ~210 to 198 Ma magmatism along the Mocoa-Garzón trend is provided by the U-Pb (zircon) data, and based upon our multi-sample database, we conclude that the majority of the Mocoa-Garzón granitoids crystallized between ca. 180 and 172 Ma.

A fourth ca. 169–152 Ma sub-episode is recorded in granitoids of the northern Ibagué (Leal-Mejía 2011; Villagómez et al. 2011; Cochrane 2013; Cochrane et al. 2014b) and Segovia batholiths (Leal-Mejía 2011). With respect to the Ibagué Batholith, zircon U-Pb ages indicate that it is a large composite intrusive comprised of at least two temporally spatially defined magmatic pulses at ca. 189–182 Ma and ca. 165–152 Ma. Unfortunately, available data does not permit the precise definition of the contact between the southern and northern sectors. The contact shown in Fig. 5.10 is an approximation based upon field and DEM observations and historic K-Ar age data. With respect to the Segovia Batholith (the Western Batholith of Bogotá and Aluja 1981 or Segovia Batholith of Ballasteros 1983), the 188.9 ± 2 Ma age presented by Cochrane (2013) pertains to a sample which is actually located within the southern Norosí Batholith (compare our Fig. 5.10 with Cochrane 2013, Fig. 5.1 on p. 88), well to the east of the mapped limits of the Segovia Batholith. The Cochrane (2013) sample is herein included in the ca. 189–180 Ma San Lucas suite and accords well with previous age dates for the Norosí Batholith.

In addition to the four major episodes of holocrystalline plutonism outlined above, three localized events comprised of hypabyssal, porphyritic-textured dikes, sills and stocks are observed (Fig. 5.10). These include (1) a cluster of porphyritic dikes and sills at Santa Cruz on the NW margin of the Serranía de San Lucas, a sample from which returned a zircon U-Pb age of 178.1 ± 5.6 Ma (Leal-Mejía 2011); (2) porphyritic stocks at Mocoa (Sillitoe et al. 1984) a sample of which returned a zircon U-Pb age of 170.2 ± 2.7 Ma with an inheritance ages ranging from ca. 184 Ma to ca. 1200 Ma (Leal-Mejía 2011); and (3) numerous porphyritic stocks emplaced along the eastern margin of the northern Ibagué Batholith. Hypabyssal porphyry from Infierno-Chilí area returned a zircon U-Pb age of 149.3 ± 2.8 Ma (Leal-Mejía 2011). Cochrane (2013) revealed a 146.8 ± 1.5 Ma zircon U-Pb age for quartz porphyry near Lérida. Numerous similar, undated porphyritic stocks outcrop along the eastern margin of the northern Ibagué Batholith, extending from Rovira south to beyond Chaparral (Fig. 5.10) and are herein assigned to the same temporal suite.

Discussion and Synthesis of Spatial Distribution of Late Triassic-Jurassic Granitoids

In the synthesis of Meso-Cenozoic granitoid magmatism presented by Aspden et al. (1987), "eastern" and "western" granitoid belts of late Triassic- Jurassic age were recognized. These authors suggested that the eastern belt (including Santander and Mocoa) may be older than the western (Ibagué-Sonsón[1]-Segovia[2]-Sierra Nevada de Santa Marta) belt.

(Footnotes: [1]The Sonsón Batholith, then thought to be of Jurassic age, has now been shown to be a composite intrusion of Permo-Triassic and Paleocene age (see Leal-Mejía et al. 2011). [2] Aspden et al. (1987) grouped the Norosí and San Martín batholiths of the Serranía de San Lucas with the Segovia Batholith of Bogotá and Aluja (1981). They provided no data for Norosí and San Martín and hence did not recognize that they represent temporally distinct batholiths).

Spikings et al. (2015), based upon new and compiled U-Pb (zircon) age dates, presented an analysis of late Triassic-Jurassic granitoid magmatism at the scale of the entire Northern Andes. In Colombia they reiterate the westward migration of granitoid magmatism from the Santander Plutonic Group to the Norosí Batholith between ca. 196 and 189 Ma; however, they do not differentiate the ca. 168–158 Ma Segovia Batholith as defined by Bogotá and Aluja (1981), Ballesteros (1983), Leal-Mejía (2011) and Leal-Mejía et al. (2011) and detailed herein. To the south, Spikings et al. (2015) group the northern and southern Ibagué and Mocoa-Garzón trend batholiths as the undifferentiated Ibagué Batholith and reveal a composite U-Pb age ranging from ca. 189 to 146 Ma. Their data demonstrate that the zircon U-Pb ages tend to cluster within restricted ranges within distinct sectors of the batholith and that the southern sector of the Ibagué Batholith returns significantly older ages when compared to the Mocoa-Garzón trend and northern Ibagué (see their Fig. 5.10 and our Figs. 5.10 and 5.11).

When our zircon U-Pb database is combined with that presented by Spikings et al. (2015), a more detailed temporal-spatial analysis of late Triassic-Jurassic granitoid magmatism is permitted. Four distinct age ranges are observed within at least six separate arc segments, and the temporal-spatial migration of late Triassic-Jurassic granitoid magmatism based upon U-Pb (zircon) ages may be visualized in the colour coding of Figs. 5.10 and 5.11. An E-W transect across the northern sector of the Colombian Cordilleras highlights an east-to-west younging trend beginning with the ca. 210–196 Ma batholiths of the Santander Massif, passing westwards through the ca. 189–180 Ma batholiths of the Serranía de San Lucas and into ca. 167–158 Ma Segovia Batholith. This tendency is accentuated if the estimated 100 kilometres of post-Jurassic sinistral displacement along the Bucaramanga–Santa

Marta fault system (Campbell 1968; Etayo-Serna and Rodríguez 1985; Cediel et al. 2003) are restored, placing the SW sector of the Pueblo Bello-Patillal Batholith and associated Guatapurí volcanics in very close proximity and immediately along trend with the Serranía de San Lucas intrusive-volcanic suite (Fig. 5.10). Notably, this restoration will not affect the position of the El Jordán Fm., which rests along the NW margin of the Santander Massif, but on the west side of the Bucaramanga–Santa Marta Fault. This suggests that the Jordán Fm. along with the Guatapurí and Noreán Fms. forms remnants of a formerly unified volcanic province.

To the south, in the northern and southern Ibagué batholiths, the Mocoa Batholith and the intrusions exposed along the Garzón Massif, the east-to-west younging trend is not clearly defined. The ca. 188–180 Ma (southern Ibagué) and ca. 180–172 Ma (Mocoa-Garzón) episodes migrate along a NNE-oriented axis, and an apparent southward and eastward migration of the magmatic arc axis is recorded. Current regional structural interpretations depict the ca. 180–172 Ma Mocoa-Garzón intrusions as tectonic slices caught up in dextral oblique basement reactivation structures responsible for Miocene uplift in the Garzón Massif (Fig. 5.2; Cediel and Cáceres 2000; Cediel et al. 2003). If restoration of an (albeit) undefined component of post-emplacement dextral translation along the Garzón Massif structures is taken into account, the apparent eastward migration of magmatism is reduced (although not completely eliminated), and slices of the Mocoa Batholith become coaxial with the ca. 189–180 Ma southern Ibagué Batholith.

Following emplacement of the Mocoa-Garzón intrusions, the ca. 166–152 Ma northern segment of the Ibagué Batholith was intruded along trend to the NNE. Thus, granitoid migration in the southern segment of the late Triassic-Jurassic arc during the ca. 189–152 Ma period was primarily along a NNE-oriented axis.

With respect to the location and timing of hypabyssal, porphyritic dikes, sills and stocks, it is observed that the three temporal suites were emplaced within and/or along the contacts with a respective major batholith of penecontemporaneous age (Fig. 5.10). Thus, the ca. 178 Ma Santa Cruz dikes and sills are located along the NW margin of the ca. 189–180 Ma Norosí Batholith, the ca. 170 Ma Mocoa stocks are located within the ca. 180–172 Ma Mocoa Batholith, and the ca. 152–146 Ma Rovira-Lerida stocks are emplaced along the eastern margin of the ca. 168–155 Ma northern Ibagué Batholith. In all cases, porphyritic magmatism was initiated within ca. 2–5 m.y. following the waning of holocrystalline plutonism. The porphyritic granitoids thus appear to represent closure-phase magmatism emplaced during the late evolution of the respective holocrystalline arc segment, prior to wholesale arc migration or cessation of active magmatism.

In summary, zircon U-Pb ages for late Triassic-Jurassic granitoid magmatism in the Colombian Andes permit the temporal definition of four major magmatic episodes including granitoid batholith emplacement within six spatially separate arc segments, in addition to three spatially temporally separate events involving late-stage, volumetrically minor hypabyssal porphyries. Each major batholith (or group of batholiths in the case of the Santander Plutonic Group) is considered to represent a temporally and spatially separate arc segment developed within the overall context of late Triassic-Jurassic subduction-related granitoid magmatism affecting much of

Northern Andean margin during this time period (Aspden et al. 1987; Litherland et al. 1994; Cediel et al. 2003; Leal-Mejía et al. 2011; Cochrane et al. 2014b; Spikings et al. 2015). In Colombia, magmatism migrates over time, both along the length of the magmatic arc axis and in a transverse sense, related to the interpreted movement vector of the subducting Pacific Plate. Tectonic setting and evolution during the late Triassic-Jurassic will be discussed in Sect. 5.4.2, following the presentation of additional lithogeochemical and isotopic information.

5.3.3.3 Lithogeochemical and Isotopic Characteristics of Late Triassic-Jurassic Granitoids

Lithogeochemistry

Figures 5.12 and 5.13 present whole-rock lithogeochemical analyses for 136 samples of late Triassic-Jurassic granitoids, including holocrystalline and hypabyssal intrusive and volcanic rocks, as compiled from Dörr et al. (1995), Bustamante et al. (2010), Leal-Mejía (2011), Bissig et al. (2014), Cochrane et al. (2014b) and Van der Lelij et al. (2016). Of the samples represented herein, some 36% (49 samples) are considered altered, based upon the criteria discussed in Sect. 5.3.1.2. Altered samples are identified by the unfilled symbols used in Figs. 5.12 and 5.13 lithogeochemical plots. No lithogeochemical data are available for the ca. 180 Ma batholiths of the Sierra Nevada de Santa Marta or their coeval volcano-sedimentary sequences (i.e. the Guatapurí Fm.).

The Colombian late Triassic to Jurassic batholiths are low-K to high-K calc-alkaline (Irvine and Baragar 1971; Peccerillo and Taylor 1976) in composition. All main phase batholiths are metaluminous, with the exception of the Santander Plutonic Group where localized peraluminous members are recorded (e.g. Bissig et al. 2014). Specific lithogeochemical features of the individual granitoid suites are reviewed below.

Santander Plutonic Group

The ca. 210–196 Ma Santander Plutonic Group produces the most differentiated trend of bulk compositions, ranging from gabbro-diorite through to granite and leucogranite ("alaskite"). Relative to the other Colombian late Triassic-Jurassic intrusive suites, the Santander Plutonic Group reveals (1) a higher degree of alkalinity, (2) a tendency towards weakly to strongly peraluminous compositions (including the leucogranites of the Vetas-California area, Bissig et al. 2014) and (3) enrichment in trace elements and REE ($\Sigma REE = 47.42$–503.96 ppm). Overall, REE patterns reveal moderate to steep decreasing slopes (($La/Yb)_N = 3.47$–34.22). The HREE define a relatively flat pattern for the intermediate members (($Gd/Yb)_N = 0.63$–2.54) and a slightly decreasing pattern for the leucogranites (($Gd/Yb)_N = 0.51$–3.35).

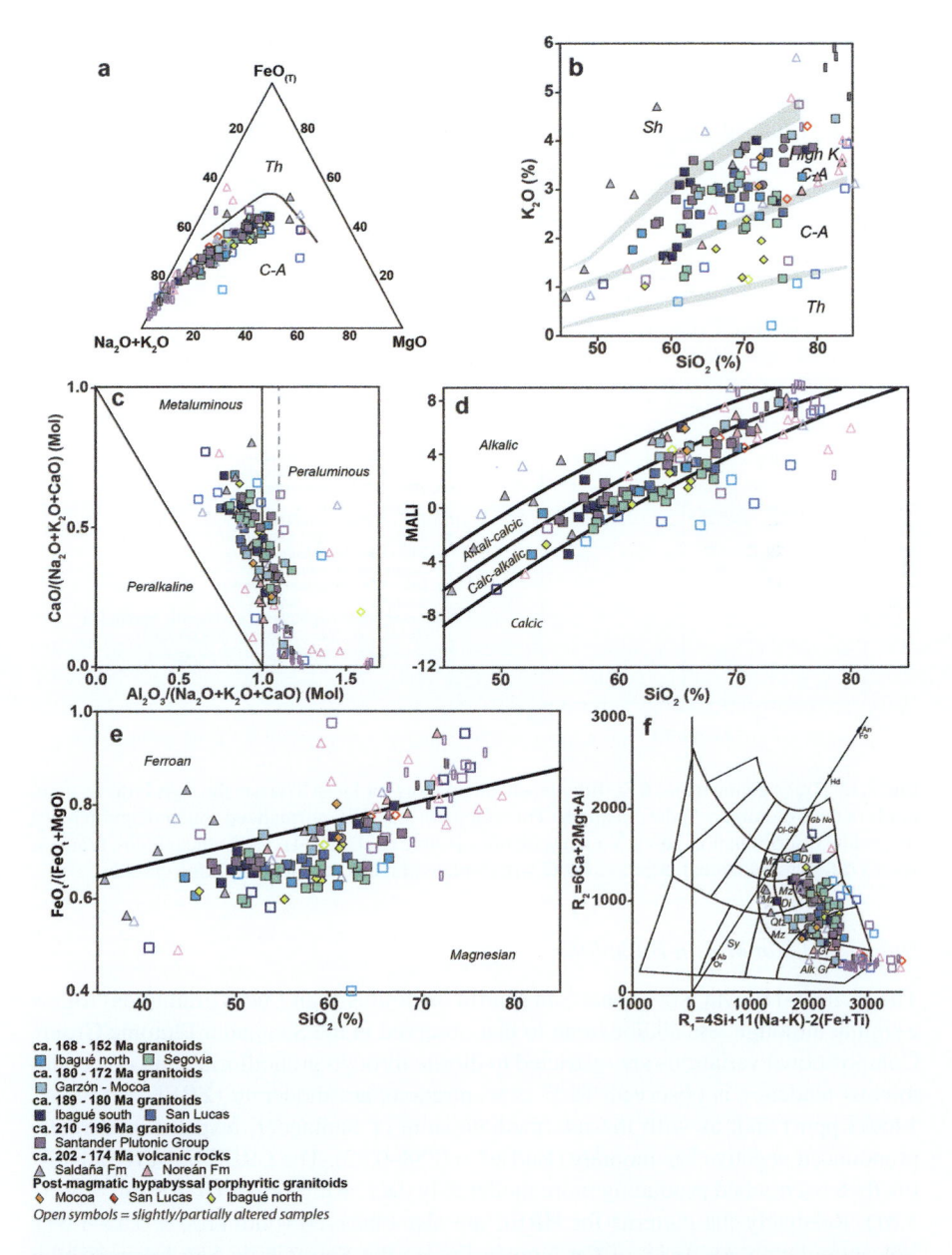

Fig. 5.12 Major element lithogeochemical plots for latest Triassic through Jurassic granitoids in the Colombian Andes. (**a**) AFM plot, curve after Irvine and Baragar (1971); (**b**) K_2O vs. SiO_2 plot, boundary fields in grey as summarized by Rickwood (1989); (**c**) alumina saturation plot after Barton and Young (2002); (**d** and **e**) MALI and Fe-index vs. SiO_2 plots, respectively, after Frost et al. (2001); (**f**) R1-R2 classification plot after De La Roche et al. (1980). *Th* tholeiite, *C-A* calc-alkaline, *Sh* Shoshonite, *Gb No* gabbro-norite, *Gb Di* gabbro-diorite, *Di* diorite, *Mz Di* monzodiorite, *Mz* monzonite, *To* tonalite, *Qtz Mz* quartz monzonite, *Gd* granodiorite, *Gr* granite, *Alk Gr* alkali granite

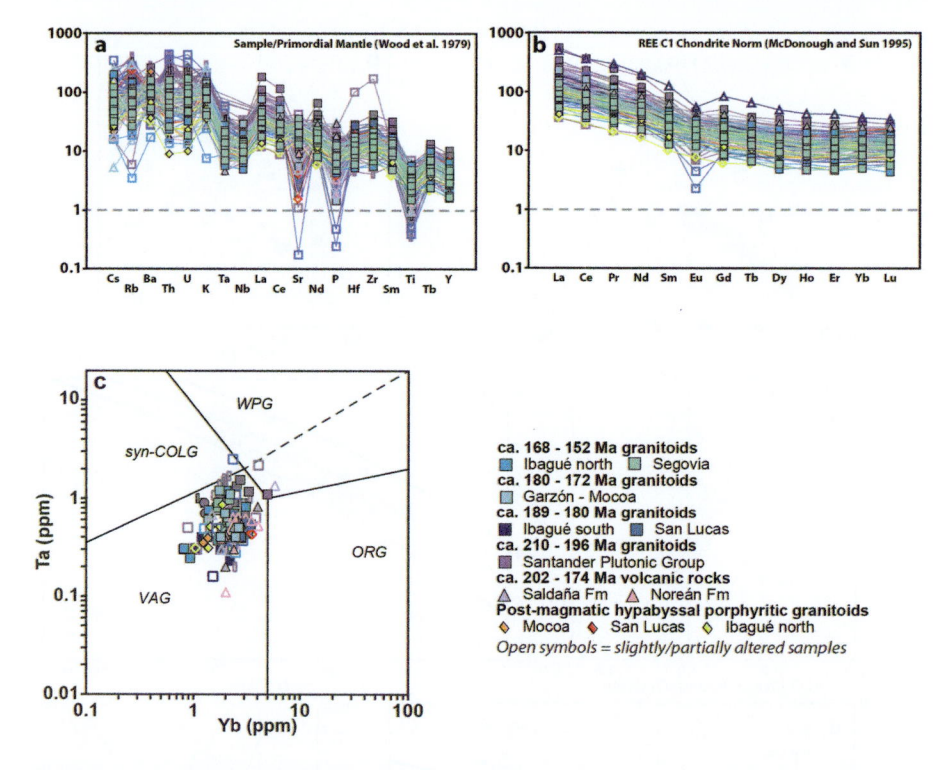

Fig. 5.13 Trace element and REE lithogeochemical plots for latest Triassic through Jurassic granitoids in the Colombian Andes. (**a** and **b**) Trace element and REE normalized spider diagram plots; (**c**) granite discrimination Ta vs. Yb diagram after Pearce et al. (1984). *VAG* volcanic-arc granites, *syn-COLG* syn-collisional granites, *WPG* within-plate granites, *ORG* ocean ridge granites

Norosí and San Martín Batholiths

The ca. 189–182 Ma Norosí and San Martín batholiths (San Lucas granitoids) follow a similar although less alkalic trend to that observed in the Santander Plutonic Group. Compositional variations are restricted to diorite through granodiorite, and no peraluminous tendency is observed. REE concentrations are moderate (ΣREE = 117.24–146.08 ppm) and, as with the intermediate suite of Santander, reveal a moderately pronounced negative Eu anomaly (Eu/Eu* = 0.58–0.82). The LREE however are distinctly less enriched producing more moderately decreasing slopes ((La/Sm)$_N$ = 2.73–3.81). Relatively flat patterns for HREE are also observed ((Gd/Yb)$_N$ = 0.83–1.31). Volcanosedimentary rocks of the Noreán Fm. in the Serrania de San Lucas exhibit similar REE patterns to the Norosi and San Martin granitoids.

Southern Ibagué Batholith

The ca. 189–180 Ma granitoids of the southern Ibagué Batholith are compositionally more variable when compared to Norosí and San Martín, ranging from metaluminous calc-alkaline gabbro and diorite through high-K (alkali-calcic) granodiorite, quartz monzonite and locally granite. The population appears to be bimodal, but this may be a reflection of the relatively small sample set. Portions of the southern Ibagué Batholith are pyroxene-dominant with lessor amounts of biotite. The REE are less enriched than both Santander and Norosí-San Martín (ΣREE = 88.07–209.80 ppm). The decreasing LREE slopes are somewhat steeper ($(La/Sm)_N$ = 2.89–5.83) than those observed in Norosí-San Martín ($(La/Sm)_N$ = 2.73–3.81), whilst Eu anomalies are only weakly negative to slightly positive (Eu/Eu^* = 0.73–1.09). The volcanic rocks of the Saldaña Formation have slightly higher REE contents (ΣREE = 118.30–240.33 ppm) and similar weak negative Eu anomalies (Eu/Eu^* = 0.77–0.95).

Mocoa-Garzón Trend Batholiths

The ca. 180–172 Ma granitoids of the Mocoa-Garzón trend show a metaluminous, high-K calc-alkaline (alkali-calcic) character associated with hornblende-biotite-bearing granodiorite to monzogranite compositions. REE contents in the phaneritic granitoids are enriched (e.g. Altamira granite; ΣREE = 223.40 ppm), when compared to the Norosí, San Martín and southern Ibagué batholiths, although not to the degree as seen in the Santander Plutonic Group. The Mocoa porphyries (ΣREE = 100.18–104.76 ppm) are distinctly less enriched in REE than the penecontemporaneous phaneritic granitoids (ΣREE = 113.95–175.06 ppm). REE patterns include relatively steep decreasing slopes ($(La/Yb)N$ = 7.97–16.77). Eu anomalies are moderately negative for the phaneritic granitoids (Eu/Eu^* = 0.67–0.83) and slightly negative to weakly positive for porphyries (Eu/Eu^* = 0.95–1.10).

Northern Ibagué and Segovia Batholiths

The ca. 168–155 Ma granitoids of the northern Ibagué and Segovia batholiths present metaluminous, low-K calc-alkaline (calc-alkalic to calcic after Frost et al. 2001) compositions dominated by hornblende with lesser biotite-bearing diorite to quartz diorite. REE patterns are flatter ($(La/Yb)_N$ = 3.58–16.3) and values are overall less enriched (ΣREE = 57.63–141.49 ppm) than those observed for Norosí-San Martín and southern Ibagué. Northern Ibagué shows slightly negative to moderately positive Eu anomalies (Eu/Eu^* = 0.77–1.47), similar to southern Ibagué, whereas very weak negative Eu anomalies are observed in samples from the Segovia Batholith (Eu/Eu^* = 0.72–1.12).

Northern Ibagué Hypabyssal Porphyry Suite

Lithogeochemical data are limited for the ca. 152–145 Ma porphyritic granitoids of the northern Ibagué Batholith, and the data set contains analyses for various undated hypabyssal porphyry intrusions of inferred latest Jurassic age. The suite includes metaluminous, low-K calc-alkaline (calc-alkalic to calcic) gabbro-diorite to granodiorite with less enriched REE contents (ΣREE = 68.07–124.50 ppm) and similar slopes ($(La/Yb)_N$ = 2.97–9.67) when compared to the northern Ibague and Segovia batholiths. Eu anomalies for the porphyries are very weakly negative to slightly positive (Eu/Eu^* = 0.91–1.08).

Whole-Rock Lithogeochemistry Summary and Discussion for Late Triassic-Jurassic Granitoids

The composite lithogeochemical data set used in this study demonstrates the Colombian late Triassic to Jurassic batholiths are of low-K to high-K calc-alkaline (Irvine and Baragar 1971; Peccerillo and Taylor 1976) or magnesian, calcic to alkali-calcic (Frost et al. 2001) in composition. All main phase batholiths are metaluminous, with the exception of the Santander Plutonic Group where a clear trend towards strongly peraluminous compositions is recorded in the leucogranite suite of Bissig et al. (2014). The granitoids are dominated by biotite and/or hornblende diorite through granodiorite but include localized ranging gabbrodioritic, monzonitic and granitic phases. Consistent LILE enrichment compared to HFSE is observed, as are negative anomalies of refractory elements such as Ta, Nb and Ti. LREE enrichment compared to HREE is recorded in all suites. Eu anomalies range from markedly negative to slightly positive, being generally consistent within individual batholiths. The foregoing characteristics are consistent with classification of the Colombian granitoids as Cordilleran-type granitoids (Frost et al. 2001), volcanic arc granitoids (Pearce et al. 1984), or K-spar and amphibole-rich calc-alkaline granitoids (Barbarin 1999), generated in transitional to subduction-type settings. This conclusion is consistent with data and conclusions presented by previous workers including Alvarez (1983), Aspden et al. (1987), Dorr et al. (1995), Bustamante et al. (2010), Leal-Mejía et al. (2011), Bissig et al. (2014) and Spikings et al. (2015).

When the composite lithogeochemical data set for the late Triassic to Jurassic Colombian granitoids is considered within the spatial-temporal and geological framework for the individual batholiths presented in Fig. 5.10, the lithogeochemical plots (Figs. 5.12 and 5.13) demonstrate clear east-to-west trends towards more primitive (less alkaline, more magnesian, less enriched in both trace and REEs, less fractionated REEs, weaker to no Eu anomaly) whole-rock compositions. In northern Colombia this is recorded in the enriched, calc-alkalic to alkali-calcic (high-K calc-alkaline) and peraluminous compositions of the Santander Plutonic Group, westwards through the intermediate compositions of the San Lucas granitoids, and into the less enriched, calcic to calc-alkalic, metaluminous compositions of the Segovia Batholith. As shown in Fig. 5.10, this tendency coincides with the east-to-west younging of the major batholiths and with changes in the nature and composition of the intruded basement. In southern Colombia a similar lithogeochemical trend from the alkali-calcic compositions of the Mocoa-Garzón granitoids to the calcic to calc-alkalic compositions of the southern Ibagué Batholith to the west is observed, although the age relationships are reversed, with the Mocoa-Garzón granitoids being younger than the southern Ibagué Batholith. This suggests that the major, minor, trace and rare-earth element lithogeochemistry of the individual batholiths is less a function of age as it is of the nature of the intruded basement and, likely, the specific tectonic framework and conditions at the time of emplacement (Barbarin 1999; Frost et al. 2001). These themes will be discussed further following presentation of isotopic data for the Colombian granitoids.

When the lithogeochemistry of the hypabyssal porphyry suites (i.e. the ca. 178 Ma Santa Cruz dikes and sills, ca. 170 Ma Mocoa porphyries and the ca. 152–146 Ma northern Ibagué porphyries) is compared with the respective, spatially related, slightly older, holocrystalline batholith, in general, the porphyries tend to (1) be less enriched in K (i.e. less alkaline), (2) be less enriched in trace elements and REE and (3) have a less pronounce to neutral or even positive Eu anomaly. These trends are best observed in the unaltered porphyries of the northern Ibagué Batholith and are present but potentially modified by post-crystallization alteration and mineralization at Santa Cruz and Mocoa. Notwithstanding, the data suggest that the hypabyssal porphyry suites consistently reveal more mantelic compositions when compared to the spatially related, slightly older, holocrystalline batholith.

Sr-Nd-Pb Isotope Geochemistry

Sr-Nd Isotope Geochemistry Results

Sr-Nd isotope data for Colombian late Triassic-Jurassic granitoids, including for the Santander Plutonic Group (Restrepo-Pace, 1995; Bissig et al. 2014; Van der Lelij et al. 2016), the Mocoa Batholith porphyries (Leal-Mejía 2011), the Norosí and San Martín batholiths, Santa Cruz porphyries and the Noreán volcanics (Leal-Mejía 2011; Cochrane et al. 2014b), the southern Ibagué Batholith and the southern Ibagué volcanics (Leal-Mejía 2011), the central and northern Ibagué Batholith and hypabyssal porphyries (Cochrane et al. 2014b) and the Segovia Batholith (Leal-Mejía 2011), are presented in Fig. 5.14 and tabulated in Appendix 3. For comparative purposes, selected data sets for the Precambrian basement rocks of the Santander Massif (Cordani et al. 2005; Ordoñez-Carmona et al. 2006; Bissig et al. 2014), Garzón Massif (Restrepo-Pace et al. 1997; Cordani et al. 2005) and Chicamocha Terrane (Cuadros et al. 2014) are also presented.

Late Triassic–early Jurassic intrusives of the Santander Plutonic Group are characterized by a wide range of highly radiogenic $^{87}Sr/^{86}Sr_{(i)}$ ratios (0.70533 to 0.73660) with negative $\varepsilon Nd_{(t)}$ values (−19.34 to −3.46). Data plot within a similarly disparate field defined by samples of Precambrian and Paleozoic continental basement rocks of the Santander Massif (Cordani et al. 2005; Ordoñez-Carmona et al. 2006; Bissig et al. 2014).

No Sr-Nd data are available for the main phase Mocoa-Garzón trend batholiths. Results for the ca. 170 Ma Mocoa porphyries reveal moderate to high $^{87}Sr/^{86}Sr_{(i)}$ ratios (~0.70600) and negative $\varepsilon Nd_{(t)}$ values (−5.60 to −3.32). $\varepsilon Nd_{(t)}$ values for the Mocoa porphyries plot within the range presented by Restrepo-Pace et al. (1997) and Cordani et al. (2005); however, full comparison of the data sets is hampered by the lack of $^{87}Sr/^{86}Sr$ analyses for Garzón Massif basement.

Unlike the Santander Plutonic Group, data for the Norosí and San Martín batholiths plot within a more discrete array, characterized by moderately high $^{87}Sr/^{86}Sr_{(i)}$ ratios (0.70674–0.70826) with negative $\varepsilon Nd_{(t)}$ values (−6.65 to +0.09). Coeval volcanic rocks of the Noreán Fm. return somewhat more mantelic $\varepsilon Nd_{(t)}$

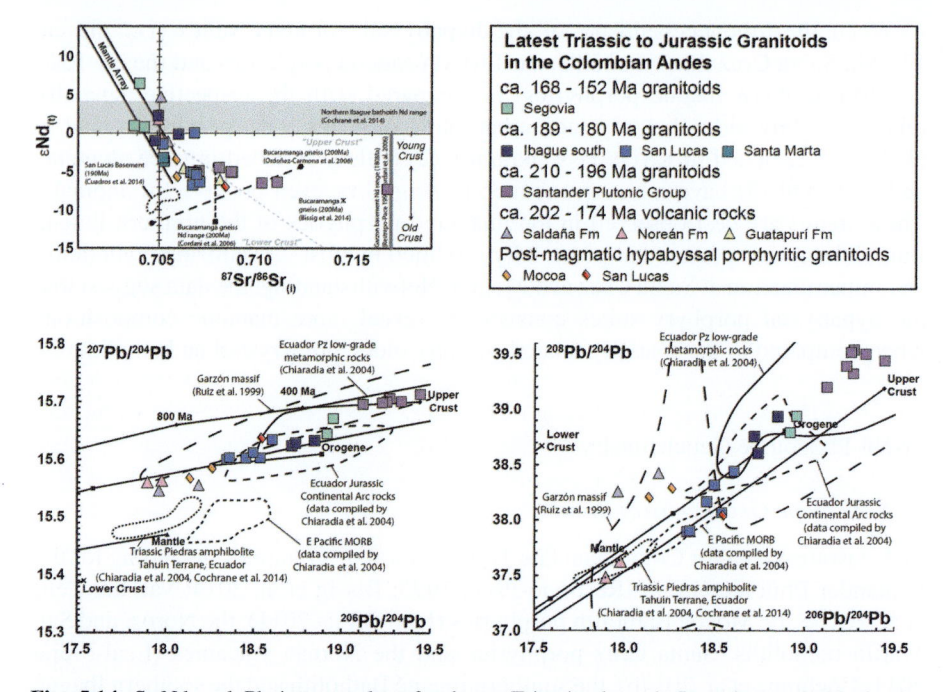

Fig. 5.14 Sr-Nd and Pb isotope plots for latest Triassic through Jurassic granitoids in the Colombian Andes. Additional data for igneous and metamorphic suites from the surrounding region are included for reference

signatures than the Norosí and San Martín batholiths. Conversely, the Santa Cruz porphyry dikes, on the western margin of the Serranía de San Lucas, record a high $^{87}Sr/^{86}Sr_{(i)}$ ratio (0.70851) and a slightly more negative $\varepsilon Nd_{(t)}$ value (−6.9).

The southern Ibagué Batholith and associated volcanic rocks yield mixed initial $^{87}Sr/^{86}Sr$ ratios around the bulk Earth composition plotting within or near the mantle array ($^{87}Sr/^{86}Sr_{(i)}$ = 0.70489 to 0.70609; $\varepsilon Nd_{(t)}$ = −0.96 to +4.83). No Sr isotope data is available for the northern Ibagué Batholith, although Nd isotope data presented by Cochrane et al. (2014b) record positive $\varepsilon Nd_{(t)}$ values (+0.32 to +3.86) similar to the $\varepsilon Nd_{(t)}$ values for the southern Ibagué Batholith.

Finally, samples from the Segovia Batholith exhibit the lowest $^{87}Sr/^{86}Sr_{(i)}$ ratios (0.70385 to 0.70434) and positive $\varepsilon Nd_{(t)}$ values (+0.86 to +6.52), generally falling along the mantle array.

Sr-Nd Isotope Geochemistry Summary and Discussion

Whole-rock Sr-Nd isotope data for the Colombian late Triassic-Jurassic granitoids plot in clusters, on a per-intrusive suite basis, with relatively little overlap between the sample sets for individual batholiths (Fig. 5.14). The overall $^{87}Sr/^{86}Sr_{(i)}$ isotope composition of the Colombian granitoids ranges from the highly evolved values of the ca. 210–196 Ma Santander Plutonic Group ($^{87}Sr/^{86}Sr_{(i)}$ = 0.70533 to 0.73660) to the juvenile values of the ca. 168–155 Ma Segovia Batholith ($^{87}Sr/^{86}Sr_{(i)}$ = 0.70385 to 0.70434). Commensurate with these data, $\varepsilon Nd_{(t)}$ values for the Santander Plutonic Group are negative ($\varepsilon Nd_{(t)}$ = −19.34 to −3.46), whilst the Segovia Batholith returns $\varepsilon Nd_{(t)}$ values up to +6.52. The Norosí-San Martín and southern and northern Ibagué batholiths return intermediate $^{87}Sr/^{86}Sr_{(i)}$ and $\varepsilon Nd_{(t)}$ values which plot in semi-discrete arrays between the above-mentioned data sets (Fig. 5.14).

As with the lithogeochemical data presented above, $^{87}Sr/^{86}Sr_{(i)}$ and $\varepsilon Nd_{(t)}$ data may be considered within a spatial-temporal and geological framework (Fig. 5.10). In the northern sector of the Colombian Cordilleras, $^{87}Sr/^{86}Sr_{(i)}$ and $\varepsilon Nd_{(t)}$ data reveal an east-to-west trend of increasingly more juvenile $^{87}Sr/^{86}Sr_{(i)}$ and $\varepsilon Nd_{(t)}$ values, extending from the highly evolved, upper crustal-influenced compositions of the ca. 210–196 Ma Santander Plutonic Group to the mixed values of the ca. 189–180 Ma Norosí and San Martín batholiths which cluster at the base and along the lower section of the mantle array, into to the primative $^{87}Sr/^{86}Sr_{(i)}$ and $\varepsilon Nd_{(t)}$ values for the Segovia Batholith (Fig. 5.14). A similar east-to-west tendency is observed in the south where the Mocoa porphyry, hosted within the Garzón Massif, returns mixed crustal values at the base of the mantle array, whilst the southern Ibagué Batholith to the west returns higher (positive) $\varepsilon Nd_{(t)}$ values which plot farther up the mantle array (Fig. 5.14).

Placed into a geological context, the above east-to-west trend is supported by changes in the nature of the intruded basement complex as shown in Fig. 5.10. In the north, Sr and Nd isotope data for the Santander Plutonic Group plot within the broad data field outlined for samples of the Bucaramanga gneiss with a distinct tendency towards upper crustal values. This suggests partial derivation and/or contamination of the Santander Plutonic Group granitoids from Precambrian and/or early Paleozoic basement rocks widely exposed in the Santander Massif (Goldsmith et al. 1971). To the west, Sr and Nd isotope data for the Norosí and San Martín batholiths of the San Lucas region are more tightly clustered mostly near the base of and extending up to the lower section of the mantle array. Sr-Nd isotope characterization of the basement rocks in the Serranía de San Lucas is restricted to analyses for the metamafic constituents which return $^{87}Sr/^{86}Sr_{(i)}$ and $\varepsilon Nd_{(t)}$ values consistent with a depleted mantle or lower crustal source, whilst $^{87}Sr/^{86}Sr_{(i)}$ values for the felsic basement components of the region are poorly constrained (Cuadros et al. 2014; Fig. 5.14). $^{87}Sr/^{86}Sr_{(i)}$ and $\varepsilon Nd_{(t)}$ data for the San Lucas Jurassic granitoids suggests evolution along the mantle array with a moderate degree of crustal input, perhaps derived from the less refractory felsic components of the basement for which $^{87}Sr/^{86}Sr_{(i)}$ ratios have yet to be well defined (Cuadros et al. 2014). The San Lucas granitoids have apparently assimilated significantly less upper crustal material than the Santander Plutonic Group granitoids. Farther west, $^{87}Sr/^{86}Sr_{(i)}$ and $\varepsilon Nd_{(t)}$ data for the Segovia Batholith reveal juvenile values with little indication of assimilation of, or contamination by, enriched continental crust. Host rocks for the Segovia Batholith include, to the west, early Paleozoic metasedimentary rocks of the Cajamarca-Valdivia island arc assemblage (Restrepo-Pace 1992; Cediel and Cáceres 2000; Cediel 2011), with quartzo-feldspathic gneisses of the San Lucas complex (González 1999) to the east. In either case contacts with the Segovia Batholith are faulted and not well exposed. No Sr-Nd isotope data are available for either of these units. Notwithstanding, the juvenile $^{87}Sr/^{86}Sr_{(i)}$ and $\varepsilon Nd_{(t)}$ signatures of the Segovia Batholith suggest little interaction with continental basement perhaps due to (1) rapid batholith emplacement in a highly extensional environment and/or (2) the absence of underlying continental basement in this region (in this context the San Lucas gneiss may be interpreted as a tectonic float of continental basement contained between the Otú and Palestina fault zones, as opposed to indicating the presence of continuous continental basement beneath the Segovia region).

To the south, a similar east-to-west pattern of diminishing crustal input is suggested between the Mocoa porphyry and the Ibagué Batholith. Mocoa is underlain by Precambrian continental basement of the Garzón Massif (Cediel and Cáceres 2000; Gómez et al. 2015a); however, actual hosts for the porphyritic intrusions analysed in this study include Jurassic holocrystalline intrusive and coeval volcanic rocks of the Mocoa-Garzón trend, for which no Sr-Nd isotope analyses are available. Our Mocoa porphyry samples plot at the base of the mantle array, within the negative $\varepsilon Nd_{(t)}$ range documented for the Garzón Massif (Fig. 5.10); however, little, if any, evolution of the Mocoa porphyry with respect to $^{87}Sr/^{86}Sr_{(i)}$ is suggested. Based upon available data, it is not possible to ascertain the influence of Garzón Massif Precambrian basement vs. Jurassic Mocoa-Garzón trend granitoids, in the Sr-Nd isotope composition of the Mocoa porphyry. To the west, the ca. 188–180 Ma southern Ibagué Batholith and coeval volcanic rocks, and the ca. 166–152 Ma northern Ibagué Batholith, return more mantelic signatures including mostly positive $\varepsilon Nd_{(t)}$ values. $^{87}Sr/^{86}Sr_{(i)}$ ratios for the southern Ibagué granitoids cluster about 0.70500 placing the composite data set along the middle mantle array. Despite their apparent age difference, the data suggest that similar Sr-Nd isotope systematics can be inferred for the southern and northern Ibagué batholiths. Both batholiths share a similar tectonic position along the Palestina fault and suture separating Precambrian Chicamocha basement from the peri-cratonic domain represented by the Cajamarca-Valdivia Terrane (Fig. 5.10). No Sr-Nd isotope data are available for basement rocks along the Ibagué trend, although the Ibagué and San Lucas batholiths share a similar structural position along their eastern margin (Fig. 5.10), and values similar to those recorded for Chicamocha basement in the San Lucas region (Cuadros et al. 2014) can be inferred. As such, when compared with the Santander, Garzón and San Lucas granitoids, we interpret the mostly mantelic $^{87}Sr/^{86}Sr_{(i)}$ and $\varepsilon Nd_{(t)}$ signatures for the southern and northern Ibagué batholiths to reflect very limited, if any, crustal assimilation or contamination. Rapid ascent of mantle-derived magmas, facilitated by extensional reactivation of the preexisting Palestina suture (see Sect. 5.4.2), could result in the mantelic $^{87}Sr/^{86}Sr_{(i)}$ and $\varepsilon Nd_{(t)}$ signatures recorded along the Ibagué trend.

Based upon the composite Sr and Nd isotope data, factors controlling the Sr-Nd isotope composition of the late Triassic-Jurassic granitoids included (1) the Sr-Nd isotope composition of the magmatic source region, which, in all cases with the possible exception of the Santander Plutonic Group, was dominated by the (depleted?) mantle, and (2) the nature and composition of the basement complex into which the granitoids were emplaced (Fig. 5.10). Undoubtedly, the tectonic and structural framework at the time of emplacement was also important. Further discussion of Sr-Nd results for the Colombian late Triassic-Jurassic granitoids will be pursued below, following presentation of lead isotope and additional data.

Pb Isotope Geochemistry Results

Figure 5.14b and Appendix 3 present available whole-rock Pb isotope data for the Colombian late Triassic-Jurassic granitoid suite, including for the Santander Plutonic Group (Bissig et al. 2014; Van del Lelij et al. 2016) and for the Mocoa porphyries, the Norosí-San Martín batholiths, Noreán volcanics and Santa Cruz porphyries and the southern Ibagué Batholith and volcanics (Leal-Mejía 2011).

No Pb isotope analyses for the basement complexes hosting the late Triassic-Jurassic granitoids were produced during the present study. For comparative purposes, Fig. 5.14 outlines available Pb isotopic data fields for (1) the Garzón Massif (Ruiz et al. 1999), applicable to the Mocoa porphyry intrusions; (2) Paleozoic metasedimentary basement of the Loja Terrane, Ecuador (Chiaradia et al. 2004), which may serve as a proxy for the unknown values of the Cajamarca-Valdivia assemblage in Colombia, given that regional correlation between the early Paleozoic metasedimentary sequences of Ecuador and Colombia has been proposed by various authors (Restrepo-Pace 1992; Cediel et al. 2003; Kennan and Pindell 2009; Spikings et al. 2015); (3) the lead isotope composition of the Piedras amphibolite, Ecuador, considered by Chiaradia et al. (2004) to represent a Triassic MORB-type mantle source reservoir which, based upon late Triassic-Jurassic tectonic models for the Northern Andes (e.g. Spikings et al. 2015; Van de Lelij et al. 2016), may provide a reasonable estimate for Triassic MORB-type mantle in Colombia; and (4) lead isotope data for Jurassic arc-related granitoids of Ecuador (Chiaradia et al. 2004), which may be compared directly with their penecontemporaneous Colombian counterparts.

With respect to the Colombian granitoids, the most radiogenic values are observed within the Santander Plutonic Group ($^{206}Pb/^{204}Pb$ = 19.12–19.44, $^{207}Pb/^{204}Pb$ = 15.70–15.71, $^{208}Pb/^{204}Pb$ = 39.20–39.54; Bissig et al. 2014; Van del Lelij et al. 2016), which straddle the upper crust lead evolution curve of Zartman and Doe (1981).

No Pb isotopic data are available for the main-stage batholiths of the Mocoa-Garzón trend. Samples of the Mocoa porphyries reveal radiogenic values, plotting just below the Orogene lead evolution curve ($^{206}Pb/^{204}Pb$ = 18.14–18.26; $^{207}Pb/^{204}Pb$ = 15.57–15.59; $^{208}Pb/^{204}Pb$ = 38.21–38.29), slightly less radiogenic than the hosting Grenvillian metamorphic basement rocks of the Garzón Massif (Ruiz et al. 1999; Fig. 5.14).

Samples from the southern Ibagué Batholith cluster above the Orogene curve ($^{206}Pb/^{204}Pb$ = 18.72–18.85; $^{207}Pb/^{204}Pb$ = 15.62–15.63; $^{208}Pb/^{204}Pb$ = 38.61–38.94) and return significantly less radiogenic values than those of the Santander Plutonic Group. The penecontemporaneous southern Ibagué volcanics return less radiogenic values than the coeval batholith ($^{206}Pb/^{204}Pb$ = 17.96–18.19; $^{207}Pb/^{204}Pb$ = 15.55–15.56; $^{208}Pb/^{204}Pb$ = 38.26–38.43). The Norosí and San Martín batholiths of the San

Lucas region are in turn somewhat less radiogenic than the coeval southern Ibagué Batholith, clustering to the left and just above the Orogene curve ($^{206}Pb/^{204}Pb = 18.35$–$18.61$; $^{207}Pb/^{204}Pb = 15.60$–$15.63$; $^{208}Pb/^{204}Pb = 37.90$–$38.45$). A similar relationship is observed between samples from the Norosí and San Martín batholiths and the coeval Noreán volcanics, with the Noreán volcanics revealing a less radiogenic Pb isotope composition than that observed for the coeval batholiths. Data for the Noreán volcanics plot just below the Orogene curve ($^{206}Pb/^{204}Pb = 17.90$–$17.98$; $^{207}Pb/^{204}Pb = 15.560$–$15.64$; $^{208}Pb/^{204}Pb = 37.48$–$37.63$).

Finally, samples from the Segovia Batholith present a radiogenic composition clustering between the Orogene and the upper crust curves ($^{206}Pb/^{204}Pb = 18.92$–$18.95$; $^{207}Pb/^{204}Pb = 15.64$–$15.67$; $^{208}Pb/^{204}Pb = 38.79$–$38.94$) and are located within the Pb isotope compositional field for the early Paleozoic metasedimentary rocks of the Loja Terrane (Fig. 5.14), which serve as proxy for the host Cajamarca-Valdivia basement. Data fall just above the Orogene curve and are somewhat more radiogenic than those for the southern Ibagué Batholith.

Late Triassic-Jurassic Pb Isotope Summary and Discussion

The overall Pb isotopic composition of the Colombian late Triassic-Jurassic granitoids is moderately to highly evolved with all $^{207}Pb/^{204}Pb$ values exceeding 15.55 and $^{206}Pb/^{204}Pb$ values extending as high as 19.44. Similar to the Sr-Nd isotope data presented earlier, whole-rock Pb isotope data for the Colombian late Triassic-Jurassic granitoids plot in discrete arrays, on a per-batholith or granitoid suite basis, with little overlap between the sample sets for individual batholiths (Fig. 5.14b). With the exception of the Santander Plutonic Group, the composite data form a linear array of clusters falling along, and at a slightly steeper slope to, the Orogene lead evolution curve of Zartman and Doe (1981). The granitoids of the Santander Plutonic Group form an isolated, highly radiogenic array disposed along and essentially parallel to the upper crust lead evolution curve.

Chiaradia et al. (2004) interpret the elongate, sloped array produced on the uranogenic diagram by the Ecuadorian Jurassic intrusions (Fig. 5.14b) to represent the mixing of magmas derived from a relatively homogenous MORB-type mantle whose lead isotope composition is approximated by the Triassic Piedras amphibolite (Fig. 5.14b), with crustal Pb derived from the basement units which host the Jurassic intrusions, in accord with the composite pre-Jurassic continental-oceanic basement recorded beneath the Ecuadorian Andes (e.g. Litherland et al. 1994). The variable character of the resulting intrusive lead isotope signatures is considered to primarily reflect variations in the composition of the basement host rocks (Chiaradia et al. 2004).

The Colombian late Triassic-Jurassic granitoid lead data plot in discrete clusters on a per-batholith basis. With the exception of the highly evolved lead isotopic compositions of the Santander Plutonic Group, the Colombia data plot coincident with the lead isotope field for Ecuadorian Jurassic arc-related granitoids as presented by Chiaradia et al. (2004). As noted, little lead isotope data is available for the host basement complexes in Colombia, although, as in Ecuador, composite, pre-Mesozoic basement architecture has also been documented in the Colombian Andes (Restrepo-Pace 1992; Cediel and Cáceres 2000; Cediel et al. 2003; Ordoñez-Carmona et al. 2006; Spikings et al. 2015). We propose, as per arguments presented in Ecuador by Chiaradia et al. (2004), that the observed variations in Pb isotope composition for the Ibagué, Norosí, San Martín and Segovia batholiths may be derived through the mixing of lead from a time-evolved MORB-type mantle source (approximated by the Piedras amphibolite), including lead derived from the Orogene, with lead inherited from basement complexes represented by the Chicamocha Terrane and the Cajamarca-Valdivia Terrane. With respect to samples from the Santander Plutonic Group, individual lead analyses plot in a linear form, essentially parallel to the upper crust lead evolution curve of Zartman and Doe (1981), suggesting that the lead isotope composition of the Santander granitoids was essentially derived from the combined Proterozoic and early Paleozoic continental basement of the Santander Massif (westernmost Guiana Shield and Grenvillian granulite belt), with little or no contribution of MORB-type mantle or Orogene lead. As recorded by the Sr-Nd isotope data, the Pb isotope composition of late Triassic-Jurassic granitoids in the Colombian Andes reflects the east-to-west changes in the composition of the intruded basement units as shown in Fig. 5.10.

Synthesis and Conclusions of Lithogeochemistry and Sr, Nd and Pb Isotope Geochemistry for Late Triassic-Jurassic Granitoids

Lithogeochemical data and Sr-Nd and Pb isotope systematics combined with U-Pb (zircon) age dating for the Colombian late Triassic-Jurassic batholiths reveal clear temporal-spatial trends and permit consistent qualitative conclusions with respect to magmatic sources and evolution and the degree of contamination through crustal anatexis or assimilation with host basement units. Lithogeochemical data indicate that all of the late Triassic-Jurassic batholiths are of the Cordilleran (Frost et al. 2001), volcanic arc (Pearce et al. 1984) or calc-alkaline (Barbarin 1999) types, typical of transitional (Barbarin 1999; in the case of the Santander Plutonic Group) and subduction-related tectonic settings. In northern Colombia, U-Pb (zircon) age dates demonstrate westward migration of the axis of magmatism from the ca. 209–196 Ma Santander Plutonic Group into the ca. 189–182 Ma main-phase batholiths of the Serranía de San Lucas and subsequently into the ca. 168–155 Ma Segovia Batholith. Lithogeochemical and Sr-Nd isotope data document diminishing crustal contamination and increasingly more juvenile melt compositions progressing from east to west. Data support an upper mantle source region and the variable mixing of mantle and crustal contributions for all batholiths, with the exception of the Santander Plutonic Group for which significant degrees of melt contamination through crustal anatexis and/or assimilation can be inferred. This is supported by the findings of Van der Lelij (2013), who, based upon Lu-Hf and Sr isotope data, concluded that Paleozoic and Mesozoic granitoids emplaced in the Santander Massif and Merida Andes between ca. 472 and 196 Ma were primarily derived through the recycling of Precambrian basement including lower to upper crustal sources with limited, if any, juvenile input from the depleted mantle. A similar conclusion can be drawn from the Pb isotope data for the Santander Plutonic Group which suggest in situ derivation and evolution of Pb with little contribution from Orogene or MORB-type mantle sources.

In the south, U-Pb (zircon) age dates for the southern Ibagué, Mocoa-Garzón trend and northern Ibagué batholiths suggest south and minor eastward migration of magmatism from the southern Ibagué to Mocoa-Garzón batholiths and subsequently along trend to the NNE into the northern Ibagué Batholith. Lithogeochemical and Sr-Nd and Pb isotope data reveal similar, predominantly upper mantle-derived compositions for the southern and northern Ibagué batholiths, despite their differences in age. The data suggest granitoid generation from a similar magmatic source region and emplacement under like tectonic conditions, in either case facilitated by the suture contact between the Chicamocha and Cajamarca-Valdivia units, which limited interaction between granitoid magmas and either basement domain. Sr-Nd and Pb isotope data are lacking for the Jurassic granitoids of the Mocoa-Garzón trend, but geological and lithogeochemical data infer greater degrees of magma interaction with the hosting Grenvillian metamorphic rocks of the western Guiana Shield as widely exposed in the Garzón Massif (Kroonenberg 1982; Ibañez-Mejía et al. 2011; Gómez et al. 2015a), although apparently not to the same degree as observed in the Santander Plutonic Group.

In conclusion, individual late Triassic to Jurassic granitoid batholiths of the Colombian Andes represent temporally and spatially separate arc segments, intruded into geologically distinct basement complexes. U-Pb (zircon) age, lithogeochemical and Sr-Nd and Pb isotope data suggest that granitoid chemical and isotopic characteristics and evolution are essentially independent of age and were primarily determined by processes within the magmatic source region for the granitoid melts and by the composition and/or degree of interaction with the hosting basement complex. Data for the individual batholiths reflect the spatial migration of late Triassic to Jurassic magmatism, combined with the unique geological conditions encountered by each granitoid arc segment at the time of emplacement. An overview and interpretation of the structural framework and tectonic evolution at the time of emplacement of the Colombian late Triassic to Jurassic granitoids are presented in the magmato-tectonic synthesis contained in Sect. 5.4.2.

5.3.4 Cretaceous to Eocene Granitoid Magmatism: Distribution, Age and Nature

Volumetrically significant granitoids of Cretaceous to Eocene age comprise much of the northwesternmost segment of the Colombian Andes, within the northern Central Cordillera and within the Chocó Arc segment of the Western Cordillera. Based upon geological exposure throughout the Colombian Andes, over 80 percent of Colombian Cretaceous to Eocene granitoid magmatism is concentrated within two composite intrusions and their satellite plutons, including the Antioquian and

Sonsón batholiths. Of these, the Antioquian Batholith (Feininger and Botero 1982) and its satellites are by far the largest, occupying an exposed area exceeding some 7800 square kilometres, more than the combined area of all the remaining Colombian Cretaceous to Eocene granitoids. The remaining granitoids, although volumetrically less significant, provide important information regarding the tectonic history of the region during the Cretaceous-Eocene.

5.3.4.1 Distribution

The distribution of major Colombian Cretaceous to Eocene granitoids including their associated volcanic sequences, where present, is shown in Fig. 5.15. Based upon geographic distribution and tectonic history, two broad groups of Cretaceous-Eocene granitoids can be recognized in Colombia, (1) an Eastern Group of autochthonous continental affinity, intruding the Cajamarca-Valdivia metamorphic basement complex, which was in situ within the Northern Andean tectonic mosaic at the time of pluton emplacement (i.e. prior to the early-mid-Cretaceous), and (2) a Western Group, including allochthonous granitoids of peri-cratonic or intra-oceanic affinity, hosted within accreted oceanic volcanic and sedimentary rocks of the Farallon Plate and CCOP/CLIP assemblage, presently underlying the cordilleran regions and coastal plains along the Colombian Pacific to the west of the Cauca and Garrapatas-Dabeiba fault and suture systems (Fig. 5.15).

The Eastern Group includes the late Cretaceous to Paleocene Antioquian Batholith and its satellite plutons (Ovejas Batholith and Altavista, La Unión and La Culebra stocks), the Paleocene Sonsón Batholith and other smaller Paleocene to Eocene intrusives such as the El Bosque Batholith and the Mariquita, Manizales, El Hatillo and Santa Bárbara stocks. The Santa Marta Batholith and Latal, Toribio and Buritáca plutons, located on the leading apex of the Sierra Nevada de Santa Marta (Tschanz et al. 1974; Mejía et al. 2008; Duque 2009; Cardona et al. 2011), are also included within this group. Notable features of many of the Eastern Group plutons, when compared to their Jurassic counterparts, include their generally sub-equant shapes (length-to-width ratios mostly less than 2:1) and the lack (or lack of preservation) of a coeval volcanic pile.

Within the Western Group, the Santa Fé, Sabanalarga and Buga batholiths and the Mistrató and other minor plutons (e.g. Jejenes Stock) are hosted within Cretaceous oceanic rocks of the Dagua and Cañas Gordas terrane assemblages. The Western Group granitoids may be considered to form components of the CCOP/CLIP assemblage, as discussed by Kerr et al. (1997) and Sinton et al. (1998). There is little published geological or geochemical information regarding some of these intrusions in Colombia, and in some cases precise radiometric age dates and lithogeochemical information have only recently been obtained (e.g. Buga, Villagómez et al. 2011; Santa Fé, Weber et al. 2015; Jejenes, Leal-Mejía 2011). An initial understanding of the origin and nature of these plutons and their relationship with their host rocks is herein presented.

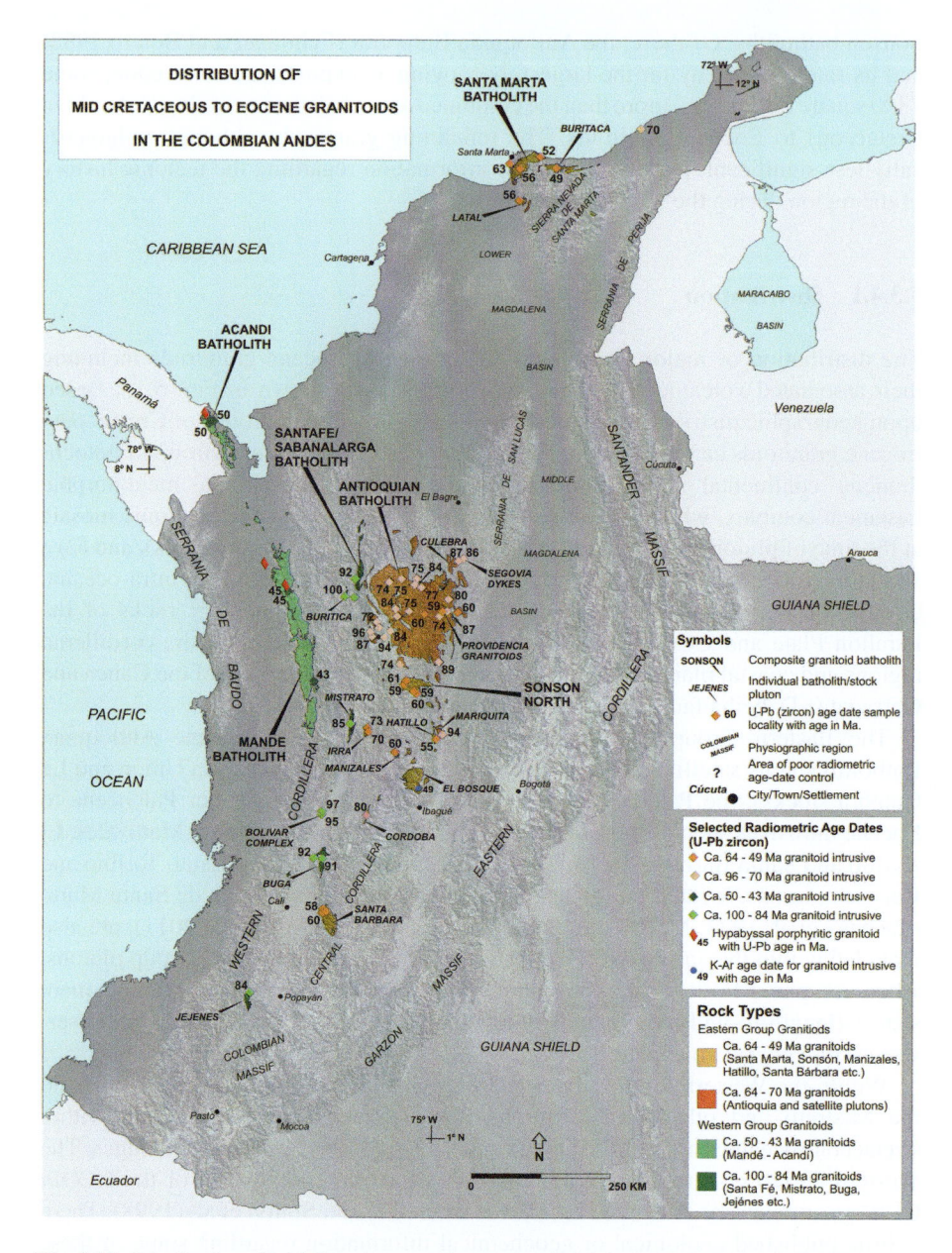

Fig. 5.15 Distribution of mid-Cretaceous to Eocene granitoids in the Colombian Andes. Principle modern-day physiographic provinces of the region are shown for reference. (Granitoid shapes modified after Cediel and Cáceres 2000; Gómez et al. 2007; Gómez et al. 2015a)

Farther to the west, the Paleocene-Eocene Mandé-Acandí batholiths, including the coeval La Equis-Santa Cecilia Formation volcanic and pyroclastic rocks (Fig. 5.15), are the most significant expression of granitoid magmatism within the Western Group. These granitoids were generated in an intra-oceanic setting upon late Cretaceous oceanic crust which forms the basement of the western segment of the Chocó Arc (Montes et al. 2012, 2015).

5.3.4.2 Age Constraints on Cretaceous-Eocene Granitoid Magmatism

Recent U-Pb (zircon) age determinations for Cretaceous to Eocene granitoids in the Colombian Andes, including intrusions from both the Eastern and Western groups, have been conducted by various authors. Results are included for works dedicated to the Antioquian Batholith and its surroundings (Correa et al. 2006; Ibañez-Mejía et al. 2007; Ordoñez-Carmona et al. 2007a; Restrepo-Moreno et al. 2007; Leal-Mejía 2011; Villagómez et al. 2011); the Sonsón Batholith (Ordoñez et al. 2001; Leal-Mejía 2011); the Mariquita Stock (Leal-Mejía 2011); the Manizales, El Hatillo and El Bosque plutons (Bayona et al. 2012; Bustamante et al. 2017); the Santa Marta Batholith (Mejía et al. 2008; Duque 2009; Cardona et al. 2011); the Buga Batholith (Villagómez et al. 2011); the Jejenes and Irra stocks (Leal-Mejía 2011); and the Mandé-Acandí batholiths (Leal-Mejía 2011; Wegner et al. 2011; Montes et al. 2012).

In total, the above data set represents over one hundred eighty-five high-precision U-Pb (zircon) magmatic crystallization ages which can be used to model Cretaceous to Paleogene magmatism in Colombia. The composite data are displayed in histogram format in Fig. 5.16. In many cases, the new data represent the first well-constrained age dates when compared to the historic largely K-Ar-based database of Maya (1992). In other cases, the data permit a much better definition of the multiple magmatic pulses which comprise large and complex intrusions, such as the Antioquian Batholith.

Within the Eastern Group of granitoids, the oldest pluton is the volumetrically minor Mariquita Stock, which produced a U-Pb (zircon) age of ca. 93.5 Ma (Leal-Mejía 2011). Large-scale, volumetrically significant and continuous plutonism begins in the mid-Cretaceous with the Antioquian Batholith, including its satellite plutons, between ca. 96 and 72 Ma. This event extends into lesser Paleocene and Eocene magmatism at ca. 62–54 Ma, recorded in the Antioquian and Sonsón batholiths and Manizales, El Bosque, El Hatillo, Santa Barbara intrusions and other minor plutons to the south.

The Antioquian Batholith is a composite poly-phase pluton emplaced in at least four pulses, spanning the late Cretaceous to Paleocene (Fig. 5.16). The earliest ca. 96–92 Ma phase is associated with more mafic to intermediate magmatism as recognized in the Altavista and San Diego stocks (Correa et al. 2006) and the mafic-intermediate xenoliths commonly embedded within the younger felsic, main-phase members of the batholith. Volumetrically, two distinct phaneritic-equigranular

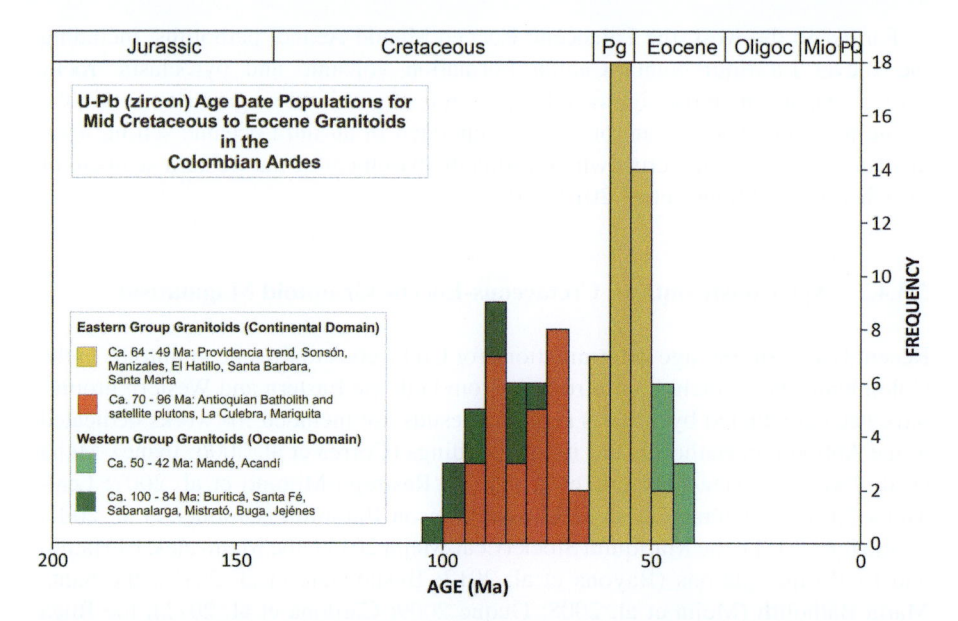

Fig. 5.16 U-Pb (zircon) age date populations for mid-Cretaceous through Eocene granitoids in the Colombian Andes. Note penecontemporaneous ages for the Western Group, allochthonous oceanic suite vs. the Eastern Group, autochthonous continental suite, representing the coeval emplacement of granitoids in distinct geotectonic environments

tonalitic to granodioritic pulses, from ca. 89 to 82 Ma and from ca. 81 to 72 Ma, account for the majority (>90%?) of the main mass of the Antioquian Batholith and satellite stocks. The Culebra Stock near Segovia returned an age of ca. 87.5 Ma. Granodiorite porphyry dikes extending to the NE into the Segovia area returned an age of ca. 86 Ma. The Ovejas Batholith returned ages ranging from ca. 76 to 72 Ma (Restrepo-Moreno et al. 2007), whilst the La Unión Stock to the south returned ca. 73.5 Ma with inheritance from ca. 82.8 Ma (Leal-Mejía 2011).

Minor Paleocene granitoid magmatism is also recorded in isolated areas within the Antioquian Batholith domain. The Caracolí Stock on the east-centre margin of the batholith returned an age of ca. 60 Ma, with inheritance from ca. 79 Ma. Medium-grained equigranular tonalite from near Providencia in the Nus River valley returned various dates ranging from ca. 60 to 58 Ma, whilst a medium- to coarse-grained quartz biotite granite porphyry stock containing distinctive euhedral bipyramidal quartz crystals, located west of Santo Domingo, revealed an age of ca. 60 Ma (Leal-Mejía 2011).

Volumetrically more significant Paleocene magmatism within the Eastern Group is documented in the Sonsón Batholith. This granitoid pluton was formerly considered to be of Jurassic age (Cediel and Cáceres 2000; González 2001; Gómez et al. 2007); however, U-Pb (zircon) age dating reveals it is a composite body, comprised of granitoid rocks of Permo-Triassic age in the south (Leal-Mejía 2011; Fig. 5.15) and of Paleocene age (ca. 61–57 Ma) in the north (Ordoñez et al. 2001; Leal-Mejía 2011).

The Sonsón Batholith is presently interpreted to include the northern sector extending around and to the east of the town of Sonsón (Leal-Mejía 2011; Fig. 5.15). The precise contact between these two ages of intrusive has yet to be cartographically defined.

Additional, recent Paleocene U-Pb (zircon) dates have also been reported for the Eastern Group Manizales Stock (ca. 59 Ma, Bayona et al. 2012), indicating that autochthonous granitoid magmatism continued to the south of Sonsón during this time period. A general southward and eastward younging trend for magmatism can be inferred to continue into the Eocene with the emplacement of the El Hatillo Stock at ca. 55 Ma (Bayona et al. 2012; Bustamante et al. 2017) and the presence of additional granitoid plutons, including the El Bosque Batholith which also provides a U-Pb (zircon) age of ca. 55 Ma (Bustamante et al. 2017)

Finally, within the Eastern Group plutons of northernmost Colombia, Paleocene to Eocene granitoid magmatism spanning the age range from ca. 64 to 47 Ma (Mejía et al. 2008; Duque 2009; Cardona et al. 2011) is recorded along the apex of the Sierra Nevada de Santa Marta (Tschanz et al. 1974). Detailed study of the Santa Marta Batholith and satellite plutons including the Latal, Toribio and Buritaca stocks, by Duque (2009), revealed emplacement of the suite in two pulses between ca. 58 and 50 Ma. The principal components of the SW Santa Marta Batholith proper and Latal pluton, including distinctive coarse- and fine-grained phases, were intruded between ca. 58 and 55 Ma, followed by emplacement of the NE sector of the Santa Marta Batholith and the Buritaca and Toribio stocks, by ca. 50 Ma. Based upon the composite U-Pb (zircon) data, Duque (2009) interprets a general NE migration of magmatism within the main-phase Santa Marta Batholith, terminating in the Buritaca pluton. Additional, early, volumetrically minor, ca. 64–62 Ma, two-mica trondhjemitic leucogranites, identified by the author (e.g. Playa Salguero), were considered unrelated to main phase batholith emplacement (see Duque-Trujillo et al. 2018).

With respect to the Cretaceous to Eocene Western Group (CCOP/CLIP) granitoids located to the west of the Cauca and Garrapatas-Dabeiba fault and suture system (Fig. 5.15), the oldest of these plutons, hosted within Cañas Gordas oceanic basement, include the Buriticá tonalite and associated Santa Fé Batholith, which have returned U-Pb (zircon) dates of ca. 100 Ma and 90 Ma, respectively (Weber et al. 2015). The Sabanalarga Batholith, located in fault contact immediately to the east (Nívia and Gómez 2005; Gómez et al. 2007), has not been dated using the U-Pb technique but appears to represent a tectonically duplicated segment of the Santa Fé Batholith. Further detailed mapping and age dating are required to better define the relationships between these intrusions and the host basement complex.

Along the trend to the south of Santa Fé and Sabanalarga, the Mistrató Batholith (Fig. 5.15) also appears within strongly tectonized Cañas Gordas volcano-sedimentary rocks, in intrusive/structural contact with the Barroso Fm. An ca. 85 Ma U-Pb (zircon) age was presented for the Mistrató Batholith by the Agencia Nacional de Hidrocarburos and Universidad de Caldas (2011). A similar age was recorded, farther south, in the southern sector of the Western Cordillera, where the previously undated Jejenes Stock returned a U-Pb age of ca. 84 Ma (Leal-Mejía 2011). In this case, intrusive relationships with the CCOP-related Dagua terrane are observed.

To the NNE of the Jejénes Stock, the Buga Batholith (Fig. 5.12) has returned a U-Pb (zircon) age of ca. 92 Ma (Villagómez et al. 2011). Buga appears to have been emplaced with pre-CCOP basement rocks of the Dagua terrane (Anaime Fm; Nívia 1992). Both the western and eastern margins of the batholith have been tectonically modified.

The youngest and by far largest intrusion of the Western Group allochonous granitoids is the Mandé-Acandí Batholith (Fig. 5.15), hosted within the El Paso-Baudo assemblage of northwesternmost Colombia (Cediel et al. 2010). Field observations and regional magnetic data (Cediel et al. 2010) indicate the Mandé Batholith is a composite body comprised of holocrystalline phaneritic and porphyritic phases ranging from diorite to granodiorite and granite. It is flanked to the east and west by the penecontemporaneous Santa Cecilia-La Equis volcanic sequence, of Paleogene age (Cediel et al. 2010). A thermal aureole is recorded within the volcanic sequence indicating the Mandé Batholith intrudes the volcanic pile. Leal-Mejía (2011) provided U-Pb (zircon) dates of ca. 46–44 Ma for quartz diorite porphyry which cuts phaneritic granodiorite within the north central sector of the batholith at Pantanos. An ca. 62 Ma (Paleocene) inheritance age, interpreted to have been donated by the volcanic pile or main batholith, was observed for these samples. Within the northern extension of the Mandé magmatic arc, including the Acandí Batholith in Panama's San Blas Range, Paleocene-Eocene U-Pb (zircon) magmatic crystallization ages are also observed. In Colombia, Montes et al. (2012) and Montes et al. (2015) record a maximum age of ca. 50 Ma for the Acandí Batholith.

Discussion of Spatial Distribution of Cretaceous-Eocene Granitoids

Based upon geographic distribution and geological setting, two major groups of Colombian Cretaceous to Eocene granitoids have been identified, including 1) Eastern Group granitoids and 2) Western Group granitoids. The Eastern Group represents autochthonous, continental granitoid magmatism of Cretaceous to Eocene age, largely dominated by two major magmatic pulses at ca. 89–82 Ma and ca. 79–72 Ma, generating the main mass of the Antioquian Batholith, its satellite plutons and the Irra Stock (ca. 70 Ma). Magmatism is rather abruptly shut down after ca. 72 Ma but reinitiates at ca. 62–58 Ma, within and to the south of the Antioquian Batholith, with the emplacement of various smaller plutons, the largest of which is the Sonsón Batholith. The available U-Pb age data demonstrate the general southward and eastward migration of autochthonous magmatic centres of the Eastern Group during post main-phase Antioquian Batholith time, from the Paleocene to the early Eocene. The Paleocene-Eocene granitoid centres can be traced from the 61 to 58 Ma Sonsón and Manizales intrusives in the north to the El Hatillo, El Bosque and Santa Bárbara plutons to the south and east, all of which produce Paleocene-Eocene U-Pb (Bayona et al. 2012; Bustamante et al. 2017) and/or K-Ar (Maya 1992) radiometric age dates.

The Western Group (CCOP/CLIP-related) granitoids may also be considered in two spatially and temporally separate groups, including an early group (Sabanalarga/

Santa Fé, Buriticá, Jejénes, Buga, etc.) dating from ca. 100 to 82 Ma, hosted within the Dagua-Cañas Gordas terranes, and the significantly younger ca. 50–42 Ma granitoids of the Mandé-Acandí arc, hosted within the El Paso-Baudó terrane. Emplacement of the early group is essentially penecontemporaneous with the development of the early phases of the continental Antioquian Batholith. The Western Group granitoids, however, are consistently hosted within oceanic terrane assemblages which have been deemed to be allochthonous (e.g. Cediel et al. 2003; Kerr et al. 2003; Kennan and Pindell 2009) and are considered to represent granitoid magmatism in an intra-oceanic environment, related to the generation and migration of the CCOP/CLIP assemblage, prior to accretion along the Colombian margin. Further temporal and spatial differentiation of the Cretaceous to Eocene granitoids of the Eastern and Western groups, within the context of the tectonic evolution of the region, will be discussed in detail following the presentation of lithogeochemical and isotopic data in the following section.

5.3.4.3 Lithogeochemical and Isotopic Characteristics of Cretaceous to Eocene Granitoids

Historically, little whole-rock lithogeochemical or isotopic data has been available for the Colombian Cretaceous to Eocene granitoid suite. Older works or compilations such as Alvarez (1983), Feininger and Botero (1982) or González (2001) contain some basic major element oxide data but include only limited or no minor, trace and rare-earth element data and virtually no Pb, Sr or Nd isotopic data, thus limiting the interpretation of petrogenetic and tectonic constraints for these rocks.

Recently, with the use of ICP-based analytical techniques, studies applying combined whole-rock major-minor-trace-rare-earth element studies, and additional isotopic analyses, to the Cretaceous-Eocene granitoid suite, have become available. Important contributions which analyse multiple plutons at a regional scale include Villagómez et al. (2011) and Leal-Mejía (2011). Additional studies involving specific intrusions include Ordoñez et al. (2001) for the Sonsón Batholith; Correa et al. (2006), Ibañez-Mejía et al. (2007), Ordoñez-Carmona et al. (2007a) and Restrepo-Moreno et al. (2007) for the Antioquian Batholith; Wegner et al. (2011) and Montes et al. (2012) for the Mandé-Acandí batholiths; and Bayona et al. (2012) and Bustamante et al. (2017) for the Manizales, El Hatillo and El Bosque plutons. Representative lithogeochemical data for the Cretaceous-Eocene Eastern Group and Western Group magmatic suite is shown in Figs. 5.17, 5.18, 5.19, and 5.20.

Lithogeochemistry

Eastern Group Granitoids

The Antioquian Batholith: The ca. 96–58 Ma Antioquian Batholith granitoid suite, including satellite bodies (e.g. the Ovejas Batholith and the Altavista, La Unión and La Culebra stocks) and the coeval Segovia dikes, is represented by 57 samples, 14 of which are altered. All of the magmatic phases show similar broad-scale lithogeochemical features such as a metaluminous nature, within a highly differentiated calc-alkaline compositional trend, which varies over time, from gabbro to granite. With respect to the classification scheme of Frost et al. (2001), the Antioquian Batholith suite demonstrates a weakly ferroan trending to magnesian composition with decreasing age, whilst most samples demonstrate a distinctly calcic tendency. Trace element spider diagram patterns for the Antioquian Batholith granitoids show magmatic arc-related signatures, with enrichment of HFSE with respect to LILE and conspicuous negative Ta-Nb anomalies (Fig. 5.18). The REE patterns show highly variable REE contents ($\Sigma REE = 21.82$–335.69) and also variable negative to positive Eu anomalies (Eu/Eu* = 0.50–2.85) (Fig. 5.18).

The ca. 62–58 Ma Providencia granitoid suite, which may be considered post main-phase batholith in age, is characterized by lower-K biotite-bearing granodiorite to granite with compositionally distinct (high-Na) plagioclase and "adakite-like" geochemical features (Richards and Kerrich 2007; e.g. high SiO_2 (≥ 56 wt%), Al_2O_3 (≥ 15 wt%) and Na_2O (≥ 3.5 wt%) contents, low K_2O (≤ 3 wt%) contents and Sr enrichment (≥ 400 ppm), accompanied by depletion of Y (≤ 18 ppm) and Yb (≤ 1.9 ppm)). Providencia suite REE trends are slightly depleted with respect to the main phases of the batholiths ($\Sigma REE = 47$–160.87). They describe gently decreasing slopes and no significant Eu anomaly.

The Irra Stock: Major, minor and trace element data for the ca. 70 Ma Irra Stock (Figs. 5.17 and 5.18) reveal characteristics which distinguish it from the main phases of the Antioquian Batholith. It is a metaluminous syenite of the shoshonite series (alkali), and it is enriched in both trace and rare-earth elements; however the mantle-normalized plot displays positive Ba and Sr anomalies and negative Nb, Ta, P and Ti anomalies similar to arc-related rocks. REE contents are relatively enriched ($\Sigma REE = 99.16$–216.5 ppm). REE plots reveal moderately fractionated chondrite-normalized patterns ($(La/Yb)_N = 18.28$–24.00). No significant Eu anomaly is observed. The Irra Stock is located within the Romeral tectonic zone and is considered to form part of an in situ phase of minor alkaline magmatism, similar to the Sucre intrusive suite located in a similar tectonic position within Romeral to the north (Vinasco 2018). Both lithogeochemical and age data for the Irra Stock are contrary to data observed for the low-K Western Group granitoids (see below).

The Sonsón Batholith: Samples from different phases of the Sonsón Batholith including phaneritic quartz-diorites, leucogranites and diorite porphyry dikes are represented by five relatively unaltered samples and one leucogranite sample with evidences of alteration (Fig. 5.17). The samples are metaluminous in nature (A/CNK <1.1) of medium- to high-K calc-alkaline affinity, although the leucogranites plot marginally peraluminous due to the partial metasomatic replacement of biotite by muscovite. An arc-magmatism signature for the Sonsón Batholith samples is revealed by the trace element spider diagram (Fig. 5.18), where higher U (12.1–16.1 ppm) and Th (21.3–24.2 ppm) contents and positive incompatible element anomalies (e.g., K, Th, U and Ta), accompanied by strong negative compatible

Eastern Group Granitoids

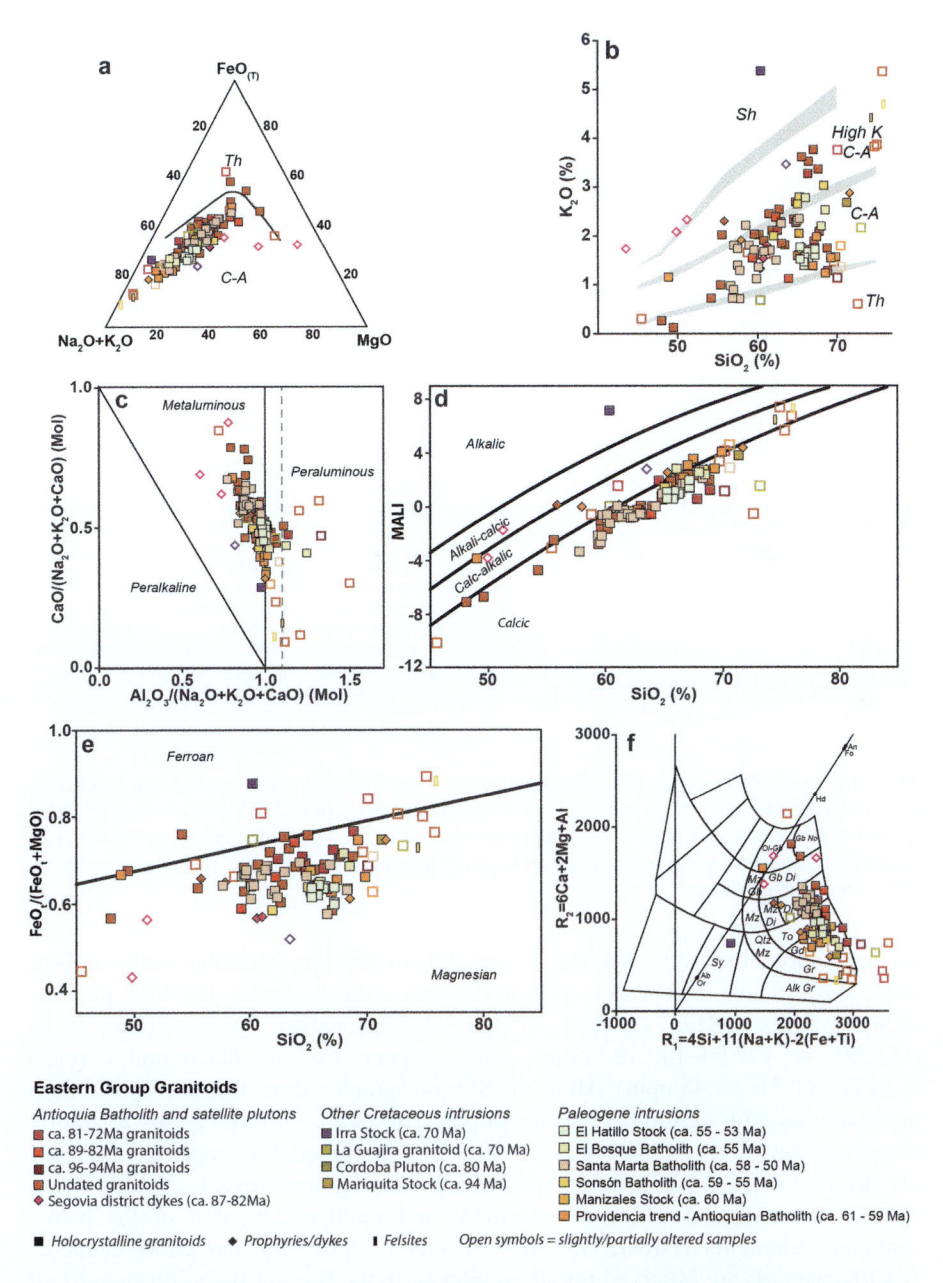

Antioquia Batholith and satellite plutons
- ca. 81-72Ma granitoids
- ca. 89-82Ma granitoids
- ca. 96-94Ma granitoids
- Undated granitoids
- Segovia district dykes (ca. 87-82Ma)

Other Cretaceous intrusions
- Irra Stock (ca. 70 Ma)
- La Guajira granitoid (ca. 70 Ma)
- Cordoba Pluton (ca. 80 Ma)
- Mariquita Stock (ca. 94 Ma)

Paleogene intrusions
- El Hatillo Stock (ca. 55 - 53 Ma)
- El Bosque Batholith (ca. 55 Ma)
- Santa Marta Batholith (ca. 58 - 50 Ma)
- Sonsón Batholith (ca. 59 - 55 Ma)
- Manizales Stock (ca. 60 Ma)
- Providencia trend - Antioquian Batholith (ca. 61 - 59 Ma)

■ *Holocrystalline granitoids*   ◆ *Prophyries/dykes* | *Felsites*   Open symbols = slightly/partially altered samples

Fig. 5.17 Major element lithogeochemical plots for mid-Cretaceous through Paleocene, Eastern Group (continental) granitoids in the Colombian Andes. (**a**) AFM plot, curve after Irvine and Baragar (1971); (**b**) K_2O vs. SiO_2 plot, boundary fields in grey as summarized by Rickwood (1989); (**c**) alumina saturation plot after Barton and Young (2002); (**d** and **e**) MALI and Fe-index vs. SiO_2 plots, respectively, after Frost et al. (2001); (**f**) R1-R2 classification plot after De La Roche et al. (1980). *Th* tholeiite, *C-A* calc-alkaline, *Sh* shoshonite, *Gb No* gabbro-norite, *Ol-Gb* (olivine-) gabbro, *Gb Di* gabbro-diorite, *Di* diorite, *Mz Di* monzodiorite, *To* tonalite, *Gd* granodiorite, *Gr* granite *Alk Gr* alkali granite, *Sy* syenite

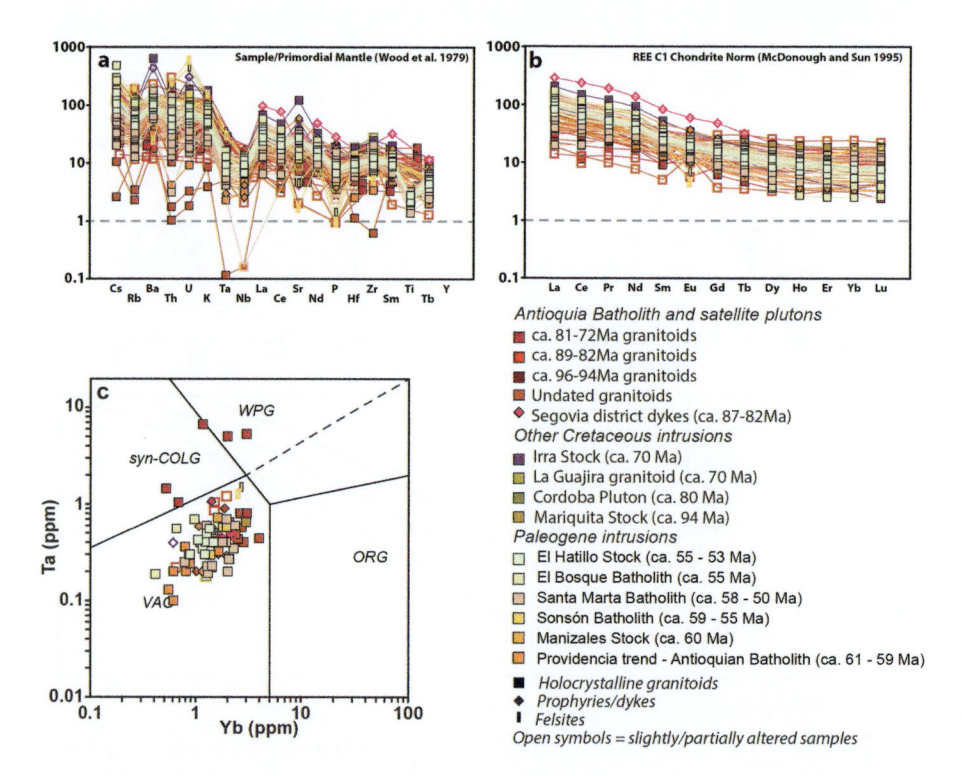

Fig. 5.18 Trace element and REE lithogeochemical plots for mid-Cretaceous through Paleocene, Eastern Group (continental) granitoids in the Colombian Andes. (**a** and **b**) Trace Element and REE normalized spider diagram plots; (**c**) granite discrimination Ta vs. Yb diagram after Pearce et al. (1984). *VAG* volcanic-arc granites, *syn-COLG* syn-collisional granites, *WPG* within-plate granites, *ORG* ocean ridge granites

element anomalies (e.g., Ba, Sr, Zr and Ti) for the leucogranites confirm their more evolved character with respect to the quartz-diorite rocks. The REE patterns indicate higher REE contents in the allanite-bearing quartz-diorite rocks ((ΣREE = 127.94–142.79 ppm) with respect to the leucogranite rocks (ΣREE = 68.56–89.48 ppm). All of the Sonsón samples show moderately fractionated patterns with gentle decreasing slopes. The quartz-diorite reveals a slightly steeper overall slope ((La/Yb)$_N$ = 10.84–11.51), with moderate negative Eu anomaly (Eu/Eu* = 0.54–0.72). By comparison, the leucogranite reveals stronger negative Eu anomalies (Eu/Eu* = 0.22–0.34) and relative depletion of the heavy rare-earth elements (HREE; La-Sm). Relatively flat light rare-earth elements (LREE) patterns are observed for all samples from the Sonsón Batholith suite ((Gd/Yb)$_N$ = 0.89–1.57; Leal-Mejía 2011). Although of similar age, none of the rock types of the Sonsón Batholith show the Na-rich "adakite-like" geochemical signature observed in the Providencia suite of the Antioquian Batholith.

The Mariquita Stock: The Mariquita Stock is a metaluminous, medium-K calc-alkaline granodiorite. It plots magnesian-calcic in the classification scheme of Frost et al. (2001). Trace element and REE volcanic reveal arc signatures with moderate REE contents (ΣREE = 80.82 ppm), decreasing slopes ($(La/Yb)_N$ = 3.28) and a moderate negative Eu anomaly (Eu/Eu* = 0.55; Fig. 5.18).

Other Paleogene Eastern Group Granitoids: Other samples of Eastern Group granitoids (Fig. 5.17) include twenty-one samples of the main phase of the Santa Marta Batholith (Mejía et al. 2008; Duque 2009), one sample of the Manizales Stock (Leal-Mejía 2011), three samples of the El Bosque Batholith (Leal-Mejía 2011; Bustamante et al. 2017) and twelve samples of the El Hatillo Stock (Bustamante et al. 2017). Samples of the Manizales Stock and the El Bosque Batholith appear altered, whilst those of the Santa Marta Batholith and the El Hatillo Stock do not. In general, samples from these plutons exhibit metaluminous to weakly peraluminous, medium- to marginally high-K calc-alkaline character, with SiO_2 content ranging from 57 to 71 wt%. A general tendency towards an increasingly magnesian calc-alkalic character, and increasing aluminity and silica content, with decreasing age is observed. A notable exception to this trend is the Eocene granitoids of the Sierra Nevada de Santa Marta, which were emplaced under differing tectonic conditions when compared with the granitoids of similar age located in the Central Cordillera. With respect to REE contents, ΣREE for samples from the Manizales Stock and El Bosque Batholith are 107.67 ppm and 112.36–169.53 ppm, respectively, whereas the El Hatillo Stock shows lower values (ΣREE = 85.18–129.54 ppm). The Manizales Stock exhibits a fractionated REE pattern ($(La/Yb)_N$ = 9.6) with very subtle negative Eu anomaly (Eu/Eu* = 0.89) and a relatively flat HREE trend ($(Gd/Yb)_N$ = 1.27). The El Bosque Batholith shows a similar decreasing slope for light rare-earth elements (La-Sm) as in the Manizales Stock, and no significant Eu anomaly (Eu/Eu* = 0.73–0.98). It records strong depletion of the HREE (Gd-Lu) where a concave upward (spoon-shaped) pattern for the HREE is observed, which may be related to hornblende fractionation in the magma source.

Western Group Granitoids

Lithogeochemical plots for the Western Group of Cretaceous to Eocene granitoids are presented in Figs. 5.19 and 5.20. Some 72 samples in total are represented. In contrast to the Eastern Group, 55% (40) of the samples from the Western Group are notably affected by some degree of post-emplacement alteration. This alteration is a reflection of the observably tectonized nature of most of the intrusions as viewed in outcrop, where abundant faulting and jointing, coupled with high rainfall conditions, have led to chemical weathering and deep oxidation. Thus, in many respects, the quality of the data set is not on par with that from the Eastern Group. Notwithstanding, the observations regarding the general lithogeochemical characteristics of the Western Group granitoids revealed herein are considered valid.

All Western Group granitoids demonstrate a low-K character (K_2O content <1%). As a reflection of this observation, Western Group granitoids are mixed magnesian-ferroan in nature, with a distinctly calcic tendency. Leal-Mejía (2011) observed the low-K behaviour for dioritic phases of the Buga Batholith and Jejenes

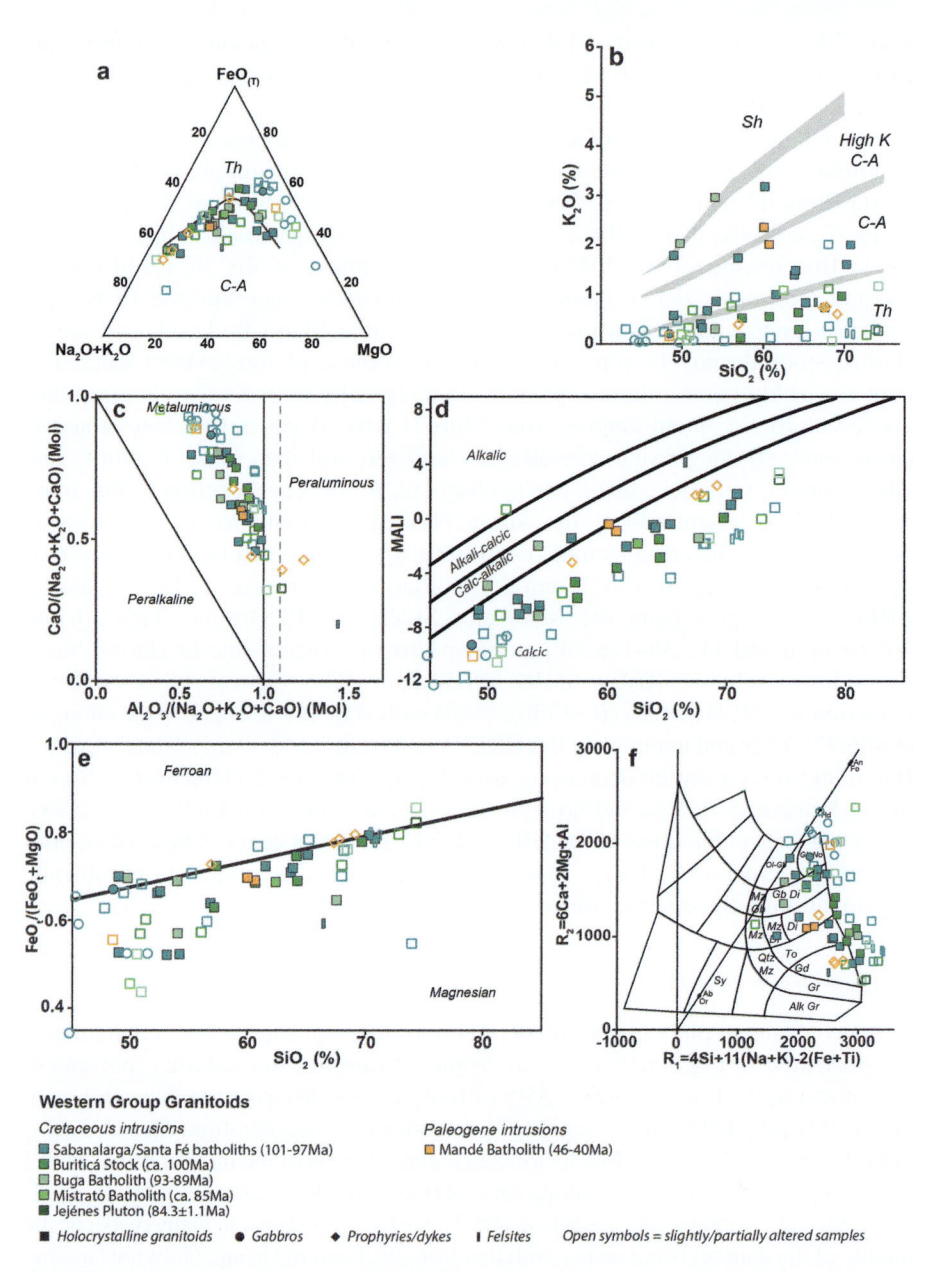

Fig. 5.19 Major element lithogeochemical plots for mid-Cretaceous through Eocene, Western Group (oceanic) granitoids in the Colombian Andes. (**a**) AFM plot, curve after Irvine and Baragar (1971); (**b**) K_2O vs. SiO_2 plot, boundary fields in grey as summarized by Rickwood (1989); (**c**) alumina saturation plot after Barton and Young (2002); (**d** and **e**) MALI and Fe-index vs. SiO_2 plots, respectively, after Frost et al. (2001); (**f**) R1-R2 classification plot after De La Roche et al. (1980). *Th* tholeiite, *C-A* calc-alkaline, *Sh* shoshonite, *Gb No* gabbro-norite, *Ol-Gb* (olivine-)gabbro, *Gb Di* gabbro-diorite, *Di* diorite, *Mz Di* monzodiorite, *Mz* monzonite, *To* tonalite, *Gd* granodiorite

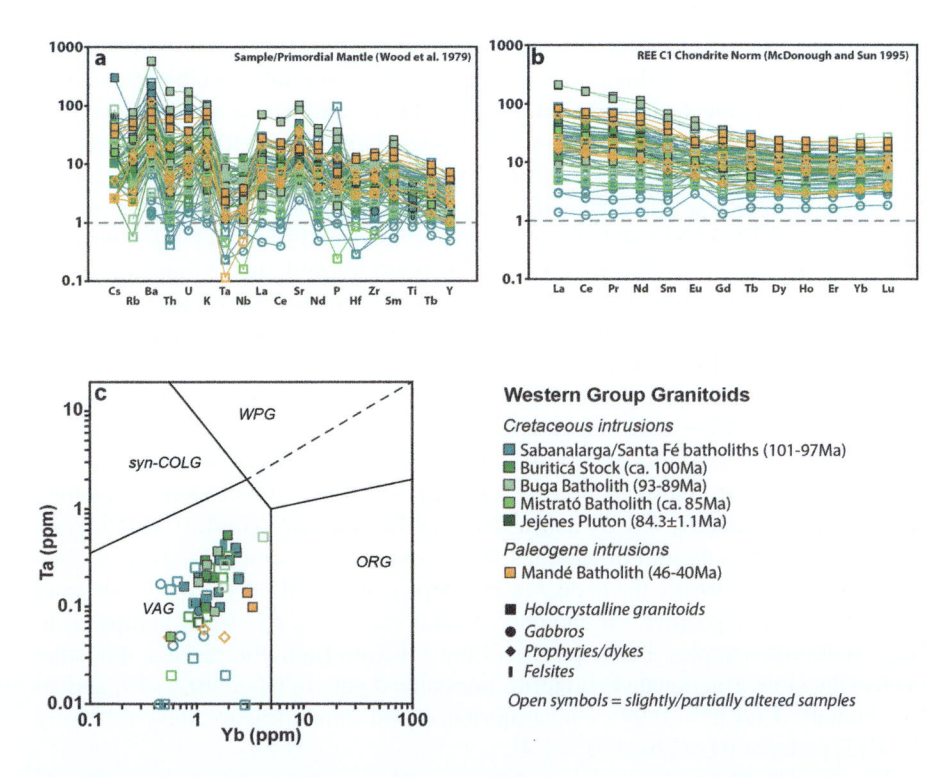

Fig. 5.20 Trace element and REE lithogeochemical plots for mid-Cretaceous through Eocene Western Group (oceanic) granitoids in the Colombian Andes. (**a** and **b**) Trace element and REE normalized spider diagram plots; (**c**) granite discrimination Ta vs. Yb diagram after Pearce et al. (1984). *VAG* volcanic-arc granites, *syn-COLG* syn-collisional granites, *WPG* within-plate granites, *ORG* ocean ridge granites

Stock, as well as for the Mandé Batholith, and noted that samples from the Jejenes Stock and Mistrató, Buga and Mandé batholiths samples plot in the tholeiite field of the Peccerillo and Taylor (1976) diagram. Exceptions to this general trend include some dioritic to granodioritic samples of the Buga and Mandé batholiths which returned K_2O contents >2% (Leal- Mejía 2011).

Buga, Santa Fé and Mistrató batholiths and Jejenes Stock: The late Cretaceous Buga, Santa Fé (including Sabanalarga) and Mistrató batholiths and Buriticá and Jejenes Stock are represented by 65 lithogeochemical samples. Of these samples, over 54% (35 samples) present evidence of hydrothermal alteration. Notwithstanding, the lithogeochemical analyses establish the metaluminous character and variable low-K tholeiitic to high-K calc-alkaline affinity of the suite (Fig. 5.19). The Jejénes Stock sample plots in the peraluminous field, likely as a result of petrographically observable hydrothermal alteration (saussuritization) in the analysed sample (Leal-Mejía 2011). Samples of the Buga, Santa Fé (Sabanalarga) and Mistrató batholiths show bimodal compositions with more melanocratic samples of gabbroic to dioritic composition and more leucocratic samples of granodioritic or tonalitic composition. The trace element spider diagram patterns of unaltered samples exhibit enrichment of HFSE with respect to LILE and negative Ta-Nb and Ti anomalies, typical of arc-magmatism (Fig. 5.20). The Santa Fé (Sabanalarga) Batholith and the Buriticá Stock return REE contents ranging from ΣREE = 3.66–130.14 ppm, with chondrite-normalized patterns exhibiting moderate to steep decreasing slopes $((La/Yb)_N = 0.77–12.88)$ and slightly negative to slightly positive Eu anomalies $(Eu/Eu^* = 0.74–1.40)$. The Buga Batholith shows variable behaviour as implied by observed compositional variations. Data presented by Leal-Mejía (2011) and Villagómez et al. (2011) indicate the more mafic phases contain variable REE contents (ΣREE = 15.42–41 ppm) with flat chondrite-normalized patterns approximating ten times chondrite value $((La/Yb)_N = 0.8–2.5)$. Other phases show higher REE contents (ΣREE = 229–241 ppm) and exhibit more fractionated patterns $((La/Yb)_N = 21–33)$, with no significant Eu anomalies $(Eu/Eu^* = 0.95–0.98)$. A third phase, with intermediate REE contents (ΣREE = 104.48 ppm), moderate fractionation $((La/Yb)_N = 2.5)$ and a negative Eu anomaly $(Eu/Eu^* = 0.52)$ are observed. Mistrató Batholith samples show lower REE contents (ΣREE = 2.6–42.5 ppm) with a variable flat to decreasing slope patterns $((La/Yb)_N = 0.71–10)$ and subtle to strongly positive Eu anomaly $(Eu/Eu^* = 1.02–2.61)$ when compared to Buga Batholith samples. Felsic phases of the Mistrató Batholith reveal a moderate decreasing slope fractionated chondrite-normalized pattern $((La/Yb)_N = 10)$ and no Eu anomaly $(Eu/Eu^* = 0.99)$, whilst diorites of the same intrusive show relatively flat REE patterns $((La/Yb)_N = 0.7–1.4)$.

The Jejénes Pluton is represented by a single sample, a saussuritized low-K tonalite which reveals relatively flat arc-related trace element patterns similar to the Buga and Mistrató batholiths. It contains relatively low concentrations of rare-earth elements (ΣREE = 34.8 ppm) which reveal a slightly fractionated pattern $((La/Yb)_N = 4.12)$ and a slight, positive Eu anomaly $(Eu/Eu^* = 1.42)$.

The Mandé Batholith: Holocrystalline and porphyritic samples of the Paleocene-Eocene Mandé Batholith are represented by seven samples. Five reveal evidence of alteration, whilst two are relatively unaltered. The samples are metaluminous with exception of two of the altered samples which plot in the peraluminous field, likely due to alteration. As with the Cretaceous granitoids of the Western Group, the Mandé suite includes low-K tholeiitic and medium- to high-K calc-alkaline members. The low-K samples record enhanced degrees of alteration. Notwithstanding, most varieties of potassic alteration involve K-enrichment (Warren et al. 2007), and hence the low-K samples are herein considered to reflect an originally low-K protolith. This is backed by apparent REE depletion (ΣREE = 23.79–88.86 ppm) in the low-K samples vs. the higher overall REE contents (ΣREE = 109.86–125.93 ppm) of the medium- to high-K samples, suggesting the low-K samples are distinctly more primitive. REE chondrite-normalized patterns for all samples show moderate decreasing slopes $((La/Yb)_N = 2.27–5.81)$ with moderate to subtle negative Eu anomalies $(Eu/Eu^* = 0.63–1.07)$.

Whole-Rock Lithogeochemistry Summary and Discussion for the Cretaceous-Eocene Granitoids

Based upon the composite lithogeochemical data set for Cretaceous to Eocene granitoids used in this study, it is possible to differentiate and characterize the Eastern and Western granitoid groups derived from the spatial analysis outlined above. Further consideration of these groups, in conjunction with U-Pb (zircon) age data, permits the identification of subgroups within each group, as follows: the Eastern Group includes (1) the ca. 96–72 Ma autochthonous, continental granitoids of the Antioquian Batholith and (2) the ca. 62–49 Ma granitoids of the Providencia suite, Sonsón Batholith and Central Cordilleran plutons to the south of the Antioquian Batholith and the Eocene intrusives of the Sierra Nevada de Santa Marta. The Providencia suite granitoids, although temporally belonging to the Paleocene suite, are spatially contained within the confines of the Antioquian Batholith and display some unique lithogeochemical characteristics which have been highlighted above. Within the allochthonous, oceanic Western Group (CCOP-/CLIP-related) granitoids, two groups are distinguished, (1) ca. 100–82 Ma, Santa Fé through Jejénes trend, hosted along the tectonized margin of the Dagua and Cañas Gordas terranes, and (2) the ca. 50–42 Ma granitoids and hypabyssal porphyries of the Mandé-Acandí arc, hosted within El Paso-Baudó oceanic basement.

The Antioquian Batholith granitoids demonstrate a differentiation trend from gabbroic through quartz-dioritic compositions, with generally increasing alkalinity and silica contents and increasing levels of REE fractionation vs. time. The later phases reveal characteristics of within-plate and syn-collisional granitoids, suggesting increased crustal input or crustal interaction/contamination (Pearce et al. 1984). Interestingly, the Paleocene Providencia suite marks a clear return to more mantelic less-fractionated compositions. The Irra Stock is a relatively enriched, alkali series shoshonite emplaced within the Romeral tectonic zone at ca. 70 Ma. It is of similar age and composition as alkali granitoids emplaced within the Romeral zone, to the north at Sucre, as recently revealed by Vinasco (2018).

The post-Antioquian granitoids to the south demonstrate similar compositional trends to those of the Antioquian suite, with a general tendency towards increased alkalinity and silica contents and greater degrees of REE fractionation. Pronounced Eu anomalies are particularly recorded for the leucocratic phases of the Sonsón Batholith. The Sierra Nevada de Santa Marta granitoids demonstrate metaluminous, magnesian-calcic lithochemistry granitoids with relatively typical arc-related trace and REE profiles.

Although interpretation is obscured to some degree by notably greater levels of post-crystallization alteration and tectonism, when compared to the Eastern Group granitoids, the whole-rock lithogeochemistry of ca. 100–82 Ma Western Group granitoids is clearly reflective of their spatial distribution within Farallon and CCOP/CLIP-related, oceanic domain basement rocks. Despite alteration patterns, all of the ca. 100–82 Ma Western Group granitoids plot within the volcanic arc granitoids field of Pearce et al. (1984). When compared to the Eastern Group, the Santa Fé through Jejénes trend plutons are decidedly more calcic in composition, with lesser degrees of alkalinity and silica, HFSE and REE enrichment, typical of more primitive granitoids developed within intra-oceanic vs. continental arcs. Hints of Fe enrichment are observed within the Western Group suites on both the AFM and $FeO_{(total)}/(FeO_{(total)} + MgO)$ vs. silica diagrams. Notwithstanding, it is noteworthy that the larger batholiths, including Santa Fé and Buga, were apparently sufficiently stable and long-lived to develop more alkali- and silica-rich fractionates, relatively enriched in HFSE.

With respect to the ca. 62–42 Ma Western Group Mandé-Acandí Batholith, it is observed that the representing sample population is small, considering that Mandé-Acandí represents a far greater volume of granitoid magma, exceeding that of all of the ca. 100–82 Ma Western Group granitoids combined. Notwithstanding, and despite additional differences in age with respect to the Santa Fé through Jejénes trend, the early, phaneritic holocrystalline phases of the Mandé-Acandí Batholith record similar major, minor and trace and rare-earth element (REE) trends, reflective of development and emplacement within similar, albeit younger, allochthonous, intra-oceanic basement. Notably, some of the younger porphyritic granitoids of the Pantanos-Pegadorcito area, and some of the phaneritic granitoid samples, return a much more evolved composition, richer in alkalies and silica, with enhanced trace and REE contents, and the incipient development of negative Eu anomalies. These observations are considered to be reflections of the longer-lived and mature development of the Mandé-Acandí arc, complete with generation of a coeval volcanic pile (Santa Cecilia-La Equis Fm.).

Based upon the forgoing data presentation and conclusions, variations in the age vs. lithogeochemical expression of the entire suite of Cretaceous through Eocene (i.e. ca. 100–ca. 40 Ma) age Eastern and Western Group granitoids of the Colombian Andes, as observed with the Jurassic-aged correlatives, are primarily a function of the nature and composition of the basement into which the granitoid magmas have been emplaced. Even more so, the specific tectonic environment and conditions at the time of granitoid emplacement, varying from continental and autochthonous to CCOP-/CLIP-related oceanic and allochthonous, are in evidence. In the case of the Cretaceous to Eocene granitoids, differences in granitoid composition are very much a reflection of the dynamic and changing tectonic conditions taking place primarily within the Pacific realm during the formulative phases of the Northern Andean Orogeny. We now present additional geochemical and isotopic data for the Cretaceous-Eocene granitoids of the Colombian Andes, in order to more precisely discuss the tectono-magmatic evolution of the region in the coming sections of this presentation.

Isotope Geochemistry

Sr-Nd Isotope Geochemistry: Results, Summary and Discussion

A relatively complete set of Sr-Nd isotope data, representative of both the Eastern Group and Western Group of Colombian Cretaceous to Eocene granitoids, is presented in Fig. 5.21. Data for both the Eastern and Western groups and their

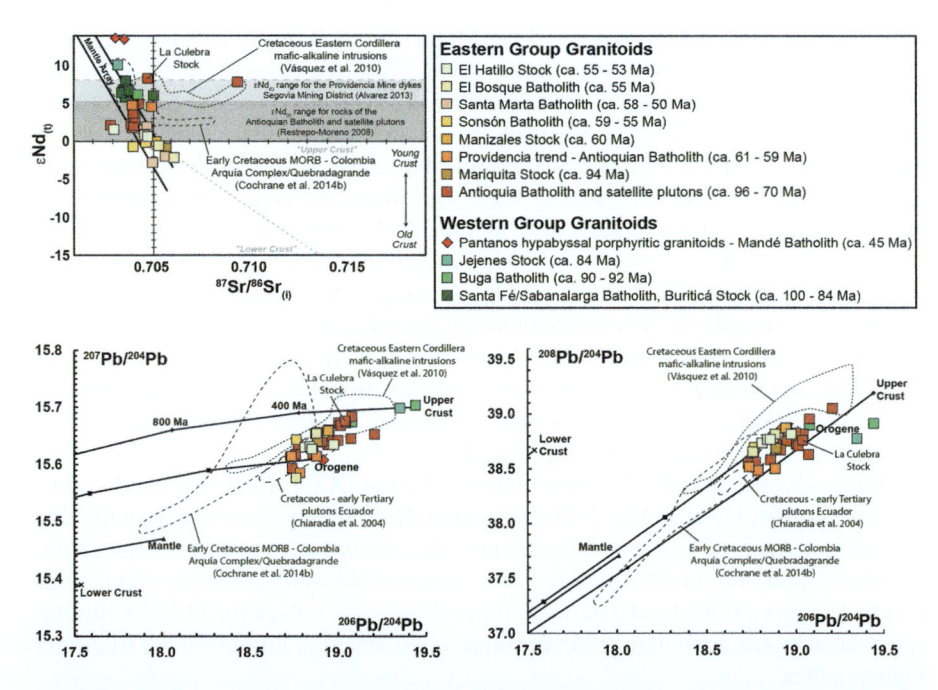

Fig. 5.21 Sr-Nd and Pb isotope plots for mid-Cretaceous through Eocene granitoids in the Colombian Andes. Additional data for igneous and metamorphic suites from the surrounding region are included for reference. Note the primitive isotopic composition of the Western Group oceanic granitoids vs. the continental granitoids of the Eastern Group. Notwithstanding, the strongly mantellic signature of the initial post-collisional Providencia suite in the eastern domain is noteworthy, compared to the evident crustal contamination observed for the remainder of the Eastern Group post-collisional suite. See text for discussion

respective subgroups, as described above, are clearly resolved, with essentially no observed Sr-Nd compositional overlap. For reference purposes, we have included two additional data sets within the context of Fig. 5.21: (1) data for early(?) Cretaceous, tectonically disrupted, continental margin MORB contained within the Romeral melange (Quebradagrande/Arquía units of Nívia et al. 1996) and (2) data for autochthonous early to mid-Cretaceous, rift-related mafic intrusions located within the Eastern Cordillera (Fabre and Delaloye 1983; Vázquez et al. 2010). In general terms, it is noted that all of the data sets represented in Fig. 5.21, including for both mafic and granitoid rock types, plot along the mantle array. Alteration of original Sr-Nd isotopic composition is suggested, particularly in terms of increasing $^{87}Sr/^{86}Sr_{(i)}$ ratios, within the Cretaceous MORB and Eastern Cordilleran mafic suites. The Cretaceous MORBs are highly tectonized and commonly demonstrate some degree of carbonitization, as do the Eastern Cordilleran intrusions, which have been exposed to post-emplacement basin dewatering events during late Cretaceous and Cenozoic basin inversion (Vázquez et al. 2010; Shaw et al. 2018).

With respect to the Eastern Group Cretaceous to Eocene granitoids shown in Fig. 5.21, Sr-Nd isotope values for the ca. 96–72 Ma Antioquian Batholith suite indicate a mantle-derived source (Leal-Mejía 2011). The Antioquian Batholith demonstrates a vertical array with similar initial $^{87}Sr/^{86}Sr_{(i)}$ ratios (0.70398–0.70455) and increasing $\varepsilon_{Nd(t)}$ values, from the ca. 87–82 Ma facies to ca. 72 Ma facies ($\varepsilon_{Nd(t)}$ = +1.74 to +4.77). The younger Paleocene-Eocene plutons of the Eastern Group similarly plot along the mantle array but show a clear shift to lower $\varepsilon_{Nd(t)}$ values and higher $^{87}Sr/^{86}Sr_{(i)}$ ratios, suggesting more evolved compositions with greater degrees of crustal interaction, as previously suggested by the whole-rock lithogeochemical data. The ca. 62–58 Ma Providencia suite, hosted within the confines of the Antioquian Batholith, provides an exception to this tendency, displaying clearly more juvenile Sr-Nd compositions. A crustal component becomes more evident in the Cretaceous Mariquita Stock, which, based upon regional geologic mapping, can be inferred to have interacted with disrupted continental basement of Mesoproterozoic(?) age, along the inferred contact between the Chicamocha and Cajamarca-Valdivia basement complexes (Fig. 5.15).

With respect to the allochthonous Western Group CCOP/CLIP granitoids, all of the analyses displayed in Fig. 5.21 plot within the mantle source region, reflecting their ubiquitous genesis and emplacement within the oceanic regime. Samples from the Mandé Batholith yielded clearly mantelic Sr and Nd isotope values, with the late porphyritic phases displaying particularly enriched $\varepsilon_{Nd(t)}$ values (to +15.26). Similar enrichment trends are noted within the more fractionated granitoids of the Buga and Jejénes suites.

Pb Isotope Geochemistry: Results and Summary and Discussion

Available lead isotope geochemical data for the Eastern and Western Group Cretaceous to Eocene granitoids of the Colombian Andes are also presented in Fig. 5.21. As with the Sr-Nd diagram, Pb data for early(?) Cretaceous Romeral melange MORB (Quebradagrande/Arquía units; Nívia et al. 1996), and for Eastern Cordillera rift-related mafic intrusions (Fabre and Delaloye 1983; Vázquez et al. 2010), have been included in the diagram. In addition, the data envelope for Paleozoic metasedimentary basement of the Loja Terrane, Ecuador (Chiaradia et al. 2004), as a proxy for the unknown values of the Cajamarca-Valdivia assemblage (Restrepo-Pace 1992; Cediel et al. 2003) which hosts the majority of the Eastern Group granitoids and formed the Colombian continental margin during early-mid-Cretaceous times, is shown.

Lead isotope geochemistry both in the ca. 96–72 Ma and the ca. 62–49 Ma subgroups of Eastern Group granitoids clusters in a relatively narrow range of values ($^{206}Pb/^{204}Pb$ = 18.74–19.21, $^{207}Pb/^{204}Pb$ = 15.58–15.68 and $^{208}Pb/^{204}Pb$ = 38.48–39.05), extending in a moderately steep array, between the Orogene and the upper crust lead evolution curves of Zartman and Doe (1981). A general tendency towards more radiogenic values, suggesting greater degrees of crustal interaction, between the earlier Antioquian Batholith suite and the Paleocene-Eocene suites of the Central Cordillera to the south, can be observed. Regardless, as with the lithogeochemical and Sr-Nd data, a clear return to less interacted, more mantelic compositions is reflected in the data from the ca. 62–58 Ma Providencia granitoid suite, contained within the confines of the Antioquian Batholith.

To the west, within the continental margin to oceanic domain, Cretaceous MORB of the Romeral mélange is observed to be notably radiogenic (Fig. 5.21b), in keeping with oceanic rocks formed along a marginal basin, receiving sediments and rifted fragments of the relatively radiogenic continental basement represented by the Cajamarca-Valdivia Terrane. The influence of continental margin sedimentation may be extrapolated to explain the only slightly less radiogenic Pb isotope composition of the ca. 100–82 Ma subgroup of the Western Group CCOP/CLIP-related granitoids (Fig. 5.21b) (the lead isotope results for which range from $^{206}Pb/^{204}Pb$ = 19.08–19.44, $^{207}Pb/^{204}Pb$ = 15.67–15.70 and $^{208}Pb/^{204}Pb$ = 38.77–38.91). This observation supports the formation of the ca. 100–82 Ma CCOP-/CLIP-related intra-oceanic arc in relatively close proximity to the Colombian continental margin, along a subduction trench which was receiving continentally derived sediment, prior to accretion. Farther west, available lead isotope analyses for samples from the Mandé Batholith (i.e. the Pantanos porphyry suite), however, depict a shift to less radiogenic compositions ($^{206}Pb/^{204}Pb$ = 18.92–18.96, $^{207}Pb/^{204}Pb$ = 15.61–15.64 and $^{208}Pb/^{204}Pb$ = 38.56–38.60). This feature can be interpreted to reflect development of the ca. 62–42 Ma Mandé(-Acandí) arc in a more distal intra-oceanic environment, isolated from the influence of significant volumes of radiogenic, continentally derived sediments.

Synthesis and Conclusions of Lithochemistry and Sr, Nd and Pb Isotope Geochemistry for Cretaceous-Eocene Granitoids

Spatial, temporal and lithogeochemical considerations, supported by whole-rock trace element, REE and Sr-Nd and Pb isotopic compositional data, for mid-Cretaceous through Eocene (ca. 100 through ca. 42 Ma) granitoids of the Colombian Andes, permit the identification and consistent characterization of two broad groups of granitoids: the Eastern and Western groups. All of the granitoids within both groups may be classified as volcanic arc-related granitoids; however, each group formed within a distinct litho-tectonic setting: the Eastern Group is comprised of autochthonous granitoids emplaced within a continental setting, whilst the Western Group is comprised of allochthonous, accreted arc granitoids, formed within an intra-oceanic setting. Each of these principal groups may be further subdivided into two individual subgroups, primarily based upon age constraints provided by the U-Pb (zircon) age database, although the lithogeochemical and isotopic evolution of each subgroup is reflective of the evolving petrogenesis and tectonic setting of the region.

The Cretaceous phases of the Eastern Group (Antioquian Batholith suite) are subduction-related continental arc (Cordilleran) granitoids with relatively limited degrees of crustal interaction, as recorded in their Sr-Nd and Pb isotope signatures. Following a ca. 10 Ma hiatus in continental granitoid magmatism, the Paleocene-Eocene Eastern subgroup granitoids to the south of the Antioquian Batholith show similar, although more evolved, lithogeochemical compositions and depict isotopic trends towards increasing crustal interaction (e.g. Bustamante et al. 2017). A marked exception to this trend remains the ca. 62–58 Ma Providencia suite, which displays a consistent return to more mantelic lithogeochemical, trace element, REE and isotopic compositions. This is an important observation from a metallogenic viewpoint, given that the largest Au resource presently outlined within the entire Cretaceous-Eocene granitoid suite of the Colombian Andes is paragenetically related to the Providencia granitoid suite (Leal-Mejía et al. 2010; Leal-Mejía 2011; Shaw et al. 2018).

In contrast to the Eastern Group granitoids, the Western Group plutons display more primitive lithogeochemical characteristics reflective of their generation within an intra-oceanic setting. Sr-Nd and Pb isotope data, however, suggest, at least locally, a significant crustal component, interpreted to represent the relative near proximity of the Colombian segment of the CCOP/CLIP intra-oceanic arc system and accreted Western Group granitoids, to the continental margin during the ca. 100–82 Ma time interval. The ca. 62–42 Ma Mandé(-Acandí) subgroup of the Western granitoids is more reflective of primitive intra-oceanic lithogeochemical compositions, albeit with a more highly evolved component, as recorded by the composition of younger (ca. 46–42 Ma) porphyritic stocks observed at Pantanos and elsewhere along the Mandé-Acandí trend. From a Sr-Nd isotope standpoint, available samples of the Mandé suite are consistently mantle-derived, and Pb-isotope compositions are significantly less radiogeneic than granitoids of the ca. 100–82 Ma subgroup. From a tectonic standpoint, the Mandé-Acandí arc was generated upon CCOP crust representing the trailing edge of the CCOP plateau, following the accretion of most of the Western Tectonic Realm (Cañas Gordas and Dagua terranes) in the late Cretaceous (Cediel et al. 1994; Cediel et al. 2003, 2010; Spikings et al. 2015). In this context, the Colombian Pacific margin would have been comprised of accreted oceanic materials during genesis of the Paleocene-Eocene Mandé-Acandí granitoids, and, unlike the ca. 100–82 Ma granitoids, the Mandé arc system would have been shielded from exposure to continentally derived sediments of the Cajamarca-Valdivia Terrane, accounting for its less radiogenic Pb isotope signatures.

5.3.5 Latest Oligocene to Pliocene Granitoid Magmatism: Distribution, Age and Nature

5.3.5.1 Introduction

Latest Oligocene to Pliocene granitoid magmatism is widely but irregularly distributed throughout the Colombian Andes. Unlike the magmatic periods described previously, large masses of holocrystalline granitoids of Oligo-Miocene age are not exposed in outcrop within the Colombian Andes, certainly due to factors related to uplift and erosional level. Notwithstanding, small batholiths and stocks of holocrystalline rocks and abundant hypabyssal porphyry clusters and volcanic deposits are widespread, especially within the western Andean ranges and intermontane valleys, and testify to the active development of granitoid magmatisn during the tectonic development of the Colombian Andes during the latest Oligocene and Miocene.

As with previous magmatic periods, latest Oligocene to Pliocene magmatism manifests as a complex distribution of spatially constrained arc segments of differing ages.

Unlike the Cretaceous to Eocene intrusions, however, all latest Oligocene to Pliocene magmatic rocks are considered to have been emplaced broadly in situ, into continental margin metamorphic and accreted oceanic basement rocks juxtaposed prior to and during the various phases of the Northern Andean Orogeny (see Sect. 5.4.4). Latest Oligocene to Pliocene magmatism in Colombia merges, both temporally and spatially, into the modern-day Pleistocene to Recent volcanic arc of the Northern Andes. The distribution, nature and constraints upon Pleistocene to Recent magmatism are well documented in various geological compilations (e.g. Cediel and Cáceres 2000; Cediel et al. 2003; Stern 2004; Gómez et al. 2015a; Marín-Cerón et al. 2018) and will not be discussed in detail herein.

Based primarily upon textural characteristics, three broad categories of latest Oligocene to Pliocene granitoid rocks can be distinguished. These include (1) holocrystalline, phaneritic granitoids forming stocks and small-scale batholiths, (2) high-level hypabyssal porphyritic rocks occurring as clusters of small-scale stocks and dikes and (3) volcanic rocks, including flows and pyroclastic sequences. Spatially and genetically related volcanic sequences are generally absent in the vicinity of the deeper-level phaneritic intrusive suites, whilst the hypabyssal porphyry stocks may at least locally intrude a penecontemporaneous volcanic edifice. Erosional factors may account for the lack of a volcanic component in the deeper-level plutons, although in some instances it is possible that a volcanic component was never developed.

5.3.5.2 Distribution

The great majority of latest Oligocene to Miocene granitoids in Colombia are concentrated within the western ranges and along the intermontane valleys of Colombian Andes, with only minor, isolated occurrences of high-level granitoids of late Miocene and Pliocene age having been documented in the Eastern Cordillera and Santander Massif.

The distribution of latest Oligocene to Miocene holocrystalline plutons is primarily observed in the physiographic Western Cordilleras of Colombia, where numerous small batholiths and stocks are observed (Fig. 5.22). In the southwest these include the Piedrancha and Cuembí batholiths and associated minor stocks observed at El Vergel, La Llanada and Cumbitara, all of which intrude CCOP-/CLIP-related oceanic volcanic and sedimentary rocks of the Dagua and Diabásico Groups (Arango and Ponce 1982). Regional mapping suggests that this trend of plutons extends northwards into the region to the west of Cali (Fig. 5.22), where little modern radiometric age or lithogeochemical data are available. Farther to the north, within the confines of the Chocó Arc, the holocrystalline Farallones Batholith, the Urrao pluton and the El Cerro Igneous Complex are observed to intrude Cañas Gordas Group basement to the west of the Middle Cauca River valley (Rodríguez and Zapata 2012; Zapata and Rodríguez 2013). The northern extension of this trend can only be inferred based upon the appearance of small plutons within the regional mapping database (González 2001; Gómez et al. 2015a).

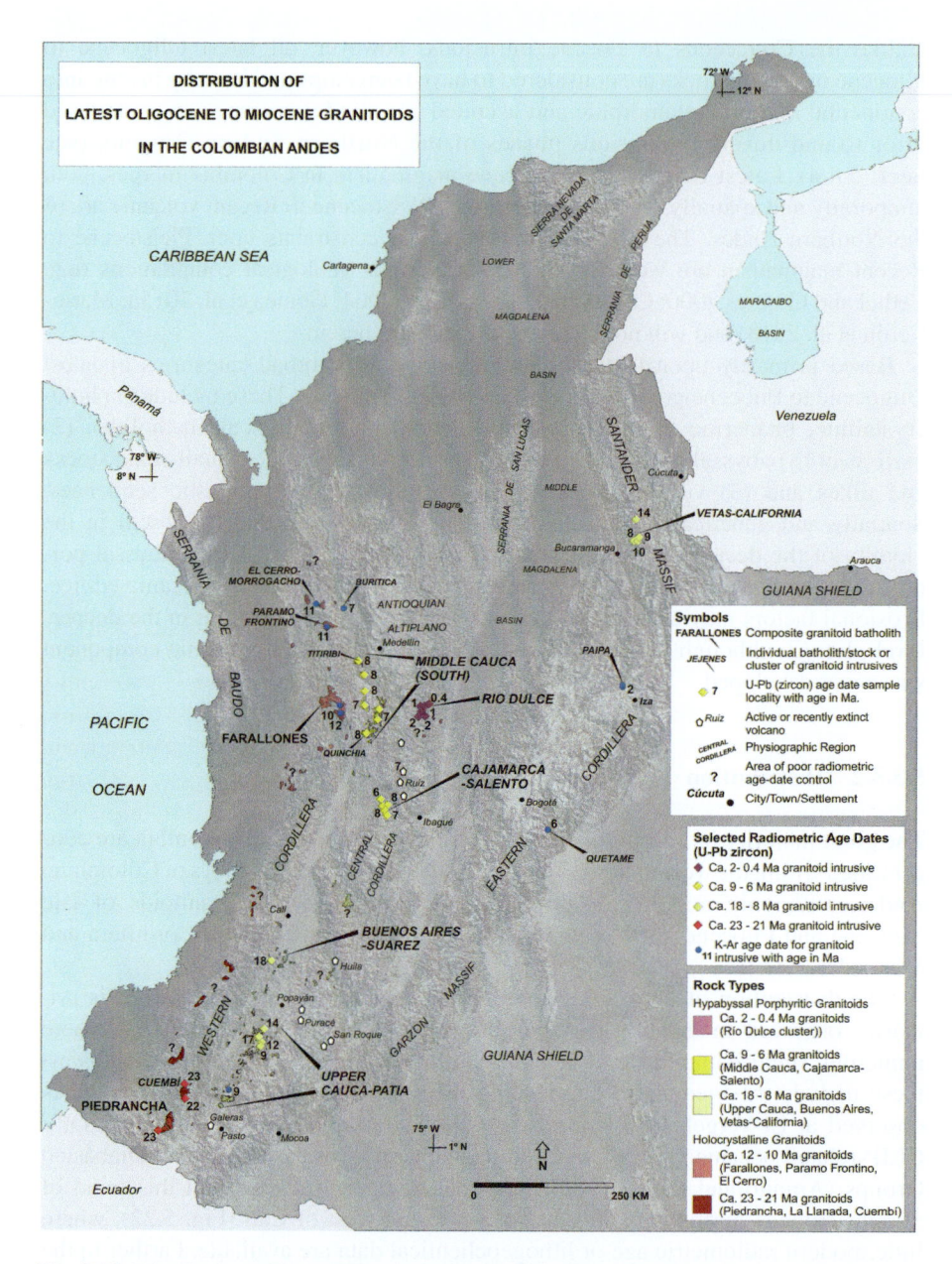

Fig. 5.22 Distribution of latest Oligocene through Mio-Pliocene granitoids in the Colombian Andes. Principal modern-day physiographic provinces of the region are shown for reference. (Granitoid shapes modified after Cediel and Cáceres 2000; Gómez et al. 2007; Gómez et al. 2015a)

With respect to the hypabyssal porphyritic granitoid stocks and dikes and associated volcanic rocks, these clusters are particularly well exposed along the eastern margins of the Western Cordilleras and the western margin of the Central Cordillera, within/along the physiographic depressions of the Patía and Upper Cauca drainage basins (Fig. 5.22). Additional, Miocene and Pliocene hypabyssal porphyry clusters are observed within the Central Cordillera, the Santander Massif and the Eastern Cordillera (including the Quetame Massif).

In the south, along the Patía and Upper Cauca drainage, a ca. 200 km long SW-NE trending series of Miocene hypabyssal porphyry clusters extends from Arboledas (Berruecos) to the area of Cerro Bolívar and Almaguer-La Vega (Betulia Igneous Complex) and northwards through Altamira, Dominical, Piedra Sentada and La Sierra. It is possible that this belt continues farther north beneath Plio-Pliestocene to Recent volcanic cover around Popayán. Another cluster of hypabyssal stocks appears in the Upper Cauca basin to the north of Popayán, at Santander de Quilichao-Buenos Aires-Suárez (París and Marín 1979).

Farther north, along both the eastern and western margins of the Middle Cauca River valley to the north of Pereira, numerous clusters of hypabyssal granitoid porphyritic stocks and dikes are observed over a ca. 100 km long N-S trending magmatic belt (González 1990, 1993, 2001). Along this belt (Fig. 5.22), porphyry intrusives are commonly observed, either as isolated plugs or volumetrically more significant clusters of stocks and dikes. From south to north, some of the more important clusters of stocks outcrop near Marsella, at Manizales-Villa María, from Ansermas to Quinchía and from Río Sucio to Supía, La Felisa, Marmato and Valparaíso-Caramanta at Támesis, around Jericó (the Quebradona cluster) and from Venecia-Fredonia north to Titiribí. Spatially coincident with the Middle Cauca River valley trend are the thick volcanic, pyroclastic and volcaniclastic sequences of the late Miocene Combia Fm., which also outcrop in the Middle Cauca River region, primarily to the west of the river between Anserma in the south and Jericó-Tarzo in the north. The Combia Fm. has also been mapped around Venecia-Fredonia and Titiribí. Combia is considered to be, in part at least, the extrusive expression of the late Miocene porphyry centres.

To the north of Titiribí, hypabyssal porphyry centres become scarce and isolated and are of generally only inferred to be late Miocene age. This observation may be in part a facet of the difficulty in recognition of these generally small intrusives in regional mapping, under heavy vegetation cover and deep tropical soil profiles. Isolated porphyritic granitoid stocks are observed to the west of Anzá and around Buriticá, where late Miocene ^{40}Ar-^{39}Ar ages have been published (Lesage et al. 2013). This northern section of the Middle Cauca porphyry belt may extend through Peque and as far north as Puerto Libertador, where isolated, small, holocrystalline and porphyritic granitoid stocks and dikes are observed at El Alacran and Montiel. The age of these last units, however, has yet to be established.

Within the Central Cordillera, a significant cluster of Miocene porphyritic granitoids is observed at Cajamarca-Salento, including the Colosa porphyry (Lodder et al. 2010) and other surrounding hypabyssal intrusives extending as far east as the Toche river. This cluster may extend to the north where it would be covered by volcanic and pyroclastic rocks of the modern-day Northern Andean volcanic arc. Farther to the north, within the Central Cordillera, an additional cluster of hypabyssal granitoids and associated pyroclastic rocks appears in the Manzanares-Samaná-Nariño (Antioquia) region. This cluster was referred to as the Río Dulce suite (Fig. 5.22) by Leal-Mejía (2011). Isolated high-level volcanic occurrences extend as far east as Norcasia.

Within Colombia's eastern cordilleran system, magmatic rocks of Miocene to Pliocene age are very scarce and only punctually developed. A localized cluster of hypabyssal granitoids is located in the Vetas-California area of the Santander Massif (Fig. 5.22; Mantilla et al. 2009; Leal-Mejía 2011; Bissig et al. 2014). To the south, minor, isolated, high-level porphyritic granitoids and associated volcanic flows and pyroclastic rocks are observed in the Eastern Cordillera at Paipa and Iza (Garzón 2003; Pardo et al. 2005a, b; Vesga and Jaramillo 2009), whilst similar occurrences are observed to the south and east of Bogotá at Quetame (Ujueta et al. 1990).

The distribution and nature of Pleistocene to Recent magmatism in Colombia is well documented and readily observed on regional geological compilations such as those presented by Cediel and Cáceres (2000) and Gómez et al. (2015a). This volcanic chain, including its extension to the south into Ecuador, is comprised of about 75 active volcanoes which are recognized within the Andean geological literature as the Northern Volcanic Zone (e.g. Stern 2004). In Colombia, this magmatic arc is primarily manifested along the Central Cordillera and the Patía-Upper Cauca River physiographic depression, where extensive volcanic and pyroclastic deposits are related to active volcanic edifices. Volcanism is dominated by lavas and pyroclastic rocks of bas-andesitic, andesitic and dacitic and occasionally basaltic composition.

Stern (2004) distinguished three separate segments comprising the active Colombian volcanic arc, including (1) the northern segment (Cerro Bravo, Santa Isabel, Nevado del Ruíz, Nevado del Tolima, Cerro Machín volcanoes), (2) the central segment (Nevado del Huila, Puracé, Sotará volcanoes) and (3) the southern segment (Cumbal, Azufral, Galeras, Doña Juana volcanoes). The northern and central segments are located in the Central Cordillera, whereas the southern segment is located in the Patía-Upper Cauca River depression and along the eastern margin of the southern Western Cordillera.

Cediel et al. (2003) recognized that the regional-scale structural architecture of the Northern Andes plays a fundamental role in the distribution and localization of volcanic edifices in Colombia. They documented the coincidence of volcanic cone and segment ("sub-chain") locations, with the trace of the various paleo-suture systems active in the tectonic assembly of the Northern Andean region since the mid-Proterozoic. In Colombia five sub-chains were defined, including, from west to east, the Cauca sub-chain (Chiles-Cumbal-Azufral-Olaya volcanoes), the inter-Andean sub-chain (Galeras-Morazurco volcanoes), the Romeral-Peltetec sub-chain (La Victoria-Chimbo-Bordoncillo-Doña Juana-Sotará-Puracé volcanoes), the Palestina sub-chain (La Horqueta-Paletará-Huila-Tolima-Ruíz-Herveo volcanoes) and the Suaza sub-chain (Guamués-Acevedo volcanoes).

5.3.5.3 Age Constraints on Miocene to Pliocene Granitoid Magmatism

Published U-Pb (zircon) data or information regarding the precise age of Miocene-Pliocene magmatism in the Colombia Andes is surprisingly limited (e.g. Maya 1992; Frantz et al. 2003; Tassinari et al. 2008; Mantilla et al. 2009; Lodder et al. 2010; Henrichs 2013; Lesage et al. 2013; Bissig et al. 2014). Based primarily upon older K-Ar (whole-rock or mineral separate) analyses, previous work and published geological maps in general refer to interpreted Miocene- to Pliocene-aged magmatic rocks as being "Neogene" in age, thus historically precluding any form of detailed analysis of the spatial appearance and migration of Miocene to Pliocene magmatic rocks with time.

More recently, Leal-Mejía (2011) presented a study containing 24 new U-Pb (zircon) age determinations for Miocene-Pliocene holocrystalline granitoid intrusions and hypabyssal porphyritic stocks, backed by various new K-Ar mineral separate and whole-rock dates. This work permits a more precise time-space analysis and understanding of the evolution of Miocene and Pliocene granitoid magmatism throughout the Colombian Andes. A histogram depicting the age distribution of Miocene to Pliocene magmatic rocks based upon recent U-Pb (zircon) age determinations is presented in Fig. 5.23.

With respect to the holocrystalline, phaneritic plutons, the U-Pb (zircon) data of Leal-Mejía (2011) indicates the oldest, ca. 23 Ma magmatic rocks are located in southwest Colombia, extending from the Piedrancha Batholith north to El Vergel, La Llanada and Cumbitara (Cuembí) plutons. These plutons record a latest Oligocene to early Miocene magmatic event spanning the ca. 24–21 Ma interval. Analysis of the Piedrancha Batholith near Piedrancha yielded an Oligo-Miocene magmatic crystallization age of ca. 23.4 Ma. To the north the El Vergel Stock returned an early Miocene magmatic crystallization age of ca. 21.9 Ma. The Cumbitara Stock returned a crystallization age of ca. 23.1 Ma. No inheritance ages were observed in any of the samples. As noted above, based upon available mapping, sporadic plutons, which could be of similar age, are mapped extending northwards along the trend of the Western Cordillera into the region to the west of Cali (Fig. 5.22).

Farther to the north, for the holocrystalline, phaneritic intrusives including the Farallones Batholith and the Urrao pluton, no new U-Pb (zircon) dates are available. These units have historically been dated by the K-Ar (hornblende) method and have returned ages including 11 ± 2 Ma (Calle et al. 1980; Zapata and Rodríguez 2013) and 11 to 12 Ma (Botero 1975), respectively. To the north of Urrao, along an approximate N-S axis, Leal-Mejía (2011) recorded a K-Ar (biotite) date of ca. 11.8 Ma for the El Cerro Igneous Complex. Regional mapping suggests this trend of plutons may extend further northwards into the region to the north of Dabeiba (González 2001).

With respect to the porphyritic granitoid suites, U-Pb (zircon) analyses to date have revealed ages ranging from Miocene to Plio-Pleistocene. Consideration of the composite data set reveals a complex time-space distribution of these rock types.

Beginning in southwestern Colombia, along the Patía and Upper Cauca drainage, the trend of porphyritic stocks and dikes, extending from Arboledas (Berruecos) in the south to Almaguer-La Vega (Betulia Igneous Complex), and Piedra Sentada and La Sierra returned ages spanning the ca. 17–9 Ma interval (Leal-Mejía 2011). The oldest magmatic crystallization age was obtained from the Dominical porphyry, which returned an age of ca. 17 Ma. The Cerro Gordo porphyry returned a magmatic crystallization age of ca. 14 Ma. Holocrystalline hornblende biotite tonalite from La Dorada and tonalite porphyry from Altamira returned similar ages of ca. 11.8 Ma and 11.6 Ma, respectively, whilst a hornblende diorite porphyry from La Dorada returned a magmatic age of ca. 9.2 Ma. To the north of Popayán, the northern Cauca Department hypabyssal porphyry cluster at Santander de Quilichao-Buenos Aires-Suárez was also dated by Leal-Mejía (2011). Diorite porphyry from near Suárez returned a magmatic crystallization age of ca. 17.7 Ma, which compares well with the age of similar porphyries from the Dominical area to the south.

The hypabyssal porphyry suites of the Middle Cauca return distinctly younger ages than those of the Patía-Upper Cauca trend. Hypabyssal granitoids along the Middle Cauca extend from Marsella in the south to Titiribí in the north. Available U-Pb (zircon) data span the range between ca. 9 and 4 Ma (Fig. 5.23). Quartz diorite porphyry from Quinchía (Dos Quebradas) returned a U-Pb (zircon) age of ca. 8.0 Ma (Leal-Mejía 2011). Diorite porphyry from the Marmato area returned a ca. 6.5 Ma age (Frantz et al. 2003). Granodiorite of the Támesis Stock, sampled near Támesis, returned a U-Pb (zircon) age of ca. 7.2 Ma (Leal-Mejía 2011), in marked contrast to historic K-Ar age data which suggested a Cretaceous age (Maya 1992). At Yarumalito, Henrichs (2013) provided U-Pb (zircon) crystallization ages of 7.0 ± 0.15 Ma and 6.95 ± 0.16 Ma for samples of andesite and diorite, respectively. North of Yarumalito, Leal-Mejía (2011) provided various additional U-Pb (zircon)

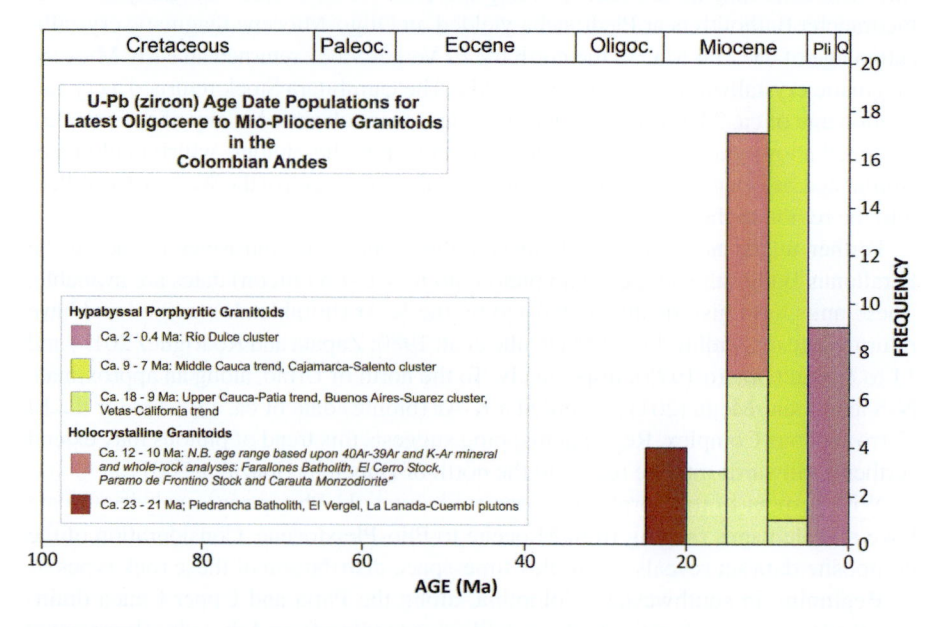

Fig. 5.23 U-Pb (zircon) and selected K-Ar and ^{40}Ar-^{39}Ar age date populations for latest Oligocene through Mio-Pliocene granitoids in the Colombian Andes

age dates: diorite porphyry at La Aurora within the Jericó (Quebradona) porphyry cluster, 8 km to the SE of the Jericó townsite, yielded a U-Pb (zircon) age of ca. 8 Ma. The La Mina diorite porphyry, located about 5 km to the south of Venecia, produced an age of ca. 7.6 Ma, and hornblende granodiorite porphyry from the El Medio creek, located a few hundred metres to the SW from the Titiribí townsite, yielded a ca. 7.6 Ma U-Pb (zircon) age.

To the north of Titiribí, hypabyssal porphyritic granitoid magmatism along the Middle Cauca becomes less apparent and has only been documented in isolated, mostly undated occurrences. Leal-Mejía (2011) provided an 11.8 ± 1.1 Ma K-Ar (magmatic hornblende) age for hornblende diorite porphyry at Buriticá. ^{40}Ar-^{39}Ar step heating analysis of similar porphyry by Lesage et al. (2013) produced a hornblende cooling age of 7.41 ± 0.4 Ma, suggesting various phases of porphyritic diorites may be present. Elsewhere, along the general northern extension of the Middle Cauca trend, porphyritic granitoids of unconfirmed late Miocene age are observed to the west of Anzá, north of Dabeiba, near Peque and to the SW of Puerto Libertador, at El Alacrán and Montiel (Teheran), where small, isolated quartz diorite stocks and porphyritic dikes intrude volcano-sedimentary basement rocks of the San Jacinto terrane. Most of these occurrences are too small and isolated to be resolved at the scale of available regional geologic maps (e.g. Cediel and Cáceres 2000; Gómez et al. 2015a). For the purposes of our analysis, these occurrences are considered to represent the northern extension of granitoid magmatism along the Middle Cauca trend, and we tentatively assign these rocks a late Miocene age of emplacement.

The volcanic and pyroclastic sequences of the Combia Fm. are exposed throughout the Middle Cauca River valley region (Fig. 5.22), and many of the Middle Cauca trend hypabyssal porphyry intrusives are hosted within the Combia Formation. Detailed geological mapping suggests the Combia Fm. records a transition from Oligo-Miocene siliciclastic sedimentation along the paleo-Cauca and Sinifaná basins (i.e. Amagá Fm.) to active granitoid magmatism during the middle to late Miocene. Historic K-Ar dating places the Combia suite at ca. 9.1 Ma (Restrepo et al. 1981 in Toro et al. 1999), although it is uncertain what level of the volcanic pile this would represent. Based upon stratigraphic and cross-cutting relationships in relation to the ca. 7 Ma hypabyssal stocks near Yarumalito, Henrichs (2013) concluded that porphyry emplacement was closely related to the final stages of Combia Fm. volcanism. Leal-Mejía (2011) dated an andesite of the upper Combia Formation outcropping near Támesis at ca. 6.1 Ma (K-Ar, whole-rock). This is in broad agreement with the ca. 6 Ma age of "Combia volcanism" suggested by Ramírez et al. (2006) and with K/Ar and ^{40}Ar-^{39}Ar mineral and whole-rock age dates produced by various studies and compiled by Rodríguez and Zapata (2014). In reality, the Combia Fm. is an extensive volcano-sedimentary unit locally exceeding 1000 meters in stratigraphic thickness. It is cut by many of the hypabyssal granitoids listed above indicating that volcanism initiated before ca. 8 Ma and may have continued to as late as ca. 4 Ma should volcanism have accompanied the full range of hypabyssal porphyritic granitoid magmatism as indicated by available U-Pb (zircon) age dates. Notwithstanding, lithogeochemical data presented below suggest that the early, mafic (bas-andesitic) phases of the Combia Fm. may be related to the ca. 12–10 Ma Farallones Batholith suite.

To the south and west of the Middle Cauca region, within the Central Cordillera, the Cajamarca-Salento hypabyssal porphyry cluster (Fig. 5.22), including various granitoid porphyry bodies outcropping near Cajamarca, Tierradentro, Montecristo and Salento, was dated by Leal-Mejía (2011). In general, results span the ca. 8.3–6.3 Ma range. Magmatic crystallization ages for the hypabyssal intrusive suite at La Colosa (see Lodder et al. 2010) yield ages spanning the ca. 8.3–7.3 Ma interval. Three early diorite porphyries returned U-Pb (zircon) ages between ca. 8.3 and 7.9 Ma, whilst paragenetically later granodiorite porphyries yielded slightly younger ages of ca. 7.6 Ma and 7.5 Ma. A latest phase of quartz porphyry returned a 7.3 Ma magmatic crystallization age. Inheritance ages obtained from zircon crystals from the La Colosa porphyries span a wide range between ca. 1060 and 13 Ma. Elsewhere, hypabyssal granitoid porphyry from La Morena and Tierradentro returned ages of ca. 8.4 Ma and 8.1 Ma, respectively. A quartz diorite porphyry from the Montecristo area returned a ca. 7.6 Ma U-Pb (zircon) age, and granodiorite porphyry collected near Salento returned ca. 6.3 Ma.

Continuing along the Central Cordillera, to the NNE of the Cajamarca-Salento hypabyssal porphyry cluster, beyond active volcanic cover provided by the Tolima-Santa Isabel-Ruíz volcanic complex, the Plio-Pleistocene Río Dulce porphyry cluster (Fig. 5.22) was also revealed by U-Pb (zircon) age dating completed by Leal-Mejía (2011). Eight U-Pb analyses were provided at Río Dulce, for the suite of hypabyssal granitoids and associated pyroclastic volcanic rocks. Magmatism spanning the 2.4 Ma to 0.4 Ma interval was recorded, in at least three distinct magmatic pulses, including at 2.4 to 2.3 Ma, 1.2 to 1.0 Ma and 0.4 Ma. The oldest magmatic ages at Río Dulce are revealed in two granodiorite porphyry samples collected to the SSE from the Nariño (Antioquia) townsite. Both returned the same age of ca. 2.4 Ma. A porphyry fragment from a nearby intrusive breccia returned an age of ca. 2.3 Ma. Diorite porphyry which seems to cut the breccia also returned a ca. 2.3 Ma age. A second magmatic pulse was recognized in the Espíritu Santo-Santa Bárbara porphyry, located about 5 km to the SE of the Nariño townsite. Quartz diorite porphyry from Espíritu Santo hill returned an age of ca.1.0 Ma. In addition, two samples collected along the Espíritu Santo Creek, about 3 km to the east, returned similar ages of ca. 1.2 Ma and 1.0 Ma. Finally, in the northern Río Dulce area, diorite porphyry collected at La Cabaña hill, about 12 km to the east of the Nariño townsite, returned a Pleistocene magmatic age of ca. 0.4 Ma.

Overall, ages for the hypabyssal intrusive rocks of the Río Dulce area porphyry suite obtained by Leal-Mejía (2011) are compared well with Pliocene to Pleistocene K-Ar ages (Maya, 1992) and geological relationships established for the nearby Tolima-Ruíz volcanic complex, suggesting that the apparently extinct Río Dulce cluster formed, in its time, the northernmost extension of the modern-day Northern Andes volcanic arc. Further north and along the trend of the active Northern Andes volcanic zone, within the Central Cordillera (including the Serranía de San Lucas), no additional manifestations of Miocene or younger granitoid magmatism have been documented or established using modern radiometric age dating techniques (Fig. 5.22).

Within Colombia's Eastern Cordilleran system, including within the Sierra Nevada de Santa Marta, Serranía de Perijá, Santander Massif, Eastern Cordillera and Garzón Massif, Miocene and younger granitoid magmatism is far scarcer than within the Central and Western Cordilleran ranges. In this context, documented granitoids or their volcanic equivalents, of confirmed Miocene or younger age, constitute less than a fraction of one percent of the geological record of the entire eastern Colombian Andes. Notwithstanding, localized occurrences of granitoid rocks have in recent years been documented, and their age and occurrence contribute to an understanding of the magmato-tectonic evolution of the region as a whole.

Within the Santander Massif, a NNE-oriented generally linear trend of late Miocene hypabyssal granitoids along the Vetas-California mining district has recently been documented in detail by various authors including Mantilla et al. (2009), Leal-Mejía (2011) and Bissig et al. (2014). Mantilla et al. (2009) presented U-Pb (zircon) data for two samples of granitoid porphyry from near the Vetas and California townsites. Results yielded magmatic crystallization ages of ca. 9.0 Ma and 8.4 Ma, respectively. Leal-Mejía (2011) dated hypabyssal granodiorite porphyry cropping at the San Celestino Mine near California, yielding a ca. 10.2 Ma U-Pb (zircon) age. Both the above cited studies noted various inheritance ages reflected in zircon grains, including at ca. 30 Ma, 50 Ma, 180 Ma and 200 Ma. Mantilla et al. (2009) related the older ages to inherited zircons from metamorphic and magmatic rocks in the area, including the granitoid rocks of the Santander Plutonic Group, and explained the younger inheritance ages as possible age mixing in zircon grains with more complex internal structure or inherited zircon grains from younger unidentified magmatic pulses in the area.

In recent studies more directly pertaining to gold mineralization along the Vetas-California trend, Rodríguez (2014) and Rodríguez et al. (2017) produced $^{40}Ar/^{39}Ar$ studies returning Pliocene to Pleistocene plateau ages ranging from ca. 3.9 to 1.2 Ma, for alteration minerals (primarily alunite) contained within and overprinting the Miocene porphyritic granitoids and Proterozoic through Jurassic basement rocks of the Santander Massif. Although U-Pb (zircon) age data for porphyritic granitoids from the sector has yet to produce such young magmatic crystallization ages, both of the foregoing studies conclude that the Plio-Pleistocene $^{40}Ar/^{39}Ar$ ages for alteration alunite reflect the presence of coeval granitoid magmatism, located at shallow depths in upper crustal levels beneath the district, which has yet to be revealed by uplift and erosion.

Some 175 km to the south of Vetas-California, within the domain of Colombia's Eastern Cordillera (sensu stricto), punctual occurrences of high-level porphyritic granitoids and associated flows and pyroclastic rocks are known at Paipa-Iza. K-Ar and $^{40}Ar/^{39}Ar$ radiometric dates at Paipa span the range between ca. 2.4 and 1.8 Ma; however, field evidence suggests volcanic activity continued into more recent times (Cepeda and Pardo-Villaveces 2004; Pardo et al. 2005a). An additional 180 km further to the south, similar granitoids to those observed at Paipa, although of more limited aerial extent, are located near the town of Quetame. A K-Ar (whole-rock) age of ca. 5.6 Ma was recorded for the Quetame granitoids by Ujueta et al. (1990).

Discussion and Synthesis of Spatial Distribution of Oligocene to Pliocene Granitoids

Unlike some of the Cretaceous through Eocene magmatic arc segments of the Colombian Andes, tectono-magmatic analyses (e.g. Apsden et al. 1987; Cediel et al. 2003; Leal-Mejía, 2011) indicate that all of the latest Oligocene through Pleistocene granitoid arc segments and isolated granitoid occurrences, summarized above, are autochthonous within the Miocene tectonic configuration and evolution of the region. In this context, the spatial vs. temporal evolution of continental granitoids during the Neogene is linked to pre- and syn-Neogene crustal architecture and intra-continental stress field evolution, coupled with the nature of oceanic vs. continental plate interactions along the Pacific and Caribbean margins. As our magmato-tectonic analysis presented in Sect. 5.4.4.2 will discuss, we consider all of the major latest Oligocene through Pliocene granitoid arc segments discussed above to be related to the westward subduction of Pacific (Nazca) Plate crust, beginning in the late Oligocene.

The spatial vs. temporal relationships of latest Oligocene through Plio-Pleistocene granitoids are revealed by regional geological mapping (e.g. Cediel and Cáceres 2000; Gómez et al. 2015a), in conjunction with relatively recent U-Pb (zircon) and other age date studies cited above. The data reveal that the oldest granitoids from this time period include holocrystalline quartz-diorite and granodiorite of the ca. 24–21 Ma Piedrancha-Cuembí trend. This arc segment was apparently relatively short-lived and extinct, prior to the eastward migration of arc axial magmatism into the hypabyssal granitoid porphyry-dominated, ca. 17–9 Ma Upper Cauca-Buenos Aires-Suarez arc segment. Pleistocene to modern-day arc magmatism currently overprints the southern portion of the Upper Cauca porphyry belt, as manifest in active volcanoes such as Galeras and Doña Juana, and continues to migrate eastwards as manifest in volcanic fields around San Roque and the Nevado del Huila.

Following closure of the Piedrancha-Cuembí arc segment and in the later phases of Upper Cauca magmatism, holocrystalline granitoid magmatism abruptly reappears in the northern Chocó Arc sector, of the physiographic Western Cordillera, intruding the Cañas Gordas Terrane basement complex. Farallones-El Cerro is a spatially and temporally distinct arc, unrelated to Piedrancha-Cuembí-Upper Cauca. It formed to the north of paleo-transform faults within the Pacific-Nazca Plate (Fig. 5.22) and represents the results of the differential interaction of a segmented subducting Nazca Plate along the Colombian Pacific margin (Sect. 5.4.4.2). The ca. 12–10 Ma Farallones-El Cerro segment was also short-lived, as arc axial magmatism migrated eastwards into Romeral melange basement, with the appearance of widespread porphyritic granitoids and associated volcanic rocks of the upper Combia Fm. along the belt-like Middle Cauca arc segment, between ca. 9 and 4 Ma. Simultaneous magmatism beginning at ca. 9 Ma was recorded to the ESE of the Middle Cauca, within the Cajamarca-Salento porphyry cluster, hosted by Cajamarca-Valdivia metamorphic basement rocks of the Central Cordillera. A similar cluster of porphyritic granitoids of Plio-Pleistocene age appears at Río Dulce, to the NNE of Cajamarca-Salento. Both Cajamarca-Salento and Río Dulce are broadly coaxial with the northernmost segment of the active Northern Andes volcanic arc (Stern 2004), and no additional manifestations of active Andean volcanism are manifest north of Río Dulce (ca. 5° 30'N). Additional discussion of the development and migration of Mio-Pliocene arc segments with respect to the evolution of the Nazca Plate along the Colombian Pacific margin will be presented in Sect. 5.4.4.2.

Miocene through Plio-Pleistocene occurrences of granitoid magmatic rocks within Colombia's eastern cordilleran system are widely spaced and volumetrically minor, and current data does not permit the temporal or spatial correlation of the granitoids within the geological context of the region or with respect to the volumetrically much more abundant Neogene granitoids which are widespread throughout the physiographic Central and Western Cordilleras. Additional discussion of the Vetas-California, Paipa and Quetame granitoids will also be supplied in the magmato-tectonic analysis of the region presented in Sect. 5.4.4.2.

5.3.5.4 Lithogeochemical and Isotopic Characteristics of Oligocene to Pliocene Granitoids

Latest Oligocene to Pliocene granitoids in the Colombian Andes include a texturally diverse suite of medium to coarse-grained, holocrystalline, phaneritic equigranular and porphyritic rock types. Detailed lithogeochemical information for the suite has only recently become available. Historic data, including basic major element analyses from limited localities, was presented by Alvarez (1983). More recently, various authors have provided more complete analysis of the Miocene and Pliocene suite, including samples from most of the important arc segments and porphyry clusters throughout the Colombian Andes. Data from Gil-Rodríguez (2010), Leal-Mejía (2011), Rodríguez and Zapata (2012), Lesage et al. (2013), Zapata and Rodríguez (2013), Bissig et al. (2014), Gil-Rodríguez (2014) and Cruz et al. (2014) are included in the present data set.

In terms of whole-rock lithogeochemistry, 150 samples of latest Oligocene to Plio-Pleistocene holocrystalline and hypabyssal granitoids are presented herein (Figs. 5.24, 5.25, 5.26, and 5.27). Some 43.3% (65 samples) of the sample set are considered altered, considering element mobility criteria described in Sect. 5.3.1.2. This is consistent with the observation that many of the altered samples were collected in areas subject to hydrothermal alteration related to porphyry-style and/or epithermal mineralization. Altered samples are represented by unfilled symbols in the lithogeochemical plots. The sample set as a whole is considered to be representative of the major Miocene to Pleistocene arc segments of the Colombian Andes.

The Colombian latest Oligocene to Plio-Pleistocene granitoids are medium- to high-K calc-alkaline in composition. Most of the main holocrystalline granitoids and hypabyssal stocks and dikes are metaluminous, with exception of samples of Miocene porphyritic granitoids of the Santander Massif, which exhibit clear

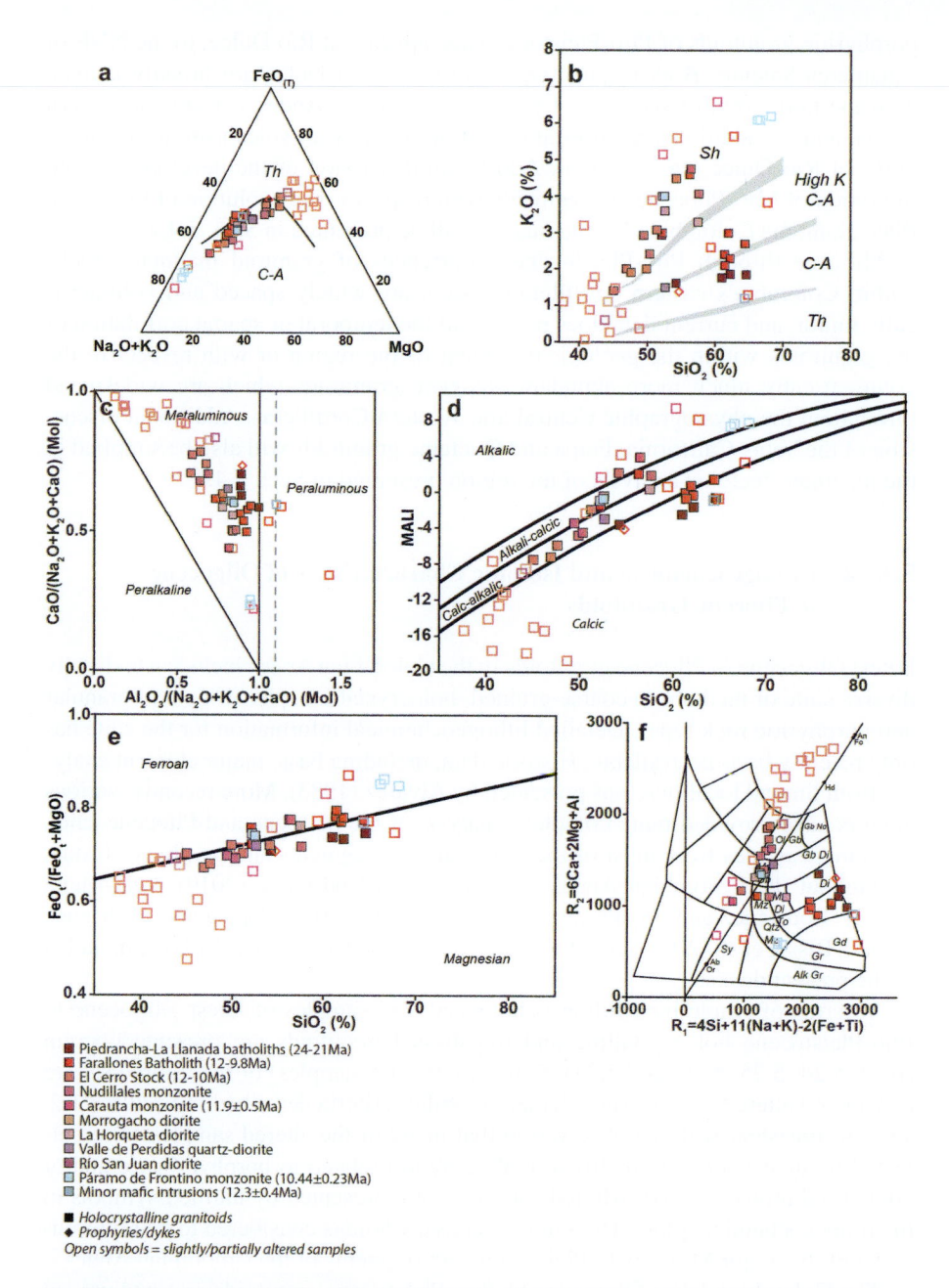

Fig. 5.24 Major element lithogeochemical plots for mid-Cretaceous through Eocene, Western Group (oceanic) granitoids in the Colombian Andes. (**a**) AFM plot, curve after Irvine and Baragar (1971); (**b**) K_2O vs. SiO_2 plot, boundary fields in grey as summarized by Rickwood (1989); (**c**) alumina saturation plot after Barton and Young (2002); (**d** and **e**) MALI and Fe-index vs. SiO_2 plots, respectively, after Frost et al. (2001); (**f**) R1-R2 classification plot after De La Roche et al. (1980). *Th* tholeiite, *C-A* calc-alkaline, *Sh* shoshonite, *Gb No* gabbro-norite, *Ol-Gb* (olivine-)gabbro, *Gb Di* gabbro-diorite, *Di* diorite, *Mz Di* monzodiorite, *Mz* monzonite, *To* tonalite, *Gd* granodiorite

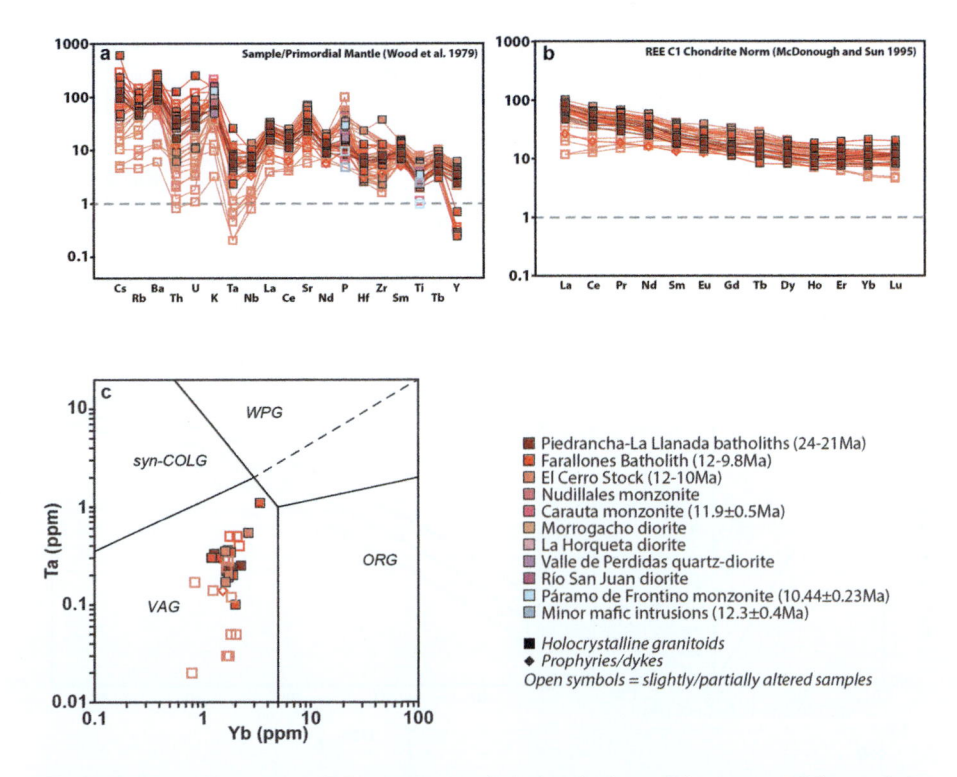

Fig. 5.25 Trace element and REE lithogeochemical plots for latest Oligocene and Miocene holocrystalline granitoids in the Colombian Andes. (**a** and **b**) Trace element and REE normalized spider diagram plots; (**c**) granite discrimination Ta vs. Yb diagram after Pearce et al. (1984). *VAG* volcanic-arc granites, *syn-COLG* syn-collisional granites, *WPG* within-plate granites, *ORG* ocean ridge granites

peraluminous affinity, and some altered samples of both, holocrystalline and porphyritic granitoids, which have been shifted into the peraluminous field due to sub-solidus hydrothermal alteration. With respect to the classification scheme of Frost et al. (2001), many of the Neogene granitoids, both holocrystalline plutonic and hypabyssal porphyritic, record a tendency towards ferroan compositions (Figs. 5.24 and 5.26). We feel this is a reflection of the tendency of many of the analysed samples to contain observable quantities (accessory to +2 modal percent) of pyrite, which would affect the Fe number, but does not signify Fe enrichment in the technical (i.e. tholeiitic) sense of the term (Irvine and Baragar 1971). This interpretation is supported by the "calc-alkaline" trend for the same samples, revealed by the AFM diagrams in Figs. 5.24a and 5.26a. Notwithstanding alteration of the Fe number, the presence of pyrite does not affect the generally calcic to calc-alkalic, metaluminous designation of these samples. A brief review of specific lithogeochemical features of individual Oligocene to Pleistocene granitoid arc segments and clusters is now presented.

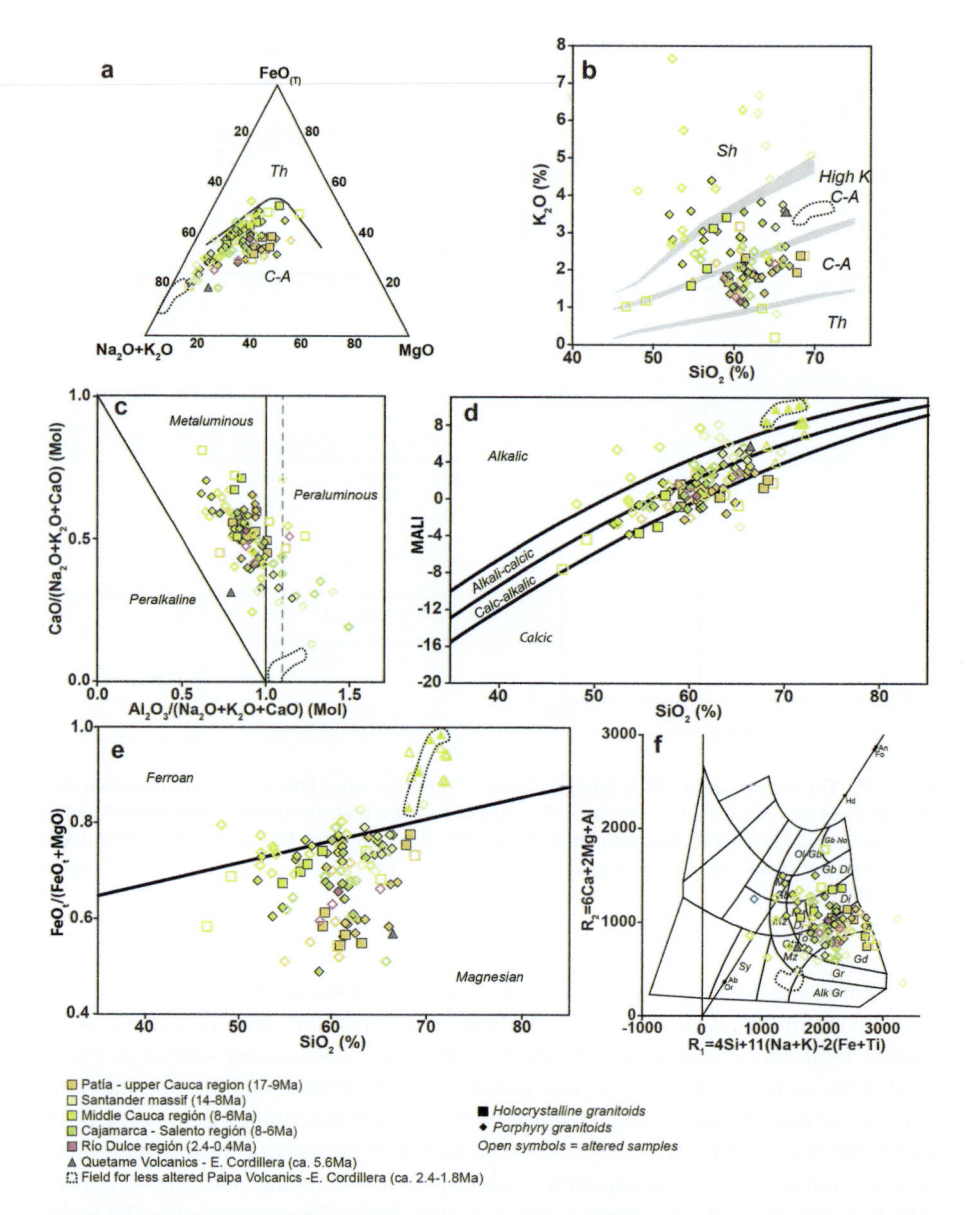

Fig. 5.26 Major element lithogeochemical plots for latest Oligocene and Mio-Pliocene porphyritic granitoids in the Colombian Andes. Major element lithogeochemical plots for mid-Cretaceous through Eocene, Western Group (oceanic) granitoids in the Colombian Andes. (**a**) AFM plot, curve after Irvine and Baragar (1971); (**b**) K_2O vs. SiO_2 plot, boundary fields in grey as summarized by Rickwood (1989); (**c**) alumina saturation plot after Barton and Young (2002); (**d** and **e**) MALI and Fe-index vs. SiO_2 plots, respectively, after Frost et al. (2001); (**f**) R1-R2 classification plot after De La Roche et al. (1980). *Th* tholeiite, *C-A* calc-alkaline, *Sh* shoshonite, *Gb No* gabbro-norite, *Ol-Gb* (olivine-)gabbro, *Gb Di* gabbro-diorite, *Di* diorite, *Mz Di* monzodiorite, *Mz* monzonite, *To* tonalite, *Gd* granodiorite

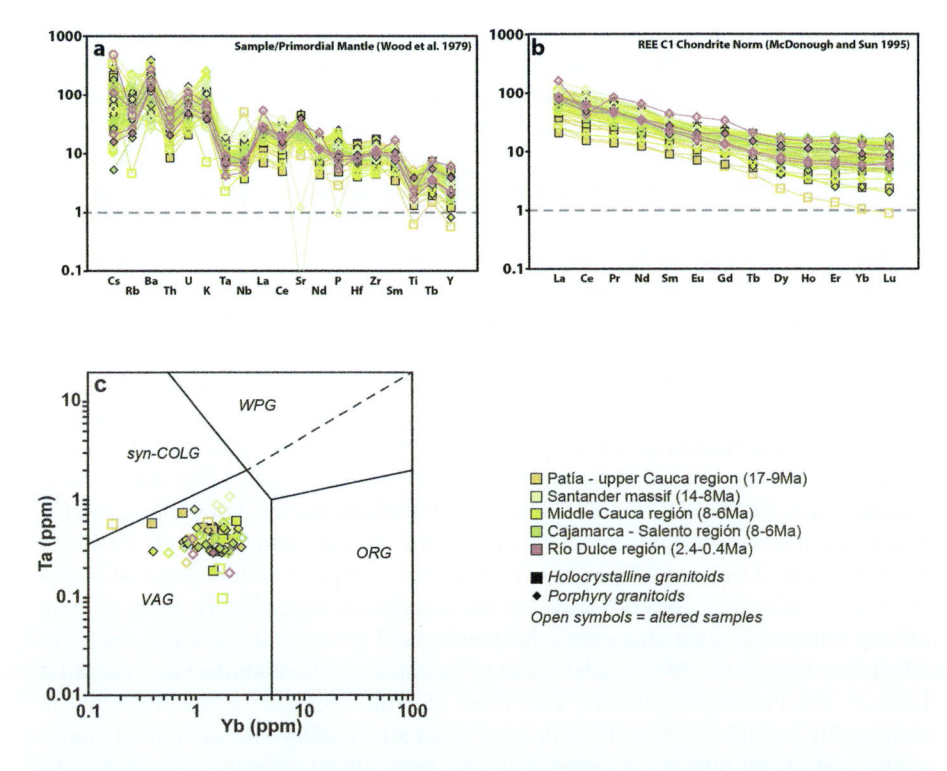

Fig. 5.27 Trace element and REE lithogeochemical plots for latest Oligocene and Miocene porphyritic granitoids in the Colombian Andes. (**a** and **b**) Trace element and REE normalized spider diagram plots; (**c**) granite discrimination Ta vs. Yb diagram after Pearce et al. (1984). *VAG* volcanic-arc granites, *syn-COLG* syn-collisional granites, *WPG* within-plate granites, *ORG* ocean ridge granites

Lithogeochemistry

Latest Oligocene-Early Miocene Holocrystalline (Phaneritic) Granitoids

The Piedrancha-La Llanada-Cuembi trend batholiths: The Piedrancha-La Llanada batholiths are represented by four unaltered samples. These samples are metaluminous, medium-K calc-alkaline rocks of diorite to tonalite composition (Fig. 5.24). Trace element spider diagram patterns for the samples of the Piedrancha-La Llanada batholiths show arc-magmatism patterns with enrichment of HFSE with respect to LILE and notable negative Ta-Nb and Ti anomalies. The REE patterns show moderate REE contents (ΣREE = 29.5–40.5 ppm) and no significant Eu anomalies (Eu/Eu* = 0.95–1.06).

The Farallones-Páramo de Frontino-El Cerro trend batholiths: The Farallones Batholith is represented by thirteen samples, five of them with evidence of alteration. Samples of the Farallones Batholith are mostly high-K calc-alkaline metaluminous with some samples straddling the limit between medium- and high-K calc-alkaline fields. Notably, some unaltered samples exhibit a more alkaline (higher K content) affinity, plotting in the Shoshonite field. Consequently, most of the unaltered samples show a tonalitic composition, and more alkaline rocks exhibit a monzo-gabbro to monzonite composition. Arc-related patterns are revealed by trace element spider diagrams. Chondrite-normalized REE diagrams show low to moderate REE contents (ΣREE = 39.8–112.3 ppm) and no significant negative or positive Eu anomalies (Eu/Eu* = 0.82–1.11).

Additional samples of Miocene holocrystalline granitoids hosted within Cañas Gordas basement of the northern Western Cordillera, compiled primarily from Rodríguez and Zapata (2012) and Zapata and Rodríguez (2013), include the El Cerro Stock (nine samples, four altered), Páramo de Frontino (five samples, two altered), Nudillales (three samples, two altered), Carauta (three samples, one altered), La Horqueta (two samples, one altered), Morrogacho (one unaltered sample), Valle de Perdidas (one altered sample) and Río San Juan (one altered sample) stocks and one altered sample of a minor mafic intrusion (Fig. 5.24). All of these samples have major and minor oxides concentrations but no trace element/REE geochemical data, with exception of one of the altered samples of the El Cerro Stock, presented by Leal-Mejía (2011), which has complete multi-elemental analysis. The samples of these granitoids are metaluminous, and, notably, most of them exhibit a more alkaline character (shoshonite field potash values; alkali-calcic to alkali after Frost et al. 2001), with respect to the calcic to calc-alkaline Piedrancha-La Llanada and Farallones plutons. This more alkaline character is also reflected in their highly variable composition, plotting from alkali gabbro through quartz monzonite and syeno-diorite, in contrast to the more dioritic-tonalitic compositions observed at Piedrancha-La Llanada and Farallones. The trace element pattern for the sample of the El Cerro Stock reflects some element mobility associated with alteration, with significant depletion in Th and U, whereas the chondrite-normalized REE pattern shows relatively low REE contents (ΣREE = 34.55 ppm) and a very subtle positive Eu anomaly (Eu/Eu* = 1.08) (Fig. 5.25).

Miocene to Plio-Pleistocene Hypabyssal Porphyry Granitoid Suites

The Patía-Upper Cauca hypabyssal porphyry suite: Middle to late Miocene Patía-Upper Cauca hypabyssal porphyry suite is well-represented by thirty-three samples of porphyritic to weakly porphyritic granitoids. Nine of these samples exhibit evidence of element mobility associated with alteration. The samples are metaluminous in character with a medium-K calc-alkaline affinity and compositions ranging from diorite and granodiorite. Arc-related geochemical signatures are revealed by patterns in the trace element spider diagram, and chondrite-normalized REE diagram patterns indicate moderate to high REE contents (ΣREE = 27.27–130.98 ppm). Slight Eu anomalies varying between slightly negative and slightly positive (Eu/Eu* = 0.87–1.18) are recorded. HREE patterns are relatively flat ((Gd/Yb)$_N$ = 0.89–4.48), with concentrations between two and twenty times chondrite (Fig. 5.27).

The Middle Cauca hypabyssal porphyry suite: Late Miocene hypabyssal porphyry granitoids of the Middle Cauca are well-represented by 38 samples, 19 of them with evidence of element mobility related to alteration. The unaltered samples define a highly differentiated compositional trend ranging from monzo-gabbro to granodiorite and tonalite. They are metaluminous in character with generally medium- to high-K calc-alkaline affinity. The suite mostly plots within the calc-alkalic field of Frost et al. (2001). A clear effect of hydrothermal alteration, pushing the suite into the alkali-calcic and alkalic fields (shoshonite), is observed (Fig. 5.26). Trace element spider diagrams of unaltered samples confirm the magmatic arc-related geochemical signature with evident negative Ta-Nb and Ti anomalies and significant depletion in Th and U. Chondrite-normalized REE diagram patterns show moderate to relatively high REE contents (ΣREE = 36.48–141.05 ppm). Eu anomalies vary between slightly negative to moderately positive (Eu/Eu* = 0.87–1.21). HREE concentrations are variable, between two and twenty times the concentration of chondrites (Fig. 5.27).

The Cajamarca-Salento hypabyssal porphyry suite: Late Miocene hypabyssal porphyry granitoids of the Cajamarca-Salento region are represented by twenty samples, ten of which show evidence of alteration. The unaltered samples are metaluminous in character with medium- to high-K calc-alkaline affinity, whereas the altered samples are shifted towards the peraluminous field, probably due to the effects of alteration. Composition ranges between diorite and granodiorite-tonalite. Trace element spider diagram patterns of unaltered samples confirm the magmatic arc-related geochemical signature with negative Ta-Nb and Ti anomalies. Chondrite-normalized REE diagram patterns show moderate to relatively high REE contents (ΣREE = 68.33–127.71 ppm) with slightly negative to slightly positive Eu anomalies (Eu/Eu* = 0.80–1.12). HREE concentrations are variable, between four and fifteen times the concentration in chondrites (Fig. 5.27).

The Río Dulce hypabyssal porphyry suite: Pliocene to Pleistocene hypabyssal porphyry granitoids of the Río Dulce region are represented by four samples, although three of them show evidences of alteration. The samples plot in the metaluminous field with exception of the most altered sample which plots in the peraluminous field and are of medium-K calc-alkaline affinity with tonalite compositions. Arc-magmatism geochemical signatures are also revealed by trace element spider diagram patterns. Chondrite-normalized REE diagram patterns show moderate to relatively high REE contents (ΣREE = 84.61–127.66 ppm) with steep decreasing slopes ($(La/Yb)_N$ = 8.33–14.97) with no significant Eu anomalies (Eu/Eu* = 0.80–1.12). HREE concentrations are around ten times the concentration in chondrites (Fig. 5.27).

The Santander Massif hypabyssal porphyry suite: Middle to late Miocene hypabyssal porphyry granitoids of the Santander Massif, especially those clustered around Vetas-California, are well-represented by twelve samples, six of them revealing element mobility due to alteration. The unaltered samples show highly evolved geochemical features, including a peraluminous character and high-K calc-alkaline affinity. Potassium enrichment is recorded in the altered samples with K_2O values >4.0 wt%. The Vetas-California suite ranges compositionally from granodiorite to tonalite, with some samples plotting close to the limit with the quartz monzonite and granite fields. Trace element spider diagram patterns of unaltered samples confirm the magmatic arc-related geochemical signature, with evidently negative Ta-Nb and Ti anomalies. Chondrite-normalized REE diagram patterns show relatively high REE contents ($\Sigma REE = 104.87-185.10$ ppm) and flat to slightly negative Eu anomalies (Eu/Eu* = 0.77–0.97). Significant enrichment in LREE at Vetas-California, with respect to other Miocene hypabyssal porphyry suites (up to 110 times chondrite), is observed, accompanied by a relative depletion of MREE and HREE (about ten times chondrite; Fig. 5.27).

Paipa-Iza and Quetame Granitoids: Available major element lithogeochemistry for Pliocene-Pleistocene volcanic rocks in the Eastern Cordillera includes one sample of a rhyodacite porphyry at the Quetame region (Ujueta et al. 1990) and fifteen samples of Paipa volcano products described as alkali (k-feldspar) rhyodacites and trachytes and calc-alkaline rhyolites (Cepeda and Pardo-Villaveces 2004; Pardo et al. 2005b). Despite evident element mobility associated to hydrothermal/volcanic alteration and/or weathering of samples from Paipa (only four out of fifteen samples seem to be relatively fresh), some general trends can be observed and compared the late Miocene-Pleistocene magmatism observed to the north in the Santander Massif region (Leal-Mejía 2011; Bissig et al. 2014; Cruz et al. 2014) (Fig. 5.26). The less altered/weathered Paipa volcanic rocks exhibit higher silica contents ($SiO_2 = 68.2-71.6$ wt%) when compared to the Santander Massif porphyries ($SiO_2 = 63.4-66.3$ wt%), whereas the Quetame volcanic rock returned 66.4 wt% SiO_2. Alumina content in samples from Paipa and the Santander Massif porphyries is similar ($Al_2O_3 = 16.2-18.1$ wt% and $Al_2O_3 = 17.2-17.7$ wt%, respectively), whilst the Quetame volcanics sample shows significantly lower values ($Al_2O_3 = 14.8$ wt%). Fe_2O_3, MgO, CaO and TiO_2 values for the Santander Massif porphyries are significantly higher ($Fe_2O_3 = 2.5-4.9$ wt%, MgO = 0.7–0.12 wt%, CaO = 2.4–4.0 wt% and $TiO_2 = 0.3-0.5$ wt%) than in the Paipa volcanics ($Fe_2O_3 = 0.8-2.0$ wt%, MgO = 0.02–0.5 wt%, CaO = 0.2–0.8 wt% and $TiO_2 = 0.1-0.3$ wt%). Fe_2O_3 and CaO values for the Quetame volcanics sample ($Fe_2O_3 = 2.9$ wt% and CaO = 3.2 wt%) are comparable to those of the Santander Massif porphyries, whereas TiO_2 value (0.27 wt%) is more in the range of the Paipa samples. The MgO value for the Quetame volcanics sample (2.0 wt%) is significantly higher than MgO values observed for both Santander Massif porphyries and Paipa volcanics. Na_2O values are slightly higher in the Paipa volcanics when compared to the Santander Massif porphyries ($Na_2O = 5.9-6.8$ wt% and $Na_2O = 4.0-5.1$ wt%, respectively), whereas K_2O values are comparable for both sample sets ($K_2O = 3.4-3.7$ wt% and $K_2O = 3.0-3.8$ wt%, respectively).

All of the Paipa-Quetame samples show a well-defined calc-alkaline trend in the AFM diagram (Fig. 5.26a), with samples from Paipa volcanics being slightly more evolved/fractionated with respect to the Santander Massif porphyries. In addition, all of the less altered samples plot in the high-K calc-alkaline field (Fig. 5.26b). The presence of abundant K-feldspar phenocrysts in both Quetame and Paipa volcanic rocks (Ujueta et al. 1990; Pardo et al. 2005b), as well as the Santander Massif porphyries (Mantilla et al. 2009; Cruz et al. 2014), suggests a more alkaline affinity for these rocks; however, major element lithogeochemistry indicates that these rocks are mostly metaluminous to weakly peraluminous in nature (Fig. 5.26c), and do not plot in the peralkaline field. At Paipa, post-crystallization hydrothermal alteration includes abundant secondary silica and pyritic sulfidation. In either case, these factors significantly reduce the confidence level of any interpretations and conclusions drawn with respect to these granitoid suites, especially those derived based soley upon major element lithogeochemistry.

Classification diagrams for feldspathic igneous rocks proposed by Frost et al. (2001) and Frost and Frost (2008) (Fig. 5.26d, e) clearly differentiate the magnesian, oxidized, calcic to calc-alkalic Santander Massif porphyries and alkali-calcic Quetame sample, from the ferroan (reduced) alkalic samples of the Paipa suite. Moreover, calculation of the alkalinity index (AI) and the feldspathoid silica-saturation index (FSSI) proposed by Frost and Frost (2008) returned positive values for both indexes (AI = 0.5–6.7, FSSI = 13.5–47.9), which confirm a silica-saturated metaluminous/peraluminous character for these quartz-bearing rocks rather than a peralkaline character. The R1-R2 classification plot for plutonic rocks (Fig. 5.26f) also demonstrates the alkalic affinity for rocks from the Paipa volcanics (alkali granite/quartz syenite) with respect to more calc-alkaline rocks of the Santander Massif porphyries (granodiorite/tonalite) and the Quetame volcanics (quartz monzonite).

Whole-Rock Lithochemistry Summary and Discussion for the Oligocene-Pliocene Granitoids

Whole-rock, trace and REE data for the majority of the latest Oligocene through Plio-Pleistocene holocrystalline and hypabyssal porphyritic granitoids of the western Colombian Andes, regardless of age, plot in consistent, narrow ranges, with variations in the lithogeochemical composition of most samples attributable to the effects of late or post-crystallization alteration of the alkali contents. The western Colombian intrusive suites are metaluminous, calcic to calc-alkaline plutons with typical arc-related trace element patterns and flat, unfractionated REE patterns lacking well-developed Eu anomalies, considered typical of relatively undifferentiated, primitive, subduction-related granitoids emplaced within oceanic crust, as represented by the Romeral mélange and Cañas Gordas and Dagua terranes, which form the basement complexes to the western Colombian Andes. The trend towards alkali enrichment observed in some of the small, isolated plutons associated with the ca. 12–10 Ma Farallones-El Cerro trend may be explained by late magmatic K-metasomatism, as described petrographically by Escobar and Tejada (1992), without invoking additional magmatic source regions or differentiation processes. Similar arc-related major, trace and REE compositions are recorded by the Cajamarca-Salento and Río Dulce suites, despite the observation that they are hosted continent-ward, within Cajamarca-Valdivia Terrane basement. Cajamarca-Valdivia, however, is not of typically "continental" composition (Restrepo-Pace 1992; Cediel et al. 2003; Cediel 2011) and in this context is considered to have preserved the more primitive bulk compositions reflected in the analyses of the Cajamarca-Salento and Río Dulce porphyry suites.

Conversely, granitoid porphyries and volcanic rocks at Vetas-California, Paipa-Iza and Quetame were emplaced within composite mid-Proterozoic-early Paleozoic continental metamorphic basement within the Santander Massif and Eastern Cordillera. All of these suites represent isolated, low-volume outliers of granitoid rocks emplaced significantly to the east and north of the principle Miocene granitoid suites of the Western and Central Cordilleras and of the Pleistocene to recent Northern Andean volcanic arc. Based upon the lithogeochemical data provided, the Vetas-California and Quetame granitoids conform to a magnesian, calcic to alkali-calcic suite, whilst the Paipa-Iza granitoids are of ferroan, alkali affinity (Frost et al. 2001), and in this context, as a whole, the Santander-Eastern Cordilleran granitoids record a bimodal distribution (Fig. 5.26). Notwithstanding, the data are limited and geographically disperse, and do not yet permit interpretation of the potential petrogenetic relationships between these outlier suites. Within the context of the compositional trends of the entire Oligo-Miocene to Pliocene granitoid suite presented herein, however, the Vetas-California, Paipa-Iza and Quetame granitoids provide the most consistently differentiated/evolved lithogeochemistry, especially in terms of alkalinity, aluminium indices and trace and REE patterns. We interpret these observations to reflect greater degrees of crustal interaction and assimilation/contamination from the thick continental basement of the Santander Massif and Chicamocha Terrane (Figs. 5.2 and 5.22) vs. the more primitive basement compositions provided by the Cajamarca-Valdivia Terrane and the oceanic terranes of the Western Tectonic Realm, which host the Oligo-Miocene-Pliocene granitoids of the Central and Western Cordilleras.

Isotope Geochemistry

Sr-Nd Isotope Geochemistry, Results, Summary and Discussion

Available Sr-Nd isotope data for the entire suite of holocrystalline phaneritic and hypabyssal porphyritic granitoid rocks of latest Oligocene to Pliocene age, with the exception of the Paipa-Iza and Quetame occurrences, are presented in Fig. 5.28. All of the arc segments and porphyry clusters documented above, albeit most with a limited number of samples, are represented within the data set. In addition, for comparative purposes, we include Sr-Nd isotope data published for mantle and crustal xenoliths contained within Plio-Pleistocene garnetiferous pyroclastic rocks occurring at Mercaderes (Weber et al. 2002; Rodríguez-Vargas et al. 2005), spatially coincident with late Miocene hypabyssal granitoid porphyry within our Upper Cauca-Patía arc segment, which we interpret to establish a representative range of values for the Sr-Nd composition of the mantle/crust beneath the SW Colombian margin during the Neogene.

As displayed in Fig. 5.28, with the exception of the Vetas-California porphyry cluster, essentially the entire latest Oligocene to Plio-Pleistocene suite of granitoids considered within this study plot within a narrow vertical field, originating within the mantle array and tending towards increasing $\varepsilon Nd_{(t)}$ values. $^{87}Sr/^{86}Sr_{(i)}$ values ubiquitously plot within a narrow range ($^{87}Sr/^{86}Sr_{(i)} = 0.70395$ to 0.70506), with very little scatter, with only two samples of tonalite-granodiorite from the Piedrancha-Cuembí suite plotting outside of this range (Fig. 5.28). The Piedrancha-Cuembí suite is in fact the most scattered of the Miocene granitoid data sets with respect to Sr-Nd isotope systematics, an observation we attribute to the widespread carbo-

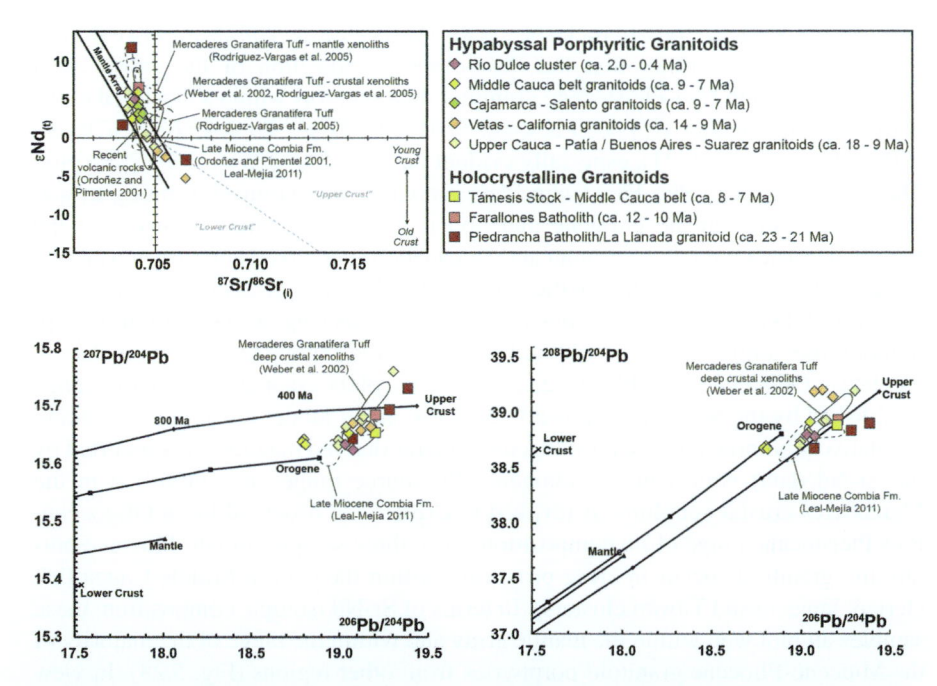

Fig. 5.28 Sr-Nd and Pb isotope plots for latest Oligocene through Mio-Pliocene holocrystalline and porphyritic granitoids in the Colombian Andes. Data for mantle and deep crustal xenoliths from the Cauca-Patía area, and volcanic rocks of the Combia Fm. from the Middle Cauca area, are plotted for reference. See text for discussion

nitization the suite has suffered, a process which could affect the $^{87}Sr/^{86}Sr_{(i)}$ ratios of these rocks. Notwithstanding, the Sr-Nd data sets for the Miocene-Plio-Pleistocene granitoids of the Cajamarca-Valdivia Terrane and Western Tectonic Realm plot essentially co-spatial with the isotopic signatures provided by the Mercaderes mantle xenoliths.

The Sr-Nd isotope composition of the Vetas-California granitoid porphyries, revealed in Fig. 5.28, is commensurate with the lithogeochemical compositions, trends and conclusions outlined in Sect. 5.3.5.4. The Vetas-California suite depicts a Sr-Nd compositional trend of decreasing $\varepsilon Nd_{(t)}$ values with increasing $^{87}Sr/^{86}Sr_{(i)}$, originating within the central mantle array and evolving towards crustally influenced values. Additional discussion of the isotopic evolution of the latest Oligocene through Plio-Pleistocene granitoid suite will be provided following presentation of Pb isotope data below.

Pb Isotope Geochemistry, Results and Summary and Discussion

Available Pb isotope geochemical results for latest Oligocene to Plio-Pleistocene granitoids of the Colombian Andes are presented in Fig. 5.28. For comparative purposes Pb isotope data published for crustal xenoliths contained within the

Mercaderes garnetiferous pyroclastic rocks (Weber et al. 2002) are also shown. Figure 5.28 demonstrates that the Pb isotope composition of the entire latest Oligocene through Plio-Pleistocene granitoid data set plots within a clustered range with very little scatter ($^{206}Pb/^{204}Pb$ = 18.79–19.39, $^{207}Pb/^{204}Pb$ = 15.62–15.76 and $^{208}Pb/^{204}Pb$ = 38.68–39.21), especially evident on the $^{207}Pb/^{206}Pb$ vs. $^{206}Pb/^{204}Pb$ plot. The latest Oligocene-Plio-Pleistocene data is notably well grouped, forming a tight, steep array between the Orogene and upper crust lead evolution curves of the Plumbotectonics model of Zartman and Doe (1981), in marked contrast to typically more shallow arrays provided by the data sets for the latest Triassic-Jurassic granitoids (Fig. 5.14) and mid-Cretaceous-Eocene granitoids (Fig. 5.21). With the exception of three samples of granitoid porphyries from the Middle Cauca region, data of the latest Oligocene-Plio-Pleistocene granitoids plots co-spatial with the range established by the Mercaderes crustal xenoliths, and a model involving the mixing of relatively homogenous, less radiogenic, mantle-derived magmas (as supported by the Sr-Nd data) with a more radiogenic Pb source range, as established in the Mercaderes crustal xenoliths, is invoked to explain the observed latest Oligocene-Plio-Pleistocene range of Pb compositions. The three samples of late Miocene porphyritic granitoids occur in close proximity within the central Middle Cauca belt (Jericó, Venecia and Titiribí clusters). In terms of Sr-Nd isotopic composition, these samples all plot well within the mantle array and within the range of the majority of the Miocene-Pliocene granitoid porphyries from other regions (Fig. 5.28). In view of this, we interpret this small population to represent the mixing of mantle-derived Pb compositions similar to those of the granitoids from other regions, with a more radiogenic, crustal-sourced Pb of a somewhat distinct composition to that defined by the Mercaderes crustal xenoliths. This is in keeping with the observation that the Romeral tectonic zone, which forms basement to the entire suite of Miocene granitoid porphyries, along both the Upper Cauca-Patía and southern Middle Cauca belts, is a heterogeneous lithotecton comprised of a mix (mélange) of rock types of differing age and continental, peri-cratonic and oceanic provenance. In this respect, the relatively homogenous appearance of Pb isotope compositions for the latest Oligocene-Plio-Pleistocene suite may well be a function of the relatively few localities for which Pb isotope analyses are available, especially given the restricted distribution of crustal xenoliths such as those documented at Mercaderes.

Synthesis and Conclusions of Lithogeochemistry and Sr, Nd and Pb Isotope Geochemistry for Oligocene-Pliocene Granitoids

Whole-rock lithogeochemistry, including major, minor, trace element and REE data, combined with analyses documenting the Sr-Nd and Pb isotope composition of latest Oligocene to Miocene and Plio-Pleistocene holocrystalline phaneritic and hypabyssal porphyritic rocks from numerous localities within the northeastern, central and western Colombian Andes permits the consistent characterization of the entire suite as subduction-related, Cordilleran (Frost et al. 2001), volcanic arc (Pearce et al. 1984) or calc-alkaline (Barbarain 1999) granitoids, formed within a continental arc setting. The principal host domains for the Neogene granitoid suite are the oceanic basement terranes of the Western Tectonic Realm (Romeral, Cañas Gordas, Dagua); however, important occurrences are also observed within the Cajamarca-Valdivia Terrane, within Colombia's physiographic Central Cordillera, essentially coaxial with the modern-day Northern Andes volcanic arc (Stern 2004). Isolated Miocene to Plio-Pleistocene granitoid outliers are also observed farther east, at Vetas-California in the Santander Massif and at Paipa-Iza and Quetame, in the Eastern Cordillera. The tectonic assembly of the region was essentially complete at the time of emplacement of each arc segment or granitoid cluster, and all of the latest Oligocene through Plio-Pleistocene granitoid suites may be considered autochthonous with respect to the tectonic evolution of the Colombian Andes.

Review of the whole-rock lithogeochemical and isotope data for the latest Oligocene to Plio-Pleistocene suite demonstrates remarkably consistent compositional trends, despite the varied nature of the basement complexes into which the granitoids were emplaced. Rare-earth element and isotopic trends for the majority of the suite suggest limited degrees of magmatic fractionation and isotopic exchange at crustal levels, consistent with the rapid emplacement of subduction-related, mantle-derived melts, facilitated by the preexisting structural architecture, which includes various paleo-sutures, as exemplified by the Palestina, Romeral and Cauca fault systems. The most evolved granitoids within the latest Oligocene-Pliocene suite include those of the Vetas-California area, which have evidently undergone somewhat greater degrees of fractionation, assimilation and/or isotopic exchange with the thick continental basement exposed within the Santander Massif. Further discussion of the nature, distribution and tectonic evolution of Neogene granitoid magmatism in the Colombian Andes is presented in Sect. 5.4.4.2.

5.4 Phanerozoic Tectono-Magmatic Evolution of the Colombian Andes

Aspden et al. (1987) presented a temporal-spatial analysis of granitoid magmatism in the Colombian Andes based upon published K-Ar and Rb-Sr radiometric age dates. They defined five episodes of subduction-related granitoid magmatism, including the Triassic, Jurassic, Cretaceous, Paleogene and Neogene, and offered a schematic interpretation of the tectonic framework for each episode within the generalized tectonic configuration of the entire Northern Andean region. These same authors identified various factors which influenced the nature, distribution and geometry of Meso-Cenozoic subduction-related granitoid arcs in Colombia. These factors included oblique plate convergence, low-angle subduction including changes in the angle of the subducting oceanic plate and the role of aseismic features and the accretion of allochthonous components contained within the oceanic domain along the Pacific margin, in the development of, and hiatuses in, the subduction process. More recently, kinematic models for the tectonic and structural evolution of the Northern Andes (e.g. Cediel et al. 2003; Kennan and Pindell 2009) and Caribbean Plate (e.g. Pindell and Kennan 2001; Nerlich et al. 2014), constructed at a similar scale to the work of Aspden et al. (1987), have independently confirmed and expanded upon many of the assertations presented by these early authors.

The increased resolution and widespread distribution of the present-day U-Pb age, lithogeochemical and isotopic database, when combined with updated concepts for the geological evolution of the Colombian Andes, permit a reassessment and more detailed reconstruction of the tectono-magmatic evolution of the region than that afforded in the Aspden et al. (1987) analysis. In the following section, we present sequential reconstructions detailing the Phanerozoic tectono-magmatic evolution of granitoids in the Colombian Andes, based upon the major magmatic episodes defined by the U-Pb (zircon) age date, lithogeochemical and isotopic database, as described in detail in the foregoing sections. Annotated schematic illustrations and time-space analyses for the early Paleozoic through middle-late Triassic, latest Triassic through Jurassic, early to middle Cretaceous, middle Cretaceous through Eocene and latest Oligocene through Miocene-Pliocene are provided in Figs. 5.29, 5.30, 5.31, 5.32, 5.33, 5.34, 5.35, and 5.36. Descriptive text pertaining to each time period highlights the temporal and spatial evolution of granitoid magmatism within the litho-tectonic and morpho-structural development of the region, prior to, leading up to and during the Meso-Cenozoic Northern Andean orogeny.

In terms of nomenclature pertaining to the various phases of tectonic development of the Colombian Andes, we have adhered to terminology used in the work of Cediel et al. (1994), Cediel and Cáceres (2000), Cediel et al. (2003), Cediel (2011) and Cediel (2018). This work provides a coherent and sufficiently detailed framework, at an appropriate temporal and spatial scale for the Colombian Andes, spanning the Proterozoic to Mio-Pliocene, within which to integrate the periods of granitoid magmatism defined herein. Table 5.2 provides a summary of tectonic events recorded within the Colombian Andes, as described in detail in the works of the previously cited authors.

5.4.1 Pre-northern Andean Orogeny Granitoids: Early Paleozoic Through Mid-Late Triassic

Our study has identified three episodes of granitoid magmatism recorded within the Colombian Andes, which were generated and emplaced within the context of pre-Northern Andean Orogeny tectono-magmatic development. These episodes include the early Paleozoic (ca. 485–439 Ma), Carboniferous (ca. 333–310 Ma) and Permo-Triassic (ca. 288–223 Ma). With respect to all three episodes, the granitoid-magmatic record is relatively sparse and punctually developed, especially when compared to wide-spread and volumetrically exponential magmatism developed during the Meso-Cenozoic. Indeed, we remind the reader that the full extent of all three early Phanerozoic magmatic events has yet to be fully defined, based upon presently available radiometric age dates vs. the resolution of existing field-based geological mapping, which doesn't yet recognize some of the important early

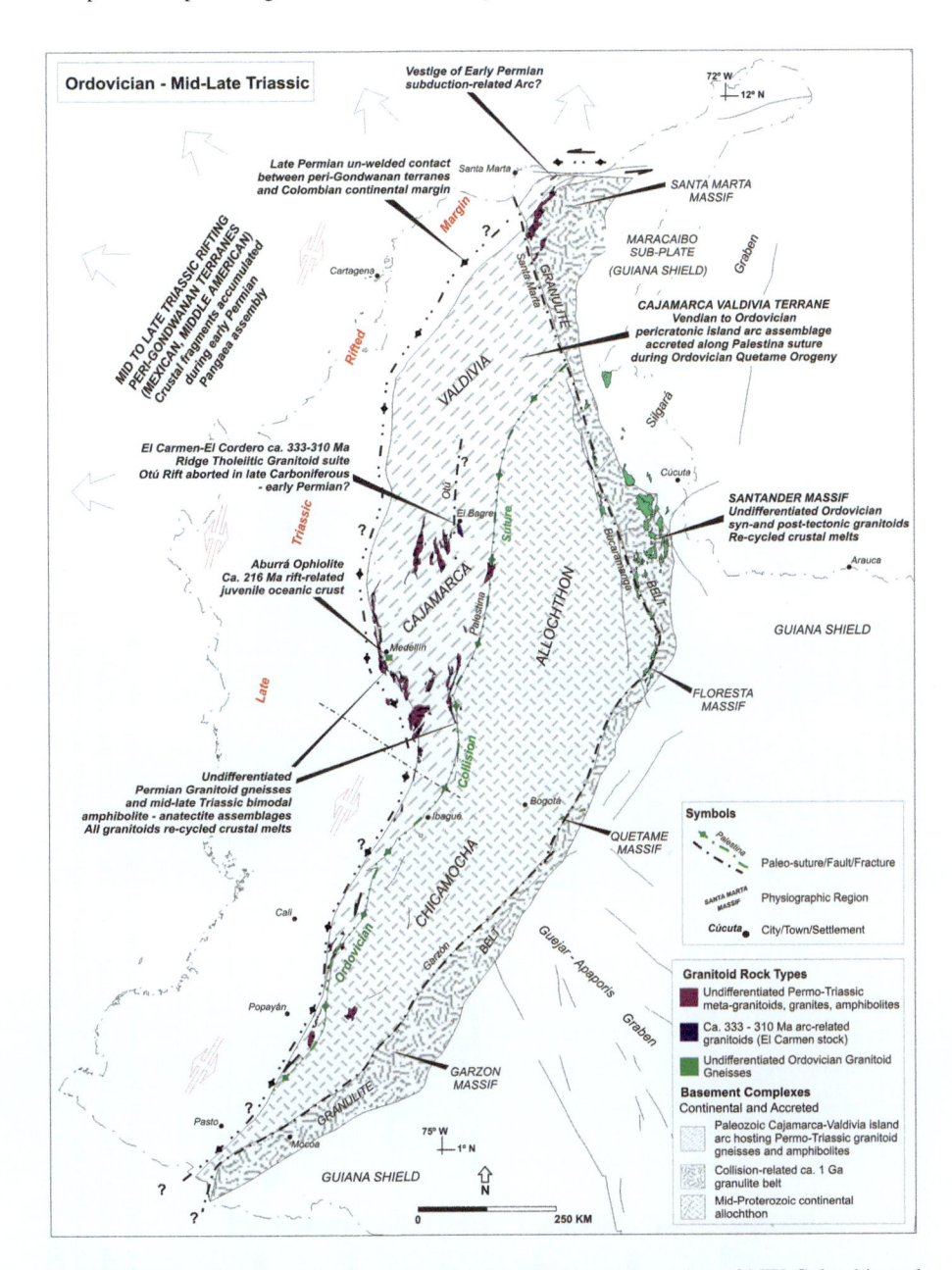

Fig. 5.29 Major litho-tectonic elements and interpreted tectonic setting of NW Colombia and surrounding area during the late Triassic. The spatial relationship between early Paleozoic, Carboniferous and Permo-Triassic granitoids exposed in the Colombian Andes is shown. (Granitoid shapes modified after Cediel and Cáceres 2000; Gómez et al. 2007; Gómez et al. 2015a. Litho-tectonic terrane and fault nomenclature modified after Cediel et al. 2003. See text for additional details)

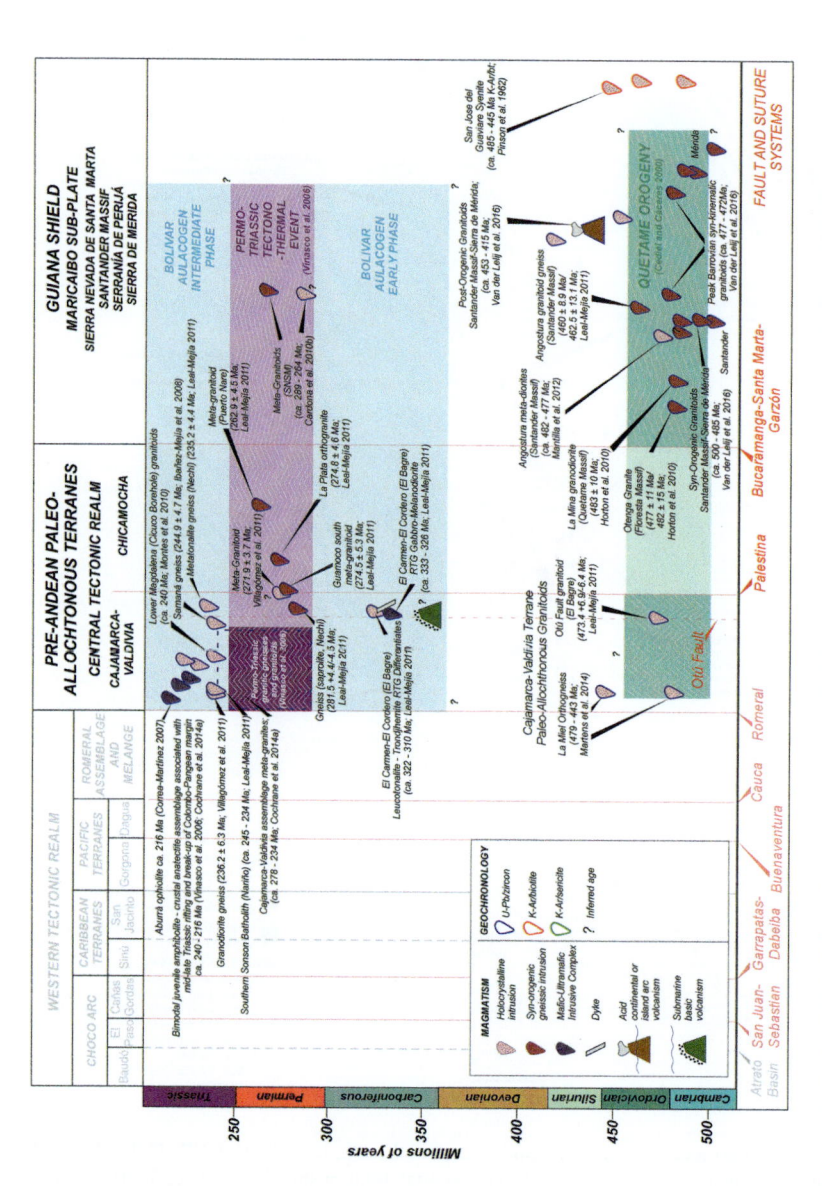

Fig. 5.30 Time-space analysis of early Paleozoic through mid-late Triassic granitoids in the Colombian Andes and surrounding region, in relation to tectonic framework, major litho-tectonic elements and orogenic events. The age and nature of individual granitoid intrusive suites of the time period are indicated. The profile contains elements projected onto a ca. NW–SE line of section through west-central Colombia. (Litho-tectonic terrane and fault nomenclature modified after Cediel et al. (2003). See text for additional details)

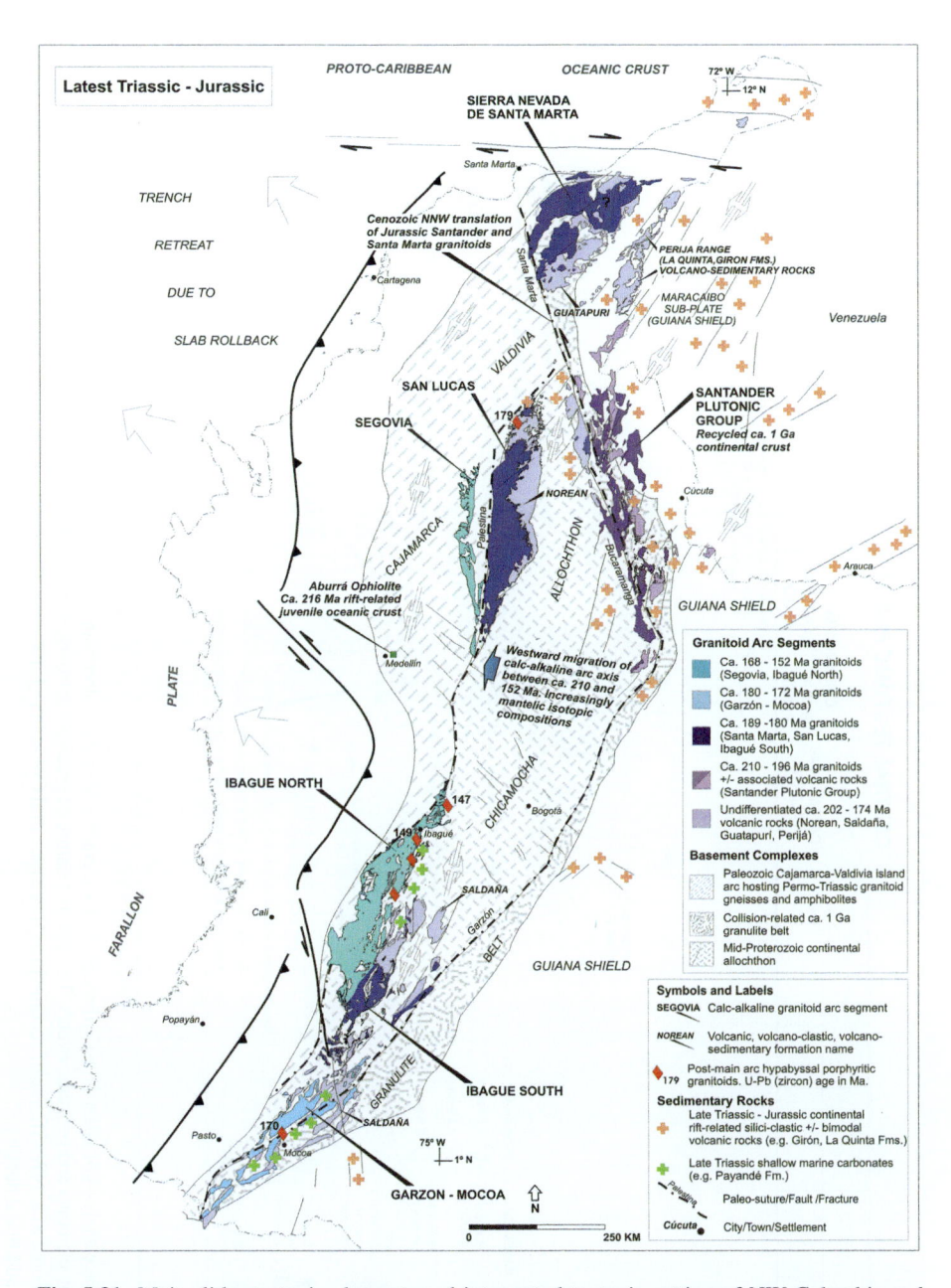

Fig. 5.31 Major litho-tectonic elements and interpreted tectonic setting of NW Colombia and surrounding area during the latest Triassic through Jurassic, highlighting the spatial-temporal relationship between the major Jurassic arc segments exposed in the Colombian Andes. (Granitoid shapes modified after Cediel and Cáceres 2000; Gómez et al. 2007; Gómez et al. 2015a. Litho-tectonic terrane and fault nomenclature modified after Cediel et al. 2003. See text for additional details)

Fig. 5.32 Time-space analysis of latest Triassic through Jurassic granitoids in the Colombian Andes and surrounding region, in relation to tectonic framework, major litho-tectonic elements and orogenic events. The age and nature of granitoid intrusive suites of the same time period are indicated. The profile contains elements projected onto a ca. NW–SE line of section through west-central Colombia. (Litho-tectonic terrane and fault nomenclature modified after Cediel et al. 2003. See text for additional details)

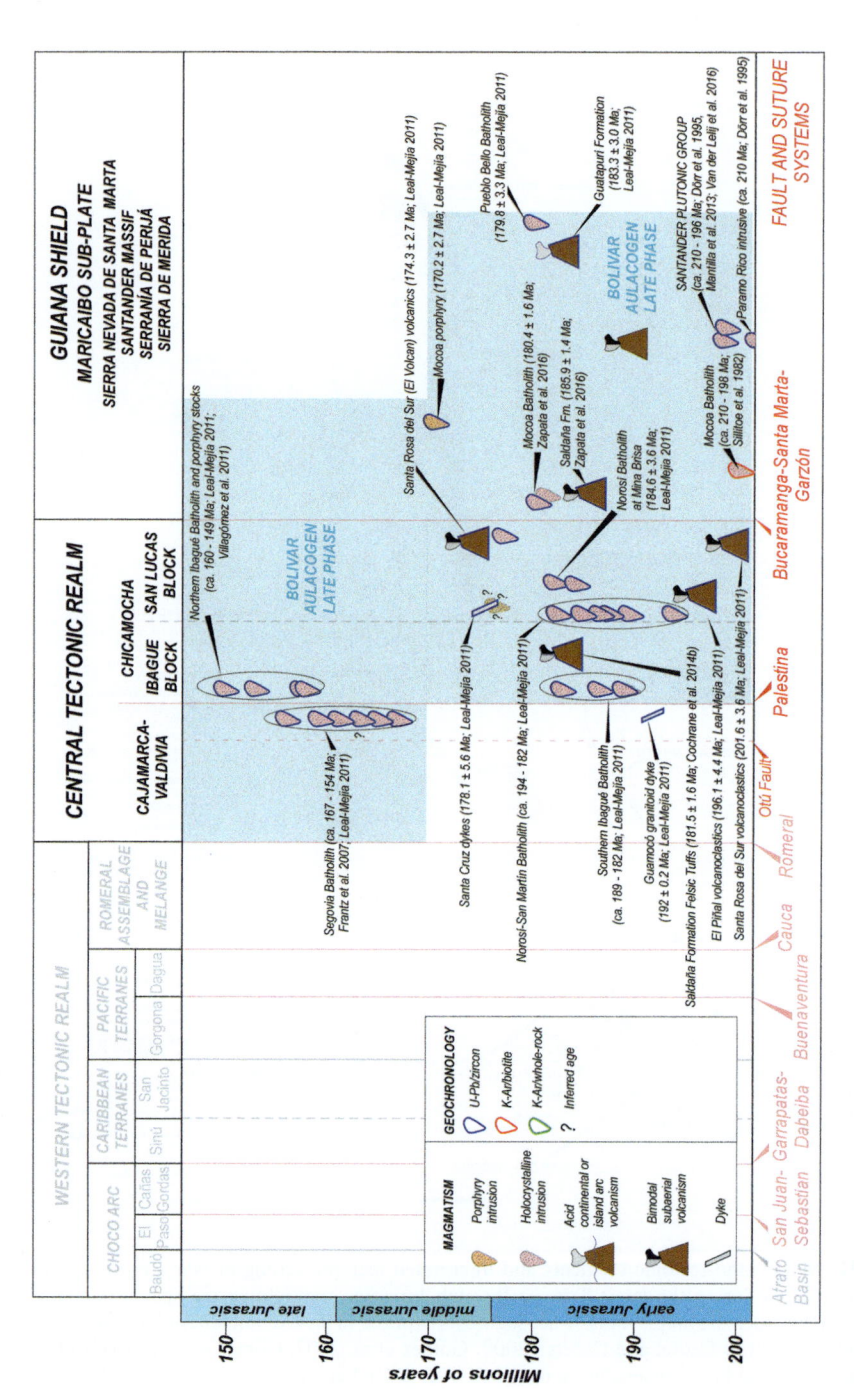

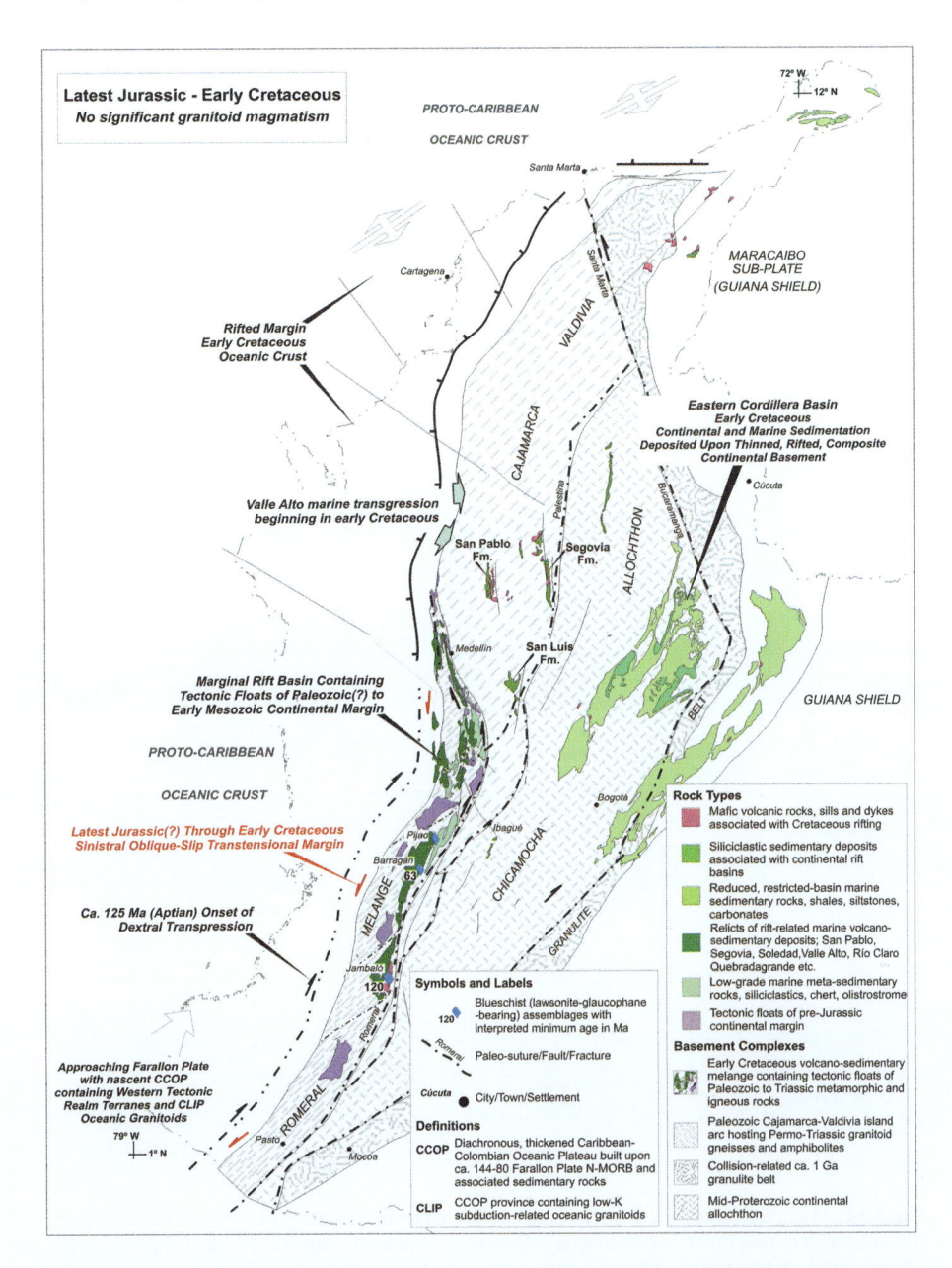

5 Spatial-Temporal Migration of Granitoid Magmatism and the Phanerozoic... 355

Fig. 5.33 Major litho-tectonic elements and interpreted tectonic setting of NW Colombia and surrounding area during the early to mid-Cretaceous. Note the absence of significant volumes of granitoid rocks within the exposed geological record of the Colombian Andes for this time period. (Lithological unit shapes modified after Cediel and Cáceres 2000; Gómez et al. 2007; Gómez et al. 2015a. Litho-tectonic terrane and fault nomenclature modified after Cediel et al. 2003. See text for additional details)

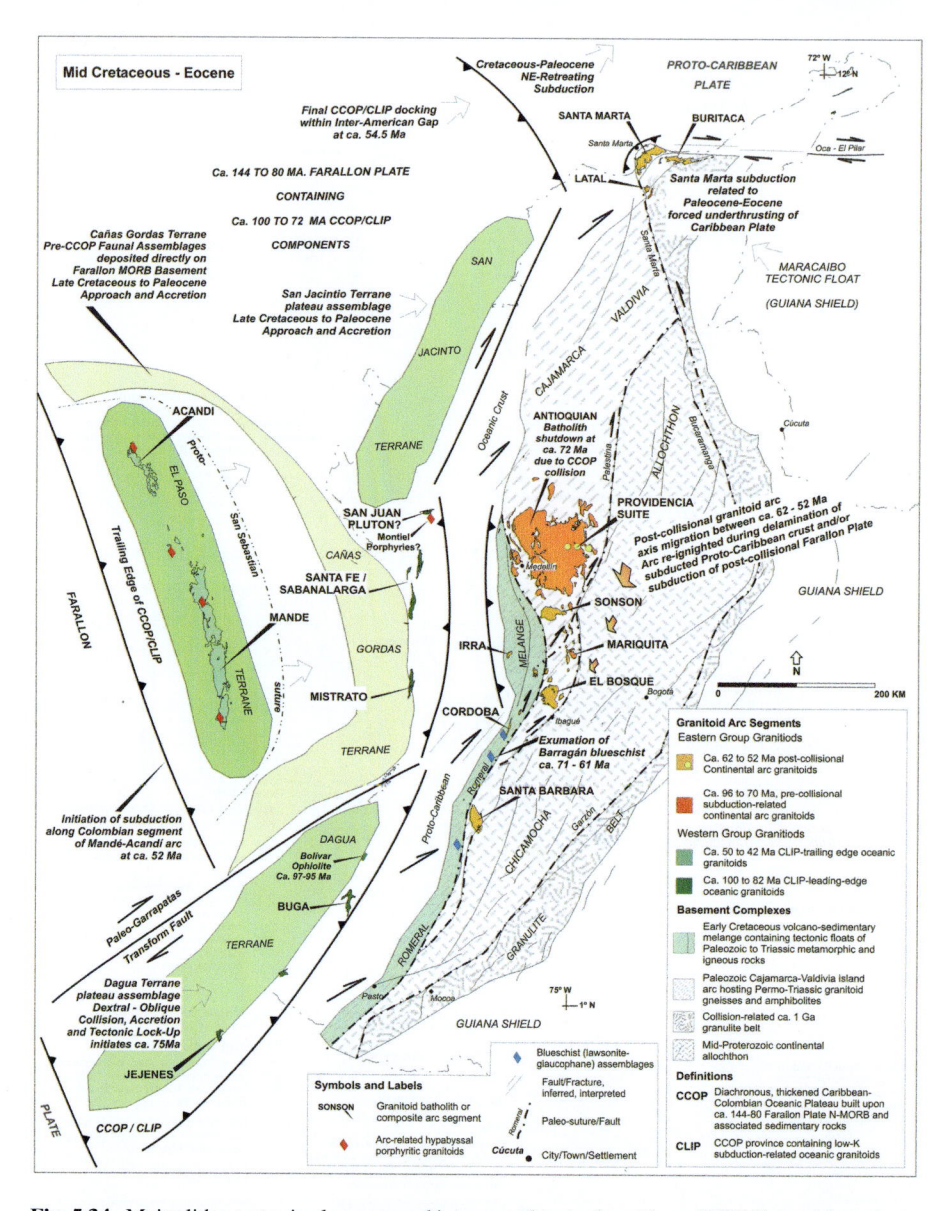

Fig. 5.34 Major litho-tectonic elements and interpreted tectonic setting of NW Colombia and surrounding area during the mid-Cretaceous to Eocene. A schematic depiction of the temporal-spatial relationship between Eastern Group (continental) granitoids and Western Group (oceanic) granitoids is presented. (Granitoid shapes modified after Cediel and Cáceres 2000; Gómez et al. 2007; Gómez et al. 2015a. Litho-tectonic terrane and fault nomenclature modified after Cediel et al. 2003. See text for additional details)

Fig. 5.35 Time-space analysis of early Cretaceous through Eocene granitoids in the Colombian Andes and surrounding region, in relation to tectonic framework, major litho-tectonic elements and orogenic events. The age and nature of granitoid intrusive suites of the same time period are indicated. The profile contains elements projected onto a ca. NW–SE line of section through west-central Colombia. (Litho-tectonic terrane and fault nomenclature modified after Cediel et al. 2003. See text for additional details)

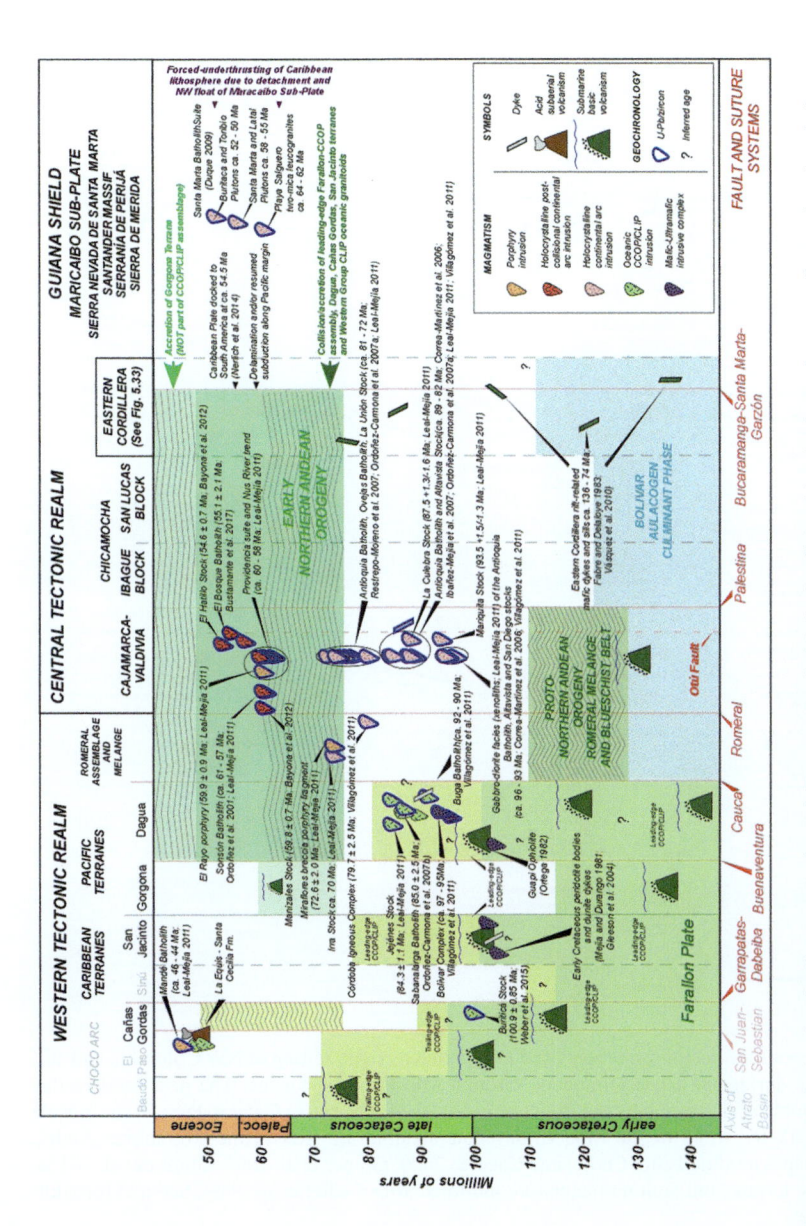

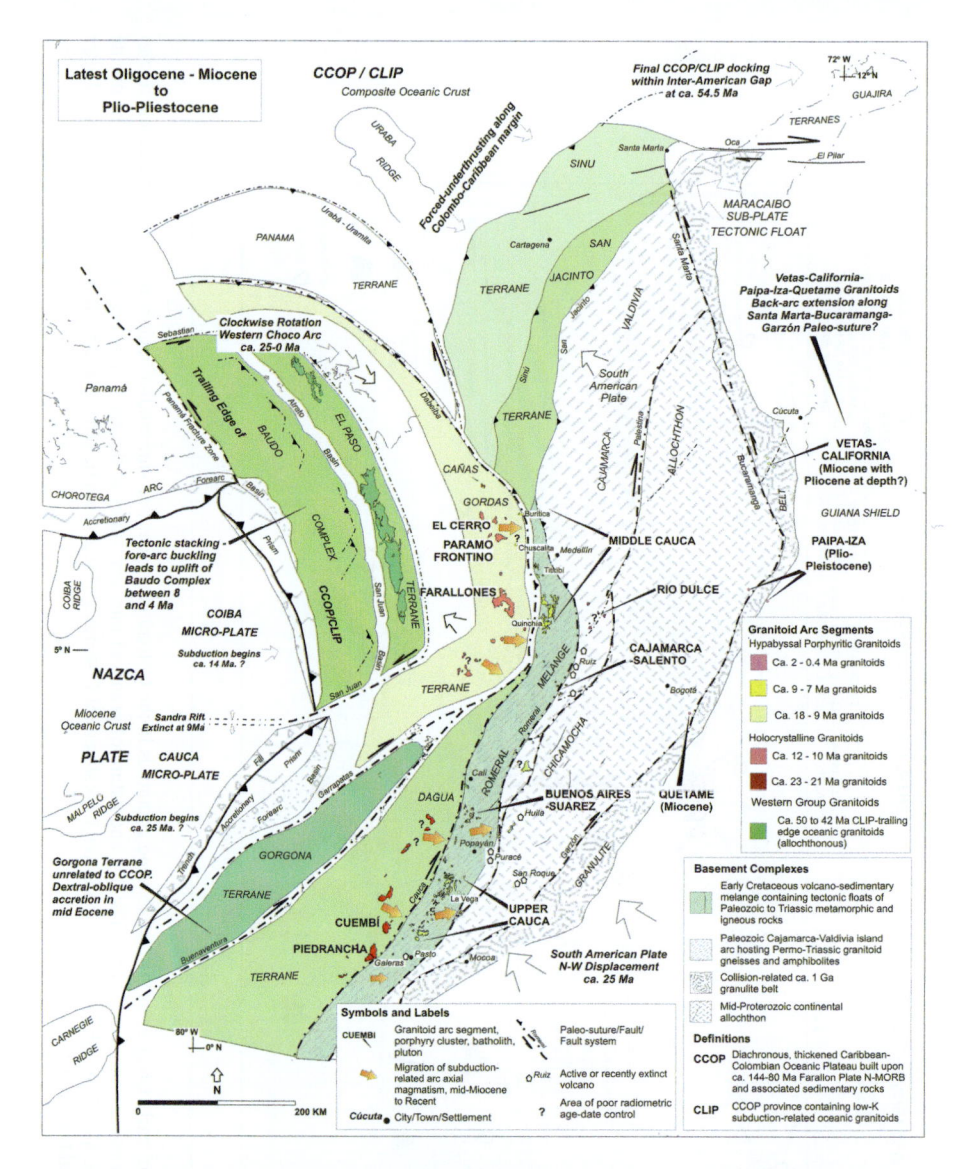

Fig. 5.36 Major litho-tectonic elements and interpreted tectonic setting of NW Colombia and surrounding area during the latest Oligocene through Mio-Plio-Pleistocene. The near modern-day tectonic assembly of the region by the Pliocene is observed. The active Galeras-Puracé-Huila-Ruíz volcanoes mark the trend of the modern-day calc-alkaline arc axis in the Colombian Andes. (Granitoid shapes modified after Cediel and Cáceres 2000; Gómez et al. 2007; Gómez et al. 2015a. Litho-tectonic terrane and fault nomenclature modified after Cediel et al. 2003. See text for additional details)

Table 5.2 Summary of Colombian tectono-magmatic episodes and regional tectonic comparisons. Litho-tectonic and morpho-structural units as defined by Cediel et al. (2003) and indicated in Fig. 5.2

Time period (magmatic episode)	Colombian tectonic phase (with age of associated granitoids)	Regional temporal comparatives (Orogenies)	Distribution of Colombian granitoids (basement domain)	Tectonic regime (Colombia)
Pre-Cambrian	Orinoco Orogeny (ca. 1.2–0.9 Ga)	Grenville Orogeny, North America	Granulite Belt in MSP (SNSM, Santander massif) and Garzón massif	Collisional, Compressional, Accretionary
Early Paleozoic	Quetame Orogeny (ca. 485–473 Ma)	Caparonesis (Venezuela), Ocloy (Ecuador-Perú), Famantinian (Argentina), Taconian-Acadian (N. America-N. Europe)	MSP (Santander massif, SNSM), Floresta and Quetame massifs, CTR (CA-VA, Central Cordillera)	Collisional, Compressional, Accretionary Extension after ca.465 Ma
Carboniferous	Bolívar Aulacogen, (early phase) (ca. 333–310 Ma)	–	CTR (CA-VA, Otú rift, Northern Central Cordillera)	Extensional, rifting (failed)
Permian-early Triassic	Permo-Triassic tec.-thermal event (ca. 290–250 Ma)	Gondwanide Orogeny, Alleghanian-Appalachian Orogeny, N. America	CTR (mostly CA-VA, Central Cordillera), SNSM	Compressional, transpressional?
Mid-late Triassic	Bolívar Aulacogen (intermediate phase) (ca. 250–216 Ma)	–	CTR (CA-VA, Central Cordillera), MSP (Santander massif, SNSM), Garzón massif	Extensional, rifting

(continued)

Table 5.2 (continued)

Time period (magmatic episode)	Colombian tectonic phase (with age of associated granitoids)	Regional temporal comparatives (Orogenies)	Distribution of Colombian granitoids (basement domain)	Tectonic regime (Colombia)
Latest Triassic-Jurassic	Bolívar Aulacogen (late phase) (ca. 210–146 Ma)	–	CTR (San Lucas range, Central Cordillera and CA-VA (Segovia Batholith), MSP (Santander massif, SNSM), Garzón massif	Extensional, slab rollback
Early cretaceous	Bolívar Aulacogen (culminant phase)	–	No significant Granitoids	Extensional, rifting (Valle Alto Rift)
Mid-cretaceous-Eocene	Early Northern Andean Orogeny (ca. 100–42 Ma)	Andean Orogeny (Peruvian and Incaic phases), Peltetec melange (Ecuador), Laramide and Sevier Orogenies, North America	CTR (CA-VA, eastern group continental granitoids), SNSM, WTR (western group CCOP/CLIP gtoids)	Transpressional, collisional, accretionary
Earliest Oligocene-Mio-Pliocene	Late Northern Andean Orogeny (ca. 24–0.4 Ma)	Late Andean Orogeny, Perú (Quecha phase), late northern Andean Orogeny, Ecuador	WTR, RM, CTR (CA-VA), MSP (Santander massif), EC	Oblique to orthogonal compression, collisional, Accretionary, Nazca plate subduction, back-arc extensión?

MSP Maracaibo Sub-plate, *SNSM* Sierra Nevada de Santa Marta, *CTR* Central Continental Realm, *CA-VA* Cajamarca-Valdivia Terrane, *WTR* Western Tectonic Realm, *CCOP/CLIP* Caribbean-Colombian Oceanic Plateau/Caribbean Large Igneous Province, *RM* Romeral Melange, *EC* Eastern Cordillera

Phanerozoic granitoid suites which constitute the region as a whole (e.g. early Paleozoic and Carboniferous granitoids of the Cajamarca-Valdivia Terrane). In addition, we emphasize that most of the early Phanerozoic (meta-)granitoids are deeply eroded, deformed and metamorphosed and have been subject to a complex series of tectono-magmatic events following their emplacement and spanning the Meso-Cenozoic. Within this framework, and notwithstanding, the early Paleozoic,

Carboniferous and Permo-Triassic granitoids constitute the (albeit limited) principle magmatic record for almost 300 million years, that is, over half of the Phanerozoic tectonic and geologic record of the Colombian Andes.

In consideration of the above, we suggest that the interpretation of detailed tectonic frameworks for the early Paleozoic, Carboniferous and Permo-Triassic granitoids in Colombia (and the Northern Andean region in general) is a complex affair. This is exemplified by the observation that published tectono-magmatic models for early Phanerozoic granitoid magmatism in the Colombian Andes, as derived from recent integrated lithogeochemical and isotopic studies (see detailed reviews and summaries presented by Cochrane (2013), Cochrane et al. (2014a), Van der Lelij (2013), Van der Lelij et al. (2016) and Spikings et al. (2015), are presented within highly schematic global-scale paleo-geographic reconstructions, which are controversial (Van der Lelij et al. op. cit.); Cochrane et al. 2014a; Spikings et al. op. cit.) and often difficult to reconcile at scales applicable to the litho-tectonic and morpho-structural units comprising the Colombian Andes.

In the following discussion of Colombian early Phanerozoic tectono-magmatic development, we have opted to forgo large-scale and schematic paleo-geographic and tectonic reconstructions, which may be found in well-versed and readily accessible sources (e.g. Weber et al. 2007; Cochrane et al. 2014a; Van der Lelij et al. 2015; Spikings et al. 2015). Figure 5.29 depicts the actual distribution of early Paleozoic, Carboniferous and Permo-Triassic granitoids, as modified from the existing regional geological map base (Cediel and Cáceres 2000; Gómez et al. 2015a). The granitoids are depicted within the context of their host basement complexes, and the figure is annotated with information pertaining to the petrogenesis, tectonic environment and tectonic evolution of the region, as derived from the information sources utilized in diagram construction. No attempt at paleo-geographic reconstruction with respect to the distribution of granitoids from the various age groupings has been initiated. The early Phanerozoic tectono-magmatic evolution of the region is additionally summarized in time-space format, presented in Fig. 5.30.

5.4.1.1 Tectonic Framework for Early Paleozoic Granitoids: The Quetame Orogeny and Early Bolívar Aulacogen

During the latest Proterozoic to early Paleozoic, the composite basement of the paleo-Andean continental region in Colombia was comprised of the western margin of the Guiana Shield (Amazon Craton), the >ca. 1.2 Ga Chicamocha Terrane paleo-continental allochthon and an intervening belt of mid-Proterozoic (ca. 1.2–0.9 Ga) granulite-grade metamorphic rocks, petrogenetically dominated by recycled early-mid-Proterozoic continental crust (Fig. 5.29; see Sect. 5.2.1). This assemblage comprised the subsiding basement to thick deposits of autochthonous marine and epicontinental sediments of Vendian and Cambrian(?), late Ordovician, Silurian, Devonian and Carboniferous to Permian age (Cediel et al. 1994; Silva et al. 2005). In Colombia, early Paleozoic supracrustal sequences underwent Cordilleran deformation and regional, Barrovian-type, sub-greenschist to amphibolite-grade metamorphism (e.g. Goldsmith et al. 1971; Ward et al. 1973; Restrepo-Pace 1995), during what has been referred to in Colombia as the Quetame Orogeny (Cediel and Cáceres 2000; Cediel et al. 2003). Within the Northern Andean region, this event may be compared with the Caparonensis Orogeny in Venezuela, the Ocloy Orogeny in Ecuador and Perú as well as the Famantinian Orogeny of northern Argentina and the Taconian-Acadian and Caledonian orogenies, of North America and Northern Europe, respectively. Within Colombia's Eastern Cordilleran system, early Paleozoic granitoids associated with this orogenic framework are located within the Santander, Floresta and Quetame massifs (Figs. 5.29 and 5.30).

The composite geologic, radiometric age date, lithogeochemical and isotopic database for the early Paleozoic (e.g. Cediel et al. 1994; Cediel and Cáceres 2000; Horton et al. 2010; Leal-Mejía 2011; Mantilla et al. 2012; Van de Lelij 2013) permits an initial understanding of granitoid magmatism within the context of the Quetame orogenic cycle. Based upon recent, detailed lithogeochemical and isotopic analysis of granitoids from the eastern Colombian and Mérida (Venezuela) Andes, Van der Lelij (2013) and Van der Lelij et al. (2016) identified three phases of granitoid magmatism within the context of early Paleozoic tectono-magmatic development, which they integrate within the interpreted geodynamic evolution of the autochthonous pre-Andean margin. These include (1) early, ca. 499–473 Ma synkinematic and peak metamorphic granitoids, which they interpret to have been generated/emplaced during a period of compression, crustal thickening, metamorphism and orogenesis; (2) ca. 472–452 Ma granitoids, emplaced during post-orogenic collapse, extension and basin formation; and (3) ca. 452–415 Ma granitoids emplaced during resumed compression, basin closure and crustal thickening. Although these authors interpret the continual subduction of Iapetus oceanic crust beneath the NW Gondwana margin during the entire ca. 499–415 Ma period (see Fig. 5.15 of Van der Lelij et al. 2015), their detailed Hf, Sr, Nd and Pb isotope data led them to conclude that all of the ca. 499–415 Ma granitoids are primarily composed of recycled crustal melts, with increasing but minor contributions of enriched and depleted mantle material during the ca. 472–452 Ma period, facilitated by active extension and crustal thinning, respectively. In this context, the early Paleozoic granitoids apparently do not represent subduction-derived melts per se, and Van der Lelij et al. (2016) invoke a process of lithospheric mantle upwelling and heat advection at the base of the crust, in the generation and partitioning of primarily crustal-derived melts.

The data of Van der Lelij (2013) and Van der Lelij et al. (2016) did not include, however, the emerging population of early Paleozoic granitoids located significantly to the west of the Santander-Floresta-Quetame massifs, hosted within Cajamarca-Valdivia Terrane metamorphic basement which underlies much of Colombia's Central Cordillera. Cajamarca-Valdivia is stratigraphically comprised of polydeformed Vendian and early Paleozoic marine meta-sedimentary and volcanic rocks, including the Cajamarca, Valdivia and Montebello Groups (Restrepo-Pace 1992; Cediel and Cáceres 2000; González 2001; Silva et al. 2005). Geochemical and geological characterization studies presented by Restrepo-Pace (1992) and paleogeographic reconstructions presented by Cediel et al. (1994) and Cediel (2011) suggest Cajamarca-Valdivia (referred to as the Central Andean Terrane by Restrepo-Pace op. cit.) represents a peri-cratonic island arc and continental margin accretionary prism assemblage, developed along the western Colombian continental margin beginning in the Vendian-Cambrian and accreted along the Palestina fault and suture system during the late Ordovician-Silurian. In this context, early Paleozoic granitoids contained with the Cajamarca-Valdivia assemblage, including the ca. 473 Ma quartz diorite outcropping along the Otú Fault (Leal-Mejía 2011) and the ca. 479–445 Ma La Miel orthogneiss (Villagómez et al. 2011; Martens et al. 2014), represent granitoids emplaced within the peri-cratonic realm, as members of the Cajamarca-Valdivia arc complex. A more complete petrogenetic and tectonic characterization of these granitoids, unfortunately, is lacking, due to the absence of lithogeochemical analyses.

In view of the tectonic framework for early Paleozoic granitoid magmatism summarized herein, we suggest the phases of syn- and post-orogenic granitoid magmatism documented in the eastern Colombian Andes (Goldsmith 1971; Restrepo-Pace 1995; Van der Lelij 2013; Van der Lelij et al. 2015), and granitoids contained within the Central Cordillera (Villagómez et al. 2011; Leal-Mejía 2011; Martens et al. 2014) were developed within the context of the Ordovician-Silurian Quetame Orogeny as described by Cediel and Cáceres (2000) and Cediel et al. (2003). This orogeny appears to have been driven by the approach and accretion of the Cajamarca-Valdivia island arc assembly and closure of the Iapetus Ocean.

The youngest early Paleozoic granitoids from the Santander Massif and Cajamarca-Valdivia Terrane date from ca. 439 to 445 Ma, respectively. Existing U-Pb (zircon) age date data indicate a paucity of granitoid occurrences throughout the Colombian Andes, spanning the period from ca. 439 to 333 Ma (Figs. 5.4, 5.5, and 5.6), indicating a general hiatus in granitoid magmatism over a ca. 100 m.y. span.

The term Bolivar Aulacogen was originally proposed by Cediel and Cáceres (2000) and Cediel et al. (2003) to describe the prolonged period of continental taphrogenesis surrounding northwestern South America, beginning in the mid-late Paleozoic and continuing through to the early Cretaceous (Cediel et al. 1994; Cediel and Cáceres 2000; Cediel et al. 2003). In eastern Colombia and western Venezuela, this extensional regime initiated with the development of an intercontinental rift and deposition of marine strata in the Pennsylvanian through Permian (Sierra de Mérida, Eastern Cordillera). The extensional regime changed briefly to transpressive in the late Permian, as recorded by tight folds associated with strike-slip faulting observed in the Sierra de Mérida (Marechal 1983). Rifting resumed during the Triassic (e.g. Payandé rift, Cediel and Cáceres 2000), continued into the Jurassic (e.g. Morrocoyal rift, Geyer 1973; Siquisique rift, Bartok et al. 1985; Perijá rift, Cediel and Cáceres 2000) and culminated in the early Cretaceous with the opening of the Valle Alto rift (Cediel and Cáceres 2000), prior to the onset of the transpressive regime characteristic of the proto-Northern Andean Orogeny.

Within the context of the Bolívar Aulacogen, granitoid development may be considered within three stages. As observed above, the initial phase of the Bolívar Aulacogen was essentially amagmatic, coinciding with a hiatus in granitoid magmatism in Colombia extending from ca. 439 to 333 Ma. The intermediate phase is demonstrated by an emerging record of magmatism beginning in the mid-Carboniferous and extending into the Permian and mid-late Triassic, dominated by granitoid anatectites and bimodal granitoid-gabbo (amphibolite) assemblages. The late phase is characterized by subduction-related volcano-plutonic arc systematics, developed within a highly extensional regime during the Jurassic (Sect. 5.4.2.1). We will now outline the development of granitoid magmatism during the intermediate phase of the Bolívar Aulacogen, from the mid-Carboniferous to the Permo-Triassic.

5.4.1.2 Tectonic Framework for Carboniferous Granitoids

Regional-scale tectonic reconstructions for the Carboniferous of the Northern Andes depict a generally passive margin. Interpreted north- to west-directed subduction was localized along the conjugate Laurentian margin prior to the final amalgamation of Pangaea in the Permian (e.g. Keppie 2008; Ramos 2009; Van der Lelij et al. 2016). In Colombia, detailed basin and facies analysis suggests the ca. 333–310 Ma period was a time of flysch-type sedimentation and of general magmatic quiescence (Cediel et al. 1994; Cediel et al. 1998; Cáceres et al. 2003). Indeed, Van der Lelij (2013) notes that there is little evidence (on a regional level) to support the existence of an active margin outboard of northwestern Gondwana between ca. 415 Ma and 290 Ma.

Notwithstanding, Leal-Mejía (2011) documented Carboniferous, ca. 330–310 Ma granitoid magmatism in the El Carmen-El Cordero Stocks, hosted within Cajamarca-Valdivia Terrane basement along the Otú Fault in Colombia's northern Central Cordillera (Figs. 5.3, 5.5, and 5.6). Available petrographic, lithogeochemical data (Leal-Mejía 2011; see Sect. 5.3.2.3, Figs. 5.7 and 5.8) denote an apparently bimodal gabbro-melanodiorite-leucotonalite assemblage at El Carmen-El Cordero. The suite is of metaluminous-weakly peraluminous, magnesian-calcic composition (Frost et al. 2001), and all samples return strongly mantelic Sr-Nd isotope signatures (Fig. 5.9). Based upon the low-K, hydrous nature of the El Carmen-El Cordero granitoids, the suite does not appear to represent an A-type (Loiselle and Wones 1979; see Frost et al. 2001) assemblage. A general calc-alkaline trend may be implied on the AFM diagram, although this is inconclusive, given gaps in intermediate compositions within the differentiation series. This may simply reflect the limited sample population upon which the present classification in based (n = 7). Overall, however, utilizing the classification scheme of Barbarin (1999), the El Carmen-El Cordero suite conforms well to mantle-derived, "tholeiitic" granitoids, of the RTG (Ridge Tholeiitic Granitoids) type. RTG suites characteristically include gabbro through tonalite, trondhjemite and plagiogranite assemblages which are Na-rich and mantelic with low $^{87}Sr/^{86}Sr$ ratios. Such suites are interpreted to be associated with tholeiitic, gabbro-dominant assemblages generated along oceanic spreading ridges. The more felsic members of the series are derived in small volumes through extreme crystal fractionation of basaltic melts and occur as dikes and plutons hosted within ophiolite complexes/oceanic crust (Barbarain 1999 and references cited therein).

In terms of age, lithochemistry and Sr-Nd isotope composition (Leal-Mejía 2011), the El Carmen-El Cordero suite presently stands unique, not only in the Colombian Andes but for the entire Northern Andean region. The age of these intrusives significantly pre-dates the age of the well-documented Permo-Triassic arc-related (e.g. Cardona et al. 2010b) and bimodal meta-granitoids, granitoid gneisses and amphibolites (Vinasco et al. 2006; Cardona et al. 2010b; Cochrane 2013; Spikings et al. 2015; see below). As mentioned, the geological context of the El Carmen-El Cordero suite is not fully understood. The granitoids are hosted within the confines of the Cajamarca-Valdivia Terrane and, based upon age constraints, were emplace at least 70 m.y. after accretion of the Cajamarca-Valdivia assemblage to continental Colombia. The granitoids are localized along the Otú Fault, a major N-S striking feature, which in the past has been interpreted as a potential plate boundary (e.g. Restrepo and Toussaint 1988; González 2001). Lithogeochemical data supplied by Leal-Mejía (2011) and summarized herein suggest the El Carmen-El Cordero granitoids represent a RTG suite (Barbarain 1999) complete with low-volume leucotonalite and trondhjemite differentiates, petrogenetically associated with oceanic spreading and ophiolite formation. We suggest that the El Carmen-El Cordero suite reflects the progressively extensional environment prevalent during the intermediate stages of the Bolívar Aulacogen. The Otú Fault could represent the longitudinal axis of a rift basin which opened to the point of at least locally producing oceanic lithosphere. The U-Pb (Zircon) ages produced to date by the El Carmen-El Cordero assemblage suggest the Otú rift was active over a > 23 m.y. period. No additional geological (sedimentological) record of the basin is known to exist within the region, although it could be contained within the poly-deformed metamorphic sequences of the Central Cordillera which have yet to be accurately dated. Alternatively, it may have been mostly removed by erosion during Meso-Cenozoic tectonic events. In either case, rifting and basin formation along the Otú Fault were short-lived and appear to have been aborted by the early Permian(?). An ensuing period of granitoid quiescence is observed, between ca. 310 and 289 Ma prior to the appearance of a new population of granitoid gneisses, granitoids and amphibolites with petrographically, lithogeochemically and isotopically distinct characteristics, during the Permo-Triassic.

5.4.1.3 Tectonic Framework for Permian to Mid-Late Triassic Granitoids

Granitoids returning Permian U-Pb (zircon) dates appear in the Colombian Andes at ca. 289 Ma, and granitoid magmatism sensu lato continued throughout the Permian and into the mid-late Triassic. In recent years, numerous workers have produced disparate, localized radiometric age date, lithogeochemical and isotopic data pertaining to the Permo-Triassic granitoid suite (e.g. Ordoñez and Pimentel 2002; Saenz 2003; Cardona et al. 2010b; Leal-Mejía 2011; Villagómez et al. 2011; Van der Lelij 2013; Rodríguez et al. 2014), whilst detailed, integrated studies focussed specifically upon these rocks have been undertaken by Vinasco (2004) and Cochrane (2013). Upon integration of these studies, a composite understanding of the tectonic framework of Permo-Triassic granitoid magmatism can be derived.

Global tectonic reconstructions proposed by various authors (Keppie and Ramos 1999; Keppie 2004; Cocks and Torsvik 2006; Weber et al. 2007; Cardona et al. 2010b; Van der Lelij 2013) suggest that during the final assembly of Pangaea in the Permian, northwestern South America was positioned along the WNW-margin of Gondwana, at a complex juncture between Gondwana, Laurensia and numerous loosely assembled pericratonic terranes, accumulated during closure of the Rheic Ocean (e.g. the Middle American and Mexican terranes) but tangential to the principle Gondwana-Laurentia suture (Ouachita-Marathon front; Keppie 2008; Cochrane et al. 2014a; Van der Lelij 2013). Following Pangaea assembly most cartoons depict the development of a west-facing subduction zone, suggesting the eastward subduction of proto-Pacific oceanic crust along much of the western Pangaean margin (e.g. Cocks and Torsvik 2006; Keppie 2008; Cardona et al. 2010b; Cochrane et al. 2014a; Van der Lelij et al. 2016).

In this context, based upon whole-rock lithogeochemical major, trace and rare-earth element analyses, Cardona et al. (2010b) and Villagómez et al. (2011) interpret Permian meta-granitoids and granitoid gneisses outcropping in the Sierra Nevada de Santa Marta (ca. 288–264 Ma) and the Central Cordillera (ca. 272 Ma), respectively, to represent vestiges of early Permian, subduction-driven continental margin magmatic arcs. We note, however, that the textural, mineralogical, structural and lithogeochemical characteristics of these occurrences share many of the features typical of the entire Permian through mid-late Triassic granitoid suite (Figs. 5.7 and 5.8), including the complexly zoned nature of contained zircons which produce multiple inheritance ages. We suggest that in the absence of more in-depth isotopic studies (Sr, Nd, Pb, Hf), it is premature to assign these meta-granitoids to a specific tectonic environment based upon lithogeochemical analyses alone, especially given the complexities which have historically been encountered in the interpretation of other Colombian granitoid suites (e.g. early Paleozoic meta-granitoids, early Jurassic Santander Plutonic Group).

Vinasco (2004) and Vinasco et al. (2006) produced a detailed and integrated petrographic, U-Pb (zircon), ^{40}Ar-^{39}Ar and Sr-Nd isotope and composite lithogeochemical study of Permo-Triassic granitoid gneisses and less deformed granitoids from several locations in Colombia's Central Cordillera, and it was these authors who were first to recognized the regional distribution and significance of this meta-granitoid suite. Vinasco et al. (2006) observed that inherited zircons from syntectonic peraluminous granitic gneisses returned ca. 280 Ma metamorphic ages, whilst ca. 250 Ma ages were returned from neoformed zircons. The less deformed crustal granitoids returned ages of ca. 230 Ma. They demonstrated that, although individual samples plot medium- to high-K calc-alkaline in composition, the suite is consistently peraluminous (S-type) and that Sr-Nd isotope data suggest high degrees of interaction, assimilation or derivation of magma from upper crustal sources. They note that isotopic data for the ca. 230 Ma suite reveals increasing contributions of juvenile mantle. Vinasco et al. (2006) suggest that the meta-granitoid suite is the product of regional Permo-Triassic tectono-thermal orogenesis associated with the assembly and break-up of the Pangaea supercontinent. A genetic model presented by Vinasco et al. (2006) suggests the Permian to mid-late Triassic suite records collision-related metamorphism at ca. 280 Ma, followed by crustal thickening and the emplacement of syn-kinematic (gneissic) peraluminous granitoids at ca. 250 Ma. Orogenic collapse led to the emplacement of late tectonic granitoid intrusions at ca. 230 Ma., marking the onset of Pangaea break-up in the Northern Andean region.

In addition to documenting the age and nature of Permo-Triassic granitoids in central Colombia, Vinasco (2004) and Vinasco et al. (2006) also observed the spatial and temporal relationship between peraluminous granitoids and amphibolite, on both the eastern flank (e.g. Padua amphibolite) and western flank (e.g. El Retiro amphibolite and Aburrá ophiolite) of the Central Cordillera. These authors suggested the amphibolites represent mantle-derived mafic melts which played a role in crustal anatexis and the overall petrogenesis of the ca. 230 Ma peraluminous granitoids, during mid-late Triassic regional extension.

In a more recent study pertaining to the late Paleozoic-Cenozoic tectonic evolution of the Northern Andean region, Cochrane (2013) and Cochrane et al. (2014a) provided additional lithogeochemical and isotopic analyses of the Permo-Triassic meta-granitoids, including a detailed analysis of the spatially related amphibolites from the Andes of both Colombia and Ecuador. The findings of these authors concur with those suggested by Vinasco et al. (2006). Cochrane (2013) notes that important lithogeochemical and isotopic features of ca. 275–240 Ma meta-granitoids include whole-rock $(La/Yb)_N$ ratios of ca. 11–16, generally magmatic zircon Th/U ratios of 0.26–1.27 and zircon εHf_i values between +2 and −12, which he interprets as consistent with anatectites generated via relatively low degrees of crustal melting, including a minimal juvenile component. Cochrane (op. cit.) concludes that Permian-early Triassic granitoid magmatism in NW South America likely occurred as a consequence of the collision and final amalgamation of western Pangaea, although he notes that the composite lithogeochemical and isotopic data do not unambiguously constrain the specific tectonic environment within which the Permian-earliest Triassic anatectites formed.

Beginning at ca. 240 Ma, Cochrane (2013) and Cochrane et al. (2014a) document the emplacement of anatectic granitoids accompanied by the appearance of tholeiitic sills and dikes (amphibolites). These authors observed that most of the post ca. 240 Ma crustal anatectites yield large intra-sample εHf_i variations and much lower $(La/Yb)_N$ and Th/U ratios than the Permian-early Triassic meta-granitoids, potentially reflecting source mixing with coeval juvenile mafic magmatism. With respect to the amphibolites emplaced between ca. 240 and 216 Ma, analyses presented by Correa (2007) and Cochrane et al. (2014a) reveal tholeiitic N-MORB to Back Arc Basin Basalt (BABB) compositions. Cochrane et al. (2014a) note that early (ca. 240–232 Ma) mantle-derived tholeiites with εHf_i values from +7.4 to +11.2 yield some zircon εHf_i values which suggest that older amphibolites assimilated continental crust, whilst the ca. 232–216 Ma amphibolites reveal diminished crustal contamination and incrementally juvenile isotopic compositions. Cochrane et al. (2014a) present a model for the ca. 240–216 Ma anatectic granitoids and juvenile amphibolites involving the thinning of continental lithosphere during Pangaea break-up. They suggest the rift stage of continental disassembly involved basaltic underplating which led to emplacement of mafic melts and, in turn, anatectic melting of the continental crust. Based upon the collective data, they conclude that rifting led to sea-floor spreading after ca. 223 Ma with ocean crust formation occurring by ca. 216 Ma (Correa 2007; Cochrane et al. 2014a).

The foregoing tectonic models for Permo-Triassic granitoids in Colombia are based primarily upon lithogeochemical, isotopic and petrogenetic arguments. They provide important temporal and spatial constraints with respect to existing models which demonstrate the taphrogenic character of the intermediate phases of the Bolívar Aulacogen during the Permo-Triassic, as derived primarily from surface geological and borehole mapping, geophysical studies and sedimentary facies and basin analysis (e.g. Cediel et al. 1994; Cediel et al. 1998; Cediel and Cáceres 2000). The magmatic vs. sedimentary-based models are particularly sympathetic beginning with the onset of mid-late Triassic continental rifting.

With respect to the Permian tectonic assembly of Pangaea, however, some degree of controversy surrounds the nature and extent of the effects of continental collision and tectono-thermal metamorphism in Colombia, and in various locations, it is difficult to reconcile early Permian subduction(?) and continental collision within the taphrogenic context of the Bolivar Aulacogen. For example, based upon the RTG assemblage observed at El Carmen-El Cordero, and discussed above, an extensional regime is observed into the late Carboniferous. Within the Quetame Massif, along the Sumapaz Range, a near-complete upper Paleozoic to early Mesozoic stratigraphic section is preserved (Cediel and Cáceres 2000), containing carbonate and evaporate sequences of Carboniferous through early to middle Permian age, which show no obvious tectono-metamorphic effects. The El Carmen-El Cordero granitoid suite also provides a case in point for this geological quandary. As documented above, the El Carmen-El Cordero suite is of mid-Carboniferous age (ca. 333–310 Ma), predating the Permian-early Triassic tectono-thermal event by over 30 million years. Detailed petrographic study of the El Carmen-El Cordero suite by Leal-Mejía (2011), however, failed to reveal significant post-crystallization penetrative deformation or metamorphic mineral assemblages beyond the pumpellyite-prehnite-chlorite-epidote grade, an assemblage which could just as easily have resulted from the low-temperature hydrothermal alteration which affects the suite (Leal-Mejía 2011; Shaw et al. 2018).

Notwithstanding, cartoons depicting the relative position of continental Colombia within Gondwana and with respect to Laurentia, the Middle American-Mexican terranes and the Ouachita-Marathon front during the late Carboniferous-early Permian are highly speculative, and the majority of the recent global-scale reconstructions suggest the region was peripheral to the principle Gondwana-Laurentia suture (e.g. Keppie 2008; Weber et al. 2007; Cadona et al. 2010b; Van der Lelij 2013; Cochrane et al. 2014a). A Permo-Triassic suture per se has yet to be clearly documented within the context of Colombia-based paleo-tectonic reconstructions (e.g. Cediel et al. 1994), and the proximity of continental Colombia to the Ouachita-Marathon front remains largely undetermined. The highly complex nature of the western Pangaea juncture is evident, and the potential role of the Middle American-Mexican terranes in stress field buffering along the collision zone has yet to be evaluated. It is intuitive that the presence of numerous small peri-cratonic crustal fragments will have a first-order effect upon the development of a well-defined or easily identifiable suture trace, especially if the region was located in a tangential position with respect to the principle collision front. From a structural standpoint, tight folds associated with late Permian strike-slip faulting in the Sierra de Mérida (Marechal 1983) may provide a record of Pangaean assembly from within the continental autochthon. Further investigations regarding the Permo-Triassic tectono-thermal event on a Colombian vs. regional scale are clearly warranted.

In conclusion, granitoid magmatism within the Colombian Andes during the Permian through mid-late Triassic is represented by widespread but generally small-volume occurrences of granitoid gneisses and anatectites, observed primarily within the Cajamarca-Valdivia Terrane underlying much of the Central Cordillera but also within the Sierra Nevada de Santa Marta and, to a lesser degree, in the Santander Massif. Based upon the data sets compiled herein, these granitoids provide a magmatic record reflecting the tectonic history of western Pangaea during the Permian and Triassic (e.g. Vinasco et al. 2006; Cochrane et al. 2014a). The granitoids are clearly characterized by their ubiquitous peraluminous (S-type) nature, contrasting markedly with the voluminous metaluminous granitoids dominating the Colombian Andes during the Meso-Cenozoic. Most authors concur that the Permian to early Triassic meta-granitoids and granitoid gneisses provide a record of crustal thickening and anatexis coincident with Pangaea amalgamation, whilst the bimodal mid-late Triassic peraluminous granite-amphibolite suite reflects continental rifting, culminating in ocean crust formation during Pangaea disassembly. Finally, we note that the role of subduction, as depicted in numerous large-scale paleo-tectonic reconstructions of the western Pangaean region (e.g. Weber et al. 2007; Cardona et al. 2010b; Cochrane et al. 2014a and references cited therein), and the contribution of subduction-derived melts (e.g. Cardona et al. 2010b; Villagómez et al. 2011), in the petrogenesis of the Colombian Permo-Triassic granitoids, have yet to be clearly demonstrated.

With the onset of oceanic rifting and advanced continental break-up along the Colombian proto-Pacific margin, the intermediate phase of the Bolívar Aulacogen, characterized by low-volume peraluminous granitoids, gave way to a regime permissive to the emplacement of voluminous, subduction-related metaluminous granitoids. The development of large-scale Jurassic batholiths accompanied by abundant volcanic rocks, emplaced within an extensional regime, during the late phase of development of the Bolívar Aulacogen will now be discussed.

5.4.2 Late Bolívar Aulacogen: Tectonic Framework for Latest Triassic-Jurassic Granitoids

Schematic models for the late Triassic-Jurassic structural and tectonic evolution of Colombia and the Northern Andes have been presented by numerous authors over the last five decades, including Bürgl (1967), Irving (1975), Sillitoe et al. (1982), Burke et al. (1984), Etayo-Serna et al. (1983), Aspden et al. (1987), Restrepo and

Toussaint (1988), Pindell et al. (1988), Cediel et al. (1994), Cediel and Cáceres (2000), Pindell and Kennan (2001), Cediel et al. (2003), Kennan and Pindell (2009), Cochrane et al. (2014b) and Spikings et al. (2015). Many of these models were imprecise, incomplete and/or overly selective with respect to the composite database of geological and cartographic information available for the region or, alternatively, were drawn at scales encompassing all of NW South America and the Caribbean, which did not permit the exposition of detailed and specific geological, stratigraphic, radiometric age date, lithogeochemical and isotopic information.

In order to update and better constrain these models, at a scale specifically representative of the Colombian Andes, we have integrated the late Triassic-Jurassic radiometric age, isotopic and lithogeochemical information presented above into the detailed paleo-facies, structural and tectonic framework provided by Cediel et al. (1994). The resulting composite late Triassic-Jurassic tectono-magmatic configuration is presented in Fig. 5.31. A summarized time-space analysis for the magmatic evolution of the region during the late Triassic-Jurassic is illustrated in Fig. 5.32.

The transition from middle to late Triassic rifting and continental break-up to the formation of late Triassic-Jurassic subduction-related magmatic arcs, marking the late phase of the Bolívar Aulacogen, is first recorded in Colombia in the ca. 210–196 Ma granitoids of the Santander Plutonic Group. Spikings et al. (2015) interpret the formation of a proto-subduction zone along the NW Colombian (Gondwana) margin beginning around this time. Late Triassic-Jurassic rift-related sedimentation in the Maracaibo and Perijá Rifts (Cediel et al. 1994; Cediel and Cáceres 2000; Cáceres et al. 2003; Cediel et al. 2003) indicates active rifting accompanied by emplacement of the Santander granitoid suite. Based upon the NNW orientation of the long axis of the Santander suite, initial subduction (if present) was broadly NE-directed. Although the granitoids are interpreted to have been emplaced in a continental arc setting, lithogeochemical and isotopic data indicate melts were primarily derived from, or mixed with, crustal sources, with a limited mantelic component (Van de Lelij 2013; Bissig et al. 2014). A crustal source is in keeping with the lithogeochemical and isotopic composition of Proterozoic and early Paleozoic metamorphic basement rocks of the Santander Massif which host the Santander Plutonic Group. Magma generation may be more specifically related to extension-related mantle upwelling and thermal-induced partial melting of lower crustal basement underlying the Santander Massif than to the subduction and partial fusion of oceanic lithosphere per se, as would be implied in typical models for arc-related, calc-alkaline granitoids. In either case, extension was insufficient to allow the wholesale entry of mantle-derived melts into the upper crust (Van der Lelij 2013).

Following ca. 196 Ma, WNW migration of the calc-alkaline magmatic arc axis is observed (Fig. 5.31). The ca. 189–180 Ma granitoids of the southern Ibagué, Norosí, San Martín and Pueblo Bello-Patillal Batholiths and the ca. 180–172 Ma Mocoa-Garzón intrusions represent extensive subduction-related magmatism with a clear metaluminous character, increasing mantelic component and diminishing degree of interaction with sialic continental basement (Alvarez 1983; Dörr et al. 1995; Leal-Mejía et al. 2011; Cochrane 2013). Arc axis migration was accompanied by ~30 degrees of clockwise rotation of the long axis of the arc into a NNE orientation, suggesting a shift to broadly SE-oriented subduction. A marked increase in magma volume is represented by the ca. 189–172 Ma granitoids, all of which include a significant explosive volcanic component (e.g. the Saldaña, Noreán, Jordán and Guatapurí Fms.). The ca. 189–172 Ma arc segments were thus emplaced under highly extensional conditions, in some instances coaxial to precursor Permo-Triassic rift-related sedimentary grabens (e.g. Payandé Rift and Ibagué Batholith; Cediel et al. 1994; Cediel and Cáceres 2000). The slight eastward migration of the Mocoa-Garzón intrusions with respect to the southern Ibagué Batholith suggests the onset of a locally compressional regime in southern Colombia at the end of this magmatic cycle, possibly related to declining rates of extension and/or shallowing of the oceanic slab subduction angle.

Continued WNW migration of the magmatic arc axis is observed with the emplacement of the ca. 168–155 Ma Segovia and northern Ibagué Batholiths (Fig. 5.31). Further shifts to more juvenile, mantle-derived compositions are observed for these batholiths, with lesser REE enrichment and Sr-Nd isotope ratios trending into the depleted mantle array. Notwithstanding, the absence of associated volcanic piles or evidence of coeval volcanism suggests these granitoids were emplaced within an increasingly neutral to compressive tectonic regime. The erosion of significant Jurassic volcanic stratigraphy during Cenozoic Northern Andean orogenic events cannot however be ruled out.

To the south, in the southern Ibagué Batholith, no granitoids dating from the ca. 168 to 155 Ma episode are observed, and based upon the available data, no additional Jurassic granitoid magmatism is recorded for this area, signifying the shutdown of subduction by ca. 172 Ma. We interpret the development of NW–SE-striking transform fault or slab tear in the Pacific Plate (Fig. 5.31), to the north where subduction continued between ca. 168 and 155 Ma, whilst to the south a complete shutdown of the Jurassic arc in Colombia is observed.

Following the final episode of holocrystalline intrusions at ca. 152 Ma, volumetrically minor hypabyssal porphyry stocks were emplaced along the eastern (back arc) margin of the northern Ibagué Batholith between ca. 152 and 145 Ma (Fig. 5.31). They record, if anything, a net eastward migration of magmatism, suggesting the (temporary) cessation of regional extension and a trend towards a more neutral to compressive tectonic conditions during closure of late Triassic-Jurassic arc-related granitoid magmatism.

The late Triassic-Jurassic granitoids of the Colombian Andes were generated within a highly complex tectonic regime involving the early rifting and break-up of western Pangaea and the separation of the Middle American terranes, followed by the continuous broadly east-directed subduction of Pacific oceanic crust beneath NW South America. The net result of this tectonic evolution was the temporal development of four major granitoid episodes, with associated volcanism and hypabyssal porphyry emplacement, manifest in at least six spatially distinct arc segments, emplaced within a highly extensional tectonic regime. Hamilton (1994) notes that continental margin magmatic arcs are extensional by nature, as recorded in the development of back-arc basins and arc-axial grabens. He cites slab-pull (rollback) due to the sinking of dense fore-arc oceanic lithosphere into the mantle as the major factor in the development of extension across a magmatic arc. Leal-Mejía (2011), Cochrane et al. (2014b) and Spikings et al. (2015) considered Pacific oceanic slab rollback an important cause of WNW granitoid arc migration in northern Colombia between ca. 210 and 152 Ma. In addition to the extensional effects caused by slab rollback, however, a net SE-directed movement vector for the South American Plate, nearly opposite that of slab rollback, has been proposed for most of the Jurassic and early Cretaceous (e.g. Kennan and Pindell 2009). Thus, we interpret the tectonic framework for reactivation of pre-Mesozoic basement structures, the development of middle to late Triassic continental rifts and the emplacement of the late Triassic-Jurassic subduction-related granitoids in Colombia to be a reflection of the extreme extensional conditions brought on by the combination of Pacific slab rollback and SE-directed migration of the South American Plate throughout the Jurassic.

5.4.2.1 Culmination of the Bolivar Aulacogen: Valle Alto Rift

Paleo-tectonic reconstructions suggest that interactions between the Pacific Plate and NW South America became highly oblique or strike-slip during the Jurassic-Cretaceous transition (Cediel et al. 1994; Cediel et al. 2003; Keppie 2004; Kennan and Pindell 2009; Cochrane 2013; Spiking et al. 2015), leading to shutdown of the subduction-driven granitoid magmatism which dominated the Jurassic. After ca. 145 Ma, the geological record confirms a rift-dominated tectonic regime, with the formation of juvenile oceanic crust along the Colombian Pacific margin (Nivia et al. 2006; Cochrane 2013; Spikings et al. 2015) and opening of the early–middle Cretaceous Valle Alto-Eastern Cordillera Basin Rift (Cediel et al. 1994; Cediel and Cáceres 2000; Cediel et al. 2003). This event was marked by deep continental rifting and subsidence, the invasion of the Cretaceous seaway and the deposition of marine and epicontinental sequences over extensive areas of the Central Tectonic Realm (including the Cajamarca-Valdivia Terrane), the Maracaibo Sub-plate and the continental platform of the Guiana Shield (e.g. see Sarmiento 2018) (Fig. 5.33). The axis of the Valle Alto rift is marked by Colombia's Eastern Cordilleran basin, which contains up to 6 km of Cretaceous marine deposits characterized by a transgressive sequence of basal, restricted marine mudstones, carbonates and evaporates overlain by progressively deeper water, reduced (carbonaceous) shales and mudstones, deposited in at least four diachronous subbasins (Sarmiento 2001). Small volumes of compositionally heterogeneous rift-related alkaline and tholeiitic mafic intrusions mark periods of maximum extension, subsidence and subbasin development (Fabre and Delaloye 1983; Vásquez et al. 2010). The mafic intrusions range in age from ca. 136 to 74 Ma (Fabre and Delaloye 1983; Vásquez et al. 2010). Lithogeochemical and isotopic data published by Vásquez et al. (2010) demonstrate the mantle-derived character and variable degrees of LREE enrichment and contribution of old crustal material to the parent melts. The oldest intrusions (Pacho, ca. 136 Ma) plot in the field of "continental basalts", reflecting the continental character of the early rifted crust beneath the Eastern Cordillera, whilst the younger intrusions reveal lithogeochemical and isotopic data which is progressively more ocean-like (Vásquez et al. 2010). Additional rift-related Cretaceous marine volcano-sedimentary deposits (e.g. San Pablo, Segovia, Valle Alto and Soledad Fms., Fig. 5.33) (Gonzalez 2001) are found as localized erosional remnants within Colombia's Central Cordillera.

Along the Colombian Pacific margin, the period spanning the latest Jurassic through ca. 124 Ma was under left lateral transtension (Cediel et al. 1994; Kennan and Pindel 2009; Fig. 5.33) and formed an active depocenter for Berriasian through Aptian and Albian sedimentary rocks of continental margin and oceanic affinity and mixed assemblages of tholeiitic and calc-alkaline basalt and andesite, with associated mafic and ultramafic intrusive rocks (e.g. Quebradagrande Complex, Nívia et al. 1996). This marginal basin also contained disjointed slivers of early Paleozoic and Permo-Triassic metamorphic rocks (e.g. Bugalagrande complex; McCourt and Feininger 1984; Arquía Complex; Nívia et al. 1996) typical of the rifted Northern Andean continental margin during the early Cretaceous (Litherland et al. 1994; Cediel et al. 2003). Plate reorganization beginning in the Aptian (Cediel et al. 1994; Maresch et al. 2000; Pindell and Kennan 2001) led to deep burial, metamorphism and tectonic reworking of the marginal basin assemblages along the Colombian margin (e.g. Orrego et al. 1980; McCourt and Feininger 1984; Maresch et al. 2000; Bustamante 2008; Maresch et al. 2009), accompanied by large-scale dextral-oblique transpressive shearing along the Romeral fault system (Ego et al. 1995). The complex tectonic architecture of the Romeral mélange (Cediel and Cáceres 2000; Cediel et al. 2003) was established at this time.

In the early Cretaceous, the Colombian Pacific was thus dominated by a rifted transtensional-transform margin and by plate movement vectors, which, from a tectono-magmatic standpoint, were not conducive to the formation of subduction-related granitoids (e.g. Aspden et al. 1987; Cediel et al. 1994; Pindell and Kennan 2001). This observation is principally supported by the absence of subduction-related, calc-alkaline granitoids in continental Colombia during the period from ca. 145 to 96 Ma (Fig. 5.4), suggesting little, if any, subduction took place beneath the Colombian continental margin during this time.

Prolonged regional extension related to the Bolivar Aulacogen and the culminant Valle Alto rift, and the ensuing ca. 50 Ma hiatus in granitoid magmatism in the Colombian Andes, is terminated in the mid- to late Cretaceous, when plate reconfiguration in the Pacific regime led to dextral oblique convergence along the Colombian margin (Figs. 5.33 and 5.34). This shift signalled the onset of the late Mesozoic-Cenozoic Northern Andean Orogeny (Cediel and Cáceres 2000; Cediel et al. 2003), comprised of a series of punctuated tectono-magmatic events, including the generation of subduction-related, calc-alkaline, continental margin and peri-cratonic volcano-magmatic arcs and the sequential approach, collision and accretion of the Western Tectonic Realm allochthonous terrane assemblages of Pacific provenance along the Colombian Pacific and Caribbean margins.

The tectonic evolution of Colombia and the Northern Andes during this time was intimately linked to the opening of the Proto-Caribbean basin and to the genesis and emplacement of the Caribbean Plate (e.g. Cediel et al. 1994; Kerr et al. 1997; Pindell and Kennan 2001; Cediel et al. 2003; Kerr et al. 2003; Kennan and Pindell. 2009). As highlighted in the following section, significant volumes of subduction-related granitoids reappear in the Colombian Andes at ca. 96 Ma, with emplacement of the precursor phases of the Antioquian Batholith (Leal-Mejía 2011).

5.4.3 Early Northern Andean Orogeny: Tectonic Framework for Cretaceous-Eocene Granitoids

Within the historical context, the Northern Andean Orogeny in Colombia has been described by various authors (e.g. Bürgl 1967; Campbell 1974; Irving 1975). General disagreement was observed, however, with respect to the timing and spatial distribution of events, especially concerning the timing of deformation and granitoid magmatism. Based upon integrated time-space analysis and considering the nature and geological history of the pre-Andean tectonic framework, Cediel et al. (2003) redefined the Northern Andean Orogeny to include orogenic events occurring since the transition from the generally extensional-transtensional regime of the Bolivar Aulacogen to the transpressive (accretionary) regime beginning in the mid-Cretaceous (Aptian-Albian) and continuing up to the present. In historic works, the driving mechanisms behind deformation and magmatism were poorly understood. In recent times, however, numerous works demonstrate the sequential tectonic evolution of the Colombian Andes and the integral relationship between the nature, composition, migration and emplacement of the Caribbean Plate and the tectonic development of the Northern Andean Block as a whole (e.g. Cediel et al. 1994; Kerr et al. 1997; Sinton et al. 1998; Pindell and Kennan 2001; Cediel et al. 2003; Kerr et al. 2003; Kennan and Pindell 2009; Nerlich et al. 2014; Cediel 2011).

With respect to the development of granitoid magmatism during the period spanning the mid-Cretaceous through Eocene, as observed in outcrop, recorded upon regional scale geological maps (Cediel and Cáceres 2000; Gómez et al. 2007; Gómez et al. 2015a) and verified by the available radiometric age dating studies highlighted above, two groups of granitoids within the Colombian Andes, including the Eastern and Western Groups, may be defined. Each of these groups has been subdivided into subgroups, based primarily upon the age vs. spatial distribution of the granitoid intrusions (Figs. 5.15 and 5.16). Thus, within the Eastern group, the ca. 96–72 Ma Antioquian Batholith and satellite plutons and the ca. 62–50 Ma intrusions to the south of the Antioquian Batholith suite, and in the Sierra Nevada de Santa Marta, may be considered. Within the Western Group, the ca. 100–84 Ma and ca. 50–42 Ma subgroups are highlighted. We emphasize that, based upon geological setting and geotectonic considerations, supported by lithogeochemical and isotopic arguments, the Eastern and Western groups reflect fundamental differences in petrogenesis and mode of emplacement: the Eastern Group is autochthonous intrusions generated in situ within the continental regime, whilst the Western Group is allochthonous in nature, generated within the intra-oceanic regime, prior to accretion to the Colombian continental margin.

5.4.3.1 Tectonic Setting for the Ca. 96–72 Ma Antioquian Batholith Arc Segment

Figure 5.34 presents a schematic representation of the composite tectonic setting of the mid-Cretaceous-Eocene, taking into account the distribution of granitoid rocks from this time period, as presently recognized within the geologic mosaic of the Colombian Andes. Figure 5.35 contains a detailed time-space analysis for the ca. 100–40 Ma granitoids within the context of the established tectonic elements of the region.

Various tectonic reconstructions for the region surrounding NW South America (e.g. Pindell and Kennan 2001; Cediel et al. 2003; Kennan and Pindell 2009; Wright and Wyld 2011; Spikings et al. 2015; Weber et al. 2015), in conjunction with U-Pb (zircon) age dating of Colombian granitoids presented herein, suggest initiation of E- to NE-directed, dextral-oblique subduction beneath the western Colombian margin, beginning at ca. 100 Ma, resulting in the appearance of metaluminous, calcic to calc-alkalic continental arc granitoids beginning at ca. 96 Ma.

With respect to the Antioquian Batholith and its suite of satellite plutons, magmatism was generated along a west-facing arc segment, within the Colombian continental block, represented by the Cajamarca-Valdivia Terrane. Three important magmatic pulses have been identified, including early calcic gabbros and diorites emplaced at ca. 96–92 Ma, followed by main phase batholith emplacement including two distinct tonalitic to granodioritic suites, in two pulses, from ca. 89 to 82 Ma and from ca. in 81 to 72 Ma, accounting for greater than 90% of batholith volume. The ca. 89–72 Ma period would coincide with the eastward subduction of Proto-Caribbean and ± marginal basin crust beneath northwestern South America. The generally dextral, transpressive regime of emplacement for the Antioquian Batholith suite has been highlighted by numerous authors (Aspden et al. 1987; Cediel et al. 1994; Pindell and Kennan 2001; Cediel et al. 2003), accounting for the relatively limited extent of the ca. 96–72 Ma continental arc segment, especially when compared with the orthogonal subduction regimes dominating the Cretaceous granitoid arcs observed in the Central Andes of Perú and Chile. At ca. 72 Ma, granitoid magmatism within the Antioquian Batholith ceases abruptly, and a hiatus of ca. 10 m.y. is recorded prior to the reinitiation of granitoid magmatism within the continental domain.

5.4.3.2 Tectonic Setting for the Ca. 100–84 Ma Western Group Arc Segment

The ca. 100–84 Ma granitoids of the Western Group, including the Buriticá, Santa Fé (Sananalarga), Mistrató, Buga and Jejénes and associated intrusive suites, form a curvilinear arc segment which extends for over 600 km, aligned along the NNW-oriented tectonized front of the Western Tectonic Realm, in sutured contact with the continental margin facies represented by the Romeral melange, immediately to the east (Fig. 5.34). Geological, lithogeochemical and isotopic considerations indicate the ca. 100–84 Ma Western Group granitoids represent the vestiges of a primitive calcic to calc-alkaline arc system, generated within the intra-oceanic domain and emplaced within the host Dagua and Cañas Gordas terrane assemblages of the Western Tectonic Realm, prior to their accretion to the continental margin.

Many recent tectonic reconstructions focus upon the oceanic domain along the NW margin of South America during the mid-Cretaceous through late Cretaceous (e.g. Kennan and Pindell 2009; Wright and Wyld 2011; Nerlich et al. 2014; Spikings et al. 2015; Weber et al. 2015). These reconstructions illustrate the appearance of intra-oceanic arcs associated with east-facing subduction of Proto-Caribbean oceanic crust beneath the approaching Caribbean-Colombian Oceanic Plateau (CCOP/CLIP; Kerr et al. 1997, 2003; Sinton et al. 1998). This system of primitive arcs, emplaced within the Farallon Plate and within overlying oceanic plateau rocks, has been variably referred to as the "Great Arc of the Caribbean" (Burke et al. 1984; Kennan and Pindell 2009; Hastie and Kerr 2010), the "Ecuador-Colombia Leeward Arc" (Wright and Wyld 2011), the "Greater Antillean Arc" (Nerlich et al. 2014) and the "Rio Cala Arc" (Spikings et al. 2015). In Colombia, the ca. 100–84 Ma metaluminous granitoids contained within the Dagua and Cañas Gordas terranes (Fig. 5.34), including the Buriticá, Santa Fé (Sananalarga?), Mistrató, Buga and Jejénes intrusives, are interpreted herein to represent accreted constituents of the Greater Arc.

Various studies address the timing and kinematics of Farallon Plate-CCOP/CLIP assemblage collision and accretion to the Colombian margin during the late Mesozoic, during what we herein refer to as the early Northern Andean Orogeny. Detailed paleo-facies and stratigraphic reconstruction, and basin analysis, at the scale of the entire Colombian Andes (Cediel et al. 1994; Cediel and Cáceres 2000; Cáceres et al. 2003), depict the continental margin tectonic response to the approach and sequential collision of the Cañas Gordas, Dagua and Gorgona terranes, beginning in the Campanian and extending progressively continent-ward, as recorded in uplift-related unconformities recorded in the physiographic Central and Eastern Cordilleras and Santander Massif, extending into the Eocene and Oligocene. This stratigraphic data is supported by the detailed study of seismic sections depicting the subsurface structure and tectonic evolution of Meso-Cenozoic sedimentary basins in Colombia (Cediel et al. 1998; Sarmiento 2018) and by thermochronological data suggesting rapid exhumation in the Central Cordillera between ca. 75 and 55 Ma (Spikings et al. 2015). Reconstruction of the evolution and trajectory of the NE-migrating CCOP/CLIP assemblage, from the Pacific realm into the inter-American gap, suggest (final?) docking of Farallon-CCOP/CLIP components along the NW margin of South America at ca. 54.5 Ma (Nerlich et al. 2014), closely followed by accretion of the Gorgona Terrane beginning in the mid-Eocene (Cediel et al. 2003; Kerr and Tarney, 2005). Thus, the early Northern Andean Orogeny is a diachronous, regional event which, in Colombia, evolved both spatially and temporally over a span exceeding 20 m.y.

With respect to the evolution of the ca. 100 through 72 Ma subduction-related granitoids within the region, we interpret the demise of the ca. 100–84 Ma Western Group arc segment to be related to the near-complete, west-directed consumption of Proto-Caribbean crust located between the Farallon-CCOP/CLIP assembly and the Colombian margin, during CCOP/CLIP migration into the peri-cratonic realm, by ca. 84 Ma. Continued NW-directed convergence between the Farallon-CCOP/CLIP assembly and the continental margin was accommodated by dextral oblique transform faulting (Fig. 5.34) and, more locally, by the east-directed subduction of remnants of marginal basin, Proto-Caribbean, possibly leading edge slivers of Farallon Plate oceanic crust beneath the continental margin, resulting in the main-phase emplacement of the Antioquian Batholith suite. The Antioquian arc segment however was generally short-lived and was rapidly extinguished at ca. 72 Ma, due to impingement of the Farallon-CCOP/CLIP assemblage upon the continental margin.

We interpret the ensuing hiatus in continental magmatism to be related to the invasion of the Antioquian segment trench by the Farallon-CCOP/CLIP assemblage. This magmatic hiatus is considered to reflect various factors/events associated with post-accretionary tectonic reorganization, prior to the reinitiation of the granitoid magmatism recorded within the post-Antioquian, ca. 62–50 Ma Eastern granitoids subgroup. Initially, "chocking-off" of the subduction zone was due to invasion by buoyant CCOP/CLIP fragments such as those represented by the Dagua terrane. Continued plate convergence was dominated by dextral-oblique transpression and offshore transform faulting (Aspden et al. 1987; Pindell and Kennan 2001; Cediel et al. 2003; Wright and Wyld 2011), within an overall regime which was not conducive to continued subduction nor to immediate slab breakoff and subduction reinitiation. Coupling stress beginning in the Maastrichtian was partitioned into various structural components of the Andean mosaic. The development of the Cauca fault and suture system (Ego et al. 1995; Cediel et al. 2003), which separates the accreted Western Tectonic Realm assemblages from the continental margin, took place at this time. Additional tectonic tightening and reactivation along preexisting structures, including the Palestina (Feininger et al. 1972) and Romeral fault systems (Ego et al. 1995; Cediel and Cáceres 2000; Cediel et al. 2003; Vinasco 2018), facilitated collision-related uplift of litho-tectonic units throughout the Central Tectonic Realm.

5.4.3.3 Tectonic Setting for the Ca. 62–50 Ma Eastern Group Post-collisional Arc Segment

Following a ca. 10 Ma hiatus, continental arc granitoids reappear within the autochthonous, continental domain, as recorded in our ca. 62–52 Ma Eastern Group intrusions (i.e. Providencia, Sonsón, Manizales, El Hatillo, El Bosque and Santa Marta plutons), contained within the physiographic Central Cordillera and Sierra Nevada de Santa Marta (Fig. 5.34). These intrusions represent the reinitiation of granitoid magmatism, albeit at a much reduced rate/volume, following the initial invasion/collision of Farallon-CCOP/CLIP components along the Colombian margin and the extinction of the Antioquian Batholith-related magmatism. In this context we have referred to the ca. 62–52 Ma Eastern Group intrusions as "post-collisional" granitoids in Figs. 5.34 and 5.35. U-Pb (zircon) age dates for these plutons illustrate the southward migration of the post-collisional arc axis, from the Providencia suite into the ancestral Central Cordillera to the south of the Antioquian Batholith. The lithogeochemical and isotopic tendencies of these intrusions are clearly distinguished on Figs. 5.17, 5.18, and 5.21, especially with respect to the increased degree of isotopic exchange through direct anatexis or contamination from crustal sources, as suggested by the available Sr-Nd data. Recent Hf isotope data supplied by Bustamante et al. (2017) additionally supports this observation. Initial εHf values presented by these authors for the El Hatillo Stock (-0.7 to $+5.6$) and the El Bosque Batholith (-4.5 to $+1.3$) suggest moderate to high degrees of crustal inheritance and recycling. Indeed, the El Bosque Batholith contains inherited Permian-aged zircons (Bustamante et al. 2017), supplying direct evidence of the recycling of Central Cordilleran basement (Cajamarca-Valdivia Terrane).

Leal-Mejía (2011) and Bustamante et al. (2017) draw attention to the strong "adakite-like" trend produced by the ca. 62 Ma Providencia suite. Bustamante et al. (op. cit.) contrast this trend with the lower Sr/Y vs. Y ratios produced by the Sonsón and El Bosque Batholiths and El Hatillo Stock. These authors suggest a petrogenetic model involving magmatic differentiation at the base of a thick lower crust, related to convergence/subduction of the CCOP lithosphere, with apparently increasing degrees of crustal contamination as magmatism migrated southwards.

Notwithstanding, Leal-Mejía (2011) observed that potentially analogous lithogeochemical and isotopic trends may be derived through a model involving delamination of subducted oceanic lithosphere and asthenospheric upwelling, following terrane collision. This author provides as example the work of Parada et al. (1999), who explain the lithogeochemical and isotopic evolution of the late Jurassic-Cretaceous Chilean Coastal Batholith of the Central Andes as the result of collision of an oceanic ridge with the continental margin. These authors interpreted pre-collisional, metaluminous, calc-alkaline magmas to be products of east-directed subduction-related arc magmatism. Following oceanic ridge collision and the cessation of subduction-related magmatism, Parada et al. (1999) invoke a model of lithospheric delamination leading to the upwelling of asthenospheric mantle and extensional deformation in the overlying continental crust, followed by the emplacement of post-collisional granitoids with "adakite-like" signatures. They note that $\varepsilon Nd_{(t)}$ values within the Coastal Batholith show a vertically increasing trend, coincident with the transition to "adakite-like" compositions. A very similar vertical increasing $\varepsilon Nd_{(t)}$ array is observed within the Antioquian Batholith suite (Fig. 5.21), prior to the emplacement of the "adakite-like" compositions reflected in the Providencia and to a lesser degree El Hatillo suites (Leal-Mejía 2011).

In this context, we suggest that the reappearance of granitoid magmatism as represented by the ca. 62–52 Ma Eastern Group arc segment does not necessarily represent the resumption of *subduction* along the Colombian Pacific margin and could alternatively be explained using a model of post-collisional lithospheric delamination, asthenospheric upwelling and thermally induced anatexis to generate post-collisional granitoid magmatism in the Central Cordillera. Such a scenario conforms well with the punctuated nature of observed magmatism, as recorded within the U-Pb age database vs. the proposed tectonic development of the region (Fig. 5.34), in addition to explaining the observed lithogeochemical and isotopic trends, including the dominantly negative εHf_i values for the El Bosque Batholith. Indeed, Vinasco (2018) interprets recent U-Pb (zircon) age and lithochemical data for the alkaline Sucre (Antioquia) intrusions, to suggest the initiation of delamination of the Caribbean assemblage as early as 70 Ma. The Sucre intrusions produce very similar age and lithochemical data to that of the Irra Stock as presented herein (Figs. 5.16 and 5.17) and suggest these plutons represent the early delamination suite, emplaced along the Romeral suture boundary.

We note that the differentiation of granitoids resulting from thermal heat transfer between the mantle and lower crust during asthenospheric upwelling, from more typical subduction-related granitoids, using basic lithogeochemical and isotopic analyses, is not necessarily straightforward, especially in the presence of tectonically thickened continental crust, where enhanced degrees of crustal anatexis, assimilation or contamination, may be intuitively suspected. A nearby example of this situation has already been revealed in the evolving petrogenetic interpretation of the early Jurassic granitoids of the Santander Plutonic Group, within the Santander Massif, where historic interpretations (e.g. Goldsmith et al. 1971; Aspden et al. 1987; Dorr et al. 1995) of relatively "typical" subduction-related petrogenesis have been supplanted by a model involving partial fusion of lower crustal source rocks by asthenospheric upwelling and heat transfer, as revealed by advanced lithogeochemical and isotopic studies, including Lu-Hf isotope analyses (e.g. Van der Lelij 2013).

In either case, Paleocene-Eocene granitoid magmatism along the Central Cordillera was short-lived and was abruptly extinguished again at ca. 52 Ma. As with the shutdown of the Antioquian Batholith, extinction of the Paleocene-Eocene arc appears to be associated with collision of another oceanic ridge, in this case represented by the Gorgona Terrane, which was accreted to the Colombian Pacific margin in the Eocene (Cediel et al. 2003; Kerr and Tarney 2005). Following emplacement of the ca. 62–52 Ma, post-collisional, Eastern Group granitoids, a resumed, ca. 30 m.y. hiatus in subduction-related granitoid magmatism is observed, as recorded by the absence of significant volumes of observable granitoids throughout central continental Colombia, dating from a period extending from the early Eocene (ca. 52 Ma) to the latest Oligocene (Figs. 5.3, 5.4, 5.16, and 5.23; Leal-Mejía 2011).

Although temporally related to the ca. 62–52 Ma Eastern post-collisional granitoid subgroup, the emplacement kinematics of the Santa Marta Batholith granitoids are clearly distinct from those of the Central Cordillera. The detailed studies of Mejía et al. (2008), Duque (2009) and Cardona et al. (2011) provide insight into the tectono-magmatic evolution of Paleogene granitoids along the NW apex of the Sierra Nevada de Santa Marta. Duque (2009) and Duque-Trujillo et al. (2018) identified various intrusive phases within and surrounding the Santa Marta Batholith, ranging in age from ca. 64 to 50 Ma. These authors conclude that an early (ca. 64–62 Ma) suite of volumetrically minor, peraluminous leucogranites (e.g. Playa Salguero pluton) were probably derived via anatexis of local amphibolite basement and are petrogenetically unrelated to the Santa Marta Batholith suite per se. They emphasize the localized, punctuated and short-lived nature of the Santa Marta arc segment and the absence of post-ca. 50 Ma granitoid magmatism in the region, following closure of the Santa Marta arc system, and conclude that Santa Marta suite magmatism was not associated with a long-lived or well-established subduction zone. A model involving the forced underthrusting of thickened, buoyant oceanic lithosphere beneath the apex of the Santa Marta Massif was proposed. Duque (2009) and Duque-Trujillo et al. (2018) relate the emplacement of the ca. 64–62 Ma peraluminous leucogranites hosted within the Gaira Group accretionary complex (Cediel and Cáceres 2000) to the partial fusion of amphibolitic basement due to the initial interaction of the Farallon-CCOP/CLIP assemblage with the northern Colombian margin, followed by main-phase emplacement and NE migration of Santa Marta suite-related magmatism between ca. 58 and 50 Ma. This model is in keeping with previous interpretations of the kinematics and temporal development of the Gaira Group accretionary prism and Santa Marta batholith, as presented by Cediel and Cáceres (2000) and Cediel et al. (2003; see Cediel 2018), in which detachment and NW migration of the Maracaibo Sub-plate beginning in the Paleocene (Fig. 5.34) resulted in the localized forced underthrusting of CCOP/CLIP crust, metamorphism within the Gaira Group and punctual granitoid magmatism, as recorded within the Playa Salguero pluton, Santa Marta Batholith and associated plutons.

5.4.3.4 Tectonic Setting for the Ca. 62–40 Ma Mandé-Acandí Western Group Arc Segment

The Paleocene-Eocene Mandé-Acandí arc assemblage (Fig. 5.15), including the metaluminous, low-K calc-alkaline (calcic) Mandé and Acandí batholiths, volcanic and pyroclastic rocks of the La Equis-Santa Cecilia Fms. and hypabyssal porphyry centres at Pantanos-Pegadorcito, Murindó and Acandí and elsewhere, represents the most significant expression of granitoid magmatism within the Western Group of granitoids. It is the only assemblage within the ca. 100–40 Ma suite for which a coeval volcanic member is preserved. Geological, lithogeochemical and isotopic data for the holocrystalline Mandé-Acandí Batholith and associated hypabyssal porphyritic rocks is consistent with an origin within an intra-oceanic arc, emplaced within CCOP/CLIP crust as represented by the El Paso Terrane-Baudó Complex (Cediel el at. 2009; Montes et al. 2012; Cediel 2018). Figure 5.34 depicts the pre-docking, intra-oceanic configuration Mandé-Acandí arc and host terranes. These same litho-tectonic units are depicted within the detailed time-space analysis presented in Fig. 5.35.

Based upon schematic reconstructions depicting the tectonic evolution of NW South America and offshore Pacific and Caribbean domains during the Paleogene, the Mandé-Acandí arc is contained within the trailing edge of the CCOP/CLIP plateau, associated with a SW-facing (NE-verging), intra-oceanic subduction zone, contained within the Farallon Plate (e.g. Aspden et al. 1987; Pindell and Kennan 2001; Kennan and Pindell 2009; Wright and Wyld 2011; Montes et al. 2012; Nerlich et al. 2014; Weber et al. 2015). Additional intra-oceanic granitoids associated with this subduction zone, located in Central America, date from the

Cretaceous (Buchs et al. 2010) and include the Middle American arc series, as depicted, for example, by Pindel and Kennan (2001) and Wright and Wyld (2011).

Magmatism related to the northern (Panamanian) segment of the Mandé-Acandí arc may have initiated as early as 62 to 59 Ma (Wegner et al. 2011; Montes et al. 2012) however published U-Pb (zircon) crystallization ages for granitoids from the Acandí Batholith in Colombia range from ca. 50 Ma (Montes et al. 2012, 2015). To the south, holocrystalline and porphyritic granitoids from the Pantanos-Pegadorcito area return U-Pb dates of ca. 45 Ma, whilst granitoids collected on the southern margin of the Mandé Batholith returned ages of ca. 43 Ma. Thus, U-Pb (zircon) crystallization ages suggest Mandé-Acandí is a multiphase arc, emplaced over a period of ca. 20 m.y., with an overall younging trend from north to south (Fig. 5.15). We note that the flare-up of the Mande-Acandí arc segment is penecontemporaneous with the onset of strong dextral-oblique transpression, tectonic tightening and uplift observed within the Colombian continental block, brought on by collision of litho-tectonic components of the Farallon-CCOP/CLIP assembly (Cañas Gordas, Dagua terranes) along the Colombian margin beginning at ca. 75 Ma. The ensuing "tectonic lock-up" along the NW margin of South America during the late Cretaceous-Paleocene may have played a role in the decoupling of the Farallon Plate from the trailing edge of the CCOP/CLIP plateau and the development of the Paleocene-Eocene segment of the Middle American Trench, along which east-directed subduction of Farallon crust beneath the trailing edge of the CCOP/CLIP plateau resulted in emplacement of the Mandé-Acandí arc.

Nerlich et al. (2014) reconstruct the genesis, evolution and migration of the Caribbean plate/basin into the inter-American gap, based upon the Pacific hotspot reference frame (Wessel and Kroenke 2008) and the Global Moving Hotspot Reference Frame (Doubrovine et al. 2012). Nerlich et al. (2014) conclude that the Caribbean Plate docks with the South America by ca. 54.5 Ma, roughly coincident with the switch from divergence to convergence between North and South America (Müller et al. 1999). Docking at 54.5 Ma is in good agreement with schematic tectonic models depicting the Caribbean Plate reaching its near-final resting place during the Eocene (e.g. Pindell and Kennan 2001; Kennan and Pindel 2009).

Following docking of the Caribbean Plate, granitoid magmatism does not reappear in the Colombian Andes until the Oligo-Miocene. The re-establishment of subduction along the Colombian Trench and the continued convergence between the South American Plate and the trailing edge of the CCOP/CLIP plateau are aspects of the late Northern Andean Orogeny, described forthwith in Sect. 5.4.4.

5.4.4 The Late Northern Andean Orogeny: Tectonic Framework and Evolution for Latest Oligocene to Plio-Pleistocene Granitoids

As documented above, following emplacement of the ca. 62–50 Ma, post-collisional, Eastern Group granitoids, the observable early Eocene to latest Oligocene geological record of the continental Colombian Andes is marked by the absence of significant volumes of granitoids (e.g. Cediel and Cáceres 2000; Gómez et al. 2015a), suggesting a hiatus in subduction-related granitoid magmatism or continental arc development during this period. It is feasible that, should such magmatism have existed, it could have been erased by subsequent uplift and erosion during the later phases of the Northern Andean Orogeny. Such a contention, however, is not supported by the limited available, albeit localized, detrital zircon studies from the Colombian Andes (e.g. Nie et al. 2012; Saylor et al. 2012), which, conversely, reveal the absence of 50 to 20 Ma detrital zircon populations. Based upon the foregoing, we conclude that granitoid magmatism associated with subduction along the Colombian Pacific margin effectively terminated with the emplacement of the ca. 62–52 Ma post-collisional arc and accretion of the Gorgona Terrane. Subsequent tectonic development during the ensuing ca. 30 m.y. magmatic hiatus is characterized by continued dextral compression, transform faulting and plate reorganization along the Pacific margin and structural tightening throughout continental Colombia (Cediel et al. 1994; Cediel et al. 2003; Kennan and Pindell 2009; Cediel 2018).

Beginning at ca. 24 Ma, granitoid magmatism reappears in the south–westernmost Colombian Andes (Fig. 5.36), hosted within CCOP-/CLIP-related rocks of the Dagua terrane (accreted in the late Cretaceous-Paleocene) but well to the south of the location of the proposed trailing edge of the Caribbean Plate in the late Oligocene (e.g. Cediel and Cáceres 2000; Pindell and Kennan 2001; Kennan and Pindell 2009; Hastie and Kerr 2010; Montes et al. 2012; Nerlich et al. 2014). This new phase of granitoid magmatism signals the reactivation of arc development within continental and western Colombia, which dominates the Neogene tectono-magmatic development of the region during the late Northern Andean Orogeny, following the early Eocene docking of the Caribbean Plate (Nerlich et al. 2014)

Detailed time-space analysis based upon U-Pb (zircon) crystallization ages for latest Oligocene through Miocene and Plio-Pleistocene granitoids and associated volcanic rocks throughout the Colombian Andes (Figs. 5.22, 5.36, and 5.37) demonstrates that extensive, composite "Neogene" arc magmatism recorded on regional geologic maps of the physiographic Central and Western Cordilleras, along the Cauca and Patia valleys and elsewhere (e.g. Cediel and Cáceres 2000; Gómez et al. 2015a), in fact consists of a complex distribution of magmatic arc segments, the location of which is observed to migrate in time and space, in both an overall south-to-north and west-to-east pattern (Cediel et al. 2003; Leal-Mejía 2011). The genesis and spatial evolution of these arc segments may in turn be attributed to (1) complexities in the late Oligocene-Miocene collision between continental South America and the trailing edge of the Caribbean Plate, resulting in accretion of the El Paso-Baudó Terrane, and (2) the penecontemporaneous evolution of the eastern Farallon (Nazca-Cocos) Plate along the Colombian Pacific margin. In view of these factors, we present observations pertaining to the convergence and collision of the South American Plate with the Caribbean plateau and to the Miocene evolution of the easternmost Farallon Plate, as they pertain to the magmatic evolution of continental Colombia, prior to presenting conclusions pertaining to the temporal-spatial evolution of granitoid arc segments in the onshore realm during the latest Oligocene, Miocene and Plio-Pleistocene.

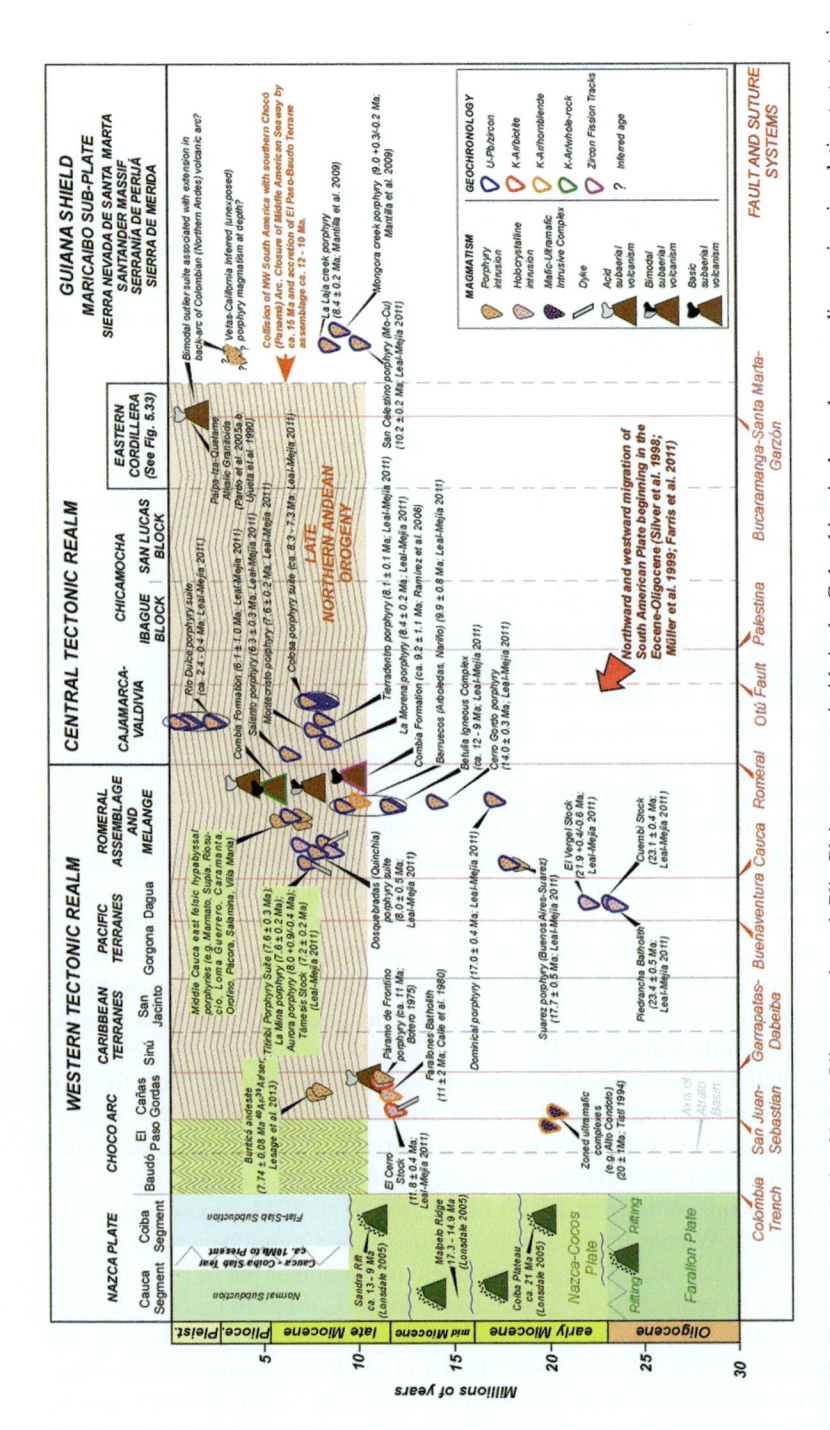

Fig. 5.37 Time-space analysis of latest Oligocene through Plio-Pleistocene granitoids in the Colombian Andes and surrounding region, in relation to tectonic framework, major litho-tectonic elements and orogenic events. The age and nature of granitoid intrusive suites of the same time period are indicated. The profile contains elements projected onto a ca. NW–SE line of section through west-central Colombia. (Litho-tectonic terrane and fault nomenclature modified after Cediel et al. 2003. See text for additional details)

5.4.4.1 Convergence and Collision Between South America and the Trailing Edge of the Caribbean Plateau

Tectonic reconstructions demonstrate that most of the CCOP/CLIP components of the Western Tectonic Realm, including the Dagua, Cañas Gordas, Gorgona and San Jacinto terranes, were loosely in place within the near-shore realm by the late Oligocene to early Miocene (e.g. Cediel et al. 1994; Pindell and Kennan 2001; Cediel et al. 2003; Kennan and Pindell 2009; Cediel 2018) (Figs. 5.34 and 5.36). To the west, the El Paso-Baudó segment of the Chocó Arc (including the Mandé-Acandí Batholith) was located within the peri-cratonic realm hosted upon the trailing edge of the Caribbean Plate.

Detailed analysis of the structural and kinematic evolution of the eastern Panamá Arc (i.e. Chocó Arc) for the late Eocene through Miocene, presented by Farris et al. (2011) and Montes et al. (2012), depicts the WNW-ESE orientation of the El Paso-Mandé-Acandí-Baudó assemblage along the trailing edge of the CCOP/CLIP plateau, essentially colinear to the trend of the Middle American arc, in consort with NE-directed subduction of Farallon crust, as universally depicted in regional tectonic reconstructions spanning the mid-Cretaceous and Paleogene (e.g. Aspden et al. 1987; Pindell and Kennan 2001; Cediel et al. 2003; Kerr et al. 2003; Kennan and Pindell, 2009; Farris et al. 2011; Wright and Wyld 2011; Montes et al. 2012; Nerlich et al. 2014; Weber et al. 2015).

Convergence between the El Paso-Baudó Terrane and the NW Colombian margin took place in the late Oligocene (Duque-Caro 1990; Pindell and Kennan 2001; Cediel et al. 2003; Cediel et al. 2010). Based upon structural, lithogeochemical and thermochronological data, Farris et al. (2011) present a model depicting the collision between South America and the southern Panama (i.e. Chocó) Arc. Their model proposes the N and W convergence of the South American block upon the Chocó Arc (as opposed to the continued N and E migration of the Caribbean plateau), prior to collision beginning at 23–25 Ma. This proposal is in agreement with early structural data and conclusions regarding the vergence and evolution of the Panamá thrust and fold belts (Silver et al. 1990), plate motion data for South America (Silver et al. 1998; Müller et al. 1999) and early Eocene docking constraints for the Caribbean Plate (Nerlich et al. 2014).

Additional lithostratigraphic, radiometric, paleomagnetic, structural and thermochronological data presented by Montes et al. (2012) demonstrate initial NW-vergent thrusting and clockwise rotation due to W-E convergence of the southwestern margin of the Chocó (El Paso-Baudó) assembly with South America during the Oligocene. Following collision at 25–23 Ma (Farris et al. 2011), continued clockwise rotation and E-W convergence led to closure of the Central American seaway by ca. 15 Ma (Montes et al. 2012). Accretion and obduction of the El Paso Terrane, including severance of the Mandé-Acandí Arc from its Caribbean roots, took place between 15 and 12 Ma, within the broad context of the development of the Panamanian orocline (Silver et al. 1990; Farris et al. 2011), resulting in development of the San Juan-Sebastian (Uramita) suture system and apparent NW vergence of the Panamá thrust and fold belt (Figs. 5.34 and 5.36), during the mid- through late

Miocene (Duque-Caro 1990; Cediel et al. 2010; Montes et al. 2012). Uplift of the western El Paso Terrane (Baudó Range) and development of the Atrato Basin took place between ca. 8 and 4 Ma (Cediel et al. 2010) and appear to be a feature associated with the evolution of the Farallon and Nazca-Cocos plates and the reestablishment of subduction along the Colombian Pacific margin (Sect. 5.4.4.2).

It is important to note that convergence between the El Paso segment of the Chocó Arc and the NW South American margin during the Oligocene to Miocene left no apparent record of granitoid arc development within continental Colombia (Central Tectonic Realm, Maracaibo Sub-plate), as may be inferred by the absence of mapped Oligo-Miocene granitoids (e.g. Cediel and Cáceres 2000; Cediel et al. 2003; Leal-Mejía 2011; Gómez et al. 2015a), or significant Oligo-Miocene detrital zircon populations (Nie et al. 2012; Saylor et al. 2012). We interpret this observation to reflect the near in situ emplacement of the Mandé-Acandí Arc, riding passively within/upon the trailing edge (El Paso Terrane segment) of the Caribbean Plate. This assemblage developed as an intact member of the CCOP/CLIP plateau as a whole and was little transported following docking of the Caribbean Plate at ca. 54.5 Ma (Nerlich et al. 2014). We conclude that Oligocene-Miocene convergence between the El Paso assemblage and NW South America did not involve development of a significant intra-plate subduction zone between the Caribbean plateau and the Colombian continental margin. This conclusion is in keeping with the arguments involving the buoyancy of CCOP/CLIP lithosphere (Molnar and Atwater 1978) and with the interpretation of various authors limiting interaction of the Caribbean-NW Colombian margin to a model involving the amagmatic, limited, SE-directed forced underthrusting of thick CCOP/CLIP lithosphere beneath the South American margin during this time period (Van der Hilst 1990; Van der Hilst and Mann 1994; Cediel et al. 2003; Kerr et al. 2003; Farris et al. 2011).

5.4.4.2 Evolution of the Farallon-Nazca-Cocos Plate System and Neogene Reinitiation of Subduction in the Colombian Andes

Coeval with Oligo-Miocene tectonic development in NW Colombia, east-directed subduction of the Farallon Plate beneath the Colombian Pacific margin was reestablished by ca. 24 Ma (Leal-Mejía 2011). Throughout the Miocene and into the Plio-Pleistocene, pulses of metaluminous, calc-alkaline magmatism in temporally and geographically distinct volcano-magmatic arc segments are revealed by U-Pb (zircon) crystallization ages for subduction-related granitoids (Figs. 5.36 and 5.37). This granitoid magmatism, which includes localized coeval volcanism, was emplaced within metamorphic rocks of the Central Tectonic Realm and Maracaibo Sub-plate and within more recently accreted oceanic rocks comprising the Western Tectonic Realm. These basement complexes are all considered to have formed part of the Colombian accretionary mosaic at the time of emplacement of their contained latest Oligocene through Plio-Pleistocence granitoids, and hence the entire Neogene granitoid suite is considered autochthonous with respect to the continental margin.

At the end of the Oligocene, the triple junction between the Farallon, South American and Caribbean plates was located in the near-shore Colombian Pacific, along the south–westernmost margin of the Panamá-Choco Arc (e.g. Pindell and Kennan 2001; Cediel et al. 2003; Lonsdale 2005). Accretion-related transform faults of the Garrapatas and San Juan-Sebastian sutures (including the southern Uramita Fault of Duque-Caro (1990) and Montes et al. (2012)), marking the southern margin of the Choco Arc terranes (i.e. the trailing edge of the Caribbean Plate), were already established as broadly NE-SW corridors of crustal-scale weakness (e.g. Barrero 1977; Duque-Caro 1991; Cediel et al. 2003) (Figs. 5.34 and 5.36). At ca. 23 Ma, the Farallon Plate splits to form the Nazca and Cocos plates along a ca. E-W rift that also extended into the Colombian Pacific, in the vicinity of the junction between the Middle American and South American subduction zones (Pindell and Kennan 2001; Lonsdale 2005). Continued plate reorganization in the Pacific realm (Lonsdale 2005) and W-directed motion of the South American Plate (Silver et al. 1998; Farris et al. 2011) resulted in near-orthogonal convergence between the Farallon Plate and the Colombian margin. Mid-Miocene rifting within the Nazca Plate is marked by the formation of the E-W-oriented Sandra Rift off the Colombian Pacific margin, which presently separates the Coiba microplate to the north from Malpelo Ridge and associated crust to the south (Lonsdale 2005) (Fig. 5.36). Ocean crust associated with seafloor spreading along the Sandra Rift dates from between ca. 14 and 9 Ma (Lonsdale 2005) and comprises the oceanic slab juxtaposed along the present-day northern Colombian Trench between ca. 5°N and 8°N. To the south, similar crust of somewhat older (ca. 14–18 Ma) age is preserved (Lonsdale 2005).

Within the Colombian onshore realm, the analysis of earthquake hypocentral solutions, gravity and magnetic data, tomographic imaging, petrogenetic data and the distribution of modern-day volcanic activity has led numerous authors in recent years to present models for Miocene to recent subduction beneath continental Colombia (Santô 1969; Dewey 1972; Pennington 1981; Van der Hilst and Mann 1994; Taboada et al. 2000; Sarmiento 2001; Zarifi et al. 2007; Vargas and Mann 2013; Bissig et al. 2014; Chiarabba et al. 2015). Many of these studies are attempts to reconcile modern-day earthquake activity observed in the Santander Massif (Bucaramanga seismic nest), with the distribution of modern-day volcanic activity, and localized late Miocene to Pliocene magmatism observed in the Vetas-California area of the Santander Massif (Mantilla et al. 2009; Bissig et al. 2014) and at Paipa-Iza in the northernmost Eastern Cordillera (Floresta Massif) (Pardo 2005a, b). All of the foregoing authors agree that available data suggests eastward to south-eastward subduction of a composite oceanic slab comprised of the segmented, Miocene-age, Nazca Plate. Some authors suggest the Nazca Plate is undergoing down-slab interaction beneath continental Colombia with CCOP/CLIP oceanic crust of Cretaceous age (e.g. Pennington 1981; Taboada et al. 2000; Zarifi et al. 2007; Vargas and Mann 2013).

Seismic tomography and additional geophysical data presented by Pennington (1981), Taboada et al. (2000), Sarmiento (2001), Zarifi et al. (2007), Vargas and Mann (2013) and Chiarabba et al. (2015) have been interpreted to reflect an E-W discontinuity or tear in the oceanic slab presently subducting beneath western

Colombia (e.g. the Caldas tear of Vargas and Mann 2013) (Fig. 5.36). This discontinuity is inferred to be located between ca. 4.8°N and 5.2 °N, broadly coincident with the southern end of the Serranía de Baudó (Taboada et al. 2000) and the ENE striking segment of the San Juan Sebastian-Uramita suture system (Cediel et al. 2010; Montes et al. 2012). Vargas and Mann (2013) and Chiarabba et al. (2015) note that the discontinuity also coincides with the interpreted western projection of features located within the subducting Nazca plate, including the Sandra Rift and the Coiba Transform fault, respectively.

Santô (1969), Dewey (1972), Pennington (1981), Van der Hilst and Mann (1994), Taboada et al. (2000), Sarmiento (2001), Cediel et al. (2003), Zarifi et al. (2007), Vargas and Mann (2013), Bissig et al. (2014) and Chiarabba et al. (2015) interpret the geometry and nature of the oceanic slab segments presently subducting beneath continental Colombia, on either side of the E-W discontinuity. All authors agree that south of ca. 5°N, the Nazca Plate is undergoing moderately steep subduction, at an angle of between ca. 30° and 40°, steepening to >50° beneath the Eastern Cordillera. This southern segment of "normally" dipping Nazca crust was referred to as the Cauca segment by Pennington (1981). Active volcanism associated with Cauca segment subduction manifests in the Colombian portion of the Northern Andean volcanic zone, which (coincidentally) terminates at about 5°N (Figs. 5.3 and 5.36).

North of 5°N, variable interpretations of the nature and geometry of subducted oceanic crust have been presented. Beneath the eastern Colombian Andes, seismic data and tomographic imaging are suggestive of a dipping slab, which has long been interpreted to be associated with abundant earthquake activity surrounding the Bucaramanga seismic nest (Santô 1969; Dewey 1972; Pennington 1981). Pennington (1981) referred to this shallowly dipping slab as the Bucaramanga segment. Some authors interpret this segment to represent CCOP/CLIP lithosphere, which in turn is interpreted to represent the down-slab prolongation of late Cretaceous CCOP/CLIP crust exposed within the Chocó Arc (El Paso-Baudó Terrane) (Pennington 1981; Taboada et al. 2000; Sarmiento 2001; Zarifi et al. 2007; Vargas and Mann, 2013; Bissig et al. 2014). However, the sparse and discontinuous nature of seismic activity recorded along the interpreted up-dip segment of the Bucaramanga slab (see profiles presented by Pennington 1981; Taboada et al. 2000; Vargas and Mann 2013; Chiarabba et al. 2015) led authors to suggest that the Bucaramanga segment is no longer connected to surface plates (Santô 1969; Dewey 1972; Sarmiento 2001; also see Plate 2C of Taboada et al. 2000; and Fig. 5.5 of Vargas and Mann 2013).

Other authors (e.g. Pennington 1981; Zarifi et al. 2007; Vargas and Mann 2013) suggest "apparent" up-dip continuity between the Bucaramanga segment and inferred, shallowly-dipping Cretaceous CCOP/CLIP lithosphere beneath the Central Cordillera, which in turn would be connected to CCOP lithosphere exposed in the Chocó Arc. These authors interpret the abrupt northward termination of the North Andes volcanic arc at ca. 5°N to reflect amagmatic, flat-slab subduction of CCOP lithosphere.

Notwithstanding, 600 km to the west of the Santander Massif, along the Colombian Trench, various authors suggest the Miocene, Coiba microplate segment of the Nazca Plate is (also) undergoing eastward subduction, between ca. 5°N and ca. 8°N (Aspden et al. 1987; Van der Hilst and Mann 1994; Taboada et al. 2000; Cediel et al. 2003; Lonsdale 2005; Vargas and Mann 2013; Chiarabba et al. 2015). Van der Hilst and Mann (1994) and Chiarabba et al. (2015) argue that the Coiba segment contains buoyant features, such as thickened volcanic ridges, and is subducting at a lower angle when compared with the Cauca segment to the south. These authors attribute the lack of modern volcanic arc development N of 5°N, to amagmatic, flat-slab subduction of the Coiba microplate. Indeed, the most recent modelling of seismic data presented by Chiarabba et al. (2015) suggests down-slab continuity of the Miocene Coiba microplate, extending from the Colombia trench into the region beneath the northeastern Colombian Andes. Thus, Chiarabba et al. (2015) present a simplified model involving massive, down-slab devolatilization of the thickened, Miocene, Coiba microplate, which can equally be invoked to explain the lack of arc-related volcanism in the up-slab section to the N of 5°N, seismic activity in the Bucaramanga nest and localized granitoid magmatism in the Santander Massif. In this context, the model of Chiarabba et al. (2015) is more in line with earlier modelling and arguments presented by Van der Hilst and Mann (1994) which suggest the Nazca, and not Caribbean Plate, is subducting beneath NW Colombia and that the presence of continuous or fragmented CCOP crust is not required to explain the observed seismic or magmatic phenomena. These and other authors (e.g. Cediel et al. 2003; Farris et al. 2011) suggest the Caribbean Plate is undergoing, rather, S- to SE-directed forced underthrusting along the western Colombian Caribbean margin (Fig. 5.36).

Interestingly, beyond observations regarding the absence vs. presence of modern-day arc-related volcanism, N and S of ca. 5°N, respectively, few proponents of the various subducting slab models have taken into account the evolution and spatial migration of subduction-related granitoid arc segments manifest within the western Colombian Andes during the latest Oligocene through Miocene. This period coincides with the birth and growth-related architectural evolution of the Nazca Plate (Lonsdale 2005) and with the reinitiation of subduction-related granitoid magmatism throughout western Colombia, leading to the conformation of the Colombian segment of the North Andes volcanic arc (Aspden et al. 1987; Cediel et al. 2003; Leal-Mejía 2011). Aspden et al. (1987) highlighted the presence of Neogene granitoids along the Western Cordillera and Cauca-Patia intermontane valley, which they considered to be associated with late Oligocene to present subduction along the entire Colombian Pacific margin. Taboada et al. (2000) related late Miocene magmatism observed along the Western Cordillera and Romeral mélange, between 5°N and 7°N, to the development of a wedge of hot asthenosphere which favoured melting and granitoid magmatism between the subducting Nazca Plate and accreted sections of the CCOP/CLIP. They suggest that the presence of CCOP/CLIP lithosphere beneath the Central Cordillera to the east acts as a shield which prevents the penetration of rising melts and as such explains the absence of active volcanism to the N of 5°N. Cediel et al. (2003) provided an explanation for the punctuated emplacement and spatial-temporal evolution of Miocene holocrystalline and porphyry-related arc segments in western Colombia, based upon the Miocene tectonic assembly of the region, involving the differential subduction of Nazca Plate crust on either side of the paleo-Garrapatas transform fault. Chiarabba et al. (2015) suggest that the Coiba segment of the Nazca Plate initially underwent normal, moderate- to high-angle subduction, which could explain the development of subduction-related granitoids between 5°N and 7°N. According to their model, the entry of buoyant material into the Colombia trench at ca. 10 Ma led to enhanced tearing of the Nazca slab along pre-established planes of weakness (e.g. the Coiba transform fault), the onset of low-angle subduction N of ca. 5°N and the consequent cessation of eastward-progressing magmatism, explaining the absence of the modern-day volcanic activity associated with the subducting Coiba segment.

As outlined in detail in Sect. 5.3.5 and within Figs. 5.22, 5.23, and 5.36, at least six granitoid arc segments/clusters of latest Oligocene through Miocene and Plio-Pleistocene age, and of significant length, continuity and outcropping area, are recorded within Colombia's physiographic Western and Central Cordilleras and along the Cauca-Patia intermontane valley. We consider the modern-day Colombian portion of the Northern Andean volcanic zone (which itself is segmented; Cediel et al. 2003; Stern 2004; Marín-Cerón et al. 2018), to represent a temporally separate arc segment, although it is cospatial with, and locally superimposed upon, Mio-Pliocene segments. In addition to the above, within Colombia's eastern cordilleran system, isolated granitoid occurrences of Miocene and Pliocene age are recorded at Vetas-California in the Santander Massif and Paipa-Iza and Quetame in the Eastern Cordillera. With respect to the generally N-S- to NNE-oriented axis of the Miocene arc segments and the NNE trend of the modern-day Northern Andean volcanic zone, the Vetas-Paipa-Quetame occurrences are located well to the east (on average over 150 km east) of the magmatic arc axis. We do not consider the Vetas-Paipa-Quetame granitoids to constitute a definable arc segment. They are low volume, localized occurrences which, based upon clear differences in lithochemistry and widely spaced, non-coaxial distribution, are considered outliers with respect to the magmatic trends of central and western Colombia.

In the south, the ca. 23–21 Ma Piedrancha-Cuembí holocrystalline suite and the ca. 18–9 Ma Upper Cauca-Patía porphyry suite are associated with subduction of the southern, Cauca segment of the Nazca plate. Continued eastward migration of the granitoid volcanic arc axis is recorded in the southern portion of the active Colombian volcanic arc (e.g. San Roque, Huila volcanoes). To the north, the ca. 12–10 Ma Farallones-El Cerro holocrystalline suite and the ca. 9–5 Ma Middle Cauca porphyry suite are associated with subduction of the Coiba segment. Granitoid magmatism related to the continued subduction of the Coiba segment records eastward migration of arc-axial magmatism, observed in the late Miocene Cajamarca-Salento hypabyssal porphyry cluster and the Plio-Pleistocene Río Dulce, both located within the Cajamarca-Valdivia basement rocks of the Central Cordillera. Both of these granitoid clusters are essentially coaxial with the northern portion of the active Colombian volcanic arc. The active Machín volcano is located on the eastern margin of the Cajamarca-Salento porphyry cluster, whilst Plio-Pleistocene to recent volcanic cover from the Nevado del Tolima limits exposure of this same porphyry cluster immediately to the north. Along trend to the NNE at Río Dulce, Plio-Pleistocene ages for hypabyssal granitoid intrusive and associated volcanic rocks compare well with similar ages for volcanic materials from the Ruíz, Santa Isabel and Tolima stratovolcanic complexes (Maya 1992). In this context, Rio Dulce (ca. 5.7°N) may be interpreted to represent the northernmost expression of volcanism associated with the modern-day Colombian volcanic arc (Fig. 5.36). We interpret the eastward migration of both the Cauca and Coiba-related arc axial magmatism to reflect progressive shallowing of the subduction angle of the respective segments of the Nazca oceanic crust, associated with the consumption of progressively younger and thermally buoyant Nazca Plate lithosphere (Lonsdale 2005) probably augmented by the entrance of buoyant aseismic features such as the Carnegie and Sandra Ridge into the (Ecuador-)Colombia trench, effectively inhibiting the subduction process (e.g. Chiarabba et al. 2015).

Composite lithogeochemical and isotopic data presented herein permit interpretation of all of granitoid suites emplaced along the western Colombian convergent margin during the Mio-Pliocene, including those located to the N of ca. 5°N (i.e. the ca. 12 Ma Farallones-El Cerro trend and the ca. 9–5 Ma Middle Cauca trend), as mantle-derived, metaluminous, calc-alkaline granitoids typical of subduction-related suites. It may be observed that, aside from differences in age, the granitoid suites comprising the various Colombian arc segments of western Colombia are very similar in major, minor, trace element, and isotopic composition. All of the suites demonstrate typical gabbro through granodiorite trends with strongly mantelic compositions and, in no instance, are enhanced levels of crustal contamination (e.g. peraluminous tendencies, anomalously high Sr isotope compositions) implicit in the petrogenetic trends demonstrated by the data set.

Based upon the above arguments, we interpret Neogene granitoid magmatism throughout western Colombia (i.e. the Western and Central Cordilleras and Cauca-Patía intermontane valley) to be the result of the subduction of composite Nazca crust beneath the composite Colombian margin since the late Oligocene. Differences in the rate and style of east-dipping subduction on either side of the Cauca-Coiba slab tear, beginning in the latest Oligocene, are reflected in the complex spatial and temporal distribution of Colombian onshore volcano-plutonic arc magmatism throughout the early Miocene and Plio-Pleistocene to recent (e.g. Cediel et al. 2003). We conclude that all of the western Colombian granitoid arc segments/clusters were emplaced following passage and docking of the trailing edge of the Caribbean Plate and do not represent the subduction of Farallon-CCOP/CLIP assemblage lithosphere per se.

Mio-Pliocene granitoids of Colombia's Eastern Cordilleran system, including those of the Vetas-California area in the Santander Massif and at Paipa-Iza and Quetame, within the Eastern Cordillera sensu stricto, form volumetrically small and isolated occurrences located over 150 km east of the subduction-related magmatic axis defined by the active Colombian volcanic arc. Available lithogeochemical data for this group of granitoids is incomplete and does not permit a full analysis of the petrogenesis of these occurrences nor a complete comparison amongst themselves.

Miocene granitoid magmatism in the Santander Massif ranges from ca. 14 to 9 Ma (Mantilla et al. 2009; Leal-Mejía 2011; Mantilla et al. 2013; Bissig et al. 2014; Cruz et al. 2014) although recent studies suggest that unexposed magmatism of

Pliocene age is likely present at shallow depth below the trend (Rodríguez, 2014; Rodríguez et al. 2017). The lithogeochemical and isotopic database for the Vetas-California granitoids is fairly complete, and the clear evolution of the suite to more evolved (siliceous, alkaline, peraluminous) compositions when compared to the Neogene porphyritic granitoids of western Colombia is evident (Figs. 5.26 and 5.27). Bissig et al. (2014) indicate that the hydrous, oxidized Vetas-California porphyries evolved from mantle-derived melts which have assimilated moderate amounts of crustal material, potentially including Guiana Shield, granulite belt and/or Paleozoic supracrustal rocks, typical of the basement assembly of the Santander Massif. Bissig et al. (2014) provide radiogenic Sr, Nd and Pb isotope data which suggest the Miocene granitoids contain juvenile material, unlike the Paleozoic and Jurassic granitoids of the area which, based upon radiogenic Lu-Hf isotope analyses, appear to primarily represent recycled ca. 1 Ga continental crust (Van der Lelij 2013; Cochrane 2013). Lu-Hf isotope analyses for the Miocene Vetas-California porphyry suite have yet to be performed.

Within the Eastern Cordillera, some 175 and 360 km the south of Vetas-California, respectively, the granitoids of Paipa-Iza and Quetame reveal additional isolated, low-volume occurrences of Mio-Pliocene granitoids situated significantly to the east of the active Colombian volcanic arc axis. Major element lithogeochemical data from Paipa (Pardo et al. 2005b) indicate ferroan, alkalic, peraluminous compositions, dissimilar to the Vetas-California suite, atypical of Cordilleran granitoids and perhaps more akin to A-type granitoids, characteristic of melts generated in extensional environments (Frost et al. 2001). Lithogeochemical data for Quetame is restricted to a single major-element analysis of ca. 5.6 Ma felsic porphyry (Ujueta et al. 1990). The analysis reveals attributes of both the Vetas-California and Paipa suites; however, it is difficult to draw any firm conclusions based upon a single major element lithogeochemical analysis. Notwithstanding, Pardo et al. (2005b) and Ujueta et al. (1990) conclude that the lithogeochemical data for both Paipa (Iza) and Quetame, respectively, is markedly distinct from the typically calc-alkaline (calcic to calc-alkalic after Frost et al. 2001) compositions revealed along the active Colombian volcanic arc to the west.

From a petrogenetic standpoint, direct comparison of the Paipa-Iza-Quetame lithogeochemical data with that of Vetas-California is difficult due to the lack of trace, REE and radiogenic isotope data at Paipa-Iza-Quetame. From a major-element standpoint, however, the ferroan, alkalic nature of the Paipa-Iza granitoids contrasts markedly with the magnesian-calc alkalic suite from Vetas-California (Fig. 5.26), and if the Eastern Colombian granitoids of Mio-Pliocene age, although relatively widely space in occurrence, are considered as a whole, the suite may be considered to provide a bimodal distribution in terms of observed major element lithochemistry.

Aside from lithogeochemical comparisons, the Vetas-California-Paipa-Iza-Quetame suites share an important relationship with respect to distribution and structural controls. The occurrences are aligned along a ca. NNE-axis, whose trace is approximately parallel with respect to, and located east of (i.e. in the back-arc), the ca. NNE-oriented axis of the active Colombian volcanic arc (Figs. 5.22 and

5.36). In addition, the Santander-Eastern Cordilleran granitoids are all located along the trace of the Bucaramanga–Santa Marta–Garzón fault and suture system (Cediel and Cáceres 2000; Cediel et al. 2003) (Figs. 5.2 and 5.36), a long-lived, active, crustal-scale feature with significant vertical continuity, which could have facilitated the emplacement of mantle-derived melts into the upper crust.

The regional tectonic setting and relationship of the isolated granitoid occurrences of Colombia's eastern cordilleran system, to Mio-Pliocene granitoid magmatism and active Andean-style volcanism related to Nazca Plate subduction in the western and central Colombian Andes, have yet to be fully established. Taken as a whole, the present geographic position of these occurrences locates them in a back-arc position and along a sub-parallel NNE trend to the magmatic axis of the active Colombian volcanic arc. The bimodal lithogeochemical composition of the Santander-Eastern Cordillera occurrences suggests the suite as a whole may be rift-related. Based upon the foregoing, we suggests the Vetas-California-Paipa-Iza-Quetame granitoids could represent indications of crustal extension, focussed along the active Bucaramanga–Santa Marta–Garzón fault system, and rift-related magmatism within the back-arc of the Northern Andean volcanic zone in Colombia.

Figures 5.36 and 5.37 demonstrate complexities of the nature, geometry and timing of latest Oligocene-Miocene to Plio-Pleistocene magmatic arc development and granitoid magmatism in Colombia. It can be observed that extensive, composite "Neogene" granitoid magmatism in Colombia is in fact composed of a series of more spatially temporally limited arc segments, including a bimodal suite of outlier occurrences located in the back-arc region. Granitoid magmatism demarcating the composite arc is observed to migrate in time and space, in both a south-to-north and west-to-east sense. The emplacement, localization and lithochemistry of the numerous arc and outlier segments were influenced by the nature and composition of various basement complexes, facilitated by the location and reactivation of paleo-fault and suture systems throughout the Colombian Andes.

5.5 Summary and Concluding Statement

Plutonic and hypabyssal porphyritic granitoids and locally their volcanic equivalents constitute important components of the geological record of the Colombian Andes, not only from a volumetric standpoint but additionally as a reflection of the complex, diverse and dynamic tectonic evolution of the region. Based upon the composite U-Pb (zircon) age date database ca. 1995–2017, the analysis presented in this chapter has identified six principle episodes of Phanerozoic granitoid magmatism including early Paleozoic (ca. 485–439 Ma), Carboniferous (ca. 333–310 Ma), Permo-Triassic (ca. 289–225 Ma), latest Triassic-Jurassic (ca. 210–146 Ma), late Cretaceous to Eocene (ca. 100–42 Ma) and latest Oligocene to Mio-Pliocene (ca. 23–1.2 Ma). A continuum of this last episode into the Plio-Pleistocene through Recent manifests in the modern-day Colombian (Northern Andean) volcanic arc. The spatial distribution and analytical resolution of the U-Pb (zircon) database permit the identification of subpopulations within the major granitoid episodes and, in turn, a detailed analysis of the spatial and temporal migration of granitoid magmatism during the entire Colombian Phanerozoic.

Our analysis has integrated the major granitoid episodes into the pre-Northern Andean Orogeny, proto-Northern Andean Orogeny and Northern Andean Orogeny phases of Colombian tectonic evolution, and, supported by lithogeochemical and radiometric isotope data for many of the granitoid suites, we have used the granitoid populations as indicators of the tectonic framework in which the granitoids were generated and emplaced. Three pre-Northern Andean Orogeny granitoid populations are identified. Early Paleozoic granitoids of the Santander, Floresta and Quetame massifs include (1) ca. 499–473 Ma syn-kinematic and peak metamorphic granitoids, which are interpreted to have been generated/emplaced during a period of compression, crustal thickening, Barrovian-style metamorphism and orogenesis; (2) ca. 472–452 Ma granitoids, emplaced during post-orogenic collapse, extension and basin formation; and (3) ca. 452–415 Ma granitoids emplaced during resumed compression, basin closure and crustal thickening. The role of subduction per se in the petrogenesis of the early Paleozoic granitoids has yet to be clearly established, given that the entire ca. 485–439 Ma suite apparently represents primarily recycled, crustal-derived melts with limited juvenile contribution. Processes as diverse as crustal thickening, Barrovian-type metamorphism, extension, crustal thinning, lithospheric mantle upwelling and heat advection at the base of the crust have been invoked in the generation of these granitoids.

Early Paleozoic granitoid magmatism in eastern Colombia may have been brought on by approach and accretion Cajamarca-Valdivia island arc complex during the Quetame Orogeny. The Cajamarca-Valdivia Terrane, underlying much of Colombia's Central Cordillera, also contains an emerging population of similarly aged early Paleozoic granitoids, which are only now beginning to be recognized. Based upon current interpretations, these granitoids are considered allochthonous or peri-cratonic with respect to their Santander-Floresta-Quetame Massif counterparts and the continental Colombian tectonic mosaic as recorded during the early Paleozoic.

Following the emplacement of the last of the Colombian early Paleozoic granitoids at ca. 439 Ma, the region entered an amagmatic phase extending through to ca. 333 Ma. This granitoid hiatus denotes the onset of the Bolívar Aulacogen, a prolonged period of continental taphrogenesis characterized by extensional tectonics, the development of intra-continental and continental margin rifts and deposition of epicontinental and marine sedimentary strata in the Carboniferous through Permian.

An initial record of rift-related magmatism beginning in the mid-Carboniferous is recorded in the El Carmen-El Cordero gabbro-leucotonalite-trondjhemite suite hosted within Cajamarca-Valdivia Terrane basement along the Otú Fault within the Central Cordillera. The Carmen-El Cordero granitoids represent a Ridge Tholeiitic Granitoid assemblage (Barbarain 1999) petrogenetically associated with oceanic spreading and ophiolite formation. We suggest that the El Carmen-El Cordero suite reflects the progressively extensional environment prevalent during of the intermediate stages of the Bolívar Aulacogen. In the first instance, however, activity along the Otú rift was short-lived, and rifting was apparently aborted during a tectono-thermal event which affected much of the Northern Andean region, beginning in the early Permian.

The early Permian tectono-thermal event, including the emplacement of ca. 289–240 Ma syn-orogenic (±subduction-related?) peraluminous granitoid gneisses, is associated with collision and crustal thickening during the amalgamation of western Pangaea. As with the early Paleozoic granitoids, the Permian gneissic granitoids appear to represent primarily recycled melts derived from S-type upper crustal sources. Following ca. 240 Ma, the resumption of rifting is registered by a widespread but low-volume bimodal suite of metaluminous tholeiitic amphibolites and peraluminous anatectic granitoids, observed to intrude the Cajamarca-Valdivia Terrane throughout much of the Central Cordillera but also recorded in the Santander Massif and Upper Magdalena Basin. Both amphibolites and granitoids record an increasingly juvenile composition over time. The emplacement of this rift-related suite culminates in seafloor spreading after ca. 223 Ma and ocean crust formation by ca. 216 Ma, as represented by the Aburrá (Santa Elena) ophiolite. As with the early Paleozoic granitoid suite, the role of subduction and the contribution of subduction-derived magmatism in the petrogenesis of the Permian and mid-late Triassic peraluminous granitoids is uncertain. Processes including crustal thickening and anatexis during continental amalgamation, and regional extension, crustal thinning and basaltic underplating during continental break-up, have been suggested as root causes for the generation of these granitoids.

In Colombia, we suggest that the understanding of Permo-Triassic granitoid magmatism remains in many respects at a preliminary stage. The Permo-Triassic granitoid suite is under-represented within the Colombian geological map base, as many of these gneissic granitoids have been historically assigned to the early Paleozoic or Precambrian or in the case of the southern Sonsón Batholith, to the Jurassic. The further use of resilient U-Pb (zircon) dating techniques and the identification of new or mis-assigned Permo-Triassic granitoids will oblige a return to field-based mapping in order to define the physical limits of the Permo-Triassic intrusive suite. The accurate representation and interpretation of this important suite on published geologic maps will in turn permit better understanding of the tectonic development of the region as a whole during this time period.

Following incipient continental break-up and the formation of oceanic crust along the Colombian proto-Pacific margin beginning in the mid-Triassic, regional extension continued. The onset of the late Bolívar Aulacogen at this time is accompanied by a brief hiatus in granitoid magmatism, extending from ca. 225 to 210 Ma. Resumption of granitoid magmatism in the latest Triassic is characterized by the appearance of a complex spatial and temporal array of voluminous, continental arc granitoids including coeval volcanic rocks, of mostly metaluminous composition, which are interpreted to represent subduction-derived melts. These latest Triassic-Jurassic granitoids and volcanic rocks are quite unlike the low-volume peraluminous granitoids characteristic of previous extensional phases. They characterize a highly extensional but subduction-related regime dominant during the late Bolívar Aulacogen.

Notwithstanding, in northern Colombia, radiogenic isotope analyses (Lu-Hf, Sr-Nd, Pb-Pb) suggest that even some metaluminous latest Triassic-early Jurassic granitoids of the Santander Plutonic Group remain primarily comprised of recycled mid-Proterozoic continental material and were generated via processes involving regional extension, asthenospheric upwelling and thermal anatexis of the lower crust. In this context the Santander granitoids may represent a highly contaminated transitional suite, and not represent subduction-related melts per se. Continued extension leads to westward migration of arc axial magmatism during the mid-Jurassic (Sierra Nevada de Santa Marta and San Lucas batholiths) and into the late Jurassic (Segovia Batholith). Related granitoids demonstrate increasingly juvenile compositions and diminishing isotopic contributions from the hosting basement rocks and are considered to represent subduction-related melts emplaced within an overall extensional regime associated with a westward-retreating trench and slab rollback within the proto-Farallon Plate. The Segovia Batholith may in fact represent an eroded peri-cratonic island arc developed upon rifted Cajamarca-Valdivia Terrane basement. In southern Colombia, mid- and late Jurassic magmatism is also clearly temporally and spatially distinguished in the southern Ibagué, Mocoa-Garzón and northern Ibagué batholiths. The westward migration of arc-axial magmatism is less well defined, however, and granitoid plutonism migrates primarily along the NNE axial trend of the granitoid arc segments. Significant volumes of Jurassic volcano-sedimentary rocks (e.g. Noreán, Guatapurí, Saldaña Fms.) and hypabyssal porphyritic granitoids are preserved within the Jurassic arc segments, especially those of middle Jurassic age, an observation we interpret to reflect the extensional environment of arc formation. All known occurrences of hypabyssal porphyritic rocks associated with Jurassic holocrystalline batholiths (e.g. Santa Cruz, Mocoa, Rovira) were emplaced within 2 to 5 m.y. of the shutdown of the associated main phase holocrystalline batholiths.

Continued extension during culmination of the Bolívar Aulacogen led to the development of a rifted continental margin floored by Proto-Caribbean oceanic crust, the opening of the culminant intercontinental Valle Alto rift and the invasion of the Cretaceous seaway over much of the region. An ensuing 50 m.y hiatus in subduction-related magmatism from between ca. 145 and 95 Ma is indicated, based upon the absence of significant granitoid occurrences of this age range throughout continental Colombia.

Plate reorganization in the Pacific during the early Cretaceous led to the onset of the early Northern Andean Orogeny, marked initially by dextral transpression and the formation of blueschist assemblages along the Colombian Pacific margin beginning prior to ca. 120 Ma and followed by the appearance of a complex assemblage of subduction-related granitoids generated within both the autochthonous continental and allochthonous oceanic realms. The Eastern Group granitoids, including primarily the ca. 96–72 Ma Antioquian Batholith and its satellite plutons, represent continental arc magmas derived via the eastward, dextral-oblique subduction of Proto-Caribbean ± leading-edge Farallon Plate crust beneath the Colombian continental margin. Antioquian Batholith magmatism was extinguished at ca. 72 Ma due to the collision and accretion of CCOP-/CLIP-related terranes of the Western

Tectonic Realm (Cañas Gordas, Dagua, San Jacinto). A low-volume, short-lived post-collisional arc was reignited within the Central Cordillera, to the south of the Antioquian Batholith, between ca. 62 and 50 Ma (Sonsón, Manizales, El Hatillo, El Bosque, Córdoba plutons), possibly marking (1) the temporal reinitiation of subduction along the Pacific margin or alternatively (2) asthenospheric upwelling and thermally induced partial melting of lower crustal materials due to delamination of recently subducted oceanic lithosphere. In either case, granitoid magmatism within the continental domain was extinguished during final approach and accretion of the Gorgona Terrane along the Pacific margin in the early Eocene.

Contemporaneous with development of the Paleocene-Eocene post-collisional arc of the Central Cordillera, granitoid magmatism was also developed in the ca. 57–50 Ma Santa Marta Batholith and associated plutons located along the apex of the Sierra Nevada de Santa Marta. This localized and short-lived arc segment was generated kinematically independently from plate interactions along the Colombian Pacific margin and is interpreted to be related to the low-angle subduction or forced underthrusting of oceanic crust and tectonic stacking along the NW margin of the Sierra Nevada de Santa Marta, due to the NW-directed migration of the continental Maracaibo tectonic float.

Located to the west of the continental granitoids of ca. 96–50 Ma age and hosted within accreted oceanic terranes of the Western Tectonic Realm, two spatially temporally separate groups of subduction-related granitoids are also encompassed within early Northern Andean Orogeny development. The first, dating from ca. 100 to 84 Ma was generated during westward subduction of Proto-Caribbean crust beneath the northward and eastward migrating Farallon-CCOP/CLIP assemblage. These primitive, allochthonous, intra-oceanic arc granitoids (Buriticá, Santa Fé, Sabanalarga, Mistrató, Buga, Jejénes), correlative to the Greater (or leading-edge) Arc of the Caribbean series, were detached from their Farallon/CCOP roots in the late Cretaceous-Paleocene during dextral-oblique collision of the Farallon-CCOP/CLIP assembly and accretion of the Western Tectonic Realm terranes. We interpret the late Cretaceous-Paleocene tectonic lock-up between the Farallon-CCOP/CLIP assemblage and the northwestern South American margin to have led to the shutdown of subduction and granitoid arc magmatism in both continental Colombia and along the Colombian segment of the Greater (leading-edge) Arc.

We suggest that the late Cretaceous-Paleocene-Eocene tectonic pile-up of buoyant, leading-edge CCOP-CLIP fragments along the NW South American margin was also a persuading factor in the detachment and initiation of east-directed subduction of the Farallon Plate beneath the trailing edge of the CCOP/CLIP assemblage, beginning in the Paleocene. In Colombia, this magmatism is represented by the ca. 50–42 Ma, intra-oceanic Mandé and Acandí batholiths including associated hypabyssal porphyritic stocks, all of which comprise the younger representatives of the allochthonous, Western Group granitoids associated with early Northern Andean orogenic development. Mandé-Acandí correlates with the Middle American Arc series of Central America. The arc emplaced into the trailing edge of the CCOP, which, in Colombia, is represented by the El Paso Terrane including the Baudó

Complex. This composite arc and oceanic basement assemblage, however, were not accreted to the Colombia margin until the Miocene.

Many investigations of the origins and spatial vs. temporal migration of the Farallon-CCOP/CLIP assemblage demonstrate Pacific provenance and N and E migration into the inter-American gap, during the mid-late Cretaceous and early Paleogene. These same investigations suggest the Caribbean Plate docked with (i.e. was fixed with respect to) the South American Plate in the early Eocene (by ca. 54.5 Ma). We contend that the magmatic record of interactions between the South American, Proto-Caribbean and Farallon plates and the Caribbean-Colombian Oceanic Plateau is duly indicated by the pre- and syn- and post-collisional granitoid arc segments within the continental domain (Eastern Group granitoids) and the leading- and trailing-edge, intra-oceanic (Western Group) granitoids, presently accreted along the Colombian Pacific margin. In this context, we suggest the re-evaluation of tectonic models which require the amagmatic, low-angle or flat-slab subduction/consumption of large volumes of oceanic lithosphere (Pacific, Farallon, CCOP) beneath continental Colombia, during time intervals in which the existence of an accompanying magmatic arc within the continental cannot be demonstrated (e.g. early-mid-Cretaceous, ca. 145–96 Ma).

Following early Northern Andean Orogeny terrane assembly and docking of the Caribbean Plate, an additional hiatus in subduction-related granitoid magmatism in continental Colombia, spanning the period from ca. 50 to 23 Ma, is recorded. This hiatus is marked by the absence of outcropping granitoids, including the lack of detrital zircon populations dating from this time interval. Autochthonous granitoid arc-related magmatism resumed along the Colombian Pacific margin at ca. 23 Ma. The following events characterize the tectonic and magmatic development of the region leading up to and during the late Northern Andean Orogeny:

(1) The N and W migration of the South American Plate, relative to the stationary Caribbean Plate, beginning as early as the Eocene. Plate interaction along the Colombo-Caribbean margin was limited to tectonic tightening, stacking, buckling and uplift of the San Jacinto and Sinú terranes and the forced-underthrusting of Caribbean lithosphere. The absence of granitoid arc magmatism throughout continental Colombia, coincident with Cenozoic Colombo-Caribbean Plate interaction, is again stressed. N and W migration of the South American Plate continued into the mid-Miocene resulting in the culmination of the late Northern Andean Orogeny, including closure of the Middle American Seaway, collision/accretion of the El Paso Terrane and uplift of the Baudo Complex along the northwesternmost Colombian margin between ca. 8 and 4 Ma.

(2) Restructuring/rifting of the Farallon Plate within the eastern Pacific domain, resulting in development of the Nazca-Cocos plate system. Continued rifting within the Nazca segment between ca. 20 and 9 Ma gave rise to the Cauca and Coiba microplates, separated by the ca. E-W striking Sandra Ridge. Granitoids associated with Nazca Plate subduction along the Colombian Pacific margin first appear within the ca. 23–21 Ma Piedrancha-Cuembí arc segment, located in south–westernmost Colombia, well to the south of the trailing edge of the CCOP. The progressively

shallowing angle of subduction of the southern (Cauca) segment of the Nazca Plate led to eastward migration of the granitoid arc axis into the Cauca-Patía region between ca. 18 and 9 Ma. Continued eastward and northward migration of the magmatic arc axis during the Mio-Plio-Pleistocene led to conformation of the modern-day Northern Andes volcanic arc in southern and central Colombia.

To the north, tectonic tightening associated with South American-CCOP plate interaction hindered initiation of subduction related to the Coiba segment of the Nazca Plate, with the first manifestation of subduction-related granitoids appearing in the Farallones-Páramo Frontino-El Cerro arc segment at ca. 12–10 Ma. Again, progressive shallowing of the subduction angle, probably due to trench clogging by buoyant aseismic material (e.g. Sandra Ridge), led to eastward migration of arc axial magmatism into the Middle Cauca valley and the Central Cordillera (Cajamarca-Salento porphyry cluster) between ca. 9 and 4 Ma, followed by emplacement of the Plio-Pleistocene Río Dulce cluster to the north and coaxial conformation of the northernmost segment of the active Colombian volcanic arc (Ruíz-Santa Isabel-Tolima volcanic complex).

Based upon the foregoing, all of the latest Oligocene through Plio-Pleistocene granitoid arc/volcanic segments in the Colombian Andes are demonstrably associated with the segmented subduction of the Nazca Plate beneath the Pacific margin, beginning in the latest Oligocene, and all of the documented Oligo-Miocene arc segments are considered autochthonous with respect to continental Colombia. In addition to these subduction-related granitoids, minor, isolated occurrences of Mio-Pliocene hypabyssal and volcanic rocks (Vetas-California, Paipa-Iza, Quetame) are observed within the Santander Massif and Eastern Cordillera, to the east of the active Colombian volcanic arc. On the basis of major element lithochemistry, these back-arc occurrences comprise a bimodal suite. They form a coaxial trend with respect to the overall NNE orientation of the Miocene through modern-day granitoid arc axis. We interpret the Vetas-California-Paipa-Iza-Quetame occurrences to represent incipient rift-related magmatism whose emplacement was facilitated by back-arc extension focussed along the deep crustal conduits provided by the Bucaramanga–Santa Marta–Garzón fault and suture system.

The age, nature and spatial vs. temporal distribution of granitoids, as presently exposed within the Colombian geologic mosaic, provide valuable clues to the deciphering of the Phanerozoic tectono-magmatic history of the Colombian Andes. Although important advances have been made in the last decade, especially with respect to the generation of high-resolution age date, lithogeochemical and isotopic data, much work remains to be done, in continued sampling within the context of high-quality field-based mapping. Data verification, integration and synthesis into the ample and evolving geological, geophysical and tectono-sedimentalogical database which exists for the region will be an essential component of this process. We consider the analysis presented within this chapter as preliminary and, beyond the advance in understanding we feel it represents, would hope it will inspire continued investigation of the less studied, polemic and unresolved details regarding Phanerozoic tectono-magmatic evolution in the Colombian Andes.

References

Agencia Nacional de Hidrocarburos (ANH), Universidad de Caldas (2011) Estudio integrado de los núcleos y registros obtenidos de los pozos someros (slim holes) perforados por la ANH. Agencia Nacional de Hidrocarburos, Unpublished Report, Manizales, 304 p

Alfonso R (2000) Catálogo de unidades ígneas de Colombia: Batolito de Mocoa. INGEOMINAS, Bogotá

Alvarez JA (1983) Geología de la Cordillera Central y el occidente colombiano y petroquímica de los intrusivos granitoides meso-cenozoicos. Boletín Geológico INGEOMINAS 26(2):1–175

Altenberger U, Concha AE (2005) Late Lower to early Middle Jurassic arc magmatism in the northern Ibagué-Batholith/Colombia. Geología Colombiana 30:87–97

Arango JL, Ponce A (1982) Mapa Geológico del Departamento de Nariño. Memoria Explicativa. INGEOMINAS, 40 p

Aspden JA, McCourt WJ, Brook M (1987) Geometrical control of subduction-related magmatism: the Mesozoic and Cenozoic plutonic history of western Colombia. J Geol Soc 144(6):893–905

Aspden JA, Fortey N, Litherland M, Viteri F, Harrison SM (1992) Regional S-type granites in the Ecuadorian Andes: possible remnants of the breakup of western Gondwana. J S Am Earth Sci 6(3):123–132

Ballesteros CI (1983) Mapa Geológico generalizado del Departamento de Bolívar. INGEOMINAS, Bogotá

Barbarin B (1999) A review of the relationships between granitoid types, their origins and their geodynamic environments. Lithos 46(3):605–626

Barrero D (1977) Geology of the Central Western Cordil- lera, west of Buga and Roldanillo, Colombia Ph.D. Thesis, Colorado School of Mines

Bartok PE, Renz O, Westermann GEG (1985) The Siquisique Ophiolites, Northern Lara State, Venezuela: a discussion on their Middle Jurassic ammonites and tectonic implications. Geol Soc Am Bull 96:1050–1055

Barton MD, Young S (2002) Non-pegmatitic deposits of beryllium: mineralogy, geology, phase equilibria and origin. Rev Mineral Geochem 50(1):591–691

Bayona G, Cardona A, Jaramillo C, Mora A, Montes C, Valencia V, Ayala C, Montenegro O, Ibañez-Mejía M (2012) Early Paleogene magmatism in the northern Andes: insights on the effects of Oceanic Plateau–continent convergence. Earth Planet Sci Lett 331-332:97–111

Bird P (2003) An updated digital model of plate boundaries. Geochem Geophys Geosyst 4(3):1027–1059

Bissig T, Mantilla LC, Hart CJR (2014) Petrochemistry of igneous rocks of the California-Vetas mining district, Santander, Colombia: implications for northern Andean tectonics and porphyry Cu (–Mo, Au) metallogeny. Lithos 200-201:355–367

Bogotá J, Aluja J (1981) Geología de la Serranía de San Lucas. Geología Norandina 4:49–55

Boinet T, Bourgois J, Bellon H, Toussaint JF (1985) Age et répartition du magmatisme Prémèsozoïque des Andes de Colombie. Comptes-rendus des séances de l'Académie des sciences Paris 300(10):445–450

Botero G (1975) Edades radiométricas de algunos plutones colombianos. Minería 27:8336–8342

Buchs DM, Arculus RJ, Baumgartner PO, Baumgartner-Mora C, Ulianov A (2010) Late Cretaceous arc development on the SW margin of the Caribbean plate: insights from the Golfito, Costa Rica, and Azuero, Panama, complexes. Geochem Geophys Geosyst 11(7):1–35

Bürgl H (1967) The orogenesis of the Andean system of Colombia. Tectonophysics 4:429–443

Burke K, Cooper C, Dewey JF, Mann P, Pindell JL (1984) Caribbean tectonics and relative plate motions. In: Bonini WE, Hargraves RB, Shagam R (eds) The Caribbean–South American Plate boundary and regional tectonics. Geological Society of America Memoir 162:31–63

Bustamante A (2008) Geotermobarometria, geoquímica, geocronologia e evolução tectônica das rochas da fácies xisto azul nas áreas de Jambaló (Cauca) e Barragán (Valle del Cauca), Colômbia. Ph.D. thesis, Universidade de São Paulo

Bustamante C, Cardona A, Bayona G, Mora A, Valencia V, Gehrels G, Vervoort J (2010) U-Pb LA-ICP-MS geochronology and regional correlation of middle Jurassic intrusive rocks from the Garzon Massif, Upper Magdalena Valley and Central Cordillera, Southern Colombia. Boletín de Geología 32(2):93–109

Bustamante C, Cardona A, Archanjo CJ, Bayona G, Lara M, Valencia V (2017) Geochemistry and isotopic signatures of Paleogene plutonic and detrital rocks of the Northern Andes of Colombia: a record of post-collisional arc magmatism. Lithos 277:199–209

Cáceres C, Cediel F, Etayo F (2003) Guía introductoria de la distribución de facies sedimentarias de Colombia, Mapas de distribución de facies sedimentarias y armazón tectónico de Colombia a través del Proterozoico y del Fanerozoico. INGEOMINAS, Bogotá

Calle B, Toussaint JF, Restrepo JJ, Linares E (1980) Edades K/Ar de dos plutones de la parte septentrional de la Cordillera Occidental de Colombia. Geología Norandina 2:17–20

Campbell CJ (1968) The Santa Marta wrench fault of Colombia and its regional setting. Paper presented at the 4th Caribbean Geological Conference, Port of Spain, 28 March – 12 Abril 1965

Campbell CJ (1974) Colombian Andes. In: Spencer AM (ed) Mesozoic-Cenozoic Orogenic Belts. Geological Society of London Special Publications (4):705–724

Cardona A, Chew D, Valencia VA, Bayona G, Miškovic A, Ibañez-Mejía M (2010a) Grenvillian remnants in the Northern Andes: Rodinian and Phanerozoic paleogeographic perspectives. J S Am Earth Sci 29(1):92–104

Cardona A, Valencia V, Garzón A, Montes C, Ojeda G, Ruiz J, Weber M (2010b) Permian to Triassic I to S-type magmatic switch in the northeast Sierra Nevada de Santa Marta and adjacent regions, Colombian Caribbean: tectonic setting and implications within Pangea paleogeography. J S Am Earth Sci 29(4):772–783

Cardona A, Valencia VA, Bayona G, Duque J, Ducea M, Gehrels G, Jaramillo C, Montes C, Ojeda G, Ruíz J (2011) Early-subduction-related orogeny in the northern Andes: Turonian to Eocene magmatic and provenance record in the Santa Marta Massif and Rancheria Basin, northern Colombia. Terra Nova 23:26–34

Cediel F (2011) Major Tecto-sedimentary events and basin development in the Phanerozoic of Colombia: In: Cediel F (ed) Petroleum Geology of Colombia. Regional Geology of Colombia, vol 1. Agencia Nacional de Hidrocarburos (ANH) – EAFIT, pp 13–108

Cediel F (2018) Phanerozoic orogens of Northwestern South America: cordilleran-type orogens, taphrogenic tectonics and orogenic float. In: Cediel F, Shaw RP (eds) Geology and tectonics of Northwestern South America: the Pacific-Caribbean-Andean junction. Springer, Cham, pp 3–89

Cediel F, Cáceres C (2000) Geological map of Colombia. Geotec, Ltd., Bogotá

Cediel F, Etayo F, Cáceres C (1994) Facies distribution and tectonic setting through the Phanerozoic of Colombia. INGEOMINAS (ed) Geotec Ltd., Bogota

Cediel F, Shaw RP, Cáceres C (2003) Tectonic assembly of the Northern Andean Block. In: Bartolin, C, Buffler RT, Blickwede J (eds) The circum-Gulf of Mexico and the Caribbean: hydrocarbon habitats, basin formation, and plate tectonics. AAPG Memoir 79:815–848

Cediel F, Barrero D, Cáceres C (1998) Seismic Atlas of Colombia: Seismic expression of structural styles in the basins of Colombia. Robertson Research International, UK (ed) Geotec Ltd., Bogotá, vol 1 to 6

Cediel F, Restrepo I, Marín-Cerón MI, Duque-Caro H, Cuartas C, Mora C, Montenegro G, García E, Tovar D, Muñoz G (2010) Geology and hydrocarbon potential, Atrato and San Juan basins, Chocó (Panamá) arc, Colombia, Tumaco Basin (Pacific realm). Colombia, Agencia Nacional de Hidrocarburos (ANH)-EAFIT, Medellín

Cepeda H, Pardo-Villaveces N (2004) Vulcanismo de Paipa. Technical Report. INGEOMINAS, Bogotá

Chiarabba C, De Gori P, Faccenna C, Speranza F, Seccia D, Dionicio V, Prieto GA (2015) Subduction system and flat slab beneath the Eastern Cordillera of Colombia. Geochem Geophys Geosyst 17(1):16–27

Chiaradia M, Fontboté L, Paladines A (2004) Metal sources in mineral deposits and crustal rocks of Ecuador (1 N–4 S): a lead isotope synthesis. Econ Geol 99(6):1085–1106

Clavijo J, Mantilla L, Pinto J, Berna L, Perez A (2008) Evolución geológica de la Serranía de San Lucas, norte del valle medio del Magdalena y noroeste de la Cordillera Oriental. Boletín de Geología 3045–62

Cochrane R (2013) U-Pb thermochronology, geochronology and geochemistry of NW South America: rift to drift transition, active margin dynamics and implications for the volume balance of continents. PhD thesis, Université de Genève

Cochrane R, Spikings R, Gerdes A, Ulianov A, Mora A, Villagómez D, Putlitz B, Chiaradia M (2014a) Permo-Triassic anatexis, continental rifting and the disassembly of western Pangaea. Lithos 190–191:383–402

Cochrane R, Spikings R, Gerdes A, Winkler W, Ulianov A, Mora A, Chiaradia M (2014b) Distinguishing between in-situ and accretionary growth of continents along active margins. Lithos 202–203:382–394

Cocks LRM, Torsvik TH (2006) European geography in a global context from the Vendian to the end of the Palaeozoic. In: Gee DG, Stephenson RA (eds) European lithosphere dynamics. Geol Soc Lond Mem 32:83–95

Cordani UG, Cardona A, Jiménez DM, Liu D, Nutman AP (2005) Geochronology of Proterozoic basement inliers in the Colombian Andes: tectonic history of remnants of a fragmented Grenville belt. Geological Society of London Special Publications 246(1):329–346

Correa AM, Pimentel M, Restrepo JJ, Nilson A, Ordoñez O, Martens U, Laux JE, Junges S (2006) U-Pb zircon ages and Nd-Sr isotopes of the Altavista stock and the San Diego gabbro: new insights on Cretaceous arc magmatism in the Colombian Andes. Abstract presented at the V South American Symposium on Isotope Geology (SSAGI), Punta del Este, 24–27 April 2006

Correa AM (2007) Petrogênese e evolução do ofiolito de Aburrá, cordilheira central dos Andes colombianos. PhD Thesis, Universidade de Brasilia

Cruz N, Carrillo JA, Mantilla LC (2014) Consideraciones petrogenéticas y geocronología de las rocas ígneas porfiríticas aflorantes en la Quebrada Ventanas (Municipio Arboledas, Norte de Santander, Colombia): implicaciones metalogenéticas. Boletín de Geología 36(1):103–118

Cuadros FA, Botelho NF, Ordóñez-Carmona O, Matteini M (2014) Mesoproterozoic crust in the San Lucas range (Colombia): an insight into the crustal evolution of the northern Andes. Precambrian Res 245:186–206

Davies JF, Whitehead RE (2006) Alkali-alumina and MgO-alumina molar ratios of altered and unaltered rhyolites. Explor Min Geol 15(1–2):75–88

Davies JF, Whitehead RE (2010) Alkali/alumina molar ratio trends in altered Granitoid rocks hosting porphyry and related deposits. Explor Min Geol 19(1–2):13–22

De La Roche H, Leterrier JT, Grandclaude P, Marchal M (1980) A classification of volcanic and plutonic rocks using R1R2-diagram and major-element analyses—its relationships with current nomenclature. Chem Geol 29(1–4):183–210

Deckart K, Godoy E, Bertens A, Saeed A (2010) Barren Miocene granitoids in the Central Andean metallogenic belt, Chile: geochemistry and Nd-Hf and U-Pb isotope systematics. Andean Geol 37(1):1–31

Dewey JW (1972) Seismicity and tectonics of western Venezuela. Bull Seismol Soc Am 62(6):1711–1751

Dörr W, Grösser JR, Rodríguez GI, Kramm U (1995) Zircon U-Pb age of the Páramo Rico tonalite-granodiorite, Santander Massif (Cordillera Oriental, Colombia) and its geotectonic significance. J S Am Earth Sci 8(2):187–194

Doubrovine PV, Steinberger B, Torsvik TH (2012) Absolute plate motions in a reference frame defined by moving hot spots in the Pacific, Atlantic, and Indian oceans. J Geophys Res 117(B09101):1–30

Duque JF (2009) Geocronología (U/Pb y $^{40}Ar/^{39}Ar$) y geoquímica de los intrusivos paleógenos de la Sierra Nevada de Santa Marta y sus relaciones con la tectónica del Caribe y el arco magmático circun-Caribeño. MSc thesis, Universidad Nacional Autónoma de México

Duque-Caro H (1990) The Chocó block in the northwestern corner of South America: structural, tectonostratigraphic, and paleogeographic implications. J S Am Earth Sci 3(1):71–84

Duque-Caro H (1991) Contributions to the geology of the Pacific and the Caribbean coastal areas of Northwestern Colombia and South America. PhD Thesis, Princeton University

Duque-Trujillo J, Sánchez J, Orozco-Esquivel T, Cárdenas A (2018) Cenozoic magmatism of the maracaibo block and its tectonic significance. In: Cediel F, Shaw RP (eds) Geology and tectonics of Northwestern South America: the Pacific-Caribbean-Andean junction. Springer, Cham, pp 551–594

Ego F, Sébrier M, Yepes H (1995) Is the Cauca-Patía and Romeral fault system left or right-lateral? Geophys Res Lett 22(1):33–36

Escobar LA, Tejada N (1992) Prospección de Platino en Piroxenitas y de Oro en Skarn en la Mina Don Diego, El Cerro, Frontino (Antioquia). B.Sc. Thesis, Universidad Nacional de Colombia, Facultad de Ciencias, Seccional de Medellín

Estrada JJ (1995) Paleomagnetism and accretion events in the Northern Andes: PhD Thesis, State University of New York

Etayo-Serna F, Barrero D, Lozano H, Espinosa A, González H, Orrego A, Zambrano F, Duque H, Vargas R, Núñez A, Álvarez J, Ropaín C, Ballesteros I, Cardozo E, Forero H, Galvis N, Ramírez C, Sarmiento L, Albers JP, Case JE, Singer DA, Bowen RW, Berger BR, Cox DP, Hodges CA (1983) Mapa de terrenos geológicos de Colombia. Publicaciones Geológicas Especiales, vol 14. INGEOMINAS, Bogotá

Etayo-Serna F, Rodríguez GI (1985) Edad de la Formación Los Santos. In: Etayo-Serna F, Laverde-Montaño F, Pava A (eds) Proyecto Cretácico: Contribuciones. Publicaciones Geológicas Especiales, vol 16. INGEOMINAS, Bogotá

Fabre A, Delaloye M (1983) Intrusiones básicas cretácicas en las sedimentitas de la parte central de la Cordillera Oriental. Geología Norandina 6:19–28

Farris DW, Jaramillo C, Bayona G, Restrepo-Moreno SA, Montes C, Cardona A, Mora A, Speakman RJ, Glascock MD, Valencia V (2011) Fracturing of the Panamanian Isthmus during initial collision with South America. Geology 39(11):1007–1010

Feininger T (1970) The Palestina fault, Colombia. Bull Geol Soc Am 81:1201–1216

Feininger T, Botero G (1982) The Antioquian Batholith, Colombia. Publicación Geológica Especial INGEOMINAS 12:1–50

Feininger T, Barrero D, Castro N (1972) Geología de parte de los departamentos de Antioquia y Caldas (sub-zona II-B). Boletín Geológico INGEOMINAS 20(2):1–173

Frantz JC, Ordoñez OC, Franco E, Groves DI, McNaughton NJ (2003) Marmato porphyry intrusion, ages and mineralization. In: Abstracts of the IX Colombian Geological Congress, Medellín, 30 July – 1 August 2003

Frost BR, Frost CD (2008) A geochemical classification for feldspathic igneous rocks. J Petrol 49(11):1955–1969

Frost BR, Barnes CG, Collins WJ, Arculus RJ, Ellis DJ, Frost CD (2001) A geochemical classification for granitic rocks. J Petrol 42(11):2033–2048

Gansser A (1955) Ein Beitrag zur Geologie und Petrographie der Sierra Nevada de Santa Marta (Kolumbien, Suedamarika). Schweiz. Mineral Petrogr Mitt 35(2):209–279

Gansser A (1973) Facts and theories on the Andes. J Geol Soc Lond 129:93–131

Garzón T (2003) Geoquímica y potencial minero asociado a cuerpos volcánicos de la región de Paipa, Departamento de Boyacá, Colombia. M.Sc. thesis, Universidad Nacional de Colombia

Geyer OF (1973) Das Praekretazische Mesozoikum von Kolumbien. Geologisches Jarbuch 5:1–56

Gil-Rodríguez J (2010) Igneous Petrology of the Colosa Gold-rich Porphyry System (Tolima, Colombia). M.Sc. thesis, The University of Arizona

Gil-Rodríguez J (2014) Petrology of the Betulia Igneous Complex, Cauca, Colombia. J S Am Earth Sci 56:339–356

Goldsmith R, Marvin RF, Mehnert HH (1971) Radiometric ages in the Santander Massif, Eastern Cordillera, Colombian Andes. US Geol Surv Prof Pap 750D:D44–D49

Gómez J, Nivia A, Montes NE, Jiménez DM, Tejada ML, Sepúlveda MJ, Osorio JA, Gaona T, Diederix H, Uribe H, Mora M (2007) Mapa Geológico de Colombia. INGEOMINAS, Bogotá

Gómez J, Montes NE, Nivia A, Diederix H (2015a) Mapa Geológico de Colombia 2015. Servicio Geológico Colombiano, Bogotá

Gómez J, Montes NE, Alcárcel FA, Ceballos JA (2015b) Catálogo de dataciones radiométricas de Colombia en ArcGIS y Google Earth. In: Gómez J, Almanza MF (eds) Compilando la geología de Colombia - Una visión a 2015. Servicio Geológico Colombiano, Publicaciones Geológicas Especiales 33:63–419

González H (1990) Mapa geológico generalizado del Departamento de Risaralda. INGEOMINAS, Escala 1:200,000, Versión digital 2010

González H (1993) Mapa geológico generalizado del Departamento de Caldas - geología y recursos minerales. Memoria explicativa, INGEOMINAS, 62p

González H (1999) Geología del Departamento de Antioquia. Escale 1:400.000. INGEOMINAS, Bogotá

González H (2001) Mapa geológico del Departamento de Antioquia: Geología, recursos minerales y amenazas potenciales. Memoria Explicativa. INGEOMINAS, Bogotá

Goossens PJ, Rose WI, Flores D (1977) Geochemistry of tholeiites of the Basic Igneous Complex of northwestern South America. Geol Soc Am Bull 88(12):1711–1720

Hamilton WB (1994) Subduction systems and magmatism. In: Smellie JL (ed) Volcanism associated with extension at consuming plate margins: Geological Society of London Special Publication 81:3–28

Hastie AR, Kerr AC (2010) Mantle plume or slab window?: physical and geochemical constraints on the origin of the Caribbean oceanic plateau. Earth Sci Rev 98(3):283–293

Henrichs I (2013) Caracterização e idade das intrusivas do sistema pórfiro Yarumalito, Magmatismo Combia, Colombia. M.Sc. Thesis, Universidade Federal do Rio Grande do Sul

Horton BK, Taylor JE, Nie J, Mora A, Parra M, Reyes-Harker A, Stockli DF (2010) Linking sedimentation in the northern Andes to basement configuration, Mesozoic extension, and Cenozoic shortening: evidence from detrital zircon U-Pb ages, Eastern Cordillera, Colombia. Geol Soc Am Bull 122(9–10):1423–1442

Ibañez-Mejia M, Tassinari CCG, Jaramillo-Mejia J (2007) U-Pb ages of the "Antioquian Batholith" – Geochronological constraints of late Cretaceous magmatism in the central Andes of Colombia. Abstract presented at the XI Congreso Colombiano de Geología, Bucaramanga, 14–17

Ibañez-Mejía M, Jaramillo-Mejía JM, Valencia VA (2008) U–Th/Pb zircon geochronology by multi-collector LA-ICP-MS of the Samaná Gneiss: a Middle Triassic syn-tectonic body in the Central Andes of Colombia, related to the latter stages of Pangea assembly. In: VI South American Symposium on Isotope Geology, San Carlos de Bariloche–Argentina, 13–17 April 2008

Ibañez-Mejía M, Ruiz J, Valencia VA, Cardona A, Gehrels GE, Mora AR (2011) The Putumayo Orogen of Amazonia and its implications for Rodinia reconstructions: new U–Pb geochronological insights into the Proterozoic tectonic evolution of northwestern South America. Precambrian Res 191(1):58–77

Irvine TN, Baragar WRA (1971) A guide to the chemical classification of the common volcanic rocks. Can J Earth Sci 8(5):523–548

Irving EM (1975) Structural evolution of the northernmost Andes, Colombia. U.S. Geological Survey, Professional Paper 846

Kellogg JN, Ogujiofor IJ, Kansakar DR (1985) Cenozoic tectonics of the Panama and North Andes blocks. In: Memoirs of the 6th Latin American Congress on Geology, vol 1. Bogotá, p. 40

Kennan L, Pindell JL (2009) Dextral shear, terrane accretion and basin formation in the Northern Andes: best explained by interaction with a Pacific-derived Caribbean plate? Geological Society of London Special Publications 328(1):487–531

Keppie JD (2004) Terranes of Mexico revisited: a 1.3 billion year odyssey. Int Geol Rev 46(9):765–794

Keppie JD (2008) Terranes of Mexico revisited: a 1.3 billion year Odyssey. In: Keppie JD, Murphy JB, Ortega-Gutierrez F, Ernst WG (eds) Middle American Terranes, potential correlatives, and Orogenic processes. CRC Press/Taylor & Francis Group, Boca Raton, 7–36

Keppie JD, Ortega-Gutierrez F (2010) 1.3–0.9 Ga Oaxaquia (Mexico): remnant of an arc/backarc on the northern margin of Amazonia. J S Am Earth Sci 29(1):21–27

Keppie JD, Ramos VA (1999) Odyssey ofterranes in the Iapetus and Rheic oceans during the Paleozoic. In: Ramos VA, Keppie JD (eds) Laurentia-Gondwana connections before Pangea. The Geological Society of America, Special Paper 336:267–276

Kerr AC, Tarney J (2005) Tectonic evolution of the Caribbean and northwestern South America: the case for accretion of two Late Cretaceous oceanic plateaus. Geology 33(4):269–272

Kerr AC, Tarney J, Marriner GF, Nivia A, Saunders AD (1997) The Caribbean–Colombian Cretaceous Igneous Province: The internal anatomy of an oceanic plateau. In: Mahoney JJ, Coffin MF (eds) Large Igneous Provinces: Continental, oceanic, and planetary flood volcanism. American Geophysical Union, Geophysical monograph 100:123–144

Kerr AC, White RV, Thompson PME, Tarney J, Saunders AD (2003) No oceanic plateau—No Caribbean plate? The seminal role of an oceanic plateau in Caribbean plate evolution. In: Bartolin, C, Buffler RT, Blickwede J (eds) The circum-Gulf of Mexico and the Carib-bean: hydrocarbon habitats, basin formation, and plate tectonics. AAPG Memoir 79:126–168

Kroonenberg SB (1982) A Grenvillian granulite belt in the Colombian Andes and its relation to the Guiana shield. Geol Mijnb 61(4):325–333

Kurhila M, Andersen T, Rämö OT (2010) Diverse sources of crustal granitic magma: Lu–Hf isotope data on zircon in three Paleoproterozoic leucogranites of southern Finland. Lithos 115(1):263–271

Leal-Mejía H (2011) Phanerozoic Gold Metallogeny in the Colombian Andes: A tectono-magmatic approach. Ph.D. thesis, Universitat de Barcelona

Leal-Mejía H, Shaw RP, Melgarejo JC (2011) Phanerozoic granitoid magmatism in Colombia and the tectono-magmatic evolution of the Colombian Andes. In: Cediel F (ed) Petroleum Geology of Colombia. Regional Geology of Colombia, vol 1. Agencia Nacional de Hidrocarburos (ANH) – EAFIT, p 109–188

Lesage G, Richards JP, Muehlenbachs K, Spell TL (2013) Geochronology, geochemistry, and fluid characterization of the Late Miocene Buriticá Gold Deposit, Antioquia Department, Colombia. Econ Geol 108:1067–1097

Litherland M, Aspden JA, Jemielita RA (1994) The metamorphic belts of Ecuador. Overseas Mem Br Geol Surv 11:1–147

Lodder C, Padilla R, Shaw RP, Garzón T, Palacio E, Jahoda R (2010) Discovery history of the La Colosa Gold Porphyry deposit, Cajamarca, Colombia. Soc Econ Geol Spec Pub 15:19–28

Loiselle MC, Wones DR (1979) Characteristics and origin of anorogenic granites. Geol Soc Am Abstr Programs 11:468

Lonsdale P (2005) Creation of the Cocos and Nazca plates by fission of the Farallon plate. Tectonophysics 404(3):237–264

Mantilla LC, Valencia VA, Barra F, Pinto J, Colegial J (2009) U-Pb geochronology of porphyry rocks in the Vetas – California gold mining area (Santander, Colombia). Boletín de Geología (UIS) 31(1):31–43

Mantilla LC, Bissig T, Cottle JM, Hart CRJ (2012) Remains of early Ordovician mantle-derived magmatism in the Santander Massif (Colombian Eastern Cordillera). J S Am Earth Sci 38:1–12

Mantilla LC, Bissig T, Valencia V, Hart CJR (2013) The magmatic history of the Vetas-California mining district, Santander Massif, Eastern Cordillera, Colombia. J S Am Earth Sci 45:235–249

Maresch WV, Stöckhert B, Baumann A, Kaiser C, Kluge R, Krückhans-Lueder G, Brix MR, Thomson M (2000) Crustal history and plate tectonic development in the southern Caribbean. Sonderheft Zeitschrift fuer Angewandte. Andean Geol 1:283–289

Maresch WV, Kluge R, Baumann A, Pindell JL, Krückhans-Lueder G, Stanek K (2009) The occurrence and timing of high-pressure metamorphism on Margarita Island, Venezuela: a constraint on Caribbean-South America interaction. Geol Soc Lond Spec Publ 328(1):705–741

Marechal P (1983) Les Temoins de Chaine Hercynienne dans le Noyau Ancien des Andes de Merida (Venezuela): Structure et evolution tectometamorphique. Ph.D. Thesis, Universite de Bretagne Occidentale

Marín-Cerón MI, Leal-Mejía H, Bernet M, Mesa-García J (2018) Late cenozoic to modern-day volcanism in the Northern Andes; A geochronological, petrographical and geochemical review. In: Cediel F, Shaw RP (eds) Geology and tectonics of Northwestern South America: the Pacific-Caribbean-Andean junction. Springer, Cham, pp 603–641

Martens U, Restrepo JJ, Ordóñez-Carmona O, Correa-Martínez AM (2014) The Tahamí and Anacona terranes of the Colombian Andes: missing links between the South American and Mexican Gondwana margins. J Geol 122(5):507–530

Maya M (1992) Catalogo de dataciones isotópicas en Colombia. Boletín Geológico INGEOMINAS 32(1–3):127–188

McCourt WJ, Feininger T (1984) High pressure metamorphic rocks of the Central Cordillera of Colombia. British Geological Survey Reprint Series 84:28–35

McDonough WF, Sun SS (1995) The composition of the earth. Chem Geol 120(3–4):223–253

McGeary S, Ben-Avraham Z (1989) The accretion of Gorgona Island, Colombia: Multichannel seismic evidence. In: Howel DG (ed) Tectonostratigraphic terranes of the Circum-Pacific Region. Circum-Pacific Council for Energy and Mineral Resources, Earth Sciences Series 1:543–554

Mejía P, Santa M, Ordoñez O, Pimentel M (2008) Consideraciones petrográficas, geoquímicas y geocronologicas de la parte occidental del Batolito de Santa Marta. Revista Dyna 15:223–236

Molnar P, Atwater T (1978) Interarc spreading and Cordilleran tectonics as alternates related to the age of subducted oceanic lithosphere. Earth Planet Sci Lett 41(3):330–340

Montes C, Guzmán G, Bayona G, Cardona A, Valencia V, Jaramillo C (2010) Clockwise rotation of the Santa Marta Massif and simultaneous Paleogene to Neogene deformation of the Plato-San Jorge and Cesar-Ranchería basins. Journal of South American Earth Sciences 29(4):832–848

Montes C, Cardona A, McFadden R, Morón SE, Silva CA, Restrepo-Moreno S, Ramírez DA, Hoyos N, Wilson J, Farris D, Bayona GA, Jaramillo CA, Valencia V, Bryan J, Flores JA (2012) Evidence for middle Eocene and younger land emergence in central Panama: implications for Isthmus closure. Geol Soc Am Bull 124(5–6):780–799

Montes C, Cardona A, Jaramillo C, Pardo A, Silva JC, Valencia V, Ayala C, Pérez-Angel LC, Rodríguez-Parra LA, Ramirez V, Niño H (2015) Middle Miocene closure of the American seaway. Science 348(6231):226–229

Müller RD, Royer JY, Cande SC, Roest WR, Maschenkov S (1999) New constraints on the Late Cretaceous/tertiary plate tectonic evolution of the Caribbean. In: Mann P (ed) Caribbean basins. Sedimentary basins of the world, vol 4. Elsevier, Amsterdam, pp 33–59

Nelson HW (1957) Contribution to the geology of the Central and Western Cordillera of Colombia in the sector between Ibagué and Cali. Leidsche Geologische Mededelingen 22:1–75

Nerlich R, Clark SR, Bunge HP (2014) Reconstructing the link between the Galapagos hotspot and the Caribbean plateau. GeoResJ 1-2:1–7

Nie J, Horton BK, Saylor JE, Mora A, Mange M, Garzione CN, Basu A, Moreno CJ, Caballero V, Parra M (2012) Integrated provenance analysis of a convergent retroarc foreland system: U–Pb ages, heavy minerals, Nd isotopes, and sandstone compositions of the Middle Magdalena Valley basin, northern Andes, Colombia. Earth Sci Rev 110(1):111–126

Nívia A, Gómez J (2005) El Gabro Santa Fé de Antioquia y la Cuarzodiorita Sabanalarga, una propuesta de nomenclatura litoestratigráfica para dos cuerpos plutónicos diferentes agrupados previamente como Batolito de Sabanalarga en el Departamento de Antioquia, Colombia. Memoirs of the X Colombian Geological Congress, Bogotá

Nívia A (1992) Mapa geológico generalizado del Departamento del Valle del Cauca. Escala 1:300.000. INGEOMINAS - British Geological Survey (BGS), Bogotá

Nivia A, Marriner G, Kerr A (1996) El Complejo Quebrada Grande: una posible cuenca marginal intracratónica del Cretáceo Inferior en la Cordillera Central de los Andes Colombianos. In: Abstracts of the VII Colombian Geological Congress, vol III, 108–123

Nivia A, Marriner GF, Kerr AC, Tarney J (2006) The Quebradagrande complex: a lower cretaceous ensialic marginal basin in the Central Cordillera of the Colombian Andes. J S Am Earth Sci 21:423–436

Núñez A (1998) Catalogo de unidades litoestratigraficas de Colombia – Batolito de Ibagué: INGEOMINAS, Bogotá

O'connor JT (1965) A classification for quartz-rich igneous rocks based on feldspar ratios. US Geological Survey Professional Paper B 525:79–84

Ordoñez O (2001) Caracterização Isotópica Rb-Sr e Sm-Nd dos Principais Eventos Magmáticos nos Andes Colombianos. Ph.D. thesis, Universidade de Brasilia

Ordoñez O, Pimentel MM (2002) Rb–Sr and Sm–Nd isotopic study of the Puquí complex, Colombian Andes. J S Am Earth Sci 15(2):173–182

Ordoñez O, Pimentel MM, Armstrong RA, Gioia SMCL, Junges S (2001) U-Pb SHRIMP and Rb-Sr ages of the Sonsón Batholith. In: III South American Symposium on Isotope Geology, Pucon - Chile, 21–24 October 2001

Ordoñez-Carmona O, Restrepo JJ, Pimentel MM (2006) Geochronological and isotopical review of pre-Devonian crustal basement of the Colombian Andes. J S Am Earth Sci 21(4):372–382

Ordoñez-Carmona O, Pimentel M, Laux JH (2007a) Edades U-Pb del Batolito Antioqueño. Boletín de Ciencias de la Tierra 22:129–130

Ordoñez-Carmona O, Pimentel MM, Frantz JC, Chemale F (2007b) Edades U-Pb convencionales de algunas intrusiones colombianas. Abstract presented at the XI Congreso Colombiano de Geología, Bucaramanga, 14–17 Aug 2007

Orrego A, Cepeda H, Rodríguez G (1980) Esquistos glaucofánicos en el área de Jambaló, Cauca (Colombia). Geología Norandina 1:5–10

Parada MA, Nyström JO, Levi B (1999) Multiple sources of the Coastal Batholith of central Chile (31-34°S) – geochemical and Sr-Nd isotopic evidence and tectonic implications. Lithos 46:505–521

Pardo N, Cepeda H, Jaramillo JM (2005a) The Paipa volcano, Eastern Cordillera of Colombia, South America – volcanic Stratigraphy. Earth Sci Res J 9(1):3–18

Pardo N, Cepeda H, Jaramillo JM (2005b) The Paipa volcano, Eastern Cordillera of Colombia, South America (Part II) – petrography and major elements petrology. Earth Sci Res J 9(2):148–164

París G, Marín P (1979) Mapa Geológico Generalizado del Departamento del Cauca. Memoria Explicativa, INGEOMINAS, 38 p

Pearce JA, Harris NB, Tindle AG (1984) Trace element discrimination diagrams for the tectonic interpretation of granitic rocks. J Petrol 25(4):956–983

Peccerillo A, Taylor SR (1976) Geochemistry of Eocene calc-alkaline volcanic rocks from the Kastamonu area, Northern Turkey. Contrib Mineral Petrol 58(1):63–81

Pennington WD (1981) Subduction of the eastern Panama Basin and seismotectonics of northwestern South America. J Geophys Res Solid Earth 86(B11):10753–10770

Pindell JL, Cande SC, Pitman WC, Rowley DB, Dewey JF, Labrecque J, Haxby W (1988) Plate kinematic framework for models of Caribbean evolution. Tectonophysics 155:121–138

Pindell J, Kennan L (2001) Processes & Events in the Terrane assembly of Trinidad and E. Venezuela. In: GCSSEPM Foundation 21st annual research conference transactions, Petroleum Systems of Deep-Water Basins, 159–192

Priem HNA, Andriessen PAM, Boelrijk NAIM, de Boorder H, Hebeda EH, Verdurmen EA, Huguett A, Verdurmen EAT, Verschure RH (1982) Geochronology of the Precambrian in the Amazonas region of southeastern Colombia (western Guiana shield). Geol Mijnb 61(3):229–242

Priem HNA, Kroonemberg SB, Boelrijk NAI, Hebeda EH (1989) Rb-Sr and K-Ar evidence of 1.6Ga basement underlying the 1.2Ga Garzón-Santa Marta granulitic belt in the Colombian Andes. Precambrian Res 42:315–324

Ramírez DA, López A, Sierra GM, Toro G (2006) Edad y proveniencia de las rocas volcanico sedimentarias de la Formación Combia en el suroccidente antioqueño – Colombia. Boletín de Ciencias de la Tierra 19:9–26

Ramos VA (1999) Plate tectonic setting of the Andean Cordillera. Episodes 22(3):183–190

Ramos VA (2009) Anatomy and global context of the Andes: main geologic features and the Andean orogenic cycle. In: Kay SM, Ramos VA, and Dickinson WR (eds) backbone of the Americas: shallow Subduction, plateau uplift, and ridge and Terrane collision. Geol Soc Am Mem 204:31–65

Redwood SD (2018) The Geology of the Panama-Chocó Arc Springer volumen – Confirm definitive reference

Restrepo JJ, Toussaint JF (1988) Terrains and continental accretion in the Northern Andes. Episodes 11:189–193

Restrepo JJ, Toussaint JF, González H (1981) Edades Mio-Pliocenas del magmatismo asociado a la Formación Combia, Departamento de antioquia y Caldas, Colombia. Geología Norandina 3:21–26

Restrepo JJ, Ordóñez-Carmona O, Armstrong R, Pimentel MM (2011) Triassic metamorphism in the northern part of the Tahamí Terrane of the central cordillera of Colombia. J S Am Earth Sci 32(4):497–507

Restrepo-Moreno SA, Foster DA, Kamenov GD (2007) Formation age and magma sources for the Antioqueño Batholith derived from LA-ICP-MS Uranium-Lead dating and Hafnium-isotope analysis of zircon grains. Geol Soc Am Abstr Programs 39(6):493

Restrepo-Pace PA (1992) Petrotectonic characterization of the Central Andean Terrane, Colombia. J S Am Earth Sci 5(1):97–116

Restrepo-Pace PA (1995) Late Precambrian to Early Mesozoic tectonic evolution of the Colombian Andes based on new geological, geochemical and isotopic data. PhD thesis, The University of Arizona

Restrepo-Pace PA, Ruíz J, Gehrels G, Cosca M (1997) Geochronology and Nd isotopic data of Grenville-age rocks in the Colombian Andes – new constraints for late Proterozoic–early Paleozoic paleocontinental reconstructions of the Americas. Earth Planet Sci Lett 150:427–441

Restrepo-Pace PA, Cediel F (2010) Northern South America basement tectonics and implications for paleocontinental reconstructions of the Americas. J S Am Earth Sci 29:764–771

Richards JP, Kerrich R (2007) Adakite-like rocks – their diverse origins and questionable role in metallogenesis. Econ Geol 102(4):537–576

Rickwood PC (1989) Boundary lines within petrologic diagrams which use oxides of major and minor elements. Lithos 22(4):247–263

Rodríguez AL (2014) Geology, Alteration, Mineralization and Hydrothermal Evolution of the La Bodega – La Mascota deposits, California-Vetas Mining District, Eastern Cordillera of Colombia, Northern Andes. M.Sc. thesis, The University of British Columbia

Rodriguez G, Arango MI (2013) Formación Barroso: arco volcanico toleitico y diabasas de San José de Urama: un prisma acrecionario T-MORB en el segmento norte de la Cordillera Occidental de Colombia. Boletín de Ciencias de la Tierra 33:17–38

Rodríguez G, Zapata G (2012) Características del plutonismo Mioceno superior en el segmento norte de la Cordillera Occidental e implicaciones tectónicas en el modelo geológico del noroccidente colombiano. Boletín de Ciencias de La Tierra 31:5–22

Rodríguez G, Zapata G (2014) Descripción de una nueva unidad de lavas denominada Andesitas basálticas de El Morito-correlación regional con eventos magmáticos de arco. Boletín de Geología 36(1):85–102

Rodríguez AL, Bissig T, Hart CJ, Mantilla LC (2017) Late Pliocene high-Sulfidation epithermal gold mineralization at the La Bodega and La Mascota Deposits, Northeastern Cordillera of Colombia. Econ Geol 112(2):347–374

Rodríguez G, Arango MI, Zapata G, Bermúdez JG (2014) Petrografía y geoquímica del Neis de Nechí. Boletín de Geología 36(1):71–84

Rodríguez-Vargas A, Koester E, Mallmann G, Conceição RV, Kawashita K, Weber MBI (2005) Mantle diversity beneath the Colombian Andes, northern volcanic zone: constraints from Sr and Nd isotopes. Lithos 82(3):471–484

Royero JM, Clavijo J (2001) Mapa geologico generalizado del Departamento de Santander. Memoria Explicativa: INGEOMINAS, Bogotá

Ruiz J, Tosdal RM, Restrepo PA, Murillo-Muñetón G (1999) Pb isotope evidence for Colombia-southern Mexico connections in the Proterozoic. In: Ramos VA, Keppie JD (eds.) Laurentia-Gondwana Connections before Pangea. Geological Society of America Special Paper 336:183–197

Saenz EA (2003) Fission track thermochronology and denudational response to tectonics in the north of The Colombian Central Cordillera. MSc Thesis, Shimane University

Santô T (1969) Characteristics of seismicity in South America. Bull Earthquake Res Inst 47:635–672

Sarmiento LF (2001) Mesozoic rifting and Cenozoic basin inversion history of the Eastern Cordillera, Colombian Andes. Inferences from tectonic models. Ph.D. Thesis, Vrije Universiteit

Sarmiento LF (2018) Cretaceous stratigraphy and paleo-facies maps of Northwestern South America. In: Cediel F, Shaw RP (eds) Geology and tectonics of Northwestern South America: the Pacific-Caribbean-Andean junction. Springer, Cham, pp 673–739

Saylor JE, Stockli DF, Horton BK, Nie J, Mora A (2012) Discriminating rapid exhumation from syndepositional volcanism using detrital zircon double dating: implications for the tectonic history of the Eastern Cordillera, Colombia. Geol Soc Am Bull 124(5–6):762–779

Schmidt-Thomé M, Feldhaus L, Salazar G, Muñoz R (1992) Explicación del mapa geológico, escala 1:250 000, del flanco oeste de la Cordillera Occidental entre los ríos Andágueda y Murindó, Departamentos Antioquia y Chocó, República de Colombia. Geologisches Jahrbuch Reihe B, Band B78

Shaw RP, Leal-Mejía H, Melgarejo JC (2018) Phanerozoic metallogeny in the Colombian Andes: a tectono-magmatic analysis in space and time. In: Cediel F, Shaw RP (eds) Geology and tectonics of Northwestern South America: the Pacific-Caribbean-Andean junction. Springer, Cham, pp 411–535

Sillitoe RH, Jaramillo L, Damon PE, Shafiqullah M, Escovar R (1982) Setting, characteristics and age of the Andean porphyry copper belt in Colombia. Econ Geol 77:1837–1850

Sillitoe RH, Jaramillo L, Castro H (1984) Geologic exploration of a molybdenum-rich porphyry copper deposit at Mocoa, Colombia. Economic Geology 79(1):106–123

Silva JC, Arenas JE, Sial AN, Ferreira VP, Jiménez D (2005) Finding the Neoproterozoic-Cambrian transition in carbonate successions from the Silgará Formation, Northeastern Colombia: an assessment from C-isotope stratigraphy. In: Memorias del X Congreso Colombiano de Geología, Bogota

Silver EA, Reed DL, Tagudin JE, Heil DJ (1990) Implications of the north and south Panama thrust belts for the origin of the Panama orocline. Tectonics 9:261–281

Silver PG, Russo RM, Lithgow-Bertelloni C (1998) Coupling of south American and African plate motion and plate deformation. Science 279:60–63

Singewald QD (1950) Mineral resources of Colombia (other than petroleum). US Geol Surv Bull 964–B:56–204

Sinton CW, Duncan RA, Storey M, Lewis J, Estrada JJ (1998) An oceanic flood basalt province within the Caribbean plate. Earth Planet Sci Lett 155(3):221–235

Spikings R, Cochrane R, Villagómez D, Van der Lelij R, Vallejo C, Winkler W, Beate B (2015) The geological history of northwestern South America – from Pangaea to the early collision of the Caribbean large Igneous Province (290–75 Ma). Gondwana Res 27:95–139

Stanley CR, Madeisky HE (1994) Lithogeochemical Exploration for Hydrothermal Ore Deposits using Pearce Element Ratio Analysis. In: Lentz DR (ed) Alteration and Alteration Processes associated with Ore-forming Systems. Geological Association of Canada, Short Course Notes 11:193–211

Stanley CR, Madeisky HE (1996) Lithogeochemical exploration for metasomatic zones associated with hydrothermal mineral deposits using Pearce Element Ratio Analysis. Short Course Notes on Pearce Element Ratio Analysis, Mineral Deposit Research Unit (MDRU), the University of British Columbia, Canada

Stern RJ (2002) Subduction zones. Reviews of geophysics 40(4):3-1-3-38

Stern CR (2004) Active Andean volcanism: its geologic and tectonic setting. Revista geológica de Chile 31(2):161–206

Stevenson RK, Patchett PJ (1990) Implications for the evolution of continental crust from Hf isotope systematics of Archean detrital zircons. Geochim Cosmochim Acta 54(6):1683–1697

Taboada A, Rivera LA, Fuenzalida A, Cisternas A, Philip H, Bijwaard H, Olaya J, Rivera C (2000) Geodynamics of the northern Andes - Subductions and intracontinental deformation (Colombia). Tectonics 19(5):787–813

Tassinari CCG, Diaz F, Buenaventura J (2008) Age and source of gold mineralization in the Marmato mining district, NW Colombia – a Miocene-Pliocene epizonal gold deposit. Ore Geol Rev 33:505–518

Tistl M (1994) Geochemistry of platinum-group elements of the zoned ultramafic Alto Condoto Complex, Northwest Colombia. Econ Geol 89(1):158–167

Tistl M, Burgath KP, Höhndorf A, Kreuzer H, Muñoz R, Salinas R (1994) Origin and emplacement of tertiary ultramafic complexes in northwest Colombia: evidence from geochemistry and K-Ar, Sm-Nd and Rb-Sr isotopes. Earth Planet Sci Lett 126(1–3):41–59

Toro G, Restrepo JJ, Poupeau G, Saenz E, Azdimousa A (1999) Datación por trazas de fisión de circones rosados asociados a la secuencia volcano-sedimentaria de Irra (Caldas). Boletín de Ciencias de la Tierra 13:28–34

Tschanz CM, Marvin RF, Cruz J, Mehnert H, Cebulla G (1974) Geologic evolution of the Sierra Nevada de Santa Marta area, Colombia. Geol Soc Am Bull 85:273–284

Trumpy D (1949) Geology of Colombia. Shell Unpublished Report No. 23323

Ujueta G, Macia C, Romero F (1990) Cuerpo Riodacítico del Terciario Superior en la Región de Quetame, Cundinamarca. Geología Colombiana 17:143–150

Van der Hilst RD (1990) Tomography with P, PP and pP delay-time data and the three-dimensional mantle structure below the Caribbean region. Ph.D. Thesis, Instituut voor Aardwetenschappen der Rijksuniversiteit Utrecht

Van der Hilst R, Mann P (1994) Tectonic implica- tions of tomography images of subducted lithosphere beneath northwestern South America. Geology 22:451–454

Van der Lelij R (2013) Reconstructing north-western Gondwana with implications for the evolution of the Iapetus and Rheic Oceans: a geochronological, thermochronological and geochemical study. PhD thesis, Université de Genève

Van der Lelij R, Spikings RA, Ulianov A, Chiaradia M, Mora A (2016) Palaeozoic to early Jurassic history of the northwestern corner of Gondwana, and implications for the evolution of the Iapetus, Rheic and Pacific Oceans. Gondwana Res 31:271–294

Vásquez M, Altenberger U, Romer RL, Sudo M, Moreno-Murillo JM (2010) Magmatic evolution of the Andean Eastern Cordillera of Colombia during the cretaceous: influence of previous tectonic processes. J S Am Earth Sci 29(2):171–186

Vargas CA, Mann P (2013) Tearing and breaking off of subducted slabs as the result of collision of the Panama arc-indenter with Northwestern South America. Bull Seismol Soc Am 103(3):2025–2046

Vesga AM, Jaramillo JM (2009) Geoquímica del domo volcánico en el Municipio de Iza, Departamento de Boyacá – Interpretación geodinámica y comparación con el vulcanismo Neógeno de la Cordillera Oriental. Boletín de Geología (UIS) 31(2):97–108

Villagómez D (2010) Thermochronology, geochronology and geochemistry of the Western and Central cordilleras and Sierra Nevada de Santa Marta, Colombia: The tectonic evolution of NW South America. PhD thesis, Université de Genève

Villagómez D, Spikings R, Magna T, Kammer A, Winkler W, Beltrán A (2011) Geochronology, geochemistry and tectonic evolution of the Western and Central cordilleras of Colombia. Lithos 125:875–896

Vinasco CJ (2004) Evolução crustal e história tectônica dos granitóides permo-triássicos dos Andes do Norte. PhD thesis, Universidade de Sao Paulo

Vinasco C (2018) The romeral shear zone. In: Cediel F, Shaw RP (eds) Geology and tectonics of Northwestern South America: the Pacific-Caribbean-Andean junction. Springer, Cham, pp 833–870

Vinasco CJ, Cordani UG, González H, Weber M, Pelaez C (2006) Geochronological, isotopic, and geochemical data from Permo-Triassic granitic gneisses and granitoids of the Colombian Central Andes. J S Am Earth Sci 21:355–371

Ward DE, Goldsmith R, Cruz J, Jaramillo C, Restrepo H (1973) Geología de los cuadrangulos H-12 Bucaramanga y H-13 Pamplona, Departamento de Santander. Boletín Geológico INGEOMINAS 21(1–3):1–132

Warren I, Simmons SF, Mauk JL (2007) Whole-rock geochemical techniques for evaluating hydrothermal alteration, mass changes, and compositional gradients associated with epithermal Au-Ag mineralization. Econ Geol 102:923–948

Weber B, Iriondo A, Premo W, Hecht L, Schaaf P (2007) New insights into the history and origin of the southern Maya block, SE México: U–Pb–SHRIMP zircon geochronology from metamorphic rocks of the Chiapas massif. Int J Earth Sci 96(2):253–269

Weber MB, Tarney J, Kempton PD, Kent RW (2002) Crustal make-up of the northern Andes – evidence based on deep crustal xenolith suites, Mercaderes, SW Colombia. Tectonophysics 345(1):49–82

Weber M, Cardona A, Valencia V, García-Casco A, Tobón M (2010) U/Pb detrital zircon provenance from late cretaceous metamorphic units of the Guajira peninsula, Colombia: tectonic implications on the collision between the Caribbean arc and the South American margin. J S Am Earth Sci 29(4):805–816

Weber M, Gómez-Tapias J, Cardona A, Duarte E, Pardo-Trujillo A, Valencia VA (2015) Geochemistry of the Santa Fé Batholith and Buriticá Tonalite in NW Colombia and evidence of subduction initiation beneath the Colombian Caribbean plateau. J S Am Earth Sci 62:257–274

Wegner W, Wörner G, Harmon RS, Jicha BR (2011) Magmatic history and evolution of the Central American land bridge in Panama since cretaceous times. Geol Soc Am Bull 123(3–4):703–724

Wessel P, Kroenke LW (2008) Pacific absolute plate motion since 145 Ma: an assessment of the fixed hot spot hypothesis. J Geophys Res 113(B06101):1–21

Wilson M (1989) Igneous petrogenesis a global tectonics approach. Chapman & Hall, London

Wood DA, Tarney J, Varet J, Saunders AD, Bougault H, Joron JL, Treuil M, Cann JR (1979) Geochemistry of basalts drilled in the North Atlantic by IPOD leg 49: implications for mantle heterogeneity. Earth Planet Sci Lett 42(1):77–97

Wright JE, Wyld SJ (2011) Late cretaceous subduction initiation on the eastern margin of the Caribbean-Colombian oceanic plateau: one great arc of the Caribbean (?). Geosphere 7:468–493

Zapata G, Rodríguez G (2013) Petrografía, Geoquímica y edad de la Granodiorita de Farallones y las rocas volcánicas asociadas. Boletín de Geología 35(1):81–96

Zapata S, Cardona A, Jaramillo C, Valencia V, Vervoort J (2016) U-Pb LA-ICP-MS geochronology and geochemistry of Jurassic volcanic and plutonic rocks from the Putumayo region (Southern Colombia):tectonic setting and regional correlations. Boletín de Geología 38(2):21–38

Zarifi Z, Havskov J, Hanyga A (2007) An insight into the Bucaramanga nest. Tectonophysics 443(1):93–105

Zartman RE, Doe SM (1981) Plumbotectonics – the model. Tectonophysics 75:135–162

Chapter 6
Phanerozoic Metallogeny in the Colombian Andes: A Tectono-magmatic Analysis in Space and Time

Robert P. Shaw, Hildebrando Leal-Mejía, and Joan Carles Melgarejo i Draper

6.1 Introduction

Unlike the highly fertile and active metalliferous domains of the central Andes, the metallogenesis sensu lato of the Northern Andes and especially Colombia remains mostly undocumented. Whereas entire issues of international journals such as *Economic Geology* and *Mineralium Deposita* have been devoted to the metallogenic provinces of Perú, Chile, Bolivia, Brasíl and Argentina (e.g. Skinner 1999), less than a handful of modern publications specifically describing the metalliferous deposits of Colombia are internationally available. Few of these are comprehensive, and most were presented over 25 years ago. This observation is confusing, given that Colombia and the Northern Andes in general comprise a highly fertile metallotectonic environment (Petersen 1979; Sillitoe 2008), as supported, for example, by extensive past gold-, silver- and platinum-group metal production, historically the most significant in all of South America (e.g. Emmons 1937; Table 6.1). In addition, Colombia remains a significant (although mostly artisanal) producer of gold, silver and much sought-after emeralds and is the largest producer of ferronickel and platinum in South America. Copper, lead, zinc and iron are produced, as principal

R. P. Shaw (✉) · J. C. Melgarejo i Draper
Departament de Mineralogia, Petrologia i Geologia Aplicada, Facultat de Ciències de la Terra, Universitat de Barcelona, Barcelona, Catalonia, Spain
e-mail: shaw6301@telus.net

H. Leal-Mejía
Departament de Mineralogia, Petrologia i Geologia Aplicada, Facultat de Ciències de la Terra, Universitat de Barcelona, Barcelona, Catalonia, Spain

Mineral Deposit Research Unit (MDRU), The University of British Columbia (UBC), Vancouver, BC, Canada

Table 6.1 Colombian historic gold production as compared to other South American countries over the period 1492 to 1934, as reported by Emmons (1937)

Country	Estimated gold production (Troy Oz)	Contribution to estimated total production (%)	Colombian production (compared to other countries)
Colombia	**48,976,465**	**37.85**	NA
Brasíl	38,732,908	29.94	1.3x
Chile	11,039,469	8.53	4.5x
Bolivia	9,849,979	7.61	5x
Perú	7,736,428	5.98	6x
Guyane (French Guyana)	4,373,337	3.38	11x
Venezuela	3,687,110	2.85	13x
Guyana	2,418,961	1.87	20x
Ecuador	1,226,831	0.95	40x
Suriname	1,130,482	0.87	44x
Argentina	208,977	0.16	238x
Total (to 1934)	**129,380,945**	**100.00**	

NA Not applicable

commodities or as byproducts, on a modest scale from a variety of geologic environments; however, neither the breath nor depth of their potential has been completely explored let alone documented. Tellingly, literally hundreds of metalliferous manifestations, occurrences, active producing mines or abandoned showings, including Au, Ag, Pb, Zn, Cd, Cu, Mo, Sb, Hg, Cr, Ni, Pt, Pd, Ti, Mn and Fe deposits, are paper-compiled in governmental catalogues and on mineral occurrence maps, mostly dating from the 1950s to 1990s. Notwithstanding, the majority of these manifestations have not been historically explored, and hence minimal empirical academic studies, such as deposit mapping, minerographic and alteration-paragenetic-isotopic-lithogeochemical studies or radiometric age dating, are available. As a consequence, few historic attempts have been made to produce an integrated temporal-spatial metallogenic framework for the Colombian Andes.

Clearly, the level of modern metallogenic understanding in the region is not on par with neighbouring Andean nations, although reasons behind this general lack of metallogenic consideration cannot be attributed to the general lack of mining history in Colombia. Pre-Colombian goldsmithing technology and craftsmanship, second to none in the Western Hemisphere, attracted extensive Conquest- and Colonial-era exploration and exploitation, and the Spanish Colonial through Independence periods produced a number of million ounce producing gold camps, which were already in demise by the late nineteenth century. The following lament, paraphrased from the invaluable treatise on Colonial-era gold mining by Restrepo (1888, p. 190), provides much insight into the state of Colombian mining in early post-Colonial times:

> It is generally believed that a mine is abandoned when its ores are exhausted and it is no longer capable of remunerating the costs of exploitation. Or so it should be, but this is not what has happened in Colombia, where the Wars of Independence and our endless civil disputes, mine inundation by subterranean waters, and the lack of method and knowledge, shortage of machinery, difficulty with transport, lawsuits etc. have in many cases caused this disastrous result.

Indeed, mine abandonment in Colombia is steeped in the complex social and ethnocultural history of the country. Regardless small-scale gold and/or platinum mining remains a very traditional activity in virtually all of the historic mining camps. Such activities, however, have in many instances been replaced by agrarian practices which take advantage of the fertile soils and ideal climatic conditions over much of the Cordilleran region, where numerous historically productive but long since abandoned mineral occurrences are presently covered by sugarcane, fruit and coffee crops, lush pastureland or tree farms. In order to explain apparent differences in metallogenic endowment between the Central and Northern Andes, Petersen (1970) highlighted the climatic differences between the humid, vegetated north and the arid altiplanos of Perú, Bolivia, Chile and Argentina. Indeed, deep tropical to semi-tropical weathering/leaching, saprolite and latosol development, and extensive vegetative cover have contributed to difficulties in the modern discovery, definition and development of ore deposits in the region. In addition, the complicated socio-political climate in Colombia over the last 50+ years has limited uninhibited field access in many regions of the country. In fact, many of the mining districts have been specifically targeted, creating an obvious obstacle for the academic, governmental and industrial sectors and thus reducing the execution, availability and scope of modern technical investigations.

6.2 Methodology and Scope of Analysis

The study of metallogeny demands the integration of genetic aspects of metalliferous mineral deposits (source of metals, source of fluids, timing of ore deposition and mineral paragenesis) with the broad-scale tectonic and magmatic style and evolution of a region. Many of the fundamental relationships between metalliferous mineral deposits, magmatism and plate tectonics are well documented, and it has long been accepted that magmatic trends and compositions and tectonic setting sensu lato exert a first-order control upon regional metallogenesis (e.g. Strong 1976; Guilbert and Park 1986; Sawkins 1990; Kirkham et al. 1995; Society of Economic Geologists 2002; Kerrich et al. 2005; Groves and Bierlein 2007; Bierlein et al. 2009). The metallogenic framework presented herein examines the age, style and distribution of metalliferous mineral occurrences in the Colombian Andes and places them into the evolving Phanerozoic tectono-magmatic framework of the region.

With respect to the underlying tectonic framework used to depict Colombian metallogeny, important advances have been made in recent years (e.g. Cediel et al.

1994; Cediel and Cáceres 2000; Cediel et al. 2003; Kennan and Pindell 2009; Cediel et al. 2010; Cediel 2011; Leal-Mejía 2011; Spikings et al. 2015; Leal-Mejía et al. 2018). The base concepts advanced by Cediel et al. (1994), Cediel and Cáceres (2000) and Cediel et al. (2003) are highly suitable to metallogenic applications. These authors describe the geology and tectonic evolution of more than 30 lithotectonic and morpho-structural domains (tectonic realms, terranes, terrane assemblages, physiographic regions) comprising the entire Northern Andean block, including regional-scale fault and suture systems. The analysis is focussed upon onshore and peri-cratonic geologic evolution and describes regional tectonic events spanning the entire Phanerozoic, which facilitates the integration of magmatic episodes and mineralizing events at a scale apt to the definition of metallogenic provinces within the context of the entire Colombia Andes.

Regarding commodity types, we restrict our analysis to precious metals (Au, Ag, PGEs) and the most important base and industrial metals, including Cu, Pb, Zn, Mo, Ni, Cr and Fe. All these metals present either significant production histories in Colombia or their occurrences are sufficiently well known with respect to location, age and geological context, that they may be confidently integrated into our time-space charts. The only non-metallic mineral we have included is emerald, for which Colombia is considered a world-class producer of high-quality gemstones and for which abundant modern studies permit temporal and spatial integration.

It must be recognized that available information pertaining to Colombian Au (±Ag) occurrences, due to their importance from a historic to modern-day perspective, far outweighs that of the other metals included in this analysis. Indeed, gold forms the principal economic commodity in more Colombian metal occurrences than in all of the known remaining metal occurrences combined. In this context, our analysis is, in many respects, primarily an analysis of Colombian gold metallogeny. Notwithstanding, given the metal associations typical of many hydrothermal mineral deposit types, such as a Au ± Cu-Mo association in porphyry-associated mineral systems (Sillitoe 2000) or a Au±Ag-Zn-Pb association in intermediate-sulphidation epithermal systems (Sillitoe and Hedenquist 2003; Simmons et al. 2005), an integral understanding of gold metallogeny in the Colombian Andes leads to an understanding of the metallogenesis of its co- or subproduct metals.

Au-dominant metallogeny, however, can be considered "typical" of the Colombian Andes, given that there are few mineral districts that are not Au-dominant, or in which Au ± Ag and Cu do not themselves form important co-products. This observation may be an artefact of the historic importance and production history Au has held in the region, combined with the relative paucity of exploration and resource development work undertaken specifically in search of other metals, and future exploration and discoveries may change this perspective. At present however, the Colombian Andes may be considered a Au-biased metallogenic province (Sillitoe 2008) in same sense that the Chilean Andes may historically be considered a Cu-dominant domain (Sillitoe and Perelló 2005) or much of the Mexican Cordillera considered Ag biased (Camprubí 2009).

A marked exception to the above observation is Colombia's Eastern Cordillera, an inverted mid-Mesozoic to Paleogene failed rift-related sedimentary basin, which purports a widespread but little documented base metal-dominant metallogeny, apparently essentially devoid of precious metals. Little metals exploration has been undertaken in Colombia's Eastern Cordillera; however, the abundance of historically recorded and exploited Zn, Pb and Cu occurrences speaks of an overlooked metallotect with interesting exploration potential. A summary of Eastern Cordilleran metallogeny is presented herein.

6.3 Principal Sources of Information

There are hundreds if not thousands of historic documents pertaining to Colombian mineral and metal occurrences and historic through modern-day metal production, spanning the pre-Columbian through Colonial and Modern eras. Volumes of publicly available metal-type, production and location data are archived in libraries in the principal Colombian cities. The most readily available collections are housed at the Bogotá headquarters of the Colombian Geological Survey (former Instituto de Investigaciones en Geociencias, Minería y Química - INGEOMINAS) and at regional offices in Medellín, Bucaramanga, Cartagena, Cali, Manizales, Popayán, Pasto and Ibagué. The Universidad Nacional in Bogotá, the Escuela de Minas in Medellín, the Universidad Industrial de Santander in Bucaramanga and the Universidad de Caldas in Manizales, among others, also host important collections, including regional data and undergraduate ± masters-level theses which are not widely circulated. Not surprising, over 95 percent of this historic technical information pertains to precious metal (Au-Ag ± Pt) occurrences and their paragenetically associated metal assemblages (Cu, Pb, Zn, Sb, Hg, etc.). Virtually all of this literature is written in the Spanish language.

Fortunately, an abundance of historic Colombian mineral occurrence and production data has been analysed and reduced to a much more manageable format in various published historical and technical compendiums, spanning the pre-Columbian to Modern eras. Pertaining to precious metals, fundamental works by Restrepo (1888), Emmons (1937), Singewald (1950), Wokittel (1960) and the Instituto de Estudios Colombianos (1987) provide excellent historical and production-focused compilations and discussions. The Publicaciones Geológicos Especiales series, published by INGEOMINAS (Mejía et al. 1986; Villegas 1987; Mutis 1993), provides thorough regional-scale compilations pertaining to all metalliferous mineral occurrences. The ACIGEMI project (INGEOMINAS 1998) includes a digital 1:500,000 compilation containing geological, geochemical and mineral occurrence and mine location data with topographic overlays. Additional joint cooperation exploration, mostly dating from the 1960s through 1980s, between INGEOMINAS and external institutions, such as the United States Geological Survey (USGS), the United Nations (UN) and the Japan International Cooperation Agency (JICA), also produced important data. These programmes included general resource inventories (e.g. Tschanz et al. 1968) but also specifically targeted porphyry Cu (Mo, Au) potential (Sillitoe et al. 1982, 1984; Japan International Cooperation Agency 1987; Gómez-Gutiérrez and Molano-Mendoza 2009). More recent generalized government-compiled information may be found in online repositories associated with mineral resources and mining in Colombia, including the Unidad de Planeación Minero Energético (www.upme.gov.co/mineria) and the Agencia Nacional de Minería (www.anm.gov.co).

As discussed above, literature regarding metallogeny throughout much of the Colombian Andes is inextricably linked to Au and its socio-economic importance spanning the pre-Columbian, Colonial-post-Colonial and early Modern eras (Restrepo 1888; Emmons 1937). Developments in the Colombian gold mining industry, stemming primarily from foreign investment in the sector between ca. 1890 and 1950, are reflected in a relatively continuous (considering the era) stream of international publications describing Colombian gold districts, deposits and mining methods (e.g. Nichols and Farrington 1899; Halse 1906; Gamba 1910; Perry 1914a, b; del Rio 1930; Hoffmann 1931; Rundall 1931; Grosse 1932; Emmons 1937 and references contained therein; Wilson and Darnell 1942a, b). This international flow of information ceased after ca. 1950, and, with the exception of the government-level mineral occurrence compilations outlined above, only a few publications describing gold deposits in Colombia are available (e.g. Rodríguez and Warden 1993; Rossetti and Colombo 1999; Felder et al. 2005; Gallego and Akasaka 2007, 2010; Sillitoe 2008; Tassinari et al. 2008; Lesage et al. 2013; Bissig et al. 2014, 2017; Rodríguez et al. 2017). These works are primarily descriptive, and although they address the genetic aspects of specific Au (+co-metal) occurrences, none attempted to place the numerous gold districts of varying age and style into an integrated tectono-magmatic (i.e. metallogenic) framework.

Based primarily upon intimate and consistent spatial association between hydrothermal Au (Ag-Cu-Mo-Zn-Pb-Sb-As) occurrences and granitoid intrusive and volcanic rocks in the Colombian Andes, a close genetic relationship has been inferred by many past workers (e.g. Restrepo 1888; Grosse 1932; Emmons 1937; Singewald 1950; Wokittel 1960; Sillitoe et al. 1982; Mejía et al. 1986; Rodríguez and Warden 1993; Rossetti and Colombo 1999; Shaw 2000a, b, 2003a, b; Sillitoe 2008). This inference is well supported to date by the available publications which have applied radiometric dating techniques to Colombian mineral occurrences (e.g. Sillitoe et al. 1982, 1984; Tassinari et al. 2008; Leal-Mejía 2011; Leal-Mejía et al. 2010; Lesage et al. 2013; Mantilla et al. 2013; Rodríguez et al. 2017; Bissig et al. 2017).

It should be observed, however, that the radiometric age database sensu lato for Colombia is malnourished, especially when available radiometric age data specifically applicable to Colombian mineral deposits are considered. In general, historic Colombian radiometric age data consist primarily of pre-1985 K-Ar and Rb-Sr isochron dates (Maya 1992), many of which demonstrate poor repeatability and large margins of error. Furthermore, prior to 2006, less than a handful of published U-Pb (zircon) ages for the dozens of Phanerozoic granitoid batholiths and stocks and associated volcanic units of the Colombian Andes were available. The age constraints for most granitoids were defined primarily based upon field relationships combined, where applicable, with the historic K-Ar or Rb-Sr data, and numerous uncertainties remained with respect to the timing of emplacement of individual batholiths and stocks. Little of the combined data could be applied to the spatial-temporal dynamics of Colombian metallogenesis, and early attempts at an integrated Colombian tectono-metallogenic framework (e.g. Shaw 2000a, b, 2003a, b) were badgered by poor age constraints.

Between 2005 and 2011, industry-sponsored mineral exploration grants funded studies which addressed the paucity of empirical data specifically applicable to the understanding of Au + co-metal metallogenesis at the scale of the entire Colombian Andes (Lodder et al. 2010; Leal-Mejía 2011; Leal-Mejía et al. 2011a). These studies were guided by previous metallogenic analyses (Shaw 2000a, b, 2003a, b) and included field review and sampling of all of the important Colombian primary (hard rock) producing Au districts and many lesser-known historic Au and base metal occurrences. Special attention was paid to the field relationships between mineral occurrences and their host basement, host and proximal granitoid plutons, hypabyssal porphyry intrusives and coeval volcanic and volcanoclastic sequences. Sampling transects provided regional coverage of plutonic, hypabyssal and volcanic suites. In total, 107 new U-Pb (zircon) age dates and 282 whole-rock major-minor-trace-RE element lithogeochemical analyses for granitoids throughout the Colombian Andes were obtained. Investigations were supported by new K-Ar, Ar-Ar and Re-Os ages, as well as radiogenic (Sr-Nd-Pb) and stable (S) isotopic data. This information, when combined with detailed petrographic, metallographic, paragenetic and alteration studies, in conjunction with tectono-magmatic analysis, permitted construction of a gold + co-metal metallogenic framework for the Colombian Andes which spans the entire Phanerozoic (Leal-Mejía and Melgarejo 2008, 2010; Leal-Mejía et al. 2006, 2009, 2010, 2011a, b, 2015; Lodder et al. 2010). The Leal-Mejía (2011) Au metallogenic framework forms the basis of the time-space analysis presented herein.

In addition to the U-Pb (zircon) dates of Leal-Mejía (2011) and co-workers, various researchers over the past decade have also provided important contributions to the Colombian radiometric age, isotopic and lithogeochemical database for granitoid intrusive and volcanic rocks. These studies include works by Vinasco et al. (2006), Ibañez-Mejía et al. (2007), Restrepo-Moreno et al. (2007), Mantilla et al. (2009, 2012, 2013), Villagómez et al. (2011), Bayona et al. (2012), Montes et al. (2012, 2015), Van der Lelij (2013), Cochrane (2013), Cochrane et al. (2014a, b), Weber et al. (2015), Van der Lelij et al. (2016), Zapata et al. (2016) and Bustamante et al. (2017), among others. As a result, there are presently more than 290 well-located U-Pb (zircon) age dates backed by lithogeochemical studies upon which to analyse the Phanerozoic evolution and migration of granitoid magmatism throughout the Colombian Andes.

Finally, aside from the plethora of Au-dominated literature, important data pertaining to Colombian base and platinum-group metal occurrences have been integrated into our time-space charts. These include works by Ortiz (1990) and

Jaramillo (2000) for Cu (Zn, Pb, Au, Ag)-bearing volcanogenic massive sulphide deposits, Mejía and Durango (1981) and Gleeson et al. (2004) for the Cerro Matoso Ni deposit and Tistl (1994) for PGEs of the San Juan and Atrato basins. With respect to the base metal occurrences of the Eastern Cordillera, contributions by Kimberley (1980) for oolitic Fe and Fabre and Delaloye (1983) for Pb, Zn, Cu and Fe are noteworthy. With respect to the emerald deposits of the Eastern Cordillera, numerous studies pertaining to their mineralogy, paragenesis, age, structural evolution and uplift history have been completed (e.g. Cheilletz et al. 1994; Ottaway et al. 1994; Branquet et al. 1999; Banks et al. 2000; Giuliani et al. 2000), which permit their placement within the regional metallogenic scheme.

6.4 Tectono-magmatic Framework of the Colombian Andes

The tectonic evolution of the Northern Andes including much of Colombia has long been recognized as highly complex (e.g. Bürgl 1967; Gansser 1973; Irving 1975; Shagam 1975; Etayo-Serna et al. 1983; Aspden et al. 1987; Restrepo and Toussaint 1988; Cediel et al. 1994; Ramos 1999). Tectonic solutions for the region must take into account a variety of factors atypical of, for example, the classical subduction-driven tectono-magmatic and metallogenic models applicable to the central Andes of Perú and Chile. In this context, the Northern Andes has been the focus of more recent tectonic and magmatic analyses (e.g. Cediel and Cáceres 2000; Taboada et al. 2000; Pindell and Kennan 2001; Cediel et al. 2003; Keppie 2008; Kennan and Pindell 2009; Leal-Mejía et al. 2011b; Cediel 2011; Montes et al. 2012; Spikings et al. 2015; Van der Lelij et al. 2016; Leal-Mejía et al. 2018). Based upon these works, important observations and controls upon tectono-magmatic and metallogenic model construction in Colombia must be taken into account. Some of these observations include the following:

1. An understanding of Colombian tectonic evolution necessitates an understanding of the integrated evolution of the entire Northern Andean region, from northern Perú through Venezuela and Panamá, including evolution of both the Pacific and Caribbean domains. Acceptable tectonic models cannot be obtained through the imposition of geopolitical limits upon tectonic model construction.
2. An understanding of pre- and proto-Andean tectonic configurations in Colombia is critical to the understanding of tectono-magmatic evolution during Meso-Cenozoic events leading up to, and during, the Northern Andean orogeny. The presence of Proterozoic and Paleozoic paleo-allochthonous crustal components and fault and suture systems underlying much of the cordilleran region of Colombia has had a marked influence upon Meso-Cenozoic through recent orogenic events (including the localization of continental rifts, the emplacement and migration of volcano-magmatic arcs, morpho-structural expressions, structural style, control of uplift, sedimentation patterns, etc.).
3. The Meso-Cenozoic Northern Andean orogeny (Cediel et al. 2003) is essentially accretionary in nature. It is intimately linked to the evolution and kinematics of

the Pacific, Caribbean and South American Plates and the arrival and emplacement of associated allochthonous oceanic terranes along the Pacific and Caribbean margins. These factors control the development and evolution of subduction zones along the NW Colombian margin and the genesis, timing and spatial migration of onshore magmatic arcs and associated volcano-sedimentary basins.
4. Regional kinematic models describing the arrival and emplacement of allochthonous oceanic terranes during the Northern Andean orogeny emphasize highly oblique terrane approach and collision with low-angle to flat subduction on intervening segments of Pacific/Caribbean oceanic crust. Dextral and sinistral transpression and transtension have influenced the nature, geometry and migration of subduction-related arc segments and regional deformation, uplift, unroofing and erosion.
5. The present-day Colombian Andes are underlain by a mosaic of mid-late Proterozoic through Cenozoic autochthonous, parautochthonous and allochthonous crustal fragments, separated by crustal-scale fault and suture systems, upon which at least seven major episodes of granitoid magmatism have been sequentially superimposed (Leal-Mejía et al. 2018). Volcano-magmatic activity during the Northern Andean orogeny culminated in the development of the active, modern-day Northern Andean volcanic arc (Stern 2004).

The following section provides brief descriptions of the principal tectonic elements of the Colombian Andes, and the kinematics and sequencing of Colombian tectonic assembly based primarily upon litho-tectonic elements described by Cediel and Cáceres (2000) and Cediel et al. (2003). This information forms the backdrop to the litho-tectonic and morpho-structural framework and the integration of the mineral districts and deposits subsequently presented in our metallogenic time-space charts.

6.4.1 Colombian Litho-tectonic and Morpho-structural Elements

Figure 6.1 depicts the geo-tectonic framework of Colombia and the adjacent region. Based upon detailed geological, geochemical and geophysical analysis (Cediel et al. 1994; Cediel et al. 1998; Cediel and Cáceres 2000; Cediel et al. 2003; Cediel 2011, 2018), over 30 litho-tectonic and morpho-structural entities including their delimiting crustal-scale fault and suture systems are outlined. These entities may be grouped into various terranes and terrane assemblages, which in turn may be grouped into four major tectonic realms. These realms include the Guiana Shield, the Maracaibo Sub-plate, the Central Tectonic and the Western Tectonic Realms. Of these four tectonic realms, the Central Tectonic Realm and the Western Tectonic Realm host the great majority of the historic and documented mineral occurrences in Colombia. A brief description and development history for each tectonic realm, as it pertains to this study, is given below.

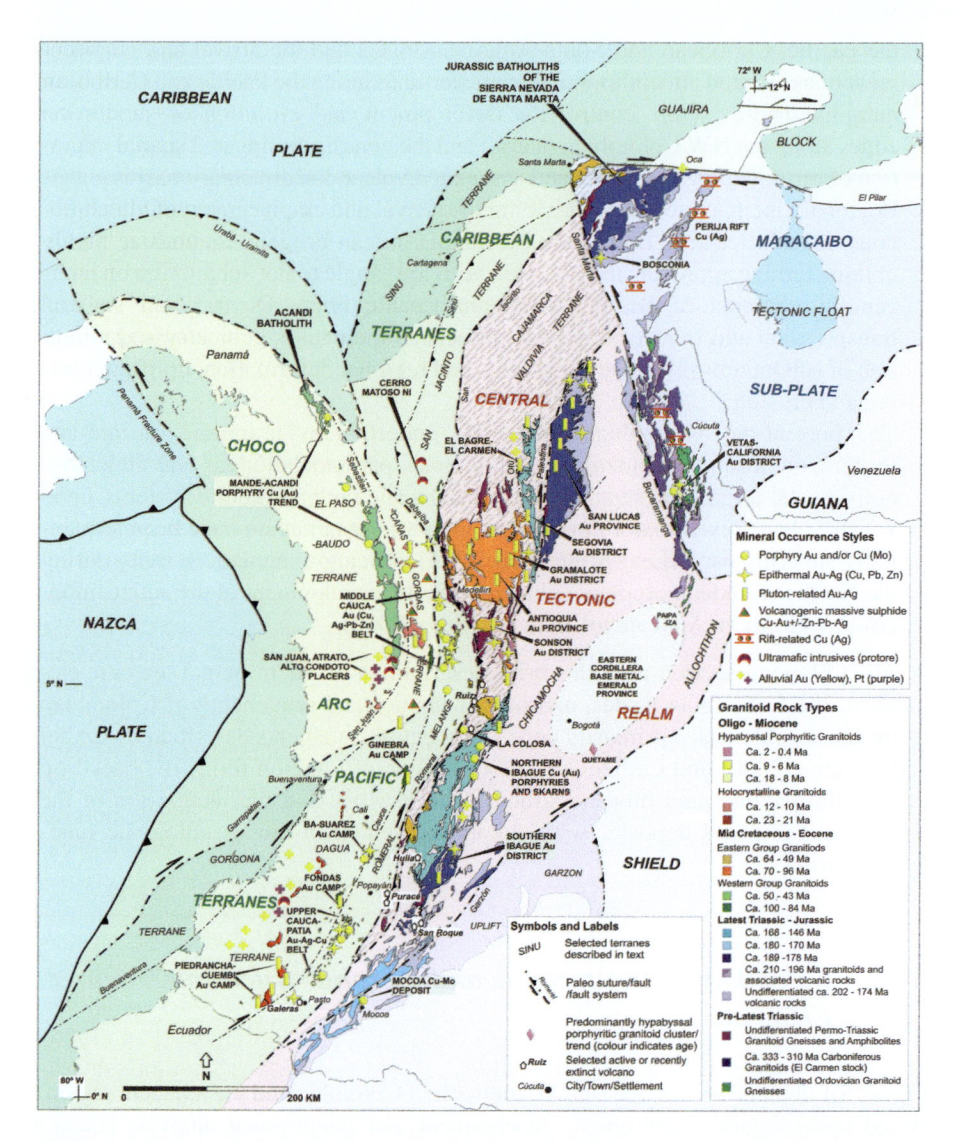

Fig. 6.1 Selected mineral occurrences and historic through modern mining districts in the Colombian Andes, in relation to major Phanerozoic granitoid arc segments and regional litho-tectonic and morpho-structural elements, as described in text

6.4.1.1 Guiana Shield Realm (GS)

This litho-tectonic realm consists of the autochthonous mass of the Precambrian Guiana Shield (Amazon Craton), which formed the backstop for the progressive continental growth of northwestern South America beginning in the middle to upper Proterozoic. Exhumed suture-related granulites exposed in the Garzón massif, the

Santander Massif and the Sierra Nevada de Santa Marta demonstrate continental collision and high-grade metamorphism during the final assembly of Rodinia (Cordani et al. 2005; Ibañez-Mejía et al. 2011) and the ensuing ca. 1 Ga Orinoquiense Orogen (Kroonenberg 1982; Restrepo-Pace 1995; Cediel and Cáceres 2000; Restrepo-Pace and Cediel 2010; Cediel 2011). An interpreted collisional remnant, sutured to the western margin of the Guiana Shield, was denominated the Chicamocha terrane by Cediel and Cáceres (2000). This paleo-allochthon is presently interpreted to comprise the basement for the eastern half of the Central Tectonic Realm (Fig. 6.1; described below). The suture zone along which the mid-Proterozoic collision took place coincides with the present-day Bucaramanga-Santa Marta-Garzón fault system.

6.4.1.2 Maracaibo Sub-plate Realm (MSP)

The MSP is a composite tectonic realm also underlain by the Guiana Shield, but much of its uplift history is linked to the Meso-Cenozoic tectonic evolution of the region. Its northern limit, in contact with the Caribbean Plate, is defined by the dextral Oca-El Pilar fault system (Fig. 6.1), whilst its west margin, in contact with the Central Tectonic Realm, is defined by the reactivated Bucaramanga-Santa-Marta fault. Topographic relief is provided by the Santander and Quetame Massifs, the Sierra de Mérida (Venezuela), the Serrania de Perijá and the Sierra Nevada de Santa Marta, the uplift history of which are linked to detachment and NW-vergent tectonic float during the Meso-Cenozoic (Cediel and Cáceres 2000; Cediel et al. 2003; Cediel 2011, 2018). The MSP contains numerous litho-tectonic and morpho-structural components, including exhumed Proterozoic and early Paleozoic basement massifs (Santa Marta, Santander, Floresta). Late Triassic-Jurassic ensialic extensional volcano-sedimentary basins are exposed along the Santander Massif, Sierra Nevada de Santa Marta and Serranía de Perijá, whilst uplift in the Santander Massif and Sierra Nevada de Santa Marta has unroofed major latest Triassic-Jurassic batholiths. Known metal occurrences of any significance within the Maracaibo Sub-plate Realm are actually quite scarce (e.g. Mejía et al. 1986). Notable exceptions include the Bailadores volcanogenic massive sulphide deposit (Sierra de Mérida, Venezuela), the Jurassic rift-related Cu (Ag) occurrences of the Girón-La Quinta Formation in the Serranía de Perijá and Au-Ag deposits associated with localized felsic magmatism of Mio-Pliocene age in the Vetas-California district of the Santander Massif.

6.4.1.3 Central Tectonic Realm (CTR)

The CTR is a composite, temporally and compositionally heterogeneous realm which occupies a wedge between the Guiana Shield Realm, the Maracaibo Sub-Plate Realm and the Western Tectonic Realm (Fig. 6.1). It forms the basement complex which underlies the entire central portion of the Colombian Andes. The CTR forms host to important metallogenic events during the Paleozoic and Meso-Cenozoic.

In this context, the description of its salient litho-tectonic components and geological evolution is herein left purposely detailed.

The CTR is considered part of the South American continental plate. It is comprised of a variety of litho-tectonic and morpho-structural entities. Its composite metamorphic basement is comprised of the Proterozoic Chicamocha terrane and the Paleozoic to early Mesozoic Cajamarca-Valdivia terrane (CA-VA). Superimposed upon these core components are Jurassic magmatic arc segments including the San Lucas, Ibagué and Segovia blocks and the late Cretaceous Antioquian Batholith, all of which dominate Colombia's physiographic Central Cordillera. The Lower, Middle and Upper Magdalena basins and Colombia's geologic Eastern Cordillera (EC) were also developed upon CTR metamorphic basement (Fig. 6.1). The Chicamocha and Cajamarca-Valdivia constituents of the CTR are allochthonous to parautochthonous with respect to the Guiana Shield autochthon, having been sutured to the region in pre-Andean times, whilst the Mesozoic to Recent components are considered to be autochthonous with respect to Chicamocha-CA-VA metamorphic basement.

The oldest constituent of the CTR is the Precambrian Chicamocha terrane (Cediel and Cáceres 2000; Cediel et al. 2003; Cediel 2011), containing relict fragments of Rodinia (Ramos 2009) and/or Oaxaquia basement (Keppie and Ortega-Gutierrez 2010), welded to the Amazon Craton during a 1.2 Ga to 0.95 (Grenvillian age) metamorphic event locally known as the Orinoquian Orogeny (Restrepo-Pace 1995; Cediel and Cáceres 2000; Restrepo-Pace and Cediel 2010). The terrane is represented by fragmented granulite-grade bodies of migmatite and quartz-feldspar gneiss, mostly outcropping along the eastern margin of the Ibagué and San Lucas blocks (Cediel and Cáceres 2000; Cordani et al. 2005; Cediel 2011; Leal-Mejía 2011; Cuadros et al. 2014; Gómez et al. 2015a).

Chicamocha is bound to the west by the composite Cajamarca-Valdivia terrane (Cediel and Cáceres 2000; Cediel et al. 2003; Cediel 2011) which broadly coincides with the Central Andean Terrane as described by Restrepo-Pace (1992). Cajamarca-Valdivia contains amphibolitic, graphitic and semi-pelitic schists and marbles, metamorphosed to greenschist through epidote amphibolite grade and generally assigned a Neoproterozoic to early Paleozoic age (Feininger et al. 1972; Restrepo-Pace 1992; Cediel et al. 1994; Ordoñez-Carmona et al. 2006; Cediel 2011; Spikings et al. 2015). This is in agreement with Ediacaran to Cambrian C-isotope stratigraphy ages for contained carbonates (Silva et al. 2005). Based upon geochemical and geological characterization presented by Restrepo-Pace (1992), Cajamarca-Valdivia represents a peri-cratonic island arc and continental margin accretionary prism assemblage, accreted along the western Chicamocha terrane in the early Paleozoic (Restrepo-Pace 1992; Cediel and Cáceres 2000; Cediel 2011). Cajamarca-Valdivia forms the basement to Permian through mid-late Triassic gneissic granitoids, meta-amphibolites and peraluminous granites representing the assembly and break-up of Pangaea in the Northern Andean region (Vinasco et al. 2006; Cochrane et al. 2014a; Spikings et al. 2015). The Palestina fault system (Feininger 1970), including associated structures such as the Chapeton and Pericos faults, represents the suture between the Chicamocha and Cajamarca-Valdivia assemblages (Cediel and Cáceres 2000; Cediel et al. 2003).

It is important to note that present-day geological mapping does not permit understanding of the precise distribution of Proterozoic, early Paleozoic and Permo-Triassic constituents of CA-VA (Cediel and Cáceres 2000; Gómez et al. 2015a), and zircon-based U-Pb age dating is only beginning to reveal the complexity of the CA-VA assemblage. For example, various units which were formerly thought to be of Proterozoic or early Paleozoic age are now known to belong to the Permo-Triassic assemblage (Restrepo-Pace 1995; Ordoñez-Carmona et al. 2006; Vinasco et al. 2006; Leal-Mejía 2011; Spikings et al. 2015).

The timing of CA-VA assemblage accretion along the western Chicamocha margin and the origins of the Palestina fault and suture system have been examined by various authors (Feininger 1970; Cediel et al. 1994, 2003; Restrepo-Pace 1995; Cediel and Cáceres 2000; Restrepo-Pace and Cediel 2010; Cediel 2011). Regional tectono-sedimentary analysis (Cediel et al. 1994), arc-related magmatic patterns and available metamorphic age dates (Restrepo-Pace 1995; Van der Lelij et al. 2016) suggest early Paleozoic accretion of CA-VA during the Quetame orogeny (Cediel and Cáceres 2000; Cediel 2011). Peak Barrovian conditions, including the emplacement of syn-kinematic granitoids, were attained at ca. 477–472 Ma (Restrepo-Pace 1995; Van der Lelij et al. 2016).

Following amalgamation, important constituents were superimposed upon the composite metamorphic basement of the Central Tectonic Realm. During the Carboniferous, oceanic rifting led to emplacement of the El Carmen-El Cordero ridge tholeiitic granitoid (RTG; Barbarin 1999) suite, whilst during the Late Triassic-Jurassic, the San Lucas, Ibagué and Segovia granitoid arc segments were generated. These composite Mesozoic lithotectons represent temporally/geographically constrained, ensialic extensional basin – continental margin magmatic arc couplets. Volcano-sedimentary basin development and subsequent emplacement of the calc-alkaline San Lucas and Ibagué Batholiths appears to have been localized by extensional reactivation of the Palestina suture. The general east-to-west younging trend of major Jurassic arc-related batholiths, from the Santander Plutonic Group through the San Lucas Batholiths and into the Segovia Batholith (Fig. 6.1), has been interpreted by various authors (Leal-Mejía 2011; Cochrane 2013; Spikings et al. 2015; Leal-Mejía et al. 2018) to reflect regional extension due to slab rollback. Cediel and Cáceres (2000), Cediel et al. (2003) and Leal-Mejía (2011) interpreted the development of the Jurassic volcano-sedimentary basins and associated arc segments within the context of the Bolivar Aulacogen, a taphrogenic framework dominating Colombian and NW South American tectonics during the mid- and late Paleozoic and Mesozoic. Events associated with the Bolivar Aulacogen and affecting the CTR include Carboniferous oceanic rifting (aborted), the break-up of Pangaea and the opening of the proto-Caribbean basin during the late Mesozoic.

From a mineral deposit standpoint, the rifts and volcano-sedimentary arc segments of Carboniferous and Jurassic age, constructed upon CTR basement, form important metallogenic provinces, especially with respect to rift-related Cu, pluton-related and epithermal Au (Ag) and porphyry-related Cu (Mo) occurrences in Colombia.

The Late Mesozoic history of the Central Tectonic Realm is dominated by continued extension and opening of the Valle Alto rift (Cediel et al. 1994, 2003; Cediel and Cáceres 2000). This period involves deep crustal rifting, the emplacement of mafic sill and dike suites (Fabre and Delaloye 1983; Vásquez et al. 2010), the invasion of the Cretaceous seaway and the deposition of deep to shallow marine sequences over extensive areas of the CTR, the Maracaibo Sub-plate, and the continental platform of the Guiana Shield (Cediel et al. 1994; Sarmiento 2018). The latest Jurassic through Cretaceous record for this period has been exhumed and exposed throughout Colombia's Eastern Cordillera and within erosional relicts such as the San Pablo and Segovia Fms. preserved within the physiographic Central Cordillera. The Eastern Cordilleran basin hosts a poorly documented assemblage of syngenetic and epigenetic base metal occurrences and precious mineral deposits and forms an extensive metallogenic province which will be outlined in more detail below. The San Pablo Formation also hosts syngenetic base metal occurrences which have been integrated into our time-space charts.

Plate reorganization leading to opening of the proto-Caribbean basin and the evolution and passage of the Caribbean Plate along the NW South American margin beginning in the early to mid-Cretaceous signalled the end of the taphrogenic regime characteristic of the Bolivar Aulacogen (Cediel et al. 1994, 2003; Cediel 2018). In this context the late Mesozoic and Cenozoic to Recent components of the Central Tectonic Realm are dominated by metaluminous, subduction-related arc granitoids, related to the assembly and accretion of the Western Tectonic Realm (detailed below) and the conformation of the present-day Northern Volcanic zone (Stern 2004). Of these granitoid assemblages, the most important by far is the Antioquian Batholith, a mid- to late Cretaceous, polyphase calc-alkaline, plutonic complex which intrudes the Cajamarca-Valdivia terrane and dominates the geology of the entire present-day northernmost Central Cordillera (Feininger and Botero 1982; González 2001; Leal-Mejía 2011). Granitoid suites extending into the Paleocene and early Eocene are observed to the south, represented by the Manizales, El Hatillo and Córdoba stocks and the El Bosque Batholith (Cediel and Cáceres 2000; Gómez et al. 2015a; Leal-Mejía et al. 2018). Some of these plutons exhibit important Au (Ag, Cu, Mo, W) metallogeny. Miocene-Pliocene and Pleistocene to Recent Andean-type volcanism forms a NNE-trending belt of hypabyssal porphyry intrusives, partially eroded volcanic edifices and active stratovolcanic cones stretching along the western margin of the CTR. Late Miocene and Pliocene hypabyssal porphyry clusters (e.g. Cajamarca-Salento, Rio Dulce; Lodder et al. 2010; Leal-Mejía 2011; Leal-Mejía et al. 2018) are associated with important porphyry-style Au and epithermal Au (Ag-Pb-Zn, Cu, Mo) mineralization hosted within Cajamarca-Valdivia metamorphic basement.

6.4.1.4 Western Tectonic Realm (WTR)

The approach, assembly and accretion of the allochthonous Western Tectonic Realm (Fig. 6.1) provided the driving mechanism for arc-related magmatism and metallogeny during the late Meso-Cenozoic Northern Andean orogeny (Cediel et al. 2003; Leal-Mejía 2011; Leal-Mejía et al. 2018). Within the WTR, three composite terrane assemblages are recognized, including the Pacific (PAT), Caribbean (CAT) and the Chocó Arc (CHO). The Romeral and Dagua terranes of the PAT assemblage, and the San Jacinto and Sinú terranes of the CAT assemblage, roughly correspond to litho-tectonic units recognized by Etayo-Serna et al. (1983). The combined PAT and CHO assemblages approximate the "Provincia Litosférica Oceánica Cretácica del Occidente de Colombia" or "PLOCO" of Nivia et al. (1996) and form the geographic "Western Cordillera" of Colombia. All of these terrane assemblages contain fragments of Pacific oceanic crust, oceanic plateaus, aseismic ridges and/or ophiolite with associated marine sedimentary rocks. All developed within/upon oceanic basement and based upon faunal assemblages (e.g. Etayo-Serna and Rodríguez 1985), paleomagnetic data (e.g. Estrada 1995) and recent paleogeographic reconstructions (e.g. Cediel et al. 1994, 2003; Kennan and Pindell 2009; Montes et al. 2012), all are allochthonous to parautochthonous with respect to continental South America. The composite terrane assemblages of the Western Tectonic Realm are characterized as follows.

Pacific (PAT) Terrane Assemblage

The PAT consists of the Romeral, Dagua, and Gorgona terranes (Fig. 6.1). The Romeral assemblage has been interpreted as a regional-scale tectonic mélange (Cediel et al. 2003) containing intensely deformed and fragmented blocks (tectonic floats?) of amphibolite and carbonaceous schist, high-pressure metamorphic rocks (eclogite, blueschist), layered mafic and ultramafic complexes, marine and pericratonic arc-related volcanic rocks, ophiolite and meta-sediments, dating from the Paleozoic, Jurassic and lowermost Cretaceous. The suite was assembled along the Pacific margin during the early Cretaceous and is in direct tectonic contact with the CTR to the east, along the Romeral fault system (Ego et al. 1995; Cediel et al. 2003; Vinasco 2018). The Romeral mélange underlies much of the Cauca-Patía intermontane valley, including the northern inter-Andean depression to the north and south of the city of Pasto. It forms the basement to numerous important Au (Cu, Ag-Zn-Pb) districts associated with felsic to intermediate volcanism and hypabyssal porphyry emplacement during the Miocene (e.g. the Upper and Middle Cauca belts).

To the west of the Romeral mélange, the Dagua terrane is comprised of an assemblage of oceanic mafic and ultramafic rocks (Diabasico Gp.) which forms the basement for important thicknesses of flyschoid siliciclastic sedimentary rocks including chert, siltstone, marlstone and greywacke (Dagua Gp.). Lithogeochemical studies indicate that the mafic and ultramafic volcanic and intrusive rocks are of oceanic tholeiitic N- and E-MORB affinity, interpreted to represent accreted fragments of Farallon oceanic crust and to include ophiolite, aseismic ridges and/or oceanic plateaus belonging to the Caribbean-Colombian oceanic plateau (CCOP) or Caribbean large igneous province (CLIP), as described by Kerr et al. (1997) and Sinton et al. (1998), respectively. A mantle plume-hotspot-oceanic flood basalt origin for this assemblage has been proposed by these authors. A summary of radiometric (mostly $^{40}Ar/^{39}Ar$) age dates for accreted mafic-ultramafic rocks in northern South America and the Caribbean suggests that CCOP/CLIP-related mafic-ultramafic magmatism may be considered in three stages: a volumetrically restricted phase initiated at ca. 100 Ma and followed by the widespread eruption of oceanic plateau rocks dating from ca. 92 and 87 Ma (Kerr et al. 1997; Sinton et al. 1998; Kerr et al. 2003; Hastie and Kerr 2010). Lesser but still widespread basaltic magmatism is subsequently recorded between ca. 77 and 72 Ma (Kerr et al. 1997; Sinton et al. 1998). Within the Dagua terrane, mid-Cretaceous $^{40}Ar/^{39}Ar$ ages for oceanic plateau rocks (Kerr et al. 1997; Sinton et al. 1998) are in broad agreement with mid- to late Cretaceous biostratigraphic ages for oceanic sedimentary rocks contained within the Dagua Gp. (Etayo-Serna and Rodríguez 1985).

Approach and collision of the CCOP/CLIP and accretion of the Dagua terrane took place in the late Cretaceous-Paleocene, along the Cauca fault and suture system (Ego et al. 1995). Related deformation and uplift generated a regional unconformity throughout much of the CTR (Cediel et al. 1994; Cediel and Cáceres 2000). Hydrothermal Au (Ag) mineralization is contained within mid-Cretaceous intra-oceanic granitoids originally emplaced into the CCOP/CLIP and accreted to the Colombian margin at this time. Following accretion, in the latest Oligocene-early Miocene, the Dagua terrane was additionally intruded by subduction-related holocrystalline granitoids, which in turn host important Au (Ag, Cu, Mo) mineralization.

Further west, the Gorgona terrane is located mostly offshore, on the southwestern margin of the Colombian Pacific. Gorgona also represents an accreted oceanic plateau of mantle plume affinity, containing massive basaltic and spinifex-textured komatiitic lava flows, pillow lavas and a peridotite-gabbro complex. Radiometric ages provided by Sinton et al. (1998) range from ca. 87 to 83 Ma; however, paleomagnetic data and paleogeographic reconstructions presented by Estrada (1995), Kerr and Tarney (2005) and Kennan and Pindell (2009) suggest Gorgona has no clear correlation with the CCOP/CLIP. Gorgona accretion to the western margin of the Dagua terrane along the Buenaventura fault took place during the Eocene (McGeary and Ben-Avraham 1989; Cediel et al. 2003; Kerr and Tarney 2005; Leal-Mejía et al. 2018).

Caribbean Terrane Assemblage

Two principal terranes are contained within this assemblage, the San Jacinto and Sinu (Fig. 6.1). San Jacinto includes a MORB-type tholeiitic basement considered a fragment of the CCOP/CLIP intercalated with deep marine carbonaceous cherts, mudstones and marlstones of upper Cretaceous age. It was accreted to the northern CTR along the San Jacinto fault during the late Cretaceous-Paleocene (Cediel et al. 2003). It forms basement to important Ni laterite deposits at Cerro Matoso and is cut by porphyritic dikes and stocks of late Cretaceous (?) age, associated with Au-Cu mineralization at El Alacrán-San Matías (Montiel), Córdoba. The Sinú terrane is comprised of similar basement to San Jacinto, overlain by turbidite sequences of Oligocene age. It was juxtaposed along the San Jacinto margin in the Miocene.

The Chocó (Panamá) Arc

The Chocó Arc assemblage in Colombia (Duque-Caro 1990; Schmidt-Thomé et al. 1992; Cediel et al. 2003; Cediel et al. 2010; Redwood 2018), together with Campanian to Eocene mafic oceanic and intermediate arc-related plutonic rocks of the Darién (San Blas) Range and Azuero Peninsula in Panamá (Wegner et al. 2011; Montes et al. 2012), represents the eastern segments of the Panamá double arc. In Colombia, the basement of the composite Chocó Arc is comprised of two distinct litho-tectonic assemblages, the Cañas Gordas terrane and the El Paso-Baudó terrane which includes the Mandé-Acandí arc (Fig. 6.1) (Cediel et al. 2010).

Cañas Gordas consists of mixed volcanic rocks of the Barroso Fm. with overlying fine-grained sedimentary rocks of the Penderisco Fm. The Barroso Fm. is dominated by tholeiitic to calc-alkaline, massive, porphyritic and amygdaloidal basaltic to andesitic flows, tuffs and agglomerates (Rodriguez and Arango 2013). Sedimentary interbeds within the Barroso Fm. mapped near the town of Buriticá contain Barremian through middle Albian fossil assemblages (González 2001 and references cited therein). Weber et al. (2015) consider gabbros hosted within the Barroso Fm. to belong to the CCOP/CLIP plateau assemblage, although biostatigraphic data suggests that Barroso also contains sections of older, pre-CCOP/CLIP, Farallon oceanic basement. The Penderisco Fm. includes thinly bedded, mudstone, siltstone, marlstone, greywacke and chert. Two members, including Urrao and Nutibara, contain marine fossil assemblages dating from the Aptian-Albian to the Upper Cretaceous (González 2001 and references cited therein). The eastern margin of the Cañas Gordas terrane was intruded by the Buriticá Tonalite and the Santa Fé Batholith at ca. 100 Ma and ca. 90 Ma, respectively (Weber et al. 2015). The terrane assemblage was accreted to the continental margin during the late Cretaceous to Paleocene.

The El Paso-Baudó terrane of the Chocó Arc are comprised of late Cretaceous to Paleogene sections of tholeiitic basalt of E-MORB affinity (Goossens et al. 1977; Kerr et al. 1997), overlain by minor pyroclastic rocks, chert and turbidite. El Paso-Baudó is considered to represent a late Cretaceous fragment of the trailing edge of the CCOP/CLIP assemblage. The Mandé-Acandí arc with associated plutonic, hypabyssal porphyritic stocks and pyroclastic volcanic sequences (La Equis – Santa Cecilia Fms; Sillitoe et al. 1982; Schmidt-Thomé et al. 1992; Leal-Mejía 2011; Montes et al. 2012) was developed upon El Paso-Baudó oceanic basement between ca. 60 and 42 Ma (Leal-Mejía 2011; Montes et al. 2012, 2015). Ural-Alaskan-type zoned ultramafic complexes at Alto Condoto and Mumbú were also intruded into El Paso-Baudó terrane basement at ca. 20 Ma (Tistl 1994; Cediel et al. 2010). Development of the San Juan and Atrato basins began in the Paleogene, with final docking and uplift of the El Paso-Baudó-Mande assemblage along the western Cañas Gordas margin in the late Miocene (Cediel et al. 2010; Montes et al. 2012, 2015). Faults related to the assembly and accretion of the Chocó Arc, including the Garrapatas-Dabeiba and San Juan-Sebastian systems (Cediel and Cáceres 2000; Cediel et al. 2003, 2010) and Uramita system (Duque-Caro 1990; Montes et al.

2012), reactivate, deform and/or truncate earlier structures associated with the Romeral, Cauca and San Jacinto fault systems.

The assembly and accretion of the composite Chocó Arc had important metallogenic consequences. Various Cu (Zn, Pb, Au, Ag)-rich volcanogenic massive sulphide occurrences are hosted within the Barroso Fm., and significant porphyry-associated Cu (Au) and Au + base metal epithermal mineralization is associated with calc-alkaline arc magmatism recorded along the Mandé-Acandí Batholith (Sillitoe et al. 1982). Uplift and erosion of the Alto Condoto and similar ultramafic complexes have been suggested as a source for the Pt (Pd-Au) placer deposits in the San Juan and Atrato basins (Tistl 1994). In addition, the Cañas Gordas terrane hosts numerous mid- to late Miocene (syn- to post-accretionary) metaluminous, calc-alkaline ± alkaline plutonic, hypabyssal porphyritic and volcanic rocks which contain pluton- and porphyry-related and volcanic-hosted Au (Cu) and Au-Ag (base metal) mineralization.

6.4.1.5 Structural Evolution of the Colombian Andes

Faulting in the Colombian Andes is abundant and complex. The importance of large-scale strike-slip faulting in particular must be recognized, not only in terms of a dominant structural style but as a key element in the tectono-magmatic evolution of the region (Aspden et al. 1987; Cediel et al. 1994, 2003; Kennan and Pindell 2009; Colmenares et al. 2018; Leal-Mejía et al. 2018). Many of the major fault systems in the Colombian Andes have prolonged, polyphase histories. Some mark paleo-rifts and/or subduction zones which actively accompanied the evolution and emplacement of granitoid arc segments during the Paleozoic and Meso-Cenozoic. The influence of these ancient structures as crustal-scale plumbing systems facilitating the posterior localization of additional volcanic-plutonic arc segments has exerted a pre-determinative role on the spatial appearance and temporal migration of many of the important metallogenic provinces throughout the region (Leal-Mejía 2011). Based upon the preceding description of tectonic realms and litho-tectonic-morpho-structural provinces, as displayed on our time-space charts, a summary of the evolution of the principal unit-bounding structures based upon and updated from more detailed explanations provided by Cediel et al. (2003) and Colmenares et al. (2018) is now presented. The role of these structures in the tectono-magmatic development of important Colombian mineral districts is emphasized herein. The location and nature of all of the fault systems outlined below are revealed in Fig. 6.1.

Bucaramanga-Santa Marta-Garzón

This fault system forms a paleo-suture which welded the Precambrian Chicamocha terrane (eastern Central Tectonic Realm) to the Guiana Shield and influenced the emplacement of Jurassic calc-alkaline batholiths exposed along the Santander and Garzón massifs. Sinistral reactivation of the Bucaramanga-Santa Marta segment during the Aptian-Albian defined the western boundary of the Maracaibo Sub-plate. Bucaramanga-Santa Marta acted as a massive sidewall or lateral ramp during episodic Paleogene through Mio-Pliocene exhumation of the Santander Massif, Sierra Nevada de Santa Marta and Serranía de Perijá. Differential movement between Bucaramanga-Santa Marta and the Santander fault (Cediel et al. 2003) to the east resulted in the development of NE-striking transtensional fault and fracture arrays which controlled the emplacement of late Miocene to Pliocene porphyry-associated and epithermal Au (Ag, Cu, Mo) mineralization in the Vetas-California district. Plio-Pleistocene magmatism manifests above the buried trace of the Bucaramanga segment, as a series of rhyodacitic to rhyolitic plugs which outcrop at Paipa, Iza and Quetame in Colombia's Eastern Cordillera.

Palestina

This fault system forms the eastern limit of the Cajamarca-Valdivia terrane and constitutes an early Paleozoic suture between CA-VA and Chicamocha. Reactivated during the late Triassic, the fault facilitated emplacement of Jurassic magmatic arcs of the San Lucas and Ibagué blocks during development of the late Bolívar Aulocogen (Cediel and Cáceres 2000; Cediel et al. 2003). These volcano-plutonic arc segments host numerous cogenetic pluton-related and epithermal Au-Ag districts as well as Cu (Mo) porphyry and Cu (Au) skarn occurrences. Component faults of the Palestina system verge and connect towards the south with the Romeral fault system (the paleo-continental margin). Dextral reactivation of the Palestina fault is recorded in the Aptian-Albian and continued in the late Cretaceous (Feininger 1970). Reactivation appears linked to activity along the Romeral fault system (see below).

Otú

The role of the Otú Fault in the Paleozoic to early Mesozoic conformation of the Cajamarca-Valdivia terrane and CCSP has also been called into question by various authors (e.g. Toussaint 1993). The fault appears internal to the composite CA-VA assemblage, and regional maps (e.g. Cediel and Cáceres 2000; Gómez et al. 2015a) depict similar lower Paleozoic metamorphic assemblages distributed on either side. Notwithstanding, recent U-Pb (zircon) age dating and lithogeochemical analyses from assorted granitoid plutons outcropping along Otú reveal oceanic, rift-related magmatism of Carboniferous age and arc-related magmatism of Jurassic and Cretaceous age (Leal-Mejía et al. 2010), suggesting Otú forms a deep crustal conduit and indicating it's possible origin along an aborted rift (Leal-Mejía et al. 2018). The regional significance and possible influence of the Otú fault with respect to the distribution and genesis of numerous vein-type Au-Ag occurrences along the Segovia-Remedios-Nechí trend (Leal-Mejía et al. 2010) have been discussed by various authors (Alvarez et al. 2007; Londoño et al. 2009; Mendoza and Giraldo 2012; see below).

Romeral Fault System

The Romeral fault marks the suture trace of accreted Jurassic and early Cretaceous oceanic assemblages along the paleo-continental margin of the Central Tectonic Realm. Existing data and interpretations for this complex suture, fault and accompanying tectonic mélange demonstrate various phases of movement and reactivation (Vinasco 2018). Early left-lateral transtension associated with the final phases of the Bolívar Aulocogen dominated movements in the early Cretaceous (Cediel et al. 1994; Cediel 2011). Beginning in the Aptian compression and deep burial of continental margin assemblages, followed by large-scale dextral-oblique transpressive shearing (Ego et al. 1995; Cediel and Cáceres 2000; Maresch et al. 2009; Cediel 2011) is recorded in the generation and exhumation of greenschist-amphibolite, blueschist and eclogite-bearing metamorphic assemblages (Orrego et al. 1980; McCourt and Feininger 1984; Bustamante 2008; Maresch et al. 2009), and in steep to vertical N-S-striking tectonic fabrics exposed within the present-day Romeral mélange. Transcurrent motion dominated movements along the Romeral system during the early through mid-Cretaceous. The lack of significant volumes of granitoid magmatism along/within the Romeral mélange suggests limited subduction of oceanic lithosphere beneath the continental margin from ca. 145 to 95 Ma (Aspden et al. 1987; Leal-Mejía 2011; Leal-Mejía et al. 2018).

The steep N-S tectonic fabrics characteristic of Romeral mélange basement provided important controls to the emplacement of late Miocene hypabyssal porphyry stocks and dikes along the Middle Cauca Porphyry Au (Cu) Belt (Leal-Mejía 2011; Bissig et al. 2017). Dextral shear transfer from the Romeral system across the western segment of the CTR and into the pre-existing Palestina fault was instrumental in the development of the structural architecture of the CTR (Restrepo-Pace 1992; Cediel 2011). This interplay between the Romeral and Palestina fault systems influenced the development of transpressive-transtensional pull-apart basins and late Cenozoic to modern-day continental arc-related magmatism contained within the CTR. Late Miocene and Pliocene-Pleistocene calc-alkaline hypabyssal porphyry clusters and volcanic centres within CA-VA basement host Au (Ag, Cu)-rich porphyries and associated Au-Ag-base metal epithermal vein and breccia deposits.

Cauca Fault System

The Cauca fault system forms a suture between the Dagua and Romeral oceanic assemblage. The generally right-lateral strike-slip character of the Cauca system varies along strike, and the dextral component can only be inferred at some localities (Ego et al. 1995; Cediel and Cáceres 2000; Cediel et al. 2003). Movement on the Cauca system and reactivation of the Romeral and Palestina faults register Dagua terrane docking during the late Cretaceous-Paleocene.

Buenaventura Fault

This fault is easily recognized on regional magnetic and gravity maps where it manifests as a NE-trending rectilinear lineament (Cediel et al. 1998). The fault coincides with the suture trace delimiting the Gorgona and Dagua terranes. Cenozoic movement along the Buenaventura fault is interpreted as dextral transpressive (Cediel and Cáceres 2000; Cediel et al. 2003). Reactivation of the Cauca and Romeral fault systems, including west-vergent Miocene thrusting along the Cauca-Patia interandean valley, is associated with docking of the Gorgona terrane (Cediel et al. 2003). The resulting structural architecture of the Romeral mélange and Dagua terrane basement rocks was influential in the superposition of contained early Miocene pluton-related Au (Ag, Cu) and mid- to late Miocene porphyry-related Au-Cu and epithermal Au-Ag (Sb) occurrences.

Garrapatas-Dabeiba Fault System

The Garrapatas fault represents a paleo-transform within the Farallón plate (Barrero 1977) which behaved in a strike-slip manner during the late Meso-Cenozoic (e.g. Aspden et al. 1987; Pindell and Kennan 2001; Cediel et al. 2003). Presently, this major break in the oceanic crust forms the principal boundary fault between the PAT assemblage to the south and the Chocó Arc assemblage to the north (Mountney and Westbrook 1997; Cediel et al. 2003, 2010). The early Garrapatas fault permitted the kinematically and temporally independent interaction of the Pacific and Chocó Arc assemblages with continental South America during the Northern Andean orogeny. It served as the southern lateral ramp which, in combination with the Dabeiba fault to the north, facilitated the accretion of the Cañas Gordas terrane in the late Cretaceous-Paleocene.

San Juan-Sebastian Fault System

This fault system is related to collision of the El Paso-Baudó terrane (including the Mandé-Acandí arc) along the NW Colombian margin (Cañas Gordas terrane) during the mid- to late Miocene. Silver et al. (1990), Farris et al. (2011) and Montes et al. 2012 present models involving the N and W migration of the South American Plate upon the fixed trailing edge of the Caribbean-Colombian oceanic plateau, resulting in accretion of the El Paso-Baudó assemblage. Continued compression and rotation led to the development of east-verging en echelon thrust faults and rotated NE-trending anticlinal fold axis within the Atrato Basin (Cediel et al. 1998, 2010). The evolution of the Nazca Plate and development of a subduction zone along the Colombian Pacific margin during the mid-Miocene produced a positive flexure in the oceanic plate and emergence of the Baudo Range in the late Miocene.

6.5 Metallotects and Metallogenic Epochs in the Colombian Andes

Metallic mineral deposits preserved in the Colombian Andes are hosted in rocks ranging from Precambrian to Recent in age. A wide variety of deposit types are distributed throughout the region, indicative of the predominant tectono-magmatic setting and processes prevalent on a temporal vs. spatial basis during the Phanerozoic. Of greatest economic significance, historically and at present, are epigenetic, mesothermal pluton- and porphyry- ± contact metamorphic (hornfels, skarn)-related deposits and epithermal volcano-sedimentary-hosted precious metals occurrences, spatially and temporally related to subduction-related metaluminous, calc-alkaline granitoid magmatism generated within continental margin or peri-cratonic fringing magmatic arc settings. Notwithstanding, deposits intimately associated with or derived from mafic-ultramafic magmas, characteristic of both ocean floor extensional (e.g. ophiolite) and convergent margin (e.g. Ural-Alaskan-type zoned ultramafic complexes) settings, are also represented. Syngenetic and syn-diagenetic deposits of the VMS (volcanogenic massive sulphide), SEDEX (sedimentary-exhalative), sediment-hosted Cu and oolitic Fe formation types are also present and are considered to have been generated in both oceanic and continental rift-related settings. Additionally, numerous metal and mineral occurrences associated with the migration/escape of brines from the root zone of the Eastern Cordilleran basin during Meso-Cenozoic structural inversion are documented. These include structurally controlled base metal- and precious mineral-bearing veins, vein arrays and tectonic breccias, and mantos and carbonate replacement – MVT (Mississippi Valley-type) – occurrences, generated in an essentially amagmatic setting. Ultimately, the wide variety of genetic models applicable to Colombian metal and mineral deposit types is a clear reflection of the complex and dynamic tectonic evolution underlying the Northern Andean region.

Figure 6.1 presents an overview of some of the most important metalliferous districts and deposits in the Colombian Andes, categorized by interpreted deposit type. Detailed tectono-magmatic analyses and time-space charts, placing the most representative deposit types into a Phanerozoic temporal vs. tectono-magmatic framework, are presented for the (1) pre-Jurassic (Figs. 6.2 and 6.3), (2) latest Triassic-Jurassic (Figs. 6.4 and 6.5), (3) early Cretaceous (Figs. 6.7 and 6.10), (4) middle Cretaceous through Eocene (Figs. 6.9 and 6.10) and (5) latest Oligocene through Plio-Pleistocene (Figs. 6.12 and 6.13). In addition, detailed location maps outlining the numerous mineral occurrences and deposit types found within some of most important and historic mineral districts are offered.

With respect to the time-space charts, the vertical axis depicts time, spanning the Phanerozoic, from the early Cambrian to the Pleistocene. The time period depicted on each chart is broadly dictated by dynamic shifts in regional tectono-magmatic evolution. The horizontal axis essentially represents an E-W composite tectono-structural cross section across the Colombian Andes. The section is idealized and schematic, as due to the overall tectonic configuration of the region (e.g. Fig. 6.1), the tectono-stratigraphic units and fault and suture systems cut by any one E-W line of section vary according to latitude. In this context, some of the major litho-tectonic and morpho-structural units represented on the time-space charts are projected over greater or lesser distances and at times oblique to the line of section, either to the north or south (e.g. the Sierra Nevada de Santa Marta, Serrania de Perijá, Southern Ibagué block, Dagua terrane, San Jacinto terrane, etc.).

Notwithstanding, from a tectono-stratigraphic standpoint, our attempt has been to focus upon the basement complexes (including their bounding crustal-scale fault systems) which are host to the most important and representative of the historic through active Colombian metal/mineral districts. These diagrams integrate the appearance, over time, of the great majority of said districts into a coherent regional magmatic and litho-tectonic framework. It is observed that the majority of Colombian metallogenic development is Meso-Cenozoic in age and follows the evolution of volcano-magmatic arcs along the regional NNE Northern Andean trend. As such, we feel the time-space charts, in combination with the accompanying plan maps, provide good two-dimensional representation of the temporal-spatial evolution of Colombian metallogeny. The age, style and principal metal associations for each district are shown within the context of their host lithotecton, and the temporal and spatial appearance of metal/mineral deposits is established in accord with progressive regional tectono-magmatic assembly. In essence, our analysis permits the definition of metallogenic provinces (which we consider synonymous with metallotects in the sense of Laffitte (1966) and Routhier (1983)) and metallogenic epochs, at the scale of the Colombian Andes.

6.5.1 Pre-Jurassic Metallogeny: Quetame Orogeny and Initiation of the Bolívar Aulacogen

The prolonged and complex tectonic evolution of the Colombian Andes during pre-Jurassic times (Cediel et al. 1994; Cediel et al. 2003; Vinasco et al. 2006; Cochrane et al. 2014a; Van der Lelij et al. 2016) led to the amalgamation of the basement components of the Maracaibo Sub-plate and Central Tectonic Realm, dominated by assemblages of middle greenschist through amphibolite- and granulite-grade meta-igneous and sedimentary rocks. Although these assemblages are host to numerous mineral and metal occurrences, radiometric age dating and general geological arguments indicate that the great majority of the mineral deposits contained within the MSP and CTR significantly post-date the age of their pre-Jurassic host basement.

In this context, there are few mineral districts/occurrences for which a pre-Jurassic age may be confidently assigned (Figs. 6.2 and 6.3). These include the Caño Negro-Quetame red bed-hosted Cu occurrences outcropping along the eastern margin of the Quetame Massif (Rodríguez 1984; Rodríguez and Warden 1993), the Bailadores volcanogenic massive sulphide (VMS) deposit located in the Sierra de

Mérida (Venezuela; Carlson 1977), mesothermal vein-type Au (Ag) mineralization hosted within the El Carmen-El Cordero stock at El Bagre (Leal-Mejía 2011) and cumulate chromitite layers hosted within the Aburrá ophiolite (Alvarez 1987; Correa-Martínez 2007). Of these, only the El Carmen-El Bagre mesothermal Au veins (and their associated residual and placer deposits) have seen any significant historic exploitation (albeit at an artisanal level) and more recently resource-level exploration and development. Bailadores contains a qualified resource (1.6 MMT grading 26% Zn, 7% Pb, 1.5% Cu; Carlson 1977; Staargaard and Carlson 2000) which remains undeveloped, whilst the Quetame Cu and Santa Elena Cr occurrences may be qualified as prospects.

Tectonic models for the pre-Mesozoic of the Northern Andes are controversial and remain the subject of active debate, especially for the early Paleozoic to late Permian (see Cochrane et al. 2014a; Van der Lelij et al. 2016; Cediel 2018; Leal-Mejía et al. 2018), where the geological and geochronological database in various sectors remains deficient. In this context, and given the sparse and disparate temporal-spatial distribution of pre-Jurassic mineral deposits, it is difficult to define coherent metallogenic provinces or epochs in the context of the time-space analyses presented in Figs. 6.2 and 6.3. Based upon available data, however, the mineral deposits presented may be considered reflective of the interpreted tectonic settings during the various stages of pre-Jurassic evolution proposed for the region.

6.5.1.1 Bailadores

This occurrence is located within the Sierra de Mérida (Venezuela) (Fig. 6.2) but is included here as this region exposes an important section of Northern Andean geology, especially representative during the Paleozoic. Bailadores is interpreted as a Kuroko-type volcanogenic massive sulphide deposit (Carlson 1977; Staargaard and Carlson 2000). Mineralization is hosted within a localized section of siliceous pyroclastic meta-volcanic rocks interfingered with marine meta-pelitic sedimentary rocks of the Mucuchachí Fm. Van der Lelij et al. (2016) provide a U-Pb (zircon) age of 452.6 ± 2.7 Ma for the Mucuchachí meta-tuff. Sulphide mineralization is considered penecontemporaneous with tuff deposition and marine pelitic sedimentation (Carlson 1977; Staargaard and Carlson 2000). Kuroko-type deposits are commonly generated in extensional settings within continental margin or volcanic island arcs. Preferred sites include arc-axial grabens and arc-proximal normal faults along the margins of back-arc basins. Van der Lelij et al. (2016) suggest that felsic volcanism of the Mucuchachí Fm. is related to intra-arc extension associated with the emplacement of granitoids during the mid-Ordovician.

6.5.1.2 Caño Negro-Quetame-Cerro de Cobre

These Cu (U) occurrences are associated with argillaceous and arenaceous red beds of early Carboniferous age outcropping over 100 km of strike length along the eastern margin of the Quetame Massif. Rodríguez and Warden (1993) consider the occurrences to present geological and geochemical similarities to mineralization in the Central African Cu Belt and the Polish Kupferschiefer. Cox et al. (2007) indicate that such deposits may be hosted in marine or lacustrine argillaceous rocks. Mineralization is of diagenetic origin, forming prior to lithification of the host rock and being generally independent of igneous processes. Paleo-facies maps for the early Carboniferous (Cediel et al. 1994) depict mixed shallow marine and intertidal sedimentation throughout the Quetame area and a period of general magmatic quiescence. This is supported by U-Pb (zircon) data (e.g. Horton et al. 2010; Leal-Mejía et al. 2018) indicating a general hiatus in magmatism throughout the eastern Colombian Andes during the Carboniferous.

6.5.1.3 El Carmen-El Bagre Au District

Leal-Mejía (2011) described auriferous quartz-sulphide veins hosted with the El Carmen-El Cordero stock. The El Carmen-El Bagre district (Fig. 6.2) is comprised of numerous NNW- to NNE-striking veins, the most important of which include the El Carmen and La Ye systems, which can be traced in artisanal and more formalized exploitations for over 5 km along strike. Host rock to the veins includes low-K leucotonalite and pyroxene-hornblende-bearing gabbro-diorites of the El Carmen-El Cordero suite (Leal-Mejía 2011; Leal-Mejía et al. 2018). Exploited veins, ranging from 0.5 to 4 m and averaging ~1 m thick, consist of massive milky quartz containing native gold and up to 20% mixed sulphides, dominated by pyrite with occasional galena, chalcopyrite and rare sphalerite. Sulphide and native gold distribution within the veins is patchy, and some sections of the veins can be devoid of mineralization. Wallrock alteration related to the veins includes m-scale haloes of moderate to pervasive sericite ± chlorite and carbonate replacing feldspar within the host intrusive. The La Ye vein is observed to cut a phaneritic leucotonalite dike of somewhat more felsic composition to that of the host leucotonalite. Additional feldspar porphyry dikes observed at various localities clearly cut both the El Carmen suite and the veins at La Ye.

The El Carmen-El Cordero stock has historically been mapped within the limits of the late Jurassic Segovia Batholith (Feininger et al. 1972; Aspden et al. 1987; González 2001; Gómez et al. 2007; Leal-Mejía et al. 2010), and El Carmen-El Cordero mineralization was considered to belong to the same trend of mesothermal vein occurrences hosted within the Segovia Batholith at Segovia-Remedios, 60 km to the south (e.g. Londoño et al. 2009; Leal-Mejía et al. 2011a). Notwithstanding, Leal-Mejía (2011) provided U-Pb (zircon) age dates ranging from ca. 333 to 310 Ma for the El Carmen-El Cordero igneous assemblage, establishing that El Carmen-El Cordero is significantly older than the Segovia Batholith. K-Ar (sericite) age dating of pervasively altered fragments of the El Carmen stock encapsulated within the La Ye vein returned an age of 280 ± 6 Ma, whilst an unaltered-unmineralized, cross-cutting porphyritic dike returned a K-Ar whole-rock age of 167 ± 5 Ma (Leal-Mejía 2011).

The Jurassic date for the cross-cutting dike coincides with the age of the Segovia Batholith proper and confirms that the El Carmen veins are pre-Segovia Batholith in age. With respect to the age of the alteration sericite, it is possible that ca. 280 Ma represents the age of hydrothermal alteration associated with vein formation. Alternatively, it is noted that ca. 280 Ma coincides with interpreted tectono-thermal metamorphism in Colombia, associated with the early-mid-Permian assembly of Pangaea (Vinasco et al. 2006; Cochrane et al. 2014a). Thus, ca. 280 may record metamorphic resetting of the age of the hydrothermal sericite.

Based upon the foregoing hydrothermal alteration and vein filling at El Carmen-La Ye is constrained to the interval between ca. 310 and 280 Ma. We observe, however, that ca. 290 to 250 Ma tectono-thermal meta-granitoids are widespread constituents within the Precambrian-Paleozoic basement complex at numerous locations within the Colombian Andes (Vinasco et al. 2006; Cochrane et al. 2014a; Leal-Mejía et al. 2018 and references cited therein), including within the Maracaibo Sub-plate and Central Tectonic Realm (Fig. 6.2). In the context of regional granitoid magmatism vs. gold metallogeny, however, the Permo-Triassic suite is generally unmineralized, and Leal-Mejía et al. (2011a) note that in no instance have such granitoids been genetically linked to gold mineralization. As such, we interpret mineralization at El Carmen-El Bagre to be genetically related to the cooling history of the El Carmen-El Cordero stock and to have been emplaced at ca. 310 Ma.

In terms of age, lithogeochemistry and Sr-Nd isotope composition (Leal-Mejía 2011), the El Carmen-El Cordero suite presently stands unique, not only in the Colombian Andes but for the entire Northern Andean region, and the geological context of these intrusions not fully understood. The age of these intrusives significantly predates well-documented Permo-Triassic granitoid gneisses and peraluminous anatectites and amphibolites (Vinasco et al. 2006; Cardona et al. 2010; Cochrane et al. 2014a). The granitoids are hosted within the confines of the Cajamarca-Valdivia terrane and, based upon age constraints, were emplace at least 70 m.y. after accretion of the Cajamarca-Valdivia assemblage to continental Colombia, and there is no evidence that they are subduction-related. The granitoids are localized along the Otú fault, a major N-S-striking feature, which has been interpreted as a potential plate boundary (e.g. Toussaint 1993; González 2001). Lithogeochemical data supplied by Leal-Mejía et al. (2018) suggest the El Carmen-El Cordero granitoids represent a ridge tholeiitic granitoid (RTG) suite (Barbarin 1999), comprised of tholeiitic gabbro-diorite with low-volume leucotonalite and trondhjemite differentiates, petrogenetically associated with oceanic spreading and ophiolite formation. The Otú fault could represent the longitudinal axis of an aborted rift basin, which opened to the point of at least locally producing oceanic lithosphere. Such a tectonic setting would be in broad agreement with tectonic models for the early Bolívar Aulacogen and the Carboniferous in Colombia (Cediel et al. 1994; Cediel and Cáceres 2000; Leal-Mejía et al. 2018), which is considered a time of regional extension and shallow marine sedimentation characterized by the absence of subduction-related magmatic arcs.

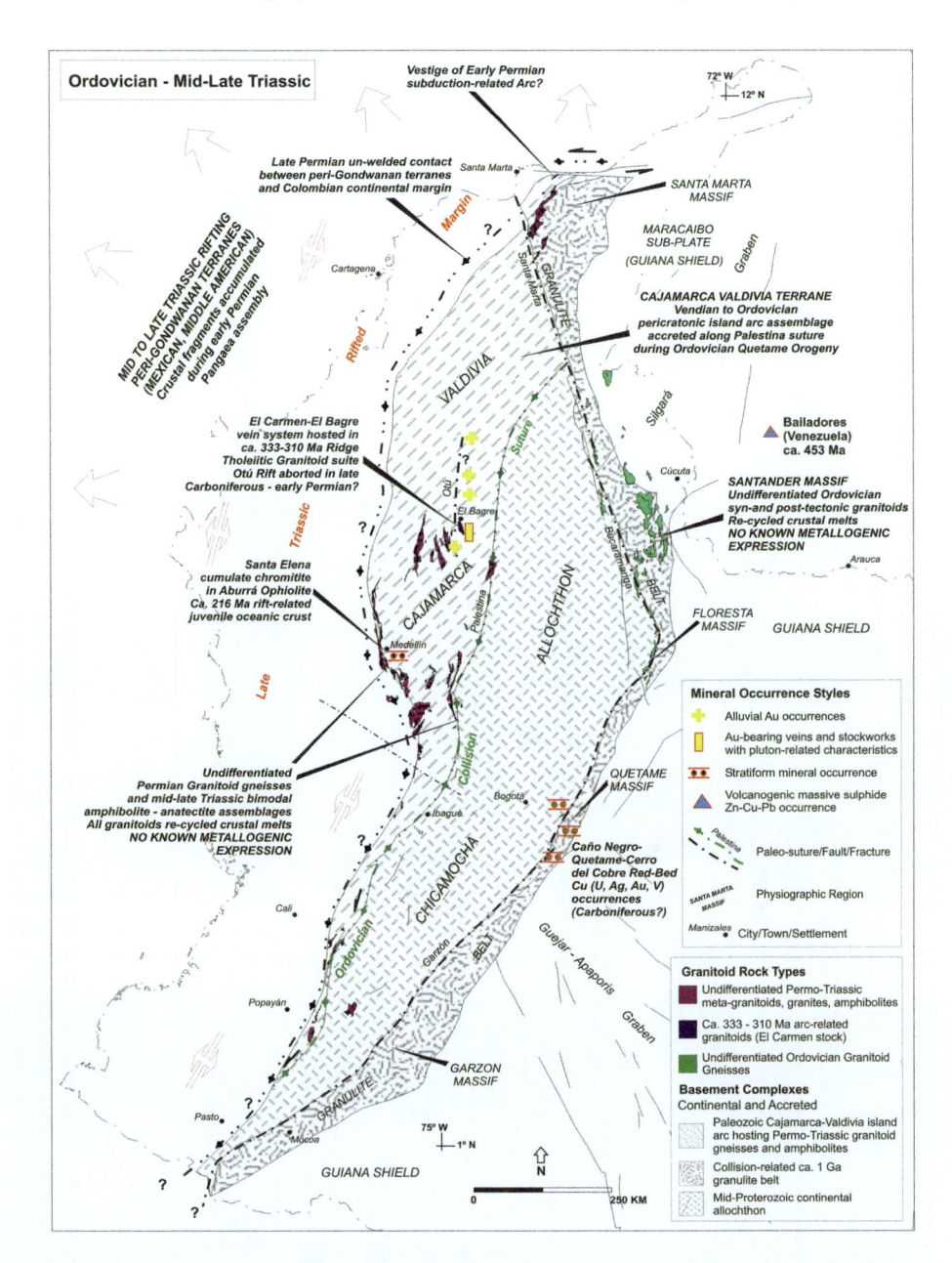

Fig. 6.2 Mineral occurrences of interpreted Ordovician through mid-late Triassic age in the Colombian Andes, in relation to tectonic setting, major litho-tectonic elements and granitoid intrusive suites of the same time period

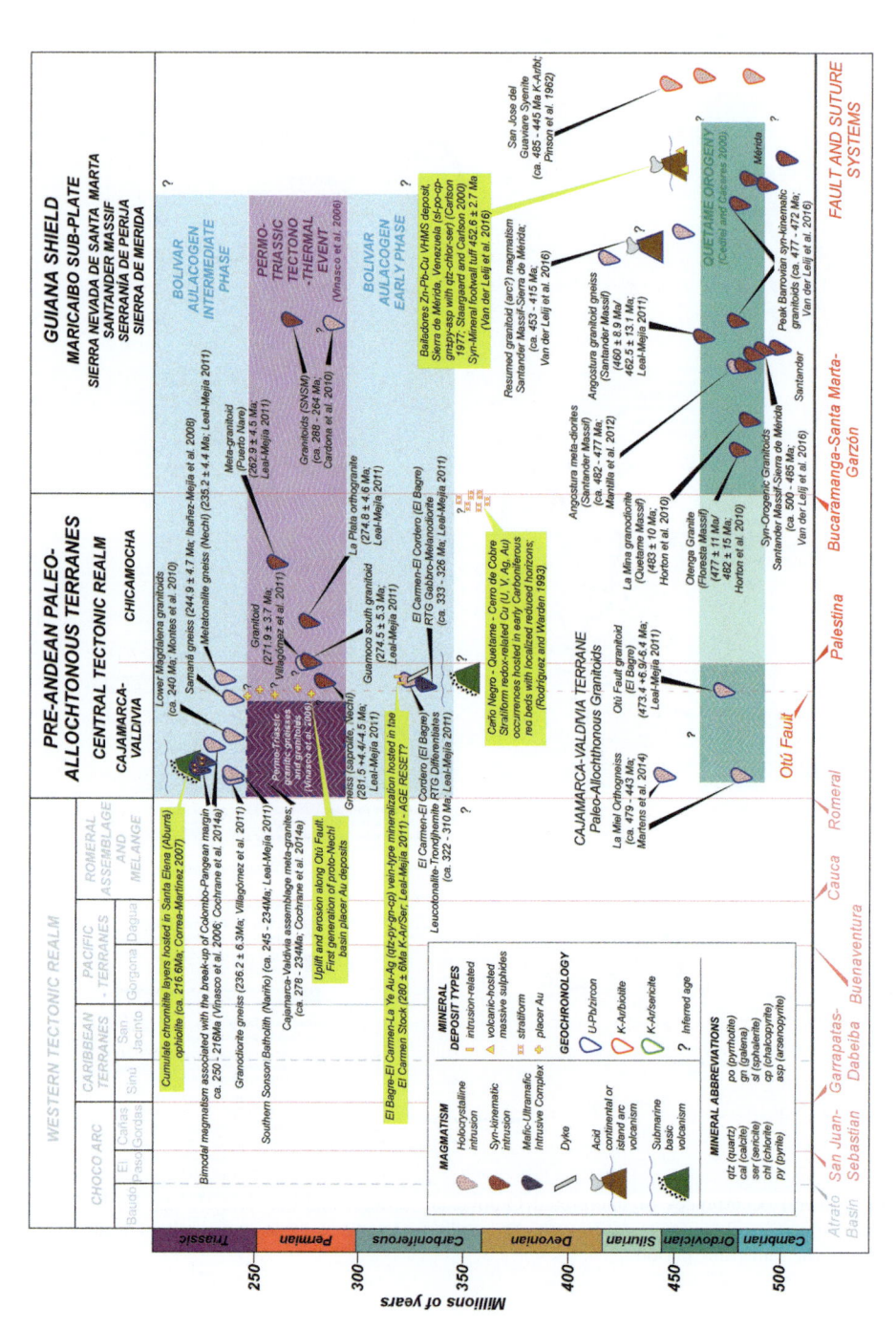

Fig. 6.3 Time-space analysis of mineral occurrences of interpreted Ordovician through mid-late Triassic age in the Colombian Andes and surrounding region, in relation to tectonic framework, major litho-tectonic elements and orogenic events and the age and nature of granitoid intrusive suites of the same time period. The profile contains elements projected onto an ca. NW-SE line of section through west-central Colombia

6.5.1.4 Santa Elena Chromitite

These occurrences (Fig. 6.2) are hosted within a tectonized belt of serpentinized gabbro, dunite and peridotite known as the Aburrá ophiolite (Correa-Martínez 2007). Disseminated to massive chromitite occurs within disjointed pods, the largest of which (Patio Bonito) contained some 30,000 T (Alvarez 1987), which has since been exploited. Cumulate podiform chromite deposits are common magmatic co-products of ophiolite petrogenesis (Mosier et al. 2012). Correa-Martínez (2007) provided a U-Pb (zircon) age of ca. 216 Ma for the Aburrá ophiolite, accompanied by lithogeochemical data suggesting that the highly depleted ultramafic rocks were emplaced within a back-arc basin to N-MORB oceanic setting. Cochrane et al. (2014a) presented a tectonic model depicting the progressive extension of continental crust leading to rifting and primitive ocean crust development during the breakup of Pangaea and separation of the Middle American and Mexican terranes from NW South America (Gondwana) during the latest Triassic.

6.5.2 Jurassic-Early Cretaceous Metallogeny: The Late and Culminant Bolivar Aulacogen

Regional extension, driven by slab detachment and rollback (Leal-Mejía 2011; Cochrane et al. 2014b; Spikings et al. 2015), continued into the Jurassic, and the onset of subduction resulted in the generation of voluminous metaluminous, calc-alkaline volcano-plutonic arcs throughout the Northern Andes (Aspden et al. 1987; Cediel et al. 2003; Spikings et al. 2015; Leal-Mejía et al. 2018). This shift in tectonic regime was accompanied by the first of the regionally defineable Colombian metallogenic epochs (Fig. 6.4). Jurassic metallogeny in Colombia is dominated by epigenetic volcano-plutonic arc-related precious and base metal occurrences formed in the mesothermal pluton, porphyry, skarn and epithermal environments, genetically related to the cooling history of spatially associated, penecontemporaneous granitoid magmatism generated within the context of the Bolivar Aulacogen. From an economic and production standpoint, vein-type (sensu lato) Au (±Ag) deposits, and their associated residual and alluvial derivatives, are by far the most important, and exploitation of these deposits from a multitude of generally artisanal operations remains very active even today.

Aside from Au-Ag, various non-precious metal occurrences of Jurassic age have been documented. These include predominantly sediment-hosted Cu (Ag) and porphyry Cu (Mo) occurrences from which production has not been recorded, neither historically nor in modern times. Notwithstanding, due to their scale and economic potential, these occurrences have been the subject of exploration programmes and academic studies (e.g. Maze 1980; Viteri 1980; Sillitoe et al. 1982; Sillitoe et al. 1984), permitting their integration into the time-space analysis presented in Fig. 6.5.

In addition to genetically related mineralization, Jurassic-aged granitoids form the host rocks to epigenetic mineralization, superimposed during subsequent

Cretaceous through Pliocene metallogenic events. New radiometric age dates and geological and geochemical relationships (e.g. Leal-Mejía et al. 2010; Leal-Mejía 2011) demonstrate that what was considered "Jurassic age" mineralization hosted within granitoids of, for example, the Santander Plutonic Group and the Segovia Batholith (e.g. Shaw 2000a, b; Sillitoe 2008; Londoño et al. 2009) is now known to significantly post-date the age of the host pluton, to the point where it is not possible for mineralization to be genetically related to the cooling history of the host Jurassic granitoids.

In concert with Northern Andean tectono-magmatic evolution during the Bolivar Aulacogen, we now provide descriptions of some of the most important Jurassic-aged mineral occurrences, representative of the Colombian Jurassic metallotects depicted upon our time-space charts.

6.5.2.1 Sedimentary Cu Occurrences of the Serranía de Perijá and Santander Massif

The initial phases of tectonic development of the Bolívar Aulacogen during the late Triassic-Jurassic involved regional extension, continental rifting, bimodal magmatism and the deposition of extensive siliciclastic/volcanoclastic deposits upon the composite basement comprised of the Maracaibo Sub-plate and Central Tectonic Realm. Rift-related deposits are contained within the Triassic Payandé rift (Cediel and Cáceres 2000; Cediel et al. 2003), the Lower Jurassic Morrocoyal rift (Geyer 1973) and the Middle Jurassic Siquisique rift (Bartok et al. 1985). The Maracaibo-Perijá rift hosts the latest Triassic-Jurassic Girón-La Quinta, Jordán and Bocas Formations (Cediel 1969; Maze 1980, 1984, Cediel and Cáceres 2000). Jurassic rift-related deposits in eastern Colombia incorporate Jurassic zircons with an age peak at ca. 185–200 Ma (Horton et al. 2010).

Maze (1980) and Viteri (1980) describe volcano-sedimentary-hosted Cu occurrences within the La Quinta-Girón Fms. of the Serranía de Perijíá and Santander Massif in northwestern Venezuela and Colombia. Variable assemblages of native Cu, hematite, pyrite, chalcopyrite, bornite, magnetite and chalcocite are contained as interstitial disseminations within red beds, arkose and conglomerate, vesicular mafic flows, localized latite flows and along the margins of mafic dikes. Maze (1980) considers the age of the mineralization to be consistent with a syn-diagenetic origin. In Colombia, Cu mineralization hosted within the Girón Fm. outcrops in a discontinuous belt extending for over 100 km along the NW flanks of the Serranía de Perijá, from Curumaní to San Diego and El Molino. Maze (1980) emphasizes the widespread nature of the La Quinta-Girón Cu province and suggests that regional-scale processes were responsible for its formation. An association between aulacogens and sediment-hosted Cu mineralization has long been recognized (e.g. Burke and Dewey 1973; Cox et al. 2007).

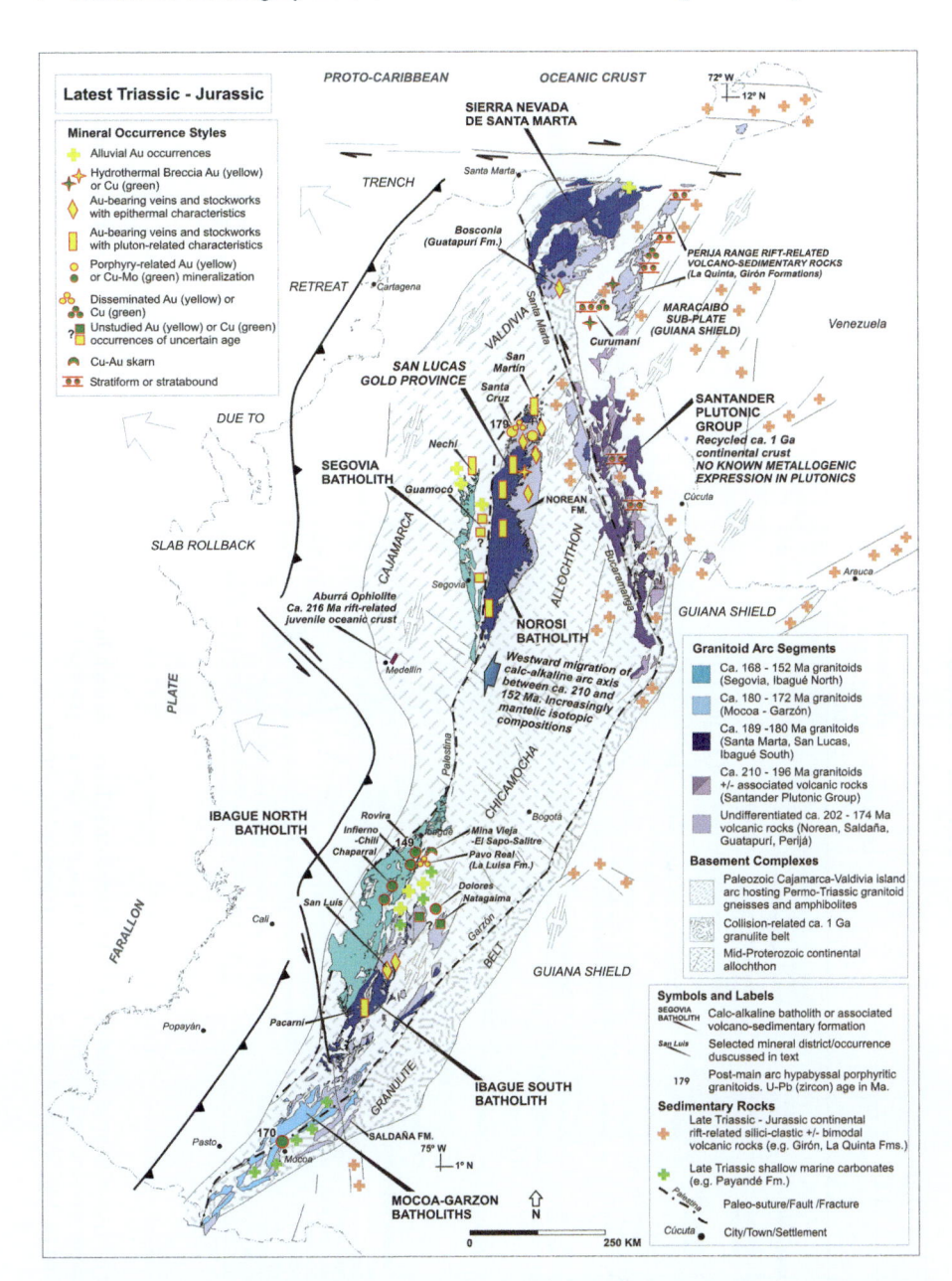

Fig. 6.4 Mineral occurrences of interpreted latest Triassic through Jurassic age in the Colombian Andes, in relation to tectonic setting, major litho-tectonic elements and granitoid intrusive and volcanic suites of the same time period. Note the highly extensional regime into which Jurassic volcanism, granitoid magmatism and metallogeny are interpreted to have been emplaced

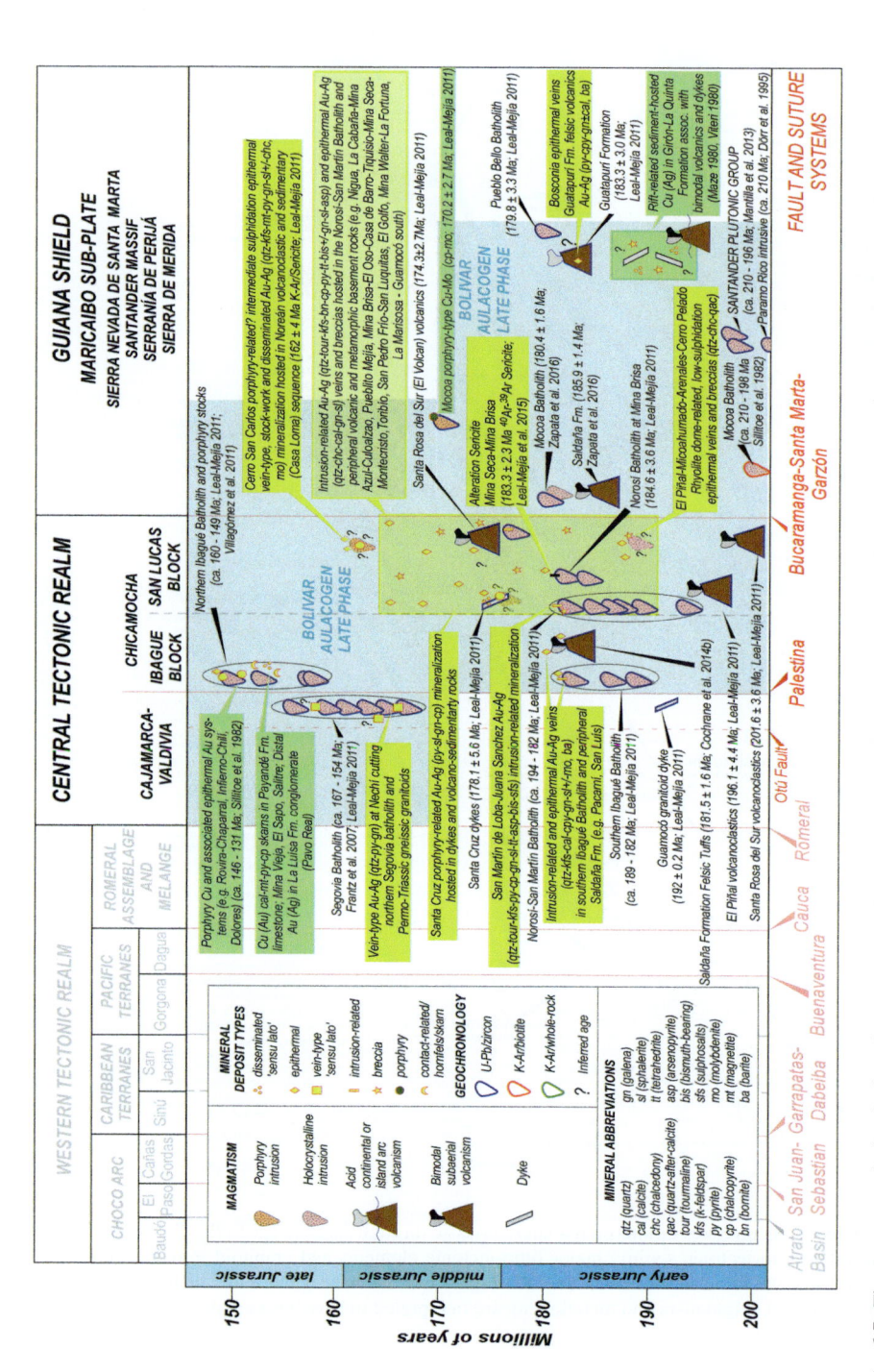

Fig. 6.5 Time-space analysis of mineral occurrences of interpreted latest Triassic through Jurassic age in the Colombian Andes, in relation to tectonic framework, major litho-tectonic elements and orogenic events and the age and nature of granitoid intrusive suites of the same time period. The profile contains elements projected onto an ca. NW-SE line of section through west-central Colombia

6.5.2.2 Jurassic Volcano-Plutonic Arc-Related Au-Ag and Cu-Mo Metallogeny

Subduction-related latest Triassic-Jurassic granitoids, including major composite holocrystalline batholiths, hypabyssal porphyry stocks and penecontemporaneous volcano-sedimentary rocks, represent the most extensive period of magmatic activity recorded within the present-day geological exposure of the Colombian Andes. Jurassic granitoids form a SSW-NNE-oriented array of arc segments extending from the Ecuador border to the Sierra Nevada de Santa Marta on the Caribbean coast (Aspden et al. 1987; Cediel and Cáceres 2000; Gómez et al. 2015a).

Based upon the analysis of U-Pb (zircon) crystallization age and lithogeochemical and isotopic data for latest Triassic-Jurassic granitoids, Leal-Mejía (2011) and Leal-Mejía et al. (2018) identified four major magmatic episodes involving granitoid batholith emplacement within six spatially separate arc segments. In addition, they identified three distinct, volumetrically minor hypabyssal porphyry suites of Jurassic age. The principal magmatic episodes/arc segments include the ca. 210–196 Ma Santander Plutonic Group, the ca. 189–182 Ma southern Ibagué-Norosí-San Martín Batholiths, the ca. 180–173 Ma Mocoa-Garzón and Sierra Nevada de Santa Marta Batholith suites and the ca. 170–152 Ma northern Ibagué and Segovia Batholiths. Hypabyssal porphyritic stocks and/or dike swarms are associated with the Mocoa, Norosí and northern Ibagué Batholiths and in all cases tend to post-date main phase batholith emplacement by 3–5 million years. Various authors have described the E to W younging trend of the Colombian Jurassic granitoids (Aspden et al. 1987; Cediel et al. 2003; Leal-Mejía 2011; Spikings et al. 2015), from the Santander Plutonic Group, passing westward through the Serranía de San Lucas and southern Ibagué granitoids, into the northern Ibagué and Segovia Batholiths (Fig. 6.1). Leal-Mejía et al. (2018) detail the spatial-temporal migration of active Jurassic arc segments both along the length of the magmatic arc axis as well as in a transverse sense, related to E to W rollback of the subducting Pacific plate.

Figures 6.4 and 6.5 summarize the appearance and distribution of Au-Ag and Cu (Mo) mineralization associated with Jurassic holocrystalline batholiths, coeval volcanic sequences and hypabyssal porphyry suites. Temporal-spatial analysis reveals that all of the composite Jurassic granitoid suites host important Au-Ag and/or Cu (Mo) mineralization. However, as observed above, recent radiometric age dating confirms that mineralization in some cases significantly post-dates the cooling history of the spatially associated Jurassic suite (e.g. Santander Plutonic Group, Segovia Batholith). Figs. 6.4 and 6.5 include only deposits we consider to be genetically related to the cooling history of Jurassic plutonic, volcanic or hypabyssal rocks. The style, age and tectonic framework of mineralization, on a regional, district and deposit scale, are indicated.

Jurassic Volcano-Plutonic Arc-Related Deposit Types

Mineralogical, textural and geochemical attributes and lithological associations, including a spatial relationship with granitoid magmatism, permit classification of the mineral deposits discussed below within the epigenetic, hydrothermal, igneous-related category. This includes deposit types more strictly related to holocrystalline plutonic suites (e.g. intrusion-related or pluton-related gold deposits; Sillitoe 1991; Thompson et al. 1999; Lang and Baker 2001; Hart 2007), to hypabyssal porphyritic intrusions (e.g. Sillitoe 2000; Sillitoe 2010) as well as those of an epithermal nature (e.g. Simmons et al. 2005), hosted within penecontemporaneous volcano-sedimentary rocks above or lateral to plutonic or porphyritic intrusions. The foregoing deposit types are considered to have a fundamental genetic relationship with magmatic fluids derived from a host and/or a nearby parental intrusion. Intrusion-related gold deposits and their associated metal assemblages are often classified with respect to the redox state of the source pluton(s) (Thompson et al. 1999; Hart 2007), as recorded mineralogically in the presence of modal magnetite (oxidized) vs. ilmenite (reduced) and lithogeochemically as recorded in analysed whole-rock ferric/ferrous ratios (Ishihara 1981; Sillitoe 1991; Thompson et al. 1999).

Within the general intrusion-related category, mineralization styles in Colombia are quite varied. Pluton (stock and batholith)-hosted veins, sheeted vein systems and localized stockworks containing Au±Ag and base metal assemblages are observed in Serranía de San Lucas and southern Ibagué Batholiths, whilst Au (Cu) skarn deposits are hosted within the northern Ibagué Batholith. Jurassic porphyry-related mineralization includes Cu porphyry-style occurrences with peripheral epithermal Au-Ag mineralization, associated with stocks in the northern Ibagué Batholith (Sillitoe et al. 1982). Epithermal mineralization associated with Jurassic volcanic and/or intrusive rocks is observed throughout the Serrania de San Lucas, at Bosconia and Aracataca in the southwestern Sierra Nevada de Santa Marta and in the southern Ibagué Batholith. Based upon textural, mineralogical and alteration criteria for epithermal deposits (Sillitoe and Hedenquist 2003; Simmons et al. 2005), mineralization is dominated by intermediate- and, more locally, low-sulphidation, quartz-sericite-illite±adularia±calcite–Au-Ag-base metal-bearing veins, stockworks, mantos, hydrothermal breccias and disseminations, in some cases localized within or marginal to felsic domes. Epithermal mineralization hosted within volcanic rocks in San Lucas and southern Ibagué forms kilometre-scale linear trends, which when followed along strike in some cases are observed to be rooted within penecontemporaneous plutonic rocks. Multiphase and overprinting mineralization and alteration assemblages are recorded along these trends, suggesting a broad continuum between pluton-related and epithermal mineralization types. Au-Ag-base metal mineralization associated with porphyritic stocks, dikes, sills and felsic domes in the Serranía de San Lucas is of a more epithermal nature. Jurassic Cu-Mo porphyry mineralization at Mocoa (Sillitoe et al. 1984) has no apparent precious metal expression.

We now provide brief descriptions of the most important mineral provinces, districts, occurrences and deposit styles, associated with Colombian Jurassic volcano-plutonic arcs. Information is presented on a per arc segment basis, from oldest to youngest, focussing upon occurrences where Au (Ag) is the principal economic commodity. Most of these districts have never been documented in readily accessible international literature. The principal source of information pertaining to Jurassic gold occurrences in Colombia is the doctoral thesis of Leal-Mejía (2011), which contains detailed descriptions and geologic, petrographic, trace element, minerographic, paragenetic and isotopic data. We have integrated more recent unpublished observations herein. With respect to Jurassic porphyry-related Cu and Mo occurrences (Fig. 6.4), we note that these deposits have been the subject of readily accessible, detailed investigations by Sillitoe et al. (1982, 1984) and Sillitoe and Hart (1984), and in this context, they will not be reiterated herein. Available information suggests the Santander Plutonic Group and the Mocoa-Garzón and the northern Ibagué and Segovia Batholiths have no apparent in situ genetically related Au (Ag) expression, and these units will not be discussed further herein. In situ Au (Ag) mineralization hosted within the Segovia Batholith and the Santander Plutonic Group will be detailed in the Cretaceous-Eocene and Oligocene-Pliocene section, presented below.

ca. 189 to 182 Ma Norosí and San Martín Batholith Suites

The Norosí and San Martin Batholiths and coeval Noreán Fm. volcano-sedimentary pile host widespread and abundant gold mineralization referred to herein as the San Lucas Gold Province (Figs. 6.4 and 6.5). Although exploitation of alluvial and residual gold concentrations along the margins of the San Lucas range dates from pre-Colombian times (Restrepo 1888), the region remains an active generative exploration target, and large tracts, especially in the south, remain essentially unexplored. Notwithstanding, artisanal mine workings in the Serrania de San Lucas reveal dozens of gold occurrences, clustered into kilometre-scale concentrations of occurrences (districts or camps; Fig. 6.6) or as isolated manifestations, distributed along the entire length of the +300 km N-S-trending San Lucas arc segment. Manifestations are mostly hosted within the Norosí and San Martin Batholiths and Noreán Fm. but also occur within the Mesoproterozoic and Paleozoic metamorphic rocks which form basement to the San Lucas arc region. Notably, mineralization does not cut early Cretaceous sedimentary strata outcropping along the eastern margin and southwest flank of the San Lucas range. An apparent concentration of gold occurrences is observed in the north; however, this may be a reflection of superior access facilitated by the historic Magdalena river system and recessed topography, combined with a drier microclimate which enhances outcrop exposure.

Leal-Mejía (2011) provided detailed characterization studies of various gold occurrences in the San Lucas region. The following summary is based upon these and subsequent investigations completed by the present authors. Gold deposit types are dominated by pluton-related occurrences and by epithermal deposits. A spatial continuum between these two deposit types is observed, both vertically, where mineralization passes from epizonal plutonic environments of the Norosí and San Martin Batholiths into isolated volcano-sedimentary roof pendants and laterally

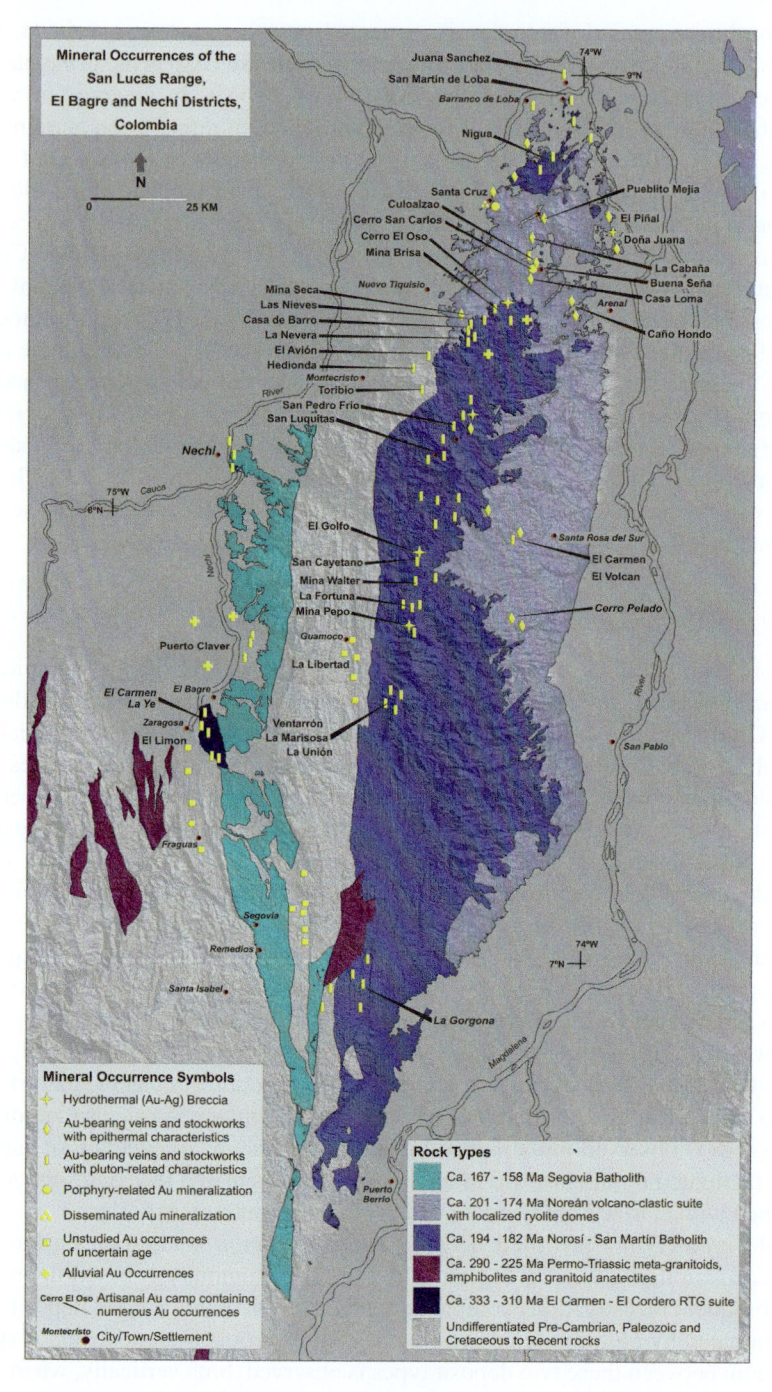

Fig. 6.6 Selected mineral occurrences of interpreted Carboniferous and latest Triassic through Jurassic age in the San Lucas Range and surrounding area of the Colombian Andes, in relation to granitoid intrusive and volcanic rocks of the same approximate time period. Physiographic features of the map area are revealed by the 30 m digital elevation model (DEM) base image

where mineralization rooted within the batholiths can be traced along hydrothermal conduits and through lateral intrusive contacts, into the adjacent volcano-sedimentary sequence. Additional epithermal deposits are related to localized felsic domes, whilst in some instances, such as at Pueblito Mejía and Santa Cruz, mineralization is related to deeper-seated intermediate to felsic porphyritic dikes. Deposit types manifest in numerous styles of Au-Ag ± Cu, Pb, Zn (As, Bi, Sb) mineralization, including as veins, vein swarms, stockworks and breccias in plutonic rocks and as contact zone replacements, veins, mantos and stratiform replacements in volcano-sedimentary strata. Vein arrays, breccias, mantos and replacements within kilometre-scale alteration haloes are related to felsic domes and porphyritic dike swarms. The gold occurrences and mineralization styles in the Serrania de San Lucas are too numerous to describe individually herein. In the following paragraphs, we present a generalized summary of deposit characteristics representative of the major deposit types mentioned above. The most important gold clusters of the Serranía de San Lucas region are shown in Figs. 6.4 and 6.6. Schematic interpretations of the geological, spatial, temporal and paragenetic development of the deposit types of the region are revealed in Fig. 6.5.

San Lucas Pluton-Related Au Occurrences The Norosí and San Martín Batholiths host numerous gold occurrence clusters, including, from north to south, Juana Sanchez, San Martín-Barranco de Loba, Nigua-La Mota, La Estrella-Culoalzao, Cerro El Oso-Mina Brisa, Mina Seca-Casa de Barro, San Pedro Frío-San Luquitas, Mina Walter-La Fortuna, La Marisosa and Ventarrón (Fig. 6.6). Gold occurrences are generally bound within broad corridors containing veins, vein swarms and breccias, hosted within metasomatized and hydrothermally altered intrusive. Individual corridors locally attain true widths of up to 50 metres and in some cases are traceable for various kilometres along strike (e.g. San Martín de Loba, Casa de Barro, Mina Seca). Individual veins within these corridors can attain true widths of up to 10 metres, containing gold concentrations exceeding 10 ppm, although vein widths are highly variable and probably average in the 0.5 to 2 metre range. Wallrock alteration associated with, and mineralization contained within, individual veins exhibits a complex and prolonged multistage paragenesis (Leal-Mejía 2011; Leal-Mejía et al. 2015). Considering a composite of the pluton-hosted gold clusters mentioned above, at least four vein development stages are observed.

Stage 1 consists of infilling with crystalline and saccharoidal quartz+tourmaline (schorl)+magnetite±pyrite and calcite, associated with metre-scale haloes of wallrock replacement by tourmaline and potassium feldspar. Secondary biotite is observed along vein selvages. Only minor amounts of gold, if any, are considered to have been deposited during this stage.

Stage 2 involves brecciation of the earlier quartz-tourmaline assemblage and further vein development, with the deposition of abundant massive pyrite and chalcopyrite+crystalline quartz. Locally, abundant bornite± chalcopyrite, sphalerite, galena and minor arsenopyrite are observed (e.g. Culoalzado). Secondary chalcocite and covellite are commonly recorded in the near-surface supergene environment. Greater than 70% of the global Au (Ag) budget is considered to have been introduced during this stage, as native gold accompanying the massive pyrite-chalcopyrite assemblage. Ag-Au ratios are generally low, ranging from ca. 0.5:1 to 3:1. Hand samples of the massive ore can contain well in excess of 1% Cu. The mixed sulphide component of individual Stage 2 veins commonly exceeds 50% and can attain 80% by volume. In large structures such as La Puerta and El Caño (San Martín de Loba), Casa de Barro, El Piojo (San Pedro Frio) and La Marisosa, multimeter thicknesses of massive sulphide+quartz infilling, containing high-grade Au (Ag) mineralization, can be observed. Stage 2 alteration is limited to varying degrees of silicification and pyritic sulphidation along vein margins, extending for tens of cm into the intrusive wallrock.

Stage 3 involves the reactivation and localized brecciation of Stage 1 and Stage 2 vein infillings and the development of new hydrothermal conduits, commonly at moderate (10 to 40 degree) angles to the early formed veins. The margins of these new conduits are not well defined, being gradational through decimetre- to metre-scale alteration zones into fresh intrusive. Stage 3 infillings are dominated by finer-grained, crystalline, grey and banded quartz with local cockade textures and abundant granular pyrite. The ore mineral assemblage includes sphalerite and galena with native Au-Ag admixtures and a host of Cu-Pb-Ag-Bi-Sb-As-bearing sulphosalts, including emplectite, tetrahedrite, polybasite and matildite among others (Leal-Mejía 2011). Ag-Au ratios can increase to in excess of 10:1. Brecciated and altered fragments of the host intrusive are commonly incorporated into the vein assemblage. Stage 3 alteration is dominated by the strong to pervasive replacement of intrusive wallrock by fine greisen-like muscovite with disseminated coarse cubic pyrite, overprinting, where present, the Stage 1 K-spar-tourmaline-biotite assemblage. Muscovite-pyrite haloes extend for various decimetres to metres on either side of individual veins and coalesce to form composite zones of pervasive wallrock replacement, tens of metres in thickness, in cases where multiple veins are present (e.g. Mina Brisa, Casa de Barro, San Martín de Loba). Decimetre-scale silicification is observed along the margins of individual veins and in zones of brecciation and stockworking. The muscovite-pyrite assemblage is weakly auriferous, carrying up to 1 ppm gold in proximity to mineralized veins.

Stage 4 involves the crackle brecciation of Stage 1, 2 and 3 infillings and the injection of fine-grained banded chalcedony and opaline silica. Stage 4 infillings are accompanied by little or no sulphide phases and contain no economic mineralization.

In order to better constrain the age of mineralization/hydrothermal alteration for the pluton-related Au occurrences described above, $^{40}Ar/^{39}Ar$ age dating of a sample of well-developed Stage 3 alteration muscovite from the Mina Seca-Mina Brisa sector was undertaken. The resulting age of 183.3 ± 2.3Ma (Leal-Mejía et al. 2015) falls within ca. 189 to 182 Ma range of U-Pb (zircon) magmatic crystallization ages provided for the Norosí Batholith in general and compares particularly well with the 184.6 ± 3.6 Ma age for the Norosí Batholith at Mina Brisa (Leal-Mejía 2011), near the $^{40}Ar/^{39}Ar$ sample locality.

San Lucas Basement-Hosted Mineralization Pluton-related mineralization hosted within Meso-Proterozoic to early Paleozoic metamorphic basement in the

San Lucas region is observed at Juana Sanchez, La Cabaña, Montecristo and Guamoco. Varying styles of gold mineralization are observed at these localities. At Juana Sanchez, batholith-rooted mineralization as described above extends into isolated xenoliths or roof pendants of intermediate to mafic orthogneiss without any significant variations in style or composition. At La Cabaña, located in the Pueblito Mejía mining sector (Fig. 6.6), however, mineralization hosted within an intermediate to mafic orthogneiss roof pendant or xenolith takes on a more epithermal aspect. Metre-wide zones of brecciation and veinlet formation were developed in at least two paragenetic stages. Stage 1 involved hydraulic brecciation of gneissic wallrock and the deposition of multi-centimetre, euhedral, prismatic quartz±ankerite crystals in symmetrical open-space infillings perpendicular to fracture selvages. Overgrowths of abundant coarse-grained crystalline pyrite, Fe-rich sphalerite with fine-grained chalcopyrite inclusions ("chalcopyrite disease"), galena and native Au (probably electrum based upon the pale yellow colour in polished section) form the Stage 1 ore mineral assemblage. Stage 2 involved continued infilling of open spaces by prismatic quartz and coarse crystalline white calcite, accompanied by localized aggregates of coarse-grained Fe-poor sphalerite. Stage 2 infills lack a precious metal component. Incomplete filling of vein and breccia cavities at La Cabaña resulted in the preservation of open spaces containing well-terminated euhedral quartz and calcite crystals. Minor post-mineral crackle brecciation and the infusion of greenish opaline silica complete the vein paragenetic assemblage. With respect to alteration, centimetre- to decimetre-scale haloes encompass the La Cabaña veins and breccias. They are marked by silica with moderate to strong sericite and patchy adularia, replacing the gneissic wallrock. Late kaolin-lined fractures cut the silica-sericite-adularia assemblage.

The Guamoco district forms a 25 km N-S elongate trend of auriferous vein occurrences, hosted within intercalated Proterozoic quartzo-feldspathic and Permian peraluminous granitoid gneisses along the west-central margin of the Norosí Batholith (Fig. 6.6). Leal-Mejía (2011) revealed a 1048 ± 23.5 Ma U-Pb (zircon) age for felsic neosome in quartzo-feldspathic gneiss hosting the La Libertad vein at the northern end of the Guamoco trend. The contact between the Norosí Batholith and metamorphic basement is structural, broadly coincident with the N-S-striking rectilinear Palestina Fault (Feininger 1970; Cediel et al. 2003; Fig. 6.4). Vein orientation along the Guamoco trend ranges from NNW- through NNE-striking and is strongly influenced by the N-S tectonic grain of metamorphic basement throughout the region (see DEM Fig. 6.6). Most basement-hosted gold occurrences along the N-S trend are located 0.5 to 5 km west of the contact with the Norosí Batholith, but mineralization is not known to be developed along the contact. The Norosí Batholith adjacent to the trend also hosts numerous gold occurrences, and the southern end of the Guamoco trend cuts the Norosí Batholith in the Marisosa sector; however, the batholith-hosted occurrences do not share the N-S vein tendency observed to the north. Mineralization along the Guamoco trend is only exposed in the surface regime. It is mostly weathered and oxidized. The following description is based upon field data and hand sample observations only. The veins at Guamoco range from 20 cm up to 5 m and average about 1 m in thickness. Individual veins can be traced for up to 500 metres along strike. Massive milky to greyish quartz provides >90% of the vein filling; the remainder consists mostly of pyrite with minor, spotty occurrences of galena, sphalerite and chalcopyrite ± native gold. The mineralogic and paragenetic simplicity of the Guamoco veins contrasts markedly with nearby mineralization hosted within the Norosí Batholith (e.g. La Marisosa, Ventarrón, La Unión) which is typical of the four-stage Pluton-related paragenesis outlined above. No additional constraint on the age of gold mineralization at Guamoco is presently available. Observation that the trend cuts into the Norosí Batholith near La Marisosa places the maximum age of the occurrences between ca. 189 and 180 Ma; however, a genetic relationship with the Norosí Batholith and basement-hosted occurrences such as La Libertad can only be suggested.

San Lucas Epithermal Volcano-Sedimentary-Hosted Occurrences Vein- and breccia-type mineralization of an epithermal character is also widespread within the San Lucas range, as exposed in artisanal gold workings at Cerro San Carlos, El Piñal-Doña Juana, Pueblito Mejía, Santa Cruz, Mina Brisa, Micoahumado, Mina Totumo and Cerro Pelado (Fig. 6.6). Vein and breccia infillings consist of abundant sulphides including pyrite, sphalerite, galena, chalcopyrite and locally arsenopyrite and tetrahedrite, accompanied by quartz ± carbonate. Textures suggestive of epithermal levels of mineralization, including cockscomb and druzy quartz terminations, colloform-crustiform banding in quartz with chalcedony and quartz-after-calcite replacements (e.g. lattice/bladed textures), are recorded. Wallrock alteration in the volcanic sequences is dominated by strong sericitization proximal to mineralized structures, within tens-of-metre haloes containing illite+pyrite and locally kaolinite. Late chlorite and epidote are recorded more distally (Leal-Mejía 2011).

Gold mineralization at Cerro San Carlos, El Piñal, Doña Juana, Micoahumado and Cerro Pelado is associated with Jurassic felsic domes along eastern margin of the San Lucas range. Diapiric doming of Norean Formation volcano-sedimentary strata and the peripheral development of radial vein sets and mantiform veining (e.g. El Piñal, Casa Loma) are observed. U-Pb (zircon) dates for volcanic and hypabyssal rocks associated with the dome centres range from ca. 201 to 172 Ma (Leal-Mejía 2011). At Cerro San Carlos, Leal-Mejía (2011) documented the development of early sodic-calcic and potassic alteration assemblages in felsic pyroclastic rocks, containing secondary albite-actinolite-quartz-K-spar and biotite, with magnetite and minor molybdenite, attributed to the presence of a weakly mineralized porphyry system. The early porphyry-related alteration assemblage is strongly overprinted by widespread sericite-illite-dominant assemblages which introduce or redistribute (e.g. Sillitoe 2000) the majority of the gold mineralization. Mineralization is contained within 20 cm to 2 m wide, NE-striking feeder structures containing quartz and up to 80 percent coarse-grained pyrite. Low-grade silicified crackle breccias contain auriferous pyrite infillings. Alteration sericite from San Carlos drill core returned a 164 ± 4 Ma K-Ar date (Leal-Mejía 2011).

At Doña Juana-El Piñal numerous occurrences of minor epithermal Au-Ag veins, veinlet clusters and breccias are contained within a three by six kilometre N-S area, hosted within the margins of flow-banded rhyolite domes (e.g. Cerro El Piñal, Cerro Pan de Azucar) and associated crystal-lithic tuffs and agglomerates, for which Leal-Mejía (2011) recorded a U-Pb (zircon) age of 196.1 ± 4.4 Ma. Mineral assemblages include primarily quartz±calcite and minor pyrite. Lattice or bladed quartz-after-calcite textures, indicative of the replacement of platy calcite by quartz, are common within the veins and breccias. Open spaces within the breccias contain fine druzy quartz. Late crackle brecciation is filled with finely banded chalcedony. Wallrock alteration is dominated by silicification and greenish sericitization proximal to mineralized structures, within broader illite±chlorite-rich haloes. Larger veins measure from 10 to 50 cm in thickness and display typical colloform-crustiform banding. Geological reconnaissance suggests that these structures have limited strike extent.

At Pueblito Mejía, high-grade Au (Ag-Pb-Zn-Cu) veins are hosted within an approximately 400 m-thick section of medium to thickly bedded andesite, dacite and rhyolite crystal-lithic tuff and agglomerate of the Noreán Fm. The volcanic sequence rests unconformably upon gneissic metamorphic basement similar to that seen at La Cabaña and is cut by fine-grained diorite and granodiorite porphyry dikes. The mineralized corridor is exposed in numerous artisanal underground workings. Mineralization is hosted within NE-striking vein sets which can be traced discontinuously for almost 2 km along strike. A close spatial relationship between the veins and the diorite-granodiorite dikes is observed. The principal veins vary from 20 cm to 1 m in thickness and are commonly accompanied by centimetre-scale veinlet development in wallrock, for up to 1 m on either side of the main vein. The largest veins have been mined over a vertical range of ca. 300 m. Vein filling is dominated by a mixed sulphide assemblage containing coarse crystalline aggregates of pyrite>galena>sphalerite>chalcopyrite, which can comprise from ca. 10 up to 90% of the vein filling by volume. Gangue mineralogy is dominated by comb-textured and colloform quartz, calcite-ankerite±rhodochrosite and late chalcedony. Commercial laboratory analysis of selected ore samples reveals that high sulphide concentrations correlate well with gold grades, with individual samples containing up to 146 ppm Au. Multi-ppm Au mineralization is commonly accompanied by 0.05 to 2% Pb, 0.3 to 0.7% Zn and 0.02 to 0.2% Cu, with between 200 and 2000 ppm As. Ag-Au ratios range from 0.5:1 to 5:1 but average close to 1:1. Galena is the best visual indicator of enhanced gold grades.

Wallrock alteration at Pueblito Mejía ranges from intense and pervasive silicification + sericite-pyrite replacement of volcanic rock in close proximity to the veins, grading to illite+sericite+pyrite+chlorite over a distance of decimetres to metres, depending upon the size of the vein. Veining and alteration display a close spatial relationship to 1 to 2 m in thick diorite-granodiorite dikes. The dikes display sinuous contacts with the hosting volcanic sequence. Where exposed in underground workings, they are pervasively altered to a sericite-illite-pyrite assemblage and host quartz and pyrite veinlets and stringers. Part-per-million-level gold grades are recorded in well-altered dikes. Similar dikes, with a similar mineralization style and alteration signature, are observed at the Santa Cruz Au-Ag occurrence (Fig. 6.6; see below) where they have been dated using the U-Pb (zircon) method at ca. 178 Ma (Leal-Mejía 2011).

At Santa Cruz, Au-Ag (Zn-Pb-Cu) mineralization is intimately associated with pervasively altered and mineralized diorite and granodiorite porphyry dikes and sills containing up to 10% pyrite as disseminations and fine fracture fillings within a strongly sericite-illite altered groundmass. Peripheral to the dikes and sills, a broad zone of mineralized joint and fracture fillings, bedding plane replacements and mantos and pyritic disseminations affects sandstones and siltstones of the late Triassic to early Jurassic Sudan and Morrocoyal Fms. and felsic volcanic rocks of the overlying Noreán Fm. Fracture fillings within the volcano-sedimentary sequence are dominated by pyrite±sphalerite-galena and quartz. Sericite-illite and pyrite are the dominant alteration minerals. Individual mineralized dike samples return values as high as 3 ppm Au, whilst fracture fillings and mantos in the volcano-sedimentary sequence can contain tens to locally hundreds of ppm Au. Ag-Au ratios range from <1:1 to 10:1, whilst mineralized materials consistently contain anomalous values of Zn, Pb, Cu and As, generally in the 100 to 1000 ppm range. U-Pb (zircon) dating of granodiorite porphyry at Santa Cruz returned a 178.1 ± 5.6 Ma magmatic crystallization age (Leal-Mejía 2011). Mineralization is considered to be genetically related to the emplacement and cooling history of the porphyry dikes.

ca. 189–182 Ma Southern Ibagué Batholith

Auriferous veins are observed in two distinct districts along the southern and eastern margin of the Ibagué Batholith, at Pacarní and San Luís (Fig. 6.4). Mineralization is hosted in ca. 189 to 184 Ma quartz diorite to quartz monzonite of the southern Ibagué Batholith and within penecontemporaneous intermediate to felsic volcanic flows, breccias and volcanoclastic rocks of the Saldaña Fm. A district-scale association between Au (Ag) mineralization and spatially separate calcite-barite-bearing veins (Mutis 1993) and breccias is observed. The Ibagué Batholith and Saldaña Fm. are unconformably overlain by continuous sequences and isolated roof pendants of late Aptian-early Albian marine sedimentary rocks. In no instance within the southern Ibagué region is hydrothermal gold mineralization documented to cut these sequences. Although vein-type mineralization tends to cluster at Pacarní and San Luis, various additional occurrences are observed along the ~50 km trend which separates the two districts (Fig. 6.4).

Pacarní District At Pacarní, quartz+sulphide veins are exposed in artisanal underground mine workings, distributed along a NE-elongated ca. 8 by 4 km trend. Mineralization is hosted within medium-grained, holocrystalline biotite quartz diorite of the southern Ibagué Batholith for which Leal-Mejía (2011) provided a U-Pb (zircon) age of 188.4 ± 2.4 Ma. A spatial association with late magmatic segregations, including hornblende syenite, aplite and pegmatite dikes, is observed. Individual veins at Pacarní range from 10 to 50 cm in width and can be traced for up to 200 m along strike. They consist of massive, crystalline milky to translucent quartz and contain base metal sulphides up to 10% by volume, including pyrite, chalcopyrite and galena±sphalerite, and local isolated inclusions of molybdenite. Laboratory analysis of mineralized vein material indicates gold concentrations ranging from 5 to 50 ppm. Ag-Au ratios are generally between 1:1 and 2:1, whilst Cu values range from 0.1 to 1.7% and Pb values from 0.1 to 1%. Trace metals such as As and Bi each attain sporadic values as high as 1000 ppm, suggesting the presence of As- and Bi-bearing minerals which have not been identified in hand specimen. Wallrock alteration is generally weak and includes fine potassium feldspar in fractures, overprinted by patchy sericite replacing feldspar and biotite in the host intrusive.

San Luís District At San Luís, Au-Ag mineralization is contained within ca. 10 by 5 km NE-trending area, centred some 50km to the NE of Pacarní (Fig. 6.4). Mineralization is hosted mostly within Saldaña Fm. volcanic and volcanoclastic rocks, although it locally occurs within quartz diorite of the Ibagué Batholith (locally referred to as the San Luís stock). The Saldaña Fm. in and around the district is dominated by medium- to thinly bedded volcanic tuffs and agglomerates and lesser flows of dacitic to rhyolitic composition. Auto-brecciated flows and spherulitic textures in rhyolite are observed. Minor andesite flows and localized sedimentary interbeds are intercalated with the felsic volcanics. Crackle brecciation and infilling in the volcanic sequence by chalcedonic quartz and opaline silica is recorded throughout the district.

Gold-silver mineralization at San Luis is revealed as veins, veinlet stockworks and breccias, exposed in historic and active artisanal underground and open-cut exploitations. Individual veins are generally narrow, ranging from 10 cm to 1 m in thickness and can be traced for up to 300 m along strike. Localized zones of loose veinlet stockworking and brecciation, developed around well-mineralized feeder veins, can attain widths of up to 20 m but have limited apparent strike extent. Veins are dominantly comprised of quartz±coarsely crystalline calcite. Quartz is massive and milky, grading to banded fine-grained chalcedonic silica. Crustiform-coloform textures are developed locally. Druzy quartz infillings are observed in voids and open spaces in breccias. Gold-silver values are confined to veins and breccias, where they are correlated with the coarse-grained base metal sulphide assemblage which includes pyrite>galena>chalcopyrite>sphalerite. The appearance of 1 to 3 mm grains of visible native gold within the massive quartz is not uncommon. Mineralized veins and breccias record gold concentrations within hand specimens ranging from 2 to greater than 300 g/t. Observed Ag-Au ratios are generally between 0.5:1 and 3:1, but locally ratios are as high as 40:1. A strong correlation is recorded between galena and gold. Lead values associated with mineralized samples range from 0.2 to 2%. The presence of chalcopyrite also correlates well with gold, with Cu values in analysed samples ranging from 0.02 to 2.5%. Wallrock alteration within the volcanic sequence is variably developed as a function of wallrock composition and proportional to the scale of veining and brecciation. In individual veins, weak to strong silicification occurs in centimetric to decimetric haloes along vein margins, often accompanied by minor chalcedonic veinlet development. Decimetre- to metre-scale argillic zones containing illite and chlorite±adularia, pyrite and calcite encompass the silicified haloes. Veinlet stockworking and brecciation are accompanied by broader zones of alteration of similar composition to those noted above. Fine crackle veinlets, void fillings and replacements of opaline silica represent the final, post-mineral phase of vein and breccia filling.

ca. 180 Ma Sierra Nevada de Santa Marta Batholiths

Bosconia Au-Ag Occurrences Despite the widespread distribution of Jurassic granitoids throughout much of the Sierra Nevada de Santa Marta (Tschanz et al. 1974), documented metalliferous mineral occurrences are actually scarce (Tschanz et al. 1968). At Bosconia, on the southwestern corner of the Sierra Nevada de Santa Marta (Fig. 6.4), Au (Ag) mineralization is hosted within thick-bedded dacite and rhyolite flows and lithic-crystal tuffs of the early-middle Jurassic Guatapurí Fm. (183.3 ± 0.3 Ma, U-Pb (zircon); Leal-Mejía 2011). The northern margin of the Guatapurí Fm. is intruded by the Pueblo Bello-Patillal Batholith (179.8 ± 3.3 Ma U-Pb (zircon); Leal-Mejía 2011). Localized mineralization is developed along a discontinuous northeast trend, as auriferous quartz veins and quartz-filled breccias averaging between 10 and 20 cm thick which can be traced for tens of metres along strike. Quartz is crystalline to sacchroidal. Well-developed crystal terminations are observed in centimetre-scale voids and remnant open spaces. The veins contain coarse-grained pyrite±chalcopyrite±galena aggregates, comprising up to 4% of the total vein-filling phases by volume. Alteration within the volcanic wallrocks is only weakly developed and includes decimetre-scale weak argillic haloes containing illite and minor pyrite±chlorite, which grade quickly to a propylitic assemblage containing chlorite, calcite, epidote and minor pyrite.

In addition to gold mineralization, barite-calcite±fluorite with minor pyrite-chalcopyrite veins and breccias are observed along the Bosconia trend. These structures are associated with strong epidotization of the host Guatapurí volcanics. A broad spatial association between the auriferous quartz-sulphide and barite-calcite structures is observed at the district scale, but cross-cutting relationships or paragenetic link between them has not been established. In general, the Bosconia Au-Ag occurrences exhibit characteristics of epithermal mineralization exposed in the higher temperature root zone of the hydrothermal system. Notably, a similar combination of auriferous and barite-bearing structures hosted within similar aged plutonic rocks and associated volcano-sedimentary strata is observed in the gold districts of the southern Ibagué Batholith.

ca. 166 to 155 Ma Northern Ibagué Batholith

In the northern Ibagué Batholith, Cu (Au-Ag) mineralization associated with late Jurassic magmatism manifests as a discontinuous 15 km N-S belt of skarn deposits located to the east of the town of Rovira (Fig. 6.4). The best known occurrences, Mina Vieja, Salitre and El Sapo (Villegas 1987), are located along the contact between the easternmost margin of the northern Ibagué Batholith and Triassic basal conglomerates and marine limestones of the Luisa and Payandé Fms., respectively. The Luisa Fm. contains broad areas of hornfelsing, silicification and disseminated pyrite. The Payandé limestones are recrystallized and host patchy white, coarse-grained crystalline calcite. At Mina Vieja and El Sapo, mineralization is developed as erratic bodies containing a coarse-grained assemblage of calcite, magnetite and hematite with garnet, diopside and minor epidote. Chalcopyrite and pyrite±sphalerite are the ore minerals. Mina Vieja, the largest of the known mined bodies, contained an estimated 400,000 t resource grading 1.7% Cu, 1 g/t Au and 33 g/t Ag (Villegas 1987). Three kilometres west of El Sapo, the Pavo Real Au occurrence contains disseminated and fracture-controlled gold-pyrite mineralization hosted within silicified and hornfelsed conglomerates and sandstones of the Triassic Luisa Fm. Meinert et al. (2005) note that metal-rich skarns are most commonly the product of interaction between magmatic and crustal rocks. Metal and mineralogical criteria vs. lithogeochemical data presented by these authors suggest that the northern Ibagué skarns fall within the Cu (Au) classification. Based upon available information, the skarn bodies are spatially constrained to the intrusive contact between the northern Ibagué Batholith and late Triassic Payandé Fm. Although this sector of the Ibagué Batholith has not specifically been dated, radiometric age dates (U-Pb, zircon) of samples to the west and north cluster in the ca. 157 to 152 Ma range (Leal-Mejía et al. 2018). Early Cretaceous marine sedimentary rocks unconformably overlay late Triassic-Jurassic volcano-plutonic rocks throughout the northern Ibagué region. The northern Ibagué skarns are herein assigned a late Jurassic age and are considered coeval with the emplacement of the northern Ibagué Batholith.

6.5.2.3 Metallogeny of the Culminant Bolivar Aulacogen and the Valle Alto Rift

The extensional regime of the Bolívar aulacogen culminated in the latest Jurassic to Cretaceous with the cessation of subduction-related metaluminous calc-alkaline arc-related magmatism and the opening of the Valle Alto rift (Cediel and Cáceres 2000). This event was marked by deep continental rifting and subsidence, the invasion of the Cretaceous seaway and the deposition of marine and epicontinental sequences over extensive areas of the Central Tectonic Realm (including the Cajamarca-Valdivia terrane), the Maracaibo Sub-plate and the continental platform of the Guiana Shield. A brief hiatus in the extensional regime is recorded as a regional Lower Aptian erosional gap (Cediel et al. 1994; Cediel and Cáceres 2000; Sarmiento 2018). Resumed regional extension and subsidence terminated in the late Cretaceous with a shift of tectonic regime to transpressional during the onset of the Northern Andean orogeny (Cediel et al. 2003).

The axis of the Valle Alto rift is marked by Colombia's Eastern Cordilleran basin, which contains up to 6 km of Cretaceous marine deposits characterized by a transgressive sequence of basal, restricted marine mudstones, carbonates and evaporates overlain by progressively deeper water, reduced (carbonaceous) shales and mudstones, deposited in at least four diachronous sub-basins (Sarmiento 2001). Small volumes of compositionally heterogeneous rift-related alkaline and tholeiitic mafic intrusions mark periods of maximum extension, subsidence and sub-basin development (Fabre and Delaloye 1983; Vásquez et al. 2010). The mafic intrusions range in age from ca. 136 to 74 Ma (Fabre and Delaloye 1983; Vásquez et al. 2010). Lithogeochemical and isotopic data published by Vásquez et al. (2010) demonstrate the mantle-derived character and variable degrees of LREE enrichment and contribution of old crustal material to the parent melts. The oldest intrusions (Pacho, ca. 136 Ma) plot in the field of "continental basalts", reflecting the continental character of the early rifted crust beneath the Eastern Cordillera, whilst the younger intrusions reveal lithogeochemical and isotopic data which is progressively more ocean like (Vásquez et al. 2010). The Cenozoic history of the Eastern Cordillera is marked by regressive marine and increasing continental-derived and freshwater deposits. Punctuated uplift-related unconformities are recorded in the Eocene, Oligocene and Miocene, marking various phases of basin inversion during the Northern Andean orogeny (Cediel and Cáceres 2000; Cediel et al. 2003). Elsewhere in Colombia, additional Cretaceous marine volcano-sedimentary deposits are found as localized erosional remnants (e.g. San Pablo, Segovia Soledad Fms.), in the Central Cordillera (Fig. 6.7).

Eastern Cordillera Mineralization

Three groups of mineralization hosted within the Cretaceous through Eocene strata of the Eastern Cordillera are included in our time-space analysis (Figs. 6.7, 6.8 and 6.9). These include (1) emerald mineralization, (2) oolitic oxide facies Fe formation deposits and (3) Zn-Pb-Cu-Fe (Ba) base metal sulphide occurrences.

Eastern Cordillera Emerald Deposits

The emerald deposits of Colombia's Eastern Cordillera have been mined since pre-Colombian times. The deposits are of world-class calibre and are considered the source of the world's finest gems (Banks et al. 2000). As such, Colombian emeralds have been studied from the gemstone to district scale, and numerous modern technical publications addressing their geological, chemical, isotopic and structural evolution, ore mineralogy and hydrothermal paragenesis and age and origin are readily available (see Ottaway 1991; Cheilletz et al. 1994, 1997; Giuliani et al. 2000; Banks et al. 2000 and Branquet et al. 2015, and references cited therein). The deposits are hosted within two distinct belts along the eastern and western margins of the Eastern Cordillera, each bound by a polyphase zone of thrust faulting. Host rocks include siliceous and carbonated black shales and dolomitic limestones of Lower Cretaceous (Berriasian through Hauterivian) age. The deposits are of hydrothermal origin and are epigenetic with respect to their host rocks. Emerald + pyrite-carbonate-albite±quartz-fluorite-parisite-sphalerite and bitumen are contained within mineralized pockets within stratiform tectonic breccias, associated with zones of faulting,

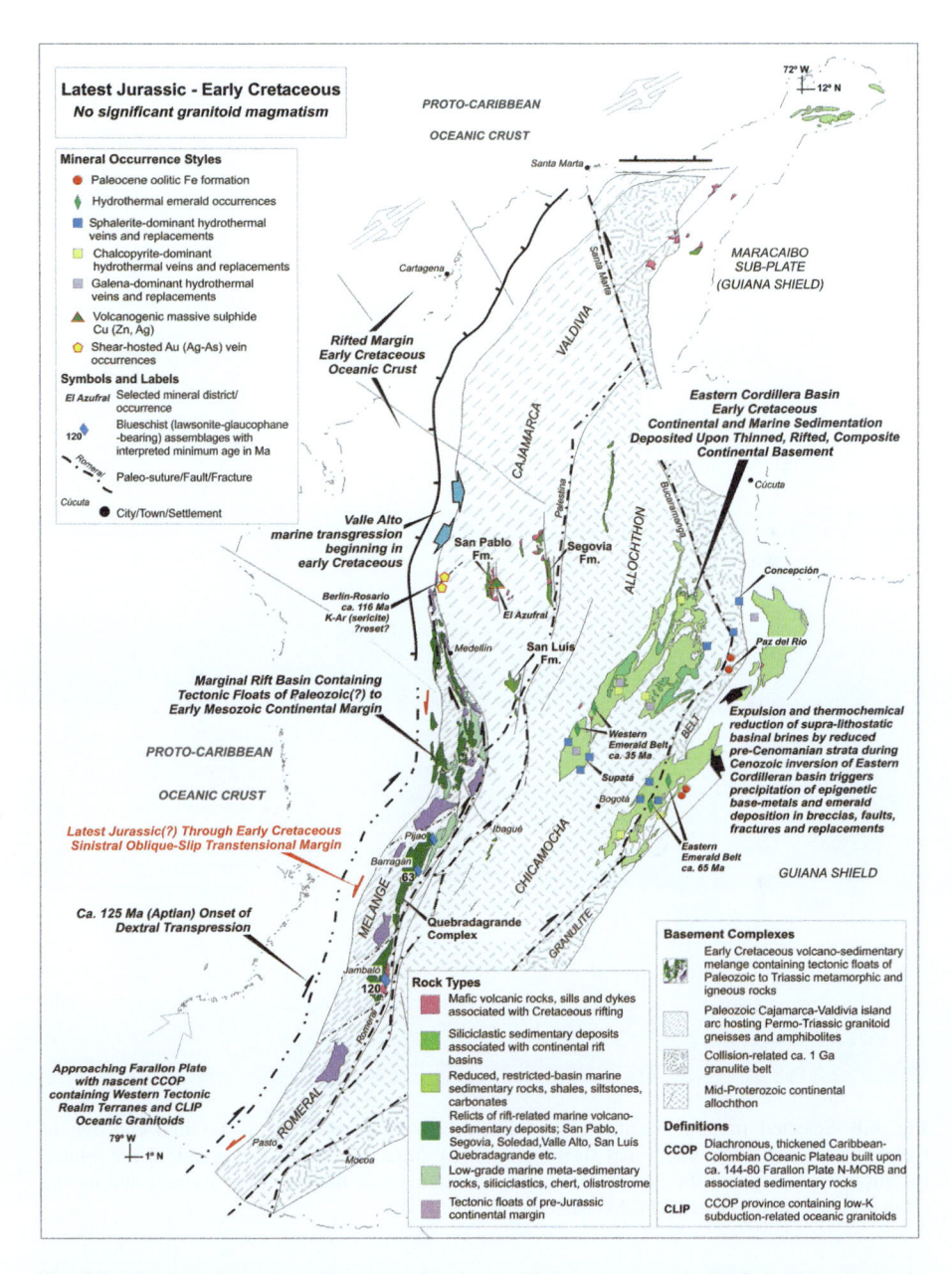

Fig. 6.7 Mineral occurrences of interpreted early to mid-Cretaceous age in the Colombian Andes, in relation to tectonic setting and selected major litho-tectonic elements of the same time period. Note the general hiatus in granitoid magmatism throughout the region during the early-mid-Cretaceous

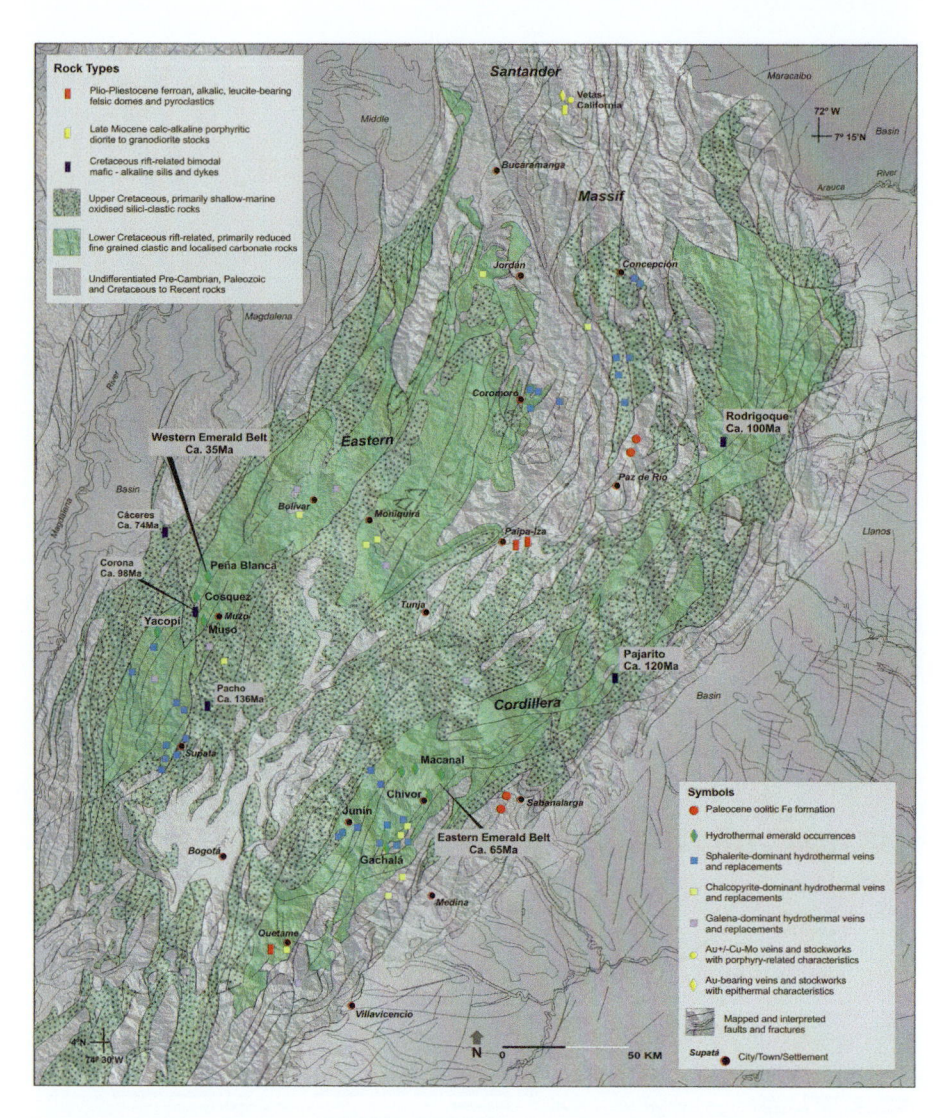

Fig. 6.8 Selected mineral occurrences of interpreted Cretaceous through Oligocene age in Colombia's Eastern Cordillera and surrounding area. Note the absence of granitoid arc-related metallogeny throughout the region. Physiographic features of the map area are revealed by the 30 m digital elevation model (DEM) base image

brecciation and intense fluid-rock interaction, including metasomatic alteration and the development of albitites with epigenetic calcite, dolomite, pyrite, micas and quartz (Cheilletz and Giuliani 1996). $^{40}Ar/^{39}Ar$ (mica) dating indicates the eastern belt (Chivor, Macanal, Gachalá) formed at ca. 65 Ma, whilst the western belt returns ages ranging from ca. 35–38 Ma (Muzo) to 31–33 Ma (Cosquez). A complex and evolving model involving the migration and mixing of deep (5–6 km), hot (+250 °C), sulphate-bearing, evaporite-derived brines from the root zone of the Eastern Cordillera has been proposed (Giuliani et al. 2000; Banks et al. 2000). Expulsion of supra-lithostatic fluid caused fracturing and brecciation of the host black shale (Branquet et al. 2015). Thermochemical reduction of sulphate during interaction with organic matter released beryllium, chromium and vanadium into solution and led to wallrock alteration and the growth of mineral infillings in veins and breccias (Giuliani et al. 2000). Temperature and pressure at the time of mineralization have been estimated at 290–360 °C and 1.12–1.06 kbar (Cheilletz et al. 1994).

Oolitic Fe Formation, Paz de Río

Kimberley (1980) described oolitic shallow-inland sea iron formation of Eocene to Miocene age which occurs in at least four areas of northwestern South America, including Paz de Río, Sabanalarga, Cúcuta and Lagunillas (Venezuela). At Paz de Río and Sabanalarga, in Colombia's Eastern Cordillera (Fig. 6.8), commercially exploited oolitic iron formation is found near the base of the 1,400 m-thick, late Eocene, Concentración Fm. The Fe-rich beds vary from 0.5 to about 8.0 m in thickness. They strike ca. N30E although the structural orientation of the beds varies considerably due to post-depositional block faulting. Maximum east-west outcrop width of the iron formation near Paz de Río is 8 km. Kimberley (1980) notes that the iron formation is thickest near a faulted edge of the outcrop belt, and he postulates that the original extent was probably significantly greater than that preserved in outcrop. Typical iron formation contains variable admixtures of hematite, goethite, siderite and chamosite ± pyrite, containing from ca. 30 to 50% total Fe. The Paz de Río iron formations are interpreted to have formed through the precipitation of Fe within transgressive, oxygenated nearshore bar and beach sediments, deposited in a landlocked or shallow-inland sea (Kimberley 1980).

Eastern Cordilleran Zn-Pb-Cu-Fe (Ba) Base Metal Sulphide Occurrences

Widespread and numerous and base metal sulphide occurrences are known within the Eastern Cordillera (Fig. 6.8). Although some of these have been historically exploited (e.g. Wokittel 1960), few have received modern-day exploration, evaluations regarding their economic potential or academic studies pertaining to their origin and paragenesis. The primary purpose of this brief review is to draw attention to the Eastern Cordillera as a potentially overlooked base metal province. Occurrence location data presented in Figs. 6.7 and 6.8 is taken from the compilation works of Fabre and Delaloye (1983), Mejía et al. (1987), Alvarez (1987) and Mutis (1993).

As noted, the Eastern Cordilleran is comprised of a rift-related sedimentary basin containing thick sequences of Cretaceous transgressive marine sandstones, evaporates and carbonates, carbonaceous siltstones, shales and mudstones, overlain by lesser transitional to continental Cenozoic sediments. Localized ca. 136 to 74 Ma alkaline and tholeiitic mafic intrusions intrude the Valanginian to Campanian section of the basin, marking periods of maximum extension and basin subsidence (Fabre and Delaloye 1983; Vásquez et al. 2010). The basin was structurally inverted during the Cenozoic via the reactivation of pre-existing structural discontinuities (e.g. Sarmiento 2001). Uplift, mostly during the Miocene, was the result of dual northeast-directed and northwest-directed transpressive stresses, resulting in the development of divergent thrust fronts on either side of the Eastern Cordillera (Geotec Ltd 1996; Cediel et al. 1998, 2003; Cediel and Cáceres 2000).

The generalized distribution of base metal sulphide occurrences within the Eastern Cordillera is shown in Figs. 6.7 and 6.8. Based upon geologic-tectonic setting, field observations and literature descriptions, the Zn-Pb-Cu-Fe (Ba) sulphide occurrences throughout the region fall within the broad sediment-hosted base metal class of deposits, with demonstrable attributes of the shale-hosted, sedimentary-exhalative (SEDEX) and Mississippi Valley-type varieties (Leach et al. 2005). The majority of the occurrences are of an epigenetic nature with respect to the host strata, forming replacements and breccia bodies hosted in carbonate rocks, and mantos, replacements, structurally/stratigraphically controlled breccias, fault-controlled vein sets and tectonic vein arrays within agillaceous and siliciclastic sedimentary rocks. Local occurrences of stratiform lenses of sulphide finely intercalated with argillaceous sediment are also observed.

Figure 6.8 reveals the stratigraphic distribution of base metal sulphide occurrences in the Eastern Cordillera modified from the seminal work of Fabre and Delaloye (1983). Based upon numerous known manifestations, these authors note that base metal sulphide occurrences are abundant within uppermost Jurassic (Titonian) through Cenomanian strata throughout the entire Eastern Cordilleran basin but in no instance are strata younger that the Cenomanian known to host base metal sulphide mineralization. They observe that, although limestone is the preferred host to mineralization, occurrences are also found within sandstones and the voluminous carbonaceous shales and siltstones that dominate the Lower Cretaceous stratigraphy of the Eastern Cordillera. They demonstrate a similitude between the structural/stratigraphic distribution of the eastern and western belt emerald occurrences and base metal accumulations along the eastern and western margins of the Eastern Cordillera (Figs. 6.7 and 6.8). Fabre and Delaloye (1983) document a broad temporal and district-scale spatial relationship between mineralization and the ca. 136 to 74 Ma alkaline and tholeiitic mafic intrusions (Fig. 6.8), observing that the cessation of rift-related magmatism closely coincides with the apparent cessation of base metal mineralization. They suggest a genetic model invoking hydrothermal activity related to mafic magmatism in the remobilization of metals contained within the Lower Cretaceous sediments and the deposition of base metal sulphides within epigenetic structures within Cenomanian and older host rocks.

Supatá Zn (Cu) Occurrences Given the paucity of recent published information, we now describe mineralization located near the town of Supatá, in order to demonstrate some of the salient features of sediment-hosted sulphide mineralization in the Eastern Cordillera.

Stratigraphy at Supatá consists of two informal members of the Villeta Group of broadly Barremian-Aptian age. These members include (1) a lower sequence of black (carbonaceous), locally cherty, shale and mudstone with minor siltstone which regionally attains stratigraphic thicknesses of over 2000 m and (2) an upper, generally oxidized series of more thinly bedded siltstones, laminated fine-grained wackes and calcareous and bioturbated sandstones with local shell beds. ca. 135 Ma (Vásquez et al. 2010) mafic dikes and sills are observed to cut at least the lower member 5 to 10 km to the north of Supatá, near the town of Pacho. The structural setting at Supatá is dominated by a N-S-oriented, south-plunging, open antiform. Penetrative deformation is registered as an S1 axial plane foliation within the lower member shales, whilst a brittle spaced cleavage is observed in the upper member. Local N-S shearing and the formation of mineral-filled tension joints are also recorded within the lower member. Sediment-hosted Zn (Cu) is presently known only within the lower Villeta Gp. member at Supatá. It is observed at two locations: in an abandoned mine area, some 2 km north of the Supatá townsite, and along the La Batea Creek, some 3 km south of the townsite.

At the abandoned Supatá mine, mineralization does not outcrop, and underground access is now inhibited by collapse. Mineralization is contained in at least one N-S-striking manto and associated fracture fillings. The manto, hosted within lower member Villeta Gp. black shale, ranges up to 4 metres in thickness. Strike continuity is unknown. Abundant mineralization sampled from an abandoned ore pile consists of massive, coarsely crystalline sphalerite with minor inclusions of chalcopyrite and pyrite. The sphalerite is loosely brecciated and cut by minor veinlets of druzy quartz, calcite and fine pyrite. Analysis of a representative sample from the ore pile returned 57.9% Zn, 0.4% Cu, 4.0% Fe and 185 ppm Pb.

In the La Batea Creek, sulphide mineralization is well exposed along the course of the stream cut. Two in situ varieties are observed. Type 1 includes a series of discontinuous stratiform lenses of fine-grained, recrystallized sphalerite intercalated with carbonaceous shale, containing minor pyrite, calcite, quartz and possibly fine-grained galena. The lenses are oblique to and are cut by the S1 foliation and are interpreted to represent an S0 surface, concordant with original bedding. We interpret the lenses to represent SEDEX mineralization deposited contemporaneously with lower Villeta Gp. shale sedimentation. Type 2 mineralization manifests as foliation parallel to cross-cutting brittle, conjugate shear and A-C-type joint fillings up to 5 cm in thickness. The joints are filled with pure fibre-crystalline, low-iron (yellow) sphalerite and minor quartz. Type 2 mineralization clearly post-dates the S1 foliation. Additional mineralized float fragments observed along the creek bed include breccias containing pyritized black shale fragments with sphalerite, calcite and quartz and additional fragments of massive coarsely crystalline siderite.

An absence of technical studies limits interpretations regarding the genesis of base metal sulphide mineralization in the Eastern Cordillera. Notwithstanding, field observations recording the attributes of occurrences at Supatá and elsewhere combined with the abundance of detailed investigations regarding the tectonic, structural and thermal evolution of the Valle Alto rift and Eastern Cordilleran basin (e.g. Fabre 1987; Sarmiento 2001) and the nature and genesis of its contained emerald deposits (references previously cited) permit speculation regarding the metallogenic evolution of Colombia's Eastern Cordilleran base metal province.

The latest Jurassic through Cretaceous and Cenozoic tectonic evolution of the Eastern Cordilleran rift basin is key to the understanding of its observed metal/mineral deposits and its metallogenic potential. Fabre and Delaloye (1983) observed the widespread distribution of base metal sulphide occurrences and spatially related mafic magmatism and argued that large-scale processes were responsible for the magmatic and metallogenic evolution of the basin. Diachronous mafic magmatism and thermal subsidence in the Eastern Cordillera is considered to mark periods of maximum rifting and mantle melting beneath the most subsiding segments of individual sub-basins (Vásquez et al. 2010). Basin subsidence was accompanied by active syn-sedimentary normal (growth) faulting (Sarminento 2001). Early basin evolution was characterized by the deposition of transitional and shallow marine siliciclastics and carbonates, followed by rapid subsidence and the deposition of thick sequences of carbonaceous siltstones, shales and mudstones (Cediel et al. 1994; Cediel and Cáceres 2000). The carbonaceous nature of much of the Valanginian through Albian argillaceous sediments suggests anoxic conditions and limited circulation in the Eastern Cordillera sub-basins (Sarmiento 2018). Basin inversion and the migration of supra-lithostatic fluid from the root zone of the Eastern Cordillera were facilitated by reactivation of rift-phase, syn-sedimentary growth faults, in various phases during the Cenozoic (Cheilletz et al. 1994, 1997; Sarmiento 2001; Branquet et al. 2015).

Many of the features associated with the shale-hosted base metal sulphide occurrences in Colombia's Eastern Cordillera are represented within models for SEDEX base metal sulphide occurrences as reviewed by Goodfellow et al. (1993) and Leach et al. (2005). SEDEX deposits are characteristic of rifted margins and, more specifically, failed intracontinental rifts. High heat flow and hydrothermal circulation can be linked to contemporaneous magmatic activity, particularly within subsiding basins which contain spatially and temporally associated igneous rocks, as seen in the Eastern Cordillera. Syn-sedimentary faults form important pathways for the ascent of metal-bearing brines from deeper basin aquifers, whilst restricted depressions or sub-basins form important traps for exhaled brines. Syngenetic SEDEX deposits are hosted by reduced, fine-grained siltstones, shales and mudstones and/or carbonate units contained within reduced sediments. The stratiform lenses of fine-grained recrystallized sphalerite intercalated with carbonaceous shale seen in the La Batea Creek at Supatá may be interpreted to represent SEDEX-style mineralization contemporaneous with the deposition of the reduced shales of the lower Villerta Gp. Alternatively, the lenses may be of diagenetic origin. Sediment-hosted deposits are commonly accompanied by disseminated, stratiform Fe (pyrite) and barite mineralization and Fe-rich carbonate (siderite, ankerite) horizons or veining, features which are observed in numerous locations in the Eastern Cordillera (Fabre and Delaloye 1983; Villegas 1987).

Genetic models proposed for the eastern and western emerald belts (e.g. Giuliani et al. 2000; Branquet et al. 2015) may also be evoked to explain many of the attributes the epigenetic sulphide-bearing mineralization seen at Supatá and elsewhere. Indeed, Giuliani et al. (2000) note that except for the presence of accessory emerald, the Eastern Cordillera emerald occurrences are similar to sediment-hosted, stratabound and stratiform base metal deposits. Fracturing and brecciation of host carbonaceous shale and siltstone during basin inversion and the expulsion of mature, supra-lithostatic metal-charged brines could have been accompanied by thermo-chemical sulphate reduction and sulphide deposition in tectonic vein arrays, mantos and breccias, due to interaction with reduced, organic-rich wallrocks. Notably, both emerald deposits and base metal sulphide deposits of the Eastern Cordillera share a common structural-stratigraphic setting, and pyrite and carbonates are abundant in the paragenetic assemblage of the emerald occurrences.

These observations imply that the epigenetic base metal sulphide occurrences of the Eastern Cordillera could significantly post-date the age of their host strata. As such, a temporal/genetic link between early Cretaceous to Cenomanian host strata, the cessation of mafic magmatism and the location of base metal sulphide occurrences, as observed by Fabre and Delaloye (1983), may be largely coincidental. Emerald mineralization took place at ca. 65 Ma and 38–31 Ma (eastern belt and western belt, respectively; Cheilletz et al. 1994, 1997) and post-dates the rift-related emplacement of mafic magmatism (ca. 136–74 Ma, Fabre and Delaloye 1983; Vásquez et al. 2010). The perturbance of K-Ar and $^{40}Ar/^{39}Ar$ systematics for some mafic intrusives at ca. 66 Ma is recorded as "alteration" by both, Fabre and Delaloye (1983) and Vásquez et al. (2010), and suggests that a basin-wide dewatering event took place at about this time. The localization of base metal sulphide (and emerald) deposits in pre-Cenomanian strata appears more an artefact of the structural evolution of the basin and of the affinity for mineral deposition triggered by the reduced carbonaceous composition of the pre-Cenomanian host rocks than of the actual age of the host strata or of a direct link with mafic magmatism.

San Pablo Fm. Cu (Ag, Zn) Occurrences

Also developed within the context of the Valle Alto rift are the Santa Elena Cu (Zn, Ag) massive sulphide occurrences hosted within the San Pablo Fm., located near the town of Guadalupe, Antioquia (Figs. 6.7 and 6.10). The San Pablo Fm. is constrained to a N-S-trending, ca. 33 km-long by 8 km-wide erosional relict of mixed lower Cretaceous rocks, dominated by basalt and bas-andesite to the west and siliciclastic rocks, including sandstones, siltstones, shales and minor cherts, to the east (Hall et al. 1972; González 2001). Gabbro through peridotite sills and dikes (González 2001) within the mafic portions of the volcano-sedimentary package suggests ophi-olitic affinities. The eastern and southern contacts of the San Pablo Fm. are intruded by the mid-Cretaceous Antioquia Batholith (Feininger et al. 1972; Leal-Mejía et al. 2018), whilst to the north it rests conformably upon metamorphic basement of the Cajamarca-Valdivia terrane (González 2001). Both the San Pablo Fm. and the intrusive rocks are cut by subvertical, NE- through E-W-striking shear zones.

Cu (Ag, Zn) mineralization at Santa Elena outcrops in three localities, in the El Azufral (Ortiz 1990), El Arroyo and San Julian creeks. Mineralization is best exposed at El Azufral, in ENE-striking structural zones containing massive to locally laminated (sheared), fine-grained mixtures of pyrite, pyrrhotite and chalcopyrite with minor bornite (supergene?), quartz and magnetite. Mineralization at El Azufral sustains a thickness of 12 m over ca. 80 m of strike length in outcrop. Analyses of representative hand specimens and core samples of massive sulphide indicate mineralization averages in the 2 to 3% Cu range, from 5 to 25 ppm Ag and 0.01 to 0.02% Zn.

Some controversy exists over the orientation of the El Azufral massive sulphide occurrences, given that they appear to strike obliquely to the regional NNE stratigraphic trend of the San Pablo Fm. and hence not necessarily be of a stratiform or stratabound nature. Notwithstanding, contacts at El Azufral appear structural, and detailed geological mapping is greatly inhibited by dense vegetation cover, deep tropical weathering and latosol development, steep topography with thick colluvial cover and the lack of sub-surface exploration, and hence, the local understanding of the geometry of the El Azufral occurrences with respect to the regional stratigraphic and structural setting has yet to be established. To our knowledge, no detailed technical investigations pertaining to El Azufral mineralization have been published. The fine-grained, massive nature of pyrite-pyrrhotite-chalcopyrite mineralization at El Azufral is typical of volcanogenic massive sulphide mineralization deposited in submarine oceanic environments. Considering the pyrite-pyrrhotite-chalcopyrite mineral assemblage and Cu (Ag, Zn) metal associations at El Azufral, and the siliciclastic-marine mafic volcanic lithologic association of the host San Pablo Fm., within the context of the extensional tectonic environment of the Cretaceous Valle Alto rift, we interpret the El Azufral occurrences to belong to the Besshi-type volcanogenic massive sulphide class of deposits (Slack 1993; Franklin et al. 2005; Morgan and Schulz 2010). Siliciclastic-mafic volcanic suite-hosted subclasses of these deposits, as recorded at El Azufral, are typically formed along rifted continental margins or within intracontinental rifts, at the early stage of separation when a supply of siliciclastic sediment is readily available (Slack 1993; Morgan and Schulz 2010). It is clear that the application of such a model at El Azufral must take into account post-depositional tectonism associated with regional Meso-Cenozoic deformation of the San Pablo Fm.

6.5.3 Cretaceous-Eocene Metallogeny: The Early Northern Andean Orogeny

Prolonged regional extension related to the Bolivar Aulacogen terminated in the mid- to late Cretaceous (Fig. 6.9). Tectonic plate reconfiguration in the Pacific regime led to oblique convergence along the Colombian margin and closure of the Bolívar Aulacogen. This shift signalled the onset of the late Mesozoic-Cenozoic Northern Andean orogeny, comprised of a series of punctuated tectono-magmatic events characterized by the generation of subduction-related, calc-alkaline, continental margin and peri-cratonic volcano-magmatic arcs and the sequential approach, collision and accretion of the Western Tectonic Realm allochthonous terrane assemblages of Pacific provenance along the Colombian Pacific and Caribbean margins. The tectonic evolution of Colombia and the Northern Andes during this time was intimately linked to the evolution of the proto-Caribbean basin and to the genesis and emplacement of the Caribbean Plate (e.g. Pindell and Kennan 2001; Cediel et al. 2003; Kerr et al. 2003; Nerlich et al. 2014).

As in the Jurassic, metallogeny during the Northern Andean orogeny is dominated by epigenetic, hydrothermal, volcano-plutonic granitoid arc-related precious ± base metal occurrences formed in the mesothermal pluton, porphyry and epithermal environments, genetically related to the cooling history of spatially associated granitoid magmatism. From an historic production standpoint, vein-type Au (±Ag) deposits, and their associated residual and alluvial derivatives, are the most important deposit types, and as for the Jurassic examples, artisanal exploitation remains active today.

Aside from epigenetic volcano-plutonic arc-related deposits, various important mineral occurrences, including three producing mines, are associated with accreted oceanic volcanic and intrusive rocks contained within terranes of the Western Tectonic Realm. The El Roble-Santa Anita, El Dovio and Anzá volcanogenic massive sulphide deposits are hosted within the Cañas Gordas terrane, whilst the ultramafic bodies which served as protore for the nickeliferous laterites at Cerro Matoso are hosted within MORB basalt of the San Jacinto terrane.

We now review Colombian metallogeny generated within the context of the Northern Andean orogeny, spanning the period from the early-mid-Cretaceous to Pleistocene. Considering the tectono-magmatic assembly of the region during this period, we have opted to present two sets of time-space charts and plan-view maps: for the early Cretaceous through Eocene (the proto- and early Northern Andean orogeny; Figs. 6.9 and 6.10) and the Oligocene through Pleistocene (late Northern Andean orogeny; Figs. 6.12 and 6.13). The first of these periods covers the transition from the Bolívar Aulacogen to the re-establishment of continental arc magmatism and the development of peri-cratonic oceanic island arcs associated with the evolution and NW migration of the Caribbean Plate along the Northern Andean margin. The second covers the final accretionary assembly of the mosaic of terranes comprising the modern-day Colombian Andes and follows the temporal-spatial development of onshore, subduction-related granitoid arc segments during the Neogene, each with its own unique assembly of epigenetic precious ± base metal-rich mineral occurrences.

6.5.3.1 Early Cretaceous Hiatus in Granitoid Magmatism and the Proto-Northern Andean Orogeny

The terminal phase of the Bolívar Aulacogen was marked by culminant extension, marine sedimentation and mafic to intermediate magmatism along the rifted margin of NW South America. During the period, spanning the latest Jurassic through ca. 124 Ma, the Pacific margin of Colombia was under left-lateral transtension (Cediel et al. 1994; Kennan and Pindell 2009) and formed an active depocentre for Berriasian

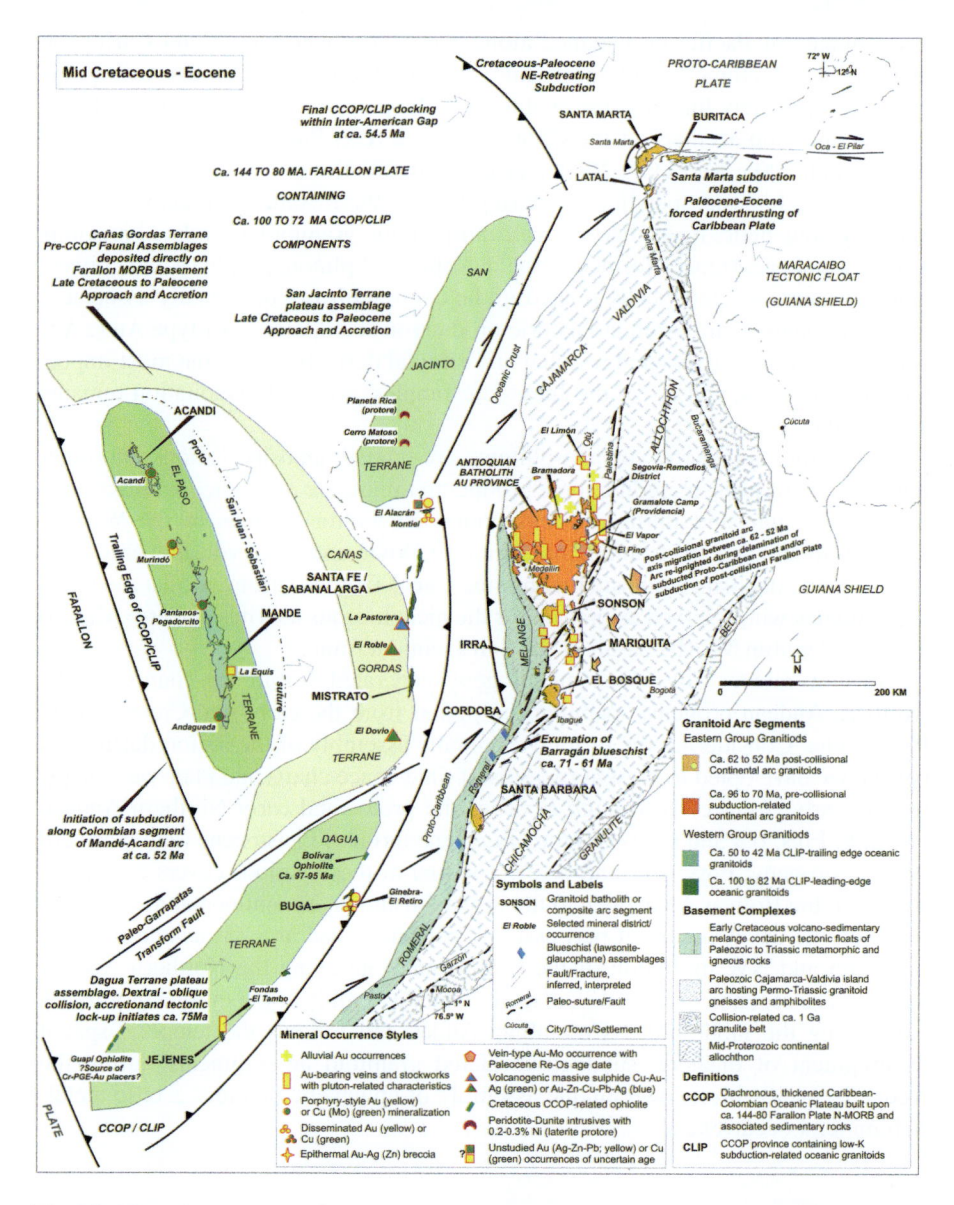

Fig. 6.9 Mineral occurrences of interpreted mid-Cretaceous through Eocene age in the Colombian Andes, in relation to tectonic setting, major litho-tectonic elements and autocthonous vs. allochthonous granitoid intrusive suites of the same time period. Note the onset of transpression and segmented oblique subduction, as well as the appearance of accreted intra-oceanic metallotects along the Colombian Pacific margin

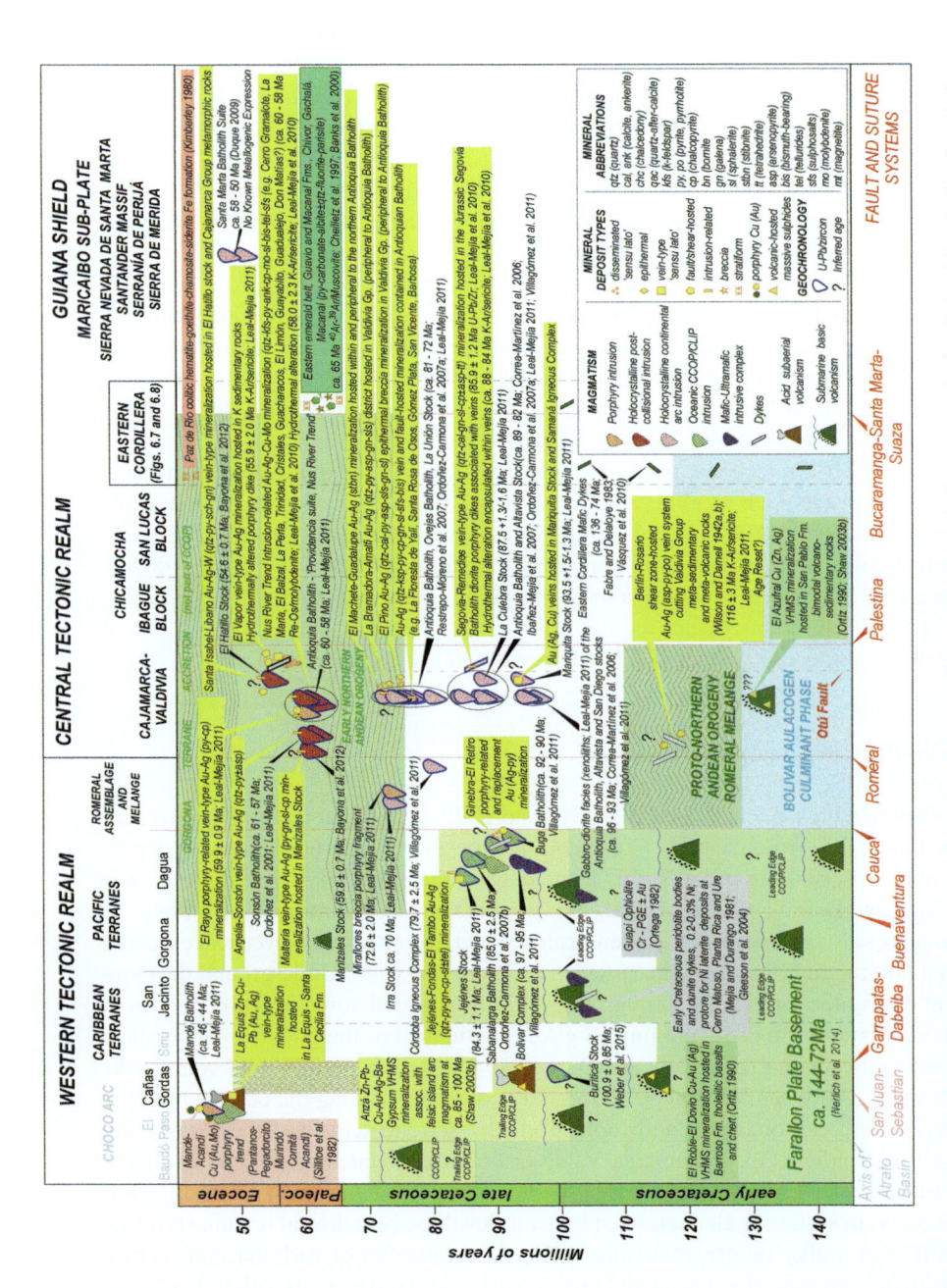

Fig. 6.10 Time-space analysis of mineral occurrences of interpreted early Cretaceous through Eocene age in the Colombian Andes, in relation to tectonic framework, major litho-tectonic elements and orogenic events and the age and nature of granitoid intrusive suites of the same time period. The profile contains elements projected onto an ca. NW-SE line of section through west-central Colombia

through Aptian and Albian sedimentary rocks of continental margin and oceanic affinity, and mixed assemblages of tholeiitic and calc-alkaline basalt and andesite, with associated mafic and ultramafic intrusive rocks (e.g. Quebradagrande Complex, Nivia et al. 1996). This marginal basin also contained disjointed slivers of early Paleozoic and Permo-Triassic metamorphic rocks (e.g. Bugalagrande complex, McCourt and Feininger 1984; Arquia Complex, Nivia et al. 1996) typical of the rifted Northern Andean continental margin during the early Cretaceous (Litherland et al. 1994; Cediel et al. 2003; Vinasco 2018). Plate reorganization associated with the proto-Northern Andean orogeny began in the Aptian (Cediel et al. 1994; Kennan and Pindell 2009), with deep burial, metamorphism and tectonic reworking of the marginal basin assemblages along the Colombian margin (e.g. Orrego et al. 1980; McCourt and Feininger 1984; Maresch et al. 2000; Bustamante 2008; Maresch et al. 2009), accompanied by large-scale dextral-oblique transpressive shearing along the Romeral fault system (Ego et al. 1995). The complex tectonic architecture of the Romeral mélange was established at this time (Cediel and Cáceres 2000; Cediel et al. 2003; Vinasco 2018).

In the early Cretaceous, the Colombian Pacific was thus dominated by a transform margin (Aspden et al. 1987; Cediel et al. 1994; Kennan and Pindell 2009; Wright and Wyld 2011; Spikings et al. 2015). The result is the general absence of subduction-related calc-alkaline continental arc granitoids from ca. 145–95 Ma, suggesting little if any subduction took place beneath the continental margin during this period. This transcurrent regime also seems to manifest in an overall lack of metalliferous deposits which can be temporally linked to this period.

Berlin-Rosario Au (Ag) Vein System

As shown in Figs. 6.7 and 6.10, the only historically significant mineral occurrences which may date from the early Cretaceous are the quartz lode-hosted Au-Ag deposits of the Berlin-Rosario district, located near the town of Briceño, Antioquia (Wilson and Darnell 1942a, b). The vein system extends discontinuously for over 13 kilometres along strike and has been explored and exploited over a vertical extent of some 800 metres. The vein measures up to 25 metres wide. Underground development during the late 1920s through early 1940s recorded some 350,000 ounces of Au production from ore averaging 18 g Au\t, with most of the gold being contained as free grains in quartz (Wilson and Darnell 1942a, b). The mineralized veins strike N-S and dip between 50° and 80° E, broadly constrained along the contact between hanging wall carbonaceous and footwall quartz-sericite schists of the lower early Paleozoic Valdivia Gp. (Cajamarca-Valdivia terrane). Undated diorite bodies cut the schists, and the vein system is cut by undated felsic and mafic dikes. The Berlin-Rosario vein system is characterized by well-developed crack-seal texture (Ramsey 1980), with milky quartz enclosing multiple laminations of carbonaceous schist. Pyrrhotite, arsenopyrite, pyrite and chalcopyrite are the dominant sulphide phases, occurring in fractures in quartz and commonly replacing fragments and laminations of included schist.

Access to the Berlin-Rosario district has been limited to very brief visits over the last few decades, and we are not aware of any modern technical studies addressing the age and paragenesis of this vein system. Thus the age and timing of vein formation vs. gold introduction and the role of the spatially associated granitoids, if any, have yet to be established. Leal-Mejía (2011) provided a K-Ar (sericite) date of 116 ± 3Ma for a sample of sericite-altered schist from a crack-seal lamellae hosted in milky quartz. It is uncertain if this date represents the age of hydrothermal alteration associated with vein formation and/or the introduction of mineralizing fluids or alternatively if it is a reset age associated with tectonic reworking of the Colombian margin during the Aptian.

6.5.3.2 Mid-Cretaceous to Eocene Continental and Intra-oceanic Arc-Related Metallogeny

Metallogeny in the Colombian Andes during the Cretaceous to Eocene demonstrates a strong spatial and temporal relationship with the complex distribution of mid-Cretaceous to Eocene metaluminous, calc-alkaline granitoids contained within the physiographic Central and Western Cordilleras. Leal-Mejía et al. (2018) informally assigned the subduction-related granitoids of this period to two groups: "Eastern" and "Western".

The Eastern group granitoids represent metaluminous, calc-alkaline arc magmatism generated during east-dipping subduction of oceanic (proto-Caribbean, Farallon/CCOP) crust beneath the mid-Cretaceous western Colombian margin. They were emplaced into autochthonous metamorphic basement rocks of the Central Tectonic Realm (mostly the Cajamarca-Valdivia terrane) underlying Colombia's physiographic Central Cordillera (Cediel and Cáceres 2000; Gómez et al. 2015a). Eastern group plutons may be subdivided into pre- and post-collisional granitoids (Leal-Mejía et al. 2018), based upon age, lithogeochemical considerations and timing of intrusion with respect to approach and collision of the Caribbean-Colombian oceanic plateau (CCOP) and accretion of the Dagua, Cañas Gordas and San Jacinto terranes in the late Cretaceous-Paleocene. Epigenetic mineralization of various styles and at least four distinct ages is observed within or peripheral to Eastern group granitoids (Fig. 6.10).

The Western group granitoids were generated in an intra-oceanic environment, and emplaced within oceanic crust, associated with the subduction of the proto-Caribbean and Farallon Plates (i.e. in both cases, oceanic lithosphere of normal thickness) beneath the margins of the thick, buoyant CCOP. The Western group granitoids were subsequently accreted during the impingement of CCOP/CLIP lithosphere along the Colombian continental margin, during at least two related accretionary events. In all cases, the Western Group granitoids may be considered allochthonous with respect to continental Colombia.

Thus, Colombian arc-related metallogeny during the mid-Cretaceous to Eocene presents an accordingly complex time-space distribution of mineral occurrences, in parallel with the age and distribution of arc-related granitoids. Epigenetic, hydrothermal metalliferous deposits, including precious and base metal mineralization, formed in the pluton-related, porphyry and epithermal environments, are associated with both Eastern and Western group granitoids. The important historic producing and known high potential manifestations shown in Figs. 6.9 and 6.10 are now discussed in more detail.

Cretaceous to Eocene Eastern (Continental) Group Granitoid Metallogeny

Recent tectonic reconstructions of the region surrounding NW South America (e.g. Aspden et al. 1987; Cediel et al. 2003; Kennan and Pindell 2009; Wright and Wyld 2011; Spikings et al. 2015; Weber et al. 2015), in conjunction with U-Pb (zircon) age dating of Colombian granitoids (reviewed by Leal-Mejía 2011; Leal-Mejía et al. 2018), suggest initiation of E- to NE-directed, dextral-oblique subduction of the proto-Caribbean and/or Farallon Plate crust beneath the Colombian margin beginning at ca. 100 Ma, resulting in the appearance of metaluminous, calc-alkaline continental arc granitoids beginning at ca. 95 Ma. Within the Eastern group (continental) granitoids, the most important plutonic suites include the pre-collisional Antioquian Batholith and Mariquita stock and the post-collisional Sonsón and El Bosque Batholiths and Manizales and El Hatillo stocks. Hydrothermal Au-Ag (±base metal) mineralization spatially-temporally associated with these plutons is now reviewed.

ca. 96 to 72 Ma Pre-collisional Antioquian Batholith Suite

Within the Eastern group granitoids, the most extensive and important gold province from both a historical and modern-day perspective is hosted within and peripheral to the Antioquian Batholith (Feininger and Botero 1982), and its suite of satellite plutons (Fig. 6.11). Leal-Mejía (2011) and Leal-Mejía et al. (2018) identified four magmatic pulses contributing to the formation of these composite batholiths. Three of these pulses are considered pre-collisional. They include calc-alkaline gabbros and diorites emplaced at ca. 96 to 92 Ma, and two distinct tonalitic to granodioritic suites emplaced at ca. 89 to 82 Ma and ca. 81 to 72 Ma. A final ca. 61 to 58 Ma tonalite to granodiorite pulse, which produced minor stocks contained within the Antioquian Batholith, is considered post-collisional.

Dozens of historically productive and presently active artisanal gold occurrences are spatially related to the Antioquian Batholith and satellite plutons (Mejía et al. 1986; Villegas 1987; Mutis 1993). Three gold metallogenic events related to the Antioquian Batholith suite, including at ca. 89–85 Ma, ca. 81–72 Ma and ca. 62–58 Ma, have been interpreted by Leal-Mejia et al. (2010) and Leal-Mejía (2011).

ca. 89–82 Ma Pre-collisional Phase Granitoids of this age are genetically related to district-scale vein-type gold mineralization in the Segovia-Remedios district, where numerous auriferous veins are recorded in an ca. 25 km belt extending from south of the town of Remedios to north of the town of Segovia (Figs. 6.9 and 6.11).

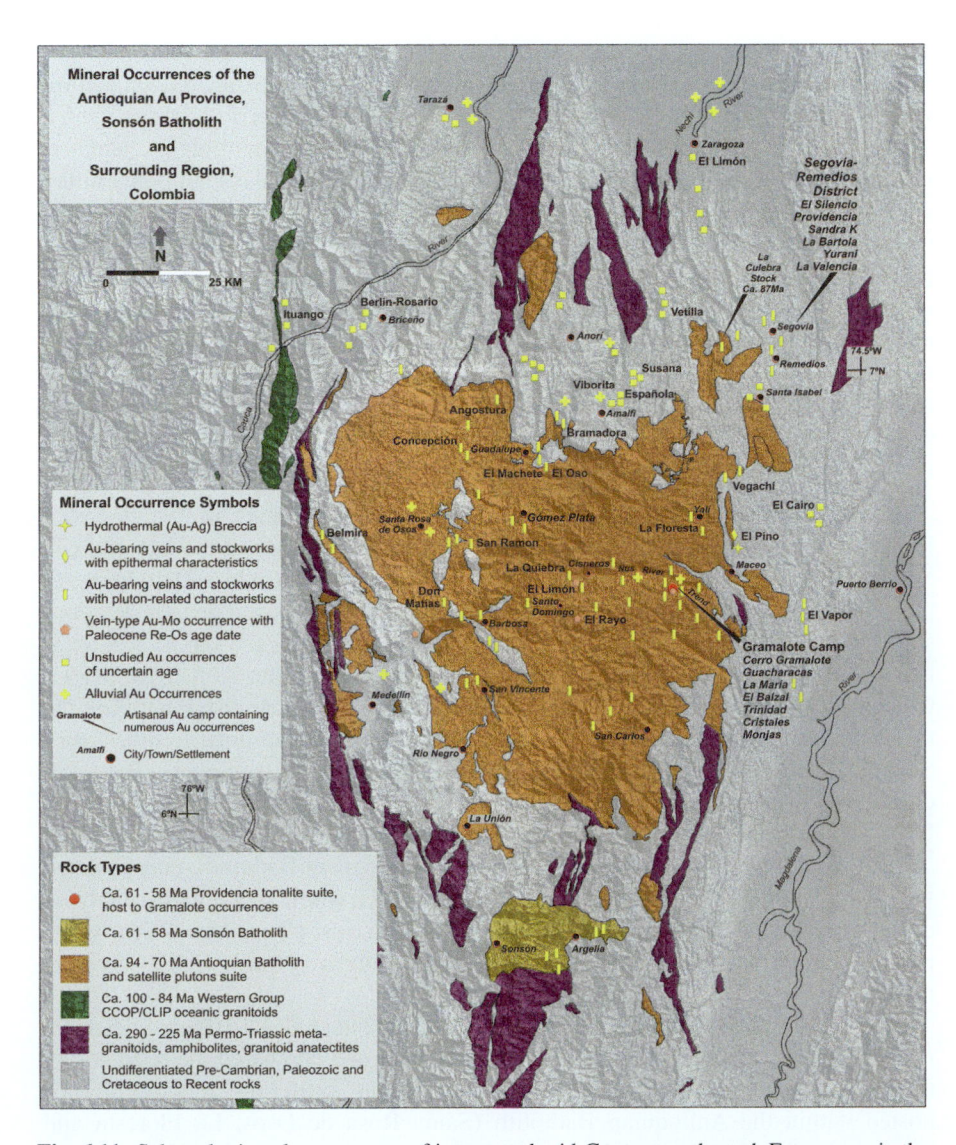

Fig. 6.11 Selected mineral occurrences of interpreted mid-Cretaceous through Eocene age in the Antioquian Batholith Au province and surrounding area of the Colombian Andes, in relation to granitoid intrusive rocks of the same approximate time period. Physiographic features of the map area are revealed by the 30 m digital elevation model (DEM) base image

Veins are mostly hosted within hornblende-biotite diorite of the Jurassic Segovia Batholith; however, some cut meta-pelitic schist of the Cajamarca-Valdivia Group to the west, whilst others are hosted within Cretaceous volcano-sedimentary rocks of the Segovia Fm. to the east. The most important veins in the district, located at the town of Segovia, include El Silencio, Providencia, La Castellana, La Pomarrosa and Sandra K (historic production >100 metric T Au). The veins are relatively narrow, averaging between 0.5 and 1.5 metres in thickness, however, are recognized for their strong continuity both along strike and down dip. Echeverri (2006) presented a paragenetic scheme for vein formation at Segovia. Stage 1 vein filling is characterized by abundant pyrite and sphalerite within a gangue of massive, commonly milky quartz. Pyrite contains small inclusions of pyrrhotite, galena, sphalerite and electrum. Stage 2 involves the fracturing of Stage 1 phases, the replacement of pyrite by galena and additional sphalerite and the deposition of subhedral to euhedral cubes of pyrite. Stage 3 includes additional fracturing and open-space filling with galena and chalcopyrite ± tetrahedrite and argentite. Minor calcite gangue was also deposited at this time. Wallrock alteration associated with vein development includes intense sericitization ± carbonitization and disseminated pyrite, in haloes locally extending for 1 to 2 metres from the vein margin and abruptly giving way to a propylitic assemblage including epidote, chlorite ± calcite and pyrite.

Previous authors have suggested that the auriferous veins of the Segovia-Remedios district are genetically related to the cooling history of the late Jurassic Segovia Batholith (Shaw 2000a, b; Sillitoe 2008). Leal-Mejía et al. (2010) however presented radiometric age and lead isotope data which support a genetic relationship between gold mineralization at Segovia-Remedios and ca. 89 to 82 Ma magmatism in the Antioquia Batholith suite, including the satellite 87.5 ± 1.6 Ma La Culebra stock. These authors observed an intimate spatial relationship between the mineralized veins and granodiorite porphyry and fine-grained dolerite dikes at El Silencio, Providencia and elsewhere in the district. Samples of the Segovia Batholith diorite and a granodiorite porphyry dike collected in the Providencia mine returned U-Pb (zircon) ages of 158.7 ± 2.0 Ma and 85.9 ± 1.2 Ma, respectively. Hydrothermal sericite from a pervasively altered enclave of the Segovia Batholith, encapsulated within the Providencia vein, returned 88 ± 2 Ma (K/Ar, whole rock), whilst analyses of the sericite-altered Providencia porphyry dike returned 88 ± 3 Ma (K/Ar, whole rock). A similarly altered dolerite dike at Sandra K returned 84 ± 3 Ma (K/Ar, whole rock).

In addition, Leal-Mejía et al. (2009) and Leal-Mejía et al. (2010) completed Pb isotopic analyses on pyrite from Segovia-Remedios and compared the results with the Pb isotopic composition of pyrite from various auriferous vein occurrence hosted within the Antioquian Batholith (Santa Rosa de Osos, La Floresta and Gramalote). Results revealed that the Segovia-Remedios and Antioquian Batholith samples plot within the same narrow $^{206}Pb/^{204}Pb$ array, suggesting a similar Pb isotopic source for mineralization in both the Segovia-Remedios and Antioquian Batholith Au occurrences. They noted that the Pb isotopic composition of pyrites from samples from other more distant Au districts plots in clearly distinct arrays (Leal-Mejía et al. 2009).

ca. 81 to 72 Ma Pre-collisional Phase U-Pb (zircon) dates for granitoid collected over much of the main body of the Antioquian Batholith return ca. 81 to 72 Ma ages (Leal-Mejía et al. 2018). Gold mineralization within and along the contacts of the Antioquian Batholith is widespread, with important historic districts located at

Santa Rosa de Osos, Gómez Plata, Guadalupe, Yali, Amalfi and La Bramadora (Fig. 6.11). This region was referred to as the Central Antioquian gold district by Leal-Mejía et al. (2010). At Santa Rosa de Osos (San Ramon), Gómez Plata and Yali, auriferous, sulphide-rich vein systems are hosted entirely within holocrystalline biotite-hornblende granodiorite. At Guadalupe, Cerro El Oso, Amalfi and La Bramadora, mineralization is hosted partially within the batholith but mostly within country rocks including Valdivia Gp. schists and within Aptian-Albian marine sedimentary rocks of the San Pablo Fm. Leal-Mejía (2011) provided descriptions of mineralization from various deposits in the region. Veins hosted within the batholith are rich in coarse-grained base metal sulphides, dominated by pyrite, including galena, sphalerite, chalcopyrite and cubanite and locally containing stibnite, native bismuth and silver-copper-bearing sulphosalts such as polybasite, all deposited in at least two paragenetic stages. Gangue mineralogy is dominated by massive quartz and local bladed calcite. Wallrock alteration includes early silicification and K-spar+pyrite along the immediate vein margins, overprinted by metre-scale haloes containing sericite and disseminated pyrite. Vein-type mineralization hosted within the Valdivia Gp. is observed at La Bramadora and Amalfi (the La Italia, La Susana, La Matilde, El Topacio and La Española workings). These occurrences are dominated by infillings of massive milky quartz, calcite, pyrite and base metal sulphides with arsenopyrite and lead-antimony sulphosalts including boulangerite. Felsic porphyry dikes are spatially related to the veins at La Bramadora although the relationship with mineralization has yet to be established. At El Machete, near Guadalupe, gold (Sb, As) mineralization is contained within quartz veinlets filling widespread centimetre-scale joints and fractures within early Cretaceous quartz-arenite of the San Pablo Fm. Sulphide mineralogy is dominated by pyrite with arsenopyrite and stibnite, as fracture fillings and disseminations. The lead isotopic composition of sulphide minerals from Yali, Santa Rosa de Osos and La Bramadora falls within a narrow range and compares well with that of mineralization from other areas of the Antioquia Batholith (Leal-Mejía et al. 2009; Leal-Mejía 2011). Based upon field relationships, mineral and alteration assemblages and geological and geochemical observations, precious metal mineralization contained within the Central Antioquian gold district is interpreted to be genetically related to the cooling history of ca. 81 to 72 Ma granitoid magmatism comprising the main mass of the Antioquian Batholith.

ca. 62 to 58 Ma Post-collisional Phase

Epigenetic Au (Ag, Cu, Mo) mineralization associated with ca. 62 to 58 Ma (post-collisional) granitoid magmatism hosted within the Antioquian Batholith was documented by Leal-Mejía and Melgarejo (2008), Leal-Mejía et al. (2010) and Leal-Mejía (2011). Mineralization is observed along an E-W elongate corridor transecting the central portion of the main batholith, extending from El Vapor in the east to just east of Medellin in the west (the Nus River trend; Figs. 6.9 and 6.11). Mineralization forms discrete veins and zones of sheeted centimetre-scale veinlets and stockworking, hosted within the ca. 62 to 58 Ma Providencia tonalite or within older phases of the batholith, often associated with aplite, porphyry and pegmatite dikes, as observed at Cerro Gramalote, Cristales, Guadualejo, Santo Domingo and Guayabito. Regardless of the age of the host intrusive, ca. 60 to 58 Ma magmatism exhibits lithogeochemical, alteration and mineralogical features which distinguish it from the earlier metallogenic phases of the batholith (Leal-Mejía et al. 2018). Leal-Mejía (2011) demonstrated a Na-rich "adakite-like" tendency for the ca. 62 to 58 Ma tonalite which hosts Au (Ag, Cu, Mo) mineralization at Cerro Gramalote near Providencia. This historic occurrence has a century-long artisanal production history and currently hosts a multimillion ounce gold resource (AngloGold Ashanti 2015). The tonalite is biotite-rich and contains distinctive mm-scale, clove brown titanite crystals, distinguishing the Providencia tonalite from older biotite-rich phases found throughout the batholith. Mineralization at Cerro Gramalote is associated with sheeted and stockwork quartz and quartz+ankerite veining which was emplaced in two paragenetic stages (Leal-Mejía and Melgarejo (2008); Leal-Mejía 2011). The first stage is dominated by quartz, potassium feldspar, pyrite, molybdenite, chalcopyrite and minor gold, in veinlets which commonly exhibit centimetre-scale potassium feldspar alteration haloes. The second stage of vein filling is again dominated by quartz and pyrite; however it includes variable quantities of sphalerite, galena and chalcopyrite and is distinguished by the presence of a complex assemblage of tellurides and bismuth sulphides and sulphosalts. Most of the gold at Cerro Gramalote was introduced during the second stage, associated with cm- to dm-scale wallrock alteration haloes containing coarse-grained sericite replacing magmatic biotite, and pyrite with ankeritic carbonate, in many parts of the deposit overprinting the Stage 1 potassic assemblage. Sericite-pyrite haloes coalesce to form tens-of-metre-scale altered zones in areas of high quartz vein density. Leal-Mejía (2011) provided radiometric age dates for the tonalite (U-Pb, zircon), Stage 1 molybdenite (Re-Os) and Stage 2 alteration sericite (K-Ar) at Cerro Gramalote, returning ages of 60.7 ± 1.0 Ma, 58.7 ± 0.3 Ma and 58 ± 2 Ma, respectively.

At various localities along the Nus River trend, including Cristales, El Limon cascade, Guadualejo, La Quiebra, Santo Domingo and El Vapor, Leal-Mejia (2011) described vein-type Au-Ag-Cu-Mo mineralization, containing assemblages of native Au, molybdenite, Ag-bearing tellurides and Bi-bearing sulphosalts, associated with potassium feldspar and sericite-pyrite alteration. Molydenite separates from El Limon Cascade, and Santo Domingo returned Re-Os ages of 60.0 ± 0.3 Ma and 59.1 ± 0.3 Ma, respectively. Biotite granodiorite porphyry and quartz porphyry dikes at Cristales and Santo Domingo returned U-Pb (zircon) ages of 61.8 ± 1.3 and 59.9 ± 0.9 Ma, respectively, broadly contemporaneous with U-Pb (zircon) crystallization ages obtained for the Gramalote tonalite, and Re-Os (molybdenite) separates from all of the above-mentioned prospects.

At El Vapor, on the east end of the Nus River trend, Au-pyrite-sphalerite-galena-chalcopyrite-bearing quartz vein arrays are hosted within early Cretaceous clastic sedimentary rocks of the Segovia Fm. Sericite-pyrite-altered granodiorite porphyry dikes intimately associated with mineralization returned a K-Ar (sericite) age of 55.9 ± 2.0 Ma (Leal-Mejia 2011). The close temporal coincidence between U-Pb (zircon) magmatic crystallization ages, molybdenite mineralization and Au+base metal sulphide-/sulphosalt-associated alteration assemblages, backed by field relationships and isotope geochemical data, support a close genetic link between ca. 62 to 58 Ma granitoid magmatism, hydrothermal alteration and Au (Ag-Cu-Mo) mineralization along the Nus River trend.

ca. 62–52 Ma Post-collisional Granitoids to the South of the Antioquian Batholith

Following cessation of subduction-related magmatism in the main phase of the Antioquian Batholith at ca. 72 Ma, and a hiatus of ca. 10 Ma, granitoid magmatism was reinitiated at a greatly reduced rate between ca. 62 and 52 Ma, as recorded in the Providencia tonalite suite, and in the Sonsón, Manizales, El Hatillo, El Bosque and other small, unnamed plutons located in the Central Cordillera to the south of the Antioquian Batholith (Fig. 6.9). This magmatism demonstrates a general southward and eastward migration of the Eastern group continental arc axis. These plutons record a significant reduction in magma volume following the late Cretaceous-Paleocene arrival of the Caribbean-Colombian oceanic plateau and accretion of the Dagua-Cañas Gordas-San Jacinto terranes. Resumption of granitoid magmatism at ca. 62 Ma may be related to delamination of previously subducted proto-Caribbean margin and/or a brief period of subduction of Farallon/CCOP lithosphere (Bustamante et al. 2017; Leal-Mejía et al. 2018), prior to arrival and accretion of the Gorgona terrane in the mid-Eocene (Cediel et al. 2003; Kerr and Tarney 2005). Gorgona terrane arrival ultimately led to a resumed hiatus in subduction-related magmatism in the continental domain extending from the Eocene to the latest Oligocene.

Epigenetic, vein-type Au (Ag) mineralization is observed within and peripheral to the Sonsón Batholith and the Manizales, El Hatillo and other unnamed stocks (Fig. 6.9). Within the ca. 61 to 57 Ma Sonsón Batholith (Ordoñez et al. 2001; Leal-Mejía 2011), veins containing high-grade Au + Ag are exposed in oxidized and mostly abandoned artisanal tunnels located to the NE of the town of Argelia. The veins strike NE, dip steeply and measure from ca. 5 to 25 cm in thickness. They are comprised of over 80% mixed sulphides, including, in approximate order of abundance, pyrite and arsenopyrite with lesser amounts of galena, sphalerite and chalcopyrite and on the order of 10% quartz. Strong sericite alteration is observed in wallrock granodiorite along the vein margins. Similar vein-type mineralization is seen in the Maltería camp near Manizales, hosted within the ca. 59 Ma Manizales stock (Bayona et al. 2012). Sulphide-rich veins cut both the stock and metamorphic basement rocks of the Cajamarca-Valdivia terrane. The age of the mineralization is not well constrained.

Within the El Hatillo stock (ca. 55 Ma; Bayona et al. 2012; Bustamante et al. 2017), near Santa Isabel, milky quartz veins contain Au-Ag, minor base metal sulphides and scheelite. Mineralization is hosted within the stock and extends into Cajamarca Gp. metamorphic rocks. To the north, along the NE-oriented Libano-Falan trend, similar auriferous milky quartz veins contain pyrite, sphalerite and minor galena and chalcopyrite. The age of mineralization in these districts is not well constrained. Maximum ages are dictated by the age of the host plutons, and the spatial relationship between gold mineralization and Paleocene to early Eocene granitoid magmatism in these areas is established.

Cretaceous to Eocene Western (Oceanic) Group Granitoid Metallogeny

Recent tectonic reconstructions for the mid-Cretaceous through Eocene (e.g. Kennan and Pindell 2009; Wright and Wyld 2011; Spikings et al. 2015; Weber et al. 2015) illustrate the appearance of intra-oceanic arcs associated with west-facing subduction of Farallon oceanic crust beneath the approaching Caribbean-Colombian oceanic plateau (CCOP) (Kerr et al. 1996, 1997, 2003). This system of primitive arcs, built upon oceanic plateau basement, has been variably referred to as the "Great Arc" (Burke et al. 1984; Kennan and Pindell 2009), the "Ecuador-Colombia Leeward Arc" (Wright and Wyld 2011) and the "Rio Cala Arc" (Spikings et al. 2015), whilst the composite of CCOP basement containing primitive arc granitoids has been referred to as the Caribbean large igneous province or CLIP (Sinton et al. 1998; Spikings et al. 2015) In Colombia, metaluminous calc-alkaline granitoids belonging to the CLIP are hosted within the Dagua and Cañas Gordas terranes (Cediel and Cáceres 2000; Cediel et al. 2003). Based upon available age dates, these granitoids include the Sabanalarga, Buriticá and Santa Fé (Weber et al. 2015) and Buga (Villagómez et al. 2011) Batholiths and the Mistrato and Jejénes stocks (Leal-Mejía et al. 2018). These plutons return U-Pb (zircon) crystallization ages ranging from ca. 100 to 84 Ma and were accreted to the Colombian margin along with slivers of Farallon-CCOP oceanic lithosphere in the late Cretaceous-Paleocene, during collision of the leading or lateral edge of the CCOP, along the continental margin. The Buga Batholith and Jejénes stock host epigenetic Au-Ag mineralization (Figs. 6.9 and 6.10), which if penecontemporaneous with the host intrusions would have formed in an inter-oceanic to peri-cratonic environment prior to final accretion to the Colombian margin.

ca. 92–90 Ma Buga Batholith

The Buga Batholith is a polyphase pluton (Leal-Mejía et al. 2018) for which U-Pb (zircon) age dating has produced ca. 92–90 Ma crystallization ages (Villagómez et al. 2011). It is in mostly structural contact with meta-tholeiite of the Amaime Fm. and gabbro of the Ginebra ophiolite, although numerous dikes cutting Amaime Fm. along the contact suggest the relation was initially intrusive (Nivia 2001). Gold occurrences within the Buga Batholith are located to the NE of the town of Ginebra. Artisanal Au production is derived from a discontinuous 10 km N-S-trending belt extending north from the principal mining centre of El Retiro. At El Retiro, Au (Ag) mineralization is hosted within a medium-grained biotite tonalite stock and an associated set of felsic dikes, located near the western contact between the Buga Batholith and the Anaime Fm. Neither the mineralized biotite tonalite nor the felsic dikes have specifically been dated, but based upon available mapping, both are included within the confines of the Buga Batholith (Nivia 2001). Strong to pervasive mineralization and alteration are exposed in open-cut artisanal workings as widespread quartz-pyrite veins, veinlets, and stockworks, associated with intense to pervasive sericite-pyrite-carbonate alteration observed almost continuously over an area of some 400 by 400 metres, with various isolated vein occurrences within biotite granodiorite being recorded for up to one kilometre from the main open-cut workings. Mineralization and strong alteration additionally extend into shallowly dipping layers of the Amaime Fm. along steeply dipping to vertical fault-vein feeders and related fractures and along extensive, low-angle, layer-parallel replacements of host Amaime amphibolite by quartz, sericite and pyrite. Pyrite is very abundant as fracture fillings and disseminations throughout the deposit. Minor amounts of galena, sphalerite and chalcopyrite are recorded in some of the larger quartz veins (Pulido 2005). Geological field relationships indicate that the productive quartz-sericite-pyrite-carbonate-altered granodiorite is in intrusive contact to the east with unmineralized, propylitically (epidote-chlorite-quartz)-altered, coarse-grained granodiorite. Radiometric age dating of the biotite tonalite stock and felsic dikes and of the syn-mineral alteration sericite would better constrain the timing relationships between hydrothermal mineralization at El Retiro and the cooling history of the main phases of the Buga Batholith.

ca. 84 Ma Jejénes stock

The Jejénes stock is comprised of a cluster of coarse-grained, low-K, biotite±hornlende tonalite plutons which intrude Diabasico Fm. (Dagua terrane) mafic volcanic rocks along the eastern margin of the physiographic Western Cordillera to the west of the city of Popayán (Orrego and Acevedo 1993). Leal-Mejía (2011) presents an 84.3 ± 1.1 Ma U-Pb (zircon) crystallization age for tonalite hosting Au (Ag) mineralization in the Fondas artisanal mining camp located some 12 km west of the town of El Tambo. Mineralization at Fondas is contained within a NE-trending, 4–5 km-long by ca. 600 m-wide corridor containing anastomosing veins and stockworks, hosted mostly within the Jejénes stock but also within proximal mafic volcanic rocks of the Diabasico Fm. Mineralized veins contain quartz and up to 3% mixed sulphides dominated by medium- to coarse-grained pyrite with minor galena, sphalerite and chalcopyrite and traces of molybdenite. Structures ranging from clusters of cm-scale veinlets to more massive veins measuring up to 70 cm are contained within broad multimeter haloes of strong to pervasive sericite-pyrite alteration replacing the original tonalite. Radiometric age dating of the syn-mineral alteration sericite would better constrain the timing of hydrothermal mineralization at Fondas with respect to the U-Pb (zircon) crystallization age of the Jejénes stock.

ca. 62 to 40 Ma Mandé-Acandí Arc

Also included within the Western group of intra-oceanic granitoids is the Paleocene-Eocene Mandé-Acandí arc assemblage. These rocks represent the most significant expression of granitoid magmatism within the Western group of granitoids.

In Colombia, the assemblage includes the metaluminous, low-K calc-alkaline Mandé and Acandí Batholiths, coeval volcanic and pyroclastic rocks of the La Equis-Santa Cecilia Fms. and hypabyssal porphyry centres located at Pantanos-Pegadorcito, Murindó, Rio Andagueda, Comitá, Acandí, Rio Pito (Panamá) and elsewhere. Recent tectonic models (e.g. Pindell and Kennan 2001; Kennan and Pindell 2009; Montes et al. 2012; Wright and Wyld 2011; Weber et al. 2015), and age and lithogeochemical considerations presented by Montes et al. (2012), suggest that the Mandé-Acandí arc developed as a response to NE-directed subduction of Farallon oceanic crust beneath the trailing edge of the CCOP, which by this time included the thick sequences of oceanic basalt exposed within the El Paso-Baudó terrane. Based upon tectonic reconstructions presented by Cediel et al. (2003), Montes et al. (2012) and Leal-Mejía et al. (2018), the Mandé Batholith, including a suite of penecontemporaneous metal occurrences and slices of CCOP basement, was accreted to the Colombian margin during the mid-late Miocene.

Figs. 6.9 and 6.10 outline the most significant metalliferous mineral occurrences spatially associated with granitoids of the Mandé-Acandí arc. These include a broadly arcuate NNW-oriented trend of porphyry-related Cu (Au, Mo) prospects extending discontinuously for almost 400 km from Río Pito (Panamá) in the north to Río Andagueda in the south and a cluster of volcanic-hosted vein deposits including the historic La Equis Au (Ag, Zn, Pb, Cu) prospect.

Mandé-Acandí Porphyry Trend With respect to the Mandé-Acandi porphyry-related Cu (Au, Mo) trend, some of these prospects were studied and described by Sillitoe et al. (1982). The overall trend of occurrences, which is spatially related to holocrystalline tonalite and granodiorite of the Mandé-Acandí Batholith, was referred to as the "Western sub-belt" by Sillitoe et al. (1982) in relation to other Colombian porphyry-related occurrences of various ages located to the east. Porphyry-style Cu (Au, Mo) manifestations have been recorded along the belt at (from north to south) Río Pito (Panamá), Acandí, Murindó, Pantanos-Pegadorcito, Comitá and Río Andagueda (Figs. 6.9). Sillitoe et al. (1982) suggest a genetic model for the Mandé-Acandí porphyry occurrences involving the generation of subduction-related volcanic arc magmatism in an intra-oceanic setting with arc construction upon oceanic crust.

Published U-Pb (zircon) crystallization ages for Mandé-Acandí Batholith presented by Montes et al. (2015) range from ca. 59 Ma for samples of granitoids from the San Blas Range to the north of Acandí to ca. 50 Ma for granitoids of the Acandí Batholith and to ca. 43Ma for granitoids collected on the southern margin of the Mandé Batholith. The crystallization ages along with petrographic data and field observations reviewed by González (2001) suggest that Mandé-Acandí is a multi-phase batholith and was emplaced over a period of ca. 20 Ma, with an overall younging trend from north to south.

Within the porphyry occurrences, U-Pb (zircon) ages are only available from Pantanos-Pegadorcito. This porphyry complex intrudes holocrystalline granodiorite of the main Mandé Batholith. Pre-mineral porphyritic tonalite returned a crystallization age of ca. 45 Ma, with inheritance ages ranging from ca. 59 to 67 Ma

(Leal-Mejía 2011). This crystallization age compares well with the K/Ar (alteration sericite) age of 42.7 ± 0.9 Ma for mineralized (chalcopyrite-bornite) dacite porphyry from the same occurrence published by Sillitoe et al. (1982). Additionally, the 59–67 Ma inheritance age supports observations recorded by Sillitoe et al. (1982) that an appreciable time interval may have separated the emplacement of certain phases of the Mandé-Acandí Batholith and the generation of some of the mineralized porphyry Cu (Au, Mo) systems. Additional K/Ar ages provided by Sillitoe et al. (1982) include 54.7 ± 1.3 Ma for magmatic hornblende from late mineral tonalite porphyry at Murindó and 48.1 ± 1.0 Ma for sericite-altered tonalite cut by porphyry at Acandí. This last age compares well with magmatic crystallization ages of ca. 50 Ma presented by Montes et al. (2012) for the Acandí Batholith.

Based upon available radiometric age dating, good overall spatial-temporal correlation between ca. 60 to 43 Ma emplacement of the holocrystalline phases of the Mandé-Acandí Batholith and the dated mineralized porphyry centres is established. In reality, however, many details regarding the metallogenetic links between the batholith and penecontemporaneous porphyry centres remain to be established. The Mandé Batholith segment remains especially remote and incipiently explored, and little additional work has been undertaken on this important trend since the investigations summarized by Sillitoe et al. (1982).

La Equis Zn-Pb-Cu (Au, Ag) Prospect Arias and Jaramillo (1987) summarize information regarding the La Equis Zn-Pb-Cu (Au, Ag) prospect. La Equis is hosted within the Paleocene La Equis Fm., about 2 km to the west of the intrusive contact with the Mandé Batholith. The La Equis Fm. is comprised of a series of felsic to intermediate pyroclastics and flows considered to represent volcanism coeval with granitoid magmatism along the Mandé-Acandí arc. At the La Equis prospect, NNW-striking fracture zones and breccias contain base metal sulphides, including, in order of abundance, pyrite, sphalerite, chalcopyrite and galena, hosted within a gangue assemblage containing quartz and barite. Significant values in Au and Ag are recorded. Alteration of the volcanic host is widespread and includes sericite and pyrite as disseminations and fine fracture fillings. To the east, the Mandé Batholith is also altered with disseminations and fine fractures hosting pyrite ± sericite, chlorite and epidote. Surface, diamond drill and underground exploration undertaken at the Progreso and Capoteros occurrences in the 1970s apparently revealed a close relationship between volcanism, intrusion, hydrothermal alteration and tectonism (Arias and Jaramillo 1987). Two genetic models were suggested for the occurrences, the first involving the development of hydrothermal veins related to emplacement of the Mandé Batholith and the second as Kuroko-type mineralization formed in a submarine exhalative environment. The lack of a subaqueous sedimentary component, however, suggests that the La Equis Fm. represents a predominantly subaerial volcanic pile. This is supported by the presence of columnar jointing associated with the subaerial cooling of andesite flows, observed to the east of El Progreso. Based upon field relationships and information presented by Arias and Jaramillo (1987), we interpret the La Equis Zn-Pb-Cu (Au, Ag) occurrences to represent epigenetic vein- and/or manto-type mineralization formed in an intra-oceanic volcanic arc environment, associated with the emplacement and cooling history of the Mandé Batholith.

Metallogeny Related to Cretaceous Oceanic Basement Terranes

Cretaceous CCOP/CLIP volcanic, intrusive and sedimentary rocks, which form basement to the Western group Cretaceous to Eocene granitoids described above, also host important metal and mineral occurrences. These include syn-volcanic Cu-Zn (Pb-Au-Ag) deposits of the massive sulphide type and orthomagmatic Ni and Cr (±PGE) occurrences associated with mafic-ultramafic intrusive complexes. In Colombia, these deposits all share certain characteristics. All were formed in an intra-oceanic environment, associated with CCOP/CLIP magmatism. All were accreted to the Colombian margin at different times during the Northern Andean orogeny (Figs. 6.7 and 6.9) and hence are allochthonous with respect to the Cretaceous Northern Andean margin. Brief descriptions of the most representative examples of these deposit types are now presented.

Guapí Ophiolite, Dagua Terrane Few metal occurrences syngenetic to the mafic-ultramafic rocks of the Diabasico Gp. of the Dagua terrane are known (Figs. 6.9 and 6.10). Ortega (1982) described the Guapí ophiolite striking NNE for over 75 km along the westernmost margin of the physiographic Western Cordillera, between the Iscuandé and Micay rivers. The ophiolite is comprised of layers of basalt, gabbro, orthopyroxenite, dunite and sepentinite with minor fine-grained marine shales. Chromite and magnetite occur both as cumulate layers but also as disseminations and fracture fillings in the mafic and ultramafic lithologies. More importantly, the Guapí ophiolite and similar units form the headwaters to important Au ± PGE-bearing alluvial districts distributed along the Colombian Pacific margin, which have been exploited since pre-Colombian times, the most important of which include (from S to N) Barbacoas, Napí and Timbiquí. The source of the precious metals contained within these alluvial has never been established. Based upon the presence of Au sourced from other circum-Pacific ophiolite complexes considered to belong to the CCOP/CLIP assemblage (e.g. Nicoya-Osa Peninsula, Costa Rica; Berrangé 1992), the Guapí ophiolite may be considered a potential source for these metals.

VMS Occurrences of the Cañas Gordas Terrane Volcanogenic massive sulphide occurrences have been identified at various localities within the mixed volcano-sedimentary basement of the Cañas Gordas terrane (Fig. 6.9). Ortíz (1990) and Jaramillo (2000) described Cu (Au, Zn, Ag) mineralization at El Roble and El Dovio, whilst more recently identified Zn-Pb-Cu (Au-Ag) occurrences at Anza have yet to be well documented and will be described in more detail herein. All of these occurrences are considered to have formed coeval with mafic magmatism in the Pacific domain. All are characterized by massive-textured sulphide mineralization, and all display significant degrees of deformation and dismemberment brought on during accretion of their hosting basement terranes.

Cu (Au, Zn, Ag) mineralization at El Roble (including Santa Anita) and El Dovio (Figs. 6.9 and 6.10) occurs as strata- and fault-bound lenses hosted within altered and tectonized carbonaceous cherts along faulted contacts with tholeiitic basalt. The largest known lens, at El Roble, measures ca. 200 m long by 100 m deep by 45 m thick and has been mined since the 1990s. Host rocks belong to the Cañas Gordas Gp. and have been assigned a middle to upper Cretaceous age. Mineralization consists of predominantly fine-grained, massive pyrite, chalcopyrite and pyrrhotite with recoverable Au±Ag and minor sphalerite. Very fine-grained alternating laminations of sulphides with chert at El Roble suggest an exhalative origin. Localized stringer and breccia (feeder and vent?) zones contain abundant quartz with sulphide, chlorite and minor calcite. The basalts are altered to fine-grained amphibole with overprinting chlorite and calcite. Ortíz (1990) suggests that the deposits were formed in starved euxenic marine basins in which syn-volcanic hydrothermal circulation expelled sulphide- and Au-Ag-bearing hydrothermal solutions at or near the seafloor. He considers the occurrences to demonstrate attributes akin to those documented for volcanogenic-exhalative Cyprus-type Cu-Au deposits (e.g. Franklin et al. 2005).

At the La Pastorera gypsum mine, located some 6 km west of the town of Anzá (Figs. 6.9 and 6.10), Zn-Pb-Cu (Au, Ag) sulphide mineralization is exposed in open-cut and underground workings. As at El Roble and El Dovio, mineralization consists of tectonized lenses hosted within Cretaceous marine volcanic and sedimentary rocks of the Cañas Gordas terrane. Notably, however, in the vicinity of the La Pastorera mine, abrupt changes in the lithologic composition of the Barroso Fm. are observed, from a typically tholeiitic basalt- and chert-dominated assemblage to a series of andesitic and dacitic pyroclastic rocks, including agglomerates, tuffs and volcano-sedimentary breccias, with subordinate siliceous to cherty exhalites and calcareous mudstone layers.

The La Pastorera intermediate-felsic volcanic and pyroclastic package and its contained massive sulphide occurrences have been examined in some detail in the vicinity of the La Pastorera gypsum mine. Three informal stratigraphic units have been recognized. From base to top, these include (1) a series of agglomerates and crystal-lithic tuffs with minor intercalations of chert, calcareous mudstone and basalt. This sequence contains tectonized layers of gypsum/anhydrite with strata- and fault-bound lenses of siliceous, banded and massive polymetallic sulphides and chert. This unit is overlain by (2) crystal-lithic tuffs of intermediate composition which in the immediate (faulted) hanging wall of the gypsum-sulphide package are intensely pyritized and contain pyritic beds or replacement zones up to 3 m thick. Additional intercalations of chert and minor calcareous mudstone also are observed. Finally, (3) a thick sequence of fine tuffs with intercalations of massive to pillowed basalt and minor chert and calcareous mudstone is recorded. This upper unit is in fault contact to the west with more typical basalt-dominated sequences of the Barroso Fm.

Structural geology in the La Pastorera mine area is complex and not well understood. Based upon observed structural-stratigraphic relationships, the Barroso Fm. including the mineralized intermediate pyroclastic sequence is tightly folded and contained within east-vergent fault panels. Stratigraphy near the gypsum mine strikes generally N-S; however, this orientation changes to almost E-W to the north of the mine area. Generally steep dips to both the east and west are suggestive of a series of upright anticlines and synclines which plunge to the south. Field observations and mapping within open-cut workings at La Pastorera suggest that the gypsum and sulphide mineralization is contained within or flanking a N-S-trending, S-plunging fold hinge.

La Pastorera sulphide mineralization is intimately associated with gypsum/anhydrite. In the late 1990s, gypsum and sulphides were exposed within a >300 m longitudinal section along the La Pastorera open pit, however continued mining activities and collapse mostly obscured exposure. Notwithstanding, within this zone, the stratified nature of the volcanic-pyroclastic sequences and gypsum and sulphide mineralization is evident. Both gypsum and sulphides are localized within the intermediate pyroclastic sequence, which exhibits fine disseminated pyrite and strong to intense chlorite-sericite alteration. Semi-continuous caps of pyritized mudstone-tuff and exhalative chert ± barite overlie the gypsum and sulphides. Sulphide bodies attain up to 12 m, averaging 4 to 5 m in thickness. They are lensoid and demonstrate complex structural-stratigraphic relationships with gypsum/anhydrite. Sulphide mineralization is observed to be thickest where the gypsum horizons are thickest, and both gypsum and sulphide horizons are observed to thicken towards the north.

Gypsum mineralization at La Pastorera is considered to be of exhalative origin. It is finely crystalline and contains inclusions and laminations of carbonaceous argillite and pyrite. Sulphide mineralization is comprised of massive to semi-massive and brecciated, finely crystalline to medium- and coarse-grained aggregates of mixed base metal sulphide minerals, including sphalerite, chalcopyrite and galena. Sulphide phases are contained within a finely siliceous and/or gypsum-rich and argillaceous matrix. Sphalerite is Fe-rich. It occurs as more coarsely crystalline masses or within finely laminated sulphide intercalations. Galena generally occurs mixed with sphalerite, colloidal silica and argillaceous materials in laminated finely crystalline exhalite. Chalcopyrite occurs with sphalerite and pyrite, as finely crystalline and medium-grained aggregates, within a highly siliceous matrix. Complex and contorted silica-sphalerite-chalcopyrite laminations are suggestive of soft-sediment deformation. In some instances, gypsum/anhydrite appears to be injected into the sulphide assemblage. Both Au and Ag contents are significant at La Pastorera. Au shows a broad correlation with the presence of galena in hand samples; however, no studies addressing the mode of occurrence of the precious metals are available.

Some 200 m to the south of La Pastorera, in the Aragon, open-cut, gypsum, abundant pyrite-rich mineralization and hydrothermal alteration is observed in black argillaceous sediments. Exposure is incomplete however, and the presence of polymetallic sulphides can only be inferred. Prospecting and exploration to the north of the La Pastorera pit have encountered additional isolated pyritic sulphide blocks and gypsum in outcrop and demonstrate that the mineral system may extend for up to 2 km along strike.

Based upon geological setting, field observations and mineral assemblage and textural considerations, the gypsum-polymetallic sulphide occurrences at La

Pastorera are considered to represent tectonically disrupted precious metal-rich volcanogenic massive sulphide deposits which developed penecontemporaneously with localized intermediate to felsic marine volcanism during the mid-Cretaceous. Regional geological mapping places the occurrences along the axial trend to the south of the Santa Fé Batholith. Related granitoid stocks intrude the Barroso Fm. within 4 km to both the north and south of La Pastorera (Fig. 6.9), and hence a spatial relationship between mineralization and intra-oceanic calc-alkaline arc magmatism at ca. 90 Ma is established. Lithogeochemical and age dating studies of these plutons and of the mineralized pyroclastic sequences at La Pastorera would aid in clarifying potential temporal and/or genetic relationships.

Ni Laterites of the San Jacinto Terrane The nickeliferous laterite deposits at Cerro Matoso (Mejía and Durango 1981; Gleeson et al. 2004), located along the Caribbean margin near the town of Montelíbano, are hosted within MORB-type tholeiitic basalts of the San Jacinto terrane (Fig. 6.9 and 6.10). San Jacinto is considered a fragment of the Cretaceous CCOP and has been obducted along the Colombian margin in the Eocene (Cediel et al. 2003; Kennan and Pindell 2009). Ni concentrations within the Cerro Matoso laterites are residual in origin, having formed by deep tropical weathering, leaching and reprecipitation during subaerial exposure, possibly since the mid-late Eocene (Gleeson et al. 2004). Notwithstanding, the Cerro Matoso deposits constitute a world-class nickel resource. They have been mined since 1981 and as of 2005 have produced 55,000 metric tonnes of high purity, low carbon ferronickel granules per annum, from proven reserves of ca. 40 M tonnes grading 2.4% Ni (Gleeson et al. 2004). The source rock at Cerro Matoso is a pre-late Cretaceous (Gleeson et al. 2004) enstatite-bearing peridotite (harzburgite) (Mejía and Durango 1981). It is cut by small dunite dikes and hosts lenses of serpentinized peridotite containing abundant magnesite veinlets (Mejía and Durango 1981). The peridotite body measures some 2.5 km by 1.7 km and is elongated in a NW-SE direction. The margins of the body are marked by steeply inclined faults (Mejía and Durango 1981) suggesting tectonic emplacement into its present position. Nickel in the protore peridotite averages between 0.2 and 0.3% and is contained principally within olivine (Mejía and Durango 1981). Intense tropical weathering led to decomposition of primary silicate phases and the liberation of nickel into the saprolite profile where it recombined to form secondary hydrous magnesium-nickel silicates (Mejía and Durango 1981). Pre-existing joint and fracture sets formed important conduits for the penetration of meteoric waters during weathering and the redeposition of supergene ore minerals within the saprolite profile. Multiple cycles of profile collapse, residual concentration and secondary enrichment led to local concentrations of Ni up to 9 wt. % (Mejía and Durango 1981; Gleeson et al. 2004).

Pluton and Porphyry-Related Cu-Au Occurrences at Montiel-El Alacrán These prospects are located in NW Colombia to the south of the Cerro Matoso Ni deposits (Fig. 6.9). No published radiometric age dates are available for the granitoids at Montiel-El Alacrán. Porphyry-style Cu-Au mineralization at Montiel-Teheran is associated with hypabyssal porphyry dikes and is in many respects similar to mineralization of Miocene age observed along the Middle Cauca trend (see below). Granitoids at Montiel-El Alacrán intrude mid-late Cretaceous oceanic basement of the San Jacinto terrane, but beyond this, present data does not better constrain the age of magmatism or Cu-Au mineralization. It is difficult to accurately place these occurrences within our time-space analysis. Our tectonic reconstruction supplied in Fig. 6.9 interprets them as having formed within the intra-oceanic to peri-cratonic environment, prior to or during accretion of the San Jacinto terrane.

At Montiel and nearby Teheran, porphyry-style Au-Cu mineralization is observed in open-cut artisanal workings, where deeply weathered, weakly porphyritic, biotite-altered diorite dikes or small plugs containing moderate to intense sheeted and stockworked quartz+magnetite veining are hosted within mafic volcanic basement of the San Jacinto terrane. Mineralization, veining and potassic alteration extend for tens of metres into the mafic volcanic basement rocks.

2.5 km to the SW of Montiel, at El Alacrán, an undated chloritized phaneritic diorite stock intrudes a late Cretaceous volcano-sedimentary succession comprised of meta-greywacke, chert, marlstone, magnetite-rich Fe formation and quartzite intercalated with porphyritic and amygdaloidal andesite, volcanic tuff, breccia and agglomerate (Vargas 2002). The volcano-sedimentary succession is contained within the eastern, W-dipping limb of an open, N-plunging syncline. On the southern flank of the diorite stock, extensive open-cut and underground artisanal Au workings are developed within and proximal to a series of ca. E-W-striking fault- and fracture-controlled veins and breccias. Mineralization extends south for >500 m along the strike of the volcanoclastic sequence, as a series of structural and stratigraphically controlled replacements, disseminations and fault and fracture fillings containing mixed sulphides, magnetite and native Au. Mineralization is best developed where E-W-striking breccias and faults intersect reactive, N-S-striking volcano-sedimentary units, especially those containing magnetite-rich Fe formation, where strong to intense mantiform replacement of magnetite by chalcopyrite and pyrite extends for tens of metres into the volcano-sedimentary sequence. Mineralization gradually diminishes along strike to the south of the diorite stock, where outcropping, ductily deformed and boudinaged layers containing magnetite-rich Fe formation are devoid of sulphide mineralization. Au-Cu mineralization at El Alacrán is spatially related to, and diminishes with distance from, the chloritized diorite stock. The age of the El Alacrán diorite, Au-Cu mineralization and its relationship to the nearby porphyry-related Au-Cu mineralization at Montiel and Teheran has yet to be precisely established.

6.5.4 Late Oligocene-Pleistocene Metallogeny: The Late Northern Andean Orogeny

Following the emplacement of ca. 62–52 Ma Eastern (continental) group post-collisional granitoids, an ca. 30 m.y. hiatus in subduction-related granitoid magmatism is recorded within the continental domain, extending from the early Eocene to the late Oligocene (Leal-Mejía 2011; Leal-Mejía et al. 2018). In this context, there is a general paucity of arc-related metallogeny in continental Colombia, from the mid-Eocene to Oligocene. This prolonged magmatic/metallogenic hiatus is considered to reflect various factors/events associated with tectonic reorganization prior to the reinitiation of east-directed subduction along the Pacific margin. Initially, the "chocking off" of subduction was due to the invasion of the subduction zone by buoyant oceanic CCOP/CLIP (e.g. Dagua terrane) and Gorgona terrane fragments. Continued plate convergence was dominated by dextral-oblique transpression (Aspden et al. 1987; Pindell and Kennan 2001; Cediel et al. 2003) in an overall regime which was not conducive to continued subduction nor to immediate slab break-off and subduction reinitiation. Coupling stress beginning in the Maastrichtian was partitioned into tectonic tightening and reactivation along pre-existing structures, including the Palestina (Feininger 1970), Romeral and Cauca (Ego et al. 1995; Cediel and Cáceres 2000; Cediel et al. 2003) fault systems, and into collision-related uplift of litho-tectonic units throughout the Central Tectonic Realm, as revealed by the development of regional unconformities, the deposition of continental uplift-related epiclastic sequences (Cediel et al. 1994; Cediel and Cáceres 2000) and thermochronological data supporting rapid exhumation in the Central Cordillera between ca. 75 and 55 Ma (Spikings et al. 2015) and extending into the Oligocene (Cediel et al. 1994). From an economic standpoint, this period of uplift led to the unroofing and erosion of hypogene gold occurrences hosted within and surrounding the Antioquian Batholith and to the generation of important alluvial and colluvial deposits contained within perched Paleogene gravel terraces, as historically exploited via hydraulic mining methods near Santa Rosa de Osos, Amalfi (e.g. La Viborita) and along the Nus River valley (Fig. 6.11).

6.5.4.1 Tectonic Framework for the Late Oligocene Through Pliocene

Paleo-tectonic reconstructions following the early Northern Andean orogeny demonstrate that most of the CCOP/CLIP components of the Western Tectonic Realm, including the Dagua, Cañas Gordas, San Jacinto and Gorgona terranes, were loosely in place by the Eocene (e.g. Pindell and Kennan 2001; Cediel et al. 2003; Kennan and Pindell 2009; Leal-Mejía et al. 2018). To the west, the El Paso-Baudó segment of the Chocó Arc (including the Mandé-Acandí assemblages) was located within the peri-cratonic realm along the trailing edge of the Caribbean Plate, but it is important to recognize that reconstructions of the origin, evolution and spatial migration of the Caribbean Plate (including the CCOP/CLIP assemblage) suggest that it had docked within the inter-American gap along the northern South American Plate by ca. 54.5 Ma (Müller et al. 1999; Nerlich et al. 2014).

In this context, the final approach and accretion of the El Paso-Baudó terrane along the NW Colombian margin in the late Oligocene-early Miocene (Duque-Caro 1990; Pindell and Kennan 2001; Cediel et al. 2003, 2010; Farris et al. 2011; Montes et al. 2012) are associated with the N and W migration of the South American Plate, beginning at ca. 25 Ma (Silver et al. 1990; Müller et al. 1999; Farris et al. 2011).

Lithostratigraphic, radiometric, paleomagnetic, structural and thermochronological data presented by Farris et al. (2011) and Montes et al. (2012) demonstrate the S-and-E to N-and-W translation of South America, inciting forced underthrusting of the buoyant CCOP margin, initial clockwise rotation of the Chocó (El Paso-Baudó) segment of the eastern Panamá Arc and closure of the Central American seaway by ca. 15 Ma. Continued westward translation and clockwise rotation led to decoupling of the Chocó segment (including the Mandé-Acandí granitoid arc) from its CCOP root and terrane obduction along the San Juan-Sebastian and Uramita-Urabá fault systems (Fig. 6.12) during the late Miocene (Cediel et al. 2003; Duque-Caro 1990; Montes et al. 2012). Uplift of the Baudó Range and coeval closure of the Atrato Basin are recorded between ca. 8 and 4 Ma (Cediel et al. 2010; Montes et al. 2012).

East-directed subduction of the Farallón Plate beneath the Colombian Pacific margin was not reestablished until the late Oligocene (Pindell and Kennan 2001; Cediel et al. 2003). Leal-Mejía et al. (2018) provide a detailed analysis of the restructuring of the eastern Farallon Plate and the reinitiation of subduction along the Colombian margin during the latest Oligocene-Miocene. Rifting within the Farallon Plate led to formation of the Nazca and Cocos plates at ca. 23 Ma (Lonsdale 2005). Continued rifting within the Nazca Plate along the Sandra Ridge, between ca. 14 and 12 Ma (Lonsdale 2005), formed the Cauca and Coiba microplates, both of which are currently subducting along the Colombian Pacific margin (Fig. 6.12).

Within the continental domain, time-space analysis of U-Pb (zircon) crystallization ages for latest Oligocene through Miocene, and Plio-Pleistocene granitoids and associated volcanic rocks (Figs. 6.12 and 6.13) demonstrate that extensive, composite "Neogene" arc magmatism emplaced within autochthonous metamorphic rocks of the Central Tectonic Realm and accreted oceanic rocks of the Romeral mélange and Western Tectonic Realm (i.e. within the physiographic Central and Western Cordilleras and along the Cauca and Patia valleys) in fact consists of a complex distribution of magmatic arc segments, the location of which is observed to migrate in time and space, in both an overall south-to-north and west-to-east pattern (Cediel et al. 2003; Leal-Mejía 2011; Leal-Mejía et al. 2018). Based upon time-space analysis, at least five distinct calc-alkaline granitoid arc segments associated with latest Oligocene through Miocene subduction along the Colombian trench are identified. Beginning in the south, the ca. 23 to 21 Ma Piedrancha-Cuembi holocrystalline suite and the ca. 17 to 9 Ma Piedrasentada-Berruecos-Buenos Aires-Suarez porphyry suite are associated with subduction of the Cauca segment of the Nazca plate (Fig. 6.12). To the north, the ca. 12–10 Ma Farallones-El Cerro-Dabeiba holocrystalline suite and the ca. 9–5 Ma Middle Cauca porphyry suite are associated with subduction of the Coiba segment. Granitoid magmatism related to the continued subduction of the Cauca segment records eastward migration of arc-axial magmatism and conformation of the modern-day Colombian volcanic arc from ca. 8.5 Ma to the present. Similarly, in the north, eastward migration of the Middle Cauca arc axis into the Central Cordillera resulted in porphyritic magmatism at Cajamarca-Salento. The northernmost manifestation of granitoid magmatism within the Central Cordillera appears at ca. 5.7°N, in the Plio-Pleistocene Río Dulce porphyry cluster,

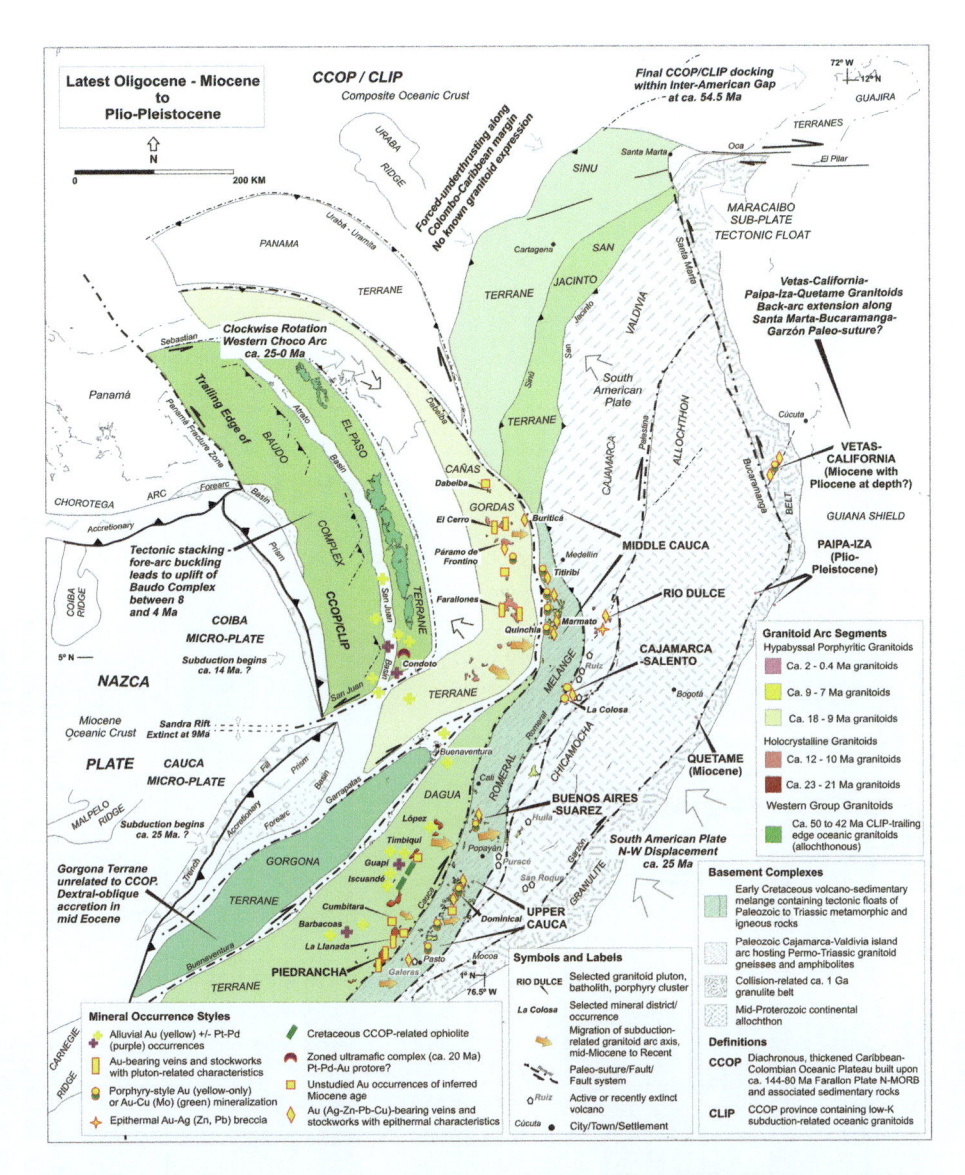

Fig. 6.12 Mineral occurrences of interpreted latest Oligocene through Plio-Pleistocene age in the Colombian Andes, in relation to tectonic setting, major litho-tectonic elements and granitoid intrusive and volcanic suites of the same time period. Note the near modern-day tectonic configuration for the region. All Oligocene through Plio-Pleistocene metallogeny is considered autochthonous to the region, although basement complexes hosting the individual mineral districts are of highly variable composition

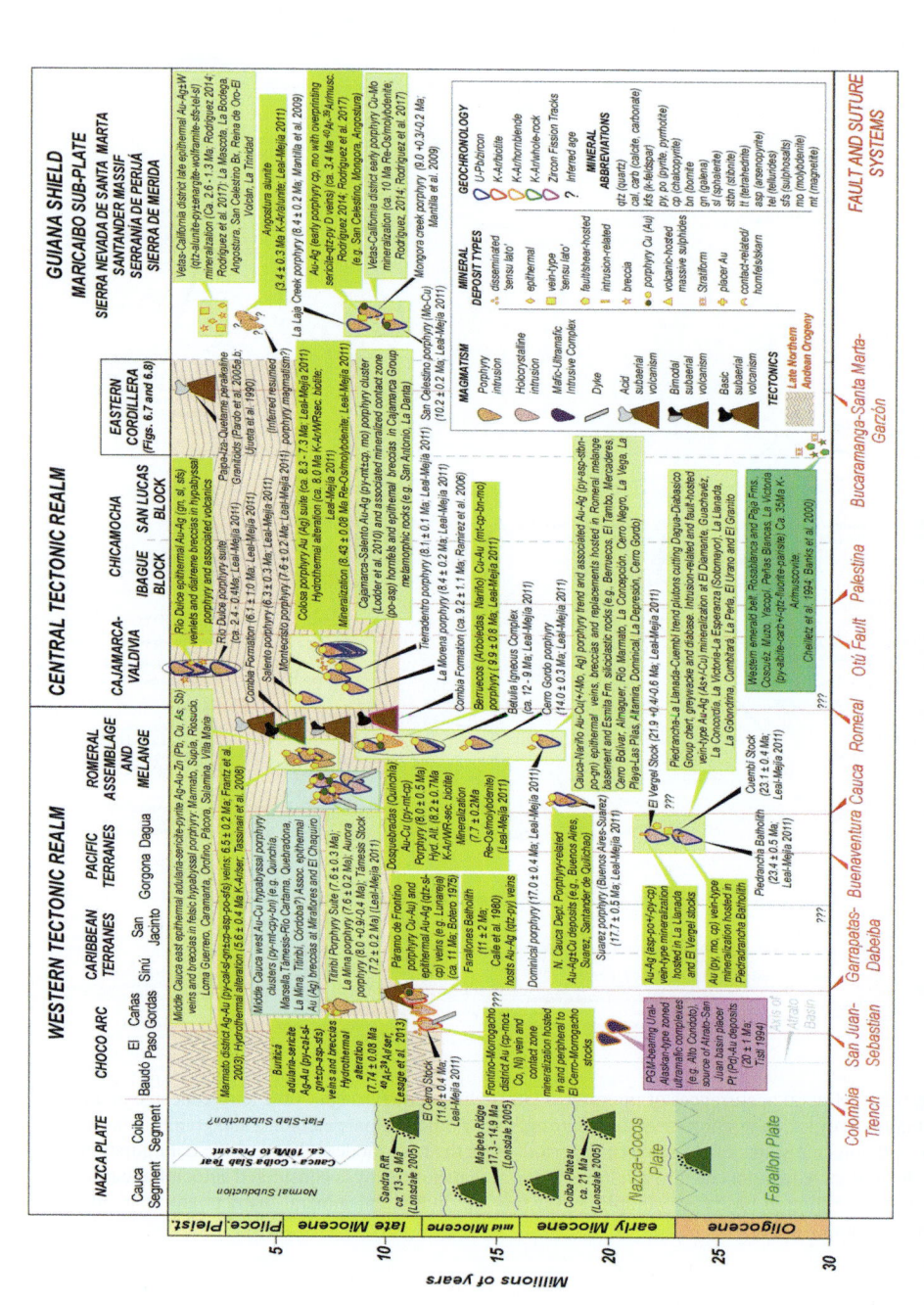

Fig. 6.13 Time-space analysis of mineral occurrences of latest Oligocene through Plio-Pleistocene age in the Colombian Andes, in relation to tectonic framework, major litho-tectonic elements and orogenic events and the age and nature of granitoid intrusive and volcanic suites of the same time period. The profile contains elements projected onto an ca. NW-SE line of section through west-central Colombia

which is essentially coaxial with the northernmost segment of the modern-day Colombian volcanic arc (Leal-Mejía 2011; Leal-Mejía et al. 2018) (Fig. 6.12).

Lithogeochemical and isotopic data (Leal-Mejía 2011; Leal-Mejía et al. 2018) permit interpretation of all of the above-mentioned suites as predominantly mantle-derived, metaluminous, calc-alkaline granitoids typical of subduction-related melts. It may be observed that, aside from differences in age, the granitoid suites comprising the various western Colombian arc segments are very similar in major-, minor- trace element and isotopic composition. All of the suites demonstrate typical gabbro through granodiorite trends with strongly mantelic isotopic compositions and in no instance are enhanced levels of crustal contamination (e.g. peraluminous tendencies, anomalously high Sr isotope compositions) implicit in the petrogenesis revealed by the data set.

In addition to the latest Oligocene-Miocene arc segments of the Western and Central cordilleras, Leal-Mejía et al. (2018) note three additional, isolated occurrences of Mio-Pliocene granitoids in Colombia's Eastern Cordilleran system, including at Vetas-California in the Santander Massif and at Paipa-Iza and Quetame, within the Eastern Cordillera (Figs. 6.12 and 6.13). These manifestations form volumetrically small and isolated occurrences located over 150 km east of the subduction-related magmatic arc axis defined by the active Colombian volcanic arc. Miocene granitoid magmatism in the Santander Massif ranges from ca. 14 to 9 Ma (Mantilla et al. 2009; Leal-Mejía 2011; Mantilla et al. 2012; Bissig et al. 2013; Bissig et al. 2014; Cruz et al. 2014) although recent studies suggest that unexposed magmatism of Pliocene age is likely present at shallow depth below the trend (Rodriguez 2014; Rodríguez et al. 2017). Bissig et al. (2014) suggest that the Vetas-California granitoids are related to the detachment and devolatilization of subducted CCOP lithosphere although plate configuration is conceptually difficult to establish. Within the Eastern Cordillera, some 175 and 360 km south of Vetas-California, respectively, the granitoids of Paipa-Iza and Quetame reveal additional isolated, low-volume granitoid occurrences of Mio-Pliocene age. The lithogeochemistry of the Pliocene Paipa-Iza occurrences is distinct from Vetas-California and is more suggestive of A-type rift-related granitoids. The lithogeochemical and isotopic characterization at Paipa-Iza and Quetame, however, has yet to be established, and petrogenetic relationships with Vetas-California remain unclear. The Vetas-California-Paipa-Iza-Quetame suites do however share important relationships with respect to distribution and structural controls. The occurrences are aligned along an ca. NNE-axis, in a back-arc position with respect to the active Colombian arc, and all are located along the trace of the Bucaramanga-Santa Marta-Garzón fault and suture system (Cediel and Cáceres 2000; Cediel et al. 2003) (Fig. 6.12). Leal-Mejía et al. (2018) suggest that these granitoids represent manifestations of back-arc magmatism facilitated by extension along the deep crustal Bucaramanga-Santa Marta-Garzón suture.

Leal-Mejía et al. (2018) interpret Neogene granitoid magmatism throughout the Colombian Andes, during the late Northern Andean orogeny, to be the result of the subduction of composite Nazca crust beneath the composite Colombian margin, beginning in the latest Oligocene. Differences in the rate and style of east-dipping subduction on either side of the Cauca-Coiba slab tear, beginning in the latest Oligocene, are reflected in the complex spatial and temporal distribution of Colombian onshore volcano-plutonic arc and back-arc magmatism during the early Miocene and Pliocene. Localization of individual arc segments and granitoid occurrences was influenced by the pre-existing structural and tectonic architecture of the region. All of these factors provided first-order controls upon the metallogeny which accompanied latest Oligocene through Pliocene granitoid magmatism in Colombia.

6.5.4.2 Metallogeny of Latest Oligocene Through Pliocene Granitoid Arc Segments

Metallogeny of the late Northern Andean orogeny is intimately and exclusively related to the emplacement and cooling history of latest Oligocene through Plio-Pleistocene metaluminous arc granitoids. As such, mineral occurrences of Miocene-Pleistocene age manifest within a similarly complex, parallel, spatial-temporal framework to that depicted by the hosting granitoid arc segments (Figs. 6.12 and 6.13). From a historic production standpoint, some important and productive mineral districts are associated with Au (Ag) production from primary source deposits related to Miocene granitoid centres, including, for example, at Titiribí, Marmato and Vetas-California. Regardless, the total production from these young primary source deposits remains relatively minor, when compared to historic production from secondary, alluvial-colluvial concentrations associated with older, more deeply eroded primary source regions (Fig. 6.14). Notwithstanding, active exploration in Colombia over the last decade has been focused more specifically upon young, low-grade, large-tonnage Au-Cu and Ag-Au-Zn (Pb-Cu) targets, such as those provided in the porphyry environment. As a result, mineral occurrences related to Miocene-aged granitoid centres currently contain, by far, the largest explored but undeveloped resources of Au, Ag and Cu in Colombia, when compared with the published resource base related to pre-Triassic through Eocene metallogenic events (Fig. 6.14).

Figure 6.13 outlines the temporal-spatial development of the most significant metalliferous districts and occurrences associated with latest Oligocene through Pleistocene-age granitoid arc segments in the Colombian Andes. As in previous

---

Fig. 6.14 (continued) geological domains and mixed genetic Au sources. (**b**) Production+resources vs. interpreted age of source mineralization. We have left the majority of the historic alluvial production from older geological domains unclassified with respect to age. Regardless, in either graph, the trends are clear. The great majority of historic/recent production was/is captured from artisanally mined, alluvial and colluvial deposits derived from older (pre-Cenozoic), eroded source regions, whilst the present qualified resources reflect the tendency in modern exploration, to search for large-tonnage, low-grade occurrences which coincide with geologically young (Miocene-Pliocene), high-level porphyry-related and epithermal environments. Modern exploration has confirmed the potential for large-tonnage/low-grade deposits in the Colombian Andes; however the data do not accurately reflect the exploration potential of numerous high-grade/low-tonnage Au districts hosted within older rocks, which have received only cursory exploration coverage

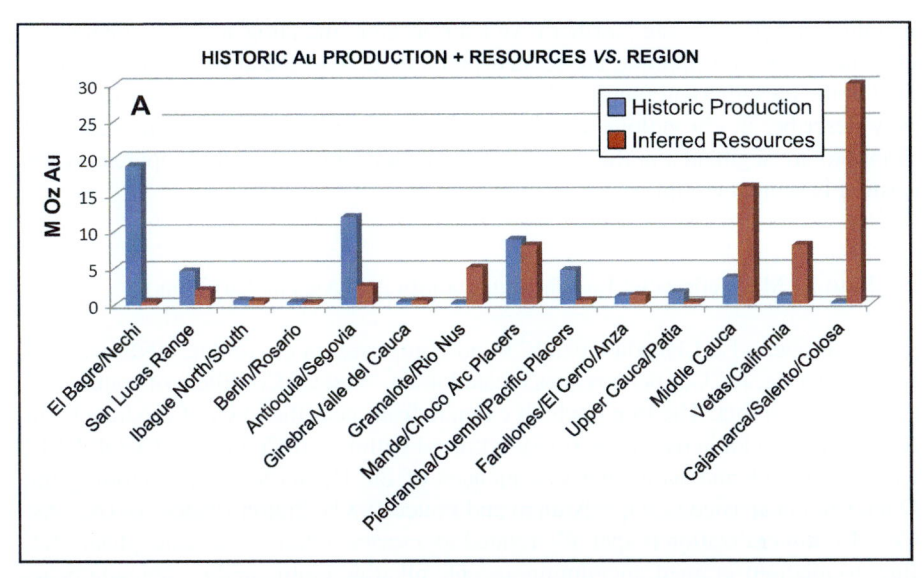

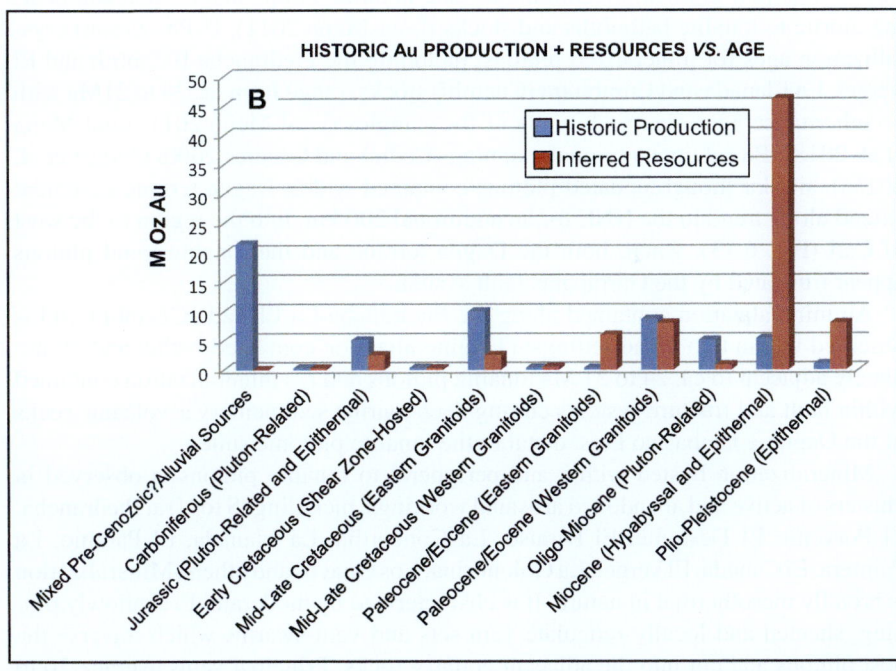

Fig. 6.14 Representative gold production in the Colombian Andes from pre-Colombian through modern times, as compiled from Restrepo (1883), Emmons (1936), AngloGold Ashanti (2007), UPME (2017), Banco de la Republica and other sources. AngloGold Ashanti considered the confidence level of pre-1900 production data to be ca. 50%, whilst that of post-1900 data is ca. 90%. Resources compiled from published corporate exploration data do not include Cu, Ag or potential base metal credits. (**a**) Production+resources vs. geographic producing region. Some regions with abundant alluvial production, such as El Bagre-Nechi and Antioquia-Segovia, will reflect mixed

metallogenic phases, late Northern Andean orogeny metallogeny is dominated by epigenetic volcano-plutonic arc-related precious ± base metal occurrences generated in the mesothermal pluton-related, porphyry and epithermal environments. We now describe, in chronological order, mineralization contained within individual granitoid arc segments, in the context of the tectono-magmatic development of the hosting basement complex.

ca. 24 to 21 Ma Piedrancha-La Llanada-Cuembí Au (Ag, As, Cu) Trend

The Piedrancha-La Llanada-Cuembí trend comprises an ca. 85 km NNE-oriented belt of Au (Ag, As, Cu) occurrences located in the Western Cordillera of southwesternmost Colombia. The trend, which extends from near the town of Piedrancha in the south to the Patia river (Fig. 6.15), is hosted within mid-Cretaceous CCOP/CLIP assemblage volcano-sedimentary sequences of the Dagua terrane, (including the Dagua and Diabasico Groups; Arango and Ponce 1982). Pluton-related Au (Ag, As, Cu, Mo) mineralization is spatially related to a series of holocrystalline, phaneritic, fine- to medium-grained, metaluminous calc-alkaline biotite and hornblende-bearing diorite to tonalite batholiths and stocks (Leal-Mejía 2011). U-Pb (zircon) crystallization ages for mineralized plutons, including the Piedrancha Batholith and El Vergel, La Llanada and Cumbitara (Cuembí) stocks, range from ca. 24 to 21Ma with no inheritance ages observed in any of the samples (Leal-Mejía 2011; Leal-Mejía et al. 2018). Based upon regional mapping (Cediel and Cáceres 2000; Gómez et al. 2015a), similar though undated plutons contained within Dagua terrane basement extend along trend to the NNE for an additional 200 km, into the region to the west of Cali (Fig. 6.13), where both the Dagua terrane and the tonalite trend plutons appear truncated by the Garrapatas fault system.

Au mineralization contained along the Piedrancha-La Llanada-Cuembí trend is observed within two broad settings: (1) mineralization contained within and immediately adjacent to ca. 24 to 21 Ma tonalite plutons and (2) mineralization contained within fault and fracture systems cutting deep marine sedimentary ± volcanic rocks of the Dagua ± Diabasico Fms. distal to the tonalite plutonic suite.

Mineralization hosted within and peripheral to tonalite plutons is observed in clusters of active and abandoned artisanal workings, including (S to N) at Piedrancha, El Porvenir, El Desquite, El Paraíso, La Concordia, La Llanada, El Páramo, La Palmera, El Canadá, El Vergel, La Golondrina, Los Guavos and others. Mineralization is broadly mesothermal in nature. It is characterized by moderate- to shallowly dipping, sheeted and locally reticulate vein sets and vein swarms which traverse the host plutons and cut into the adjacent country rocks. Principal veins average 20 to 30 cm thick and are commonly accompanied by stockworks of cm-scale veinlets. Economic mineralization is generally confined to the vein sets. It consists of up to 60% mixed sulphides, dominated by approximately equal portions of pyrrhotite and arsenopyrite ± pyrite with local accumulations of chalcopyrite, minor galena and sphalerite and native Au. The veins at El Porvenir (Piedrancha Batholith) contain abundant molybdenite (Mutis 1993) with arsenopyrite and minor base metal sulphides.

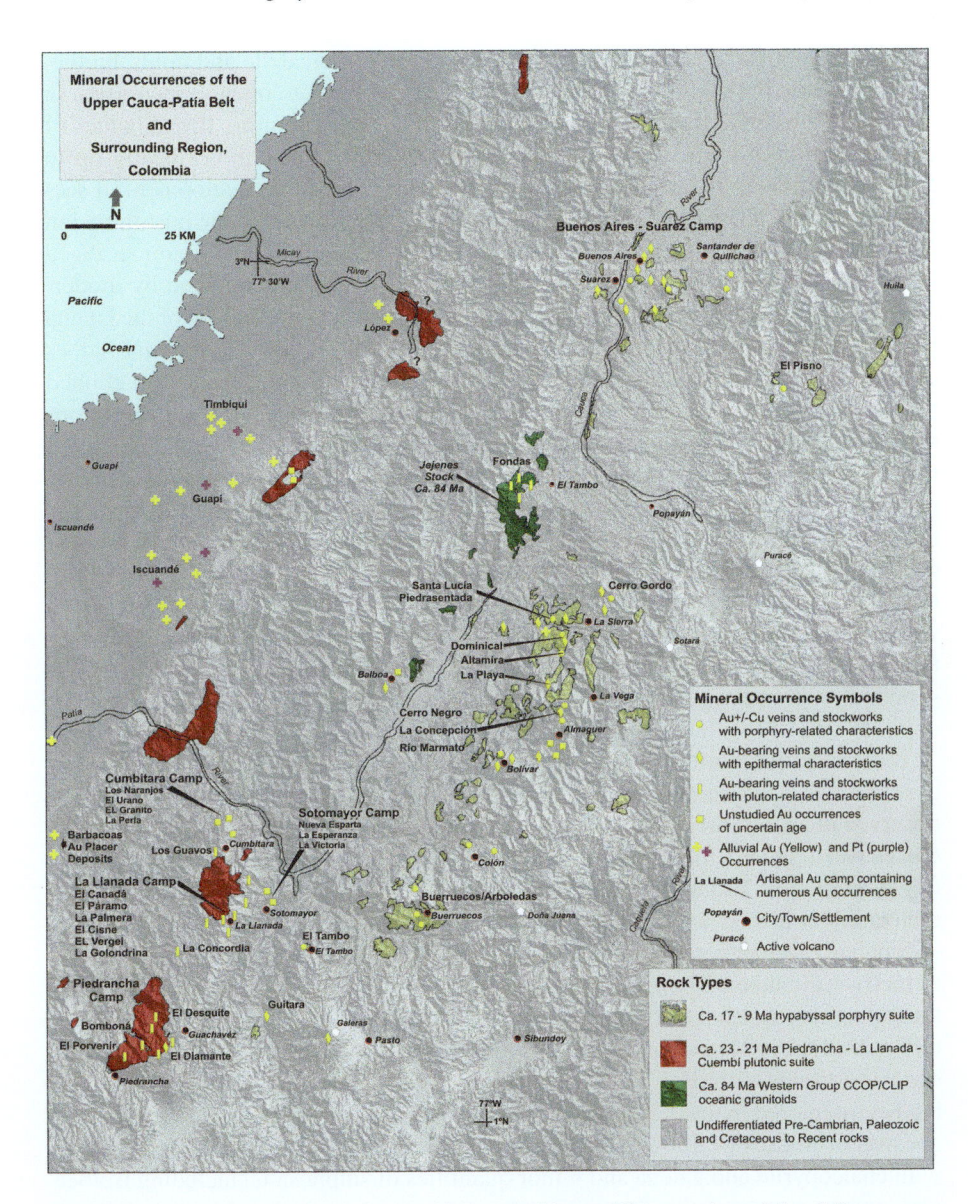

Fig. 6.15 Selected mineral occurrences of interpreted latest Oligocene through late Miocene age in Upper Cauca-Patía region and surrounding area of the Colombian Andes, in relation to granitoid intrusive rocks of the same approximate time periods. Physiographic features of the map area are revealed by the 30 m digital elevation model (DEM) base image

Gangue is dominated by quartz with subordinate ankerite/calcite and minor biotite. Wallrock alteration within the intrusives varies from weak to intense, increasing in relation to proximity and intensity/density of mineralized veining. Broad haloes of carbonitization and fine pyrrhotite-arsenopyrite-pyrite sulphidation (locally up to 5% by volume) envelop mineralized structures. Notable is the development of secondary biotite, replacing magmatic biotite and hornblende within the host intrusion (Leal-Mejía 2011). Increasing silicification is observed in proximity to mineralized veins. The wallrocks to most of the mineralized plutons are dominated by moderately to thinly bedded, carbonaceous chert, cherty shale and greywacke of the Dagua Fm. The intrusive-country rock contact is marked by metre-scale thermal haloes comprised of compact, siliceous biotite hornfels containing abundant fine-grained pyrrhotite-arsenopyrite±pyrite, with localized faulting and siliceous gouge development. Within the sedimentary package, vein sets generally terminate within tens of metres of the intrusive contact. Detailed radiometric age date, isotopic, mineral paragenetic and structural studies are not available for the pluton-related deposits of the Piedrancha-La Llanada-Cuembí trend. A maximum age of ca. 24 to 21 Ma is established by the crystallization age of the host plutons. Based upon the generally mesothermal mineral and alteration assemblages, the intimate spatial relationship with the tonalite plutons and the apparent lack of pervasive retrograde alteration minerals (e.g. sericite-illite, chlorite) associated with mineralization, we suggest a genetic link between gold mineralization and the emplacement and cooling history of ca. 24 to 21 Ma granitoids.

Mineralization hosted within carbonaceous chert, shale and greywacke ± pyroclastic rocks, diabase and basalt of the Dagua ± Diabasico Fms., respectively, is observed in active and abandoned artisanal exploitations near the towns of (S to N) Guachavéz (e.g. El Diamante), Sotomayor (e.g. La Nueva Esparta) and Cumbitara (e.g. La Perla, El Urano, El Granito) (Fig. 6.15). In these locations, vein-type mineralization is hosted within fault and fracture systems rooted within the volcano-sedimentary sequences, where granitoid igneous rocks are locally absent or limited to metre-scale dikes. The El Diamante prospect was investigated in some detail by the Japan International Cooperation Agency and the Metal Mining Agency of Japan (JICA-MMAJ 1984) and by Molano and Shimazaki (2003). Au is contained in ore shoots and veins within ca. 25 m-wide, N50-60W-striking structural corridor, which has been explored for ca. 1,200 m along strike and to a depth of 200 m. Mineralization consists of masses of fine-grained sulphides dominated by pyrite and arsenopyrite with chalcopyrite and galena and minor quantities of sulphosalts, including freibergite, pyrargyrite, proustite, argentite and polybasite, deposited within three paragenetic stages (Molano and Shimazaki 2003). Au mostly occurs as fine disseminations and replacements within arsenopyrite and arsenical pyrite. The predominant gangue mineral is quartz, which is commonly brecciated and milled by post-mineral faulting. Based upon fluid inclusion and O, D and S stable isotope data, Molano and Shimazaki (2003) concluded that the fluids responsible for Au mineralization at El Diamante were of mesothermal character and predominantly magmatic ± meteoric in origin and possibly related to emplacement of the nearby Piedrancha Batholith (23.4 ± 0.5 Ma; Leal-Mejía 2011) (Fig. 6.15).

To the N-NNE of Cumbitara, artisanal mining along ca.10 km discontinuous trend reveals Au mineralization hosted within Dagua Fm. sedimentary rocks (e.g. La Perla, El Urano and El Granito workings) and diabase and basalt of the Diabasico Fm. (e.g. Los Naranjos and San Martín workings). Mineralization is contained within auriferous veins and vein networks localized along steeply dipping (E or W) to vertical reverse faults. The general mineralized trend strikes NNW through NNE. Mineralized structures are characterized by the presence of one or more central veins contained within or lying parallel to the main fault plane. Individual high-angle veins average ca. 20 cm thick and may occur in zones up to 3 m thick, hosting two or three individual veins. Proximal to the main structure, numerous conjugate and antithetic veinlets are often observed. Structural flexure along the high-angle fault planes has generated low-angle, dilatant jogs resulting in marked local thickening of the mineralized structure, in places up to 4 or 5 metres. These zones are characterized by bifurcation of the main fault-parallel vein, strong antithetic fracturing and the development of an intense stockwork of mineralized veinlets with strong hydrothermal alteration affecting the surrounding host rock. Mineralogically, veining is dominated by milky quartz ± calcite gangue, containing native Au and up to 4% mixed sulphides, including principally pyrite, minor arsenopyrite and lesser galena ± sphalerite. Within the principal vein, sulphides are mostly concentrated within vein-parallel laminations, which may include fine fragments of the local country rock. Wallrock alteration proximal to mineralized structures includes silicification and up to 5% finely disseminated euhedral pyrite. Pervasive disseminated carbonate alteration and the development of fine quartz-pyrite veinlets exhibiting bleached alteration haloes are observed somewhat more distal to the main structure.

ca. 17.5 to 9 Ma Piedrasentada-La Vega-Berruecos Porphyry Au (Ag, Cu) Trend

Following the emplacement of ca. 24 to 21 Ma Piedrancha-La Llanada-Cuembí arc segment, eastward migration of the calc-alkaline magmatic arc axis (Figs. 6.12 and 6.15) has been interpreted to represent shallowing of the subduction angle of the Nazca plate, possibly associated with the splitting of the Farallon Plate at ca. 23 Ma (Lonsdale 2005), and/or the subduction of young, thick and buoyant (e.g. Malpelo Ridge or Carnegie Ridge-like) oceanic crust (e.g. Cediel et al. 2003). ca. 17.5 to 9 Ma Piedrasentada-La Vega-Berruecos granitoids extend semi-continuously for ca. 200 km along a NNE trend, exposed along the Cauca-Patía intermontane valley and within the Colombian Massif. The trend is comprised of predominantly hypabyssal porphyritic, with lesser medium-grained equigranular, phaneritic, felsic stocks, dikes and sills, observed in polyphase clusters at (from S to N) El Tambo (Nariño), Berruecos-Arboleda, San Pablo-Colón, Cerro Bolivar, Cerro Negro-La Concepción, La Vega (Betulia igneous complex), Altamira, Dominical-Piedrasentada and Cerro Gordo-La Sierra. These granitoids are located to the east of the Dagua terrane and Cauca fault and suture and are hosted mostly within mixed metamorphic and oceanic volcanic, ultramafic and sedimentary rocks of the Romeral mélange and Oligo-Miocene-aged continental siliciclastic rocks of the Esmita Fm. deposited upon the Romeral basement along the evolving Cauca-Patía basin (Orrego and Paris 1990; Orrego and Acevedo 1993; Orrego et al. 1999; Cediel et al. 2003; Leal-Mejía 2011; Marín-Cerón et al. 2018). Along trend to both the south and the north, the granitoid belt may continue beneath extensive Plio-Pliestocene to Recent volcanic cover. Beyond volcanic cover to the north of Popayán, the trend reappears in the Buenos Aires-Suárez-Santander de Quilichao region, where similar porphyritic granitoids of similar age are observed (París and Marín 1979; Leal-Mejía et al. 2018, see below).

Metallogeny along ca. 17.5 to 9 Ma Piedrasentada-La Vega-Berruecos trend is Au (Cu) and Au-Ag±As±Sb biased and is spatially and temporally related to many of the hypabyssal porphyritic granitoid clusters noted above. Mineralization commonly extends into the immediate country rocks to the porphyries, as alteration zones and hornfelsing within the basement complex, and more distally manifests as epithermal disseminations, manto-style replacements, veins, joint and fracture fillings and breccias within the siliciclastic sequences of the Esmita Fm.

Sillitoe et al. (1982) referred to the Miocene hypabyssal porphyry occurrences along the Cauca-Patia valley as the "central sub-belt" relative to porphyry trends located to the east (Jurassic Rovira-Infierno-Chilí trend) and the west (Eocene Mandé-Acandí trend). They provided a 17.4 ± 0.4 Ma K-Ar (magmatic biotite) age for propylitically altered dacite (granodiorite) porphyry from Piedrasentada (Santa Lucía). Gómez-Gutierrez and Molano-Mendoza (2009) provided more detail regarding porphyry-style Au-Cu mineralization at this locality. They note an early andesite (diorite) intrusive phase containing chalcopyrite and pyrite, with dominant secondary biotite alteration, and M-, A-, EB- and B-type veinlets (Sillitoe 2000), typical of high-temperature magmato-hydrothermal formation. A second intrusive phase contains dominantly propylitic alteration (Sillitoe et al. 1982). Mineralization and alteration locally affect the chloritized mafic volcanic basement suite. Auriferous, epithermal-textured veins, veinlets and localized breccias containing banded and colloform quartz, disseminated pyrite and coarse radiating aggregates of stibnite, hosted within the Esmita Fm., are observed up to 2 km to the north and south of the porphyry centre.

Some 5 km to the SSW, near Dominical, similar porphyry-related Au (Cu) mineralization is more intensely developed (JICA 1987). Early A-type quartz veining and associated secondary biotite alteration in porphyritic diorite is overprinted by a moderate to locally dense stockwork of D-type quartz-pyrite veinlets with associated pervasive sericite-pyrite alteration. Propylitic alteration, including abundant epidote+pyrite, is dominant peripheral to the stockwork zones. Leal-Mejía et al. (2018) provided a 17.0 ± 0.4 Ma (U-Pb, zircon) age for propylitically altered porphyry from Dominical. Similar quartz-pyrite stockworks are hosted within diorite porphyry outcropping 5 km to the south near the village of Altamira. Numerous narrow veins and localized breccias containing epithermal-textured quartz and stibnite are hosted within the Esmita Fm. to the east of Dominical (Mutis 1993).

Ten kilometres to the NE of Dominical, at Cerro Gordo, numerous dikes and sills of coarse-grained biotite diorite porphyry cut domed and weakly deformed, thinly to moderately bedded siltstone and sandstone of the Esmita Fm. Au (Ag) mineralization are present as widespread, pyritic disseminations, joint and fracture fillings, veinlets and bedding contact replacements. A broad halo of weak to locally moderate argillic (illite) alteration encompasses the mineralized area. Leal-Mejía et al. (2018) present a 14.0 ± 0.3 Ma (U-Pb, zircon) age for diorite porphyry at Cerro Gordo.

East of La Vega, porphyry-style Au mineralization is associated with numerous hypabyssal porphyritic and medium-grained phaneritic stocks, sills and dikes of the Betulia igneous complex (Orrego et al. 1999; Gil-Rodríguez 2014). Leal-Mejía et al. (2018) reveal U-Pb (zircon) ages of 11.6 ± 0.2 Ma and 9.2 ± 0.2 Ma for diorite and granodiorite porphyries containing pervasive secondary biotite + pyrite alteration. At La Concepción, near Cerro Negro, exposures in artisanal mine workings and nearby outcrop demonstrate the complex field relationships between geological structure, various phases of hypabyssal porphyry, basement rocks of the Romeral mélange and sedimentary rocks of the Esmita Fm.

La Concepcíon Au-Ag Prospect Au-Ag mineralization exposed in historic mine workings at La Concepción is hosted within biotite diorite porphyry, Esmita Fm. siltstone and semi-pelite and poly-deformed quartz-chlorite-muscovite (sericite) schists of the Romeral mélange basement. Mineralization has not been studied in detail, and the following description is based upon field observations. Historic mining has mostly been developed within a 3 m-thick, low-angle (dip sub-horizontal to 25° NW) zone of dilatency exposed for ca. 500 m along strike. In the north, the low-angle structure cuts highly altered and silicified biotite-plagioclase porphyry, whilst in the south, it separates basement schist in the hanging wall from siltstone and semi-pelite of the Esmita Fm. in the footwall and may therefore be interpreted to represent a thrust-style detachment. Mineralization along the low-angle dilatency is characterized by multiple phases of moderate to intense brecciation and massive quartz-pyrite-arsenopyrite-pyrrhotite infilling. Mineralization intensity diminishes towards both the northern- and southern-exposed margins, where the low-angle dilatency is seen to close into narrow, more discrete, shallowly dipping veinlets. Transecting the low-angle dilatency near the mid-point of the exposed mineralization is a N65°W-striking, moderately to steeply NE-dipping fracture (fault?) zone, characterized by brecciation, quartz stockworking and intense silicification, extending dominantly into the footwall of the low-angle structure. In the distal (50 to 400 m) footwall, mineralization is observed in Esmita Fm. silty pelite and biotite plagioclase porphyry, as a series of irregular quartz-pyrite-arsenopyrite veinlets and stockworks and mm-scale fractures lined with mixed sulphides. The proximal hanging wall of the low-angle structure is topographically near vertical and not easily accessed. Unlike the footwall, hanging wall mineralization appears diminished, characterized by abundant fine pyrite veinlets and fracture fills, extending for some tens of metres into the hanging wall, but lacking the intense silicification observed in the footwall zone.

At least three distinct intrusive phases cut the metamorphic and sedimentary rocks at La Concepción: (1) a fine- to medium-grained equigranular biotite-hornblende diorite observed to the east and southeast of the mine area, (2) a fine-grained biotite ± hornblende-plagioclase porphyry, hosting phenocrysts of either phases to 3 mm size, exposed in the footwall of the mine area and in the hanging wall up to the crest of the overlying ridge, and (3) a late, fine-grained, crowded hornblende ± biotite-quartz-plagioclase porphyry, containing cm-size bi-pyramidal quartz phenocrysts, observed as dikes cutting the plagioclase porphyry. The fine-grained biotite ± hornblende-plagioclase porphyry appears most directly related to mineralization.

Hydrothermal alteration at La Concepción, including silicification, sericitization and disseminated pyrite sulphidation, is widespread, pervasive and locally intense. Compact siliceous biotite hornfels developed within Esmita Fm. siltstone extends for tens of metres in the footwall of the low-angle structure. Silicification and hornfelsing is also strong along the N65W fracture, extending well into the footwall of the low-angle dilatency, where zones of brecciation contain horfelsed clasts and have undergone silica flooding. Adjacent to individual quartz-sulphide veinlets, finely disseminated pyrite is observed; however, the overall content of disseminated sulphide is relatively low (<2%). Fine-grained sericite development is pervasive within the biotite plagioclase porphyry. Argillitization, probably associated with sulphide weathering, is observed in outcrop.

The structural geology of the mine area is dominated by the mineralized low-angle thrust dilatency (average dip 10 to 15°W) and the high-angle, N65°W fracture. The Esmita Fm. in the vicinity of the mine area is viewed, in plan, through an erosional window which has penetrated the hanging wall basement schists. The high-angle, N65°W fracture may represent a feeder structure to low-angle mineralization and to mineralization developed along the contact between the biotite plagioclase porphyry and Esmita Fm. The asymmetrical distribution of alteration/mineralization on the hanging wall vs. footwall of the low-angle structure suggests the role of the hanging wall schists as a cap rock, forcing hydrothermal fluids laterally along the low-angle dilation and into the footwall of the mineral system. The timing of mineralization at La Concepción appears syn-kinematic with respect to the structural development of the area. Mineralization is spatially focussed within the biotite plagioclase porphyry and associated thermal/hydrothermal alteration halo. This porphyry unit has not been specifically dated at La Concepción but is interpreted to belong to ca. 11.6 ± 0.2 Ma and 9.2 ± 0.2 Ma Betulia igneous complex suite.

Continuing 15 km SSW from La Concepción, at Cerro Bolívar, numerous narrow quartz-sulphide veins and breccias are exposed in artisanal workings along the contact between hypabyssal porphyritic diorite and granodiorite and poly-deformed graphitic micaceous schist of the Romeral mélange. At the time of the authors' visit, mineralization was only exposed within the schistose basement rocks where at least two generations of quartz veining are observed; an early, unmineralized, folded and boudinaged variety which appears to have developed during the metamorphic history of the schistose basement protolith and a narrow, more planer-rectilinear variety which forms weak, discontinuous stockworks cutting the earlier metamorphic veining. The second vein type is auriferous, containing local concentrations of pyrite, galena and stibnite. Similar veining can be traced along a NNE-trending axis into the region to the S of the town of Almaguer and to the SSW around the towns of Colón and San Pablo where numerous porphyritic diorite to granodiorite stocks and plugs are observed to intrude graphitic quartz-mica schist basement and Esmita Fm. silty sandstone. Alvarez et al. (1979) provided a 13.3 ± 3 Ma K-Ar age date for granodiorite porphyry at nearby Cerro San Cristobal.

At Berruecos (Arboledas), near the southern limit of the exposed Piedrasentada-La Vega-Berruecos Porphyry trend, porphyry-style Cu-Au mineralization is observed in at least three separate locations, along Mazamorras, Olaya and Manjoy Creeks. In all cases, the host to mineralization is a suite of hypabyssal biotite ± hornblende-plagioclase porphyritic rocks of dioritic ± granodioritic composition hosted within mixed mafic volcanic and schistose metamorphic rocks of the Romeral mélange basement. Leal-Mejía (2011) provided a K-Ar (magmatic hornblende) age of 9.9 ± 0.8 Ma for granodiorite porphyry of the Berruecos suite. Mineralization at Berruecos is characterized by Au-rich porphyry-style alteration (Sillitoe 2000), including early pervasive secondary biotite + magnetite as disseminations and fine veinlets, both with associated fine-grained hypogene sulphides including pyrite, chalcopyrite, bornite and molybdenite. Total hypogene sulphide content associated with potassic alteration averages between 1 and 3%. Potassic alteration is locally overprinted along throughgoing fractures and minor faults by sericite±pyrite, whilst alteration passes rapidly along the margins of the porphyry complex to propylitic, with disseminated pyrite, chloritization of mafic phases and the partial replacement of plagioclase by epidote. Supergene enrichment has led to localized, near-surface concentrations of cuprite, neotosite, chalcocite and native Cu. Secondary Cu-oxides, including admixtures of malachite, brocanthite, chalcanthite and azurite, are observed in outcrops along the Mazamorras creek. In artisanal Au workings at Santa Ana, 4 km to the NNW of Berruecos, N-trending clusters of veinlets, weak stockworks and breccias containing druzy quartz and abundant Fe-oxides after pyrite are developed in quartzose sandstones of the Esmita Fm.

ca. 17 Ma Buenos Aires-Suárez Porphyry Au (Cu) Cluster

Along trend to the north of the Piedrasentada-La Vega-Berruecos porphyry suite, beyond Plio-Pleistocene volcanic cover to the north of Popayán, the Buenos Aíres-Suárez and Santander de Quilichao clusters of hypabyssal porphyritic dioritic to granodioritic stocks, sills and dikes (Fig. 6.15) intrude mafic volcanic basement rocks of the Dagua and/or Romeral terrane and moderately to thinly bedded siliciclastic sedimentary rocks of the Oligo-Miocene Esmita Fm. (París and Marín 1979). Au mineralization is exposed in numerous active open-cut and underground artisanal workings, mostly located within Esmita Fm. sedimentary rocks but also locally exposed within diorite and granodiorite porphyry and mafic volcanic basement. The Esmita Fm. in the vicinity of the porphyry clusters has been diapirically domed and is cut by radiating fracture and joint sets associated with doming during porphyry emplacement. Mineralization manifests as pyritic joint and fracture fillings, manto-style replacements along bedding planes, localized breccias and broad disseminations within receptive Esmita Fm. sandstones and sandy siltstones. Alteration associated with these rock types is generally argillic, comprised of illite ± fine sericite and pyrite, with localized silicification. Vein and fracture fillings are dominated by pyrite and druzy quartz. Mineralized diorite and granodiorite porphyry is also exposed in some artisanal workings. At La Toma and El Molino, porphyritic hornblende granodiorite hosts patchy potassic alteration comprised of localized high-temperature A-type quartz veinlets (Sillitoe 2000), disseminated secondary biotite+/-magnetite and minor chalcopyrite +/- molybdenite. Potassic alteration is overprinted by widespread weak to pervasive sericite-pyrite (phyllic alteration) and auriferous quartz-pyrite stockwork veining. Leal-Mejía et al. (2018) present a U-Pb (zircon) crystallization age of 17.7 ± 0.5 Ma for hornblende diorite porphyry containing secondary biotite, collected at La Toma. This age compares well with K-Ar and U-Pb age dates presented by Sillitoe et al. (1982) and Leal-Mejía et al. (2018), respectively, for similar porphyry-related and epithermal mineralization along the Piedrasentada-La Vega-Berruecos porphyry belt to the south.

Based upon geological mapping (París and Marín 1979; Orrego and Paris 1990; Orrego and Acevedo 1993; Orrego et al. 1999; Cediel et al. 2003; Gómez et al. 2015a) and available U-Pb (zircon) age data (Leal-Mejía et al. 2018), granitoid magmatism in southern Colombia, to the south of the Caldas-Coiba tear (Vargas and Mann 2013; Chiarabba et al. 2016) temporarily ceased at ca. 9 Ma. To the north of the Caldas-Coiba tear, active mid-Miocene accretion of the Panamá-Choco arc was still underway, and subduction initiation was delayed with respect to the south, inhibited by the thick, buoyant nature of the subducting CCOP and the greater degree of coupling between the CCOP and overlying continental lithosphere. In this context, the first appearance of holocrystalline arc-related plutonism to the north of the Caldas-Coiba tear, analogous to that observed in ca. 24 to 21 Ma Piedrancha-Cuembí arc segment to the south, manifests at ca. 12 Ma, along the Farallones-El Cerro trend of phaneritic granitoid stocks and batholiths. Additional porphyritic granitoids ± related volcanic rocks are observed along the Middle Cauca belt, to the east of Farallones-El Cerro (Figs. 6.12 and 6.16). Both of these calc-alkaline arc segments host significant pluton- and/or porphyry-related and epithermal precious ± base metal deposits, which we continue to discuss in chronological order.

ca. 12 Ma Farallones-El Cerro Au (Ag, Cu) Trend

The Farallones-Frontino trend of holocrystalline phaneritic plutons includes the Farallones Batholith, the Urrao (Pàramo de Frontion) Pluton and the El Cerro igneous complex. This ca. 150 km-long, NNW-trending arc segment extends along the topographic apex of the northern Western Cordillera, where it is observed to intrude Penderisco Fm. oceanic sedimentary rocks which form basement to the eastern

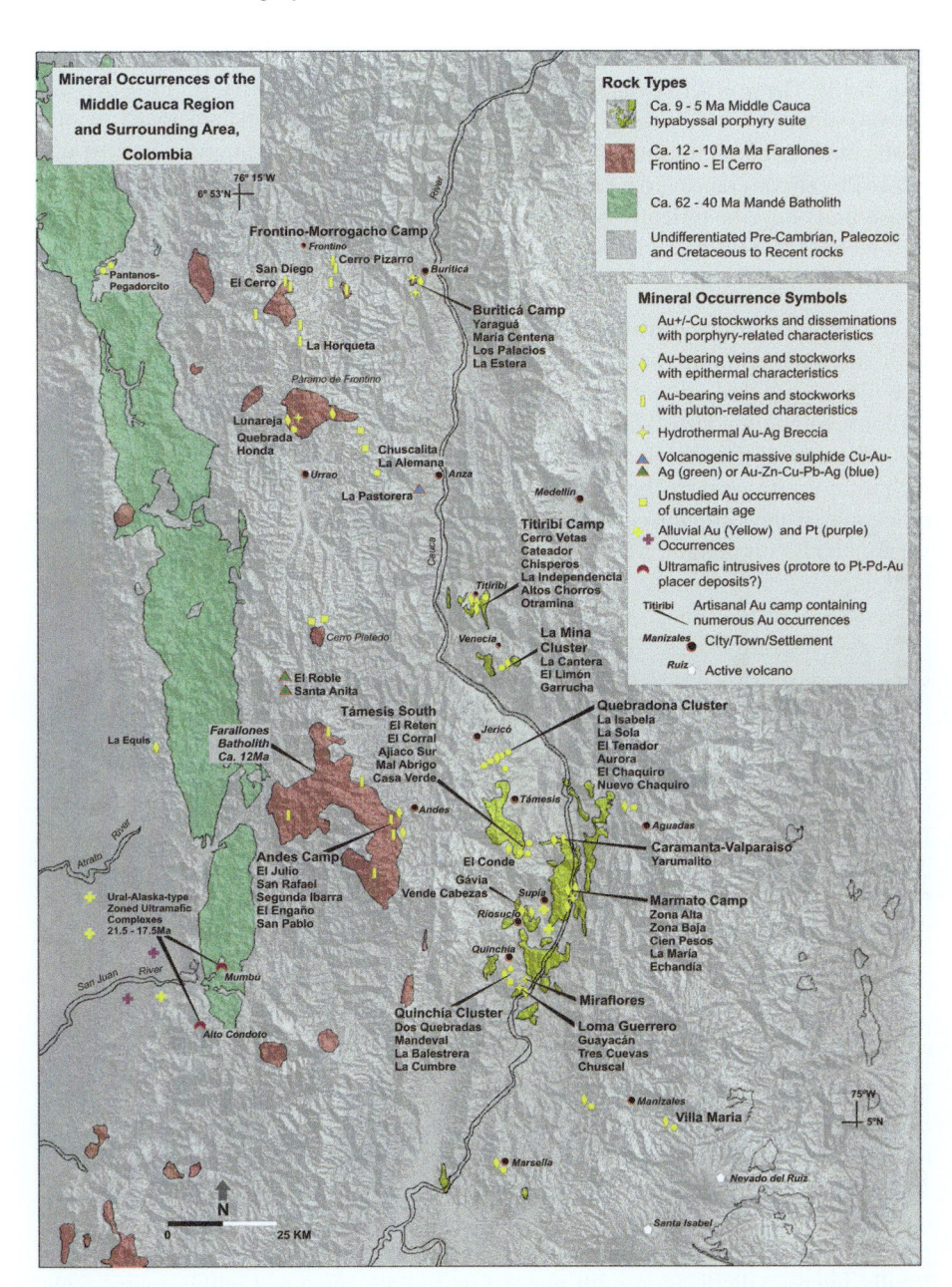

Fig. 6.16 Selected mineral occurrences of interpreted Cretaceous and late Miocene age in the Middle Cauca and surrounding area of the Colombian Andes, in relation to granitoid intrusive rocks throughout the region. Physiographic features of the map area are revealed by the 30 m digital elevation model (DEM) base image

segment of the Cañas Gordas terrane, to the north of the Garrapatas transform (Fig. 6.12). The northern extension of this plutonic trend can be inferred based upon the appearance of additional granitoid plutons in the region to the north of Dabeiba (Calle et al. 1980; González 2001; Cediel and Cáceres 2000; Gómez et al. 2015a). No U-Pb (zircon) dates are available for the Farallones, Urrao or El Cerro plutons. These units have historically been dated by the K-Ar (magmatic hornblende, biotite) method, which has consistently returned ages ranging from ca. 12 to 11 Ma (Botero 1975; Calle et al. 1980; Leal-Mejía 2011).

ca. 12 Ma Farallones Batholith Au (Ag) mineralization spatially associated with the Farallones Batholith occurs as discrete pluton-hosted veins and as veins, fracture fillings, sulphide-rich gouge zones and breccias hosted within Penderisco Fm. carbonaceous siliciclastic rocks proximal to the batholith contact. Numerous active and abandoned artisanal workings are clustered along the eastern margin of the batholith, to the west of the towns of Andes and Betania, although this may be a reflection of the difficult access and extreme topography encountered along the western batholith margin. Mineralization, even within the active workings, is never well exposed, as the entire zone is heavily vegetated and artisanal exploitation is restricted to narrow underground tunnels often developed along deeply oxidized fault and fracture zones. Batholith-hosted mineralization (e.g. La Reina, Santa Cecilia) consists of discrete, ca. E-W-striking fractures and veins averaging less than 20 cm thick, containing mostly fine pyrite with arsenopyrite ± molybdenite, quartz and locally clay-rich gouge. Alteration within the batholith is closely restricted to the vein margins and includes the early replacement of magmatic hornblende by biotite followed by the chloritization of mafic phases and the locally intense development of fine veinlets and fractures filled with epidote, pyrite and chlorite. A zone of variably siliceous hornfels is developed along the margin of the batholith, extending for at least 500 m into thinly bedded, folded Penderisco Fm. carbonaceous greywacke, siltstone, shale and mudstone. Mineralization within the sediments (e.g. San Rafael, El Julio, Segunda Ibarra, El Engaño, San Pablo workings) overprints both the folding and the hornfels. It manifests as structurally controlled fault breccias, joint and fracture fillings and thin bedding plane replacements. Faults breccias measure up to 2 metres wide and contain hornfelsed/silicified fragments of Penderisco Fm. host rock, with lesser massive milky and druzy quartz infilling mixed with clayey gouge and very fine ("sooty-textured") sulphides dominated by pyrite and arsenopyrite ± galena, sphalerite and silver-rich sulphosalts. Joint and fracture fillings are narrow, ranging from 2 to 20 cm wide, and contain clay and sulphide-rich gouge with generally minor quartz. Wallrock alteration is generally restricted to the immediate margins of the structures and includes weak to locally intense silicification with minor fine sericite contained within metre-scale haloes of finely disseminated and fracture-controlled pyritic sulphidation.

The precise timing of Au (Ag) mineralization contained within and along the eastern margin of the Farallones Batholith has not been determined. The 11 ± 2Ma K-Ar (magmatic hornblende) age provided by Calle et al. (1980) is considered to represent a maximum age for mineralization, and based upon close spatial relationships, we interpret mineralization to be genetically related to the cooling history of the Farallones Batholith.

ca. 12 to 11 Ma Paramo de Frontino Volcanic Complex The Páramo de Frontino (Páramo de Urrao) (Fig. 6.16) represents the best preserved mid-Miocene high-level volcano-plutonic complex recognized along the northern segment of the Western Cordillera. A variety of mineral deposit types are hosted within this complex, including volcanic-hosted epithermal Ag-Au-base metal veins and porphyry-style Cu-Au occurrences. Au (Ag)-rich fault veins, breccias and fracture fillings, similar to those recorded along the margins of the Farallones Batholith, are hosted within Cañas Gordas (Penderisco Fm.) carbonaceous siliciclastic basement rocks around the periphery of the volcano-plutonic complex.

The geology of the Páramo de Frontino is sparsely documented, and the following generalizations of the geology and mineral occurrences of the area are based upon historic reporting by Wokittel (1954) and González (2001) and field observations by the present authors. The Páramo de Frontino stock is comprised of a phaneritic biotite and augite-bearing diorite to monzodiorite (González 2001) and was dated at 11 Ma (K-Ar, biotite) by Botero (1975). The age relationship between the stock, hypabyssal porphyry plugs and dikes and the felsic to intermediate volcanic rocks observed in the higher elevations of the Páramo has not been definitively established. González (2001) assigns the volcanoclastic sequences to the Pliocene; however, the intimate spatial relationship between the Páramo de Frontino stock and the volcanic pile suggests they may be cogenetic. The volcanic portions of the complex consist of sub-horizontal, bedded, coarse volcanic agglomerates and breccias and finer-grained lithic, crystal, lapilli and ashflow tuffs which range from andesitic to dacitic in composition and from fresh to strongly hydrothermally altered. Based upon the interpretation of satellite imagery in conjunction with field observations, the Páramo de Frontino is interpreted as a partially eroded caldera complex intruded by high-level hypabyssal plutons.

Wokittel (1954) described various mineral occurrences located within the Páramo de Frontino. Ascending the SW slope páramo along the Quebrada Honda stream cut, he noted various zones of "auriferous-argentiferous-cupriferous pyrites" disseminated in the andesitic wallrocks, between ca. 2,200 and 3,100 m elevation. Upon the páramo, the most significant mineralization includes the Lunareja structure, exposed between 3,400 and 3,600 m elevation. Lunareja consists of a 30 m-wide zone of fracturing, brecciation, sheeted and stockwork-style veining and epithermal-textured, colloform-banded and druzy quartz and chalcedony infilling developed within andesitic and dacitic pyroclastic rocks. The zone strikes N105°E, dips 85°S to vertical and can be traced for over 3 km in outcrop. Within the core structure, mineralization consists primarily of massive and fragmented/brecciated mixed sulphides, dominated by brown sphalerite and chalcopyrite, with localized concentrations of pyrite, argentite and tetrahedrite and minor stibnite, galena and native gold, set within a gangue matrix of banded and brecciated quartz, amethyst, agate, jasper and chalcedony. Additional, paragenetically late quartz, siderite, calcite, barite and hematite are also observed, primarily filling or lining vugs and open spaces within the sulphide-silica-dominated phases. Sheeted and stockwork veining on the margins of the core structure contain quartz and amethyst, however, based upon observed boxworks, may have included carbonate, barite and/or mixed sulphides, which have been leached from surface outcrop. Wokittel (1954) suggested that mineralization at Lunareja was deposited in at least three paragenetic stages: (1) early base metals with quartz, (2) quartz-chalcedony with sulphosalts and precious metals, and (3) late gangue minerals including calcite, siderite, barite, hematite and Mn-bearing phases. Based upon exploration drifting at Caja de Oro, Los Rusos and El Fierro, Wokittel (1954) notes that the distribution of sulphide phases within the core structure is highly erratic and that the greatest concentration of sulphides occur within a 3 m section on the southern margin of the fracture zone. Based upon textural and mineralogical attributes, paragenesis and geological setting, Lunareja demonstrates many of the features common to intermediate-sulphidation epithermal vein deposits formed coeval with metaluminous, calc-alkaline andesitic-dacitic granitoid arc magmatism (Sillitoe and Hedenquist 2003).

ca. 11.8 Ma El Cerro (Cerro Frontino, La Horqueta, Morrogacho) Igneous Complex Along trend to the north of the Páramo de Frontino, the El Cerro igneous complex is comprised of a series of small, semi-isolated, petrogenetically related stocks, located between the towns of Frontino, Cañas Gordas and Abriaquí, including at Cerro Frontino (not to be confused with the Páramo de Frontino), La Horqueta and Morrogacho. Au (Cu, Zn, Ag, As, W, Co ± PGE) mineralization is hosted within dikes, sheeted veinlets and contact zones within the intrusives, within hornfels and skarn developed along the intrusive margins and within sheeted veinlets and stockworks, replacement-style mantos and breccias zones developed within Cañasgordas (Penderisco Fm.) sedimentary rocks peripheral to the individual stocks. Based upon petrographic analyses, Escobar and Tejada (1992) and González (2001) observe that the individual plutons of the El Cerro complex range from melanodiorite to pyroxenite, dominated by cpx±hornblende-biotite-bearing monzodiorite.

Mineralization hosted within intrusive rocks and associated hornfels is best developed at Cerro Frontino (Mina San Diego) where fine- to medium-grained phaneritic hornblende- and biotite-bearing pyroxenite, pyroxene gabbro and melanodiorite are cut by dikes of pegmatite containing multi-cm crystals of clinopyroxene, hornblende, biotite and plagioclase. Au mineralization is exposed in numerous artisanal workings where it manifests within at least 10, widely spaced (20 to > 50m), N80 to 110E-striking, 45 to 80 degree S-dipping corridors, each containing sparse sheeted veinlets. Individual veinlets, including visible alteration haloes, are narrow, ranging from less than one to 10 cm in width. Two to three sub-parallel veinlets are commonly clustered within individual 1 to 2 m-wide corridors, and individual corridors have been mined for up to 300 m along strike. The core of individual veinlets measures only a few mm to 1–3cm thick. Ore petrography studies by Molina and Molina (1984) and Escobar and Tejada (1992) indicate ore deposition in two paragenetic phases, including (1) Au-molybdenite-scheelite-cobaltite ± lollingite and quartz followed by (2) chalcopyrite-pyrrhotite-sphalerite-quartz ± scheelite.

Accumulations of coarse-grained native Au within the veinlets are common and locally spectacular. Gangue phases include minor quartz and calcite ± felted masses of biotite, tremolite, chlorite, diopside and wollastonite. Detailed petrographic studies by Escobar and Tejada (1992) document late magmatic K metasomatism along the mineralized corridors and, in general, within the Cerro Frontino pyroxenite, with the sequential replacement of augite and diopside by hornblende and, in turn, by euhedral Fe-rich biotite, which occupies 20 or more modal percent of the rock. K/Ar analysis of late magmatic biotite by Leal-Mejía (2011) returned an age of 11.8 ± 0.4 Ma (Fig. 6.12). Late hydrothermal effects include the local replacement of biotite by chlorite and epidote. Alteration haloes along the veinlet margins are marked by the coarsening of biotite and by the appearance of a calc-silicate mineral assemblage including tremolite, scapolite, wollastonite, scheelite, magnetite, idocrase, apatite, calcite and epidote. Escobar and Tejada (1992) note the presence of anomalous Pt to 118 ppb, associated with pyroxenitic wallrocks. The pegmatite dikes are weakly auriferous. They contain individual crystals of clinopyroxene, hornblende, biotite and plagioclase up to 5 cm, with finer-grained interstitial phases including quartz, calcite, epidote, chlorite, pyrite, chalcopyrite, sphene and magnetite. The margins of the Cerro Frontino stock are marked by a tens-of-metre zone of strong hornfelsing and calc-silicate development within siltstone and arenaceous sandstone of the Penderisco Fm. The hornfels is highly siliceous, compact and massive to weakly banded. It is comprised of saccharoidal quartz, diopside and tremolite-actinolite with lesser amounts of scapolite and minor calcite, idocrase, garnet, apatite, epidote, sphene, magnetite and disseminated pyrite (Escobar and Tejada 1992).

Due E of Cerro Frontino, at Morrogacho-Cerro Pizarro, Au mineralization exposed in numerous artisanal tunnels is contained within narrow sheeted veins, stockworks, breccias and fracture and fault zones hosted within weakly hornfelsed carbonaceous siltstone, sandstone and shale of the Penderisco Fm., in the vicinity of the Morrogacho stock. Mineralization is similar to that observed around the Farallones Batholith. Individual veinlets and fractures contain pyrite, pyrrhotite, arsenopyrite and chalcopyrite±sphalerite within a gangue dominated by quartz and calcite. Wallrock alteration associated with mineralization includes general hornfelsing with silicification, disseminated pyrite and minor sericite forming haloes along the margins of the mineralized structures.

Gold mineralization spatially associated with the El Cerro igneous complex is intimately associated with individual stocks and dikes and their thermal contact haloes, including at Cerro Frontino, Morrogacho and La Horqueta. At Cerro Frontino, Au mineralization appears to represent late magmato-hydrothermal solutions associated with intense late magmatic K (biotite) metasomatism of pyroxenite and melanodiorte, accompanied by calc-silicate alteration and mineral deposition. The 11.8 ± 0.4 Ma K/Ar (biotite) age (Leal-Mejía 2011) is considered to represent a maximum age for mineralization. More distal occurrences (e.g. Cerro Pizarro) are interpreted to be linked to the cooling history of the individual plutons.

ca. 9 to 4 Ma Middle Cauca Porphyry-Related Au-Cu
and Ag-Au-Zn (Pb, Cu) Trends

Based upon albeit limited K-Ar age dating, granitoid plutonism along ca. 12 Ma Farallones-El Cerro trend, north of the Caldas-Coiba tear, ceased at ca. 11 Ma, and eastward migration of the metaluminous calc-alkaline granitoid arc axis into the Middle Cauca region (Shaw 2003a, 2003b; Sillitoe 2008; Leal-Mejía 2011) is observed (Fig. 6.12). Based upon available radiometric age dates and lithogeochemical data (Leal-Mejía et al. 2018), subduction-related granitoid magmatism first appears along the Middle Cauca at ca. 9 Ma.

The Middle Cauca hypabyssal porphyry belt forms an ca. 120 km-long, N-S granitoid arc segment extending on either side of the Cauca River valley from Pereida in the south to Buriticá in the north (Fig. 6.16). The majority of the Middle Cauca porphyry belt, from N of Pereida to Titiribí, intrudes Romeral mélange basement and probable Dagua terrane basement to the W of the Cauca River. Within this region, the oceanic basement terranes are unconformably overlain by early-mid-Miocene siliciclastic sequences of the Amaga Fm. (González 2001) and mid- to late Miocene volcanic and volcanoclastic rocks of the Combia Fm., both of which are important hosts to mineralization in the porphyry-proximal environment. Between Titiribí and Buriticá, the belt is mostly hosted within mixed Cañas Gordas terrane oceanic volcano-sedimentary rocks.

From a historic to modern-day perspective, the Middle Cauca porphyry belt is perhaps the best documented (e.g. Sillitoe 2008) and most explored of the Colombian precious metals provinces, and historic camps dating from pre-Colonial and Colonial times, such as Marmato, Titiribí and Buriticá, have received a certain degree of attention within more recent, readily accessible international literature. In this context, we will forego detailed descriptions of individual camps of the Middle Cauca, in favour of providing a generalized composite summary of deposit styles. In addition, we will provide citations of the most relevant geological studies pertaining to individual deposits or districts.

From a metallogenic standpoint, two broad hypabyssal porphyry-associated mineralization styles are observed along the Middle Cauca: (1) porphyry-related Au-Cu mineralization sensu stricto (Sillitoe 2000), comprised of multiphase quartz+magnetite+sulphide stockwork veining and disseminations and paragenetically associated calcic to potassic alteration assemblages centred upon clusters of weakly to moderately porphyritic hornblende-biotite plagioclase diorite ± granodiorite dikes and stocks and extending into the adjacent wallrocks, and (2) intermediate- to low-sulphidation, epithermal Ag-Au-Zn (Pb-Cu) deposits associated with phyllic/argillic (sericite, illite±chlorite, quartz, pyrite) alteration assemblages, hosted within and peripheral to hypabyssal granodiorite to quartz monzonite porphyry stocks, commonly restricted to structural corridors, breccias, fault-vein arrays and stratigraphic discontinuities, transecting the porphyritic stocks and mineralizing country rocks in the circum-porphyry environment. In various instances, the two styles of mineralization coexist within a single mineral camp, as spatially separate or overprinting assemblages. We note, however, that the best developed examples of each style are contained within spatially separate deposits.

Middle Cauca Porphyry-Style Au-Cu Occurrences

Porphyry-style Au-Cu mineralization associated with porphyritic diorite ± granodiorite intrusions, hosted within Romeral and Dagua terrane basement and cutting Miocene Amagá and/or Combia Fm. cover rocks along the southern Middle Cauca belt, is observed at (from S to N) Marsella, Villamaría, Quinchía, south Támesis, La Quebradona, La Mina and Titiribí (Fig. 6.16). Farther north, similar porphyry-related Au-Cu mineralization of unconstrained age, hosted within Cañas Gordas terrane volcano-sedimentary basement, outcrops at Chuscalito-Mina Alemana. The age of these porphyry-style manifestations to the N of Titiribí can presently only be inferred based upon similarities with well-constrained late Miocene Au-Cu occurrences to the south and based upon location within the approximately N-trending linear projection of the southern Middle Cauca belt.

Focusing upon the southern Middle Cauca belt, the most intensely mineralized porphyry centres are characterized by the presence of multiple phases of generally finer-grained, sparsely to moderately crowded, hornblende-biotite plagioclase ± quartz porphyry producing two or more superimposed mineralizing events. Mineralization is characterized by multiple and overprinting alteration assemblages typical of Au-rich porphyries worldwide (Sillitoe 2000). Early, high-temperature, calcic and potassic alteration phases include calcic amphibole+magnetite+sulphide in veinlets and veinlets and disseminations of quartz+magnetite+secondary biotite+sulphides, respectively. K-feldspar, occurring as haloes about magnetite and/or quartz veinlets, is common but generally subordinate to biotite as a potassic alteration phase. Sodic alteration, recorded as albite haloes along early veinlets, is locally observed but is neither common nor strongly developed. Magnetite (+amphibole) and quartz+magnetite veinlets are generally sinuous in nature, forming weak to moderate and, in the case of multiple generations, intense, multi-directional stockworks within the porphyry host and in some cases extending for tens of metres into the surrounding basement rock, where consistent grade Au and Cu mineralization is commonly maintained. Pyrite and chalcopyrite are the principal sulphide phases, occurring in veinlets accompanying calcic and potassic assemblages and as disseminations. Bornite is locally abundant, mostly as a supergene phase. Small amounts of molybdenite in veinlets are observed in most deposits. Total sulphide in most cases is generally low, averaging <2%, and total magnetite often exceeds total sulphide content. Veinlets of anhydrite and gypsum are locally well developed in some deposits (e.g. La Mina, Titiribí) whilst apparently absent in others. In the cases where hypogene alteration and mineralization are not overprinted by lower temperature phyllic-argillic assemblages, porphyry-related calcic-potassic alteration and Au-Cu mineralization pass abruptly into well-developed, regionally distributed, generally barren propylitic assemblages, recorded within spatially associated, unmineralized (intra-mineral, post-mineral) porphyry, mafic volcanic basement rocks or bas-andesitic to andesitic volcanoclastic rocks of the

Combia Fm. Propylitic alteration throughout the Middle Cauca is dominated by abundant chlorite, epidote, pyrite and calcite±quartz, distributed as disseminations and within fine fractures and veinlets. Notwithstanding, virtually all of the high-temperature, calcic-potassic Au-Cu cores along the Middle Cauca are in fact to some degree overprinted, by phyllic-argillic assemblages associated with continued, post-calcic+potassic+Cu-Au, porphyritic magmatism or volcanism along the Middle Cauca belt. Such overprints range from discrete and minor to pervasive and extreme and are associated with the development of various styles of overprinting and peripheral epithermal Ag-Au-Zn (Pb, Cu) mineralization, spatially related to the Au-Cu systems at the camp or porphyry cluster scale. The most important localities dominated by epithermal styles of mineralization will be outlined in the following section. We now provide additional details and age constraints pertaining to the most prominent porphyry-related Au-Cu centres along the Middle Cauca. It must be noted that, historically, and to date, little Au (Cu, Ag) production has been derived directly from hypogene porphyry-style Au-Cu ores. Additionally, due to their relatively recent discovery, little has been published regarding porphyry-related Au-Cu mineralization along the Middle Cauca. Unpublished data generated by private industry is not considered herein.

Quinchía Three calcic-potassic Au±Cu porphyry core zones, aligned along an ca. 3 km N-S trend have been identified at Quinchía, including (S to N) La Cumbre, Mandeval and Dosquebradas (Leal-Mejía 2011). Mineralization at La Cumbre may be characterized as weakly to moderately developed and is locally overprinted by late, discrete, auriferous pyrite-sericite veins hosted within minor gouge-rich fault zones (e.g. La Balastrera). At Mandeval, early calcic-potassic assemblages and intense quartz veinlet stockworking are overprinted by pervasive sericite-chlorite-pyrite-clay (intermediate argillic) alteration leading to widespread leaching of Au-Cu mineralization. Spaced auriferous pyrite-sericite veins containing minor quartz and sphalerite, localized along minor faults cut both alteration assemblages. At Dosquebradas, intense early quartz stockworking and pervasive biotite alteration are developed within mafic volcanic basement rocks of the Dagua Fm.(?) along the margin of similarly altered diorite porphyry. Weakly mineralized (intra-mineral) diorite porphyry from the Dosquebradas suite returned a magmatic crystallization age of 8.0 ± 0.5 Ma, whilst K-Ar (whole-rock) dating of early diorite porphyry containing pervasive secondary biotite alteration returned a weighted average of 8.2 ± 0.7 Ma (Leal-Mejía 2011; Leal-Mejía et al. 2018). Additionally, molybdenite from a monomineralic veinlet in sericitized diorite porphyry returned a Re-Os age of 7.7 ± 0.2 Ma (Leal-Mejía 2011).

South Támesis ca. 30 km north of Quinchía, recent exploration has identified a 3 km, NNE-oriented trend of Au (Cu±Mo) porphyry centres hosted within and along the SE margin of the Támesis stock. Well-developed mineralization has been identified at (S-N) El Reten, El Corral, Ajiaco Sur, Malabrigo and Casa Verde, whilst artisanal workings hosted within the Támesis stock to the west reveal additional Au (Cu) mineralization at El Conde. Mineralization is best developed at El Reten, where intense quartz-sulphide stockworking with associated calcic-potassic alteration is hosted within early diorite porphyry and extends for tens of metres into bas-andesitic Combia Fm. wallrocks. K-feldspar forms the dominant potassic phase within the porphyry but is replaced by dominant secondary biotite in the volcanic wallrocks. Quartz stockworking and alteration are less intense at El Corral and Casa Verde. At Ajiaco Sur, mineralization is hosted directly within a medium-grained dioritic phase of the Támesis stock. At El Conde, mineralization is also hosted within phaneritic diorite and granodiorite of the Támesis stock. NE-striking veins up to 30 cm thick with associated sheeted veinlets contain quartz, magnetite, pyrite and chalcopyrite. Various phases of barren, post-mineral porphyry are also observed along the trend where early phases, including mineralization are commonly overprinted by late propylitic (chlorite-epidote) alteration. No radiometric age dates have been published for the Támesis South porphyry Au-Cu trend. Leal-Mejía et al. (2018)) provide U-Pb (zircon) ages for the Támesis stock ranging from ca. 7.8 to 7.2 Ma. We interpret the south Támesis porphyry centres to be related to the late magmatic evolution of the Támesis stock.

La Quebradona Located some 22 km to the NNW of south Támesis, the La Quebradona porphyry cluster (Fig. 6.16) is located within ca. 10 km diametre circular structure interpreted as a partially eroded, nested caldera. To date, five spatially separate calcic-potassic centres have been identified within the confines of the circular structure, including La Isabella, La Sola, El Tenedor, La Aurora and El Chaquiro (Leal-Mejía 2011). Mineralization outcropping in road cuts and stream cuts at La Isabella, La Sola and El Tenedor is only weakly developed and appears to involve only one or two phases of early diorite porphyry. Au-rich mineralization is best developed at La Aurora where a well-mineralized, originally contiguous polyphase diorite cluster measuring ca. 400 m in diametre is divided by a 100 m dike of post-mineral diorite. Mineralization and alteration extend for tens of m into Combia Fm. volcanic host rock. Early mineral diorite was dated at 7.6 ± 0.2 Ma U-Pb (zircon) (Leal-Mejía et al. 2018). At El Chaquiro, weakly developed potassic alteration and magnetite-quartz stockworking are developed within early mineral diorite outcropping in La Quebradona creek. Adjacent to and topographically above the diorite, a large phreatomagmatic breccia is developed in volcanic agglomerate and tuff of the Combia Fm. The breccia measures some 400 m along a NNE axis and can be traced for over 100 m vertically. It is characterized by intense breccia development accompanied by pervasive illite-sericite-pyrite-tourmaline alteration (Leal-Mejía 2011) and may be related to porphyry-style Au-Cu at depth (Sillitoe 2000). Distal to the breccia, at La Coqueta, intermediate-sulphidation epithermal veins and veinlets contain the assemblage pyrite-sphalerite-chalcopyrite-electrum ± galena-acanthite-hessite-cervelleite-hedleyite-tetrahedrite-matildite-stromeyerite-freibergite within a gangue matrix comprised of quartz and ankerite (Pujol et al. 2012).

La Mina 20 km to the NNE of La Quebradona, at La Mina, various porphyry centres, aligned along a N-S axis, outcrop over an ca. 1.5 km trend. Mineralization is best exposed at La Cantera (a roadside quarry), where moderate to strong calcic-potassic assemblages associated with at least two phases of early mineral porphyry contain strong Au-Cu mineralization. Alteration and mineralization extend for tens of metres into Combia Fm. bas-andesitic wallrocks. Diorite porphyry from La Cantera returned a U-Pb (zircon) age of 7.6 ± 0.2 Ma (Leal-Mejía et al. 2018). Six hundred metres to the north, less well-developed mineralization is exposed in a series of historic hydraulic open-cut trenches. Farther north at El Limón, the Au-Cu trend appears to be truncated by courser-grained, post-calcic-potassic, granodiorite porphyry containing strong to pervasive illite-sericite-pyrite with overprinting propylitic alteration assemblages and hosting localized intermediate-sulphidation Ag-Au-Zn (Pb, Cu) veins and breccias (Grosse 1926; Botsford 1926).

Titiribí Titiribí is an important historic Au-Ag-producing district with a formal and artisanal mining history extending over at least 240 years. Excellent detailed documentation of the geology and gold mineralization at Titiribí is provided in the works of Grosse (1926,1932; summarized by Emmons 1937) and by Botsford (1926), and, specifically, the work of Grosse (1926) must be considered among the finest of studies pertaining to the geology and mineralization of any mineral district in Colombia. The historic Titiribí mining district was mostly developed along highgrade epithermal veins (reviewed below) localized along the lower eastern flank of Cerro Vetas, a polyphase monzodiorite, diorite-granodiorite to quartz monzonite porphyry complex outcropping at higher elevations to the west of Titiribí townsite. The porphyry complex intrudes and mineralizes pre-Cenozoic Romeral mélange metamorphic basement, comprised of amphibolite and graphitic quartz-mica schist and mafic volcanic rocks, including overlying Miocene Amagá Fm. siliciclastic sedimentary rocks. Porphyry-style Au-Cu mineralization at Cerro Vetas was not historically exploited and was only identified during recent exploration activities (Meldrum 1998; Leal-Mejía 2011). Mineralization is associated with multiple phases of diorite to monzodiorite and monzonite porphyry. Early alteration is dominated by a potassic assemblage containing secondary biotite with veinlets of quartz+magnetite ± a generally subordinate calcic amphibole component. Biotite is replaced by K-feldspar as the potassic phase in late syn-mineral monzodiorite porphyry. Chalcopyrite provides the hypogene Cu-bearing phase, whilst bornite is absent. Strong to intense and widespread secondary biotite development is observed within schistose basement rocks along the NW margin of the porphyry system (Meldrum 1998). With respect to age, Leal-Mejía et al. (2018)) provide a U-Pb (zircon) age of 7.6 ± 0.3 Ma for propylitically altered late mineral diorite from Cerro Vetas.

Chuscalita-Mina Alemana The southern extension of the Middle Cauca porphyry Au-Cu belt, hosted within Romeral-Dagua basement, terminates just to the north of Titiribí. Farther to the NW, located ca. 14 km WNW of Anzá townsite (Figs. 6.12 and 6.16), porphyry-related Au-Cu mineralization is hosted within mafic to intermediate volcanic rocks of the Barroso Fm. (Cañas Gordas terrane), exposed in outcrop and artisanal Au workings along the Quebrada Chuscalita and at Mina Alemana. Along Quebrada Chuscalita, mineralization is hosted within weakly porphyritic, fine- to medium-grained quartz diorite to granodiorite, containing sparse veinlets but pervasive secondary biotite alteration accompanied by fine disseminated chalcopyrite, bornite, pyrite and magnetite. Compact, siliceous biotite-pyrrhotite hornfels is well developed along the contact with the Barroso Fm. 250 m to the west, beyond the hornfels aureole, at Mina Alemana, steeply dipping to vertical, cm-scale sheeted veinlets, containing quartz, pyrite, pyrrhotite, chalcopyrite, bornite, magnetite and native Au, outcrop within an ca. N110E-trending corridor cutting mixed volcano-sedimentary strata of the Barroso Fm. No radiometric age dates are available to better constrain the porphyry-related mineralization at Chuscalita-Mina Alemana.

Middle Cauca Epithermal Ag-Au-Zn (Pb-Cu) Trend

Intermediate- and low-sulphidation epithermal deposits hosted within and peripheral to Middle Cauca suite dioritic to granodiotitic hypabyssal porphyry intrusions include some of the best documented and historically productive hypogene Au (Ag) occurrences in Colombia. In some instances, the epithermal manifestations are related to composite porphyry districts such as at Quinchía, La Mina and Titiribí, where they may form spatially separate mineralized centres or, more commonly, are adjacent to and/or partially overprint calcic- and potassic-altered Au-Cu porphyry cores. In other instances, such as at Supía-Riosucio, Marmato and Buriticá, significant porphyry-style Au-Cu mineralization, or related calcic-potassic alteration, has not been identified. Epithermal mineralization along the Middle Cauca takes on various styles and variations, many of which are a reflection of the nature of host rocks and host basement. Typical porphyry-hosted epithermal mineralization, however, as seen at Villamaría, Supía-Riosucio, Marmato, Pácora, Yarumalito and Chisperos (Titiribí) (Fig. 6.16), is generally characterized by the presence of structurally controlled, syn-tectonic fault-veins with associated sheeted and weak stockwork-style or joint-controlled veining and localized breccias, hosted within hundreds of metres- to km-scale, weak to strong and pervasive phyllic grading to argillic alteration haloes. Affected porphyries are dominated by coarser-grained, moderately crowded, hornblende-biotite plagioclase ± K-feldspar quartz diorite to granodiorite and quartz monzonite, which, where observable, often intrude the finer-grained, sparsely porphyritic early diorites commonly associated with Au-Cu mineralization. The epithermal veins and breccias are sulphide-rich and characterized by multiple stages of mineral infilling, dominated abundant pyrite±pyrrhotite, Fe-rich and Fe-poor sphalerite with lesser chalcopyrite, arsenopyrite and galena ± localized tetrahedrite, stibnite, molybdenite and assorted Ag-Pb-sulphosalts. Both native Au and electrum are observed. Gangue phases consist of calcite±ankerite and generally subordinate quartz±adularia and marcasite. Core mineralization within individual structures generally measures in the range of 5 to 30 cm; however, in exceptional cases, it may attain thicknesses of >1 m. Core structures however tend to anatomose and pinch and swell abruptly along strike and down dip, and veins contain a ubiquitous quantity of fault gouge comprised of mixed clays, milled porphyry wallrock and ground mixed sulphide. In exceptional cases, larger structures, including gouge and gangue phases, may measure multiple metres thick over limited distances and contain various lenses of core sulphide mineralization, with internal layers of gouge, suggesting vein formation was syn-tectonic and took place during repeated phases of structural disruption and hydrothermal infilling. Wallrock alteration proximal to veins is dominated by pervasive, medium- to course-grained sericite, which in turn is enveloped by widespread disseminated argillic (illite-pyrite) alteration and, commonly, by fine pyrite±quartz-filled joint sets. Epithermal mineralization and alteration overprint a regional propylitic (epidote-calcite-albite-pyrite±chlorite-quartz) assemblage, clearly evident in distal and unmineralized penecontemporaneous porphyry species. Principal mineralized core veins may be clustered in sheeted and anastamosed arrays in metric proximity between one structure and another, but more commonly, core structures are spaced at tens to hundreds of metres or more. High-grade mineralization in most cases is closely confined to the mineralized fault veins and breccias, and little or no economic disseminated mineralization is to be found within the widespread argillic haloes between structures. In the case of widely spaced core structures, and a resultant lack of coalescence between alteration haloes, wallrock alteration will pass from argillic to the regional propylitic assemblage. In this context, from the standpoint of bulk mineability, economic grades within these large-scale, porphyry-hosted epithermal systems remain entirely dependent upon the structural density provided by the mineralized core structures ± spatially associated veinlets.

Based upon geological setting, ore mineral and paragenetic assemblages and alteration styles, epithermal mineralization along the Middle Cauca demonstrates features characteristic of epigenetic, intermediate-sulphidation, adularia-sericite precious+base metal deposits associated with porphyritic stocks and related volcanic rocks within calc-alkaline granitoid arcs (Hayba et al. 1985; Rossetti and Colombo 1999; Sillitoe and Hedenquist 2003; Sillitoe 2008). The structurally controlled nature and the apparent lack of a syn-mineral magmatic component (dikes, magmatic injections) within or accompanying vein formation suggest epithermal mineralization is late or generally post-dates emplacement with respect to the hosting porphyry stocks. Available radiometric age data support this in part, however, suggests mineralization closely follows (e.g. Tassinari et al. 2008; Henrichs 2013) or is penecontemporaneous with (e.g. Lesage et al. 2013) the cooling of the host pluton. We now provide additional details pertaining to the most prominent epithermal centres along the Middle Cauca.

Quinchia Two distinct styles of epithermal mineralization, revealed by extensive artisanal workings, are highlighted within the Quinchía district, including at Loma Guerrero (Chuscal) and Miraflores (Leal-Mejía 2011). At Loma Guerrero, mineralization is localized along the contact between the late Cretaceous Irra stock (Leal-Mejía 2011; Leal-Mejía et al. 2018) and altered late Miocene granodiorite porphyry. On the northern flank of the hill, at Tres Cuevas, Au mineralization is developed in brecciated, silicified and pervasively sericitized porphyry, containing quartz and pyrite in veinlets and pyrite as disseminations and nests filling open spaces. To the south at El Chuscal, underground workings on the Guayacán vein system reveal narrow, high-grade auriferous veinlets cutting pervasively argillitized granodiorite porphyry. Vein fillings consist of early quartz and pyrite. A second phase of sulphide deposition, consisting of galena, sphalerite, chalcopyrite and tetrahedrite with native Au, replaces early pyrite (Leal-Mejía 2011).

Two kilometres to the north of Loma Guerrero, at Miraflores (Rodríguez and Warden 1993; Rodríguez et al. 2000), an auriferous, sub-cylindrical magmato-hydrothermal breccia body measuring some 250 by 280 m in plan cuts Cretaceous mafic and ultramafic oceanic volcanic and intrusive basement rocks. The breccia body contains widespread but erratic low-grade gold values and is cut by narrow, NNW-striking, high-grade feeder veins which have been exploited by artisanal miners to over 100 m depth. The breccia varies from clast to matrix supported (Rodríguez et al. 2000; Ceballos and Castañeda 2008) and is characterized by abundant interstitial porosity permitting the development of well-terminated hydrothermal infilling phases. Lithic fragments are dominated by angular to sub-rounded clasts of mafic and ultramafic basement rocks but include felsic porphyry clasts of both late Cretaceous and late Miocene age (Leal-Mejía 2011). Hydrothermal alteration and infilling is dominated by a calcic assemblage including (in approximate paragenetic order) epidote, quartz and calcite with late, well-terminated zeolites, including heulandite and chabazite (Ceballos and Castañeda 2008; Leal-Mejía 2011). Lattice-textured calcite was interpreted by Ceballos and Castañeda (2008) to indicate fluid boiling. The ore mineral assemblage includes early pyrite with a later phase of pyrite accompanied by lesser galena, sphalerite, chalcopyrite and hessite (Ceballos and Castañeda 2008; Leal-Mejía 2011). Native gold accompanies the late sulphide assemblage and also occurs locally as spectacular dendritic and crystalline infillings within breccia cavities. Neither the Miraflores breccia nor the epithermal occurrences at Loma Guerrero have been dated precisely by radiogenic means. Mineralization in all cases cuts, alters or contains clasts of late Miocene porphyry and is interpreted herein to be genetically related to the cooling history of the Quinchía late Miocene hypabyssal porphyry cluster (Fig. 6.16).

Supía-Riosucio ca. 17 km due north of Miraflores, epithermal Au mineralization hosted within hypabyssal diorite and granodiorite porphyry and associated felsic pyroclastic rocks of the Combia Fm. outcrops along highway between the towns of Supía and Riosucio. Historically, the great majority of the Au production from this sector has been recovered from alluvial deposits along the Supía River to the south of the town of Supía. Notwithstanding, narrow sulphide-rich intermediate-sulphidation veins have also been widely exploited by artisanal means within the sector known as Gavia and Vende Cabezas. The veins are widely spaced and contain the typical pyrite-sphalerite±galena and chalcopyrite sulphide assemblage, hosted within a broad area of intense argillic alteration containing abundant pyrite as disseminations and joint and fracture fillings.

Marmato The Marmato camp is an important Au-Ag producer with a production history spanning more than five centuries (Restrepo 1888). Numerous academic works (e.g. Rodriguez 1987; Warden and Colley 1990; Rodríguez and Warden 1993; Rossetti and Colombo 1999; Díaz 2002; Vargas 2005; Tassinari et al. 2008;

Leal-Mejía 2011; Santacruz 2011,2016; Santacruz et al. 2012, 2014) document the intermediate-sulphidation, hypabyssal porphyry-hosted, fault-vein and breccia-style epithermal Ag-Zn-Au (Pb, Cu) mineralization, which continues to be exploited in dozens of artisanal and more formalized underground developments in the Zona Alta, Cien Pesos, Zona Baja, Echandía and La María zones and in other more isolated vein systems contained within the overall 10 km^2 camp. Marmato, the type locality for the high-Fe sphalerite mineral "marmatite", is also the type locality for porphyry-hosted fault-vein mineralization along the Middle Cauca. Epithermal mineralization conforms to the previously related generalized description, and the reader is referred to the above cited references for additional details. The Marmato porphyry cluster is comprised of multiple individual porphyry phases, yet surprisingly few radiometric age dates have been published for these rocks. Frantz et al. (2003) provided a 6.5 ± 0.2 Ma U-Pb (zircon) crystallization age for diorite porphyry hosting mineralized fault veins in the Zona Alta. This coincides well with a 6.7 ± 0.6 Ma $^{40}Ar/^{39}Ar$ age for magmatic biotite from the Marmato stock, provided by Vinasco (2001). Tassinari et al. (2008) published a 5.6 ± 0.6 Ma K-Ar (sericite) age for granodiorite porphyry hosting mineralized veins in the Zona Baja and interpreted this to represent the age of mineralization. Interestingly, Vinasco (2001), in parallel studies pertaining to faulting along the Middle Cauca, provided a 5.6 ± 0.4 Ma $^{40}Ar/^{39}Ar$ (biotite) step heating age for the Marmato stock, which he interpreted to represent tectonic reactivation along the Cauca-Romeral fault systems. The age is indistinguishable from the Tassinari et al. (2008) interpreted age for mineralization and strongly supports field observations described above regarding the post-host porphyry, syn-tectonic emplacement and evolution of Marmato-style vein systems along the Middle Cauca. Recently, Santacruz (2016) provided ten LA-ICP-MS U-Pb (zircon) ages for multiple phases of the "Marmato-Aguas Claras Suite – MACS" between ca. 6.6–6.3 Ma and ca. 5.7 Ma for pre-mineralization and post-mineralization porphyry phases, respectively. Moreover, the age of mineralization at Marmato was also constrained by two $^{40}Ar/^{39}Ar$ plateau ages obtained from adularia in veins and veinlets of the upper (5.96 ± 0.02 Ma) and lower (6.05 ± 0.02 Ma) mineralized zones of the deposit (Santacruz 2016).

Caramanta-Valparaiso Centred ca. 12 km to the NNW of Marmato, between the towns of Caramanta and Valparaiso, a broadly E-W-striking corridor of widely spaced fault-veins transects late Miocene diorite to granodiorite porphyry, Amagá Fm. sedimentary and Combía Fm. volcanic rocks. Auriferous veins, observed in artisanal workings along the Quebrada Honda and at Yarumalito, are characterized by high total sulphide content, with late quartz, calcite and sericite gangue-rich selvages, hosted within broad zones of illite-pyrite alteration affecting both porphyries and the Combia Fm. volcanic rocks. At Yarumalito, localized secondary biotite and magnetite, in disseminations and veinlets, are observed within diorite porphyry, suggesting that argillic alteration associated with epithermal mineralization overprints an early potassic event. Similar mineralization and alteration are exposed to the east at Orofino and Bermejal, in outcrop and highway cuts along the western margin of the Cauca River. At Yarumalito, Henrichs (2013) provided U-Pb (zircon) crystallization ages of 7.00 ± 0.15 Ma and 6.95 ± 0.16 Ma for samples of andesite and diorite, respectively, both host to mineralization. She concluded that porphyritic magmatism was closely related to the final stages of Combia Fm. volcanism and that the Yarumalito epithermal deposits were emplaced in fault-veins shortly thereafter.

Titiribí As indicated above, Titiribí was an important late Colonial and post-Colonial Ag-Au (with by-product Zn, Cu and Pb) district, with pre-1930 production estimated at between 1.5 and 2.5 million ounces Au equivalent (Botsford 1926). Production was derived from a complex and geometrically diverse series of precious and base metal-rich deposits, localized along faults, unconformities, bedding plane discontinuities, mantiform replacements and disseminations and intrusive contact zones, hosted within Romeral mélange basement schist and Amagá Fm. siliciclastic sedimentary rocks, intruded by the Cerro Vetas late Miocene diorite to granodiorite and quartz monzonite porphyry. The principal mining centres within the district, located to the N and NW of Titiribí townsite, included Altos Chorros, La Independencia-Zancudo, Cateador-Chisperos and Otramina, each containing numerous individual deposits. Grosse (1926, 1932) provided detailed descriptions of the occurrences, including observations regarding geological structure, host rocks, ore, gangue and alteration mineral paragenesis and grade (Au-Ag-Pb-Zn-Cu-Sb-As) of typical ore from many of the individual structures. Grosse (op. cit.) noted that diapiric doming and reverse faulting within the basement complex and Amagá Fm., around the Cerro Vetas porphyry complex, provided a first-order control to the distribution of structural and stratigraphic dilatencies and traps, which host mineralization throughout the district. He concluded that the veins, mantos, replacements and impregnations at Titiribí were derived from hydrothermal segregations associated with the Cerro Vetas "laccolith". The deposits formed more or less simultaneously, following emplacement of the intrusion (Grosse 1926).

More recent studies pertaining to the epithermal deposits at Titiribí have focussed upon ore mineralogy and mineral paragenesis, geothermometry and sulphur and Pb-isotope studies (e.g. Leal-Mejía et al. 2006; Gallego and Akasaka 2007, 2010; Leal-Mejía 2011; Uribe 2013) of ore samples collected from underground exposures within basement metamorphic rocks (Sabaletas schist) and permeable quartz-rich sandstones and conglomerates of the Amagá Fm. Results generally confirm the findings of Grosse (1926, 1932). Structural, stratigraphic and contact-controlled mineralization is characterized by an assemblage consisting predominantly of admixtures of massive and granular sulphides, with limited amounts of gangue and, in structurally controlled cases, clay-rich fault gouge. Sulphides are dominated by abundant arsenopyrite, pyrite and sphalerite, lesser amounts of galena and chalcopyrite and numerous Ag-bearing Pb-Cu-Sb sulphosalts, deposited in at least two paragenetic stages, with native gold, native silver and electrum being introduced late in the ore mineral paragenesis (Gallego and Akasaka 2007; Leal-Mejía 2011). Gangue minerals include quartz and dolomite±calcite. Geothermometric studies by Gallego and Akasaka (2007, 2010) suggest overall sulphide and quartz gangue deposition took place between ca. 420 and 235°C. Studies pertaining specifically to late tetrahedrite deposition within the Independencia tunnel veins, however, suggest ore-stage deposition took place at lower temperatures, between ca. 220 and 170°C (Leal-Mejía 2011). Sulphur isotope values for chalcopyrite, arsenopyrite, sphalerite, galena and boulangerite presented by Leal-Mejía (2011) cluster in a narrow $\delta^{34}S$ range between ca. -2.6 and +3.3 per mil, consistent with a mantle-derived source for S. Alteration associated with the high-sulphide infillings is predominantly phyllic-argillic, including sericite-illite, accompanied by lesser quartz with carbonate and pyrite, developed along margins of mineralized zones. Low-grade haloes containing pyritic disseminations and tectonic arrays of quartz-carbonate-pyrite veinlets may extend for various metres on either side of the high-grade mineralized structures.

Based upon overall geologic setting, field observations and mineralogical, paragenetic, alteration and geothermometric parameters observed across the entire district, high-grade Ag-Au mineralization at Titiribí is consistent with the intermediate-sulphidation epithermal class of deposits (Sillitoe and Hedenquist 2003). No radiometric age dates specifically pertaining to these deposits have been published. Notwithstanding, field, stratigraphic and cross-cutting relationships indicate they are late Miocene in age and they are interpreted to be spatially and temporally related to the emplacement and cooling history of the ca. 7.6 Ma (Leal-Mejía et al. 2018) Cerro Vetas porphyry cluster (ca. Grosse 1926).

Buriticá Located ca. 90 km NNW of Titiribí, the Ag-Au-Zn (Pb, Cu) mineralization at Buriticá is hosted within and peripheral to late Miocene porphyritic diorite to granodiorite which intrude CCOP/CLIP basement comprised of oceanic mafic volcanic and sedimentary rocks of the early Cretaceous Cañas Gordas terrane and the ca. 100 Ma Buriticá stock. Like Marmato and Titiribí, historic production from numerous deposits within the Buriticá camp, including Yaraguá, Los Palacios, María Centena and La Estera, dates from pre-Colombian times (Restrepo 1888), and extensive artisanal and semi-formalized exploitation continues at present. Mineralization at Buriticá is dominated by ENE and ESE striking, steeply dipping fault-veins and localized breccia bodies, hosted within late Miocene porphyry but also within Cañas Gordas Gp. sediments and the Buriticá stock. The porphyry-hosted fault-veins and breccias are in many respects similar to those observed at Supía, Marmato, Yarumalito and elsewhere along the southern Middle Cauca belt. Recent studies (Lesage 2011; Lesage et al. 2013) indicate that the Buriticá vein system overprints early weak potassic and propylitic alterations within the host porphyry stocks. Mineralization manifests as structurally controlled, Ag-Au-sulphide-rich veins and breccias bodies, characterized by (1) the early deposition of sulphide (pyrite>sphalerite>chalcopyrite+galena) with minor tetrahedrite, native Au and electrum, followed by (2) abundant quartz with minor sulphides and (3) the brecciation and the deposition of lesser pyrite+sphalerite+tetrahedrite+stibnite+native Au/electrum and late, abundant calcite (Lesage et al. 2013). Vein-proximal wallrock alteration associated with mineralization is dominated by a phyllic assemblage which includes sericite/muscovite+adularia+quartz+calcite+pyrite and which grades rapidly, in the absence of additional veining, to an epidote-dominant propylitic assemblage. Based upon mineralogical, paragenetic, fluid inclusion and stable isotope data, Lesage et al. (2013) characterized porphyry-hosted Ag-Au-Zn mineralization at Buriticá as intermediate-sulphidation, epithermal, in nature. Leal-Mejía (2011) provided an 11.8 ± 1.1 Ma K-Ar (magmatic hornblende) age for pre-mineral hornblende diorite porphyry from Buriticá. $^{40}Ar/^{39}Ar$ step heating analysis of similar porphyry by Lesage et al. (2013) produced a hornblende cooling age of 7.41 ± 0.4 Ma, suggesting various phases of early (pre-mineral) diorite may be present. Additional $^{40}Ar/^{39}Ar$ step heating analysis of alteration muscovite by Lesage et al. (2013) produced a cooling age of 7.74 ± 0.08 Ma, which these authors interpret as the age of mineralization. Overlap in the Ar-Ar magmatic hornblende vs. the alteration muscovite cooling ages suggests that epithermal mineralization at Buriticá is related to the cooling history of the host late Miocene porphyry complex.

ca. 8.3 to 7.3 Ma Cajamarca-Salento Porphyry Au Province

The late Miocene Cajamarca-Salento porphyry province includes numerous calc-alkaline, hypabyssal diorite to granodiorite porphyry stocks and clusters of stocks contained within a sub-equant, ca. 400 km² area extending between the towns of Cajamarca (Tolima Department) and Salento (Quindio Department) (Núñez 2001) (Fig. 6.17). The province contains the recently discovered La Colosa Au-porphyry (Lodder et al. 2010) and related occurrences at Montecristo, Tierradentro and Salento (Leal-Mejía 2011). Based upon current mineral resource estimates exceeding 28M oz (>870 metric tonnes) of contained Au (AngloGold Ashanti 2015), the La Colosa deposit alone represents the most important modern-day Au discovery in the Colombian Andes. Early stream sediment geochemistry and prospecting (Lozano 1984; Pulido 1988a, b) suggested the potential for "disseminated" Au occurrences in the region, and localized, sporadic exploitation of minor alluvial occurrences and epithermal deposits within fringing drainages is recorded. The Au-rich porphyry occurrences sensu stricto, however, show no evidence of historic or recent artisanal exploitation. In a preliminary analysis, Sillitoe (2008) included the La Colosa deposit within the general trend of the Middle Cauca Belt (Shaw 2003b). Such inclusion, however, requires significant southward extension and eastward deformation of the general N-S trend and belt-like geometry of the Middle Cauca, and based upon location, basement composition and architecture, geochemical arguments and attributes related to the mineralogy, alteration and scale of Au-porphyry mineralization at La Colosa, we consider the Cajamarca-Salento cluster as a separate mineral province.

The Cajamarca-Salento porphyry province is located within Colombia's physiographic Central Cordillera, limited by the Romeral tectonic zone to the west and the Quindio, Bermellón and Toche Rivers to the N, S and E, respectively. Late Miocene porphyritic stocks and dikes intrude greenschist- to amphibolite-grade carbonaceous and quartz-chlorite-mica schists of the Cajamarca-Valdivia terrane. To the N and E, thick, unconsolidated volcanoclastic deposits inhibit identification of possible extensions of the province under Mio-Pliocene to Recent volcanic cover.

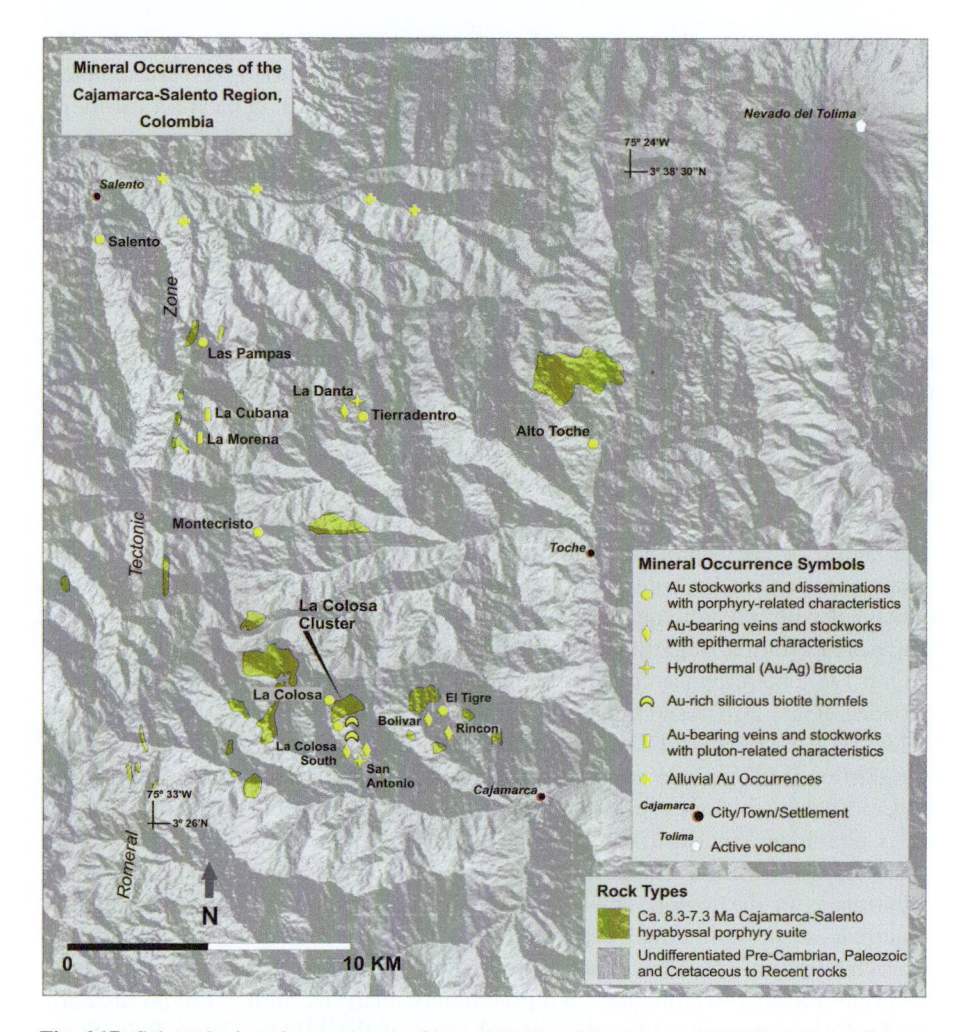

Fig. 6.17 Selected mineral occurrences of interpreted late Miocene age in the Cajamarca-Salento porphyry cluster and surrounding area of the Colombian Andes, in relation to granitoid intrusive rocks of the same approximate time period. Physiographic features of the map area are revealed by the 30 m digital elevation model (DEM) base image

The province is located to the NE of the point of divergence of two crustal-scale fault systems, Romeral and Palestina. Analysis of digital elevation images and Miocene through neotectonic movement vectors along these bounding faults (Figs. 6.12 and 6.17) permits the interpretation of porphyry emplacement within a lozenge-shape zone of dextral transtension. Subsequent exhumation associated with west-vergent thrusting exposes the basement complex and Cajamarca-Salento cluster through an erosional window in the Mio-Pliocene to Recent volcanic cover.

Descriptions of porphyry-style mineralization within the Cajamarca-Salento Au province have been published only for the La Colosa deposit (Lodder et al. 2010; Leal-Mejía 2011). These descriptions remain preliminary and will certainly be refined by ongoing investigations. Gil-Rodríguez (2010) and Leal-Mejía (2011) investigated at least 12 texturally and paragenetically separate phases of diorite and granodiorite porphyry associated with mineralization at La Colosa. Three phases of early mineral, fine-grained diorite and porphyritic diorite and two phases of intrusive breccias exhibit pervasive early potassic and later sodic-calcic alteration. The potassic assemblage includes minor quartz, magnetite and early biotite veinlets, K-feldspar replacements around plagioclase phenocrysts and, most notably, widespread to intense disseminations of fine-grained secondary biotite, which impart an overall reddish tone to the affected porphyry host. Potassic alteration is overprinted by widespread and patchy but locally intense sodic-calcic assemblages which include fine-grained pseudomorphic albite replacements accompanied by fibrous aggregates of dark green actinolite and calcic epidote, commonly dispersed along pyrite-rich veinlets (Leal-Mejía 2011). The sulphide assemblage is dominated by abundant pyrite, occurring in veinlets, as massive replacements and as intermineral disseminations, deposited in at least two paragenetic stages (Leal-Mejía 2011). Additional sulphide phases are observed to replace Stage 1 pyrite, including minor chalcopyrite>pyrrhotite>arsenopyrite>>galena, sphalerite and molybdenite>> very fine-grained Au-Ag-Bi-bearing tellurides (Leal-Mejía 2011). Stage 2 pyrite appears to post-date this assemblage; however, a complete paragenetic sequence for the late-stage sulphides has yet to be established. Native Au is widespread at La Colosa. Andedral, rounded grains generally measure less than 20 microns and occur as isolated blebs or replace Stage 1 pyrite and accompany the chalcopyrite-pyrrhotite-arsenopyrite-galena assemblage. The highest overall porphyry-related grades, commonly exceeding 1 g/t Au, are associated with the early-phase diorites containing strong potassic-sodic-calcic alteration assemblages (Gil-Rodríguez 2010; Lodder et al. 2010; Leal-Mejía 2011). Conversely, intermineral diorites demonstrate weak intermediate argillic (sericite + chlorite + illite), and propylitic (chlorite + epidote ± calcite) alteration that locally overprints higher temperature potassic and sodic-calcic alteration types. Mineralization includes pyrite ± minor chalcopyrite and pyrrhotite, in veinlets and as disseminations, and gold grades are, on average, <0.4 g/t Au (Lodder et al. 2010). Late mineral intrusions and dikes at La Colosa are quartz dioritic in composition. They are affected by weak to moderate phyllic and propylitic alteration and contain modest amounts of pyrite as disseminations and in fine veinlets but are essentially barren with respect to Au (Gil-Rodríguez 2010; Lodder et al. 2010).

The fertile, early diorites of the La Colosa complex were emplaced along a NNW trend, marked by the topographic axis of the La Guala ridge. The NE margin of the early mineral phases is marked by abrupt intrusive contacts, principally with barren, late mineral quartz diorite to the east. To the south, along La Colosa Creek, carbonaceous Cajamarca schists in contact with early and intermineral diorite porphyry form a broad zone of brecciation and strong to intense hydrothermal metamorphism, with replacement of schists by compact, auriferous, siliceous hornfels containing abundant secondary biotite, pyrrhotite and pyrite. To the south and west (upslope), along La Guala ridge, NNE-striking fracture zones cut hydrothermally altered carbonaceous schist. The fractures contain narrow, high-grade Au, druzy quartz and chalcedony veinlets, with very fine-grained aggregates of pyrite-marcasite (melnikovite pyrite), and are suggestive of lower temperature, epithermal exsolutions associated with the late cooling history of the underlying porphyry complex. The near rectilinear, NNW-striking, western margin of the porphyry complex is marked by the E-dipping Belgica fault zone, which separates the porphyry complex from Cajamarca Gp. schist to the west. The movement history along this structure has yet to be deciphered. Mineralization and alteration extends into the footwall schists, and abundant early mineral intrusive breccia is observed along the hangingwall (eastern) margin of the fault zone, suggesting the structures were originally extensional in nature, permissive to the emplacement of multiple early porphyry phases and to the penetration of hydrothermal fluids into the footwall schists. Notwithstanding, in outcrop, the footwall schists are presently highly sheared, and kinematic indicators can be interpreted to suggest reverse movement. The abundant late mineral quartz diorites and intrusive contacts observed on the eastern margin of the intrusive complex are essentially absent on the western margin and along the Belgica fault zone, suggesting the footwall structures passed rapidly from extensional to compressional following emplacement of the early mineral diorite complex.

Concerning the age of the La Colosa porphyry complex and contained Au mineralization, Leal-Mejía et al. (2018) provide U-Pb (zircon) crystallization ages of 8.3 ± 0.2 Ma and 8.1 ± 0.3 Ma for early mineral (high-grade) porphyritic diorites, whilst early mineral granodiorite porphyry returned a crystallization age of 7.9 ± 0.3 Ma. Intermineral (low-grade) granodiorite and diorite porphyries returned crystallization ages of 7.6 ± 0.2 Ma to 7.5 ± 0.2 Ma, respectively, and a late mineral (barren) granodiorite porphyry dike returned a crystallization age of 7.3 ± 0.2 Ma. With respect to alteration, Leal-Mejía (2011) provided a K-Ar age of 8.0 ± 0.8 Ma for secondary biotite extracted from a sample of early mineral diorite porphyry, whilst K-Ar (whole-rock) ages for early mineral breccia with pervasive secondary biotite alteration averaged 7.9 ± 0.8 Ma. Finally, Re-Os analysis of veinlet molybdenite returned an age of 8.43 ± 0.08 Ma. In this context, we interpret Au mineralization at La Colosa to have been emplaced relatively early in the magmatic evolution of the porphyry complex, mostly between ca. 8.4 and 7.9 Ma.

With respect to granitoids contained elsewhere within Cajamarca-Salento Province, Leal-Mejía et al. (2018) provide U-Pb (zircon) crystallization ages for porphyritic stocks associated with Au-porphyry-style mineralization. At La Morena and Tierradentro, the crystallization age of porphyritic diorites is 8.4 ± 0.2 Ma and 8.1 ± 0.1 Ma, respectively. At Montecristo quartz diorite porphyry returned an age of 7.6 ± 0.2 Ma, whist at Salento granodiorite porphyry returned an age of 6.3 ± 0.3 Ma. The U-Pb crystallization ages provide a maximum age limit for porphyry-related mineralization contained within the stocks, although additional information regarding these mineral occurrences has yet to be published.

Approximately 8.3 to 7.3 Ma porphyritic magmatism contained within the Cajamarca-Salento Province is nearly coaxial with the NNE-trending, modern-day Colombian volcanic arc. The active Machin volcano is located 10 km ENE of the La Colosa porphyry complex, whilst Plio-Pleistocene to Recent volcanic cover from the Nevado del Tolima complex limits identification of potential extensions of late Miocene porphyry-style mineralization to the NNE. Plio-Pleistocene to Recent volcanic deposits continue uninterrupted along trend to the NNE, where thick cover associated with the active Santa Isabel and Nevado del Ruiz volcanic complex is encountered. Available radiometric dates for these volcanic complexes reveal latest Pliocene and Pleistocene to recent ages (see compilation in Maya 1992; Gómez et al. 2015b); however, evidence for NNE continuation of the late Miocene porphyry belt beneath this volcanic cover has yet to be revealed. Farther along trend to the NNE, beyond recent volcanic cover, a cluster of Plio-Pleistocene porphyritic granitoids with associated Au-Ag mineralization is observed at Río Dulce (Leal-Mejía 2011).

ca. 2.4 to 0.4 Ma Río Dulce Epithermal Cluster

The Río Dulce cluster, located to the east of the town of Nariño (Antioquia Department) and extending farther east into the area surrounding Samaná and Norcasia (Caldas Department), includes numerous small hypabyssal diorite to granodiorite porphyry stocks with associated high-level diatreme breccias and localized deposits of pyroclastic rocks, emplaced within/upon upper greenschist- and amphibolite-grade metamorphic rocks of the Cajamarca-Valdivia terrane. Leal-Mejía (2011) noted Au-Ag (Zn, Pb, Cu) mineralization associated with Plio-Pleistocene hypabyssal porphyritic rocks and diatreme breccias at various localities, including Arboledas, Espíritu Santo, Santa Rita, La Cabaña, La Torre and Guyaquil-La Morena (Fig. 6.12). Mineralization observed at the most representative of these localities is described below.

ca. 2.4 Ma Arboledas Diatreme The Au-Ag-bearing phreatomagmatic diatreme breccia at Arboledas is exposed in artisanal underground workings. It is comprised of coarse-grained, sub-angular to sub-rounded clasts of diorite to granodiorite porphyry, pumice and tuff and graphitic schist basement, with a matrix containing rock flour and fine-grained juvenile igneous material. The breccia has been identified within an area measuring some 350 by 150 m, coincident with hydrothermal silicification, subordinate argillic (illite-smectite) alteration and the deposition of abundant hydrothermal pyrite as disseminations and minor veinlets. Minor amounts of sphalerite, galena and chalcopyrite are observed. Hydrothermal alteration and mineralization affect the fragments, matrix and immediate Cajamarca Gp. wallrocks hosting the diatreme breccia. Leal-Mejía (2011) and Leal-Mejía et al. (2018) presented U-Pb (zircon) age dates for the Arboleda area porphyries and diatreme breccia. Samples of granodiorite porphyry from two separate stocks returned identical crystallization ages of 2.4 ± 0.1 Ma. An altered diorite porphyry fragment from the

Arboledas diatreme breccia returned an age of ca. 2.3 ± 0.1 Ma, whilst a diorite porphyry dike which intrudes the diatreme breccia returned an identical age of 2.3 ± 0.1 Ma age.

ca. 1.1 Ma Espíritu Santo Porphyry and Diatreme Breccia 10 km NNE of Arboledas, at Espiritu Santo, coarse-grained polymictic diatreme breccias with fine-grained dacite porphyry injections are localized along the contact with a porphyritic diorite-granodiorite stock. The complex is in turn localized along the contact between schistose metamorphic rocks of the Cajamarca Gp. to the east and peraluminous Permo-Triassic granitoids of the Nariño Batholith to the west. Within the Espíritu Santo diatreme, Au-Ag mineralization is associated with intense hydrothermal alteration and fine-grained disseminations of pyrite with traces of sphalerite, galena, chalcopyrite and molybdenite(?). Hydrothermal alteration is characterized by near pervasive replacement of the breccia by silica and sericite-illite±carbonate. Within the coarse-grained diorite-granodiorite stock, localized zones of vertical flow banding with moderate to strong quartz-sericite-pyrite alteration are enveloped by broader zones of propylitic alteration (epidote-chlorite-carbonate). Leal-Mejía (2011) and Leal-Mejía et al. (2018) provided U-Pb (zircon) crystallization ages for four separate locations in and around the Espíritu Santo porphyry stock. Quartz diorite porphyry from Espíritu Santo hill returned ages 1.3 ± 0.1 Ma and 1.0 ± 0.2 Ma. Additional diorite porphyry samples collected ca. 3 km to the east, along the Espíritu Santo Creek, returned similar ages of ca. 1.2 ± 0.1 and 1.0 ± 0.1 Ma.

Santa Rita Sector Mineralization similar to that observed at Espíritu Santo continues for over 5 km along an approximate NNE trend into the area around Santa Rita, where additional small stocks and dikes of granodiorite porphyry and phreatomagmatic diatreme breccias contain locally abundant disseminated pyrite±chalcopyrite. The porphyries intrude Cajamarca Gp. schists and quartzites which exhibit contact metamorphic aureoles containing fine secondary biotite. Hydrothermal effects developed along the immediate contact zones overprint the hornfels and include weak to moderate silicification, chloritization and millimetric pyrite veinlets. No radiometric age dates are available for the Santa Rita porphyries. Some 5 km to the NNE, however, along the Espíritu Santo-Santa Rita trend, Leal-Mejía (2011) and Leal-Mejía et al. (2018) present a 0.4 ± 0.1 Ma U-Pb (zircon) crystallization age for unmineralized diorite porphyry at La Cabaña. Mineralization at Santa Rita is considered to be related to the ca. 1.1 to 0.4 Ma Espíritu Santo-La Cabaña trend.

Based upon geological setting, lithogeochemical considerations (Leal-Mejía et al. 2018) and radiomentric age constraints, the Rio Dulce area is interpreted to host a shallowly eroded, high-level, volcanic dome-diatreme complex associated with the emplacement of hypabyssal, metaluminous, calc-alkaline diorite and granodiorite porphyry stocks and dikes of Plio-Pleistocene age. The presence of abundant pyrite with minor sphalerite and galena and associated argillic (sericite-illite ± smectite) ±quartz and carbonate alteration suggests Au-Ag mineralization at Río Dulce is of the intermediate-sulphidation type (Sillitoe and Hedenquist 2003) and associated with the emplacement and cooling history of the Plio-Pleistocene igneous complex.

The Plio-Pleistocene Río Dulce cluster of hypabyssal porphyritic granitoids and associated epithermal Au-Ag occurrences represent the youngest expression of hypabyssal porphyry magmatism with spatially-temporally associated epigenetic precious metals mineralization, as presently identified within western Colombia. From a geologic, lithogeochemical and metallogenic standpoint, the Río Dulce epithermal cluster has many similarities to the late Miocene Cajamarca-Salento porphyry province, albeit exposed at a shallower erosional level.

As at Cajamarca-Salento, the Río Dulce cluster is located essentially coaxial to the Colombian segment of the active, NNE-trending North Andes volcanic zone (Fig. 6.12). The Plio-Pleistocene age dates revealed at Río Dulce compare well with similar ages for volcanic materials from the Ruíz, Santa Isabel and Tolima stratovolcanic complexes (Maya 1992; Gómez et al. 2015b), located along trend to the SSW, and in this context, Rio Dulce is interpreted to represent the northernmost expression of volcanism associated with the modern-day Colombian (and indeed, Andean) magmatic arc.

Thus, metaluminous, calc-alkaline granitoid magmatism along the North Andes volcanic axis ends near the latitude of Río Dulce, and no evidence for the northward continuation of the arc is observed within the regional geological (e.g. Cediel and Cáceres 2000; Gómez et al. 2015a) or radiometric age (e.g. Maya 1992; Gómez et al. 2015b; Leal-Mejía et al. 2018) databases. This observation is in agreement with the concept of the E-W-trending Caldas-Coiba tear (Vargas and Mann 2013; Chiarabba et al. 2016), the surface projection of which would pass somewhat to the south of Rio Dulce (Fig. 6.12). According to these authors, the abrupt northern termination of the North Andes volcanic axis may be associated with shallow to subhorizontal subduction of CCOP lithosphere to the north of the Caldas tear, as opposed to normal subduction of the Nazca Plate to the south.

Notwithstanding, one final Miocene-Pliocene trend of hypabyssal, porphyritic granitoids with important associated precious and base metal mineralization is located significantly to the N and E of Río Dulce, at Vetas-California within the Santander Massif. This historic precious metal district is located well off axis with respect to the Northern Andean volcanic zone and N of the Caldas tear as proposed by Vargas and Mann (2013) and Chiarabba et al. (2016). A description of the porphyritic granitoid-related Cu-Mo (Au) and Au-Ag (W) mineralization in the Vetas-California district is now provided.

ca. 10 to 8.5 Ma and 2.6 to 1.3 Ma Vetas-California Porphyry Cu-Mo (Au) and Au-Ag Trend

The prolonged tectono-magmatic history of the Santander Massif has become increasingly well documented and understood in recent years (e.g. Ward et al. 1973; Dörr et al. 1995; Restrepo-Pace 1995; Cediel et al. 2003; Zarifi 2006; Cediel 2011; Mantilla et al. 2012, 2013; Bissig et al. 2014; Van der Lelij et al. 2016) and provides an excellent example of a region which has been affected by overprinting phases of granitoid magmatism accompanied by complex tectonic reworking and structural evolution. Known metalliferous deposits in the region are sparse (e.g. Ward et al. 1970; Mejía et al. 1986), relative to the quantity and variety of occurrences observed in the Central and Western Colombian Cordilleras. From a historical production and present-day economic standpoint, the most important deposits of the region are restricted to the metallogenically isolated Vetas-California Au-Ag (Cu, Mo) district (Figs. 6.12 and 6.13), where historic evaluations combined with recent exploration and development have outlined mineral resources from three semi-contiguous deposits, Angostura, La Bodega and La Mascota, totalling >7M oz (218 metric tonnes) Au, 19 M oz (591 metric tonnes) Ag and 84M lbs (41,852 metric tonnes) Cu (Bissig et al. 2014; Rodriguez 2014; Rodríguez et al. 2017).

The Vetas-California district is located ca. 35 km NE of the city of Bucaramanga (Figs. 6.12 and 6.18). The district consists of two principal mineralized trends; (1) La Baja-La Alta, which extends in a NE direction for >7 km, from near the town of California into the Angostura-La Alta area, roughly coinciding with the NE-trending, structurally controlled valley of the Río La Baja, and (2) Vetas, located some 10 km to the SE of California, where numerous mineral occurrences are located in the vicinity of the Vetas townsite. The Au-Ag deposits of the Vetas-California district have been exploited since pre-Colombian and early Colonial times (e.g. Restrepo 1888), and the apparent potential for modern, large-scale development is such that numerous publications pertaining to the area are readily available.

Ward et al. (1970, 1973) mapped the mineral district and surrounding area in detail and provided a geological framework and geochemical analyses for most of the important underground workings along the La Baja trend and for various deposits at Vetas. These authors observed the spatial relationship between mineralization and numerous high-level, hypabyssal porphyritic granitoid dikes and irregular-shaped stocks, characteristic of the Vetas-California district. They noted that the altered porphyries cut lower Cretaceous stratigraphy on the SW margin of the La Baja trend and concluded that mineralization was post-early Cretaceous in age. Additional mapping and geological compilation for the district was subsequently provided by Mendoza and Jaramillo (1975) and Royero and Clavijo (2001). Sillitoe et al. (1982) discussed porphyry-associated mineralization at Angosturas and included the district within their Eastern sub-belt of Colombian porphyry-related deposits. Felder et al. (2005) provided a more detailed description of mineralization, alteration and structural controls in the Angosturas sector, highlighting the abundance of hypogene alunite, overprinting sericitic alteration within the Angosturas ore mineral assemblage. In recent years, research projects addressing the mineralogy, alteration, paragenesis, fluid geochemistry, stable isotopic composition and age of mineralization in the district have been provided by Diaz and Guerrero (2006), Leal-Mejía (2011), Mendoza, (2011), Raley (2012), Rodriguez (2014) and Rodríguez et al. (2017). Mantilla et al. (2009, 2012, 2013) and Bissig et al. (2012a, b, 2014) presented detailed geologic and petrogenetic studies of the various ages of granitoids outcropping within and around Vetas-California. Bissig et al. (2014) presented a composite model relating petrogenetic aspects of the late Miocene porphyritic granitoids at Vetas-California to the Cu (-Mo, Au) metallogeny of the California-Vetas Mining District and to the late Miocene to Recent tectonic configuration of proposed low-angle subduction of CCOP lithosphere beneath the Santander Massif. Models involving subduction of the Caribbean Plate, however, are difficult to reconcile in light of paleo-tectonic configurations suggesting that (1) large volumes of the CCOP assemblage were likely never subducted beneath NW South America (e.g. Aspden et al. 1987; Cediel et al. 1994; Cediel et al. 2003; Nerlich et al. 2014; Weber et al. 2015; Leal-Mejía et al. 2018), (2) the CCOP assemblage has largely been fixed with respect to the South American Plate since ca. 54.5 Ma (Müller et al. 1999; Nerlich et al. 2014) and (3) tectono-magmatic analyses suggesting the complete absence of continental arc development and granitoid magmatism throughout the central and eastern Colombian Andes between ca. 52 and 10 Ma (e.g. Leal-Mejía et al. 2018).

Mineralization throughout the Vetas-California district is characterized by closely to widely spaced, generally vertical to steeply dipping, quartz+pyrite-rich veins, veinlets, massive replacement zones and polyphase breccias. Mineralization often follows fractures related to narrow shear zones, which themselves are mineralized. Sulphide concentrations tend to be lenticular, and narrow seams of gouge are common along many of the veins (Ward et al. 1970). Mineralization may be hosted within, along the margins of, or in the vicinity of the altered±mineralized porphyritic granodiorites which occur throughout the district; however, it is not constrained to any one rock type, and the hydrothermal alteration zones with associated veins and breccias extend well beyond the margins of the porphyritic rocks, to affect the Precambrian, Paleozoic and Mesozoic basement rocks at the district scale. The degree of wallrock alteration varies from localized and structurally controlled along vein margins, to widespread, intense and pervasive. Composite, intense alteration along the upper La Baja trend (e.g. La Bodega, Angosturas), for example, is such that it is difficult to accurately identify host lithology due to textural and mineralogical destruction and overprinting brought on by hydrothermal alteration and pyrite replacement (Ward et al. 1970; Sillitoe et al. 1982; Felder et al. 2005; Leal-Mejía 2011). In this context, the paragenesis of the district is complex and yet to be fully documented. Evidence for repeated shearing, brecciation, replacement and recrystallization is widely seen, and several generations of quartz and sulphides are present (e.g. La Mascota-La Bodega; Mendoza 2011; Rodriguez 2014, Rodríguez et al. 2017). High-grade, vertically plunging ore shoots are often developed at vein and fracture intersections (Ward et al. 1970; Felder et al. 2005). Radiometric age dating (e.g. Leal-Mejía 2011; Mantilla et al. 2013; Rodriguez 2014; Rodríguez et al. 2017; reviewed below) also suggests the superposition of spatially coincident but temporally separate mineralizing events (Fig. 6.13).

Mineralized veins along the La Baja trend cluster in groups, with each group containing numerous veins. Ward et al. (1970) highlight three main composite groups, including, from SW to NE, (1) La Baja through San Cristobal, (2) La Mascota through Angosturas to La Alta and (3) El Silencio-La Picota (Fig. 6.18). The majority of the veins along the La Baja trend individually strike NE and ca. E-W. They are arranged in en echelon fashion in a northeast-stepping fashion, with

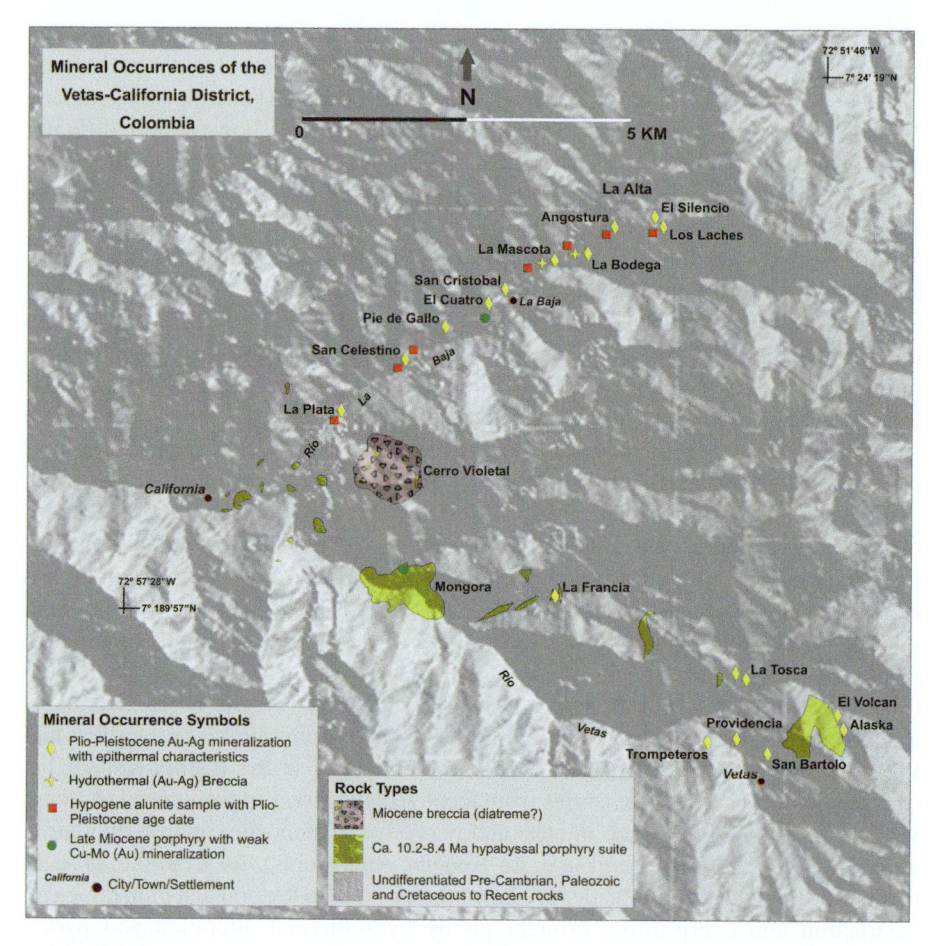

Fig. 6.18 Selected mineral occurrences of interpreted late Miocene through Pliocene age in the Vetas-California Au district of the eastern Colombian Andes, in relation to granitoid intrusive rocks of the same approximate time period, after Ward et al. (1970) and Rodríguez et al. (2017). Physiographic features of the map area are revealed by the 30 m digital elevation model (DEM) base image

the axis roughly parallel to Río La Baja, and Ward et al. (1970) consider them to represent a transtensional array developed between major N- to NE-striking faults which bound the district. The extensive tectono-hydrothermal breccias which form host to mineralization at La Mascota trending into La Bodega (Mendoza 2011; Rodriguez 2014; Rodríguez et al. 2017) form a steeply dipping tabular body which follows a NE strike. Host rocks along the La Baja trend include primarily the Precambrian Bucaramanga gneiss and quartz monzonite and porphyritic granodiorite of Jurassic and late Miocene age, respectively, whilst, in addition, at Angosturas at least, a portion of the resource is hosted within highly altered and mineralized Ordovician granitoids (Leal-Mejía 2011).

At Vetas, mineralization is primarily hosted within the Bucaramanga gneiss, cut by localized dikes and small stocks of late Miocene porphyry ± Jurassic monzogranite. Alteration, including primarily silicification and sericitization accompanied by disseminated pyrite, is less pervasive than along the La Baja trend, and more apt to be confined to narrow discrete zones associated with mineralized veining. Ward et al. (1970) outline various groupings of mineralized veins at Vetas, including (1) San Bartolo-Trompeteros, (2) La Tosca and (3) El Volcan-Alaska (Fig. 6.18). These authors note that near Vetas the strike of veins is N to NNW, whilst the veins near El Volcan strike NNE to NE.

District-scale field observations presented by Ward et al. (1970), Mendoza and Jaramillo (1975) and Mantilla et al. (2009), in association with detailed petrographic, paragenetic and alteration studies along the La Baja trend at Angosturas (Felder et al. 2005; Diaz and Guerrero 2006), La Plata (Raley 2012; Barbosa 2016), La Mascota, La Bodega, El Cuatro (Mendoza 2011; Rodriguez 2014; Rodríguez et al. 2017) and at Vetas (Bissig et al. 2012a; Reyes 2013; Sánchez 2013), permit the interpretation of a prolonged and complex tectono-magmatic and hydrothermal history for the Vetas-California district. When combined with radiometric age dates provided by Mantilla et al. (2009, 2013), Leal-Mejía (2011), Bissig et al. (2012b), Rodriguez (2014) and Rodríguez et al. (2017), multiple stages of granitoid magmatism + mineralization + alteration, spanning almost 10 m.y. period, can be postulated.

Altered and mineralized late Miocene porphyritic rocks are intimately associated with mineralization at Vetas-California (Ward et al. 1970), and cross-cutting relationships with respect to the porphyries establish a maximum age for mineralization throughout the district. Mantilla et al. (2009) presented U-Pb (zircon) data for two separate granitoid porphyries, collected near the Vetas and California townsites. Results yielded magmatic crystallization ages of ca. 9.0 Ma and 8.4 Ma, respectively. Leal-Mejía (2011) dated hypabyssal granodiorite porphyry out cropping at the San Celestino Mine, obtaining a 10.2 ± 0.2 Ma U-Pb (zircon) crystallization age. The San Celestino porphyry contains early, weakly developed porphyry-style Cu-Mo mineralization which is cut by sheeted, auriferous, pyrite-rich fractures with associated phyllic to argillic alteration haloes, interpreted to represent an intermediate-sulphidation, epithermal overprint upon the Cu-Mo mineralization. A 10.14 ± 0.04 Ma Re-Os age for a molybdenite from El Cuatro (Bissig et al. 2012b; Rodríguez et al. 2017) coincides well with the nearby San Celestino U-Pb (zircon) age of Leal-Mejía (2011) and is considered to represent the age of early Cu-Mo mineralization along the La Baja trend. Rodriguez (2014) and Rodríguez et al. (2017) present a poorly constrained $^{40}Ar/^{39}Ar$ (hydrothermal sericite) age of <10 Ma at La Bodega. This age may coincide with the age of early phyllic alteration accompanying weakly auriferous intermediate-sulphidation veining associated with the cooling history of the ca. 10.2 to 8.4 Ma granodiorite porphyries. The erratic analytical results presented for this sample, however, may also reflect disturbance of the $^{40}Ar/^{39}Ar$ systematics by subsequent mineralizing events (Rodríguez op.cit.).

Following emplacement (and cooling?) of the evident (outcropping) ca. 10.2 to 8.4 Ma porphyries along the La Baja trend, superimposed ore mineral assemblages combined with radiometric age dates for related alteration minerals suggest the juxtaposition of a complex series of epithermal events which are considered to be responsible for the introduction of the majority of the Au, Ag, Cu and associated ore minerals of economic interest at Vetas-California. At Angosturas, quartz veins and replacements containing sulphides and sulphosalts (including pyrite, chalcopyrite, tetrahedrite, bornite, digenite, enargite, native Au, electrum, sylvanite and hessite; Diaz and Guerrero 2006) associated with hypogene alunite+pyrite and sericite+pyrite alteration (Felder et al. 2005) are commensurate with high- to intermediate-sulphidation epithermal mineralization (e.g. Sillitoe and Hedenquist 2003; Simmons et al. 2005), the detailed paragenesis for which has yet to be established. Leal-Mejía (2011) provided a K-Ar (alunite) age of 3.4 ± 0.3 Ma for hypogene alunite containing disseminated pyrite, collected from a vein in Angosturas drill core. Bissig et al. (2012b) and Rodríguez et al. (2017) provided $^{40}Ar/^{39}Ar$ ages of ca. 3.90 and 2.48 Ma for hydrothermal sericite and alunite, respectively, collected from underground artisanal workings (La Perezosa tunnel) at the same prospect. Along trend to the SW of Angosturas, in contiguous artisanal workings at La Bodega, Rodriguez (2014) supplied a 3.54 ± 0.13 Ma $^{40}Ar/^{39}Ar$ plateau age for alteration sericite. Further to the SW, at San Celestino, Bissig et al. (2012b) and Rodríguez et al. (2017) recorded a 3.23 ± 0.06 Ma $^{40}Ar/^{39}Ar$ age for hypogene alunite overprinting porphyry-style mineralization, and at La Plata, near the SW margin of the La Baja trend, these same authors provided a similar 3.43 ± 0.07 Ma age for alunite contained within altered Triassic leucogranite. Hence, an ca. 3.9-3.4 Ma phase of advanced- (alunite) to intermediate- (sericite) argillic alteration is recorded along the entire La Baja trend.

Further detailed studies pertaining to alteration geochronology and paragenetic sequencing for the La Mascota and La Bodega deposits, however, were provided by Rodriguez (2014) and Rodríguez et al. (2017). At these deposits, additional $^{40}Ar/^{39}Ar$ plateau ages for alteration alunite cluster in ca. 2.6 to 2.2 Ma and ca. 1.9 to 1.27 Ma ranges. The earlier range is considered to be the principal period of ore introduction at both deposits, which according to Rodriguez (2014) took place in three paragenetic sub-stages: (1) quartz-alunite+copper sulphides+Au+electrum, (2) quartz-wolframite+Au+electrum and (3) quartz-alunite+enargite+Au+electrum. All three sub-stages additionally include pyrite, Au-Ag tellurides±proustite and tennentite-tetrahedrite series minerals, among others. During the subsequent ca. 1.9 to 1.27 Ma event, alteration/mineralization was dominated by quartz-alunite-pyrite+sphalerite, with the localized development of porous ("vuggy") silica; however, no important Au, Ag or Cu deposition took place (Rodriguez 2014).

Based upon fluid inclusion and stable isotope studies, Rodriguez (2014) concluded that the various stages of epithermal mineralization at La Mascota and La Bodega took place under conditions of decreasing temperature and decreasing pH. The main ore fluid was considered to be magmatic in origin, with minor, late-stage mixing with meteoric waters. Notably, based upon available U-Pb (crystallization) ages, the youngest recorded magmatic rocks in the Vetas-California district include ca. 10.2 to 8.4 Ma porphyries outcropping along the La Baja trend and elsewhere. In this context, it is unlikely that the ore-bearing stages of high- to intermediate-sulphidation epithermal mineralization observed along the La Baja trend, which began at ca. 3.9 Ma, are related to the cooling history of the 10.4 to 8.5 Ma porphyry bodies. Rodriguez (2014) and Rodríguez et al. (2017 interpret the data to suggest that additional pulses of (porphyritic?) magmatism, including at ca. 3.9 to 3.4 Ma, 2.6 to 2.2 Ma and 1.9 to 1.3 Ma, which have yet to be mapped and dated or, more likely, unroofed by erosion, may be responsible for the multiple stages of advanced argillic alteration and Au-Ag-Cu-W-Zn mineralization in the Vetas-California district.

Late Pleistocene to Recent Volcanism, Au Degassing and Deposition at Galeras Volcano

One final study of mineralizing processes in the Colombian Andes is worthy of mention: that of the very young (late Pleistocene to Recent) degassing and deposition of gold associated with active volcanism within the composite Galeras andesitic stratovolcano of southwestern Colombia (Fig. 6.12). Galeras is historically the most active volcano within the Colombian calc-alkaline volcanic arc, having erupted many times since the mid-sixteenth century. ^{40}Ar/^{39}Ar dating of andesite flows and rhyolite tuff from the flanks of the volcano suggests that most of the present edifice was formed since ca. 0.7 Ma (Goff et al. 1994). The history and nature of recent Galeras eruptions were studied in detail by Stix et al. (1993) and Goff et al. (1994). Analyses of hydrothermally altered rocks, vein "ore", recent andesite, fumarole discharges and fumerole sublimates by Goff et al. (1994) demonstrate that Galeras has deposited high-grade Au in Pleistocene hydrothermal events and that recent (1992–1993) andesite and magmatic volatiles contain Au at levels of ca. 0.015 mg/kg and 0.04 mg/kg, respectively. These authors note that at estimated and postulated consistent flux/depositional rates of ca. 20 kg/yr, a precious metal deposit exceeding 200 t Au could form at shallow levels within the Galeras edifice in only 10,000 years. Evidence of past mineral deposition is observed in proto-Galeras flows in the Río Guaitara valley ca. 16 km west of the Galeras summit, where an intermediate- to high-sulphidation vein up to 3 m wide contains early dolomite with replacement quartz and kaolin, pyrite, tetrahedrite, chalcopyrite, enargite and sphalerite, with bonanza-grade accumulations of Au as late-stage electrum. This mineralization was dated at between ca. 630 and 460 Ka, using the ^{234}U-^{230}Th method on dolomite leachate (Goff et al. 1994). Goff et al. (op.cit.) note that magmatic S contained within Galeras magma is relatively reduced and attribute the formation of SO_2 gas and associated H_2SO_4 to shallow decompression within the magmatic conduit. They characterize the hydrothermal environment within the country rocks surrounding the magmatic conduit as "high-sulphidation", with alteration assemblages overprinting recent andesite breccias collected within the active cone containing quartz, gypsum, anhydrite, alunite, 5-10% pyrite and up to 2.5 ppm Au. Porphyritic dacite lavas distal to the active conduit are intensely altered to intermediate pH assemblages containing quartz, sericite, chlorite and pyrite. Goff et al. (1994) note that SO_4 and Cl in acid spring waters which drain from Galeras are derived from magmatic volatiles, even though waters are dominantly meteoric. Gold mineralization associated with active volcanism at Galeras provides important insight and a first-order, modern-day example of the link between subduction-driven granitoid magmatism and hydrothermal precious and base metal mineralization in the Colombian Andes. We have postulated a similar genetic relationship between hydrothermal alteration and mineralization for many of the granitoid arc segments emplaced during the tectono-magmatic evolution of much of the Colombian Meso-Cenozoic.

6.6 Summary and Concluding Statement

Based upon our analysis, at the level of the Colombian Andes, we have identified numerous fertile metallotects containing a wide variety of mineral deposits, framed within five broad metallogenic epochs, which we have defined based upon important changes in the tectonic and magmatic evolution of the Northern Andes during the Phanerozoic. Mineral deposit types and styles, of both syngenetic and epigenetic origin, are manifold and range from the products of orthomagmatic cumulate segregation and hydrothermal processes genetically related to marine and continental volcanism/magmatism to accumulations related to uplift, basin development, chemical sedimentation and the amagmatic migration of mineral-rich brines. Important secondary and residual deposits are also recorded, related to the complex uplift, weathering and erosion history of the Northern Andean region. Our analysis shows that many significant deposits and mineral districts presently residing in Colombia were formed in the peri-cratonic or intra-oceanic realm and are allochthonous to the Colombian cordilleras, a result of the complex Meso-Cenozoic accretionary tectonic history of the region. Overall, our analysis emphasizes the sympathetic relationship observed between mineral/metal occurrences in the Colombian Andes and the tectonic development of the region as a whole, i.e. many deposit types may be considered typical of specific tectono-magmatic settings, such as podiform Cr deposits associated with ophiolite sequences, volcanogenic sulphide occurrences generated within the intra-oceanic realm or porphyry-related Cu occurrences associated with Andean-style volcanic arcs. In this context, the temporal-spatial analysis of mineral occurrences throughout the Colombian Phanerozoic provides a reaffirmation of the complex tectonic evolution proposed for the region, and the appearance of certain mineral deposits at specific time(s) and place(s) provides support for the occurrence of unrecognized or controversial tectonic process or events. At the same time the analysis highlights many of the atypical aspects of Colombian and Northern Andean tectonism and metallogeny, when compared with the better-documented provinces of the Central Andes of, for example, Perú, Chile, Bolivia and Argentina.

With respect to our Colombian metallogenic epochs, as recorded within our time-space charts (Figs. 6.3, 6.5, 6.10 and 6.13), we note that the prolonged period represented by the pre-Jurassic chart is the least well supported, not so much due to a lack of technical information regarding tectonic processes and developments during the Paleozoic-early Mesozoic but more so with respect to, in absolute terms, the lack of mineral occurrences which can confidently be linked to pre-Jurassic times. We offer various lines of reasoning to explain this observation. Firstly, the great majority of the early Paleozoic through Permo-Triassic rocks outcropping in the Colombian Andes are moderate- to high-grade metamorphic rocks which have been subjected to burial, tectonic reworking, uplift and erosion during prolonged and aggressive accretionary-driven tectonism, in accordance with regional prograde metamorphic events registered in the early Paleozoic and Permo-Triassic. In this context, the likelihood of mineral deposit preservation, in general terms, is low. Fertile metallogenic regimes clearly existed however, at least locally, during this extended time period, as demonstrated by the Bailadores (Sierra de Mérida, Venezuela) VMS and the El Carmen-El Cordero RTG-related Au-vein deposits. In either case, one of the keys to deposit preservation may reside in the extensional, rift-related setting in which both were formed, which could have contributed to their burial and protection from subsequent tectonic events.

Notwithstanding, the close genetic relationship between granitoid magmas and metalliferous deposits sensu lato, as observed for the great majority of Jurassic through recent metals occurrences throughout the Colombian Andes, is not well sustained during pre-Jurassic times. Albeit, in the first instance, pre-Jurassic granitoids (including the early Paleozoic, Carboniferous and Permo-Triassic suites) are volumetrically minor when compared with younger granitoid suites. In addition, they are deeply eroded, perhaps beyond a level conducive to the preservation of associated hydrothermal mineral deposits, which, even at the pluton-related level, tend to accumulate in the higher levels and copulas of associated intrusions. Alternatively, perhaps a more fundamental, first-order aspect of metallogenesis, involving metals source(s) is at play: lithogeochemical and isotopic observations indicate that the early Paleozoic (ca. 485–439 Ma) granitoids, Permian-early Triassic (ca. 289–250 Ma) granitoid gneisses and mid-Triassic (ca. 240–225 Ma) anatectic granitoids all represent recycled lower crustal melts. Coincidentally, not a single metal occurrence has been genetically linked to these predominantly peraluminous granitoids. Conversely, the well-mineralized El Carmen-El Cordero granitoids comprise the only pre-Jurassic suite for which a strongly mantelic isotopic signature is recorded. These observations provide a first-order testament to the metallogenic fertility of mantle-derived magmas vs. crustal-derived anatectites.

Following the pre-Jurassic, the second Colombian metallogenic epoch essentially spans the latest Triassic through Jurassic. It is reflective of regional taphrogenesis throughout NW South America (the Bolívar Aulacogen) prevalent during and following the break-up of Pangea, and included the early formation of volcano-sedimentary Cu (Ag) occurrences associated with continental rift-related bimodal volcanism and sedimentation, followed by the superposition of regionally extensive, autochthonous, subduction-driven, metaluminous volcano-plutonic arc magmatism and the formation of back-arc basins, with associated epithermal and pluton-related Au-Ag-base metal, porphyry-related Cu (Mo, Au) and skarn-type Cu (Au) mineralization. The sheer volume of the Jurassic batholiths and associated volcanic rocks attests to the highly extensional environment, driven primarily by slab rollback, into which Jurassic granitoid magmatism was emplaced. As in the pre-Jurassic, the role of mantle vs. crustal-derived fertility may be called into question: recent lithogeochemical and isotopic data suggests that the latest Triassic to early Jurassic Santander Plutonic Complex represents primarily recycled continental crust, and again, no significant metal deposits are known to be genetically associated with the cooling history of this granitoid suite. In contrast, as arc development migrated westward, and granitoid magmatism became predominantly mantle-derived, significant and widespread cogenetic Au, Ag, Cu, Pb, Zo, Mo, etc. deposits appear, in the Norosí, San Martín, Ibagué south, Ibagué north and Mocoa Batholiths and associated volcanic rocks. Preservation of these deposits during subsequent Northern Andean orogenesis in the Meso-Cenozoic is attributed to the originally extensional environment in which the deposits were formed and to the pre-existence and role of crustal-scale suture and fault systems (including Santa Marta – Bucaramanga - Garzón, Palestina and Otú) within the pre-Jurassic-post-Jurassic architecture of the Northern Andean Block, which provided important buffers and shock absorbers for the localization of stress during regional Meso-Cenozoic tectonism and uplift.

By the end of the Jurassic, continued trench retreat led to the demise of arc magmatism in continental Colombia, after ca. 146 Ma, and to the development of an early Cretaceous rifted margin floored by proto-Caribbean oceanic crust, to the N and W. This tectonic framework accompanied the beginning of the third metallogenic epoch (ca. 145–100 Ma), which is characterized by rift-related marine volcanism with clastic and carbonate sedimentation during the culminant phase of the Bolívar Aulacogen. Opening of the Valle Alto rift (Eastern Cordilleran basin) took place at this time, and rift-related volcanogenic Cu (Zn, Ag) occurrences such as El Azufral are evident. In addition, accumulations of metal-rich brines and exhalative Pb, Zn ± Cu occurrences within third- and fourth-order marine depocentres of the Valle Alto rift and Eastern Cordilleran basin were deposited at this time. These metals would become protore for the widespread, high-grade, epigenetic, replacement-style and fault- and breccias-hosted base metal and emerald occurrences emplaced during Meso-Cenozoic dewatering and structural inversion of the Eastern Cordilleran basin. This epoch is equally characterized by the absence of significant volumes of arc-related granitoids and/or their associated hydrothermal metal occurrences.

The tectonic quiescence of the Colombo-proto-Caribbean margin terminates abruptly at ca. 125 Ma, during onset of the proto-Northern Andean orogeny, marked by strong dextral shear, greenschist- with locally blueschist-grade metamorphism and tectonic reconfiguration of the Pacific-Farallon margin. Within the context of this event, only one mineral occurrence, the Berlín-Rosario shear zone-hosted Au (Ag) vein system, has been integrated in our time-space charts. Available data pertaining to the genesis of this significant gold occurrence, however, is sparse and does not permit understanding of the genetic aspects of the occurrence or of the metallogenic implications of the Romeral event.

The fourth metallogenic epoch in Colombia is the most complex and varied, marked by the return of hydrothermal Au (Ag, Cu, Mo, base metals) deposits related to voluminous subduction-related granitoid arc magmatism in the continental domain, in addition to the arrival of various styles of mineralization hosted within allochthonous terranes accreted during the early Northern Andean orogeny. Within the autochthonous continental domain, pluton-related Au (Ag) districts formed between ca. 87 and 72 Ma are associated with repeated pulses of metaluminous granitoid magmatism comprising the Antioquian Batholith and its satellite plutons. Associated districts include important historic through modern-day Au producers at Segovia-Remedios, Bramadora and Santa Rosa de Osos, among numerous others, emplaced prior to shut down of the Antioquian Batholith and associated magmatism during collision and accretion of the Dagua and Cañas Gordas terranes of the Western Tectonic Realm, beginning at ca. 75 Ma. Following an ca. 10 m.y. hiatus, granitoid magmatism was reignited in the continental domain. The multimillion ounce ca. 60–58 Ma Gramalote Au (Ag, Cu, Mo) deposit, hosted within the Providencia granitoid suite along the Nus River Valley, is associated with this post-collisional magmatism, as are locally important Au deposits hosted within the ca. 61–57 Ma Sonsón Batholith, the ca. 59 Ma Manizales stock and the ca. 55 Ma El Hatillo stock. The strongly mantelic isotope compositions of the Providencia suite are noteworthy, whilst significant degrees of crustal contamination are observed as post-collisional arc magmatism migrated south, reflected in the geochemistry and W-rich metal chemistry of the El Hatillo stock. Collision and accretion of the Gorgona terrane along the Colombian Pacific margin led to extinction of granitoid magmatism in the continental domain in the Eocene. A hiatus in granitoid magmatism and paucity of granitoid-related metallogeny within continental Colombia, associated with plate reorganization along the Colombian Pacific extending from the Eocene into the late Oligocene, is recorded.

Metal occurrences formed in the allochthonous realm and accreted to the Colombian margin along with the components of the Western Tectonic Realm during the early Northern Andean orogeny are typical of mineral deposits associated with intra-oceanic and peri-cratonic settings. They include orthomagmatic segregations and differentiates which formed the protore for important PGE (Au) placers (Guapí ophiolite) and laterite Ni (Cerro Matoso) deposits, Cu- and Au-rich VMS occurrences of the Cañas Gordas terrane and pluton- and porphyry-related Au (Ag, Cu) occurrences related to primitive (low-K), metaluminous intra-oceanic arc granitoids emplaced along the leading-edge of the Caribbean-Colombian oceanic plateau (e.g. El Retiro, El Tambo-Fondas, El Alacrán-Montiel). The highly prospective Cu (Au) porphyry occurrences of the Mandé-Acandí arc (e.g. Acandí, Murindó, Pantanos-Pegadorcito, Río Andagueda) were emplaced along the trailing edge of the CCOP during the mid-late Eocene, however, were not accreted to the Colombian margin, along with the host El Paso-Baudó terrane assemblage, until the Miocene, during the late Northern Andean orogeny. With respect to the assortment of allochthonous mineral occurrences hosted within accreted terranes of the Western Tectonic Realm, we note the clear distinction between metal assemblages and deposit types within terranes dominated by anhydrous, Fe-enriched, tholeiitic (ferroan after Frost et al. 2001) N-MORB and E-MORB suites (e.g. Dagua and San Jacinto terranes) vs. those containing hydrous calc-alkaline (magnesian after Frost et al. 2001) granitoid differentiates (e.g. Cañas Gordas terrane). The mineral potential inherent to tholeiitic vs. calc-alkaline magmatism should be borne in mind when considering metals exploration in the various basement domains of the Western Tectonic Realm.

The fifth Colombian metallogenic epoch, spanning the latest Oligocene through Recent, is linked with the reinitiation of subduction of Nazca Plate oceanic lithosphere beneath the Colombian Pacific margin and the resultant emplacement of metaluminous, medium through high-K calc-alkaline (calcic through alkali-calcic after Frost et al. 2001) holocrystalline plutonic and hypabyssal porphyritic granitoid arc segments, into the previously accreted terranes forming the basement complexes of the emerging Central and Western Colombian Cordilleras. In this context, fifth epoch metallogeny in Colombia is comprised exclusively of autochthonous, arc-related, hydrothermal pluton- and porphyry-associated Au-Ag-Cu-Pb-Zn (Mo, As, Sb) mineral occurrences. The spatial and temporal evolution of these occurrences is intimately related to the spatial-temporal evolution of granitoid arc segments in an overall southwest-to-northeast sense, within the tectono-magmatic development of the late Northern Andean orogeny, which is in turn related to the configuration of the Cauca vs. Coiba segments of the Nazca Plate, as it ultimately led to conformation of the Colombian segment(s) of the modern-day Northern Andean volcanic arc. The rapid unroofing of the earliest holocrystalline intrusives exposed Au (Ag)-rich quartz-sulphide±carbonate vein deposits hosted by ca. 23–21 Ma and 12–10 Ma plutons within with the Piedrancha-Cuembí and Farallones-El Cerro arc segments, respectively, and within tectonic vein arrays spatially associated with the host plutons. The sequential eastward migration of arc-axial magmatism, as recorded in the ca. 18–9 Ma Upper Cauca-Patia belt and Buenos Aires-Suárez cluster and the ca. 9–5 Ma Middle Cauca belt, was accompanied by the emplacement of felsic hypabyssal porphyritic granitoids with associated porphyry-related Au (Cu) deposits and a diverse suite of spatially associated, porphyry, volcanic and sediment-hosted, low- and intermediate-sulphidation epithermal Au-Ag-base metal occurrences. Reactivation of the deep crustal architecture of the Romeral and Cauca fault and suture systems facilitated emplacement of Upper Cauca-Patía, Buenos Aires-Suárez and Middle Cauca magmatism. Continued eastward migration of hypabyssal porphyry magmatism during the late Miocene-Pleistocene led to emplacement of the ca. 8.3–7.3 Ma Cajamarca-Salento porphyry cluster and the ca. 2.4–0.4 Ma Río Dulce high-level porphyry and diatreme complex, both within Cajamarca-Valdivia terrane metamorphic basement underlying the Central Cordillera. Cajamarca-Salento and Río Dulce are contained within regional-scale transpressional lozenges, bound by the Romeral and Palestina sutures, related to tectonic accommodation within the continental block during the final stages of the Northern Andean orogeny. These Miocene and Pleistocene clusters are essentially coaxial with the NNE-trending, modern-day, Northern Volcanic Zone. The Cajamarca-Salento cluster hosts the La Colosa porphyry Au deposit, with a documented resource exceeding 28 M oz (880 metric tonnes) of Au, as well as numerous additional porphyry Au and related epithermal occurrences. Epithermal, precious metal-rich diatremes and breccias associated with the Plio-Pleistocene Río Dulce cluster are perhaps reflective of the upper levels at Cajamarca-Salento, eroded in the late Miocene.

400 km to the NE of the Cajamarca-Salento cluster, in the Vetas-California district of the Santander Massif, significant Plio-Pleistocene high-sulphidation Au (Ag-Cu-Zn-W) epithermal mineralization cuts Precambrian basement gneiss, Paleozoic peraluminous and Jurassic metaluminous granitoids and late Miocene

hypabyssal porphyries containing Miocene porphyry-related Cu-Mo (Au) mineralization. Although unexposed, Pleistocene granitoid magmatism is suspected to be present beneath the Vetas-California district. The relationship of Mio-Pliocene granitoid magmatism at Vetas-California with other (unmineralized) late Miocene-Pliocene felsic magmatic centres located elsewhere within Colombia's Eastern Cordilleran system, in the back-arc of the active Colombian volcanic arc, has yet to be clearly established.

Regardless of age, the Mio-Pliocene porphyry arc segments and clusters of the Colombian Andes reveal clearly mantelic isotope compositions and similar, relatively well-constrained lithogeochemical and isotopic trends. Individual arc segments were superimposed, however, upon basement complexes of highly variable age and composition, ranging from (ca. west-to-east) Cretaceous oceanic volcano-sedimentary and E-MORB plateau rocks in the Western Tectonic Realm and mixed Paleozoic through Cretaceous marginal basin assemblages of the Romeral melange to tectonized Paleozoic carbonaceous schist and meta-volcanic assemblages of the Cajamarca-Valdivia terrane and high-grade Precambrian metamorphic rocks and Paleo-Mesozoic granitoids, of continental affinity in the Santander Massif. Observed variations in the lithogeochemical, trace and rare Earth element and isotopic evolution of the individual Mio-Pliocene and Pleistocene porphyry suites and their inherent metallogenic variations all interpreted to have been derived from a relatively homogenous mantle source and may be attributable to the nature and composition, including oxidation state and tectonic architecture, of the basement complex(es) which host the individual porphyry suite(s) and their associated mineral occurrences.

Despite advances in tectono-magmatic understanding, and the exponential augmentation of the radiometric age, and high-quality lithogeochemical and isotopic geochemical database available for Colombian igneous rocks, the study of Phanerozoic metallogeny in the Colombian Andes remains in its infancy. A deeper and more integrated understanding of the complex interrelationships between regional and local tectonic evolution, magmatism and metallogenesis must await the results of detailed studies from numerous individual deposits and mineral districts, of various ages, in a variety of litho-tectonic settings, ideally using modern, combined geochemical, thermochronological and isotopic analytical techniques, combined with essential field-based mapping, well-controlled sampling and petrographic and paragenetic studies. The magmato-tectonic vs. metallogenic analysis presented herein may be considered a framework, guide and point of departure for such studies.

References

Alvarez J (1987) Minerales de Cromo. In: Villegas A (ed) Recursos minerales de Colombia, 2nd edn, INGEOMINAS, Publicaciones Geológicas Especiales 1, pp 185–199
Alvarez J, Marulana N, Botero G, Linares E (1979) Edad K-Ar de stock de San Cristobal, Nariño. Publicación Geológica Especial, Faculdad de Ciencias, Medellín 11:1–4
Alvarez M, Ordóñez O, Valencia M, Romero A (2007) Geología de la zona de influencia de la falla Otú en el distrito minero Segovia-Remedios. Dyna 74(153):41–51

AngloGold Ashanti (2007) Producción histórica de Au en Colombia, Revisión. Oral presentation presented at the XI Congreso Geológico Colombiano, Bucaramanga, 14–17 Aug 2007
AngloGold Ashanti (2015) Mineral resource and ore reserve report. http://www.aga-reports.com/15/download/AGA-RR15.pdf. Accessed 13 Sept 2016
Arango JL, Ponce A (1982) Mapa Geológico del Departamento de Nariño. Memoria Explicativa. INGEOMINAS,. 40 p
Arias A, Jaramillo L (1987) Minerales de Cobre. In: Villegas A (ed) Recursos minerales de Colombia, vol 1, 2nd edn. INGEOMINAS, Publicaciones Geológicas Especiales, Bogota, pp 122–184
Aspden JA, McCourt WJ, Brook M (1987) Geometrical control of subduction-related magmatism: the Mesozoic and Cenozoic plutonic history of western Colombia. J Geol Soc 144(6):893–905
Bayona G, Cardona A, Jaramillo C, Mora A, Montes C, Valencia V, Ayala C, Montenegro O, Ibañez-Mejía M (2012) Early Paleogene magmatism in the northern Andes: insights on the effects of Oceanic Plateau–continent convergence. Earth Planet Sci Lett 331–332:97–111
Banks DA, Giuliani G, Yardley BWD, Cheilletz A (2000) Emerald mineralisation in Colombia: fluid chemistry and the role of brine mixing. Miner Deposita 35(8):699–713
Barbarin B (1999) A review of the relationships between granitoid types, their origins and their geodynamic environments. Lithos 46(3):605–626
Barbosa S (2016) Evolucion del Sistema Hidrotermal del Prospecto La Plata, Distrito Minero de Vetas-California (DMVC), Santander, Colombia. BSc thesis, Universidad Industrial de Santander
Barrero D (1977) Geology of the Central Western Cordil- lera, west of Buga and Roldanillo, Colombia. PhD thesis, Colorado School of Mines
Bartok PE, Renz O, Westermann GEG (1985) The Siquisique ophiolites, Northern Lara State, Venezuela: a discussion on their Middle Jurassic ammonites and tectonic implications. Geol Soc Am Bull 96(8):1050–1055
Berrangé JP (1992) Gold from the Golfo Dulce placer province, southern Costa Rica. Revista Geológica de América Central 14:13–37
Bierlein FP, Groves DI, Cawood PA (2009) Metallogeny of accretionary orogens: the connection between lithospheric processes and metal endowment. Ore Geol Rev 36(4):282–292
Bissig T, Mantilla LC, Rodriguez A, Raley, C, Hart CJR (2012a) The California-Vetas district, eastern Cordillera, Santander, Colombia: Late Miocene porphyry and epithermal mineralization hosted in Proterozoic gneisses and Late Triassic-Early Jurassic intrusions. Poster presented at the Society of Economic Geologists SEG2012 conference, Westin Hotel, Lima, Peru, 23–26 Sept 2012
Bissig T, Rodriguez A, Mantilla LC, Hart CJR (2012b) Hydrothermal evolution of the California-Vetas District, Santander Colombia: new age constraints on hydrothermal minerals. In: Bissig T, Hart CJR (eds) Colombia gold and porphyry project year 1 technical report. Mineral Deposit Research Unit, The University of British Columbia, Vancouver, BC, Canada
Bissig T, Mantilla LC, Hart CJR (2013) Petrochemistry of igneous rocks in the California-Vetas District, Santander, Colombia: relationship to late Miocene porphyry Cu-Mo (-Au) mineralization and tectonic implications. In: Bissig T, Hart CJR (eds) Colombia gold and porphyry project year 1 technical report. Mineral Deposit Research Unit, The University of British Columbia, Vancouver, BC, Canada
Bissig T, Mantilla LC, Hart CJR (2014) Petrochemistry of igneous rocks of the California-Vetas mining district, Santander, Colombia: implications for northern Andean tectonics and porphyry Cu (–Mo, Au) metallogeny. Lithos 200–201:355–367
Bissig T, Leal-Mejía H, Stevens RB, Hart CJR (2017) High Sr/Y magma petrogenesis and the link to porphyry mineralization as revealed by Garnet-Bearing I-type granodiorite porphyries of the Middle Cauca Au-Cu Belt, Colombia. Econ Geol 112(3):551–568
Botero G (1975) Edades radiométricas de algunos plutones colombianos. Revista Minería 27(169–179):8336–8342
Botsford RS (1926) The Zancudo Mining District. Memorias del Ing. R.S. Botsford, Sociedad del Zancudo, 94p

Branquet Y, Giuliani G, Cheilletz A, Laumonier B (2015) Colombian emeralds and evaporites: tectono-stratigraphic significance of a regional emerald-bearing evaporitic breccia level. Abstract presented in the 13th biennal meeting of the Society for Geology Applied to Mineral Deposits, Nancy (France), 24–27 Aug 2015

Branquet Y, Laumonier B, Cheilletz A, Giuliani G (1999) Emeralds in the Eastern Cordillera of Colombia: two tectonic settings for one mineralization. Geology 27(7):597–600

Burke K, Dewey JF (1973) Plume-generated triple junctions: key indicators in applying plate tectonics to old rocks. J Geol 81:406–433

Burke K, Cooper C, Dewey JF, Mann P, Pindell JL (1984) Caribbean tectonics and relative plate motions. In: Bonini WE, Hargraves RB, Shagam R (eds) The Caribbean–South American Plate boundary and regional tectonics, vol 162. Geological Society of America Memoir, Boulder, pp 31–63

Bürgl H (1967) The orogenesis of the Andean system of Colombia. Tectonophysics 4:429–443

Bustamante A (2008) Geotermobarometria, geoquímica, geocronologia e evolução tectônica das rochas da fácies xisto azul nas áreas de Jambaló (Cauca) e Barragán (Valle del Cauca), Colômbia. PhD thesis, Universidade de São Paulo

Bustamante C, Cardona A, Archanjo CJ, Bayona G, Lara M, Valencia V (2017) Geochemistry and isotopic signatures of Paleogene plutonic and detrital rocks of the Northern Andes of Colombia: a record of post-collisional arc magmatism. Lithos 277:199–209

Calle B, Toussaint JF, Restrepo JJ, Linares E (1980) Edades K/Ar de dos plutones de la parte septentrional de la Cordillera Occidental de Colombia. Geología Norandina 2:17–20

Camprubí A (2009) Major metallogenic provinces and epochs of Mexico. SGA News 25 June 2009

Cardona A, Valencia V, Garzón A, Montes C, Ojeda G, Ruiz J, Weber M (2010) Permian to Triassic I to S-type magmatic switch in the northeast Sierra Nevada de Santa Marta and adjacent regions, Colombian Caribbean: tectonic setting and implications within Pangea paleogeography. J South Am Earth Sci 29(4):772–783

Carlson GG (1977) Geology of the Bailadores, Venezuela, massive sulfide deposit. Econ Geol 72:1131–1141

Ceballos L, Castañeda DM (2008) Aspectos metalogenicos de la Brecha Lítica de Miraflores, Vereda Miraflores – La Cumbre, sector sureste del Municipio de Quinchía, Departamento de Risaralda. BSc thesis, Universidad de Caldas

Cediel F (1969) Die Girón-Gruppe: eine frueh-mesozoische Molasse der Ostkordillere Kolumbiens. Neues Jahrbuch Geologie und Paläontologie, Abhandlungen, Munchen 133(2):111–162

Cediel F (2011) Major Tecto-sedimentary events and basin development in the Phanerozoic of Colombia. In: Cediel F (ed) Petroleum geology of Colombia. Regional geology of Colombia, vol 1. Agencia Nacional de Hidrocarburos (ANH) – EAFIT, pp 13–108

Cediel F (2018) Phanerozoic orogens of Northwestern South America: cordilleran-type orogens, taphrogenic tectonics and orogenic float. In: Cediel F, Shaw RP (eds) Geology and tectonics of Northwestern South America: the Pacific-Caribbean-Andean junction. Springer, Cham, pp 3–89

Cediel F, Etayo F, Cáceres C (1994) In: Ingeominas (ed) Facies distribution and tectonic setting through the Phanerozoic of Colombia. Geotec Ltd., Bogota

Cediel F, Cáceres C (2000) Geological map of Colombia. Geotec, Ltd., Bogotá

Cediel F, Barrero D, Cáceres C (1998) In: Robertson Research International, UK (ed) Seismic atlas of Colombia: seismic expression of structural styles in the basins of Colombia, vol 1–6. Geotec Ltd., Bogotá

Cediel F, Shaw RP, Cáceres C (2003) Tectonic assembly of the Northern Andean block. In: Bartolini C, Buffler RT, Blickwede J (eds) The circum-Gulf of Mexico and the Caribbean: hydrocarbon habitats, basin formation, and plate tectonics, AAPG memoir 79. American Association of Petroleum Geologists, Tulsa, OK, pp 815–848

Cediel F, Restrepo I, Marín-Cerón MI, Duque-Caro H, Cuartas C, Mora C, Montenegro G, García E, Tovar D, Muñoz G (2010) Geology and Hydrocarbon Potential, Atrato and San Juan Basins, Chocó (Panamá) Arc, Colombia, Tumaco Basin (Pacific Realm). Colombia, Agencia Nacional de Hidrocarburos (ANH)-EAFIT, Medellín

Cheilletz A, Giuliani G (1996) The genesis of Colombian emeralds: a restatement. Miner Deposita 31(5):359–364
Cheilletz A, Féraud G, Giuliani G, Rodriguez CT (1994) Time-pressure and temperature constraints on the formation of Colombian emeralds: an $^{40}Ar/^{39}Ar$ laser microprobe and fluid inclusion study. Econ Geol 89(2):361–380
Cheilletz A, Giuliani G, Branquet Y, Laumonier B, Sanchez AJ, Féraud G, Arhan T (1997) Datation K-Ar et $^{40}Ar/^{39}Ar$ à 65±3 Ma des gisements d'émeraude du district de Chivor Macanal: argument en faveur d'une déformation précoce dans la Cordillère orientale de Colombie. Comptes Rendus de l'Académie des Sciences Paris 324:369–377
Chiarabba C, De Gori P, Faccenna C, Speranza F, Seccia D, Dionicio V, Prieto GA (2016) Subduction system and flat slab beneath the Eastern Cordillera of Colombia. Geochem Geophys Geosyst 17(1):16–27
Cochrane R (2013) U-Pb thermochronology, geochronology and geochemistry of NW South America: rift to drift transition, active margin dynamics and implications for the volume balance of continents. PhD thesis, Université de Genève
Cochrane R, Spikings R, Gerdes A, Ulianov A, Mora A, Villagómez D, Putlitz B, Chiaradia M (2014a) Permo-Triassic anatexis, continental rifting and the disassembly of western Pangaea. Lithos 190–191:383–402
Cochrane R, Spikings R, Gerdes A, Winkler W, Ulianov A, Mora A, Chiaradia M (2014b) Distinguishing between in-situ and accretionary growth of continents along active margins. Lithos 202–203:382–394
Colmenares F, Román-García L, Sánchez JM, Ramirez JC (2018) Diagnostic structural features of NW South America: Structural crosssections based upon detail field transects. In: Cediel F and Shaw RP (eds) Geology and Tectonics of Northwestern South America: The Pacific-Caribbean-Andean Junction
Cordani UG, Cardona A, Jiménez DM, Liu D, Nutman AP (2005) Geochronology of Proterozoic basement inliers in the Colombian Andes: tectonic history of remnants of a fragmented Grenville belt. Geol Soc Lond Spec Publ 246(1):329–346
Correa-Martínez AM (2007) Petrogênese e evolução do ofiolito de Aburrá, cordilheira central dos Andes colombianos. PhD thesis, Universidade de Brasilia
Cox D, Lindsey D, Singer D, Moring B, Diggles M (2007) Sediment-hosted copper deposits of the world: deposit models and database. USGS open-file report 03-107, Version 1.3 (2003), revised 2007, available online at https://pubs.usgs.gov/of/2003/of03-107/of03-107.pdf
Cruz N, Carrillo JA, Mantilla LC (2014) Consideraciones petrogenéticas y geocronología de las rocas ígneas porfiríticas aflorantes en la Quebrada Ventanas (Municipio Arboledas, Norte de Santander, Colombia) – implicaciones metalogenéticas. Boletín de Geología 36(1):103–118
Cuadros FA, Botelho NF, Ordóñez-Carmona O, Matteini M (2014) Mesoproterozoic crust in the San Lucas Range (Colombia): an insight into the crustal evolution of the northern Andes. Precambrian Res 245:186–206
del Rio S (1930) Placer mining in Colombia. Engineering and Mining Journal 129(7):354–356
Díaz F (2002) Composição isotópica e idade das mineralizações de Au epitermal do distrito mineiro de Marmato, noroeste de Colômbia. MSc thesis, Universidade De São Paulo
Diaz LA, Guerrero M (2006) Asociaciones Mineralógicas de las Menas Auroargentíferas y su Distribución en el Yacimiento Angostura (California, Santander). BSc thesis, Universidad Industrial de Santander
Dörr W, Grösser JR, Rodríguez GI, Kramm U (1995) Zircon U-Pb age of the Páramo Rico tonalite-granodiorite, Santander Massif (Cordillera Oriental, Colombia) and its geotectonic significance. J South Am Earth Sci 8(2):187–194
Duque-Caro H (1990) The Chocó block in the northwestern corner of South America: structural, tectonostratigraphic, and paleogeographic implications. J South Am Earth Sci 3(1):71–84
Echeverri B (2006) Genesis and thermal history of gold mineralization in the Segovia-Remedios mining district of northern Colombia. MSc thesis, Shimane University
Ego F, Sébrier M, Yepes H (1995) Is the Cauca-Patía and Romeral fault system left or right-lateral? Geophys Res Lett 22(1):33–36

Emmons WH (1937) Gold Deposits of the World: with a section on prospecting. McGraw-Hill, New York

Escobar LA, Tejada N (1992) Prospección de Platino en Piroxenitas y de Oro en Skarn en la Mina Dan Diego, El Cerro, Frontino (Antioquia). BSc thesis, Universidad Nacional de Colombia (Medellín)

Etayo-Serna F, Rodríguez GI (1985) Edad de la Formación Los Santos. In: Etayo-Serna F, Laverde-Montaño F, Pava A (eds) Proyecto Cretácico: contribuciones. Publicaciones geológicas especiales, vol 16. INGEOMINAS, Bogotá

Etayo-Serna F, Barrero D, Lozano H, Espinosa A, González H, Orrego A, Zambrano F, Duque H, Vargas R, Núñez A, Álvarez J, Ropaín C, Ballesteros I, Cardozo E, Forero H, Galvis N, Ramírez C, Sarmiento L, Albers JP, Case JE, Singer DA, Bowen RW, Berger BR, Cox DP, Hodges CA (1983) Mapa de terrenos geológicos de Colombia. Publicaciones Geológicas Especiales INGEOMINAS 14(1):1–235

Estrada JJ (1995) Paleomagnetism and accretion events in the Northern Andes. PhD thesis, State University of New York

Fabre A (1987) Tectonique et géneration d'hydrocarbures: un modèle de l'evolution de la Cordillère Orientale de Colombie et du bassin de Llanos pendant le Crétacé et le Tertiaire. Archives des Sciences Genève 40(2):145–190

Fabre A, Delaloye M (1983) Intrusiones básicas cretácicas en las sedimentitas de la parte central de la Cordillera Oriental. Geología Norandina 6:19–28

Farris DW, Jaramillo C, Bayona G, Restrepo-Moreno SA, Montes C, Cardona A, Mora A, Speakman RJ, Glascock MD, Valencia V (2011) Fracturing of the Panamanian Isthmus during initial collision with South America. Geology 39(11):1007–1010

Feininger T (1970) The Palestina fault, Colombia. Bull Geol Soc Am 81:1201–1216

Feininger T, Botero G (1982) The Antioquian Batholith, Colombia. Publicación Geológica Especial INGEOMINAS 12:1–50

Feininger T, Barrero D, Castro N (1972) Geología de parte de los departamentos de Antioquia y Caldas (sub-zona II-B). Boletín Geológico INGEOMINAS 20(2):1–173

Felder F, Ortiz G, Campos C, Monsalve I, Silva A (2005) Angostura project: a high sulfidation Gold-Silver deposit located in the Santander complex of the North Eastern Colombia. Paper presented at the ProExplo Conference, Lima (Peru), 24–27 May 2005

Franklin JM, Gibson HL, Jonasson IR, Galley AG (2005) Volcanogenic massive sulfide deposits. In: Hedenquist JW, Thompson JFH, Goldfarb RJ, Richards JP (eds) Economic geology 100th anniversary volume, vol 1905–2005, pp 523–560

Frantz JC, Ordoñez OC, Franco E, Groves DI, McNaughton NJ (2003) Marmato porphyry intrusion, ages and mineralization. Abstract presented at the IX Congreso colombiano de Geología, Medellín, 30 July–01 Aug 2003

Frost BR, Barnes CG, Collins WJ, Arculus RJ, Ellis DJ, Frost CD (2001) A geochemical classification for granitic rocks. J Petrol 42(11):2033–2048

Gallego AN, Akasaka M (2007) Silver-bearing and associated minerals in El Zancudo deposit, Antioquia, Colombia. Resource Geology 57(4):386–399

Gallego AN, Akasaka M (2010) Ag-rich tetrahedrite in the El Zancudo deposit, Colombia – occurrence, chemical compositions and genetic temperatures. Resour Geol 60(3):218–233

Gamba FP (1910) Mining in the Department of Pasto, Colombia. Eng Min J 89:1104–1105

Gansser A (1973) Facts and theories on the Andes. J Geol Soc Lond 129:93–131

Geotec Ltd. (1996) The Foothills of the Eastern Cordillera, Colombia, Geological Map, Scale 1:100.000, five sheets

Geyer O (1973) Das prakretazlsche Mesozoikum von Kolumbien. Geol Jahrb B5:1–156

Gil-Rodríguez J (2010) Igneous petrology of the Colosa gold-rich porphyry system (Tolima, Colombia). MSc thesis, University of Arizona

Gil-Rodríguez J (2014) Petrology of the Betulia Igneous Complex, Cauca, Colombia. J South Am Earth Sci 56:339–356

Giuliani G, France-Lanord C, Cheilletz A, Coget P, Branquet Y, Laumomnier B (2000) Sulfate reduction by organic matter in Colombian emerald deposits: chemical and stable isotope (C, O, H) evidence. Econ Geol 95(5):1129–1153

Gleeson SA, Herrington RJ, Durango J, Velásquez CA, Koll G (2004) The mineralogy and geochemistry of the Cerro Matoso S.A. Ni laterite deposit, Montelíbano, Colombia. Econ Geol 99:1197–1213

Goff F, Stimac J, Larocque C, Hulen J, McMurty G, Adams A, Roldán A, Trujillo P, Counce D, Chipera S, Mann D, Heizler M (1994) Gold degassing and deposition at Galeras volcano, Colombia. GSA Today 4(10):243–247

Gómez J, Montes NE, Nivia A, Diederix H (2015a) Mapa geológico de Colombia 2015. Servicio Geológico Colombiano, Bogotá

Gómez J, Montes NE, Alcárcel FA, Ceballos JA (2015b) Catálogo de dataciones radiométricas de Colombia en ArcGIS y Google Earth. In: Gómez J, Almanza MF (eds) Compilando la geología de Colombia – Una visión a 2015, Publicaciones Geológicas Especiales 33. Servicio Geológico Colombiano, Bogotá, pp 63–419

Gómez J, Nivia A, Montes NE, Jiménez DM, Tejada ML, Sepúlveda MJ, Osorio JA, Gaona T, Diederix H, Uribe H, Mora M (2007) Mapa Geológico de Colombia. INGEOMINAS, Bogotá

Gómez-Gutiérrez DF, Molano-Mendoza JC (2009) Evaluación de zonas de alteración hidrotermal y fases intrusivas, para el prospecto "Stock Porfirítico de Piedra Sentada" (Vereda Santa Lucía) Cauca, Colombia. Geologia Colombiana 34:75–94

González H (2001) Mapa geológico del Departamento de Antioquia: geología, recursos minerales y amenazas potenciales. Memoria Explicativa. INGEOMINAS, Bogotá

Goossens PJ, Rose WI, Flores D (1977) Geochemistry of tholeiites of the Basic Igneous Complex of northwestern South America. Geol Soc Am Bull 88(12):1711–1720

Goodfellow WD, Lydon JW, Turner RJW (1993) Geology and genesis of stratiform sedmienthosted (SEDEX) zinc-lead-silver sulphidedeposits. In: Kirkham RV, Sinclair WV, Thorpe RI, Duke JD (eds) Mineral Deposit Modelling, pp 201–251

Grosse E (1926) Estudio Geológico del Terciario Carbonífero de Antioquia en la parte occidental de la Cordillera Central de Colombia entre el Rio Arma y Sacaojal ejecutado en los anos de 1920 – 1923 para el Gobierno del Departamento de Antioquia (Ferrocarril de Antioquia). Dietrich Reimer (Ernst Vohsen), Berlin

Grosse E (1932) Zur Kenntnis der Gold-Silberlagerstätten von Titiribi. Zeitschrift für praktische Geologie 40:44–45

Groves DI, Bierlein FP (2007) Geodynamic settings of mineral deposit systems. J Geol Soc London 164(1):19–30

Guilbert JM, Park CF Jr (1986) The geology of ore deposits. Freeman & Co, New York

Hall RB, Alvarez AJ, Rico HH (1972) Geología de los Departamentos de Antioquia y Caldas (Subzona II – A). Boletín Geológico (Ingeominas) 20(1):1–85

Halse E (1906) The occurrence of pebbles, concretions and conglomerate in metalliferous veins. Am Inst Min Eng Trans 36:154–177

Hart CJR (2007) Reduced intrusion-related gold systems. In: Goodfellow WD (ed) Mineral deposits of Canada: a synthesis of major deposit types, district metallogeny, the evolution of geological provinces, and exploration methods, vol 5. Geological Association of Canada, Mineral Deposits Division, Special Publication, pp 95–112

Hastie AR, Kerr AC (2010) Mantle plume or slab window?: physical and geochemical constraints on the origin of the Caribbean oceanic plateau. Earth Sci Rev 98(3):283–293

Hayba D, Bethke P, Heald P, Foley N (1985) Geologic, mineralogic, and geochemical characteristics of volcanic-hosted epithermal precious-metal deposits. Rev Econ Geol 2:129–167

Henrichs I (2013) Caracterização e idade das intrusivas do sistema pórfiro Yarumalito, Magmatismo Combia, Colombia. MSc thesis, Universidade Federal do Rio Grande do Sul

Hoffmann F (1931) Die Gold-Silberlagerstätte von Titiribi (Kolumbien). Zeitschrift für praktische Geologie 39(1–13):19–26

Horton BK, Taylor JE, Nie J, Mora A, Parra M, Reyes-Harker A, Stockli DF (2010) Linking sedimentation in the northern Andes to basement configuration, Mesozoic extension, and Cenozoic shortening: evidence from detrital zircon U-Pb ages, Eastern Cordillera, Colombia. Geol Soc Am Bull 122(9–10):1423–1442

Ibañez-Mejía M, Tassinari CCG, Jaramillo-Mejía J (2007) U-Pb ages of the "Antioquian Batholith" – geochronological constraints of late Cretaceous magmatism in the central Andes of Colombia. Abstract presented at the XI Congreso Colombiano de Geología, Bucaramanga, 14–17 August 2007

Ibañez-Mejía M, Ruiz J, Valencia VA, Cardona A, Gehrels GE, Mora AR (2011) The Putumayo Orogen of Amazonia and its implications for Rodinia reconstructions: new U–Pb geochronological insights into the Proterozoic tectonic evolution of northwestern South America. Precambrian Res 191(1):58–77

INGEOMINAS (1998) Atlas colombiano de información geológico-minera para inversión (ACIGEMI). Escala 1:500,000, digital format.

Instituto de Estudios Colombianos (1987) El Oro en Colombia. Universidad INCCA de Colombia, Bogotá, p 290

Irving EM (1975) Structural evolution of the northernmost Andes, Colombia, vol 846. U.S. Geological Survey, Professional Paper, Washington, DC

Ishihara S (1981) The granitoid series and mineralization. Economic Geology 75th Anniversary Volume, pp 458–484

Japan International Cooperation Agency (JICA) (1987) Informe sobre Exploración de minerales del área de Almaguer, Departamento de Cauca, Colombia. Compilación Fases I, II y III.

Japan International Cooperation Agency of Japan (JICA), Metal Mining Agency of Japan (MMAJ) (1984) Estudio preliminar de factibilidad del desarrollodel area el Diamante-Paraiso-Bombona, Departamento de Narino. Internal report, pp 1–183

Jaramillo L (2000) Geological setting and potential of VMS deposits in Colombia. In: Sherlock R, Logan MAV (eds) VMS deposits of Latin America, vol 2. Geological Association of Canada, Mineral Deposits Division, Special Publication, Vancouver, pp 325–332

Kennan L, Pindell JL (2009) Dextral shear, terrane accretion and basin formation in the Northern Andes: best explained by interaction with a Pacific-derived Caribbean Plate? Geol Soc Lond Spec Publ 328(1):487–531

Kerr AC, Tarney J (2005) Tectonic evolution of the Caribbean and northwestern South America: the case for accretion of two Late Cretaceous oceanic plateaus. Geology 33(4):269–272

Kerr AC, Tarney J, Marriner GF, Nivia A, Saunders AD (1997) The Caribbean–Colombian Cretaceous igneous province: the internal anatomy of an oceanic plateau. In: Mahoney JJ, Coffin MF (eds) Large igneous provinces: continental, oceanic, and planetary flood volcanism, Geophysical monograph, vol 100. American Geophysical Union, Washinghton, pp 123–144

Kerr AC, Tarney J, Marriner GF, Nivia A, Saunders AD, Klaver GT (1996) The geochemistry and tectonic setting of late Cretaceous Caribbean and Colombian volcanism. J South Am Earth Sci 9:111–120

Kerr AC, White RV, Thompson PME, Tarney J, Saunders AD (2003) No oceanic plateau—no Caribbean plate? The seminal role of an oceanic plateau in Caribbean plate evolution. In: Bartolin C, Buffler RT, Blickwede J (eds) The circum-Gulf of Mexico and the Carib-bean: hydrocarbon habitats, basin formation, and plate tectonics, vol 79. AAPG Memoir, pp 126–168

Kerrich R, Goldfarb RJ, Richards JP (2005) Metallogenic provinces in an evolving geodynamic framework. Econ Geol 100:1097–1136

Keppie JD (2008) Terranes of Mexico revisited: a 1.3 billion year odyssey. In: Keppie JD, Murphy JB, Ortega-Gutierrez F, Ernst WG (eds) Middle American terranes, potential correlatives, and orogenic processes. CRC Press – Taylor & Francis Group, Boca Raton, pp 7–36

Keppie JD, Ortega-Gutierrez F (2010) 1.3–0.9 Ga Oaxaquia (Mexico): remnant of an arc/backarc on the northern margin of Amazonia. J South Am Earth Sci 29(1):21–27

Kimberley MJ (1980) The Paz de Rio oolitic inland-sea iron formation. Econ Geol 75:97–106

Kirkham RV, Sinclair WD, Thorpe RI, Duke JM (1995) Mineral deposit modeling, vol 40. Geological Association of Canada, Special Paper, St. John's

Kroonenberg SB (1982) A Grenvillian granulite belt in the Colombian Andes and its relation to the Guiana shield. Geol Mijnb 61(4):325–333

Lang JR, Baker T (2001) Intrusion-related gold systems: the present level of understanding. Miner Deposita 36:477–489

Laffitte P (1966) La métallogénie de la France. Bulletin de la Société Géologique de France 7(1):53–72

Leach DL, Sangster DF, Kelley KD, Large RR, Garven G, Allen CR, Gutzmer J, Walters S (2005) Sediment-hosted lead-zinc deposits; a global perspective. In: Hedenquist JW, Thompson JF, Goldfarb RJ, Richards JP (eds) Economic geology one hundredth anniversary volume 1905–2005. Littleton Society of Economic Geologists, Littleton, pp 561–607

Leal-Mejía H (2011) Phanerozoic gold metallogeny in the Colombian Andes: a tectono-magmatic approach. PhD thesis, Universitat de Barcelona

Leal-Mejía H, Melgarejo JC (2008) Mineralogy of "Independecia" gold deposit, Titiribi Mining District, Colombia. Poster presentation and abstract presented at the XXVIII Spanish Society of Mineralogy (SEM) meeting (SEM2008), Zaragoza (Spain), 16–19 Sept 2008

Leal-Mejía H, Melgarejo JC (2010) Mineralogy, mineral chemistry and isotope geochemistry of gold mineralization at San Martin de Loba, San Lucas Range, Colombia. Poster presentation and abstract presented at the 20th general meeting of the International Mineralogical Association (IMA2010), Budapest (Hungary), 21–27 Aug 2010

Leal-Mejía H, Castañeda M, Shaw RP, Melgarejo JC, Sepulveda OI (2006) Mineralogy of "Independecia" gold deposit, Titiribi Mining District, Colombia: poster presentation and abstract presented at the XXVI Spanish Society of Mineralogy (SEM) Meeting (SEM2006), Oviedo, Spain, 11–14 Sept 2006

Leal-Mejía H, Tassinari CCG, Melgarejo JC (2009) Pb-Pb systematics on sulfides from Andean Colombian gold deposits. Oral presentation and abstract presented at the Circum-Caribbean and North Andean tectonomagmatic evolution: impacts on palaeoclimate and resource formation Workshop, Cardiff (Wales, UK), 2–3 Sept 2009

Leal-Mejía H, Shaw RP, Padilla R, Valencia VA (2010) Magmatism vs. Mineralization in the Segovia-Remedios and Central Antioquia Au Districts, Colombia. Poster and abstract presented at the Society of Economic Geologists conference (SEG2010), Keystone (Colorado), 2–5 Oct 2010

Leal-Mejía H, Melgarejo i Draper JC, Shaw RP (2011a) Phanerozoic gold metallogeny in the Colombian Andes. Oral presentation and abstract presented at the 11th Society of Geology Applied to mineral deposits biennial meeting (SGA 2011), Antofagasta (Chile), 26–29 Sept 2011

Leal-Mejía H, Shaw RP, Melgarejo JC (2011b) Phanerozoic granitoid magmatism in Colombia and the tectono-magmatic evolution of the Colombian Andes. In: Cediel F (ed) Petroleum geology of Colombia, vol. 1, Regional geology of Colombia. Agencia Nacional de Hidrocarburos (ANH) – EAFIT, pp 109–188

Leal-Mejía H, Shaw RP, Hart CJR, Bissig T (2015) Jurassic pluton-related Au-Ag mineralization at Mina Uno, Mina Seca – Mina Brisa Trend, Serranía de San Lucas Au Province, Colombia. Poster presented at the MDRU 25th anniversary celebration, Vancouver (BC), 10–11 Apr 2015

Leal-Mejía H, Shaw RP, Melgarejo JC (2018) Spatial-temporal migration of granitoid magmatism and the tectono-magmatic evolution of the Colombian Andes. In: Cediel F, Shaw RP (eds) Geology and Tectonics of Northwestern South America: The Pacific-Caribbean-Andean Junction, Springer, Cham, pp 253–398

Lesage G (2011) Geochronology, petrography, geochemical constraints, and fluid characterization of the Buriticá gold deposit, Antioquia Department, Colombia. MSc thesis, University of Alberta

Lesage G, Richards JP, Muehlenbachs K, Spell TL (2013) Geochronology, geochemistry, and fluid characterization of the late Miocene Buriticá gold deposit, Antioquia Department, Colombia. Econ Geol 108(5):1067–1097

Litherland M, Aspden JA, Jemielita RA (1994) The metamorphic belts of Ecuador. Br Geol Surv Overseas Mem 11:1–47

Lodder C, Padilla R, Shaw RP, Garzón T, Palacio E, Jahoda R (2010) Discovery history of the La Colosa Gold Porphyry deposit, Cajamarca, Colombia. Society of Economic Geologists, Special Publication, vol 15, pp 19–28

Londoño C, Montoya JC, Ordóñez O, Restrepo JJ (2009) Características de mineralizaciones vetiformes en el distrito El Bagre-Nechi, Antioquia. Bóletin de Ciencias de la Tierra 26:29–38

Lonsdale P (2005) Creation of the Cocos and Nazca plates by fission of the Farallon plate. Tectonophysics 404:237–264

Lozano O (1984) Prospección geoquímica para oro, plata, antimonio y mercurio en los municipios de Salento, Quindío y Cajamarca, Tolima. Boletín Geológico (INGEOMINAS) 27:4–76

Mantilla LC, Valencia VA, Barra F, Pinto J, Colegial J (2009) U-Pb geochronology of porphyry rocks in the Vetas – California gold mining area (Santander, Colombia). Boletín de Geología (UIS) 31(1):31–43

Mantilla LC, Bissig T, Cottle JM, Hart CRJ (2012) Remains of early Ordovician mantle-derived magmatism in the Santander Massif (Colombian Eastern Cordillera). J South Am Earth Sci 38:1–12

Mantilla LC, Bissig T, Valencia V, Hart CJR (2013) The magmatic history of the Vetas-California mining district, Santander Massif, Eastern Cordillera, Colombia. J South Am Earth Sci 45:235–249

Maresch WV, Stoeckhert B, Baumann A, Kaiser C, Kluge R, Krueckhans-Lueder G, Brix MR, Thomson M (2000) Crustal history and plate tectonic development in the southern Caribbean, Zeitschrift fuer Angewandte Geologie. Sonderheft Zeitschrift Angewandete Geologie 1:283–289

Maresch WV, Kluge R, Baumann A, Pindell JL, Krückhans-Lueder G, Stanek K (2009) The occurrence and timing of high-pressure metamorphism on Margarita Island, Venezuela: a constraint on Caribbean-South America interaction. Geol Soc Lond Spec Publ 328(1):705–741

Marín-Cerón MI, Leal-Mejía H, Bernet M, Mesa-García J (2018) Late Cenozoic to modern-day volcanism in the Northern Andes: a geochronological, petrographical and geochemical review. In: Cediel F, Shaw RP (eds) Geology and tectonics of Northwestern South America. Springer, Cham, pp 603–641

Maya M (1992) Catalogo de dataciones isotópicas en Colombia. Boletín Geológico (INGEOMINAS) 32(1–3):127–188

Maze WB (1980) Geology and copper mineralization of the Jurassic La Quinta formation in the Sierra de Perijá, northwestern Venezuela. Paper presented at the IX Caribbean geological conference, Santo Domingo (Dominican Republic), 15–26 Aug 1980

Maze WB (1984) Jurassic La Quinta Formation in the Sierra de Perijá, northwestern Venezuela: geology and tectonic environment of red beds and volcanic rocks: the Caribbean–South American plate boundary and regional tectonics. Geol Soc Am Mem 162:263–282

McCourt WJ, Feininger T (1984) High pressure metamorphic rocks of the Central Cordillera of Colombia. In: British geological survey reprint series, vol 84, pp 28–35

McGeary S, Ben-Avraham S (1989) The accretion of Gorgona Island, Colombia: multichannel seismic evidence. In: Howel DG (ed) Tectonostratigraphic terranes of the Circum-Pacific region, Circum-Pacific Council for energy and mineral resources, earth sciences series, vol 1, pp 543–554

Meinert LD, Dipple GM, Nicolescu S (2005) World skarn deposits. Econ Geol 100th Anniv Vol:299–336

Mejía LJ, Pulido O, Angarita L, Buenaventura J (1986) Mapa de ocurrencias minerales de Colombia. INGEOMINAS, Preliminary Edition, Bogota, 4 sheets

Mejía V, Durango J (1981) Geología de las lateritas niquelíferas de Cerro Matoso S.A. Boletin de Geología (Universidad Industrial de Santander) 15(29):99–116

Meldrum S (1998) Titiribí porphyry copper project, Antioquia, Colombia: data compilation and porphyry model. Unpublished report, 21 p

Mendoza M (2011) Estudio textural de las brechas del sector La Mascota, Proyecto La Bodega (Municipio de California, Departamento de Santander), BSc thesis, Universidad Industrial de Santander

Mendoza F, Giraldo K (2012) Definition of structural controls on formation of lode deposits of associated to major Out Shear Zone Fault System at El Limón Au vein, Zaragoza – Antioquia, Colombia. Poster presented at the 2012 Geological Society of America annual meeting, Charlotte Convention Center, Charlotte, NC, 4–7 Nov 2012

Mendoza H, Jaramillo L (1975) Geología y Geoquímica del área de California, Santander. Boletín Geológico (INGEOMINAS) 22(2):1–98

Molano JC, Shimazaki H (2003) Mineralogía, Geoquímica y algunos aspectos genéticos de la Mina El Diamante – Nariño (Colombia). Boletín de Geología 25(40):105–116

Molina C, Molina A (1984) Principales características geológicas y mineralógicas de la mina El Cerro, Frontino, Antioquia. BSc thesis, Universidad Nacional de Colombia

Montes C, Cardona A, McFadden R, Morón SE, Silva CA, Restrepo-Moreno S, Ramírez DA, Hoyos N, Wilson J, Farris D, Bayona GA, Jaramillo CA, Valencia V, Bryan J, Flores JA (2012) Evidence for middle Eocene and younger land emergence in central Panama: implications for Isth-mus closure. Geol Soc Am Bull 124(5–6):780–799

Montes C, Cardona A, Jaramillo C, Pardo A, Silva JC, Valencia V, Ayala C, Pérez-Angel LC, Rodríguez-Parra LA, Ramirez V, Niño H (2015) Middle Miocene closure of the American Seaway. Science 348(6231):226–229

Morgan LA, Schulz KJ (2010) Physical volcanology of volcanogenic massive sulfide deposits. In: Shanks WCP III and Thurston R (eds) Volcanogenic massive sulfide occurrence model. U.S. Department of the Interior and U.S. Geological Survey Scientific Investigations Report 2010–5070–C, pp 65–104

Mosier DL, Singer DA, Moring BC, Galloway JP (2012) Podiform chromite deposits – database and grade and tonnage models. U.S. Geological Survey Scientific Investigations Report 2012–5157:1–45

Mountney NP, Westbrook GK (1997) Quantitative analysis of Miocene to Recent forearc basin evolution along the Colombian convergent margin. Basin Res 9:177–196

Müller RD, Royer JY, Cande SC, Roest WR, Maschenkov S (1999) New constraints on the Late Cretaceous/Tertiary plate tectonic evolution of the Caribbean. In: Mann P (ed) Caribbean basins. Sedimentary basins of the world, vol 4. Elsevier, Amsterdam, pp 33–59

Mutis V (1993) Catálogo de los yacimientos, prospectos y manifestaciones minerales de Colombia, 2nd edn. INGEOMINAS, Publicaciones Geológicas Especiales 13:1–536

Nerlich R, Clark S, Bunge H (2014) Reconstructing the link between the Galapagos hotspot and the Caribbean Platea. Geo Res J 1 & 2:1–7

Nichols HW, Farrington OC (1899) The ores in Colombia – from mines in operation in 1892. Field Columbian Museum, Publication 33, Geological Series 1(3):123–177

Nivia A (2001) Mapa Geológico del Departamento del Valle del Cauca, Memoria Explicativa. INGEOMINAS, Bogotá

Nivia A, Marriner G, Kerr A (1996) El Complejo Quebrada Grande: una posible cuenca marginal intracratónica del Cretáceo Inferior en la Cordillera Central de los Andes Colombianos. In: Abstracts of the VII Colombian Geological Congress, vol III, pp 108–123

Núñez A (2001) Mapa geológico generalizado del Departamento del Tolima – Geología, recursos geológicos y amenazas geológicas, Memoria Explicativa. INGEOMINAS, Escala 1:250.000, Bogotá

Ordoñez O, Pimentel MM, Armstrong RA, Gioia SMCL, Junges S (2001) U-Pb SHRIMP and Rb-Sr ages of the Sonsón Batholith. In: III South American Symposium on Isotope Geology, Pucon – Chile, 21–24 Oct 2001

Orrego A, Acevedo A (1993) Geología de la Plancha 364 – Timbío, Informe No. 2169. Ingeominas, Bogotá

Ordoñez-Carmona O, Restrepo JJ, Pimentel MM (2006) Geochronological and isotopical review of pre-Devonian crustal basement of the Colombia Andes. Journal of South American Earth Sciences 21(4):372–382

Orrego A, Paris G (1990) Cuadrángulo N-6, Popayán, Geología, Geoquímica y ocurrencias minerales. INGEOMINAS, Bogotá

Orrego A, Cepeda H, Rodríguez G (1980) Esquistos glaucofánicos en el área de Jambaló, Cauca (Colombia). Informe 1729. INGEOMINAS, Bogotá

Orrego A, Paris G, Ibañez D, Vásquez E (1999) Geología y geoquímica de la Plancha 387 – Bolívar. Publicaciones Especiales de Ingeominas 22:55–114

Ortega CR (1982) Complejo ofiolítico en la Cuenca del río Guapí. Boletín de Geología (Universidad Industrial de Santander) 15(29):117–123

Ortiz F (1990) Massive sulfides in Colombia. In: Fontboté L, Amstutz GC, Cardozo M, Cedilllo E, Frutos J (eds) Stratabound ore deposits in the Andes. Society of Geology Applied to Mineral Deposits, Special Publication, vol 8, pp 379–387

Ottaway TL (1991) The geochemistry of the Muzo emerald deposit, Colombia. MSc thesis, University of Toronto

Ottaway TL, Wicks FJ, Bryndzia LT, Kyser TK, Spooner ETC (1994) Formation of the Muzo hydrothermal emerald deposit in Colombia. Nature 369:552–554

París G, Marín P (1979) Mapa Geológico Generalizado del Departamento del Cauca. Memoria Explicativa. INGEOMINAS, 38 p

Perry RW (1914a) Quartz Mining in Colombia. Engineering and Mining Journal 97(18):889–892

Perry RW (1914b) Quartz mining in Colombia II. Engineering and Mining Journal 97(18):945–948

Petersen U (1970) Metallogenic provinces in South America. Geologische Rundschau 59(3):834–897

Petersen U (1979) Metallogenesis in South America: progress and problems. Episodes 4:3–11

Pindell J, Kennan L (2001) Processes & events in the terrane assembly of Trinidad and E. Venezuela. In: GCSSEPM foundation 21st annual research conference transactions, petroleum systems of deep-water basins, pp 159–192

Pujol N, Schamuells S, Melgarejo JC, Leal-Mejía H (2012) Mineralogy of Au mineralization at the Quebradona Creek, Jericó (Antioquia, Colombia). Revista de la SEM 16:256–257

Pulido OH (1988a) Reconocimiento regional para mineralizaciones de oro diseminado en tres zonas de los departamentos de Caldas, Quindio y Tolima. Boletín Geológico (INGEOMINAS) 27:21–36

Pulido OH (1988b) Geología y geoquímica del área de San Antonio, Cajamarca, Tolima. Boletín Geológico (INGEOMINAS) 27:37–84

Pulido W (2005) Caracterización geológico-minera del depósito aurífero de El Retiro, Ginebra (Valle del Cauca). BSc thesis, Universidad Nacional de Colombia

Raley CA (2012) Mineralogical characterization of sulfide mineralization, alteration and microthermometry of related fluid inclusions of the La Plata Prospect, Colombia. BSc thesis, The University of British Columbia

Ramos VA (1999) Plate tectonic setting of the Andean Cordillera. Episodes 22(3):183–190

Ramos VA (2009) Anatomy and global context of the Andes: main geologic features and the Andean orogenic cycle. In: Kay SM, Ramos VA, and Dickinson WR (eds) Backbone of the Americas: shallow subduction, plateau uplift, and ridge and terrane collision. Geological Society of America, Memoir 204:31–65

Ramsey JG (1980) The crack-seal mechanism of rock deformation. Nature 284:135–139

Redwood SD (2018) The geology of the Panama-Chocó Arc. In: Cediel F, Shaw RP (eds) Geology and tectonics of Northwestern South America. Springer, Cham, pp 901–926

Restrepo JJ, Toussaint JF (1988) Terrains and continental accretion in the Northern Andes. Episodes 11:189–193

Restrepo V (1888) Estudio sobre las minas de oro y plata de Colombia. Imprenta de Silvestre y Compañía, Bogotá

Restrepo-Moreno SA, Foster DA, Kamenov GD (2007) Formation age and magma sources for the Antioqueño Batholith derived from LA-ICP-MS Uranium-Lead dating and Hafnium-isotope analysis of zircon grains. Geological Society of America Abstracts with Programs 39(6):493

Restrepo-Pace PA (1992) Petrotectonic characterization of the Central Andean Terrane, Colombia. J South Am Earth Sci 5(1):97–116

Restrepo-Pace PA (1995) Late Precambrian to Early Mesozoic tectonic evolution of the Colombian Andes based on new geological, geochemical and isotopic data. PhD thesis, The University of Arizona

Restrepo-Pace PA, Cediel F (2010) Northern South America basement tectonics and implications for paleocontinental reconstructions of the Americas. J South Am Earth Sci 29:764–771

Reyes JJ (2013) Identificación de los estilos de mineralización en la Mina Reina de Oro (Municipio de Vetas, Departamento de Santander) a partir de estudios de alteración hidrotermal y mineralogía de mena. BSc thesis, Universidad Industrial de Santander

Rodriguez AL (2014) Geology, Alteration, Mineralization and Hydrothermal Evolution of the La Bodega-LaMascota deposits, California-Vetas Mining District, Eastern Cordillera of Colombia, Northern Andes. MSc thesis, The University of British Columbia

Rodríguez C (1984) Les indices mineralises du massif de quetame Colombie. Rapport de fin de cycle CESEV, Université de Nancy (unpublished)

Rodriguez C (1987) lnforme final de exploración, Etapa 1. Proyecto zonas aledañas a Marmato. Ecominas, Bogota (unpublished report)

Rodriguez G, Arango MI (2013) Formación Barroso: arco volcanico toleitico y diabasas de San José de Urama: un prisma acrecionario T-MORB en el segmento norte de la Cordillera Occidental de Colombia. Boletín de Ciencias de la Tierra 33:17–38

Rodríguez C, Warden AJ (1993) An overview of some Colombian gold deposits. Miner Deposita 28:47–57

Rodríguez AL, Bissig T, Hart CJR, Mantilla LC (2017) Late pliocene high-sulfidation epithermal gold mineralization at the La Bodega and La Mascota Deposits, Northeastern Cordillera of Colombia. Econ Geol 112(2):347–374

Rodríguez G, Celada CM, Cossio U, Munoz R, Balcero G (2000) Evaluación Geológico – Minera del Yacimiento de Miraflores, Distrito Minero de Quinchía, Risaralda (Colombia). Informe 2467, Ingeominas, Medellín

Rossetti P, Colombo F (1999) Adularia-sericite gold deposits of Marmato (Caldas, Colombia): field and petrographical data. Geol Soc Lond Spec Publ 155:167–182

Routhier P (1983) Where are the metals for the future? Editions du BRGM, Orleans

Royero JM, Clavijo J (2001) Mapa geologico generalizado del Departamento de Santander, Memoria Explicativa. INGEOMINAS, scale 1:400.000, Bogotá

Rundall WH (1931) The gold mines of the Frontino & Bolivia and associated companies. Min Mag 44:73–81

Sánchez SM (2013) Estudio de las litologías y alteraciones hidrotermales asociadas, presentes en el tramo comprendido entre el casco urbano del Municipio de Vetas y la Mina 'Reina de Oro' (Macizo de Santander, Colombia). BSc thesis, Universidad Industrial de Santander

Santacruz L (2011) Microtermometría de inclusiones fluidas aplicada al depósito de Marmato. BSc thesis, Universidad Nacional de Colombia

Santacruz L, Redwood S, Molano JC, Cecchi A (2012) Evolution of sulfidation states in the Marmato gold deposit, Colombia. Poster presented at the Society of Economic Geologists SEG2012 conference, Westin Hotel, Lima, Peru, 23–26 Sept 2012

Santacruz L, Redwood S, Molano JC, Cecchi A (2014) Affinity between bismuth and gold in the Marmato gold deposit, Colombia: a probable case of the liquid bismuth collector model. Poster presented at the Society of Economic Geologists SEG2014 conference, Keystone, CO, 27–30 Sept 2014

Sarmiento LF (2001) Mesozoic rifting and Cenozoic basin inversion history of the Eastern Cordillera, Colombian Andes. Inferences from tectonic models. PhD thesis, Vrije Universiteit

Sarmiento LF (2018) Cretaceous stratigraphy and paleo-facies maps of Northwestern South America. In: Cediel F, Shaw RP (eds) Geology and tectonics of Northwestern South America: the Pacific-Caribbean-Andean junction. Springer, Cham, pp 673–739

Sawkins FJ (1990) Metal deposits in relation to plate tectonics, 2nd edn. Springer-Verlag, Heildelberg
Schmidt-Thomé M, Feldhaus L, Salazar G, Muñoz R (1992) Explicación del mapa geológico, escala 1:250 000, del flanco oeste de la Cordillera Occidental entre los ríos Andágueda y Murindó, Departamentos Antioquia y Chocó, República de Colombia. Geologisches Jahrbuch Reihe B, Band B78
Shagam R (1975) The northern termination of the Andes. In: Nairn AEM, Stehli FG (eds) The ocean basins and margins volume 3: the Gulf of Mexico and the Caribbean. Plenum Press, New York, pp 325–420
Shaw RP (2000a) Gold Mineralisation in Colombia: geologic setting, metallogeny, production history, and exploration update. Paper presented at the Peru's 4th international gold symposium, Lima, Peru, May 2000
Shaw RP (2000b) Gold mineralization in the Northern Andes: magmatic setting vs. metallogeny: abstract presented at the 11th International mining congress, Bogota, 2000
Shaw RP (2003a) Hacia un análisis litotectónico de metalogénesis en Colombia. Keynote speech and abstract presented at the 9th Colombian geological congress, Medellín, Colombia, 2003
Shaw RP (2003b) Colombia: a "Ten Best Picks" analysis of gold metallogeny and project potential. Unpublished report prepared for AngloGold Ltd., Lima, Perú, 204 p
Sillitoe RH (1991) Intrusion-related gold deposits. In: Foster RP (ed) Gold metallogeny and exploration. Chapman & Hall, Glasgow, pp 165–209
Sillitoe RH (2000) Gold-rich porphyry deposits: descriptive and genetic models and their role in exploration and discovery. Reviews in Economic Geology 13:315–345
Sillitoe RH (2008) Major gold deposits and belts of the North and South American Cordillera: distribution, tectonomagmatic settings and metallogenetic considerations. Econ Geol 103(4):633–687
Sillitoe RH (2010) Porphyry copper systems. Econ Geol 105:3–41
Sillitoe RH, Hart SR (1984) Lead-isotopic signatures of porphyry copper deposits in oceanic and continental settings. Geoquinica et Cosmoquimica Acta 48:2135–2142
Sillitoe RH, Hedenquist JW (2003) Linkages between volcanotectonic settings, ore-fluid compositions, and epithermal precious metal deposits. Soc Econ Geol Spec Publ 10:315–343
Sillitoe RH, Perelló J (2005) Andean copper province: tectonomagmatic settings, deposit types, metallogeny, exploration, and discovery. Economic Geology 100th Anniversary Volume 1905–2005:845–890
Sillitoe RH, Jaramillo L, Damon PE, Shafiqullah M, Escovar R (1982) Setting, characteristics and age of the Andean porphyry copper belt in Colombia. Econ Geol 77:1837–1850
Sillitoe RH, Jaramillo L, Castro H (1984) Geologic exploration of a molybdenum-rich porphyry copper deposit at Mocoa, Colombia. Econ Geol 79:106–123
Silva JC, Arenas JE, Sial AN, Ferreira VP, Jiménez D (2005) Finding the Neoproterozoic-Cambrian transition in carbonate successions from the Silgará Formation, Northeastern Colombia: an assessment from C-isotope stratigraphy. In: Memorias del X Congreso Colombiano de Geología, Bogota
Silver EA, Reed DL, Tagudin JE, Heil DJ (1990) Implications of the north and south Panama thrust belts for the origin of the Panama orocline. Tectonics 9:261–281
Simmons SF, White NC, John DA (2005) Geologic characteristics of epithermal precious and base metal deposits. Economic Geology 100th Anniversary Volume:485–522
Singewald QD (1950) Mineral Resources of Colombia (other than Petroleum). U S Geol Surv Bull 964-B:56–204
Sinton CW, Duncan RA, Storey M, Lewis J, Estrada JJ (1998) An oceanic flood basalt province within the Caribbean plate. Earth Planet Sci Lett 155(3):221–235
Skinner BJ (1999) Preface. In: Skinner BJ (ed) Geology and ore deposits of the Central Andes. Economic geology special publication, vol 7, p iii
Slack JF (1993) Descriptive and grade-tonnage models for Besshi-type massive sulfide deposits. In: Kirkham RV, Sinclair WD, Thorpe RI, Duke JM (eds) Mineral deposit models, vol 40. Geological Association of Canada, Special Paper, Denver, pp 343–371

Society of Economic Geologists (2002) Society of Economic Geologists special session: the global tectonic setting of ore deposits – present understanding and new advances, pp 11–14. Geol Soc Am Abstr Programs 34(6):11–14

Spikings R, Cochrane R, Villagómez D, Van der Lelij R, Vallejo C, Winkler W, Beate B (2015) The geological history of northwestern South America – from Pangaea to the early collision of the Caribbean Large Igneous Province (290–75 Ma). Gondw Res 27:95–139

Staargaard CF, Carlson GC (2000) The Bailadores volcanogenic massive sulphide deposit, Venezuela. In: Sherlock RL, Logan MA (eds) Volcanogenic massive sulphide deposits of Latin America. Geological Association of Canada Mineral Deposits Division, St. John's, pp 315–323

Stern CR (2004) Active Andean volcanism: its geologic and tectonic setting. Revista Geológica de Chile 31(2):161–206

Stix J, Calvache M, Fischer TP, Gómez D, Narvaez L, Ordoñez M, Ortega A, Torres R, Williams SN (1993) A model of degassing at Galeras Volcano, Colombia, 1988–1993. Geology 21(11):963–967

Strong DF (ed) (1976) Metallogeny and plate tectonics. Geological Association of Canada, Special Paper 14, Ontario

Taboada A, Rivera LA, Fuenzalida A, Cisternas A, Philip H, Bijwaard H, Olaya J, Rivera C (2000) Geodynamics of the northern Andes – subductions and intracontinental deformation (Colombia). Tectonics 19(5):787–813

Tassinari CCG, Diaz F, Buenaventura J (2008) Age and source of gold mineralization in the Marmato mining district, NW Colombia: a Miocene-Pliocene epizonal gold deposit. Ore Geol Rev 33:505–518

Thompson JFH, Sillitoe RH, Baker T, Lang JR, Mortensen JK (1999) Intrusion-related gold deposits associated with tungsten-tin provinces. Miner Deposita 34:323–334

Tistl M (1994) Geochemistry of platinum-group elements of the zoned ultramafic Alto Condoto complex, northwest Colombia. Econ Geol 89:158–167

Toussaint JF (1993) Evolución geológica de Colombia, Precámbrico–Paleozóico. Universidad Nacional de Colombia, Medellín

Tschanz CM, Hall RB, Feininger T, Ward DE, Goldsmith R, MacLaughlin DH, Maugham EK (1968) Summary of mineral resources in four selected areas of Colombia. United States geological survey, open-file report 68-279, pp 1–58

Tschanz CM, Marvin RF, Cruz J, Mehnert H, Cebulla G (1974) Geologic Evolution of the Sierra Nevada de Santa Marta area, Colombia. Geol Soc Am Bull 85:273–284

UPME (2017) Producción de minerales, producción de oro, histórico anual.. http://www.upme.gov.co/generadorconsultas/Consulta_Series.aspx?idModulo=4&tipoSerie=117&grupo=358. Accessed 08 Mar 2017

Uribe C (2013) Hydrothermal evolution of the Titiribí Mining District. Unplublished report, MDRU Colombia Gold and Porphyry Project,. 126p.

Van der Lelij R (2013) Reconstructing north-western Gondwana with implications for the evolution of the Iapetus and Rheic Oceans: a geochronological, thermochronological and geochemical study. PhD thesis, Université de Genève

Van der Lelij R, Spikings RA, Ulianov A, Chiaradia M, Mora A (2016) Palaeozoic to Early Jurassic history of the northwestern corner of Gondwana, and implications for the evolution of the Iapetus, Rheic and Pacific Oceans. Gondw Res 31:271–294

Vargas H (2002) El Alacran Skarn, Deposito de Cobre-Oro-Plata, Puerto Libertador, San Juan de Asis, Cordoba, Licencia No. 022-23, Colombia, 48 p., with digital database in CD-ROM format.

Vargas R (2005) Evaluación geológica, geoquímica y génesis de la Zona De Exclusión en Marmato, Caldas, Colombia. MSc thesis, Universidad Nacional de Colombia

Vargas CA, Mann P (2013) Tearing and breaking off of subducted slabs as the result of collision of the Panama arc-indenter with northwestern South America. Bull Seismol Soc Am 103(3):2025–2046

Vásquez M, Altenberger U, Romer RL, Sudo M, Moreno-Murillo JM (2010) Magmatic evolution of the Andean Eastern Cordillera of Colombia during the Cretaceous: influence of previous tectonic processes. J South Am Earth Sci 29(2):171–186

Villagómez D, Spikings R, Magna T, Kammer A, Winkler W, Beltrán A (2011) Geochronology, geochemistry and tectonic evolution of the Western and Central cordilleras of Colombia. Lithos 125:875–896

Villegas A (ed) (1987) Recursos Minerales de Colombia, vol 1, 2nd edn. INGEOMINAS, Publicaciones Geológicas Especiales, pp 1–564

Vinasco C (2018) The Romeral shear zone. In: Cediel F, Shaw RP (eds) Geology and tectonics of Northwestern South America: the Pacific-Caribbean-Andean junction. Springer, Cham, pp 833–870

Vinasco CJ (2001) A Utilização da Metodologia ^{40}Ar-^{39}Ar para o Estudo de Reativações Tectônicas em Zonas de Cisalhamento, Paradigma-O Falhamento de Romeral nos Andes Centrais da Colombia. MSc thesis, Universidade de São Paulo

Vinasco CJ, Cordani UG, González H, Weber M, Pelaez C (2006) Geochronological, isotopic, and geochemical data from Permo-Triassic granitic gneisses and granitoids of the Colombian Central Andes. J South Am Earth Sci 21:355–371

Viteri E (1980) Aspectos Geologicos de las exploraciones cupriferas en la Sierra de Perija, Venezuela. Paper presented at the IX Caribbean geological conference, Santo Domingo, Dominican Republic, 15–26 Aug 1980

Ward DE, Goldsmith R, Cruz J, Restrepo H (1970) Mineral resources of the southern half of Zone III, Santander, Norte de Santander and Boyacá: USGS Open File Report 70-359, pp 1–171

Ward DE, Goldsmith R, Cruz J, Restrepo H (1973) Geología de los cuadrangulos H-12 Bucaramanga y H-13 Pamplona, Departamento de Santander. Boletín Geologico (INGEOMINAS) 21(1–3):1–131

Warden A, Colley H (1990) Field investigations of the Marmato gold deposits of Caldas and Risaralda, Colombia. Trans Inst Min Metall Tech Note 99B:52–54

Weber M, Gómez-Tapias J, Cardona A, Duarte E, Pardo-Trujillo A, Valencia VA (2015) Geochemistry of the Santa Fé Batholith and Buriticá Tonalite in NW Colombia e Evidence of subduction initiation beneath the Colombian Caribbean Plateau. J South Am Earth Sci 62:257–274

Wegner W, Wörner G, Harmon RS, Jicha BR (2011) Magmatic history and evolution of the Central American Land Bridge in Panama since Cretaceous times. Geol Soc Am Bull 123(3–4):703–724

Wilson FK, Darnell BF (1942a) A lode gold mine in Colombia. Engineering and Mining Journal 143(4):62–66

Wilson FK, Darnell BF (1942b) A lode gold mine in Colombia II. Engineering and Mining Journal 143(5):58–62

Wokittel R (1954) Informe preliminar sobre los yacimientos de minerales del Mpio. de Urrao, Depto. de Antioquia. Informe 1025, Investigaciones de Geología Minera, Instituto de Geología Nacional, Ministerio de Minas y Petroleos, Bogotá, Colombia

Wokittel R (1960) Recursos minerales en Colombia. Servicio Geológico Nacional, Compilación de los estudios geológicos oficiales en Colombia 10:1–393

Wright JE, Wyld SJ (2011) Late Cretaceous subduction initiation on the eastern margin of the Caribbean-Colombian Oceanic Plateau: one great arc of the Caribbean (?). Geosphere 7:468–493

Zapata S, Cardona A, Jaramillo C, Valencia V, Vervoort J (2016) U-Pb LA-ICP-MS Geochronology and geochemistry of Jurassic volcanic and plutonic rocks from the Putumayo Region (Southern Colombia): tectonic setting and regional correlations. Boletín de Geología 38(2):21–38

Zarifi Z (2006) Unusual subduction zones: case studies in Colombia and Iran. PhD thesis. University of Bergen

Chapter 7
Paleogene Magmatism of the Maracaibo Block and Its Tectonic Significance

José F. Duque-Trujillo, Teresa Orozco-Esquivel, Carlos Javier Sánchez, and Andrés L. Cárdenas-Rozo

7.1 Introduction

The northwestern region of South America includes several tectonic provinces such as the southern Caribbean Plate, the Guajira, and the North Andean and Maracaibo blocks. This complexity is enhanced by the tectonism related to the highly dynamic interaction between the Caribbean and South American plates. Velocity vectors calculated from GPS data (GEODVEL 2010 by Argus et al. 2010) indicate that the Maracaibo block (MB) is being transported at ~60 mm/yr. with a current northwest relative motion with respect to the South American Plate and a general eastward motion for the North Andean block and the Southern Caribbean margin (Trenkamp et al. 2002; Colmenares and Zoback 2003).

The basement of the southern Caribbean Plate (Fig. 7.1) corresponds to an ca. 90 Ma oceanic plateau (Burke 1988; Pindell and Kennan 2009), with crustal thickness between 10 and 20 km. This basement is overlain by a late Cretaceous to Cenozoic sedimentary sequence ca. 3–8 km thick (Mauffret and Leroy 1997). The boundary between the Caribbean and South American plates is defined by the South Caribbean deformed belt (SCDB; Fig. 7.1), a fold- and thrust-belt structure. In that region, the thickness of the deformed sedimentary sequence is ca. 10–18 km, suggesting high sedimentation rates during the late Miocene to Recent (Bernal-Olaya et al. 2015). Furthermore, this tectonic setting has triggered mud diapirism, which has been identified in several sites within the fold and thrust belt (Quintero 2012).

J. F. Duque-Trujillo (✉) · C. J. Sánchez · A. L. Cárdenas-Rozo
Earth Sciences Department, EAFIT University, Medellín, Colombia
e-mail: jduquetr@eafit.edu.co

T. Orozco-Esquivel
Centro de Geociencias, Universidad Nacional Autónoma de México, Querétaro, Qro., Mexico

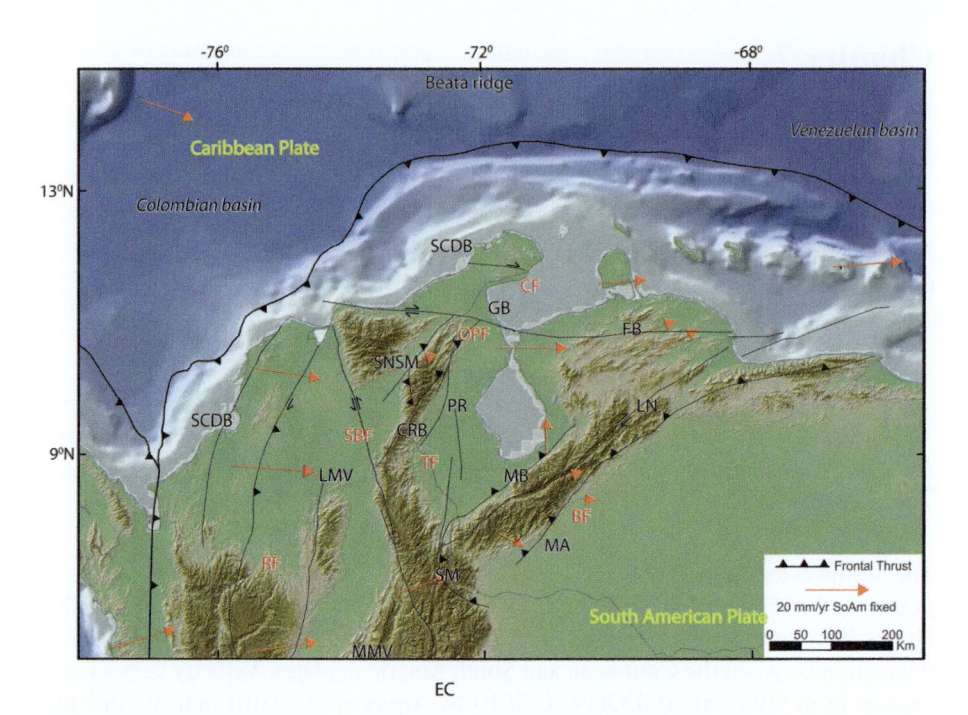

Fig. 7.1 Regional map showing the main morphotectonic elements in the Maracaibo block: Cesar-Rancheria Basin (CRB), Lara Nappes (LN), Maracaibo Basin (MB), Merida Andes (MA), Perijá Range (PR), Santander Massif (SM), and Sierra Nevada de Santa Marta (SNSM) massif. Neighboring elements to the Maracaibo block include the Eastern Cordillera (EC), Falcon Basin (FB), Guajira Basin (GB), Lower Magdalena Valley (LMV) Basin, Middle Magdalena Valley (MMV) Basin, and South Caribbean deformed belt (SCDB). The main fault systems in the region include the Bocono (BF), Cuiza (CF), Oca (OF), Romeral (RF), and Santa Marta-Bucaramanga (SBF) faults. Orange arrows represent current GPS velocity vectors after Colmenares and Zoback (2003)

In the northeastern part of the South American continental plate, the SCDB includes the Maracaibo block (MB). Morphologically, the MB shows a conspicuous triangular shape, with its northwestern apex corresponding to the Sierra Nevada de Santa Marta (SNSM), which is bounded to the southeast by the Cesar-Rancheria Basin that separates the SNSM from the Perijá range and the Santander Massif. Further to the southeast, the Maracaibo Basin is bounded to the south by the Merida Andes (Fig. 7.1). The northern boundary of MB corresponds to the Oca-El Pilar right-lateral fault system with displacements of ca. 90 km (Kellogg 1984); this fault constitutes the boundary to the Guajira block and the Falcon Basin (Fig. 7.1). In the southern Guajira Block, an ca. 4–7 km thick Jurassic to Cenozoic sedimentary sequence is found in the Baja Guajira Basin (Gomez 2001; INGEOMINAS 2007), whereas its northern counterpart contains metamorphic and igneous bodies associated with a late Cretaceous to Paleogene arc-continental collision event (Weber et al. 2009). The Cuiza fault, the most significant structure within the Guajira Block, is an east-west striking right-lateral fault system, which extends from western to eastern offshore areas of the Guajira Peninsula. The displacement along this fault has been estimated at 15–25 km (Gomez 2001). The sedimentary sequence of the Falcon Basin (Figs. 7.1 and 7.2) overlies a Cretaceous basement of metamorphosed island-arc lithologies accreted to South America during the Paleocene-early Eocene (Blanco et al. 2015). The basal sedimentary sequence deposited in the Falcon Basin during Eocene-Middle Miocene times records a marine setting related to extensional episodes (Fig. 7.2). On the other hand, the overlying sequence reflects transition to continental facies as well as sedimentation hiatuses that resulted from compression and inversion events (Macellari 1995; Ostos et al. 2005; Bezada et al. 2008).

The southwestern boundary of the MB is defined by the left-lateral Santa Marta-Bucaramanga fault system, for which controversial estimates pertaining to the amount of displacement, range from ca. 60 and 120 km (Campbell 1965; Florez-Niño 2001). This fault system separates the MB from the Lower and Middle Magdalena Valley Basins (Fig. 7.1). Close to this boundary, the Lower Magdalena Valley Basin is filled by a ca 8 km thick sedimentary sequence unconformably deposited on an oceanic basement during Oligocene to recent times in an extensional regime (Flinch et al. 2003; Cerón et al. 2007). To the south, the sedimentary sequence in the Middle Magdalena Valley Basin overlies a basement of continental affinity deposited during Mesozoic to recent times. The Cenozoic sedimentary sequence consists of an ca. 7 km thick non-marine succession, deposited as a result of uplift in both the Central and Eastern Cordilleras (Gómez et al. 2005).

On the basis of these observations, it is possible to establish that rigid motion has produced significant shortening, uplift, and exhumation inside the MB, which is evidenced by the building of topographically important mountain ranges (i.e., Sierra Nevada de Santa Marta (SNSM) massif, Perijá Range, and Mérida Andes), characterized by the exposure of both metamorphic and igneous basement rocks (Fig. 7.3). Additionally, the geological history of this block is critical in the establishment of the interaction between the northern South American and the Caribbean plates, and subsequent tectonic-influenced events (e.g., Cenozoic-Caribbean geology, long-term paleoceanographic changes in the Atlantic and Pacific oceans, Cenozoic faunal distribution in northern South America, and spatial distribution of energy resources, among others).

To date, two tectonic models have been proposed to explain the geological history of MB: (i) northwestward overthrusting of crystalline rocks of the Andes on a thrust fault extending into the mantle and overriding the Maracaibo Basin (Kellogg and Bonini 1982; Cediel et al. 2003) and (ii) incipient low-angle subduction of a buoyant, young, and hot Caribbean Plate (Pindell et al. 2005).

In this paper, we examine the petrogenesis of sporadic Cenozoic magmatic events recorded in the westernmost SNSM (Tschanz et al. 1974; Cardona et al. 2010b, 2011a; Duque 2009) and elsewhere in the Colombian Andes, with the aim of evaluating subduction processes and their importance in the Late Cretaceous-Cenozoic tectonic evolution of the MB.

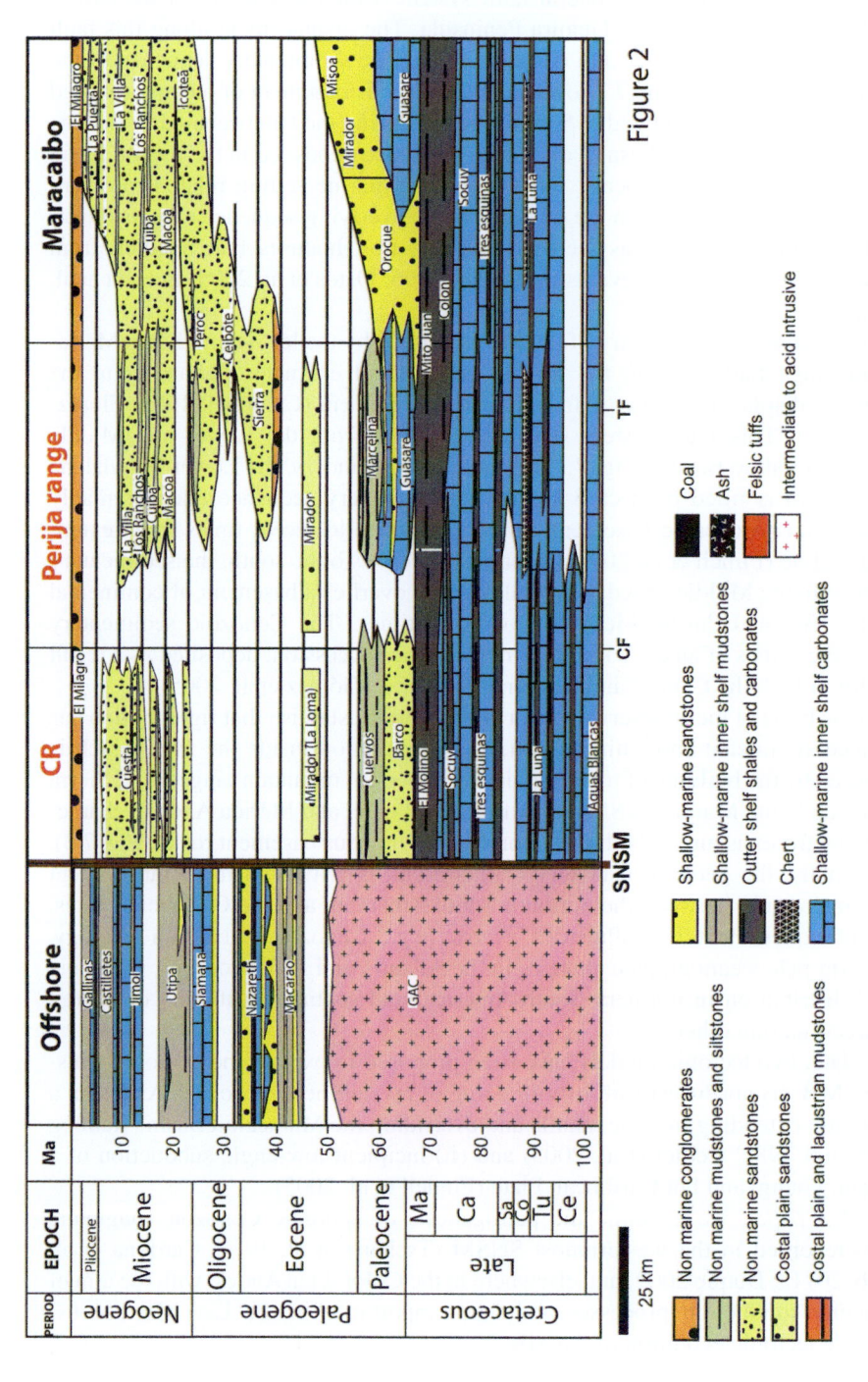

Fig. 7.2 General stratigraphic chart for the Colombian Caribbean offshore, Cesar-Rancheria Basin (CRB), Perijá Range (PR), and Maracaibo Basin (MB) (Fig. 7.1). Compiled from Miller (1962), Forero (1970), Cáceres et al. (1980), Kellogg (1984), Maze (1984), Parnaud et al. (1995), Mann et al. (2006), Vence (2008), and Ayala (2009)

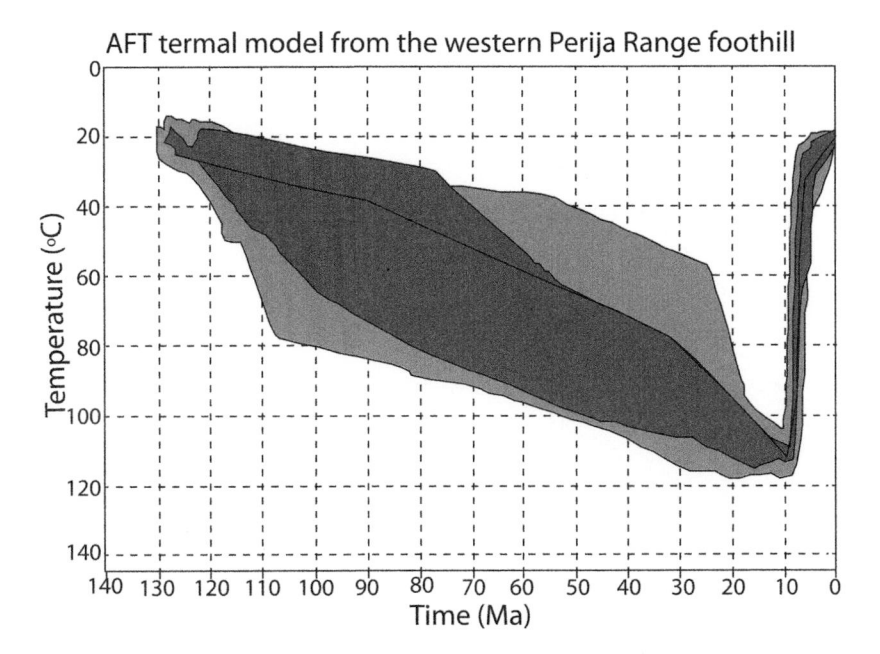

AFT termal model from the western Perija Range foothill

Fig. 7.3 Time-temperature modeling modified from Hernández and Jaramillo (2009) using A-FT for an early Cretaceous sample collected in the western foothills of the Perijá Range. Notice the accelerated exhumation during the late Miocene

7.2 Paleogene Magmatism in the Maracaibo Block

7.2.1 General Settings

Paleogene magmatismin the MB is mainly represented by the Santa Marta Batholith (SMB), located in the northwestern tip of SNSM (Fig. 7.4). Tschanz et al. (1969, 1974) first described this magmatic body which intrudes the two most external metamorphic belts of SNSM (i.e., Sevilla and Santa Marta belts) (Fig. 7.4). As a result, this magmatism has been divided into two separated belts of different ages and compositions (Tschanz et al. 1974).

The first magmatic belt, represented by the Latal and Toribio Plutons, is constituted by Paleocene to early Eocene intrusive bodies, with compositions ranging from diorite to hornblende diorite, and pegmatitic diorite. The Latal Pluton, which intrudes Precambrian gneisses of Los Mangos Granulite unit, is the southernmost intrusion representing this magmatic event (Tschanz et al. 1974). The Toribio Pluton, which is located within the SMB, was considered by Tschanz et al. (1974) as a different intrusion, related to the Latal Pluton by similar composition (in lieu of the lack of geochronological data) (Fig. 7.4). The Latal Pluton, dated at 54.15 ± 2.3 Ma using K/Ar in hbl (the date was recalculated after Steiger and Jäger 1977 and corresponds to sample 10 from Tschanz et al. 1974), is more complex than

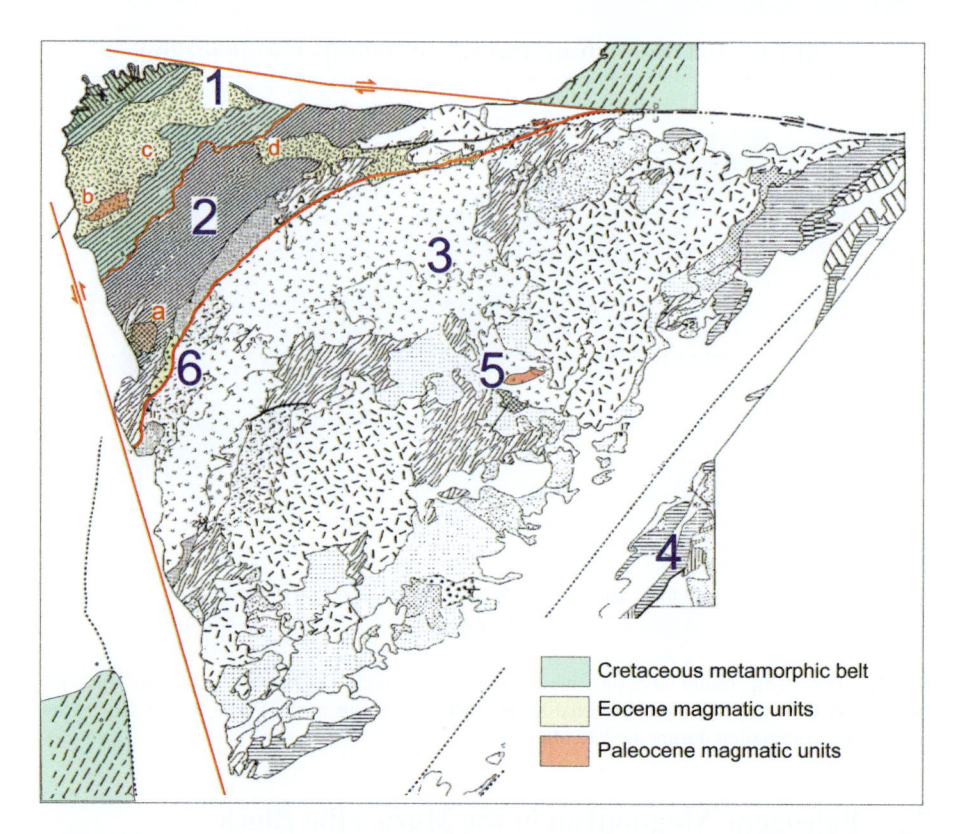

Fig. 7.4 Simplified geological map from SNSM. (1) Santa Marta Province, (2) Sevilla Province, (3) Sierra Nevada Province, (4) Perijá Range, (5) Atanques laccolith, (6) Socorro stock. Paleocene intrusive rocks in red: (a) Latal Pluton and (b) Toribio Pluton. Eocene intrusive rocks in yellow: (c) Santa Marta Batholith, (d) Buritaca Pluton. Modified after Tschanz et al. (1974)

the Toribio Pluton, because it is composed of at least three different lithologies with complex interelationships. The main lithologies in the Latal Pluton include hornblende diorite and hornblendite that discordantly cut diorite bodies. Spectacular pegmatitic dikes composed of hornblende and plagioclase are also present and seem to be the result of a latter intrusive phase into the hornblendites and probably also into the diorite (Tschanz et al. 1974).

The second magmatic belt—the Santa Marta Intrusive Complex—is includes the SMB (45.52 ± 2.0 Ma, K/Ar in hbl and 41.32 ± 1.2 Ma K/Ar in bt) and the Buritaca Pluton (45.88 ± 2.9 Ma, K/Ar in hbl and 45.26 ± 1.4 Ma K/Ar in bt) (Tschanz et al. 1974) (Fig. 7.4). These age dates were recalculated after Steiger and Jäger (1977), and correspond to samples 8 and 7 from Tschanz et al. (1974). These magmatic bodies consist of highly homogeneous bodies of hornblende-biotite granodioritic composition, commonly foliated and marked by a mafic mineral lineation, especially visible in the eastern SMB. Tschanz et al. (1974) interpreted the foliation and lineation as border zone features, generated during strong interaction between the magma and the wall rock, which also produced metasomatic alteration and granitization of hosting metamorphic rocks. As part of this magmatic belt, Tschanz et al. (1974) described a series of small leucocratic bodies of muscovite granite outcropping in the northeastern SMB along the Mendihuaca River and to the west of the Buritaca Pluton. These bodies were not dated; however, Tschanz et al. (1974) assigned them an age of ca. 50.5 Ma, as they appeared to be younger than the SMB and Buritaca Plutons, but older than a series of aplitic dikes which cut the SMB which were dated at 44.69 ± 1.5 Ma (K/Ar in muscovite (date recalculated after Steiger and Jäger (1977) and corresponds to sample 9 from Tschanz et al. (1974)).

Tschanz et al. (1974) reported two other magmatic rocks yielding Paleocene-Eocene ages in the SNSM: the Atanques laccolith and the Socorro stock. These plutons remain poorly studied, and their relation with Santa Marta magmatism during the Cenozoic is still unconstrained. The Atanques laccolith is a magmatic body which crops out in the southern part of the SNSM (Fig. 7.4). It is described as a very coarse-grained porphyritic rock that contains plagioclase phenocrysts in a fine- to medium-grained gray groundmass. The age of this body is inferred from an anomalous biotite K/Ar age of 57.3 ± 1.7 Ma from the Atanques batholith, which should have been partially reset during the intrusion of the laccolith (Tschanz et al. 1974). This age is doubtful because this laccolith does not seem to be related with other Paleogene bodies on the southeastern side of the SNSM, and a relationship with the magmatic belts of the northwestern tip of SNSM is not easy to sustain. The Socorro stock is located to the southeast of the Latal Pluton and seems to intrude the Sevilla lineament (Fig. 7.4). Tschanz et al. (1974) classified this body as a biotite granodiorite and reported two discordant K/Ar ages of 37.8 ± 1.1 Ma and 131 ± 4 Ma. Tschanz et al. (1974) selected the 131 Ma age as the cooling age of the Socorro stock, and the 37.4 Ma age is interpreted as an altered age, possible due to Ar loss during protoclastic deformation. More recently, Cardona et al. (2010a) reported an age of 54.3 ± 1 Ma (Zr, U/Pb LA-ICP-MS) for rocks of this body supporting emplacement in the Eocene, and suggested the existence of at least one other Cretaceous (~131 Ma) magmatic body in the area.

7.2.1.1 Field Geology and Petrology

Santa Marta Batholith (SMB)

The SMB constitutes a continuous unit ca. 40 km long and 15 km wide, which extends from the southern part of the Santa Marta area to the northern side of Tayrona National Nature Park. It is composed of a very homogenous mass of amphibole-biotite granodiorite to tonalite, with punctual variation due to mafic enclaves or cumulitic rocks (Fig. 7.5). The major constituent of the rock is plagioclase, followed by orthoclase, amphibole, biotite, apatite, zircon, ilmenite, and titanite. Texturally, this body varies from medium-fine to coarse-grained. This feature is used to characterize the different magmatic facies, as the mineralogical composition is almost invariant. Although four different facies were identified during

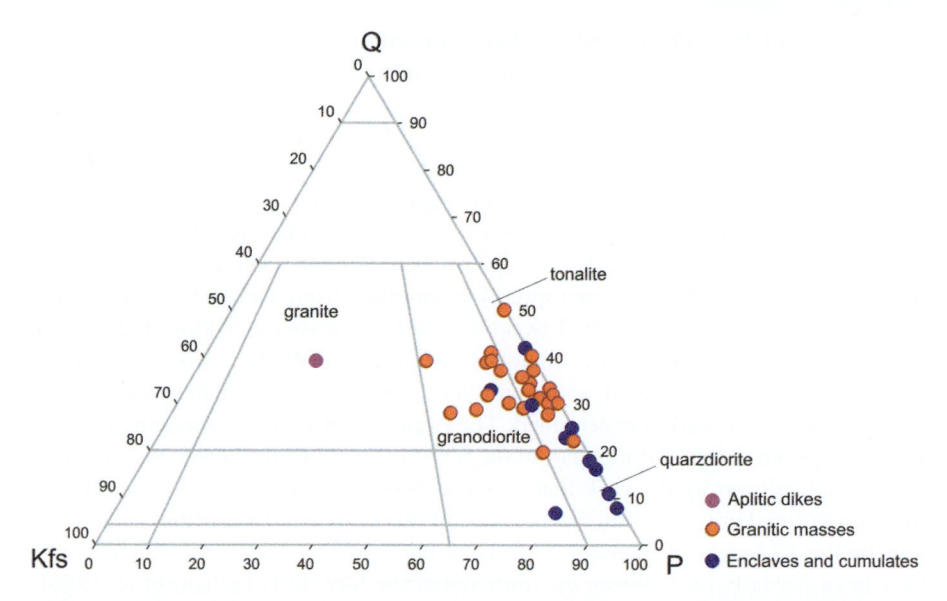

Fig. 7.5 QAP Streckeisen classification diagram from the Santa Marta Batholith (SMB) rocks

our fieldwork, only three are represented on the map: (1) the main magmatic facies, (2) a fine-grained facies, and (3) a poikilitic facies (Fig. 7.6). A fourth, cumulitic facies, was found only in restricted units or as big boulders in creeks and therefore, is not represented on the map.

As illustrated in the magmatic facies map (Fig. 7.6), the main facies is the dominate composition of the batholith, recorded in ~80% of the area. This facies is composed of amphibole-biotite granodiorite to tonalite (Fig. 7.7), consisting of plagioclase, quartz, orthoclase, hornblende, and biotite, with apatite, zircon, and opaque minerals as accessories. Enclaves of mafic minerals and poikilitic megacrysts of orthoclase are common textural features of the entire magmatic body. The general texture of this magmatic body is characterized by a gentle mafic mineral lineation (Fig. 7.8a). It is well developed, especially in the eastern border, where compositional banding developed, possibly due to tectonically controlled magmatic flow (Fig. 7.8b). A plastic-state cataclastic deformation, formed between 300 and 400 °C, is overimposed on the magmatic flow texture. This deformation has affected the most refractory minerals, such as plagioclase (Fig. 7.8c–e). We interpret this deformation to have been localized along the margin (lateral or basal) of the magmatic body, and to have been caused by shear friction between the magma and the host rock. This explanation concurs in part with the interpretation of Tschanz et al. (1969), who also identified this exturally distinct zone, and described it as a "border zone", which they mapped along the western border of the batholith. These authors interpreted these characteristics as the result of granitization of the metamorphic host rock and contamination of the intruded granitoid with the metamorphic wall rock.

The fine-grained facies is represented by a small and elongated (NE-SW) stripe, located at southwestern SMB, close to the Playa Salguero leucogranite. This facies

Fig. 7.6 Distribution of magmatic facies in the Santa Marta Batholith. (1) Main magmatic facies, (2) poikilitic facies, (3) fine-grained facies. PNN Tayrona: Tayrona National Nature Park. (Modified after Tschanz et al. (1974))

represents less than 5% of the total batholith in outcrop area (Fig. 7.6). It has the same mineralogical composition of the main magmatic facies and is recognized by its fine-grained texture and highly marked mineral lineation. Mafic enclaves are abundant, and most of them are aligned in the same orientation as the mineral lineation.

Rocks of the poikilitic facies outcrop at the northeastern SMB, constituting ca.15% of the exposed batholith area. The best exposures include road cuts along the Santa Marta-Riohacha highway, near the Piedras River (Fig. 7.6). The poikilitic facies has the same mineralogy as the main magmatic facies, but its distinctive characteristic is the poikilitic character of amphibole, which includes crystals of plagioclase, quartz, biotite, and opaque minerals, indicating that the amphibole had a more rapid growth rate than the other minerals (Fig. 7.9). The general texture of this magmatic facies is similar to the main magmatic facies texture, although mineral lineation is weaker and the grain size tends to be coarser. Mafic enclaves are less common than in the rest of the unit, while xenoliths are more common, especially fragments of amphibolites with quartz-feldspar segregations. Due to the scarce rock exposures, cross-cutting relationships between the poikilitic and main magmatic facies are only observed in the northern part of the batholith, at Los Naranjos beach, where the two facies appear to have interacted as a mixture of two partially molten

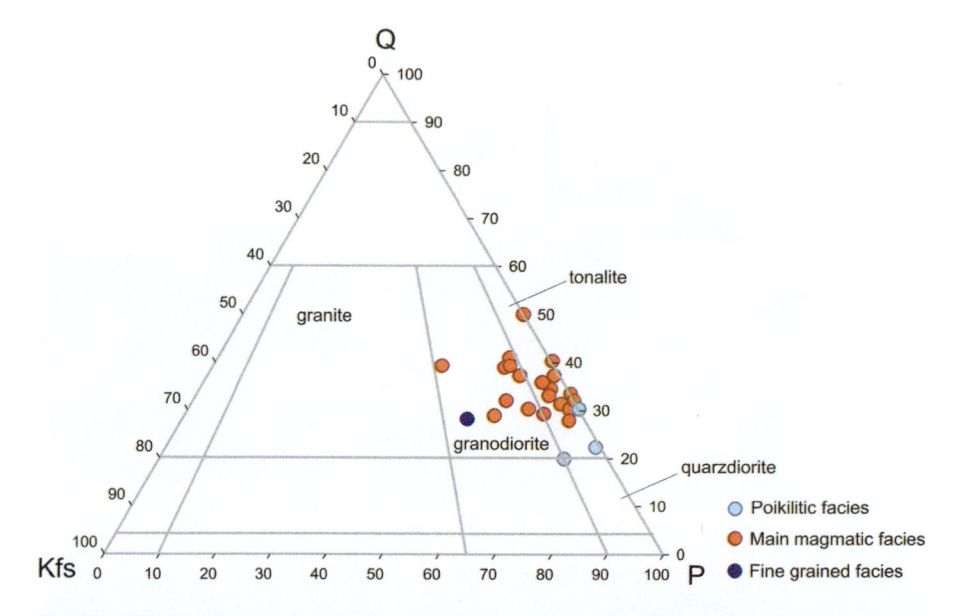

Fig. 7.7 QAP Streckeisen classification diagram for rocks of the different magmatic facies found in the Santa Marta Batholith

Fig. 7.8 Textural variations in the main magmatic facies of the Santa Marta Batholith. (**a**) Absence of mineral lineation and deformed mafic enclaves. (**b**) Highly deformed rocks with well-defined mineral lineation and mafic-rich bands. (**c–e**) photomicrographs from site (**b**) showing plastic-state deformation. (**c, d**) Crossed nicols, (**e**) Parallel nicols

Fig. 7.9 Photomicrographs from a poikilitic hornblende crystal, main characteristic of the poikilitic facies within the Santa Marta Batholith. Notice the plagioclase, quartz, biotite, and opaque inclusions. Crossed nicols (1), Parallel nicols (2)

Fig. 7.10 Evidence for magma mixing in the Santa Marta Batholith. Mixing occurred between the main magmatic facies (ppal) and the poikilitic facies (poi). Amphibolite xenoliths (Xe) and mafic magmatic enclaves (MME) are present

masses, producing a complex relationship between the poikilitic facies intruded into the main facies. In this context, the borders between these facies are transitional and diffuse, suggesting some assimilation in the contact zone between these two magmas, which may have been in a ductile state at the time of the emplacement (Fig. 7.10). Large xenoliths of amphibolites are also recorded in the Los Naranjos outcrop.

The cumulitic facies is not mapped in Fig. 7.6 because of its restricted occurrence. In most of the cases, it was only observed as large boulders with well developed cumulate characteristics, located along the Gaira River (Fig. 7.11). Larger outcrops of similar rocks were found in the Latal Pluton. These cumulitic rocks are composed of ca. 95% pegmatitic amphiboles and 5% of plagioclase> pyroxene >olivine. These have been interpreted as large accumulations of amphibole due to solid phase segregation during fractional crystallization of a water-rich magma.

Fig. 7.11 Pegmatitic cumulate composed by ~95% of amphibole. (**a**) Amphibole-plagioclase cumulate texture, (**b**) transitional contact between SMB rock and cumulate

This segregation could have taken place along the walls or bottom of the magmatic chamber. Bottom segregation and later remobilization to upper parts of the magmatic chamber (to be partially assimilated at the edges) would constitute a plausible hypothesis for the formation of the pegmatitic crystals. The occurrence of cumulate rocks in the Latal Pluton and SMB strongly suggests an intimate relationship between these two plutons.

A widespread and common characteristic throughout the SMB is the presence of mafic magmatic enclaves. In general, these enclaves have the same aspect and same mineralogical composition (amphibole, biotite, plagioclase, quartz, and opaque minerals). Although some enclaves are constituted by >80% of mafic minerals, intercrystalline relationships are the same as in the granitic mass. Notwithstanding, high textural variability is observed in the different enclaves, including fine-grained, coarse-grained, porphyritic enclaves, with cumulate textures, deformed, with diffuse and sharp edges, among others. In this context, it is difficult to suggest a simple and single genesis for these enclaves. Nevertheless, some field relationships allow us to infer that the enclaves could be related to disaggregation and mingling processes of mafic intrusions or remobilized cumulitic rocks, as described by Barbarin (2005) and Tobisch et al. (1997).

The latest magmatic activity registered in the SMB seems to be represented by a series of aplitic dikes which cut the entire SMB suite and its host rocks. The contact between this late-phase magmatism and the main granitoid varies from diffuse (where the aplitic magma drags some mafic enclaves) to sharp, indicating that these dykes likely intruded at different stages and not only at the end of SMB magmatic activity.

Buritaca Pluton (BP)

The Buritaca Pluton is an intrusive magmatic unit, located on the northeastern side of SMB (Fig. 7.4). It was described by Tschanz et al. (1974). Petrological similarities (i.e., mineralogy, mineral lineation, and abundance of mafic magmatic enclaves and xenoliths), between the SMB and BP suggest they may share common processes with respect to magma genesis and evolution. However, these two units have distinct regional orientations (Fig. 7.4). The SMB shows a NE-elongated shape (concordant with the regional structures), whereas the BP has an E-W elongated shape. Such a difference may be related to changes in the regional paleo-stress distribution at the time of emplacement. The E-W trend of the BP contrasts not only with the SMB but also with other regional structures, suggesting that E-W deformation may have been overimposed by the time of BP intrusion. This deformation is possibly been related to the Oca fault, displacement along which would have facilitated the intrusion of magmas. Furthermore, intense post-crystallization deformation is evident in BP rocks, mainly recorded as cataclastic deformation in mylonitic zones, possibly related to minor faults within the Oca Fault System.

Latal Pluton (LP)

The Latal Pluton was first described by Tschanz et al. (1974), as an intrusive complex formed by at least three different types of rocks with different ages. Although this intrusive is detached from the Santa Marta magmatic complex, these authors associated the LP with Paleogene magmatism based upon age and petrological similarities. The LP is mainly composed of hornblende diorites. Minor hornblendites intrude the diorites, forming segregations and dikes (Tschanz et al. 1969).

Because of the poor exposure and complex relationships between the different rocks, the LP unit is not well understood. With respect to the entire Paleocene magmatic complex, the LP has the greatest variation in mafic magmatic rocks. The unit is composed of quartz diorite, tonalite, and diorite bodies, with a strong mineral lineation, defined by mafic minerals (Figs. 7.12 and 7.13). As described by Tschanz et al. (1969), the suite has been subjected to moderate to intense deformation at over 300 °C, which is evidenced by the bending of plagioclase crystals (Figs. 7.14 and 7.15).

As found in the magmatic bodies mentioned above, mafic magmatic enclaves are common in the LP as well. These have the same grain size as the main mass—dominantly composed of hornblende, biotite, and plagioclase—with strong mineral lineation oriented with the rock foliation. It is common to find fine-grained dikes cutting the granitoid in different directions. Although a genetic relationship is difficult to prove, it may be inferred that the mafic enclaves were formed by disruption of mafic dikes as occurred in the SMB.

Within the LP magmatic bodies composed of pegmatitic hornblendites and pyroxenites, consisting mainly of euhedral mafics minerals (>90%) and some plagioclase (Figs. 7.12 and 7.13b), are observed in out crop. The field relationships

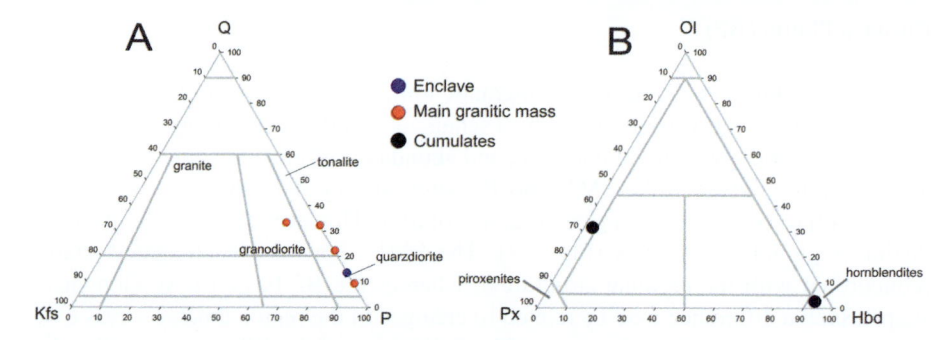

Fig. 7.12 Streckeisen classification diagram for the Latal Pluton rocks. (**a**) Main granitic mass, (**b**) mafic magmatic enclaves in the Latal Pluton

Fig. 7.13 (**a**) Latal Pluton main granitic mass cut by a fine-grained dike and mafic enclaves. (**b**) Cumulate hornblendite from the Latal Pluton

between these bodies and the main intrusive amss of the LP remain unclear. Two hornblende types (tremolite-actinolite and hornblende) were found in these rocks, both as primary crystals and as product of uralitization from pyroxene. One of these cumulate bodies is composed of clinopyroxene, orthopyroxene, olivine, and amphibole. These are interpreted as cumulate masses formed by the precipitation of crystals during early stages of fractional crystallization. The cumulates found in the LP may be related with the cumulate blocks found in the SMB, strengthening the suggestion of a genetic relationship between these two magmatic bodies.

Toribio Pluton (TP)

The ca. 20 km^2, NE-SW elongated TP is included within the SMB. The contact relationships between the two intrusives however, is not observed in the field. Tschanz et al. (1969) described the TP as having formed from the same hornblende diorite as the LP. However, pegmatitic hornblendites are absent in TP. Petrographically, the studied samples herein indicate that, as described by

Fig. 7.14 Photomicrographs of the Toribio Pluton. (1 and 2) Typical texture and mineral association from the main granitic mass of the Toribio Pluton. Crossed and parallel nicols, respectively. (3 and 4) Semi-plastic deformation rocks of the Toribio Pluton rocks. Crossed and parallel nicols, respectively

Tschanz et al. (1969), the TP is mainly composed of medium-grained diorites and tonalities (Fig. 7.16), with a weak mineral lineation. The mineralogical composition is the same as that of LP and SMB: hornblende, plagioclase, biotite, quartz, and some K-feldspar filling fractures. Although some evidence of brittle deformation was found in one sample, intense plastic deformation is more common, similar to that observed in rocks from the border zone of SMB, which also reveals a superimposed brittle deformation. Tschanz et al. (1969) considered the TP and the LP to belong to a small stock, based on their similarities and proximity. Nevertheless, some evidence suggests that the TP may be a block of granitized amphibole schist or a portion of magma highly contaminated by amphibole schist.

Leucocratic Granitoids

A series of leucocratic granitoids intruding the northwestern tip of the SNSM are described by Tschanz et al. (1974). These are located in the western part of BP and also in the northeastern SMB (along the Mendihuaca River). Recently, Duque-Trujillo et al. (2010) reported a previously unknown leucocratic granite outcrop near the town of Gaira and named it the Playa Salguero leucogranite (Fig. 7.17).

Fig. 7.15 Photomicrographs of the Latal Pluton. (1 and 2) General appearance from the Latal Pluton main magmatic facies. (3 and 4) Cumulate hornblendite. (5 and 6) Cumulate pyroxenite with olivine. Crossed nicols (1, 3, and 5), parallel nicols (2, 4, and 6)

Tschanz et al. (1969) did not report this particular granite but propose correlation of the leucogranites in general, with leucocratic dikes and intrusions found in the Gaira Schists. The Playa Salguero leucogranite comprises an ca. 10 km^2 magmatic body with a NE-SW elongation, concordant with the encompassing geological units. The best exposures of this unit are found along the road between Playa Salguero and Pozos Colorados (south of Santa Marta city), where it intrudes the Santa Marta Schists (Fig. 7.18).

In hand sample the rock is fine-grained, white to gray in color, and mainly composed of plagioclase (~30%), K-feldspar (~25%), quartz (~35%), muscovite (~5%), biotite (~5%), and garnet (up to 3%) (Figs. 7.19 and 7.20). The rock has a well-defined mineral lineation marked by biotite and muscovite and is commonly cut by a series of aplitic (sometimes pegmatitic) dikes in several directions

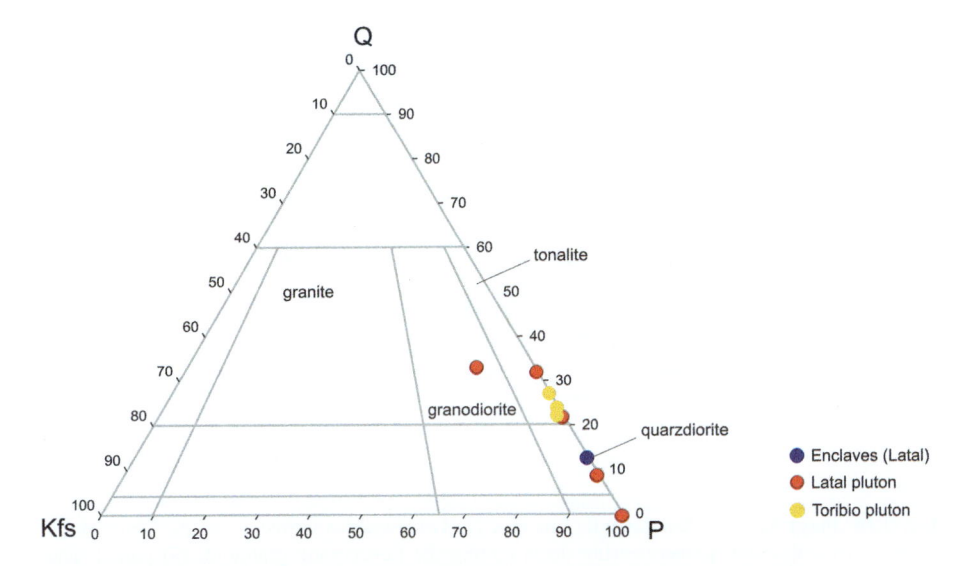

7 Paleogene Magmatism of the Maracaibo Block and Its Tectonic Significance 567

Fig. 7.16 QAP Streckeisen classification diagram from Latal and Toribio Plutons

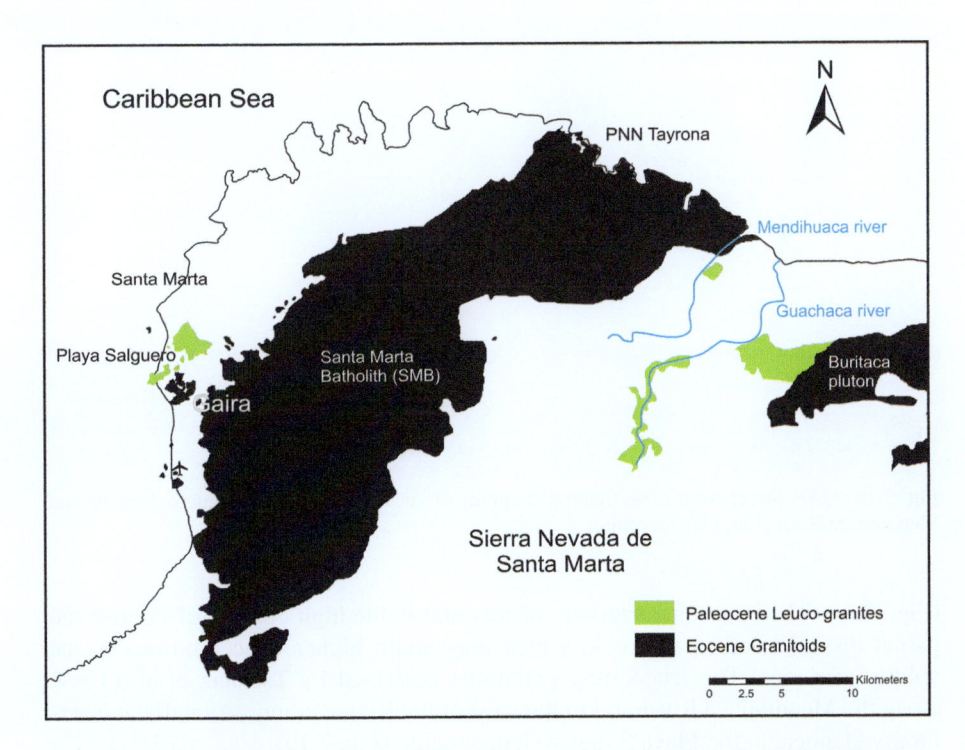

Fig. 7.17 Distribution of the Paleocene-Eocene leucogranitic rocks of the Santa Marta Province. (Modified after Tschanz et al. (1974))

Fig. 7.18 Playa Salguero leucogranite outcrop. (1) Leucogranite intrusion on the Santa Marta Schists, (2) leucocratic garnet-bearing dikes cutting the Leucocratic granitoid, (3) garnet- and muscovite-rich leucocratic dike

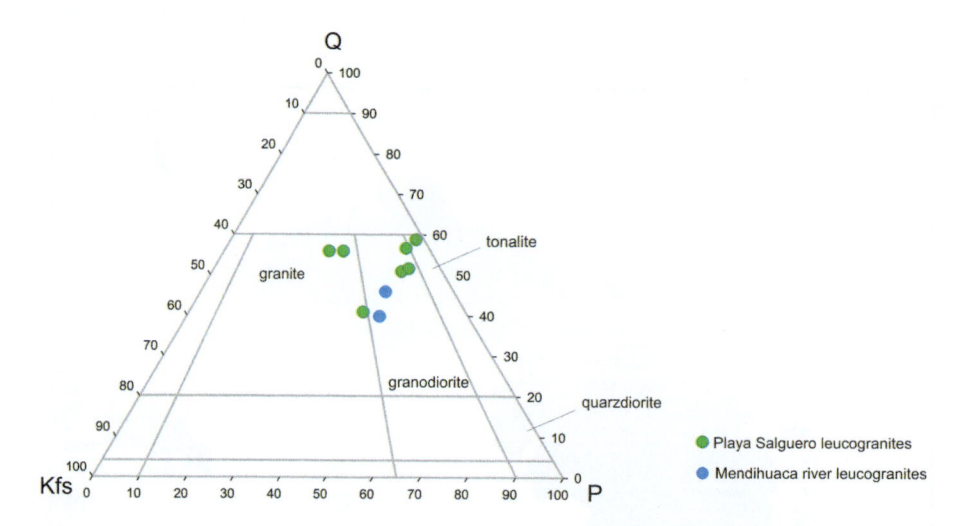

Fig. 7.19 QAP Streckeisen classification diagram of rocks from Playa Salguero (green) and Mendihuaca River (blue) leucogranites

(Fig. 7.18). The main characteristic of this unit is the high content of intense-red garnet disseminated in the rock, which may attain higher concentrations in the aplitic intrusions. The leucocratic granitoids described by Tschanz et al. (1969) along the Mendihuaca River and to the west of the BP are compositionally and texturally identical to the Playa Salguero leucogranite (Fig. 7.19).

Fig. 7.20 Photomicrographs of the Playa Salguero leucogranite. (1 and 2) Typical mineral association on the Playa Salguero leucogranite, (3) deformed crystals of garnet, plagioclase, and biotite, (4) K-feldspar-bearing leucogranite.

7.2.1.2 Geochronological Data

Abundant geochronological data has been reported for the Paleogene intrusive units of SNSM. Tschanz et al. (1969) obtained K/Ar dates for most of the units. Mejía-Herrera et al. (2008) reported the first U/Pb ages, obtained from the main magmatic facies of the SMB. Later, Cardona et al. (2011a) presented a detailed U/Pb geochronologic study of the plutonic rocks from the northwestern part of SNSM. Duque-Trujillo (2009) also presented a detailed $^{40}Ar/^{39}Ar$ and U/Pb geochronologic study of this magmatism, with special emphasis on the petrogenetic evolution of the SMB. These authors also report three $^{40}Ar/^{39}Ar$ ages from some previously dated samples. Salazar et al. (2016) presented the first U-Pb (SHRIMP) data, and undertook a magnetic fabric and shear deformation study of the Santa Marta Pluton. Furthermore, low-temperature (<250 °C) geochronologic data have been also obtained for these plutons. Villagómez et al. (2011) reported thermochronological (apatite fission tracks) data from transects across the SNSM. One of the transects cut the NW tip of the massif, and included some samples of the SMB. Also, Cardona et al. (2011b) obtained zircon and apatite U-Th/He data on samples from the NW margin of the Santa Marta Massif, including samples from the SMB.

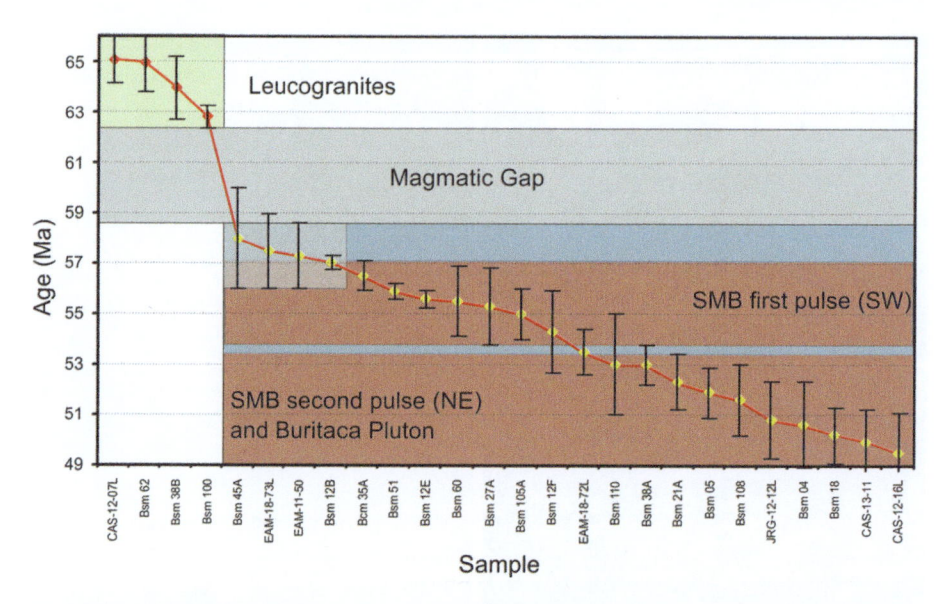

Fig. 7.21 U/Pb age distribution of granitoid samples from the Santa Marta Province highlighting two main magmatic events during the Paleocene and Eocene, separated by a magmatic gap

This information is essential, not only to reconstruct the timing and evolution of this magmatism within a regional context, but also to allow us to understand the geological evolution of SMB, which represents the main magmatic unit representing Paleogene SNSM magmatism. Moreover, medium- to low-temperature geochronological data also allow us to quantify and constrain the timing of uplift and exhumation of the northwestern tip of the SNSM, a keystone for the understanding of the interaction between the South American and Caribbean Plates and the complex tectonic evolution of the northwestern corner of South America.

Paleocene Magmatism

Paleocene crystallization ages are scarce in the Paleogene magmatic record of the SNSM. Cardona et al. (2011a) reported two zircon U-Pb ages of ~64 and ~63 Ma (Fig. 7.21) for the Playa Salguero leucogranite, on samples taken in Pozos Colorados and Playa Salguero, respectively. Cardona et al. (2011a) also reported a Paleocene age (65.1 ± 0.9 Ma) obtained from a leucogranite found in the western part of the BP. This age confirms that all leucogranites were emplaced in a short time interval and may be genetically related to a Paleocene magmatic arc intruding late Cretaceous metamorphic rocks of the Santa Marta Province. The Paleocene ages obtained for the Playa Salguero leucogranite are concordant with the 70.3 ± 1.0 Ma age obtained by Cardona et al. (2009) from a biotite tonalite recovered at 2.1 km depth in the Rancheria 2 drill hole, located in the Lower Guajira Basin.

Also, Tschanz et al. (1969) reported a Paleocene age for the Atanques laccolith, on the southeastern part of the SNSM—suggesting that Paleocene magmatism affected the whole massif. Nevertheless, this age is doubtful, because it corresponds to a biotite K/Ar age of 57.3 ± 1.7 Ma, which also yielded an age of 162 ± 11 Ma in hornblende from the same sample. This sample was collected 1 km from the lower contact between the Atanques laccolith and the SMB, which may explain the difference between these ages due to a partial Ar lost from the biotite K/Ar isotopic system during the intrusion of the laccolith.

Eocene Magmatism

Excluding the Paleocene leucogranites, the entire Paleogene magmatic suite of the SNSM returns Eocene crystallization ages. After a 5 m.y. magmatic gap, magmatism occurs continuously on the northwestern tip of the SNSM from 58 to 49 Ma (Cardona et al. 2011a) (Fig. 7.21). The evolution of this Eocene magmatism can be divided in two major magmatic phases which are petrographically similar but geographically separated. The first magmatic phase, with a homogeneous age of ca. 56 Ma, comprises the southwestern part of SMB (Toribio and Latal Plutons; Fig. 7.21). The second magmatic phase, which yields a homogenous age of ca. 51 Ma, comprises the northeastern part of SMB and the BP (Fig. 7.21). Thus, a magmatic migration towards northeast is recorded during the Early Eocene, which is supported by U-Pb zircón ages (Fig. 7.22).

Different magmatic processes occurring before and after the main magmatic pulses are recognized (Fig. 7.21). With ages between ca. 58 and 57 Ma, shortly before the first magmatic pulse, mafic rocks from Toribio Pluton and mafic magmatic enclaves from the main SMB are identifiable (Fig. 7.21), which imply that the enclaves and the mafic rocks from Toribio could have crystallized in an earlier event related to cumulate rock formation. Furthermore, the cumulates represent the most primitive rocks of the magmatic system and may have been later remobilized from depth by magma convection currents or new magmatic pulses.

After the first SMB magmatic pulse and partially before the second pulse (ca. 55–52 Ma), a series of aplitic dikes were emplaced (Fig. 7.21). These dikes commonly cut the SMB sharply, indicating that the SMB was more or less crystallized by the time of dike intrusion. Field relationships of different magmatic facies—mingling, partial overlapping of magmatic pulses, and a general northeastward migration of the magmatism—indicate that the SMB is a composite batholith formed as a result of long-lived (ca. 10 m.y.) and complex magmatic processes.

Medium- and Low-Temperature Geochronology

A classical method to address the time-temperature evolution of a rock is through thermochronology. This method is based upon the application of different geochronological systems, with different closure temperatures, on the same rock unit, in

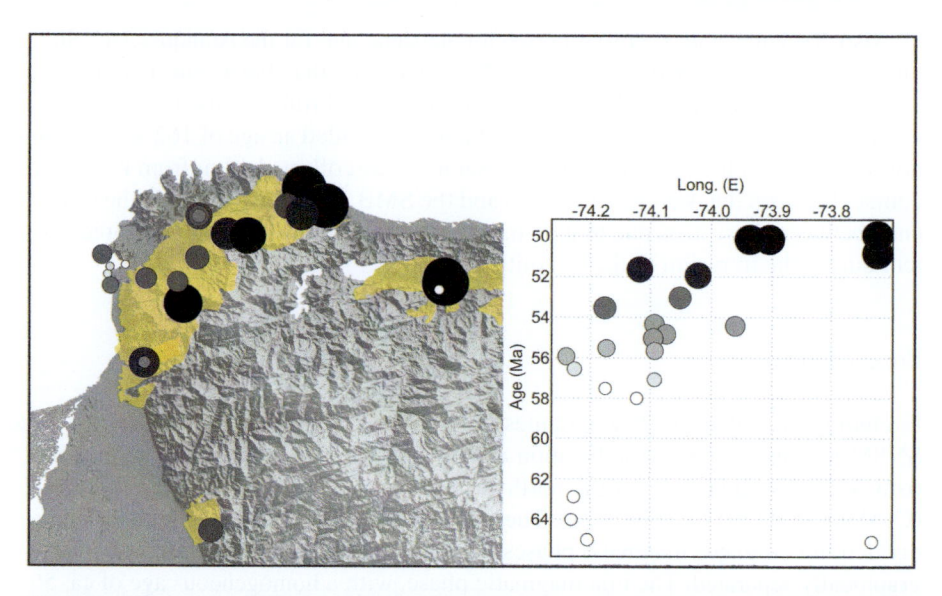

Fig. 7.22 Geographic distribution of U/Pb ages of magmatic rocks from the Santa Marta Province. Age of samples is identified by spot size and color. Small white spots are the youngest samples, and larger black sports are the oldest samples. The inserted graphic shows the distribution of age with longitude, showing that the Paleocene event is widespread and that the Eocene event displays a strong migration from southwest to northeast with time

order to estimate the age at which the rock passed through a predetermined temperature (s) (Farley 2002; Reiners and Shuster 2009). Extensive geochronologic studies have been conducted on the SMB and the SNSM, including U-Pb in zircon, $^{40}Ar/^{39}Ar$ in hornblende, biotite and K-feldspar, fission tracks in apatite, and U-Th/He in zircon and apatite. These methods cover the temperature range from ca. 900 to 70 °C (Reiners et al. 2005). This temperature range, considering a mean medium to upper crust geothermal gradient of 30 °C/km, would correspond to a depth range of 2 to >16 km (Farley 2002).

As indicated, the SMB has been extensively dated using the U-Pb method (Cardona et al. 2011a; Duque-Trujillo et al. 2010), obtaining a Paleocene-Eocene (59–49 Ma) age for the magmatic activity. The U-Pb (zircon) system has high closure temperatures, around >900 °C (Cherniak and Watson 2001; Reiners et al. 2005). These ages are commonly considered to represent crystallization or emplacement of a magmatic body. Cardona et al. (2011b), using the Al-in-hornblende calibration of Schmidt (1992) in samples from the SMB, calculated an emplacement pressure between 4.9 ± 0.6 and 6.4 ± 0.6 kbar. These pressure values correspond to emplacement depths in the range of 15–19 km (Fig. 7.23) and are consistent with the nature of amphibolite facies rocks and peak pressures of ca. 6.6 ± 0.8 kbar (17.7–19.2 km) of the host rock (Bustamante et al. 2009; Cardona et al. 2010b). The calculated depths of emplacement imply that the northwestern tip of the SNSM massif has been subjected to ca. 16 km of unroofing during the last ca. 55 Ma (Cardona et al. 2011b).

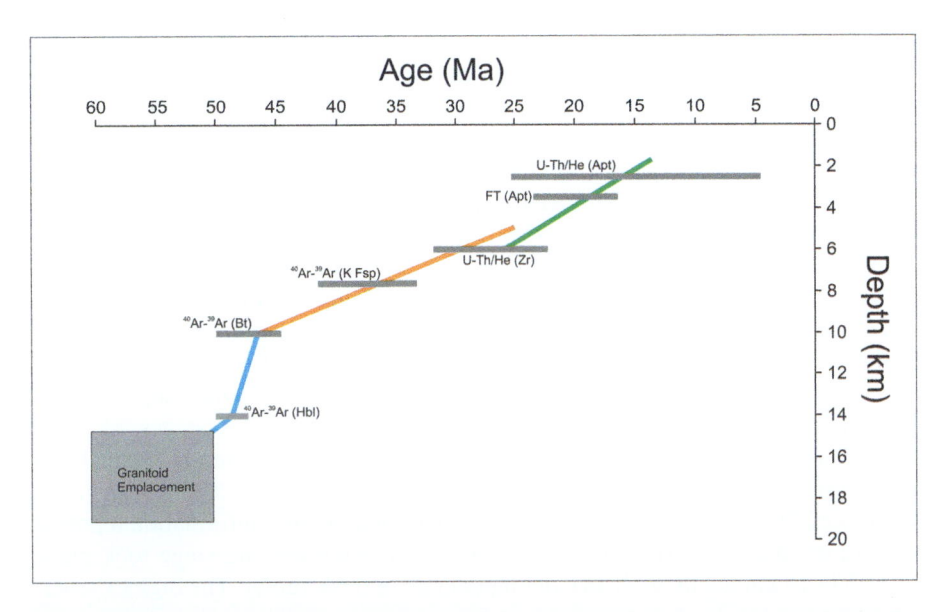

7 Paleogene Magmatism of the Maracaibo Block and Its Tectonic Significance

Fig. 7.23 Age vs. depth diagram with data obtained by different geochronological methods for samples from the Santa Marta Batholith. Lower temperature ages were obtained for diverse geological units from the Santa Marta Province. See text for references

An important temperature range between 550 and 250 °C is covered by $^{40}Ar/^{39}Ar$ systematics in hornblende (550 ± 100 °C), biotite (300 ± 50 °C), and K-feldspar (250 ± 100 °C) (see review in Reiners et al. 2005). As a result of the different closure temperatures, in a crystallized magmatic body, older ages are expected for hornblende than biotite and the youngest ages for K-feldspar. Duque-Trujillo et al. (2010) reported $^{40}Ar/^{39}Ar$ ages for samples previously dated by the U-Pb method. All ages obtained in the same sample fulfill the principles of age order depending on the system closure temperature, suggesting the absence of later tectono-thermal events that would have modified the isotopic systematics. Consequently, $^{40}Ar/^{39}Ar$ ages are interpreted as cooling ages and can be used to calculate the cooling rate of the SMB. Reported $^{40}Ar/^{39}Ar$ ages in hornblende yielded ages between 50 and 47.7 Ma, ages in biotite are in the range of 49.5–44 Ma, and ages in K-feldspar are in the range of 41.8–33.7 Ma (Fig. 7.23).

The low-temperature thermochronometers U-Th/He in zircon (Z-He) and apatite (A-He) and fission tracks in apatite (A-FT) have also been applied to rocks from the Santa Marta Province, including the SMB, by Cardona et al. (2011b), in order to constrain exhumation rates on the upper continental crust (ca. 6 km depth). The suggested closure temperatures for the U-Th/He system in zircon and apatite are in the range of 160–200 °C and 70 °C, respectively, and the retention temperature for fission tracks in apatite is in the range of 90–120 °C (Reiners et al. 2005; Farley 2002). Cardona et al. (2011b) reported Z-He ages in the range from 18.7 to 26.2 Ma for nine samples, with poor correlation relative to elevation. The same authors report

A-He from 24.6 to 5.5 Ma, which show a well-defined pattern of increasing age relative to elevation (Fig. 7.23). Apatite fission-track data are reported by Villagómez et al. (2011), who obtained ages ranging from 16.3 to 24 Ma, on samples from the SMB and host rocks (Fig. 7.23).

A diagram of depth vs. age, containing all available thermochronological data for the SMB and host rocks (Fig. 7.23) permits analysis of the cooling and exhumation history of the SMB and the northwestern tip of the SNSM. Due to the absence of public domain heat flow data for the SNSM, a thermal gradient throughout the Cenozoic has been assumed. In this context, we concurr with Villagómez et al. (2011), Mora et al. (2008), and Spikings et al. (2000), who assumed a 30 °C/km geothermal gradient for their thermochronological studies along the SNSM massif, and the foreland basins in the Colombian and Ecuadorian cordilleras.

The resulting thermal history can be divided into three cooling stages. The first stage spans emplacement (50–60 Ma) to 49.5–44 Ma, when the pluton reached 300 °C. This stage is characterized by high cooling rates of around 70 °C/my (Fig. 7.23). Assuming a 30 °C/km thermal gradient and thermobarometric information reported by Cardona et al. (2011b), the emplacement and initial cooling stage took place from 16 to 10 km depth, with an exhumation rate of 0.66 mm/yr. The high exhumation rate was related to highly active tectonism, as is also indicated by high temperature deformation in several exposures of the SMB, suggested by the strong crystal deformation which dominates the SMB (Salazar et al. 2016). Tschanz et al. (1969) described a superimposed deformation in greenschist to amphibolite facies in some exposures of the SMB. Such deformation may be associated with this early stage of exhumation.

The second cooling stage, defined for the range between 300 and 180 °C (Fig. 7.23), is characterized by slower (although still high) cooling rates of around 5.7 °C/my, occurring between 45 and ca. 25 Ma. Assuming a 30 °C/km thermal gradient, this cooling stage would correspond to SMB exhumation from 10 to 6 km depth, with an exhumation rate of 0.19 mm/yr.

The third exhumation stage at the northwestern tip of the SNSM (Santa Marta Province) corresponds to the 180–70 °C range (Fig. 7.23), defined by A-FT (Villagómez et al. 2011) and A-He analysis (Cardona et al. 2011b) along elevation profiles. Based on the results reported by these authors, two high exhumation pulses are proposed around 25 and 16 Ma. Calculated rates are in the range of 0.5–0.09 mm/yr., with an extremely fast rate of 0.8 mm/y between 30 and 25 Ma, calculated on the basis of A-FT by Villagómez et al. (2011).

7.2.1.3 Geochemical Characteristics of the Santa Marta Province Magmatism

This section is based upon whole-rock major and trace element analysis of different magmatic products belonging to the Paleocene-Eocene magmatic suite of the SNSM massif, as presented by Duque (2009) and Cardona et al. (2010b).

Major Elements

The Paleocene magmatic facies of the SNSM span a wide compositional range, from ultramafic rocks to granites. Volumetrically, however, the dominant compositions are tonalite and diorite (Fig. 7.24) with SiO_2 contents ranging from 40% to 80% (Fig. 7.24). On the basis of the SiO_2 content, this magmatic suite is divided into three compositional groups, which correspond to rocks with similar characteristics.

The most primitive rocks in the series constitute the first compositional group, that includes rocks classified as cumulates and cumulitic enclaves, including part of the main intrusive mass of the Latal Pluton. Rocks of this group are classified as gabbros and ultramafic rocks and display a restricted variation in SiO_2 content at relatively low values between 45 and 55 wt% (Fig. 7.24). These low Si rocks have variable K contents and plot from the low K to the shoshonitic series in the SiO_2 vs. K_2O diagram (Fig. 7.24). This variation is mainly defined by the composition of enclaves and cumulates and could be explained by the cumulate character of these rocks.

The next compositional group consists of more evolved rocks, with SiO_2 content ranging from 55 to 70 wt%, classified as diorites and tonalities with minor granodiorites and gabbrodiorites. They constitute a coherent group within the medium-K rock series (Fig. 7.24). This group is dominated by rocks belonging to the main magmatic masses of the SMB, Buritaca and Toribio Plutons, with only one sample from the Latal Pluton belonging to this group. This sample plots close to a mafic enclave found within the SMB.

The most evolved rocks have SiO_2 contents which range from 70 to 80 wt% (Fig. 7.24). This group is includes aplite dikes and the leucogranites of the Playa Salguero facies, Mendihuaca River, and west of the Buritaca Pluton. These rocks have a color index <5%. However, geochemically, most of the analyzed samples classify as medium- to low-K granodiorites (Fig. 7.24), indicating an anomalous composition, as also indicated by the high garnet content in most of these rocks. It is important to point out that the Leucocratic granitoids are notably older than the SMB suite, and correspond to a different magmatic event.

Regarding the alumina saturation index, all the intermediate samples of the second group fall in the metaluminous field (Fig. 7.25), whereas the more evolved samples of the third group (leucogranitic) plot inside the felsic peraluminous and low peraluminous fields (Fig. 7.25), which precludes the process of alumina enrichment that is usual in the "S-type" granites.

In order to determine other compositional characteristics of these rocks, the analyzed samples were plotted in the An-Ab-Or diagram for silicic plutonic rocks (Fig. 7.26), where the samples of the SMB and Buritaca and Latal Plutons fall in the tonalite and granodiorite fields. The aplite dikes fall in the granite field, following the evolution trend of granitic melts. On the other hand, leucogranitic samples fall in the trondhjemite and in part, the tonalite field, following the proposed evolution of trondhjemitic melts (Fig. 7.26). Other reported geochemical characteristics classify these leucogranites as high-Al trondhjemites, sharing some compositional fea-

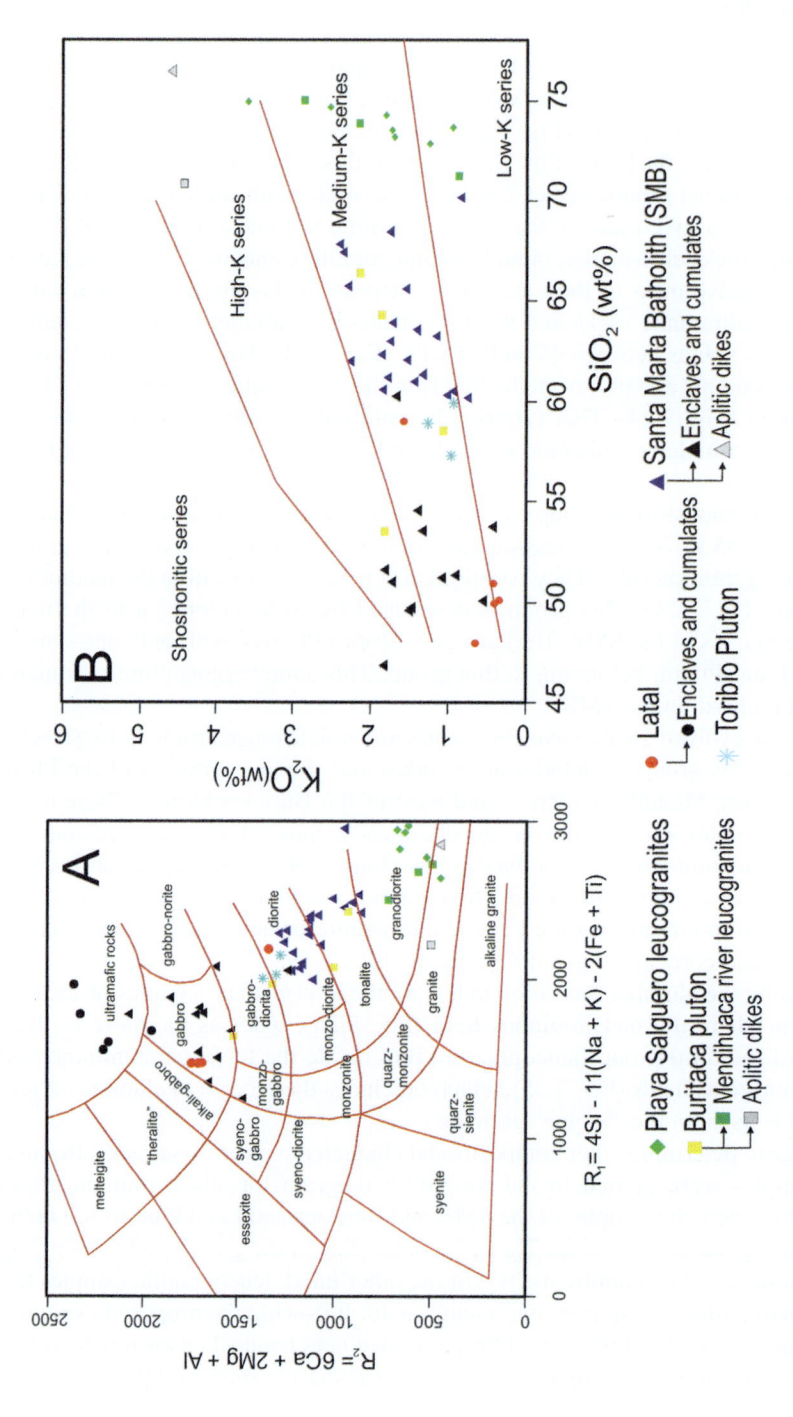

Fig. 7.24 Geochemical classification diagrams: (**a**) R1–R2 major elements diagram (De la Roche et al. 1980), (**b**) magmatic series discrimination diagram based on the SiO_2 vs. K_2O content (Peccerillo and Taylor 1976)

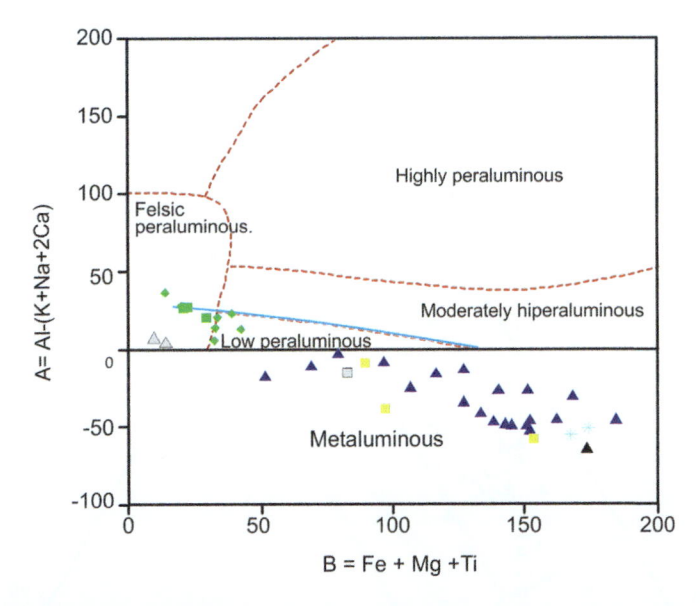

Fig. 7.25 Rock classification diagram according to the degree of alumina saturation from Villaseca et al. (1998), after Le Fort (1981)

tures with high-Si adakites and experimental melts resulting from melting of amphibolite and basalt (Martin et al. 2005; Moyen et al. 2003).

Trace Elements

In general, all analyzed samples from the SMB and associated plutons exhibit an "arc signature", pointing to a role involving slab-derived fluid magma genesis (Winter 2001; Wilson 1989). The main expression of this characteristic is the enrichment in large-ion lithophile elements (LILE), Pb, and Sr with respect to the high-field-strength elements (HFSE) and rare earth elements (REE). On the basis of the observed geochemical characteristics, samples were separated into four different groups.

The first group comprises the rocks with the lowest SiO_2 contents (45–55 wt%) and includes cumulates, enclaves, and part of the main intrusive mass of the Latal Pluton. In a normalized multielemental diagram, this group is characterized by almost flat patterns, irregular behavior of the most incompatible elements, small to absent Ti anomaly, and lower contents in elements from La to Sm (Fig. 7.27a). This suggests that the rocks from this group were formed by accumulation of minerals such as biotite, amphibole, plagioclase, and less olivine and pyroxene. The patterns in chondrite-normalized REE diagrams of samples from this group are relatively flat with a slight (enclaves and main mass) to absent (cumulate) enrichment in the light REE relative to the heavy REE (Fig. 7.27b). These samples have weak negative Eu anomalies reflected in Eu/Eu* < 1 values (Fig. 7.28).

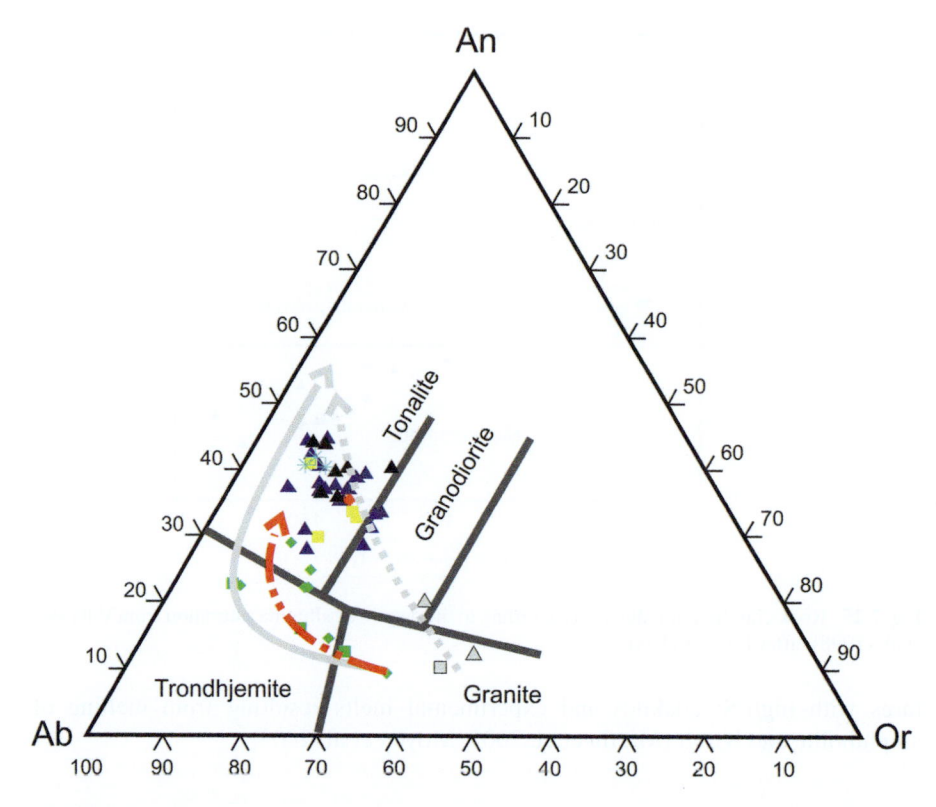

Fig. 7.26 Classification diagram based on normative values of albite (Ab), orthoclase (Or), and anorthite (An) from O'Connor (1965) modified by Barker (1979). Solid gray line shows the evolution path of trondhjemitic liquids, and dotted gray line the evolution path of granitic liquids. Same legend as used in Fig. 7.24

The second group consists of rocks with SiO_2 content between 49 and 64 wt% from the main intrusive masses of all magmatic units (SMB, Buritaca, Toribio, and Latal). All samples from this group have very similar patterns in a normalized multielement diagram (Fig. 7.27a), despite the wide range of SiO_2 values. They are characterized by an enrichment in LILE over HFSE, positive Pb and Sr anomalies, and negative Th, Nb-Ta, and Ti anomalies (Fig. 7.27a). Samples of this group have more fractionated REE patterns between light REE and heavy REE when compared to first group, as well as lower abundances of the middle and heavy REE. Negative Eu anomalies are absent in most samples with only a few samples showing very weak positive or negative anomalies (Fig. 7.28).

The third group is exclusively composed of SMB rocks with SiO_2 contents ranging from 60 to 76 wt%. Geochemical features of this group of samples resemble those of the second group. However, this group has more prominent negative Nb-Ta and positive Pb anomalies (Fig. 7.27b). Due to the higher degree of differentiation, these samples have strongly fractionated REE patterns, showing depletion in the

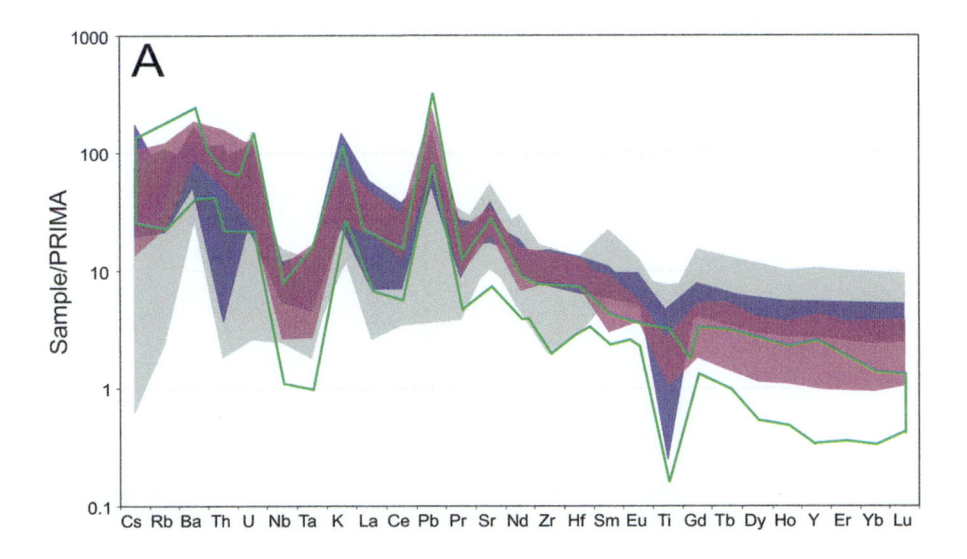

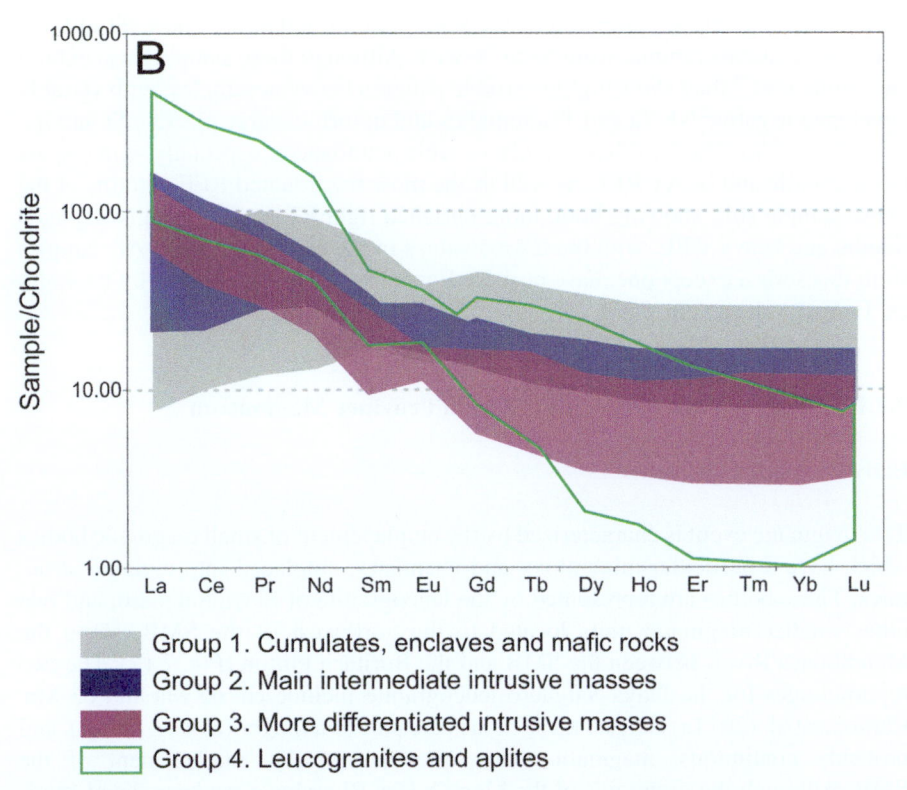

Fig. 7.27 Normalized multielement diagrams for the Paleocene-Eocene intrusive rocks of the Santa Marta batholith. (**a**) Trace elements normalized to primitive mantle (PRIMA) values of Sun and McDonough (1989). (**b**) Rare earth elements normalized to chondrite values of Anders and Grevesse (1989)

Fig. 7.28 Variation of Eu/Eu* (Eu* = EuN/(SmN + GdN)$^{1/2}$) as function of SiO$_2$. Values of Eu/Eu* higher than unity indicate positive Eu anomalies, and those lower reflect negative Eu anomalies

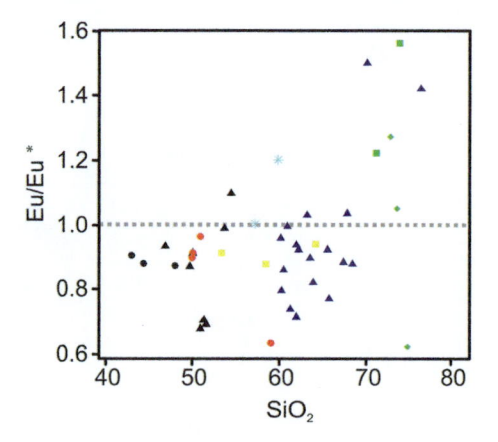

MREE and HREE. Less differentiated samples from this group have weak negative to absent Eu anomalies, whereas the more differentiated samples have positive ones (Fig. 7.28).

The fourth group is composed of the leucogranites and aplite dikes, all of them with SiO$_2$ contents ranging from 68 to 76 wt%. Although these samples also exhibit "arc signatures," they show highly variable patterns between samples, with variably developed negative Nb-Ta and Ti anomalies and notorious large positive Sr anomalies (Fig. 7.27a). The REE have highly variable abundances, especially with respect to the middle and heavy REE, as well as the most fractionated REE patterns of the entire sample suit, showing both, more enriched light REE, and strongly depleted middle and heavy REE, with the lowest values in the elements Er-Yb. All samples from this series, except one, have positive Eu anomalies, characterized by the highest Eu/Eu* values (Fig. 7.28).

7.2.1.4 Petrogenesis of the Santa Marta Province Magmatism

Early Paleocene Magmatism

This magmatic event is characterized by the emplacement of small magmatic bodies which have a particular mineralogy and chemistry, emplaced into regional structures. These bodies are represented by the leucogranite of Playa Salguero, and two other smaller magmatic units located to the northwest of the SMB, along the Mendihuaca River, between the SMB and the Buritaca Pluton (Fig. 7.17). The two reported ages for the Playa Salguero leucogranite include ca. 65 Ma and 63 Ma. Cardona et al. (2011a) suggest these bodies are part of a longer (at least 8 m.y.), and probably continuous, magmatic event, previous to the emplacement of the SMB. Although the granitoids of the Mendihuaca River have not been dated, similarities described by Tschanz et al. (1969) suggest that these bodies form part of the same event. Cardona et al. (2011a) also reported a 64.04 ± 0.36 Ma age (U/Pb in zircon) for a tonalite collected on the western part of the Buritaca Pluton, which strengthens the existence of an early Paleocene magmatic event in this part of the Santa Marta Province.

Mineralogical and geochemical characteristics of the leucogranites permit their classification as trondhjemites, with geochemical similarities to high-silica adakites (Martin and Moyen 2002). The geochemical characteristics strongly suggest that melting of a mafic protolith under the stability of amphibole, garnet, and rutile played a significant role in the generation of the leucogranite. These characteristics agree with partial melting of a garnet-bearing amphibolite between 650 and 950 °C. Similar characteristics have been reported for the origin of trondhjemitic melts from hydrous amphibolite protoliths (Winther 1996; Getsinger et al. 2009; Condie 2005; Prouteau et al. 2001; Garcia-Casco et al. 2008). In the Santa Marta Province, the leucocratic granitoids are closely related to rocks metamorphosed under conditions of the greenschist and amphibolite facies, which constitute a stacked subduction complex accreted to the South American Plate during Great Caribbean Arc collision (Cardona et al. 2010b). Similar relations have been described for trondhjemite of the Catalina Schist in California (Sorensen and Barton 1987) and Sierra del Convento, in Cuba (Lázaro et al. 2009). We propose that partial melting of MORB-like rocks, interbedded with metasediments, of the subduction complex during peak metamorphic conditions (amphibolite facies), is a viable process for the formation of the trondhjemites.

Trondhjemitic melt formation in tectonic settings characterized by collision and subsequent subduction initiation requires a high geothermal gradient in order to reach the necessary P-T conditions for partial melting (Martin et al. 2005). A high geothermal gradient is plausible for the Santa Marta region during early Paleocene interaction between the Caribbean and South American due to (1) the absence of subduction processes predating the collision of the Great Caribbean Arc against South American (Pindell and Kennan 2009) that may have depressed the mantle isotherms; (2) overthrusting of the South American Plate over the Caribbean Plate after the collision (Van der Hilst and Mann 1994; Miller et al. 2009; Pindell and Kennan 2009); (3) the hot and buoyant nature of the thick and young Caribbean Plate, which would have led to a very low underthrusting angle and to the accretion of oceanic and sedimentary materials over the continent; and (4) the slow plate convergence rate due to the highly oblique nature of the underthrusting process. Geochemical features and tectonic conditions indicate that the Santa Marta Province leucogranites may be considered as markers for the early Paleocene collision between the Caribbean and South American plates in the northern part of Colombia, previous to the initiation of a proper underthrusting between these plates.

Later Paleocene-Early Eocene magmatism

After an ca. 5 m.y. magmatic gap, magmatism in the Santa Marta Province resumed for ca. 10 m.y., between ca. 58 and 49 Ma, forming the main magmatic masses of the Santa Marta Batholith and the Buritaca, Toribio, and Latal Plutons (Figs. 7.4,

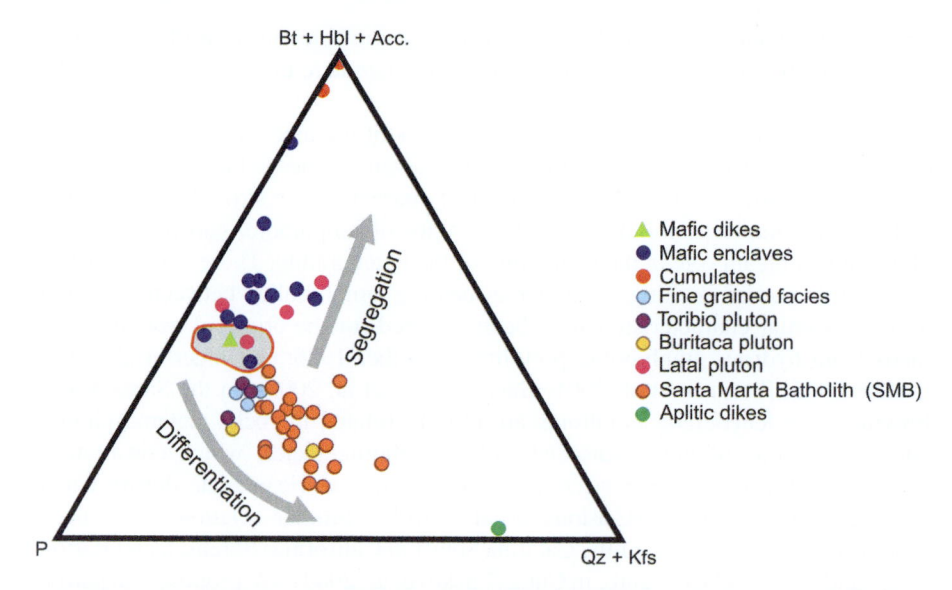

Fig. 7.29 Evolution model proposed for the SMB samples and associated rocks showing the trends of crystal segregation and crystal fractionation (differentiation). The probable composition of the parental magma is circled with a red line. Model constructed after Barbarin (2005). Bt, biotite; Hbl, hornblende; Acc, accessory minerals; P, plagioclase; Qz, quartz; Kfs, alkali feldspar

7.6, and 7.17). Although continuous, this magmatism may be separated, at least geographically, into two separate events. The first magmatic event, around 56 Ma, formed the Latal Pluton and the southwestern part of the SMB. During the second event, between 52 and 50 Ma, magmatism migrated toward the northeast, as indicated by the ages of the northwestern part of the SMB and the Buritaca Pluton (Fig. 7.21).

Although the Eocene intrusive rocks of the Santa Marta Province constitute a classical calc-alkaline magmatic suite, they are lithologically highly variable. At least three different rock groups may be identified: (1) low SiO_2 cumulates and mafic enclaves, (2) typical tonalities and granodiorites that form the granitic masses of the larger magmatic bodies, and (3) high SiO_2 rocks (>70 SiO_2 wt%) and aplite dikes. These lithological groups may also be identified using a basic petrogenetic model, constructed on the basis of modal rock mineralogy (Fig. 7.29), and based upon the premises proposed by Barbarin (2005) for the Sierra Nevada, California.

The cumulate rocks, mainly found in the Latal Pluton, are composed of ca. 90% of amphibole with some plagioclase and other mafics. On the other hand, the mafic enclaves are of highly variable composition. This can be related to the fact that most of them have a cumulate character and demonstrate different degrees of hybridization and mingling with the main intrusive mass. However, crystal segregation from a parental magma and differential accumulation in a magma chamber is the most probable explanation for the composition of the mafic and ultramafic enclaves and cumulates (Fig. 7.29). Cumulate rocks tend to be the most mafic and earliest crystallized

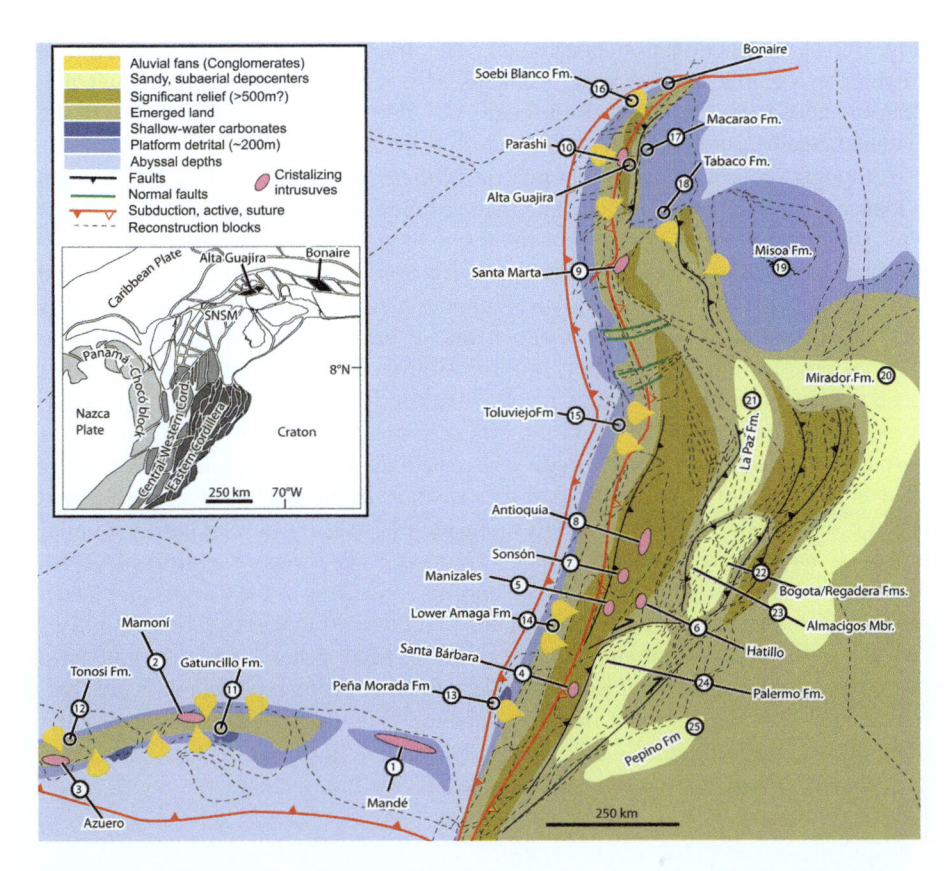

Fig. 7.30 Paleogeographic reconstruction of the Northern Andes and Southern Caribbean at ca. 50 Ma, taken from Cardona et al. (2014). This reconstruction is based on Montes et al. (2012) and Pindell and Kennan (2009), modified for Maracaibo from Escalona and Mann (2011); Central America from Meschede and Frisch (1998); Leeward Antilles from Muessig (1984), Priem et al. (1986), Stearns et al. (1982). References for plutonic bodies, sedimentary environments (Cardona et al. 2014 and references therein). Numbers indicate reference to specific plutonic bodies and stratigraphic sections in Cardona et al. (2014)

in a magmatic complex, leading the evolution of the rest of the magmatic complex (Barbarin 1999, 2005). This is consistent with the observation that all dated mafic enclaves from the SMB yielded systematically older ages when compared with the rest of the Eocene magmatic suite (Fig. 7.30).

The probable parental magma composition can be located around the less mafic enclaves and most mafic rocks of the main intrusive masses (Fig. 7.29). From this composition the magmas may have evolved by crystal fractionation and differentiation processes to generate the compositions of the different main intrusive bodies. The fractionation of large amounts of amphibole from the parental magma is evidenced not only by the presence of the cumulate bodies but also by the decrease of Nb/Ta and K/Rb, the increase of La/Yb ratios, and the depletion of the middle REE

(Fig. 7.27), as fractionation progressed. The fractionation of large amounts of amphibole requires magma water contents of 6 wt%, implying a high-fluid flux (Moore and Carmichael 1998; Rodríguez et al. 2007) in the magma genesis of the Eocene Santa Marta Province.

Although an "arc signature" is present in the SMB magmas (Fig. 7.27), some evidence suggests that the Eocene magmatic event was not associated to an Andean-type subduction zone. Under scoring the evidence is the fact that Paleocene-Eocene magmatism in the SNSM was a relatively low-volume, localized and short-term event (ca. 10 m.y.), and no other magmatic units have been recognized in the nearby region. Nevertheless, a geochemical "arc signature" is not exclusive to magmas produced in a suprasubduction-metasomatized mantle wedge, and some of those geochemical characteristics may be acquire in the absence of a fully developed subduction system. Considering this scenario, plausible magmaic processes for the generation of a compositionally intermediate parental magma can be envisaged in the melting of the lower part of a thickened crust or of the underthrusted oceanic slab at relatively low pressures, outside the stability field of garnet. Similar processes have been proposed for the origin of TTG series (Karsli et al. 2010; Martin et al. 2005). The young age and abnormal thickness of the Caribbean Plateau imply high buoyancy of this oceanic plate, which would make its subduction difficult beneath the continental plate. In addition, high water content would have favored the partial melting of igneous and sedimentary rock types at relatively shallow depths. This suggests that incipient subduction could have favored the production of water-rich magmas and their emplacement into the recently obducted accretionary prism and oceanic crust which forms basement to the Santa Marta Province.

7.3 Tectonic Implications of the Santa Marta Province Magmatism

Late Cretaceous to Eocene circum-Caribbean tectonics were dominated by the oblique interaction of the Caribbean Plate with North and South America (Burke 1988; Pindell 1993; Spikings et al. 2001, 2005; Pindell et al. 2005; Vallejo et al. 2006). In northern Colombia (Santa Marta Province and Guajira), this process has been well documented by the identification of intraoceanic fragments accreted between 96 and 76 Ma and late Cretaceous-Paleocene middle to high pressure metamorphic complexes (Weber et al. 2009, 2011; Cardona et al. 2010b). The Paleocene (65–63 Ma) trondhjemitic leucogranites of the Santa Marta Province reinforce the idea of a convergent interaction between the Caribbean and South American plates during this time. Thickness and age of the Caribbean Plate would have led to a very low-angle underthrusting of the Caribbean Plate under the South American Plate. In addition, a high thermal regime would have led to partial melting of the oceanic plate to generate the leucogranitoids. Within the circum-Caribbean realm, similar

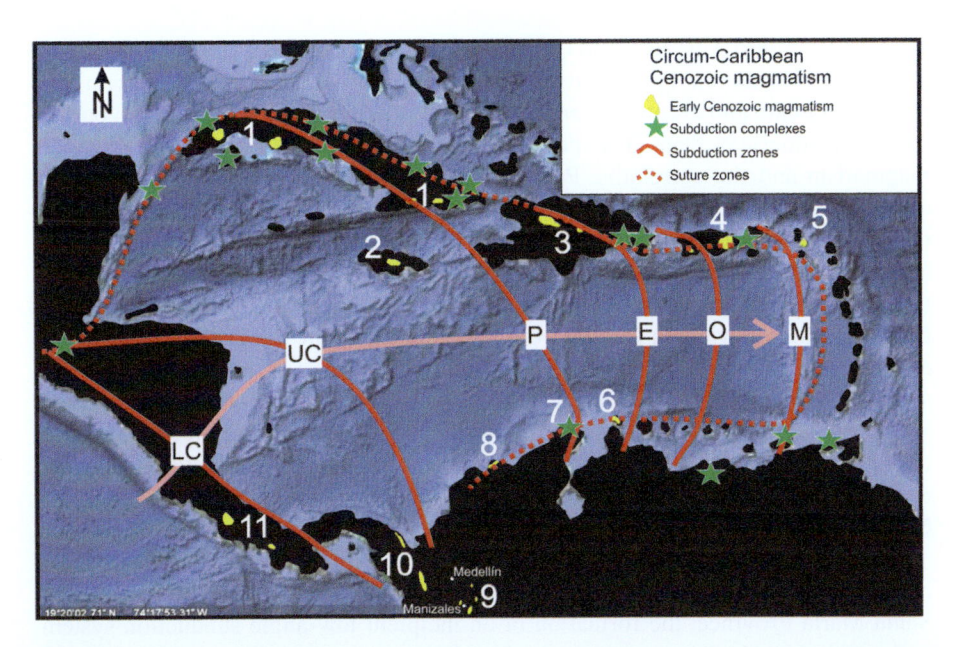

Fig. 7.31 Diagram showing the location of plutonic rocks and subduction complexes associated with the late Cretaceous-Miocene migration of the Caribbean Plate. Late Cretaceous (LC), early Cretaceous (UC), Paleocene (P), Eocene (E), Oligocene (O), Miocene (M). (1) Cuba, (2) Jamaica, (3) La Española, (4) Puerto Rico, (5) San Martín, (6) Aruba, (7) Guajira, (8) Santa Marta, (9) Cordillera Central, (10) Cordillera Occidental, (11) Panamá. (Modified from French and Schenk (2004); image from Google Earth. Location of the subduction zone and complexes modified from Garcia-Casco (2009)

trondhjemitic rocks found in Sierra del Convento, Cuba, have been interpreted by Garcia-Casco et al. (2008) as the product of partial melting of amphibolite lithologies in a young oceanic slab during the initial phases of subduction.

Further, intrusive rocks with TTG characteristics, mostly associated with subduction complexes, are common all around the Caribbean, including in the Sierra Maestra (Cuba), Hispaniola, Puerto Rico, Virgin Islands, Margarita, Aruba, Tobago, Parashi, Santa Marta, Antioquia, and Buga (Fig. 7.31) (Lidiak and Jolly 1996; Maresch et al. 2009; Pindell and Kennan 2009; Cardona et al. 2011a, 2014). These intrusives are found on both margins of the Caribbean Plateau with ages becoming younger to the east, a feature that has been used to track the "tectonic escape" of the Caribbean Plate during the Cenozoic (Müller et al. 1999).

Müller et al. (1999), based upon mantle-referenced palinspastic reconstructions, documented a space reduction between North and South America starting in the early Eocene, when the Caribbean Plate collided with the North American Bahamas platform, inciting a change in the relative motion of the Caribbean Plate relative to North and South America from a northeastward to a mainly eastward motion (Mann 1997). North and South America subsequently converged in a more orthogonal direction (Pindell et al. 1988; Müller et al. 1999), leading to a space reduction between the South American and Caribbean plates (Müller et al. 1999; Nerlich et al. 2014). This situation could have produced highly oblique collision with the Caribbean Plate, and eventually, an incipient subduction zone in the northern part of the South American Plate. This process is evidenced by the Santa Marta Province magmatism and the correlative Parashi Pluton, emplaced, active between 51 and 47 Ma (Cardona et al. 2014), and other plutons located in the Cordillera Central, including the Manizales Stock (59.8 ± 0.7 Ma), El Hatillo Stock (54.6 ± 0.7 Ma), Santa Barbara Batholith (58.9 ± 0.4 Ma and 58.4 ± 0.4 Ma), Stock of Sonsón (60.7 ± 1.4 Ma), and Antioquia Batholith (59.2 ± 1.2 Ma, 59.9 ± 0.9 Ma, and 60.7 ± 1.0 Ma) (review in Bayona et al. 2012; Bustamante et al. 2016; Leal-Mejía et al. 2018). In central and northern Colombia, the onset of Eocene magmatism seems to be diachronous, starting around 60 Ma in the Cordillera Central, becoming younger to the north where the Santa Marta Batholith and related plutons intruded (58–49 Ma), and ending at 50–47 Ma in the northern Guajira with the Parashi Pluton (Fig. 7.30). This trend is also observed at a smaller scale in the Sierra de Santa Marta Province (Fig. 7.22).

Following early Paleocene low-angle underthrusting of the Caribbean Plate beneath the South American Plate and associated trondhjemitic magmatism in the Santa Marta Province, the formation of an incipient low-angle subduction system without the development of a proper mantle wedge would have progressed as a result of the continuous convergence between the North and South American plates. In this scenario, a high thermal regime and high water content of the subducting Caribbean Plateau may have promoted the partial melting of the down-going slab or lower crust. This could have generated water-rich intermediate magmas with TTG-like geochemical characteristics, which then evolved to form the SMB, Toribio, Latal, and Buritaca Plutons by continuous processes of magma differentiation through crystal fractionation and segregation, as well as the assimilation of host rocks.

Evidence found in the sedimentary record supports the underthrusting of the Caribbean Plate beneath South America in the Paleocene. In the Cesar-Rancheria Basin, located to the east of the SNSM massif (Fig. 7.1), the Paleocene-early Eocene section consists of a thick sedimentary package of shallow-marine sandstones and sporadic limestone (especially at northern basin), which grade upward to transitional coal-bearing mudstones and continental sandstones. This sedimentary package is particularly well preserved along the eastern flank of the basin (Mora and Garcia 2006; Sanchez and Mann 2015), whereas it was mostly eroded along the western margin, implying a more active exhumation process close to the SNSM subsequent to the deposition of this section (Ayala et al. 2012). The sedimentary sequence in the Baja Guajira Basin, to the north of the SNSM within the northern block of the Oca-El Pilar fault (Fig. 7.1), shows a significant depositional/erosional hiatus from late Cretaceous to middle Eocene (García et al. 2010), which may be associated with the exhumation of the SNSM driven by the Caribbean-South American interaction. In the Maracaibo Basin, located to the east of the Perijá range (Fig. 7.1), the Paleocene–middle Eocene peak of subsidence is possibly indicating development of a foredeep caused by tectonic loading and subsidence during arc accretion (Escalona and Mann 2011). To the west of the Southern Caribbean deformed belt (Fig. 7.1), onlapping stratigraphic terminations trnucated against Caribbean Plate basement in a pre-Oligocene sedimentary package suggest initiation of a flexure driven by underthrusting of the Caribbean Plate beneath the South American margin (Bernal-Olaya et al. 2015).

Subsequent to Eocene-Oligocene magmatic events documented in different localities along the central and northern Colombian Andes, a regional magmatic gap is observed (Bayona et al. 2012; Leal-Mejía et al. 2018). This hiatus is commonly associated with tectonic margin segmentation, block rotation, basin opening, basin in filling, and basin deformation, documented in Colombia north of 7.5°N (Macellari 1995; Montes et al. 2005, 2010; Beardsley and Avé Lallemant 2007). A clockwise rotation of the SNSM block by ca. 30° was propsed by Montes et al. (2010). The petrogenetic models presented in this paper support such a model (Fig. 7.32).

Thermobarometric calculations suggest that the SMB intruded at mid-crustal depth of ca.16 km, implying that the northwestern tip of the SNSM massif has undergone unroofing of this magnitude since 57 Ma (Cardona et al. 2011b).

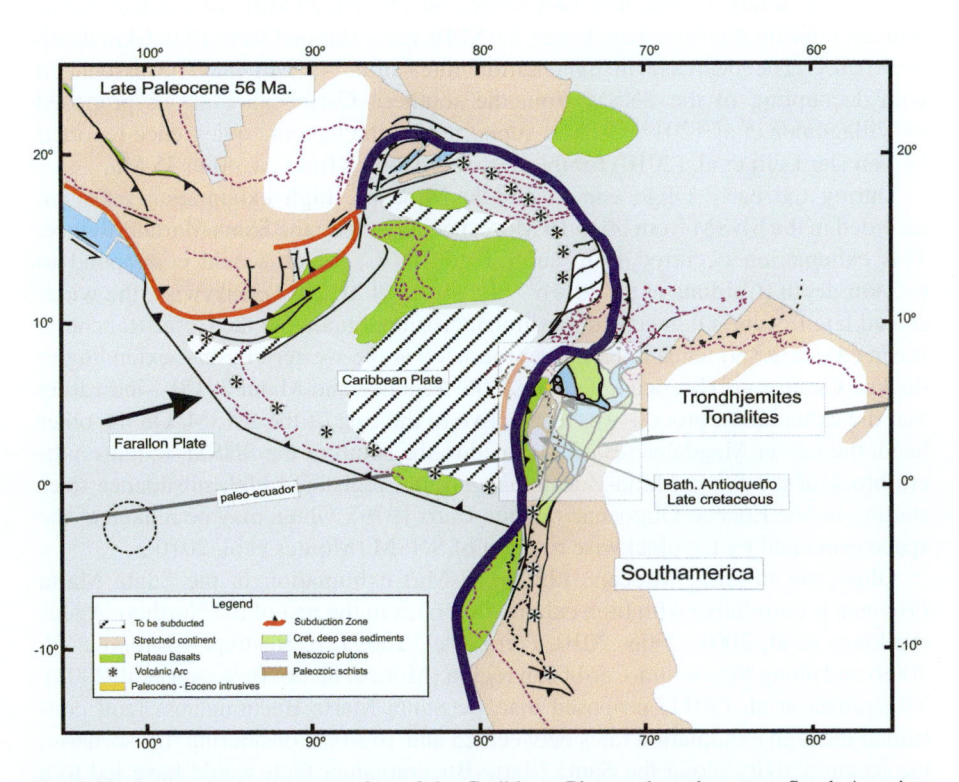

Fig. 7.32 Paleotectonic reconstruction of the Caribbean Plate and northwestern South America for the late Paleocene (56 Ma) indicating the main tectonic features. Compiled and modified from Pindell and Kennan (2007) and Montes et al. (2005). The Sierra Nevada de Santa Marta block is shown in the position proposed by Montes et al. (2005)

Following emplacement, the plutons cooled at a high cooling rate, reaching a temperature of 300 °C by ca. 47 Ma (^{40}Ar–^{39}Ar biotite age), which corresponds to ca. 10 km depth (Fig. 7.23). This translates into an exhumation of about 6 km between emplacement and up to 46 Ma. Salazar et al. (2016) found a strong northeast-southwest magnetic fabric in the SMB by using anisotropy of low-field magnetic susceptibility (AMS). A co-axial fabric is also evident in the strong mineral lineation in the SMB and Buritaca Plutons. These authors propose that this fabric was developed during greenschist to amphibolite facies (~500 °C) metamorphism from 57 to 46 Ma, in which the principle shear vector was oriented east-west (based on the actual position of the batholith), and induced the generation of a northeast-southwest strain (S-surface). High cooling rates and the strong crystal fabric observed in the SMB are thus interpreted to be related to the northwest-southeast convergence between the Caribbean and South American plates during the Eocene. This convergence would have developed a high regional northwest-southeast strain and high exhumation rates due to the flat subduction of the Caribbean Plateau under a rigid crustal block as represented by the SNSM.

After the early Eocene tectonic event, exhumation seems to have slowed down between the middle Eocene and early Oligocene (45 and 25 Ma). Although exhumation rates during this time were lower, the SMB was exhumed from 10 to 6 km depth in 10 m.y. The decrease in exhumation rates after 45 Ma may be associated with decoupling of the SNSM from the southern Caribbean Plate as proposed by Villagómez et al. (2011). This is supported by the tectonic quiescence reported by Van Der Lelij et al. (2010) for the Leeward Antilles from ca. 40 to 35 Ma.

During the early Oligocene to middle Miocene, high exhumation rates are recorded in the SNSM from 30 to 16 Ma, especially along the Santa Marta Province. This exhumation occurred from ca. 180 to 70 °C, which would correspond to 6–2 km depth (Cardona et al. 2011b; Villagómez et al. 2011). Likewise, the widespread late Eocene-Oligocene depositional/erosional hiatus in the Cesar-Rancheria Basin (Ayala 2009) tends to be more protracted at the western basin—extending to the late Cretaceous (Mora and Garcia 2006; Sanchez and Mann 2015)—indicating that the exhumation process was also active to the east of the SNSM. On the other hand, the Lower Magdalena Basin, located to the south of the SNSM, in the southern block of the Santa Marta-Bucaramanga fault, registered high subsidence rates starting in late Eocene-Oligocene (Duque-Caro 1979), which may be related to the space generated by the clockwise rotation of SNSM (Montes et al. 2010).

Oligocene to early Miocene (25 to 16 Ma) exhumation in the Santa Marta Province is correlative with high exhumation rates in the rest of the Northern Andes (Spikings et al. 2001, 2005, 2010, Parra et al. 2009a, b; Restrepo-Moreno et al. 2009) and along the circum-Caribbean region (Mora et al. 2010; Sisson et al. 2008). Villagómez et al. (2011) proposed that the Santa Marta-Bucaramanga fault controlled the high exhumation rates between 25 and 16 Ma. Considering this scenario, the recent activity along the Santa Marta-Bucaramanga fault would have led to a differential exhumation of the SMB in a differential way, resulting in the exposure of deeper crustal levels in the southwestern SMB. This could explain the difference in age and crustal affinity between the southwestern and northeastern SMB. The

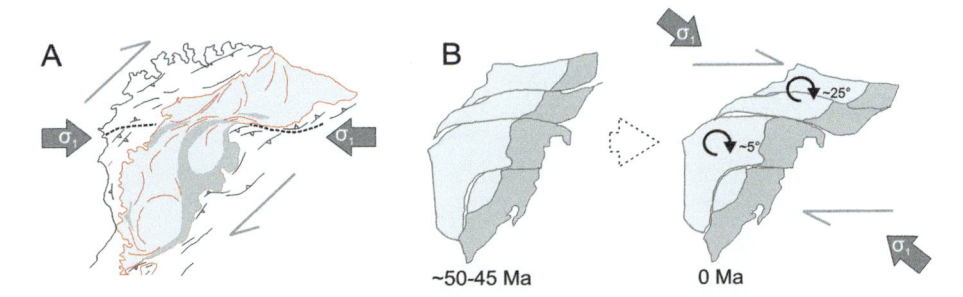

Fig. 7.33 (a) Tectonic model proposed by Salazar et al. (2016) using the magnetic foliation (red lines) measured on the Santa Marta Batholith (SMB). This foliation is proposed to have formed under as a result of right-lateral shear stress (Modified from Salazar et al. 2016). (b) Tectonic reconstruction of the SMB at ~50–45 Ma based on the restoration of translated and rotated rigid blocks along major faults

northeastern SMB reveals older ages and cumullate rocks are absent. Conversely, in the similar-age southwestern SMB, cumulate rocks, interpreted to be related to deeper crustal levels, are abundant.

The driving forces controlling the collision, obduction, underthrusting, and exhumation along the northwestern corner of South America during the Paleocene to Oligocene were apparently the rate and vector of convergence between South and North America (Müller et al. 1999; Nerlich et al. 2014). Conversely, during the early Miocene, uplift and exhumation of crustal blocks would have been driven by changes in underthrusting rates due to variations in thickness and superficial topography of the oceanic subducting plate, as, during this time, clear evidence of plate collision is lacking (Pardo-Casas and Molnar 1987; Silver et al. 1998; Spikings et al. 2001). Salazar et al. (2016) propose that convergence between the Americas, with the Caribbean Plate trapped between them, has led to the tectonic escape of the latter. This, coupled with the compressional deformation observed in recent tectonic elements, beginning in the latest Oligocene - early Miocene, evidence continued oblique collision along the South American-Caribbean margin after this time (Müller et al. 1999; Pindell et al. 2005; Nerlich et al. 2014). Salazar et al. (2016), based upon the magnetic foliation defined by AMS and the metamorphic foliation of batholith-hosting wall-rocks, proposed that the emplacement of the SMB took place when the magmatic arc was being deformed by dextral shear, as a result of the oblique convergence of the Caribbean and South American plates (Fig. 7.33a). This deformation regime was active since the late Eocene, as indicated by crystallization and cooling ages obtained for the SMB. Continued dextral shear led to the deformation patterns presently observed in the SNSM.

The faulting pattern of the SNSM has been divided into six different regions (Fig. 7.34) by INGEOMINAS (2007). Regions I to IV are characterized by large E-W-striking - NW-verging right-lateral transpressive structures which transect the entire massif. An exception to this is present in region V (Cesar-Rancheria Valley), where the structural configuration portrays NW-verging reverse faults (Fig. 7.34). Nevertheless, the overall configuration is consistent with a regional NW-SE compressive stress regime (Fig. 7.34).

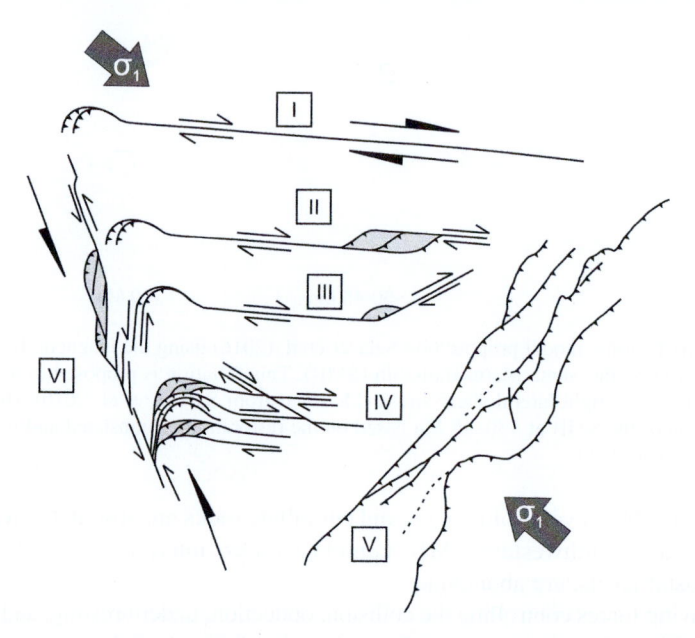

Fig. 7.34 Structural patterns for the tectonic regions of the SNSM (Modified from INGEOMINAS 2007)

INGEOMINAS (2007) noticed that the degree of deformation and fault density in the SNSM decrease toward to the northern part of the massif, suggesting that the main compressive stress is located to the south and deformation is propagating from the SE to NW.

In order to reconstruct the pre-deformational shape of the SMB, restoration of the right-lateral movement on the northern structures of the SNSM massif (Jordán, El Carmen, and San Lorenzo faults) is required. Figure 7.33b depicts a palinspastic reconstruction of the SMB along both, the eastern and western borders, with the San Lorenzo and Santa Marta metamorphic units, respectively. To make this restoration plausible, a counterclockwise rotation of about 25° is applied to the northernmost block, such that it may fit with the southern part of the batholith. Restoration results in a single NE-trending magmatic unit, dislocated by an E-W-oriented shear zone that transects the middle of the batholith. This shear zone is interpreted to have been active at the time when the magmatic rock was still in a plastic state, allowing for the development of both the E-W-oriented mineral and magnetic foliations within the SMB and along the El Carmen fault, as described by Salazar et al. (2016).

The clockwise rotation of rigid blocks along right-lateral strike-slip fault systems is a commonly observed feature, in this particular case, ratifying the right-lateral movement along the Oca, El Jordan, and El Carmen faults. The observed kinematics are produced by the relative NW-verging displacement of the SNSM massif with respect to the Caribbean Plate, which is also evidenced by the east ward displacement of the Guajira Peninsula with respect to the SNSM massif.

7.4 Basin Filling as a Tectonic Response

Lower Cretaceous basins located around and within the Maracaibo Basin (MB) record growth and deformation processes in response to continent-oceanic arc collision (i.e., subduction, terrane accretion, tilting, and uplift) (Bayona et al. 2011).

The Rio Negro Fm, a succession of siliciclastic sediments, deposited in Lower Cretaceous extensional basins, is overlain by vast deposits of Aptian-Albian carbonates (Cogollo Fm) which cover Triassic-Jurassic plutonic and volcanic rocks from the SNSM, the basement of the Perijá Range, and the Paleozoic basement of the Maracaibo Basin. These units are covered by bituminous limestones, shales, and cherts of Turonian and Santonian age (La Luna Fm), which record anoxic deepwater deposition which continued up to the Campanian (Bayona et al. 2011 and references therein). The deepest marine conditions of deposition were reached during the formation of the lowermost beds of the Colón Fm (~85 Ma), marked by a period of very low depositional rates. At this time, a first-order marine regression is recorded by a series of shallow-marine to continental sediments of including the Hato Nuevo, Manantial, and Guasare Fms (Bayona et al. 2011). This shallowing of the depositional systems is also recorded in the upper member of the Colón Fm (Campanian to Paleocene) by the vertical variation of planktic/benthic foraminifera fauna (Martínez and Hernández 1992). An integrated petrographic and geochronologic analysis performed for the Upper Cretaceous sediments indicates supply of detritus from the Colombian Central Cordillera, and the absence of clear sediment input from the SNSM (Bayona et al. 2011).

During early Paleocene, a period of continent-ward migration of the shoreline and lower siliciclastic input into the Maracaibo block is marked by the deposition of the calcareous rocks of Hato Nuevo and Manantial Fms. These are interpreted formed to have been deposited in a marginal basin to shallow-marine platform environment (Etayo-Serna 1979), which evolved to a siliciclastic depocenter with coal-bearing strata, as represented in the eastern part of the MB by the Guasare Fm. The shallowing of the marine environment and installation of a siliciclastic-dominated environment was clearly diachronic, being older in the RB and younger to the eastern side of the MB. During the Late Paleocene, siliciclastic deposition dominated the MB. Continental depositional settings were located to the west (Cerrejón Fm) and marginal depositional settings to the east (Marcelina Fm). Then, during the early to middle Eocene depositional settings are characterized by fluvial channels and floodplains as recorded in the Tabaco and Palmito Fms (early Eocene) and Misoa Fm (early to middle Eocene) (Bayona et al. 2011 and references therein). Provenance and paleocurrent studies performed on the Manantial and Cerrejón Fms suggest the important input of sediments from the SNSM, which by this time was being eroded, especially along the northwestern tip of the Massif, from where, it has been proposed, the majority of the sediment contained within these units, was supplied (Bayona et al. 2007, 2011).

After the early Eocene, as series of regional discontinuities are recorded as angular erosional surfaces, reverse faults, folds, and uplifted highs (e.g., Perijá range and

El Palmar high), are recorded along from the Cesar-Rancheria Basin and extending to the eastern part of the Maracaibo Basin (Kellogg 1984).

From Lower Cretaceous to Eocene, the highest subsidence rates in the Maracaibo block are observed from the late Paleocene to the early Eocene. During this interval, coal-bearing strata were deposited all along the MB. Nevertheless, this deposition seems to have been diachronous, beginning in the Rancheria Basin, and migrating eastward to cover the Guasare Platform by the late Paleocene. By the early Eocene, sedimentation rates at the Rancheria Basin decreased, whilst, in the western Maracaibo Basin, thick continental sedimentary sequences were still being deposited (Misoa Fm). Sediment provenance for the late Paleocene to early Eocene strata (Cerrejón Fm), and tectonic subsidence rates, correlate with the documented timing of deformation along the western collisional margin of the Sierra Nevada de Santa Marta massif and Central Cordillera. These features represent topographic relief along the South American Plate margin resulting from the low-angle interaction and forced-underthrusting/subduction of the Caribbean Plate during this time (Cardona et al. 2009, 2010b, 2011a). Nevertheless, during the late Eocene to Miocene, high subsidence rates associated with crustal stretching, are recorded in the 260 km wide Plato-San Jorge Basin (PSJB), including its flat-lying, 2 to 8 km thick, fine-grained sedimentary sequence, which is as old as Oligocene in age (Montes et al. 2010). A complex array of faults cut the basement and the overlying strata which are as young as the Upper Miocene in age (Montes et al. 2010). These faults are mostly NE-SW and ENE-WSW striking. The first group, was probably inherited from Jurassic widespread rifting, while the second group is possibly associated with localized extension related to late Cretaceous to Eocene strike-slip faulting, as widely recognized along the western margin of the South American Plate (Mora-Bohórquez et al. 2017).

The configuration of the RB and PSJB, tectonic subsidence, and the westward migration of continental sedimentation away from the Paleocene collisional margin, from the Rancheria Basin into the Guasare Basin, is consistent with the proposed vertical tilting of the Sierra Nevada de Santa Marta as a rigid crustal block during the low-angle underthrusting/subduction of the Caribbean Plate beneath the South American Plate.

7.5 Cenozoic of the Caribbean-South American Plate Interaction Model

Models for the evolution of the Caribbean province envision two end members. In the first one, the Caribbean Plate was formed in its current location. This "in situ model" proposes the formation of the Caribbean Plate by extension during the separation between North America and South America (Meschede and Frisch 1998; Müller et al. 1999). On the other hand, the second model proposes an origin for the Caribbean Plate as oceanic crust in the Pacific region—the "Pacific model"— with or without the participation of the Galapagos hotspot (Pindell and Dewey 1982; Burke 1988; Pindell et al. 1988; Pindell and Kennan 2009). Proponents of this second model have reached a general consensus about the northward and eastward migration of the Caribbean Plate (Freymueller et al. 1993; Kellogg et al. 1995; Pindell and Kennan 2009; Altamira and Burke 2015) facilitated by an over 100 km wide dextral shear zone (Audemard and Audemard 2002).

The northeastward motion of the Caribbean Plate has been inferred via the documentation of transpressional and transtensional features at the plate margins. Ave'lallemant and Gordon (1999) describe Oligocene transtensional deformational styles on Roatan Island, Honduras. Ratschbacher et al. (2009) interpret several events of Eocene to recent transpression and transtension during translation of the eastern Chortis block, and Galindo and Lonergan (2013) propose complex strain partitioning along the South Caribbean deformed belt at northern Colombia, consisting of right-lateral, strike-slip faulting and linked compressional structures at the rear of the accretionary prism.

Several geophysical studies hypothesize subduction of the Caribbean Plate beneath South America. Using P-wave travel-time residuals and non-isostatic topography, Yarce et al. (2014) suggested slab-associated upper mantle flow in northern Colombia, which may be interpreted to indicate Caribbean flat slab subduction that becomes steeper toward the region of the Bucaramanga "nest." Tomographic studies also have imaged a high velocity zone beneath the continental crust which also may be interpreted to represent a shallowly (<15°) subducting slab and for some authors may explain Bucaramanga "nest" seismicity, related to the hypothetical dehydration and eclogitization of thickened oceanic crust (Van der Hilst and Mann 1994; Benthem et al. 2013; Bernal-Olaya et al. 2015; Chiarabba et al. 2015). Finally, gravity and magnetic studies suggest that the SNSM should be dynamically supported by the underthrust Caribbean slab (Cerón et al. 2007). Considering the composite geophysical studies noted herein, and documentation regarding the timing and nature of magmatim along the NW apex of the SNSM, most studies are in agreement that Caribbean-South American plate interaction records arc collision, followed by low-angle subduction and the underthrusting of oceanic crust.

The migration of depocenters documented in sedimentary basins along northern South America has been interpreted to be associated with the transient oblique convergence, collision, and tectonic load of the Caribbean Plate and the South American Plate margin. In the Maracaibo Basin (Fig. 7.1), late Eocene depocenters are preserved to the northwest, overlying an early Paleocene-late Eocene unconformity, whereas lower Miocene depocenters are preserved to the southeast, overlying a more extended unconformity (Fig. 7.2) (Parnaud et al. 1995). This regional unconformity has been associated with foreland uplift conditioned by isostatic rebound after the Caribbean arc overthrust the South American margin (Escalona and Mann 2011). In the Cesar-Rancheria Basin (Fig. 7.1), a lower Miocene-Pliocene depocenter is preserved on the western margin of the basin, underlying by an Oligocene-late Cretaceous unconformity (Mora and Garcia 2006; Ayala 2009). This unconformity may have been caused by the uplift of the SNSM during Caribbean arc-South American Plate collision. The unconformity extends from the Pliocene to the late Cretaceous in the eastern basin, denoting a younger uplift in the western foothills of the Perijá Range.

An alternative model for the Caribbean-South American plate interaction is the "Orogenic float model" (Oldow et al. 1990; Cediel et al. 2003; Cediel 2018). In this model a detached block of attenuated continental crust overrides— along a northwest-directed vector —the Caribbean Plate at a low angle, and includes the detachment of Oligocene-lower Miocene rocks that may have accommodated deformation generated during Caribbean-South American plate convergence (Kellogg 1984; Audemard and Audemard 2002; Cediel et al. 2003; Corredor et al. 2003). The absence of present-day subduction is apparent as supported by the lack of recent magmatism and Benioff zone development in northwestern Colombia (Cerón et al. 2007; Leal-Mejía et al. 2018).

The results and information presented in this paper support the "Pacific model" for the evolution of the Caribbean Plate and its emplacement along the northwest South American margin. In addition, the characterization of magmatic bodies related in detail herein help constrain the timing and processes related to any proposed arc-continental collision between the Caribbean and South American plates. With respect to the Colombian Caribbean, this process was initited by the Paleocene, with the possible initiation of flat subduction during the late Paleocene-early Eocene. This timing is in accordance with the processes of uplift and subsidence, as documented in the sedimentary basins along the northwest South American margin. We suggest that divergent models that attempt to explain the petrogenesis and tectonic evolution of the trondjemitic and calc-alkaline series magmatic suites along the apex of the SNSM may be difficult to sustain.

Acknowledgments This research was partially supported by IGCP, geophysics commission graduate scholarship program (2008) to Duque-Trujillo, J.; Project #2289 from Fundación para la promoción de la Investigación y la Tecnología del Banco de la Republica (BANREP), and PAPIIT project #IN-114508. Authors would like to thank Susana María Hoyos Muñoz for assistance in the editing with text and figure edition.

References

Altamira A, Burke K (2015) The ribbon continent of South America in Ecuador, Colombia, and Venezuela. AAPG Special Volumes. Memoir 108: Petroleum Geology and Potential of the Colombian Caribbean Margin, pp 39–84

Anders E, Grevesse N (1989) Abundances of the elements: meteoritic and solar. Geochimica et Cosmochimica Acta 53:197–214

Argus DF, Gordon RG, Demets C (2010) GEODVEL, MORVEL, and the velocity of Earth's center. In AGU Fall Meeting Abstracts

Audemard FE, Audemard FA (2002) Structure of the Mérida Andes, Venezuela: relations with the South America-Caribbean geodynamic interaction. Tectonophysics 345:299–327. https://doi.org/10.1016/S0040-1951(01)00218-9

Ave'lallemant HG, Gordon MB (1999) Deformation history of Roatán Island: implications for the origin of the Tela basin (Honduras). Sedimentary Basins of the World, 4, pp 197–218

Ayala RC (2009) Análisis tectonoestratigráfico y de procedencia en la subcuenca de cesar: relación con los sistemas petroleros [Master of Science]. Universidad Simón Bolívar, pp 183–198

Ayala RC, Bayona G, Cardona A, Ojeda C, Montenegro OC, Montes C, Valencia V, Jaramillo C (2012) The paleogene synorogenic succession in the northwestern Maracaibo block: tracking intraplate uplifts and changes in sediment delivery systems. J S Am Earth Sci 39:93–111. https://doi.org/10.1016/j.jsames.2012.04.005

Barbarin B (1999) A review of the relationships between granitoid types, their origins and their geodynamic environments. Lithos 46:605–626. https://doi.org/10.1016/S0024-4937(98)00085-1

Barbarin B (2005) Mafic magmatic enclaves and mafic rocks associated with some granitoids of the central Sierra Nevada batholith, California: nature, origin, and relations with the hosts. Lithos 80:155–177. https://doi.org/10.1016/j.lithos.2004.05.010

Barker F (1979) Trondhjemites: definition, environment and hypotheses of origin. In: Barker F (ed) Trondhjemites, dacites and related rocks. Elsevier, Amsterdam, pp 1–12

Bayona G, Lamus-Ochoa F, Cardona A, Jaramillo CA, Montes C, Tchegliakova N (2007) Provenance analysis and paleocene orogenic processes in the Rancheria Basin (Guajira, Colombia) and surrounding areas. Geologia Colombiana 32:21–46

Bayona G, Montes C, Cardona A, Jaramillo C, Ojeda G, Valencia V, Ayala-Calvo C (2011) Intraplate subsidence and basin filling adjacent to an oceanic arc-continent collision: a case from the southern Caribbean-South America plate margin. Basin Res 23:403–422. https://doi.org/10.1111/j.1365-2117.2010.00495.x

Bayona G, Cardona A, Jaramillo C, Mora A, Montes C, Valencia V, Ayala C, Montenegro O, Ibañez-Mejia M (2012) Early paleogene magmatism in the northern Andes: insights on the effects of oceanic plateau-continent convergence. Earth Planet Sci Lett 331–332:97–111. https://doi.org/10.1016/j.epsl.2012.03.015

Beardsley AG, Avé Lallemant HG (2007) Oblique collision and accretion of the Netherlands leeward antilles to South America. Tectonics. https://doi.org/10.1029/2006TC002028

Benthem S, Govers R, Spakman W, Wortel R (2013) Tectonic evolution and mantle structure of the Caribbean. J Geophys Res Solid Earth 118(6):3019–3036

Bernal-Olaya R, Mann P, Vargas CA (2015) Earthquake, Tomographic, Seismic Reflection, and Gravity Evidence for a Shallowly Dipping Subduction Zone beneath the Caribbean Margin of Northwestern Colombia. In: Memoir 108: Petroleum Geology and Potential of the Colombian Caribbean Margin, pp 247–270

Bezada MJ, Schmitz M, Jácome MI, Rodríguez J, Audemard F, Izarra C, BOLIVAR Active Seismic Working Group (2008) Crustal structure in the Falcón Basin area, northwestern Venezuela, from seismic and gravimetric evidence. J Geodyn 45(4):191–200

Blanco JM, Mann P, Nguyen L (2015) Location of the suture zone separating the great arc of the Caribbean from continental crust of northwestern South America inferred from regional gravity and magnetic data

Burke K (1988) Tectonic evolution of the Caribbean. Annu Rev Earth Planet Sci 16:201–230. https://doi.org/10.1146/annurev.earth.16.1.201

Bustamante C, Cardona A, Saldarriaga M, García-Casco A, Valencia V, Weber M (2009) Metamorfismo de los esquistos verdes y anfibolitas pertenecientes a los Esquistos de Santa Marta, Sierra Nevada de Santa Marta (Colombia): ¿registro de la colisión entre el Arco Caribe y la margen Suramericana? Boletín de Ciencias de la Tierra, Universidad Nacional de Colombia

Bustamante C, Archanjo CJ, Cardona A, Vervoort JD (2016) Late jurassic to early cretaceous plutonism in the Colombian Andes: a record of long-term arc maturity. Bull Geol Soc Am 128:1762–1779. https://doi.org/10.1130/B31307.1

Cáceres H, Camacho R, Reyes J (1980) The Geology of the Rancheria Basin. In: Geotec (ed) Geological Field-Trips, Colombia 1980–1989. Asociación Colombiana de Geólogos y Geofísicos del Petróleo, Bogotá, pp 1–31

Campbell CJ (1965) The Santa Marta wrench fault of Colombia and its regional setting. Transactions of the Caribbean Geology Conference, 4, pp 247–261

Cardona A, Valencia V, Bayona G, Jaramillo C, Ojeda G, Ruiz J (2009) U/Pb LA-MC-ICP-MS zircon geochronology and geochemistry from a post-collisional biotite granite of the Baja Guajira Basin, Colombia: implications for late cretaceous and neogene Caribbean–South American Tectonics. J Geol 117(6):685–692

Cardona A, Valencia V, Garzón A, Montes C, Ojeda G, Ruiz J, Weber M (2010a) Permian to triassic I to S-type magmatic switch in the northeast Sierra Nevada de Santa Marta and adjacent regions, Colombian Caribbean: tectonic setting and implications within Pangea paleogeography. J S Am Earth Sci 29:772–783. https://doi.org/10.1016/j.jsames.2009.12.005

Cardona A, Valencia V, Bustamante C, Garcia-Casco A, Ojeda G, Ruiz J, Saldarriaga M, Weber M (2010b) Tectonomagmatic setting and provenance of the Santa Marta Schists, northern Colombia: insights on the growth and approach of cretaceous Caribbean oceanic terranes to the South American continent. J S Am Earth Sci 29:784–804. https://doi.org/10.1016/j.jsames.2009.08.012

Cardona A, Valencia VA, Bayona G, Duque J, Ducea M, Gehrels G, Jaramillo C, Montes C, Ojeda G, Ruiz J (2011a) Early-subduction-related orogeny in the northern Andes: turonian to eocene magmatic and provenance record in the Santa Marta Massif and Rancheria Basin, northern Colombia. Terra Nov 23:26–34. https://doi.org/10.1111/j.1365-3121.2010.00979.x

Cardona A, Valencia V, Weber M, Duque J, Montes C, Ojeda G, Villagomez D (2011b) Transient Cenozoic tectonic stages in the southern margin of the Caribbean plate: U-Th/He thermochronological constraints from eocene plutonic rocks in the Santa Marta massif and Serranía de Jarara, northern Colombia. Geol Acta 9(3–4):445–469

Cardona A, Weber M, Valencia V, Bustamante C, Montes C, Cordani U, Muñoz CM (2014) Geochronology and geochemistry of the Parashi granitoid, NE Colombia: tectonic implication of short-lived early eocene plutonism along the SE Caribbean margin. J S Am Earth Sci 50:75–92. https://doi.org/10.1016/j.jsames.2013.12.006

Cediel F (2018) Phanerozoic orogens of Northwestern South America: cordilleran-type orogens, taphrogenic tectonics and orogenic float. Springer, Cham, pp. 3–89

Cediel F, Shaw RP, Caceres C (2003) Tectonic assembly of the Northern Andean Block. AAPG Mem 79:815–848

Cerón JF, Kellogg JN, Ojeda GY (2007) Basement configuration of the Northwestern South America – Caribbean margin from recent geophysical data. Ciencia, Tecnol y Futur 3:25–49

Cherniak D, Watson E (2001) Pb diffusion in zircon. Chemical Geology 172(1–2):5–24

Chiarabba C, De Gori P, Mele FM (2015) Recent seismicity of Italy: active tectonics of the central Mediterranean region and seismicity rate changes after the Mw 6.3 L'Aquila earthquake. Tectonophysics 638:82–93. https://doi.org/10.1016/j.tecto.2014.10.016

Colmenares L, Zoback MD (2003) Stress field and seismotectonics of northern South America. Geology 31:721–724. https://doi.org/10.1130/G19409.1

Condie K (2005) TTGs and adakites: are they both slab melts? Lithos 80:33–44

Corredor J, Morell J, Armstrong R et al (2003) Remote continental forcing of phytoplankton biogeochemistry: observations across the "Caribbean-Atlantic front". Geophys Res Lett 30:1–4. https://doi.org/10.1029/2003GL018193

De La Roche H, Leterrier J, Grandclaude P, Marchal M (1980) A classification of volcanic and plutonic rocks using R1-R2. Diagrams and major elements analysis – its relationships whit current nomenclature. Chem Geol 29:183–210

Duque J (2009) Geocronología (U-Pb y 40Ar/39Ar) y geoquímica de los intrusivos paleogenos de la Sierra Nevada de Santa Marta y sus relaciones con la tectonica del Caribe y el arco magmatico circun-Caibe~no. Universidad Nacional Autonoma de Mexico, p 189, MSc Dissertation

Duque-Caro H (1979) Major structural elements and evolution of northwestern Colombia. Geological and geophysical investigations of continental margins: AAPG Memoir, 29, pp 329–351

Duque-Trujillo J, Orozco-Esquivel T, Cardona A, Ferrari L, López-Martinez M, Solaro L, Valencia V (2010) The Paleogene intrusive of the Sierra Nevada de Santa Marta, Colombia: short-lived magmatism related to the collision of the caribbean plate with South América. In 2010 GSA Denver Annual Meeting

Escalona A, Mann P (2011) Tectonics, basin subsidence mechanisms, and paleogeography of the Caribbean-South American plate boundary zone. Mar Pet Geol 28:8–39. https://doi.org/10.1016/j.marpetgeo.2010.01.016

Etayo-Serna F (1979) Moluscos de una capa del Paleoceno de Manantial (Guajira). Bol Geol Univ Ind Santander 13(27):5–55

Farley KA (2002) (U-Th)/He dating: techniques, calibrations, and applications. Rev Mineral Geochem 47(1):819–844

Flinch J, Amaral J, Doulcet A, et al (2003) Structure of the offshore sinu accretionary wedge. Northern Colombia VIII Simp Boliv Cuencas Subandinas, pp 76–83

Florez-Niño J (2001) Elastic geomechanical model of Bucaramanga and Oca faults and the origin of the Sierra Nevada de Santa Marta, Northern Andes, Colombia. In: Proceedings AGU Fall Meeting Abstracts Volume 1, p 834

Forero A (1970) Estartigrafia del Pre-Cretacico en el flanco occidental de la Serrania de Perijá. Geologia Colombiana 7:7–78

French C, Schenk C (2004) Map showing geology, oli and gas fields, and geologic provinces of the Caribbean region. U.S. Geological Survey. Open file report 97–470- k

Freymueller JT, Kellogg JN, Vega V (1993) Plate motions in the north Andean region. J Geophys Res 98:21853. https://doi.org/10.1029/93JB00520

Galindo P, Lonergan L (2013) Evolution of the Bahia Basin: evidence for vertical-Axis block rotation and basin inversion at the Caribbean plate margin offshore northern Colombia. In AAPG International Conference and Exhibition, p 11

García M, Mier R, Cruz G (2010) Reconstrucción de la historia paleo-thermal de la subcuenca de la baja Guajira, Colombia. Boletín de Geología 32(2):55–71

Garcia-Casco A (2009) Subduction complex associated with the Caribbean Plate migration from late Cretaceous. http://www.ugr.es/~agcasco/personal/. Accessed 20 March of 2017

Garcia-Casco A, Lazaro C, Rojas-Agramonte Y et al (2008) Partial melting and counterclockwise P T path of subducted oceanic crust (Sierra del Convento Melange, Cuba). J Petrol 49:129–161. https://doi.org/10.1093/petrology/egm074

Getsinger A, Rushmer T, Jackson MD, Baker D (2009) Generating high Mg-numbers and chemical diversity in tonalite-trondhjemite-granodiorite (TTG) magmas during melting and melt segregation in the continental crust. J Petrol 50:1935–1954. https://doi.org/10.1093/petrology/egp060

Gomez I (2001) Structural style and evolution of the Cuisa fault system. Guajira, Colombia

Gómez E, Jordan TE, Allmendinger RW, Hegarty K, Kelley S (2005) Syntectonic cenozoic sedimentation in the northern middle Magdalena Valley basin of Colombia and implications for exhumation of the Northern Andes. Geol Soc Am Bull 117(5–6):547–569

Hernández O, Jaramillo JM (2009) Reconstrucción de la historia termal en los sectores de Luruaco y Cerro Cansona–cuenca del Sinú-San Jacinto y en el piedemonte occidental de la Serranía del Perijá entre Codazzi y la Jagua de Ibirico–cuenca de Cesar Ranchería. Informe Final Cuenca Cesar-Ranchería. Agencia Nacional de Hidrocarburos (Bogotá), p 58

Herrera PM, Santa Escobar M, Ordoñez Carmona O, Pimentel M (2008) Consideraciones Petrograficas, Geoquimicas y Geocronologicas de la parte Occidental del Batolito de Santa Marta. Dyna 75:223–236

INGEOMINAS (2007) Geología de las planchas 11, 12, 13, 14, 18, 19, 20, 21, 25, 26, 27, 33, 34 y 40. Proyecto: "Evolución geohistórcia de la Sierra Nevada de Santa Marta" Informe Contrato de prestación de servicios No. PS 025–06

Karsli O, Dokuz A, Uysal I et al (2010) Generation of the early cenozoic adakitic volcanism by partial melting of mafic lower crust, Eastern Turkey: implications for crustal thickening to delamination. Lithos 114:109–120. https://doi.org/10.1016/j.lithos.2009.08.003

Kellogg JN (1984) Cenozoic tectonic history of the Sierra de Perijá, Venezuela-Colombia, and adjacent basins. In: The Caribbean-South American Plate Boundary and Regional Tectonics, pp 239–262

Kellogg JN, Bonini WE (1982) Subduction of the Caribbean Plate and basement uplifts in the overriding South American Plate. Tectonics 1:251–276. https://doi.org/10.1029/TC001i003p00251

Kellogg JN, Vega V, Stailings TC et al (1995) Tectonic development of Panama, Costa Rica, and the Colombian Andes: constraints from Global Positioning System geodetic studies and gravity. Geol Soc Am Spec Pap 295:75–90. https://doi.org/10.1130/SPE295-p75

Lázaro C, García-Casco A, Rojas Agramonte Y, Kröner A, Neubauer F, Iturralde-Vinent M (2009) Fifty five million year history of oceanic subduction and exhumation at the northern edge of the Caribbean plate (Sierra del Convento mélange, Cuba). J Metamorph Geol 27(1):19–40

Leal-Mejia H, Shaw RP, Melgarejo JC (2018) Spatial/temporal migration of granitoid magmatisn and the phanerozoic tectono-magmatic evolution of the Colombian Andes. In: Cediel F and Shaw RP (eds). Geology and Tectonics of Northwestern South America: The Pacific-Caribbean-Andean Junction, Springer, pp 253–397

Le Fort P (1981) Manaslu leucogranite: a collision signature of the Himalaya, a model for its génesis and emplacemenr. J Geophys Res 86:10545–10568

Lidiak EG, Jolly WT (1996) Circum-Caribbean granitoids: characteristics and origin. Int Geol Rev 38:1098–1133

Macellari CE (1995) Cenozoic sedimentation and tectonics of the southwestern Caribbean pull-apart basin, Venezuela and Colombia. Pet Basins South Am 62:757–780

Mann P (1997) Model for the formation of large, transtensional basins in zones of tectonic escape. Geology 25:211–214. https://doi.org/10.1130/0091-7613(1997)025<0211:MFTFOL>2.3.CO;2

Mann P, Rogers R, Gahagan L (2006) Overview of plate tectonic history and its unresolved tectonic problems. In: Bundschuh J, Alvarado GE (eds) Central america: geology, resources and hazards, vol 1. Leiden, Taylor and Francis/Balkema, pp 201–237

Maresch WV, Kluge R, Baumann A, Pindell JL, Krückhans-Lueder G, Stanek K (2009) The occurrence and timing of high-pressure metamorphism on Margarita Island, Venezuela: a constraint on Caribbean-South America interaction. Geol Soc Lond, Spec Publ 328(1):705–741

Martin H, Moyen JF (2002) Secular changes in tonalite-trondhjemite-granodiorite composition as markers of the progressive cooling of Earth. Geology 30:319–322. https://doi.org/10.1130/0091-7613(2002)030<0319:SCITTG>2.0.CO

Martin H, Smithies RH, Rapp R, Moyen JF, Champion D (2005) An overview of adakite, tonalite^trondhjemite^granodiorite (TTG), and sanukitoid: relationships and some implications for crustal evolution. Lithos 79:1–24

Martínez J, Hernández R (1992) Evolution and drowning of the late Cretaceous Venezuelan carbonate platform. J S Am Earth Sci 5:197–210

Mauffret A, Leroy S (1997) Seismic stratigraphy and structure of the Caribbean igneous province. Tectonophysics 283:61–104. https://doi.org/10.1016/S0040-1951(97)00103-0

Maze WB (1984) Jurassic La Quinta Formation in the Sierra de Perijá, northwestern Venezuela: geology and tectonic environment of red beds and volcanic rocks. In: Bonini WE, Hargraves RB, Shagam R (eds) The Caribbean-South American plate boundary and regional tectonics, vol 162. Geological Society of America, Boulder, pp 263–282

Meschede M, Frisch W (1998) A plate-tectonic model for the mesozoic and early cenozoic history of the Caribbean plate. Tectonophysics 296:269–291. https://doi.org/10.1016/S0040-1951(98)00157-7

Miller JB (1962) Tectonic trends in the Sierra de Perijá and adjacent parts of Venezuela and Colombia. AAPG Bull 46:1565–1595

Miller MS, Levander A, Niu FL, Li AB (2009) Upper mantle structure beneath the Caribbean-South American plate boundary from surface wave tomography. J Geophys Res Earth:B01312. https://doi.org/10.1029/2007jb005507

Montes C, Hatcher RD, Restrepo-Pace PA (2005) Tectonic reconstruction of the northern Andean blocks: oblique convergence and rotations derived from the kinematics of the Piedras-Girardot area, Colombia. Tectonophysics 399:221–250

Montes C, Guzman G, Bayona G et al (2010) Clockwise rotation of the Santa Marta massif and simultaneous Paleogene to Neogene deformation of the Plato-San Jorge and Cesar-Ranchería basins. J S Am Earth Sci 29:832–848. https://doi.org/10.1016/j.jsames.2009.07.010

Montes C, Bayona G, Cardona A, Buchs D, Silva C, Moron SE, Hoyos N, Ramirez DA, Jaramillo C, Valencia V (2012) Arc-continent collision and Orocline formation: closing of the Central American Seaway. J Geophys Res 117(B4):B04105

Moore G, Carmichael I (1998) The hydrous phase equilibria (to 3 kbar) of an andesite and basaltic andesite from western Mexico: constraints on water content and conditions of phenocryst growth. Contrib Mineral Petrol 130:304. https://doi.org/10.1007/s004100050367

Mora A, Garcia A (2006) Cenozoic tectono-stratigraphic relationships between the Cesar Sub-Basin and the southeastern lower Magdalena Valley Basin of Northern Colombia. Search and Discovery 30046:1–11

Mora A, Parra M, Strecker MR et al (2008) Climatic forcing of asymmetric orogenic evolution in the Eastern Cordillera of Colombia. Bull Geol Soc Am 120:930–949. https://doi.org/10.1130/B26186.1

Mora A, Horton BK, Mesa A, Rubiano J, Ketcham RA, Parra M, Blanco V, Garcia D, Stockli DF (2010) Migration of Cenozoic deformation in the Eastern Cordillera of Colombia interpreted from fission track results and structural relationships: implications for petroleum systems. AAPG Bull 94(10):1543–1580

Mora-Bohórquez JA, Ibáñez-Mejia M, Oncken O et al (2017) Structure and age of the Lower Magdalena Valley basin basement, northern Colombia: new reflection-seismic and U-Pb-Hf insights into the termination of the central andes against the Caribbean basin. J S Am Earth Sci 74:1–26. https://doi.org/10.1016/j.jsames.2017.01.001

Moyen JF, Martin H, Jayananda M, Auvray B (2003) Late Archaean granites: a typology based on the Dharwar Craton (India). In: Precambrian Research, pp 103–123

Muessig KW (1984) Structure and Cenozoic tectonics of the Falcón Basin, Venezuela and adjacent areas. In: Bonini WE, Hargraves RB, Shagam R (eds) The Caribbean-South American plate boundary and regional tectonics, vol 162. Geological Society of America, Boulder

Müller RD, Royer JY, Cande SC, Roest WR, Maschenkov S (1999) New constraints on the Late Cretaceous/Tertiary plate tectonic evolution of the Caribbean, Caribbean basins. Elsevier, Amsterdam, pp 33–59

Nerlich R, Clark SR, Bunge HP (2014) Reconstructing the link between the Galapagos hotspot and the Caribbean plateau. Geo Res J 1–2:1–7. https://doi.org/10.1016/j.grj.2014.02.001

O'Connor J (1965) A classification for quartz-rich igneous rocks based on feldspar ratios. US Geological Survey

Oldow JS, Bally AW, Ave Lallemant HG (1990) Transpression, orogenic float, and lithospheric balance. Geology 18:991–994

Ostos M, Yoris F, Lallemant HGA (2005) Overview of the southeast Caribbean–South American plate boundary zone. Geol Soc Am Spec Pap 394:53–89

Pardo-Casas F, Molnar P (1987) Relative motion of the Nazca (Farallon) and South American Plates since Late Cretaceous time. Tectonics 6:233–248. https://doi.org/10.1029/TC006i003p00233

Parnaud F, Gou Y, Pascual J, Capello MA, Truskowski I, Passalacqua H (1995) Stratigraphic synthesis of western Venezuela. Pet Basins South Am AAPG Mem no 62, pp 681–698

Parra M, Mora A, Sobel ER et al (2009a) Episodic orogenic front migration in the northern Andes: constraints from low-temperature thermochronology in the Eastern Cordillera, Colombia. Tectonics. https://doi.org/10.1029/2008TC002423

Parra M, Mora A, Jaramillo C et al (2009b) Orogenic wedge advance in the northern Andes: Evidence from the Oligocene-Miocene sedimentary record of the Medina Basin, Eastern Cordillera, Colombia. Bull Geol Soc Am 121:780–800. https://doi.org/10.1130/B26257.1

Pecerillo R, Taylor S (1976) Geochemistry of Eocene calk-alkaline volcanic rocks from the kastamonu area, Northem Turkey. Contrib Mineral Petrol 58:63–81

Pindell JL (1993) Regional synopsis of Gulf of Mexico and Caribbean Evolution. Trans. GCSSEPM 13th Annu. Res. Conf, pp 251–274

Pindell J, Dewey JF (1982) Permo-triassic reconstruction of western Pangea and the evolution of the Gulf of Mexico/Caribbean region. Tectonics 1(2):179–211

Pindell J, Kennan L (2007) Cenozoic kinematics and dynamics of oblique collision between two convergent plate margins: the Caribbean-South America collision in Eastern Venezuela, Trinidad and Barbados, Transactions of GCSSEPM, pp 458–553

Pindell J, Kennan L (2009) Tectonic evolution of the Gulf of Mexico, Caribbean and northern South America in the mantle reference frame: an update. Geol Soc London 328:1–55. https://doi.org/10.1144/SP328.1

Pindell JL, Cande SC, Pitman WC et al (1988) A plate-kinematic framework for models of Caribbean evolution. Tectonophysics 155:121–138. https://doi.org/10.1016/0040-1951(88)90262-4

Pindell J, Kennan L, Maresch WV, Draper G (2005) Plate-kinematics and crustal dynamics of circum-Caribbean arc-continent interactions: tectonic controls on basin development in Proto-Caribbean margins. Geol Soc Am Spec Pap 394:7–52. https://doi.org/10.1130/2005.2394(01)

Priem HNA, Beets DJ, Verdurmen ET (1986) Precambrian rocks in an early Tertiary conglomerate on Bonaire, Netherlands Antilles (southern Caribbean borderland): evidence for a 300 km eastward displacement relative to the South American mainland. Geol Mijnb 65(1):35–40

Prouteau G, Scaillet B, Pichavant M, Maury R (2001) Evidence for mantle metasomatism by hydrous silicic melts derived from subducted oceanic crust. Nature 410:197–200

Quintero JD (2012) Interpretación sísmica de volcanes de lodo en la zona occidental del abanico del abanico del delta del río Magdalena, Caribe Colombiano, Tesis (pregrado), Universidad Eafit, 64 p

Ratschbacher L, Franz L, Min M et al (2009) The North American-Caribbean Plate boundary in Mexico-Guatemala-Honduras. Geol Soc London, Spec Publ 328:219–293. https://doi.org/10.1144/SP328.11

Reiners PW, Shuster DL (2009) Thermochronology and landscape evolution. Phys Today 62:31–36. https://doi.org/10.1063/1.3226750

Reiners PW, Ehlers TA, Zeitler PK (2005) Past, present, and future of thermochronology. Rev Mineral Geochem 58:1–18. https://doi.org/10.1016/j.jmig.2010.03.005

Restrepo-Moreno SA, Foster DA, Stockli DF, Parra-Sánchez LN (2009) Long-term erosion and exhumation of the "Altiplano Antioqueño", Northern Andes (Colombia) from apatite (U-Th)/He thermochronology. Earth Planet Sci Lett 278:1–2

Rodríguez C, Sellers D, Dungan M et al (2007) Adakitic dacites formed by intracrustal crystal fractionation of water-rich parent magmas at Nevado de Longav?? Volcano (36.2??S; Andean Southern Volcanic Zone, Central Chile). J Petrol 48:2033–2061. https://doi.org/10.1093/petrology/egm049

Salazar CA, Bustamante C, Archanjo CJ (2016) Magnetic fabric (AMS, AAR) of the Santa Marta batholith (northern Colombia) and the shear deformation along the Caribbean Plate margin. J S Am Earth Sci 70:55–68. https://doi.org/10.1016/j.jsames.2016.04.011

Sanchez J, Mann P (2015) Integrated structural and basinal analysis of the Cesar–Rancheria Basin, Colombia: implications for its tectonic history and petroleum systems. In: Bartolini C, Mann, P (Eds), Petroleum Geology and Potential of the Colombian Caribbean Margin, vol 108

Schmidt MW (1992) Amphibole composition in tonalite as a function of pressure: an experimental calibration of the Al-in-hornblende barometer. Contrib Mineral Petrol 110:304–310. https://doi.org/10.1007/BF00310745

Silver PG, Russo RM, Lithgow-Bertelloni C (1998) Coupling of South American and African Plate motion and plate deformation. Science 80(279):60–63. https://doi.org/10.1126/science.279.5347.60

Sisson VB, Avé Lallemant HG, Sorensen SS (2008) Correlation of Eocene-Oligocene Exhumation around the Caribbean. Venezuela, Dominican Republic, Honduras, and Guatemala, Geological Society of America, Abstracts, pp 292–298

Sorensen SS, Barton MD (1987) Metasomatism and partial melting in a subduction complex Catalina Schist, southern California. Geology 15(2):115–118

Spikings RA, Seward D, Winkler W, Ruiz GM (2000) Low-temperature thermochronology of the Northern Cordillera Real, Ecuador: tectonic insights from zircom and apatite fission track analysis. Tectonics 19:649–668. https://doi.org/10.1029/2000TC900010

Spikings RA, Winkler W, Seward D, Handler R (2001) Along-strike variations in the thermal and tectonic response of the continental Ecuadorian Andes to the collision with heterogeneous oceanic crust. Earth Planet Sci Lett 186:57–73. https://doi.org/10.1016/S0012-821X(01)00225-4

Spikings RA, Winkler W, Hughes RA, Handler R (2005) Thermochronology of allochthonous terranes in Ecuador: unravelling the accretionary and post-accretionary history of the Northern Andes. Tectonophysics 399:195–220. https://doi.org/10.1016/j.tecto.2004.12.023

Spikings RA, Crowhurst PV, Winkler W, Villagomez D (2010) Syn- and post-accretionary cooling history of the Ecuadorian Andes constrained by their in-situ and detrital thermochronometric record. J S Am Earth Sci 30:121–133. https://doi.org/10.1016/j.jsames.2010.04.002

Stearns C, Mauk FJ, Van der Voo R (1982) Late cretaceous-early tertiary paleomagnetism of Aruba and Bonaire (Netherlands Leeward Antilles). J Geophys Res Solid Earth 87(B2):1127–1141

Steiger RH, Jäger E (1977) Subcommission on geochronology: convention on the use of decay constants in geo- and cosmochronology. Earth Planet Sci Lett 36:359–362. https://doi.org/10.1016/0012-821X(77)90060-7

Sun S, McDonough W (1989) Chemical and isotopic systematics of oceanic basalts: implications for mantle composition and processes. Geol Soc Lond (Special Publication) 42:313–345

Tobisch OT, McNulty BA, Vernon RH (1997) Microgranitoid enclave swarms in granitic plutons, central Sierra Nevada, California. Lithos 40:321–339. https://doi.org/10.1016/S0024-4937(97)00004-2

Trenkamp R, Kellogg JN, Freymueller JT, Mora HP (2002) Wide plate margin deformation, southern Central America and northwestern South America, CASA GPS observations. J S Am Earth Sci 15:157–171. https://doi.org/10.1016/S0895-9811(02)00018-4

Tschanz CM, Jimeno A, Cruz J (1969) Geology of the Santa Marta area (Colombia). Instituto Nacional de Investigaciones Geológico Mineras, Informe 1829, p 288

Tschanz CM, Marvin RF, Cruz BJ, Menhert HH, Cebula GT (1974) Geologic Evolution of the Sierra Nevada de Santa Marta, Northeastern Colombia. Geol Soc Am Bull 85:273–284. https://doi.org/10.1130/0016-7606(1974)85<273:GEOTSN>2.0.CO;2

Vallejo C, Spikings RA, Luzieux L et al (2006) The early interaction between the Caribbean Plateau and the NW South American Plate. Terra Nov 18:264–269. https://doi.org/10.1111/j.1365-3121.2006.00688.x

Van der Hilst R, Mann P (1994) Tectonic implications of tomographic images of subducted lithosphere beneath northwestern South America. Geology 22:451–454. https://doi.org/10.1130/0091-7613(1994)022<0451:TIOTIO>2.3.CO;2

Van Der Lelij R, Spikings RA, Kerr AC et al (2010) Thermochronology and tectonics of the Leeward Antilles: evolution of the southern Caribbean Plate boundary zone. Tectonics. https://doi.org/10.1029/2009TC002654

Vence E (2008) Subsurface structure, stratigraphy, and regional tectonic controls of the Guajira Margin of Northern Colombia [Master of Science]. The University of Texas at Austin, p 128

Villagómez D, Spikings R, Mora A et al (2011) Vertical tectonics at a continental crust-oceanic plateau plate boundary zone: fission track thermochronology of the Sierra Nevada de Santa Marta, Colombia. Tectonics. https://doi.org/10.1029/2010TC002835

Villaseca C, Barbero L, Rogers G (1998) Crustal origin of Hercynian peraluminous granitic batholiths of Central Spain: petrological, geochemical and isotopic (Sr, Nd) constraints. Lithos 43(2):55–79

Weber MBI, Cardona A, Paniagua F et al (2009) The Cabo de la Vela Mafic-Ultramafic Complex, Northeastern Colombian Caribbean region: a record of multistage evolution of a Late Cretaceous intra-oceanic arc. Geol Soc London (Spec Publ) 328:549–568. https://doi.org/10.1144/SP328.22

Weber M, Cardona A, Valencia V et al (2011) Geochemistry and geochronology of the guajira eclogites, northern Colombia: evidence of a metamorphosed primitive cretaceous caribbean Island-arc. Geol Acta 9:425–443. https://doi.org/10.1344/105.000001740

Wilson M (1989) Igneous petrogénesis. A global tectonic approach. Chapman and Hall, London, p 421

Winter JD (2001) An introduction to igneous and metamorphic petrology. Prentice Hall, London, p 697

Winther KT (1996) An experimentally based model for the origin of tonalitic and trondhjemitic melts. Chem Geol 127:43–59. https://doi.org/10.1016/0009-2541(95)00087-9

Yarce J, Monsalve G, Becker TW et al (2014) Seismological observations in Northwestern South America: evidence for two subduction segments, contrasting crustal thicknesses and upper mantle flow. Tectonophysics 637:57–57. https://doi.org/10.1016/j.tecto.2014.09.006

Chapter 8
Late Cenozoic to Modern-Day Volcanism in the Northern Andes: A Geochronological, Petrographical, and Geochemical Review

M. I. Marín-Cerón, H. Leal-Mejía, M. Bernet, and J. Mesa-García

Abbreviations

ACPB	Amagá-Cauca-Patía
AFC	Assimilations by fractional crystallization
AFM	Diagram alkali-iron-magnesium
AFS	Algeciras fault system
AOC	Altered oceanic crust
AVZ	Austral Volcanic Zone
BAU	Baudó terrane
Ca	(circa) Approximately
CAM	Caribbean mountain terrane
CA-VA	Cajamarca-Valdivia terrane
CC	Central Cordillera
CET	Colombia-Ecuador trench
CG	Cañasgordas terrane
CS	Carbonate-rich sediments

M. I. Marín-Cerón (✉)
Departamento de Ciencias de la Tierra, Universidad EAFIT, Medellín, Colombia
e-mail: mmarince@eafit.edu.co

H. Leal-Mejía
Mineral Deposit Research Unit (MDRU), The University of British Columbia (UBC), Vancouver, Canada

M. Bernet
Institut des Sciences de la Terre, Université Grenoble Alpes, Grenoble, France

J. Mesa-García
Departamento de Ciencias de la Tierra, Universidad EAFIT, Medellín, Colombia

Geology Department, University of Michigan, Ann Arbor, MI, USA

CV	Cauca Valley
CVZ	Central Volcanic Zone
DA	Dagua terrane
E	East
E.g.	For example (exempli gratia)
EC	Eastern Cordillera
ENE	East northeast
Et al	And others (et alia)
Fig	Figure
Fm.	Formation
GA	Macizo de Garzón
GOR	Gorgona terrane
GPa	GigaPascal
GU-FA	Guajira-Falcón terranes
HFS	High-field strength
HFSE	High-field-strength elements
HIMU	High U/Pb mantle
HREE	Heavy rare earth elements
HS	Hemipelagic sediments
I.E	In essence (id est)
K	Cretaceous
Km	Kilometers
LCC	Lower continental crust
LIL	Large-ion lithophile
LILE	Large-ion lithophile elements
LREE	Light rare earth elements
Ma	Mega-annum
MMB	Middle Magdalena Basin
MORB	Mid-ocean-ridge basalt
MSP	Maracaibo subplate
N	Neogene
NAB	Northern Andean Block
NE	Northeast
NNE	North northeast
NVZ	Northern Volcanic Zone
NW	Northwest
OIB	Ocean island basalt
P	Paleogene
P	Pressure
PA	Panamá terrane
Pa	Pascal
PCB	Panamá-Chocó Block
Pz	Paleozoic
RA	Rear arc
REE	Rare earth element
RO	Romeral terrane

SE	Southeast
SJ	San Jacinto terrane
SL	San Lucas Block
SM	Sierra Nevada de Santa Marta
SN	Sinú terrane
SS	Subducted sediments
SVZ	South Volcanic Zone
SW	Southwest
T	Temperature
TAS	Total alkali-silica
UCC	Upper continental crust
VF	Volcanic front
W	West
WBZ	Wadati-Benioff zone
WC	Western Cordillera
WSW	West southwest

8.1 Introduction

The geodynamic setting of the Northern Andean Block (NAB) is controlled by the interaction between the Nazca, South American, and Caribbean Plates. This configuration includes the development of a subduction zone along the northwestern South American continental margin (Fig. 8.1). This margin was previously subjected to collision and accretion of terranes of oceanic affinity (e.g., Marriner and Millward 1984; McCourt et al. 1984; Aspden and McCourt 1986; Etayo-Serna et al. 1986; Aspden et al. 1987; Restrepo and Toussaint 1988; Kellogg and Vega 1995; Pindell et al. 1998; Trenkamp et al. 2002; Cediel et al. 2003; Kennan and Pindell 2009; Cediel et al. 2011).

The processes recorded through time in subduction zones represent the interaction between the fluids/melts coming from the subducted slab (including altered oceanic crust (AOC) and subducted sediments (SS)), mantle metasomatism and partial melting, interaction of mantle-derived magmas with the lower and upper crust, and the emplacement parameters (e.g., Tatsumi 2003; Tatsumi 2005).

In the NAB, during the late Paleogene to Neogene, tectonic reorganization, including the breakup of the Farallón Plate into the Nazca and Cocos plates (~26 Ma), and the onset of collision of the Panamá-Chocó Block (PCB) (23–25 Ma) resulted in a reconfiguration of subduction and arc-related magmatism in the Colombian Andes (e.g., Marriner and Millward 1984; Aspden et al. 1987; Pardo-Casas and Molnar 1987; Kellogg and Vega 1995; Cediel et al. 2003; Wilder 2003; Restrepo-Moreno et al. 2010; Farris et al. 2011; Montes et al. 2012; Leal-Mejía et al. 2018).

Magmatism recorded during this period is represented by the emplacement of holocrystalline and hypabyssal porphyritic plutons and the deposition of thick sequences of volcanic flows and pyroclastic rocks (Figs. 8.1 and 8.2) (e.g.,

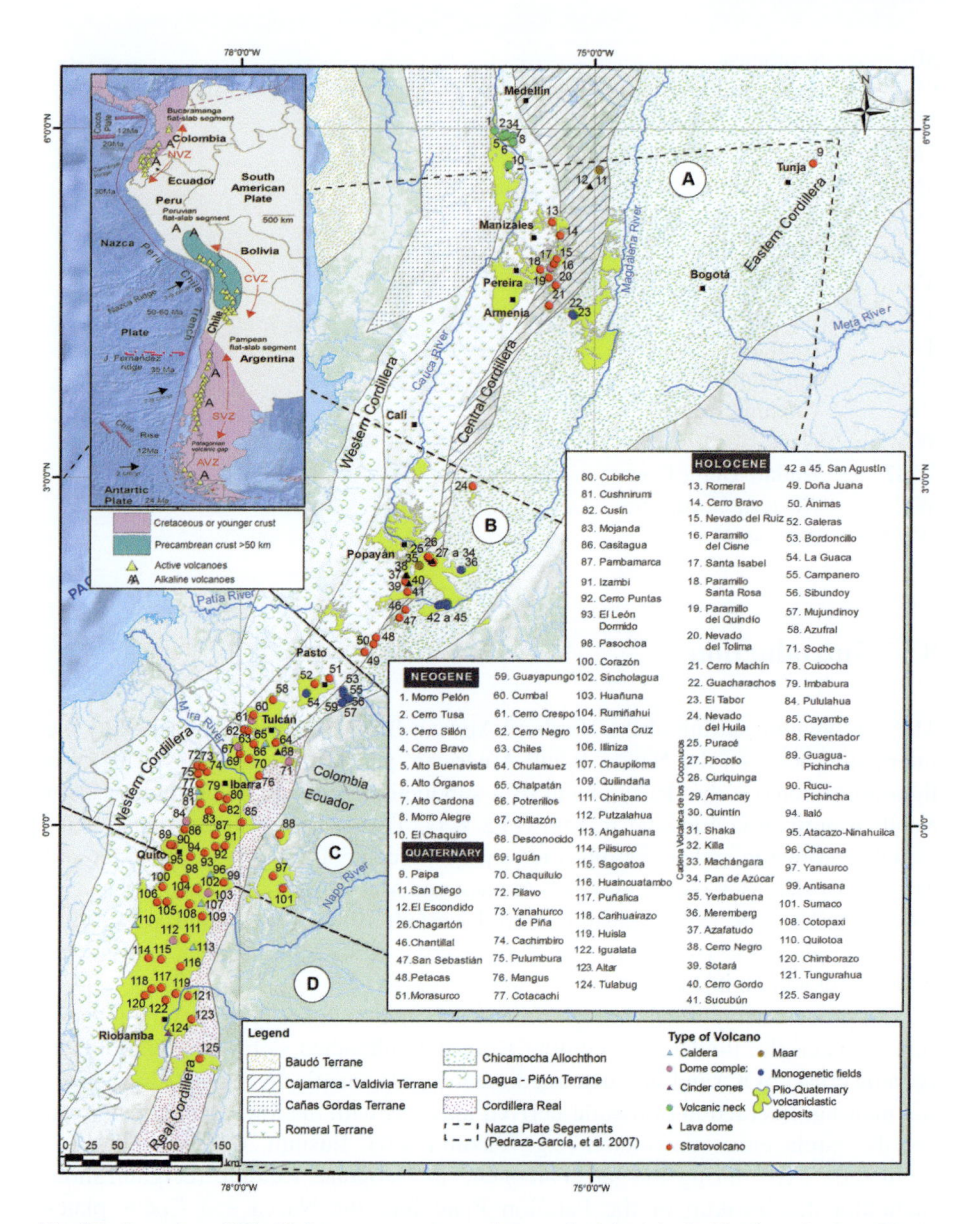

Fig. 8.1 Location of Plio-Pleistocene to recent volcanoes (red dots) in the Northern Andean Block in relation to the geometric model for subduction of the segmented Nazca Plate as presented by Pedraza-García et al. (2007). Characteristics of volcanoes within zones A, B, C, and D are discussed in text. Pre-Andean basement domains after Ramos and Aleman (2000). NVZ Northern Volcanic Zone, CVZ Central Volcanic Zone, SVZ South Volcanic Zone, AVZ Austral Volcanic Zone. Flat slab segments Bucaramanga, Peruvian, and Pampian after Thorpe and Francis (1979), Thorpe et al. (1982) in R.S. Thorpe (ed) (1982). Andesite. Orogenic Andesite and related rocks. Harmon et al. (1984). Winter (2001) An introduction to igneous and metamorphic petrology

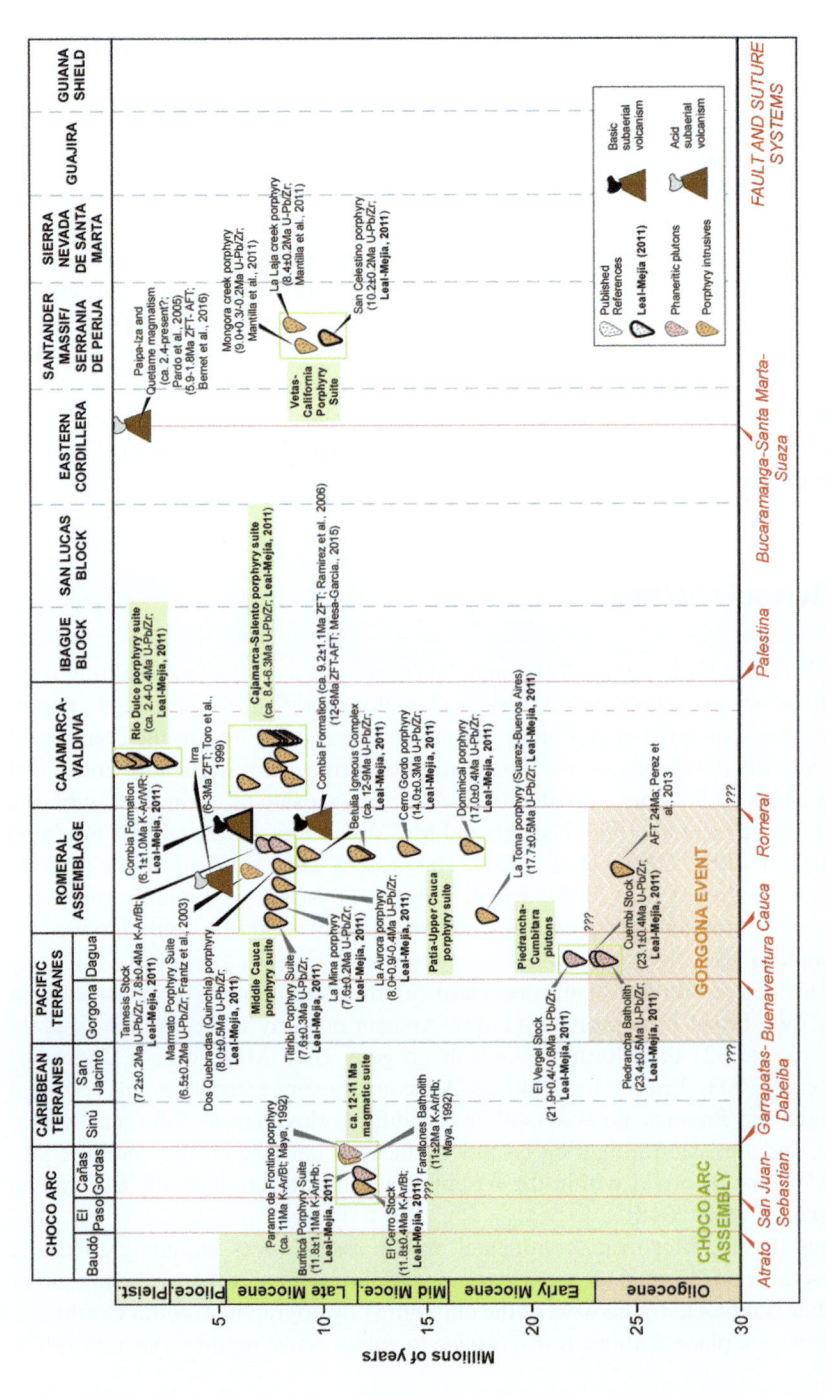

Fig. 8.2 Chronostratigraphic diagram of Late Paleogene to Neogene granitoids in the Colombian Andes (after Leal-Mejía 2011; Pérez et al. 2013; Mesa-García 2015; Bernet et al. 2016). The main lithotectonic and morphostructural units are modified from Cediel et al. (2003) and Cediel et al. (2011)

Campbell 1974; Marriner and Millward 1984; McCourt et al. 1984; Aspden et al. 1987; Calvache et al. 1997; Estrada et al. 2001; González 2001; Marín-Cerón 2007; Marín-Cerón et al. 2010; Leal-Mejía 2011; Cediel et al. 2011), primarily outcropping within the physiographic provinces of the Western Cordillera (WC), Interandean Depression or Cauca Valley (CV), Central Cordillera (CC), and Eastern Cordillera (EC). A detailed description and synthesis of late Cenozoic magmatism within the Colombian Andes is presented herein, based upon previous work and including new data from our ongoing studies. Based upon age determinations and geographic distribution, the episodes of magmatism considered herein include (1) emplacement of hypabyssal porphyritic plutons along the Cauca and Patía valleys (CV) dating from ca. 17 to 6 Ma, (2) deposition of Combia Fm. volcanic rocks within the Amagá paleo-basin from ca. 12 to 6 Ma, (3) deposition of volcanic rocks of the Irra Fm. along the middle CV and at Paipa-Iza within the Eastern Cordillera between ca. 6 and 3 Ma, and (4) volcanism contained within the Colombian segment of the active NVZ, extending from ca. 3 Ma to present.

8.2 Tectonic Setting

The NAB (sensu Cediel et al. 2003) may be considered a microplate resulting from interactions between the South American, Farallon, and Caribbean tectonic plates, including the sequential accretion of terranes of oceanic affinity to the continental margin and the development and reactivation of terrane-limiting suture zones and faults. Various authors have divided the NAB into a mosaic of terrane assemblages (e.g., Etayo-Serna et al. 1986; Aspden and McCourt 1986; Álvarez 1987; Restrepo and Toussaint 1988; Cediel et al. 2003). The present work is based on the tectonic analysis of Cediel et al. (2003), which uses geophysical, geological, lithostratigraphic, structural, and geochemical studies to delimit, describe, and classify suspect terranes.

The tectonic evolution of northwestern South America may be considered in terms of two broad time periods: (1) pre-Andean orogeny (Precambrian to early Mesozoic) and (2) the Northern Andean orogeny (late Mesozoic to Cenozoic) (Cediel et al. 2003). Prior to the Northern Andean Orogeny, tectonic events recorded in the mid-late Proterozoic (Grenvillian), middle Ordovician to Silurian, Permo-Triassic, and Jurassic through early-mid Cretaceous underscore the complex pre-Andean history recorded within the basement terranes throughout the Colombian Andes (e.g., Cediel 2018).

During the late Mesozoic, continental margin assemblages including epiclastic basinal sediments, oceanic arc terranes, and ophiolite were accreted to the margin of the continental block, to the west of the emerging physiographic Central Cordillera. Tectonism took place within a transpressive to compressive regime (Álvarez 1987), characteristic of the Northern Andean Orogeny in Colombia. Subduction-related magmatism within the continental block during this time period (e.g., Antioquian Batholith and satellite plutons) was related to the northward and eastward migration of allochthonous blocks contained within the Farallon Plate. Additional intra-oceanic and continental margin granitoids (e.g., Sabanalarga-Santa Fé Batholith, Heliconia Diorite, the Pueblito Diorite, the Pueblito Gabbro, the Altamira Gabbro, the Cambumbia Stock, Buga Batholith, Jejenes Stock) formed part of the accreted assemblage. Early Northern Andean orogeny tectonism culminated with collision and accretion of the Gorgona Terrane to the Colombian margin beginning in the Eocene (Kerr et al. 1998).

Rifting of the Farallón Plate, into Nazca and Cocos plates, at ~26 Ma (Lonsdale 2005) had a major impact on the regional scale geodynamic setting of northwestern South America during the Neogene. Plate reorientation and changes in convergence direction along the Colombian Pacific margin are observed (e.g., Aspden et al. 1987; Pardo-Casas and Molnar 1987; Cediel et al. 2003; Lonsdale 2005). The onset of the late Northern Andean orogeny, brought on by the compressive/transpressive collision between the PCB and northwestern South America at 23–25 Ma (e.g., Farris et al. 2011), was accompanied by reconfiguration of subduction along the Pacific margin and an increase in the rate of uplift and exhumation throughout much of the Colombian Cordilleran system (e.g., Restrepo-Moreno et al. 2010; Farris et al. 2011; Montes et al. 2012). Reinitiation of magmatism within the continental domain (e.g., Toussaint and Restrepo 1982; Marriner and Millward 1984; Aspden et al. 1987; Cediel et al. 2003) accompanied the Neogene tectonic reconfiguration of the region and led to the conformation of the Colombian segment of the NVZ (Stern 2004). Leal-Mejía et al. (2018) present a detailed tectonic analysis of subduction-related magmatism during this period. Additionally, within the context of Cenozoic tectonic evolution, dextral transpression-transtension along the Romeral and Cauca fault systems led to the generation of a system of pull-apart basins called referred to as Amagá-Cauca-Patía (ACPB) (Sierra 1994; Sierra and Marín-Cerón 2011), located along the Cauca-Patia Valley between the physiographic Western and Central Cordilleras (Fig. 8.3c, d).

Present-day geodynamic processes are dominated by interaction of the Nazca and Caribbean Plates with the South American plate (e.g., Mann and Burke 1984; Thorpe 1984; Pardo-Casas and Molnar 1987; Duque-Caro 1990; Mann and Corrigan 1990; Russo et al. 1992; van der Hilst and Mann 1994; Cediel et al. 1997; Trenkamp et al. 2002; Cediel et al. 2003; Kennan and Pindell 2009). Recent volcanism is related to the geometry and evolution of the Wadati-Benioff zone (WBZ). The latest geophysical studies (Pedraza-García et al. 2007) indicate that the WBZ changes dip and maximum depth along strike, permitting the definition of four Nazca Plate segments, referred to herein, from north to south, as arc segments A to D (Fig. 8.4).

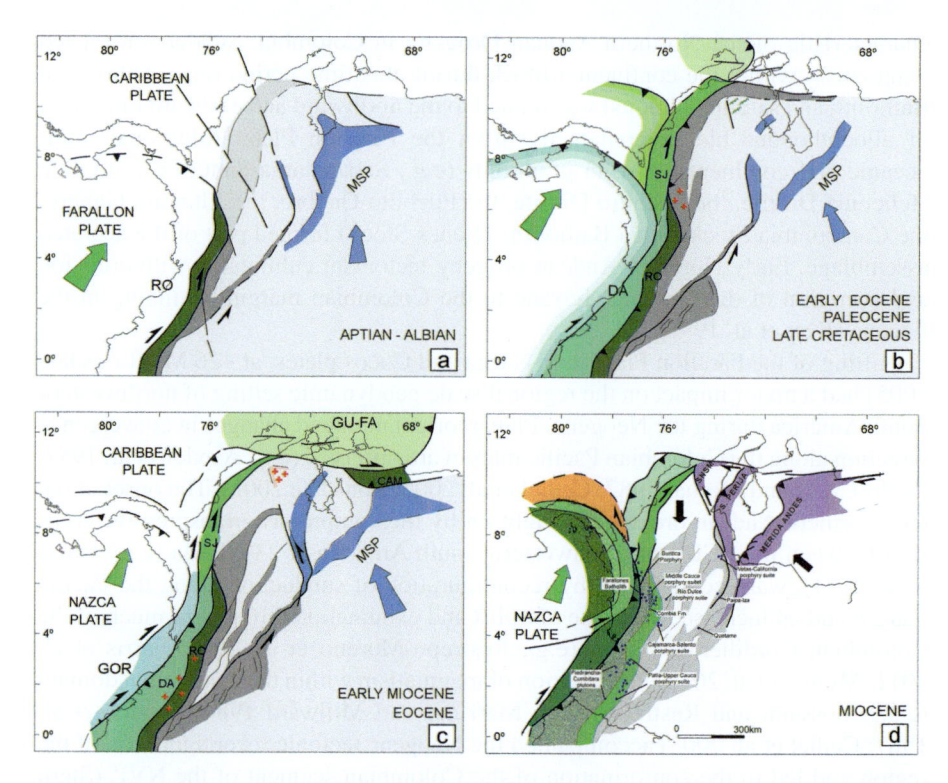

Fig. 8.3 Tectonic evolution of northwestern South America during four important periods in the geodynamic evolution of Colombia through geological time (After Cediel et al. 2003). No modifications are made to this Fig. RO Romeral terrane, MSP Maracaibo subplate realm, DA Dagua terrane, GU-FA Guajira-Falcón terranes, GOR Gorgona terrane, BAU Baudó terrane, PA Panamá terrane, CG Cañasgordas terrane, GA Macizo de Garzón, EC Eastern Cordillera, SN Sinú terrane, SJ San Jacinto terrane, SM Sierra Nevada de Santa Marta, red crosses represent magmatism

8.3 Previous Work

Based upon the composite radiometric age date database available at the time, Aspden et al. (1987) defined five major magmatic episodes in the Colombian Andes, including during the Triassic, Jurassic, Cretaceous, Paleogene, and Neogene. In general, the magmatic episodes were correlated with periods of subduction of Farallón Plate beneath continental South America (e.g., Marriner and Millward 1984; McCourt et al. 1984; Cediel et al. 2003; Saenz 2003; Restrepo-Moreno et al. 2009; Rodríguez et al. 2012).

Detailed time-space analysis of published U-Pb (zircon) age data for Phanerozoic granitoids, supported by lithogeochemical and isotope data, Leal-Mejía et al. (2018) identified six principle periods of Phanerozoic granitoid magmatism in the Colombian Andes, including early Paleozoic (ca. 485–439 Ma), Carboniferous (ca.

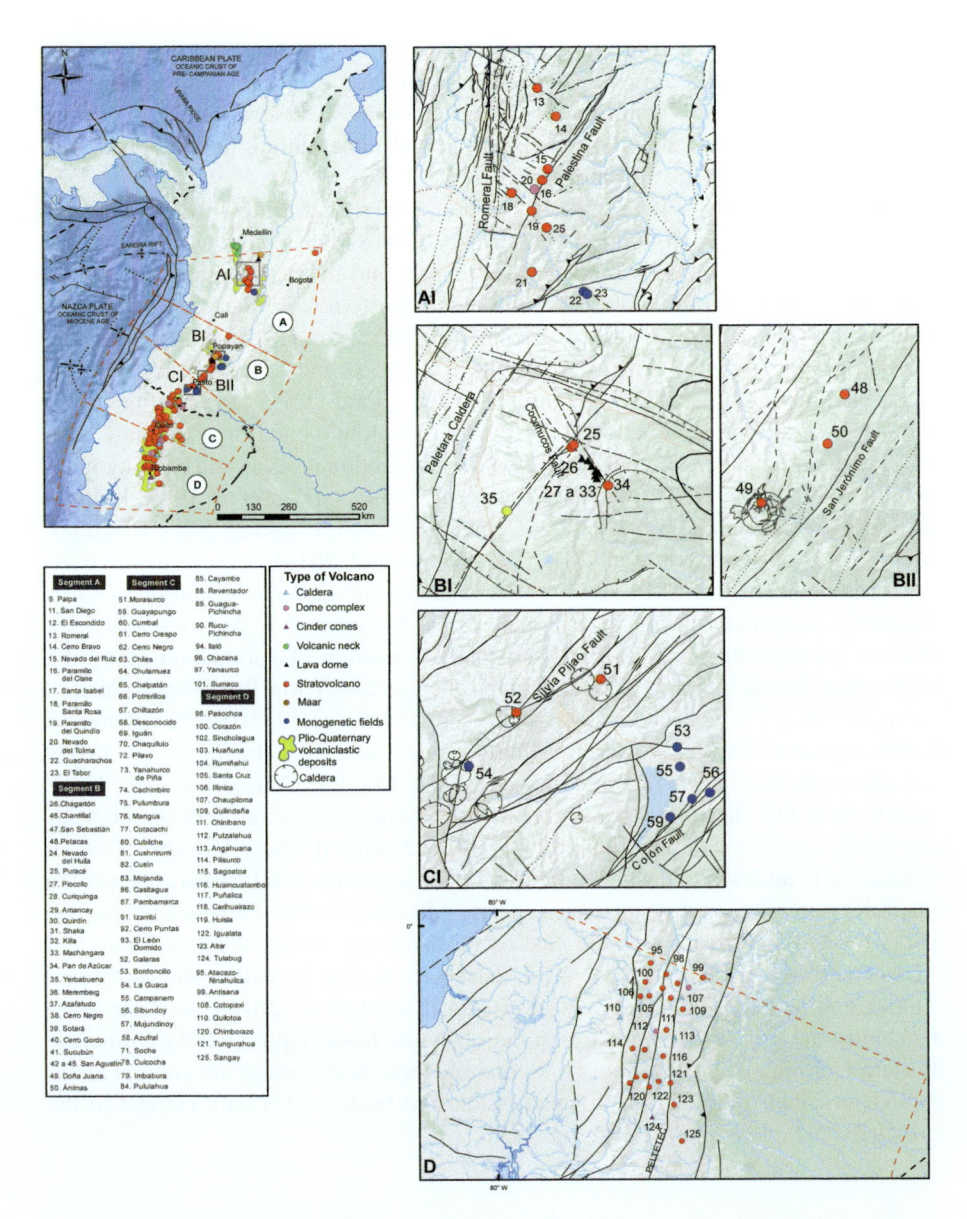

8 Late Cenozoic to Modern-Day Volcanism in the Northern Andes... 611

Fig. 8.4 Location of Plio-Pleistocene to recent volcanoes (red triangles) in segments A, B, C, and D. (Volcano location and structural data compiled after Gómez et al. 2007; Torres-Hernández 2010; Hall et al. 2008; Mejía et al. 2012; Bohórquez et al. 2005; Pardo et al. 2005)

333–310 Ma), Permo-Triassic (ca. 289–225 Ma), latest Triassic-Jurassic (ca. 210–146 Ma), late Cretaceous to Eocene (ca. 100–42 Ma), and latest Oligocene to Mio-Pliocene (ca. 23–1.2 Ma), all emplaced prior to conformation of the present-day Northern Andean Volcanic Zone. Leal-Mejía et al. (2018) highlighted sub-periods within each magmatic period. Importantly, these authors demonstrate that not all of the recorded granitoids are associated with subduction of oceanic lithosphere and suggest that granitoids resulting from processes as diverse as oceanic rifting, regional taphrogenesis, mantle upwelling, crustal underplating and anatexis, and tectonic delamination are all represented.

During the late Paleogene to Neogene (i.e., after ca. 26 Ma), rifting within the Farallon Plate and reconfiguration of the Nazca-Cocos Plate led to the reinitiation of subduction beneath the Colombia Pacific margin and the development of a magmatic arc within the physiographic Central and Western Cordilleras and along the intervening Cauca Valley. Related volcanism formed part of the lithostratigraphic record of the Amagá-Cauca-Patía Basin (Sierra and Marín-Cerón 2011). Reorientation of slab convergence directions (e.g., Pardo-Casas and Molnar 1987) and the onset of collision between northwestern Colombia and the Panamá-Chocó Block beginning at ca. 23–25 Ma affected the subduction process and ensuing magmatic activity in the NAB (e.g., McCourt et al. 1984; Aspden et al. 1987; Kellogg and Vega 1995; Trenkamp et al. 2002; Cediel et al. 2003; Lonsdale 2005; Restrepo-Moreno et al. 2010; Farris et al. 2011; Leal-Mejía et al. 2018).

Based upon existing U-Pb (zircon) age data and the spatial distribution of magmatism in Colombia during the Neogene, Leal-Mejía et al. (2018) identified six distinct arc segments (Fig. 8.5) related to the evolution of subduction along the Colombian Pacific margin. These included the ca. 23–21 Ma Piedrancha-Cuembí segment, the ca. 17–9 Ma upper Cauca-Patía segment, the ca. 12–10 Ma Farallones-El Cerro segment, the ca. 9–4 Ma Middle Cauca segment, the ca. 8.2–7.6 Ma Cajamarca-Salento segment, and the ca. 2–0.4 Ma Rio Dulce segment. These authors note that these last two segments are essentially coaxial with the present-day NNE-trending, calc-alkaline arc axis of the Northern Andes Volcanic Zone in Colombia. In addition to the abovementioned arc segments, Leal-Mejía et al. (2018) identified three isolated occurrences of late Miocene to Plio-Pleistocene igneous rocks located significantly to the east of the Northern Andes Volcanic Zone and Vetas-California (Mantilla et al. 2011; Bissig et al. 2012), within the Santander Massif and at Paipa-Iza (Pardo et al. 2005); Bernet et al. 2016), and intrusives located to the east of Bogotá City, in the Quetame town (Ujueta 1991) within the Eastern Cordillera. These authors note that the three occurrences are coaxial along a NNE-oriented trend and are located within the back-arc (?) with respect to the principle calc-alkaline arc axis to the west.

8.4 Present Study

Our study compares and contrasts petrographic and lithogeochemical data for mid-Miocene through recent volcanic and hypabyssal porphyritic rocks from various localities in the Colombian Andes. In addition, we present information pertaining to the geometry and segmented nature of the present-day Colombian volcanic arc, and, supported by Pb isotope data, contrast the lithogeochemical composition of modern

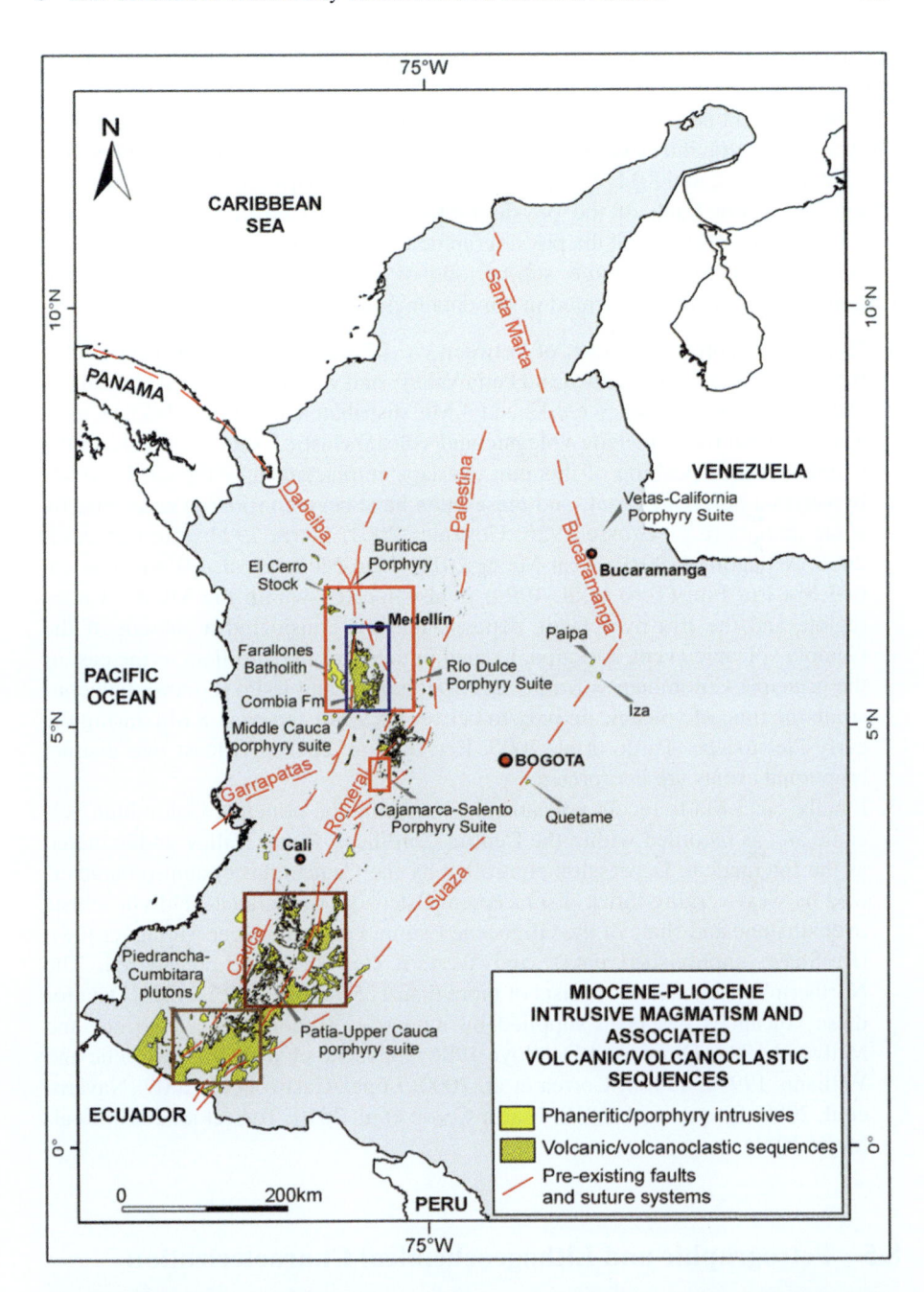

Fig. 8.5 Distribution of mid-Miocene to Pliocene magmatic rocks in the Colombian Andes including the 17–6 Ma Upper Patía-Cauca hypabyssal porphyry suite (dark red square), the 12–6 Ma Middle Cauca and Cajamarca-Salento hypabyssal porphyry suites (blue square), and the 12–3 Ma Combia and Irra Formations (after Sierra et al. 1995; Leal-Mejía 2011, orange square)

volcanism in Colombia with that of other locations in the Central and Southern Andes.

Compiled and new lithogeochemical data for mid-Miocene through recent hypabyssal porphyritic intrusive and volcanic rocks in Colombia are shown at Figs. 8.6, 8.7, 8.8, 8.9, 8.10, and 8.11. The principle geographic distribution of these rocks is along the eastern flank of the physiographic Western Cordillera and along the Interandean Depression and the physiographic Central and Eastern Cordilleras.

With respect to age, three sub-periods of Neogene magmatism from the Colombian Andes are represented in our data including:

1. Hypabyssal porphyry bodies of between ca. 17 and 6 Ma. The distribution of these rocks is along the Cauca and Patía Valleys and within the Central Cordillera.
2. Volcanic rocks of between ca. 12 and 4 Ma, distributed along the Middle Cauca Valley. These rocks include volcanic and volcanoclastic rocks belonging to the Combia Fm. Deposition of this unit overlaps emplacement of the ca. 17–6 Ma hypabyssal porphyry event, and these units have been considered cogenetic by some authors (e.g., Grosse 1926; González 2001; Sierra 1994; Ramírez et al. 2006; Villagómez 2010; Leal-Mejía 2011; Leal-Mejía et al. 2018). The ca. 6–3 Ma Irra Fm. (Toro et al. 1999) is also located within the Middle Cauca region, and the Irra pyroclastic deposits may be considered a subunit of the Combia volcanic event. At Paipa, located in the Eastern Cordillera to the east of the principle Colombian volcanic arc, fission track and U-Pb (zircon) data constrain the time of volcanic activity to between 5.9 Ma and ca. 1.8 Ma during the early Pleistocene (Pardo et al. 2005; Bernet et al. 2016). At least two distinct eruptional events are interpreted.
3. Finally, ca. 3 Ma to recent volcanism, comprising the principle Colombian volcanic arc, as recorded within the Central Cordillera, Cauca Valley understudied as the Interandean Depression controlled by the Cauca fault system (characterized by west-verging thrust displacement Paleozoic Cajamarca-Valdivia schists over siltstone and shale of the Oligocene Esmita Formation near Almaguer town (R. Shaw, unpublished data)) and Western Cordillera are considered. The Northern Volcanic Zone consists of more than 125 volcanoes (Fig. 8.1). Data for these volcanoes has been supplied by numerous authors (e.g., Marriner and Millward 1984; Droux and Delaloye 1996; Calvache et al. 1997; Calvache and Williams 1997a, 1997b; Correa et al. 2000; López-Castro et al. 2009; Navarro et al. 2009; Duque et al. 2010; Marín-Cerón et al. 2010; Toro et al. 2010; Leal-Mejía 2011).

8.5 Petrographic and Lithogeochemical Characterization

A summary of the petrographic characteristics of our dataset for Neogene volcanic and hypabyssal porphyritic rocks (Fig. 8.5) is presented in Table 8.1. This table includes characteristic mineral assemblages and the classification for the various

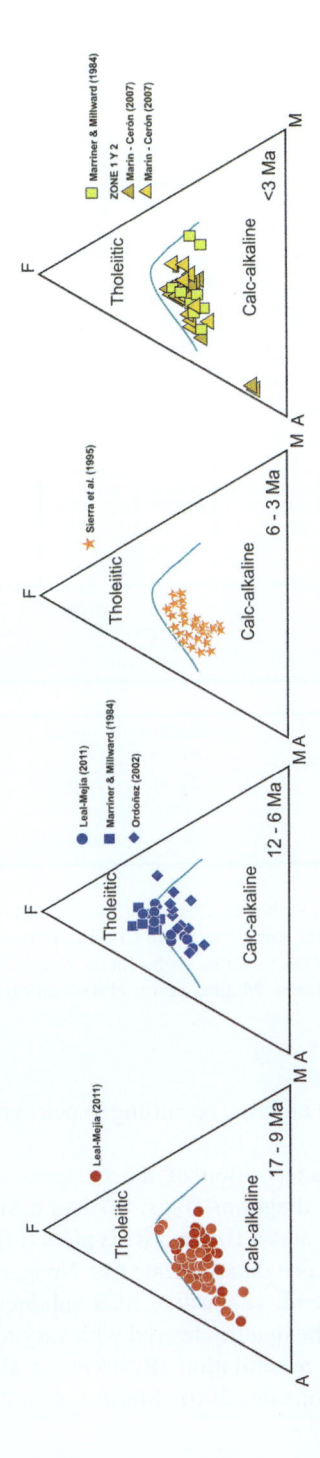

Fig. 8.6 AFM diagram plot (after Irvine and Baragar 1971) for Neogene igneous rocks from Colombian Andes; compiled after Leal-Mejía (2011), Marriner and Millward (1984), Ordoñez (2002), Sierra et al. (1995), and Marín-Cerón et al. (2010)

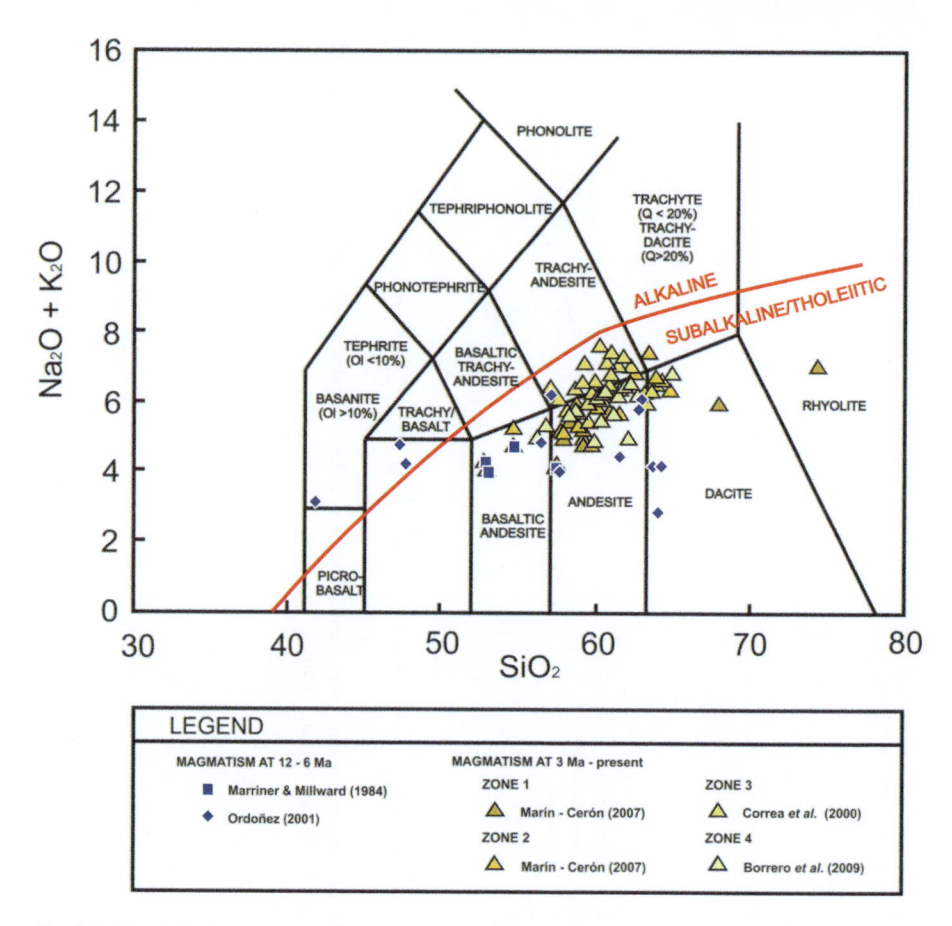

Fig. 8.7 Total alkalis vs. silica (TAS) with rock classification diagram (after LeMaitre et al. 1989) for samples of volcanic rocks from the ca. 12 to 6 Ma Combia Formation, and active volcanos, including the Chiles, Cumbales, Azufral, Galeras, Doña Juana–Sotará, Puracé-Coconucos, Huila, Machín-Ruiz-Tolima volcanic complexes. Magma series classification (red line) after Irvine and Baragar (1971)

suites as given by the respective authors, according to petrographic and/or lithogeochemical analyses.

The lithogeochemical characterization of the Neogene suites is presented in AFM diagrams (Fig. 8.6), TAS diagrams (Figs. 8.7 and 8.8), trace and minor element spider diagrams (Figs. 8.9 and 8.10), and REE plots (Fig. 8.11); all references are cited in the figure captions. The data indicate that Neogene magmatism is characterized by a calc-alkaline trend, suggestive of a subduction-related signature. Magma source is considered to be mantle-derived with varying degrees of continental crust contamination and/or assimilation (Restrepo et al. 1981; Marriner and Millward 1984; Sierra 1994; González 2001; Marín-Cerón 2007; Toro et al. 2010;

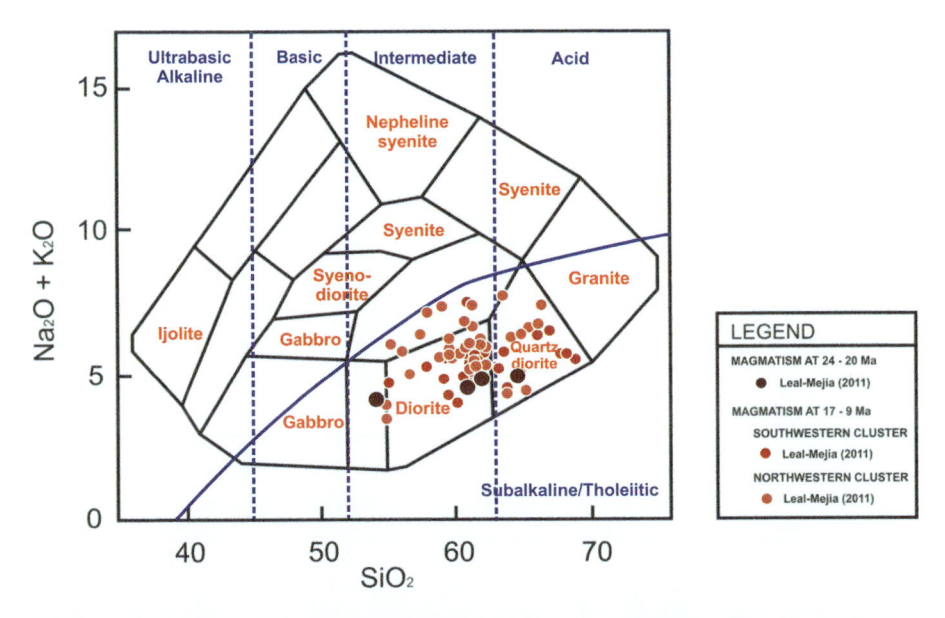

Fig. 8.8 Total alkalis versus silica (TAS) diagram (Cox et al. 1979) for samples of ca. 17–6 Ma hypabyssal porphyritic intrusions of the Cauca-Patía, Middle Cauca, and Cajamarca-Salento regions as compiled after Leal–Mejía (2011)

Leal-Mejía 2011). Hydrothermal alteration may also have played a role in major and trace elemental variations (Leal-Mejía et al. 2018).

In general, the sample suites for Neogene magmatism considered herein plot within the medium to high-K fields on the TAS diagram. They are enriched in high-field-strength (HFS) elements and light rare earth elements (LREE) and depleted in large-ion lithophile (LIL) elements and heavy rare earth elements (HREE) (Figs. 8.9, 8.10, and 8.11). The main geochemical characteristics of each magmatic event are presented below.

8.5.1 Ca. 17–6 Ma Porphyry Intrusions

The porphyry intrusions magmatism (17–6 Ma), represented by hypabyssal porphyritic bodies, is the most extensive intrusive episode that took place during the Neogene. Leal-Mejía (2011) and Leal-Mejía et al. (2018) consider these porphyry bodies in terms of three distinct arc segments (belts or trends), including the ca. 17–9 Ma Cauca-Patía segment in SW Colombia, the ca. 8–6 Ma Middle Cauca segment in west-central Colombia, and the ca. 8.2–7.6 Ma Cajamarca-Salento cluster, located to the east in the Central Cordillera. The Middle Cauca belt is spatially, and at least in part temporally, associated with Combia Fm. volcanism, which is thought to extend from ca. 12 to 6 Ma. No time-equivalent volcanic units have been identified for the Cauca-Patía and Cajamarca-Salento trends.

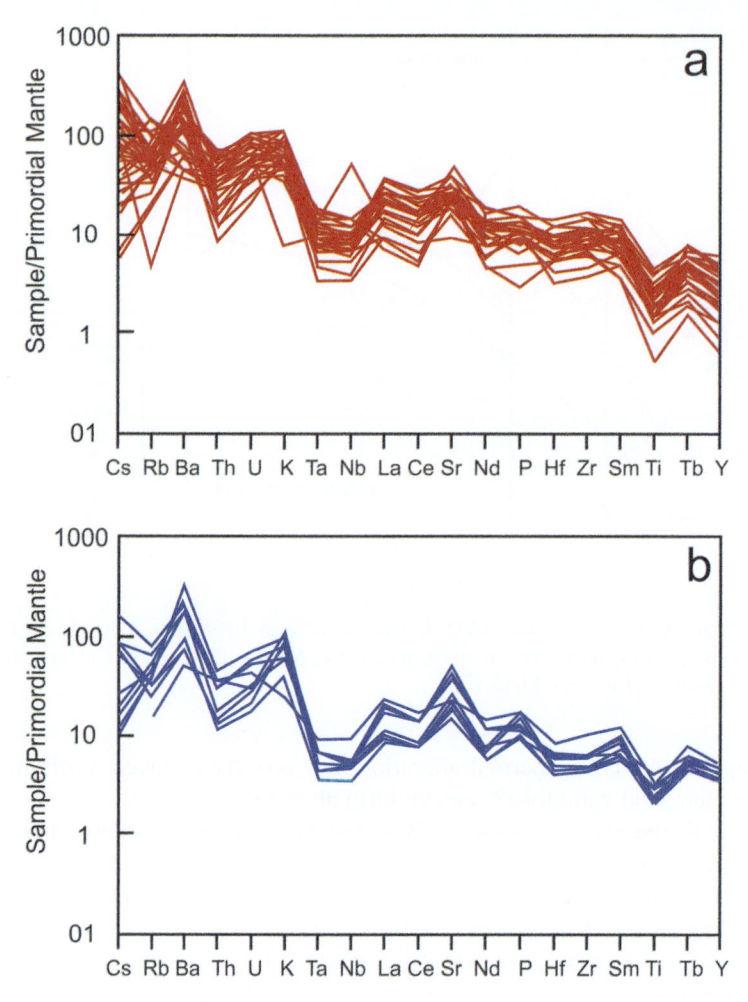

Fig. 8.9 Trace element spider diagrams normalized to primordial mantle (after Wood et al. 1979) for (**a**) 17–6 Ma hypabyssal porphyritic intrusions from the Cauca-Patía, Middle Cauca, and Cajamarca-Salento regions and (**b**) 12–6 Ma Combia Fm. volcanic rocks. (Data compiled after Leal–Mejía 2011)

The SW- to NE-trending Cauca-Patía belt, located along the eastern flank of the Western Cordillera and the western flank of the Central Cordillera (Fig. 8.3a), includes the Betulia igneous complex and the Dominical and Cerro Gordo hypabyssal intrusive complexes, among numerous additional isolated stocks and plugs. In general, the belt is comprised of small diorite to granodiorite stocks belonging to the calc-alkaline series (Figs. 8.6 and 8.7).

The Middle Cauca porphyritic intrusives range from diorites to granodiorites, with an SiO_2 content between 54.71 wt% and 66.41 wt%. Plagioclase and quartz are the main phenocryst phases in the rocks (Leal-Mejía 2011). These hypabyssal

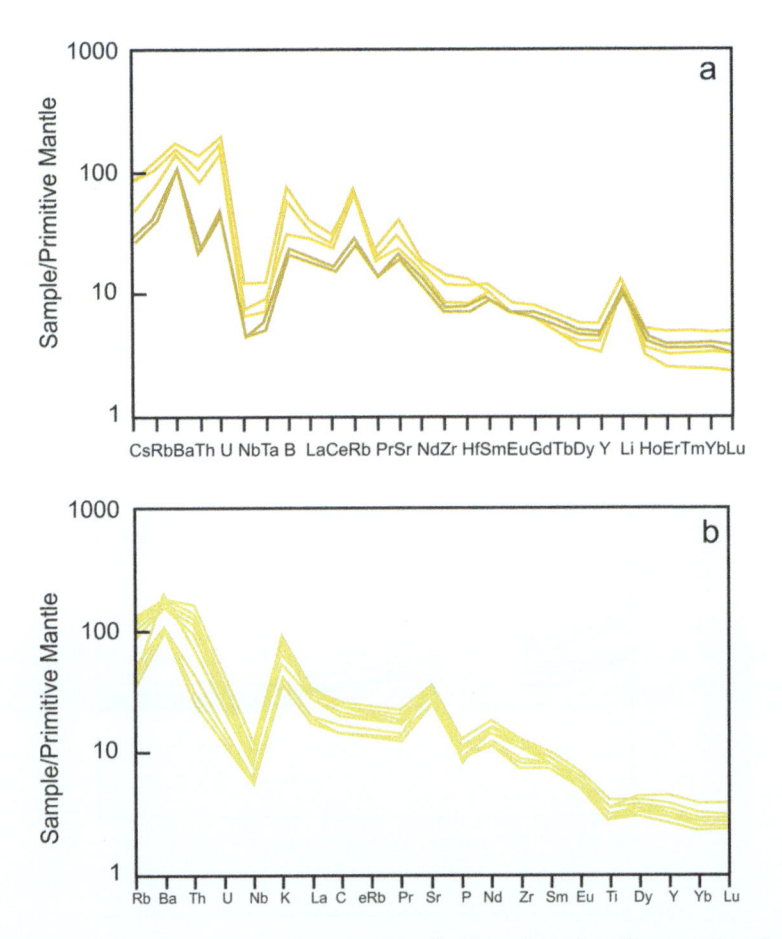

Fig. 8.10 Trace element spider diagrams normalized to Primitive Mantle (after Sun and McDonough 1989) for ca. 3 Ma to recent volcanic rocks from (**a**) B–C Segments (SW) of the present-day volcanic arc, and (**b**) Segment A (NW) of the present-day volcanic arc (see Fig. 8.1 for location of arc segments. Data from Marín-Cerón et al. 2010; Toro et al. 2010)

porphyries intrude basement assemblages of the Romeral melange and the Cañas Gordas terrane, which were accreted to the continental margin during the early phases of the Northern Andean orogeny.

To the south and west, the Cajamarca-Salento suite is comprised of porphyritic hornblende diorite and granodiorite, with 55.20–66.7 wt% SiO_2. This suite intrudes Paleozoic basement rocks of the Cajamarca-Valdivia group, to the east of the Romeral fault system (Leal-Mejía 2011).

Both the Middle Cauca and Cajamarca-Salento suites demonstrate medium-K calc-alkaline compositions. Based upon preliminary isotope data (0.70398–0.70511 for $^{87}Sr/^{86}Sr$; 0.51276–0.51294 for $^{143}Nd/^{144}Nd$; Leal-Mejía 2011), both suites are interpreted to be of mantle derivation, with low upper crustal contamination and/or assimilation.

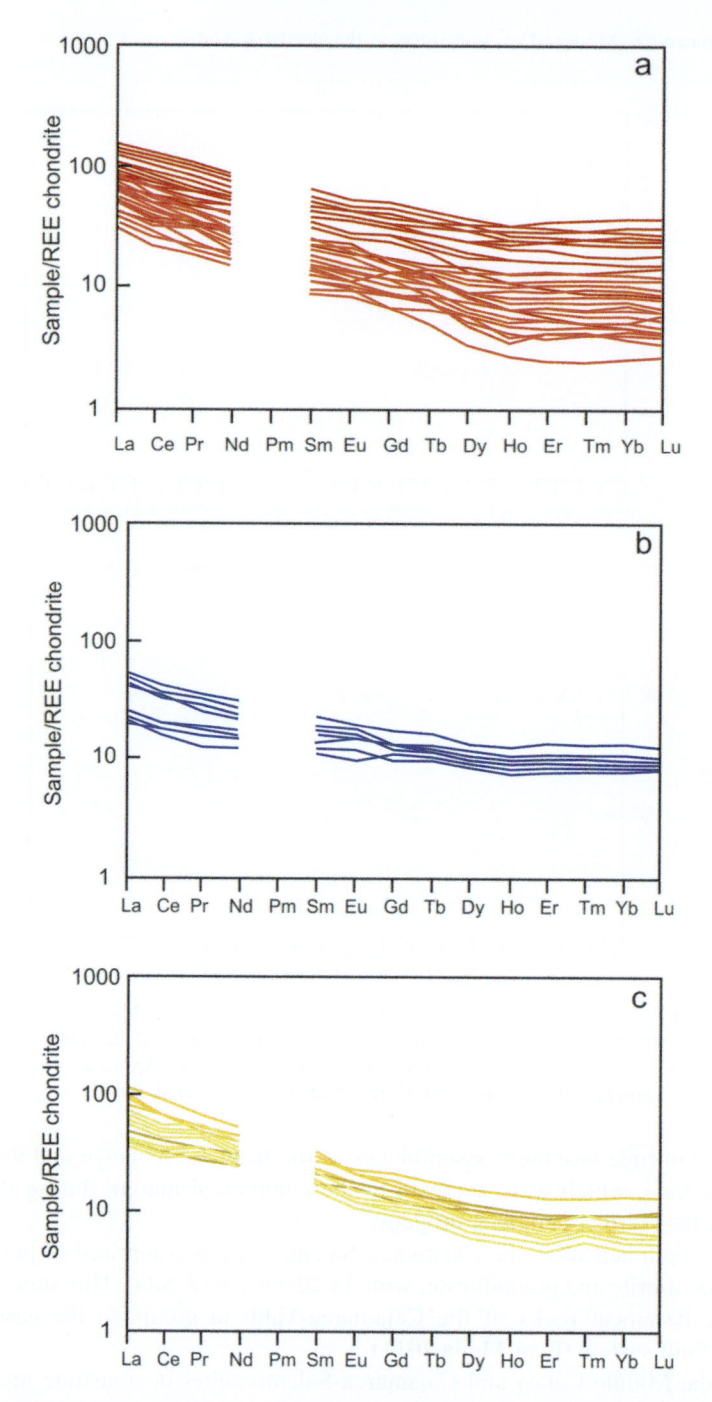

Fig. 8.11 Chondrite-normalized rare earth elements (REE) plots for (**a**) 17–6 Ma hypabyssal porphyritic intrusives from the Cauca-Patía, Middle Cauca, and Cajamarca–Salento regions, (**b**) volcanic rocks of the Combia Formation, and (**c**) 3 Ma to recent volcanic rocks of the Colombian Andes, compiled from Leal-Mejía (2011), Marín-Cerón et al. (2010) and Toro et al. (2010). Normalized after Sun and McDonough (1989)

Table 8.1 Petrographic characterization of Late Paleogene to Neogene magmatism along the CC, CG, and eastern flank of the WC. The data is presented in chronostratigraphic order with the older magmatism at the bottom and the youngest at the top

Unit or suite		Lithology	Mineral assemblage	Author
Magmatism 3 Ma to present				
Southwestern volcanic arc	Chiles, Cumbal Azufral, Galeras volcanoes Doña Juana-Sotará and Puracé-Coconucos volcanic complexes	Andesites ± rhyolites ± dacites Andesites ± rhyolites ± dacites	Pl + Cpx + Opx ± Amp ± Ms. ± Ilm ± Ap Pl + Cpx + Opx ± Amp ± Ol ± Qtz ± Op	Calvache et al. (1997), Calvache and Williams (1997a), Marín-Cerón (2007), López-Castro (2009), Duque et al. (2010)
Northwestern volcanic arc	Huila volcano Machín-Ruiz-Tolima volcanic complex	Andesites Porphyritic andesite	Pl + Qtz + Or + Px ± Ilm ± Ap ± Mag Pl + Cpx + Hb ± Opx ± Op ± Ap	Correa et al. (2000), Toro et al. (2010)
Magmatism 6–3 Ma				
Irra Formation		Andesitic tuff-tuffaceous conglomerate	Hb + Pl + Qtz ± Zrn ± Ap	Sierra (1994), Estrada et al. (2001), Toro et al. (1999)
Magmatism 12–6 Ma				
Combia Formation	Lava flows La Taparo section	Andesites-Basalts Tuff-tuffaceous breccia-agglomerates	Pl + Hb + Opx ± Ol ± Cpx Pl + Amp + Px + Qtz + Cal ± Grt ± Ms. ± Bt ± Op ± Or ± Ep ± Chl	Grosse (1926), González (2001), Rios and Sierra (2004), López and Ramírez (2006)
	Bolombolo-Peñalisa section Guineales-Peñalisa section	Ash-lapilli tuff Tuff-tuffaceous sandstone	Cpx + Pl + Hb + Bt ± Ol ± Cal ± Ap Px + Hb + Pl + Qtz ± Chl ± Zrn ± Ap	

(continued)

Table 8.1 (continued)

Unit or suite	Lithology	Mineral assemblage	Author
Magmatism 17–6 Ma			
Northern cluster (9–6 Ma)			
Cajamarca-Salento porphyry suite	Hornblende porphyritic dacites	Qtz + Kfs + Pl + Amp ± Bt ± Ep ± Ap ± Zr ± Mag ± Ilm	Leal-Mejía (2011)
	Hornblende porphyritic andesites	Qtz + Pl + Amp ± Ep ± Ap ± Mag ± Ilm	
Middle Cauca porphyry suite	Hornblende porphyritic diorite	Pl + Hb + Ep ± Ap ± Mag ± Ilm	Restrepo et al. (1981), González (2001), Leal-Mejía (2011)
	Biotite porphyritic monzonite	Pl + Amp + Bt ± Ap ± Mag	
	Pyroxene hornblende porphyritic andesite-basalts	Pl + Amp + Px ± Ep ± Ap ± Zrn ± Mag ± Ilm	
Titiribí facies	Amphibole porphyritic andesite	Pl + Kfs + Qtz + Amp ± Mag	
La Mina facies		Qtz + Pl + Kfs + Bt + Cpx ± Amp ± Ap ± Ttn ± Zrn ± Ep ± Mag ± Ilm ± Py	
La Quebradona-La Aurora porphyry suite	Hornblende biotite quartz micromontzo diorite	Qtz + Kfs + Pl + Aug + Hb ± Bt ± Ap	
Tamesis facies	Granodiorite-diorite	Qtz + Pl + Kfs + Bt ± Amp ± Ap ± Mag	
Quebrada San Pedro facies	Biotite rhyodacite		
The Oro Fino facies	Hornblende porphyritic dacite	Pl + Amp + Qtz ± Ep ± Aln ± Ap ± Zrn ± Mag ±	
Marmato facies	Hornblende biotite dacite	Pl + Qtz ± Kfs ± Bt + Amp ± Ap ± Mag ± Ep ± Chl	
Dos quebradas facies	Hornblende porphyritic andesite	Pl + Amp + Bt ± Ap ± Zrn ± Mag	
Mandeval facies	Early porphyritic diorites	Pl + Qtz + Cpx + Amp ± Zrn ± Ap ± Mag	
	Porphyritic quartz diorite	Qtz + Pl + Amp ± Ap ± Mag	

Southern cluster (17–9 Ma)				
Patia – Upper Cauca porphyry suite	La Dorada facies	Hornblende biotite tonalities	Pl + Qtz ± Bt ± Amp ± Ep ± Ap ± Zrn ± Mag	Leal-Mejía (2011)
	Garrapatas facies	Porphyritic tonalities	Pl + Qtz + Amph ± Ep ± Ap ± Py	
	San Jerónimo facies	Hornblende porphyritic diorites	Pl + Bt + Cpx + Amp ± Ap ± Zrn ± Ilm	
	Cajibro facies	Porphyritic dacite	Qtz + Kfs + Pl + Amp ± Bt ± Grt ± Ap ± Ilm	
		Hornblende biotite porphyritic granodiorite	Qtz + Pl + Or ± Bt ± Amp ± Ap ± Zrn	
		Felsitic rhyodacite	Qtz + Pl + Amp ± Bt ± Ilm	

Note: *Amp* amphibole, *Ap* apatite, *Aug* augite, *Bt* biotite, *Cal* cal, *Chl* chlorite, *Cpx* clinopyroxene, *Ep* epidote, *Grt* garnet, *Hb* hornblende, *Ilm* ilmenite, *Kfs* K-feldspar, *Mag* magnetite, *Ms* muscovite, *Ol* olivine, *Op* opaque minerals, *Opx* orthopyroxene, *Or* orthoclase, *Pl* plagioclase, *Px* pyroxene, *Py* pyrite, *Qtz* quartz, *Tnt* titanite

The magmatic event spanning 17 to 6 Ma is especially significant from a metallogenic standpoint given the fact that many of the intrusive bodies emplaced during this period of time contain Cu and/or Au mineralization (see Shaw et al. 2018). Chiaradia et al. (2004) suggest the presence of cupriferous porphyry and hydrothermal type gold deposits related to arc magmatism during the Cenozoic in Ecuador. Taking into consideration the time-space, genetic and tectonic similarities between the Ecuadorian Andes and the Colombian Andes, it is possible to find late Miocene porphyry-like deposits in Colombia. One particular case is the Marmato stock (porphyritic volcanic rocks) of andesitic to dacitic composition generated by mantelic magmas of calc-alkaline to tholeiitic character, which upwelled to the upper crust controlled by fractures and faults of the Cauca-Romeral fault system (Tassinari et al. 2008). This stock has gold mineralization in veins generated by low to intermediate sulphidation hydrothermal fluids (Tassinari et al. 2008). Recent studies have identified Au/Cu deposits in hypabyssal porphyries correlated with the Combia Formation (Uribe-Mogollón 2013).

8.5.2 Ca. 12–6 Ma Combia Fm. Volcanism

Volcanism represented by the Combia Fm., outcropping within the Amagá Basin along the Middle Cauca Valley, is thought to extend from ca. 12 to 6 Ma (e.g., Ramírez et al. 2006; Leal Mejía et al. 2011) (Fig. 8.3). This event is through to represent, at least in part, the extrusive component of the Middle Cauca trend of subvolcanic porphyry intrusives (e.g., Cerro Corcovado, El Cangrejo latibasalt, y Popala Diabase; e.g., López et al. 2006). The later phases of Combia volcanism were dominated by explosive volcanic and pyroclastic sequences, likely deposited within complex, localized, transpressional-transtentional pull-apart basins (e.g., Sierra 1994; Ramírez et al. 2006; Piedrahita et al. 2017).

Lithogeochemical data for the Combia Formation reveal tholeiitic through calc-alkaline compositions, which define an overall calc-alkaline trend, suggestive of arc magmatism with a relatively primitive character (Fig. 8.5) (Marriner and Millward 1984; Jaramillo 1976; Ordoñez 2002, Mesa-García, 2015). SiO_2 contents are relatively low (47–57%), with K_2O ranging from 0.83 to 1.82 wt.% in basaltic to bas-andesitic flow samples (Fig. 8.6). Negative Eu anomalies imply plagioclase fractionation, and the low Ba/Ta ratio (Gill 1981 in Marriner and Millward 1984; Mesa-García 2015) also suggests a magmatic arc signature (Marriner and Millward 1984; Mesa-García 2015). These data combined with Nb–Ta values, and LILE enrichment (Fig. 8.8) suggest subduction-related magma genesis for the Combia Fm. (Marriner and Millward 1984; Mesa-García 2015).

Other studies (e.g., Jaramillo 1976; Álvarez 1983; Marriner and Millward 1984; Ordoñez 2002; Leal-Mejía 2011, Mesa-García 2015) also show the volcanic rocks of the Combia Fm. have a tholeiitic through calc-alkaline composition, with a random distribution of the magma series within the Amagá Basin (Marín-Cerón et al.

unpublished data). εNd values ranging from -1 to 11 and low radiogenic Sr isotope values suggest low-crustal contamination (Leal-Mejía 2011; Ordoñez 2002). However, the Pb isotopic data (Mesa-García 2015); Leal-Mejía et al. (2018) indicate a radiogenic Pb signature similar to the recent volcanism along the Central Cordillera, as reported by Marín-Cerón et al. (2010).

8.5.3 Ca. 6–3 Ma Irra Fm. Volcanism

This late Miocene to Pliocene explosive volcanic event is represented by the Irra Formation (e.g., Restrepo and Toussaint 1990; Toro et al. 1999) (Fig. 8.3b), which is considered herein to have been deposited as a subunit of the ca. 12–6 Ma Combia Fm., prior to the development of ca. 3 Ma to recent volcanism within the active Colombian volcanic arc to the east. Early deposits (B Member) of the Irra Fm. were emplaced at ca. 6.3 ± 0.2 (FT, zircon; Toro et al. 1999). The Irra Formation has been divided into three members. Member C is the bottom unit, composed of conglomerates and conglomeratic sandstones formed in alluvial fan environments. The source area for member C is considered to be the Central Cordillera according to the volcanic and metamorphic rock fragments probably related to the Quebradagrande and Arquía Complexes and E to SE paleocurrent results (Sierra 1994; Sierra et al. 1995). Member B consists of epiclastic flood flow deposits and minor ash flows. Member A, which contains lahars, tuff, and pyroclastic rocks, overlies these units. The geochemical signature of the Irra Fm. indicates an andesitic composition belonging to the calc-alkaline series (Fig. 8.6) (Sierra 1994; Sierra et al. 1995).

Both the A and B Members of the Irra Fm. have been correlated with volcanic activity in the Central Cordillera; the Ruiz-Cerro Bravo volcanic complex has been suggested as the most possible source area (Sierra et al. 1995). Notwithstanding, based upon age constraints and geographic location, Toro et al. (1999) correlate the Irra Fm. with volcanism along the Middle Cauca Valley. Further studies are proposed to provide more accurate constraints upon the age distribution of this volcanic activity.

8.5.4 Ca. 5.9–1.8 Ma Paipa-Iza Volcanism

Paipa-Izá volcanism in the Eastern Cordillera is temporally constrained between 5.9 Ma and 1.8 Ma (late Miocene-Pleistocene) (Pardo et al. 2005; Bernet et al. 2016). The volcanic products of this spatially isolated event include pumice and ash flow tuffs, lava domes and pyroclastic block, and ash flow tuffs. Lithogeochemical analyses indicate Papa volcanism is comprised of alkaline rhyolites and trachytes and high-K calc-alkaline rhyolites. SiO_2 values range between 68% and 72%, while $Na_2O + K_2O$ contents range from 7% to 10%. Essential minerals include phenocrysts and glomerocrysts of anorthoclase-, sanidine-, and anorthoclase-mantled plagioclase. Accessory minerals include reddish (Mg-rich?) biotite and hastingsite, whereas accessory minerals include augite, zircon, sphene, and magnetite. Petrographically, the mineral assemblage demonstrates disequilibrium textures, such as dissolution embayments, corrosion and reabsorption borders, and normal, inverse, oscillating, and patchy zonation, together with fibrous borders intercalated with euhedral borders. Lithogeochemical correlation with published data for volcanic rocks at nearby Iza confirms that acid and alkaline magmas locally erupted in the Eastern Cordillera during the late Neogene and that the resulting volcanic rocks are of markedly different composition than the ca. 3 Ma to recent volcanic rocks of calc-alkaline affinity which are characteristics of the present-day Colombian segment of the Northern Volcanic Zone located within the Central Cordillera and Cauca Valley to the west. At Paipa-Iza, structural controls upon magma emplacement have been inferred. NE-striking reverse faults, which transect the area, are parallel to regional structures of the EC. Additional NW-striking normal faults are parallel to lineaments described by Ujueta (1991). The main NW-striking structures include the Cerro Plateado fault (normal) and a tensional fracture that links the Paipa volcanic complex to the Iza volcano. Together with the E-W-striking Las Peñas fault and the NE-striking Agua Tibia fault, these four structures form a caldera margin (Cepeda et al. 2004; Cepeda and Pardo 2004).

8.5.5 3 Ma to Recent Volcanism

Plio-Pleistocene to recent volcanoes forming the Northern Volcanic Zone in Colombia and in the Northern Andean Block in general, are comprised of dacite and rhyolite to basaltic andesites and andesite lavas, with a wide variety of associated pyroclastic and ignimbrite flows. In general, the volcanic products range from tholeiitic to calc-alkaline in nature and produce a clear calc-alkaline trend on the lithogeochemical diagrams (e.g., Correa et al. 2000; Droux and Delaloye 1996; Calvache and Williams 1997a; Calvache and Williams 1997b; Correa et al. 2000; Marín-Cerón 2007; Borrero et al. 2009; López-Castro 2009; Duque et al. 2010; Toro et al. 2010; Monzier et al. 1997; Barragan et al. 1998; Bourdon et al. 2003; Bryant et al. 2006) (Figs. 8.1 and 8.4). Notably, alkali-basaltic volcanism is observed in a few volcanoes located along the eastern (back-arc?) margin of the main calc-alkaline arc (e.g., La Horqueta, San Roque; Kroonenberg et al. 1982). SiO2 content for the calc-alkaline trend volcanoes ranges from ca. 50 to 70 wt% and enrichment of LILE and depletion of HFSE are observed for these samples (Fig. 8.9) (Droux and Delaloye 1996; Calvache and Williams 1997a; Marín-Cerón 2007; López-Castro 2009; Duque et al. 2010; Toro et al. 2010; Leal-Mejía 2011).

The dominance of positive anomalies of the fluid-mobile elements B, Pb, Sr, and Li and depletion of fluid-immobile elements such as Nb and Ta have been called as an indicative of a component released from the subducted slab and consistent with the general features of trace elements in slab-derived fluid related arc volcanics (e.g.,

Nakamura et al. 1985; Sakuyama and Nesbrit 1986; Ryan and Langmuir 1987). This trace element signature is interpreted to be produced by fluids released from the subducted slab. It implies that the thermal conditions beneath the SW Colombian arc did not reach the temperatures required for melting of this young (~14 Ma, Hardy 1991) and, therefore, warm subducted slab. However, taking into account that the primary signatures of those andesitic rocks may be obscured by significant lower crustal involvement, we propose a multi-isotopic approach to trace the involvement of subducted components – altered oceanic crust (AOC) – and subducted sediments: hemipelagic sediments (HS) and carbonate-rich sediments (CS). Some volcanic products have adakite-like signatures (i.e., high Sr and LREE contents, low Y and HREE contents, high Sr/Y and La/Yb, 87Sr/86Sr <0.7045, Defant and Drummond 1990), such as at Puracé and Huila Volcanic complexes (Monsalve and Arcila 2015), and may be related to mantle-derived magmatism and H2O-rich lower crustal interaction, as we will discuss later on at the last section. Petrographically, several disequilibrium features (e.g., sieve-textured plagioclase, complex zoning, coexistence of olivine with quartz) may indicate a complex crystallization process, mainly acquired at the upper crustal levels (Marín-Cerón et al. 2010).

With respect to the Volcanic Zones throughout the South American Andes, the isotopic signatures along the Northern Andean Volcanic Zone are (1) the lowest in $^{87}Sr/^{86}Sr$ ratio (0.704167–0.704530), (2) the highest in $^{143}Nd/^{144}Nd$ ratio (0.512741–0.512975) (see Fig. 8.12a), (3) the highest in $^{208}Pb/^{204}Pb$ ratio (38.68–38.89) (see Fig. 8.12b) (Droux and Delaloye 1996; Calvache et al. 1997; Marín-Cerón et al. 2010), and (4) primitive with respect to $^{176}Hf/^{177}Hf$ ratio (0.282944–0.283066). These systematic isotopic comparisons call attention to the role of different basement types of along the Andean Cordillera. In this context we can conclude that isotopic contributions from the lower crust are the norm in mature continental arcs, not only related to the thickness of the crust but also to temperature and pressure gradients, H_2O content, and the melt fraction developed at the upper-mantle - lower crust interface (Annen et al. 2006). Notwithstanding, differing subduction components together with variable thermal regimes along the subducting slab, will also contribute to the compositional variability of primary magmas (Fig. 8.10).

With respect to crustal-level structural controls on magma emplacement, there is a clear coincidence of volcanic cone and subchain location with the trace of the various paleosuture systems along which we have herein reconstructed the Northern Andean Block (Fig. 8.1). This coincidence emphasizes the importance of underlying paleo-structure in the evolving tectonics of NAB (Cediel et al. 2003) from south to north.

1. Cauca-Pujilí subchain: Illiniza-Atacazo-Pichincha-Cotacachi (Ecuador)-Chiles-Cumbal-Azufral-Olaya (Colombia)
2. Interandean subchain (located between the Cauca-Pujili and Romeral-Peltetec sutures): Chimborazo-Igualata-Sagoatoa-Cotopaxi-Mojanda (Ecuador)-Galeras-Morazurco (Colombia)

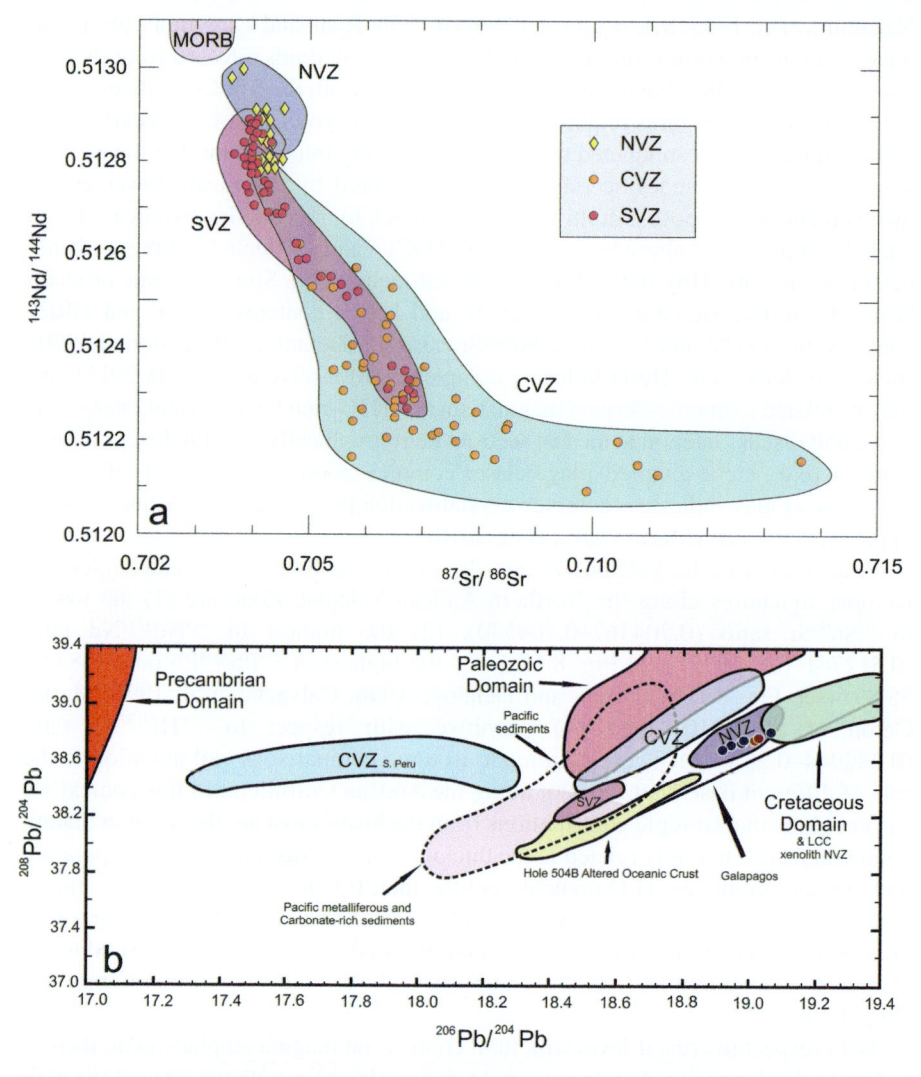

Fig. 8.12 (**a**) Sr vs. Nd isotopic ratios for the three zones of the Andes. (Data from James et al. (1976), Hawkesworth et al. (1979), James (1982), Harmon et al. (1984), Frey et al. (1984), Thorpe (1984), Hickey et al. (1986), Hildreth and Moorbath (1988), Wörner et al. (1988), Walker et al. (1991), de Silva (1991), Kay et al. (1991), Davidson and de Silva (1992). Winter (2001) An introduction to igneous and Metamorphic Petrolog. Prentice Hall; and data from this study). (**b**) Pb isotope data (207Pb/204Pb vs 206Pb/204Pb) for Northern Volcanic Zone (NVZ) volcanic rocks of the Northern Andean Block (after Marín-Cerón 2007). Unpublished data for the Combia Formation (Marin-Cerón unpublished data) are also included. Data for Central (CVZ) and Southern (SVZ) Andean Volcanic Zones, Pre-Andean basement, and Pacific oceanic volcanic rocks and sediments are shown for reference. Data set includes basement gneisses of southern Peru from Tilton and Barreiro (1980), Pacific sediments after Dasch (1981) and White et al. (1985), Paleozoic basement from Chiaradia et al. (2004), Metalliferous sediments from DSDP leg 92 from Barret et al. (1987), Cretaceous oceanic rocks after Kerr et al. (1998), Colombian lower crust xenoliths from Weber et al. (2002), ACC from Hole 504 after Pedersen and Furnes 2001)

3. Romeral-Peltetec subchain: Altar-Tungurahua-Antisanas-Cayambe-Mangus (Ecuador)-La Victoria-Chimbo-Bordoncillo-Doña Juana Sotará-Puracé (Colombia)
4. Palestina subchain: La Horqueta-Paletará-Huila-Tolima-Ruiz-Herveo (Colombia)

This final subchain is not well-defined in Ecuador. It may include Sangay, which straddles the Cordillera Real along the Llanganates fault (see "Regional Tectonics in the Northern Andes," subsection "Palestina fault system" above) in a geologic position similar to that of the Tolima-Ruiz cluster in the north-central Central Cordillera of Colombia. Suaza subchain: Sumaco-Pande Azúcar Reventador (Ecuador)-Guamuez-Acevedo (Colombia).

Based upon the analysis of earthquake hypocenter data, Pedraza-García et al. (2007) highlight changes in the dip of the Nazca Plate subducting beneath NW South America, which influences the geometry of the Benioff zone beneath the NAB. These authors define four segments with variable dips within the Nazca Plate. They refer to these segments, from north to south as segments A (Cali), B (Popayán), C (Nariño), and D (Quito) (see Figs. 8.1 and 8.4).

In addition, apparent amagmatic zones, or volcanic gaps, are observed within the Northern Volcanic Zone. The most northerly is located between 5.2°N and 7°N and could be related to an interpreted overlap zone between the Caribbean and Nazca Plates (Taboada et al. 2000). A second gap coincides within the southern A (Cali) segment between 3.5°N and 5°N, as proposed by Pedraza-García et al. (2007), where a pronounced curve appears in the isodepth contours of the subducting plate. A seismic gap at intermediate depths is also recorded in the north of Ecuador (between 0°N and 1°S), where subduction of the Carnegie Ridge is interpreted to cause changes in the dip angle in the transition zones and to result in tearing of the Nazca Plate (Gutscher et al. 1999a, b). In addition, the maximum depth of the seismicity of the subducted slab in the CET shows two regions of increasing seismicity: the first one, from 4.5°N to 5°N, and the second one, from 2°S to 1°S. Gutscher et al. (1999a, b) suggested that the presence of the Malpelo and Carnegie Ridges generates differential blockage along the Colombia-Ecuador trench.

We now describe in more detail the location, structural setting, and geochemical characteristics of Plio-Pleistocene to recent volcanoes in the NAB, within the context of the configuration of the four segments of the subducting Nazca Plate (segments A–D) as proposed by Pedraza-García et al. (2007) (Fig. 8.4). The location and name of individual volcanic complexes within each segment are shown in greater detail in Figs. 8.1 and 8.4.

8.5.5.1 Arc Segment A (Cali)

This volcanic zone contains, from south to north, the Cerro Machín, Tolima, Quindío, Santa Rosa, Santa Isabel, Cisne, Ruiz and Cerro Bravo Volcanic Complexes, San Diego, and Escondido volcanoes (Fig. 8.4), describing a very narrow volcanic arc, located above a seismic Benioff zone that clearly defines a steep subducting slab (ca. 45° dipping, Pedraza-García et al. 2007).

Structural studies coinciding with this segment (Mejía et al. 2012) indicate the region is transacted by longitudinal NE-SW and N-S structures, including the Palestina, Santa Rosa, and San Jerónimo faults. In addition, a transverse system of NW-SE- to E-W-striking faults, including Villamaría, Termales, Campoalegrito, and San Ramón, is recorded. Quaternary stress indicators indicate a WSW-ENE direction of compression and dextral strike slip deformation by simple shear along the main longitudinal faults. Left lateral displacement is transmitted to most of the NW transverse structures. A WSW-ENE direction of compression was also calculated from striated planes of Cretaceous-Paleocene and older rocks, suggesting previous right lateral displacements along longitudinal faults such as Palestina and San Jerónimo (Mejía et al. 2012). Geochemically, the major, trace and REE compositions, Sr, Nd, and Pb systematics are very similar to those reported for the NVZ (our unpublished data).

8.5.5.2 Arc Segments B (Popayán) and C (Nariño)

Over 70 volcanoes have been identified within segments B and C (Fig. 8.4), describing a less narrow volcanic arc, compared to arc segment A, that clearly defines a steep subducting slab (ca. 30° dipping, Pedraza-García et al. 2007). Studies undertaken by Marín-Cerón et al. (2010) provide insight into subduction-related and petrogenetic process along this segment (see the latest section of this chapter for detailed geochemical modeling) and classify segments B–C volcanism in Colombia in terms of two groups of volcanoes including (1) volcanic front volcanoes (VF) (Cumbal, Azufral, and Galeras) and (2) rear-arc volcanoes (RA) (Doña Juana, Puracé, Coconucos, and Huila).

There is an evident structural control on volcano location in western segments B–C, along a NE-oriented axis, coincident with numerous important faults, including Silvia-Pijao, El Crucero, Popayán, Cauca-Almaguer, Julumito, La Tetilla, Mosquerillo, and Cauca-Patía. Additional NW-oriented fractures (e.g., Paso de Bobo, Piendamó, and Paletará faults) are also important. All of these faults have been reactivated during the Quaternary (e.g., Hall et al. 2008; París et al. 1992; Velandia et al. 2005; Rovida and Tibaldi 2005). The complex structural control on volcano location in segments B–C may be related to the formation of giant calderas of more than 3 km diameter (e.g., Gabriel López, Paletará, Chagartón, Santa Elena) (Fig. 8.4).

Alkaline volcanism has been reported within eastern segment C (Fig. 8.13), in the Sibundoy Valley and San Agustin area of southern Colombia (e.g., Kroonenberg et al., 1982; Monsalve and Arcila, 2015; Borrero and Castillo 2006). The volcanism is located to the east of the principle calc-alkaline arc axis, on the eastern side of the Central Cordillera, and manifests as generally small, monogenetic alkaline cinder cones (e.g., Kroonenberg et al. 1982; Monsalve and Arcila 2015; Borrero and Castillo 2006). A tectonic model involving a slab window within segment C in

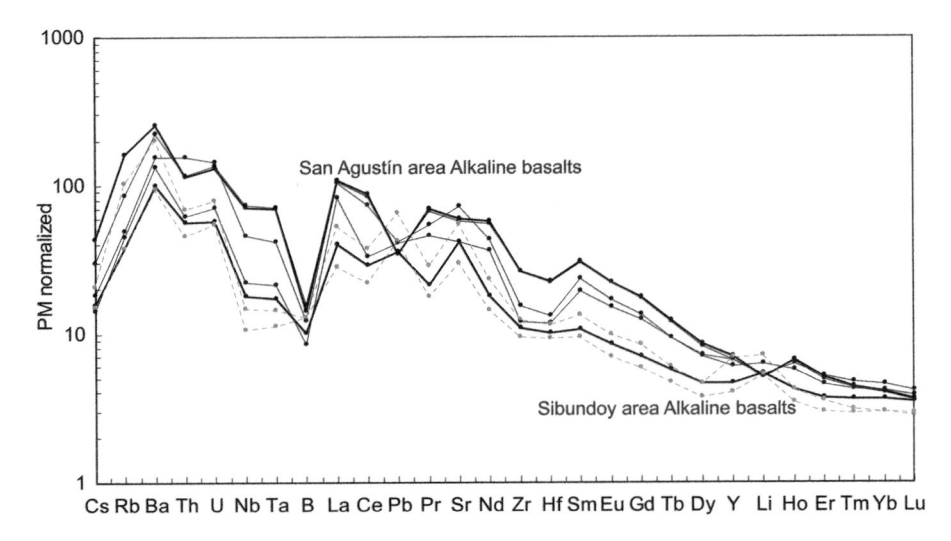

Fig. 8.13 Trace element data for alkaline volcanic rocks from the Sibundoy and San Agustin areas, normalized to Primordial Mantle (after Wood et al. 1979), data from this study

Colombia has been proposed related to asthenospheric input in this volcanism, with the subduction of the Carnegie Ridge and Malpelo Rift (Monsalve and Arcila 2015; Borrero and Castillo 2006). However, based upon our lithogeochemical studies, including major, trace element and REE data, (Fig. 8.13) together with petrography, we favor the idea of an alkaline volcanic arc with a clear subduction-related signature emplaced to the east (in the back-arc?), where magma emplacement may have been controlled by the Colon fault parallel to the main trace of the Algeciras Fault System (AFS), and related to Riedel-type synthetic and antithetic faults that form a pull-apart basin at the Sibundoy area (Fig. 8.4, Velandia et al. 2005).

8.5.5.3 Arc Segment D (Quito)

Segment D is located mostly in Ecuador (Fig. 8.4). Definition of this arc segment is more complex as the lack of intermediate seismicity in the subduction zone does not permit clear localization of the subducting slab beneath the Ecuadorian margin. At surface, the corresponding active volcanic arc is considerably wider than in Colombia and can be as broad as 110 km. The subducting slab is interpreted to dip E from 9° to 49.5° (Pedraza-García et al. 2007). Gutscher et al. (1999a, b) interpreted the lack of intermediate seismicity as the consequence of flat subduction of the part of Nazca oceanic crust supported by the buoyant Carnegie Ridge.

In general, there is a marked arc-normal lithogeochemical zonation involving enrichment in K_2O and other incompatible elements from west to east. This zonation may be related to the nature of the fluids and melts supplied by the subducting slab at increasing depths and the degree of partial melting of the mantle wedge (Barragan et al. 1998; Bourdon et al. 2003). Additionally, limited assimilation of crustal rocks by ascending mantle-derived magmas and subsequent crystal fractionation may be responsible for the rhyolitic province of the Ecuador's Eastern Cordillera (Hammersley 2003; Garrison et al. 2006).

A geochemical and isotopic study of lavas from Pichincha, Antisana and Sumaco volcanoes in the Northern Volcanic Zone (NVZ) in Ecuador shows magma genesis to be strongly influenced by slab melts. Pichincha lavas (in a fore arc position) display all the characteristics of adakites (or slab melts) and were found in association with magnesian andesites. In the main arc, adakite-like lavas from Antisana volcano could be produced by the destabilization of pargasite in a garnet-rich mantle. In the back-arc, high-niobium basalts found at Sumaco volcano could be produced in a phlogopite-rich mantle. The obvious variation in spatial distribution (and geochemical characteristics) of the volcanism in the NVZ between Colombia and Ecuador clearly suggests that the subduction of the Carnegie Ridge beneath the Ecuadorian margin strongly influences the subduction-related volcanism (Bourdon et al. 2003).

8.6 Metamorphic Decarbonation of Carbonate Sediments and Arc-Magma-Pb Radiogenic Lower Crust Interaction

In arc magmatism, the fluids/melts liberated by dehydration of subducting oceanic crust recycle lithophile elements back into the mantle wedge, triggering the production of subduction zone arc volcanism. Although there is enough evidence for the existence and mobility of fluids during subduction metamorphism, their effect on decarbonation reactions in subduction lithologies remains a topic of debate (Gorman et al. 2006). The study of such carbonate-rich sediment bearing volcanic-arcs can help to constrain the influence of carbonate breakdown in the sub-arc region and the CO_2 cycle in the subduction zone.

The SW Colombian arc (arc segments B and C) is an ideal place to study andesite magma genesis; it is dominantly composed of andesites with lesser dacites and rhyolites; phenocrysts of plagioclase, orthopyroxene, clinopyroxene, and hornblende are the major crystal components in the andesites. Titanomagnetite, ilmenite, and apatite are always present as minor phases. Petrographically, they are generally porphyritic with several disequilibrium features (e.g., reverse and complex zoning profiles in plagioclase and pyroxenes, sieve textures in plagioclase, the ubiquity of amphibole phenocrysts with reaction rims or completely replaced by oxides, and the presence of silica-rich melt inclusions in plagioclase). The abovementioned characteristics, the lack of basalts in the study area, and the clear binary mixing trends in Pb isotopic systematics (Fig. 8.14) between primary magmas and lower continental crust (LCC) (based on lower crustal xenoliths data from this region, Weber et al. 2002) suggest the importance of assimilation of LCC materials by mantle-derived arc magmas (e.g., Marín-Cerón et al. 2010).

Likewise, the available data from mantle xenoliths of the lithospheric mantle in this region indicate a depleted mantle (MORB-like) source, on the basis of HFSE concentrations and Sr-Nd isotopic systematics (Rodriguez-Vargas et al. 2005). A possible HIMU-type component in mantle beneath the arc, influenced by the Galapagos plume, is ruled out on the basis of age and distance from the spreading center (up to 450 km, Pedersen and Furnes 2001).

Based upon the primitive normalized pattern diagrams (Figs. 8.9 and 8.10), we can observe that all the studied volcanoes display the typical arc-magma signature with positive anomalies of Rb, Ba, B, and Li in addition to Th, U, and Pb and negative anomalies in the high-field-strength elements (HFSE), almost without variations across the arc. We can conclude that no OIB-type signatures are observed. Instead, a homogenous across-arc mantle wedge prior to slab-fluid interaction can be assumed within the study area (Marín-Cerón et al. 2010).

Using a multiple cross-arc isotopic variation approach, Marín-Cerón et al. (2010), we were able to determine the nature and the origin of the metazomatizing fluids, the isotopic compositions of primary magmas along the volcanic front (VF), and rear-arc (RA), as well as the degree of mantle-crust interaction, by virtue of the fact that the radiogenic isotopic compositions of Pb, Nd, and Hf are not affected by magmatic processes or by time-integrated effects in the case of Quaternary arc volcanics (Shibata and Nakamura 1997). The combination of these isotopic systematics clearly indicates the contribution of the following five end-members: multiple subduction slab components formed from altered oceanic crust (AOC) and subducted sediments (HS-CS), depleted mantle (MORB-like), and LCC materials (Fig. 8.14). The linear trends of Pb isotopes suggest that the magma of each volcanic center was formed by mixing of two homogenous end-members: lower crustal materials and primary magmas (Fig. 8.14, Marín-Cerón et al. 2010). The Pb isotope composition of likely primary magmas at the VF is less radiogenic than that at the RA (Fig. 8.14), which is opposite to the relationship observed in some typical island arcs, e.g., Izu and NE Japan arcs (Shibata and Nakamura 1997; Shibata and Nakamura 1997). Such behavior is only possible if we consider the involvement of a relatively large amount of CS as the third subduction component in the formation of primary magma at the VF in the SW Colombian volcanic arc.

Consistent with such an idea, the Hf-Pb systematics show positive linear correlations for the VF volcanoes and negative and more scattered patterns for those along the RA (Fig. 8.14). Two different patterns are also distinguishable in the Hf-Nd plot: (1) a horizontal array along the VF, with almost constant εHf and variable εNd, indicating a larger influence of CS in the mantle source and (2) a mantle-terrigenous array along the RA, indicating a lesser influence of CS in the mantle source and normal dehydration processes (Fig. 8.14). Based upon the previous discussed diagrams, together with petrographic and the trace element data, we conclude that it is necessary to invoke multiple magma mixings in the generation of SW Colombian andesite. Firstly, metamorphic decarbonation of the subducted slab and carbonate breakdown within the sub-arc region generated homogenous fluids from subduction components (AOC-HS-CS). Secondly, these fluids metasomatized the

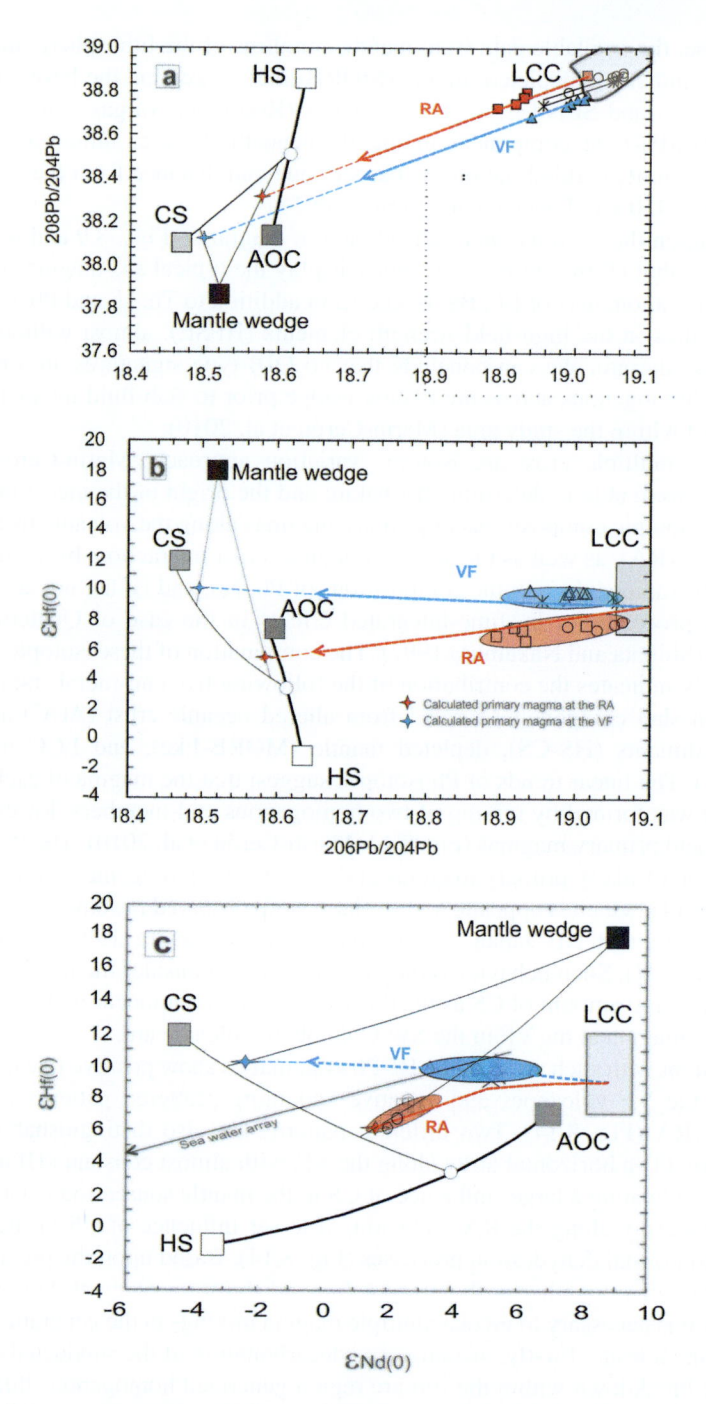

Fig. 8.14 Schematic illustration of proposed model in (**a**) ^{207}Pb/204Pb vs. ^{206}Pb/204Pb; (**b**) εHf(0) vs ^{206}Pb/^{204}Pb (**c**) εHf(0) vs εNd(0) for the SW Colombian volcanic arc (same figure from Marín-Cerón et al. 2010). Multiple isotope systems clearly show the contribution of the following

homogenous depleted mantle wedge, decreasing in amount with depth to the WBZ. Thirdly and most importantly, different degrees of interaction of arc magmas and lower crustal materials produced the intermediate to silicic magmas with radiogenic Pb signatures of the studied andesites and dacites (Marín-Cerón et al. 2010).

Our data, contrary to the results of recent empirical and modeled studies of the metamorphic decarbonation of subducted sediment at a slab (e.g., Gorman et al. 2006; Kerrick and Connolly 2001), strongly indicate the influence of carbonate-rich sediments in the chemistry of the arc-related magmas. We interpret the disparity in the depth of decarbonation, when compared with previous studies, to be related to the type of carbonate material entering the subduction zones, such as pure or impure carbonates. Recent studies indicate that the main carbonate minerals entering subduction zones are calcite and dolomite (Molina and Poli 2000). In addition, it was observed that dolomite replaces calcite/aragonite as the main carbonate phase at increasing pressure under conditions relevant to subduction zones (Molina and Poli 2000; Morlidge et al. 2006; Hammouda 2003). Thus, using experimental results in impure carbonates – dolomite (Ogasawara et al. 2000), we can predict that such CS may break down within the sub-arc region in a moderately warm subduction zone, such as observed along the SW Colombian volcanic arc (van Keken et al. 2002). Figure 8.15 indicates that subducted carbonate-rich sediments can break down at the top of the subducting slab at T~900 °C and P~4 GPa, corresponding to 120~140 km depth, that is, equivalent to the depth below the VF in the study area. Finally, the generated fluids/melts in the slab may display initially extreme trace element patterns, but compositions must be at least partially equilibrated with bulk eclogite (Feineman et al. 2007). Subsequently, the slab-derived component is decreasingly added to the depleted MORB-source cross-arc mantle with increasing depth to WBZ, as is supported by Fig. 8.14. Strong enrichment in Ba in CS (Plank et al. 2004) can also explain the high Ba/Nb ratios at the volcanic front. Similar enrichment has also been proposed as evidence of carbonate involvement in the formation of primary basaltic magmas in Nicaraguan volcanoes along the Central America arc (Plank et al. 2004; Patiño et al. 2000).

◀────────────────────────

Fig. 8.14 (continued) end-members to the volcanic front magmas: altered oceanic crust (AOC), hemipelagic sediments (HS), carbonate-rich sediments (CS), depleted mantle (MORB-like), and lower continental crust (LCC). The mantle wedge composition was assumed to be MORB-like prior to the introduction of slab-related fluids because influence of the Galapagos plume does not extend to the Colombian arc (>450 km away). The most plausible percentage of components of aqueous fluids coming from subducted slab at the volcanic front is 90% AOC, 9% HS, and 1% CS. On the other hand, 97% AOC and ~3% HS with negligible CS are required at the RA. The results indicate that dehydration and decarbonation processes at subduction zones control the chemical composition of mantle-derived magmas across the arc, and that the composition of intermediate to silicic lavas is strongly influenced by LCC contamination (up to 10% LCC). The role of upper continental crust (UCC) assimilation of magmas on the way to the surface is not easily explained by Pb, Hf, and Nd isotope systematics, and it appears to be weak or negligible. (For analytical methods, standard reference values, and estimated isotopic compositions of end-members, see supplementary information (Tables 1 and 2 from Marín-Cerón 2007)

Fig. 8.15 Carbonate stability field, dashed line from experimental results in impure carbonates – dolomite (Ogasawara et al. 2000). Blue thick line corresponds with the top of the oceanic crust from Van Keken et al. (2002), for subduction zones v = 6 cm/yr., age: 20My

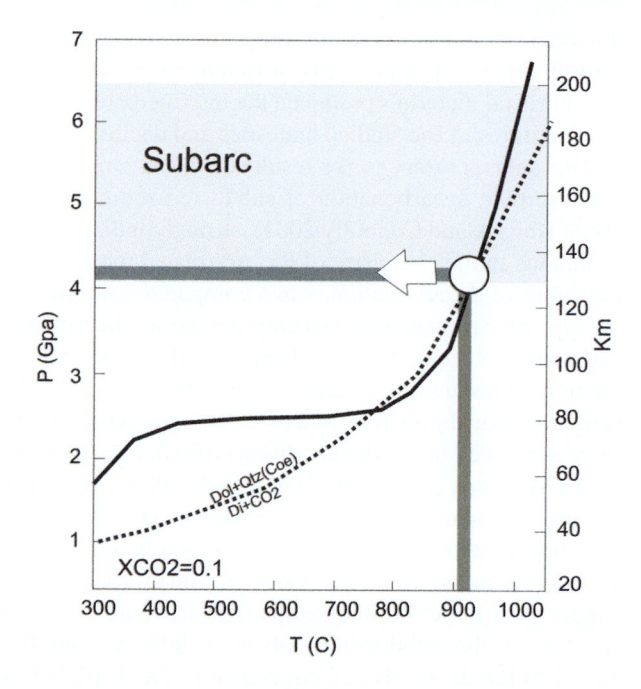

Different mantle-derived basaltic magmas are generated with increasing depth to the WBZ (Marín-Cerón et al. 2010). Once the primary magmas are formed, several post-physicochemical processes are involved in the generation of andesite magmas. The likely presence of lower crustal xenoliths in the studied area permits us to explore the mantle-derived magmas and LCC interaction. An additional advantage in this region is that the LCC has a remarkably Galapagos-like signature (Weber et al. 2002) which may make it easier to identify as an end-member component (Fig. 8.14). The current model of deep lower crust formation in the SW Colombian arc on the basis of geophysical, petrographical, and geochemical characteristics of LCC xenoliths (Weber et al. 2002) has been linked to the accretion of the mid-Cretaceous Oceanic Plateau to the NW margin of South America (Kerr et al. 1998) which may have triggered the generation of LCC with Galapagos-like isotopic signatures underplating beneath the Precambrian-Paleozoic basement in the Colombian Andes. Thus, variable degrees of interaction of mantle metasomatized magmas with materials from the deep lower continental crust can explain the Pb-radiogenic isotopic signature of the entire volcanic suite. The H_2O content of primary basaltic magmas, as well as differing melt fractions of LCC (Annen et al. 2006), may be major factors in the production of the geochemical variations of these arc magmas (Fig. 8.16a). Other factors include the assimilation rates of LCC and the rate of crystallizing phases (AFC, De Paolo 1981) and/or mixing, assimilation, storage, and hybridization (MASH, Hildreth, and Moorbath 1988) at the upper-mantle/lower crust boundary.

Fig. 8.16 (**a**) Model for the formation of intermediate to silicic magmas at convergent margins within the "hot deep zone". (Modified from Annen et al. 2006) and (**b**) pressure-temperature diagram showing liquidus surface for a silicic andesite contoured for 5, 7, and 10 wt% dissolved water (summarized by Annen et al. 2006)

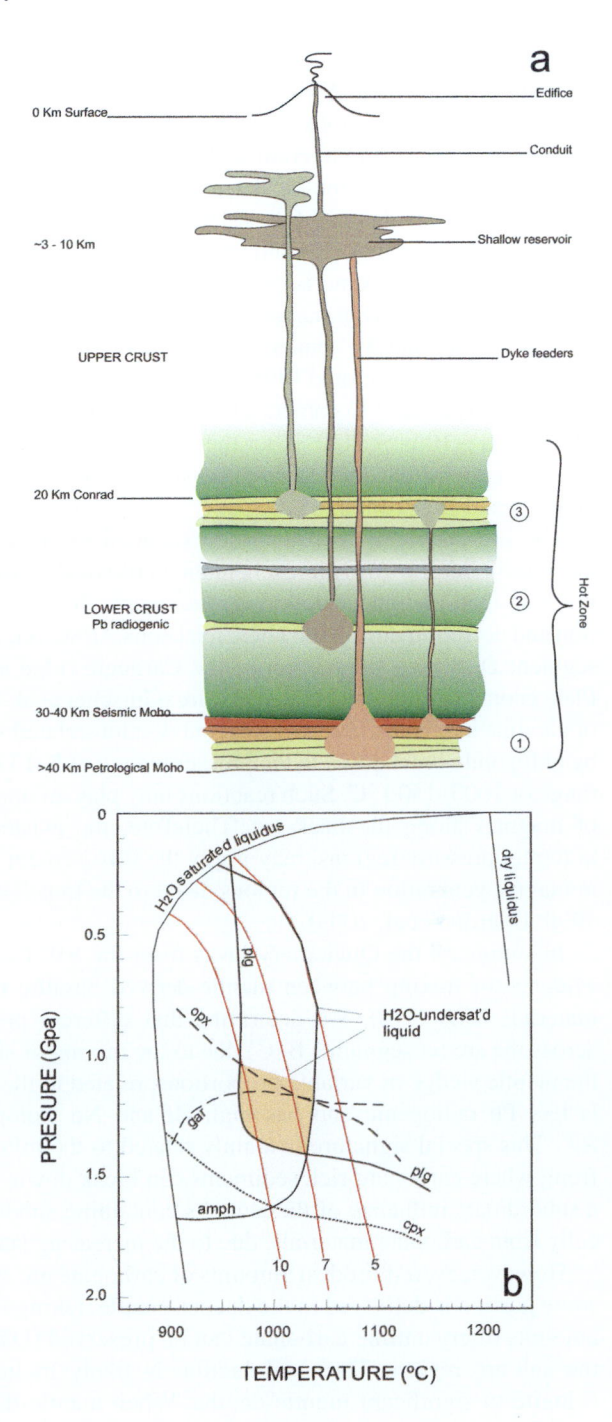

The main petrological constraints (Fig. 8.16b) with respect to the present-day andesitic volcanism in SW Colombia include (1) the presence of clinopyroxene under high-H_2O conditions in the lower crustal environment and at lower-pressure conditions. When the temperature of the melts drops, clinopyroxene becomes unstable and reacts with the melt to form amphibole, resulting in the evolved melt being more siliceous (Foden and Green 1992); (2) amphibole is stable only for H_2O contents ≥4 wt% (Eggler 1972) and temperatures below ~1050 °C (e.g., Müntener et al. 2001); (3) orthopyroxene is confined to relatively low pressures and temperatures over 920 °C, whereas garnet appears only above ~1.1GPa; and (4) plagioclase stability decreases and An content increases with increasing H_2O. Based on the experimental data of Kawamoto (1996) and Pichavant et al. (2002b), it is possible to infer that the maximum H_2O content of an andesite melt in equilibrium with plagioclase (An>80) is ~ 10 wt% H_2O. After the main mineral phases start to crystallize within the lower crust, the intermediate to silicic magma evolves according to the P-T gradient and H_2O content.

The present model can also be expanded to volcanic arc segments A and D in Ecuador, as shown within schematic geological profiles (Fig. 8.17a–c), where it is clear that the presence of Pb-radiogenic lower crust is related to subduction and accretion during the late Cretaceous. However, caution must be taken in arc segment D, due to subduction of the Carnegie ridge and the variations of Nazca Plate geometry along and across the arc (Gutscher et al. 1999a, b). The development of alkaline magmatism with a clear subduction-related signature could be triggered by phlogopite breakdown in the pressure range of 5–13 GPa, and in the temperature range of 1000–1300 °C. Such reactions may play an important role in the formation of magmas along the back-arc. Therefore, the existence of potassic amphibole in higher pressure regions, may imply the involvement of a subduction component in magma generation in the regions distal to the trench axis (e.g., Sudo and Tatsumi 1990; Bourdon et al. 2003).

In summary, the Quaternary lavas from the SW Colombian volcanic arc show evidence of mixing between mantle-derived basaltic magmas with lower crustal materials (Fig. 8.14). We determine that different primary magmas are formed across the arc (at segments B–C) due to the mixing of slab-derived fluids/melts and the mantle wedge in variable proportions, related to the depth to the WBZ. The VF is less Pb radiogenic and has high Hf and Nd isotopic ratios compared to the RA. This special signature is mainly related to the influence of CS at the volcanic front, where carbonate-rich sediments can break down. The RA volcanoes indicate a subordinate influence of fluids/melts containing subduction components, specifically from carbonate materials, due to the increasing stability field of carbonates.

However, even if modest amounts of carbonate are removed during subduction, some portion is delivered to the deeper mantle, taking into account that significant amounts of crystalline carbonate can be preserved (Dasgupta et al. 2004) beyond the sub-arc region. Thus, subduction is likely to introduce carbonate-bearing eclogite to significant mantle depths. When mantle-derived magmas intrude the lower crust, heat, volatiles and H2O are transferred to the surrounding crust, which can lead to partial melting of LCC. Silicic residual melts are generated by

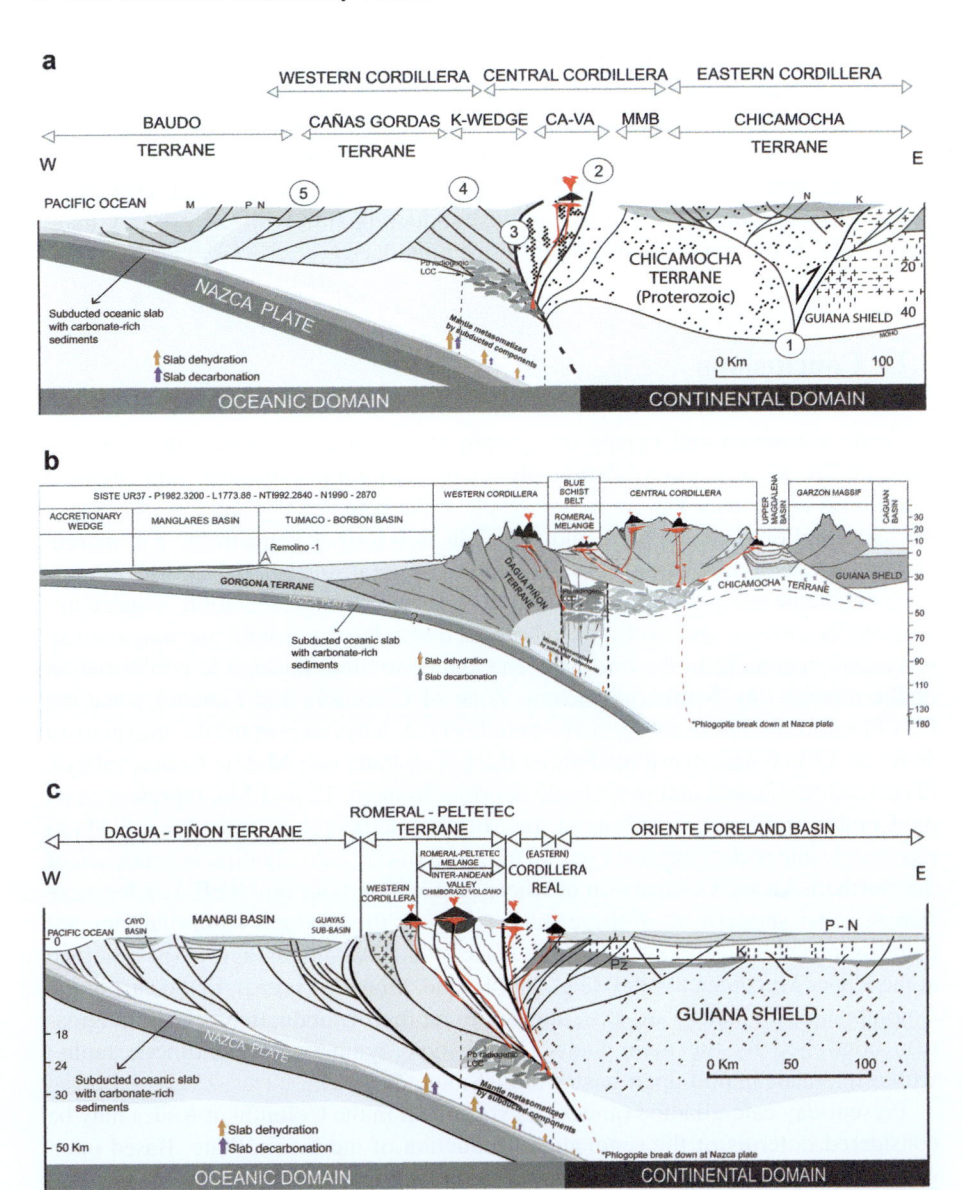

8 Late Cenozoic to Modern-Day Volcanism in the Northern Andes…

Fig. 8.17 (**a**) Schematic geological transect across segment A. Modified after Cediel et al. (2003). Principal sutures: 1 = Grenville (Orinoco) Santa Marta-Bucaramanga-Suaza faults; 2 = Palestina fault system; 3 = Romeral-Peltetec fault system; 4 = Garrapatas-Dabeiba fault system; 5 = Atrato fault system. Abbreviations: K-wedge Cretaceous wedge, CA-VA Cajamarca-Valdivia terrane, MMB Middle Magdalena Basin, sl San Lucas Block; (meta-)Sedimentary Rocks: Pz Paleozoic, K Cretaceous, P Paleogene, N Neogene. (**b**) Schematic geological transect across segments B and C. (Modified after Cediel et al. 2003; Weber et al. 2002; Marín-Cerón et al. 2010). (**c**) Schematic geological transect across segment D. (Modified after Cediel et al. 2003)

incomplete crystallization of newly arrived basalt, with some contribution from remelting of earlier intrusions and partial melting of older crustal rocks, as have been proposed by Annen et al. (2006). Moreover, our model likely explains the occurrence of scapolite in the garnet-bearing LCC xenoliths (Weber et al. 2002) within this region, which can be produced by the interaction of CO_2-rich fluids derived from decarbonation of CS in the subducting slab with the primary magmas – LCC at the upper-mantle/lower crust boundary.

8.7 Conclusions

Neogene volcanism and hypabyssal porphyritic magmatism within the Northern Andean Block, and more specifically along Colombia's Central and Western Cordilleras and Cauca Valley, are the result of the complex evolution of the Nazca (former Farallón) oceanic plate and its subduction history beneath the SW margin of continental South America. Based upon temporal and spatial considerations, three magmatic sub-periods have been outlined within our presentation, which demonstrate the history and evolution of subduction-related volcanic arc magmatism/volcanism beginning in the mid-Miocene and continuing through to conformation of the modern-day Northern Volcanic Zone of Colombia and Ecuador since the Plio-Pliestocene. These sub-periods include (1) hypabyssal porphyritic magmatism from ca. 17 to 6 Ma, distributed along the Cauca-Patía and Middle Cauca Valleys; (2) extrusive volcanic and pyroclastic deposits from ca. 12 to 3 Ma, represented by the Combia and Irra Formations of the Middle Cauca region; and (3) ca. 3 Ma to recent volcanic rocks contained within the modern-day calc-alkaline volcanic arc of the Northern Andes. Comparison of whole-rock major, trace and REE data for these sample suites suggests all of these suites represent the products of similar magmatism associated with subduction of the Nazca Plate. Alkaline magmatism observed at the Upper Magdalena Valley (e.g., San Roque, Sibundoy areas), to the east of the Colombian calc-alkaline arc axis, appears to represent subduction-related magmatism controlled by the main faults (Riedel-type synthetic and antithetic faults) within Interandean pull-apart basins (e.g., Sibundoy area).

Present-day calc-alkaline (andesitic) volcanism in the Colombian Andes may be considered in terms of the segmented subduction of the Nazca Plate. Based upon geophysical studies, four segments have been identified, including (N to S) segment A (Cali), segment B (Popayán), segment C (Nariño), and segment D (Quito). Changes in the geometry of the WBZ beneath each segment may account for variations in the composition of resulting volcanic products, while the structural configuration of each segment influenced the location, emplacement, and size of volcanic edifices (Fig. 8.17a–c).

Based upon the presence of adakite-like characteristics in some Neogene volcanic rocks of the NAB, we favor a petrogenetic model for the generation of primary subduction-related magmas which involves dehydration/decarbonation fluxing and melting of the subducted Nazca Plate. Pb isotope data for NAB volcanic rocks, however, indicates that the lower crust exercises an important influence upon magma composition during magma ascent. Based upon radiogenic Pb values, lithogeochemical and petrographic evidence, we suggest that primary mantle-derived magmas, including slab fluids/melts, were retained within lower crustal magma chambers, where they underwent Pb isotope exchange, crustal assimilation/mixing, and crystal fractionation. We consider these processes to have influenced the final major, trace, and REE composition of the resulting products of NAB volcanism. We conclude that variations in the composition of the lower crust have likely influenced the composition of present-day volcanism along the length of the entire Andean volcanic arc.

References

Annen C, Blundy JD, Sparks RS (2006) The genesis of intermediate and silicic magmas in deep crustal hot zones. J Petrol 47(3):505–539

Álvarez A (1983) Geología de la cordillera Central y el Occidente colombiano y petroquímica de los intrusivos granitoides Mesocenozoicos. Boletín Geológico, vol 26, p 175, Bogotá

Álvarez J (1987) Mapa metalogénico de las fajas ofiolíticas de la zona occidental de Colombia. Ingeominas, Bogotá, Informe No. 2024, p 41

Aspden JA, McCourt WJ (1986) Mesozoic oceanic terrane in the Central Andes of Colombia. Geology 14:415–418

Aspden JA, McCourt WJ, Brook M (1987) Geometrical control of subduction-related magmatism: the Mesozoic and Cenozoic plutonic history of Western Colombia. J Geol Soc Lond 144:893–905

Barragan R, Geist D, Hall M, Larson P, Kurz M (1998) Subduction controls on the compositions of lavas from the Ecuadorian Andes. Earth Planet Sci Lett 154(1–4):153–166

Barret TJ, Taylor PN, Lugowski J (1987) Metalliferous sediments from DSDP leg 92: The East Pacific Rise transects. Geochim Cosmochim Acta 46:651–666

Bernet M, Urueña C, Amaya S, Peña ML (2016) New thermo and geochronological constraints on the Pliocene-Pleistocene eruption history of the Paipa-Iza volcanic complex, Eastern Cordillera, Colombia. J Volcanol Geotherm Res 327:299–309. https://doi.org/10.1016/j.jvolgeores.2016.08.013

Bissig T, Mantilla FL, Rodríguez A, Raley Ch, Hart C. (2012) The Vetas-California district, eastern Cordillera, Santander, Colombia: Late Miocene porphyry and epithermal mineralization hosted in Proterozoic gneisses and Late Triassic intrusions. Abstract. XVI Peruvian Geological Congress & SEG Conference. Lima, Perú

Bohórquez O, Monsalve ML, Velandia F, Gil-Cruz F, Mora H (2005) Determinación del Marco Tectónico Regional para la Cadena Volcánica más Septentrional de la Cordillera Central de Colombia. Boletín de Geología, UIS, vol 27/44, p 55–79

Bourdon E, Eissen JP, Gutscher MA, Monzier M, Hall M, Cotten J (2003) Magmatic response to early aseismic ridge subduction: the Ecuadorian margin case (South America). Earth Planet Sci Lett 205(3–4). https://doi.org/10.1016/S0012-821X(02)01024-5

Borrero C, Castillo H (2006) Vulcanitas del S-SE de Colombia: retro-arco Alcalino y su posible relacion con una ventana Astenosferica. Boletín de Geología vol 28/2, p 24–34

Borrero C, Toro LM, Alvarán M, Castillo H (2009) Geochemistry and tectonic controls of the effusive activity related with the ancestral Nevado del Ruiz volcano, Colombia. Geofísica Internacional, vol 48/1, p 149–169

Bryant JA, Yogodzinski GM, Hall ML, Lewicki JL, Bailey DG (2006) Geochemical constraints on the origin of volcanic rocks from the Andean Northern Volcanic Zone, Ecuador. J Petrol 47(6):1147–1175

Calvache M, Cortés P, Williams S (1997) Stratigraphy and chronology of the Galeras Volcanic Complex, Colombia. J Volcanol Geotherm Res 77:5–19

Calvache ML, Williams S (1997a) Geochemistry and petrology of the Galeras Volcanic Complex, Colombia. J Volcanol Geotherm Res 77:21–38

Calvache ML, Williams S (1997b) Emplacement and petrological evolution of the andesitic dome of Galeras. J Volcanol Geotherm Res 77:57–59

Campbell CJ (1974) Colombian Andes. In: Spencer AM (ed.) Mesozoic and Cenozoic Orogenic Belts. Special Publication of the Geological Society, London, vol 4, p 705–771

Cediel F (2018) Phanerozoic orogens of Northwestern South America: cordilleran-type orogens, taphrogenic tectonics and orogenic float. Springer, Cham, pp. 3–89

Cediel F, Etayo F, Cáceres C (1997) Distribución de facies sedimentarias y su marco tectónico durante el Fanerozoico en Colombia. VI Simposio Bolivariano, Exploración petrolera en las cuencas Subandinas, Cartagena

Cediel F, Shaw RP, Cáceres C (2003) Tectonic assembly of the Northern Andean Block. In Bartolini C, Buffler RT, Blickwede J (eds) The Circum-Gulf of Mexico and the Caribbean: hydrocarbon habitats, basin formation, and plate tectonics.– AAPG Memoir 79, p 815–848

Cediel F, Leal-Mejía H, Shaw RP, Melgarego JC, Restrepo-Pace PA (2011) Petroleum geology of Colombia: regional geology of Colombia. ANH – Colombia. 1: 220

Cepeda H, Pardo N, Jaramillo J (2004) The Paipa volcano, Colombia, South America. IAVCEI meeting Poster Session. November 14–19, Pucón, Chile

Cepeda H, Pardo N (2004) Vulcanismo de Paipa. Informe técnico. INGEOMINAS-Bogota. Colombia. 140 p

Chiaradia M, Fontboté L, Beate B (2004) Cenozoic continental arc magmatism and associated mineralization in Ecuador. Mineral Deposita 39:204–222

Correa AM, Cepeda H, Pulgarín B, Ancoches E (2000) El volcán Nevado del Huila (Colombia): rasgos generales y caracterización composicional. Geogaceta 27:51–54

Cox KG, Bell JD, Pankhurst RJ (1979) The Interpretation of Igneous Rocks: London, George Allen & Unwin, p 464

Dasch EJ (1981) Lead isotopic composition of metalliferous sediments from the Nazca Plate. Mem Geol Soc Am 154:199–209

Dasgupta R, Hirschmann MM, Withers AC (2004) Deep global cycling of carbon constrained by the solidus of anhydrous, carbonated eclogite under upper mantle conditions. Earth Planet Sci Lett 227:73–85

Davidson JP, de Silva SL (1992) Volcanic rocks from the Bolivian Altiplano: Insights into crustal structure, contamination, and magma genesis in the central Andes. Geology 20:1127–1130

Defant M, Drummond MS (1990) Derivation of some modern arc magmas by melting of young subducted lithosphere. Nature 347:662–665

De Paolo DJ (1981) Trace-element and isotopic effects of combined wallrock assimilation and fractional crystallization. Earth Planet Sci Lett 53:189–202

de Silva S (1991) Styles of zoning in Central Andean ignimbrites-insights into magma chamber processes: In: Andean magmatism and its tectonic setting. Special paper 265, pp 217–232

Droux A, Delaloye M (1996) Petrography and Geochemistry of Plio-Quaternary Calc-Alkaline volcanoes of Southwestern Colombia. J S Am Earth Sci 1–2:27–41

Duque-Caro H (1990) The Choco Block in the northwestern corner of South America: Structural, tectonostratigraphic, and paleogeographic implications. J S Am Earth Sci 3:71–84

Duque JF, Toro GE, Cardona A, Calvache M (2010) Geología, geocronología y geoquímica del volcán Morasurco, Pasto, Colombia. Boletín de Ciencias de la Tierra 27:25–36

Estrada JJ, Viana R, González H (2001) Memoria Explicativa del Mapa Geológico de la Plancha 205 – Chinchiná. Escala 1:100.000. INGEOMINAS, p 92

Etayo-Serna F, Barrero D, Lozano HQ, Espinosa A, Gonzalez H, Orrego A, Ballesteros IT, Forero,HO, Ramirez CQ, Zambrano-Ortiz F, Duque-Caro H, Vargas RH, Nuñez A, Alvarez,

JA, Ropain UC, Cardozo EP, Galvis N, Sarmiento LR, Albers JP, Case JE, Singer DA, Bowen RW, Berger BR, Cox DP, Hodges CA (1986) Mapa de terrenos geológicos de Colombia. Bogotá. Publicaciones Geológicas Especiales del INGEOMINAS, vol 14/1, p 235

Eggler DH (1972) Amphibole stability in H2O-undersaturated calc- alkaline melts. Earth Planet Sci Lett 15:38–44

Farris DW, Jaramillo C, Bayona G, Restrepo-Moreno SA, Montes C, Cardona A, Mora A, Speakman RJ, Glascock MD, Valencia V (2011) Fracturing of the Panamanian Isthmus during initial collision with South America. Geology 39(11):1007–1010

Feineman MD, Ryerson FJ, DePaolo DJ, Plank T (2007) Zoisite-aqueous fluid trace element partitioning with implications for subduction zone fluid composition. Chem Geol. https://doi.org/10.1016/j.chemgeo.2007.01.008

Frey FA, Gerlach DC, Hickey RL, López-Escobar L, Minizaga-Villavicencio F (1984) Petrogenesis of the Laguna del Maule volcanic complex, Chile (36°S). Contrib. Mineral Petrol 88:133–149

Foden JD, Green TH (1992) Possible role of amphibole in the origin of andesite: some experimental and natural evidence. Contrib Mineral Petrol 109:479–493

Garrison J, Davidson J, Reid M, Turner S (2006) Source versus differentiation controls on U-series disequilibria: Insights from Cotopaxi Volcano, Ecuador. Earth Planet Sci Lett 244:548–565

Gill JB (1981) Orogenic Andesites and Plate Tectonics, vol 16. Springer, Berlin Heidelberg

Gómez J, Nivia A, Montes NE, Tejada ML, Jiménez DM, Sepúlveda MJ, Mora MP (2007) Mapa Geológico de Colombia escala 1: 1.000.000. Ingeomina, Bogotá

González H (2001) Memoria Explicativa del Mapa Geológico del Departamento de Antioquia. Escala 1:400.000. Medellín, Ingeomina, Bogotá, p 240

Gorman P J,. Kerrick DM, Connolly JAD (2006) Modeling open system metamorphic decarbonation of subductingslabs, Geochem. Geophys. Geosyst.,7(4): 21, doi:10.1029/2005GC001125

Grosse E (1926) Estudio Geológico del Terciario carbonífero de Antioquia en la parte occidental de la Cordillera Central de Colombia: Berlín, Verlag Von Dietrich Reimer, p 361

Gutscher MA, Malavielle J, Lallemand S, Collot JY (1999a) Tectonic segmentation of the North Andean margin: impact of the Carnegie Ridge collision. Earth Planet Sci Lett 170(1–2):155–156

Gutscher R, Malavieielle J, Lallemend S, Collot JY (1999b) Tectonic segmentation of the North Andean margin: Impact of the Carnegie ridge collision. Earth Planet Sci Lett 168:255–270

Hall ML, Samaniego P, Le Pennec JL, Johnson JB (2008) Ecuadorian Andes volcanism: A review of Late Pliocene to present activity. J Volcanol Geotherm Res 176:1–6

Hammersley L (2003) The Chalupas caldera. PhD Dissertation. Univ. California, Berkeley

Hammouda T (2003) High-pressure melting of carbonated eclogite and experimental constraints on carbon recycling and storage in the mantle. Earth Planet Sci Lett 214:357–368

Harmon RS, Barreiro BA, Moorbarth S, Hoefs J, Francis PW, Thorpe RS, Deruelle B, McHugh J, Virglino JA (1984) Regional O, Sr and Pb-isotope relationships in late Cenozoic calc-alkaline lavas of the Andean Cordillera. J Geol Soc Lond 141:803–822

Hardy N (1991) Tectonic evolution of the easternmost Panama Basin. J S Am Earth Sci 4:261–270

Hawkesworth CJ, Norry MJ, Roddick JC, Baker PE, Francis PW, Thorpe RS (1979) ^{143}Nd/^{144}Nd, ^{87}Sr/^{86}Sr, and incompatible trace element variations in calc-alkaline andesitic and plateau lavas from South America. Earth Planet Sci Lett 42:45–57

Hickey RL, Frey FA, Gerlach DC, López-Escobar L (1986) Multiple sources for basaltic arc rocks from the southern volcanic zone of the Andes (34° – 41° S): Trace element and isotopic evidence for contributions from subducted oceanic crust, mantle, and continental crust. J Geophys Res 91(B6):5963–5983

Hildreth W, Moorbath S (1988) Crustal contribution to arc magmatism in the Andes of central Chile. Contrib Mineral Petrol 98:455–489

Irvine TN, Baragar WRA (1971) A guide to the chemical classification of the common volcanic rocks. Can J Earth Sci 8(5):523–548

James DE (1982) A combined O, Sr, Nd, and Pb isotopic and trace element study of crustal contamination in central Andean lavas: I. Local geochemical variations. Earth Planet Sci Lett 57:47–62

James DE, Brooks C, Cuyubamba A (1976) Andean Cenozoic volcanism: Magma genesis in the light of strontium isotopic composition and trace-element geochemistry. Geol Soc Amer Bull 87(p):592–600

Jaramillo JM (1976) Volcanic rocks of the Río Cauca valley, Colombia S.A. Thesis Degree of Master of Arts, Rice University, Houston

Kawamoto T (1996) Experimental constraints on differentiation and H2O abundance of calcalkaline magmas. Earth Planet Sci Lett 144:577–589

Kay S, Mpodozis C, Ramos VA, Munizaga F (1991) Magma source variations for mid-Tertiary magmatic rocks associated with a shallowing subduction zone and a thickening crust in the Central Andes (28–33°S). In: Andean Magmatism and its Tectonic Setting, Boulder, Colorado. Harmon, R.S., Rapela, C.W., eds. Spec. Pap. Geol. Soc. Am., vol 265, p 113–137

Kellogg J, Vega V (1995) Tectonic development of Panamá, Costa Rica, and the Colombian Andes: Constraints from global positioning system geodetic studies and gravity. Geological Society of America. Special Paper vol 295, pp 75–90

Kennan L, Pindell J (2009) Dextral shear, terrane accretion and basin formation in the Northern Andes: Best explained by interaction with a Pacific-derived Caribbean Plate. The geology and evolution of the region between North and South America. Geological Society of London, Special Publication, p 58

Kerr AC, Tarney J, Nivia A, Marriner GF, Saunders AD (1998) The structure of an oceanic plateau: evidence from obducted Cretaceous terranes in western Colombia. Tectonophysics 292:173–188

Kerrick DM, Connolly JAD (2001) Metamorphic devolatilization of subducted marine sediments and the transport of volatiles into the Earth's Interior. Nature, 411:293–296

Kroonenberg S, Pichler H, Diederix H (1982) Cenozoic alkalibasaltic to ultrabasic volcanism in the uppermost magdalena valley Southern Huila department, Colombia. Geología Norandina 5:19–26

Leal-Mejía H (2011) Phanerozoic Gold Metallogeny in the Colombian Andes – A tectonomagmatic approach: Ph.D. thesis, Barcelona (Catalonia), Spain, University of Barcelona, p 1000

Leal-Mejia H, Shaw RP, Melgarejo JC (2018) Spatial/temporal migration of granitoid magmatisn and the phanerozoic tectono-magmatic evolution of the Colombian Andes. In: Cediel F and Shaw RP (eds). Geology and Tectonics of Northwestern South America: The Pacific-Caribbean-Andean Junction, Springer, pp 253–397

LeMaitre RW, Bateman P, Dudek A, Keller J, MJ L-LB, Sabine PA, Schmid R, Sorensen H, Streckeisen A, Woolley AR, Zanettin BA (1989) Classification of Igneous Rocks and Glossary of Terms. Blackwell, Oxford

Lonsdale P (2005) Creation of the Cocos and Nazca plates by fission of the Farallón plate. Tectonophysics 404(3–4):237–264

López A, Ramírez S (2006) Registro del Vulcanismo Neógeno en el suroccidente antioqueño y sus implicaciones tectónicas. Undergraduate thesis. EAFIT University. 122p

López A, Sierra GM, Ramírez S (2006) Vulcanismo Neógeno en el suroccidente antioqueño y sus implicaciones tectónicas. Boletín Ciencias de la Tierra 19:27–41

López-Castro SM (2009) Estratigrafía, petrología y geoquímica de las rocas volcánicas del flanco occidental del volcán Puracé, alrededores de Coconuco. Master thesis on Earth Sciences. EAFIT University, p 9

Mann P, Burke K (1984) Cenozoic rift formation in the northern Caribbean. Geology 12:732–736

Mann P, Corrigan J (1990) Model for late Neogene deformation in Panama. Geology 18:558–562

Mantilla FL, Mendoza H, Bissig T, Hart C (2011) Nuevas evidencias sobre el magmatismo Miocenico en el distrito minero de Vetas-California (Macizo de Santander, Cordillera Oriental, Colombia). Boletín de Geología vol 33/1, p 41–56

Marín-Cerón MI (2007) Major, trace element and multi-isotopic systematics of SW Colombian volcanic arc, northern Andes: Contributions of slab fluid, mantle wedge and lower crust to the origin of Quaternary andesites. Doctoral thesis. Okayama University, Japan, p 133

Marín-Cerón MI, Moriguti T, Makishima A, Nakamura E (2010) Slab decarbonation and CO_2 recycling in the Southwestern Colombian volcanic arc. Geochim Cosmochim Acta 74:1104–1121

Marriner GF, Millward D (1984) Petrochemistry of Cretaceous to recent Vulcanism in Colombia. J Geol Soc Lond 141:473–486

McCourt WJ, Aspden JA, Brook M (1984) New geological and geochronological data from the Colombian Andes: continental growth by multiple accretion. J Geol Soc Lond 141:831–845

Mejía EL, Velandia F, Zuluaga CA, López JA, Cramer T (2012) Análisis estructural al noreste del Volcán Nevado del Ruíz, Colombia–aporte a la exploración geotérmica. Boletín de Geología 34:27–41

Mesa-García J (2015) Combia Formation: a Miocene immature volcanic arc?. Master Thesis, EAFIT Univeristy, Medellín, Colombia. p 249

Molina JF, Poli S (2000) Carbonate stability and fluid composition in subducted oceanic crust: an experimental study on H2O-CO2-bearing basalts. Earth Planet Sci Lett 176:295–231

Monsalve ML, Arcila M (2015) Firma Adakítica en los productos recientes de los volcanes Nevado del Huila y Puracé, Colombia. Boletín Geológico 43:23–40

Montes C, Cardona A, McFadden R, Morón SE, Silva CA, Restrepo-Moreno S, Ramírez DA, Hoyos N, Wilson J, Farris D, Bayona GA, Jaramillo CA, Valencia V, Bryan J, Flores JA (2012) Evidence for middle Eocene and younger land emergence in central Panama: Implications for Isthmus closure. Science 124(5–6):780–799

Monzier M, Robin C, Hall ML, Cotten J, Mothes P, Eissen JP, Samaniego P (1997) Les adakites d'Equateur : Modèle préliminaire. Comptes Rendus Acad. Sci Paris 324:545–552

Morlidge M, Pawley A, Giles D (2006) Double carbonates breakdown reactions at high pressures: an experimental study in the system CaO-MgO-FeO-MnO-CO2. Contrib Mineral Petrol 152:365–373

Müntener O, Kelemen PB, Grove TL (2001) The role of H2O during crystallisation of primitive arc magmas under upper-most mantle conditions and genesis of igneous pyroxenites: an experimental study. Contrib Mineral Petrol 141:643–658

Nakamura E, Campbell IH, Sun SS (1985) The influence of subduction processes on the geochemistry of Japanese alkaline basalt. Nature 316:55–58

Navarro S, Pulgarín B, Monsalve ML, Cortés GP, Calvache ML, Pardo N, Murcia H (2009) Geología e historia eruptiva del complejo volcánico Doña Juana (CVDJ) Nariño. Boletín de geología 31(2)

Ordoñez O (2002) Caracterizacao isotopica Rb-Sr E Sm-Nd dos principais eventos magmáticos nos Andes Colombianos. Doctoral thesis. Brazilia University, p 165

Ogasawara Y, Ohta M, Fuksawa K, Katayama I, Maruyama S (2000) Diamond-bearing and diamond-free metacarbonate rocks from Kumdy-Kol in the Kokchetav Massif, northern Kazakhstan. The Island Arc, 9:400–416

Pardo N, Cepeda H, Jaramillo J (2005) The Paipa volcano, Eastern Cordillera of Colombia, South America: volcanic stratigraphy. Earth Sci Res J 9:3–18

Pardo-Casas F, Molnar P (1987) Relative motion of the Nazca (Farallón) and South American plates since Late Cretaceous time. Tectonics 6(3):223–248

París G, Marín W, Sauret B, Vergara H, Bles J L (1992) Neotectónica. En: Microzonificación Sismogeotécnica de Popayán. CEE-INGEOMINAS. INGEOMINAS 2, pp 28–49

Patiño LC, Carr M, Feigenson M (2000) Local and regional variations in Central American arc lavas controlled by variations in subducted sediment input. Contrib Mineral Petrol 138:265–283

Pedersen R, Furnes H (2001) Nd- and Pb-isotopic variations through the upper oceanic crust in DSDP/ODP Hole 504B, Costa Rica Rift. Earth Planet Sci Lett 189:221–235

Pedraza-García P, Vargas CA, Monsalve H (2007) Geometric Model of the Nazca Plate Subduction in Southwest Colombia. Earth Sci Res J 11(2):117–130

Pérez AM, Marín-Cerón MI, Bernet M, Sierra G, Moreno N (2013) Resultados preliminares de AFT en la Formación Amagá, Pozos el Cinco-1B y Venecia-1. In: Colombia. Event: XIV Congreso Colombiano de Geología Libro: XIV Congreso Colombiano de Geología. Resúmenes

Pichavant M, Martel C, Bourdier JL, Scaillet B (2002). Physical conditions, structure, and dynamics of a zoned magma chamber: Mount Pelée (Martinique, Lesser Antilles Arc). J Geophys Res 107, article number 2093

Piedrahita VA, Bernet M, Chadima M, Sierra GS, Marín-Cerón MI, Toro GE (2017) Detrital zircon fission-track thermochronology and magnetic fabric of the Amagá Formation (Colombia): Intracontinental deformation and exhumation events in the northwestern Andes. Sediment Geol 356:26–42. https://doi.org/10.1016/j.sedgeo.2017.05.003

Pindell JL, Higgs R, Dewey JF (1998) Cenozoic palinspatic reconstruction, paleogeographic evolution and hydrocarbon setting of the northern margin of the northern margin of South America. In: Pindell JL, Drake CL (eds) Paleogeographic evolution and non-glacial eustasy, northern South America: Society of economic and petroleum mineralogists special publications, 58:45–85

Plank T, Balzer V, Carr M (2004) Nicaraguan volcanoes record paleoceanographic changes accompanying closure of the Panama gateway. Geology 30(12):1087–1090

Ramírez DA, López A, Sierra GM, Toro GE (2006) Edad y provenincia de las rocas volcánico sedimentarias de la Formación Combia en el suroccidente Antioqueño- Colombia. Boletin Ciencias de la Tierra. 19:9–26

Ramos V, Aleman A (2000) Tectonic evolution of the Andes U. Cordani, E.J. Milani, A. Thomaz Filho, M.C. Campos Neto (Eds.), Tectonic Evolution of South America, p. 635–685 (31st Int. Geol. Congr., Rio de Janeiro)

Restrepo-Moreno SA, Foster DA, Stockli DF, Parra-Sánchez LN (2009) Long-term erosion and exhumation of the 'Altiplano Antioqueño', Northern Andes (Colombia) from apatite (U-Th)/He thermochronology. Earth Planet Sci Lett 278:1–12

Restrepo-Moreno SA, Cardona A, Jaramillo C, Bayona G, Montes C, Farris DW (2010) Constraining Cenozoic uplift/exhumation of the Panamá-Chocó Block by apatite and zircon low-temperature thermochronology: insights on the onset of collision and the morphotectonic history of the region. Abstract. GSA Denver Annual Meeting. Geol Soc Am Abstr Programs 42(5):521

Restrepo JJ, Toussaint JF, González H, (1981) Edades MioPliocenas del magmatismo asociado a la Formación Combia. Departamentos de Antioquia y Caldas, Colombia. Geología Norandina. 3: 2126

Restrepo JJ, Toussaint JF (1988) Terranes and continental accretion in the Colombian Andes. Episodes 11(3):189–193

Restrepo J, Toussaint J (1990) Cenozoic arc magmatism of northwestern Colombia: Geological Society of America Special paper, vol 41, pp 205–212

Rios AM, Sierra MI (2004) La Formación Combia: Registro de la relación entre el volcanismo Neógeno y la sedimentación fluvial, sección Guineales – Bolombolo, suroeste antioqueño. Undergraduate thesis. EAFIT University. 122 pp.

Rovida A, Tibaldi A (2005) Propagation of strike-slip faults across Holocene volcano-sedimentary deposits, Pasto, Colombia. J Struct Geol 27:1838–1855

Rodríguez G, Arango MI, Bermúdez JG (2012) Batolito de Sabanalarga, plutonismo de arco en la zona de sutura entre las cortezas oceánica y continental de los Andes del Norte. Boletín Ciencias de la Tierra 32:81–98

Rodriguez-Vargas KE, Mallmann G, Conceicao RV, Kawashita K, Weber MBI (2005) Mantle diversity beneath the Colombian Andes, Northern Volcanic Zone: Constraints from Sr and Nd isotopes. Lithos 82:471–484

Russo RM, Okal EA, Rowley K (1992) Historical seismicity of the southeastern Caribbean and tectonic implications. Pageoph 139(1):87–120

Ryan JG, Langmuir CH (1987) The systematics of lithium abundances in young volcanic rocks. Geochim Cosmochim Acta 51:1727–1741

Saenz EA (2003) Fission track thermochronology and denudational response to tectonics in the north of the Colombian Central Cordillera. Master thesis. Shimane University. 131 pp

Sakuyama M, Nesbitt RW (1986) Geochemistry of the quaternary volcanic rocks of the northeast Japan arc. J Volcanol Geotherm Res 29(1–4):413–450

Shibata T, Nakamura E (1997) Across-arc variations of isotope and trace element compositions from Quaternary basaltic volcanic rocks in northeastern Japan: Implications for interaction between subducted oceanic slab and mantle wedge. J Geophys Res 102(B4):8051–8064

Sierra G (1994) Structural and sedimentary evolution of the Irra Basin, northern Colombian Andes. Master thesis, Department of Geological Science, State University of New York, Binghamton, NY. 102 pp.

Sierra GM, MacDonald W, Estrada JJ (1995) Young rotations inferred from paleomagnetic evidence in late Tertiary strata: slip reversals along the Romeral Strike –Slip fault zone, Northern Andes. In: Estados Unidos Eos, Transactions, American Geophysical Union

Sierra GM, Marín MI (2011) Amagá, Cauca and Patía Basins at: Petroleum Geology of Colombia, vol 2, ANH-EAFIT, p 104

Stern CR (2004) Active Andean volcanism: its geologic and tectonic setting. Andean Geol 31(2):161–206

Sudo A, Tatsumi Y (1990) Phlogopite and K-amphibole in the upper mantle: Implication for magma genesis in subduction zones. Geophys Res Lett 17(1):29–32

Sun SS, McDonough W (1989) Chemical and isotopic systematics of oceanic basalts: implications for mantle composition and processes. In: Saunders AD, Norry MJ (eds.) Magmatism in ocean basins, Geol. Soc. Spec. Pub., vol 42, pp 313–345

Taboada A, Rivera LA, Fuenzalida A, Cisternas A, Philip H, Bijwaard H, Olaya J, Rivera C (2000) Geodynamics of the northern Andes: Subductions and intracontinental deformation (Colombia). Tectonics 19(5):787–813

Tassinari CCG, Díaz F, Buena J (2008) Age and sources of gold mineralization in the Marmato mining district, NW Colombia: A Miocene – Pliocene epizonal gold deposit. Ore Geol Rev 33:505–518

Tatsumi Y (2003) Some constraints on arc magma genesis. In: Eiler J (ed) Inside the Subduction Factory. American Geophysical Union Monographs, 138, pp 277–292

Tatsumi Y (2005) The subduction factory: how it operates in the evolving earth. GSA Today 15(7):4–10

Thorpe RS, Francis PW (1979) Variations in Andean andesite composition and their petrogenetic significance. Tectonophysics Bd. 57 pp 53–70 Amsterdam

Thorpe RS, Francis PW, Hammill M, Baker MCW (1982).The Andes. Andesites. Ed Thorpe, R.S. 187–205

Thorpe RS (1984) The tectonic setting of active Andean volcanism. In: Andean magmatism: Chemical and Isotopic Constraints (Harmon, R.S.; Barreiro, B.A.; editors). Shiva Geological Series, Shiva Publications, Nantwich, U.K, pp 4–8

Tilton GR, Barreiro BA (1980) Origin of lead in Andean calc-alkaline lavas, southern Peru. Science 210:1245–1247

Toro G, Restrepo JJ, Poupeau G, Saenz E, Azdimousa A (1999) Datación por trazas de fisión de circones rosados asociados a la secuencia volcano – sedimentaria de Irra (Caldas). Boletín de Ciencias de la Tierra 13:28–34

Toro LM, Borrero-Peña CA, Ayala LF (2010) Petrografía y geoquímica de las rocas ancestrales del volcán Nevado del Ruiz. Boletín de Geología 32(1):95–105

Torres-Hernández MP (2010) Petrografía, geocronología y geoquímica de las ignimbritas de la formación Popayán, en el contexto del vulcanismo del suroccidente de Colombia, pp 35–132

Toussaint JF, Restrepo JJ (1982) Magmatic evolution of the northwestern Andes of Colombia. Earth Sci Rev 18:205–213

Trenkamp R, Kellogg JN, Freymueller JT, Mora HP (2002) Wide plate margin deformation, southern Central America and northwestern South America, CASA GPS observations. J S Am Earth Sci 15:157–171

Uribe-Mogollón CA (2013) Hydrothermal evolution of the Titiribí mining district. Undergraduate thesis. EAFIT University. 127 pp

Ujueta G (1991) Tectónica y actividad ígnea en la Cordillera Oriental de Colombia (Sector Girardot-Cúcuta). En Simposio sobre Magmatismo Andino y su Marco Tectónico, memorias Tomo I, 151–192

Van der Hilst R, Mann P (1994) Tectonic implications of tomographic images of subducted lithospher beneath northwestern South America. Geology 22(5):451–454

van Keken PE, Kiefer B, Peacock SM (2002) High-resolution models of subduction zones: Implications for mineral dehydration reactions and the transport of water into the deep mantle, Geochem Geophys Geosyst, 3(10):1056. https://doi.org/10.1029/2001GC000256

Velandia F, Acosta J, Terraza R, Villegas H (2005) The current tectonic motion of the Northern Andes along the Algeciras Fault System in SW Colombia. Tectonophysics 399:313–329

Villagómez D (2010) Thermochronology, geochronology and geochemistry of the Western and Central cordilleras and Sierra Nevada de Santa Marta, Colombia: The tectonic evolution of NW South America. Doctoral thesis. University of Geneve. 166 pp

Walker GPL, Wilson CJN, Froggat PC (1991) An ignimbrite veneer deposits; the trail marker of a pyroclastic flow. J Volcanol Geotherm Res 9:409–421

White WM, Dupre B, Vidal P (1985) Isotope and trace element geochemistry of sediments from the Barbados Ridge — Demerara Plain region, Atla

Wilder DT (2003) Relative motion history of the Pacific-Nazca (Farallon) plates since 30 million years ago. Graduate Thesis. University of South Florida. 106 pp.

Winter John D (2001) An Introduction to Igneous and Metamorphic Petrology. Prentice Hall, Upper Saddle River. ISBN 0-13-240342-0

Weber MBI, Tarney J, Kempton PD, Kent RW (2002) Crustal make-up of the northern Andes: evidence based on deep crustal xenolith suites, Mercaderes, SW Colombia. Tectonophysics 345:49–82

Wood DA, Joron JL, Treuil M (1979) A re-appraisal of use of trace elements to classify and discriminate between magma series erupted in different tectonic setting. Earth Planet Sci Lett 45:326–336. 122

Wörner G, Davidson J, Moorbath S, Turner TL, McMillan N, Nye C, López-Escobar L, Moreno H (1988) The Nevados de Payachata Volcanic Region 18°S/69°W, Northern Chile. I. Geological, geochemical and isotopic observations. Bull Volcanol 30:287–303

Part V
The Northern Andean Orogen

Chapter 9
Diagnostic Structural Features of NW South America: Structural Cross Sections Based Upon Detailed Field Transects

Fabio Colmenares, Laura Román García, Johan M. Sánchez, and Juan C. Ramirez

This chapter illustrates the structural architecture of the Colombian Andes, highlighting the geometric and temporal relationships between lithologic units, faulting, and folding, as derived from field transects across key areas of the complex Colombian Cordilleran system. The study permits a better understanding of the tectonic and physiographic evolution and temporal genetic linkages contained within the Northern Andes. It is primarily a visual tool, intended to aid in the interpretation of less well-known segments of the Colombian Andes and hence in the construction of structural models in sectors where insufficient geological data is available. The study is depicted within selected cross sections, which were constructed using detailed field data collected by the authors, during structural and stratigraphic surveys produced for the petroleum, coal, and mineral exploration and development industries. The structural interpretations presented herein are supported by subsurface information derived from seismic data and/or borehole mapping. Most of the sections are balanced or, in the cases where balancing was not possible, drawn with emphasis upon detailed lithologic mapping, observing lateral facies changes, unit thicknesses variations, and strict biostratigraphic control. Structural sections belonging to six regional transects are presented in printed format. An extended digital appendix containing 15 additional figures illustrates more specific structural case studies.

The following regional transects are portrayed herein in printed format:

1. Guajira Allochthon
2. Maracaibo block
3. Eastern Cordillera
4. Central Cordillera

Electronic supplementary material: The online version of this chapter (https://doi.org/10.1007/978-3-319-76132-9_9) contains supplementary material, which is available to authorized users.

F. Colmenares (✉) · L. Román García · J. M. Sánchez · J. C. Ramirez
Geosearch Ltd., Bogotá, Colombia

5. Western Cordillera
6. Chocó indenter

The following detailed sections are appended in digital format:

1. A1. Northern Eastern Cordillera, Chinácota Graben
2. A2. Northern Eastern Cordillera, Tamá Strike-slip tectonics
3. A3. Middle Magdalena Valley, La Salina-Infantas
4. A4. Eastern Cordillera, Central Sector 3D Models
5. A5. Central Cordillera, Ibagué-Cajamarca
6. A6. Western Cordillera, Bolívar Ultramafic Zoned Complex
7. A7. Upper Magdalena Valley, Ortega
8. A8. Upper Magdalena Valley, Pechúi
9. A9. Upper Magdalena Valley, San Antonio TEA
10. A10. Eastern Cordillera, Western Foothills
11. A11. Upper Magdalena Valley, Upar
12. A12. Upper Magdalena Valley, El Pensil-La Plata
13. A13. Upper Magdalena Valley, Iskana
14. A14. Putumayo, San Juan Norte
15. A15. Northern Eastern Cordillera – Rio Nevado Canyon (schematic illustration)

9.1 Guajira Allochthon

The Guajira Peninsula, constitutes the northern most feature of the South American continent. Its current tectonic setting results from the coalescence of the eastwards transported Caribbean crust against the South American Plate. Main lithofacies differ compositionally and genetically from its time-equivalents occurring in neighboring areas, such dissimilarities (noted from the Oca Fault northwards), strongly point to an allochton origin. Moreover, the structural features of the Guajira region, far from being comprehended, portray the fragmented and overprinted testimony of the complex generation and emplacement of the Caribbean oceanic crust. In this context, La Guajira region can be divided as follows:

- Maracaibo platform: conformed by crystalline basement, including Paleozoic medium- to high-grade metamorphic rocks, intruded by latest Triassic-Jurassic granitoid plutons with associated volcanic rocks, unconformably overlain by Cretaceous littoral and marine successions, which are in turn unconformably overlain by mainly Eocene conglomerates (Renz 1960; Rollins 1965; Zuluaga et al. 2007).
- Cosinas Trough: represents a rift basin containing a thick pile of essentially Middle(?) to Upper Jurassic marine sedimentary rocks, including reef carbonates, overlain by Lower Cretaceous littoral and shallow marine siliciclastics (Burgl 1960; Renz 1960; Rollins 1965).
- Central Guajira Massif: cored by migmatized gneiss in amphibolite- and granulite facies, intruded by Jurassic granitoids. Low-grade Cretaceous metamorphic rocks have been also recorded (MacDonald 1964; Lockwood 1965; Álvarez 1967).

- Caribbean Platform: includes marine sedimentary rocks deposited under shallow water to continental slope conditions. Imbricate slivers of marine sedimentary rocks and oceanic mafic and ultramafic rocks are intruded by the Eocene quartz diorite Parashi Pluton (MacDonald 1964; Lockwood 1965; Álvarez 1967; Weber et al. 2010; Cardona et al. 2014).

Internal architectural features related to the main Cuiza and Oca faults are illustrated in Fig. 9.1. The E-W-striking right lateral Cuiza fault separates the Alta Guajira Basin to the north, from the Baja Guajira Basin to the south. The parallel right-lateral Oca Fault, divides the Lower Guajira Basin from the Cesar-Ranchería Basin.

The ENE-trending Cosinas Trough (Fig. 9.2(1.1)) is bounded by two main faults: The intersecting Puralapo and Cuiza faults in the northwest and the Cosinas Fault in the south. These faults separate the Middle to Upper Jurassic-Cretaceous infill of the Cosinas Trough from outcropping Paleozoic metamorphic rocks in the north and from Mesoproterozoic to Paleozoic high-grade metamorphic rocks covered by Upper Triassic?-Jurassic sedimentary units in the south.

The Cosinas Trough basin fill was deposited upon a distended basement consisting of the Paleozoic Macuira gneiss and Triassic rocks of the Uitpana Formation. The Triassic succession is mainly conformed by sandstone and conglomerate, with minor marine shales and limestones. Sedimentation was accompanied by the emplacement of sills and dikes related to rifting and extension of the lower crust, which continued into the Middle Jurassic. Overlying the tilted Triassic strata, in clear angular unconformity, Jurassic and Lower Cretaceous units were deposited. The Middle to Upper Jurassic units include the shale-rich shallow marine deposits of the Cheterló and Cajú formations, followed by deposition of quarzitic sandstones and conglomerates of the Chinapa Formation. The basal contact of the Jurassic units is not exposed but seems to be resting in angular unconformity over Triassic or older basement, based upon the marked lateral variability in the thickness of the Chinapa Formation, which ranges from 664 to 1177 m (Rollins 1965). The former is supported by the reported facies variations within the units, which register the development of coalescent sedimentary environments related to differential subsidence rates along the graben. The Upper Jurassic formations, including the Jipi and Cuisa shales, coarsen from E to W. This facies change has impeded detailed differentiation of the Upper Jurassic units (Rollins 1965).

Cretaceous sedimentary rocks of the Guajira Peninsula are correlated with formations defined in the Maracaibo Basin. Within the Cosinas graben, the preserved succession begins with a thick basal sequence containing coarse-grained continental siliciclastics and shallow marine reefal limestones, including the upper sandstone of the Palanz Formation. Early Cretaceous deposition surpassed the limits of the graben to the S, onlapping the rift-shoulder (foot-wall of the Cosinas Fault), and even resting unconformably upon Triassic basement (Rollins 1965). During the Valanginian to Aptian, the deposition of near-shore limestones and black shales (Moina and Yuruma formations) took place under reducing conditions and is interpreted to indicate a new pulse of subsidence, probably related to the culmination of the extensional regime. Subsequently, the Albian to Cenomanian Maraca Formation is characterized by deposition of massive fossiliferous limestones and thin shale interbeds. Cenozoic

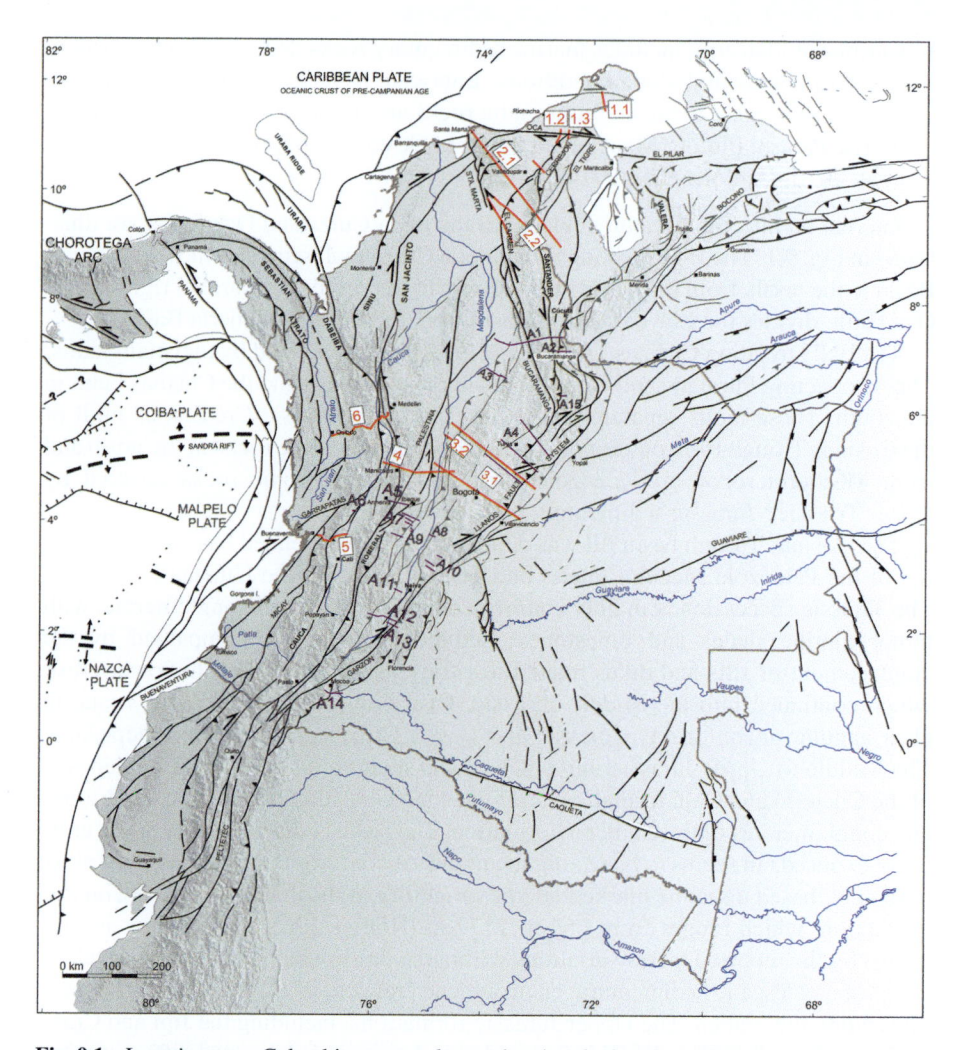

Fig. 9.1 Location map. Colombia structural map showing the locations of the figures included in this chapter: 1. Guajira (1.1 Cosinas Basin 1.2 and 1.3 Oca Fault sections); 2. Maracaibo block (2.1 Sierra Nevada de Santa Marta 2.2 Cesar-Ranchería Valley); 3. Eastern Cordillera (3.1 Cross section and restored sections in its central sector 3.2 Western Foothills of the Eastern Cordillera); 4. Central Cordillera (Bituima-Anserma); 5. Western Cordillera, Buga-Buenaventura Section; 6. Chocó Indenter, Quibdo-Caldas section. The figures included in the digital format correspond to: A1, Toledo, Chinácota Graben; A2, Tamá strike-slip tectonics; A3, Salina-Infantas Section; A4, Eastern Cordillera, central sector 3D models; A5, Central Cordillera, Ibagué-Calarcá section; A6, Bolivar Ultramafic Zoned Complex; A7, Upper Magdalena Valley, Ortega sections; A8, Upper Magdalena Valley, Pechui section; A9, Upper Magdalena Valley, San Antonio; A10, Colombia sections, Eastern Cordillera western Foothills; A11, Upar sections, Upper Magdalena Valley; A12, El Pensil-La Plata, Upper Magdalena Valley; A13, Iskana Section, Upper Magdalena Valley; A14, Putumayo, San Juan Norte; A15, Rio Nevado Canyon, Eastern Cordillera

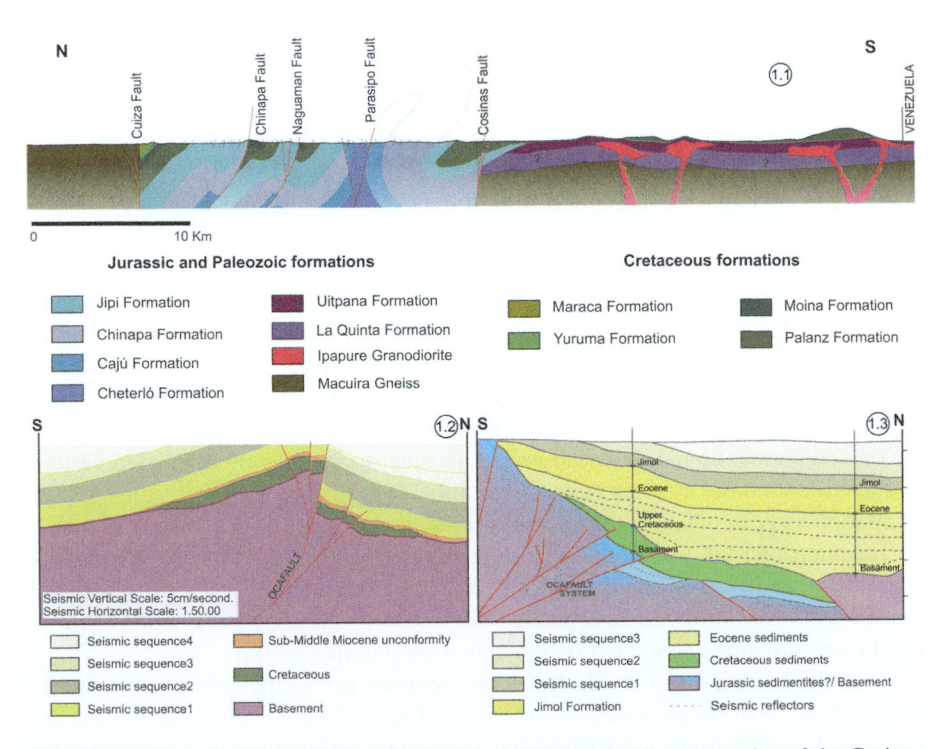

Fig. 9.2 Guajira Allochthon (Guajira Peninsula). *1.1* N-S regional cross section of the Cosinas Basin, Alta Guajira. (Modified after Rollins 1960 and Quintero 2017). Surface geology modified from Renz (1960), Rollins (1965), Irving (1972), and Zuluaga et al. (2007). *1.2* N-S profile showing the geometry of the Oca Fault in the Lower Guajira Basin. Interpreted from seismic lines CV-1300 and HUMA90–09 in TWT (Cediel et al. 1998). *1.3* Geometry of the Oca Fault further to the E in relation to 1.2. The cross section is controlled by two wells. Note the preservation of Eocene sedimentites in the northern (hanging wall) of the fault

sedimentary units are preserved only in isolated patches, within the Cosinas Trough, to the south along the Cosinas platform and to the north within the Macuira Range, where Eocene conglomerates of the Maraca Formation crop out (Rollins 1965).

In the present structural configuration, the Cuiza Fault truncates to the north the Cosinas Trough, right laterally dislocating the original graben. The prolongation of the trough can be traced to the NE in the Cocinetas area, partially outcropping beneath the Cenozoic cover.

The internal geometry and general structural and crosscutting relationships of the basin fill sequences were characterized in three main patterns by Rollins (1965). These include (1) an older set of longitudinal E-W faults and folds, (2) a group of NW-SE strike-slip faults that crosscut the E-W longitudinal set, and (3) a minor set of NE-SW strike-slip faults that intersect the E-W set and are truncated by the NW fault set. The major faults illustrated in the N-S section of the Cosinas Trough include the E-W Chinapa, and the NE-SW oblique right-lateral Naguaman and Parasipo faults.

The foldings affecting the Mesozoic Basin fill exhibits two main axial orientations: (1) E-W longitudinal folds subparallel to the older faults, which may additionally represent compression during inversion of the frontal flank of the basin, and (2) NE-trending folds that display a consistent oblique geometric relationship to the main E-W strike-slip faults. The integrated array of faults and folds is interpreted to represent transpressive deformation of the graben infill (Bartok 1993), beginning at the end of the Cretaceous.

The Fig. 9.2(1.1) was compiled from the most consistent and verifiable surface geological information (Irving 1972; Rollins 1965). The principle features within the Cosinas Trough are (1) in the north, the near-complete, steeply dipping Cretaceous succession, piggybacked upon SE-verging thrust faults; (2) in the central segment, verticalized Jurassic strata, including the basal Cheterló Formation, exposed in the core of a major overturned anticline; and (3) to the south, an overturned, north-verging syncline with Lower Cretaceous units exposed in its core. Along the common flank of these folds, the Chinapa Formation crops out, in contrast, it is absent in the other aforementioned areas. Jurassic units outcrop exclusively in the depression. The Cretaceous basal unit, the Palanz Formation, rests unconformably on the Uitpana Formation and extends over the southern rift shoulder.

The main internal structural features include the steeply dipping, inverse Chinapa, Naguaman, and Parasipo faults, which have been interpreted as compressional or transpressive structures. An alternative interpretation considers these faults as positive flower structures generated by transpression (Gómez 2001).

The Fig. 9.2 (1.2 and 1.3) illustrates two sections across the E-W Oca Fault and the progression of its geometry from W to E. The fault shows vergence toward the north, with an important sense of reverse movement and transposition of strike-slip failure. Broadly, the Oca Fault affects the Neogene sedimentary sequence, which at its base has been interpreted as a restricted carbonate buildup unconformably supported on the basement (Jurassic-Triassic sedimentary or crystalline units). Wells drilled in the northern side of Oca Fault evidenced the occurrence of Lower-Middle Miocene sedimentites, which are absent on the southern side. Within the depicted seismic section, at least four Neogene sequences are discernable.

The Oca Fault, should be understood as well, as a right-lateral mega-shear zone that represents the tectonic boundary between the South American and Caribbean plates, along which the latter has moved obliquely with respect to the continental margin, thus, accumulating a lateral displacement of more than 75 km. Besides, this tectonic zone would have played a major role in the conjoined uplifting of the SNSM block, as recorded by its enormous vertical elevation (see Fig. 9.3).

9.2 Maracaibo Block

9.2.1 Sierra Nevada de Santa Marta

The roughly tetrahedral coastal Sierra Nevada de Santa Marta (SNSM), is the northernmost prominent topographic feature (5770m-high) in the South American Plate. It is mainly composed by a wide variety of igneous and metamorphic rocks, ranging from Precambrian to Paleogene in age. A fringe of Jurassic volcano-sedimentary

9 Diagnostic Structural Features of NW South America: Structural Cross Sections… 657

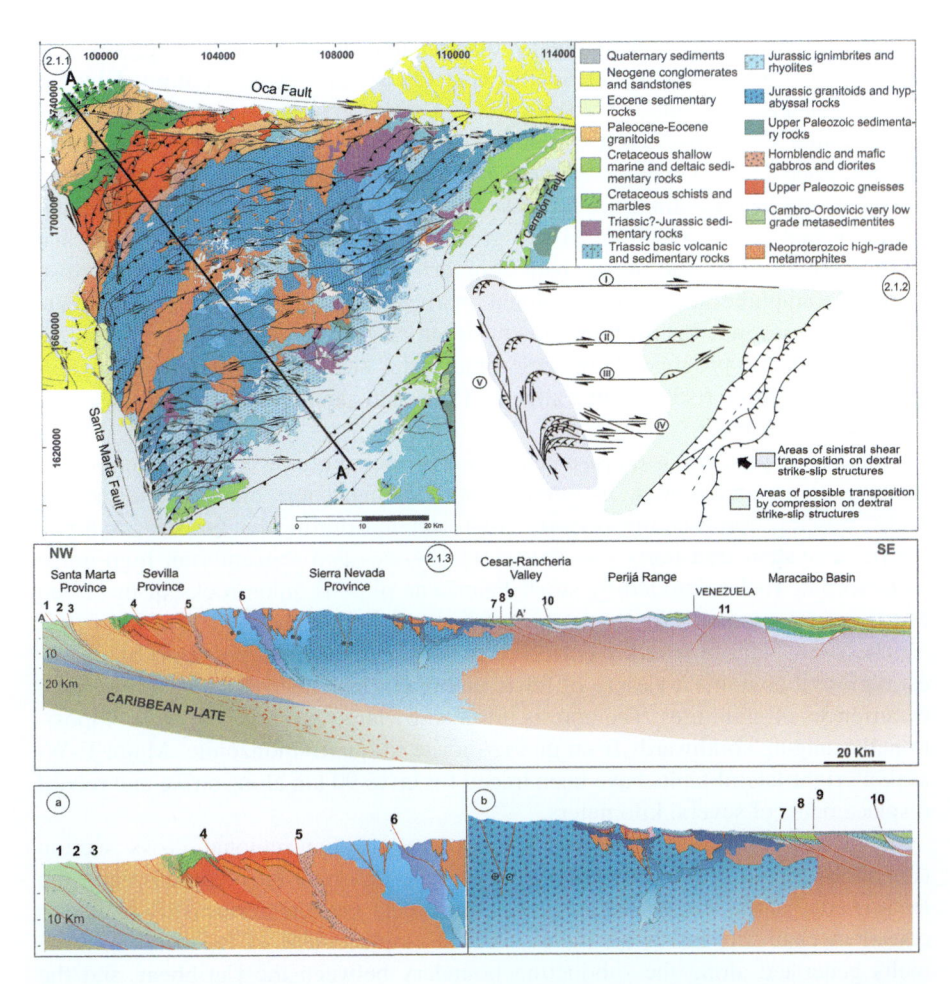

Fig. 9.3 Sierra Nevada Santa Marta. *2.1.1.* Generalized regional map of the SNSM. *2.1.2* Tectonic transposition. Generalized structural map of the SNSM and localization of the structural patterns associated with a tectonic transposition interpretation. The highlighted areas show the deformation fields configured by the associated structural patterns (Geosearch Ltda 2007). *2.1.3* Regional cross section of the Sierra Nevada de Santa Marta, Cesar-Ranchería Valley and Maracaibo Basin. 1, Taganga Fault; 2, Florían Fault; 3, Jordán Fault; 4, Orihueca Fault; 5, Manitza Fault; 6, Mindigua Fault; 7, Rio Cesarito Fault; 8, Cesarito Fault; 9, San Diego Fault; 10, Media Luna Fault; 11, El Tigre Fault; a) NW extreme in detail (Mesa and Rengifo 2011; Sanchez and Kammer 2017) of the SNSM section illustrating the NW-verging imbricate systems. *b)* Detailed cross section of the SE slope of the SNSM comparing the inclination of the top of the exposed basement in the highlands and underlying Cesar-Ranchería Valley sedimentary cover. Here illustrated as well the successive onlapping of the Jurassic to Cretaceous sedimentary units and its northwestward-thinning

cover, constituted by continental clastic and felsic volcanic rocks, outcrops along the southeastern margin of the SNSM, where it is unconformably covered by essentially marine Cretaceous deposits. The regional cross section (Fig. 9.3(2.1.3)) illustrates the structural framework of the SNSM, the Cesar-Ranchería Basin, the Perijá Range, and the Maracaibo Basin.

The SNSM has been descriptively divided into three geological "provinces", including (1) Santa Marta, (2) Sevilla, and (3) Sierra Nevada, delimited by the Orihueca and Sevilla faults (Tschanz 1969).

The Santa Marta province, accreted during the Late Cretaceous to Paleocene time, is conformed by two low-grade metamorphic belts, intruded by the Paleocene-Eocene Santa Marta Batholith. The coastal belt consist of a suit of low- to moderate-grade metamorphic rocks including greenschists, phyllites, mica schists, and amphibolites, while the inner belt contains greenschist and graphitic schist, mylonite, amphibolites, and occasional marble lenses. NNW-verging imbricate fault slices constitute the most prominent structural feature. The intervening reverse faults disrupt the progressive metamorphic facies-succesion, imbricating higher-grade metamorphic rocks against slivers of lower metamorphic grade, as observed along the Florín Fault.

The Sevilla province is constructed by a poly-metamorphic complex comprising mafic gneisses and schists, including granulites and some blocks of ultramafic rocks, possibly representing oceanic crust accreted during the Jurassic. A thrust imbricate system transports late Paleozoic gneisses and Precambrian high-grade metamorphic rocks intruded by small Paleocene plutons, upon rocks of the Santa Marta province to the NW.

The Sierra Nevada province corresponds to the central and topographically highest region. It contains a variety of felsic and mafic igneous rocks, as well as local occurrences of granulites. The core is intruded by large Jurassic plutons, compositionally ranging southwards from quartz-diorite to quartz-monzonite. Major E-W-striking right-lateral faults have been traced for up to 90 km along strike and record displacements of several kilometers.

A detailed illustration of the NW extreme of the regional cross section (Fig. 9.3(2.1.3a)), shows how the brittle-ductile deformation recorded along the leading edge of the over-thrusting continental block, must be intrinsically related to coeval mechanisms involving the rising and emplacement of Paleocene-Eocene melts generated along the subducting boundary between the Caribbean and the South American plates. In this context, the elongated hinterland-dipping root of the Santa Marta Batholith appears parallel to the associated deep mega-shear from which the thrust system branches.

On the other hand, along the southeastern slope of the SNSM, (Fig. 9.3(2.1.3b)), the exposed, tilted surface marking the top of uplifted crystalline basement, is notoriously coincident and continuous with the basement surface beneath the sedimentary cover in the Cesar-Ranchería Basin, as revealed by seismic studies. Moreover, the progressive northwestward onlapping of the Paleozoic, Jurassic and Cretaceous continental, and marine sedimentary strata, onto the basement of the SNSM, as well as the absence of Cretaceous or younger sedimentary units overlying the middle or topographically high areas of the SNSM, suggests that the tilting of the entire SNSM block toward the ESE was probably already initiated by the late Paleozoic and that an important portion of the SNSM was already exhumed by this time.

Tectonic Transposition

Six distinct patterns have been distinguished within the structural framework of the SNSM (Fig. 9.3(2.1.2)).

The anomalous mechanical and geometric character of the western terminations of the dextral strike-slip fault set, and especially the marked northwestward convexity observed within an overall compressive domain, strongly suggests the transposition of tectonic fabrics. Such geometric features could indicate interference between the left-lateral shear generated by the Santa Marta Fault System, on the right-lateral structures along and intersecting 30 km-wide corridor which occurs parallel to the western limit of the SNSM.

Along the southeastern margin of the SNSM, geometric features indicating right-lateral stress patterns appear to be over-imprinted by the compressive effect of the Perijá fold and thrust front, resulting in the observed anomalous arrangement of structures marking the easternmost terminations, as shown in the highlighted areas in Fig. 9.3(2.1.2).

From the central segment of the SNSM to the NE, a complex array of sigmoidal transpressive duplex structures are recorded, backed by the Rio Barcino Fault. This fault leads, along its 65 km length, to the successive transference of displacement between E-W right-lateral structures in the central SNSM and the Oca Fault megashear (Geosearch 2007).

9.2.2 Cesar-Ranchería Basin

The Cesar-Ranchería Basin (CR) is an elongated, asymemetric, N30E-oriented depression located between the southeastern border of the SNSM and the Serranía de Perijá (SP). The Mesozoic to Cenozoic sedimentary cover of the SP and CR rests on a SE-tilted basement consisting of Precambrian to early Paleozoic crystalline rocks. The present axis of the SP roughly corresponds with the depocenter axis of a Triassic-Jurassic rift, which distended the possibly already tilted basement surface. This process generated accommodation space for the deposition of a succession of late Triassic-Jurassic to early Cretaceous sedimentary and volcanic rocks, including the La Quinta Formation red beds and the Girón Group and equivalents as well as the Lower Cretaceous Rio Negro Formation. Rifting waned in the early Cretaceous and was followed by passive margin sedimentation during the Aptian, which buried the overall prism-shape of the Triassic-Jurassic sedimentary succession. Thin marine deposits, including sandstones, shales, and carbonates (Lagunitas, Aguas Blancas, La Luna, and Molino Shales formations), were deposited on a SE-tilted broad platform during the thermal sagging in the Late Cretaceous (Mesa et al. 2011).

The structural character of the CR began to be shaped at the end of the Cretaceous (Maastrichtian-Paleocene), and ultimately in the Early Eocene, during the climax of inversion of the rift-bounding faults of the Perijá graben due to NNW-SSE compressive forces, producing the uplift of the asymmetrically bivergent Perijá Range. During the SP uplift, the CR transitioned to an incipient foreland basin, parallelly assimilating deformation from northwest-verging reverse faults propagating form the inversion front. Uplift along the Perijá front was accompanied by deposition within the CR, of Paleocene fluvio-lacustrine sediments (Barco-Los Cuervos formations), inner platform shallow marine carbonates and littoral sediments (Hato Nuevo, Manantial formations), the Cerrejón sandstones, shales and coal horizons, and braided coarse fluvial sediments of the Tabaco Formation. Sub-Eocene and sub-Oligocene stratigraphic hiatuses are marked by regional unconformities, indicative of successive pulses of tectonic uplift. Deltaic deposits and rare reef structures ocurring in a shallow marine platform with tidal influence are assigned an Eocene-Oligocene age.

The Late Miocene to Pliocene were characterized by molassic, alluvial, and fluvial sedimentation (Cuesta and San Antonio formations), related to renewed deformation and uplift along the SP, especially well marked by a sub Late Miocene unconformity.

The three cross sections of the CR (Fig. 9.4(2.2)) illustrate the structural geometry of an advanced foreland basin that closes toward the north. It is emphasized how the main faults along the Perijá thrust front placed older units, and even crystalline basement, upon Cenozoic sedimentary deposits of the CR.

Deformation of the CR advanced to the NNW, involving primarily fold propagation along thrust planes of the shortcut system related to the main inversion faults of the SP.

The progressive NNW onlap and thinning of Cenozoic units within the CR evidences the continuity of deformation during much of the history of the basin. A pronounced pulse of deformation during the early Eocene is best registered in the sedimentary record.

Since the end of the Cretaceous, the Verdesia high represented a positive topographic barrier within the central CR that persists to the present day. This feature separated isochrone-heterotopic lithofacies in the resulting northern and southern sub-basins. Right-lateral activity along the Ariguanicito Fault, combined with a vertical component, seems to be related to the structural subdivision of the CR (Geosearch 2007).

9.3 Eastern Cordillera

The northernmost Andean ranges are physiographically divided in three cordilleras. In Colombia, the Eastern Cordillera (EC), elongated *ca.* 1000 km, exhibits a roughly sigmoidal shape. The southern segment trends NE and then bends from the Sierra Nevada del Cocuy to the NNW. Parallel to this direction, and around 90 km farther away from the Cocuy deflection, the Cordillera branches to the NE in the Sierra de Merida of the Venezuelan Andes. The structural architecture of the orogen is, in part, shaped by the presence of three large basement massifs, including the Garzón Massif to the SW, the central Quetame and Floresta massifs, and the Santander Massif to the NW.

The western and eastern boundaries of the Eastern Cordillera are defined by discontinuous en echelon fault segments that limit the cordillera along the Magdalena Basin to the W and the Putumayo, Eastern Llanos and Maracaibo Basins, to the E. These foreland basins contain deposits derived from the continuous deformation and distinct phases of uplift that affected the EC region, mainly during the early Eocene, early Oligocene, and early Miocene.

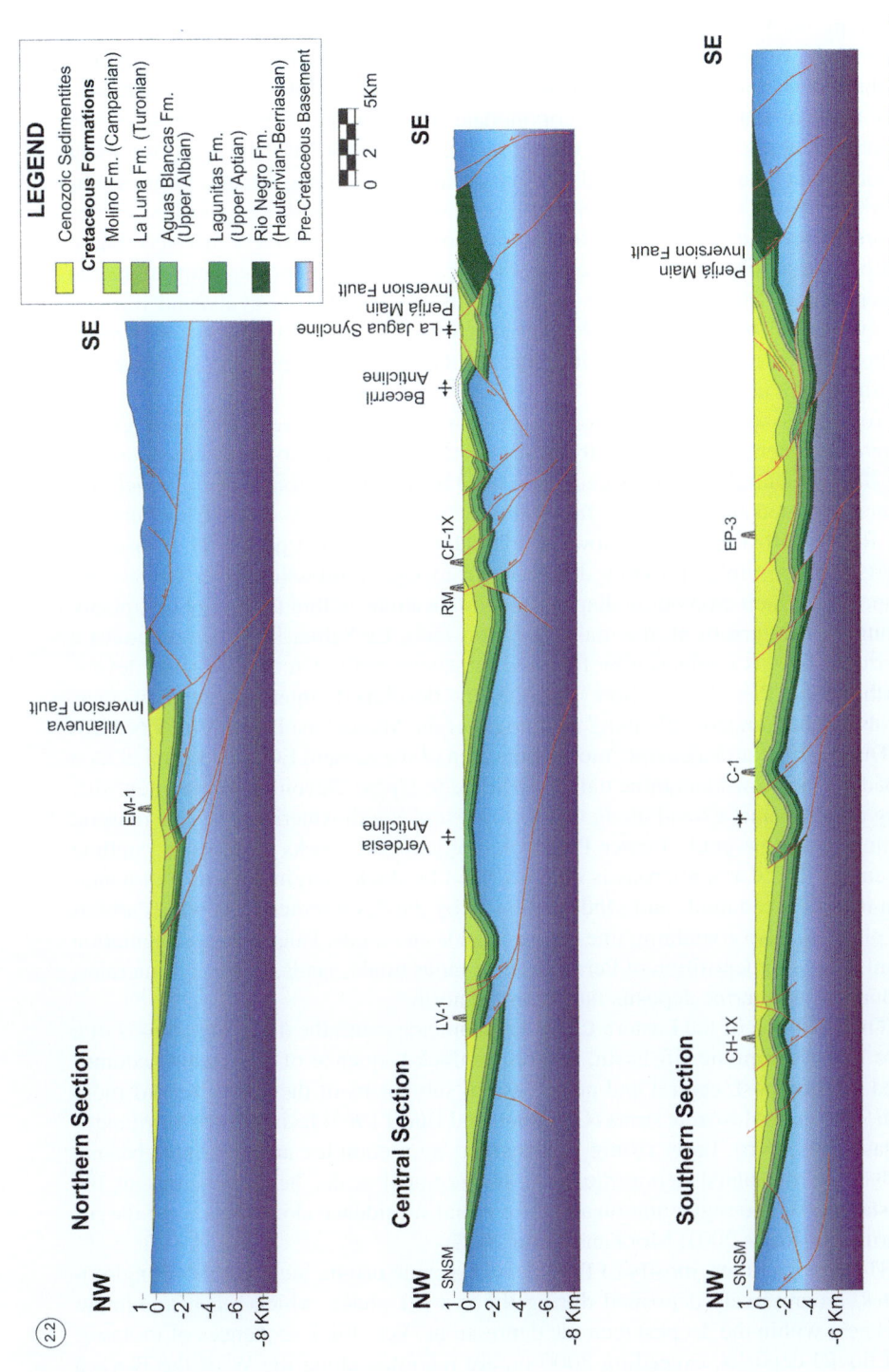

Fig. 9.4 Cross sections across the Cesar-Ranchería Basin. PMIF, Perijá main inversion fault. Oil wells. EM, El Molino-1 (proyected); EP-3, El Paso-3 (proyected); C-1, Compae-1; CH-1X, CesarH-1X; LV-1, Los Venados-1 (proyected); RM, Rio Maracas (proyected); CF-1X, Cesar F 1X (proyected) (Mesa and Rengifo 2011)

Along the Eastern Foothills of the EC, deformation is characterized by broad synclines developed in the hanging wall of the Guaicáramo and Yopal faults, cored by thick successions of Cretaceous marine and Cenozoic Molasse deposits. A complex array of faults and folds accommodate inversion and exhumation along the Eastern Foothills. Within the illustrated sections (Fig. 9.5(3.1)), the most important structures include the Tesalia and the Santa Maria faults, along which a large, asymmetrical, SE-verging propagation fold is developed. The Farallones anticline exhibits pre-Cretaceous rocks in its core, and its frontal flank is strongly inverted. These complex structures accommodate major vertical displacement (inversion) along the eastern front. The NE-plunging of the Farallones anticline is evidenced through comparison of the northern and southern sections. The included restoration demonstrates how extensional faults with large displacements may provide a mechanism for significant uplift.

Toward the west, the Cretaceous succession dips progressively westward, exhibiting detachment-style asymmetric folds rooted at Jurassic-Lower Cretaceous levels. This structural pattern extends westward to the Don Alfonso Fault, which has been interpreted as an inversion fault. The central region, between the Chocontá and the Rio Minero faults, also shows a dominant thin-skinned pattern comprised of short amplitude folds possibly detaching from Upper Jurassic-Lower Cretaceous evaporites, which crop out in diapirs at many locations within the Bogotá altiplano. Along the western front, the main inversion fault, La Salina-Bituima, generates a shortcut complex composed by fold and thrusts rooted in Cretaceous shaly levels. Furthermore the deformation mechanism developed intercutaneous wedges involving the Cenozoic fill along the margin of the Magdalena Basin (Fig. 9.6(3.2)).

During the Late Paleozoic, most of the area of the present EC was occupied by a broad and shallow epicontinental sea. Middle to Upper Devonian marine deposits, very rich in fauna, extend along the northwestern South American passive margin, resting upon low-grade Lower Paleozoic metamorphic rocks or older crystalline basement. The Carboniferous is characterized by thick marginal basin sedimentation including red muds and sands, followed by the development of a wide carbonate platform, also containing fine-grained sands and muds. Paleozoic sedimentation terminates with deposition of Permian terrigenous muds, sands, and conglomerates, including thin marine deposits farther to the north.

The axial line of the Eastern Cordillera coincides with the depositional axis of a Late Triassic-Jurassic rift basin, wherein a thick sequence of sediments accumulated. The deepest, central and northwestern sub-basins of the region record more than 9000 m of Mesozoic strata (Campbell and Bürgl 1965; Irving 1975; Sarmiento-Rojas et al. 2006). Early rifting was marked by a complex array of intra-basinal faults and associated depocenters. At the regional scale, heterogeneities at the crustal level led to segmentation and differential subsidence along the axis of the rift (Sarmiento-Rojas 2001; Mora and Parra 2008).

The Jurassic units, mostly of fluvial and volcanic origin, include thick conglomerates, progressively deposited during the syn-rift phase, which reach maximum thickness within the deepest tectonic depressions. Very thick sequences of molassic and fluvial deposits, exceeding 8000 m, are recorded along the W of the Boyacá

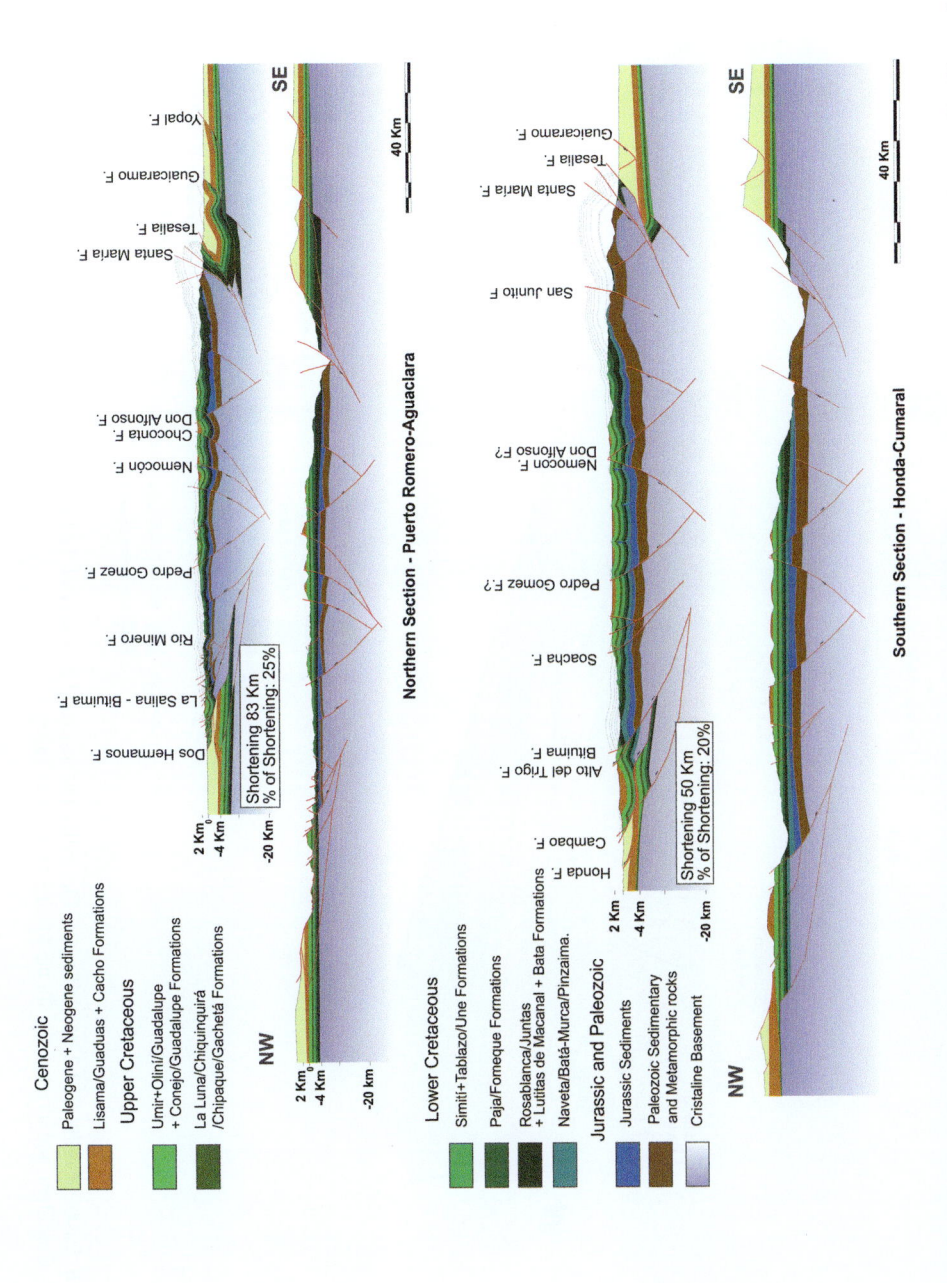

Fig. 9.5 Eastern Cordillera. Cross section and restored sections of the Eastern Cordillera in its middle sector. Restoration reference, Top Upper Cretaceous. *3.1* Cross section and restored sections of the Eastern Cordillera in its central sector. Restoration reference level: Top Upper Cretaceous

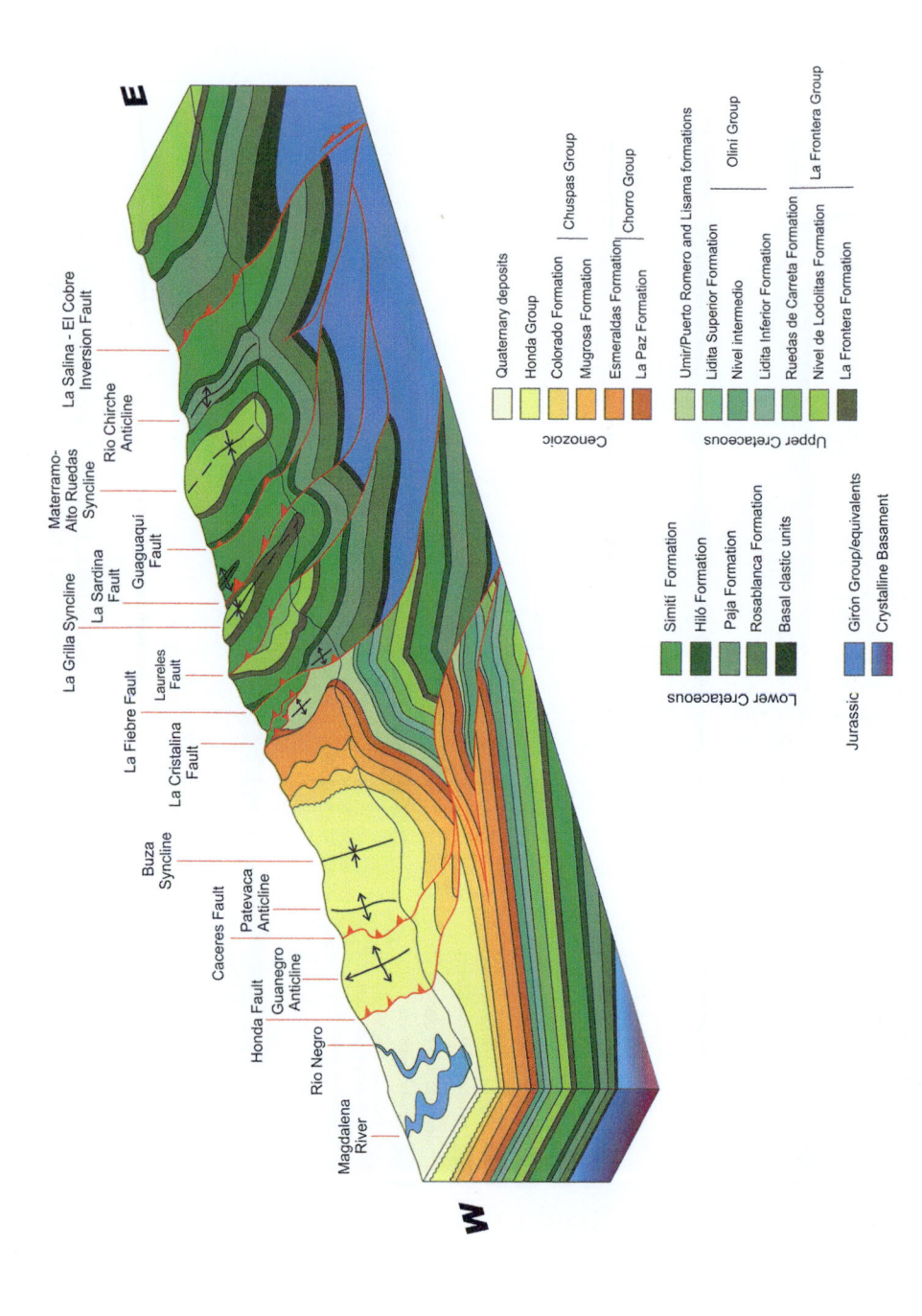

Fig. 9.6 Eastern Cordillera 3.2 Block diagram illustrating the structural geometry of the footwall of La Salina Inversión Fault in the western foothills of the Eastern Cordillera, and its relation with the Middle Magdalena Valley

Fault, one of the master rift-related faults (Fig. A4, Digital appendix). Such deposits do not occur to the E of the fault.

Regional extension prevailed into the very early Cretaceous, as recorded by the concomitant accumulation of thick, irregular, and discontinuous coarse-sand bodies, deposited along fault escarpments and locally extending onto the shoulders of the rift. Subsequently, regional deposition of mainly shallow marine and littoral sedimentary deposits throughout most of the Cretaceous, evidences the progressive marine invasion, and indirectly marks an eventual and prolonged phase of thermal subsidence, which leaded to the colmatation of the basin. Sedimentation during the latest Cretaceous marks a basinal "overflow," extending for some distance outside the basin limits, reaching areas of the tilted rift shoulders. Accordingly, areas that accumulated and preserved greater thicknesses of sedimentary deposits are located to the E of the Gámeza and equivalent faults (Fig. A4, Digital appendix), while in the central region (Floresta and Tibasosa), the accumulated thicknesses are significantly thinner. Apparently, the Arcabuco and Floresta areas were uplifted to some degree during this time.

Punctuated compression related to the approach, collision, and accretion of oceanic terranes along the Colombian Pacific margin during the Paleogene drove the final expulsion of the sea, initial reactivation of the Mesozoic Basin margin, and the development of a foreland basin to east. Sedimentation changed to marginal and fluvial conditions, reflecting the cessation of regional subsidence.

During the late Paleocene and into the early Eocene, increasing episodes of compression accelerated basin inversion, generating asymmetrical bivergent uplift and the incipient development of the eastern and western foreland basins. Thick successions of coarse clastic fluvial sediments were deposited as a consequence of the initiation and acceleration of uplifting pulses.

During the Miocene, the process of inversion along most of the Mesozoic rift-related normal faults became a generalized feature. Sedimentation along the foreland basins, coeval with exhumation, continued throughout the Cenozoic. Clastic deposition was also recorded within the nascent intermontane basins of the central portion of the EC. The accumulation of coarse-grained materials during the Pliocene, in both the intermontane and foothill basins of the EC, marks the end point of compressive deformation and uplift, responsible for the present geomorphology of the range.

The highest summit of the EC, in the Sierra Nevada del Cocuy (5410 m), coincides also with the deepest region of the basin where the thickest infill of Mesozoic sedimentation was accumulated. This particular site, could represent an ancient triple junction of the Mesozoic rift system, that nowadays exhibits one of its branches (Mérida), dislocated more than 90 km to the NNW, along the left-lateral Chitagá Fault System (Fig. A2, Digital Format).

9.4 Central Cordillera

The Central Cordillera of Colombia is constituted by fragmented Neoproterozoic basement, overlain by low- to medium-grade metamorphic rocks of the early Paleozoic Cajamarca Complex, including the Padua gneiss, and intruded by late

Permian to early Triassic meta-granitoids (*e.g.*., La Línea, Manizales), as well as by Jurassic granitoids. Along the western margin of the cordillera, the Arquía and the Quebradagrande complexes, outcrop as slivers in the Romeral tectonic zone (mélange).

Cretaceous to early Eocene granitoids (*e.g.*, Antioquia, Sonsón, El Hatillo, El Bosque), intruding the older rocks, were generated during plate convergence, oblique subduction and delamination of oceanic crust along the western Colombian paleo-margin. Miocene-Pliocene volcanoclastic, material derives from the active volcanic arc which marks the crest of the range.

The included regional section (Fig. 9.7) encompasses the complex litho-tectonic and structural frame of the Central Cordillera and the Magdalena and Cauca valleys. Multiple episodes of basin formation, closure and collision, metamorphism, and uplift can be interpreted. The oldest recognized event was identified and dated by Restrepo-Pace et al. (1997), as the emplacement of Mesoproterozoic continental basement in the region of the present Central and Eastern cordilleras. Dating (ca. 1.14 and 0.94 Ga), correlates the high-grade metamorphic rocks with the Grenvillian event, which generated the collisional belt during the assembly of Rodinia. The protoliths of these rocks correspond in part to a sedimentary sequence deposited between *ca.* 1.25 and 1.14 Ga, with source components dating from *ca.* 1.6 to 1.25 Ga (Restrepo-Pace et al. 1997).

In the Central Cordillera, early Paleozoic rocks correspond to an assemblage of meta-sedimentary and meta-volcanic rocks, including green and graphitic schists, gneisses, quartzites, and locally marbles, comprising the Cajamarca and Valdivia Groups, often referred to as the Cajamarca Complex. Protoliths of these rocks were deposited between the Neoproterozoic and the Early Paleozoic and metamorphosed during the Late Ordovician to Early Devonian (ca. Caledonian). Age and facies equivalents of the Cajamarca Complex are documented within the basement of the Eastern Cordillera, including schists and gneisses recorded at Silgará, Busbanzá, Quetame, and Güejar.

During the Late Triassic to Late Jurassic occurred the emplacement of large plutons along the entire axis of the Central Cordillera including the Mocoa, La Plata, Ibagué, Segovia, and Norosí Batholiths. Penecontemporaneous volcano-sedimentary sequences of intermediate to felsic composition located along the eastern margin of the batholiths (*e.g.*, Trumpy 1943) could be related to back-arc extension possibly combined with slab rollback (Bartok 1993; Sarmiento-Rojas et al. 2006; Cochrane et al. 2014). The age of the adjoining magmatic phases successively indicates the shifting path of the magmatic arc towards the continent.

The La Salina Fault is an important inversion fault located along the western flank of the Eastern Cordillera, as evidenced by seismic data, geological mapping and stratigraphic surveys. During the consolidation of the Jurassic rifting, it controlled subsidence and subsequent differential sedimentation on the basement of the present-day Central and Eastern cordilleras. To the east, a thick Cretaceous marine succession is recorded, while in the Central Cordillera region, significantly thinner, conglomeratic, and marginal marine clastics were deposited (*e.g.*, San Félix, Valle Alto, Abejorral formations).

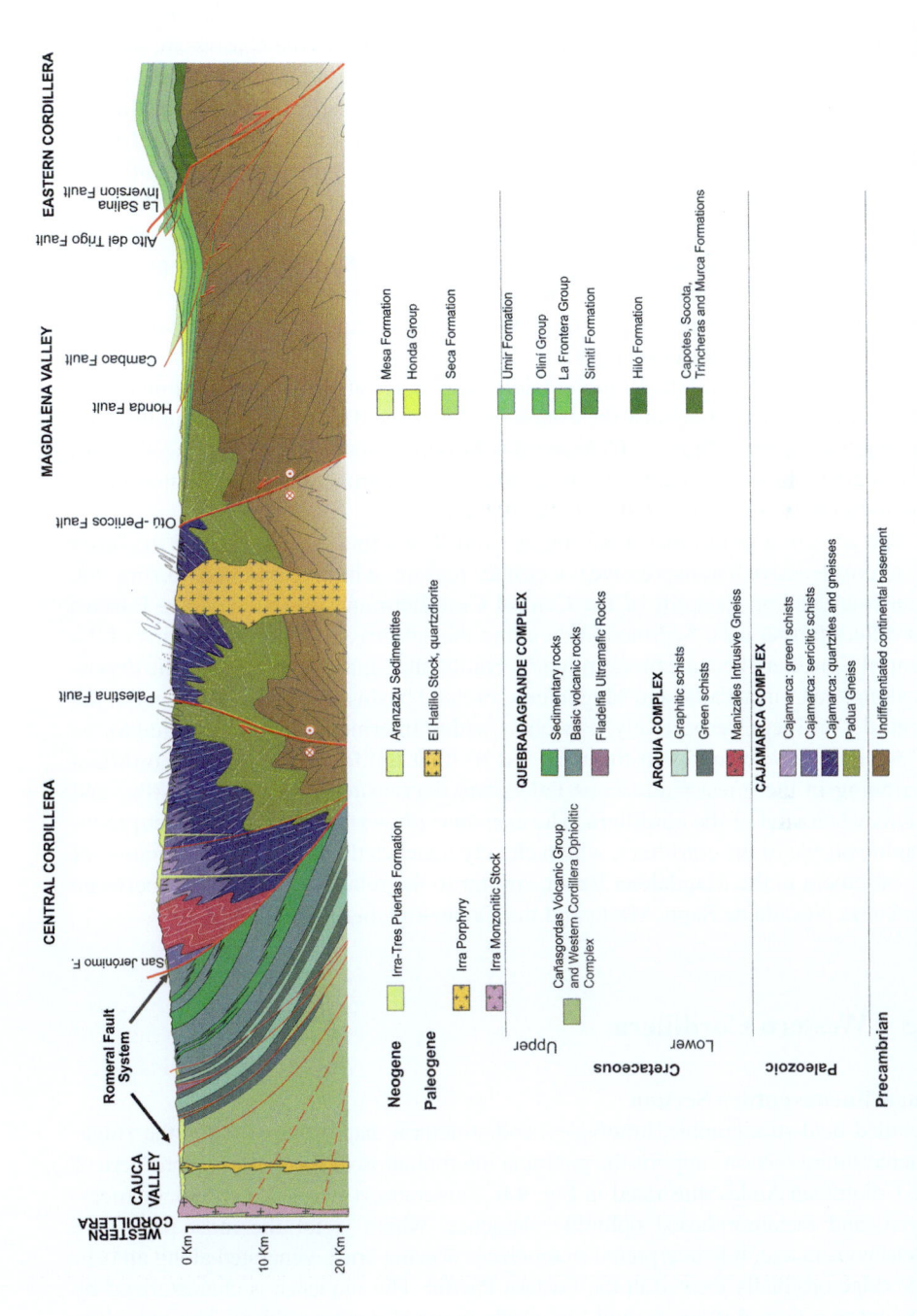

Fig. 9.7 Central Cordillera. Regional section (E-W), showing the relation of Central Cordillera, Magdalena and Cauca valleys, and Eastern and Western Cordilleras Foothills

The most relevant tectonic feature in the western flank of the Central Cordillera is the Romeral Fault System or Romeral Tectonic Zone. It is constituted by numerous sub-parallel regional faults, extending from Ecuador to the Caribbean, along a fringe of some 6 km wide. This tectonic zone separates continental basement to the East, from a mosaic of oceanic affinity fragments and associated marine sedimentites to the west. Westwards of the San Jerónimo Fault, within the Romeral Fault System, a tectonic *mélange* conformed by a thick succession of mafic volcanic and marine sedimentary rocks, along with detached slivers of the paleo-continental margin, was identified as the Arquía and Quebradagrande complexes). This mélange comprises in some regions (Génova, Jambaló), slices of high-pressure metamorphic rocks. The mafic volcanic rocks correspond to tholeiitic to calc-alkaline magmas, associated with development of oceanic crust and immature island-arc volcanism along the continental paleo-margin.

Farther to the west of the Romeral Fault System, terranes of oceanic affinity were tectonically emplaced against the continental margin. Thick, internally imbricated tectonic slices, including the Cañasgordas Group (Chocó Block) and the Western Cordillera Ophiolitic Complex to the south, represent punctuated accretionary continental growth ocurring in Late Cretaceous-Paleocene.

The accretion of terranes along the western Colombia margin marked the onset of a compressive (transpressive) tectonic regime which continued during the Cenozoic, leading to uplift of the Central Cordillera and inversion of the Eastern Cordillera farther east. Sedimentation along the eastern and western margins of the Central Cordillera became predominately continental, marked by siliciclastic deposits along the Cauca-Patía, and Magdalena foreland basins, to the west and east of the Central Cordillera, respectively, as well as within intermontane basins (Amagá).

Accretionary tectonics in the west led to final uplift of the Central Cordillera beginning in the latest Cretaceous-Paleocene, increasing the eastward tilting and extensive erosion of the cordillera. The eastward tilt is revealed in the present topographic profile of the cordillera, which closely matches the subsurface inclination of the basement in the Magdalena Basin, similar to the relationship recorded between the Sierra Nevada de Santa Marta and the Cesar-Ranchería Basin.

9.5 Western Cordillera

Buga-Buenaventura Section
Detailed field stratigraphic, lithofacies, and structural data gathered along the Buga-Buenaventura section supports the geologic interpretation of the Western Cordillera of the Colombian Andes illustrated in Fig. 9.8. This transect exposes a partial, dismembered, and metamorphosed ophiolite sequence, which is for the most part, Late Cretaceous in age. It is interpreted to represent oceanic crust, generated along an oceanic ridge originally located in the Eastern Pacific. The sequence is characterized by the superposition of dynamic and low-grade regional metamorphism. The ophiolitic suite was transported eastward and was partially consumed beneath the continental

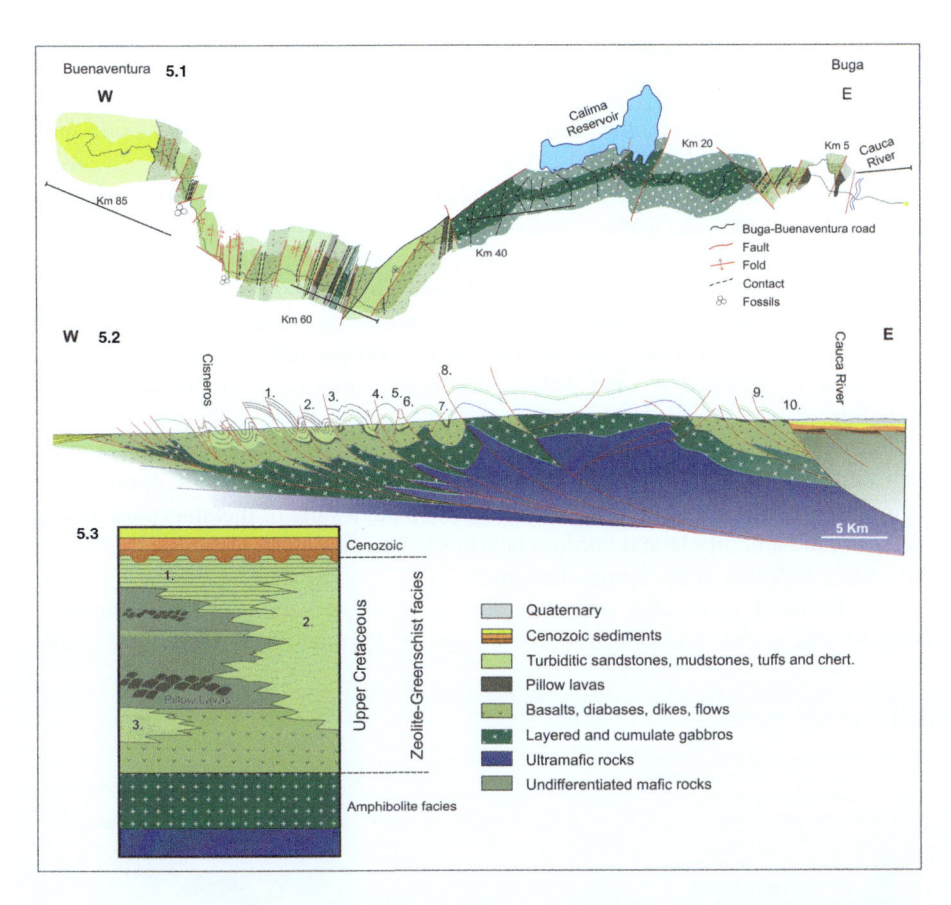

9 Diagnostic Structural Features of NW South America: Structural Cross Sections...   669

Fig. 9.8 Western Cordillera (Buga-Buenaventura). *5.1* Regional section (E-W) of the Western Cordillera between Buga and Buenaventura. Faults from W to E. 1, El Naranjo Fault; 2, Puerto Nuevo Fault; 3, Rio Bravo Fault; 4, Dagua Fault; 5, Calima Fault; 6, Zabaleta Fault; 7, Tragedias Fault; 8, Santa Bárbara Fault; 9, Media Canoa Fault; 10, Cauca Fault. *5.2* Detailed geological map along the Buga-Buenaventura road in the Western Cordillera. *5.3* Stratigraphic scheme showing the general relations of the units depicted in the regional cross section. 1, Mudstones, cherts, and tuffs; 2, sandstones, mudstones, cherts, and tuffs; 3, turbiditic sandstones (see section legend)

margin, within a Late Mesozoic subduction zone, represented in surface by the present-day Romeral Fault System. Tectonic stacking of these oceanic crust fragments influenced the repositioning of the subduction zone by shifting it gradually to the West. The Buga-Buenaventura section reveals a series of allochthonous imbricated slices, elongated along N-S to NNE-SSW axis. Tectonic stacking verges predominantly to the west. Individual slices contain gabbro, diabase, basalt, pillow lava, turbidite, chert, and pelagic sediments in complex tectonic juxtaposition. Ultramafic rocks, including the Bolívar zoned ultramafic complex (see Fig. A6, Digital Appendix), located some 45 km NNE of the Buga-Buenaventura section, are interpreted as the basal part of the ophiolite suite.

9.6 Chocó Indenter

A transect across the northern sector of the Western Cordillera (Fig. 9.9), extending from the town of Caldas (Central Cordillera) to Quibdó (Atrato Valley), reveals a system of imbricated west-verging reverse faults and localized back-thrusts, affecting for the most part, the Cañas Gordas Group.

This unit, consists of marine sediments, pillow lavas, and mafic to intermediate volcanic rocks, of Early Cretaceous age (Botero 1963). To the west of the El Toro Fault, outcrop volcanic and marine rocks of Cenozoic age (*e.g.*, Santa Cecilia and La Equis formations), which rest unconformably on the ophiolitic sequence. The succession is intruded by the Paleocene-Eocene, dioritic-granodioritic Mandé Batholith.

To the east of the Cauca Fault, within the Romeral Tectonic Zone, the exposed rock units show affinity with the basement of the Central Cordillera. Among the main units appears the Pueblito dioritic complex and a suite of metamorphic rocks, including the amphibole and epidote-bearing Sabaletas schist, considered to form the eastern segment of the Arquía Complex.

The Romeral zone, also understood as a *mélange* zone, exposes Paleozoic metamorphic rocks, gneissic granitoid bodies of Permian(?) age, and low-grade

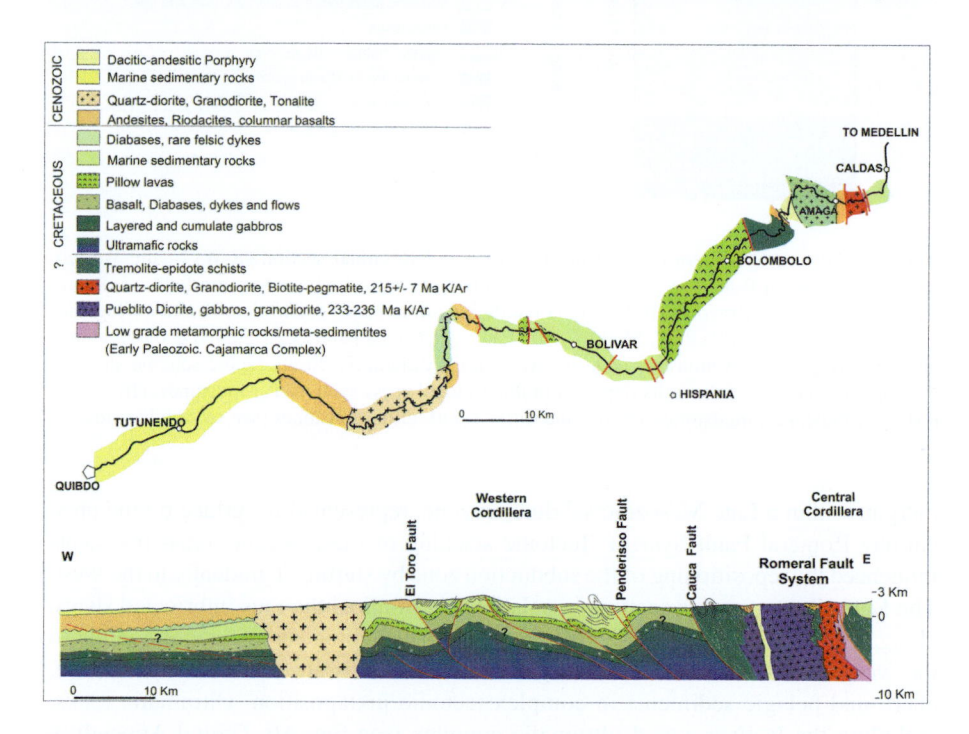

Fig. 9.9 Chocó Indenter (Caldas-Quibdo section and map). Partial, regional section (E-W), of the Chocó indenter. Geological sketch map depicts a transect along the Caldas (Antioquia)-Quibdó (Chocó) road

meta-volcano-sedimentary rocks considered to be Early Cretaceous in age (Quebradagrande Complex). Unconformably overlying these units, is preserved the Cenozoic volcano-sedimentary cover (Amagá and Combia formations).

Rock units located to the west of the Romeral and Cauca Fault Systems are interpreted to represent fragments of the accretion of oceanic crust, emplaced during the NE-wards migration and oblique collision of the Caribbean Plate along the continental margin of NW South America during the Late Mesozoic. Following accretion, the reconfiguration of the Farallon Plate leaded to subduction and the formation of a volcanic arc with the emplacement of suprasubduction volcanic rocks (Taboada et al. 2000), as recorded by the basaltic-andesitic to intermediate volcanic Combia Formation (Maya 1992). The arc, includes a belt of sub-volcanic hypabyssal porphyry intrusive clusters extending from Marsella in the south, northwards through Quinchía, Marmato, Támesis, and Titiribí. Petro-tectonic features of this section are fairly similar to those found in the Buga-Buenaventura section (Fig. 9.8).

References

Alvarez W (1967) Geology of the Simarua and Carpintero areas, Guajira peninsula, Colombia. Princeton University, Department of Geology
Bartok P (1993) Prebreakup geology of the Gulf of Mexico-Caribbean: its relation to Triassic and Jurassic rift systems of the region. Tectonics 12(2):441–459
Botero G (1963) Contribución al conocimiento de la Geología de la Zona Central de Antioquia. An Fac Minas 57:101
Bürlg H (1960) Geología de la Península de La Guajira: Serv. Geol. Nac. (Colombia). Bol Geol 6(1–3):129–168
Campbell CJ, Burgl H (1965) Section through the eastern cordillera of Colombia. Bull Geol Soc Am 76(5):567–589
Cardona A, Weber M, Valencia V, Bustamante C, Montes C, Cordani U, Muñoz CM (2014) Geochronology and geochemistry of the Parashi granitoid, NE Colombia: Tectonic implication of short-lived Early Eocene plutonism along the SE Caribbean margin. J S Am Earth Sci 50:75–92
Cediel F, Barrero D, Cáceres C (1998) Seismic atlas of Colombia. Seismic expression of structural styles in the basins of Colombia in six Vol. Robertson Research-Geotec-Ecopetrol SA, Bogotá
Cochrane R, Spikings R, Gerdes A, Ulianov A, Mora A, Villagómez D, Chiaradia M (2014) Permo-Triassic anatexis, continental rifting and the disassembly of western Pangaea. Lithos 190:383–402
Geosearch Ltda (2007) Geología de las planchas 11, 12, 13, 14, 18, 19, 20, 21, 25, 26, 27, 33 y 34 Proyecto "Evolución Geohistórica de la Sierra Nevada de Santa Marta". Invemar, Ecopetrol, Ingeominas, Bogotá
Gómez I (2001) Structural style and evolution of the Cuiza fault system. Guajira, Colombia: Master Thesis, University of Houston, Houston
Irving EM (1972) Mapa Geológico de la Península de La Guajira, Colombia (Compilación). Escala 1:100.000. Ingeominas
Irving EM (1975) Structural evolution of the northernmost Andes, Colombia (No. 846). US Govt. Print. Off
Lockwood JP (1965) Geology of the Serranía de Jarara Area. Guajira Peninsula, Colombia. Ph.D. Thesis, Princeton University, New Jersey, p 167
MacDonald WD (1964) Geology of the Serranía de Macuira Area. Guajira Peninsula, Colombia. Ph.D Thesis, Princeton Univ. New Jersey, p 237

Maya M (1992) Catálogo de dataciones isotópicas en Colombia. Ingeominas, Bol Geol 32(1–3):127–187

Mesa AM, Rengifo S (2011) Petroleum geology of Colombia, Cesar-Rancheria Basin, vol 6. ANH- Fondo Editorial Universidad Eafit, Medellín

Mora A, Parra M (2008) The structural style of footwall shortcuts along the eastern foothills of the Colombian eastern cordillera: differences with other inversion related structures. Revista Ciencia Tecnologia & Futuro 3:7–21

Quintero CA (2017) Tectónica transpresiva en el margen septentrional de la Serranía de Cosinas en la Alta Guajira (Colombia). Master Thesis. Universidad Nacional de Colombia, Facultad de Ciencias, Departamento de Geociencias

Renz O (1960) Geología de la parte Sureste de la Península de La Guajira (República de Colombia). Bol Geol (Venezuela). Spec Pub (3), 1:317–349

Restrepo-Pace PA, Ruiz J, Gehrels G, Cosca M (1997) Geochronology and Nd isotopic data for Grenville-age rocks in the Colombian Andes: new constraints for late Proterozoic–early Paleozoic paleocontinental reconstructions of the Americas. Earth Planet Sci Lett 150:427–441

Rollins JF (1960) Stratigraphy and structure of the Goajira unpublished Ph.D. Thesis, Dept. Geol., Univ. Of Nebraska, Lincoln, Nebraska (Revised version in press, Univ. of Nebraska)

Rollins JF (1965) Stratigraphy and structure of the Guajira Peninsula, northwestern Venezuela and northeastern Colombia. Univ. Nebraska Studies, New Ser., 30:1–1102

Sánchez JMS, Kammer A (2017) Marco Geológico y rasgos estructurales del Batolito de Santa Marta y sus rocas encajantes, flanco Noroccidental de la Sierra Nevada de Santa Marta. In: XVI Congreso Colombiano de Geología, III Simposio de Exploradores, Santa Marta

Sarmiento-Rojas L (2001) Mesozoic rifting and Cenozoic Basin inversion history of the eastern cordillera Colombian Andes-inferences from tectonic models. Netherlands Research School of Sedimentary Geology (NSG) Publication number 2002.01.01, p 295

Sarmiento-Rojas LF, Van Wess JD, Cloetingh S (2006) Mesozoic transtensional basin history of the eastern cordillera, Colombian Andes: inferences from tectonic models. J S Am Earth Sci 21(4):383–411

Taboada A, Rivera LA, Fuenzalida A, Cisternas A, Philip H, Bijwaard H, Rivera C (2000) Geodynamics of the northern Andes: subductions and intracontinental deformation (Colombia). Tectonics 19(5):787–813

Trumpy D (1943) Pre-cretaceous of Colombia. Geol Soc Am Bull 54:1281–1304

Tschanz Ch M, Jimeno A, Vesga C (1969) Geology of the Sierra Nevada de Santa Marta Area (Colombia). Informe 1829. Ingeominas. Bogotá

Weber M, Cardona A, Valencia V, García-Casco A, Tobón M, Zapata S (2010) U/Pb detrital zircon provenance from late cretaceous metamorphic units of the Guajira peninsula, Colombia: tectonic implications on the collision between the Caribbean arc and the south American margin. J S Am Earth Sci 29(4):805–816

Zuluaga C, Ochoa A, Muñoz C, Guerrero N, Martinez A M, Medina P, Pinilla A, Ríos P, Rodríguez P, Salazar E, Zapata E (2007) Proyecto de Investigación: Cartografía e historia geológica de la Alta Guajira, implicaciones en la búsqueda de recursos minerales. Memoria de las planchas 2, 3,5 y 6 (con parte de las planchas 4, 10 y 10BIS). Informe Interno Ingeominas, p 535

Chapter 10
Cretaceous Stratigraphy and Paleo-Facies Maps of Northwestern South America

Luis Fernando Sarmiento-Rojas

10.1 Introduction

The northwestern corner of South America (Fig. 10.1) had a complex geological history during the Cretaceous time, with interaction of at least three major tectonic plates along the active continental margin extending from Peru to Colombia and including a passive margin along the northern margin, extending into Venezuela. Some generalized paleogeographic reconstructions have been presented for the region (e.g., Etayo et al. 1976; Macellari 1988; Geotec 1992; Sarmiento-Rojas 2001; Cediel et al. 2003a, 2011; Villamil 1994, 1999, 2012; and Villamil and Pindell 2012, among others), but an integrated map set encompassing Venezuela, Colombia, Ecuador, and northernmost Peru, and including several regional stratigraphic sections and paleo-facies reconstructions spanning the Cretaceous period, has yet to be attempted. The purpose of this paper is to present 12 paleo-facies maps covering various ages of the Cretaceous period, supported by a regional sequence stratigraphic framework, and to summarize the Cretaceous tectonic/geological history of the northwestern South American region. I refer to "paleo-facies" maps as opposed to "paleogeographic" maps, because no attempt has been undertaken herein to do palinspastic restorations. Notwithstanding, in order to facilitate understanding of the complex tectonic history involving the accretion of oceanic terranes to the continental margin, schematic tectonic reconstructions are integrated within the paleo-facies maps. One fundamental condition in the construction of paleogeographic interpretations is the construction of accurate and correlatable stratigraphic sections. In the study region, there are many applied lithostratigraphic names, with varying local synonyms, and few biostratigraphic studies to make accurate regional stratigraphic correlations. In order to overcome this difficulty, I attempt to integrate

L. F. Sarmiento-Rojas (✉)
Independent Consultant, Calle 96 N° 45A-40 Interior 6 Apto 402., Bogotá 111211, Colombia

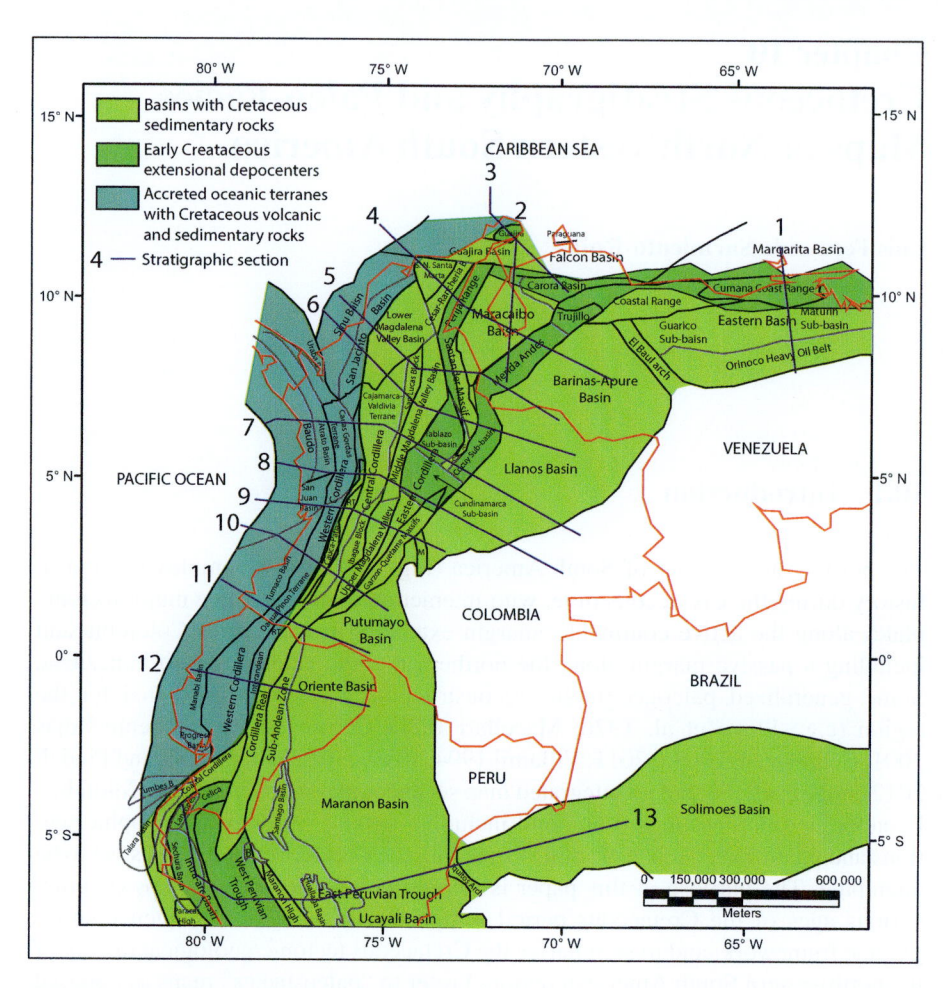

Fig. 10.1 Map of northwestern South America showing location of sedimentary basins and stratigraphic sections discussed in text. Basins indicating Cretaceous sedimentary rocks include those presently containing Cretaceous deposits, as well as those where the existence of such deposits in the past may be inferred. East-west trending Early Cretaceous depocenter in northern Venezuela is inferred from the rift system generated during separation of the Yucatan block from northern South America, based upon some plate tectonic interpretations (e.g., Pindell and Keenan 2009). F: Floresta Massif and Floresta High. M, Sierra de la Macarena; RT, Romeral Terrane as defined by Cediel et al. (2003b, 2011); B, Bagua Basin of Peru. The east-west trending Early Cretaceous depocenter in Northern Venezuela is inferred to be present below oceanic terranes accreted to the northern Venezuelan margin during the Cenozoic (not shown in this map). The maximum thicknesses for the Cretaceous sedimentary record occur in the Trujillo depocenter of northwestern Venezuela, the Tablazo, Cocuy, and Cundinamarca sub-basins of the Eastern Cordillera of Colombia and the Western Peruvian Trough of western Peru

the stratigraphy of the region into a sequence stratigraphic framework, based partially on ages reported by biostratigraphic studies, previous sequence stratigraphic studies, and in good part, using lithological and facies data from available literature and sequence stratigraphy concepts.

10.2 Methodology

I prepared several stratigraphic cross sections perpendicular to the structural grain of the Northern Andean region, including several paleo-facies maps. In order to do this, I applied the following methodology:

(a) Compilation from available literature of (a) more than 100 stratigraphic columns and stratigraphic cross sections and (b) previous paleogeographic maps
(b) Preparation of stratigraphic sections and paleo-facies maps for the studied region

10.3 Geological Setting

The northwestern corner of South America (northern Peru, Venezuela, Colombia, and Ecuador) includes litho-tectonic and morpho-structural features (Fig. 10.1) summarized below.

1. *The pre-Cambrian shields.* In south of Venezuela and east of Colombia and Ecuador lies the Guiana Shield. In east of Peru lies the Brazilian Shield. Between them, in northern Brazil, the Solimoes Basin represents an intracontinental rift developed during the Paleozoic. The Solimoes Basin trends EW along the course of the Amazon River. During the Late Cretaceous and/or earliest (?) Cenozoic, it received sediments from a west-flowing river system.
2. *The sub-Andean foreland basins bounding the Northern Andes.* These include from NE to SW: in Venezuela, the eastern Venezuela Basin, including the eastern Maturin sub-basin and the Western Guárico sub-basin, to the south of which the Orinoco Heavy Oil Belt is located. The Barinas-Apure Basin is separated from the eastern Venezuela Basin by the El Baul Arch. The Barinas-Apure Basin continues into Colombia in the Llanos Orientales Basin, south of which is located the Caguán-Putumayo Basin, which is separated from the Llanos Orientales Basin by the Sierra de La Macarena high. The Putumayo Basin of southern Colombia continues into Ecuador in the Oriente Basin and further south into Peru in the eastern part of the East Peruvian trough which contains the Marañón, Ucayali, and Madre de Dios basins. All these basins contain local Paleozoic fill and craton-ward thinning Upper Cretaceous fill, covered by Cenozoic fill which also thins toward the craton. It is possible to consider all of these basins as a single sub-Andean foreland basin with craton-ward thinning Cretaceous and Cenozoic sediments, resulting from flexural subsidence of the lithosphere due to the weight of the growing Andean orogen during the Late Cretaceous-Cenozoic time and, partially, due to Cretaceous thermal subsidence of back-arc rift basins, subsequently captured by the Andean uplift.
3. *The Northern Andes including mountainous relief and intermontane basins.* At the present time the Northern Andes may be considered a single mountain range, in Peru, consisting of the Eastern and Western cordilleras separated by the Altiplano and in Ecuador consisting of the Real and Western cordilleras

separated by the Interandean Valley. In Colombia the Andes diverge to the north of the Colombian Massif to form three distinct mountain ranges: the Eastern, Central, and Western cordilleras. The Magdalena Valley separates the Eastern and Central cordilleras, and the Cauca Valley separates the Central and Western cordilleras. The Colombian Eastern Cordillera diverges northward forming two mountain ranges: the Sierra de Perijá along the northern Colombia-Venezuela border and the Mérida Andes of Venezuela. There are also lesser coastal ranges: the Coast Range of Venezuela, the Sierra Nevada de Santa Marta and the Baudó Range in Colombia, and the Coastal Cordillera of Peru. Several intermontane basins throughout the region contain Cretaceous sediments, such as the Maracaibo Basin of Venezuela, the Middle and Upper Magdalena basins of Colombia, and several basins in Peru (Santiago, Bagua, and Huallaga, among others). Many of these basins had different Paleozoic histories or were part of different tectonic blocks. In Colombia, Etayo et al. (1983) proposed the existence of several tectonic terranes.

During the Early Cretaceous, prior to uplift of the continental-rooted Anden ranges, basin development initiated as local rifts or extensional back-arc basins bounded by normal faults, inherited from Early Mesozoic structures. These subbasins began to coalesce due to flexural thermal subsidence, into a regional, interconnected basin by the Aptian time. The Early Cretaceous sub-basins that constituted important, rapidly subsiding depocenters included Trujillo, Machiques (Perijá Range), Uribante (Mérida Andes), Cundinamarca (Eastern Cordillera of Colombia), and the Western Peruvian Trough (Western Cordillera of Peru; Macellari, 1988). Some of these extensional basins contain grabens separated by horst blocks, for example, the Western Peruvian Trough and the Eastern Peruvian Trough separated by the Marañón High, the Machiques (Perijá) and Uribante (Mérida Andes) rifts separated by the Maracaibo high, and the Cocuy and Tablazo sub-basins separated by the Floresta paleo-high. However, since the Aptian and during most of the Cretaceous, sedimentation covered most of the continental margin due to an elevated sea level combined with thermal subsidence. In this context, Etayo et al. (1983) proposed the name Cretaceous supraterrane in Colombia, encompassing the whole of the Cretaceous basin resting upon continental crust. This idea can be extrapolated over northwestern South America, from Venezuela to northern Peru. During the Cenozoic, the extensive basin was segmented into several compartments separated by structural highs, some of them resulting from inversion of Mesozoic normal faults which limited the original grabens. These Cenozoic compartments are now independent intermontane basins with differing and variable Cenozoic sedimentary fill.

Within the Andes of northern Peru, Ecuador, and Colombia, subduction-related magmatic arcs were developed along/within the margin of the continental plate during the Cretaceous. In Colombia and Ecuador, such magmatic arcs, although poorly developed, were located in the area of the Colombian Central Cordillera and its continuation to the south in Ecuador. In southernmost Ecuador and Peru, however, the Cretaceous magmatic arc was very well developed in the intra-arc Celica basin and in the intra-arc Casma basin developed in the western part of the Western Peruvian Trough. In Colombia, Ecuador, and Peru, most of the Cretaceous basins located craton-ward (to the east) of the magmatic arc can be considered "back-arc basins." In southernmost Ecuador and northern Peru, some basins (Tumbes, Talara, and Sechura, among others offshore of Peru) developed ocean-ward (west of) the magmatic arc and can be considered fore-arc basins.

4. *Terranes containing pre-Cretaceous rocks* representing the continental margin such as sections of the Central Cordillera of Colombia, the Real Cordillera of Ecuador, and portions of the Peruvian Andes and the coastal Cordillera of Peru.
5. *Cretaceous oceanic terranes in the Northern Andes (Venezuela, Colombia, and Ecuador)* which are petrologically and lithochemically similar to the Caribbean Plate. These are interpreted as fragments of the Caribbean/Farallon Plate, accreted to the continental margin during Cretaceous (Ecuador and Colombia) and Early Cenozoic (Venezuela), including the Coastal Range of northern Venezuela and Western cordilleras of Colombia and Ecuador.

10.3.1 Note on Southernmost Ecuador and Northern Peru

There is an important difference between the Andes of southern Ecuador and northernmost Peru and the Ecuadorian and Colombian Andes to the north. During the Meso-Cenozoic, the Ecuadorian-Colombian Andes were subject to collision and accretion of oceanic terranes of the Pacific and Caribbean Plate affinity. By contrast, the Peruvian Andes do not contain accreted oceanic terranes. During Cretaceous times, the Peruvian Andes were dominated by subduction of the Farallon and Phoenix plates beneath the western margin of the continent, which itself had undergone important deformation.

10.4 Summary of Mesozoic Plate Tectonic Interpretations

10.4.1 Jurassic

According to some Jurassic plate tectonic reconstructions (e.g., Ross and Scotese 1988; Pindell and Kennan 2009), the western margin of Colombia, Ecuador, and northern Peru constituted an active convergent margin related to subduction of Pacific (i.e., Farallon Plate) lithosphere and the development of a subduction-related magmatic arc in western Colombia, Ecuador, and northern Peru (Pindell and Dewey 1982; Bourgois et al. 1982a, b; McCourt et al. 1984; Pindell 1990, 1993; Pindell and Barret 1990; Pindell and Erikson 1993; Pindell and Tabbut 1995; Pindell et al. 2005; Toussaint 1995a, b; Toussaint and Restrepo 1989, 1994; Meschede and Frisch 1998). In Colombia and Ecuador, continental arc magmatism spanned most of the

Jurassic and is preserved along the length of the Central Cordillera (e.g., Ibague Batholith, Villagómez 2010; Villagómez et al. 2011) and elsewhere (see detailed discussion of Jurassic are related magmatism by Leal-Mejía et al. 2018). The youngest pulse of Mesozoic continental arc magmatism occurred at 145 Ma (Villagomez, 2010). Such an interpretation explains at least some of the Jurassic rift basins from western Colombia to Peru as back-arc basins.

In addition to continued subduction of the Farallon Plate along the western margin of South America, some tectonic plate interpretations (e.g., Pindell and Kennan 2009) suggest separation of North and South America started during the Middle to Late Jurassic, along the northern margin of South America (Venezuela and northern Colombia). As a result, a new, "Proto-Caribbean" oceanic basin began to open between northern and western Colombia, the Chortis block and Venezuela and the Yucatan block (Fig. 10.2). Such an interpretation explains Jurassic rift basins in northeastern Colombia and Venezuela.

10.4.2 Cretaceous

10.4.2.1 Venezuela

Despite the arrival of Pacific-Caribbean terranes in NW Colombia during the Late Cretaceous, generated some degree of tectonic loading and flexure in the western Maracaibo region, Venezuela, since the Jurassic and during the Early Cretaceous, much of northern South America behaved as a passive continental margin, as evidenced by subsidence curves, the age of synorogenic flysch units, stratigraphic and structural studies, and plate motion histories (e.g., Villamil and Pindell 2012). During the Meso-Cenozoic, west-to-east-directed Caribbean plate motion caused allochthonous terranes of Jurassic-Cretaceous rocks, many of which were deformed and metamorphosed during the Cretaceous, to be abducted onto the Venezuelan passive margin.

10.4.2.2 Western Colombia

Villagomez (2010) and Villagómez et al. (2011) studied the thermochronology, geochronology, and geochemistry of the Western and Central cordilleras of Colombia and proposed a model for the tectonic evolution of western Colombia from the Jurassic to Paleocene (Figs. 10.2 and 10.3):

- From ca. 145 Ma (Jurassic-Cretaceous boundary) to 130 Ma (Hauterivian) uplift and exhumation occurred due to rebound of the continental margin (Central Cordillera of Colombia) and retreat of subduction west of Colombia, including backstepping of the subduction zone to a more westerly position. According to Nivia (2001) and Nivia et al. (1996, 2006), this corresponds to the opening of the Quebradagrande marginal basin to the west of the Central Cordillera. An oceanic

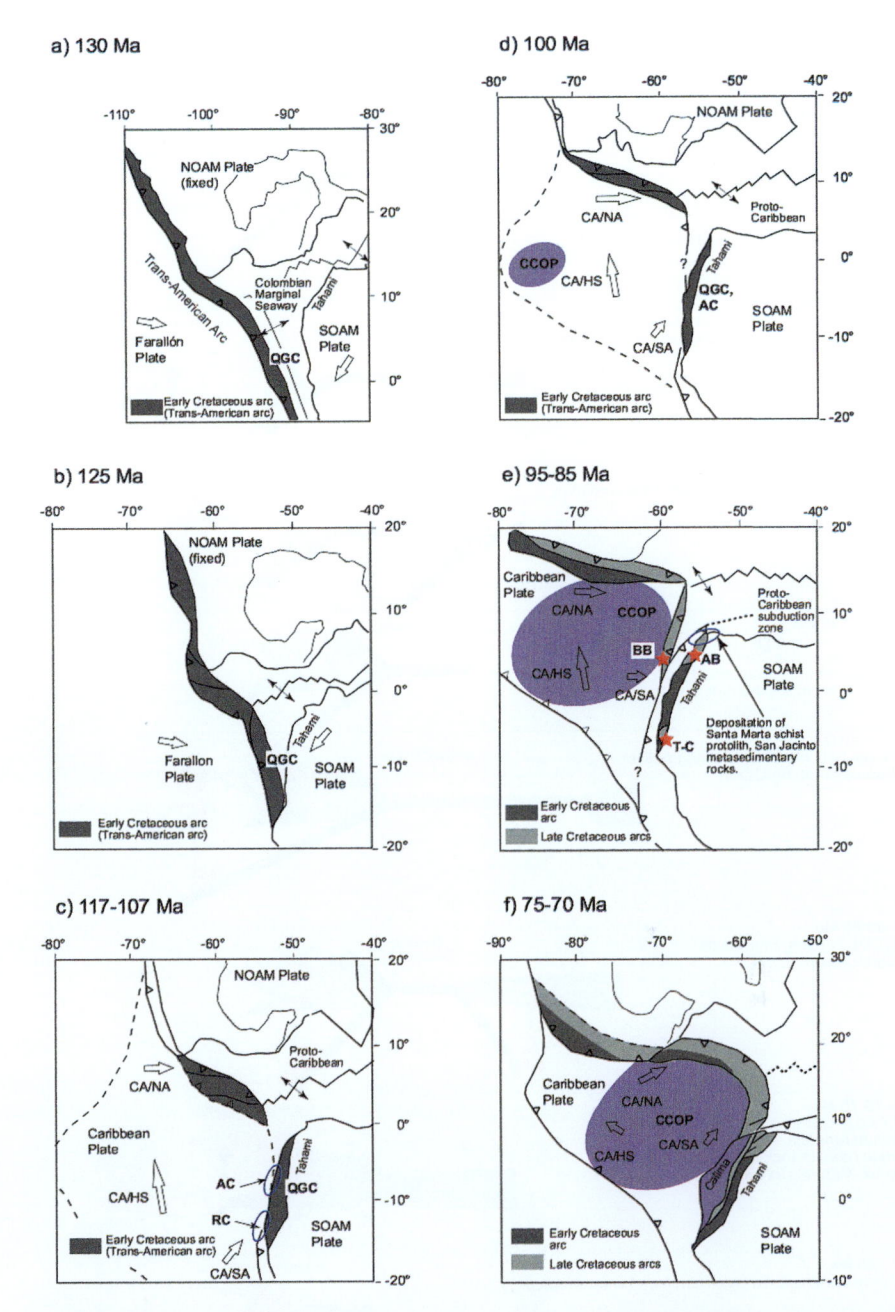

Fig. 10.2 Paleotectonic reconstructions during the Cretaceous including relative paleo-positions of North and South America, modified and simplified from Pindell and Kennan (2009). Reference frames: (**a, b**) North America, (**c–f**) Indo-Atlantic using hotspot reference frame of Müller et al., in Villagomez et al. (2011). Relative convergence direction: *CA/HS* Caribbean Plate/hotspot, *CA/NA* Caribbean Plate/North America, *CA/SA* Caribbean Plate/South America. Abbreviations: *AB* Antioquian Batholith, *AC* Arquía Complex, *BB* Buga Batholith, *CCOP* Caribbean-Colombian Oceanic Province, *NOAM* North American Plate, *QGC* Quebradagrande Complex, *RC* Raspas Complex in Ecuador, *SOAM* South American Plate, *T-C* Tangula-Curiplaya intrusions. The Early Cretaceous Trans-American Arc is shown in dark gray, Late Cretaceous arc is shown in medium gray, and the CCOP is shown in purple. (From Villagomez et al., 2011)

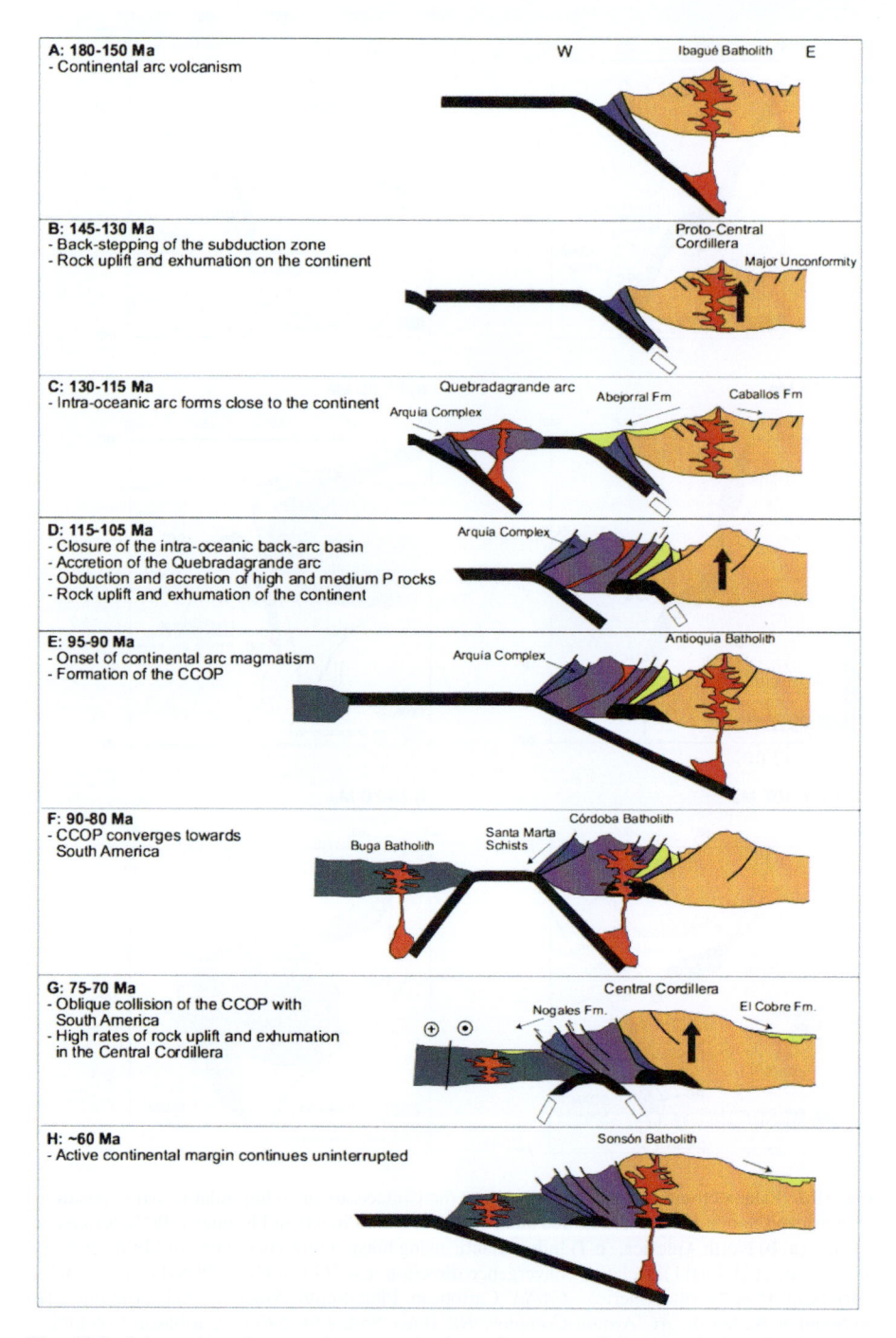

Fig. 10.3 Schematic paleotectonic reconstruction of western Colombia during the Jurassic, Cretaceous, and Paleocene (ca. 150 to 60 Ma). Black arrows indicate exhumation periods. From Villagomez et al. (2011)

marginal basin and intraoceanic arc, represented by the Quebradagrande Complex, formed during the Early Cretaceous, and its inception may have been caused by backstepping of the Jurassic slab due to the introduction of buoyant seamounts (Villagomez, 2010). The coexistence of both MORB-like gabbros and basalts in close association with pillowed arc basalts, locally covered by marine sediments with both an oceanic and continental provenance, suggests an oceanic arc origin for the Quebradagrande Complex, with a back arc located proximal to the continent. The Quebradagrande marginal basin corresponds to the Colombian marginal seaway basin of Pindell and Kennan (2009). Subduction of oceanic crust beneath the Quebradagrande marginal basin may have approached the continent. Alkali-feldspar 40Ar/39Ar cooling ages obtained from crystalline rocks located to the south of the laterally extensive Ibagué Fault yield ages of ca. 138–130 Ma (Valanginian-Hauterivian), suggesting uplift and exhumation of the southern part of the Central Cordillera, contemporaneous with the cessation of Jurassic arc magmatism.

- From 130 Ma (Hauterivian) to 115 Ma (Aptian), an intraoceanic arc forms at the western border of the Quebradagrande marginal basin, and it migrates toward the continent.
- From 115 Ma (Aptian) to 105 Ma (Albian), closure of the intraoceanic back-arc basin (Quebradagrande marginal basin of Nivia, 2001 and Nivia et al., 1996, 2006, or Colombia seaway of Pindell and Kennan, 2009). Remnants of this basin and the associated arc constitute the Quebradagrande Complex along the Romeral-Peltetec suture zone. The Quebradagrande Complex accreted against metamporphic basement of the Tahami Terrane (sensu Restrepo and Toussaint, 1988; Toussaint and Restrepo, 1989, 1994; Toussaint, 1995a, b) or Cajamarca-Valdivia Terrane (sensu Cediel et al., 2003b, 2011) in Colombia's Central Cordillera during the Late Aptian. That event was accompanied by the obduction of medium-high P-T metamorphic rocks of the Arquía Complex onto the Cretaceous fore-arc to the west. These rocks may represent abducted remnants of subduction channel sediments. In the Central Cordillera north of the Ibagué Fault, medium-temperature thermochronometers reveal the presence of a younger cooling event at 107–117 Ma (Aptian-Albian).
- From 95 (Cenomanian) to 90 Ma (Turonian), the onset of continental arc magmatism in central Colombia is observed (see discussion by Leal-Mejía et al. 2018). To the west, in the Pacific domain, the Caribbean oceanic plateau was formed (*e.g.*, Kerr et al. 1997a, b; Sinton et al. 1998). Accreted fragments of this plateau were referred to as the PLOCO (i.e., *Provincia Litosferica Oceanica Cretacica Occidental: Western Cretaceous Lithospheric Province*) by Nivia (2001). Geochronological analyses of plateau rocks from throughout NW South America and elsewhere range from ca. 105 to 72 Ma (see summary of radiometric and biostratigraphic ages in Kerr et al. 2003).
- From 90 Ma (Turonian) to 80 Ma (Campanian), the Caribbean Plate (CCOP sensu Kerr et al. 1997a, b) converged upon NW South America. The remnant ocean basin located between South America and the Caribbean Plate was consumed via double-vergent subduction, giving rise to continental- and oceanic-rooted

arcs (see summary and reconstruction by Leal-Mejía et al. 2018). The Antioquian and Cordoba batholiths are representative continental arc segments, while the Sabanalarga, Mistrato, Buga, and Jejenes plutons are representative of the oceanic arc formed along the leading edge of the CCOP.
- From 75 Ma (Campanian) to 70 Ma (Maastrichtian), the Caribbean Plate collided against the NW South American margin, along the dextral compressive Cauca-Almaguer fault system, and fragments of the CCOP, including the leading-edge arcs, were accreted to the continental margin, forming the basement of the Western Cordillera of Colombia and Ecuador. This resulted in the extinction of both magmatic arcs. High rates (ca. 1.6 km/My) of uplift and exhumation of the Central Cordillera were observed at this time there, driven by collision and accretion of the CCOP.
- From ca. 60 to 50 Ma (Paleocene), subduction was briefly reestablished beneath the western continental margin, as revealed by limited occurrences of granitoid plutons to the south of the Antioquian Batholith (Sonsón, Manizales, El Hatillo, El Bosque; Leal-Mejía et al. 2018). Along the northern continental margin, in the Sierra Nevada de Santa Marta Province, Paleo-Eocene subduction-related magmatism appears to have been driven by separate and distinct tectonic mechanisms (Duque-Trujillo et al. 2018; Cediel 2018). At the same time (ca. 65–58 Ma), the southernmost Sierra Nevada de Santa Marta was exhumed at elevated rates (≥ 0.2 Km/My), in response to the interaction between the CCOP and northern South America.

10.4.2.3 Western Ecuador

Based upon geological fieldwork and thermochronological, geochronological, and lithochemical data, Vallejo (2007) and Vallejo et al. (2009) proposed the following tectonic history of the Western Cordillera of Ecuador (Fig. 10.4).

- The ca. 123 Ma (Aptian) age reported from a set of basalts, basaltic andesites, and gabbros in western Ecuador suggest they may represent fragments of Jurassic to Early Cretaceous oceanic crust, which accreted against South America during the Early Cretaceous (*e.g.*, Litherland et al. 1994). This could represent the southern continuation of the Quebradagrande marginal basin in Colombia, as interpreted by Nivia (2001) and Nivia et al. (1996, 2006).
- From approximately 85 Ma to 83 Ma (Santonian), subduction of oceanic crust occurred west of the continental margin (as represented by the Cordillera Real of Ecuador). This oceanic plate (proto-Caribbean? Farallon?) separated the Caribbean plateau farther west from the South American margin. A remnant oceanic basin with pelagic sedimentation developed and was consumed via a double-vergent subduction system, giving rise to continental and oceanic arcs, similar to the case interpreted for western Colombia. Granitoid rocks of the La Portada and Paujíl plutons are interpreted to represent the oceanic magmatic arc developed on the leading edge of the Caribbean Plate.

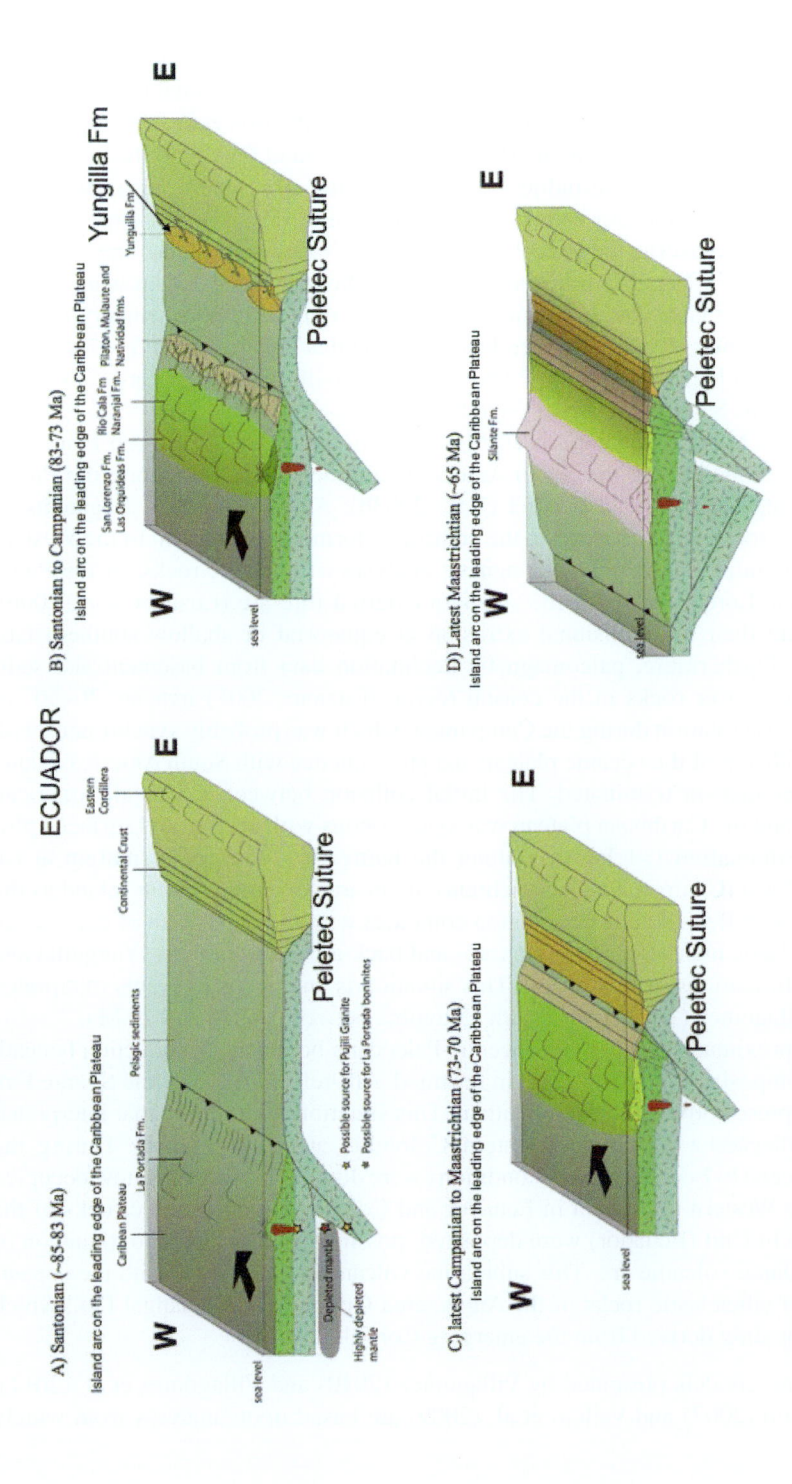

Fig. 10.4 Schematic Late Cretaceous paleotectonic reconstruction of the Western Cordillera of Ecuador and neighboring areas. From Vallejo (2007)

- From 83 Ma (Santonian) to 73 Ma (Campanian), the marginal basin was narrowing. The magmatic arc on the leading edge of the Caribbean Plate continued. Volcanic rocks of the San Lorenzo, Las Orquideas, Rio Cala, and Naranjal Fms. are representatives of this arc. Between this arc and the subduction zone, trench fill turbidites were deposited (Pilaton, Mulaute, and Natividad Fms.). To the west of the continental margin, turbidites of the Yungilla Fm. were also deposited. Intraoceanic island arc sequences (Pujilí Granite, Rio Cala Group, Naranjal Unit) intrude/overlie the CCOP and yield crystallization ages that range between ca. 85 and 72 Ma. The lithochemistry and radiometric ages of lavas associated with the Rio Cala Arc, combined with the age range and geochemistry of their turbiditic and volcanoclastic products, indicate that the arc initiated by westward subduction beneath the CCOP and are coeval with other island arc rocks (Las Orquideas, San Lorenzo, and Cayo formations).
- From 73 Ma (Campanian) to 70 Ma (Maastrichtian), the Caribbean Plate dextrally collided along the South American continental margin, along the Pujilí-Pallatanga suture zone (Cediel et al. 2003b). As in Colombia, fragments of plateau rocks were accreted to the continent, forming a basement to the Western Cordillera of Ecuador. Paleomagnetic analyses of volcanic rocks, of the Piñon and San Lorenzo units of the southern external fore-arc (Luzieux et al. 2006), indicate their pre-collisional extrusion at equatorial or shallow southern latitudes. Furthermore, paleomagnetic declination data from basement and sedimentary cover rocks in the coastal region (Luzieux 2007) indicate 20–50° of clockwise rotation during the Campanian, which was probably synchronous with the collision of the oceanic plateau and arc sequence with South America. Island arc magmatism terminated. The initial collision between the South American Plate and the Caribbean plateau was synchronous with accelerated surface uplift and exhumation (>1 km/my) along the buttressing continental margin at ca. 75–65 Ma (Campanian-Maastrichtian), in an area extending as far inland as the Cordillera Real. Rapid exhumation coincides with the deposition of continental siliciclastic material in both the fore- and back-arc environments (Yunguilla and Tena formations, respectively). This situation is analogous to events interpreted by Villagomez (2010) for western Colombia between ca. 75 and 70 Ma.
- At approximately 65 Ma (Cretaceous-Paleocene boundary), subduction beneath the composite continental margin resumed. Arc-related rocks of the Silante Fm. are representative of this magmatism. This situation is similar to that interpreted by Villagomez (2010) for western Colombia at ca. 60–50 Ma. During the Paleocene to Eocene, marine conditions were dominant in the area now occupied by the Western Cordillera in Ecuador and Colombia, and volcanic rocks of the Macuchi Unit (Ecuador) were deposited, possibly as a temporal continuation of the Silante volcanic arc. This submarine volcanism was coeval with the deposition of siliciclastic rocks of the Angamarca Gp. and the Saguangal Fm., which were mainly derived from the emerging Cordillera Real.

Tectonic models presented by Villagomez (2010) and Villagómez et al. (2011), and Vallejo (2007) and Vallejo et al. (2009) are based upon analyses from widely spaced rock samples of the Western and Central cordilleras of Colombia and the Western Cordillera of Ecuador, respectively. Due to the spaced location of the samples, and regional nature of the studies, some geological domains remained unsampled and hence are not accounted for in the above tectonic interpretations. Two examples of this include the Cretaceous tectonic evolution of southernmost coastal Ecuador and the Guajira Peninsula in northeasternmost Colombia/Venezuela.

In southernmost western Ecuador, Jaillard et al. (1995) proposed that "an accreted terrane underlain by oceanic crust formed during the Aptian-Albian." To the southeast, the oceanic crust is overlain by Cenomanian-Coniacian fine-grained pelagic deposits, coarse-grained volcanoclastic turbidites of Santonian-Campanian age, and Maastrichtian-Middle Paleocene tuffaceous shales. To the northwest, Late Campanian-Paleocene volcanoclastic beds and lava flows of island arc composition resting upon oceanic crust resulted from the opening of a marginal basin between an Early Late Cretaceous island arc (Cayo arc) and a latest Cretaceous-Paleocene island arc (San Lorenzo arc). In the latest Paleocene-Earliest Eocene, the accretion of the remnant arc to the Andean continental margin caused a major but localized deformation phase that affected the southern part of coastal Ecuador (Fig. 10.5, after Jaillard et al. 1995).

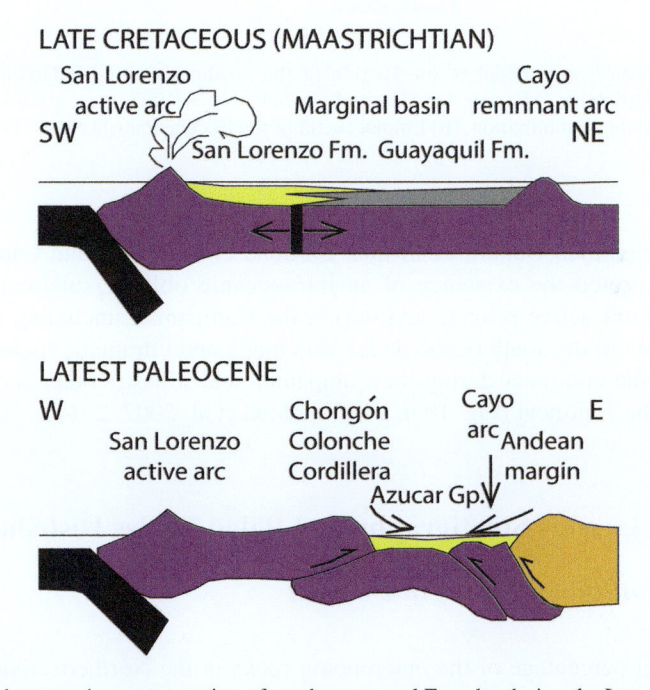

Fig. 10.5 Paleotectonic reconstruction of southern coastal Ecuador during the Late Cretaceous to Paleocene. (**a**) In the Late Cretaceous, a marginal basin opened between the Early Late Cretaceous Cayo arc and the latest Cretaceous-Paleocene San Lorenzo island arc. (**b**) In the Late Paleocene–Earliest Eocene, the Cayo remnant arc collided with the Andean continental margin and caused intense deformation of the Santa Elena Peninsula, emergence of the Chongón Cordillera, and infilling of the Santa Elena basin by coarse-grained turbidites. Modified after Jaillard et al. (1995)

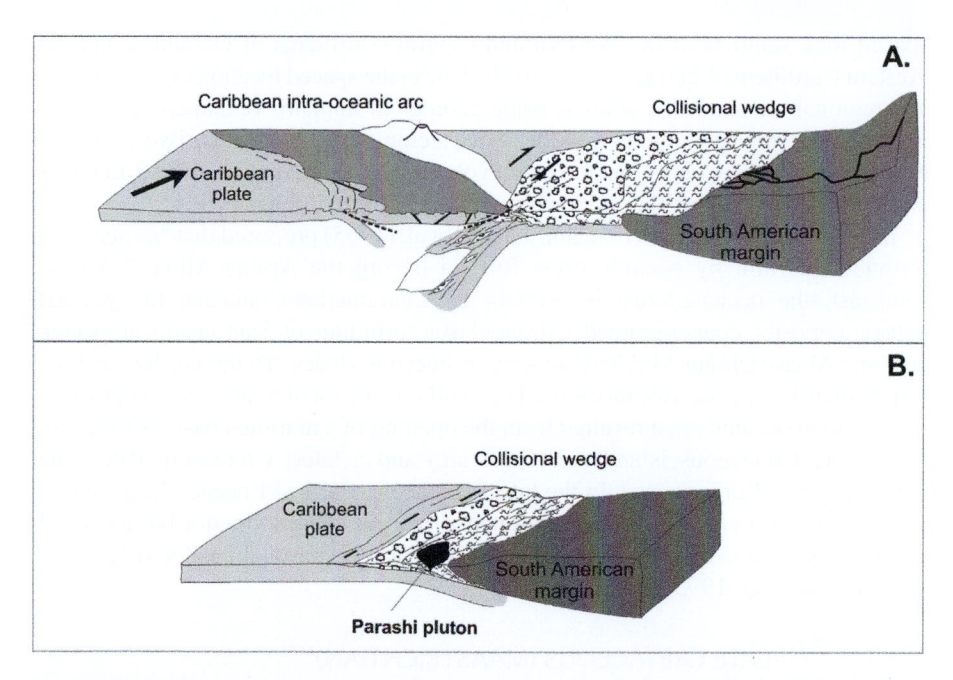

Fig. 10.6 Paleotectonic model reconstruction of the northern Colombia Guajira peninsula during the Maastrichtian to Eocene. (**a**) Maastrichtian to Paleocene oblique arc-continent collision, followed by subduction initiation. (**b**) Emplacement of the Eocene Parashi Stock. From Cardona et al. (2014)

In the Colombian Guajira Peninsula, Cardona et al. (2007) and Cardona et al. (2014) interpreted the existence of an intraoceanic oblique subduction-related arc (Jarara Fm.) active prior to and during the Campanian, including an oceanic back-arc basin to the south (Cabo de La Vela mafic and ultramafic rocks). This arc approached the continent during the Campanian and Maastrichtian and was later accreted to the continent (Fig. 10.6, after Cardona et al. 2007, 2014).

10.5 Cretaceous Stratigraphy and Paleo-Facies Distribution

10.5.1 Continental Margin Domain

An important percentage of the outcropping rocks in the Northern Andean region are of Cretaceous age. Figure 10.1 shows the sedimentary basins considered within this study, including the location of the stratigraphic sections constructed herein. Figures 10.7, 10.8, 10.9, 10.10, 10.11, 10.12, 10.13, 10.14, 10.15, 10.16, 10.17, 10.18, 10.19 and 10.20 include 13 traverse time-stratigraphic cross sections of these basins. I constructed the sections from available literature, including but not

10 Cretaceous Stratigraphy and Paleo-Facies Maps of Northwestern South America

■	Alluvial fan coarse clastics	X X	Continental crust with no sedimentation
■	Fluvial and coastal plain sandstones and mudstones	□	Oceanic crust
■	Coastal plain mainly mudstones	■	Caribbean type oceanic crust mainly basalts
■	Littoral mainly sandstones	■	Caribbean type oceanic accreted terranes
■	Inner shelf sandstones	■	Subduction related magmatic arc
■	Inner shelf limestones	■	Jurassic magmatic arc
■	Inner shelf mudstones	■	Metamorphic rocks in suture zones
■	Outer shelf shales and pelagic limestones commonly organic-rich		
■	Middle shelf bituminous limestones and shales		
■	Marine upwelling related chert		
■	Phosfatic marine shelf facies		
■	Turbidite sandstones and shales		

STRATIGRAPHIC SECTIONS AND PALEOFACIES MAPS LEGEND

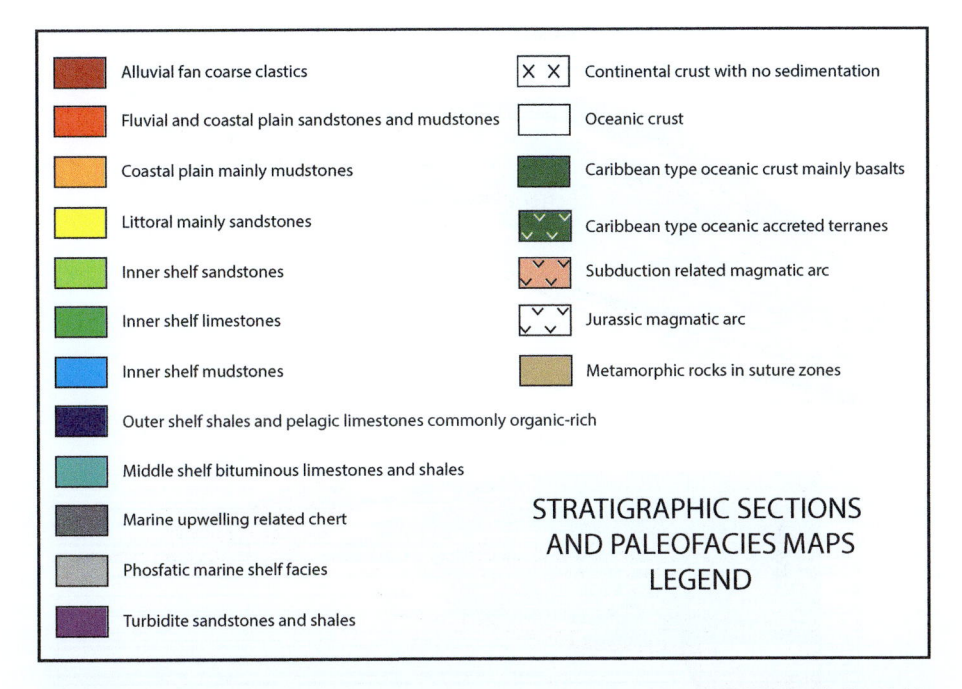

Fig. 10.7 Color legend for the stratigraphic sections depicted in Figs. 10.8, 10.9, 10.10, 10.11, 10.12, 10.13, 10.14, 10.15, 10.16, 10.17, 10.18, 10.19 and 10.20 and for the paleo-facies maps depicted in Figs. 10.21, 10.22, 10.23, 10.24, 10.25, 10.26, 10.27, 10.28, 10.29, 10.30, 10.31 and 10.32. In the paleo-facies maps, the color-filled "x"s for some facies signify these areas were covered by the sedimentary environments corresponding to the respective color, but the lack of subsidence or very reduced subsidence did not permit enough accommodation space for sediment accumulation, or if sediment accumulated it was later eroded

restricted to Fabre (1985, 1986, 1987), Macellari (1988), Cooper et al. (1995), Sarmiento-Rojas (2001), Lopez-Ramos (2005), Villamil (2012), Villamil and Arango (2012), Guerrero (2002), and Sarmiento (2015) in Colombia; Rod and Maync (1954), González de Juana et al. (1980), Lugo and Mann (1995), Parnaud et al. (1995), Mann et al. (2006), and Villamil and Pindell (2012) in Venezuela; Barragan et al. (2004) in Ecuador; and Macellari (1988), Jaillard and Sempere (1989), Jaillard (1993), and Jaillard et al. (1990, 2000, 2005) in southern Ecuador and Peru. In these stratigraphic sections, I identify sequences bounded by unconformities recognizable in some parts of the basin (usually proximal areas) or correlative conformities in other areas of the basin (usually distal areas) which were generated during times of relative (tectono-eustatic) low sea level. These unconformities are proposed sequence boundaries (SB) and are shown in the stratigraphic sections as red horizontal lines and labeled with odd numbers. I also identify maximum flooding surfaces (MFS) within these sequences generated during times of relative high sea level. Maximum flooding surfaces are shown in the stratigraphic sections as blue horizontal lines and labeled with even numbers. I assume that SB and MFS are surfaces generated regionally at specific times (time surfaces).

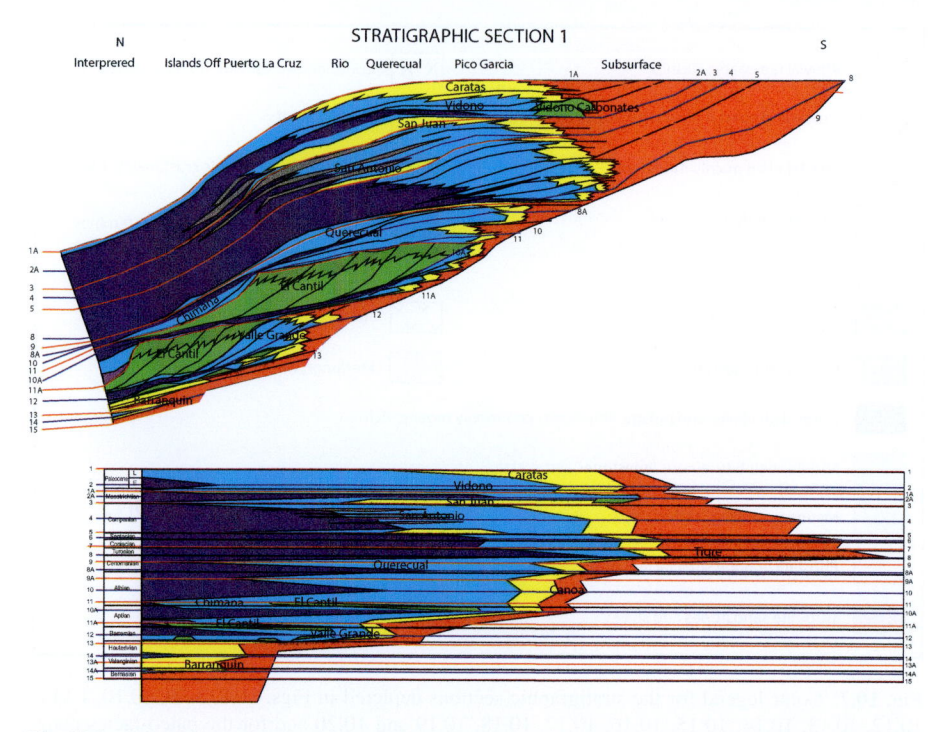

Fig. 10.8 Stratigraphic Section 1 through eastern Venezuela. See Fig. 10.1 for section location and Fig. 10.7 for facies color legend. (**a**) Section modified from Villamil and Pindell (2012). (**b**) Section with geological time in vertical axis. Horizontal red lines indicate proposed sequence boundaries (SB); horizontal blue lines indicate maximum flooding surfaces (MFS) for the proposed stratigraphic sequences. See additional explanation in text

Although this assumption probably is only a first approximation due to differential tectonic subsidence of graben blocks, especially during the earliest Cretaceous and the initial inversion of these grabens at the end of Cretaceous, this assumption is a useful tool for regional correlation.

Cretaceous rocks, including (locally) uppermost Jurassic and Paleocene deposits, form a mega-sequence bounded by regional unconformities that are at least locally angular. On a broad scale, Cretaceous rocks represent a major transgressive-regressive cycle with the maximum flooding surface close to the Cenomanian-Turonian boundary (MFS 8), corresponding to the maximum Cretaceous, and even Mesozoic, eustatic level (Fabre 1985; Villamil 1994, 2012; Sarmiento-Rojas 2001; Figs. 10.7, 10.8, 10.9, 10.10, 10.11, 10.12, 10.13, 10.14, 10.15, 10.16, 10.17, 10.18, 10.19 and 10.20). Superimposed on this large-scale trend, several smaller transgressive-regressive cycles are present, suggesting an oscillating relative tectono-eustatic level. These minor cycles correspond to the several proposed stratigraphic sequences.

Particularly in Colombia, Venezuela, and Ecuador, I have identified several of these small transgressive-regressive cycles, which can approximately be correlated throughout the region and could possibly be extended further into southern Ecuador and northern Peru, although in this southern region, the ages of the surfaces have minor

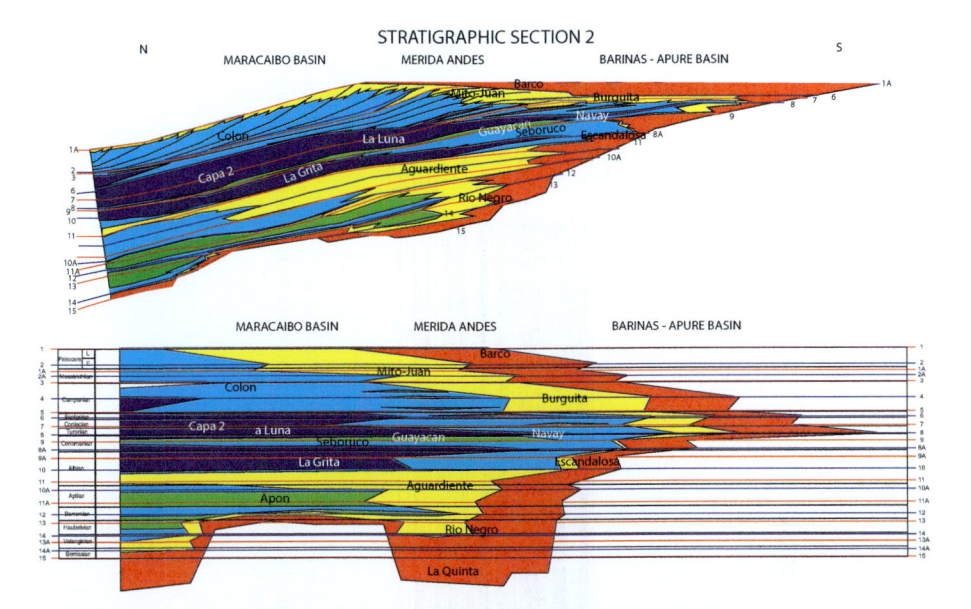

Fig. 10.9 Stratigraphic Section 2 through western Venezuela. See Fig. 10.1 for section location and Fig. 10.7 for facies color legend. (**a**) Section modified from Villamil and Pindell (2012). (**b**) Section with geological time in vertical axis. Horizontal red lines indicate proposed sequence boundaries (SB); horizontal blue lines indicate maximum flooding surfaces (MFS) for the proposed stratigraphic sequences. See additional explanation in text

changes possibly resulting from more active tectonics and basin inversion at the end of Cretaceous, as compared to the northern region. The earliest Cretaceous surfaces are more difficult to correlate regionally because differential vertical subsidence of different blocks limited by normal faults. In the stratigraphic sections (Figs. 10.7, 10.8, 10.9, 10.10, 10.11, 10.12, 10.13, 10.14, 10.15, 10.16, 10.17, 10.18, 10.19 and 10.20), these surfaces are numbered from top to bottom. In the following sections the stratigraphic description is mainly based on Villamil and Pindell (2012) for Venezuela, Sarmiento-Rojas (2001) for Colombia, and Barragan et al. (2004) for Ecuador.

10.5.1.1 Venezuela

Jurassic red beds are present in the extensional grabens of the Mérida Andes (the Uribante and Trujillo rifts) and Serranía de Perijá (Machiques rift) and in several subsurface grabens in the Maracaibo, Barinas-Apure, and eastern Venezuelan Basins (Espino and Apure-Mantecal Grabens).

According to plate tectonic interpretations, a Jurassic marine shelf section is expected in northernmost South America and particularly Venezuela. Some occurrences are apparent in northern salients of the continent, in the Colombian Guajira (Cocinas Gp.) and Caracas area. Villamil and Pindell (2012) speculated that a major (<2 km thick) section of Jurassic marine shelf section probably exists beneath the Serranía del Interior and northernmost Venezuela.

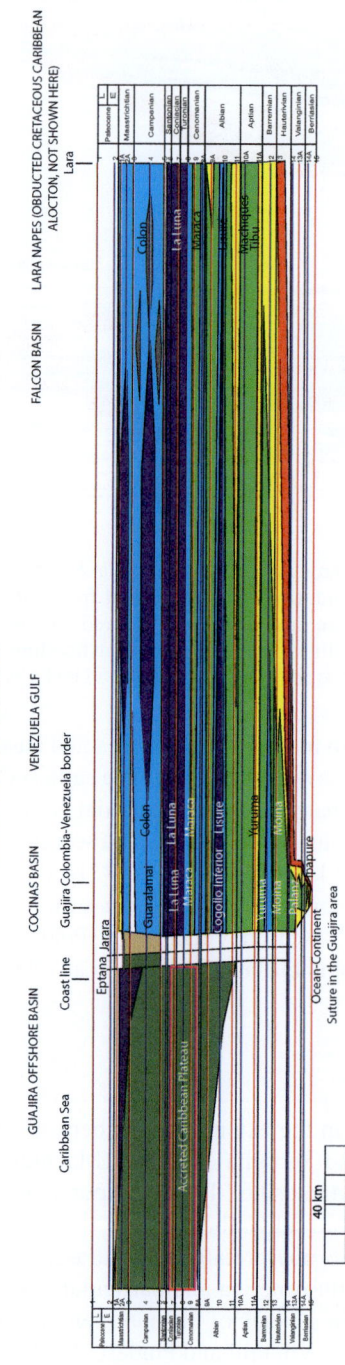

Fig. 10.10 Stratigraphic Section 3 through northernmost Colombia and western Venezuela. See Fig. 10.1 for section location and Fig. 10.7 for facies color legend. Section with geological time in vertical axis. Horizontal red lines indicate proposed sequence boundaries (SB); horizontal blue lines indicate maximum flooding surfaces (MFS) for the proposed stratigraphic sequences. See additional explanation in text

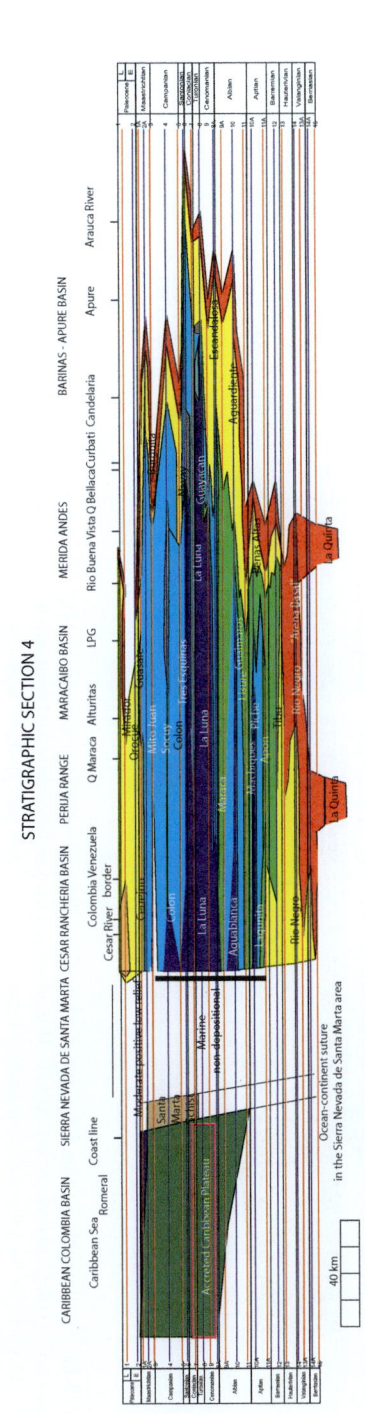

Fig. 10.11 Stratigraphic Section 4 through northern Colombia and western Venezuela. See Fig. 10.1 for section location and Fig. 10.7 for facies color legend. Section with geological time in vertical axis. Horizontal red lines indicate proposed sequence boundaries (SB); horizontal blue lines indicate maximum flooding surfaces (MFS) for the proposed stratigraphic sequences. See additional explanation in text

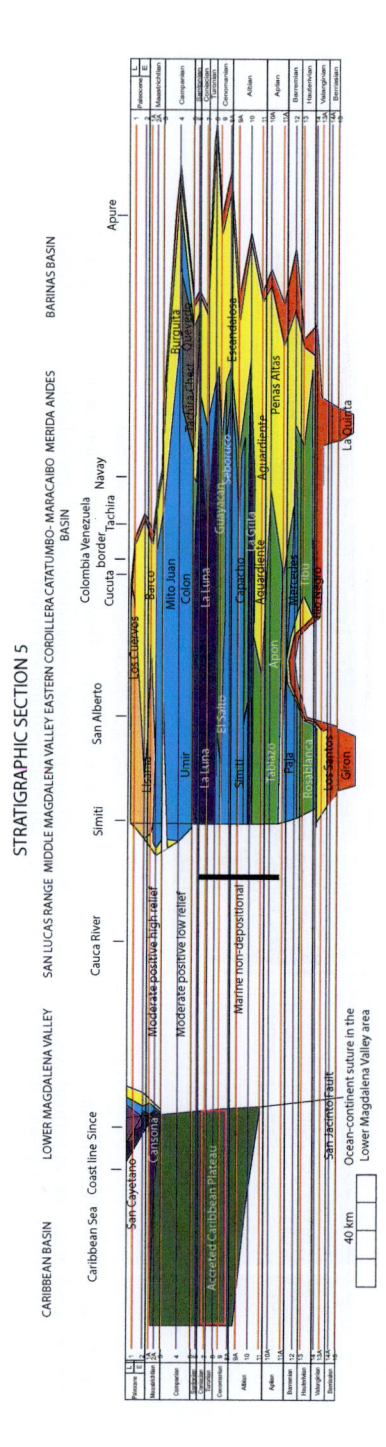

Fig. 10.12 Stratigraphic Section 5 through northern Colombia and western Venezuela. See Fig. 10.1 for section location and Fig. 10.7 for facies color legend. Section with geological time in vertical axis. Horizontal red lines indicate proposed sequence boundaries (SB); horizontal blue lines indicate maximum flooding surfaces (MFS) for the proposed stratigraphic sequences. See additional explanation in text

Fig. 10.13 Stratigraphic Section 6 through northern Colombia. See Fig. 10.1 for section location and Fig. 10.7 for facies color legend. Section with geological time in vertical axis. Horizontal red lines indicate proposed sequence boundaries (SB); horizontal blue lines indicate maximum flooding surfaces (MFS) for the proposed stratigraphic sequences. See additional explanation in text

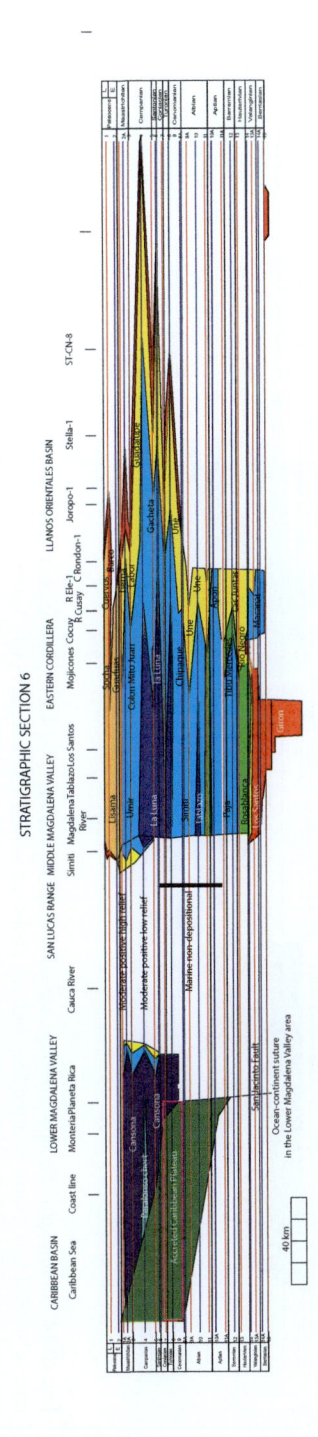

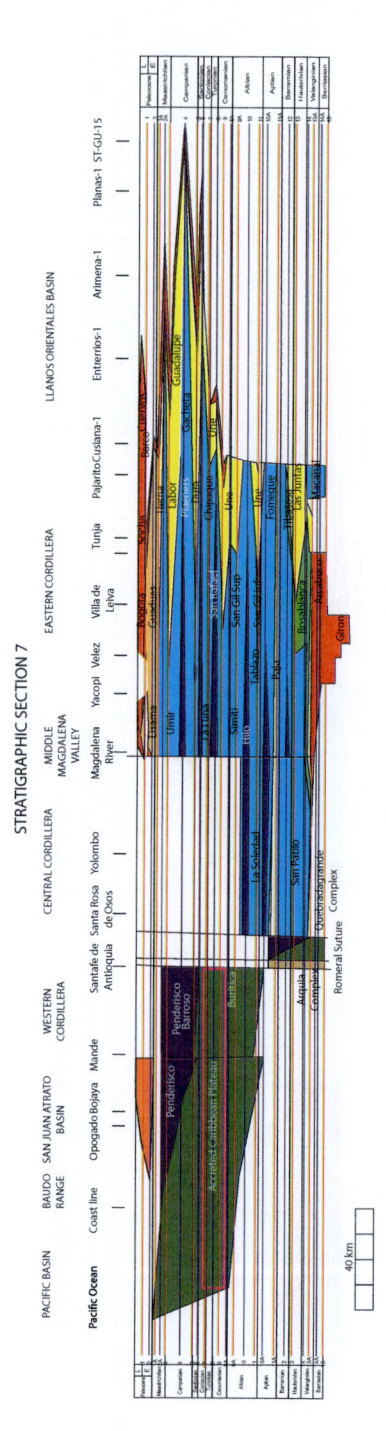

Fig. 10.14 Stratigraphic Section 7 through central Colombia. See Fig. 10.1 for section location and Fig. 10.7 for facies color legend. Section with geological time in vertical axis. Horizontal red lines indicate proposed sequence boundaries (SB); horizontal blue lines indicate maximum flooding surfaces (MFS) for the proposed stratigraphic sequences. See additional explanation in text

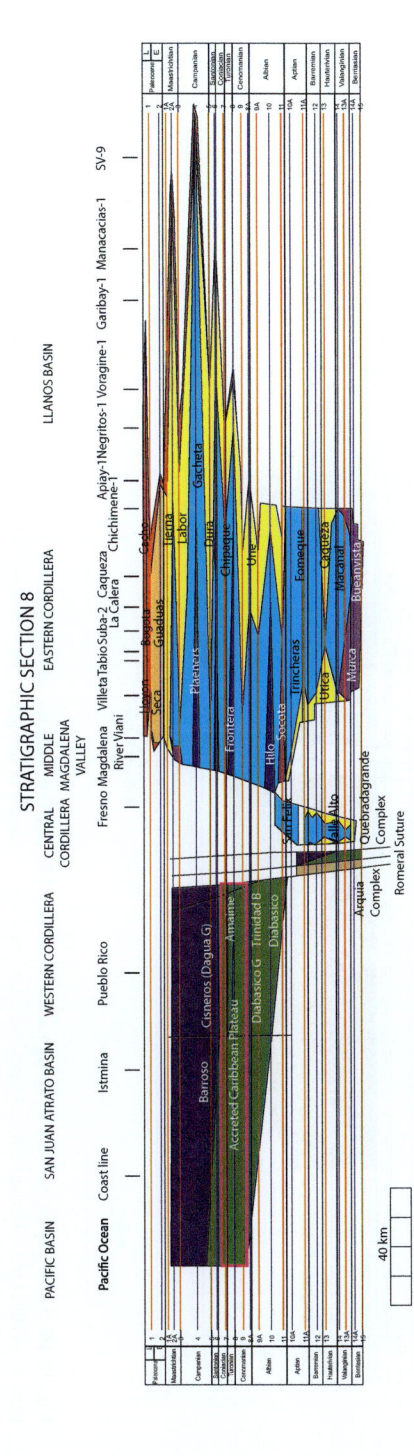

Fig. 10.15 Stratigraphic Section 8 through central Colombia. See Fig. 10.1 for section location and Fig. 10.7 for facies color legend. Section with geological time in vertical axis. Horizontal red lines indicate proposed sequence boundaries (SB); horizontal blue lines indicate maximum flooding surfaces (MFS) for the proposed stratigraphic sequences. See additional explanation in text

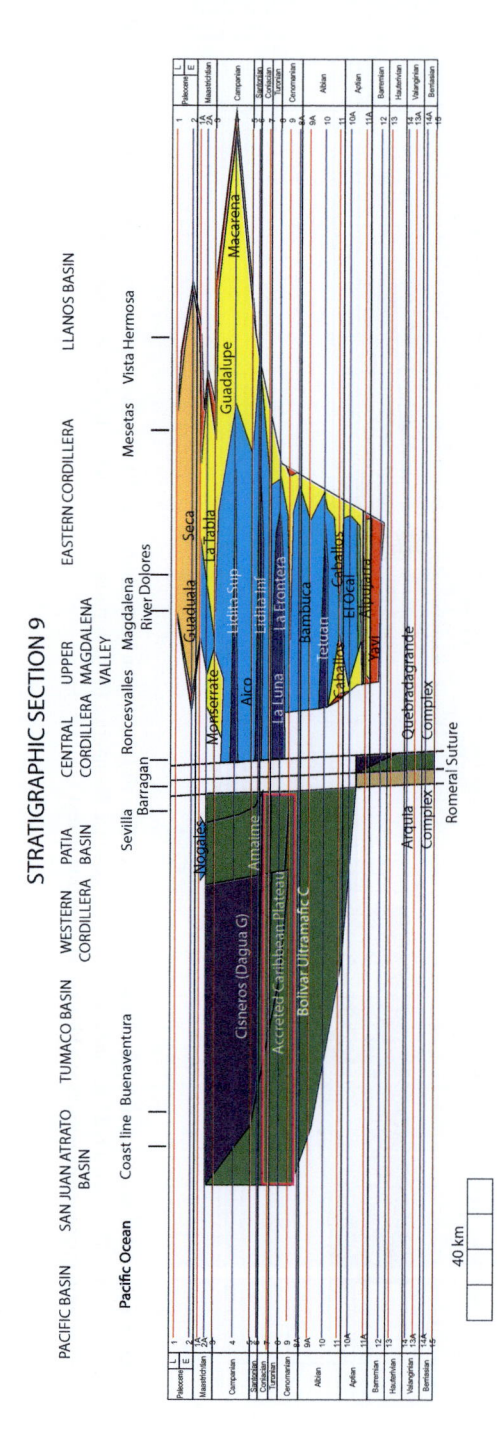

Fig. 10.16 Stratigraphic Section 9 through southern Colombia. See Fig. 10.1 for section location and Fig. 10.7 for facies color legend. Section with geological time in vertical axis. Horizontal red lines indicate proposed sequence boundaries (SB); horizontal blue lines indicate maximum flooding surfaces (MFS) for the proposed stratigraphic sequences. See additional explanation in text

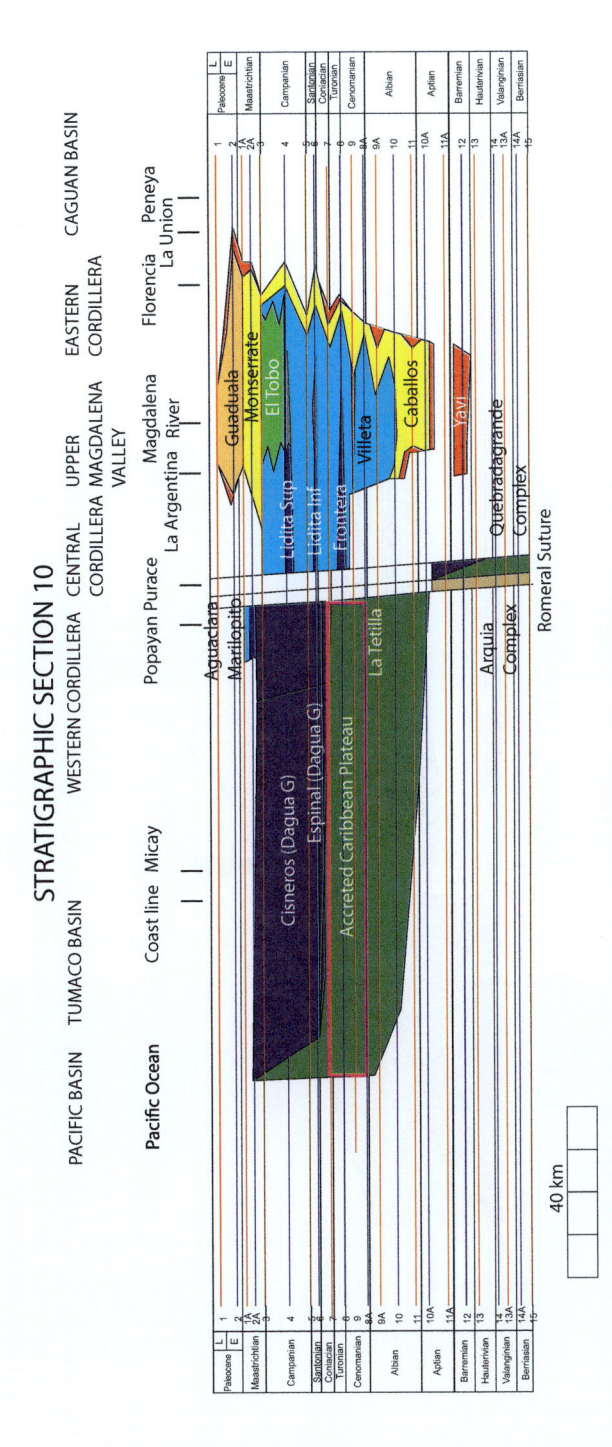

Fig. 10.17 Stratigraphic Section 10 through southern Colombia. See Fig. 10.1 for section location and Fig. 10.7 for facies color legend. Section with geological time in vertical axis. Horizontal red lines indicate proposed sequence boundaries (SB); horizontal blue lines indicate maximum flooding surfaces (MFS) for the proposed stratigraphic sequences. See additional explanation in text

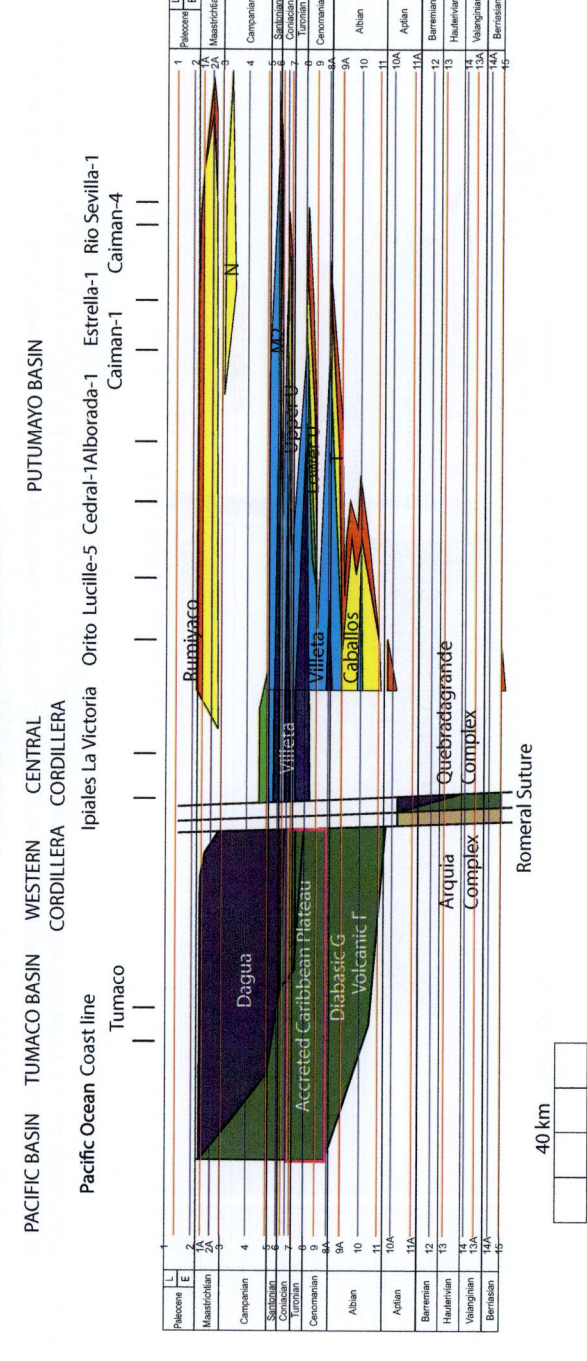

Fig. 10.18 Stratigraphic Section 11 through southern Colombia. See Fig. 10.1 for section location and Fig. 10.7 for facies color legend. Section with geological time in vertical axis. Horizontal red lines indicate proposed sequence boundaries (SB); horizontal blue lines indicate maximum flooding surfaces (MFS) for the proposed stratigraphic sequences. See additional explanation in text

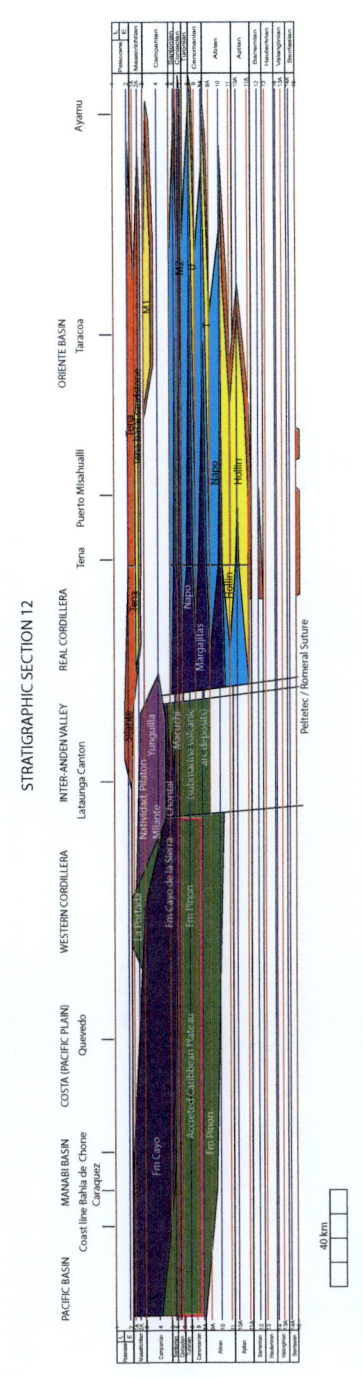

Fig. 10.19 Stratigraphic Section 12 through Ecuador. See Fig. 10.1 for section location and Fig. 10.7 for facies color legend. Section with geological time in vertical axis. Horizontal red lines indicate proposed sequence boundaries (SB); horizontal blue lines indicate maximum flooding surfaces (MFS) for the proposed stratigraphic sequences. See additional explanation in text

Fig. 10.20 Stratigraphic Section 13 through northern Peru and northwestern Brazil. See Fig. 10.1 for section location and Fig. 10.7 for facies color legend. Section with geological time in vertical axis. Horizontal red lines indicate proposed sequence boundaries (SB); horizontal blue lines indicate maximum flooding surfaces (MFS) for the proposed stratigraphic sequences. See additional explanation in text

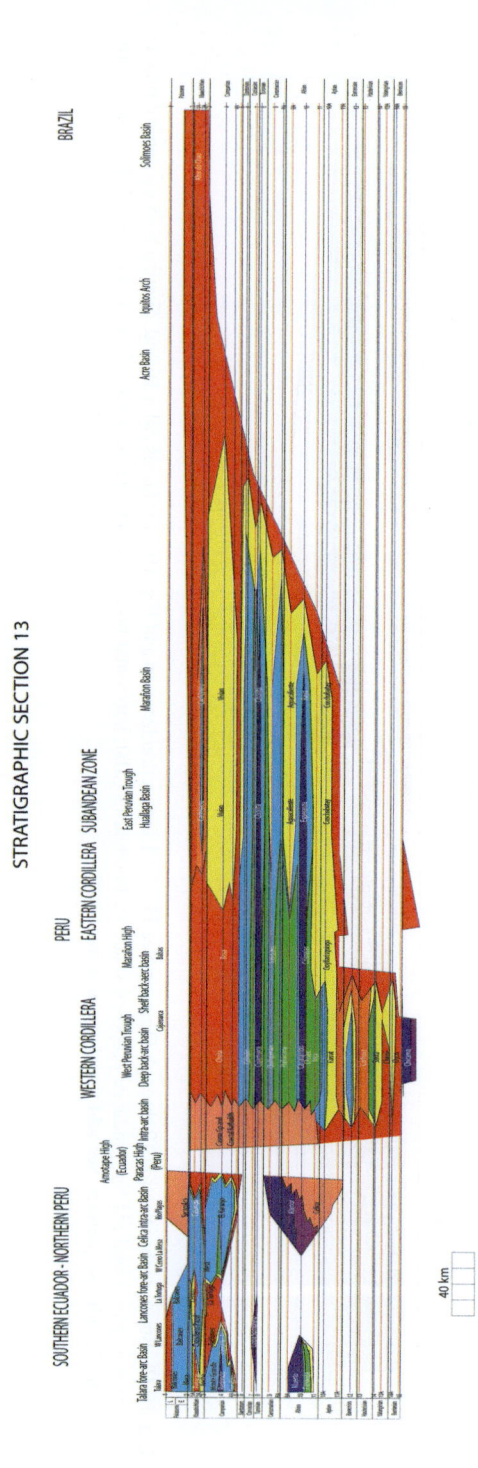

Early Cretaceous Sedimentation

In northern Venezuela, Early Cretaceous sedimentation (Figs. 10.8, 10.9, and 10.21) took place in grabens inherited from Jurassic times, localized within the Machiques and Uribante rifts. During the earliest Cretaceous (Berriasian? Valanginian?), these rift basins were still active or initiated subsidence by thermal cooling following active Jurassic extension. In the Guajira Peninsula, shallow marine sedimentation (basal clastics followed by carbonates of the Palanz Fm.; Figs. 10.10 and 10.21) is recorded; however, with the exception of northernmost Venezuela, where it is possible to predict deposition of shallow marine facies, sedimentation was dominantly

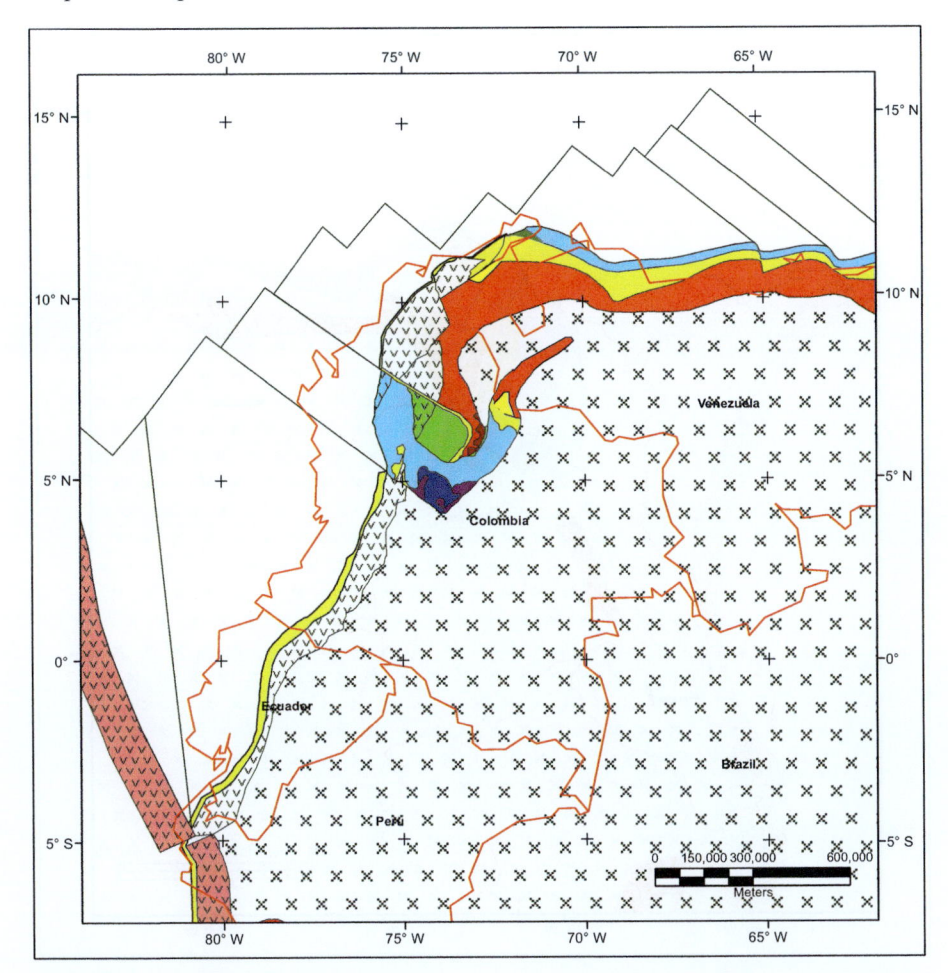

Fig. 10.21 Schematic paleo-facies map of northwestern South America during the Berriasian-Valanginian. See Fig. 10.1 for location and Fig. 10.7 for facies color legend. The schematic tectonic map reconstruction has been integrated to aid the reader in understanding the complex tectonic history of oceanic terranes accreted to the continental margin. Compiled after Etayo et al. (1976), Macellari (1988), Geotec (1992), Sarmiento-Rojas (2001), Cediel et al. (2003a, 2011), Villamil (1994, 1999, 2012), and Villamil and Pindell (2012)

continental. In the Machiques and Uribante grabens, Cretaceous sedimentation initiated within fluvial environments (*e.g.*, Rio Negro Fm. in western Venezuela and Barranquín Fm. in eastern Venezuela). During the Hauterivian (Figs. 10.8, 10.9, and 10.22), this fluvial sedimentation gradually onlapped onto the craton, while in the grabens, littoral or shallow marine sedimentation was initiated (littoral portions of the Rio Negro Fm. in western Venezuela and Barranquin Fm. in eastern Venezuela). Due to post-Jurassic tectonic or thermal subsidence, thicker sections were accumulated in the Uribante, Machiques, and Barquisimeto troughs. In general, Cretaceous rocks are separated from Jurassic rocks by an unconformity (SB 15). During the Barremian (Figs. 10.8, 10.9, and 10.22), marine transgression in general

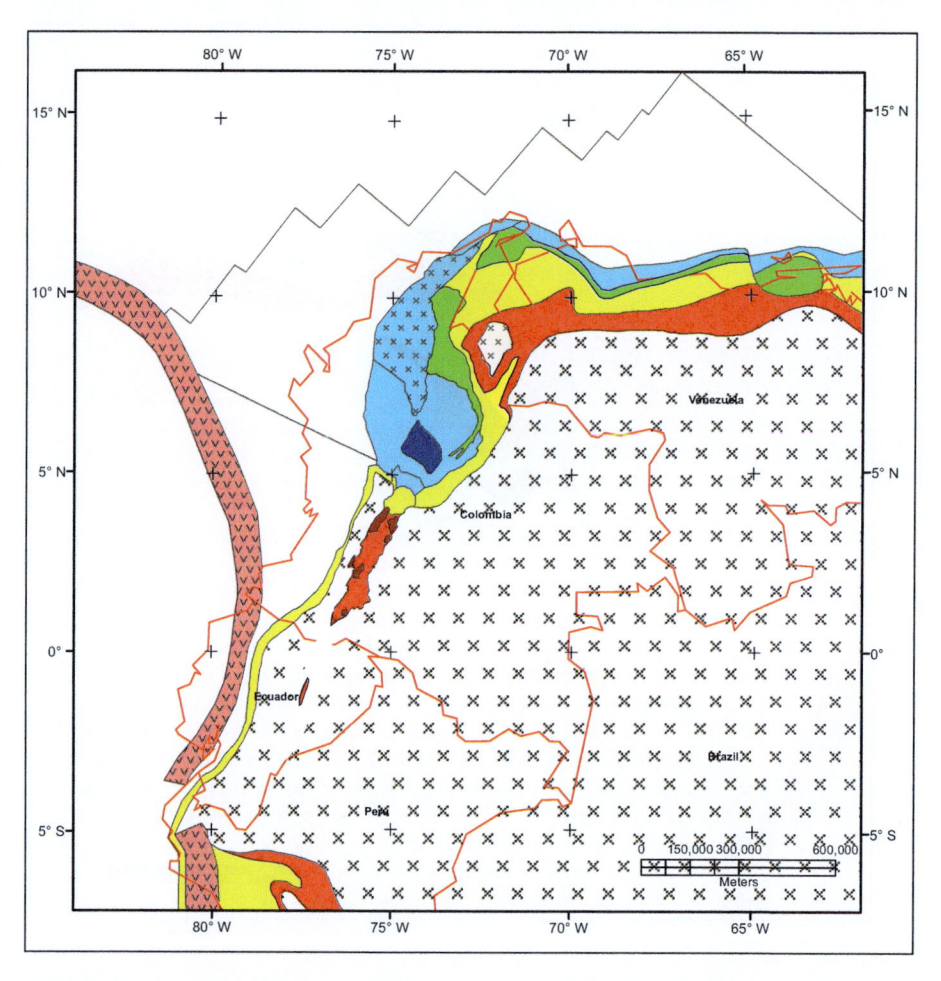

Fig. 10.22 Schematic paleo-facies map of northwestern South America during the Hauterivian-Barremian. See Fig. 10.1 for location and Fig. 10.7 for facies color legend. The schematic tectonic map reconstruction has been integrated to aid the reader in understanding the complex tectonic history of oceanic terranes accreted to the continental margin. Compiled after Etayo et al. (1976), Macellari (1988), Geotec (1992), Sarmiento-Rojas (2001), Cediel et al. (2003a, 2011), Villamil (1994, 1999, 2012), and Villamil and Pindell (2012)

advanced southward, and continental facies were covered by littoral siliciclastic marine facies (Barranquin Fm. and Valle Grande Fm. of eastern Venezuela and upper part of Rio Negro Fm. of western Venezuela), and locally some carbonate marine shelf facies were deposited, some with isolated carbonate buildups (Villamil and Pindell 2012; Morro Blanco and Taguarumo members of Barranquín Fm. of eastern Venezuela and lower part of Apón Fm. in western Venezuela and Yuruma Fm. in Guajira northernmost Colombia; Figs. 10.10 and 10.22). In general, since Berriasian to Barremian times, a general marine transgression occurred.

During Aptian (Figs. 10.8, 10.9, and 10.23), open marine inner shelf carbonate platforms were established (Apón Fm. in western Venezuela and El Cantil Fm. of

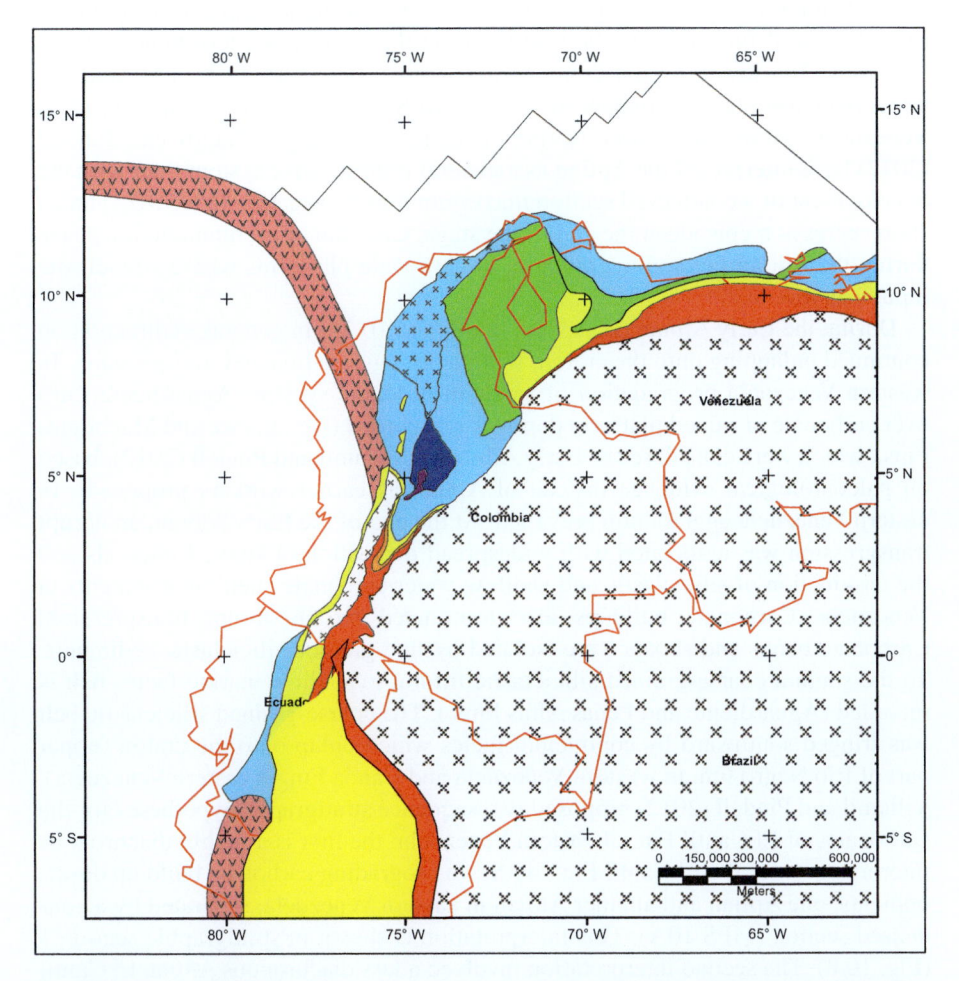

Fig. 10.23 Schematic paleo-facies map of northwestern South America during the Aptian. See Fig. 10.1 for location and Fig. 10.7 for facies color legend. The schematic tectonic map reconstruction has been integrated to aid the reader in understanding the complex tectonic history of the oceanic terranes accreted to the continental margin. Compiled after Etayo et al. (1976), Macellari (1988), Geotec (1992), Sarmiento-Rojas (2001), Cediel et al. (2003a, 2011), Villamil (1994, 1999, 2012), and Villamil and Pindell (2012)

eastern Venezuela), with local carbonate buildup developments dominated by rudist bivalves and corals. In western Venezuela carbonate shelf facies interfinger with clastics as mixed carbonate platforms (Villamil and Pindell, 2012). In other places clastic supply inhibited carbonate development. To the south, continental littoral and continental clastics continue to onlap onto Jurassic rocks or the craton (Valle Grande Fm. in eastern Venezuela and Peñas Altas Fm. and lower part of Aguardiente Fm. in western Venezuela). Fine-grained clastic shelf facies were deposited over carbonate shelf facies (shales of Valle Grande Fm. and García Fm. over Barranquín Fm. in eastern Venezuela, Apón Fm. in western Venezuela). As a consequence of the transgression, general retrogradation of siliciclastic (*e.g.*, Aguardiente Fm. over Rio Negro Fm. in western Venezuela) and carbonate facies (*e.g.*, El Cantil Fm. older in the north than in the south in eastern Venezuela) is observed. The lower Aptian contains a sequence boundary (SB 11A), and the upper Aptian contains a condensed section (MFS 10A, Guaimaro shale of western Venezuela). Above the lower Aptian sequence boundary, Villamil and Pindell (2012) have interpreted the Aptian as a general transgressive system tract with the development of a condensed section maximum flooding surface (MFS 10A) and a short regression episode at the end of the stage. Creation of accommodation space during the Aptian favored aggradation of carbonate platforms where siliciclastic input was low.

During the Early Albian (Figs. 10.8, 10.9, and 10.24) in general, sedimentation continued onlapping onto the craton, with facies aggradation and transgression. In western Venezuela progradation of sands of a deltaic system (Aguardiente Fm.) over carbonate or mixed platform deposits is recorded (*e.g.*, Lisure and Machiques Fms., which were considered as Early Albian by Villamil and Pindell (2012), based on paleontological evidence instead of Aptian as earlier workers proposed). In eastern Venezuela aggradation prevailed. At the end of the Early Albian, an abrupt transgression was associated with widespread deposition of starved intervals and the termination of siliciclastic and shallow-water carbonate shelf environments in Venezuela. Carbonate buildups also terminated by an abrupt transgression. Carbonate banks and shoals were isolated by fine-grained siliciclastic sediments. To the south a coarse-grained siliciclastic littoral to shallow marine facies belt is recorded (Aguardiente and Peñas Altas Fms.). The coarse-grained siliciclastic belt was fringed southward by continental facies which onlap onto the craton (upper part of Rio Negro Fm. in western Venezuela and Canóa Fm. in eastern Venezuela). Villamil and Pindell (2012) proposed two sequence stratigraphy hypotheses for the carbonates of El Cantil Fm. of eastern Venezuela: the first is a highly diachronous interpretation for the El Cantíl Fm., with two prograding carbonate build up developments, one Aptian and another Albian in eastern Venezuela, separated by a condensed section (MFS 10A). This interpretation is shown in stratigraphic section 1 (Fig. 10.8). The second interpretation involved a less diachronous Aptian El Cantíl Fm., lacking the two carbonate bodies separated by a shale.

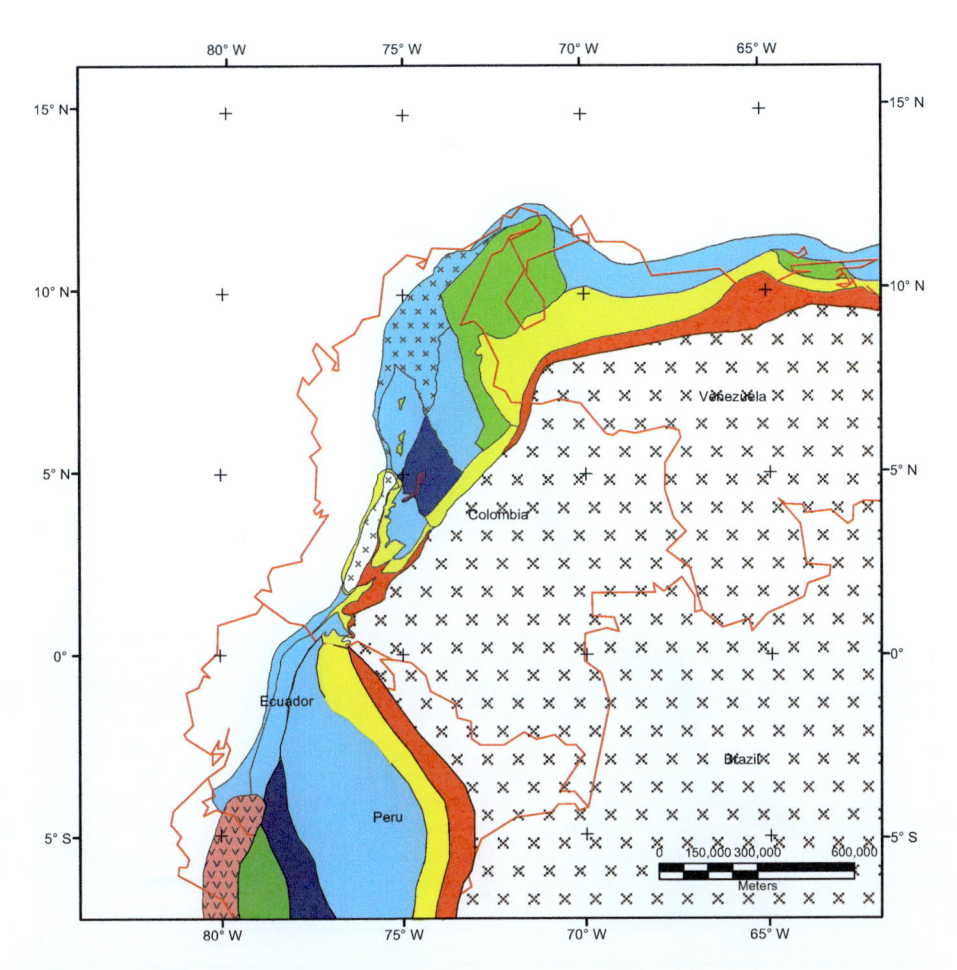

Fig. 10.24 Schematic paleo-facies map of northwestern South America during the latest Aptian-Early Albian. See Fig. 10.1 for location and Fig. 10.7 for facies color legend. The schematic tectonic map reconstruction has been integrated to aid the reader in understanding the complex tectonic history of oceanic terranes accreted to the continental margin. Compiled after Etayo et al. (1976), Macellari (1988), Geotec (1992), Sarmiento-Rojas (2001), Cediel et al. (2003a, 2011), Villamil (1994, 1999, 2012), and Villamil and Pindell (2012)

Middle Albian and Late Cretaceous Sedimentation

An abrupt, widespread transgression is observed during the Middle to Late Albian (Figs. 10.8, 10.9, and 10.25), resulting in the development of condensed sections and maximum flooding surfaces, including an important petroleum source rock interval. Two maximum flooding surfaces include:

1. A lower Middle Albian surface (MFS 10) that marks the termination of the platform carbonates and records an abrupt landward shift of facies. In eastern

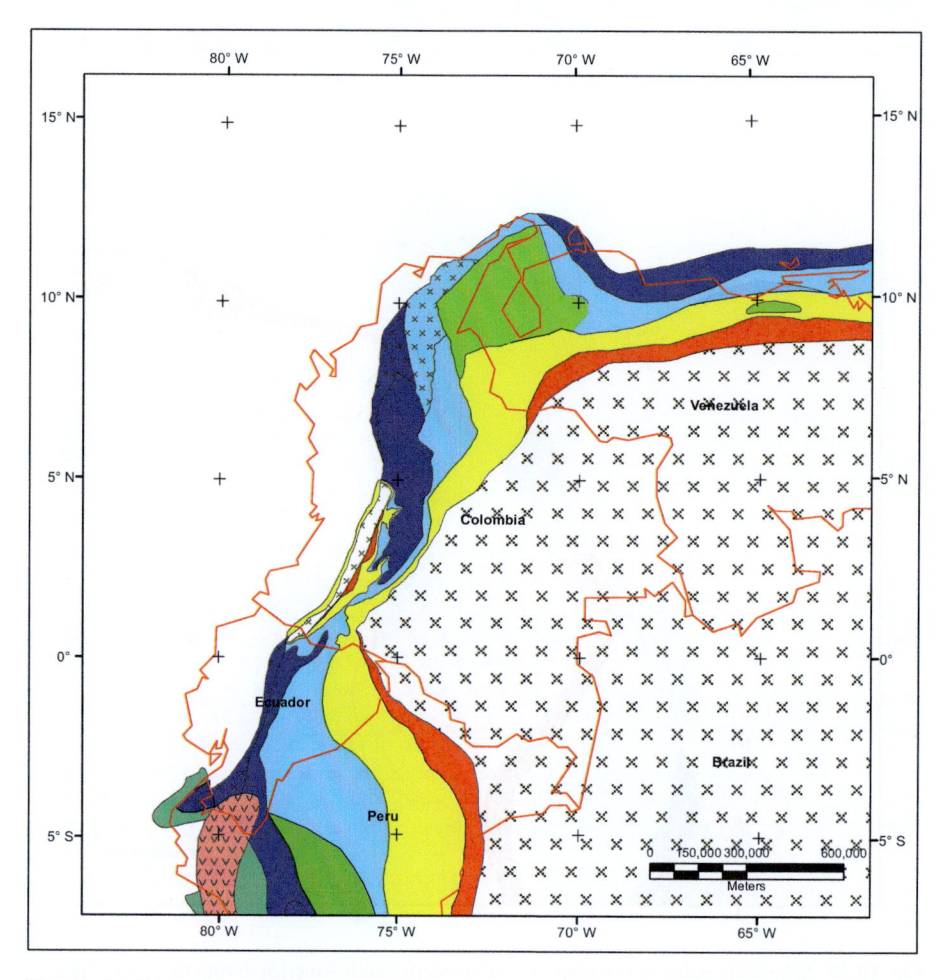

Fig. 10.25 Schematic paleo-facies map of northwestern South America during the Late Albian. See Fig. 10.1 for location and Fig. 10.7 for facies color legend. The schematic tectonic map reconstruction has been integrated to aid the reader in understanding the complex tectonic history of oceanic terranes accreted to the continental margin. Compiled after Etayo et al. (1976), Macellari (1988), Geotec (1992), Sarmiento-Rojas (2001), Cediel et al. (2003a, 2011), Villamil (1994, 1999, 2012), and Villamil and Pindell (2012)

Venezuela, MFS 10 is within a condensed section of glauconite-rich greensand at the base of Chimana Fm. In western Venezuela, the transgressive surface and maximum flooding surface (MFS 10) are condensed at the top of the Aguardiente Fm. and the middle greensand of the Escandalosa Fm. of the Barinas Basin.

2. A Late Albian maximum flooding surface (MFS 8A). The interval between the transgressive surface and the MFS 8A maximum flooding surface is condensed at the base of the Querecual Fm. in eastern Venezuela and the base of the isopic facies of the La Grita member of the Capacho Fm. in western Venezuela. In northern

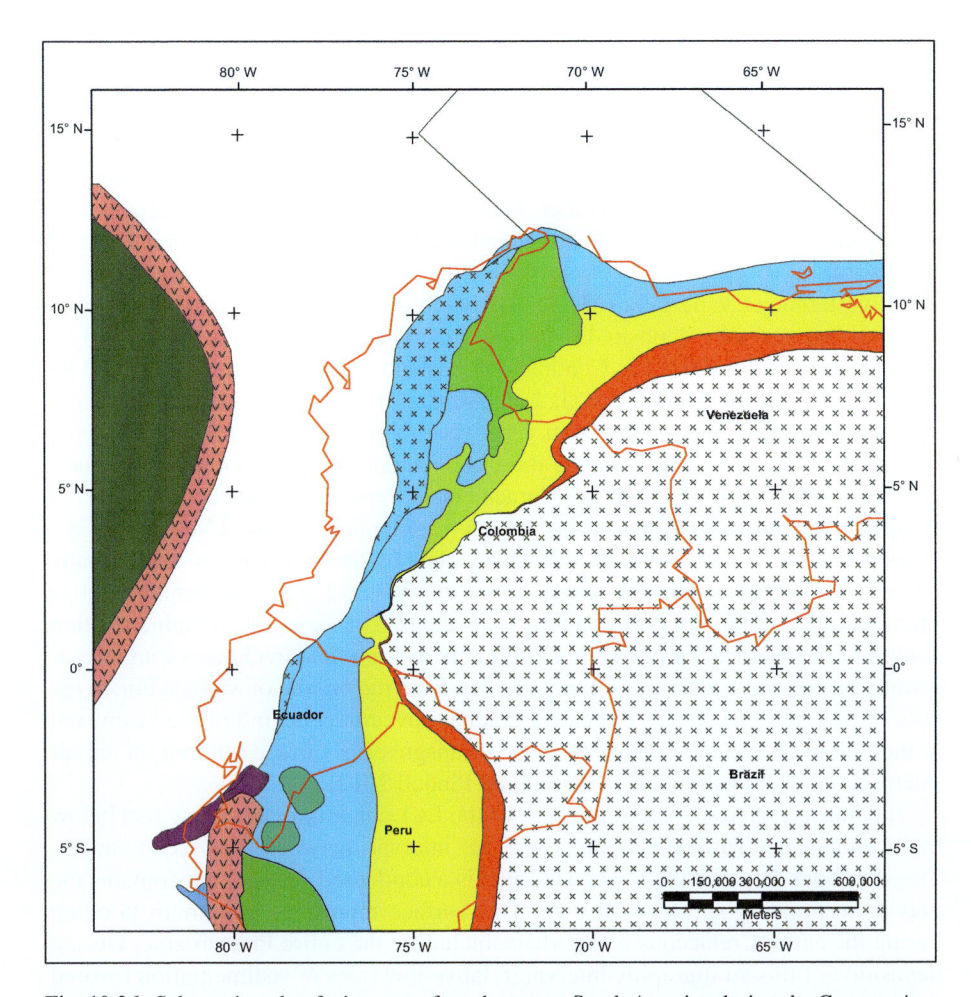

Fig. 10.26 Schematic paleo-facies map of northwestern South America during the Cenomanian. See Fig. 10.1 for location and Fig. 10.7 for facies color legend. The schematic tectonic map reconstruction has been integrated to aid the reader in understanding the complex tectonic history of oceanic terranes accreted to the continental margin. Compiled after Etayo et al. (1976), Macellari (1988), Geotec (1992), Sarmiento-Rojas (2001), Cediel et al. (2003a, 2011), Villamil (1994, 1999, 2012), and Villamil and Pindell (2012)

Venezuela (Isla La Borracha, offshore Puerto La Cruz), the MFS 8A event occurs at the contact between the Chimana Fm. and the base of the Querecual Fm. The base of the Querecual Fm is diachronous in eastern Venezuela (Late Middle to Early Late Albian in the north and Late Albian in the south).

During the Cenomanian (Figs. 10.8, 10.9, and 10.26), in eastern Venezuela, noncalcareous shales were deposited on an inner to outer marine shelf (Querecual Fm.). In western Venezuela noncalcareous shales were deposited on a marine shelf (Seboruco Fm.), and sands were deposited in littoral to shallow marine environments to the south (Escandalosa Fm.). In western Venezuela the noncalcareous Seboruco Shale represents high stand deposition with a high sedimentation rate. Above the Seboruco Shale rests the shallow-water carbonate of the Guayacán Member, and between the two, there is a sequence boundary unconformity (SB 9). This unconformity represents an abrupt shift in facies, from distal shale (Seboruco) to proximal shallow-water limestone (Guayacán Member) or even sandstone (Tocuy), as a consequence of a relative sea level drop that forced shallow-water depositional system basinward.

In westernmost Venezuela (Maracaibo Basin and Serranía de Perijá), shallow-water carbonates were deposited (Maraca Fm.). The Maraca Fm. is considered by Villamil and Pindell (2012) as Cenomanian, instead of Late Albian as proposed by earlier studies. If the Maraca Fm. is Late Albian, there has to be an unconformity with a Cenomanian hiatus of ca. 4 m.y. between the top of the Maraca Fm. and the base of La Luna Fm., which marks the Cenomanian-Turonian boundary. Villamil and Pindell (2012) propose that the Maraca Fm. is Cenomanian and correlates to the Guayacán Member, based on the observation that in other regions of Venezuela (i.e., Mérida Andes), the Guayacán Member is overlain by the La Luna Fm. and in many regions of Colombia, shallow-water carbonates of the Late Cenomanian (Caliza Mermeti) are overlain by La Luna Fm. equivalents. The same stratigraphic relation applies to the Maraca-La Luna. In addition, there is no field evidence of an unconformity representing a hiatus of 4 m.y. Rather, from the middle of Maraca Fm. to the base of La Luna Fm., facies indicate a gradual upward deepening of the basin, and at the Maraca-La Luna contact, there is a transgressive surface without an abrupt interruption of sedimentation (Villamil and Pindell 2012).

The transgressive surface at the base of the La Luna Fm. is just a few feet below the Cenomanian-Turonian boundary. The interval between the Cenomanian-Turonian boundary and the lower Turonian is a condensed section that contains the maximum flooding surface (MFS 8). This surface represents maximum flooding during the entire Cretaceous and perhaps including the entire Phanerozoic. During deposition of this stratigraphic interval, relative low rates of sedimentation favored accumulation of organic matter during times of a global marine anoxic event, allowing deposition of the best petroleum source rock in northwestern South America. The MFS 8 surface lies within a 10 m section of basal La Luna Fm., and it is characterized by a widely distributed concretion interval in western Venezuela (Villamil and Pindell 2012). Similarly, in eastern Venezuela, the Cenomanian–Turonian boundary lies within the lowest 100 m of the Querecual Fm., with similar facies.

During the Middle Turonian to Coniacian (Figs. 10.8, 10.9, 10.27, and 10.28), calcareous shales, hemipelagic limestones with micro-calcite concretions with local development of cherts, were deposited in an outer marine shelf representing a prograding high stand. This interval (La Luna Fm. in western Venezuela and Querecual Fm. in eastern Venezuela) contains cherts, indicative of vigorous upwelling conditions.

An additional transgressive event occurred during the Early Coniacian time. This event generated a highly fossiliferous laterally continuous concretion level containing abundant ammonites, planktic foraminifera, and *inoceramus* bivalves. In western

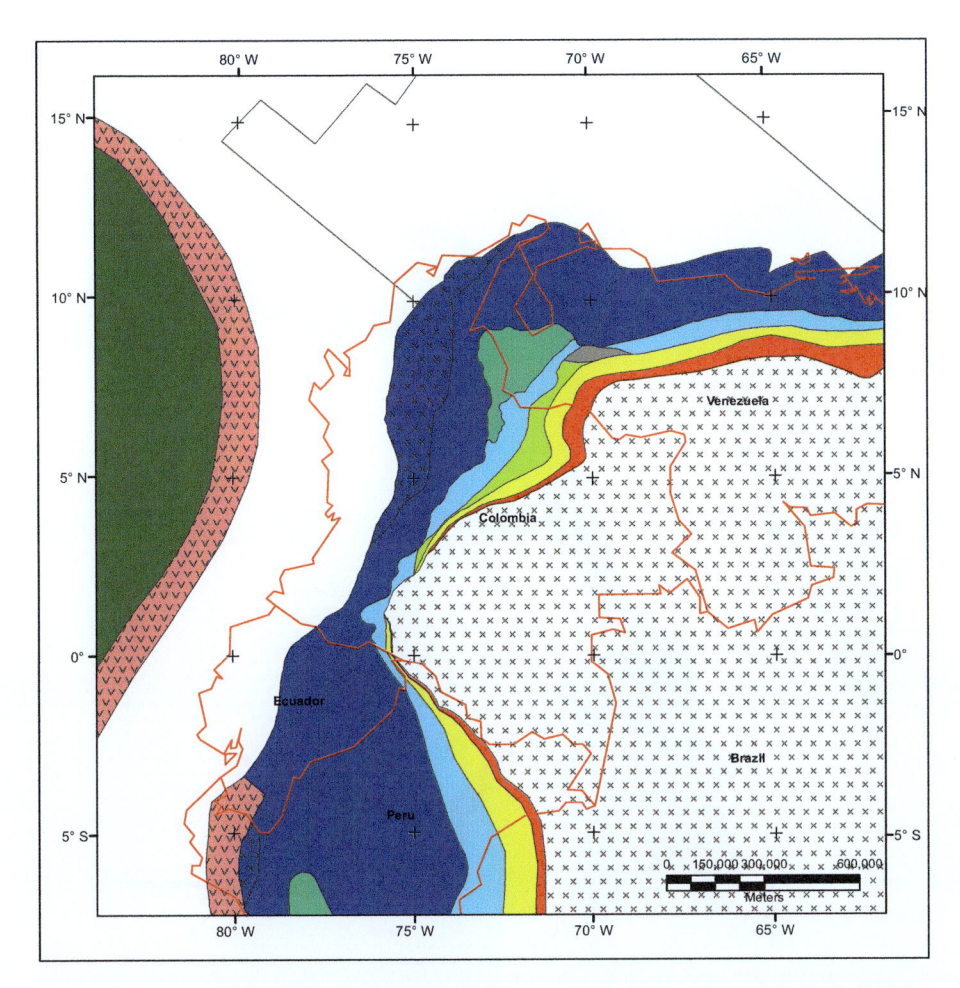

Fig. 10.27 Schematic paleo-facies map of northwestern South America during the Turonian. See Fig. 10.1 for location and Fig. 10.7 for facies color legend. The schematic tectonic map reconstruction has been integrated to aid the reader in understanding the complex tectonic history of oceanic terranes accreted to the continental margin. Compiled after Etayo et al. (1976), Macellari (1988), Geotec (1992), Sarmiento-Rojas (2001), Cediel et al. (2003a, 2011), Villamil (1994, 1999, 2012), and Villamil and Pindell (2012)

Venezuela this event lies within the La Luna Fm. while in eastern Venezuela, within Querecual Fm. (Villamil and Pindell 2012).

During the Santonian to Campanian (Figs. 10.8, 10.9, 10.29, and 10.30), intensive upwelling conditions were established over northwestern South America with deposition of chert (upper part of La Luna Fm., Ftanitas del Táchira, Tres Esquinas phosphorites in western Venezuela, and cherts of the San Antonio Fm. in eastern Venezuela). The location of active chert deposition shifted from specific locations to a regional distribution over the entire open marine shelf. During this time,

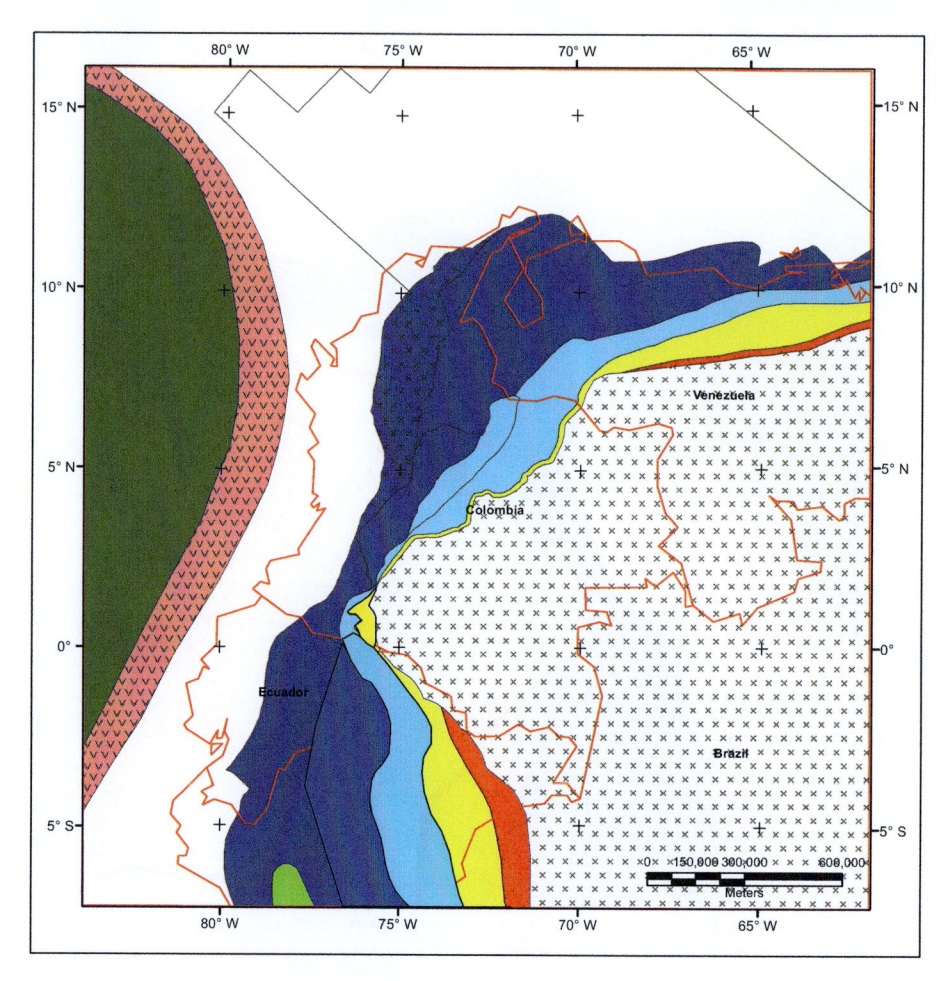

Fig. 10.28 Schematic paleo-facies map of northwestern South America during the Coniacian. See Fig. 10.1 for location and Fig. 10.7 for facies color legend. The schematic tectonic map reconstruction has been integrated to aid the reader in understanding the complex tectonic history of oceanic terranes accreted to the continental margin. Compiled after Etayo et al. (1976), Macellari (1988), Geotec (1992), Sarmiento-Rojas (2001), Cediel et al. (2003a, 2011), Villamil (1994, 1999, 2012), and Villamil and Pindell (2012)

shallow-water facies were deposited for the first time in the Barinas Basin. These facies prograded over Early Turonian maximum flooding shallow-water cherts of the Navay Fm. The upper part of the La Luna Fm. contains a phosphate-rich level called "Capa 2" in phosphate mines. Capa 2 overlies, in abrupt contact, cherts and calcareous shales of the La Luna Fm. and represents a drop in relative sea level (SB 7), followed by a rise and winnowing of siliceous fine-grained sediment. Capa 2 is overlain by cherts of the Ftanitas del Táchira, which cover large-scale ripples of the Capa 2. Another condensed section which recorded transgression and maximum

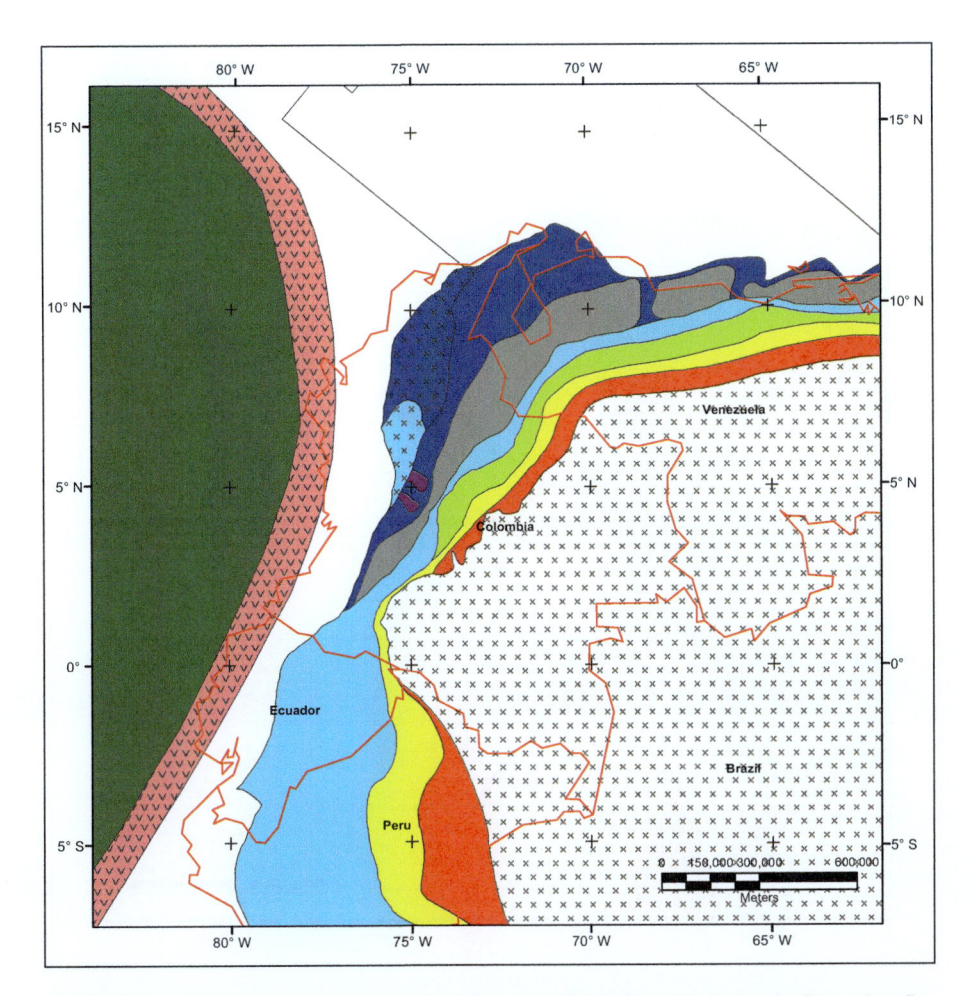

Fig. 10.29 Schematic paleo-facies map of northwestern South America during the Santonian. See Fig. 10.1 for location and Fig. 10.7 for facies color legend. The schematic tectonic map reconstruction has been integrated to aid the reader in understanding the complex tectonic history of oceanic terranes accreted to the continental margin. Compiled after Etayo et al. (1976), Macellari (1988), Geotec (1992), Sarmiento-Rojas (2001), Cediel et al. (2003a, 2011), Villamil (1994, 1999, 2012), and Villamil and Pindell (2012)

flooding (MFS 6) is the Tres Esquinas Fm. (Ghosh 1984). Downlapping of the overlying Colon shale is interpreted from outcrop and seen in seismic sections (Villamil and Pindell 2012). The condensed section of the Tres Esquinas Fm. contains glauconite and pyrite and phosphate-rich shales, abundant mosasaur and other marine reptile bones, and abundant fish debris. In eastern Venezuela, a drop in relative sea level (SB 5) is interpreted at the base of San Antonio Fm. by Villamil and Pindell (2012), with sand input in the distal basin previously dominated by fine-grained sediments of the Querecual Fm. The sequence boundary 5 at the base of the San

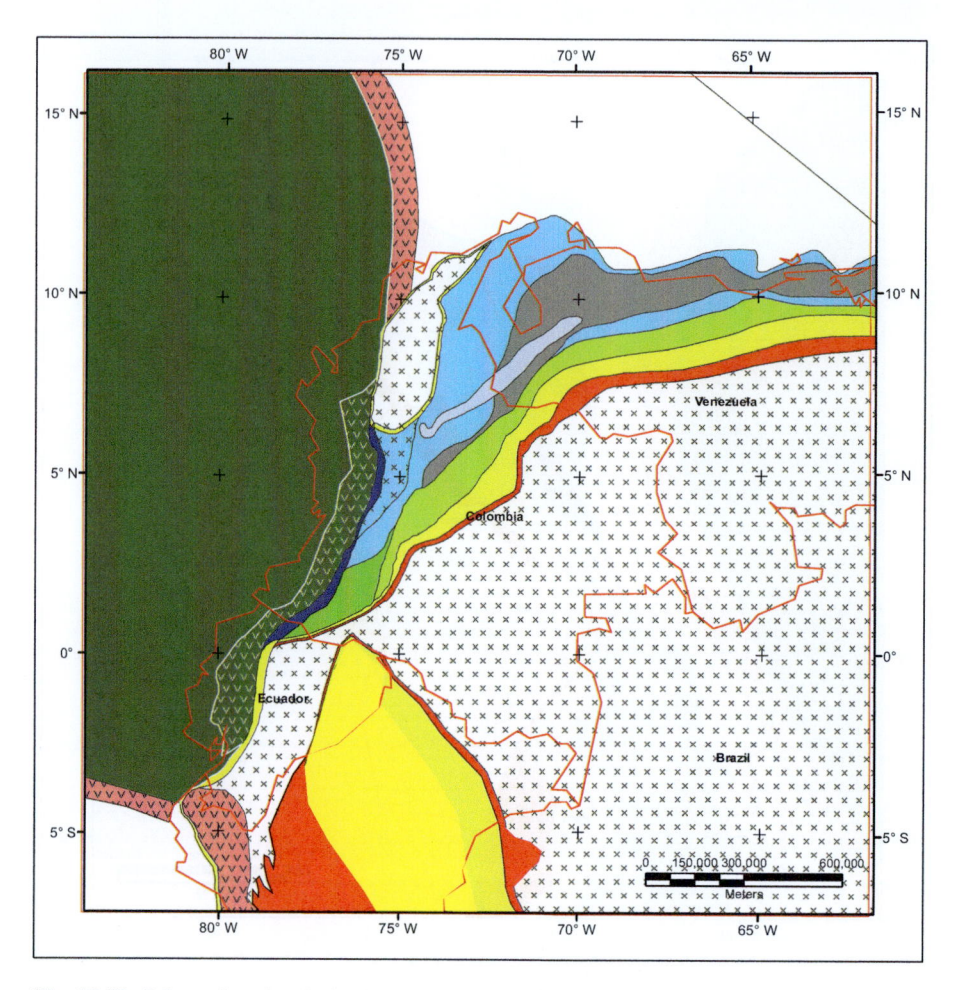

Fig. 10.30 Schematic paleo-facies map of northwestern South America during the Campanian. See Fig. 10.1 for location and Fig. 10.7 for facies color legend. The schematic tectonic map reconstruction has been integrated to aid the reader in understanding the complex tectonic history of oceanic terranes accreted to the continental margin. Compiled after Etayo et al. (1976), Macellari (1988), Geotec (1992), Sarmiento-Rojas (2001), Cediel et al. (2003a, 2011), Villamil (1994, 1999, 2012), and Villamil and Pindell (2012)

Antonio Fm. correlates with a biostratigraphic hiatus of lower Campanian plankton foraminiferal biozones, evidenced by Conney and Lorente (2009), in some basins of western Venezuela (Lake Maracaibo, Trujillo-Lara, and Barinas-Apure).

During the Maastrichtian (Figs. 10.8, 10.9, 10.31, and 10.32), sediment distribution changed in Venezuela. In western Venezuela, progradation of shallowing upward and coarsening upward shales of the Colón and Mito-Juan Fms. occurred, downlapping over the Tres Esquinas condensed section (MFS 6). In western Venezuela and Colombia, an Upper Maastrichtian and Paleocene regression and

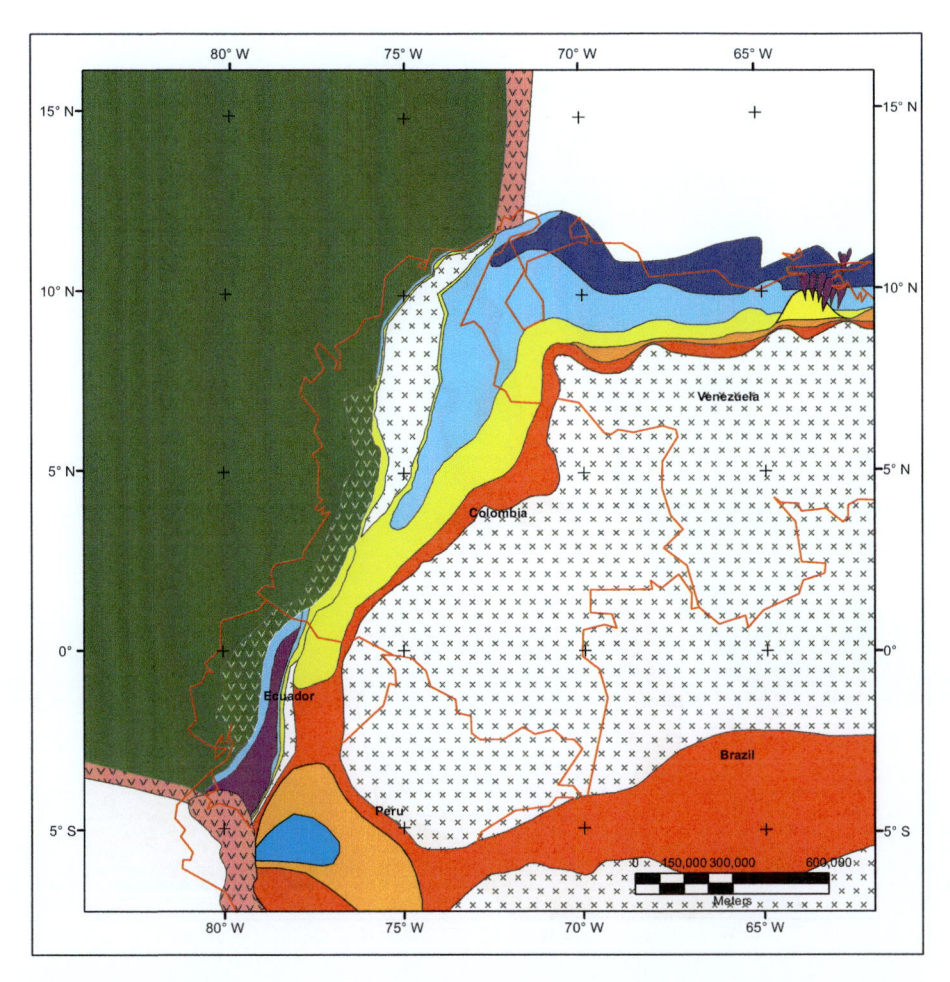

Fig. 10.31 Schematic paleo-facies map of northwestern South America during the Early Maastrichtian. See Fig. 10.1 for location and Fig. 10.7 for facies color legend. The schematic tectonic map reconstruction has been integrated to aid the reader in understanding the complex tectonic history of oceanic terranes accreted to the continental margin. Compiled after Etayo et al. (1976), Macellari (1988), Geotec (1992), Sarmiento-Rojas (2001), Cediel et al. (2003a, 2011), Villamil (1994, 1999, 2012), and Villamil and Pindell (2012)

basinward shift of facies are recorded. Inner marine shelf shales (Colón Fm.) were followed by coarser shallow marine shales (Mito Juan Fm.) ending with progradation of littoral sands (Barco Fm.) The progradation and regression filled the accommodation space, ending open marine sedimentation. The Colón and Mito Juan sedimentation high stand infilled the basin at high sedimentation rates. However, in the distal part of the basin, in western Venezuela (Lake Maracaibo and Trujillo-Lara), downlapping and condensation at the bottom of the Colón Fm. precluded deposition of several Campanian plankton foraminiferal biozones recognized by

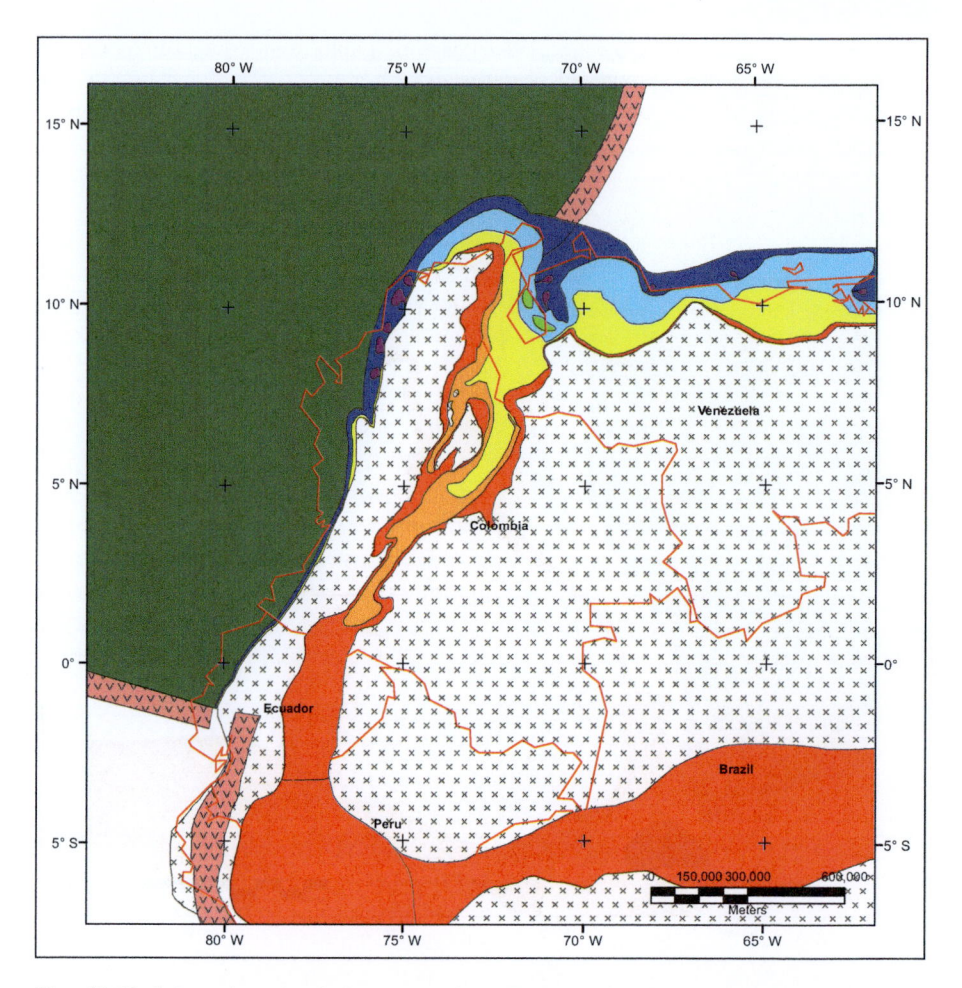

Fig. 10.32 Schematic paleo-facies map of northwestern South America during the Late Maastrichtian-Early Paleocene. See Fig. 10.1 for location and Fig. 10.7 for facies color legend. The schematic tectonic map reconstruction has been integrated to aid the reader in understanding the complex tectonic history of oceanic terranes accreted to the continental margin. Compiled after Etayo et al. (1976), Macellari (1988), Geotec (1992), Sarmiento-Rojas (2001), Cediel et al. (2003a, 2011), Villamil (1994, 1999, 2012), and Villamil and Pindell (2012)

Conney and Lorente (2009). In these areas, the stratigraphic interval between the surfaces represented by sequence boundaries 5 and 3 is condensed. Sediment input was derived from uplift of the Colombian Central Cordillera, probably increased by subtle uplift of the peripheral bulge associated with flexural subsidence, generated by the approaching Caribbean plateau in western Venezuela. In eastern Venezuela the unconformity between the chert-rich San Antonio Fm. (below) and the basal sand-rich San Juan Fm. (above) records a relative drop in sea level (SB 3), which terminated chert deposition in the basin. Sequence boundary 3, at the bottom of the

San Juan Fm. of eastern Venezuela, correlates with a hiatus evidenced by the lack of some Upper Campanian planktonic foraminiferal zones, as recognized by Conney and Lorente (2009) in western Venezuela. The San Juan Fm. of eastern Venezuela terminates with a transgressive surface that defines the base of the Vidoño Fm. The stratigraphic interval between this transgressive surface and the maximum flooding surface (MFS 2A) records relatively deep-water facies deposition of calcareous shales rich in planktonic radiolarians and foraminifera. The structureless San Juan Fm. basal sandstones could represent basin floor fans or alternatively shallow marine sands. If the basal San Juan sandstones are basin floor fans, the contact with the overlying Vidoño Fm. would be a progradational surface of slope deposits; if they are shallow marine sands, the contact with Vidoño Fm. would be a transgressive surface. In both cases the Vidoño Fm. represents a transgressive portion and a regressive progradational portion of marine deposits (Villamil and Pindell 2012).

The base of the San Juan Fm. records a fall in relative sea level (SB 3; Erikson and Pindell 1993). This was probably caused by tectonic uplift due to compressive stress on the plate during a time when the motion between North and South America changed (Villamil and Pindell 2012). In eastern Venezuela, the transgressive system tract of the lower Vidoño Fm. possibly records tectonic loading to the north. The Caratas Fm. records prograding sand during the final regression at the end of the Cretaceous. During deposition of the upper Vidoño and Caratas Fms., high stand possibly terminated with the arrival of the Caribbean forebulge in the Late Eocene, at the top of the Caratas Fm. (Erikson and Pindell 1993).

10.5.1.2 Colombia

Early Cretaceous Syn-Rift Sedimentation

Jurassic red beds are present in extensional grabens in the Guajira Peninsula, Serranía de Perijá, Serranía de San Lucas, Eastern Cordillera, and Upper Magdalena Valley and possibly in the Arauca Graben in the Llanos Basin, as a continuation of the Espino Graben of the Venezuelan Barinas Basin. In the Guajira Peninsula, shallow marine sediments of the Cocinas Gp. are also observed.

The Early Cretaceous sedimentary history in Colombia is illustrated in Figs. 10.10, 10.11, 10.12, 10.13, 10.14, 10.15, 10.16, 10.17, 10.18, 10.19, 10.20, 10.21, 10.22 and 10.23. Sedimentation started in the Tablazo sub-basin in Jurassic times and continued during the Early Cretaceous (Figs. 10.13, 10.14, 10.15 and 10.21), locally without a tectonic-related angular unconformity (*e.g.*, at the Rio Lebrija section; Cediel 1968). In other areas, Cretaceous sedimentary rocks rest along an angular unconformity (SB 15) on earlier Mesozoic, Paleozoic, or even Pre-Cambrian rocks. In the Tablazo sub-basin, the first facies were mainly sandstones (Los Santos, Tambor, and Arcabuco Fms.) deposited in fluvial environments (Renzoni 1985a, b, c; Clavijo 1985; Vargas et al. 1985; Laverde and Clavijo 1985; Galvis and Rubiano 1985; Etayo-Serna and Rodríguez 1985). Bürgl (1960, 1964, 1967) suggested that an initial marine incursion in the Cundinamarca sub-basin flooded a continental area with an arid climate, permitting evaporite formation during the early stages of marine transgression. McLaughlin (1972) cited paleontological evidence from the Berriasian-Valanginian for some evaporite occurrences. During the Berriasian, the sea flooded the basin from the northern part of the Central Cordillera to the west, toward the Cundinamarca sub-basin (Etayo-Serna et al. 1976). Subsequently, the sea proceeded from the Cundinamarca sub-basin northward filling two sub-basins, while the Santander-Floresta paleo-Massif remained emergent (Etayo-Serna et al. 1976; Fabre 1985, 1987; Sarmiento 1989; Cooper et al. 1995).

Early Cretaceous Sedimentation on the Tablazo Sub-Basin Latest Jurassic to Valanginian fluvial sedimentation In the Tablazo sub-basin was followed by mudstone deposition in marginal marine environments, recording a marine transgression (Cumbre Fm. of Mendoza 1985 and Renzoni 1985c, and Ritoque Fm. of Ballesteros and Nivia 1985 and Rolón and Carrero 1995). Later, tidal and shallow-water marine shelf carbonates (Rosablanca Fm. of Cardozo and Ramirez 1985) were deposited during the Valanginian-Hauterivian, followed by shallow marine shales (Paja Fm.) during the Hauterivian-Barremian (*c.f.* Etayo-Serna et al. 1976). Although transgression was progressive from the center of the basin, two periods of relative retreat occurred, during the Hauterivian (SB 13) and Aptian (SB 11A) (Rolón and Carrero 1995; Ecopetrol et al. 1994; Figs. 10.13, 10.14, 10.22, and 10.23). Near Villa de Leiva, the Paja Fm. records an abrupt change from organic-rich marine shelf facies in the lower (Barremian) part of the Paja Fm. (the Arcillolitas Negras Inferiores Member of Etayo 1968) changing to proximal tidal shabka and calcareous algal facies of the middle (Aptian) part of the unit (Forero and Sarmiento 1985; the Arcillolitas Abigarradas Member of Etayo-Serna 1968). This abrupt change represents a regression and a relative sea level drop (SB 11A).

Later, during the Aptian (Figs. 10.13, 10.14 and 10.23), a relative tectono-eustatic sea level rise (MFS 10A) is suggested by deeper marine facies in the upper part of the Paja Fm (Forero and Sarmiento, 1985; Ecopetrol et al. 1994; Rolón and Carrero, 1995). Near Curití, a condensed section compresses the entire Aptian into a thickness of 15 m (MFS 10A; Villamil and Pindell, 2012).

The Upper Aptian-Lower Albian Tablazo Fm. presents a problem. In the Middle Magdalena Valley, the lower part of the Tablazo Fm. consists of calcareous shale, while it contains massive fossiliferous limestone and marls in its upper part (Morales et al. 1956). The calcareous shales are rich in organic matter and are proven petroleum source rocks, interpreted to have been deposited along an anoxic outer marine shelf (i.e., La Luna-1 well; Sarmiento 2011, 2015). In contrast, along the western flank of the Eastern Cordillera near Barichara, the lowermost part of the Tablazo Fm. includes fine-grained limestones and sandy bioclastic limestones, while coarse fossils are found in the remainder of the unit. The Tablazo Fm. of the Middle Magdalena Valley, at least in its lower section, is a distal shaly organic-rich facies, while most of the Tablazo Fm. in the Eastern Cordillera represents a sandy, bioclastic, proximal shallow marine shelf facies. As a working hypothesis, I propose that the lower organic-rich part of the Tablazo Fm. in the Middle Magdalena Valley is a distal facies, correlatable with the lowermost fine-grained part of the Tablazo Fm. of the Eastern Cordillera or, alternatively, with the uppermost part of the Paja Fm. of the Eastern Cordillera. This distal interval was deposited during maximum marine flooding (MFS 10A). It is also possible that the lower part of the Tablazo Fm. of the Eastern Cordillera is coeval with a condensed section in the Middle Magdalena Valley. The massive, bedded fossiliferous limestone of the Tablazo Fm. in the Middle Magdalena Valley possibly correlates to the sandy bioclastic limestones of the Tablazo Fm. of the Eastern Cordillera.

Berriasian to Aptian Sedimentation on the Cocuy Sub-basin. During the latest Jurassic to earliest Cretaceous, marine transgression in the Cocuy sub-basin initiated in the south, as recorded by the Brechas de Buenavista Fm. (Dorado 1984) and the Calizas del Guavio Fm. (Ulloa and Rodríguez 1976; Fabre 1985; Mojica et al. 1996). To the north, facies changes record the transition from continental to shallow marine sedimentation (Lutitas de Macanal Fm.), during the Berriasian to Valanginian (Fabre 1985). During the Hauterivian to Barremian, wave-dominated deltaic sandy environments are recorded (Arenisca de Las Juntas Fm; Fabre 1985). In the Hauterivian, deposition of prograding sands in a rapidly subsiding basin (Fabre 1985) was probably facilitated by a fall in relative tectono-eustatic base level (SB 13). (Figs. 10.13, 10.14, 10.21, 10.22 and 10.23).

Early Cretaceous Sedimentation Over the Santander-Floresta Paleo-Massif. The Santander- Floresta paleo-Massif remained emergent until the Hauterivian, at which time deposition of continental sandstones, followed by progradation of deltaic sandstones (Rionegro Fm. and lower part of Tibasosa Fm.) and, in turn, by deposition of shallow marine carbonates, took place. The two sub-basins coalesced into a single basin during the Hauterivian flooding and base level rise over the paleo-Massif (Fabre 1985; Moreno 1990a, b, 1991). Santander-Floresta formed an intrabasinal high and significant barrier to sediment movement until the Aptian time (Cooper et al. 1995). The succession of sandstone (Tambor and Los Santos Fms.), limestone (Rosablanca Fm.), and dark shale (Paja Fm.) facies, recorded in the Tablazo sub-basin, is laterally younger toward the east over the Santander-Floresta paleo-Massif (sandstone, Rionegro Fm.; limestone and shale, Tibú and Mercedes Fms.) and in the Cocuy sub-basin (sandstone, Arenisca de Las Juntas Fm.; limestone and shale, Apón Fm; Fabre 1985). This lateral change in age of facies occurred as a result of the oscillating and progressive marine transgression toward the east during Valanginian to Aptian times. (Figs. 10.12, 10.13, 10.14, 10.21, 10.22 and 10.23).

Berriasian to Aptian Sedimentation in the Cundinamarca Sub-basin. Toward the south, both the Tablazo and Cocuy sub-basins record a gradual increase in dark shale content, deposited in poorly oxygenated, shallow marine environments (Caqueza Gp., Villeta Gp; Fabre 1985; Rubiano 1989; Sarmiento 1989). In the Cundinamarca sub-basin, Cretaceous sedimentation started in the Tithonian?-Berriasian-Valanginian, with turbidite deposits along both the eastern (lower

Caqueza Gp; Pimpirev et al., 1992) and western (lower part of Utica Sandstone, Murca Fm; Sarmiento 1989; Moreno 1990b, 1991) flanks. Turbidite deposition along the eastern border of the basin prevailed into the Hauterivian (Caqueza Gp; Pimpirev et al. 1992). (Figs. 10.14, 10.15, 10.21, 10.22 and 10.23).

During the earliest Cretaceous, basin subsidence exceeded sediment supply, resulting in retrogradation of the turbidite fan system, such that distal fan sediments covered middle fan mouth channel deposits. In post-Berriasian time, sediment supply increase overwhelmed basin subsidence, resulting in progradation of the turbidite fan system (Pimpirev et al. 1992) and locally by progradation of deltaic sands during the Hauterivian (upper part of Utica Sandstone; Sarmiento 1989; Moreno 1990b). This reflects a relative sea level drop (SB 13). Toward the south, the shallow marine sandstones and limestones of the Naveta Fm. mark the development of a shoreline during the Hauterivian-Barremian (Cáceres and Etayo-Serna 1969; Sarmiento 1989). Differential subsidence related to syn-sedimentary normal faulting generated unstable slopes on basin margins. These processes favored turbidite deposition from the Early Cretaceous to the Aptian (lower part of Utica Sandstone, Murca Fm., Socotá Fm; Polanía and Rodríguez 1978; Sarmiento 1989; Moreno 1990b, 1991; Caqueza Gp, Pimpirev et al., 1992).

Aptian Sedimentation in the Upper Magdalena Valley Cretaceous sedimentation is recorded for the Aptian (Vergara and Prössl 1994), although probably initiated in an extensional basin during the Late Jurassic. Feldspathic and lithic sandstones, conglomerates, and red mudstones (Yaví Fm.) were deposited as alluvial fans on valley slopes, while finer sandstones and mudstones (Alpujarra Fm. sensu Florez and Carillo 1994, Etayo-Serna 1994 and Etayo-Serna and Florez 1994) and/or Lower Sandstone member of the Caballos Fm. (sensu Guerrero et al. 2000) accumulated within a flowing northward fluvial system. (Figs. 10.16, 10.17, 10.22 and 10.23).

Abrupt lateral thickness changes and ubiquitous turbidite deposition throughout the Eastern Cordillera basin (Figs. 10.13, 10.14, 10.15 10.16, 10.21, 10.22 and 10.23) attest to local tectonic/differential subsidence depositional conditions in the Cretaceous. Regional correlation of relative Early Cretaceous tectono-eustatic cycles is difficult to establish due to active localized extensional tectonics. Since the Aptian, however, these relative tectono-eustatic level cycles become increasingly traceable.

An important transgression followed a relative sea level rise during the Late Aptian (MFS 10A), as the sea flooded all the area of the present Eastern Cordillera, including south of the Cundinamarca sub-basin (Etayo-Serna et al. 1976; Etayo-Serna 1994). During the Late Aptian, the sea gradually flooded the Upper Magdalena Valley, and dark mudstone and limestone (El Ocal Fm. sensu Florez and Carillo 1994, Etayo-Serna, 1994 and Etayo-Serna and Florez, 1994, or Middle mudstone and biomicrite member of the Caballos Fm. sensu Guerrero et al., 2000), were deposited in a shallow marine environment (Etayo-Serna 1994).

Dark-gray to black mudstone was deposited regionally upon a dysoxic shallow marine shelf which included the upper part of the Paja Fm. in the former Tablazo sub-basin, the Fómeque Fm. in the former Cocuy sub-basin, the Villeta Gp., the upper part Socotá Fm. in the former Cundinamarca sub-basin, and the El Ocal Fm. in Upper Magdalena Valley.

Cretaceous Post-Rift Sedimentation

Cretaceous post-rift sedimentation is illustrated in Figs. 10.10, 10.11, 10.12, 10.13, 10.14, 10.15, 10.16, 10.17, 10.18, 10.25, 10.26, 10.27, 10.28, 10.29, 10.30, 10.31 and 10.32. Villamil (1994) interpreted limestone-shale or chert-shale rhythmic beds as Milankovitch cycles. Using high-resolution graphical stratigraphic correlation, he showed that distal pelagic limestone-shale cycles are coeval to proximal parasequences. Assuming these cycles all have the same duration, and that subsidence was constant through time, Villamil (1994) plotted the thickness of all cycles in a modified Fisher plot (a stacking plot for cyclic rhythmic sedimentation) to obtain a curve of changes in relative accommodation space or relative tectono-eustatic base level.

Based on facies analysis, macrofossil biostratigraphy, high-resolution event, and cycle chronostratigraphy, together with the modified Fisher plots, Villamil (1994) proposed a sequence stratigraphic interpretation and a relative tectono-eustatic level history.

During the Albian (Figs. 10.10, 10.11, 10.12, 10.13, 10.14, 10.15, 10.16, 10.17, 10.18, 10.24 and 10.25), a relative base level fall (SB 11) favored progradation of deltaic and littoral sands (Caballos Fm.; Flórez and Carrillo 1994; Etayo-Serna 1994) in the area of the Upper Magdalena Valley and along the eastern border of the basin (lower part of Une Fm; Fabre 1985).

During the Middle to Late Albian (Figs. 10.10, 10.11, 10.12, 10.13, 10.14, 10.15, 10.16, 10.17, 10.18 and 10.25), the transition from near-shore marine facies of the Caballos Fm. to the deepening upward, lower part of Villeta Gp. in the Upper Magdalena Valley recorded a rise in relative tectono-eustatic level (MFS 10; Villamil 1994; Etayo-Serna 1994). This tectono-eustatic level rise was also recorded by the upward deepening trend from the shallow-water San Gil Inferior Fm. to the deeper San Gil Superior Fm., the Socotá Fm. to the Hiló Fm., and within the Une Fm. (Villamil 1994). During the Middle Albian, regionally distal, organic-rich outer shelf facies were deposited over most of the basin, during a global anoxic event responsible for the petroleum generation potential of these rocks (i.e., Tetuán Fm. in the Upper Magdalena Valley). However, there are no petroleum source rocks of Middle Albian age in the Middle Magdalena Valley. A possible explanation for this anomaly is that in the Middle Magdalena Valley, the Middle Albian maximum flooding surface (MFS 10) occurs in a condensed section with a very reduced sedimentation rate, which is not favorable for the accumulation of organic matter.

During the Late Albian-Early Cenomanian (Figs. 10.10, 10.11, 10.12, 10.13, 10.14, 10.15, 10.16, 10.17, 10.18, 10.25, and 10.26), a relative tectono-eustatic level fall (SB 9A) was recorded by progradation of the upper part of Une Fm., and a generalized shallowing upward facies trend is recorded. In the Late Albian-

Earliest Cenomanian, there is a sequence boundary (SB 9A) expressed as a forced regression (unnamed shale overlying the cherts of the Hiló Fm., shallow-water sandstone of Churuvita Fm. over deeper shale of the San Gil Superior Fm; Villamil 1994). In the Upper Cenomanian, Villamil (1994) interpreted the next marked sequence boundary (SB 9, including first sandstone in the shales of the Villeta Gp., upper sandstone part of Churuvita Fm., and uppermost sandstone of the Une Fm.).

During the Late Cenomanian, Turonian, and Coniacian (Figs. 10.10, 10.11, 10.12, 10.13, 10.14, 10.15, 10.16, 10.17, 10.18, 10.26, 10.27 and 10.28), the tectono-eustatic base level reached its maximum level for the Mesozoic. The sea flooded the entire northwestern corner of South America, and dark-gray shale was deposited from Venezuela to northern Peru. Villamil (1994) recognized smaller relative tectono-eustatic level cycles during this time interval.

A relative tectono-eustatic base level rise during the Late Cenomanian (Figs. 10.10, 10.11, 10.12, 10.13, 10.14, 10.15, 10.16, 10.17, 10.18 and 10.26) induced a slight deepening of the basin and a notorious decrease of the detrital supply. This led to basin starvation and the slow deposition of black laminated shale grading to distal micritic limestone pelagic facies. The maximum flooding surface located at the Cenomanian-Turonian boundary (MFS 8, Figs. 10.10, 10.11, 10.12, 10.13, 10.14, 10.15, 10.16, 10.17 and 10.18) is characterized by a highly fossiliferous concretion horizon within the Frontera Fm. and the lower part of San Rafael Fm. (Villamil 1994). During the Turonian-Coniacian, the present-day foothills of the Eastern Cordillera close to the Llanos Basin were flooded (Cooper et al. 1995), but not the entire Llanos Basin area. From the Middle Turonian to Late Coniacian, a gradual progradation and shallowing upward occurred during deposition of the upper part of the San Rafael Fm. and the Villeta Gp. in the Upper Magdalena Valley was related to a relative tectono-eustatic sea level fall (SB 7, Villamil 1994).

In the Upper Magdalena Valley, during the Late Coniacian to Santonian (Figs. 10.10, 10.11, 10.12, 10.13, 10.14, 10.15, 10.16, 10.17, 10.18, 10.28 and 10.29), the transition from the uppermost Villeta Gp., deposited upon an inner shelf, to the Lidita Inferior unit of the Olini Gp., deposited on a deeper middle shelf (Jaramillo and Yepez 1994; Ramirez and Ramirez 1994), points to a deepening of the basin and relative tectono-eustatic level rise (MFS 6; *c.f.* Fig. 2 of Etayo-Serna 1994).

During the Santonian, Campanian, Maastrichtian, and Paleocene (Figs. 10.10, 10.11, 10.12, 10.13, 10.14, 10.15, 10.16, 10.17, 10.18, 10.29, 10.30, 10.31 and 10.32), a general regression and progradation were recorded by littoral to transitional coastal plain facies (Guadalupe Gp., Guaduas Fm.). Guadalupe Gp. sands represent two cycles of westward shoreline progradation, aggradation, and retrogradation, dominated by high-energy quartz-rich shoreface sandstones derived from the Guiana Shield (Cooper et al. 1995). Regression did not occur continuously but with minor transgressive events recorded by fine-grained siliceous and phosphatic facies (Föllmi et al. 1992; Plaeners Fm., Olini Gp., upper part of La Luna Fm.).

A sequence boundary (SB 5) occurs at the base of the medium shale unit of the Olini Gp. (Aico Shale; Lower-Middle Santonian according to Villamil, 1994, or Late Santonian-Early Campanian according to palinostratigraphy by Jaramillo and

Yepez, 1994) and Etayo-Serna, 1994) and the shallow-water El Cobre Sandstones of Barrio and Coffield (1992) in the Upper Magdalena Valley (Villamil 1994). The shallow-water marine sands of the Arenisca Dura Fm. represent a lower forced regression system tract (sensu Posamentier et al. 1992; Cooper et al. 1995), caused by a relative sea level drop (SB 5).

Mudstones of the upper part of the Arenisca Dura Fm. and shales of the Plaeners Fm. represent a transgressive system tract (Cooper et al. 1995). During the Santonian-Early Campanian, a maximum flooding surface (MFS 4) and a relative tectono-eustatic level rise from the medium shale unit of the Olini Gp. to the Upper Chert unit has been interpreted by Villamil (1994). In contrast to the Eastern Cordillera, where the Cretaceous maximum flooding surface occurred at the Cenomanian-Turonian boundary (MFS 8), the maximum flooding in the Llanos Basin occurred during the Campanian (MFS 4 at the top of Gachetá Fm; Fajardo et al., 1993; Fig. 10.4 of Cooper et al. 1995).

During the Late Campanian (Figs. 10.10, 10.11, 10.12, 10.13, 10.14, 10.15, 10.16, 10.17, 10.18 and 10.30), relative sea level continued to drop (SB 3), and shallow marine oxygenated environments prevailed in the Eastern Cordillera. A shale stratigraphic interval between the Labor and Tierna Fms., informally called "Upper Plaeners," records a relative sea level rise (MFS 4). The Labor Fm. represents a sand-dominated, forced regression system tract (Cooper et al. 1995), induced by a relative sea level drop (SB 2A).

The regional regression and long-term relative tectono-eustatic level fall were interrupted by two small cycles of relative tectono-eustatic base level rise during Late Campanian (MFS 4), during deposition of the "Upper Plaeners" and Early Maastrichtian (MFS 2A), as suggested by Föllmi et al. (1992) and Villamil (1994). According to Cooper et al. (1995), the Upper Plaeners unit represents a condensed marine mudstone deposited during a relative tectono-eustatic level rise (MFS 4).

During the Early (?) Maastrichtian (Figs. 10.10, 10.11, 10.12, 10.13, 10.14, 10.15, 10.16, 10.17, 10.18 and 10.31), the eastern part of the basin was filled by the littoral quartz sands of the Arenisca Tierna Fm. (Fabre 1985). According to Cooper et al. (1995), the latter represents the transgressive systems tract of the next sequence that reached a maximum flooding surface (MFS 2) at the base of Guaduas Fm. (Figs. 10.10, 10.11, 10.12, 10.13, 10.14, 10.15, 10.16, 10.17, 10.18, 10.31 and 10.32). The gradual uplift of the western margin of the Upper Magdalena Valley supplied clasts of metamorphic rocks that were accumulated by fluvial systems close to the sea in a braided delta (Cimarrona Fm; Gómez and Pedraza, 1994). Sands were dispersed along a littoral belt (Monserate Fm; Ramírez and Ramírez, 1994), while in the more distal areas, carbonate silt (Díaz, 1994a, b) or mud (Umir Fm.) accumulated (Etayo-Serna 1994).

Southern Cauca Valley The Campanian basement of the southern Cauca Valley, herein termed the "Amaime Heterolithic Complex," represents tectonic stacking of oceanic basalts and sediments, trench deposits, and slivers of dismembered ophiolites (Barrero and Laverde, 1998), accreted to the continental margin. Over this accreted basement, the Nogales Fm. was deposited during the Maastrichtian.

The lower Nogales Fm. consists of matrix-supported conglomerates, calcareous sandstones, mudstone, and claystone sourced from the emergent Central Cordillera and deposited as debris flow, most probably in the proximal part of a fan delta (fan delta-front facies). The middle part of the section consists of turbiditic sandstones, siliceous claystone, and chert beds. Maximum flooding probably occurred during deposition of the cherts and claystones (MFS 2). The upper section consists of claystone and cross-bedded sandstone. Conglomerate and claystone of red and green color dominate the upper part and suggest transition to continental deposits. This segment is interpreted to be deposited in the fan delta transition zone front to alluvial fan facies (Pardo et al. 1993; Blau et al. 1995). In the southern areas of the Cauca Valley near Vijes, Cali-Timba, and El Dinde-Chimborazo, chert has been reported (Marilopito Fm.) and marine mudstones (Aguaclara Fm.) similar to those deposited during maximum flooding (MFS 2) in the Nogales Fm. These units have been considered Maastrichtian in age. In the Patía sub-basin, the uppermost conglomerates (lower part of the Rio Guabas unit) suggest transition to continental deposits. The Paleocene is represented by a stratigraphic hiatus (SB 1). (Figs. 10.16, 10.17, 10.30 and 10.31).

Northern Sinu-San Jacinto Basin. In this basin the Cansona Fm. was deposited during the Late Cretaceous (Coniacian-Santonian-Maastrichtian; Figs. 10.12, 10.13, 10.28, 10.29, 10.30 and 10.31). Its lower part was deposited unconformably (SB 5?) over oceanic basement (Guzmán et al. 1994, 2004). The Cansona Fm. is composed of foraminifera-rich calcareous mudstone, thin limestone, chert, and locally sandstone and conglomerate, within a transgressive-regressive sequence marked by a Campanian maximum flooding surface (MFS 4). During the Late Cretaceous and Early Paleogene, a coastline developed close to the present western boundary of the Lower Magdalena Valley, a region that formed positive relief and a source area for the sediments of the Cansona Fm. Conglomerates and sandstones are interpreted to record a proximal setting close to the shore line. Environments of deposition include proximal, low energy and relatively inner to middle shelf (Alfonso et al. 2009), and middle neritic to deep bathyal pelagic, as marked by planktonic foraminifera, cocolithophorids, and radiolarian (Guzmán et al. 1994). Palinofacies indicate a sub-oxic to anoxic environment (Juliao et al. 2011). The Early Paleocene corresponds to a hiatus represented by the unconformity surface between the Cansona Fm. and overlying rocks. This unconformity (SB 1) implies that the Cansona Fm. was exposed to erosion as a result of Early Paleocene uplift, as recorded by the AFTA data analyzed by Alfonso et al. (2009).

Catatumbo Sub-basin (Colombian Portion of the Maracaibo Basin) Sedimentation started as continental fluvial facies (Rionegro Fm. in Maracaibo and sandy lower part of Tibú Fm.) unconformably deposited upon (SB 15) over a pre-Cretaceous peneplanized surface. Marine transgression during the Aptian initially deposited littoral facies followed by shallow marine shelf limestones (Tibú and Mercedes Fms.). In distal areas, inner shelf mudstones were deposited and prevailed during a sea level rise (MFS 12). During the Albian, deltaic sands prograded to the NW (Aguardiente Fm.). Progradation was favored by a relative sea level fall (SB 11). Sandstones are dominantly quartz arenites sourced from the craton (Figs. 10.12, 10.21, 10.22 and 10.23).

During the Late Cretaceous (Figs. 10.12, 10.24, 10.25, 10.26, 10.27, 10.28, 10.29, 10.30, 10.31 and 10.32), sedimentation was controlled by eustatic changes, and mudstone deposition prevailed along an inner marine shelf (mudstones of the Capacho, La Luna, Colón, and Mito-Juan Fms.). During the Middle Albian and Lower Turonian, marine flooding events (MFS 10 and MFS 8 respectively) deposited organic-rich fine-grained pelagic limestones and shales (Albian Capacho Fm. and Turonian La Luna Fm.) during maximum sea levels, coinciding with global anoxic events. Subsequently, inner shelf mudstones coarsening upward to siltstones record a gradual shallowing of the basin (Colón and Mito-Juan Fms.). By the end of Cretaceous and beginning of the Paleocene, uplift in the Central Cordillera supplied sand and silt to the basin, and a final regression occurred marked by coarsening shallowing upward units (Catatumbo, Barco, and Los Cuervos Fms.).

Perijá Range and the Cesar-Ranchería Basin In these areas, Cretaceous stratigraphy is very similar to that of the northern Maracaibo Basin (Fig. 10.11). Extensional episodes in the Perijá Range controlled sedimentation during the Berriasian to Early Aptian (Figs. 10.11, 10.21, 10.22 and 10.23), until the Early Cretaceous (pre-late Aptian), and a clastic succession (Rio Negro Fm.) was deposited unconformable over Jurassic red beds (La Quinta Fm., Maze 1984) of the Perijá rift. The Rio Negro Fm., containing sandstones and conglomerates with occasional claystone and siltstone interbeds, was deposited in a fluvial to transitional environment. In the Perijá depocenter, the Rio Negro Fm. attains thickness up to 1500 m, while in the Cesar-Rancheria Basin, thickness decreases to 200 m and to a minimum of 10 m on the NW border of the basin, along the southeastern flank of the Sierra Nevada de Santa Marta.

During the Late Aptian (Figs. 10.11 and 10.24), thick-bedded fossiliferous limestones and very thin-bedded calcareous shales of the Lagunitas Fm. were deposited upon an inner carbonate shelf. During the Albian to Cenomanian (Figs. 10.11, 10.25, and 10.26), biomicritic limestones and fossiliferous, organic-rich shales, glauconitic sandstones, mudstones, and sandy limestones of the Aguas Blancas Fm. (Colombia) and Maraca Fm. (Venezuela) were deposited on an inner to middle marine shelf. Vertical changes from carbonate inner shelf facies to argillaceous middle to outer shelf facies were controlled by relative sea level changes.

During the Cenomanian to Coniacian (Figs. 10.11, 10.26, 10.27 and 10.28), bituminous biomicrites (pelagic limestones), organic-rich calcareous shales with abundant concretions, chert, including planktonic and benthonic foraminifera and ammonites contained within the La Luna Fm., were deposited on an outer to middle shelf controlled by relative sea level changes. Similarly, in western Venezuela, the interval between the Cenomanian-Turonian boundary and the lower Turonian is a condensed section that contains the maximum flooding surface (MFS 8) of the Cretaceous. During deposition of this stratigraphic interval, relatively low rates of sedimentation favored accumulation of organic matter under global marine anoxic conditions, allowing deposition of what is considered the best petroleum source rock in northwestern South America. The Tres Esquinas glauconitic mudstone member, only a few meters thick, widely recognized in Venezuela, has been also recorded in most of the wells drilled in the Cesar-Ranchería Basin (Intera Information Technologies-Bioss 1995). This member represents a condensed section recording transgression and maximum flooding (MFS 6).

Overlying the Tres Esquinas Member, biomicrites and calcareous mudstones of the lowermost part of Colón Fm. (the Molino Fm. of the Cesar-Ranchería Basin) were deposited upon an inner to middle shelf. The rest of the Molino Fm. is comprised of calcareous and carbonaceous shales with abundant benthonic and planktonic foraminifera, also deposited upon an inner to middle marine shelf (Martínez 1989). In its uppermost segment, thin interbedded sandstones mark transition to an intertidal environment.

Central Cordillera and the Lower Magdalena Valley (Plato-San Jorge Area) In most of these areas, there is no Cretaceous sedimentary record (Figs. 10.12 and 10.13). I suggest that, although these areas were covered by the sea, the lack of subsidence, or very reduced subsidence, did not provide accommodation space for sediment accumulation (indicated in the paleo-facies maps with color for sedimentary environment). In the absence of a stratigraphic record (indicated by a pattern of "x"s in Fig. 10.7), it may be postulated that accumulated sediment would have been eroded during Maastrichtian to Paleocene and Eocene uplift of these areas, as suggested by Cediel et al. (2003a, b).

10.5.1.3 Southern Colombia (Putumayo Basin) and Ecuador (Oriente Basin)

Sedimentation in southern Colombia and Ecuador is illustrated in Figs. 10.18, 10.19, 10.23, 10.24, 10.25, 10.26, 10.27, 10.28, 10.29, 10.30, 10.31 and 10.32. A slight reactivation of Jurassic normal faults during the Early Cretaceous time is recorded. Continental sedimentation was followed by development of a coastal plain and marine transgression. Deposition of shallow marine sandstone facies in the southern Upper Magdalena Valley of Colombia and Aptian in the sub-Andean Putumayo and Oriente Basins (Caballos Fm. in Colombia, and Hollín Fm. in Ecuador) during the Albian was followed by deposition of inner shelf mudstones (lower part of Napo Fm. in Ecuador, and lower part of Villeta Fm. in Colombia). Since the Turonian, in Ecuador and possibly southernmost Colombia, regional transpression leading to tectonic inversion of the Andean Ranges converted the basin to a back-arc foreland. During this time, marine sedimentation continued, and the sea reached its maximum extension toward the craton during the Cenomanian-Turonian boundary (MFS 8; upper part of Napo Fm. in Ecuador, and middle part of Villeta Fm. in Colombia). Later, depositional environments became shallower and included a proximal marine sandstone facies (Rumiyaco Fm. in Colombia, and Tena Fm. in Ecuador). Several stratigraphic sequences generated by tectono-eustatic changes have been identified in southern Colombia and Ecuador in the Aptian to Upper Cretaceous section (Hollín, Napo and basal Tena Fms. in Ecuador, and Caballos, Villeta and basal Rumiyaco Fms. in Colombia), Barragán et al., 2004). Each stratigraphic sequence includes an erosive sequence boundary corresponding to fluvial incised valleys during a sea level drop. These valleys were filled by fluvial and estuarine deposits followed by marine mudstones during a transgressive phase and deposited organic-rich mudstones during maximum flooding events. The upper part of each cycle contains distal shallow marine limestones and mudstones and proximal prograding deltaic sandstones, and most of these cycles show a transition from fluvial to estuarine and marine shelf (Barragán et al. 2004). Such sequences include the following from base to top: (1) Aptian, dominated by sandstones (lower part of Hollín Fm. in Ecuador), deposited on a surface incised by fluvial erosion of valleys generated during a relative sea level drop (SB 11A). Later, during sea level rise, sand deposition filled these valleys, (2) Lower and Middle Albian, including the fluvial to estuarine sandstones of the Caballos Fm. in Colombia, and the upper part of the Hollín Fm. in Ecuador. This sequence is bounded at the bottom by SB 11 and contains marine mudstones (base of Napo and Villeta Fms.). This configuration represents a transgressive sequence wherein the maximum flooding surface (MFS 10) is within the outer shelf distal and organic-rich facies, (3) Upper Albian to Lower Cenomanian. The base is the SB 9A surface at the base of the "T" sand. It contains inner shelf shale facies, and its maximum flooding surface is MFS 8A, in the distal outer shelf facies, (4) Cenomanian to Turonian. The base is the SB 9 surface at the base of the "lower U" sand. The maximum flooding surface is MFS 8, with organic-rich, distal shales, (5) Coniacian to Santonian. The base is the SB 7 surface, which occurs at the base of the "M2" sand in Ecuador and the "upper U" sand in Colombia. The maximum flooding surface is MFS 6. In the upper part, a minor Santonian sequence can be recognized as a subdivision of this sequence. Sequences 3 to 5 are dominated by marine shelf facies in the Napo and Villeta Fms. (6) The Upper Campanian, SB 5, at the bottom, represents an unconformity with a hiatus spanning the lower and Middle Campanian, suggesting regional uplift of the basin. This sequence is dominated by sandstone, including the "N" sand in Colombia, and "M1" sand in Ecuador. It only occurs in the eastern segment of the basin, probably due to continued uplift in the western segment.

In Ecuador, in the Sub-Andean Zone east of the Real Cordillera and west of the Oriente Basin, between the Sub-Andean and Cosanga Faults, the Margajitas Fm. (Fig. 10.19) measures more than 1 km in thickness. Pratt et al. (2005) describe the unit as a dark-gray, pyrite-bearing, noncalcareous, strongly cleaved, silty mudstone to mudstone with occasional interbeds up to 15 m thick of massive, well-sorted quartz arenite. The arenites are dark gray due to interstitial organic material and pyrite and contain about 1% distinctive blue quartz grains. Bioturbation, including horizontal traces on bases, is widespread. The Margajitas Fm. was considered Paleozoic in age by Tschopp (1948) and was included in the Upano unit, a meta-andesite-dominated Jurassic sequence, by Litherland et al. (1994). Notwithstanding, the Margajitas mudstones include conformable, non-tectonized sequences of cleaved limestone with typical Cretaceous bivalves and echinoid fauna, which measure up to

25 m thick. The Margajitas quartz arenites are petrographically identical to Hollín arenites and contain a similar proportion of blue quartz. Based on these considerations, Pratt et al. (2005) suggest a Cretaceous age for this unit. If the Margajitas Fm. is of Cretaceous age, the dark-gray pyritiferous mudstones would have been affected by high degrees of diagenesis and hence may represent organic-rich facies deposited in a distal part of the Oriente basin. Such mudstones could represent source rocks for petroleum found in the Oriente and Putumayo basins and may help explain the volume of petroleum discovered therein.

A Campanian to lower Maastrichtian stratigraphic hiatus has been described in the western part of the basin in both Colombia and Ecuador, which was generated by basin uplift and erosion in the foreland area, followed by development of a ravinement surface formed by flooding due to the subsequent sea level rise (Barragan et al. 2004). The Campanian to lower Maastrichtian hiatus is related to early phases of compressional tectonism in the sub-Andean zone (Baby et al. 2004).

During the Late Maastrichtian to Early Paleocene, shallow marine and fluvial sandstones were deposited (Tena Fm. in Ecuador and Rumiyaco Fm. in Colombia).

10.5.1.4 Southernmost Ecuador, Northern Peru, and Northwestern Brazil

In southern Ecuador and Peru, the following paleogeographical zones are recognized from east to west (Jaillard, 1993, Jaillard et al.,1990, 1995, 2000; Fig. 10.1):

1. The pre-Cambrian Guiana and Brazilian Shields.
2. The East Peruvian Trough, an eastern, westward sloping, moderately subsiding, marine shelf in along which sedimentation initiated during the Aptian. The eastern continuation constitutes the eastern basins of Ecuador (Oriente Basin) and Peru (Santiago, Huallaga, Marañón, and Ucayali Basins), containing marine and continental fill. This zone corresponds to the easternmost Cordillera Real-Eastern Cordillera (Ecuador and Peru, respectively), the sub-Andean Zone, and the eastern basins.
3. The Marañón High, an axial swell which constitutes a positive (lesser subsiding) horst block with a thin Cretaceous section, deposited during the Aptian. It is contained within the Eastern Cordillera of northern Peru (Marañón geanticline) and the southwestern Altiplano of Southern Peru.
4. The West Peruvian Trough, which formed a deep, rapidly subsiding rift basin, filled with a thick, dominantly marine section, during Mesozoic. It contains a complete section representing the whole of the Cretaceous. The Western Peruvian Trough became an incipient Western Cordillera during the Senonian. Northward its sedimentary fill thins and continues north into the southern part of the Cordillera Real and sub-Andean Zone of Ecuador. The West Peruvian Trough, the Marañon High, and the East Peruvian Trough constitute part of a back-arc basin complex.
5. The Coastal Trough, or intra-arc basin, is filled with very thick sections of latest Jurassic and mid-Cretaceous volcanic and volcanoclastic rocks which outcrop

along the present-day coast. In northwestern Peru and southwestern Ecuador, the Celica intra-arc basin is the northward equivalent of the coastal trough of Peru. In northwestern Peru, the Lancones and Talará Basins are fore-arc marginal basins, comparmentalized during the Albian and Campanian, respectively.

6. The Coastal Cordillera contains mainly Paleozoic and older rocks, because uplift and erosion during the Cenozoic removed most of the Mesozoic section. It outcrops in southern Peru. In northern Peru and southern Ecuador, the Amotape Massif is regarded as an allochthonous terrane accreted during the latest Jurassic (Mourier et al. 1988). During the Cretaceous, it was a morphological equivalent of the Coastal Cordillera in southern Peru.

Cretaceous Sedimentation in Northern Peru and Northwestern Brazil

In Peru, Cretaceous sedimentation spanning the latest Jurassic to Early Aptian is characterized by mainly deltaic siliciclastic deposits, followed by carbonate platform sedimentation up to the Campanian, and subsequently by deposition of continental red beds. Cretaceous sedimentation in northern Peru and northwestern Brazil is illustrated in Figs. 10.20, 10.21, 10.22, 10.23, 10.24, 10.25, 10.26, 10.27, 10.28, 10.29, 10.30, 10.31 and 10.32.

Latest Jurassic to Early Aptian Sedimentation

West Peruvian Trough Well-documented earliest Cretaceous rocks are restricted to the West Peruvian Trough (Dalmayrac et al. 1980, Figs. 10.20 and 10.21). To the north, lagoon deposits of probable Early Tithonian age (Simbal Fm.) are overlain, in sharp contact, by a Late Tithonian to Berriasian aggradational section of proximal turbidites, slope deposits, and basinal black shales (Chicama Gp; Jaillard and Sempere 1989), measuring up to 1500 m thick. In the West Peruvian Trough, subsequent deposition of the Oyón Fm., comprised of continental mudstones and sandstones containing some coal beds (Wilson 1963), is recorded. Rapid subsidence of the basin permitted deposition of various hundreds of meters of sediment. The abrupt vertical upward change, from deep marine sediments of the Chicama Gp. to continental facies of the Oyón Fm., is interpreted to represent a sequence boundary (SB 15) at the base of Oyón Fm. Although the Oyón Fm. is mostly continental in nature, it contains shallow marine interbeds with Berriasian ammonites. This marine interval contains a maximum flooding surface (MFS 14A). To the southwest, in Lima province, Berriasian deposits include thick andesite flows with limestone and shale interbeds (Rivera 1979; Macellari 1988).

During the Valanginian (Figs. 10.20 and 10.21), sedimentation was also restricted to the West Peruvian Trough, albeit covering a larger area than area during the Berriasian. Valanginian sedimentation extended to the north of the trough into the northern Cajamarca area, where quartz sandstones and minor interbeds of carbonaceous shales containing plant remnants and minor coal beds (Chimú Fm. Benavides 1956; Wilson 1963) record a fluvio-deltaic transitional environment (Scherrenberg et al. 2012). Over the Chimú Fm., brackish to shallow marine limestones and shales of the Santa Fm. (Benavides 1956) record Valanginian transgression, with a maximum flooding surface (MFS 14; Scherrenberg et al. 2012). Deeper water environments occurred to the southwest. This transgression was followed by deposition of regressive, varicolored nonmarine mudstones, siltstones, and quartz-sandstones of the Carhuáz Fm. during the Hauterivian-Barremian(?) (Figs. 10.20 and 10.22). A brackish interval within the Carhuáz Fm. probably contains a maximum flooding surface (MFS 12). In central Peru, Scherrenberg et al. (2012) report gypsum and oolitiic and bioclastic limestones, interpreted as shallow marine incursions (MFS 12). Overlying the Carhuáz Fm., the quartz sandstones of the Aptian(?) Farrat Fm. are recorded (Figs. 10.20 and 10.23; Wilson 1963). In central Peru, the Farrat Fm. contains plant remains, ripple marks, and other evidence of a fluvial-deltaic depositional environment (Scherrenberg et al. 2012). During deposition of Farrat Fm., progradation and maximum regression, were probably associated with a drop in sea level (SB 11 A). In conclusion, during the Berriasian to Early Aptian, several cycles of continental to shallow marine sedimentation are interpreted to be associated with the development of deltaic systems, controlled by relative tectono-eustatic changes.

Late Aptian to Campanian Sedimentation

During the Aptian, sedimentation continued in the West Peruvian Trough but also extended to the Marañón high and the Eastern Peruvian Trough (Figs. 10.20 and 10.23).

A Late Aptian to Early Albian transgression is documented over most of Peru (Figs. 10.20, 10.24, and 10.25). The basal transgressive unit in the West Peruvian Trough is the Inca Fm., which consists of interbedded brownish-gray oolitic, sandy, and ferruginous limestone, green fossiliferous shale, minor quartz-sandstone, and ferruginous siltstone (Benavides 1956). To the east, in the Marañón High, this unit is possibly replaced by the Goyllaarisquizga Fm. (Macellari 1988). To the south, the unit becomes less ferruginous and more calcareous, and it passes into the Pariahuaca Fm. (Wilson 1963). The base of the Inca Fm. and the Pariahuaca Fm. represents a transgressive surface. To the southwest in the Lima province, widespread volcanic activity is recorded at this time and extended into the Cenomanian in the western intra-arc basin (Casma Gp; Atherton et al. 1983).

Over the Marañón High, quartz sandstones of the Goyllarisquisga Fm. were deposited unconformably (SB 11A?) over metamorphic basement. It is possible that this formation is also coeval with older units of the West Peruvian Trough, but a reduced thickness accumulated over the less subsiding Marañón horst block.

In the northern part of the East Peruvian Trough, overlying the Inca Fm., Early to Middle to Albian open marine shelf fossiliferous calcareous shales and limestones of the Chulec Fm. were recorded by Benavides (1956). During the Middle Albian (Figs. 10.20, 10.24, and 10.25), maximum flooding of the West Peruvian

Trough resulted in the deposition of black bituminous shales interbedded with fetid limestones and cherts of the Pariatambo Fm. (Cobbing et al. 1981). This unit was deposited upon an outer marine shelf at a time of global anoxic conditions and contains an important maximum flooding surface (MFS 10).

In the northern part of the Marañón High, the lateral equivalents of the Chulec and Pariatambo Fms. are represented by the Crisnejas Fm., consisting of calcareous shale and sandstone with limestone interbeds (Benavides 1956). This unit probably represents a more proximal facies of the marine shelf.

In the West Peruvian Trough, during the Late Albian to Cenomanian (Figs. 10.20, 10.25, and 10.26), a normal oxygenated environment was re-established, and regressive massive limestones (Yumagual Fm. lower part of Pulluicana Gp.), followed by wavy-bedded, nodular argillaceous limestone with interbedded sandstone and shale (Mujarrun Fm. upper part of Pulluicana Gp.), were deposited upon a shallow marine shelf (Benavides, 1956). During the Cenomanian, the margins of the trough emerged (Cobbing et al. 1981), possibly due to a relative drop in sea level. The Mujarrun Fm. is a shallowing upward carbonate shelf (Jaillard and Sempere 1989); near its top there is a sequence boundary (SB 9; Macellari 1988).

In the East Peruvian Trough (Figs. 10.20, 10.24, and 10.25), sedimentation initiated on a basal unconformity (SB 11A), with cross-bedded white quartz sandstones, micaceous mudstones, and siltstones comprising the Aptian-Albian Cushabatay Fm. (Kummel, 1984). The lower part of the unit was deposited in fluvial environments, but the upper part represents deltaic or coastal barrier sands (Huerta-Kohler 1982). The Esperanza Fm. overlies the Cushabatay Fm. It is comprised of black fossiliferous shale, with occasionally glauconitic sandstone and limestone interbeds. The shales contain ostracods, foraminifera, bivalves, and gastropods, indicative of a shallow marine environment (Soto 1979). The Albian Esperanza Fm. contains a maximum flooding surface (MFS 10). To the east, in the Marañón Basin, the lateral equivalent of the Esperanza Fm. is recorded in the sandstones and shales of the Raya Fm., deposited in a more proximal shallow marine environment. Renewed regression and progradation of a delta system is recorded by coarse-grained (conglomeratic), massive to cross-bedded quartz sandstones containing minor black shale interbeds and plant remains, comprising the Agua Caliente Fm. (Soto 1979). It may be possible to interpret a sequence boundary at the top of the unit (SB 9A?).

During the Late Cenomanian and Early Turonian (Figs. 10.20, 10.26, and 10.27), transgression is recorded by the deposition of brown shales and marls interbedded with limestones (Quillquiñan Gp; Benavides 1956). This unit was deposited in a neritic environment and recorded a deepening-upward trend between the shallower deposits of the Pulluicana Gp. and the deeper sediments of the overlying Cajamarca Fm. The Cajamarca Fm. consists of fine-grained limestones, well-stratified marls, and very thin shales deposited upon an outer marine shelf during the Middle to Late Turonian (Benavides 1956). Deepening of the basin is related to a relative sea level rise (MFS 8). Over the Marañon High, the Pulluicana and Quillquiñan Gps. were replaced by the Jumasha Fm. This unit is composed of bioclastic and pelletal limestone and dolomite, with interbeds of very fine-grained limestone and siltstone

(Wilson, 1963), deposited in a shallow marine environment. In central Peru Scherrenberg et al. (2012) interpreted the depositional environment of the Jumasha Fm. in terms of a shallow, open marine epicontinental carbonate platform, with associated bars of oolite sand, distinct layers of ammonites, and moderate to strong bioturbation, containing intercalations of hydrocarbon source rocks. They interpreted these features to indicate eustatic sea levels ranging to maximum sea level stands (MFS 8). Ammonite assemblages indicate Middle-Late Albian to Late Turonian age (Scherrenberg et al. 2012).

In the West Peruvian Trough, during the Coniacian to Santonian (Figs. 10.20, 10.28, and 10.29), calcareous shale and siltstone with interbeds of nodular limestone of the Celedín Fm. were deposited in a shallow marine environment, shallower than that of the Cajamarca Fm. (Benavides 1956). In central Peru, the unit consists of limestone, marl, and gypsum, interpreted locally as a sabkha depositional environment, possibly representing basin shallowing due to a drop in sea level (SB 7). Subsequently, slightly deeper water environments to the west were possibly related to a rise in sea level (MFS 6; c.f. Scherrenberg et al. 2012).

In the East Peruvian Trough (Figs. 10.20, 10.25, 10.26, 10.27, 10.28 and 10.29), above the sequence boundary at the top of the Agua Caliente Fm., a new transgression was recorded by deposition of dark-gray marine shales with interbeds of sandstone, siltstone, and limestone (Chonta Fm.). This unit varies from a thin (100 m), predominantly sandy proximal in the southeast to a thick (1600 m), more distal, predominantly shale and limestone unit to the northwest. It is interpreted as a deltaic unit in its proximal area (Soto 1979). The lower part of the Chonta Fm. is basal sandstone followed by a shale interval, while the middle section starts with sandstone followed by an organic-rich micritic limestone. The upper part is gray shale with micritic limestone interbeds. The lower dark-gray shales possibly records a sea level rise (MFS 8A), while the sandstone at the base of the middle part may represent a proximal influx of sand during a drop in sea (SB 9). According to Jaillard and Sempere (1989), the organic-rich micritic limestone is Early Turonian in age, interpreted to record outer shelf deposition during times of maximum sea level and global anoxic conditions (MFS 8). The upper part of the unit was deposited on a shallow marine shelf. The Chonta Fm. contains micro- and macrofauna indicative of an Albian to Santonian age (Knechtel et al. 1947; Huerta-Kohler 1982).

Late Campanian to Paleocene Sedimentation

During the Campanian (Figs. 10.20, 10.30, 10.31 and 10.32), in the West Peruvian Trough and Marañón High, the appearance of widespread red beds marks the transition to completely continental sedimentation resulting from uplift of detrital source areas marking initiation of the Andean Orogeny. By the end of the Santonian, the western part of the West Peruvian Trough had been uplifted and provided the source for continental red beds of the Chota Fm. (northern Peru) and the Casapalca Fm. (central Peru; Benavides 1956). In central Peru, the upper part of this continental sequence contains sporadic floodplain and abundant alluvial fan deposits, including thick fluvial intervals demarcating braided channels. The Casapalca Fm. unconformably overlies the Celedín Fm. (SB 5). It is only preserved in the cores of synclines adjacent to thrust faults. The map pattern indicates foreland depocenters located east of the parallel fault traces (Scherrenberg et al. 2012).

In the East Peruvian Trough (Figs. 10.20, 10.30, 10.31 and 10.32), the Vivian Fm. was deposited during Campanian. It is the most important petroleum reservoir of the Marañón Basin. The unit consists of cross-bedded sandstone with minor interbeds of black shale containing plant remains. The lower section represents an extensive braided fluvial channel system (Del Solar 1982). Vertical change in depositional environment from a marine shelf in the underlying Chonta Fm. to the fluvial system of the Vivian Fm. was the result of a relative sea level drop. The base of the unit represents a sequence boundary (SB 5). The upper part of the unit represents coastal barriers (Del Solar 1982), overlain by Maastrichtian to Early Paleocene (?) black shales, claystone, and siltstones of the Cachiyacu Fm., which contains brackish to marine fauna (Kohl and Blissenchach 1962). The Cachiyacu Fm. represents the next transgression resulting from a relative sea level rise (MFS 2A). It is overlain by continental red beds of the Huchpayacu Fm., representing the first clastic input resulting from Andean uplift (Huerta-Kohler 1982).

In the western part of the Solimoes Basin of northwestern Brazil, Triassic and Jurassic red beds are observed (IHS 2008a, b). The Alter do Chao Fm. crops out in the basin and overlies the red bed deposits. Alter do Chao Fm. consists of cross-bedded sandstone, conglomerate, and mudstone, deposited in a meandering fluvial system flowing to the SW as indicated by paleocurrent data (Mendes et al. 2012; Franzinelli and Igreja 2011). The formation contains few fossils. Some authors (Price 1960; Daemon 1975; Dino et al. 1999; Mendes et al. 2012) consider it Late Cretaceous, but Caputo (2009, 2011a, b) considers it Cenozoic in age. A Late Cretaceous age for the formation is more consistent than a Cenozoic age, based on the fact that the Alter do Chao SW flowing fluvial system drained toward during low elevation basins during Cretaceous, which were completely inverted and uplifted during Cenozoic to form the Andes mountains, which began to supply detrital sediments towards the east into the Amazonas basin. In the present work, I assume an uppermost Cretaceous to Paleocene age to the Alter do Chao Fm. (Figs. 10.20, 10.31, and 10.32).

Cretaceous Sedimentation in the Fore-Arc Region of Southern Ecuador and Northwestern Peru

In southern Ecuador and northwestern Peru, the fore-arc Talara and Lancones Basins and the intra-arc Celica Basin contain an incomplete Cretaceous sedimentary and volcanic record, due to uplift during some Cretaceous time intervals (Fig. 10.20). In all of these basins, Cretaceous rocks rest unconformably upon Paleozoic Amotape Terrane basement, and pre-Albian stratigraphy is not recorded. In the intra-arc Celica Basin, volcanic rocks of the Albian Celica Fm. inter-finger with turbidites and deep marine shales of the Alamor Fm. These sediments were deposited up to Early Cenomanian. In the Talara Basin, an initial transgression during Albian times, with deposition of basal sandstone followed by shallow marine limestone and shale of the Pananga Fm., was followed by deposition of organic-rich marine shales of the Muerto Fm., which is the petroleum source rock of the Talara Basin. This unit was likely deposited during a time of relative high sea level (MFS 10), under global anoxic conditions.

During the Late Cenomanian to Santonian (Fig. 10.20), Albian to Lower Cenomanian rocks were deformed and possibly partly eroded during uplift, which corresponds to a major sequence boundary resulting from lumping other sequence boundaries (SB 5 and 7) in the Lancones and Celica Basins. In the Talara Basin and western part of the Lancones Basin, this Late Cenomanian to Santonian hiatus is also recorded but interrupted during Late Turonian-Early Coniacian times, when the Copa-Sombrero Fm. was deposited. This formation consists of deep marine turbidites and shales. According to Jaillard et al. (2005), the Copa-Sombrero Fm. represents a transgression resulting from a relative sea level rise (MFS 8). The event is also recognized in the Peruvian back-arc basins and correlatable with the lower part of Celendín Fm. in the Rentema area, the upper part of Chonta Fm. in the Marañón Basin, and the upper part of the Napo Fm. in the Oriente Basin of Ecuador. The maximum flooding surface in southern Ecuador and northern Peru: Late Turonian-Early Coniacian in age (MFS 8) is slightly different from Colombia and Venezuela. This observation is attributed to tectonic activity in southern Ecuador and Peru which has yet to be recognized in Colombia and Venezuela. In the Talara Basin and the western part of Lancones Basin, the Copa-Sombrero Fm. is bound by sequence boundaries (SB 9 and SB 7, respectively) at both the base and the top.

During the Campanian and Maastrichtian (Fig. 10.20), sedimentation in the Talara, Lancones, and Celica Basins was controlled by compartmentalized basin segments, separated by local highs. Rapid lateral changes in facies and thickness suggest sedimentation coeval with deformation and uplift, which provided detrital sources. In all the stratigraphic columns described by Jaillard et al. (2005), sedimentation apparently initiated upon unconformities, at different times in different areas, with sandstones followed by marine shale deposits.

During the Early Campanian (Fig. 10.20) in the Rio Playas area of the Celica basin, the El Naranjo Fm. records a transgression, which reached maximum flooding (MFS 4 and 4A). In the Lancones Basin, sedimentation started during the Late Campanian, with deposition of a sandstone and shale unit (Mesa Fm.). Marine transgression reached maximum flooding (MFS 4) during deposition of a middle marine limestone interval in the middle part of the Mesa Fm. In the same basin, in the La Tortuga area, Late Campanian deposits of alluvial fan breccias suggest sedimentation near an uplifted detrital source. During the Campanian in the Talara Basin, a marine transgression was recorded by the Sandino Fm. This transgression reached maximum flooding (MFS 10)when the anoxic marine shales of the Muerto Fm. were deposited.

During the Maastrichtian, in the northern Peru fore-arc basins, several marine shale units were deposited (Fig. 10.20), which according to Jaillard et al. (2005), represent marine transgressions that reached maximum flooding events during the

Early Maastrichtian (MFS 2A) and Late Maastrichtian (MFS 2). Interlayered with these shale units are coarse detrital fluvial or alluvial fan deposits and littoral sandstones; detrital fluvial or alluvial fan deposits include the Maastrichtian Casanga Fm. in the Rio Playas area of the Celica Basin, the Maastrichtian upper part of the La Tortuga Fm., in the La Tortuga area of the Lancones Basin, and the Petacas Fm. in the Talara Basin; littoral sandstones include the Maastrichtian Cenizo Fm. in the Tortuga area of the Lancones Basin and within the Cazaderos Fm. in the western part of Lancones Basin. Some of these coarse detrital units are associated with unconformities related to tectonic uplift and possible sea level drops.

10.5.2 Accreted Oceanic Terranes of the Pacific Domain

The oceanic basement of the Colombian Western Cordillera and the Cauca-Patía Valley includes the Calima Terrane (sensu Toussaint and Restrepo 1989, 1994 and Toussaint 1995a, b) or Cañas Gordas, Dagua-Piñón, Sinú and San Jacinto Terranes (sensu Cediel et al. 2003b, 2011), and the Baudó coastal range (Cuna Terrane sensu Toussaint and Restrepo 1989, 1994 and Toussaint 1995a, b, or Baudó Terrane sensu Cediel et al. 2003b, 2011). The Western Cordillera of Ecuador has been referred to as the Dagua-Piñón Terrane (Cediel et al. 2003b, 2011). The mafic to ultramafic igneous components of these composite terrane assemblages have been interpreted to have formed above an oceanic hotspot or mantle plume (Kerr et al. 1997a, b; Sinton et al. 1998). Recent studies suggest the Galapagos hotspot represents the point of origin for this magmatic activity (Nerlich et al. 2014). According to prevailing tectonic models, flair-up of the Galapagos hotspot during the mid-Cretaceous led to extrusion of large volumes of oceanic plateau lavas, which were emplaced within/upon an oceanic substrate provided by N-MORB basalts +/− deep ocean sediments comprising the Farallon Plate.

The composite assemblage of Farallon crust + mantle plume-derived volcanics has been referred to as the Caribbean-Colombian Oceanic Province (CCOP) or Caribbean large igneous province (CLIP; see summary and discussion in Nerlich et al. (2014) and Leal-Mejía et al. 2018). It is generally accepted that the ultramafic and mafic basement rocks of the Calima and Cuna Terranes (after Toussaint and Restrepo, 1989, 1994, ; Toussaint 1995a, b) or Cañas Gordas, Dagua-Piñón, Baudó, Sinú, and San Jacinto Terranes (after Cediel et al. 2003b, 2011), exposed in western Colombia and Ecuador, form part of the CCOP/CLIP assemblage. Several authors have provided lithochemical, radiometric, and biostratigraphic evidence to show that the plateau rocks exposed in the Caribbean, Colombia, and Ecuador (and elsewhere) are geochemically and temporally equivalent and range in age from ca. 100 to 88 Ma with a lesser pulse between 76 and 72 Ma (e.g., Kerr et al. 1997a, b, 1998, 1999, 2003, 2004); Sinton 1997; Sinton et al. 1998; Spikings et al., 2001; Luzieux et al. 2006; Vallejo et al. 2009; Nerlich et al. 2014).

Thus, several oceanic terranes containing mafic/ultramafic assemblages have been accreted to the continental margin north of Peru. Tectonic models presented by numerous authors (*e.g.*, Kerr et al. 1997a, b, 1998, 1999, 2002a, b, 2003, 2004; Kerr and Tarney, 2005; Pindell 1990, 1993; Cediel et al. 2003b; Pindell and Kennan 2009, among others) depict the accretion of fragments of the CCOP/CLIP assemblage (Caribbean Plate) to the continental margin, during its northward and eastward migration from the Galapagos hotspot toward its present position, beginning in the mid-Cretaceous. Interstratified with, and on top of, the CCOP/CLIP volcanic rocks, deep marine sediments also accumulated. These volcanic and sedimentary rocks are also shown in the stratigraphic sections presented in Figs. 10.10, 10.11, 10.12, 10.13, 10.14, 10.15, 10.16, 10.17, 10.18 and 10.19. However, because these rocks are highly deformed and locally metamorphosed, it is difficult to distinguish and separate conventional stratigraphic units or sequences. Except for local detailed studies (mainly in Ecuador), only generic or regional lithostratigraphic assemblages have been defined, and the level of stratigraphic knowledge of these rocks is much lower than the sedimentary rocks deposited upon/within the continental margin. No attempt to identify or correlate stratigraphic sequences in this domain has been made herein. In northern Venezuela, Caribbean oceanic terranes accreted to the continental margin have also been identified (*e.g.*, Caribbean Mountain Terrane of Cediel et al. 2003b). This accretion, however, occurred later during the Cenozoic and is not included in this paper.

10.6 Discussion and Conclusions

The present work attempts to decipher, integrate, and graphically depict the Cretaceous sedimentary record, basin development, and the dynamic interaction of paleo-facies, within the context of global eustacy and the geological and tectonic history of northwestern South America during the Late Mesozoic. One of the underpinning factors behind the complex, compartmentalized basin evolution in the region is the observation that, from the latest Jurassic to Paleocene, tectonic models record kinematic interplay involving no less than four distinct tectonic plates, including (1) South America, (2) Pacific/Farallon, (3) Proto-Caribbean, and (4) Caribbean (CCOP/CLIP), resulting in varied configurations of a triple-junction, and generating numerous tectono-sedimentary environments along extensional-transtensional, compressive-transpressive, and transcurrent margins, in marine, continental, and transitional settings. Due to this complex tectonic configuration, penecontemporaneous, active vs. passive margin and marine vs. continental sedimentation is recorded in separate basin-filling events during the Cretaceous, over the region spanning northern Peru through Ecuador, Colombia, and Venezuela.

Notwithstanding, Cenozoic reactivation of Mesozoic structure and tectonic inversion of many of the Late Mesozoic basins during Paleogene and Neogene orogenesis have provided abundant natural outcrop for the field-based study of sedimentation and paleo-facies development at the regional level, throughout the Northern Andes. This information is supported by abundant borehole and geophysical data generated during over a half-century of fossil fuel exploration and development in the region, a process which has delineated world-class petroleum and coal resources, and created a detailed and functional database for ongoing exploration and study.

The exercise of stratigraphic correlation is a preliminary step in any attempt to build paleogeographic maps. In northwestern South America, there are many names for lithostratigraphic units, and in "too many" cases, the same unit has been given different local names, or the same name has been applied to different or heterochronous stratigraphical units. As a result, and until such inconsistencies in stratigraphic nomenclature are sorted out, any attempt to build truly accurate paleogeographic maps is rendered next to impossible. Regardless, in such a case, the well-founded application of the concepts and methods of sequence stratigraphy, as applied herein, provides an essential tool for regional and local stratigraphic correlation and paleo-facies interpretation.

During Jurassic rifting throughout northwestern South America, a passive continental margin developed along the northern and northwestern edge of the continental block (Guiana Shield, Venezuela), while active subduction with a continuous and well-developed magmatic arc and craton-ward back-arc extensional basins developed along the western margin of the continent (Colombia, Ecuador, and Peru). The localization of many of these basins was controlled by normal faulting related to Jurassic and earlier rifting. Continental clastic sedimentation dominated in most of these rapidly subsiding basins, while in those undergoing extreme subsidence (*e.g.*, the West Peruvian Trough, the northern Venezuela rift basin, the Cundinamarca sub-basin), marine sediments also accumulated during the Late Jurassic.

Cretaceous sedimentary rocks, including locally, uppermost Jurassic and Paleocene deposits, form a mega-sequence bounded by regional unconformities that are at least locally angular. On a broad scale, Cretaceous sedimentary rocks represent a major transgressive-regressive cycle with the maximum flooding surface close to the Cenomanian-Turonian boundary (MFS 8), corresponding to the maximum Cretaceous, and even Mesozoic, eustatic level. Superimposed on this large-scale trend, several smaller transgressive-regressive cycles are present, suggesting an oscillating relative tectono-eustatic level. These minor cycles correspond to the several stratigraphic sequences defined herein. I have attempted to correlate these stratigraphic sequences throughout the whole of the Northern Andean study area. Most of these stratigraphic sequences have been defined by earlier workers at specific locations, but not over the region as a whole. Within the Cretaceous, I propose 10 stratigraphic sequences and correlate 21 surfaces, including sequence boundaries (SB) and maximum flooding surfaces (MFS), throughout the studied area, based upon the critical assumption that they represent time surfaces. Most of the proposed surfaces appear to be correct; however, others remain speculative and in need of better biostratigraphic data to test their validity. In this sense, I have not attempted to use new biostratigraphic data for this purpose but have integrated ages proposed in the cited literature. However, it remains clear that, in places where sedimentation occurred coeval with structural deformation and uplift, we cannot expect to correlate stratigraphic sequences over distances greater than the wavelength of the structures being generated at that time. For example, in the area of the northern

Peruvian Andes, the end of the Cretaceous (Campanian-Maastrichtian), it has been observed that syntectonic sedimentation occurred only in synclinal troughs, while structural uplift at the crest of related anticlines reduced local subsidence or even generated exhumation and erosion. Thus, correlation of stratigraphic sequences over great distances in this region is inhibited.

In theory, eustacy and tectonics are responsible for relative tectono-eustatic changes of relative sea level or base level changes. Therefore, differential vertical movements of the earth's surface generated by active tectonics will generate differential vertical changes of relative sea level or base level, controlling the development of stratigraphic sequences. In addition, at a regional scale, the wavelength of structures generated by tectonism reflects the thermal age of the lithosphere. In general, at the scale of northwestern South America, during the Cretaceous, the wavelength of structural folding seems to decrease toward the border of the continent. As a conclusion then, we may say that the validity of the proposed stratigraphic correlations may only be applied to distances equal to or shorter than the wavelength of syn-tectonic structures, which seem to decrease toward the continent margin.

Cretaceous sedimentary history can be summarized in four episodes: (1) Berriasian to Aptian, (2) Aptian to Cenomanian, (3) Cenomanian to Santonian, and (4) Campanian to Early Paleocene.

During the Berriasian to Early (?) Aptian, sedimentation was restricted to rapidly subsiding extensional basins inherited from Jurassic rifts. Because of differential tectonic subsidence between graben and horst blocks, it is difficult to recognize the eustatic signal and correlate stratigraphic sequences. Notwithstanding, great thicknesses of sediment accumulated. Sedimentation initiated in continental environments (alluvial fan and fluvial), followed by marine transgression. An exception to this occurred in the Cundinamarca sub-basin presently exposed in the Colombian Eastern Cordillera, where Berriasian sedimentation occurred in marine environments. Important lateral changes of facies and thickness, due to differential tectonic subsidence of tectonic blocks, are recorded. In Venezuela and northeastern Colombia, shallow shelf carbonate platforms developed. During times of maximum marine flooding (MFS) in some stratigraphic sequences, carbonates were replaced by shales deposited upon inner marine shelves. In Central Colombia, mud-dominated sedimentation occurred. A great thickness of inner to outer marine shales accumulated, initially with some turbidites, probably resulting from tectonically induced sea bottom instability. In Peru, a great thickness of alternating, littoral deltaic to shallow marine sandstones and shales accumulated. The marine shales were preferentially deposited during times of maximum marine flooding.

During the Late (?) Aptian to Cenomanian, regional thermal subsidence (following earlier active rifting) resulted in a regional increase in the area of marine sedimentation, and previously isolated basins coalesced into a major regional basin extending along the margin of the continent from Venezuela to Peru. During this time, marine incursion into new areas such as the Barinas-Apure basin in Venezuela, the Llanos, Upper Magdalena Valley, Putumayo basins in Colombia, the Oriente basin in Ecuador, and the Marañón and East Peruvian Trough basins in Peru is well documented. Regional thermal subsidence permitted the sedimentary record to become more sensitive to eustatic changes, and therefore, it is easier to recognize the eustatic signal and correlate stratigraphic sequences at the regional level. In Venezuela and northwestern Colombia, carbonate-dominated platforms were replaced by marine shale sedimentation, interrupted in western Venezuela by a progradation of deltaic sandstones (Aguardiente Fm.). In Colombia, mud-dominated sedimentation prevailed in the marine shelf, also interrupted by progradation of deltaic sands (Aguardiente and Une Fms.). In Peru, earlier siliciclastic deltaic sedimentation restricted to the West Peruvian Trough was replaced by regional carbonate platform sedimentation upon a shallow shelf. Regionally, from Venezuela to Peru, sea level rise coinciding with an oceanic anoxic event during the Middle Albian favored accumulation of petroleum source rocks.

During the latest Cenomanian to Santonian, regional subsidence continued, and, due to a global rise in sea level, craton-ward marine incursion into previously unaffected areas is recorded. During the latest Cenomanian to Early Turonian, the sea reached its maximum Cretaceous level, representing the sea level high for the entire Phanerozoic. Sedimentation was controlled by eustatic changes. Coincident with maximum Cretaceous flooding, oceanic anoxic events favored accumulation of organic matter at the sea bottom, resulting in deposition of the best petroleum source rocks in northern South America (La Luna Fm. and equivalents). From Venezuela to Ecuador, pelagic shale and pelagic fine-grained limestone sedimentation prevailed. During this time interval, but specifically during the Coniacian-Santonian in Venezuela and Colombia, marine upwelling conditions favored development of siliceous plankton and chert deposition. In Peru, shallow marine carbonate shelves were replaced by mud-dominated marine shelves. Due to the somewhat earlier initiation of tectonic activity (deformation, uplift) in Peru when compared to the rest of the Northern Andes from Venezuela to Ecuador, some sequence boundaries and maximum flooding surfaces of the stratigraphic sequences recognized in Peru are slightly temporally shifted, with respect to the corresponding stratigraphic equivalents in the northern Andes.

Beginning in the Santonian, and into the Campanian to Paleocene, marine regression, a general shallowing of sedimentary environments, and penecontemporaneous compressional deformation are recorded throughout northwestern South America. As noted, these processes initiated somewhat earlier in Peru. In Venezuela, at the end of the Cretaceous (Maastrichtian) and Paleocene, passive margin conditions shifted to an active margin, with obduction of oceanic terranes of Caribbean affinity, and to the west, uplift in the Sierra Nevada de Santa Marta. Along the Pacific margin of Colombia-Ecuador, during the Campanian and Maastrichtian, the collision and accretion of fragments of the Caribbean Plate (CCOP/CLIP assemblage) to the continental margin generated uplift in the Central Cordillera and its northern prolongation into the Lower Magdalena Valley (Plato-San Jorge area), the Cordillera Real, and the sub-Andean zone, including the western part of the Oriente basin. Regional Late Maastrichtian and Paleocene regression led to a transition from marine to dominantly continental environments extending from Venezuela to Ecuador. In Peru, Andean uplift beginning on the Campanian provided a source of detrital sediments, accumulated in active synforms as continental and alluvial fan deposits. By the

Paleocene, continental sedimentation dominated throughout the Northern Andean region, and basins began to compartmentalize due to active regional transpressive-transtensional deformation within the context of the South American-Pacific-Caribbean plate triple junction.

In southernmost Ecuador and northwestern Peru, sedimentation within fore-arc basins, active since the Albian, is characterized by strong lateral changes of facies and thickness, marked by periods of uplift and localized sedimentation during the entire Late Cretaceous. Enhanced tectonic activity during this time in Peru resulted in punctual time shifts with respect to the sequence boundaries and maximum flooding surfaces recognized in Venezuela, Colombia, and Ecuador.

I have herein compiled a Cretaceous sedimentary and tectonic history for the northwestern corner of South America, attempting to define and correlate a framework for several stratigraphic sequences across the entire Northern Andean region, as permitted by their sequence boundaries and maximum flooding surfaces. This methodology has proven useful in the correlation of the Cretaceous sedimentary record of basins located upon/within the continental margin.

However, I have not attempted to define and correlate similar sequences in the Cretaceous record which was deposited over the oceanic crust, and associated with accreted terranes incorporated into the Northern Andean mosaic to the north of Peru, because these units are highly deformed, and much less precisely known. Regardless, some of the proposed sequence boundaries discussed herein are certainly the result of active tectonism in the region, and many of these sequences are related to tectonic events in Peru, as proposed by several workers (*e.g.*, Jaillard, 1993; Jaillard et al., 1990, 1995, 2000). In Ecuador, Colombia, and Venezuela, Cretaceous tectonic activity affecting the stratigraphic record deposited on the continent seems to have been less intense than that recorded in Peru.

Notwithstanding, some unconformities can be correlated in time and location with known or proposed tectonic events. An example of this is the unconformity at the beginning of the Aptian, recognized in southern and central Colombia. This unconformity correlates in time and location with the closure of the Quebradagrande oceanic margin basin and the accretion of the Quebradagrande arc to southern Colombia and Ecuador. Although it is likely that this Lower Aptian unconformity is related to some degree of structural deformation, the driving mechanisms behind the unconformity remain uncertain. An alternative explanation for this unconformity may involve changes in the stress regime due to the chage from active syn-rift to thermal flexural subsidence, recognized in other parts of the world as a breakup unconformity.

In several seismic lines in south central Colombia (Upper Magdalena Valley and southern part of the Eastern Cordillera), Jaimes and De Freitas (2006) recognized an unconformity associated to some degree of structural deformation in rocks of middle Cretaceous age. It is possible that this unconformity corresponds to the aforementioned Lower Aptian unconformity, associated with the Quebradagrande closure event. Based upon biostratigraphic data from petroleum wells, however, these authors proposed an Albian-Cenomanian age for the unconformity and correlated it with changes in subduction/deformation along the Colombian and Peruvian segments of the Andes. Jaimes and De Freitas (2006) suggest that such changes may be related to the opening of the South Atlantic Ocean at equatorial latitudes.

Another example of a tectonic-related unconformity has been documented, based upon biostratigraphic and seismic data in western Venezuela, by Cooney and Lorente (2009). These authors demonstrated the absence of several Campanian foraminiferal biozones and illustrate unconformable relationships in some seismic lines, mainly within the Maracaibo Basin. When horizontalized to a seismic reflector near the top of the Cretaceous, the seismic lines reveal compressional structures developed below the unconformity. This Upper Campanian unconformity correlates temporally and spatially with the the initial collision of the Antilles magmatic arc with northern Colombia. Following collision, during the Paleocene, continental sedimentation became dominant in most of the basins of Colombia and Venezuela, as they became progressively compartmentalized during active Northern Andean deformation and uplift.

References

Alfonso M, Herrera JM, Navarrete RE, Bermúdez HD, Calderón JE, Parra FJ, Sarmiento G, Vega F, Perrilliat M de C (2009) Cartografía geológica, levantamiento de columnas estratigráficas, toma de muestras y análisis bioestratigráficos. Sector de Chalán (Cuenca Sinú–San Jacinto). ANH-ATG, Bogotá

Atherton MP, Pitcher WS, Warden V (1983) The Mesozoic marginal basin of central Peru. Nature 305:303–306

Baby P, Rivdeneira M, Barragan R (eds) (2004) La Cuenca Oriente: Geología y Petróleo. IFEA, IRD, PETROECUADOR, Quito, 296 p

Ballesteros I, Nivia A (1985) La Formación Ritoque: Registro sedimentario de una albufera de comienzos del Cretácico. In: Etayo-Serna F, Laverde-Montaño F (eds) Proyecto Crétacico, contribuciones. Chapter XIV. Ingeominas Publicación Geológica Especial 16, Bogotá, 17 p

Barragán R, Christophoul F, White H, Baby P, Rivadeneira M, Ramírez F, Rods J (2004) Estratigrafía secuencial del Cretácico en la Cuenca Oriente del Ecuador. In: Baby P, Rivadeneira M, Barragán R (eds) La Cuenca Oriente: Geología y Petróleo. IFEA, IRD, PETROECUADOR, Quito, pp 45–68

Barrero D (1979) Geology of the central western cordillera, west of Buga and Roldanillo, Colombia, vol 4. Publicaciones Geológicas Especiales del Ingeominas, Bogota, pp 1–75

Barrero D, Laverde F (1998) Estudio Integral de evaluación geológica y potencial de hidrocarburos de la cuenca "intramontana" Cauca- Patía, ILEX- Ecopetrol report, Inf. N0. 4977

Barrio CA, Coffield DQ (1992) Late cretaceous stratigraphy of the upper Magdalena Basin in the Payandé-chaparral segment (western Girardot Sub-Basin), Colombia. J S Am Earth Sci 5(2):132–139

Benavides C, Victor E (1956) Cretaceous system in northern Peru. Am Mus Nat Hist (New York) Bull 108(4):359–493

Blau J, Moreno M, Senff M (1995) *Plaxius Caucadensis n. Sp.*, a crustacean microcoprolite from the basal Nogales formation (Campanian to Maastrichtian of Colombia). Micropaleontology 41:85–88

Bourgois J, Azéma J, Tournon J, Bellon Calle B, Parra E, Toussaint JF, Glaçon G, Feinberg H, De Wever P, Origlia I (1982a) Ages et structures des complexes basiques et ultrabasiques de la facade Pacifique entre 3° N et 12° N (Colombie, Panamá et Costa-Rica). Bull Soc Géol France 7 t XXIV(3):545–554

Bourgois J, Calle B, Tournon J, Toussaint JF (1982b) The Andean ophiolitic megastructures on the Buga-Buenaventura transverse (western cordillera – Valle Colombia). Tectonophysics 82:207–229

Bürgl H (1960) El Jurásico e Infracretáceo del Río Batá, Boyacá: Servicio Geológico Nacional, Informe 1319, Boletín Geológico, Bogotá, 1–3: 169–211

Bürgl H (1964) El Jura-Triásico de Colombia. Boletín Geológico Servicio Geológico Nacional, Bogotá, 12(1–3):5–31

Bürgl H (1967) The orogenesis of the Andean system of Colombia. Tectonophysics 4(4–6):429–443

Bustamante A (2008) Geobarometria, geoquimica, geocronologia e evolução tectônica das rochas de fácies xisto azul nas áreas de Jambaló (Cauca) e Barragán (Valle del Cauca), Colômbia. PhD Thesis. Instituto de Geociências, Universidad de São Paulo, Brasil, 179

Cáceres C, Etayo-Serna F (1969) Bosquejo geológico de la región del Tequendama, Opúsculo guía de la Excursión Pre-Congreso. 1er Congreso Colombiano de Geología, Opúsculo, Bogotá, 23 p

Caputo MV (2009) Discussão sobre a Formação Alter do Chão e o Alto de Monte Alegre. In: Simpósio de Geologia da Amazônia, Boletim de Resumos, vol 11, Manaus

Caputo MV (2011a) Discussão sobre a Formação Alter do Chão e o Alto de Monte Alegre. In: Nascimento RSC, Horbe AMC, Almeida CM (eds) Contribuições à Geologia da Amazônia, Simpósio de Geologia da Amazônia, vol 7, pp 7–23

Caputo MV (2011b) Reposicionamento estratigráfico da Formação Alter do Chão e evolução tectosedimentar da Bacia do Amazonas no Mesozoico e Cenozoico. In: XIV Congreso Latinoamericano de Geología, Medellin, Colombia, Memórias, pp 173–174

Cardona A, Weber M, Wilson R, Cordani U, Muñoz CM, Paniagua F (2007) Evolución tectonomagmática de las rocas máficas-ultramáficas del Cabo de La Vela y el Stock de Parashi, Península de la Guajira: registro de la evolución orogénica Cretácica-Eocena del norte de Suramérica y el Caribe. XI Congreso Colombiano de Geologia, Bucaramanga, Agosto 14–17. 6 p

Cardona A, Weber M, Valencia V, Bustamante C, Montes C, Cordani U, Muñoz CM (2014) Geochronology and geochemistry of the Parashi granitoid, NE Colombia: tectonic implication of short-lived early Eocene plutonism along the SE Caribbean margin. J South Am Earth Sci 50:75–92

Cardozo E, Ramírez C (1985) Ambientes de depósito de la Formación Rosablanca: Área de Villa de Leiva. In: Etayo-Serna F, Laverde-Montaño F (eds) Proyecto Crétacico, contribuciones. Chapter XIII. Ingeominas Publicación Geológica Especial 16, Bogotá, 13 p

Cediel F (1968) El Grupo Girón, una molasa Mesozóica de la Cordillera Oriental. Servicio Geológico Nacional Boletín Geológico, Bogotá 16(1–3): 5–96

Cediel F (2018) Phanerozoic orogens of Northwestern South America: cordilleran-type orogens, taphrogenic tectonics and orogenic float. Springer, Cham, pp. 3–89

Cediel F, Etayo-Serna F, Cáceres C (2003a) Facies distribution and tectonic setting during the Proterozoic of Colombia (scale 1:2′000.000, map). Instituto Colombiano de Geología y Minería (Ingeominas), Bogotá

Cediel F, Shaw RP, Caceres C (2003b) Tectonic assembly of the northern Andean block. In: Bartolini C, Buffler RT, Blickwede J (eds) The Circum-Gulf of Mexico and the Caribbean: hydrocarbon habitats, basin formation, and plate tectonics, vol 79. American Association of Petroleum Geologists Memoir, pp 815–848

Cediel F, Leal-Mejia H, Shaw RP, Melgarejo JC, Restrepo-Pace PA (2011) Geology and hydrocarbon potential regional geology of Colombia. In: Cediel F (ed) Petoleum geology of Colombia, vol 1. Univeristy EAFIT, Agencia Nacional de Hidrocarburos, Bogota, 224 p

Clavijo J (1985) La secuencia de la Formación Los Santos en la Quebrada Piedra Azul: registro de una hoya fluvial evanescente. In: Etayo-Serna F, Laverde-Montaño F (eds) Proyecto Crétacico, contribuciones. Chapter IV. Ingeominas Publicación Geológica Especial 16, Bogotá, 18 p

Cobbing EJ, Pitcher WS, Wilson JJ, Baldock JW, Taylor WP, McCourt W, Snelling NJ (1981) The geology of the western cordillera of northern Peru. Institute of Geological Sciences, London. Overseas Memoir, 5, 143 p

Cooney PM, Lorente MA (2009) A structuring event of Campanian age in western Venezuela, interpreted from seismic and palaeontological data, vol 328. Geological Society, London., Special Publications, pp 687–703. https://doi.org/10.1144/SP328.27

Cooper MA, Addison FT, Alvarez R, Coral M, Graham RH, Hayward AB, Howe S, Martinez J, Naar J, Peñas R, Pulham AJ, Taborda A (1995) Basin development and tectonic history of the Llanos Basin, eastern cordillera, and middle Magdalena Valley, Colombia. AAPG Bull 79(10):1421–1443

Daemon RF (1975) Contribuição à datação da Formação Alter do Chão, bacia do Amazonas. Rev Bras Geo 5:58–84

Dalmayrac C, Laubacher G, Marocco R (1980) Geologie des Andes Peruvienses. Travaux et Documents de l'ORSTOM (Paris) 122, 501 p

Del Solar C (1982) Ocurrencia de hidrocarburos en la Formación Vivian, nororiente Peruano. In: I Simposio Exploracion Petrolera en las Cuencas Subandinas, vol 1. Association Columbiana de Geologist y Geofísicos del Petroleum, Bogotá

Diaz L (1994a) Distribution de las facies siliciclastic correspondents a la Formation Arenisca Tierna y equivalents end el Valle Superior del Magdalena. In: Etayo Serna F (ed) Studios geological del Valle Superior del Magdalena. Chapter IV. Univ. Nacional de Colombia, Ecopetrol, Bogotá, 15 p

Diaz L (1994b) Reconstruction de la Cuenca del Valle Superior del Magdalena, a finale del Creation. In: Etayo Serna F (ed) Studios geological del Valle Superior del Magdalena. Chapter XI. Univ. Nacional de Colombia, Ecopetrol, Bogotá, 13 p

Dino R, Silva OB, Abrahao D (1999) Paleontological and stratigraphic characterization of the cretaceous strata from the Alter do Choo formation, Amazonas basin. In: UNESP, Simpsio sobre o Cretaceo do Brasil and Simpósio sobre el Cretácico de América del Sur, vol 5., *Anais*, pp 557–565

Dorado J (1984) Contribución al conocimiento de la estratigrafía de la Formación Brechas de Buenavista (límite Jurásico – Cretácico), región noroeste de Villavicencio (Meta). Geología Colombiana, Bogotá, 17: 7–40

Duque-Trujillo J, Sánchez J, Orozco-Esquivel T, Cárdenas A (2018) Cenozoic magmatism of the maracaibo block and its tectonic significance. In: Cediel F, Shaw RP (eds) Geology and tectonics of Northwestern South America: the Pacific-Caribbean-Andean junction. Springer, Cham, pp 551–594

Ecopetrol, Esso and Exxon Exploration Company (1994) Integrated technical evaluation Santander sector Colombia., 1991–1994. Report, text and figures. Final Report, Houston, 38 p., 49 figs

Erikson JP, Pindell JL (1993) Analysis of subduction in northeastern Venezuela as a discriminator of tectonic models of northern South America. Geology 21:945–948

Etayo F, Renzoni G, Barrero D (1976) Contornos sucesivos del mar Cretácico en Colombia. Primer Congreso Colombiano de Geología, Mem., Bogotá, pp 217–252

Etayo-Serna F (1968) El sistema Cretáceo en la región de Villa de Leiva y zonas próximas. Geología Colombiana, Bogotá, 5:5–74

Etayo-Serna F (1994) Epílogo: A modo de historia geológica del Cretácico del Valle Superior del Magdalena. In: Etayo Serna F (ed) Estudios geológicos del Valle Superior del Magdalena. Chapter XX. Univ. Nacional de Colombia, Ecopetrol, Bogotá, 6 p

Etayo-Serna F, Florez JM (1994) Estratigrafía y estructura de la Quebrada Calambe y el cerro El Azucar, Olaya Herrera Tolima. In: Etayo Serna F (ed) Estudios geológicos del Valle Superior del Magdalena. Chapter XII. Univ. Nacional de Colombia, Ecopetrol, Bogotá, 23 p

Etayo-Serna F, Rodríguez G (1985) Edad de la Formación Los Santos. In: Etayo-Serna F, Laverde-Montaño F (eds) Proyecto Crétacico, contribuciones. Chapter XXVI. Ingeominas Publicación Geológica Especial 16, Bogotá, 13 p

Etayo-Serna F, Barrero D, Lozano H, Espinosa A, González H, Orrego A, Zambrano F, Duque H, Vargas R, Núñez A, Álvarez J, Ropaín C, Ballesteros I, Cardozo E, Forero H, Galvis N, Ramírez C, Sarmiento L (1983) Mapa de terrenos geológicos de Colombia. Ingeominas Publicación Geológica Especial 14, Bogotá, 235 p

Fabre A (1985) Dinámica de la sedimentación Cretácica en la región de la Sierra Nevada del Cocuy (Cordillera Oriental de Colombia). In: Etayo-Serna F, Laverde-Montaño F (eds) Proyecto Crétacico, contribuciones. Chapter XIX. Ingeominas Publicación Geológica Especial 16, Bogotá, 20 p

Fabre A (1986) Géologie de la Sierra Nevada del Cocuy (Cordillère Orientale de Colombie). Thèse Université de Genève de Docteur es Sciences de la Terre, No 2217, Genève, 394 p

Fabre A (1987) Tectonique et géneration d'hydrocarbures: Un modèle de l'évolution de la Cordillère Orientale de Colombie et du Bassin des Llanos pendant le Crétacé et le Tertiaire. Arch Sc Genève 40(Fasc. 2):145–190

Fajardo A, Rubiano JL, Reyes A (1993) Estratigrafía de secuencias de las rocas del Cretáceo Tardío al Eoceno Tardío en el sector central de la cuenca de los Llanos Orientales, Departamento del Casanare, Report ECP-ICP-001-93, Piedecuesta, Santander, 69 p

Flórez JM, Carrillo GA (1994) Estratigrafía de la sucesión litológica basal del Cretácico del Valle Superior del Magdalena. In: Etayo Serna F (ed) Estudios geológicos del Valle Superior del Magdalena. Chapter II. Univ. Nacional de Colombia, Ecopetrol, Bogotá, 26 p

Föllmi KB, Garrison RE, Ramirez PC, Zabrano-Ortíz, Kennedy WJ, Lehner BL (1992) Cyclic phosphate-rich successions in the upper cretaceous of Colombia. Palaeogeogr Palaeoclimatol Palaeoecol 93:151–182

Forero H, Sarmiento L (1985) La facies evaporítica de la Formación Paja en la region de Villa de Leiva. In: Etayo-Serna F (ed) Proyecto Cretacico Contribuciones. Ingeominas, Bogota., Publicacion Especial XVII, pp 1–16

Franzinelli E, Igreja H (2011) Ponta das Lajes e o Encontro das Águas, AM - A Formação Alter do Chão como moldura geológica do espetacular Encontro das Águas Manauara. In: Winge M, Schobbenhaus C, Souza CRG, Fernandes ACS, Berbert-Born M, Sallun filho W, Queiroz ET (eds) Sítios Geológicos e Paleontológicos do Brasil. Publicado na Internet em 29/11/2011 no endereço http://sigep.cprm.gov.br/sitio054/sitio054.pdf

Galvis N, Rubiano J (1985) Redefinición estratigráfica de la Formación Arcabuco, con base en análisis facial. In: Etayo-Serna F, Laverde-Montaño F (eds) Proyecto Crétacico, contribuciones. Chapter VII. Ingeominas Publicación Geológica Especial 16, Bogotá, 16 p

Geotec (1992) Facies distribution and tectonic setting through the Phanerozoic of Colombia. A regional synthesis combining outcrop and subsurface data presented in 17 consecutive rock-time slices

Ghosh SK (1984) Late Cretaceous condensed sequence, Venezuelan Andes. In: Bonini W E, Hargraves R B, Shagam R (Eds.), The Caribbean–South American Plate boundary and regional tectonics. Geol. Soc. Am. Mem. 162: 317–324

Gómez E, Pedraza P (1994) El Maastrichtiano de la región Honda Guaduas, límite N del Valle Superior del Magdalena: Registro sedimentario de un delta dominado por ríos trenzados. In: Etayo Serna F (ed) Estudios geológicos del Valle Superior del Magdalena. Chapter III. Univ. Nacional de Colombia, Ecopetrol, Bogotá, 20 p

González de Juana C, Iturralde J, Picard X (1980) Geología de Venezuela y de sus Cuencas Petrolíferas: Caracas, Ediciones Foninves, Tomos I y II, 1031 p

Guerrero J (2002) A Proposal on the Classification of Systems Tracts: Application to the Allostratigraphy and Sequence Stratigraphy of the Cretaceous Colombian Basin. Part 2: Barremian to Maastrichtian. Geologia Colombiana 27, pp 27–49, 2 Figs, 1 Table

Guerrero J, Sarmiento G, Navarrete R (2000) The Stratigraphy of the W Side of the Cretaceous Colombian Basin in the Upper Magdalena Valley. Reevaluation of selected areas and type localities ioncluding Aipe, Guaduas, Ortega and Piedras. Geoogia Colombiana 25, pp 45–110, 12 Pl., 5 Figs., 6 Tabl., 1 Microp. App.: 6 Pl. Bogotá.100 p

Guzmán G, Clavijo J, Barrera R (1994) Geología Bloque Santero, Secciones estratigráficas. Informe inédito. INGEOMINAS, Bogotá

Guzmán G, Gómez E, Serrano B (2004) Geología de los Cinturones del Sinú, San Jacinto y Borde Occidental del Valle Inferior del Magdalena, Caribe Colombiano. Escala 1:300.000. Ingeominas. Bogotá, 134p

Huerta-Kohler T (1982) Exploración petrolíera en la Cuenca Ucayali, Oriente Peruano. In: II Simposio Exploración Petrolera en las Cuencas Subandinas, V. 1. Asociación Colombiana de Geólogos y Geofísicos del Petróleo, Bogotá

IHS (2008a) Amazonas Basin. Basin Monitor. IHS, 50 p

IHS (2008b) Solimoes Basin. Basin Monitor. IHS, 50 p

Intera Information Technologies-Bioss (1995) Evaluación geológica regional de la cuenca Cesar-Ranchería. Bogotá: Empresa Colombiana de Petróleos. ECOPETROL

Jaillard E (1993) Kimmeridgian to Paleocene tectonic and geodynamic evolution of the Peruvian (and Ecuadorian) margin. In: Salfity JA (ed) Cretaceous tectonics of the Andes. Reprint. Fonds Documentaire ORSTOM. Vieweg, pp 101–167

Jaillard E, Sempere T (1989) Cretaceous sequence stratigraphy of Peru and Bolivia. Fonds Documentaire ORSTOM, 010019766, 20 p. 1. Paleontological Appendix

Jaillard EP, Solar P, Carlier G, Mourier T (1990) Geodynamic evolution of the northern and central Andes during early to middle Mesozoic times: a Tethyan model. J Geol Soc Lond 147:1009–1022

Jaillard É, Ordoñez M, Benitez S, Berrones G, Jiménez N, Montenegro G, Zambrano I (1995) Basin development in an accretionary, oceanic-floored fore-arc setting: southern coastal Ecuador during late cretaceous–late Eocene time. In: Tankard AJ, Suárez S R, Welsink HJ (eds) Petroleum basins of South America, vol 62. AAPG Memoir, pp 615–631

Jaillard É, Hérail G, Monfret T, Díaz-Martínez E, Baby P, Lavenu A, Dumont J-F (2000) Tectonic evolution of the Andes of Ecuador, Peru, Bolivia and northernmost Chile. In: Cordani UG, Milani EJ, Thomaz-Filho A, Campos DA (eds) Tectonic evolution of South America. Publication of the 31st international geological congress, Rio de Janeiro, pp 481–559

Jaillard E, Bengtson P, Dhondt A (2005) Late cretaceous marine transgressions in Ecuador and northern Peru: a refined stratigraphic framework. J S Am Earth Sci 19:307–323

Jaimes E, De Freitas M (2006) An Albian–Cenomanian unconformity in the northern Andes: evidence and tectonic significance. J S Am Earth Sci. https://doi.org/10.1016/j.jsames.2006.07.011

Jaramillo C, Yépez O (1994) Palinostratigrafía del Grupo Olini (Coniaciano-Campaniano), Valle Superior del Magdalena, Colombia. In: Etayo Serna F (ed) Estudios geológicos del Valle Superior del Magdalena. Chapter XIII, Univ. Nacional de Colombia, Ecopetrol, Bogotá, 20 p

Juliao T, Carvalho M, Torres D, Plata A, Parra C (2011) Definición de paleoambientes de la Formación Cansona a partir de palinofacies, cuenca Sinú – San Jacinto. Resumen 14 Congreso Latinoamericano de Geología, Medellín, Colombia

Kerr AC, Tarney J (2005) Tectonic evolution of the Caribbean and northwestern South America: the case for accretion of two late cretaceous oceanic plateaus. Geology 33:269–272

Kerr AC, Marriner GF, Tarney J, Nivia A, Saunders AD, Thirlwall MF, Sinton CW (1997a) Cretaceous basaltic terranes in western Colombia: elemental, chronological and Sr-Nd isotopic constraints on petrogenesis. J Petrol 38:677–702

Kerr AC, Tarney J, Marriner GF, Nivia A, Saunders AD (1997b) The Caribbean–Colombian cretaceous Igneous Province: the internal anatomy of an oceanic plateau. In: Mahoney JJ, Coffin MF (eds) Large igneous provinces: continental, oceanic, and planetary flood volcanism, vol 100. American Geophysical Union, Geophysical monograph, pp 123–144

Kerr AC, Tarney J, Nivia A, Marriner GF, Saunders AD (1998) The internal structure of oceanic plateaus: inferences from obducted cretaceous blocks in western Colombia and the Caribbean. Tectonophysics 292:173–188

Kerr AC, Iturralde-Vinent MA, Saunders AD, Babbs TL, Tarney J (1999) A new plate tectonic model of the Caribbean: implications from a geochemical reconnaissance of Cuban Mesozoic volcanic rocks. Geol Soc Am Bull 111:1581–1599

Kerr AC, Aspden JA, Tarney J, Pilatasig LF (2002a) The nature and provenance of accreted oceanic blocks in western Ecuador: geochemical and tectonic constraints. J Geol Soc 159:577–594

Kerr AC, Tarney J, Kempton PD, Spadea P, Nivia A, Marriner GF, Duncan RA (2002b) Pervasive mantle plume head heterogeneity: evidence from the late cretaceous Caribbean-Colombian oceanic plateau. J Geophys Res 107(7). https://doi.org/10.1029/2001JB000790

Kerr AC, White RV, Thompson PM, Tarney J, Saunders AD (2003) No oceanic plateau- no Caribbean plate? The seminal role of an oceanic plateau in Caribbean plate evolution. In: Bartolini C, Bufer RT, Blickwede J (eds) The Circum-Gulf of Mexico and the Caribbean: hydrocarbon habitats, basin formation, and plate tectonics, vol 79. AAPG Memoir, pp 126–168

Kerr AC, Tarney J, Kempton PD, Pringle M, Nivia A (2004) Mafic pegmatites intruding oceanic plateau gabbros and ultramafic cumulates from bolivar, Colombia: evidence for a 'wet' mantle plume? J Petrol 45:1877–1906

Knechtel MM, Richards E, Rathbun MV (1947) Mesozoic fossils of Peruvian Andes. Johns Hpokins University (Baltimore), Studies in Geology, 15, 150 p

Kohl E, Blissenchach E (1962) Las capas rojas del Cretáceo Superior-Terciario en la region del curso medio del Rio Ucayali, Oriente del Perú. Boletin de la Sociedad Geológica del Perú 39:1–37

Kummel B Jr (1984) Geological reconnaissance of the Contamana region, Peru. Bull Geol Soc Am 59:1217–1266

Laverde F, Clavijo J (1985) Análisis facial de la Formación Los Santos, según el corte Yo y Tu (Zapatoca). In: Etayo-Serna F, Laverde-Montaño F (eds) Proyecto Crétacico, contribuciones. Chapter VI. Ingeominas Publicación Geológica Especial 16, Bogotá, 9 p

Leal-Mejia H, Shaw RP, Melgarejo JC (2018) Spatial-Temporal Migration of Granitoid Magmatisn and the Phanerozoic Tectono-Magmatic Evolution of the Colombian Andes. In: Cediel F and Shaw RP (eds). Geology and Tectonics of Northwestern South America: The Pacific-Caribbean-Andean Junction, Springer, Cham, pp 253–397

Litherland M, Aspden JA, Jemielita RA (1994) The metamorphic belts of Ecuador. Overseas Geology and Mineral Resources, 11, 147 pp. 2 map enclosures at 1:500,000 scale. British Geological Survey, Nottingham

Lopez-Ramos E (2005) Chronostratigraphic correlation chart of Colombia. Ingeominas, Bogota, 56 p

Lugo J, Mann P (1995) Jurassic – Eocene tectonic evolution of Maracaibo Basin, Venezuela. In: Tankard A, Suarez S, Welsink H (eds) Petroleum basins of South America, vol 62. AAPG Memoir, pp 699–725

Luzieux L (2007) Origin and Late Cretaceous-Tertiary evolution of the Ecuadorian forearc. PhD Thesis, ETH, Zurich

Luzieux LDA, Heller F, Spikings R, Vallejo CF, Winkler W (2006) Origin and cretaceous tectonic history of the coastal Ecuadorian forearc between 1°N and 3°S: paleomagnetic, radiometric and fossil evidence. Earth Planet Sci Lett 249:400–414

Macellari CE (1988) Cretaceous paleogeography and depositional cycles of western South America. Jour. South American earth. Sciences 1(4):376–418

Mann P, Escalona A, Castillo MV (2006) Regional geologic and tectonic setting of the Maracaibo supergiant basin, western Venezuela. AAPG Bull 90(4):445–477

Martínez JI (1989) Foraminiferal biostratigraphy and paleoenvironments of the Maastrichtian colon mudstones of northern South America. Micropaleontology 35(2):97–113

Maze WB (1984) Jurassic la Quinta formation in the sierra de Perijá, northwestern Venezuela: geology and tectonic environment of red beds and volcanic rocks. In: Bonini WE, Hargraves RB, Shagam R (eds) The Caribbean-south American plate boundary and regional tectonics. Geol Soc am mem, vol 162, pp 263–282

McCourt WJ, Feininger T, Brook M (1984) New geological and geochronological data from the Colombian Andes: continental growth by multiple accretion. J Geol Soc Lond 141:831–845

McLaughlin DH Jr (1972) Evaporite deposits of Bogotá area, cordillera oriental, Colombia. AAPG Bull 56(11):2240–2259

Mendes AC, Truckenbrod W, Rodriguez Nogueira AC (2012) Análise faciológica da Formação Alter do Chão (Cretáceo, Bacia do Amazonas), próximo à cidade de Óbidos, Pará, Brasil. Revista Brasileira de Geociências 42(1):39–57

Mendoza H (1985) La Formación Cumbre, modelo de transgresión marina rítmica, de comienzos del Cretácico, Proyecto Crétacico, contribuciones. Chapter IX. In: Etayo-Serna F, Laverde-Montaño F (eds) Ingeominas Publicación Geológica Especial 16, Bogotá, 17 p

Meschede M, Frisch W (1998) A plate tectonic model the Mesozoic and early Cenozoic history of the Caribbean plate. Tectonophysics 296(3–4):269–291

Mojica J, Kammer A, Ujueta G (1996) El Jurásico del sector noroccidental de Suramerica y guía de la excursión al Valle Superior del Magdalena (Nov. 1–4/95), Regiones de Payandé y Prado, Departamento del Tolima, Colombia. Geología Colombiana, Bogotá

Morales LG, the Colombian Petroleum Industry (1956) General geology and oil occurrences of the middle Magdalena Valley, Colombia. In: Weeks LG (ed) Habitat of the middle and upper Magdalena basins, Colombia. Oil - a symposium. A.A.P.G, pp 641–695

Moreno JM (1990a) Stratigraphy of the lower cretaceous Rosablanca formation, west flank, eastern cordillera, Colombia, vol 17. Geología Colombiana, Bogotá, pp 65–86

Moreno JM (1990b) Stratigraphy of the Lower Cretaceous units central part Eastern Cordillera, Colombia, 13th International Sedimentological Congress, Nothingham, Aug 22–31, 1990. Abstract

Moreno JM (1991) Provenance of the lower cretaceous sedimentary sequences, central part, eastern cordillera, Colombia. Revista Academia Colombiana Ciencias Exactas Físicas y Naturales 18(69):159–173

Mourier T, Megard F, Pardo A, Reyes L (1988) L'evolution Mesozoique des Andes de Huancabamba et l'hypothèse de l'accretion du microcontinent Amotape-Tahuin. Bulletin de la Société Géologique de France 8(4):69–79

Nerlich R, Clark SR, Bunge HP (2014) Reconstructing the link between the Galapagos hotspot and the Caribbean plateau. Geo Res J 1-2:1–7

Nivia A (2001) Mapa Geológico del Departamento del Valle del Cauca. Memoria Explicativa. Ingeominas, Bogota, 148 p

Nivia A, Marriner G, Kerr A (1996) El Complejo Quebradagrande, una posible Cuenca marginal intracratónica del Cretaceo Inferior en la Cordillera Central de los Andes colombianos: VII Cong. Col Geol Mem 3:108–123

Nivia A, Marriner GF, Kerr AC, Tarney J (2006) The Quebradagrande complex: a lower cretaceous ensialic marginal basin in the central cordillera of the Colombian Andes. J S Am Earth Sci 21:423–436

Pardo A, Moreno M, Gómez A (1993) La Formación Nogales: una unidad sedimentaria fosilifera del Campaniano-Maastrictiano, aflorante en el flanco occidental de la Cordillera Central colombiana: VI Congreso Colombiano de Geología, memorias tomo I, p 248–261

Parnaud Y, Gou Y, Pascual J, Truskowski I, Gallango O, Passalacqua H (1995) Petroleum geology of the central part of the eastern Venezuela Basin. In: Tankard A, Suarez S, Welsink H (eds) Petroleum basins of South America, vol 62. AAPG Memoir, pp 741–756

Pimpirev CT, Patarroyo P, Sarmiento G (1992) Stratigraphy and facies analysis of the Caqueza group, a sequence of lower cretaceous turbidites in the cordillera oriental of the Colombian Andes, vol 17. Geología Colombiana, Bogotá, pp 297–308

Pindell JL (1990) Geological arguments suggesting a Pacific origin for the Caribbean plate. In: Larue DK, Draper G (eds) Transactions of the 12th Caribbean geological conference. Miami Geological Society, St. Croix, pp 1–4

Pindell JL (1993) Regional synopsis of Gulf of Mexico and Caribbean evolution. In: Pindell JL, Perkins RF (eds) Mesozoic and early Cenozoic development of the Gulf of Mexico and Caribbean region — a context for hydrocarbon exploration. Foundation thirteenth annual research conference. Gulf Coast section SEPM, pp 251–274

Pindell JL, Barret SF (1990) Geological evolution of the Caribbean region: a plate tectonic perspective. In: Dengo G, Case J (eds) The Caribbean region, vol. H, the geology of North America. Geol. Soc. am, Boulder, pp 405–432

Pindell J, Dewey J (1982) Permo-Triassic reconstruction of western Pangaea and the evolution of the Gulf of Mexico-Caribbean region. Tectonics 1:179–211

Pindell J, Erikson J (1993) The Mesozoic margin of northern South America. In: Salfity J (ed) Cretaceous tectonics of the Andes, Vieweg, pp 1–60

Pindell J, Kennan L (2009) Tectonic evolution of the Gulf of Mexico, Caribbean and northern South America in the mantle reference frame: an update. In: James K, Lorente MA, Pindell J (eds) The geology and evolution of the region between north and South America, vol 328. Geological Society of London, Special Publication, pp 1–60

Pindell JL, Tabbutt KD (1995) Mesozoic-Cenozoic Andean paleogeography and regional controls on hydrocarbon systems. In: Tankard AJ, Suárez S R, Welsink HJ (eds) Petroleum basins of South America, vol 62. AAPG Memoir, pp 101–128

Pindell J, Kennan L, Maresch W, Stanek K, Draper G, Higgs R (2005) Plate kinematics and crustal dynamics of circum-Caribbean arc-continent interactions; tectonic controls on basin development in proto-Caribbean margins. In: Ave-Lallemant-Hans-G, Sisson-Virginia-B (eds) Caribbean-south American plate interactions, Venezuela, vol 394. Special Paper Geological Society of America, pp 7–52

Polanía JH, Rodríguez OG (1978) Posibles turbiditas del Cretáceo Inferior (Miembro Socotá) en el área de Anapoima (Cundinamarca); una investigación sedimentológica basada en registros gráficos, vol 10. Geología Colombiana, Bogotá, pp 87–91

Posamentier HW, Allen GP, James DP, Tesson M (1992) Forced regressions in a sequence stratigraphic framework: concepts, examples, and exploration significance. AAPG Bull 76:1687–1709

Pratt WT, Duque P, Ponce M (2005) An autochthonous geological model for the eastern Andes of Ecuador. Terctonophysics 399:251–278

Price LI (1960) Dentes de *Theropoda* num testemunho de sonda no estado do Amazonas. Anais da Acad Bras Ciên 32:79–84

Ramírez N, Ramírez H (1994) Estratigrafía y origen de los carbonatos del Cretácico Superior en el Valle Superior del Magdalena, Departamento del Huila, Colombia. In: Etayo Serna F (ed) Estudios geológicos del Valle Superior del Magdalena. Chapter V. Univ. Nacional de Colombia, Ecopetrol, Bogotá, 15 p

Renzoni G (1985a) Paleoambientes de la Formación Tambor en la Quebrada Pujamanes. In: Etayo-Serna F, Laverde-Montano F (eds) Proyecto Crétacico, contribuciones. Chapter III. Ingeominas Publicación Geológica Especial 16, Bogotá, 18 p

Renzoni G (1985b) La secuencia facial de la Formación Los Santos por la Quebrada Piedra Azul: Registro de una hoya fluvial evanescente. In: Etayo-Serna F, Laverde-Montaño F (eds) Proyecto Crétacico, contribuciones. Chapter IV. Ingeominas Publicación Geológica Especial 16, Bogotá, 18 p

Renzoni G (1985c) Paleoambientes de la Formación Arcabuco y Cumbre de la Cordillera de Los Cobardes. In: Etayo-Serna F, Laverde-Montaño F (eds) Proyecto Crétacico, contribuciones. Chapter X. Ingeominas Publicación Geológica Especial 16, Bogotá, 14 p

Rivera R (1979) Zonas faunísticas del Cretáceo de Lima. Boletin de la Sociedad Geológica del Peru, 62-19-24

Rod E, Maync W (1954) Revision of lower cretaceous stratigraphy of Venezuela. AAPG Bull 38:193–283

Rolón LF, Carrero MM (1995) Análisis estratigráfico de la sección Cretácica aflorante al oriente del anticlinal de Los Cobardes entre los Municipios de Guadalupe-Chima-Contratación, Departamento de Santander. Tesis pregrado Geología, Univ. Nacional de Colombia, Bogotá, 80

Ross MI, Scotese CR (1988) A hierarchical tectonic model of the Gulf of Mexico and Caribbean region. Tectonophysics 155:139–168

Rubiano JL (1989) Petrography and stratigraphy of the Villeta Group, Cordillera Oriental, Colombia, South America. M.Sc. Thesis, Univ. South Carolina, Columbia, SC., 96 p

Sarmiento LF (1989) Stratigraphy of the Cordillera Oriental west of Bogotá, Colombia. M.Sc. Thesis University of South Carolina, Columbia, SC.,102 p

Sarmiento LF (2011) Middle Magdalena Basin. Geology and hydrocarbon potential. In: Cediel F (ed) Petroleum geology of Colombia, vol 11. Universidad Eafit for ANH, Medellín, p 192

Sarmiento LF (2015) Correlacion estratigrafica pozos con datos de geoquimica de roca del Valle Medio del Magdalena, parte norte del Valle Superior del Magdalena y parte Suroccidental de la Cordillera Oriental. Presentacion para Equipo de Hidrocarburos no convencionales Ecopetrol, Bogota, 110 p

Sarmiento-Rojas LF (2001) Mesozoic rifting and Cenozoic basin inversion history of the Eastern Cordillera, Colombian Andes. Inferences from tectonic models. Ph.D. Thesis, Vrije Universiteit, Amsterdam, 297 p

Scherrenberg AF, Jacay J, Holcombe RJ, Rosenbaum G (2012) Stratigraphic variations across the Marañon fold-thrust belt, Peru: implications for the basin architecture of the west Peruvian trough. Jour. South American earth. Sciences 30:147–158

Sinton CW (1997) The internal structure of oceanic plateaus: inferences from obducted cretaceous terranes in western Colombia and the Caribbean. Tectonophysics 292:173–188

Sinton CW, Duncan RA, Storey M, Lewis J, Estrada JJ (1998) An oceanic flood basalt province within the Caribbean plate. Earth Planet Sci Lett 155:221–235

Soto FV (1979) Facies y ambientes deposicionales cretácicos, area central sur de la Cuenca Marañón. Boletin de la Sociedad Geológica del Perú 60:233–250

Spikings RA, Winkler W, Seward D, Handler R (2001) Along- strike variations in the thermal and tectonic response of the continental Ecuadorian Andes to the collision with heterogeneous oceanic crust. Earth Planet Sci Lett 186:57–73

Toussaint JF (1995a) Hipótesis sobre el marco geodinámico de Colombia durante el Mesozóico temprano, Contribution to IGCP 322 Jurassic events in South America, vol 20. Geología Colombiana, Bogotá, pp 150–155

Toussaint JF (1995b) Evolución geológica de Colombia 2. Triásico Jurásico. Contribución al IGCP 322 "Correlation of Jurassic events in South America" International Geological Correlation Programme Unesco IUGS. Univ. Nacional de Colombia. Medellín, 94 p

Toussaint JF, Restrepo JJ (1989) Acresiones sucesivas en Colombia; un nuevo modelo de evolución geológica. V Congreso Colombiano de Geología, Bucaramanga, I: 127–146

Toussaint JF, Restrepo JJ (1994) The Colombian Andes during cretaceous times. In: Salfity JA (ed) Cretaceous tectonics of the Andes, Verlag Braunschweig, Wiesbaden, pp 61–100

Tschopp HJ (1948) Geologische Skizze von Ekuador. Bulletin de l'Association Suisse de Géologie Ingénieur et Pétrologie 15:14–45

Ulloa C, Rodríguez E (1976) Geología del cuadrángulo K-12, Guateque. Boletín Geológico, Ingeominas, Bogotá, 22(1): 4–55

Vallejo CV (2007) Evolution of the Wester Cordillera in the Andes of Ecuador (Late Cretaceous-Paleocene). Ph.D. Dissertation, Swiss Federal Institute of Technology Zürich. 158 p. 4 Appendixes 159–208

Vallejo C, Winkler W, Spikings RA, Luzieux L, Heller F, Bussy F (2009) Mode and timing of terrane accretion in the forearc of the Andes in Ecuador, vol 204. Geological Society of America Memoir, pp 197–216

Vargas R, Etayo RG, Téllez N (1985) Corte estratigráfico panorámico de la Formación Los Santos, Carretera Quebrada El Medio El Boquerón Santander. In: Etayo-Serna F, Laverde-Montaño F (eds) Proyecto Crétacico, contribuciones. Chapter V. Ingeominas Publicación Geológica Especial 16, Bogotá, 12 p

Vergara L, Prössl KF (1994) Dating the Yaví formation (Aptian upper Magdalena Valley, Colombia). In: Etayo Serna F (ed) Estudios geológicos del Valle Superior del Magdalena. Chapter XVIII. Univ. Nacional de Colombia, Ecopetrol, Bogotá, 14 p

Villagomez DR (2010) Thermochronology, geochronology and geochemistry of the Western and Central cordilleras and Sierra Nevada de Santa Marta, Colombia: The tectonic evolu on of NW South America. Thèse présentée à la Faculté des sciences de l'Université de Genève pour obtenir le grade de Docteur ès sciences, mention sciences de la Terre. Geneve, 125 p. Apendixes

Villagómez D, Spikings R, Magna T, Kammer A, Winkler W, Alejandro Beltrán A (2011) Geochronology, geochemistry and tectonic evolution of the western and central cordilleras of Colombia. Lithos 125(2011):875–896

Villamil T (1994) High-resolultion stratigraphy, chronology and relaitve sea level of the Albian-Santonian (Cretaceous) of Colombia. Ph.D. Thesis Univ. Of Colorado at Boulder, 446 p

Villamil T (1999) Campanian–Miocene tectonostratigraphy, depocenter evolution and basin development of Colombia and western Venezuela. Palaeogeogr Palaeoclimatol Palaeoecol 153(1999):239–275

Villamil T (2012) Chronology, relative sea level, and a new sequence stratigraphy model for distal offshore facies, Albian to Santonian, Colombia. In: Paleogeographic evolution and non-glacial Eustasy, northern South America. SEPM Special Publication No 58, pp 161–216

Villamil T, Arango C (2012) Integrated stratigraphy of latest Cenomanian and early Turonian facies of Colombia. Paleogeographic evolution and non-glacial eustasy, Northern South America. SEPM Spec Publ 58:129–159

Villamil T, Pindell JL (2012) Mesozoic paleogeographic evolution of northern South America: foundations for sequence stratigraphic studies in passive margin strata deposited during non-glacial times. In: Paleogeographic evolution and non-glacial Eustasy, northern South America. SEPM Special Publication No 58, pp 283–318

Wilson JJ (1963) Cretaceous stratigraphy of central Andes of Peru. AAPG Bull 47:1–3

Chapter 11
Morphotectonic and Orogenic Development of the Northern Andes of Colombia: A Low-Temperature Thermochronology Perspective

Sergio A. Restrepo-Moreno, David A. Foster, Matthias Bernet, Kyoungwon Min, and Santiago Noriega

Abbreviations

AER	Age-elevation relationship for LTTC data
AFT	Apatite fission-track dating
A-PAZ	Apatite partial annealing zone for fission tracks in FT dating LTTC
AHe	Apatite uranium-thorium/helium dating
AP	Antioqueño Plateau
CC	Central Cordillera
CRFS	Cauca-Romeral Fault System
EC	Eastern Cordillera
FT	Fission-track dating
GOF	Goodness of fit in LTTC modeling
LA-ICP-MS	Laser ablation inductively coupled plasma mass spectrometry
LTTC	Low-temperature thermochronology

S. A. Restrepo-Moreno (✉)
Universidad Nacional de Colombia, Facultad de Minas, Departamento de Geociencias y Medio Ambiente, Medellín, Colombia

Department of Geological Sciences, University of Florida, Gainesville, FL, USA
e-mail: sarestrepom@unal.edu.co; sergiorm@ufl.edu

D. A. Foster · K. Min
Department of Geological Sciences, University of Florida, Gainesville, FL, USA

M. Bernet
ISTerre, Université Grenoble Alps, Grenoble, France

S. Noriega
Universidad Nacional de Colombia, Facultad de Minas, Departamento de Geociencias y Medio Ambiente, Medellín, Colombia

Ma	Mega-annum
MTL	Mean track length in fission-track analyses
NAB	Northern Andes Block or NorAndean Block
PAZ	Partial annealing zone for fission tracks in FT dating LTTC
PCB	Panama-Chocó Block
PLOCO	From its definition in Spanish as "Provincia Litosférica Oceánica Cretácica Occidental"
PLCMG	From its definition in Spanish as "Provincia Litosférica Continental Mesoproterozoica Grenvilliana"
PRZ	Partial retention zone for helium in (U-Th)/He dating LTTC
SNSM	Sierra Nevada de Santa Marta
SNC	Sierra Nevada del Cocuy
WC	Western Cordillera
ZHe	Zircon uranium-thorium/helium dating
ZFT	Zircon fission-track dating
Zr-PAZ	Zircon partial annealing zone for fission tracks in FT dating LTTC

11.1 Introduction

The Andes are one of the most representative active orogens in the world. Orogenic processes along the western margin of South America, operating primarily during the Meso-Cenozoic, resulted in the construction of this massive mountain belt. Major insight into the morphotectonic events (surface uplift, erosional and tectonic exhumation, sediment production-routing and deposition, basin development, etc.) that have shaped the northern segment of the Andes has come from recent advances in the application of low-temperature thermochronology techniques such as fission-track and (U-Th)/He dating in apatite and zircon. These findings are crucial for enhancing our understanding of lithosphere-atmosphere-hydrosphere coupled systems (e.g., internal-external process interactions and feedback mechanisms) as they control orogenic development. Complex geodynamic and climatic regimes including the interaction of multiple tectonic plates and crustal blocks in subduction and collision settings, combine to produce the morphotectonic domain in the tropical Northern Andes.

The geologic and geographic importance of the Andes at the local, regional, and global scales is indisputable. The Andes constitute one of the major "factories" for the construction of continental crust, making it the foremost modern orogenic system developed by subduction of oceanic crust along a continental margin (Ramos 2009). Extending over 4000 km in a roughly N-S direction along the western margin of South America, and reaching almost 7000 m in elevation, the Andes rank as the longest and second highest mountain belt in the world. Such a topographic structure plays a pivotal role in atmospheric circulation and hence in regional and global climatic, biotic, and geomorphic patterns (Anderson 2015; Zuluaga and Houze 2015; Ruddiman 2013; Ochoa et al. 2012; Hoorn et al. 1995, 2010; Ehlers and Poulsen 2009; Weng et al. 2007; Aalto et al. 2006; Molnar 2004; Hartley 2003; Jiang et al. 2002; Montgomery et al. 2001; Poveda and Mesa 2000; Kennan 2000; Gregory-Wodzicki 2000; Mittermeier et al. 1999; Villamil 1999; Ruddiman et al. 1997; Laubacher and Naeser 1994; Andriessen et al. 1993; Hooghiemstra 1989; Van der Hammen 1989; Van der Hammen et al. 1973).

Geoscientists have reached a consensus with respect to the fact that there are complex interrelations between internal and external processes during morphotectonic development (Fig. 11.1), far beyond the old paradigm of constructive (tectonic)-destructive (erosion) forces and/or the passive isostatic response to erosional unloading (Bishop 2007; Summerfield 2005; Zeitler et al. 2001; Ollier and Pain 2000; Burbank and Pinter 1999; Koons 1998; Pinter and Brandon 1997; Avouac and Burov 1996; Molnar et al. 1993; Koons 1990). As a coherent body of knowledge, tectonic geomorphology attempts to explain orogeny in the context of both tectonic (internal) and geomorphic (external) processes (Burbank and Anderson 2011) and seeks to shed light on the long-debated topic of whether exhumation in mountain ranges is controlled primarily by climate-induced erosion or tectonics, or by a complex interplay of both (Bishop 2007; Clift 2010; Grujic et al. 2006; Harris and Mix 2002; Kennan 2000; Koons 1998; Lamb and Davis 2003; Molnar and England 1990; Montgomery et al. 2001; Raymo and Ruddiman 1992; Reiners and Shuster 2009; Summerfield and Hulton 1994; Whipple 2009).

Orogenic belts are curvilinear regions of positive topography that result from key geologic processes and complex feedback mechanisms of the coupled lithosphere-atmosphere system (Fig. 11.1). Some of these processes are evident, such as surface uplift,[1] erosion, evolution of fluvial networks, topographic construction-destruction, isostatic accommodation, lithospheric flexure, crustal fluxes (e.g., accretionary, erosional) of matter and energy, etc. (Reiners and Shuster 2009; Bishop 2007; Whipple and Meade 2006; Willett et al. 2001, 2006; Zeitler et al. 2001). Deep-seated geological processes may appear to be unrelated to surface processes, but it has been shown that even tectonic deformation, structural evolution, and metamorphism/magmatism can, to some extent, be controlled by climate and erosion (Whipple 2009; Beaumont et al. 1992, 2001; Willett 1999; Norris and Cooper 1997; Pavlis et al. 1997; Koons 1990). In this realm, too, the consensus is that the combination of erosion and exhumation is "… a significant agent in active tectonic systems,

---

[1] We use the term uplift (and other associated terminology) in the sense defined by England and Molnar (1990) as follows: Surface uplift is the displacement of the average elevation of the landscape with respect to mean sea level. Rock uplift is the net displacement of a rock parcel with respect to sea level. Rock uplift is equal to surface uplift under no-erosion and no-deposition conditions, i.e., where no exhumation/burial takes place. Exhumation implies the approximation of the rock parcel to the topographic surface. In that sense, exhumation (also referred to as denudation) can be defined as the difference between rock uplift and surface uplift. A topographic steady-state condition is then achieved when rock uplift and erosion proceed at the same rate to inhibit surface uplift.

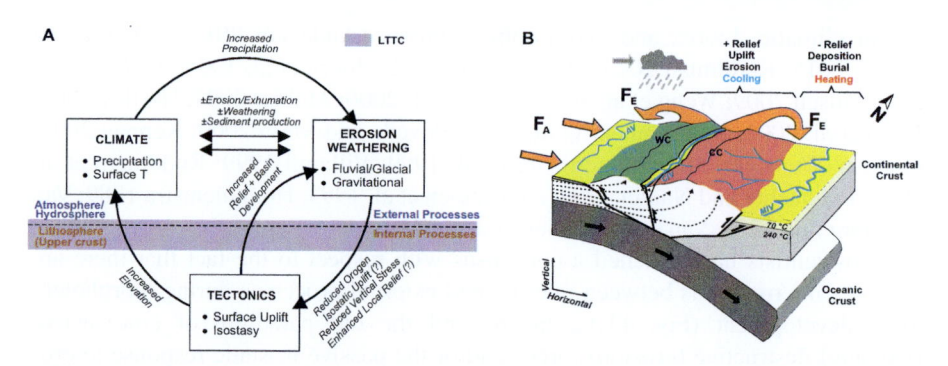

Fig. 11.1 Highly coupled internal-external process system. (**A**) The dynamic atmosphere-hydrosphere-lithosphere interface where external (erosion and climate) and internal (tectonics) processes interact through complex feedback mechanisms. For example, during compressional episodes at an active margin (e.g., Northern Andes), surface uplift can occur, increasing elevation, and potentially augmenting local relief, if fluvial or glacial incision take place, thereby increasing erosion and exhumation rates leading to further uplift via isostatic rebound. Erosion produces crustal cooling that can be recorded by LTTC. Increased detrital production is accommodated in sedimentary basins where burial is manifested in LTTC systems as heating-reseating. Climate can change either through tectonic factors (e.g., increased elevation, establishment of orographic barriers, development of glaciation) or through other forcing mechanisms (e.g., solar, ocean circulation, etc.), altering the erosional regime and producing modifications in the system via isostatic compensations, slope adjustment, fault reactivation. The range of sensitivity of LTTC (horizontal, purple band) sits in the upper crust (1–5 km), right across the lithosphere-atmosphere boundary, making it a powerful tool to address morphotectonic processes (e.g., erosional exhumation, uplift, changes in topography). (**B**) Schematic representation of mass and energy fluxes though a convergent setting such as the Northern Andes of Colombia. Accretionary (F_A) and erosional (F_E) fluxes are represented by orange arrows. Dashed lines show material transport with vertical and horizontal components represented as vectors. Uplift and exhumation focused on cordilleran massifs (red and green) are facilitated by major crustal structures (faults) enhancing erosional fluxes that are measurable though LTTC as cooling. Sediments from cordilleran massifs accumulate in interandean and/or marginal basins (yellow) producing burial and heating also detectable by LTTC. Orographic precipitation as a climatic input is shown as a blue cloud moving from west to east indicating moisture transfer from the Eastern Equatorial Pacific that "feed" geomorphic process as well as the hyper-pluvial Chocó Biogeographic Region, the rainiest place on earth. CC and WC mark the Central and Western cordilleras respectively. MIV and CIV show the position of the interandean valleys of the Magdalena and Cauca rivers, respectively. AV shows the Atrato Valley. (Modified from Stolar et al. 2006)

particularly at larger spatial scales" so that a better understanding of orogenic belts entails special attention to erosional exhumation (Zeitler et al. 2001).

Orogenic development at geodynamically complex convergent margins such as the Northern Andes is a widely debated subject. Several issues remain largely unresolved, including the timing and magnitude of main morphotectonic phases, the relative contributions to crustal thickening (e.g., magmatic addition vs. crustal shortening), the role of mantle vs. crustal processes in driving surface uplift (e.g., removal of a dense lithospheric mantle root), the function of inherited crustal structures and geodynamic shifts in facilitating (or blocking) surface uplift, preservation (or obliteration) of elevated plateaus, changes in paleotopography, and the comparative effects of subduction and/or collision on surface and rock uplift (Montgomery et al. 2001; Taboada et al. 2000; Ramos 1999; Dalziel 1986; Jordán et al. 1983). Convergent margins featuring multiple plate boundaries and a combination of subduction and accretionary dynamics, such as the Northern Andes, are prone to incessant and rapid changes in both plate convergence dynamics (angles, rates, etc.) and the varying character of subducted/accreted crustal elements (Taboada et al. 2000; Gutscher et al. 1999; Aspden et al. 1987; Pardo-Casas and Molnar 1987; Pilger 1984; Cross and Pilger 1982). Changes in these key processes can have significant effects on magmatic and metamorphic dynamics, but also on deformation, morphotectonic responses, and sedimentological and paleogeographic dynamics within the upper plate (Reyes-Harker et al. 2015; Montes et al. 2012b; Bayona et al. 2011; Cardona et al. 2011; Farris et al. 2011; Horton et al. 2010b; Restrepo-Moreno et al. 2009b, c; Villagomez-Díaz 2010; Aspden et al. 1987). It has been shown, for instance, that morphotectonic processes such as surface uplift and erosional exhumation are highly responsive to plate convergence parameters (velocity and obliquity) and the thickness of the subducted oceanic plate (Espurt et al. 2010; Sobolev and Babeyko 2010; Gerya et al. 2009; Spotila et al. 1998; Thompson et al. 1997).

Uplift of the Northern Andes throughout the Cenozoic, as well as the formation of other mountain ranges in the region (e.g., Panamá Arc/Isthmus), is a fundamental element of the environmental and biotic history of northwestern South America, the Caribbean, and the eastern equatorial Pacific (Fig. 11.2). The evolving topographic configuration modified the drainage patterns of rivers including the Amazon, Orinoco, Magdalena-Cauca, Atrato, and San Juan, on both regional and continental scales (Anderson et al. 2016; Reyes-Harker et al. 2015; Hoorn et al. 1995, 2010; Mora et al. 2010a; Villamil 1999). In addition, it promoted a more complex biodiversity setting in terms of ecological niches and species migration through varying topography (Ochoa et al. 2012; Hoorn et al. 2010; Weng et al. 2007; Andriessen et al. 1993; Hooghiemstra 1989; Van der Hammen 1989; Van der Hammen et al. 1973) and new connections of previously isolated continental masses e.g., The Great American Biotic Interchange via Isthmus establishment (Marshall et al. 1979; O'Dea et al. 2016; Woodburne 2010).

Among the processes that are relevant for understanding orogeny (crustal thickening, metamorphism, magmatism, deformation, uplift, exhumation, etc.), the present contribution focuses on the cooling histories of rocks, as they reveal specific patterns of erosional exhumation associated with surface uplift (Reiners and Shuster 2009; Reiners and Brandon 2006; Spotila 2005; Gleadow et al. 2002; Brown et al. 1994). Specifically, we review the spatial distribution of apatite and zircon (U-Th)/He and fission-track cooling ages as well as thermochronological models in the Colombian Andes to address orogenic processes in this Andean segment.

In the first part of this contribution, we introduce the main physiographic and litho-structural features of the Northern Andes in Colombia, including the principal mountain ranges (Western, Central, and Eastern cordilleras) and "isolated" topographic structures (e.g., Sierra Nevada de Santa Marta, Serranía de Baudó, Serranía de La Macarena, Serranía de Chiribiquete, etc.), as well as the subdivision of litho-structural domains or crustal blocks proposed for the region (e.g., Cediel et al. 2003). We then present fundamental concepts for the generation and interpretation of thermochronological data. Finally, existing low-temperature thermochronology (LTTC) datasets and thermotectonic models are discussed, both in a segmented fashion (i.e., by mountain ranges) and by integrating cooling ages/histories into the regional geodynamic context, with the aim of reconstructing the morphotectonic history of the Northern Andes of Colombia during the Cenozoic.

We emphasize that LTTC provides direct, explicit information on cooling of rocks (Brown et al. 1994; Gleadow and Brown 1999; Reiners and Shuster 2009), primarily related to erosional (uplift-driven erosion) and tectonic exhumation (normal fault, crustal thinning), and/or to post-magmatic cooling (Ring et al. 1999). LTTC also provides useful information on paleo-relief[2] (Flowers et al. 2008; House et al. 1998), but not strictly on surface uplift and paleoelevation (Reiners and Brandon 2006; Molnar 2004). In the words of Ghosh et al. (2006): "Elevation of the earth's surface is among the most difficult environmental variables to reconstruct from the geologic record." This remains the case for the Andes, despite efforts to address the issue (Anderson et al. 2016; Ochoa et al. 2012; Weng et al. 2007; Ghosh et al. 2006; Hooghiemstra et al. 2006; Gregory-Wodzicki 2000; Van der Hammen et al. 1973). Therefore, we focus here on the tectonothermal history of the Northern Andes, because tectonic-driven surface uplift is often followed by concomitant topographic buildup and increased erosional exhumation (Burbank and Anderson 2011; Reiners and Shuster 2009; Ollier 2006; Ollier and Pain 2000; Burbank and Pinter 1999; Pinter and Brandon 1997). It is important to note that erosion is triggered not only by changes in topographic configuration (i.e., local relief) but also by climatic factors (e.g., increased precipitation, frequency of storm events, etc.), and the controversy related to this topic has been discussed in the literature for several decades (Clift 2010; Bierman 2004; Lamb and Davis 2003; Ruddiman et al. 1997; Sugden et al. 1997; Molnar and England 1990; Summerfield and Hulton 1994).

---

[2] In this contribution, the term relief is equivalent to topographic relief and refers to the difference between the highest and lowest point in a particular area. In that sense the Andean Region of Colombia is characterized by relief in excess of 5000 m, as the highest points in the Central Cordillera reach elevations close to 5.5 km, whereas the bottoms of valleys such as the Middle Cauca and Middle Magdalena are at ~500 m. Conversely, the Caribbean, (except for the Sierra Nevada de Santa Marta with peaks reaching elevations of more than 5.5 km) Amazonian, and Orinoquia regions possess low topographic relief, i.e., <200 m on average. Local and relative relief are more specific terms, indicating the difference in elevation measured over a specified area. Figure 11.7 illustrates differences in relative relief over the Northern Andes. (For more detail on these definitions, see Summerfield, 2001, and Montgomery 2003).

11.2 The Northern Andes of Colombia

After the establishment of the global tectonics paradigm (Summerfield 2000; Wyllie 1976) and the classification of orogenic systems as either "cordilleran" or "collisional" (Dewey and Bird 1970), the Andes have been traditionally interpreted as an "archetypal" example of a cordilleran-type domain (Ramos 1999; Summerfield 2000; Dalziel 1986; Zeil 1979). For Cediel et al. (2003), such a classification entails a fundamental inaccuracy because "it views the Andes as a single, homogeneous geotectonic unit – a supposition that could not be farther from the truth." In this section, we will briefly address the geologic differences that make the Northern Andes (and the Colombian Andes in particular) unique from the litho-structural, geodynamic, and physiographic points of view, so its morphotectonic history can be better understood.

It has been pointed out that the Northern Andes, i.e., the portion of the Andean range present in Ecuador, Colombia, and Venezuela, deviate from the classic orogenic models used to explain the Central and/or Southern Andes (Cediel et al. 2003; Ramos 1999; Van der Hilst and Mann 1994; Jordán et al. 1983; Shagam 1975; Gansser 1973). Some of the features that make the Northern Andes distinct are: (i) The nature of the continental margin, (ii) the types and age of the underlying basement and the subducted/accreted crust, (iii) the prevalence of mixed subduction/accretion dynamics involving complex convergence of major plates (Nazca-Farallon, Cocos, Caribbean, and South American), continental blocks (e.g., NAB), young intra-oceanic arc complexes (e.g., PCB), oceanic plateaus (e.g., CLIP), and/or aseismic oceanic ridges, (iv) the more oblique nature of convergence leading to transpression dynamics that affect both subduction and accretion, (v) the orientation, geometry, spatial distribution, and segmentation of pre-orogenic anisotropies, (vi) the style of magmatism and deformation/metamorphism, (vii) the style of stress field regimes during uplift and orogenesis (Cortés et al. 2005; Colmenares and Zoback 2003; Cediel et al. 2003; Trenkamp et al. 2002; Taboada et al. 2000; McCourt et al. 1984; Ramos 1999; Aspden et al. 1987).

Each of the differences listed above comprises an intricate geodynamic regime that cannot be accurately explained by the traditional "collision" or "cordilleran" models. Furthermore, it is based on some of these differences that the Andean range is subdivided into the Southern, Central, and Northern Andes, each segment with its own peculiarities in terms of tectonic styles, types of magmatism, and geomorphic expression (Cediel et al. 2003; Montgomery et al. 2001; Kennan 2000; Gregory-Wodzicki 2000; Summerfield 2000; Ramos 1999; Pilger 1984; Jordán et al. 1983). Acknowledging such differences is essential to better understand the morphotectonic history of the Northern Andes, where the complicated stress fields resulting from the interactions of major plates and smaller crustal blocks cause significant crustal shortening that can be accommodated differentially by a convoluted mosaic of tectonic blocks, each with a potentially different morphotectonic history. In the following sections, we describe the general tectono-structural, lithologic, and geomorphic framework of the main mountain belts and sedimentary basins of the Northern Andes in Colombia.

11.2.1 Tectono- and Litho-Structural Framework

The Northern Andes are dominated geodynamically by the convergence of the Caribbean, Nazca-Farallon, Cocos, and South American tectonic plates (Ramos 2009; Taboada et al. 2000). In addition, the Northern Andes are detached from the South American Plate by major strike-slip faults (Cediel et al. 2003; Taboada et al. 2000; Mann and Burke 1984; Pennington 1981). This complex convergence setting is marked by subduction and accretion dynamics, whereby multiple crustal fragments have been docked to the western South American margin since the Mesozoic (Branket et al. 2012; Vinasco and Cordani 2012; Chicangana 2005; Cediel et al. 2003; Taboada et al. 2000; Burke et al. 1984).

The nature of subduction of the Nazca-Farallon plate system under NW South America has changed considerably through time. Some of the varying parameters/processes include the rate and direction of convergence, the presence-absence of subduction resisters (seamounts, ridges, buoyant oceanic crust, etc.), and the prevalence of collision/accretion over subduction (Sobolev and Babeyko 2010; Wilder 2003; Somoza 1998; Pardo-Casas and Molnar 1987; Pilger 1984; Jordán et al. 1983). The modes in which subduction and/or accretion take place affect orogenic processes. For example, collision of the Caribbean Plateau with the South American margin triggered tectonic uplift and concomitant erosional exhumation in the Late Cretaceous-Paleocene (Amaya et al. 2017; Barbosa-Espitia et al. 2013; Villagomez and Spikings 2013). Docking of the PCB to the South American margin since the Oligocene (Farris et al. 2011) resulted in the segmentation of subduction, with shallow and/or blocked subduction of the Nazca Plate in the north and steeper subduction in the south (Mann and Corrigan 1990). Such contrasting modes of subduction produced the apparent eastward migration of the deformation front that is evidenced by overall broadening of the Andean chain and absence of volcanism in the northern segments, as opposed to a narrower deformation belt in the southern segment with volcanism focused in the Central Cordillera (Taboada et al. 2000; Vargas and Mann 2013). Orogenetic changes not only affected inboard South America as PCB's morphotectonic processes were deeply influenced by its own history of collision and oroclinal development (Ramírez et al. 2016; Montes et al. 2012a, 2012b; Farris et al. 2011; Coates et al. 2004; Duque-Caro 1990a; Mann and Corrigan 1990).

Transpressional accretion and strain partition define at least two main crustal blocks: the Northern Andean Block (NAB) and the Panama-Chocó Block (PCB) (Suter et al. 2008; Cediel et al. 2003; Cortes and Angelier 2005; Taboada et al. 2000). Both crustal domains act as complex litho-structural "interfaces" between the major tectonic plates and constitute the mountainous realm of NW South America (Cediel et al. 2003). The NAB and the PCB are further partitioned into numerous lithotectonic and morphostructural blocks. Nomenclature of these blocks is variable and based primarily on faulted boundaries, continental vs. oceanic affinity, lithology, age, and rheological differences (Cediel et al. 2003; Montes et al. 2005; Restrepo and Toussaint 1988; Etayo-Serna 1986). Cediel et al. (2003) proposed a coherent subdivision of the Northern Andes into specific lithotectonic and morphostructural domains, shedding light on the distribution of tectonic plates, subplates, terranes, and composite terranes. We employ their approach here because it precludes the incorrect perception that the major physiographic features today (e.g., cordilleras/serranías and/or valleys/basins) must correspond to single lithotectonic units or specific geotectonic events, so as to constitute a valid "geologic" reference framework (Cediel et al. 2003), a notion used frequently in the past (Aspden et al. 1987; Álvarez 1983; Botero 1963).

The Northern Andean Block (NAB, sensu Cediel et al. 2003) is delimited to the east by the Boconó and Guaicáramo faults in the Colombia and Venezuela foothills, which separate the NAB from cratonic rocks of the South American Plate to the north/northwest by the Oca and El Pilar fault systems and the Caribbean deformation/subduction zone and to the west by the Nazca Subduction zone (Fig. 11.2). It is important to note that this definition of NAB includes lithotectonic and morphostructural features of both the Colombian Andes and the PCB realms (see Fig. 1- Cediel et al. 2003). In Colombia, both NAB and PCB imply a set of complex, diverse, and minor lithotectonic blocks, which include provinces of continental affinity (Pre Cambrian to Palaeozoic basement rocks with Mesozoic sedimentary and intrusive rocks of the Central and Eastern cordilleras), situated to the east of the Cauca-Romeral Fault System (CRFS), as well as provinces of oceanic affinity (Western Cordillera and PCB). The CRFS is, therefore, a regional structure extending from Ecuador to northern Colombia (i.e., part of the Dolores-Guayaquil megashear zone) and separating rocks of continental affinity to the east from units of oceanic affinity to the west (Gomez et al. 2015b; Chicangana 2005; Ego et al. 1995). This structure records activity since at least Triassic times (Vinasco and Cordani 2012).[3] To the west, the NAB includes the southernmost portion of the Panama-Chocó Block (PCB). All of the lithotectonic blocks mentioned before occur as N-S elongated ribbons separated by regional faults with predominantly inverse and/or lateral kinematics (Restrepo et al. 2009; Chicangana 2005; Cediel et al. 2003; Ego et al. 1995; Etayo-Serna 1986).

In this contribution, we consider the PCB separately, as this crustal block played a pivotal role in the latest collisional phases of western Colombia, particularly during the Oligocene to present morphotectonic history of both the PCB and the Colombian Andes (Ramírez et al. 2016; Vargas and Mann 2013; Montes et al. 2012b; Farris et al. 2011; Restrepo-Moreno et al. 2010; Suter et al. 2008; Coates et al. 2004; Taboada et al. 2000; Duque-Caro 1990a, b). The PCB's litho-structural boundaries are as follows: the Uramita-Dabeiba fault system marks the eastern boundary, the Garrapatas fault defines the southern limit, and the western edge is defined by the subduction zone with the Nazca Plate (Fig. 11.2). The PCB is considered as a rigid indenter that is part of the Panamanian Arc that was accreted onto NW South America (Case et al. 1971; Coates et al. 2004; Kellogg and Vega 1995; Montes et al. 2012a, b, 2015; O'Dea et al. 2016; Suter et al. 2008; Taboada et al. 2000) as early as the Oligocene (Farris et al. 2011). Topographically, the PCB exhibits two virtually parallel, arcuate ranges: the Darien-Cuchillo-Mandé Serranía to the north and the Majé-Baudó Serranía to the south. These dual serranías are built

---

[3] For details on the Romeral Shear Zone, see Chap. 5, Contribution No. 12, in this volume.

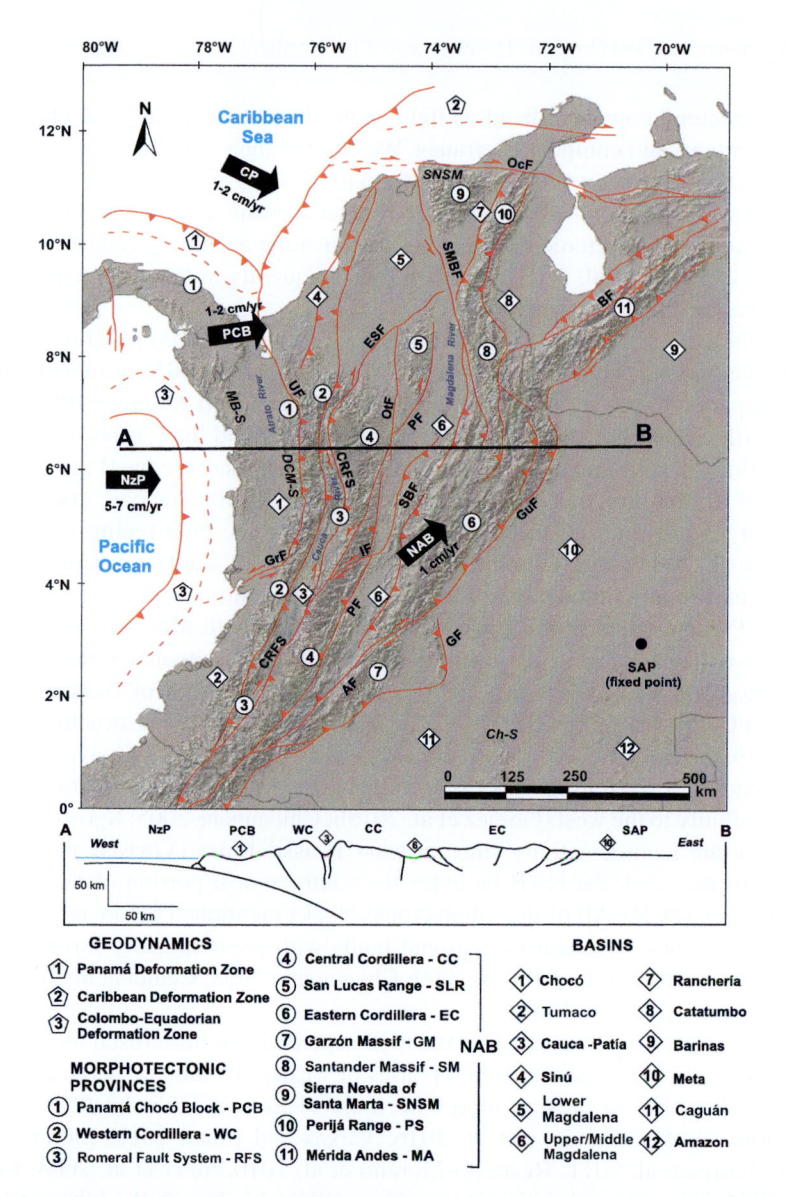

Fig. 11.2 Litho-structural and physiographic domains of the Northern Andes in Colombia and Venezuela. Three fundamental structural domains are recognized in Colombia: the South American Plate (SAP), the NorAndean Block (NAB), and the Panama-Chocó Block (PCB), the Nazca Plate (NzP) and the Caribbean Plate (CP) are key players in the geodynamic history of the Region. The Main physiographic features include the Western, Central and Eastern cordillera (part of the NAB). Important isolated topographic features correspond to the Sierra Nevada de Santa Marta and the Serranía de Chiribiquete (Ch-S). The Majé-Baudó Serranía (MB-S) and the Darién-Cuchillo-Mandé Serranía (DCM-S) are part of the PCB. Main faults and faults systems are: Boconó Fault (BF), Guaicáramo Fault (GuF), Oca Fault (OcF), Garzón Fault (GF), Santa Marta Bucaramanga Fault (SMBF), Salinas-Bituima Fault (SBF), Palestina Fault (PF), Otú Fault (OtF), Ibagué Fault (IF), Algeciras Fault (AF), Espíritu Santo Fault (ESF), Cauca-Romeral Fault System (CRFS), Uramita Fault (UF), Garrapatas Fault (GrF). Block nomenclature from Cediel et al. (2003). Basin nomenclature from Barrero et al. (2007). Regional faults from (Gomez et al 2015). Shaded relief map developed from SRTM-90 m DEM

on Cretaceous–Paleogene igneous basement rocks of oceanic affinity (Coates et al. 2004; Montes et al. 2012b). Structurally and geomorphically, both ranges are continuous along the PCB and enclose Cenozoic sedimentary formations such as the Atrato-Chucunaque Basin (Buchs et al. 2010; Case et al. 1971; Coates et al. 2004; Duque-Caro 1990b; Garzon-Varon 2012; Ramírez et al. 2016). It is often assumed that the southern reaches of the Darién-Cuchillo-Mandé Serranía in the PCB are part of the Western Cordillera in the NAB. However, the PCB is separated both topographically and structurally from the rest of the cordilleran realm of Colombia, as the locus of docking between the Panamanian and the Colombian realms is located upon the suture separating these domains along the Uramita-Dabeiba Fault (Suter et al. 2008; Montes et al. 2012a, b).

In a litho-structural sense, the triple cordilleran system of the Colombian Andes also exhibits several significant differences among the various ranges. The Western Cordillera falls into what is known as the oceanic realm (PLOCO,[4] sensu Gomez et al. 2015b) situated to the west of the CRFS. Basement rocks of the Western Cordillera encompass a series of accreted elements of oceanic affinity (magmatic arcs, volcano-sedimentary sequences, etc.), most of which are Cretaceous in age (e.g., Mistrató, Buga, and Sabanalarga batholiths). Volcano-sedimentary sequences were intruded by calc-alkaline plutonic bodies in Miocene times (e.g., Farallones del Citará, Tatamá, and Horqueta plutons) (Gomez et al. 2015b; Restrepo-Moreno et al. 2013c; Rodríguez and Zapata 2012; González and Londoño 2002a, b). In contrast, the Central Cordillera (at least its axis and eastern flank) and the Eastern Cordillera represent a geologic province of continental affinity (PLCMG,[5] sensu Gomez et al. 2015b). Tectono-magmatic activity on the Central Cordillera is by no means a recent feature, having been a key component of the cordillera's orogenetic evolution during the Mesozoic and Paleogene. The Central Cordillera is primarily composed of old (Precambrian(?) to Paleozoic), medium-to-high-grade metamorphic rocks (Restrepo 2008; Vinasco et al. 2006; Restrepo and Toussaint 1982; Feininger and Botero 1982; Botero 1963) intruded by large plutons[6] (Leal-Mejía et al. 2011) of primarily Jurassic, Cretaceous, and Paleocene-Early Eocene age (Aspden et al. 1987; Bayona et al. 2012; Botero 1963; Bustamante et al. 2016; Feininger and Botero 1982; Leal-Mejía et al. 2011; Restrepo-Moreno et al. 2007, 2009a). Modern volcanic/magmatic arcs are prevalent in the middle portion of this range (Cochrane 2013; Leal-Mejía et al. 2011; Maya and González 1995; Pilger 1984; Restrepo et al. 1981; Restrepo and Toussaint 1990). The Eastern Cordillera combines crystalline basement rocks that include old, Precambrian to Paleozoic, metamorphic massifs in the Garzón and Santander massifs, plutonic units varying in age from Mesozoic to Tertiary, and sedimentary sequences that range from the Paleozoic to the Neogene (Velandia-Patiño

---

[4] PLOCO, from its definition in Spanish as "Provincia Litosférica Oceánica Cretácica Occidental" (Gomez et al. 2015b).
[5] PLCMG, from its definition in Spanish as "Provincia Litosférica Continental Mesoproterozoica Grenvilliana" (Gomez et al. 2015b).
[6] For a detailed account of magmatism in the Northern Andes, see Chap. 4, contribution No. 5, this volume.

2018; Amaya-Ferreira 2016; Anderson et al. 2016; Anderson 2015; Horton et al. 2010b; Bayona et al. 2008; Cortés et al. 2006; Dengo and Covey 1993). The Central Cordillera constitutes the latest Cretaceous-Paleogene thrust front, whereas the Eastern Cordillera is considered a double-verging, fold-and-thrust belt with intermontane basins of various ages located along the axial zone (Bayona et al. 2008; Mora et al. 2010b; Parra et al. 2012).

Recent geophysical investigations of crustal structure-thickness in the Colombian Andes have advanced models that show a complex multiplate subduction domain (Syracuse et al. 2016; Vargas and Mann 2013) and marked isostatic anomalies (Poveda et al. 2015; Yarce et al. 2014). Both phenomena are important in terms of understanding and interpreting LTTC datasets. We point to crustal thickness because high elevations can be sustained via isostatic compensation in convergent settings through crustal thickening (e.g., shortening due to compression, magmatic addition, crustal underplating, and ductile flow of the lower crust), thinning of the mantle lithosphere (e.g., delamination and/or tectonic erosion), and/or dynamic crustal support (e.g., thermal anomalies due to magmatism). With regard to isostasy, it is clear that high elevations in the NAB of Colombia do not necessarily coincide with well-established crustal roots, i.e., some Andean segments in the region possess anomalous, isostatically uncompensated topography (Poveda et al. 2015; Yarce et al. 2014). Residual topography in the WC indicates that, regardless of its high elevation (~4000 m), this range is underlain by relatively thin crust (~30 km) so that the high topographic expression of the WC has to be supported by a dynamic process such as a hot anomaly (related to potential upwelling(?)), sustained compression linked to the ongoing collision of the PCB, and/or a combination of both. Ranges in the PCB, particularly the Majé-Baudó Serranía, show even higher topographic residuals between +1 and +2. The CC on the other hand, which displays sierra-like topography with peaks above 5000 m and extensive remnants of planation surfaces such as the AP (~2500 m), is characterized by crustal thicknesses in excess of ~55 km showing segments that are isostatically balanced (residual topography = 0 km) and others, such as the AP and San Lucas Serranía, that are slightly overcompensated (residual topography <0 km). The EC exhibits a rather asymmetric isostatic behavior. Southern segments around the Garzón Massif are overcompensated with residual topography between 0 and 1.5 km, whereas the areas at the Floresta Massif and southern Santander Massif are uncompensated with values of residual topography sometimes in excess of +1.5 km. A condition of "dynamic topography" is proposed for the positive residuals found to the east of Bucaramanga Fault near the Floresta/Santander massifs where steepening of the subducting slab produces a hot anomaly due to upwelling related to asthenospheric return flow (Syracuse et al. 2016; Yarce et al. 2014; Vargas and Mann 2013). Marked contrasts between topography and crustal thickness found between the WC, CC, and EC, i.e., positive residuals (thin crust supporting relatively high topography) for the WC and northern EC while the CC and southern EC exhibit negative residuals (thick crust supporting relatively low topography), can be at least partially justified by mantle downwelling triggered by the Nazca Slab. Such an explanation seems to be consistent with the presence of an asthenospheric wedge beneath the Central and Southern Andes of Colombia (Syracuse et al. 2016; Vargas and Mann 2013), a cordilleran segment where active volcanoes are located. The Sierra Nevada del Cocuy in the EC (~5200 m of elevation) and the isolated SNSM (maximum elevation ~5700 m) also coincide with high positive residuals. The Sierra Nevada del Cocuy in the Floresta/Santender massifs area can be explained by localized asthenopsheric flow (Yarce et al. 2014), but the SNSM, positioned over a zone of flat subduction of the Caribbean Place beneath South America, needs to be explained through alternative mechanisms. An orogenic float model (Oldow et al. 1990; English and Johnston 2004) has been proposed for the SNSM (Piraquive et al. 2017; Cardona et al. 2011; Cediel et al. 2003), where a major through-going, deep-crustal basal detachment allows the whole crustal section to "float" on the underlying lithosphere. This is typical of flat subduction regimes (an in some cases in accretionary margins), where stress coupling of the upper plate with the flat slab may transmit stresses progressively in the direction of subduction producing basement-cored block uplifts in the foreland, such as in the US Cordillera (Bird 1988). Finally, from the LTTC standpoint, orogenic floats and flat subduction are important because, in addition to cooling due to uplift and erosional exhumation, shifts from "normal" to shallow subduction can trigger removal of the hot asthenospheric replacing it by cold subducting slab leading in turn to crustal refrigeration such as documented through LTTC studies in the western USA (Dumitru et al. 1991). Under such scenario cooling ages may not reflect intervals of major regional uplift and erosional exhumation but crustal refrigeration itself. In that sense, it is possible that some of the cooling ages recorded for the SNSM and Guajira are due to crustal refrigeration under flat subduction rather than unroofing due to uplift-driven erosion.

The high geodynamic complexity of the Northern Andes results in an intricate litho-structural mosaic and in an equally complicated stress field (Ramos 2009; Cortés et al. 2005; Cediel et al. 2003; Taboada et al. 2000). Considerable crustal shortening over the Cenozoic affected discrete tectonic blocks, with various degrees of mechanical coupling, so that each block can exhibit a different morphotectonic response in terms of uplift and erosional exhumation. Variations in crustal thickness and subcrustal processes (flat vs. steep subduction, asthenospheric upwelling, etc.) also play a potential role in the distribution of topography and uplift/exhumation. While several authors have attempted to reconstruct the tectonic history of Colombia or to advance a unified taxonomy of litho-structural units (Cediel et al. 2003; Pindell et al. 2005; Restrepo and Toussaint 1988; Etayo-Serna 1986; Bonini et al. 1984; McCourt et al. 1984; Pennington 1981; Irving 1975), the task has proven as complicated as the subject under consideration. In this contribution, we use the litho-structural framework proposed by Cediel et al. (2003) for the Colombian Andes.

11.2.2 Physiographic Configuration

Immediately north of Ecuador, at latitude ~1.5° N, the Andean chain is approximately 60 km wide and forms two mountainous massifs known as the Macizo Colombiano and Nudo de los Pastos. At this location, the Andes range splits into three distinct north- to northeast-trending ranges known as the Western, Central, and Eastern cordilleras (hereafter WC, CC, and EC, respectively). The three ranges fan out to the north such that the entire cordilleran system at ~7° N exceeds 500 km in width. The CC and EC are separated by the Magdalena Valley, whereas the Cauca Valley divides the CC and WC (Figs. 11.2 and 11.3).

All three cordilleras share some general features. For example, all are bounded by major crustal structures, follow a roughly parallel orientation, are separated by major fluvial basins, and attain altitudes of 4000 m or more. In addition, the WC, CC and EC display a combination of both sierra- and plateau-like topography. The presence of high-elevation remnants of erosional surfaces in the Colombian Andes has been reported since the 1960s (Arias 1996; Kroonenberg et al. 1990; Feininger and Botero 1982; Page and James 1981; Feininger et al. 1972; Botero 1963). One of the largest and best preserved is the Antioqueño Plateau (AP) in the CC (Noriega-Londoño 2016; Restrepo-Moreno et al. 2009c; Arias 1996; Soeters 1981), although smaller plateaus can be found in the WC (Páramo de Frontino) and the EC (Páramo de Berlín and Sierra Nevada del Cocuy) (Kroonenberg et al. 1990).

When examined in detail, however, the three cordilleras exhibit certain peculiarities. The WC is built on Cretaceous volcano-sedimentary basement rocks of oceanic affinity (PLOCO) that were intruded by plutonic rocks during the Middle-to-Late Miocene (Gomez et al. 2015b; Rodríguez and Zapata 2012; González and Londoño 2002b, c; Salazar et al. 1991). With an average width of <50 km, the WC is the narrowest among the three ranges and exhibits a more striking sierra-like topography, i.e., highly sinusoidal topographic profile along its axis, with summit elevations in excess of 4000 m (e.g., Cerro Tatamá and Farallones del Citará) and saddles below 500 m (e.g., Bajo Mastrató). Only a few and very localized erosional surface remnants are found on the WC's axis (e.g., Páramo de Frontino in its northern segment). This high-relief sinusoidal topography, coupled with the WC's position adjacent the eastern equatorial Pacific coast, produces complex atmosphere-lithosphere interactions such as the general orographic precipitation effect (Zuluaga and Houze 2015). These interactions not only influence the position of low-level jets (Poveda and Mesa 2000) but also help maintain one of the wettest regions on Earth, the Chocó Biogeographic region (Mittermeier et al. 1999).

With a length of ~1000 km and average width of ~80 km, the CC runs roughly parallel to the WC (Figs. 11.2, 11.3, and 11.6). This range is associated with crystalline basement rocks of continental affinity (PLCMG) that include medium- to high-grade metamorphic rocks (Precambrian(?), Paleozoic, and Mesozoic) and large intrusive masses of Jurassic–Cretaceous age (Gomez et al. 2015b; Restrepo et al. 2009; Villagomez-Díaz 2010; Restrepo and Toussaint 1982; Feininger and Botero 1982; Feininger et al. 1972; Botero 1963). Minor remnants of Cretaceous formations that developed in shallow marine environments can also be found in the CC

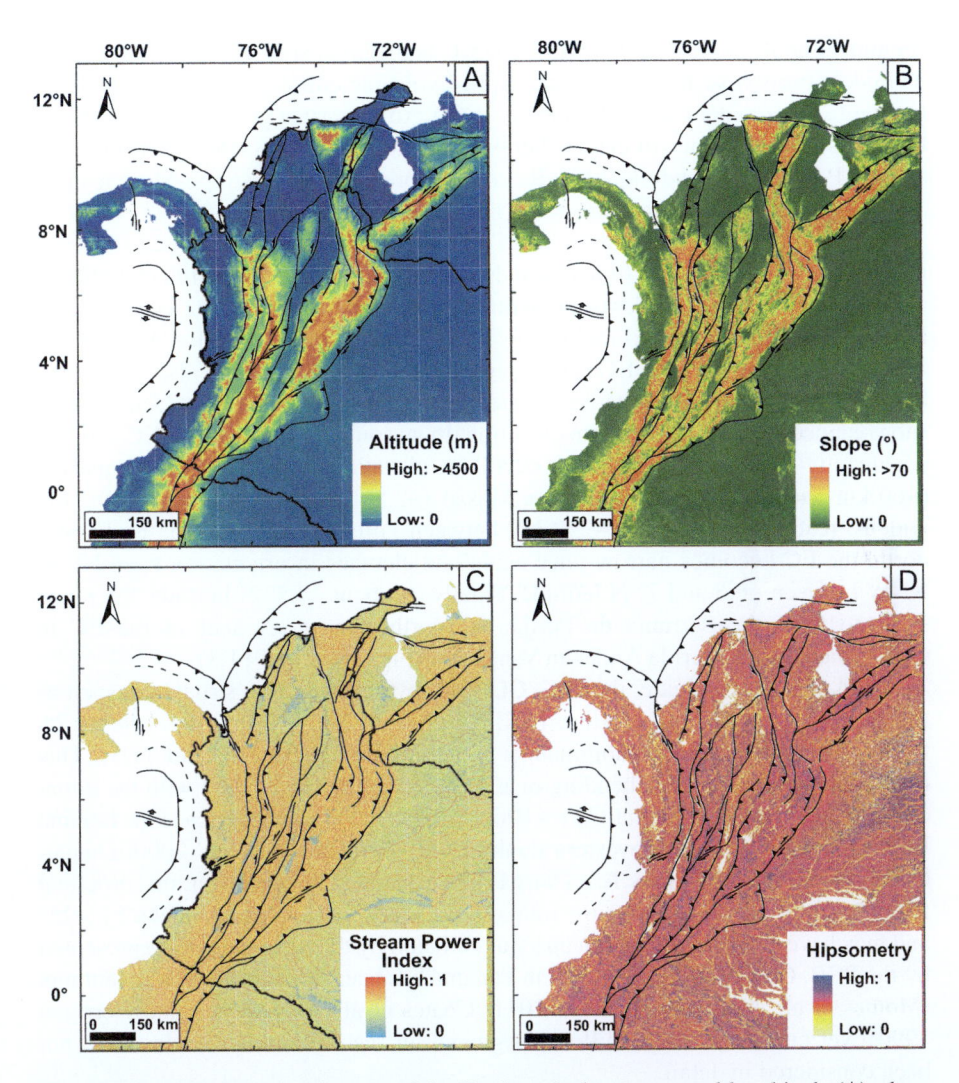

Fig. 11.3 Geomorphological features of the Northern Andes represented by altitude (**A**), slope (**B**), stream power index (**C**) and hypsometry (**D**). These variables are valuable proxies for identifying and quantifying surficial processes related to erosion and sedimentation rates and hence for differentiating relict from active landscapes and evaluating topographic equilibrium (ratio of relative area and relative elevation) in tectonically active settings (Hengl and Reuter 2008). Note the low-relief surfaces (low slope and intermediate/low hypsometry) in the northern portion of the WC and CC, the San Lucas Range (SLR), and the middle EC. These elevated, planar features are used as morphotectonic markers of surface uplift and erosion rates. Enclosed intermontane basins are markedly flat areas corresponding to the Cauca and the Middle/Upper Magdalena Valleys, all of which are related to, and/or bound by, regional faults

(Feininger et al. 1972; Botero 1963). The CC is the highest of the three ranges, and its middle portion sustains active ice-capped volcanic edifices up to 5300 m elevation (e.g., Nevado del Huila, Nevado del Ruiz). An important geomorphic feature of the CC is the prevalence of elevated erosional surfaces in its northern portion, such as the AP (Noriega-Londoño 2016; Restrepo-Moreno 2009; Restrepo-Moreno et al. 2009c Arias 1995; Page and James 1981).

The mountain range known as the EC also constitutes part of the PLCMG, since it comprises old crystalline (igneous and metamorphic) rocks of continental affinity as well as Palaeozoic-to-Cenozoic sedimentary formations (Gomez et al. 2015b; Bayona et al. 2008; Dengo and Covey 1993; Forero 1990; Van der Wiel 1989). At ~1400 km in length, the EC is the longest of the three ranges and also the second highest. The highest summits are located in the Sierra Nevada del Cocuy, where snow-capped peaks reach 5400 m elevation. In its southern reaches, the EC exhibits a generally subdued topography (<2700 m elevation) and is relatively narrow (~60 km wide). North of the Garzón Massif (~2.5° N latitude), however, the EC widens, reaching a width of ~250 km between latitudes 6.0 and 6.5° N. Farther north, the EC becomes narrow once again and changes in orientation from NE to NNW between ~6.5 and 7° N latitude. Finally, at about ~7.5° N latitude, the range splits to form the Serranía de Perijá, the northernmost segment of the EC in Colombia, and the Merida Andes in Venezuela (Figs. 11.2 and 11.3).

At ~4.5–5.5° N latitude, the WC, CC, and several bounding structures, such as the Cauca-Romeral Fault System (CRFS), change their longitudinal orientation from a NNE-SSW to a predominantly N-S direction (Figs. 11.2 and 11.3). This inflection reflects oroclinal bending in a location roughly coincident with the initial collision between the Panama-Chocó Block (PCB) and South America at the Istmina Deformation Zone, on the western flank of the WC (Taboada et al. 2000; Duque-Caro 1990a), and the Caldas Tear (Vargas and Mann 2013). Moreover, this oroclinal feature implies counterclockwise rotations of lithospheric/crustal blocks by ~25° and may have been caused by the onset of collision by the rigid indenter represented by the PCB. Oroclinal development in Panamá has been addressed by other authors (Montes et al. 2012a; Farris et al. 2011; Coates et al. 2004; Mann and Corrigan 1990), yet curvatures of cordilleran segments in the Colombian Andes have not been considered in detail.

In addition to the main cordilleras, Colombia contains several "isolated" topographic structures, such as the Sierra Nevada de Santa Marta (SNSM). Crowned by Pico Cristóbal Colón (~5800 m), the SNSM is the highest and steepest isolated coastal massif in the world.[7] Colombia's other isolated structures include the Darién and Baudó serranías, in the Chocó Biogeographic region, and the Serranía de la Macarena, which projects into the Amazonian plains, as well as minor topographic features such as the Serranía de Chiribiquete (Amazonas region), the Serranía de San Jacinto (Caribbean Plains), and the Macuira and Jarará serranías (Guajira Peninsula) (Figs. 11.2 and 11.3).

---

[7] Worldwide, the SNSM (5750 m) ranks second in elevation, behind the much broader and longer Saint Elias Range (5959 m) in the USA (Alaska) and Canada.

Along with this intricate mountainous mosaic, Colombia is characterized by a broad arrangement of fluvial networks draining into the Caribbean (e.g., Magdalena-Cauca, Atrato, and Sinú rivers), the Pacific (e.g., San Juan River), and the Atlantic (e.g., Orinoco and Amazonas rivers). Cenozoic topographic development of the Northern Andes has been central to the evolution of these rivers by driving profound changes in the distribution and extent of the fluvial network (Anderson et al. 2016; Pardo-Trujillo et al. 2015; Reyes-Harker et al. 2015; Hoorn et al. 2010; Silva et al. 2008; Suter et al. 2008; Gomez et al. 2005a, b; Gomez 2001; Villamil 1999; Hoorn et al. 1995). Yet, rivers are not simply passive "observers" of landscape change but major drivers of morphotectonic activity (Summerfield and Hulton 1994). Representing less than ~0.005% of the global water budget, rivers are nevertheless the most powerful erosional force on Earth (Syvitski and Milliman 2007; Willett et al. 2006; Milliman 2001), and it is estimated that the global fluvial network moves ~24 Gt of particulate material to the oceans and interior basins every year (Hooke 2000). On geological time scales, rivers are responsible for the accumulation of vast volumes of detrital materials in sedimentary formations (Wilkinson et al. 2009). In the Colombian Andes, rivers have incised the crust to produce deep valleys and have transported considerable volumes of detrital material to be deposited on floodplains, alluvial cones, deltas, etc. The variety and thickness of Cenozoic sediment sequences related genetically to Andean rivers (Cediel et al. 2011; Horton et al. 2010a, b; Bayona et al. 2008; Barrero et al. 2007; Van der Hammen 1961) are testimony to the efficacy of rivers in producing, transporting, and depositing sediments in tropical mountainous regions and thus in shaping tropical landscapes (Thomas 1994). Recently, however, humans have become an even greater erosional force, responsible for mobilizing staggering amounts (~35 Gt/yr) of detrital material across the landscape and outcompeting rivers and glaciers as erosive agents (Syvitski and Milliman 2007; Restrepo and Syvitski 2006; Walling 2006; Wilkinson 2005; Hooke 2000). This human-mediated process is considered a hallmark of the Anthropocene (Syvitski 2012; Restrepo-Moreno and Restrepo-Múnera 2007), and the Colombian Andes are no exception to modern, hyper-erosive regimes caused by anthropogenic activity. The Magdalena River[8] is a major contributor of detrital material to the Caribbean (Restrepo and Syvitski 2006; Walling 2006; Restrepo and Kjerfve 2000; Milliman and Meade 1983), with a sediment yield ranking among the highest worldwide (Restrepo et al. 2006) and translating to erosion rates in excess of 10 km/Ma for some of its tributaries.

High erosion rates (whether natural or anthropogenic) in tropical mountainous setting are not unexpected. Characterized by high precipitation, warm and humid climates, steep topography, and tectonically active terrain, denudation in the tropical Andes is achieved through external processes dominated by profound weathering, mass wasting, and erosion by water (Montgomery et al. 2001; Goudie 1995; Thomas 1994; Stallard 1988). The energy input required for denudation (i.e., erosion) comes from both internal (e.g., potential energy via surface uplift and increased

---

[8] For details on erosion/fluvial process in the Magdalena River, see Chap. 8, Contribution No. 17, this volume.

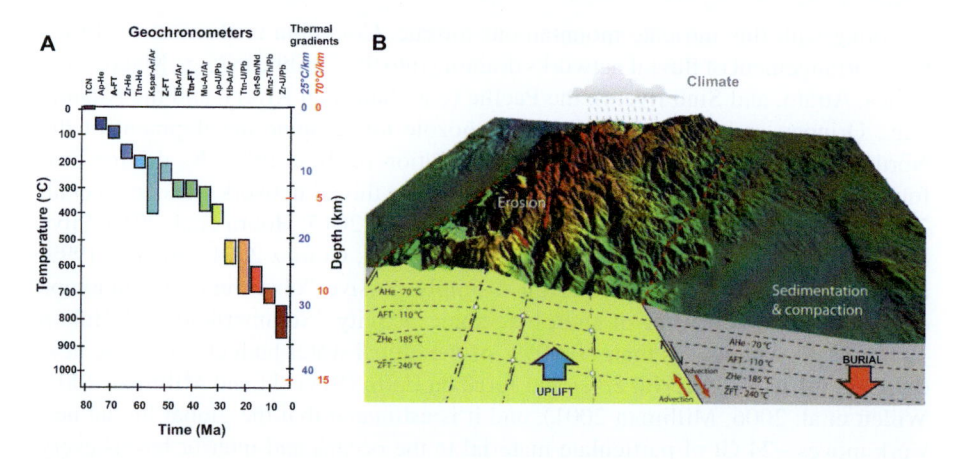

Fig. 11.4 LTTC systems and the conceptual framework. (**A**) Thermal boundaries for the most common geochronometry systems. Below ~300 °C, these tools are collectively referred as LTTC. We focus on reporting data for LTTC systems with closure temperatures below ~250 °C, such as ZFT. Terrestrial cosmogenic nuclides (TCN) are not strictly a thermochronometer but it is included as it extends the range of morphotectonic analyses to the surface of the earth allowing to directly assessing exposure dating and erosion rates at various spatiotemporal scales. Note that some minerals (e.g., apatite and zircon) are particularly useful thermochronometers as they record information in several systems, including U/Pb geochronology and FT + He LTTC, making them invaluable in morphotectonic studies both in cordilleran massifs and sedimentary basins (Modified from Peyton and Carrapa 2013). (**B**) The spatial distribution of LTTC tools relative to topography, climate, and the thermal structure of the upper crust (Courtesy of Dr. Ulrich. A. Glasmacher). Red arrow indicates heating-burial in sedimentary basins. Blue arrow marks cooling-uplift/exhumation focused on cordilleran massifs

topographic relief, earthquakes, etc.) and external (e.g., climatic shifts, variations in precipitation fluvial incision, etc.) processes. Vertical incision by rivers is a fundamental feature of Andean regions, where steep slopes and valleys with local relief in excess of 4–5 km feed the geomorphic "machine" facilitating both mass wasting and streambed erosion (Aalto et al. 2006; Montgomery et al. 2001; Restrepo-Moreno 2009, Restrepo-Moreno et al. 2009c). Topographic differences between disparate morphotectonic sectors of the Andes are therefore important as they exert direct controls on erosional exhumation and hence on lithosphere-atmosphere-hydrosphere feedbacks (Ehlers and Poulsen 2009; Hartley 2003; Insel et al. 2010b; Kennan 2000; Lamb and Davis 2003). As discussed in later sections, the erosive power of different litho-structural and physiographic domains can be approximated using slope and hypsometry data and local relief maps (Pinet and Souriau 1988) (Figs. 11.3 and 11.4). Furthermore, a simple functional relationship defining denudation velocity as directly dependent on mean local relief (Ahnert 1970) can be applied to sub-domains in the Colombian Andes. Although this relationship was developed initially for mid-latitude basins in tectonically inactive regions, it provides a first-order approximation of the magnitude of denudation rates (Montgomery 2003; Kirchner et al. 2001; Gunnell 1998; Summerfield and Hulton 1994), and we apply it here in that sense.

11.2.3 Inter-Andean and Marginal Basins

The principal sedimentary basins of the Northern Andes are delimited by the major orogenic topography (Cediel et al. 2009, 2011; Barrero et al. 2007). West of the WC, some of the sedimentary sequences within the Tumaco and Chocó Basins are associated with the geologic evolution of the PCB and the WC itself (Pardo-Trujillo et al. 2015; Cediel et al. 2009). Between the WC and CC, the Cauca-Patía and Amagá Basins occupy a deep fluvial trough that today is dominated by the Cauca River (Marín-Cerón et al. 2015; Silva et al. 2008; Barrero et al. 2007; Sierra et al. 2004; Correa and Silva 1999), while the CC and EC enclose the Upper and middle Magdalena River Basins (Bayona et al. 2008; Gomez 2001; Reyes-Harker et al. 2015). Closer to the Caribbean Sea, other important basins include the Lower Magdalena, Sinú-San Jacinto, Cesar-Rancheria, Guajira, and Catatumbo Basins (Garzon-Varon 2012; Bayona et al. 2011; Cardona et al. 2010; Montes et al. 2010; Barrero et al. 2007), while to the east, the vast plains of the Amazonas and Orinoco rivers contain the major Caguán-Putumayo, Vaupés-Amazonas, and Eastern Llanos basins (Cediel et al. 2009, 2011; Barrero et al. 2007) (Fig. 11.2). Most of these basins have received continental and/or marine sediments since at least the Mesozoic and are a clear testimony to the intensity of sediment production, routing, and deposition in the Northern Andes during the Cenozoic orogenetic phases. Many of these basins were studied in recent years to address their thermal evolution and the provenance of detrital materials (Anderson et al. 2016; Reyes-Harker et al. 2015; Horton et al. 2010a, b; Barbosa et al. 2013; Bayona et al. 2008; Marín-Cerón et al. 2015; Restrepo-Moreno et al. 2013b), though we note such topics are not covered in this contribution.

Given the lithologic, structural, morphological, and climatic differences across the three cordilleras and interspersed sedimentary basins, the Northern Andes of Colombia constitute an ideal natural laboratory to evaluate the interplay and feedback mechanisms among tectonics, surface processes, and climate (Figs. 11.1, 11.2, and 11.3). Important issues in this realm, such as the magnitude, timing, and spatial patterns of erosion, exhumation, topographic development, orogen kinematics, and/or metamorphism and magmatism, can be addressed by LTTC.

11.3 LTTC in Colombia: Some Morphotectonic Implications

In this section we discuss LTTC data available for Colombia. The information is presented for each physiographic entity (e.g., WC, CC, EC, SNSM), with a focus on those provinces with crystalline basement rocks, such as the Antioqueño Plateau, the Garzón and Santander Massifs, the SNSM, and the WC's plutons. A brief summary of LTTC datasets from the Panamanian Isthmus is also provided because such information is critical for understanding the PCB's collisional history and its morphotectonic effects on both the PCB and Colombian mountain ranges. We begin by briefly introducing the subject of LTTC and its application to tectonic geomorphology.

11.3.1 Tectonic Geomorphology and Low-Temperature Thermochronology

The last two decades have witnessed the emergence and rapid establishment of a new and multidisciplinary subfield in geology known as tectonic geomorphology (Burbank and Anderson 2011; Bishop 2007). As the appealing name suggests, this new subfield has managed to bridge two formerly distinct research areas, thereby bringing about a revolution in geological thinking and facilitating studies on the interplay between internal and external processes (Burbank and Anderson 2011; Ehlers and Farley 2003; Montgomery et al. 2001; Pinter and Brandon 1997; Reiners and Shuster 2009; Willett et al. 2006). For some scientists, this "revolution" was triggered by changes in the foundations of tectonic geomorphology and the advent of several new radiometric and isotopic analytical techniques, including (U-Th)/He, fission track, and cosmogenic isotope analyses (Summerfield 2005). In addition, refinement of in situ LA-ICP-MS techniques applied to zircons for U/Pb dating and Hf isotope analysis (Gerdes and Zeh 2006), as well as advances in apatite and zircon single-grain duble dating (Reiners et al. 2005, Shen et al. 2012) has improved our ability to accurately determine provenance and hence identify paleo-fluvial systems and geomorphically active cordilleran massifs (Carrapa 2010; Dickinson and Gehrels 2008; Gehrels 2012).

These analytical techniques, combined with field investigations, enable a better understanding of such surface and near-surface processes as burial, erosion, exhumation, sediment routing, surface exposure age, etc., which are crucial for linking tectonics and geomorphology (Burbank and Anderson 2011; Carrapa 2010; Reiners and Shuster 2009; Bishop 2007; Reiners and Ehlers 2005a, b; Summerfield 2005; Shuster et al. 2005b; Bierman 2004; Bernet and Spiegel 2004; Ehlers and Farley 2003; Farley 2002; Sugden et al. 1997; Brown et al. 1994). Complementing these morphotectonic LTTC-based studies, increased computational power and software/code development has facilitated the construction of more robust quantitative approaches to landscape evolution and orogenesis via thermotectonic modeling (Vermeesch and Tian 2014; Ketcham 2008; Willgoose 2005; Braun 2003; Gallagher et al. 2005a).

In Colombia alone, the last decade was marked by an increase in LTTC studies of both cordilleran massifs and sedimentary basins (see compilation of radiometric age by Gomez et al. 2015a, b), so that a more detailed picture of the nature of the Andean orogeny is beginning to emerge. Two new concepts dominate: specifically, the multiphase nature of the mountain-building processes and its asynchronous character. This should not be taken to mean that the surface uplift and erosional exhumation history of the region was "chaotic". Rather, the integrated analyses of several LTTC studies reveal that certain periods of enhanced uplift, topographic build-up, and erosional exhumation can be recognized. Such paroxysms can be associated with major geodynamic events at the regional to continental scale. Because this chapter will repeatedly refer to LTTC information, we briefly summarize the basic conceptual aspects of LTTC.

11.3.1.1 Processes in the Upper Crust and LTTC Approaches

The upper crust can be conceptualized as a thermal field in which isotherms tend to follow the topographic configuration of the terrain. At greater depths, however, the isotherms become flattened and eventually decouple from surface topography (Reiners and Shuster 2009; Stüwe and Hintermueller 2000; Ring et al. 1999). As a parcel of rock moves upward across the crust's thermal field (i.e., during exhumation), it cools, causing the various radiometric systems to become "closed" and thus recording when the rock parcel passed through the corresponding closure isotherm. A traditional approach to reconstructing a rock's thermal history is to determine radiometric ages for multiple systems with different closure temperatures (in the range of 70 °C to ~800 °C for available geochronometers, see Fig. 11.4). Given the relationship between temperature and depth that is implicit in the geothermal gradient, the cooling history can be translated into a record of exhumation (Restrepo-Moreno 2009; Saenz 2003), provided that certain assumptions (e.g., estimates of the geothermal gradient, absence of transient thermal episodes, etc.) are applicable.

Another approach, which is especially suitable for upper crustal sections, is to use an age-elevation relationship (hereafter AER) for one or more LTTC systems (Burbank and Anderson 2011; Brown et al. 1994; Fitzgerald 1994). To apply this method, LTTC data are produced from samples collected along topographic profiles (or boreholes) covering a large range of elevations (or depths) (Restrepo-Moreno 2009; Restrepo-Moreno et al. 2009c; Gallagher et al. 2005b; Osadetz et al. 2002; House et al. 2002; Crowhurst et al. 2002; Stockli et al. 2000; Foster and Gleadow 1996; Fitzgerald 1994; Foster et al. 1994; Brown et al. 1994). The AER commonly exhibits a positive correlation (Fig. 11.5) because samples from higher positions in the topographic profile are the first to pass through the LTTC system's closure temperature range and record older apparent cooling ages. Conversely, samples in the lower-elevation profile register young cooling ages (e.g., Fitzgerald et al. 1994). The detailed features of AER, such as inflection points and/or changes in regression slope (Fig. 11.5), can help pinpoint the timing and magnitude of specific thermotectonic events (Peyton and Carrapa 2013; Burbank and Anderson 2011; Gallagher et al. 2005b; Brown et al. 1994).

A third and highly complementary approach identifies the most likely time-temperature path(s) of a rock sample using computer models fed with a suite of thermochronologic, geochronologic, paleothermometric, spatial, and dimensional information obtained from individual samples. For instance, AFT, AHe, ZHe, ZFT, and U/Pb datasets can be routinely modeled using HeFTy® (Ketcham 2003, 2005, 2008) and QTQt (Vermeesch and Tian 2014; Gallagher 2013) computer codes. HeFTy® (a subsequent version of the AFTSolve® code) uses search algorithms to quantify the range of statistically acceptable and good-fit thermal histories for a sample adhering to user-defined constraints, and then matches the measured LTTC, and in some cases vitrinite reflectance, data (Ketcham 2005). HeFTy® and AFTSolve® are applied to individual rock samples and use a "frequentist" approach to assess the goodness of fit (GOF) between input data and thermal model predictions (Ketcham 2008; Vermeesch and Tian 2014). In contrast, the more recent code QTQt® can inte-

Age-Elevation Relationships (AER)

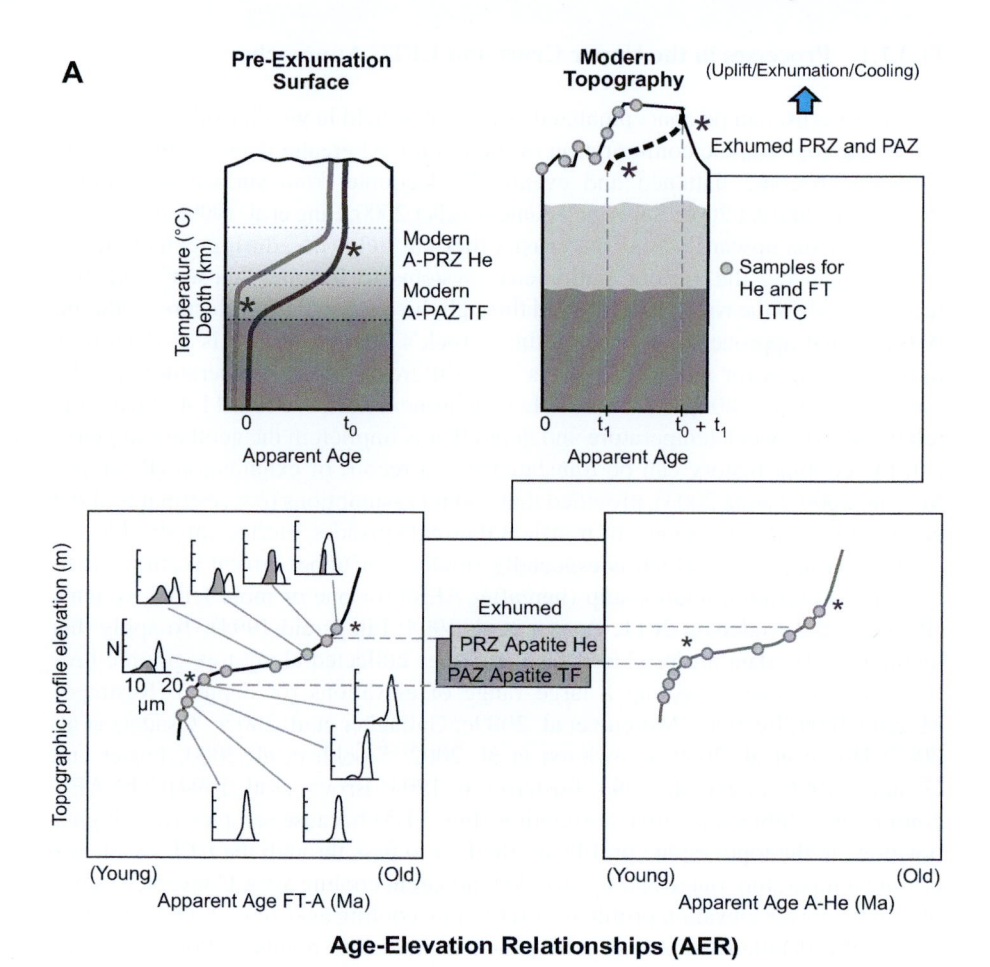

Fig. 11.5 (**A**) The concepts of the AFT-PAZ and AHe-PRZ and their use in LTTC studies along vertical profiles. Note the positive correlation in AER, the overlapping nature of both sensitivity thermal zones (AFT and AHe), and the distribution of MTL for the different segments of the AER. The exhumed PRZ/PAZ permits identification of the timing and magnitude of cooling events, in this case associated with uplift and erosional exhumation. The asterisk (*) marks inflection points in the age-elevation curve. The light-gray shading delimits the crustal regions over which He is partially retained (light gray dotted box) and tracks are partially annealed (dark gray dotted box). Dark-gray shading represents the zone of complete FT annealing and/or He diffusion. White band indicates the zone of track stability and He retention over geologic time. At the time of the pre-exhumation condition, rocks above the PAZ and PRZ possess apatite grains that register the age of first exhumation event. Subsequent exhumation pulses (t_1) are also recorded by AERs. Apatite grains in rocks below the bottom of the PAZ and PRZ record an apparent age of 0 Ma. In the modern topography situation, the formerly developed PAZ/PRZ appears as an exhumed or fossilized PAZ/PRZ that can be sampled along the valley walls, scarps, etc. In the bottom of the figure, the AER for AFT (left) and AHe (right) clearly display the positive correlation (low-young vs. high-old) expected from a vertical profile with negligible perturbation from thermal events or faulting (Adapted from Brown et al. 1994; Burbank and Anderson 2001).

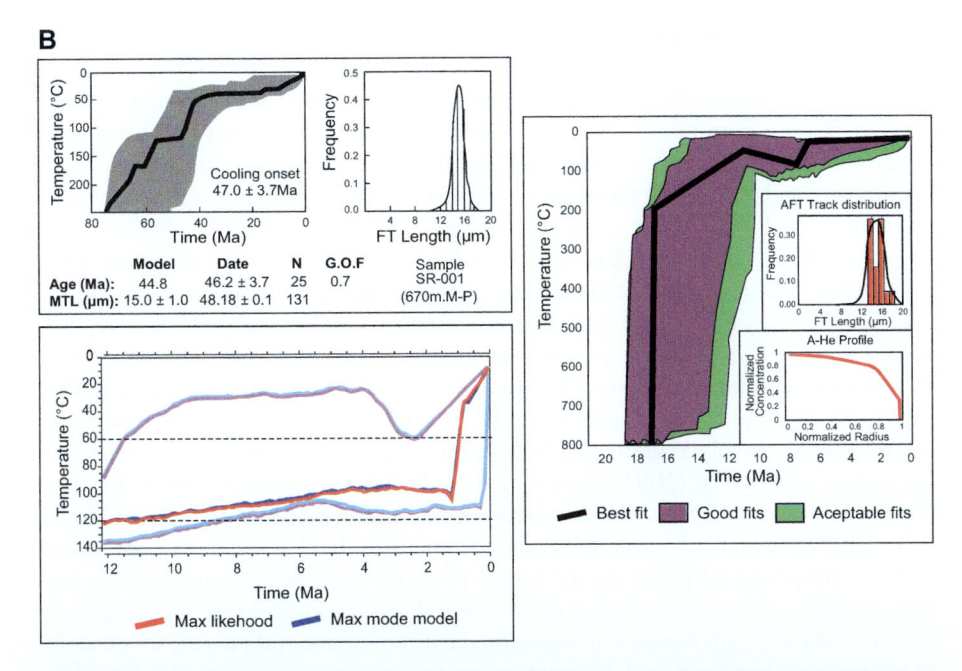

Fig. 11.5 (continued) (**B**) Examples of typical time (Ma) vs. temperature (°C) report plots (i.e., time-temperature histories) for Colombian Andean samples derived from thermal modeling using AFTSolve® (upper left, for the Antioqueño Batholith in the Central Cordillera, Restrepo-Moreno 2009), HeFTy® (right, for the Tatamá Intrusive in the Western Cordillera, Restrepo-Moreno et al. 2015), and Qt-QT® (lower left, for the Santander Massif, Amaya-Ferreira 2016; Amaya et al. 2017). In general, segments characterized by steep slopes in the time-temperature paths mark periods of rapid cooling, often attributable to erosional exhumation. AFTSolve® focuses mainly on AFT and ZFT ages and on apatite track-length distributions. Vitrinite reflectance data on maximum paleo-temperature can be used as a modeling constraint for detrital samples. Additionally, HeFTy® incorporates AHe ages and diffusion parameters. Other thermochronometers can be used to establish constraint boxes to guide the modeling exercise. Both AFTSolve® and HeFTy® work on individual samples. Qt-QT® integrates multiple parameters such as AFT, ZFT, AHe, ZHe, and vitrinite reflectance, with the possibility of including Argon data, but it allows two different approaches: multi-sample throughout topographic profiles or single sample at specific points on the landscape. For a review and comparison between HeFTy® and Qt-QT, see Vermeesch and Tian (2014)

grate several samples across vertical profiles and uses a Bayesian "Markov Chain Monte Carlo" algorithm to produce "most likely" thermal histories (Vermeesch and Tian 2014; Gallagher 2013), although the latest versions of HeFTy® also allow combining multiple samples to model common cooling histories. All of the codes mentioned above involve forward and inverse modeling functionalities. The forward simulation allows predictions of age distribution (output) for any given thermal history (input), whereas the inverse portion finds the thermal history that best matches the input data (Vermeesch and Tian 2014). Several investigations in the Colombian Andes have used either one or a combination of these approaches on multiple LTTC systems (AHe, AFT, ZHe, and ZFT) to generate valuable information on episodicity of erosional exhumation. We present several examples and will use them to pinpoint the main phases of morphotectonic upheaval in the Colombian Andes.

11.3.1.2 Basic Concepts and Applications in LTTC

Radiometric systems that are sufficiently sensitive at low-temperatures (i.e., 40–250 °C) are commonly used to constrain exhumation-related processes like normal faulting, erosion, sediment production-routing, burial, and topography formation. Such dating methods are collectively called low-temperature thermochronometers (Peyton and Carrapa 2013; Reiners and Shuster 2009; Reiners and Brandon 2006; Reiners and Ehlers 2005a, b; Spotila 2005). Two important low-temperature thermochronometric systems are (U-Th)/He and fission-track dating. The (U-Th)/He technique is based on measurements of radiogenic helium (^{4}He) that accumulates in the mineral structure via α-decay of ^{238}U, ^{235}U, ^{232}Th, and ^{147}Sm. Helium concentrations are determined by heating and degassing mineral grains, followed by quadrupole mass spectrometry. Parent isotope concentrations are measured from the dissolved grains via ICP-MS. For detailed reviews of the technique, see Ehlers and Farley (2003), Farley (2002), Farley and Stockli (2002), Reiners (2005), Reiners and Ehlers (2005b), Reiners et al. (2004), and Wolf et al. (1998).

Fission-track analysis exploits the linear structures (tracks) of microscopic damage that are produced in the crystal lattice of certain minerals (mainly apatite and zircon) by the spontaneous fission of ^{238}U. While the thickness of the original track is ~1 μm, track length varies depending on the different mineral phases, with values of approximately 16 μm and 12 μm for apatite and zircon, respectively (Gleadow et al. 2002; Brown et al. 1994). Fission-track analysis involves both grain dating and statistical treatment of track lengths (Galbraith 2005; Gleadow et al. 1986, 2002; Gallagher et al. 1998; Brown et al. 1994). First, fission-track ages are determined from the relationship between spontaneous fission-track densities (obtained by direct track counting under a high-resolution optical microscope) and concentrations of ^{238}U, the latter being determined either through the external detector method (EDM) (Donelick et al. 2005; Galbraith 2005; Hurford and Green 1983; Green 1981; Lisker et al. 2009) or by LA-ICP-MS (Chew and Donelick 2012; Hasebe et al. 2004). The LA-ICP-MS approach reduces the problems of human error associated with track counting and also permits generation of U-Pb ages in apatite (closure temperature ~ 400 °C) as a further constraint on cooling history and thermotectonic modeling (Chew and Donelick 2012; Shen et al. 2012). Finally, statistical analysis of fission-track length distributions is carried out by digital measurement and counting of confined tracks under the optical microscope. For a detailed description of the fission-track dating method, refer to (Brown et al. 1994; Chew and Donelick 2012; Donelick et al. 2005; Gallagher et al. 1998; Gleadow 1981; Tagami and O'Sullivan 2005).

The low-temperature radiometric techniques described above are now widely applied to apatite and zircon. The (U-Th)/He and fission-track analyses in apatite (AHe and AFT, respectively) are of particular interest in the field of tectonic geomorphology, and more specifically in orogenesis, due to the low closure temperatures of ~70–120 °C for AFT and ~40–80 °C for AHe. These temperatures are lower than those of any other available thermochronometers (Reiners and Shuster 2009; Donelick et al. 2005; Ehlers and Farley 2003; Farley 2002; Gallagher et al.

1998; Reiners and Ehlers 2005a, b) and, depending on geothermal gradient, correspond to the upper ~1–4 km of the crust (Figs. 11.4 and 11.5). Other commonly used low-temperature thermochronometers include (U-Th)/He and fission-track dating in zircon (hereafter ZHe and ZFT). With respective closure temperatures of ~180 and 240 °C (Bernet 2009; Reiners 2005; Tagami and O'Sullivan 2005; Reiners et al. 2004; Zaun and Wagner 1985), these techniques provide information on a complementary thermal range for assessing processes in the upper crust. Together, these methods help elucidate a wide range of processes that are intimately related to orogenesis, such as exhumation histories in the upper crust (whether via erosion or normal faulting), topographic development, sediment production and accumulation, and burial in sedimentary basins.

An important aspect of LTTC analyses applied to reconstructing morphotectonic events relates to partial resetting of both helium and FT dating. On geological timescales and at temperatures above ~80 °C, 100% of helium is lost from apatite grains of typical size via diffusion, resulting in an age of 0 Ma. In contrast, almost all helium is retained in the crystal lattice at temperatures below ~40 °C (Farley 2000, 2002; Wolf et al. 1998), thereby recording the timing of cooling events. In the temperature range between these two extremes, known as the apatite partial retention zone (A-PRZ), only partial fractions of helium are retained. For zircon, the Zr-PRZ is ~165–190 °C (Reiners 2005; Reiners et al. 2004; Ehlers and Farley 2003; Farley 2000, 2002; etc.). Similarly, fission tracks in apatite are completely annealed (i.e., self-repaired) above ~120 °C and fully preserved below ~60 °C, with the temperature interval between the two termed the apatite partial annealing zone (A-PAZ) (Brown et al. 1994; Donelick et al. 2005; Gleadow and Fitzgerald 1987; Ketcham et al. 1999; Laslett et al. 1987). The Zr-PAZ is ~200–240 °C (Reiners and Brandon 2006; Tagami and O'Sullivan 2005). It is important to note that the true thermal extents of the PRZ and PAZ are dependent on exhumation-cooling rates (Reiners and Brandon 2006), particularly when exhumation rates exceed ca. 1–5 km/Ma.

The position within the crust of the exhumed or current PRZ and PAZ can be identified via age-elevation relationships (AER), either from topographic profiles or from borehole data (Brown et al. 1994; Crowhurst et al. 2002; Gallagher et al. 1998; House et al. 2002; Osadetz et al. 2002; Stockli et al. 2000; Spotila 2005). In most cases, the AER exhibits a strong positive correlation (Fig. 11.5), such that samples that are positioned high in the topographic profile possess older ages. Conversely, samples in the lower portion of the elevation profile register younger ages. Using the AER method, both the PRZ and PAZ are characterized by significant variations in apparent age, whereas ages derived from crustal intervals above or below these zones are less variable, with some showing no significant age variation. The PAZ can be further identified by variations in the distribution of mean track length (MTL) along the AER profile. For example, in the case of AFT, samples within the PAZ show bimodal skewed distributions with high sigma values, whereas segments both above and below the PAZ yield unimodal, low sigma, often un-skewed distributions centered around 15 µm (Gallagher et al. 1998; Brown et al. 1994) (Fig. 11.5A, B). A similar situation occurs for ZFT but with the expected shorter MTL (Tagami and O'Sullivan 2005). For those sections where the PRZ and/or PAZ are well defined,

AERs can provide clues to the timing and rates of exhumation (Fig. 11.5) (Restrepo-Moreno et al. 2009c; Stockli et al. 2000; Foster et al. 1994). Once the location of the PRZ or PAZ in an exhumed rock mass is identified, the timing and magnitude of exhumation can be reconstructed (Fig. 11.5). Further details on LTTC in apatite and zircon can be found in several review papers (Peyton and Carrapa 2013; Reiners and Brandon 2006; Donelick et al. 2005; Reiners and Ehlers 2005a; Farley 2002; Gallagher et al. 1998; Brown et al. 1994).

Before we delve into morphotectonic interpretations, it is important to note that LTTC data compiled over a broad region should be treated carefully because many of those individual ages were generated for relatively large areas without the systematic sampling strategies employed by the AER approach. In doing so, some of those ages can be derived from the paleo-PRZ (or PAZ), whereas others come from above or below the paleo-PRZ. Therefore, the compiled age data cannot be interpreted in the context of a simple AER approach to truly represent a single thermal event, such as those related to erosional cooling. For example, Hoorn et al. (2010, Supporting Online Materials) report considerable variability among AFT ages from the Andes and correctly attribute such variability to a "nonhomogeneous distribution of the samples" or to different degrees of incision. Segmentation of the Colombian Andes into several discrete crustal bocks with potentially different tectonothermal histories adds to the problem of LTTC data interpretation over broader areas. These are among some of the inherent limitations of AFT and AHe LTTC (Brown and Summerfield 1997; England and Molnar 1990; Flowers and Kelley 2011; Moore and England 2001). On the other hand, the high-resolution, vertical profile approach permits constraining specific morphotectonic events. For example, AERs obtained along a vertical transect for the AP in the CC (Restrepo-Moreno et al. 2009c) suggest that uplift-driven exhumation occurred during the periods ~48–45 and ~24–22 Ma. These findings are supported by HeFTy® and AFTSolve® modeling (Restrepo-Moreno 2009) and indicate that the AP underwent two distinct pulses of uplift, the first during the Eocene and the second near the Oligo-Miocene transition. The assortment of ages in the 45–25 Ma range marks the PAZ and PRZ paleo-thermal structure, suggesting a protracted period of tectonic quiescence during which erosional surfaces developed without significant orogenic construction.

In this contribution, we show that despite the apparently diachronous nature of surface uplift/exhumation between discrete litho-structural domains, major morphotectonic pulses are identifiable in LTTC datasets for the Northern Andes of Colombia. Such pulses fall mostly within five intervals: Cretaceous-Paleocene (~69–63 Ma), Eocene (46–42 Ma), Oligocene (25–22 Ma), Miocene (16–14 Ma, 10–7 Ma), and Pliocene (3 Ma) to present. Interestingly, a similar set of intervals was proposed previously for the Colombian Andes on the basis of sediment-stratigraphic analyses, i.e., Laramic, Pre-Andina, Proto-Andina, and Eu-Andina orogenic phases (Van der Hammen 1961).[9] Nevertheless, van der Hammen's approach was

---

[9] Although the nomenclature of orogenetic phases by van der Hammen refers to orogenetic events of continental scale (in the case of the Laramic) and to the gradual consolidation of the Andean topography, we opt to maintain the same terminology in order to avoid confusion. The chronology of some of these phases in van der Hammen (1961) correspond to the Eocene (Incaic) and Miocene (Quechua) in Peru (Mégard 1984).

unable to pinpoint the exact locations and intensities of surface uplift/exhumation for the various segments of the Colombian orogen; a situation that is now resolved through the systematic use of LTTC. We provide geodynamic and morphoclimatic scenarios to explain morphotectonic phases over a broad range of spatiotemporal scales and attempt to reconstruct paleogeographic domains, particularly in terms of the development of cordilleran systems and associated inter-Andean fluvial networks/basins.

11.3.2 LTTC and Morphotectonic Reconstructions in the Colombian Andes

On a global scale, the Cenozoic era was crucial from a morphotectonic point of view. Many of the planet's present-day physiographic features were consolidated during this period, thereby influencing climatic and biotic patterns worldwide (Grujic et al. 2006; Hartley 2003; Hoorn et al. 2010; Molnar et al. 1993; Molnar and England 1990; Poulsen et al. 2010; Raymo 1994; Raymo and Ruddiman 1992; Ruddiman 2013; Summerfield 2000). The two major mountain ranges in the world, the Himalayas and the Andes, appear to be primarily tertiary features. The Himalaya and the Tibetan Plateau, for instance, developed over multiple phases of surface uplift[10] during the Eocene and Miocene, following the Eocene onset of Indian-Eurasian collision (Amano and Taira 1992; Burbank et al. 1993; Copeland et al. 1990; Harrison et al. 1993; Van der Beek et al. 2006; Zeitler et al. 2001). Similarly, the Andes, which have undergone intermittent upheavals since the Late Cretaceous (Amaya et al. 2017; Piraquive et al. 2017; Cardona et al. 2011; Hoorn et al. 2010; Villagomez-Díaz 2010; Restrepo-Moreno 2009; Restrepo-Moreno et al. 2009b, c; Saenz 2003; Van der Hammen 1961), potentially achieved their majestic appearance during the Neogene (Hoorn et al. 2010; Insel et al. 2010a; Ehlers and Poulsen 2009; Ollier 2006; Hartley 2003; Gregory-Wodzicki 2000; Kennan 2000; Kennan et al. 1997; Kroonenberg et al. 1990; Benjamin et al. 1987). The massive volumes of detrital materials accumulated in inter-Andean and marginal basins are evidence of the dynamic nature of orogenesis/mountain building as a fundamental trigger of erosional exhumation in the Colombian Andes (Pardo et al. 2012b; Cediel et al. 2011; Sierra and Marin-Ceron 2011; Horton et al. 2010a, b;

---

[10] To avoid confusion arising from the use of the terms uplift, exhumation, denudation, etc., please refer to the LTTC section of this contribution. We emphasize that LTTC techniques can be used through several approaches (vertical profiles, sample multiple dating, etc.) to constrain timing and rate(s) of cooling associated with erosional exhumation. When the term "uplift" is used to discuss LTTC datasets in various litho-structural domains of the Colombian Andes, we assume that "surface uplift" (i.e., topographic buildup) is the main trigger of erosional exhumation. In that regard, LTTC data is taken to yield only bulk erosion rates, such as the movement of a rock parcel toward the eroding topographic surface (i.e., exhumation). Therefore, LTTC does not constrain surface uplift with respect to a fixed frame of reference such as sea level. For a detailed discussion on these issues, see England and Molnar (1990), Brown and Summerfield (1997), and Reniers and Brandon (2006).

Bayona et al. 2008; Barrero et al. 2007; Toro et al. 2004; Van Houten and Travis 1968; Van der Hammen 1961).

Cenozoic orogenesis in the Andean range has been recognized from indirect evidence since the early twentieth century. Reviewing paleobotanical data from Bolivia (Berry 1919), and delving into multiple lines of evidence for several locales in the Andes of Bolivia, Peru, Ecuador, and Colombia (Steinmann 1922), changes in elevation on the order of several kilometers were proposed for the Andes. The topic was revisited in Peru by Steinmann (Steinman 1929; Steinmann et al. 1930), while, in Colombia, initial ideas on topographic buildup were tacitly addressed by Grosse (1926) to explain Cenozoic sediment production and accumulation in the Proto-Cauca Valley, west of the CC. Subsequently, the matter was grasped in a more direct fashion by Oppenheim in the 1940s when he referred to the "diastrophical cycle" (Oppenheim 1941). Almost two decades later, a review of sediment-stratigraphic data from throughout Colombia's Cenozoic basins enabled the identification of major discontinuities followed by basal conglomerates deposited over relatively short periods (Van der Hammen 1961). Prior to the extensive use of LTTC in Colombia, tectonic paroxysms were reconstructed from these coarse facies (Van der Hammen 1961) and then used to propose a chronological framework for orogenetic pulses in the Colombian Andes. Four major paroxysms are known as the Laramic, Pre-Andina, Proto-Andina, and Eu-Andina orogenetic events and occurred during the Late Cretaceous-Early Paleocene, the middle Eocene, the Late Oligocene, and the Miocene, respectively (Van der Hammen 1961). As we will see, although this chronology covers the main morphotectonic events of the Colombian Andes, there were modest advances over the rest of the twentieth century in the understanding of the timing and magnitude of morphotectonic process in cordilleran massifs.

With the establishment of the tectonic plates paradigm during the latter half of the twentieth century (Holmes 1965) and the consolidation of classic ideas on landscape evolution by Davis, Penk, and Hack (Burbank and Anderson 2011; Davis 1899, 1922; Hack 1976; Penk 1953), interest in topographic development took a new direction. However, even by the turn of the twenty-first century, relatively few studies had attempted to document orogenic developments in the Northern Andes of Colombia directly. Others, following the pioneering work of van der Hammen (1950–1970), addressed the subject of tectonic paroxysms/paleotopography/paleoelevation through indirect approaches such as sediment-stratigraphic analysis and paleofloral reconstructions based on palynology (Anderson et al. 2015; Ochoa et al. 2012; Hooghiemstra et al. 2006; Gregory-Wodzicki 2000; Andriessen et al. 1993; Kroonenberg et al. 1990). As the use of LTTC has grown more prevalent in the Northern Andes over the course of the last few decades, direct and compelling evidence of orogenesis has become increasingly available.

It was not until the 1980s that the application of LTTC facilitated a better grasp of tectonic upheavals in the Andes (Kohn et al. 1984; Nelson 1982; Shagam et al. 1984; Van der Wiel 1991). One of the first reviews to consider LTTC data in assessing orogenesis in the Colombian Andes was that of Kroonenberg et al. (1990). Spatially explicit thermochronological data were discussed in conjunction with other geological information according to discrete orogenic segments. However, the lack of sufficient LTTC information for that time (approximately 4 separate locales in the entire Colombian Andes) precluded a detailed recognition of orogenetic events. Nonetheless, even at this early stage, the episodic nature of morphotectonic activity was recognized. For example, uplift of the SNSM was placed at 6 Ma based on LTTC results by Kohn (1984) and Kellogg (1984), although the magnitude and topographic effects of this pulse were not discussed due to the absence of further details in the original datasets and interpretations. Citing Kellogg and Bonini (1982), Kroonenberg et al. (1990) reported that orogenic activity in the area led to the development of structural relief in excess of 12 km. For the Sierra Nevada del Cocuy (SNC) and Santander Massif, and based on previous AFT and ZFT results (Shagam et al. 1984), Kroonenberg et al. (1990) proposed a minor yet long-lasting pulse of initial uplift (Paleogene(?)) followed by short, high-magnitude pulses of uplift/ exhumation in the Miocene. The apparently inverted AER between the Santander Massif and the SNC, where older ages are associated with lower-elevation samples, led to the conclusion that uplift of the Santander Massif occurred when the Cocuy Basin was still in its final stage of subsidence (Kroonenberg et al. 1990). However, the scarcity of LTTC datasets along vertical profiles for that period makes it difficult to verify this claim. For the Garzón Massif, LTTC data from both crystalline massifs and detrital units (Van der Wiel 1989, 1991) show a major phase of uplift from ~12 to 9 Ma. Several interesting paleogeographic ideas were put forth proposing that, at least at this latitude, the EC was not a prominent topographic feature during the middle Miocene, thereby enabling fluvial interconnections to exist between the CC and Amazon Basin (Van der Wiel 1991; Kroonenberg et al. 1990). In the CC, and citing LTTC ages for the Manizales pluton (AFT ~10 Ma and ZFT ~62 Ma), Kroonenberg et al. (1990) concluded that uplift of this section occurred between 10 and 4 Ma ago. The lack of additional information for the CC at that time was recognized by Kroonenberg et al. (1990): "More quantitative data are needed" for better constrained morphotectonic reconstructions in the Northern Andes. In the WC, the situation was even worse, since no LTTC data existed from this portion of the Colombian Andes. Therefore, the nature of uplift and erosional exhumation in the WC had only been addressed via indirect evidence, such as crystallization time for certain Cretaceous and Miocene plutons (Aspden and McCourt 1986; Aspden et al. 1987; McCourt et al. 1984), erosional surface remnants sitting at elevations >3000 m (Case et al. 1971; Padilla 1981; Zuluaga and Mattsson 1981), evidence of Pleistocene glaciation (Zuluaga and Mattsson 1981), and cyclic leveling indicative of vertical movements >0.5 mm/y derived from gravimetric investigations (Lüschen 1986).

On the basis of scarce LTTC information, as well as indirect evidence, it was proposed that significant uplift of the Colombian Andes began in isolated locations around 25 Ma, but that broader orogenetic phases that raised the ranges above 3000 m did not occur until ~9–12 Ma to present in the CC, EC, and probably also the WC (Kroonenberg et al. 1990; Van der Hammen 1989; Van der Wiel 1989; Van Houten and Travis 1968). It was also argued that more recent morphotectonic upheavals had become focused in northeastern Colombia, in the SNSM and SNC, by 4–6 Ma. With the available data at the time suggesting accretion of the PCB by the Late Miocene–Pliocene (Duque-Caro 1990a, b; Kellogg and Vega 1995), Kroonenberg et al. (1990) proposed that this final orogenetic phase in the Colombian Andes was synchronous with PCB accretion, thereby implying potential cause-and-effect relationships.

The pioneering review on the morphotectonic evolution of Colombia based on LTTC data advanced by Kroonenberg et al. (1990) was followed by the work of Gregory-Wodzicki (2000), who, through several indirect lines of evidence (crustal deformation, volcanic deposits, marine geomorphic markers, and paleobotany), addressed the paleotopographic history of the entire Andean range with a focus on the Central Andes. In that study, indirect evidence, mostly from paleobotanical records (pollen) from Colombia's EC, was used to reconstruct floral assemblages and derive paleoelevation. On this basis it was concluded that (i) rapid uplift (mainly rock uplift due to erosion-driven isostatic rebound, as opposed to surface uplift) occurred from ca. 2 to 5 Ma at rates of ~0.6 to 3 km/Ma, (ii) initial phases of marked erosional activity (positive topography) in the EC probably occurred by ~12 Ma, and (iii) by ~5–4 Ma, the EC was still less than 40% of its current elevation. However, these paleoelevation estimates exhibit large errors, often greater than 60%. Brief mention is made of morphotectonic events in the WC and CC, suggesting that uplift there took place primarily during the Late Cretaceous-Paleocene interval but that morphotectonic constraints in both mountain chains were poor.

Following a number of preliminary studies (summarized in Kroonenberg et al. 1990; Gregory-Wodzicki 2000; Kennan 2000), hundreds of cooling ages have been reported in the literature over the last two decades (Gomez et al. 2015a). A quick review of these datasets shows that the Colombian Andes contain a record of spatio-temporally complex orogenic phases. In this section, we summarize LTTC data for the Colombian Andes, according to specific segments of the orogenic belt, in an attempt to highlight the main cooling events, most of which are attributed to erosional exhumation triggered by surface uplift. The brief discussion above on previous LTTC reviews sets the backdrop for the following segments in this contribution. As new data are reported continuously, our objective here is simply to summarize the present state of a growing database.

11.3.2.1 Western Cordillera

The WC is the least studied segment of the trident-shaped Colombian Andes. Over the last 5 years, however, several inter-institutional projects have conducted a large number of LTTC studies on cordilleran massifs and associated sedimentary basins.[11] LTTC (various systems) and zircon U/Pb data produced along vertical profiles for

---

[11] Instituto de Investigaciones en Estratigrafía (IIES) at Universidad de Caldas (Colombia), Grupo de Estudios Tectónicos (GET) at Universidad Nacional de Colombia (Medellín, Colombia), University of Florida at Gainesville (Florida, USA), University Grenoble Alpes (Grenoble, France), Agencia Nacional de Hidrocarburos-ANH (Colombia), Universidad EAFIT (Medellín, Colombia)

crystalline massifs such as the Farallones del Citará Batholith and Horqueta Stock [12] allowed the development of thermal history models showing a major thermotectonic pulse at ca. 9–10 Ma, immediately following plutonic emplacement (~12 Ma). Magmatism was proposed as a potential trigger of uplift and subsequent exhumation via crustal thickening (Barbosa-Espitia et al. 2013; Restrepo-Moreno et al. 2013a, b, c). This orogenetic phase was probably amplified by the compressive regime sustained by the continued collision of the PCB (Farris et al. 2011; Suter et al. 2008). Unroofing in excess of ~3 km was associated with uplift-driven exhumation and probably coincides with the topographic buildup that lead to the establishment of the WC in the middle Miocene as a major topographic barrier with critical orographic and environmental effects (Pardo-Trujillo et al. 2015). Interestingly, the LTTC vertical profile in excess of 2000 m at the Farallones del Citará Batholith displays virtually concordant ages in several LTTC systems (e.g., ZFT, AFT, AHe), and no inflection point is shown for an AER. These two findings indicate exhumation in excess of ~3 km and at rates of ~0.5–1 km/Ma also for the Miocene (Restrepo-Moreno et al. 2015). Similarly, in the southern WC, crystalline massifs have provided AHe and zircon U/Pb data from individual samples of Miocene plutons (Piedrancha Batholith, Danubio Stock, Tatamá Granodiorite, Támesis Stock, and Pance Stock) that reveal two cooling episodes at ca. 20–15 Ma and 12–6 Ma, likely related to exhumation (Barbosa-Espitia et al. 2013). In the Tumaco and Cauca-Patía basins, thick (~1–1.5 km) Miocene sandstone and conglomerate sequences suggest that uplift in the WC was followed by erosion and sediment transfer to the nearby basins. Concurrent inversion of sedimentary basins on both sides of the WC (e.g., Tumaco, Chocó, and Amagá basins) has also been documented (Marín-Cerón et al. 2015; Barbosa-Espitia et al. 2013; Pardo et al. 2012a; Cediel et al. 2009; Barrero et al. 2007).

Regardless of the formation ages of most WC crystalline massifs (from the Cretaceous Mistrató pluton to the middle Miocene Farallones Citará Batholith) and sedimentary formations, a marked and continued exhumation pulse occurred from ~14 to 9 Ma (Pardo-Trujillo et al. 2015). Since ~5 Ma, the WC experienced slow exhumation, with the exception of the Mandé Batholith (an Eocene pluton that is part of the PCB but which rests as a piece of subdued "plateau-like" topography accreted to the WC's western margin) for which recent AHe ages suggest significant cooling at ~2 Ma.[13] Tectonic stability apparently has dominated much of the WC since the late Miocene (Restrepo-Moreno et al. 2015), despite continued collision of the PCB with the South American margin over the same period (Coates et al. 2004; Farris et al. 2011; Kellogg and Vega 1995; Suter et al. 2008; Taboada et al. 2000). The majority of LTTC data from the WC also support the hypothesis that coexisting low-relief surfaces (Páramo de Frontino) and sierra-like topography (Farallones del Citará) are a common feature in active orogenic systems (Restrepo-Moreno et al. 2013a), both in collisional (Hetzel et al. 2011; Van Der Beek et al. 2009) and sub-

---

[12] For a detailed location and genesis of plutonic masses in the Northern Andes, see Andean Magmatism, in this volume, Chap. 4, contribution No. 4, and/or Gomez et al. (2015a, b).

[13] GET research group at Universidad Nacional de Colombia, unpublished datasets.

ductional settings (Coltorti and Ollier 1999, 2000; Gubbels et al. 1993; Kennan et al. 1997; Restrepo-Moreno et al. 2009c). These contrasting geomorphic domains, i.e., sierra and plateau, share a common denudational history in most areas of the WC. In contrast, the CC exhibits lagged geomorphic response to tectonic input despite being located at a similar latitude relative to the WC (Restrepo-Moreno 2009; Restrepo-Moreno et al. 2009c).

Comparison of LTTC and zircon U/Pb studies on several Miocene WC plutons, such as the Citará and Horqueta plutons (Restrepo-Moreno et al. 2013a), reveals that crustal thickening driven by magmatic addition, rather than by convergence, could be a major cause of surface uplift. If true, the resulting surface uplift and exhumation should be concentrated on magmatic foci, which does not appear to be the case as Miocene morphotectonic activity is also prevalent in areas of the WC lacking Neogene plutonism. Yet, for much of the WC specific mechanisms remain incompletely understood, and both surface uplift and exhumation are potentially related to (i) crustal thickening from magmatic activity and (ii) compression and fault reactivation during the collision of the PCB and South America or a combination of both. Although less plausible, the concurrence of high- and low-temperature radiochronologic data might simply reflect post-magmatic cooling into cold country rocks at shallow crustal levels. This possibility has yet to be evaluated with P-T estimates of pluton emplacement depth and thermochronology of the country rocks.

Despite the specific mechanisms, Miocene topographic buildup of the WC (Pardo-Trujillo et al. 2015; Restrepo-Moreno et al. 2013a) and concomitant development of a major orographic barrier (Poveda and Mesa 2000; Zuluaga and Houze 2015) are key elements of the geologic and environmental history of the Chocó Biogeographic region (Duque-Caro 1990b; Mittermeier et al. 1999). Such geoenvironmental evolution includes crucial geodynamic and paleogeographic processes around the equatorial circum-Pacific and circum-Caribbean region, such as docking of the PCB to the South American margin and the closure/emergence of the Panamanian Isthmus (O'Dea et al. 2016; Pardo-Trujillo et al. 2015; Garzon-Varon 2012; Montes et al. 2012a, b; Farris et al. 2011; Coates et al. 2004). Events of this magnitude have exerted a strong influence on oceanic and atmospheric circulation since the Oligocene or Miocene, although the extent of such influence remains debated (O'Dea et al. 2016; Montes et al. 2015; Molnar 2008). Some of the evidence used to support a middle-late Miocene buildup of the WC derives from changes in erosion and sedimentation rates, as well as shifts in depositional environments in the sedimentary basins on the Pacific margin (Pardo-Trujillo et al. 2015; Vallejo et al. 2015). Middle Miocene topographic buildup and the resulting orographic precipitation have strongly influenced climatic patterns and thus sediment routing systems on the western flank of the WC. For example, the Chocó Biogeographic region (Fig. 11.1), which is confined to the west of the WC, is one of the wettest (>12 m mean annual precipitation (Poveda and Mesa 2000)) and most bio-diverse geographic provinces on Earth (Mittermeier et al. 1999), two key signatures that seem related to topographic changes during middle Miocne orogenetic events (Pardo-Trujillo et al. 2015).

11.3.2.2 Central Cordillera

The CC was one of the first locations in the Northern Andes in which LTTC studies were carried out using a combined vertical profile–multiple thermochronometer approach (Restrepo-Moreno 2009; Restrepo-Moreno et al. 2009c). Extensive erosional surfaces exist today in the northern CC, where the Antioqueño Plateau (AP) maintains a mean elevation of ~2500 m (Figs. 11.1, 11.2 and 11.3) (Noriega-Londoño 2016; Arias 1996; Kroonenberg et al. 1990; Page and James 1981; Soeters 1981). These erosion surfaces are useful geomorphological markers, which, in combination with LTTC data, allow us to establish the timing and magnitude of exhumation (Villagomez-Díaz 2010; Restrepo-Moreno 2009; Restrepo-Moreno et al. 2009c; Montes-Correa 2007; Saenz 2003) (Fig. 11.6).

The earliest cooling phase in the CC occurred between the Late Cretaceous and Early Paleocene, according to AFT and ZFT ages derived largely from the Antioqueño Batholith, the Aburrá Valley area, and close to the middle Magdalena

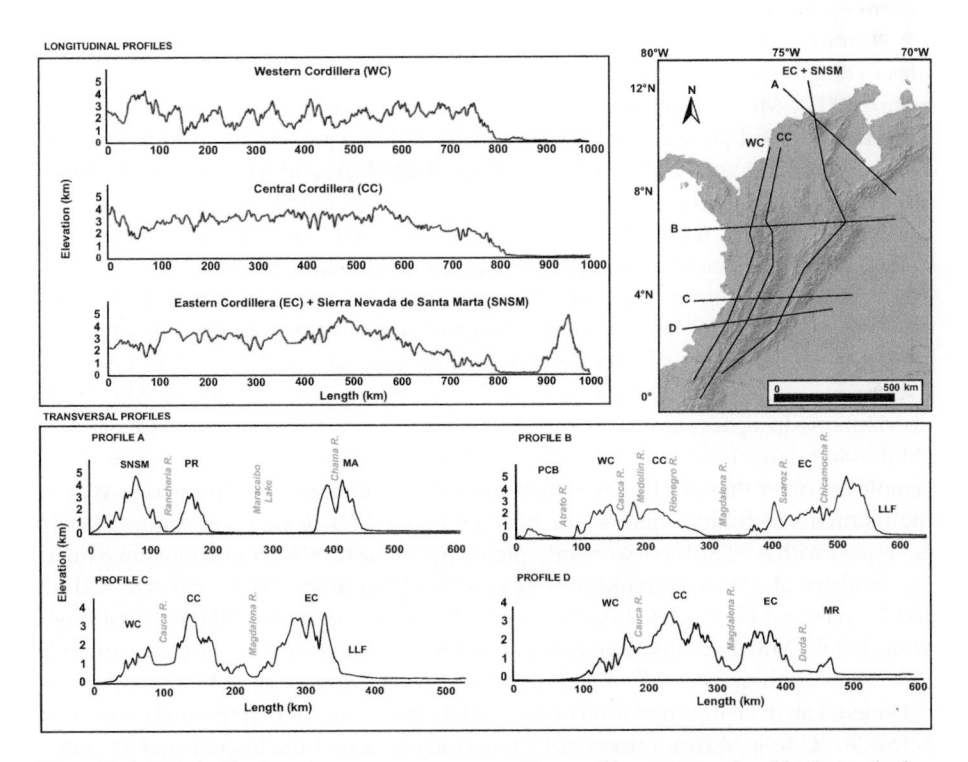

Fig. 11.6 Longitudinal and transverse topographic profiles across the Northern Andes. Longitudinal profiles of cordilleran massif display topography, maximum/minimum elevations, degree of fluvial incision (main rivers labeled in vertical blue font), and slope change in foothills. Note that the WC exhibits the highest incision, followed by the EC and CC. The latter shows a bimodal character, with high incision in the central part and low incision (i.e., AP) in the northern portion. SNSM Sierra Nevada de Santa Marta, PR Perijá Range, MA Mérida Andes, PCB Panamá-Chocó Block, LLF Llanos Foothills, MR Macarena Range, WC Western Cordillera, CC Central Cordillera, EC Eastern Cordillera

Valley (Figs. 11.7, 11.8, 11.9, and 11.10) (Noriega-Londoño 2016; Villagomez and Spikings 2013; Villagomez-Díaz 2010; Restrepo-Moreno 2009; Montes-Correa 2007; Toro et al. 2006; Gomez et al. 2005a, b; Saenz 2003). Nine ZFT ages from the Antioqueño Batholith (Saenz 2003) fall within the approximate range of 67–49 Ma. Although the sampling strategy in that study did not involve a vertical profile approach, it is possible that the age spread evident over a varied altitudinal gradient represents the zircon PAZ. If so, the extreme ages may fall close to inflection points and thus mark the onset of exhumation for two prominent orogenetic episodes in the Northern Andes, namely, the Late Cretaceous-early Paleocene and the early Eocene. Based on interpretations of multiple geochronometer (e.g., K/Ar and ZFT) data, it was suggested that erosional exhumation in the northern segments of the CC took place at rates of ~0.03–1 km/Ma and unroofing reached to approximately 2.5–3 km (Montes-Correa 2007; Saenz 2003).

Other important orogenetic phases in the CC have been reported from investigations that focused the LTTC strategy by sampling multiple plutons and high-grade metamorphic rocks from the northern end of the range and conducting AFT and ZFT through a spatially dispersed sampling array that did not involve vertical profiles (Villagomez and Spikings 2013; Villagomez-Díaz 2010; Montes-Correa 2007; Saenz 2003). Most of the intrusive bodies investigated by Saenz (2003) are in an axial position and possess Cretaceous crystallization ages (Restrepo-Moreno et al. 2009a), e.g., Antioqueño, Sonsón, and Ovejas batholiths and Altavista Stock, while the Triassic Amagá Stock (Restrepo-Moreno, U/Pb Zr unpublished data) is oriented toward the western flank of the CC. Furthermore, the Antioqueño, Sonsón, and Ovejas plutons fall close to the AP morphotectonic domain. As in the case of previous reports for the region, reported ZFT ages for all five units fall within the approximate range of 64–46 Ma, with the exception of one sample from the Amagá Stock, which yielded an older (Cretaceous) cooling age of ~157 Ma. For both the Cretaceous and Triassic lithologies, AFT ages range from ~34 to 49 Ma. MTL distributions in apatite are relatively homogenous, fluctuating around 14 µm and exhibiting relatively low dispersion of <1.5 µm. All ZFT ages reported for a given sample are older than AFT ages, which support the coherence of the dataset. Within the thermal sensitivity windows of both ZFT and AFT, time-temperature models generated with AFTSolve® were interpreted by Saenz (2003) as yielding three main tectonothermal events: (i) rapid cooling of ~50 °C/Ma from 250 °C (ZFT) to ~70 °C (AFT) in the middle-to-late Eocene, (ii) a subsequent interval of "thermal stability" until middle Miocene times, and (iii) a cooling event through surface temperature (20 °C) at cooling rates of ~4 °C/Ma. However, the last event (Miocene) may be considered an over-interpretation of the models due to the lack of thermal constrains below 70 °C (e.g., AHe). Intriguingly, this study captured the Eocene and Miocene orogenetic pulses in the CC but did not provide any information on the morphotectonic paroxysm near the Oligo-Miocene transition reported for the region by other authors (Noriega-Londoño 2016; Restrepo-Moreno 2009; Restrepo-Moreno et al. 2009c). On the basis of LTTC data and modeled t-T paths, morphotectonic upheavals were interpreted as due to uplift-driven erosional exhumation on the order of

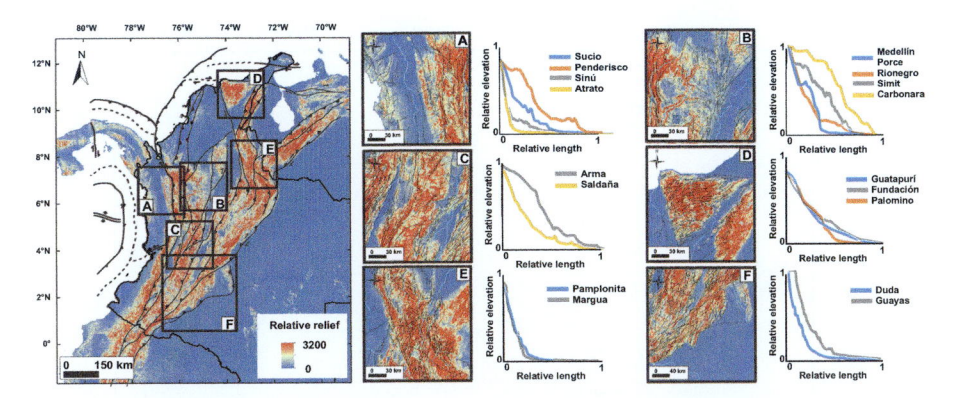

Fig. 11.7 (*Left*) Relative relief of the Northern Andes. Red denotes high and blue low relative relief. Black lines represent regional faults, after Gomez et al. (2015b). Black boxes are regions of interest and are detailed in the separate figures to the right corresponding to the main geomorphological features and morphotectonic responses of tectonothermal events discussed in the text. (**A**) Northwestern Western Cordillera and PCB's Atrato Basin, Baudó and Darien ranges. (**B**) Northern Central Cordillera with AP and San Lucas Range. (**C**) Central portion of Central Cordillera, Quindío, and Ibagué fans, segmentation of CC by Ibagué fault and intra-mountain tectonic Cauca and Magdalena Basins. (**D**) Sierra Nevada of Santa Marta, Perijá Range, and Rancheria Basin. (**E**) Santander Massif and a branch of Merida Andes. (**F**) Grazón Massif, Upper Magdalena Basin, and Macarena Range. Each detailed box on the right shows a selection of longitudinal river profiles, which indicate steady-state and/or transient stage caused by perturbation from tectonic and/or climatic input. In general, concave up-profiles indicate a steady-state behavior, in contrast to low concavity or convex profiles, which may indicate that the fluvial system is in transient state as a consequence of differential tectonic input, nick-point migrations, and/or drainage rejuvenation (for details see Hengl and Reuter 2008)

2–0.16 km/Ma for both the Eocene and Miocene pulses (Montes-Correa 2007; Saenz 2003). An alternative interpretation advanced by Saenz (2003), in response to the poor correlation in AER, is that the ZFT ages do not reflect surface process but rather the loss of heat from plutonic masses emplaced into cold, Paleozoic metamorphic rocks. We prefer to assume that the closure temperature of ZFT thermochronometers is sufficiently low to record some morphotectonic input and that age dispersion is related to the position of the samples within the exhumed Z-PAZ, combined with variable closure temperatures of individual samples due to radiation damage.

Conjecture on low-relief erosion surfaces and LTTC datasets for the time led Saenz (2003) to assign a post-Oligocene age to the AP, whereas Restrepo-Moreno et al. (2009a, b, c) proposed that the initial surfaces developed as early as the middle Eocene-Oligocene, during extended periods of tectonic quiescence lasting more than 15 Ma. Surface uplift and concomitant exhumation of the CC via erosion influenced sedimentary basins within both the proto-EC and middle Magdalena Valley. The episodic nature of uplift-driven erosional exhumation in the Colombian Andes is not only discernible from LTTC datasets but also verified at several locations, for

784   S. A. Restrepo-Moreno et al.

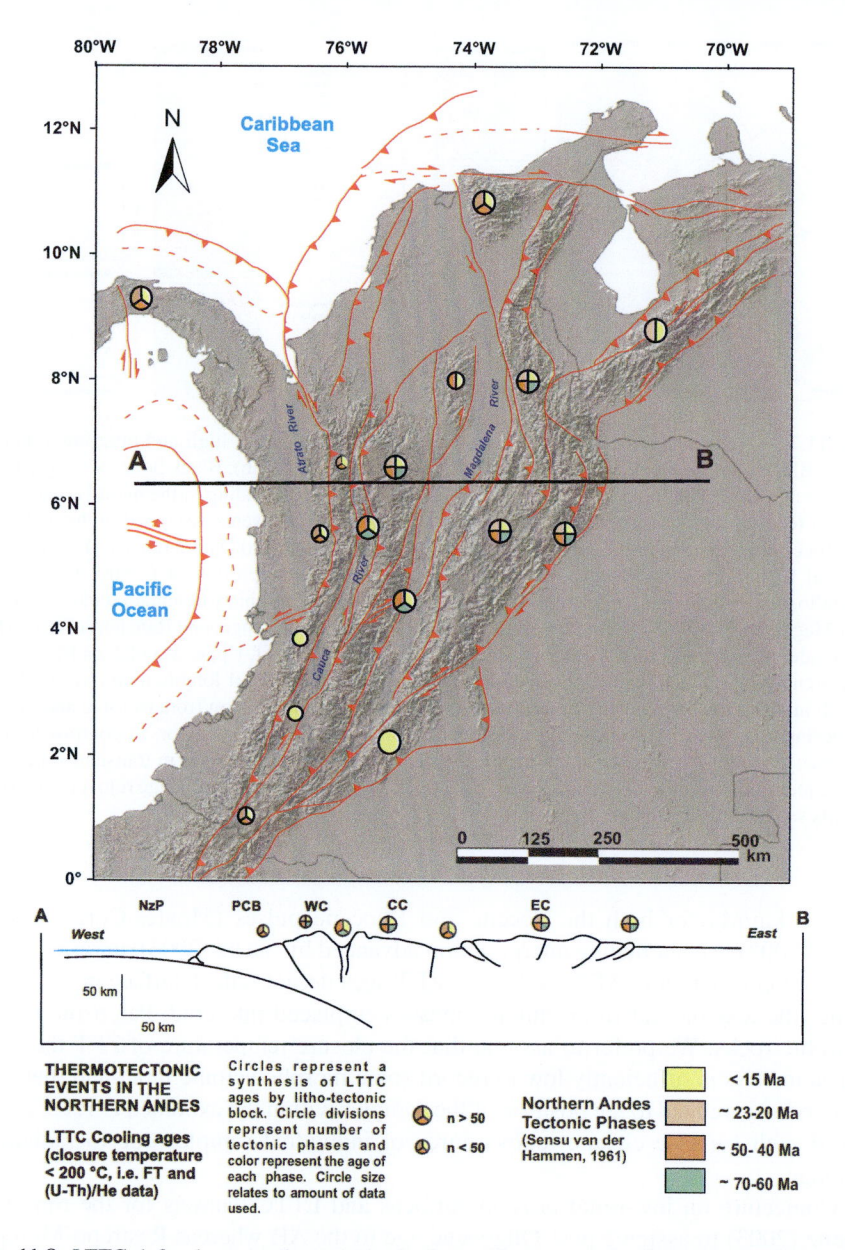

Fig. 11.8 LTTC-defined orogenetic events in the Colombian Andes. The Colombian Andes are the locus of multiple orogenetic events that have caused deformation, surface uplift, topographic buildup, erosional exhumation, sediment production-routing-deposition, and basin development.

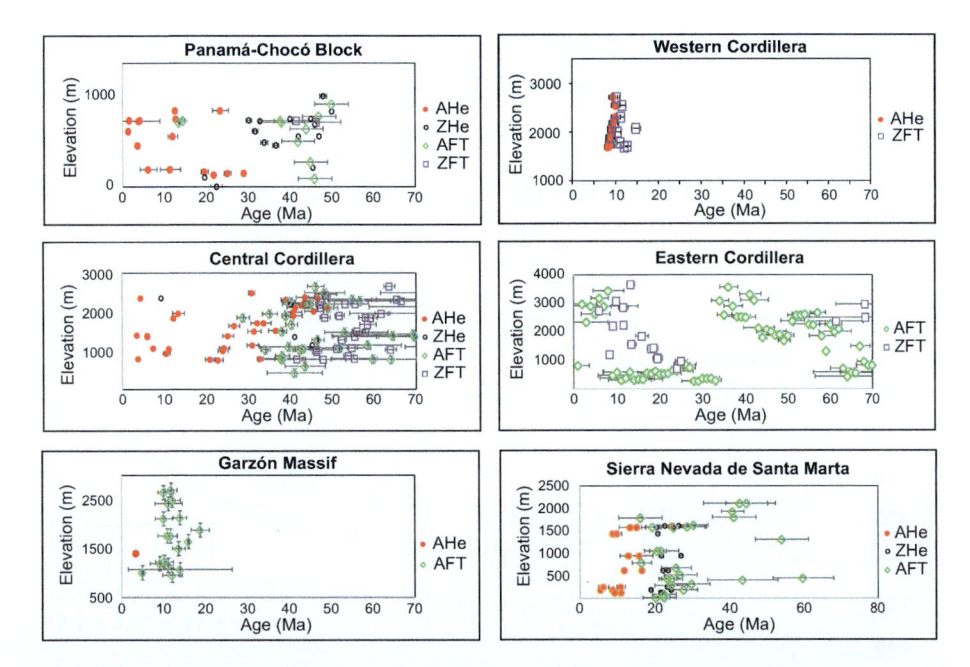

Fig. 11.9 Compilation of AERs for (U-Th)/He and fission-track data in the Northern Andes. (*Upper left*) The Panama-Chocó Block (Farris et al. 2011; Montes et al. 2012a, b, GET research group at Universidad Nacional de Colombia, unpublished datasets) registered at least three main cooling (post-magmatic) events in the Paleocene-Eocene, middle Miocene, and Mio-Pliocene. (*Upper right*) The Western Cordillera exhibits the last of three cooling of the Eu-Andina Phase pulses in the late Cenozoic. (*Middle left*) The Central Cordillera shows clear late and middle Miocene and late Oligocene cooling events but relatively dispersed AER data for the Paleocene-Eocene. (*Middle right*) The Eastern Cordillera shows AFT dispersed ages from the late Mesozoic, and ZFT-registered cooling events in middle Miocene(?). Mio-Pliocene AFT ages indicate a clear cooling event. (*Lower left*) The Garzón Massif just shows the last Eu-Andean event after 15 Ma. (*Lower right*) The Sierra Nevada de Santa Marta displays a well-defined late Cenozoic cooling event during the late and middle Miocene and dispersed data of AFT ages in the Paleocene-Eocene

Fig. 11.8 (continued) Most notable intervals are the Late Cretaceous-Early Paleocene, Early and middle Eocene, Oligo-Miocene transitions that configured a set of tectonic paroxysms first recognized in the late 1950s (van der Hammen 1961 and references therein). We direct attention to the investigations that documented such events from LTTC studies (Figure 11.11), mainly because LTTC can provide explicit information on the timing, magnitude, and location of key morphotectonic processes such as exhumation and, indirectly, surface uplift. Note the prevalence of pulses in the intervals referred to in Fig. 11.11 as the Laramic, Pre-, Proto-, and Eu-Andina tectonic phases (after Van der Hammen 1961). Circles represent (U-Th)/He and fission-track cooling ages (i.e., closure temperatures <250 °C). The number of divisions in each circle relate to described morphotectonic events/cooling phases recorded in each block

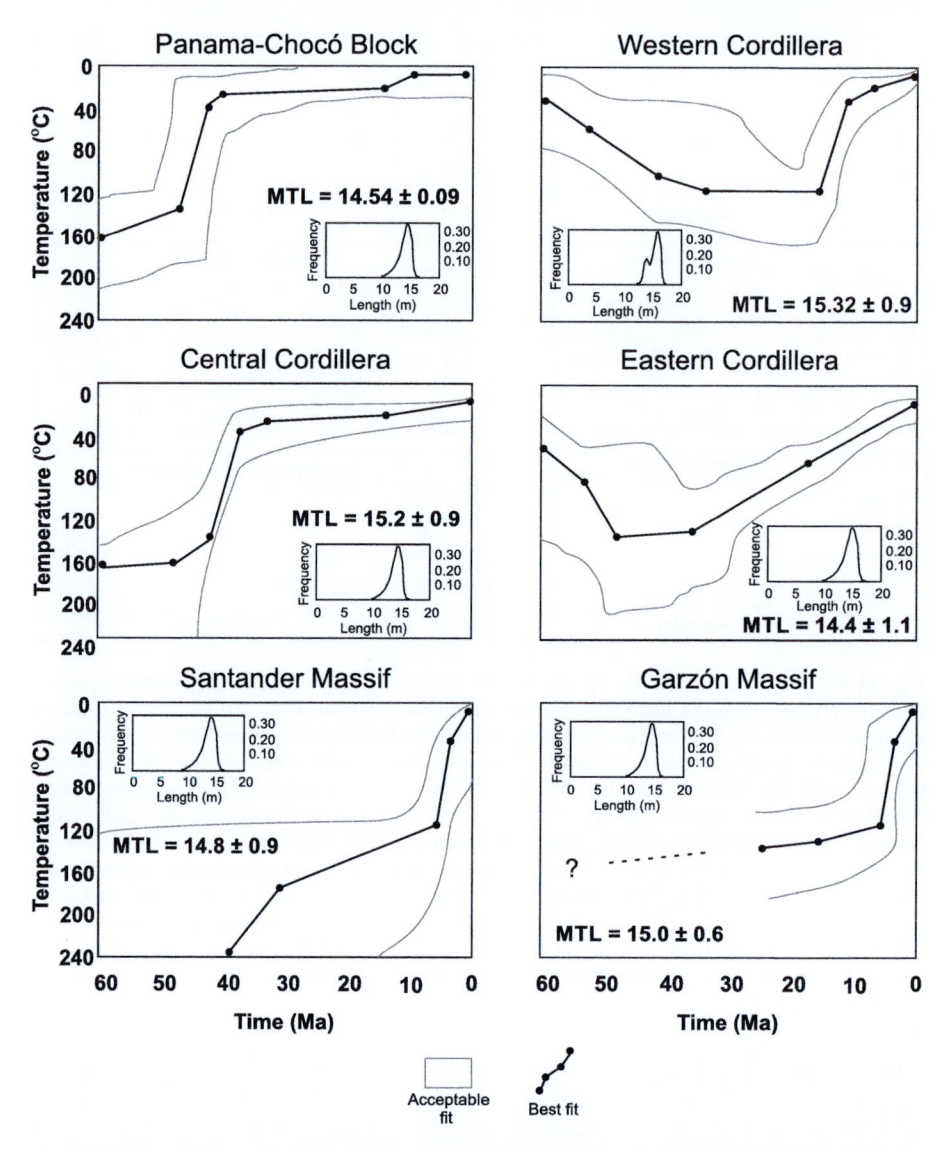

Fig. 11.10 Examples of time-temperature paths resulting from computer modeling (HeFTy®, Qt-QT®, etc.) in the Panama-Chocó Block, Western and Eastern cordilleras, and different segments of the Central Cordillera. In the PCB, several crystalline massifs (Cerro Azul, Mamoní, etc.) display a clear cooling event between 50 and 40 Ma. Cretaceous sedimentary formations at an axial position in the WC, as well some intrusive units of various ages (Cretaceous to Miocene, not shown in this figure) exhibit marked cooling between 15 and 10 Ma. Similarly, some segments of the CC reveal cooling episodes in Eocene times (from AFT and AHe). Both the Santander and Garzón Massifs in the EC record significant cooling during the Upper Miocene–Early Pliocene. All vertical (temperature) and horizontal (age) scales are the same for all boxes. Model output plots modified from Amaya et al. 2017; Amaya-Ferreira 2016; Anderson et al., 2016; Montes et al. 2012a, b; Restrepo-Moreno et al. 2009b, c, 2013a, b, c; Restrepo-Moreno 2009; Saenz 2003; Villagomez and Spikings 2013

example, in the detrital products for Paleocene sandstones of the Cocuy Basin and sandstones of the southern Bogotá Basin, both located in the EC, as well as in the middle Magdalena Basin (Gomez et al. 2005a, b; Toro et al. 2004).

On the eastern side of the Medellín River, i.e., opposite valley wall from where Restrepo-Moreno et al. (2009a, b, c) conducted their LTTC work, an additional 28 AFT and ZFT ages were generated for basement rocks consisting of high-degree metamorphic units (Montes-Correa 2007). Zircon FT ages vary between ~70 and ~47 Ma, although most converge on the middle Eocene. Apatite FT ages are consistently younger than ZFT ages and range from ~65 to 23 Ma. Apatite track-length distributions yielded various populations within the approximate range $12.8 \pm 1.6\,\mu m$ to $14.3 \pm 0.9\,\mu m$, while frequency histograms for MTL are clearly multimodal and negatively skewed and have been interpreted as indicating "…complex cooling histories" (Montes-Correa 2007). The older ages appear to reflect Late Cretaceous uplift resulting from the Cretaceous–Paleocene collision of Western Cordillera-like elements with northwestern South America (Montes-Correa 2007; Saenz 2003). Computer-based models (AFTSolve®) of cooling paths applied to 11 samples revealed two distinct thermal events. First, an Early–middle Eocene cooling pulse that brought samples from temperatures of ~250 °C to less than the AFT closure temperature. In some sectors, this pulse extended into the Oligocene. Rapid cooling is correlated with uplift-driven erosional exhumation at relatively modest rates of ~0.04 km/Ma. It is important to note that, without a proper explanation, model simulations in Montes-Correa (2007) are interpreted as yielding information down to ~30 °C, i.e., below the thermal limits of the LTTC systems. Eocene tectonothermal activity in this portion of the AP appears to have been of lower intensity than that reported from proximal locations by both AFT and AHe dating (Noriega-Londoño 2016; Restrepo-Moreno 2009; Restrepo-Moreno et al. 2009c). Second, a warming trend that started in the Late Miocene and culminated in the Late Pliocene was associated with a regional increase in the geothermal gradient (Montes-Correa 2007) driven by Miocene magmatism (e.g., Combia Formation (Ramírez et al. 2006; Restrepo et al. 1981)). Although Montes-Correa (2007) did not directly document apatite composition, mean Dpar values ($<1.7\,\mu m$) place these samples near the fluorapatite end-member, which is known to possess the lowest track stability to thermal annealing among most common apatites (Donelick et al. 1999, 2005; Duddy et al. 1988; Green et al. 1989). Thus, these apatite samples are potentially susceptible to thermal perturbation (fully or partially reset) by thermal events (magmatism, volcanism). However, because magmatism associated with the Combia Formation occupies a distal position relative to the central CC, an alternative possibility is that some of the samples in Montes-Correa's (2007) dataset were partially reset by localized hydrothermal activity (veins) related to Late Paleocene magmatism (Leal-Mejía et al. 2011). This interpretation is unsupported by the coherent behavior of LTTC datasets on the opposite side of the same valley (Restrepo-Moreno et al. 2009c) in systems such as AHe, which, although more liable to thermal resetting (Reiners and Brandon 2006), do not show any indication of thermal perturbation in any of the vertical profiles there reported. These Eocene (cooling) and Miocene-Pliocene (heating) thermal events were separated by an extended period of thermal stability, showing a similar trend when compared to other studies in nearby CC localities (Noriega-Londoño 2016; Restrepo-Moreno 2009; Restrepo-Moreno et al. 2009c).

The Eocene cooling phase associated with erosional exhumation in the CC is constrained primarily by AFT and ZFT ages (Montes-Correa 2007; Saenz 2003) and AHe and AFT ages (Restrepo-Moreno 2009; Restrepo-Moreno et al. 2009c; Villagomez and Spikings 2013). This event is recorded both in plutonic and high-grade metamorphic massifs on both flanks of the Medellín River Valley, as well as in the northwestern corner of the AP (Noriega-Londoño 2016). AHe data suggest younger pulses during the Oligo-Miocene transition (Noriega-Londoño 2016; Restrepo-Moreno 2009; Restrepo-Moreno et al. 2009c; Villagomez and Spikings 2013; Villagomez-Díaz 2010), while no Miocene events have been accurately recorded in this sector via LTTC. Elsewhere, the intermittent nature of AP exhumation is documented by AERs (Restrepo-Moreno 2009; Restrepo-Moreno et al. 2009c), while AHe data from the central portion of the Antioqueño Batholith suggest exhumation at a rate of ~0.08–0.1 km/Ma during the Eocene (~45 Ma), followed by slow exhumation (~0.02–0.008 km/Ma) during the period 40–25 Ma (Restrepo-Moreno 2009) (Figs. 11.9 and 11.10). In addition to virtually concordant Eocene AHe and AFT ages, unimodal MTL distribution histograms centered on ~14.4 μm indicate rapid Eocene cooling through the A-PAZ and A-PRZ, thus implying >3–4 km of unroofing over 4 Ma (Restrepo-Moreno 2009), i.e., exhumation rates on the order of ~1 km/Ma. But the Eocene was not the only period of enhanced morphotectonic activity in the region during the Cenozoic. Cooling events of similar magnitude near the Oligocene-Miocene transition are indicated by AHe data (Noriega-Londoño 2016; Restrepo-Moreno 2009; Restrepo-Moreno et al. 2009c; Villagomez-Díaz 2010), for example, Restrepo-Moreno (2009) derived exhumation rates of ~0.25 km/Ma from the AER of a segment beneath the AHe-PRZ. Similarly, on the CC's eastern flank, the Mariquita stock yielded AFT ages close to 32 Ma, which are interpreted to reflect exhumational cooling near the Eocene-Oligocene transition (Gomez et al. 2005a, b).

In the northwestern CC, Cenozoic tectonothermal events exhibit a marked influence of the Cauca-Romeral Fault System (CRFS). Noriega-Londoño (2016) used apatite and zircon (U-Th)/He data to perform AER analyses and model t-T paths, on the basis of which it was suggested that a rapid cooling event occurred during the Eocene. The evidence for this event was recorded simultaneously on both sides of the Espiritu Santo Fault, a poorly documented subsidiary structure of the CRFS (Figs. 11.7 and 11.8). For the Eocene event, the author estimated exhumation rates of ~0.2–0.7 km/Ma. The CRFS caused diachronous post-Eocene exhumation of lithotectonic blocks throughout the surrounding area. These processes are focused on the Espiritu Santo Fault, as evidenced by Miocene LTTC ages that are systematically younger than the majority of the AP. Vertical "extrusion" and exhumation of a small crustal block, relative to the large AP to the SE and the erosion front approaching from the NW, are probably related to increased compressional stages during the Proto-Andina orogenetic phase. This interval was characterized by both (i) ongoing collision of the PCB and (ii) orthogonalization and augmented convergence between the Nazca-Farallon and South American plates (Kennan 2000; Pardo-Casas and Molnar 1987; Pilger 1984).

The majority of LTTC data discussed for the northern portion of the CC imply that the Eocene and Oligo-Miocene events were separated by a period of tectonic quiescence that lasted for ~20 Ma, a span sufficiently long for developing low-elevation erosional surfaces such as the AP. This finding suggests a polycyclic style of peneplanation for the northern segment of the CC (Restrepo-Moreno 2009; Restrepo-Moreno et al. 2009c). The synchronicity of both the Oligocene and Eocene events with similar events in Peru and Bolivia, and with periods of faster, orthogonal convergence between the Nazca-Farallon and South American plates (Kennan 2000; Pardo-Casas and Molnar 1987; Somoza 1998), points to continental-scale controls on surface uplift and erosional exhumation in the CC (Restrepo-Moreno 2009; Restrepo-Moreno et al. 2009c).

Morphotectonic processes for southern segments of the CC have also been studied through LTTC. The Ibagué Batholith, located in the central CC, is an elongated Jurassic granitoid trending NNE between approximately 3° and 6° N latitude. In contrast to the Antioqueño Batholith in the AP, the Ibagué Batholith occupies a less rigid lithospheric segment (Montes et al. 2005). Nonetheless, the Ibagué Batholith's LTTC history exhibits similar Cretaceous–Paleocene ZFT ages and Eocene–Oligocene AFT ages, which together indicate at least two phases of rapid cooling associated with exhumation rates of ~1.6 km/Ma and 0.3 km/Ma, respectively (Villagomez-Díaz 2010). Additional (U-Th)/He ages, corresponding to the Mio-Pliocene, constrain the timing of erosion-related cooling events associated with the most recent episodes of Northern Andean orogenetic activity (Villagomez and Spikings 2013), specifically, the Eu-Andina orogenetic phase (van der Hammen 1961). Miocene morphotectonic activity in the southern portion of the CC further supports the idea that the AP functions in a stage of lagged geomorphic response to tectonic input (Restrepo-Moreno 2009).

Preliminary data from an ongoing research project in the Serranía de San Lucas, a northern segment of the CC structurally "detached" from the rest of the range, reveals distinct AHe age distributions (Fig. 11.9). For example, several plutonic units with Jurassic zircon U/Pb crystallization ages yield a cluster of AHe dates around 35 Ma. Significant LTTC age variability (~21–45 Ma) among the different samples of crystalline massifs (e.g., Ité Granite and Ité Gneiss) has been reported for samples collected at various elevations. Rather than pointing to many discrete morphotectonic events, this age spread may indicate the presence of the PRZ. Therefore, it is likely that exhumation pulses occurred toward the ends of this interval. If so, the San Lucas range would possess a morphotectonic history similar to that of the AP (Restrepo-Moreno 2009; Restrepo-Moreno et al. 2009c) despite the fact that (i) the San Lucas range is sitting ~1500–2000 m lower than the AP and (ii) both morphotectonic domains belong to discrete crustal blocks separated by the Palestina Fault System, potentially favoring the development of "pop-up" features during shifts from transpressive to compressive regimes transpression.

11.3.2.3 Eastern Cordillera

The EC contains the largest number of LTTC data in Colombia. In the Northern Andes, pioneering work using LTTC was conducted on eastern elements of both the Colombian (Van der Wiel 1991) and the Venezuelan Andes (Kohn et al. 1984; Shagam et al. 1984), while several workers have documented Cenozoic cooling driven by erosional exhumation during compressional phases (Velandia-Patiño 2018; Amaya et al. 2017; Amaya-Ferreira 2016; Mora et al. 2008, 2010b; Parra et al. 2009; Kohn et al. 1984; Shagam et al. 1984). A number of climatic controls have also been proposed. For instance, Mora et al. (2008) stressed the role of paleoclimate and increased erosion rates as major controls on asymmetry and tectonic evolution of this range. Subsequently, LTTC datasets yielded evidence of eastward-migrating Paleogene orogenesis focused primarily in the EC (Parra, et al. 2009).

To date, LTTC investigations in the EC have been undertaken primarily in the Santander and Garzón massifs.[14] In the following sections, we discuss the most relevant findings from these two locales while also alluding to other portions of the EC.

Santander Massif The Santander Massif is located in the north-central portion of the EC, where it comprises a broad, ~100 km-long region of elevated topography oriented approximately NNW–SSE. With a total width exceeding 100 km, the EC is significantly more extensive at this latitude than at the Garzón Massif farther south. Average elevation is >3000 m in the north and rises to >4000 m in the Sierra Nevada del Cocuy, potentially indicating northward tilting of the massif. The Santander Massif is part of the Maracaibo Block, along with the Sierra Nevada de Santa Marta and the Merida Andes. A principal crustal structure delimiting the Santander Massif to the west is the Bucaramanga Fault, while major faults also bound the massif to the east (Fig. 11.7). Topographic asymmetry is also apparent with a generally steeper western flank toward the Bucaramanga Fault. The massif is composed of old (Precambrian to Paleozoic) crystalline basement rocks covered by marine and terrigenous sedimentary layers of Paleozoic, Jurassic, and Cretaceous age. Jurassic and Triassic magmatism led to the development of major plutonic bodies, while volcanic activity is constrained to the Mesozoic and Mio-Pleistocene (Velandia-Patiño 2018; Amaya-Ferreira 2016; Goldsmith et al. 1971; Ward 1973).

Late Paleozoic epirogenic uplift of the Santander Massif, leading to erosional denudation of Carboniferous limestones and shales, was reported by Ward (1973), while recent work has described Cenozoic uplift and exhumation patterns from AFT and ZFT ages along topographic profiles (Velandia-Patiño 2018; Amaya-Ferreira 2016; Amaya et al. 2017). An asynchronous character is evidenced by the disparate LTTC ages for several fault-separated blocks. Relative to the Bucaramanga Fault,

---

[14] More detailed descriptions of the litho-structural characteristics of the Santander and Garzón Massifs are available in Chap. 2, Contribution No. 3 in this volume. Additional information on thermotectonic events for the Santander Massif, at deeper crustal levels and in relation to other litho-structural elements of the Maracaibo Block and the Venezuelan Andes, are addressed in Chap. 6 contribution No. 14 in this volume.

the Western and Eastern Blocks exhibit two major cooling events associated with erosional exhumation. The first occurred during the Late Cretaceous (~72 Ma) and was followed by a second pulse near the Oligocene–Miocene transition (~25 Ma). Both events are associated with average exhumation rates of ~0.2 km/Ma (Fig. 11.9). The greatest disparity between the two blocks occurred during the Pliocene, when the Western Block experienced exhumation rates of ~2.5 km/Ma compared to 0.2 km/Ma in the Eastern Block (Amaya-Ferreira 2016). Amaya et al. (2017) concluded that the Santander Massif has been exhumed through subvertical inverted structures (e.g., the Bucaramanga Fault) since the Late Cretaceous and that crustal structuring exerted a major control on that exhumation. Similar results are reported by Velandia-Patiño (2018). The AFT and ZFT data reported for the Santander Massif (Velandia-Patiño 2018; Amaya 2016), in addition to LTTC data from other EC localities (Mora et al. 2010b; Parra et al. 2009; Mora et al. 2008), suggest that the western and central portions of Cretaceous EC basins, and more specifically in the Santander Massif, experienced synchronous exhumation during the Late Cretaceous and Early Paleocene. This model contrasts with evidence for cooling during the Pliocene–Pleistocene, during which the axis of the EC exhibited differential surface uplift from north to south and the range attained its maximum elevation. Away from the Santa Marta Bucaramanga Fault, other studies of lithotectonic blocks in the Santander Massif based on LTTC systems with higher closure temperatures (i.e., 300 °C) record old (Paleozoic) cooling episodes at ~200 Ma that were associated with regional magmatism between ~205 and 196 Ma (van der Lelij et al. 2016). However, samples from closer to the Bucaramanga-Santa Marta Fault System exhibit cooling ages related to exhumation in excess of ~10 km at ~40 Ma, a pattern which led van der Lelij et al. (2016) to suggest that the Bucaramanga-Santa Marta Fault Sytem was active at least since Eocene times.

Erosional exhumation events with chronologies similar to those of the Santander Massif, but smaller degrees of unroofing, have been reported for other segments of the CC (Restrepo-Moreno 2009; Villagomez-Díaz 2010), as well as in Ecuador, Peru, Bolivia, and Argentina (Barnes et al. 2006; Coughlin et al. 1998; Laubacher and Naeser 1994; Mégard et al. 1984). The prevalence of Eocene morphotectonic activity in South America led some authors (van der Lelij et al. 2016; Restrepo-Moreno 2009; Restrepo-Moreno et al. 2009c) to propose major continental-scale controls on uplift-driven erosional exhumation associated with periods of accelerated and orthogonal convergence between Farallon and South America (Pardo-Casas and Molnar 1987; Pilger 1984; Somoza 1998). Interestingly, erosional exhumation associated with the Eocene orogenetic phase has been recognized in numerous sedimentary basins. For example, van der Hammen (1961) found several basal conglomerates from this interval in various detrital sequences throughout Colombia and advanced the idea of a prevalent pulse of mountain building he termed the Pre-Andina orogenetic phase. Similar conclusions were reached by Horton et al. (2010a, b) through analysis of provenance patterns in synorogenic basins of the EC and Santander Massif.

Van der Lelij et al. (2016) also reported Miocene–Pleistocene cooling events in the Santander Massif. For example, AFT data indicate rapid exhumation to the east of the Bucaramanga-Santa Marta Fault System from ~17 Ma. According to these authors, exhumation was controlled primarily by stress partitioning along secondary structures that allowed for continued and rapid exhumation of discrete blocks, while other structural elements remained virtually isothermal. The onset of exhumation east of the fault is not constrained by LTTC data but may have been continuous since ~40 Ma. Because the sampling approach of van der Lelij et al. (2016) and/or the actual tectonothermal histories involved do not permit the identification of inflection points in AER, the onset of rapid cooling remains unidentified. Nonetheless, the fact that samples were below the Zr-PAZ in Eocene times has been used to reconstruct erosion rates on the order of 0.1–0.2 km/Ma, assuming an average geothermal gradient of 25 °C/km. Significantly lower rates, or even a period of morphotectonic quiescence, are proposed for the interval 27–15 Ma. This is in stark contrast to other segments of the Colombian Andes such as the northern segments of the CC and the SNSM where a major pulse of erosion exhumation is reported near the Oligo-Miocene transition (Cardona et al. 2010; Piraquive et al. 2017; Restrepo-Moreno et al. 2009c; Villagómez et al. 2011).

Apatite and zircon FT datasets along topographic profiles on the western flank of the Santander Massif reveal Neogene cooling related to erosional exhumation in two distinct events, the first at ~25 and the second at 7 Ma (Amaya et al. 2017; Velandia-Patiño 2018). Average Cenozoic exhumation rates reported in the said studies are on the order of 0.12 km/Ma, although an acceleration during the Late Miocene to ~0.4 km/Ma is proposed (Amaya et al. 2017). Forward- and inverse-modeling data presented by Amaya et al. (2017) confirm both Cenozoic pulses, even though constrain boxes imposed in some of their models are relatively tight, thereby leaving little room for the model to build alternative paths. The Oligocene exhumation phase is explained in connection to the initial collision of the Panamá arc with northwestern South America, whereas the Late Miocene orogenetic events may have been associated with the collision of the PCB (Amaya et al. 2017; Velandia-Patiño 2018). However, these conclusions disregard the concomitant nature of this and other thermotectonic events (in terms of timing and magnitude) that were taking place in the CC, the SNSM, and the PCB, as well as farther south in the Bolivian and Peruvian Andes (Ramírez et al. 2016; Restrepo-Moreno et al. 2009c; Montes et al. 2012a, b; Arancibia et al. 2006; Laubacher and Naeser 1994; Mégard 1984; Mégard et al. 1984; Steinman 1929). The prevalence of late Oligocene morphotectonic events in multiple locations in the Northern and Central Andes makes it important to consider larger-scale mechanisms such as those implied in the regionally changing geodynamic scenario at the continental scale, where major plate interactions (Farallon-Nazca and South America) were fluctuating in terms of speed and reorientation of convergence.

Although we focus primarily on crystalline massifs, recent LTTC work in Cretaceous–Pliocene sedimentary sequences located on the western flank of the EC (specifically between the EC and middle Magdalena Valley; (Sánchez et al. 2012)), in the middle Magdalena Valley, and on the Eastern flank of the EC (Silva et al. 2013) reveal interesting thermotectonic patterns. AFT and AHe document hanging-wall exhumation from ~45 to 30 Ma and again from ~18 to 12 Ma, with accelerating rates between 12 and 3 Ma. Onset of the first event coincided, albeit partially, with the Eocene phase reported in numerous locations in the Northern Andes of Colombia. Miocene events are typical of other segments of the EC (Gomez et al. 2005a, b; Parra et al. 2009), the SNSM (Piraquive-Bermúdez 2016; Cardona et al. 2011), and the WC (Barbosa-Espitia et al. 2013). Partial and full resetting of several LTTC systems was controlled with vitrinite reflectance data and are also identifiable by (i) a quick examination of the high dispersion of ZFT, ZHe, and AFT data and (ii) the variability of MTL distributions (Sánchez et al. 2012). Alternatively, Sánchez et al. (2012) attributed AFT data dispersion to multiple compositional populations of apatite grains, a condition that may translate into different annealing kinetics and hence different ages and/or MTL distributions. HeFTy® model simulations for various samples confirm the Eocene and Miocene cooling phases described above. Combined paleo-thermometry (Ro%) and LTTC data permit the identification of fully reset samples for those AFT and AHe ages that recorded the most recent (Miocene) and strongest exhumation event, during which rocks were brought from depths below the PAZ or PRZ (i.e., ~2–3 km) to the upper crust. Further, Ro data show that Upper Cretaceous sections achieved burial depths of around 4–5 km during initial (Paleocene–Eocene) synorogenic sedimentation. Partially reset samples are also identified from Ro and MTL data and can be used to inform thermal history models. In such studies, clear AER (or age depth) and MTL vs. depth relationships are not sufficiently straightforward to infer the positions of exhumed PAZ or PRZ. Cooling rates derived from HeFTy® models fluctuate between 5 and 25 °C/Ma which at average paleogeothermal gradients translate to exhumation rates of ~0.2–1 km/Ma. In addition, Silva et al. (2013) used combined LTTC and zircon U/Pb provenance data to unravel multiple fault reactivation events in interrelated fault systems of the Magdalena Valley and EC since the Paleocene. Morphotectonic pulses of uplift and exhumation were asynchronous along this segment of the EC, although regional tectonic upheaval apparently clustered around the Late Miocene. These authors chose to associate this orogenetic acceleration with "superimposed collision" of the PCB, which not only caused vertical tectonics but also oroclinal bending of the EC. More importantly, climatic contrast between the humid eastern and drier western flanks of the EC promoted faster morphotectonic responses on the former (Silva et al. 2013; Mora et al. 2008).

Mechanisms invoked to explain the morphotectonic history of the Santander Massif include shifts in regional geodynamic conditions caused by the subduction of the Caribbean Plate beneath South America (Amaya-Ferreira 2016), a process that was responsible for one of the most intense phases of uplift/exhumation in the Colombian Andes during the Early Miocene (Mora et al. 2010a). Sánchez et al. (2012) posited that mechanisms associated with reactivation or onset of thrusting can account for certain pre-orogenic structures, while an alternative model invokes enhanced compression related to the onset of the PCB–South American collision (Amaya et al. 2017; Velandia-Patiño 2018; Amaya-Ferreira 2016), a process that may have commenced around the Oligocene–Miocene boundary (Farris et al. 2011; Montes et al. 2012b). As we can see, several authors have elected a remote mechanism to explain the period of rapid erosional exhumation in the Santander

Massif (Amaya-Ferreira 2016; Amaya et al. 2017; van der Lelij et al. 2016; Velandia-Patiño 2018; Sánchez et al. 2012), specifically, the continued collisional phases of the PCB since ~24 Ma. An alternative mechanism, or a combination of several, involves not only the PCB but the period of faster and orthogonal convergence between the Nazca and South American plates between ~20 and 15 Ma (van der Lelij et al. 2016), following reorganization of the converge system after fission of the Farallon Plate (e.g., Pardo-Casas and Molnar 1987).

Garzón Massif Over 250 km in length and ~30 km wide, the Garzón Massif forms the core of the southern EC (Figs. 11.6 and 11.7) and provides valuable insight into the morphotectonic processes that characterized these segments during the Late Cenozoic (Anderson et al. 2016; Van der Wiel 1991). With a maximum elevation of ~3000 m, this NE-trending segment is considerably lower topographically than both the EC to the north and the CC to the west. The Garzón Massif possesses a crystalline basement of high-grade Mesoproterozoic and Neoproterozoic metamorphic rocks and Jurassic granitoids, and delimits a structural high with a north-to-northeast-plunging configuration, in which the crystalline basement plunges beneath a northern cover of younger geologic units (Anderson 2015; Anderson et al. 2016). This segment also records long periods of dip-slip displacement as well as right lateral strike-slip deformation (Bakioglu 2014). During the Quaternary, the Garzón Massif was subjected to wrench tectonics, thereby making it the southernmost expression of major active transpression in the Colombian Andes (Bustamante et al. 2010; Chorowicz et al. 1996). Recently reported borehole and gravimetric/magnetic/seismic data (Saeid et al. 2017) indicate the occurrence of low-angle faulting/thrusting and, more importantly, periods of deformation associated with "thin-skinned" and "thick-skinned" tectonics. Such deformation affected basement rocks: between ~9 and 16 Ma, "thin-skinned" episodes resulted in shortening of ~43 km, while "thick-skinned" events between ~3 and 6 Ma produced shortening of 22 km.[15]

LTTC studies of the Garzón Massif date back to the early 1990s, when a set of apatite, zircon, and even titanite fission-track data were produced for lithologic units including granulates, gneisses, granites, granites, and skarns (Van der Wiel and Andriessen 1991; Van der Wiel 1991). Many of the samples reported in those early studies yielded reproducible ages in various LTTC systems (e.g., AFT, ZFT, and even U/Pb zircon), thereby enabling the timing of exhumation to be constrained to between ~9 and 14 Ma. Temporal proximity of ZFT and crystallization ages for Jurassic plutonic bodies is distinctive, indicating rapid post-magmatic cooling through the ZFT closure temperatures (~350–400 °C). This Jurassic thermal event, associated with magmatic activity, is overprinted on Precambrian basement units and is captured by various radiometric dating systems (Bustamante et al. 2010; Restrepo-Pace et al. 1997).

More recently, AFT data and thermal modeling of Garzón Massif crystalline rocks (Anderson et al. 2016; Anderson 2015) suggest cooling at 6–4 Ma, which, on average, is ~8 million years younger than previous interpretations from the same

---

[15] Further details on the Garzón Massif are found in Chap. 2, Contribution No. 2 of this volume.

area. All samples reported by Anderson et al. (2016) display normal (i.e., positive) age-elevation correlated values. Considering the old crystallization or metamorphic ages of host rocks, these new LTTC data are interpreted as indicating the timing of exhumational cooling from fully reset rocks that recorded "zero" ages prior to exhumation. More important, the resulting apatite MTL data exhibit unimodal distributions centered around ~14 μm, suggesting rapid cooling through the PAZ. In the same study, thermal modeling also confirms that analyzed samples from basement rock were located below the PAZ prior to exhumation. The authors concluded post-Miocene unroofing of more than 3–4 km for an assumed paleogeothermal gradient of 20–25 °C/km. Regional cooling trends appear to combine continuous exhumation since ca. 6 Ma (average cooling rate of ~20–25 °C/Ma) with the fastest exhumation occurring between 6 and 4 Ma, followed by a protracted period of slow, near-surface cooling. These trends were reported for sites west and east of the Algeciras Fault, respectively (Anderson et al. 2016).

According to Anderson et al. (2016), the contemporaneous exhumation of the Garzón Massif and the Bogotá region (Anderson et al. 2016; Anderson 2015; Hooghiemstra et al. 2006; Van der Hammen 1957, 1989; Van der Hammen et al. 1973), located >250 km north in the same range, represents a regional phase of EC surface uplift close to the Messinian–Zanclean transition. They proposed a mechanism whereby a shift to principally strike-slip motion along the Algeciras Fault was linked to the onset of more oblique convergence, higher strain partitioning, and/or the geodynamic effects of subducting anomalies such as the Carnegie Ridge (Anderson et al. 2016). This regional change, which implies broad morphotectonic responses, and the possible structural affinity and synchronous uplift/exhumational history of three separate domains (Garzón Massif, foreland structures of the Serranía de la Macarena, and the Vaupes Arch (Velandia et al. 2005; Acosta et al. 2004), led to development of the modern barrier between the Orinoco and Amazon rivers (Anderson 2015; Anderson et al. 2016). Along with the uplift of other Andean segments since the Oligocene, these pulses of surface uplift and concomitant erosional cooling play a pivotal role in the evolution of the Amazon fluvial network, with clear interactions among plate tectonics, morphotectonic activity, fluvial development, and ecological evolution in South America (Anderson et al. 2016; Hoorn et al. 1995, 2010; Mora et al. 2010a).

Other Sectors of the Eastern Cordillera An AFT age of ~26 Ma from the western flank of the EC (the Guaduas Syncline) is interpreted as indicating rapid cooling due to middle Oligocene denudation (Gomez 2001). Furthermore, FT ages from the Floresta Massif (Toro et al. 2004) and southern Santander Massif (Velandia-Patiño 2018; Amaya-Ferreira 2016; Shagam et al. 1984) segments of the EC yielded ages close to the Oligo–Miocene transition, potentially implying that surface uplift and erosional exhumation were occurring at the core of the EC by this time. If so, the relatively short MTL distributions reported from samples in some of these locales also suggest low cooling rates and residence in the PAZ (Toro et al. 2004). More recent phases of uplift-driven exhumation are apparent in the AFT data for five samples analyzed from Cretaceous and Tertiary sedimentary units in the EC's western flank at the Guaduas Syncline (Gomez 2001). AFT ages cluster between 12 and 5 Ma, suggesting that an important phase of cooling associated with erosional exhumation took place during this period. As discussed in a separate section, this event was contemporaneous with those reported for the Garzon Massif at ~12 Ma and ~6–3 Ma (Anderson et al. 2016; Van der Wiel and Andriessen 1991). According to LTTC data presented by Toro et al. (2004), the eastern flank of the EC also experienced an important phase of Andean deformation during the middle Miocene and Pliocene. Over the Bogotá region, LTTC (ZFT) was used not strictly as a means to constrain the thermal history of a sample but rather as a geochronometer to date young ash layers (outcrop and borehole samples) and thus calibrate the Neogene–Quaternary history of fluviolacustrine sediments capping the CC. At this location, the pollen record depicts major tectonic uplift of the Eastern Cordillera from ~ 5 to 3 Ma, in addition to a protracted period of climatic fluctuations that began shortly after 2.7 Ma (Hooghiemstra 1989; Van der Hammen et al. 1973).

LTTC data have been used in conjunction with structural and provenance data to unravel the tectonic inversion of symmetric rifts in the CC, EC, and Magdalena Valley (Mora et al. 2013). Early deformation and crustal cooling (mainly via erosional exhumation) were proposed for the Early Paleocene primarily in the CC and Magdalena Valley. Similar findings have been reported for various segments of the EC (Amaya et al. 2017; Gomez 2001; Shagam et al. 1984), the CC (Villagomez and Spikings 2013; Saenz 2003), the middle Magdalena Valley (Reyes-Harker et al. 2015), and several locations in the WC (Restrepo-Moreno et al. 2013b). An additional period of pronounced orogenetic activity occurred at the end of the Miocene and was responsible for maximum shortening rates and reactivation of the main inversion faults. This interval is considered a major phase of topographic development (Mora et al. 2013) and corresponds to one of the main paroxysms in the Colombian Andes, being reported in the majority of LTTC studies published so far for this Andean segment (see Fig. 11.11 and references therein). Contrary to other EC studies (Parra et al. 2009), a period of middle Eocene tectonic quiescence suggests some degree of asynchronism even over relatively short distances. Weaker morphotectonic phases during the Oligocene have not been ruled out (Mora et al. 2013). Accelerated and orthogonal convergence resulting in predominately compressive stress regimes was proposed as a mechanism to explain Andean shortening and erosional cooling, indicating that a minimum convergence rate of ~2 cm/yr. is necessary to trigger shortening in the upper plate and concomitant morphotectonic phases (Mora et al. 2013). If that is the case, it remains unclear, however, why the latter study failed to detect major orogenetic phases during the interval Late Oligocene–Early Miocene, when the fission of the Farallon Plate (into the Nazca and Cocos plates) produced rapid convergence rates and one of the most orthogonal periods of convergence in the Northern Andes (Pardo-Casas and Molnar 1987; Wilder 2003).

Reyes-Harker et al. (2015) provided new and reviewed LTTC and U/Pb geochronology data for bedrock and detrital samples collected from an axial position on the western flank of the EC, just south of the Santander Massif. Together with biostratigraphic zonations, core descriptions, sandstone petrography, facies analysis, seis-

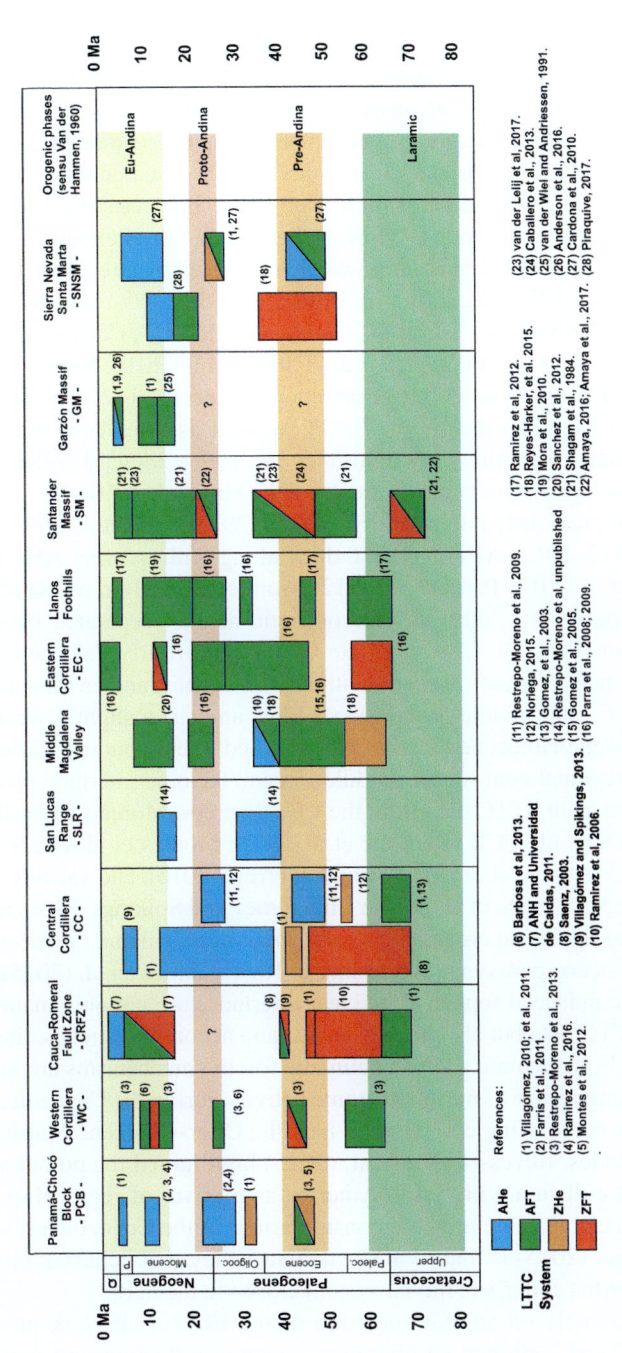

11 Morphotectonic and Orogenic Development of the Northern Andes of Colombia... 797

Fig. 11.11 Summary of reported thermotectonic/orogenetic phases in the Northern Andes of Colombia based on LTTC. Horizontal gray bands correspond to the different orogenetic phases of van der Hammen (1961)

mic information, and vitrinite reflectance (Ro%) data, their results show discrete phases of morphotectonic activity (uplift, erosional exhumation sediment production-routing-deposition) that affected not only the EC but also the proto-CC and the inter-Andean middle Magdalena Valley. A number of ZHe ages from Jurassic and Early Cretaceous units on the EC's western flank fall primarily between 30 and 55 Ma, showing a correlation between stratigraphic position and age (where lower portions record the youngest ages and vice versa). Based on these ages and on Ro % values suggesting temperatures high enough for ZHe resetting, these authors concluded that the older age group (~54 Ma) marks the onset of exhumation. As discussed previously, this period apparently was characterized by a regional morphotectonic paroxysm, with LTTC recording uplift and erosional exhumation of the CC, Santander Massif, SNSM, PCB, and WC (Amaya et al. 2017; Cardona et al. 2011; Gomez 2001; Gomez et al. 2005a, b; Pardo-Trujillo et al. 2015; Restrepo-Moreno 2009; Restrepo-Moreno et al. 2009c; Restrepo-Moreno et al. 2013b; Villagomez and Spikings 2013; Villagomez et al. 2011), as well as elsewhere in the Andes (Peru, Bolivia, and Argentina) (Coughlin et al. 1998; Mégard 1984) and the Caribbean (Rojas-Agramonte et al. 2006; Van der Lelij et al. 2010).

On the basis of previously published (Silva et al. 2013; Parra et al. 2009) and new (Reyes-Harker et al. 2015) AFT and ZFT data, along with existing AHe and ZHe analyses (Mora et al. 2013; Parra et al. 2012; Mora et al., 2010a, b), Reyes-Harker et al. (2015) proposed two intervals of orogenetic activity separated by a middle Eocene tectonic interlude: (i) a strong, Late Cretaceous–/Early Paleocene-to-Early Eocene contractional deformation event, with localized uplift and erosional exhumation in the EC, CC, and middle Magdalena Valley, and (ii) a slightly weaker, Late Eocene-to-Pleistocene/Holocene phase of renewed contractional deformation. However, in the regional context, the middle Eocene tectonic interlude proposed in that study contrasts with LTTC data from the CC (Restrepo-Moreno 2009; Restrepo-Moreno et al. 2009c), the PCB (Ramírez et al. 2016; Montes et al. 2012b), certain portions of the EC (Amaya et al. 2017; Amaya-Ferreira 2016), and various segments of the WC (Restrepo-Moreno et al. 2013a; Villagomez and Spikings 2013), as well as with the paleogeographic and orogenetic reconstructions of Villamil (1999), who suggested a middle Eocene paroxysm. According to Reyes-Harker et al. (2015), periods of morphotectonic upheaval separated by clear interludes are not only manifested in basement rock LTTC data but also in provenance and accommodation/sediment supply relationships. Both tectonic and paleoclimatic forcing mechanisms are invoked to explain the pulsating nature of morphotectonic activity during the Cenozoic. As have previous researchers reporting cooling ages near the Oligo–Miocene transition from the Colombian Andes, Reyes-Harker et al. (2015) highlighted the potential role of the onset of PCB collision with South America as a driver of regional orogenetic activity. An important paleogeographic consideration for the Cenozoic presented by Reyes-Harker et al. (2015) is that of a Magdalena Valley that behaved, albeit transiently, as a somewhat endorheic fluvial system closed to the north.

Although not strictly on an axial position of the EC, recent work provides an interesting LTTC and sediment-stratigraphic (seismic reflection and crosscutting relationships) dataset that implies deformation-induced (folding and thrusting) exhumation affecting Cretaceous-to-Miocene sedimentary rocks immediately west of the Santander Massif, between the modern Magdalena Valley and the EC's western flank (Mora et al. 2013). Eight detrital samples yielded AFT ages and MTL, while vitrinite reflectance data were utilized to identify paleogeothermal gradients and locate the A-PRZ. AFT ages vary from ~81 to 16 Ma and include detrital, partially reset, and fully reset ages due to increased temperatures following a period of Oligocene burial (4 km). Although not discussed in the main text, the majority of MTL are 12 um and exhibit high values for standard deviation, potentially implying incomplete resetting and/or extended residence periods in the PAZ. LTTC by apatite fission-track dating and HeFTy® thermal modeling define a Paleogene (~60–50 Ma) onset of rapid exhumation along the modern boundary between the Magdalena Valley hinterland basin and the EC, reflecting the shortening and cooling associated with tectonic inversion of the EC's Mesozoic rift basin. This model is supported by the deformation identified in seismic profiles (Parra et al. 2012). Although slightly younger than the exhumation ages obtained from neighboring provinces (Gomez et al. 2005a, b), discrete pulses of Early Paleocene and Early–middle Eocene deformation have been reported from LLTC studies of the CC, EC, and middle Magdalene at the same latitude (Restrepo-Moreno 2009; Restrepo-Moreno et al. 2009c; Villagomez-Díaz 2010), thereby indicating ample deformational and morphotectonic activity during that interval.

Despite the episodic nature of uplift and exhumation in several Andean locales, both in and beyond Colombia, that is evident from LTTC data, Mora et al. (2013) posited that "rather than representing discrete episodes of deformation in the CC, middle MV, and westernmost EC, [their] results suggest broad activation of an integrated structural system." Furthermore, ignoring evidence presented in many of the studies presented in this review, as well as interpretations of the importance of a pulsating tectonic input (such as that related to accelerated an orthogonal phases of convergence) (Restrepo-Moreno 2009; Restrepo-Moreno et al. 2009c; Villagomez-Díaz 2010), Parra et al. (2012) concluded that "regardless of the principal driving forces, emerging evidence for the Northern Andes highlights the departure from common Andean models of episodic shortening." Yet, from that standpoint, unless "shortening" does not involve topographic construction and erosional exhumation, explaining the extensive vertical profile LTTC data that appear to display episodicity rather than continuity in Northern Andean morphotectonic trends is problematic. Thus, for instance, recent geophysical and structural LTTC studies, however, reveal "thin-" and "thick-skinned" tectonic events in the Garzón Massif that indicate considerable shortening during the Miocene associated with uplift of up to 6 km (Saeid et al. 2017; Velandia-Patiño 2018).

11.3.2.4 Isolated Ranges and the PCB

In this section we discuss LTTC datasets for the SNSM and smaller mountain ranges in Colombian territories that are part of the PCB.

Sierra Nevada de Santa Marta Of all the "isolated" topographic structures in the Colombian Andes, the SNSM has captured the attention of the geological community since the 1950s (Piraquive-Bermúdez 2016; Cardona et al. 2011; Villagomez-Díaz 2010; Gansser 1955; Tschanz et al. 1974) thanks to both its majestic geomorphic nature and the complex geological (lithology, structures) history it entails. Geomorphically, the SNSM is a highly dissected, triangular massif with high local relief and a maximum elevation of nearly 6 km. It rises from the Caribbean Plains in northern Colombia making it the second highest coastal relief in the world. Separating the SNSM from the EC is the Cesar-Rancheria depression (Figs. 11.1, 11.6, and 11.7). Geologically, the SNSM represents one of the three main upthrusts of Precambrian rocks in the Colombian Andes, along with the Garzón and Santander Massifs in the EC.[16]

This large massif is composed of high-to-medium-grade metamorphic rocks (e.g., Los Mangos Granulites, Dibulla Gneiss, etc.) as well as large plutonic masses, such as the Pueblo Bello, Central, Santa Marta, and Patillal batholiths (Piraquive-Bermúdez 2016; Piraquive Bermúdez et al. 2016; Villagomez-Díaz 2010; Colmenares 2007; Cardona-Molina et al. 2006; Tschanz et al. 1974). Structurally, the SNSM is part of the triangular Maracaibo Block, and several authors consider the massif to be a vast promontory that has penetrated >100 km in a NW direction into the Caribbean Sea since the Tertiary (Montes et al. 2010; Colmenares 2007; Tschanz et al. 1974). The SNSM is bounded to the north by the right-lateral Oca Fault (OF) running along the coast, to the west by the left-lateral Bucaramanga-Santa Marta Fault (BSMF), and to the southeast by the right-lateral Cerrejón Fault. Internally, the massif consists of three main provinces – the Santa Marta, Sevilla, and Sierra Nevada provinces – roughly oriented in a SW–NE direction and separated by major thrust faults. A complete review of the massif's lithological and structural characteristics is provided by Piraquive-Bermúdez (2016) and Kroonenberg (Chapter 210.1007/978-3-319-76132-9_5, this volume).

Relatively few thermotectonic studies have been undertaken in the SNSM. To date, the techniques used to assess its morphotectonic history include AFT, Zr-AHe, and Al-in-Hb thermobarometry (Piraquive-Bermúdez 2016; Piraquive Bermúdez et al. 2016; Cardona et al. 2011; Villagomez-Díaz 2010). On the basis of LTTC data derived from Eocene granitoids of the SNSM and Guajira, Cardona et al. (2011) constrained the exhumation history of an important segment of northern South America and concluded that major phases of uplift/exhumation occurred in the Late Eocene, Late Oligocene, and Miocene times. Results from the Guajira region reveal the Eocene and Oligocene events, although no cooling ages were reported for the more recent Miocene phase. Thermobarometry data also yield important information for elucidating morphotectonic processes in the SNSM. Aluminum data from hornblendes of the Santa Marta Batholith (Eocene) indicate that at least 15–20 km

---

[16] For details on isolated massifs exhibiting Proterozoic and Paleozoic lithologies, see Chap. 2, Contribution No. 3 of this volume.

of unroofing has taken place since crystallization (~52 Ma). In the Guajira Peninsula, calculated pressures for the Eocene Parashi stock and stratigraphic data from Oligocene conglomerates suggest that ~8 km of crust were lost from these massifs between the Eocene and Oligocene (Cardona et al. 2011).

Similarly, Villagomez et al. (2011) identify three major orogenetic phases in the SNSM: (i) Early Paleocene, at rates of >0.2 km/Ma; (ii) Early/Mid-Eocene, at even higher rates of ~0.3–0.5 km/Ma; and (iii) Late Oligocene, when northern portions of the SNSM Province exhumed at rates approaching 0.7–0.8 km/Ma. An additional phase may have occurred during the period ~25–16 Ma. A less significant cooling event, associated with rates of ~0.1 km/Ma, is documented for positions proximal to the Bucaramanga-Santa Marta Fault System during the period 40–25 Ma. It is proposed that the high rates of uplift/erosional exhumation in the SNSM shifted to the northwest during the Paleogene/Neogene via the propagation of NW-verging thrusts. No morphotectonic cooling has been reported from the SNSM after ~16 Ma. This is a somewhat contradictory result, provided the high relief and fluvial activity that presently characterize the region, as well as the lack of erosional/relict surfaces. Villagomez et al. (2011) concluded that the rock and surface uplift responsible for the modern topography are recent phenomena, potentially ≤1 Ma, giving insufficient time for the exhumed PAZ and PRZ to become exposed. This latter issue might be resolved via a more extended vertical profile approach, using the full range of altitudes in the SNSM combined with an extensive, multiple-dating LTTC analysis of sedimentary apatite and zircon in active fluvial systems draining the SNSM.

Mechanisms proposed to explain the morphotectonic history of the SNSM/Guajira region include an increased convergence rate and more orthogonal interaction between North and South America, shallow subduction of the Caribbean Plate, and/or post-Eocene changes in plate convergence (angle and rates), which caused the South American continental margin blocks to override the Caribbean Plate (orogenic float). Initial phases are generally attributed to the collision of the Caribbean Plateau with northwestern South America, while the more recent pulses were apparently driven by under-thrusting of the Caribbean Plate beneath northern South America during onset of subduction (Cardona et al. 2011; Villagomez et al. 2011). The absence of a Miocene phase in the Gaujira segment is used to imply a separation of the Guajira from the SNMS and the EC chain due to block rotation and transtensional tectonics during post-Eocene times (Cardona et al. 2011; Montes et al. 2010). More recently, Piraquive et al. (2017) proposed a broader array of mechanisms associated with the various pulses. Permian-to-Triassic topographic buildup and erosional exhumation were attributed to slab retreat and subsequent delamination of the lower crust. Recent studies reveal that erosional exhumation in the SNSM has been episodic and exhibits contrasting magnitudes among the various portions of the massif (Piraquive et al. 2017; Piraquive Bermúdez et al. 2016; Cardona et al. 2011; Villagomez et al. 2011). Post-magmatic cooling, due to erosional exhumation at rates of ~0.8 km/Ma, initially affected the middle reaches of the SNSM's northwestern flank between 50 and 45 Ma, with increased rates of ~2 km/Ma between 45 and 40 Ma. Cooling pulses reported from the massif's western flank were similar both in timing (~49–37 Ma) and intensity (0.4–0.8 km/Ma). Subsequent Paleogene pulses (e.g., 37–35 Ma) appear to have signaled an increase in the intensity of erosional exhumation (e.g., ~3 km/Ma) in some sectors of the SNSM, while a one-to-two-order of magnitude decrease is apparent from 37 to 13 Ma. During the Miocene, apatite helium ages show contrasting uplift/exhumation rates between northern (0.09–0.03 km/Ma) and western (0.1–0.3 km/Ma) sectors of the SNSM.

In general, exhumation in the SNSM appears to have been controlled by fault activity and/or elevated geothermal gradients (Piraquive et al. 2017; Piraquive Bermúdez et al. 2016). For example, slip along major NW-verging structures since Late Paleocene-Eocene times has served to increase topographic relief, thereby favoring vertical rock exhumation in excess of 11 km. Rapid exhumation under high geothermal gradients may have occurred through formation of low-viscosity mid-crustal channels. According to Piraquive-Bermúdez (2016), this process, which can occur during pluton emplacement, is associated with intruding pods fed from a thickened continental crust and localized at dissimilar depths. Channelized crustal flow may lead to focused exhumation of crystalline rocks in the middle to low crust via a process similar to the tectonic aneurysms proposed for the Himalayan syntaxes (Zeitler et al. 2001). In the SNSM, the litho-structural configuration apparently favors erosional cooling via channelized crustal flow (Piraquive et al. 2017; Piraquive-Bermúdez 2016). The lack of Late Pliocene-to-recent erosional exhumation in the SNSM, as indicated by multiple LTTC datasets, has been used by Piraquive-Bermúdez (2016) to conclude that recent surface uplift, still unresolved by erosion, relates to isostatic rebound caused by mantle upwelling (see also Villagomez et al. 2011), which in turn resulted from convective removal of the crust beneath the SNSM. This scenario is supported by gravity and magnetic data indicating a diminished crustal thickness of <25 km beneath the SNSM (Poveda et al. 2015; Camargo 2014). Caution must be exercised when interpreting LTTC data for the SNSM and Guajira as it is plausible that some of the cooling ages there recorded are associated with crustal refrigeration under flat subduction regimes rather than to unroofing due to uplift-driven erosion.

Despite of the multiplicity of mechanisms proposed for the SNSM, we note that this pattern of Eocene and Oligocene uplift and exhumation is exhibited by other segments of Colombia's cordilleran systems, detached both structurally and geographically from the SNSM, and by the PCB and Andean elements in Ecuador, Peru, and Bolivia. Consequently, it is necessary to pose the question of whether it is reasonable to consider large, continental-scale drivers of the various orogenetic phases for the whole Andean system rather than more localized explanations. We consider it likely that plate interactions on the scale of the northern South America and Caribbean Plate boundary are linked to larger plate margin dynamics affecting the greater Andean system without disregarding the importance of more local effects (structural, isostatic, magmatic, etc.).

Mountain Ranges in the PCB The PCB encompasses two principal mountain ranges known as the Darién and Baudó serranías, which are characterized by generally low summit elevations (average of ~300 m and rarely >900 m). Both ranges, which run roughly parallel to one another, exhibit arcuate form in planar view and are more accentuated in Colombia than in Panamá (Fig. 11.2). A zone of flat topography between the two marks the position of thick sedimentary successions of the Atrato-Chucunaque basins and the modern lowlands of the Atrato River (Coates et al. 2004; Duque-Caro 1990b; Garzon-Varon 2012).

Recent work in Panamá (LTTC, U/Pb geochronology, geochemistry, etc.) has shown that major pulses of diachronous tectonism (e.g., exhumation, burial/inversion, etc.) occurred in discrete intervals during the Eocene, near the Oligo-Miocene transition, and in Early-Late Miocene (Ramírez et al. 2016; Montes et al. 2012b; Farris et al. 2011; Restrepo-Moreno et al. 2010). LTTC results from Panamá provide crucial information on erosional exhumation related to surface uplift and to the gradual establishment of the Panamanian Isthmus (O'Dea et al. 2016; Ramírez et al. 2016; Montes et al. 2012b; Farris et al. 2011; Restrepo-Moreno et al. 2010; Coates et al. 2004; Kellogg and Vega 1995; Duque-Caro 1990a, b). The possible causes of surface uplift and exhumation of the Panamá Arc are numerous and include crustal thickening due to magmatic addition, tilting of entire crustal blocks during oroclinal development, enhanced Eocene convergence of the Farallon and Caribbean plates, collisional phases of the PCB with NW South America, collision/accretion of seamounts (e.g., the Azuero Accretionary Complex), and or indenters (oceanic plateaus, Coiba Ridge) (Ramírez et al. 2016; Montes et al. 2012b; Farris et al. 2011; Buchs et al. 2010; Kerr et al. 1997).

The asynchronous nature of morphotectonic upheavals (post-crystallization plus uplift/erosional exhumation cooling) is evident from the distribution of LTTC data even in neighboring plutonic masses of similar crystallization age, such as the Cerro Azul and Mamoní plutons in the San Blas-Darien Range. In the AHe system, these units record Oligocene and middle Miocene erosional cooling at rates of ~0.2 km/Ma, respectively. Significantly higher cooling rates, on the order of 100 °C/Ma (i.e., erosional exhumation of ~4 km/Ma), are inferred from ZHe data from the same batholiths, although this value represents an average of both post-magmatic cooling and subsequent erosional cooling due to unroofing. Similarly, basement rocks in the Azuero Peninsula were exhumed at rates of 0.2 km/Ma during the Eocene (Valle Rico pluton) and Middle-to-Late Miocene (Cerro Montuoso and Parita intrusives) (Ramírez et al. 2016; Montes et al. 2012b). It is important to bear in mind that LTTC ages (systems with closure temperature below 250 °C) in Panamá are typically much younger than U/Pb zircon crystallization ages, with the exception of the Mamoní pluton, indicating that measured ages mark the onset of cooling due to erosional exhumation. The LTTC data presented by Ramírez et al. (2016) also show that sedimentary formations (e.g., Gatuncillo and Cobachón formations) experienced inversion and uplift/exhumation in the Miocene. Among the mechanisms invoked to explain erosional cooling of intrusive rocks in the Panamanian Arc are collision/accretion of seamounts (for the Azuero Peninsula), oceanic plateau collision, and the onset and continued collision of the PCB against South America (Ramírez et al. 2016; Montes et al. 2012a, b; Farris et al. 2011; Restrepo-Moreno et al. 2010).

11.4 Brief Notes on Erosion Rates and Paleogeogprahic Development in the Northern Andes

After reviewing erosion rates derived from LTTC in the Colombian Andes, it is sensible to ask just how much erosional exhumation is "a lot," so that morphotectonic and orogenetic process in the Northern Andes can be placed in the global context. Similarly, the multiple regional orogenetic paroxysms that took place during the Late Mesozoic and throughout the Cenozoic imply major changes in the paleogeographic configuration of the Northern Andes. In this section we briefly introduce the discussion on these two interrelated topics so that general ideas can be used to illuminate both subjects.

11.4.1 Erosion Rates in the Northern Andes: From a Regional to a Global Perspective

How meaningful are the magnitudes of erosional exhumation pulses in the Northern Andes, as estimated from LTTC? One could also ask whether the rates of uplift-driven erosional exhumation reconstructed from these datasets are to some extent representative of the mountainous environment of the Northern Andes. We use two simple approaches to test the overall validity of LTTC exhumation figures in this context. The first uses compilations of long- and mid-term erosion rates and compares figures with the datasets discussed in this contribution. For simplicity, we choose to represent this data in the form of a plot depicting erosion rates for various morphotectonic and morphoclimatic regimes worldwide (Fig. 11.12).

It is clear that the LTTC data discussed here fall within the normal range of magnitude for erosional exhumation trends in tectonically active, tropical-to-temperate mountainous regions (e.g., Andes, Alps, New Zealand, Taiwan). Recent studies of erosion rates in several morphoclimatic and morphotectonic regions (Montgomery 2007; Aalto et al. 2006; Heimsath 2006; Bierman 2004; Kirchner et al. 2001) and at various temporal scales (Herman et al. 2013; Matmon et al. 2003; Montgomery and Brandon 2002) indicate that rates of 0.1–1 km/Ma are typical of most orogenic systems (the Alps in Europe and the Southern Alps in New Zealand, the Andes in South America, etc). Notable exceptions are some areas of the Taiwan mountains (>5 km/Ma) and the Himalayan syntaxes (>10 km/Ma) where extremely fast erosional exhumation is driven by both structural and climatic controls, resulting in a tectonic aneurysm (Grujic et al. 2011; Willett et al. 2003; Zeitler et al. 2001). In some cases, it is possible that long-term erosion rates reflect the increased glacial activity that has characterized the last 2–3 million years of Earth's history (Shuster et al. 2005a; Peizhen et al. 2001). Our review of morphotectonic process in the Colombian Andes shows that average rates of erosional exhumation fall close to 0.1 and 1 km/Ma. Only in a few instances and specific locations, characterized by particular tectonic (structures) and/or climatic configurations (SNSM, Santander Massif, etc.), do erosion rates approach Himalayan or Taiwanese values of >5 km/Ma.

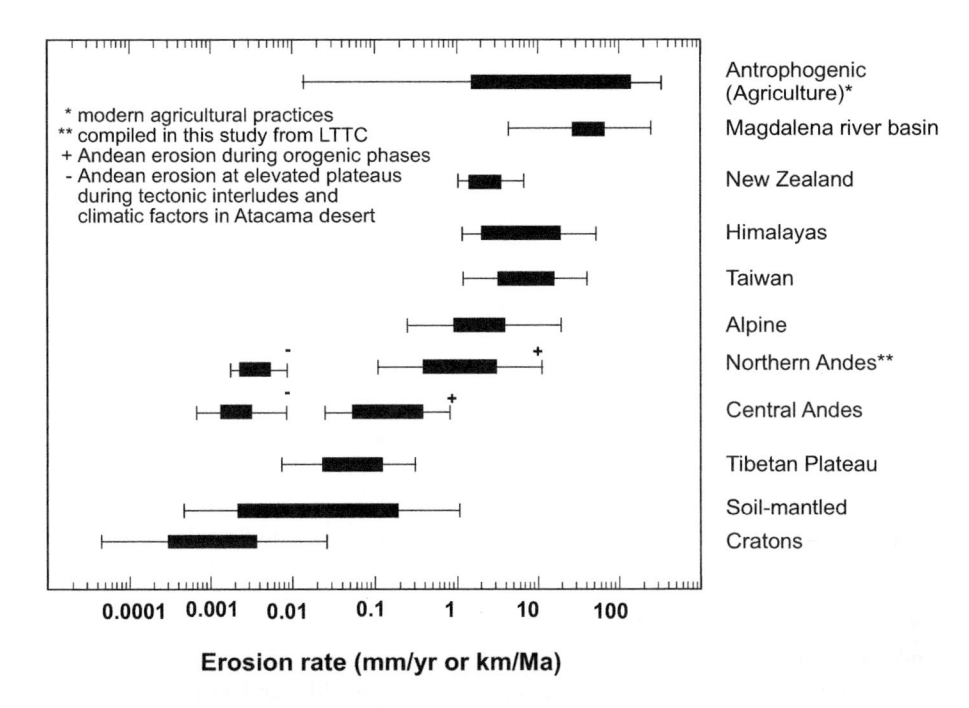

Fig. 11.12 Relative magnitudes of global erosion rates for various modern orogenic systems, cratonic environments, and anthropogenic activities (modern mechanized agriculture). (After Montgomery 2007)

The most striking feature of recent compilations of modern and geologic erosion data is the rapidity of denudation typical of industrialized agriculture, which exceeds not only background erosion in the majority of natural geomorphic environments (Montgomery 2007, 2008) but also global soil recovery (Wilkinson and Humphreys 2005). This pattern raises two important problems: (i) the issue of soil conservation and societal success (Montgomery 2008; Lal 2001) and (ii) the lack of precision implied by extrapolating modern erosion rates to reconstruct larger time spans, given the pervasive effect of humans in geomorphic (particularly fluvial) environments (Douglas 1967; Goudie 1995; Walling 2006). As an example from the Northern Andes, the modern Magdalena River basin exhibits erosion rates higher that 10 mm/yr. (i.e., >10 km/Ma!) as reconstructed for several tributary catchments using sediment transport data (Restrepo et al. 2006). According to LTTC datasets discussed in this contribution, this is an anomalous geomorphic condition that evidences the "hypererosive" regime to which tropical fluvial basins have been subjected though anthropogenic activities. One thing appears certain in terms of modern vs. long-term geomorphic patterns for the Colombian Andes: that, as in the rest of the world (Syvitski 2012; Montgomery 2008; Syvitski and Milliman 2007; Wilkinson 2005; Wilkinson and McElroy 2005; Hooke 2000), humans are the predominant geomorphic agent on Earth today, a fact that can be considered a hallmark of the "Anthropocene" (Syvitski 2012).

The second approach utilizes the functional relationship between local relief and erosion rate proposed by Ahnert (1970). Although this simple function probably incorporates significant oversimplifications, it nonetheless expresses well the intimate relationship between landscape morphology and denudation, allowing swift estimation of sediment fluxes and average erosion rates via readily available topographic information in the form of DEMs (Roering et al. 2007; Montgomery 2003; Gunnell 1998). We apply this functional relationship as a preliminary, order-of-magnitude estimate of erosion, as it is widely acknowledged that this approach was developed for short-term data in temperate mountainous systems and, thus, that tropical climate and active tectonics might modify geomorphic activity. Even under this metric, exhumation rates for the various subsegments of the Northern Andes of Colombia fall between 0.2 and 1 km/Ma, as predicted by LTTC on long-term time scales. This external check gives us confidence that the rates and magnitude of orogenetic phases predicted for the Colombian Andes on the basis of LTTC data fall within a coherent range of morphotectonic processes.

A further question helps contextualize the efficacy of erosion rates not only in sculpting the landscape but also in regulating other surficial and deep-seated orogenetic process: what would be the visible effect of something as "minimal" in denudation terms as 10 mm/yr.? Such rates are typical of the Himalayas and certain Andean segments and on modern farms. In geological terms, this short-term rate of 10 mm/yr. translates to 10 km/Ma. At such a rate, it would take less than 1 million years to fully devastate the Himalayas, assuming no additional tectonic or isostatic input (Fig. 11.13).

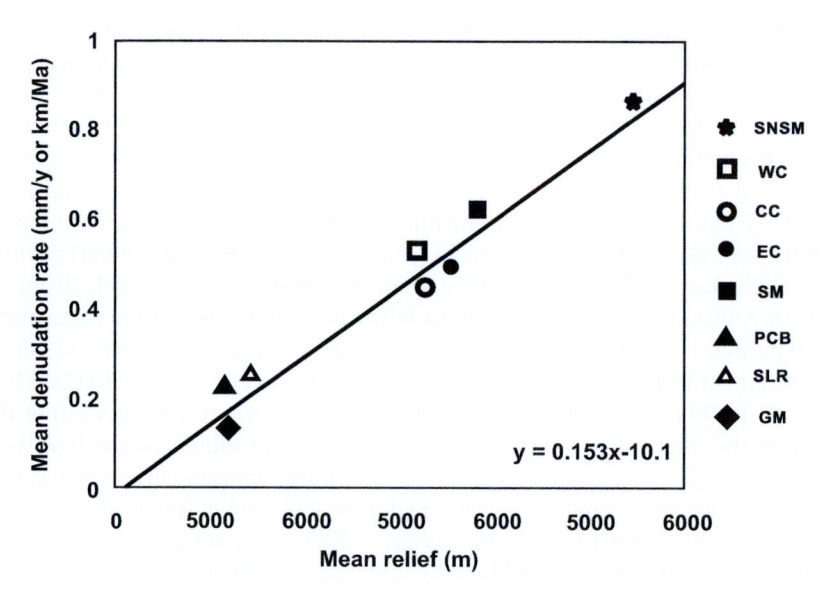

Fig. 11.13 Functional relationship between mean relief and erosion rates Ahnert (1970) applied to datasets from the Northern Andes of Colombia compiled in this contribution

11.4.2 LTTC-Derived Orogenetic Events in the Colombian Andes and Their Paleogeographic Repercussions, from the Cretaceous to the Holocene

The Cenozoic paleogeographic and paleotopographic evolution of the Northern Andes and the circum-Caribbean is still a matter of debate (Amaya et al. 2017; Anderson et al. 2016; O'Dea et al. 2016; Hoorn and Flantua 2015; Montes et al. 2015; Reyes-Harker et al. 2015; Coates and Stallard 2013; James 2006; Kroonenberg et al. 1990; Pindell et al. 2006; Restrepo-Moreno 2009; Villamil 1999). LTTC data, along with sediment stratigraphy and structural information, can be used to reconstruct the overall patterns, and Fig. 11.14 attempts to blend those data to provide a general summary of paleogeographic-paleotopographic events distributed along what we think are the major pulses of morphotectonic activity in the Late Cretaceous-Early Paleocene, Lower/Middle Eocene, Late Oligocene, and Miocene-Present.

From the Cretaceous to the present, the most significant events are plausibly as follows. The prevalence of marine sedimentary covers over a vast portion of Colombia in the Early Cretaceous reflects a major regional transgression, which came to a halt near the K-T transition. During the Late Cretaceous and Early Paleogene, LTTC shows that several crystalline massifs were experiencing erosional cooling, mainly in the northern segments of the proto-CC and proto-EC, producing the first topographic highs at that time. Deposition was localized on the submerged proto-EC, the Llanos, and in some inter-Andean positions. As uplift in the CC progressed over the course of several pulses (Paleocene, Eocene, Oligo-Miocene, etc.), the locus of deposition moved progressively eastward (Reyes-Harker et al. 2015; Mora et al. 2013; Horton et al. 2010a, b; Bayona et al. 2008; Gomez et al. 2005a, b; Gomez 2001; Villamil 1999). Compressional forces, in conjunction with profuse Late Cretaceous and Early Paleogene magmatism, probably played a key role in triggering uplift and concomitant erosional exhumation in some Northern Andean segments.

Tectonic quiescence apparently prevailed over the remainder of the Paleocene and into the Early and middle Eocene, when LTTC data from throughout the Colombian Andes, as shown in this contribution, indicate one of the strongest tectonic upheavals both for cordilleran massifs (WC, CC, EC, SNSM, etc.) and sedimentary basins (inter-Andean basin and marginal basins such as the middle Magdalena Valley and the proto-EC). Although Villamil (1999) proposed a double south-to-north and west-to-east progression of deformation, LTTC data compiled in this contribution do not support that hypothesis. In fact, regions as remote from one another as Peru and Cuba/Leeward Antilles were experiencing coeval morphotectonic activity of similar magnitude (Barnes et al. 2006; Coughlin et al. 1998; Kennan et al. 1997; Mégard 1984; Mégard et al. 1984; Rojas-Agramonte et al. 2006; Van der Lelij et al. 2010). We posit that a major Eocene perturbation of the geodynamic setting on a continental scale, such as shifts in converge speed and direction between the Farallon-South American plates, may provide a simple explanation for large-scale mountain building in the region. Obviously, differences in the litho-structural

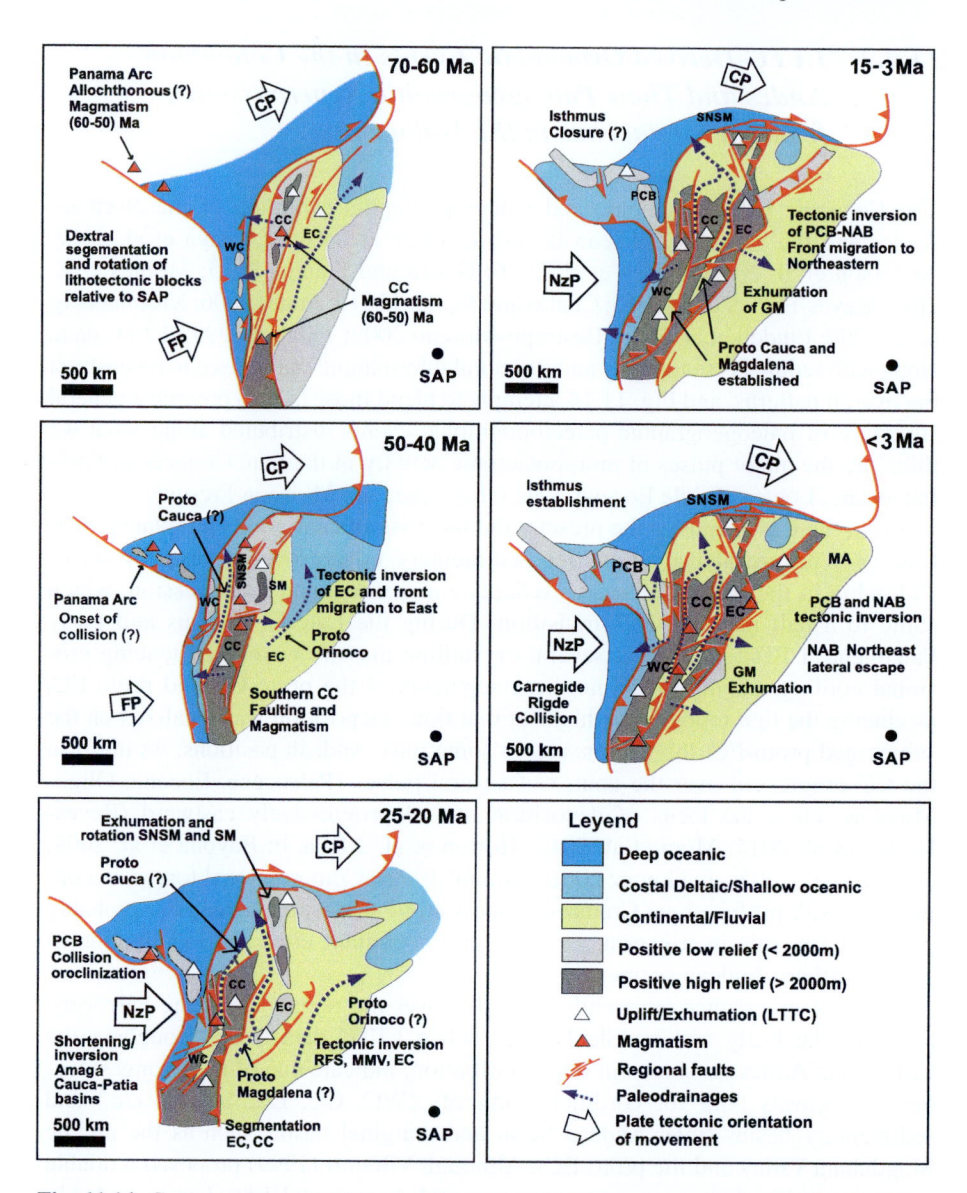

Fig. 11.14 Conceptual model of topographic and paleogeographic evolution in the Northern Andes Major plates are Farallon Plate (FP), Nazca Plate (NzP), Caribbean Plate (CP), South American Plate (SP, reference frame). The "location" of the Western, Central and Eastern cordilleras (WC, CC, and EC) and the Sierra Nevada de Santa Marta (SNSM) correspond to the relative positions of the paleo-topographic highs they represented in each period, i.e., proto-cordilleras. After Villamil (1999), Kennan and Pindel (2009), and Cediel et al. (2011)

configuration characteristic of the "multi-block" geologic milieu of the Northern Andes can exert major local controls on orogenetic activity, producing the asynchronous patterns observed in some sectors of the Colombian Andes.

The general tectonic regime, including shifts from predominantly compressive to predominantly transpressive conditions, also play a key role. The CRFS and associated structures offer a good example of this. There, major crustal structures may at times facilitate both the vertical displacement (e.g., AP, Restrepo-Moreno et al. 2009b, c) and major strike-slip motion (Vinasco and Cordani 2012; Chicangana 2005; Ego et al. 1995) necessary to develop pull-apart inter-Andean basins, such as the Amagá Formation that accommodates large amounts of detrital material being shed since the Eocene from the CC and WC into the Cauca Depression (Marín-Cerón et al. 2015; Restrepo-Moreno et al. 2013a, b, Silva et al. 2008; Sierra et al. 2004; Grosse 1926).

Another major orogenic phase in Colombia inferred from LTTC occurred near the Oligo-Miocene transition and has been reported for various sectors of the CC, WC, Santander Massif, SNSM, Serranía de Perijá, and PCB (Amaya et al. 2017; Noriega-Londoño 2016; Cardona et al. 2011; Villagomez et al. 2011; Restrepo-Moreno et al. 2009c). This event represents unroofing in excess of 2–3 km in some places, yet it goes undetected in the CRFZ, SLR, and Garzón Massif (see Fig. 11.11). According to some authors, this morphotectonic pulse modified depocenters in the EC and Magdalena Basin (Gomez 2001; Sánchez et al. 2012; Villamil 1999) and was associated with initial development of the Orinoco River as a NE-flowing fluvial system (Reyes-Harker et al. 2015). At the continental scale, major reorganization of the Amazon fluvial system also took place during this time (Hoorn et al. 1995, 2010; Mora et al. 2010a).

It is interesting to note how the evolution of LTTC studies has itself unraveled morphotectonic patterns in more detail. Take, for instance, the CC, where AFT and ZFT studies (Villagomez and Spikings 2013; Montes-Correa 2007; Saenz 2003) found no evidence for the Oligo-Miocene orogenic phase in the AP. It was not until regional AFT and AHe data were generated along vertical profiles (Restrepo-Moreno 2009; Restrepo-Moreno et al. 2009c) that the timing and magnitude of this major orogenic pulse became evident. Today, this fundamental morphotectonic event is clearly recognized in several locations in the Colombian Andes, both in cordilleran massifs and sedimentary basins.

Finally, Neogene-to-Quaternary orogenic activity prevailed in the Northern Andes. Although we have shown that the Andes are the product of multiple morphotectonic phases throughout the Late Mesozoic and Cenozoic, this period of morphotectonic paroxysms is commonly referred to as the "Andean Orogeny" (Gregory-Wodzicki 2000) or the "Eu-Andina Phase" (Van der Hammen 1961). This is probably because during this period the Northern Andes, and the Andean range in general, acquired its present configuration in terms of topography and oroclinal development, fluvial arrangement, morphoclimatic regions, and ecological niches (Reyes-Harker et al. 2015; Mora et al. 2013; Ochoa et al. 2012; Hoorn et al. 2010; Kroonenberg et al. 1990; Villamil 1999). The majority of Andean segments in Colombia for which LTTC data exist record morphotectonic activity at some point during the Miocene–Pleistocene interval. Particularly strong is the middle Miocene phase that, in the WC, was accompanied by magmatism and may have triggered the establishment of the WC as a major orographic barrier (Pardo-Trujillo et al. 2015) and the development of the hyperpluvial area we know today as the Chocó Biogeographic Region characterized by having some of the rainiest province on the planet (Poveda and Mesa 2000). Notable sites where this Miocene tectonic paroxysm is not evident from LTTC data are the San Lucas Range and the AP in the northern CC. In the case of the AP, this finding led to the proposition that the lithostructural and geomorphic configuration in the vicinity of the AP is responsible for its lagged geomorphic response to tectonic input (Noriega-Londoño 2016; Restrepo-Moreno 2009; Restrepo-Moreno et al. 2009b, c).

A key outcome of the inherited crustal grain exhibiting a predominant N–S orientation, in conjunction with continued morphotectonic activity in the Northern Andes since the Paleocene, also strongly controlled by both "thin-" and "thick-skinned" tectonics, was the gradual configuration of an inter-Andean fluvial network capable of moving tremendous amounts of detrital materials in a south-to-north fashion (e.g., Cauca and Magdalena Rivers). We note that several studies (Anderson et al. 2016; Reyes-Harker et al. 2015; Silva et al. 2013) converge on the fact that these longitudinal drainage patterns sometimes evolved into closed basins, which transiently interrupted the delivery of sediments to distal basis such as the Caribbean and LMV. This feature appears to have been a characteristic of the early stages of inversion.

11.5 Concluding Remarks

Information on the orogenic development of the Colombian Andes derived from LTTC data reveals that:

1. The large-scale topography of the Colombian Andes has evolved through multiple paroxysms during the Cenozoic. Existing LTTC studies in the Colombian Andes show a complex spatiotemporal distribution even within the same mountain ranges. However, the majority of erosional exhumation events are associated with discrete phases of tectonic-driven uplift. Such phases cluster around the intervals Early Paleocene, Eocene, Late Oligocene, and Miocene. Therefore the term Andean Orogeny used in the literature to refer to Miocene to present uplift of the Andean range is not suitable for the Northern Andes of Colombia, and perhaps not suitable for the Andes at all, because of the episodic, asynchronous, and spatially heterogeneous nature of morphotectonic activity across the range since Late Cretaceous and until present.
2. Tectonic upheavals that drove uplift and concomitant erosion-exhumation occurred, on average, at rates between 0.2 and 1 km/Ma, except for some portions of the Santander Massif and the SNSM, which experienced rates of >5 km/Ma. Unroofing proceeded at magnitudes of 2–6 km per event. These rates and the amounts of crustal removal involved are within the normal range

for tectonically active regions, such as the Andes, Alps, and Taiwan, but are, on average, lower than Himalayan rates. Surface uplift in excess of 2500 m is evident from the presence of marine sedimentary sequences now at elevations of 2000–3000 m, primarily in the EC.
3. In some locales, recognized periods of morphotectonic activity were separated by extended interludes (e.g., slow and/or oblique convergence) over 15 million years in duration. In the AP region, these interludes sustained very low rates of erosional exhumation (i.e., ~0.008 km/Ma) – basically Cratonic rates in mountainous settings (e.g., Restrepo-Moreno et al. 2009b, c)! The low rates evident in LTTC data, together with remnants of elevated plateaus found in various segments of the Colombian Andes, indicate that in some cases these interludes allowed the development of low-elevation erosional surfaces that are now integral to the geomorphic configuration of the CC, EC, and WC.
4. During extended (>10–15 Ma) intervals of tectonic quiescence, thick saprolitic covers developed. In the case of crystalline massifs, this process may have produced typical etch typography, such as that reported for the AP. Thick, deeply weathered mantles in felsic crystalline provinces served as "reservoirs" for significant volumes of quartz that subsequently were released as quartz "bursts" during the Paleogene and Neogene, which filled terrigenous sequences in various sectors of the Cauca and Magdalena fluvial basins (see, e.g., Fig. 11.14 of Reyes-Harker et al. 2015, which shows increases in quartz deposition from the Eocene to the Oligo-Miocene transition).
5. The efficacy of unroofing in the Northern Andes of Colombia is evidenced by the presence of deep-crustal materials brought to the surface during the Cenozoic (i.e., mosaics of catazonal plutonic and high-grade metamorphic lithologies found in some of the crystalline massifs reviewed above). Shallower erosional exhumation is also indicated in some regions by the proximity of crystallization and cooling ages for some plutonic masses, particularly in the WC and the PCB.
6. Some of the highest topographic features in the Andes (e.g., SNC and SNSM) do not show evidence of Neogene magmatism, thereby indicating that their Late Cenozoic orogenetic history is dominated by tectonic-driven uplift. Interestingly, these two massifs are adjacent to some of the thickest accumulations of sedimentary materials in Colombia. On the other hand, topographic highs in the WC (~4000 m) and central portions of the CC (~5000 m) are associated with Neogene magmatism, such that topographic buildup in these areas may have resulted from a combination of tectonic and magmatic mechanisms.
7. Even though phases of Andean uplift were concentrated chiefly in major cordilleran massifs separated by fluvially drained inter-Andean basins, the vast volumes of detrital material produced via Cenozoic erosion and stored in inter-Andean and marginal basins throughout Colombia signify the high degree of regional morphotectonic activity. Numerous sedimentary formations in Colombia exhibit depositional discontinuities and recurrent basal conglomeratic sequences, which are indicative of morphotectonic events also affecting the inter-Andean and marginal depressions. In effect, prior to the extensive use of

LTTC, tectonic paroxysms in the Northern Andes were reconstructed from coarse facies deposition (although conglomeratic deposits do not always represent tectonic input and/or topographic buildup). Sedimentary sequences in Colombia are important sources of LTTC information from which we can infer burial histories (heating) and inversion-exhumation (cooling), as well as provenance signatures (radio and thermochronology), to validate paleogeographic scenarios.

8. The timing of erosional exhumation in the Colombian Andes exhibits marked synchronicity with similar events across the Andean range and the circum-Caribbean (Quechua, Incaic phases). Periods of morphotectonic activity appear to coincide with major geodynamic reorganization (e.g., enhanced convergence, plate partitioning, orthogonalization, collisional episodes), giving morphotectonic activity in the Andes a continental signature. Nevertheless, more localized morphotectonic events have also been reported from several segments of the Northern Andes of Colombia indicating the dominance of local controls.

9. The mechanisms invoked to explain orogenesis in the Colombian Andes are numerous (e.g., changes in subduction parameters, compression/transpression, accretionary/collisional events, orogenic floats, crustal thickening-thinning, mantle circulation, focalized rainfall, crustal flow, etc.) and vary from region to region. Many such mechanisms are linked to complex subduction-accretion dynamics and to inherited crustal structures. Both circumstances are prevalent in the litho-structural and geodynamic realm embodied by the Northern Andes. In some locales, crustal thickening by magmatic addition or thrusting is the preferred mechanism. Localized and pre-existing weak-crust, restricted higher heat-flow, and/or orogenic floats are potentially important factors, particularly with regard to topographic residuals and isostatically uncompensated ranges such as the SNSM and the WC. A number of studies attribute Late Oligocene-to-Early Miocene orogenetic activity in multiple Andean sectors to the onset of the PCB collision. Even though the PCB is certainly an important factor, using it as a morphotectonic "culprit" makes it harder to explain spatiotemporal correlations with other, larger-scale processes, such as increased rates of convergence between the Nazca and South American plates.

10. Orogenic floats and flat subduction (as potential morphotectonic mechanisms) are important factors to consider in interpreting LTTC datasets. Changes in subduction dynamics from "normal" to shallow can force out the hot asthenosphere and replace it with the cold subducting slab leading to crustal refrigeration. Under such hypothetical scenario, it is plausible that some of the LTTC ages recorded for the SNSM and Guajira reflect onset of shallow subduction and concomitant refrigeration of the crust rather erosional exhumation.

11. Despite the diversity of mechanisms proposed as drivers of Andean uplift/exhumation in Colombia, LTTC evidence highlights the prevalence of morphotectonic paroxysms of ~3–5 Ma duration in the Paleogene, Eocene, Late Oligocene, and Miocene. The relatively diachronous nature of uplift/exhumation may be due to the differential uplift and/or strain partitioning heterogeneities common in such complex litho-structural milieux.

12. Tectonically driven uplift and associated fluvial erosion are central to defining patterns of atmospheric circulation (e.g., WC uplift and the low-level jets and rain-shadow systems, EC orographic precipitation, etc.). In some cases, positive feedbacks are established between orogenic uplift and concentration of precipitation, resulting in more efficient fluvial dissection and, ultimately, greater erosion such in the case of the SNSM and some sectors of the EC and the WC. The number of mechanisms invoked to explain erosional exhumation and uplift in the context of the complex tectonic, lithologic, and structural configuration of the Colombian Andes, and potential feedbacks established among crustal deformation/structure, geomorphology, and climate remain speculative. Many research groups continue to work in Colombia on the application of LTTC in cordilleran massifs, sedimentary basins, and modern fluvial networks to improve our understanding of these highly interconnected and spatiotemporally complex processes.
13. Fluvial erosion is key in the development of both small- (wavelength 0.1–1 km) and large-scale (wavelength 10–100 km) topographic features. In certain instances, erosion is broad and occurs over areas of $>1 \times 10^4$ km^2, potentially triggering isostatic compensation. However, the rugged topographic configuration of the Colombian Andes focuses that erosion in narrow inter-Andean valleys with local relief often exceeding 4 km. Thus, the volume of material removed is relatively small and unlikely to cause significant isostatic rebound at typical lithosphere elastic thicknesses. Nonetheless, under specific rheological conditions, focalized incision of valley bottoms might trigger increased elevation of interfluves via isostatic rebound.
14. Although several authors have suggested that the modern, trident-shaped Colombian Andes (i.e., WC, CC, EC) have been in place since the Early Tertiary, the review presented here indicates otherwise. The modern Andean topography may be strongly influenced by Miocene and younger orogenetic activity. However, various Colombian sectors record significant pulses of surface uplift and erosional exhumation since the Early Paleogene, such that the current topography is a palimpsest of Cenozoic morphotectonic (tectonics-climate) events.
15. Long-term, natural erosion rates inferred from LTTC for several segments of the Colombian Andes serve as a background to evaluate modern (i.e., anthropogenic) erosion rates. Other contributions in this volume point to sediment yields in the Magdalena Basin that can be translated to average, basin-wide erosion rates above 10 km/Ma! Such rates are similar to those in zones of tectonic aneurysm, such as the Himalayas. Increased erosion is a hallmark of the Anthropocene. Both long- and short-term erosion rate data need to be generated to better understand how the landscape evolves at different spatiotemporal scales and to assess the human impact on geomorphic processes.

The coming decades will bring a wealth of crucial LTTC information. However, our attempts to unravel key elements of the topographic evolution of the Northern Andes require multidisciplinary approaches that tackle both internal and external process (e.g., crustal deformation, mountain uplift, unroofing/exhumation, climate, paleoecology, paleogeography, etc.), as these processes are seldom operating independent of one another. Key issues for future research on Colombia's tectonic geomorphology must include better constraints on crustal structure (past and present, surface and deep), particularly concerning the chronology of fault activation and kinematics. To this end, new tools that complement LTTC studies include quantitative geomorphology supported by digital terrain analyses, reanalysis of seismic datasets, numeric modeling of topographic, structural (crust and lithosphere scale) and LTTC data, paleogeographic reconstructions, and so on. Further work in LTTC, particularly in the development of high-resolution, vertical profile studies and LTTC multiple dating in modern and old detrital material units, in combination with numerical thermomechanical computer modeling, will help provide more accurate spatiotemporal scenarios for interpreting the orogenetic history of the fascinating and geologically complex region embodied by the Northern Andes of Colombia.

References

Aalto R, Dunne T, Guyot JL (2006) Geomorphic controls on Andean denudation rates. J Geol 11:85–99

Acosta J, Lonergan L, Coward M (2004) Oblique transpression in the western thrust front of the Colombian Eastern Cordillera. J South Am Earth Sci 17(3):181–194

Ahnert F (1970) Functional relationships between denudation, relief, and uplift in large mid-latitude basins. Am J Sci 268:243–263

Álvarez J (1983) Geología de la Cordillera Central y el occidente colombiano y petroquímica de los intrusivos granitoides Meso-Cenozoicos. Boletín Geológico de Ingeominas 26(2):1–175

Amano K, Taira A (1992) Two-phase uplift of Higher Himalayas since 17 Ma. Geology 20(5):391–394

Amaya S, Zuluaga CA, Bernet M (2017) New fission-track age constraints on the exhumation of the central Santander Massif: Implications for the tectonic evolution of the Northern Andes, Colombia. Lithos 282:388–402

Amaya-Ferreira S (2016) Termocronología y Geocronología Del Basamento Metamórfico del Macizo de Santander, Departamento de Santander, Colombia. [PhD Thesis]. Universidad Nacional de Colombia, Bogotá p 172

Anderson VJ (2015) Uplift and exhumation of the Eastern Cordillera of Colombia and its interactions with climate [PhD Thesis]. University of Texas, Austin USA, p 185

Anderson VJ, Saylor JE, Shanahan TM, Horton BK (2015) Paleoelevation records from lipid biomarkers: application to the tropical Andes. Geol Soc Am Bull 127(11–12):1604–1616

Anderson VJ, Horton BK, Saylor JE, Mora A, Tesón E, Breecker DO, Ketcham RA (2016) Andean topographic growth and basement uplift in southern Colombia: Implications for the evolution of the Magdalena, Orinoco, and Amazon river systems. Geosphere 12(4):1235–1256

Andriessen PAM, Helmens KF, Hooghiemstra H, Riezebos PA, van der Hammen T (1993) Absolute chronology of the PlioceneQuaternary sediment sequence of the Bogota area, Colombia. Quat Sci Rev 12:483–501

Arancibia G, Matthews SJ, De Arce CP (2006) K-Ar and Ar-40/Ar-39 geochronology of supergene processes in the Atacama Desert, Northern Chile: tectonic and climatic relations. J Geol Soc 163:107–118

Arias LA (1995) El relieve de la zona central de Antioquia: un palimpsesto de eventos tectónicos y climáticos. Revista Faculta de Ingeniería Universidad de Antioquia 10:9–24

Arias L (1996) Altiplanos y cañones en Antioquia: una mirada genética. Revista Facultad de Ingeniería Universidad de Antioquia 8(2):84–96

Aspden J, McCourt WJ (1986) A Mesozoic oceanic terrane in the central Andes of Colombia. Geology 14:415–418

Aspden JA, McCourt WJ, Brook M (1987) Geometrical control of subduction-related magmatism – the Mesozoic and Cenozoic plutonic history of western Colombia. J Geol Soc Lond 144:893–905

Avouac JP, Burov EB (1996) Erosion as a driving mechanism of intracontinental mountain growth. J Geophys Res 17:747–717, 769

Bakioglu KB (2014) Garzón Massif basement tectonics: a geopyhysical study. Upper Magdalena Valley, Colombia: University of South Carolina

Barbosa-Espitia AA, Restrepo-Moreno SA, Pardo A, Osorio J, Ochoa D (2013) Uplift and exhumation of the southernmost segment of the Western Cordillera (Colombia) and development of the neighboring Tumaco Basin, in Proceedings 2013 GSA Annual Meeting 125th Anniversary of GSA, Denver CO USA, 2013, vol 45, Geological Society of America Abstracts with Programs, p 542

Barnes JB, Ehlers TA, McQuarrie N, O'Sullivan PB, Pelletier JD (2006) Eocene to recent variations in erosion across the central Andean foldthrust belt, northern Bolivia: implications for plateau evolution. Earth Planet Sci Lett 248(1–2):118–133

Barrero D, Pardo A, Vargas CA, Martinez JI (2007) Colombian sedimentary basins: nomenclature, boundaries and petroleum geology, a new proposal. Agencia Nacional de Hidrocarburos – A.N.H.

Bayona G, Cortés M, Jaramillo C, Ojeda G, Aristizabal JJ, Reyes-Harker A (2008) An integrated analysis of an orogen–sedimentary basin pair: Latest Cretaceous–Cenozoic evolution of the linked Eastern Cordillera orogen and the Llanos foreland basin of Colombia. GSA Bull 120(9–10):1171–1197

Bayona G, Montes C, Cardona A, Jaramillo C, Ojeda G, Valencia V (2011) Intraplate subsidence and basin filling adjacent to an oceanic arc–continental collision; a case from the Southern Caribbean–South America plate margin. Basin Res 23:403–422

Bayona G, Cardona A, Jaramillo C, Mora A, Montes C, Valencia V, Ayala C, Montenegro O, Ibañez-Mejia M (2012) Early Paleogene magmatism in the northern Andes: insights on the effects of Oceanic Plateau–continent convergence. Earth Planet Sci Lett 331–332:97–111

Beaumont C, Fullsack P, Hamilton J (1992) Erosional control of active compressional orogens. In: McClay KR (ed) Thrust tectonics. Chapman and Hall, New York, pp 1–18

Beaumont C, Jamieson RA, Nguyen M, Lee B (2001) Himalayan tectonics explained by extrusion of a low-viscosity crustal channel coupled to focused surface denudation. Nature 414(6865):738–742

Benjamin MT, Johnson NM, Naeser CW (1987) Recent rapid uplift in the Bolivian Andes: evidence from fission track dating. Geology 15:680–683

Bernet M (2009) A field-based estimate of the zircon fission-track closure temperature. Chem Geol 259:181–189

Bernet M, Spiegel C (2004) Detrital thermochronology—provenance analysis, exhumation, and landscape evolution of mountain belts. Geological Society of AmericaSpecial Paper 378. Geological Society of America, Boulder

Berry EW (1919) Fossil plants from Bolivia and their bearing upon the age of uplift of the eastern Andes. US Government Printing Office

Bierman PR (2004) Rock to sediment – slope to sea with Be-10 – rates of landscape change. Annu Rev Earth Planet Sci 32:215–255

Bird P (1988) Formation of the Rocky Mountains, western United States – a continuum computer model. Science 239(4847):1501–1507

Bishop P (2007) Long-term landscape evolution: linking tectonics and surface processes. Earth Surf Process Landf 32(3):329–365

Bonini WE, Hargraves RB, Shagam R (1984) The Caribbean-South American plate boundary and regional tectonics. Geological Society of America

Botero G (1963) Contribución al conocimiento de la geología de la zona central de Antioquia. Anales Facultad de Minas Medellín 57:7–101

Branket Y et al (2012) Andean deformation and rift inversion, eastern edge of Cordillera Oriental (Guateque–Medina area), Colombia. J South Am Earth Sci 15:391–407

Braun J (2003) Pecube: a new finite element code to solve the heat transport equation in three dimensions in the Earth's crust including the effects of a time-varying, finite amplitude surface topography. Comput Geosci 29:787–794

Brown RW, Summerfield MA (1997) Some uncertainties in the derivation of rates of denudation from thermochronologic data. Earth Surf Process Landf 22:239–248

Brown RW, Summerfield MA, Gleadow AJW (1994) Apatite fission track analysis: its potential for the estimation of denudation rates and implications for models of long-term landscape development. In: Kirby MJ (ed) Process models and theoretical geomorphology. Willey, Chichester, pp 23–53

Buchs DM, Arculus RJ, Baumgartner PO, Baumgartner-Mora C, Ulianov A (2010) Late Cretaceous arc development on the SW margin of the Caribbean Plate: insights from the Golfito, Costa Rica, and Azuero, Panama, complexes. Geochem Geophys Geosyst 11(7):1–35

Burbank DW, Anderson RS (2011) Tectonic geomorphology. Wiley, Chichester UK, 460 p

Burbank DW, Pinter N (1999) Landscape evolution: the interactions of tectonics and surface processes. Basin Res 11:1–6

Burbank DW, Derry LA, Lanord CF (1993) Reduced Himalayan sediment production 8 Ma ago despite an intensified monsoon. Nature 364:48–50

Burke K, Cooper C, Dewey JF, Mann P, Pindell JL (1984) Caribbean tectonics and relative plate motions. Geol Soc Am Mem 162:31–64

Bustamante C, Cardona A, Bayona G, Mora AR, Valencia V, Gehrels GE, Vervoort J (2010) U-Pb LA-ICP-MS geochronology and regional correlation of Middle Jurassic intrusive rocks from the Garzón Massif, Upper Magdalena Valley and Central Cordillera, southern Colombia. Boletin de Geologia 32:93–109

Bustamante C, Archanjo CJ, Cardona A, Vervoort JD (2016) Late Jurassic to Early Cretaceous plutonism in the Colombian Andes: a record of long-term arc maturity. Geol Soc Am Bull 128(11–12):1762–1779

Camargo S (2014) Estructura litosférica asociada a la Sierra Nevada de Santa Marta a partir de datos de gravimetría, magnetometría y sismología. In: Proceedings Latin-American and Caribbean Seismological Commision Bogotá Colombia, Volume Poster session, LACSC

Cardona A, Valencia VA, Bayona G, Duque J, Ducea M, Gehrels G, Jaramillo C, Montes C, Ojeda G, Ruiz J (2010) Early-subduction-related orogeny in the northern Andes: Turonian to Eocene magmatic and provenance record in the Santa Marta Massif and Rancheria Basin, northern Colombia: Terra Nova 23(1):26–34

Cardona A, Valencia V, Weber M, Duque J, Montes C, Ojeda G, Reiners P, Domanik K, Nicolescu S, Villagomez D (2011) Transient Cenozoic tectonic stages in the southern margin of the Caribbean plate: U-Th/He thermochronological constraints from Eocene plutonic rocks in the Santa Marta massif and Serranía de Jarara, northern Colombia. Geologica Acta 9(3–4):445–466

Cardona-Molina A, Cordani UG, MacDonald WD (2006) Tectonic correlations of pre-Mesozoic crust from the northern termination of the Colombian Andes, Caribbean region. J South Am Earth Sci 21(4):337–354

Carrapa B (2010) Resolving tectonic problems by dating detrital minerals. Geology 38(2):191–192

Case JE, Duran LG, Lopez A, Moore WR (1971) Tectonic Investigations in Western Colombia and Eastern Panama. GSA Bull 82(10):2685–2712

Cediel F, Shaw R, Caceres C (2003) Tectonic assembly of the northern Andean Block. In: Bartolini C, Buffler R, Blickwede J (eds) The circum-gulf of Mexico and Caribbean: hydrocarbon habitats, basin formation and plate tectonics, v. 79. AAPG Memoir, Tulsa, pp 815–848

Cediel F, Restrepo I, Marín-Cerón MI, Duque-Caro H, Cuartas C, Mora C, Montenegro G, García E, Tovar D, Muñoz G (2009) Geology and Hydrocarbon Potential, Atrato and San Juan Basins, Chocó (Panamá) Arc. Tumaco Basin (Pacific Realm), Colombia, Medellín, Colombia, Fondo Editorial Universidad EAFIT, p 172

Cediel F, Leal-Mejia H, Shaw R, Melgarego J, Restrepo-Pace P (2011) Petroleum geology of Colombia: regional geology of Colombia. Fondo Editorial U. Eafit, Ed 1:220

Chew DM, Donelick RA (2012) Combined apatite fission track and U-Pb dating by LA-ICP-MS and its application in apatite provenance analysis. Quantitative mineralogy and microanalysis of sediments and sedimentary rocks: Mineralogical Association of Canada Short Course, 42:219–247

Chicangana G (2005) The Romeral fault system: a shear and deformed extinct subduction zone between oceanic and continental lithospheres in northwestern South America. Earth Sci Res J 9(1):50–64

Chorowicz J, Chotin P, Guillande R (1996) The Garzon fault: active southwestern boundary of the Caribbean plate in Colombia. Geologische Rundschau 85(1):172–179

Clift PD (2010) Enhanced global continental erosion and exhumation driven by Oligo-Miocene climate change. Geophys Res Lett 37:1–6

Coates AG, Stallard RF (2013) How old is the Isthmus of Panama? Bull Mar Sci 89(4):801–813

Coates AG, Collins LS, Aury M-P, Berggren WA (2004) The geology of the Darien, Panama, and the late Miocene-Pliocene collision of the Panama arc with northwestern South America. GSA Bull 116(11/12):1327–1344

Cochrane R (2013) U-Pb thermochronology, geochronology and geochemistry of NW South America: rift to drift transition, active margin dynamics and implications for the volume balance of continents. University of Geneva

Colmenares F (2007) Evolución Geohistorica de la Sierra Nevada de Santa Marta. INGEOMINAS, Bogotá, p 397

Colmenares L, Zoback MD (2003) Stress field and seismotectonics of northern South America. Geology 31(8):721–724

Coltorti M, Ollier CF (1999) The significance of high planation surfaces in the Andes of Ecuador. In: Smith BJ, Whalley WB, Warke PA (eds) Uplift, erosion and stability: perspectives on long-term landscape development, vol Volume 162. Geological Society, London, pp 239–253

Coltorti M, Ollier CD (2000) Geomorphic and tectonic evolution of the Ecuadorian Andes. Geomorphology 32(1–2):1–19

Copeland P, Harrison TM, Heizler MT (1990) 40Ar/39Ar singlecrystal dating of detrital muscovite and K-feldspar from Leg 116, southern Bengal Fan: implications for the uplift and erosion of the Himalayas. Proc Ocean Drill Progr Sci 116:93–114

Correa I, Silva JC (1999) Estratigrafía y Petrografía del Miembro Superior de la Formación Amagá en la Sección El Cinco-Venecia-Quebrada La Sucia. Venecia, Antoquia, BSc: Universidad EAFIT, p 47

Cortes M, Angelier J (2005) Current states of stress in the Northern Andes as indicated by focal mechanisms of earthquakes. Tectonophysics 403:29–58

Cortés M, Angelier J, Colletta B (2005) Paleostress evolution of the northern Andes (Eastern Cordillera of Colombia): Implications on plate kinematics of the South Caribbean region. Tectonics 24(1): 1-27

Cortés M, Colletta B, Angelier J (2006) Structure and tectonics of the central segment of the Eastern Cordillera of Colombia. J South Am Earth Sci 21(4):437–465

Coughlin TJ, O'Sullivan P, Kohn BP, Holcombe RJ (1998) Apatite fission-track thermochronology of the Sierras Pampeanas, central western Argentina: implications for the mechanism of plateau uplift in the Andes. Geology 26(11):999–1002

Cross TA, Pilger RH (1982) Controls of subduction geometry, location of magmatic arcs, and tectonics of arc and back-arc regions. Geol Soc Am Bull 93(6):545–562

Crowhurst PV, Green PF, Kamp PJJ (2002) Appraisal of (U-Th)/He apatite thermochronology as a thermal history tool for hydrocarbon exploration: an example from the Taranaki Basin, New Zealand. AAPG Bull 86(10):1801–1819

Dalziel I (1986) Collision and Cordilleran orogenesis: an Andean perspective. Geol Soc (London, Special Publications) 19(1):389–404

Davis WM (1899) The geographical cycle. Geogr J 14:481–504

Davis WM (1922) Peneplains and the geographical cycle. Geol Soc Am Bull 23:587–598

Dengo CA, Covey MC (1993) Structure of the Eastern Cordillera of Colombia: implications for trap styles and regional tectonics. AAPG Bull 77(8):1315–1337
Dewey JF, Bird JM (1970) Plate tectonics and geosynclines. Tectonophysics 10(5–6):625–638
Dickinson WR, Gehrels GE (2008) U-Pb ages of detrital zircons in relation to paleogeography: triassic paleodrainage networks and sediment dispersal across southwest Laurentia. J Sediment Res 78(11–12):745–764
Donelick RA, Ketcham RA, Carlson WD (1999) Variability of apatite fission-track annealing kinetics II: crystallographic orientation effects. Am Mineral 84:1224–1234
Donelick RA, O'Sullivan PB, Ketcham RA (2005) Apatite fission-track analysis. In: Reiners PW, Ehlers TA (eds) Low-temperature thermochronology: techniques, interpretations, and applications, Reviews in mineralogy and geochemistry, vol Volume 58. Chantilly, MSA, pp 49–94
Douglas I (1967) Man, vegetation and the sediment yield of rivers. Nature 215:925–928
Duddy IR, Green PF, Laslett GM (1988) Thermal annealing of fission tracks in apatite 3. Variable temperature behavior. Chem Geol 73:25–38
Dumitru TA, Gans PB, Foster DA, Miller EL (1991) Refrigeration of the western Cordilleran lithosphere during Laramide shallow-angle subduction. Geology 19(11):1145–1148
Duque-Caro H (1990a) The Choco Block in the northwestern corner of South America: structural, tectonostratigraphic and paleogeographic implications. J South Am Earth Sci 3(1):71–84
Duque-Caro H. (1990b) Neogene stratigraphy, paleoceanography and paleobiogeography in northwest South America and the evolution of the Panama Seaway. Palaeogeography, Palaeoclimatology, Palaeoecology, 77(3-4):203–234
Ego F, Sébrier M, Yepes H (1995) Is the Cauca-Patia and Romeral fault system left or right lateral? Geophys Res Lett 22(1):33–36
Ehlers TA, Farley KA (2003) Apatite (U-Th)/He thermochronometry: methods and applications to problems in tectonic and surface processes. Earth Planet Sci Lett 206:1–14
Ehlers TA, Poulsen CJ (2009) Influence of Andean uplift on climate and paleoaltimetry estimates. Earth Planet Sci Lett 281:238–248
England P, Molnar P (1990) Surface uplift, uplift of rocks, and exhumation of rocks. Geology 18(12):1173–1177
English JM, Johnston ST (2004) The Laramide orogeny: what were the driving forces? Int Geol Rev 46(9):833–838
Espurt N, Baby P, Brusset SMR, Hermoza W, Barbarand J (2010) The Nazca Ridge and uplift of the Fitzcarrald Arch: implications for regional geology in northern South America. In: Hoorn C, Wesselingh FP (eds) Amazonia, landscape and species evolution: a look into the past. Wiley-Blackwell, London, p 464
Etayo-Serna F (1986) Mapa de Terrenos Geológicos de Colombia, Publicación Geológica Especial INGEOMINAS, vol 14. INGEOMINAS, Bogotá, pp 1–235
Farley KA (2000) Helium diffusion from apatite: general behavior as illustrated by Durango Fuorapatite. J Geophys Res 105:2903–2914
Farley KA (2002) (U-Th)/He dating: techniques, calibrations, and applications. Rev Miner Geochem 47:819–844
Farley KA, Stockli DF (2002) (U-Th)/He dating of phosphates: apatite, monazite, and xenotime. Rev Miner Geochem 48:559–577
Farris DW, Restrepo-Moreno SA, Jaramillo C, Bayona G, Montes C, Cardona A, Reiners P, Mora A, Speakman RJ, Glasscock MD (2011) Evolution of the Panamanian Isthmus. Geology 39(11):1007–1010
Feininger T, Botero G (1982) The Antioquian Batholith, Colombia, Bogotá, Publicación Geológica Especial Ingeominas, Publicaciones Geológicas Especiales, pp 1–50
Feininger T, Barrero D, Castro N (1972) Geología de parte de los departamentos de Antioquia y Caldas (Sub-zona II-B). Boletín Geológico Ingeominas 20(2):1–173
Fitzgerald PG (1994) Thermochronologic constraints on post-Paleozoic tectonic evolution of the central Transantarctic Mountains, Antarctica. Tectonics 13(4):818–836
Flowers RM, Kelley SA (2011) Interpreting data dispersion and "inverted" dates in apatite (U–Th)/He and fission-track datasets: an example from the US midcontinent. Geochimica et Cosmochimica Acta 75(18):5169–5186

Flowers RM, Wernicke P, Farley KA (2008) Unroofing, incision, and uplift history of the southwestern Colorado Plateau from apatite (U-Th)/He thermochronometry. GSA Bull 120(5/6):571–587

Forero A (1990) The basement of the Eastern Cordillera, Colombia: an allochthonous terrane in northwestern South America. J South Am Earth Sci 3:141–151

Foster DA, Gleadow AJW (1996) Structural framework and denudation history of the flanks of the Kenya and Anza Rifts, East Africa. Tectonics 15(2):258–271

Foster DA, Gleadow AJW, Mortimer G (1994) Rapid Pliocene exhumation in the Karakoram (Pakistan) revealed by fission-track thermochronology of the K2 gneiss. Geology 22:19–22

Galbraith R (2005) Statistics for fission track analysis. Chapman & Hall/CRC, Interdisciplinary Statistics, Boca Raton, p 240

Gallagher K (2013) QTQt User Guide v 5

Gallagher K, Brown RW, Johnson C (1998) Geological applications of fission-track analysis. Ann Rev Earth Planet Sci 26:519–572

Gallagher K, Stephenson J, Brown R, Holmes C, Ballester P (2005a) Exploiting 3D spatial sampling in inverse modeling of thermochronologic data. Rev Mineral Geochem 48:375–387

Gallagher K, Stephenson J, Brown R, Holmes C, Fitzgerald P (2005b) Low temperature thermochronology and modeling strategies for multiple samples 1: vertical profiles. Earth Planet Sci Lett 237:193–208

Gansser A (1955) Ein Beitrag zur Geologie und Petrographie der Sierra Nevada de Santa Marta (Kolumbien, Sudamerika). Schweizerische Mineralogische und Petrographische Mitteilungen 35(2):209–279

Gansser A (1973) Facts and theories on the Andes Twenty-sixth William Smith Lecture. J Geol Soc 129(2):93–131

Garzon-Varon F (2012) Modelamiento estructural de la zona límite entre la Microplaca de Panamá y el Bloque Norandino a partir de la interpretación de imágenes de radar, cartografía geológica, anomalías de campos potenciales y líneas sísmicas. MSc Universidad Nacional de Colombia, 102 p

Gehrels G (2012) Detrital zircon U-Pb geochronology: current methods and new opportunities. In: Busby C, Azor A (ed) Tectonics of sedimentary basins: Recent advances. Willey, pp 45–62.

Gerdes A, Zeh A (2006) Combined U–Pb and Hf isotope LA-(MC-) ICP-MS analyses of detrital zircons: comparison with SHRIMP and new constraints for the provenance and age of an Armorican metasediment in Central Germany. Earth and Planetary Sci Letters, 249(1-2):47–61

Gerya TV, Fossati D, Cantieni C, Seward D (2009) Dynamic effects of aseismic ridge subduction: numerical modelling. Europ J Mineral 21(3):649–661

Ghosh P, Garzione CN, Eiler JM (2006) Rapid uplift of the Altiplano revealed through 13C-18O bonds in paleosol carbonates. Science 311:511–514

Gleadow AJW (1981) Fission track dating methods: What are the real alternatives?. Nucl Tracks Radiat Meas 5:3–14

Gleadow AJW, Brown RW (1999) Fission track thermochronology and the long-term denudational response to tectonics. In: Summerfield MA (ed) Geomorphology and global tectonics. Wiley, Chichester, pp 57–75

Gleadow AJW, Fitzgerald PG (1987) Uplift history and structure of the transantarctic mountains: new evidence from fission track dating of basement apatites in the dry valleys area, southern Victoria Land. Earth Planet Sci Lett 82:1–14

Gleadow AJW, Duddy IR, Green PF, Lovering JF (1986) Confined fission track lengths in apatite: a diagnostic tool for thermal history analysis. Contrib Miner Petrol 94:405–415

Gleadow AJW, Belton DX, Kohn BP, Brown RW (2002) Fission track dating of phosphate minerals and the thermochronology of apatite. Rev Mineral Geochem 48:579–630

Goldsmith R, Marvina R, Mehnert HH (1971) Radiometric ages in the Santander massif, eastern Cordillera, Colombian Andes. US Geol Surv Prof Pap 750:D44–D49

Gomez E (2001) Tectonic controls on the Late Cretaceous to Cenozoic sedimentary fill of the Middle Magdalena Valley Basin, Eastern Cordillera and Llanos Basin, Colombia [Ph.D] Cornell University, p 619

Gomez E, Jordan TE, Allmendinger RW, Hegarty K, Kelley S (2005a) Syntectonic Cenozoic sedimentation in the northern middle Magdalena Valley Basin of Colombia and implications for exhumation of the Northern Andes. Geol Soc Am Bull 117(5–6):547–569

Gomez E, Jordan TE, Allmendinger RW, Cardozo N (2005b) Development of the Colombian foreland-basin system as a consequence of the diachronous exhumation of the northern Andes. Geol Soc Am Bull 117(9):1272–1292

Gomez J, Montes NE, Alcárcel FA, Ceballos JA (2015a) Catálogo de dataciones radiométricas de Colombia en ArcGIS y Google Earth. In: Gomez J, Almanza MF (eds) Compilando la geología de Colombia: Una visión a 2015, vol Volume 33. Servicio Geológico Colombiano, Bogotá, pp 63–419

Gomez J, Nivia Á, Montes NE, Almanza MF, Alcárcel F, Madrid CA (2015b) Notas explicativas: Mapa Geológico de Colombia, Bogotá, Servicio Geológico Colombiano, Compilando la geología de Colombia: Una visión a 2015, v. Publicaciones Geológicas Especiales, pp 35–60

González H, Londoño AC (2002a) Catálogo de las unidades litoestratigráficas de Colombia, Monzodiorita de Farallones (Batolito de Farallones) Nmdf: INGEOMINAS

González H, Londoño AC (2002b) Catálogo de las unidades litoestratigráficas de Colombia, Monzodiorita de la Horqueta (Stock de La Horqueta) Nmdh, Cordillera Occidental Departamento de Antioquia: INGEOMINAS

González H, Londoño AC (2002c) Catálogo de las unidades litoestratigráficas de Colombia, Tonalita de Tatamá (N1tt) Cordillera Occidental Departamento del Chocó: INGEOMINAS

Goudie AS (1995) The changing earth: rates of geomorphological processes. Wiley-Blackwell, New Jersey p 352

Green PF (1981) A new look at statistics in fission track dating. Nuclear Tracks 5(1–2):77–86

Green PF, Duddy IR, Laslett GM, Hegarty KA, Gleadow AJW, Lovering JF (1989) Thermal annealing of fission tracks in apatite 4. Quantitative modeling techniques and extension to geological time scales. Chem Geol 79:155–182

Gregory-Wodzicki KM (2000) Uplift history of the central and northern Andes: a review. GSA Bull 112:1091–1105

Grosse E (1926) El Terciario Carbonífero de Antioquia. D. Reimer-E. Vohsen, Berlin, p 361

Grujic D, Coutand I, Bookhagen B, Bonnet S, Blythe A, Duncan C (2006) Climatic forcing of erosion, landscape, and tectonics in the Bhutan Himalayas. Geology 34(10):801–804

Grujic D, Warren CJ, Wooden JL (2011) Rapid synconvergent exhumation of Miocene-aged lower orogenic crust in the eastern Himalaya. Lithosphere 3(5):346–366

Gubbels T, Isacks B, Farrar E (1993) High-level surfaces, plateau uplift, and foreland development, Bolivian central Andes. Geology 21(8):695–698

Gunnell Y (1998) Present, past and potential denudation rates: is there a link? Tentative evidence from fission-track data, river sediment loads and terrain analysis in the South Indian shield. Geomorphology 25:135–153

Gutscher MA, Malavieille J, Lallemand S, Collot J-Y (1999) Tectonic segmentation of the northern Andean margin: impact of the Carnegie Ridge collision. Earth Planet Sci Lett 168:255–270

Hack JT (1976) Dynamic equilibrium and landscape evolution. In: Melborn W, Flernal R (eds) Theories of landforms development. State University of New York Publications in Geomorphology, Binghamton, pp 87–102

Harris SE, Mix AC (2002) Climate and tectonic influences on continental erosion of tropical South America, 0-13 Ma. Geology 30(5):447–450

Harrison TM, Copeland P, Hall SA, Quade J, Burner S, Ojha TP, Kidd WSF (1993) Isotopic preservation of Himalayan/Tibetan uplift, denudation, and climatic histories of two molasse deposits. J Geol 1001(2):157–175

Hartley A J (2003) Andean uplift and climate change. J Geol Soc Lond 160:7–10

Hasebe N, Barbarand J, Jarvis K, Carter A, Hurford AJ (2004) Apatite fission-track chronometry using laser ablation ICP-MS. Chem Geol 207:135–145

Heimsath AM (2006) Eroding the land: Steady-state and stochastic rates and processes through a cosmogenic lens. Geol Soc Am Spec Pap 415:111–129

Hengl T, Reuter HI (eds) (2008) Geomorphometry: Concepts, Software, Applications. Developments in Soil Science, 33, Elsevier, p 772

Herman F, Seward D, Valla PG, Carter A, Kohn B, Willett SD, Ehlers TA (2013) Worldwide acceleration of mountain erosion under a cooling climate. Nature 504(7480):423

Hetzel R, Dunkl I, Haider V, Strobl M, von Eynatten H, Ding L, Frei D (2011) Peneplain formation in southern Tibet predates the India-Asia collision and plateau uplift. Geology 39(10):983–986

Holmes A. (1965). Principles of Physical Geology. Revised Edition. Nelson.

Hooghiemstra H (1989) Quaternary and Upper-Pliocene glaciations and forest development in the Tropical Andes: evidence from a long high-resolution pollen record from the sedimentary basin of Bogota, Colombia. Palaeogeogr Palaeoclimatol Palaeoecol 72:11–26

Hooghiemstra H, Wijninga VM, Cleef AM (2006) The paleobotanical record of Colombia: implications for biogeography and biodiversity. Ann Missouri Bot Gard 93(2):297

Hooke RL (2000) On the history of humans as geomorphic agents. Geology 28(9):843–846

Hoorn C, Flantua S (2015) An early start for the Panama land bridge. Science 348(6231):186–187

Hoorn C, Guerrero J, Sarmiento GA, Lorente MA (1995) Andean tectonics as a cause for changing drainage patterns in Miocene northern South America. Geology 23(3):237–240

Hoorn C, Wesselingh FP, Steege H, Bermudez MA, Mora A, Sevink J, Sanmartín I, Sanchez-Meseguer A, Anderson CL, Figueiredo JP, Jaramillo C, Riff D, Negri FR, Hooghiemstra H, Lundberg J, Stadler T, Särkinen T, Antonelli A (2010) Amazonia through time: Andean uplift, climate change, landscape evolution, and biodiversity. Science 330:927. https://doi.org/10.1126/science.1194585

Horton BK, Parra M, Saylor JE, Nie J, Mora A, Torres V, Stockli DF, Strecker MR (2010a) Resolving uplift of the northern Andes using detrital zircon age signatures. GSA Today 20(7):4–10

Horton BK, Saylor JE, Nie J, Mora A, Parra M, Reyes-Harker A, Stockli DF (2010b) Linking sedimentation in the northern Andes to basement configuration, Mesozoic extension, and Cenozoic shortening: Evidence from detrital zircon U-Pb ages. Eastern Cordillera, Colombia GSA Bulletin

House MA, Wernicke BP, Farley KA (1998) Dating topography of the Sierra Nevada, California, using apatite (U-Th)/He ages. Nature 396:66–69

House M, Kohn BP, Farley KA, Raza A (2002) Evaluating thermal history models for the Otway Basin, southeastern Australia, using (U-Th)/He and fission-track data from borehole apatites. Tectonophysics 349(1–4):277–295

Hurford AJ, Green PF (1983) The zeta age calibration of fission track dating. Isotope Geosci 1:285–317

Insel N, Ehlers TA, Schaller M, Barnes JB, Tawackoli S, Poulsen CJ (2010a) Spatial and temporal variability in denudation across the Bolivian Andes from multiple geochronometers. Geomorphology 122:65–77

Insel N, Poulsen CJ, Ehlers TA (2010b) Influence of the Andes Mountains on South American moisture transport, convection, and precipitation. Clim Dyn 35(7–8):1477–1492

Irving EM (1975) Structural evolution of the northernmost Andes, Colombia. U.S. Geol Surv Prof Pap 846:47

James KJ (2006) Arguments for an against the Pacific origin of the Caribbean Plate: discussion, finding for an inter-American orogin. Geologica Acta 4(1–2):279–302

Jiang JH, Wu DL, Eckermann SD (2002) Upper Atmosphere Research Satellite (UARS) MLS observation of mountain waves over the Andes. J Geophys Res Atmos 107(D20):SOL15-11–SOL15-10

Jordán TE, Isacks BL, Allmendinger RW, Brewer JA, Ramos VA, Ando CJ (1983) Andean tectonics related to geometry of subducted Nazca plate. Geol Soc Am Bull 94(3):341–361

Kellogg JN (1984) Cenozoic tectonic history of the Sierra de Perijá, Venezuela-Colombia, and adjacent basins. Geological Society of America Memoir 162:239–261

Kellogg JN, Vega V (1995) Tectonic development of Panama, Costa Rica, and the Colombian Andes: Constraints from global positioning system geodetic studies and gravity. In: Mann P (ed) Geologic and tectonic development of the Caribbean plate boundary in southern Central America: Geological Society of America Special Paper, vol 295, Geological Society of America, pp 75–90

Kennan L (2000) Large-scale geomorphology of the Andes: interrelationships of tectonics, magmatism and climate. In: Summerfield MA (ed) Geomorphology and global tectonics. John Wiley, New York, pp 167–199

Kennan L, Lamb S, Hoke L (1997) High-altitude palaeosurfaces in the Bolivian Andes: evidence for late Cenozoic surface uplift. In: Widdowson M (ed) Palaeosurfaces: recognition, reconstruction and paleoenvironmental interpretation, vol Volume 120. Geological Society of London Special Publication, London, pp 307–324

Kennan L, Pindell L (2009) Dextral shear, terrane accretion and basin formation in the Northern Andes: best explained by interaction with a Pacific-derived Caribbean Plate? Geological Society, London, Special Publications 328(1):487–531

Kerr AC, Marriner GF, Tarney J, Nivia A, Saunders AD, Thirlwall MF, Sinton CW (1997) Cretaceous basaltic terranes in Western Colombia: elemental, chronological and Sr–Nd Isotopic Constraints on petrogenesis. J Petrol 38:677–702

Ketcham RA (2003) Observations on the relationship between crystallographic orientation and biasing in apatite fission-track measurements. Am Mineral 88:817–829

Ketcham RA (2005) Forward and inverse modeling of low-temperature thermochronometry data. Rev Mineral Geochem 48:275–314

Ketcham RA (2008) HeFTy: Viola, Idaho, Apatite to Zircon, Inc

Ketcham RA, Donelick RA, Carlson WD (1999) Variability of apatite fission-track annealing kinetics III: extrapolation to geological time scales. Am Mineral 84:1235–1255

Kirchner JW, Finkel RC, Riebe CS, Granger DE, Clayton JL, King JG, Megahan WF (2001) Mountain erosion over 10 yr, 10 ky, and 10 my time scales. Geology 29(7):591–594

Kohn BR, Shagam R, Banks PO, Burkley LA (1984) Mesozoic-Pleistocene fission-track ages on rocks of the Venezuelan Andes and their tectonic implications. In: Bonini WE, Hargraves RB, Shagam R (eds) The Caribbean-South American plate boundary and regional tectonics, Geological Society of America Memoir, vol 162. Geological Society of America, Boulder, pp 365–384

Koons PO (1990) The two-sided wedge in orogeny; erosion and collision from the sand box to the Southern Alps, New Zealand. Geology 18:679–682

Koons P (1998) Big mountains, big rivers, and hot rocks. EOS (Transactions of the American Geophysical Union) 79:908

Kroonenberg SB, Bakker JGM, Van der Wiel AM (1990) Late Cenozoic uplift and paleogeography of the Colombian Andes – constraints on the development of high-Andean biota. Geologie en Mijnbouw 69(3):279–290

Lal R (2001) Soil degradation by erosion. Land Degrad Dev 12(6):519–539

Lamb S, Davis P (2003) Cenozoic climate change as a possible cause for the rise of the Andes. Nature 425:792–797

Laslett GM, Green PF, Duddy IR, Gleadow AJW (1987) Thermal annealing of fission tracks in apatite 2. A Quantitative analysis. Chem Geol 73:25–38

Laubacher G, Naeser CW (1994) Fission-track dating of granitic rocks from the Eastern Cordillera of Peru: evidence for Late Jurassic and Cenozoic cooling. J Geol Soc Lond 151(3):473–483

Leal-Mejía H, Shaw RP, Melgarejo JC (2011) Phanerozoic granitoid magmatism in Colombia and the tectono-magmatic evolution of the Colombian Andes. In: Cediel F (ed) Petroleum Geology of Colombia. Regional Geology of Colombia, vol 1. Bogotá Colombia, Agencia Nacional de Hidro-carburos (ANH) – EAFIT, pp 109–188

van der Lelij R, Spikings R, Mora A (2016) Thermochronology and tectonics of the Mérida Andes and the Santander Massif, NW South America. Lithos 248:220–239

Lisker F, Ventura B, Glasmacher AU (2009) Apatite thermochronology in modern geology. Geol Soc Lond Spec Publ 324:1–23

Lüschen E (1986) Gravity and height changes in the ocean-continent transition zone in western Colombia. Tectonophysics 130(1–4):141151–146157

MacDonald WD, Estrada JJ, Sierra GM, Gonzalez H (1996) Late Cenozoic tectonics and paleomagnetism of North Cauca Basin intrusions, Colombian Andes: dual rotation modes. Tectonophysics 261:277–289

Mann P, Burke K (1984) Neotectonics of the Caribbean. Rev Geophys 22(4):309–362

Mann P, Corrigan J (1990) Model for late Neogene deformation in Panama. Geology 18:558–562

Marín-Cerón MI, Restrepo-Moreno SA, Bernet M, Foster DA, Min K, Pardo-Trujillo A, Barbosa-Espitia A, Kamenov G (2015) The Amagá formation (Eocene?-Miocene) in the Cauca-Patía depression: A recorder of major morphotectonic and paleogeographic events between the Western and Central cordilleras. In: Proceedings GSA Annual Meeting, Baltimore, Maryland, USA, vol 47, Geological Society of America Abstracts with Programs, p 675

Marshall LG, Butler RF, Drake RE, Curtis GH, Tedford RH (1979) Calibration of the great American interchange. Science 204(4390):272–279

Matmon A, Bierman P, Larsen J, Southworth S, Pavich M, Caffee M (2003) Temporally and spatially uniform rates of erosion in the southern Appalachian Great Smoky Mountains. Geology 31(2):155–158

Maya M, González H (1995) Unidades litodémicas en la Cordillera Central de Colombia. Bol Geol INGEOMINAS 35(2):3

McCourt WJ, Aspden JA, Brook M (1984) New geological and geochronological data from the Colombian Andes: continental growth by multiple accretion. J Geol Soc Lond 141:793–802

Mégard F (1984) The Andean orogenic period and its major structures in central and northern Peru. J Geol Soc Lond 141(5):893–900

Mégard F, Noble DC, McKee EH (1984) Multiple pulses of Neogene compressive deformation in the Ayacucho intermontane basin, Andes of Central Peru. Geol Soc Am Bull 95:1108–1117

Milliman JD (2001) Delivery and fate of fluvial water and sediment to the sea: a marine geologist's view of European rivers. Scientia Marina 65:121–131

Milliman JD, Meade RH (1983) World-wide delivery of river sediment to the oceans. J Geol 91(1):1–21

Mittermeier RA, Meyer N, Mittermeier CG (1999) Hotspots: Earth's biologically richest and most endangered terrestrial ecoregions, Monterrey, Mexico, Conservation International and Agrupacion Sierra Madre, p 432

Molnar P (2004) Late Cenozoic increase in accumulation rates of terrestrial sediment: how might climate change have affected erosion rates? Ann Rev Earth Planet Sci 32:67–89

Molnar P (2008) Closing of the Central American Seaway and the Ice Age: a critical review. Paleoceanography 23(PA2201). https://doi.org/10.1029/2007PA001574

Molnar P, England PC (1990) Late cenozoic uplift of mountain-ranges and global climate change – chicken or egg? Nature 346:29–34

Molnar P, England P, Martinod J (1993) Mantle dynamics, uplift of the Tibetan Plateau, and the Indian monsoon. Rev Geophys 31(4):357–396

Montes C, Robert D, Hatcher RD, Restrepo-Pace PA (2005) Tectonic reconstruction of the northern Andean blocks: Oblique convergence and rotations derived from the kinematics of the Piedras-Girardot area, Colombia. Tectonophysics 399(1–4):221–250

Montes C, Guzman G, Bayona G, Cardona A, Valencia V, Jaramillo C (2010) Clockwise rotation of the Santa Marta massif and simultaneous Paleogene to Neogene deformation of the Plato-San Jorge and Cesar-Ranchería basins. J South Am Earth Sci 29(4):832–848

Montes C, Bayona G, Cardona A, Buchs DM, Silva C, Morón S, Hoyos N, Ramírez D, Jaramillo C, Valencia V (2012a) Arc-continent collision and orocline formation: closing of the Central American seaway. J Geophys Res Solid Earth 117(B4):1–25

Montes C, Cardona A, McFadden R, Morón S, Silva C, Restrepo-Moreno S, Ramírez D, Hoyos N, Wilson J, Farris D, Bayona GA, Jaramillo CA, Valencia V, Bryan J, Flores JA (2012b) Evidence for middle Eocene and younger land emergence in central Panama: implications for Isthmus closure. Geol Soc Am Bull 124(5–6):780–799

Montes C, Cardona A, Jaramillo C, Pardo A, Silva JC, Valencia V, Ayala C, Pérez-Angel LC, Rodriguez-Parra LA, Ramirez V, Niño H (2015) Middle Miocene closure of the Central American Seaway. Science 348(6231):226–229

Montes-Correa LF (2007) Exhumación de las rocas metamórficas de alto grado que afloran al Oriente del Valle de Aburrá, Antioquia [MSc: Universidad EAFIT], p 134

Montgomery DR (2003) Predicting landscape-scale erosion rates using digital elevation models. C. R. Geosci Surf Geosci 35:112111130

Montgomery DR (2007) Soil erosion and agricultural sustainability. Proc Natl Acad Sci 104(33):13268–13272

Montgomery DR (2008) Dirt: the Erosion of Civilizations. University of California Press, Berkeley p 296

Montgomery DR, Brandon MT (2002) Topographic controls on erosion rates in tectonically active mountain ranges. Earth Planet Sci Lett 201(3):481–489

Montgomery DR, Balco G, Willett SD (2001) Climate, tectonics, and the morphology of the Andes. Geology 29(7):579–582

Moore MA, England PC (2001) On the inference of denudation rates from cooling ages of minerals. Earth Planet Sci Lett 158:265–284

Mora A, Parra M, Strecker MR, Sobel ER, Hooghiemstra H, Torres V, Vallejo-Jaramillo J (2008) Climatic forcing of asymmetric orogenic evolution in the Eastern Cordillera of Colombia. GSA Bull 120(7–8):930–949

Mora A, Baby P, Roddaz M, Parra M, Brusset S, Hermoza W, Espurt N (2010a) Tectonic history of the Andes and sub-Andean zones: implications for the development of the Amazon drainage basin. In: Hoorn C, Wesselingh FP (eds) Amazonia, landscape and species evolution: a look into the past. Blackwell Publishing, Chichester p 464

Mora A, Horton BK, Mesa A, Rubiano J, Ketcham RA, Parra M, Blanco V, Garcia D, Stockli DF (2010b) Migration of Cenozoic deformation in the Eastern Cordillera of Colombia interpreted from fission track results and structural relationships: implications for petroleum systems. AAPG Bull 94(10):1543–1580

Mora A, Reyes-Harker A, Rodriguez G, Tesón E, Ramirez-Arias JC, Parra M, Caballero V, Mora JP, Quintero I, Valencia V (2013) Inversion tectonics under increasing rates of shortening and sedimentation: cenozoic example from the Eastern Cordillera of Colombia. Geol Soc Lond Spec Publ 377(1):411–442

Nelson E (1982) Post-tectonic uplift of the Cordillera Darwin orogenic core complex: evidence from fission track geochronology and closing temperature–time relationships. J Geol Soc 139(6):755–761

Noriega-Londoño S (2016) Geomorfología tectónica del noroccidente de la Cordillera Central, Andes del Norte-Colombia [MSc: Universidad Nacional de Colombia–Sede Medellín], 179 p

Norris RJ, Cooper AF (1997) Erosional control on the structural evolution of a transpressional thrust complex on the Alpine Fault, New Zealand. J Struct Geol 19(10):1323–1342

O'Dea A, Lessios HA, Coates AG, Eytan RI, Restrepo-Moreno SA et al (2016) Formation of the Isthmus of Panama. Sci Adv 2(8):1–11

Ochoa D, Hoorn C, Jaramillo C, Bayona G, Parra M, De la Parra F (2012) The final phase of tropical lowland conditions in the axial zone of the Eastern Cordillera of Colombia: evidence from three palynological records. J South Am Earth Sci 39:157–169

Oldow JS, Blly AW, Avé Lallemant HG (1990) Transpression, orogenic float, and lithospheric balance. Geology 18(10):991–994

Ollier CD (2006) Mountain uplift and the neotectonic period. Ann Geophys 49(1):437–450

Ollier CD, Pain CF (2000) The origin of mountains. Routledge, London, p 368

Oppenheim V (1941) Geología de la Cordillera Oriental entre los Llanos y el Magdalena. Revista Academia Colombiana de Ciencias Exactas, Físicas y Naturales 14:175–181

Osadetz K, Kohn B, Feinstein S, O'Sullivan P (2002) Thermal history of Canadian Williston basin from apatite fission-track thermochronology—implications for petroleum systems and geodynamic history. Tectonophysics 349(1):221–249

Padilla L (1981) Geomorfologia de posibles areas peneplanizadas en la Cordillera Occidental de Colombia. Revista CIAF 6:391–402

Page WD, James ME (1981) The antiquity of the erosion surfaces and late Cenozoic deposits near Medellín Colombia: implications to tectonics and erosion rates. Revista CIAF 6(1–3):421–454

Pardo A, Silva JC, Cardona A, Restrepo-Moreno SA, Osorio JA, Borrero C (2012a) Estudio geológico integrado en la Cuenca Tumaco Onshore. Síntesis cartográfica, sísmica y análisis bioestratigráfico, petrográfico, geocronológico, termocronológico y geoquímico de testigos de perforación y muestras de superficie: Agencia Nacional de Hidrocarburos

Pardo A, Silva JC, Cardona A, Restrepo-Moreno SA, Osorio JA, Borrero C (2012b) Estudio integrado de los núcleos y registros obtenidos de los pozos someros (slim holes) perforados por la ANH: Agencia Nacional de Hidrocarburos

Pardo-Casas F, Molnar P (1987) Relative motion of the Nazca (Farallon) and South American plates since Late Cretaceous time. Tectonics 6(3):233–248

Pardo-Trujillo A, Restrepo-Moreno SA, Vallejo DF, Flores JA, Trejos R, Foster D, Barbosa-Espitia AA, Bernet M, Marin-Ceron M, Kamenov G (2015) New thermotectonic and paleo-environmental constraints from the Western Cordillera and associated sedimentary basins (Northern Andes, Colombia): the birth of a major orographic barrier and the Choco Biogeographic region?. In: Proceedings 2015 GSA Annual Meeting, Baltimore, Maryland, 2015, vol 47, Geological Society of America Abstracts with Programs, p 156

Parra M, Mora A, Jaramillo C, Strecker MR, Sobel ER, Quiroz L, Rueda M, Torres V (2009) Orogenic wedge advance in the northern Andes: evidence from the Oligocene-Miocene sedimentary record of the Medina Basin, Eastern Cordillera, Colombia. GSA Bull 121(5–6):780–800

Parra M, Mora A, Lopez C, Rojas LE, Horton BK (2012) Detecting earliest shortening and deformation advance in thrust belt hinterlands: example from the Colombian Andes. Geology 40(2):175–178

Pavlis TL, Hamburger MW, Pavlis GL (1997) Erosional processes as a control on the structural evolution of an actively deforming fold thrust belt: an example from the Pamir-Tien Shan region, Central Asia. Tectonics 16:810–822

Peizhen Z, Molnar P, Downs WR (2001) Increased sedimentation rates and grain sizes 2-4 Myr ago due to the influence of climate change on erosion rates. Nature 410(6831):891

Penk W (1953) Morphological analysis of landforms. Macmillan, London, p 429

Pennington WD (1981) Subduction of the Eastern Panama Basin and Seismotectonics of Northwestern South-America. J Geophys Res 86(NB11):753–770

Peyton SL, Carrapa B (2013) An overview of low-temperature thermochronology in the Rocky Mountains and its application to petroleum system analysis

Pilger RH (1984) Cenozoic plate kinematics, subduction and magmatism: South American Andes. J Geol Soc Lond 141(5): 793-802.

Pindell J, Kennan L, Maresch WV, Stasnek K-P, Draper G, Higgs R (2005) Plate Kinematic and crustal dynamics of circum-Caribbean arc continent Interactions: tectonics controls on basin development in the Proto-Caribbean margins. In: Avé-Lallemant HG, Sisson VB (eds) Caribbean-South American Plate interactions, Venezuela, vol Volume 394. Geological Society of America, Boulder, pp 7–52

Pindell J, Kennan L, Stanek KP, Maresch WV, Draper G (2006) Foundations of Gulf of Mexico and Caribbean evolution: eight controversies resolved. Geologica Acta 4(1–2):303–341

Pinet P, Souriau M (1988) Continental Erosion and large-scale relief. Tectonics 7(3):563–582

Pinter N, Brandon M (1997) How erosion builds mountains. Sci Am 276(44):74–79

Piraquive A, Pinzón E, Kammer A, Bernet M, von Quadt A (2017) Early Neogene unroofing of the Sierra Nevada de Santa Marta, as determined from detrital geothermochronology and the petrology of clastic basin sediments. GSA Bull 130 (3-4): 355-380

Piraquive-Bermúdez A (2016) Structural Framework, deformation and exhumation of the Santa Marta Schists: accretion and deformational history of a Caribbean Terrane at the north of the Sierra Nevada de Santa Marta [Ph.D.]. Universidad Nacional de Colombia, 394 p

Piraquive Bermúdez A, Pinzón E, Bernet M, Kammer A, Von Quadt A, Sarmiento G (2016) Early Neogene unroofing of the Sierra Nevada de Santa Marta along the Bucaramanga-Santa Marta Fault. In: Proceedings EGU General Assembly Conference Abstracts 2016, vol 18, p 17214

Poulsen CJ, Ehlers TA, Insel N (2010) Onset of Convective Rainfall During Gradual Late Miocene Rise of the Central Andes. Science 328(5977):490–493

Poveda G, Mesa OJ (2000) On the existencoef Lloró (the rainies locality on Earth): enhanced ocean-land-atmosphere interaction by a Low-Level Jet. Geophys Res Lett 27(11):1675–1678

Poveda E, Monsalve G, Vargas CA (2015) Receiver functions and crustal structure of the northwestern Andean region, Colombia. J Geophys Res Solid Earth 120(4):2408–2425

Ramírez DA, López A, Sierra GM, Toro G (2006) Edad y proveniencia de las rocas volcánico sedimentarias de la Formación Combia en el Suroccidente Antioqueño, Colombia. Boletín de Ciencias de la Tierra 19:9–26

Ramírez DA, Foster DA, Min K, Montes C, Cardona A, Sadove G (2016) Exhumation of the Panama basement complex and basins: implications for the closure of the Central American seaway. Geochem Geophys Geosyst 10(1002):1–20

Ramos VA (1999) Plate tectonic setting of the Andean Cordillera. Episodes 22:183–190

Ramos VA (2009) Anatomy and global context of the Andes: main geologic features and the Andean orogenic cycle. Geol Soc Am Mem 204:31–65

Raymo ME (1994) The Himalayas, organic-carbon burial, and climate in the miocene. Paleoceanography 9(3):399–404

Raymo ME, Ruddiman WF (1992) Tectonic forcing of late cenozoic climate. Nature 359(6391):117–122

Reiners PW (2005) Zircon (U-Th)/He Thermochronometry. Rev Mineral Geochem 58:151–179

Reiners PW, Brandon MT (2006) Using thermochronology to understand orogenic erosion. Ann Rev Earth Planet Sci 34:419–466

Reiners PW, Campbell IH, Nicolescu S, Allen CM, Hourigan JK, Garver JI, Mattinson JM, Cowan DS (2005) (U-Th)/(He-Pb) double dating of detrital zircons. American Journal of Science, 305(4), 259–311

Reiners PW, Ehlers TA (2005a) In: Rosso J (ed) Low-temperature thermochronology: techniques, interpretations, and applications. Mineralogical Society of America, Chantilly p 622

Reiners PW, Ehlers TA (2005b) Past, present, and future of thermochronology. Rev Mineral Geochem 58:1–18

Reiners PW, Shuster DL (2009) Thermochronology and landscape evolution. Physics Today 62:31–36

Reiners PW, Spell TL, Nicolescu S, Zanetti KA (2004) Zircon (U-Th)/He thermochronometry: he diffusion and comparisons with 40Ar/39Ar dating. Geochim Cosmochim Acta 68:1857–1887

Restrepo JJ (2008) Obducción y metamorfismo de ofiolitas Triásicas en el flanco occidental del terreno Tahamí. Cordillera Central de Colombia: Boletín de Ciencias de la Tierra, No. 22

Restrepo JD, Kjerfve B (2000) Magdalena river: interannual variability (1975–1995) and revised water discharge and sediment load estimates. J Hydrol 235:137–149

Restrepo JD, Syvitski JPM (2006) Assessing the effect of natural controls and land use change on sediment yield in a major Andean River: the Magdalena Drainage Basin, Colombia. Ambio 35(2):65–74

Restrepo JJ, Toussaint JF (1982) Metamorfismos superpuestos en la Cordillera Central de Colombia. In: Proceedings Memorias V Congreso Latinoamericano de Geología, Buenos Aires, Argentina, pp 505–512

Restrepo JJ, Toussaint JF (1988) Terranes and continental accretion in the Colombian Andes. Episodes 11(1):189–193

Restrepo JJ, Toussaint JF (1990) Cenozoic arc magmatism of northwestern Colombia. Geol Soc Am Spec Pap 241:205–212

Restrepo J, Toussaint J, González H (1981) Edades Mio-pliocenas del magmatismo asociado a la Formación Combia, departamentos de Antioquia y Caldas, Colombia. Geología Norandina 3:21–26

Restrepo JD, Kjerfve B, Hermelin M, Restrepo JC (2006) Factors controlling sediment yield in a major South American drainage basin: the Magdalena River, Colombia. J Hydrol 316:213–232

Restrepo JJ, Carmona OO, Martens U, Correa AM (2009) Terrenos, complejos y provincias en la Cordillera Central de Colombia. Ingeniería Investigación y Desarrollo 9(2):49–56

Restrepo-Moreno SA (2009) Long-term morphotectonic Evolution and Denudation Chronology of the Antioqueño Plateau, Cordillera Central, Colombia [PhD], University of Florida, 223 p

Restrepo-Moreno SA, Restrepo-Múnera MA (2007) La Erosión: Un Problema a los Pies de la Sociedad: EOLO. Revista Ambiental-Fundación CON-VIDA 5(11):17–27

Restrepo-Moreno SA, Foster DA, Kamenov GD (2007) Formation Age and Magma Sources for the Antioqueño Batholith Derived from LA-ICP-MS Uranium–Lead Dating and Hafnium-Isotope Analysis of Zircon Grains. In: Proceedings Geological Society of America Abstracts with Programs, vol 3, no 6, GSA, pp 493–494

Restrepo-Moreno SA, Foster DA, Kamenov GD (2009a) Crystallization Age and Magma Sources of the Antioqueño And Ovejas Batholiths, Central Cordillera, Colombia: evidence From Combined LA-ICP-MS U/Pb Dating and Hf-Isotope Analysis of Zircon Grains and Whole-Rock Geochemistry. In: Proceedings Geological Society of America Annual Meeting Abstracts with Programs, Oregon, USA, vol 41, no 7, GSA, p 222

Restrepo-Moreno SA, Foster DA, O'Sullivan P, Donelick RA, Stockli DF (2009b) Cenozoic exhumation of the Antioqueño Plateau, Northern Andes, Colombia, from apatite low-temperature thermochronology. In: Proceedings AGU Fall Meeting, Volume abstract #T43B-2091

Restrepo-Moreno SA, Foster DA, Stockli DF, Parra LN (2009c) Long-term erosion and exhumation of the "Altiplano Antioqueño", Northern Andes (Colombia) from apatite (U-Th)/He thermochronology. Earth Planet Sci Lett 278(1–2):1–12

Restrepo-Moreno SA, Cardona A, Jaramillo C, Bayona G, Montes C, Farris DW (2010) Constraining Cenozoic uplift/exhumation of the Panama-Chocó block by apatite and zircon low-temperature thermochronology: insights on the onset of collision and the morphotectonic history of the region. In: Proceedings 2010 GSA Denver Annual Meeting, Denver, CO, vol 42, Geological Society of America Abstracts with Programs, p 521

Restrepo-Moreno SA, Min K, Barbosa-Espitia AA, Bernet M, Pardo A (2013a) Assessing the thermotectonic history of Colombia's Western Cordillera through thermochronology/geochronology in plutonic massifs and sedimentary basins: Tectonic, geomorphic and climatic implications. In: Proceedings 2013 GSA Annual Meeting 125th Anniversary of GSA, Denver CO USA, vol 45, Geological Society of America Abstracts with Programs, p 548

Restrepo-Moreno SA, Min K, Barbosa A, Foster DA, Bernet M, Pardo A, Marín-Cerón MI, Osorio JA, Hardwick E, Kamenov G (2013b) Thermotectonic History of Colombia's Western Cordillera and Associated Pacific and Interandean Basins: integrated application of thermochronology/geochronology/provenance analyses. In: Proceedings AGU Fall Meeting, San Francisco

Restrepo-Moreno SA, Min K, Bernet M, Barbosa A, Marín-Cerón M, Hardwick E (2013c) Thermotectonic history of the Farallones del Citará Batholith (Colombia's Western Cordillera) through multi-system, vertical profile thermochronology/geochronology: tectonic, geomorphic and climatic implications. In: Proceedings XIV Congreso Colombiano de Geología y Primer Simposio de Exploradores, Bogotá Colombia, vol Memorias v Sociedad Colombiana de Geología, p 533

Restrepo-Moreno SA, Vinasco CJ, Foster DA, Min K, Noriega S, Bernet M, Marín-Cerón MI, Bermúdez M, Botero M (2015) Cenozoic accretion and morpho-tectonic response in the northern Andes (Colombia) through low-temperature thermochronology analyses/modeling. In: Proceedings GSA Annual Meeting, Baltimore, vol 47, Geological Society of America Abstracts with Programs, p 675

Restrepo-Pace PA, Ruiz J, Gehrels G, Cosca M (1997) Geochronology and Nd isotopic data of Grenville-age rocks in the Colombian Andes: new constraints for Late Proterozoic–Early Paleozoic paleocontinental reconstructions of the Americas. Earth Planet Sci Lett 150:427–441

Reyes-Harker A, Ruiz-Valdivieso CF, Mora A, Ramírez-Arias JC, Rodriguez G, De La Parra F, Caballero V, Parra M, Moreno N, Horton BK (2015) Cenozoic paleogeography of the Andean foreland and retroarc hinterland of Colombia. AAPG Bull 99(8):1407–1453

Ring U, Brandon MT, Lister GS, Willett SD (1999) Exhumation processes: normal faulting, ductile flow and erosion, Geological Society Special Publication, vol 154. Geological Society Special Publication 378, London

Rodríguez G, Zapata G (2012) Características del plutonismo Mioceno superior en el segmento norte de la Cordillera Occidental e implicaciones tectónicas en el modelo geológico del noroccidente colombiano: Boletín de Ciencias de la Tierra, no. 31

Roering JJ, Perron JT, Kirchner JW (2007) Functional relationships between denudation and hillslope form and relief. Earth Planet Sci Lett 264(1):245–258

Rojas-Agramonte Y, Neubauer F, Bojar AV, Hejl E, Handler R, García-Delgado DE (2006) Geology, age and tectonic evolution of the Sierra Maestra mountains, southeastern Cuba. Geologica Acta 4(1–2):123–150

Ruddiman WF (2013) Tectonic uplift and climate change. Springer Science & Business Media, New York p 535

Ruddiman WF, Raymo ME, Prell WL, Kutzbach JE (1997) The uplift-climate connection: a synthesis. In: Tectonic uplift and climate change. pp 471–515, Springer, Boston, MA.

Saeid E, Bakioglu K, Kellogg J, Leier A, Martinez J, Guerrero E (2017) Garzón Massif basement tectonics: structural control on evolution of petroleum systems in upper Magdalena and Putumayo basins, Colombia. Marine Petrol Geol 88:381–401

Saenz EA (2003) Fission track thermochronology and denudational response to tectonics in the north of The Colombian Central Cordillera [Masters Thesis]. Shimane University Japan, 138 p

Salazar G, James M, Tistl M (1991) El Complejo Santa Cecilia – La Equis: Evolución y acreción de un arco magmático en el norte de la Cordillera Occidental, Colombia. In: Proceedings Simposio Sobre Magmatismo y su Marco Tectónico, Volume Memorias 2, pp 142–106

Sánchez J, Horton BK, Tesón E, Mora A, Ketcham RA, Stockli DF (2012) Kinematic evolution of Andean fold-thrust structures along the boundary between the Eastern Cordillera and Middle Magdalena Valley basin, Colombia. Tectonics 31(3): 1-24

Shagam R (1975) The northern termination of the Andes. In: The Gulf of Mexico and the Caribbean. Plenum Press, New York pp 325–420

Shagam R, Kohn BP, Banks PO, Dash LE, Vargas R, Rodriguez GI, Pimentel N (1984) Tectonic implications of Cretaceous-Pliocene fission-track ages from rocks of the circum-Maracaibo basin region of western Venezuela and eastern Colombia. In: Bonini WE, Hargraves RB, Shagam R (eds) The Caribbean-South American plate boundary and regional tectonics: Geological Society of America Memoir, vol 162. Geological Society of America, Boulder, pp 385–412

Shen CB, Donelick RA, O'Sullivan PB, Jonckheere R, Yang Z, She ZB, Miu XL, Ge X (2012) Provenance and hinterland exhumation from LA-ICP-MS zircon U–Pb and fission-track double dating of Cretaceous sediments in the Jianghan Basin, Yangtze block, central China. Sedimentary Geology, 281:194–207

Shuster DL, Ehlers TA, Rusmoren ME, Farley KA (2005a) Rapid glacial erosion at 1.8 Ma revealed by 4He/3He thermochronometry. Science 310(5754):1668–1670

Shuster DL, Vasconcelos PM, Heim JA, Farley KA (2005b) Weathering geochronology by (U-Th)/He dating of goethite. Geochimica Et Cosmochimica Acta 69(3):659–673

Sierra G, Marin-Ceron M (2011) Amagá-Cauca- Patía, Medellín Colombia, Fondo Editorial Universidad EAFIT, Petroleum Geology of Colombia, 51 p

Sierra G, Silva J, Correa L (2004) Estratigrafía Secuencias de la Formación Amagá. Boletin de Ciencias de la Tierra 15:9–22

Silva JC, Sierra GM, Correa LG (2008) Tectonic and climate driven fluctuations in the stratigraphic base level of a Cenozoic continental coal basin, northwestern Andes. J South Am Earth Sci 26:369–382

Silva A, Mora A, Caballero V, Rodriguez G, Ruiz C, Moreno N, Parra M, Ramirez-Arias J, Ibañez M, Quintero I (2013) Basin compartmentalization and drainage evolution during rift inversion: evidence from the Eastern Cordillera of Colombia. Geological Society, London, Special Publications, vol 377, pp SP377. 315

Sobolev SV, Babeyko AY (2010) What drives orogeny in the Andes? Geology 33(8):617–620

Soeters R (1981) Algunos datos sobre la edad de dos superficies de erosion en la Cordillera Central de Colombia. Revista CIAF 6:525–528

Somoza R (1998) Updated Nazca (Farallon)–South America relative motions during the last 40 My: implications for mountain building in the central Andean region. J South Am Earth Sci 11:211–215

Spotila JA (2005) Applications of low-temperature thermochronometry to quantification of recent exhumation in mountain belts. In: Reiners PW, Ehlers TA (eds) Low-temperature thermochronology: techniques, interpretations, and applications, vol Volume 58. MSA, Chantilly, VA, pp 449–466

Spotila JA, Farley KA, Sieh K (1998) Uplift and erosion of the San Bernardino Mountains associated with transpression along the San Andreas fault, California, as constrained by radiogenic helium thermochronometry. Tectonics 17(3):360–378

Stallard RF (1988) Weathering and erosion in the humid tropics. In: Lerman A, Meybeck M (eds) Physical and chemical weathering in geochemical cycles, vol Volume 251. Kluwer Academic, Dordrecht, pp 226–246

Steinman G (1929) Geologie von Peru, Heidelberg, Karl Winter, 448 p

Steinmann G (1922) Über die junge Hebung der Kordillere Südamerikas. Geologische Rundschau 13(1):1–8

Steinmann G, Stappenbeck R, Sieberg AH, Lissón CI (1930) Geología del Perú, C. Winters Universitätsbuchhandlung

Stockli DF, Farley KA, Dumitru TA (2000) Calibration of the apatite (U-Th)/He thermochronometer on an exhumed fault block, White Mountains, California. Geology 28(11):983–986

Stolar DB, Willett SD, Roe GH (2006) Climatic and tectonic forcing of a critical orogen. Spec Pap Geol Soc Am 398:241–250

Stüwe K, Hintermueller MT (2000) Topography and isotherms revisited: the influence of laterally migrating drainage divides. Earth Planet Sci Lett 184:287–303

Sugden DE, Summerfield MA, Burt TP (1997) Editorial: linking short-term geomorphic processes to landscape evolution. Earth Surf Proces Landf 22:193–194

Summerfield MA (2000) Geomorphology and global tectonics. Wiley, New York. 368 p

Summerfield MA (2005) The changing landscape of geomorphology. Earth Surf Proces Landf 30:779–781

Summerfield MA, Hulton NJ (1994) Natural controls of fluvial denudation rates in major world drainage basins. J Geophys Res Solid Earth 99(B7):13871–13883

Suter F, Sartori M, Neuwerth R, Gorin G (2008) Structural imprints at the front of the Chocó-Panamá indenter: Field data from the North Cauca Valley Basin, Central Colombia. Tectonophysics 460(1):134–157

Syracuse EM, Maceira M, Prieto GA, Zhang H, Ammon CJ (2016) Multiple plates subducting beneath Colombia, as illuminated by seismicity and velocity from the joint inversion of seismic and gravity data. Earth Planet Sci Lett 444:139–149

Syvitski J (2012) Anthropocene An epoch of our making. Global Change Newslett 78:12–15

Syvitski JPM, Milliman JD (2007) Geology, geography, and humans battle for dominance over the delivery of fluvial sediment to the coastal ocean. J Geol 115(1):1–19

Taboada A, Rivera LA, Fuenzalida A, Cisternas A, Philip H, Bijwaard H, Olaya J, Rivera C (2000) Geodynamics of the northern Andes: subductions and intracontinental deformation (Colombia). Tectonics 19(5):787–813

Tagami T, O'Sullivan PB (2005) Fundamentals of fission-track thermochronology. In: Reiners PW, Ehlers TA (eds) Low-temperature thermochronology: techniques, interpretations, and applications, Reviews in Mineralogy and Geochemistry, vol Volume 58. MSA, Chantilly, pp 19–47

Thomas MF (1994) Geomorphology in the tropics. A study of denudation and weathering in low latitudes. Wiley, New York. 482 p

Thompson AB, Schulmann K, Jezek J (1997) Thermal evolution and exhumation in obliquely convergent (transpressive) orogens. Tectonophysics 280(1):171–184

Toro J, Roure F, Bordas-Le Floch N, Le Cornec-Lance S, Sassi W (2004) Thermal and kinematic evolution of the Eastern Cordillera fold and thrust belt, Colombia. In: Swennen R, Roure F, Granath JW (eds) Deformation, fluid flow, and reservoir appraisal in foreland fold and thrust belts, vol Volume 1, pp 79–115

Toro G, Hermelin M, Schwave E, Posada B, Silva D, Poupeau G (2006) Fission-track datings and long-term stability in the Central Cordillera highlands, Colombia. In: Latrubesse E (ed) Tropical Geomorphology with Special Reference to South America, vol Volume 145. Springer, Berlin, pp 1–16

Trenkamp R, Kellogg JN, Freymueller JT, Mora HP (2002) Wide plate margin deformation, southern Central America and northwestern South America, CASA GPS observations. J South Am Earth Sci 15(2):157–171

Tschanz CM, Marvin RF, Cruzb J, Mehnert HH, Cebula GT (1974) Geologic Evolution of Sierra-Nevada-De-Santa-Marta, Northeastern Colombia. Geol Soc Am Bull 85(2):273–284

Vallejo F, Restrepo-Moreno SA, Pardo A, Trejos R, Flores JA, Plata A, Min K, Foster DA, López SA (2015) Miocene Andean uplift and its impact on planktonic communities in Eastern Equatorial Pacific basins. In: Proceedings 15th International Nannoplankton Association Meeting, Bohol, Philippines, vol 35, Special Issue Journal of Nannoplankton Research Nannoplankton Association, p 88

Van der Beek P, Robert X, Mugnier J-L, Bernet M, Huyghe P, Labrin E (2006) Late Miocene – recent denudation of the central Himalaya and recycling in the foreland basin assessed by detrital apatite fission-track thermochronology of Siwalik sediments, Nepal. Basin Res 18:413–434

Van Der Beek P, Van Melle J, Guillot S, Pêcher A, Reiners PW, Nicolescu S, Latif M (2009) Eocene Tibetan plateau remnants preserved in the northwest Himalaya. Nature Geosci 2(5):364

Van der Hammen T (1957) Palynologic stratigraphy of the Sabana de Bogota (East Cordillera of Colombia). Boletín Geológico (Bogotá) 5:187–203

Van der Hammen T (1961) Late Cretaceous and Tertiary stratigraphy and tectogenesis of the Colombian Andes. Geologie en Mijnbouw 40:181–188

Van der Hammen T (1989) History of the montane forests of the northern Andes. Plant Syst Evol 162(1-4):109–114

Van der Hammen T, Werner JH, van Dommelen H (1973) Palynological record of the upheaval of the Northern Andes: a study of the Pliocene and Lower Quaternary of the Colombian Eastern Cordillera and the early evolution of its High-Andean biota. Rev Paleobot Palynol 16:1–122

Van der Hilst R, Mann P (1994) Tectonic implication of tomographic images of subducted lithosphere beneath northwestern South América. Geology 22:451–454

Van der Lelij R, Spikings RA, Kerr AC, Kounov A, Cosca M, Chew D, Villagomez D (2010) Thermochronology and tectonics of the Leeward Antilles: evolution of the southern Caribbean Plate boundary zone. Tectonics 29:TC6003. https://doi.org/10.1029/2009TC002654

Van der Wiel A (1989) Uplift of the Precambrian Garzon Massif (Eastern Cordillera of the Colombian Andes) in relation to fluvial and vocaniclastic sedimentation in the adjacent Neiva Basin. In: Proceedings Abstract 28th Int. Geological Congr, pp 3-505

Van der Wiel AM (1991) Uplift and volcanism of the SE Colombian Andes in relation to Neogene sedimentation in the Upper Magdalena Valley [Ph.D.]. University of Wageningen, 208 p

Van der Wiel A, Andriessen P (1991) Precambrian to Recent thermotectonic history of the Garzon massif (Eastern Cordillera of the Colombian Andes) as revealed by fission-track analysis: Uplift and volcanism of the SE Colombian Andes in relation to Neogene sedimentation of the upper Magdalena Valley [Ph.D. Thesis]. Free University, Amsterdam Wageningen

Van Houten FB, Travis RB (1968) Cenozoic deposits, Upper Magdalena Valley, Colombia. Am Assoc Petrol Geol 52:675–702

Vargas CA, Mann P (2013) Tearing and breaking off of subducted slabs as the result of Collision of the Panama arc-indenter with Northwestern South America. Bull Seismol Soc Am 103(3):2025–2046

Velandia F, Acosta J, Terraza R, Villegas H (2005) The current tectonic motion of the Northern Andes along the Algeciras Fault System in SW Colombia. Tectonophysics 399(1):313–329

Velandia-Patiño FA (2018) Cinemática de las fallas mayores del Macizo de Santander – énfasis en el modelo estructural y temporalidad al sur de la Falla de Bucaramanga [PhD]. Universidad Nacional de Colombia, 229 p

Vermeesch P, Tian Y (2014) Thermal history modelling: HeFTy vs QTQt. Earth-Sci Rev 139:279–290

Villagomez-Díaz DR (2010) Thermochronology, geochronology and geochemistry of the Western and Central cordilleras and Sierra Nevada de Santa Marta, Colombia: the tectonic evolution of NW South America [PhD]. Université de Genève, 144 p

Villagomez D, Spikings R (2013) Thermochronology and tectonics of the Central and Western Cordilleras of Colombia: early Cretaceous–Tertiary evolution of the northern Andes. Lithos 160:228–249

Villagomez D, Spikings R, Mora A, Guzmán G, Ojeda G, Cortés E, van der Lelij R (2011) Vertical tectonics at a continental crust-oceanic plateau plate boundary zone: Fission track thermochronology of the Sierra Nevada de Santa Marta, Colombia. Tectonics 30(4): 1-18

Villamil T (1999) Campanian–Miocene tectonostratigraphy, depocenter evolution and basin development of Colombia and western Venezuela. Palaeogeogr Palaeoclimatol Palaeoecol 153:239–275

Vinasco CJ, Cordani U (2012) Reactivation episodes of the Romeral Fault System in the northwestern part of the Central Andes, Colombia, through 39Ar-40Ar and A-Ar results. Boletín de Ciencias de la Tierra 32:111–124

Vinasco CJ, Cordani UG, Gonzalez H, Weber M, Pelaez C (2006) Geochronological, isotopic, and geochemical data from Permo-Triassic granitic gneisses and granitoids of the Colombian Central Andes. J South Am Earth Sci 21(4):355–371

Walling DE (2006) Human impact on land-ocean sediment transfer by the world's rivers. Geomorphology 79(3–4):192–216

Ward D (1973) Geología de los cuadrángulos H-12, Bucaramanga y H-13, Pamplona, Departamento de Santander: Boletín Geológico Ingeominas

Weng CY, Hooghiemstra H, Duivenvoorden JF (2007) Response of pollen diversity to the climate-driven altitudinal shift of vegetation in the Colombian Andes. Philos Trans Royal Soc B Biol Sci 362(1478):253–262

Whipple KX (2009) The influence of climate on the tectonic evolution of mountain belts. Nature Geosci 2(2):97–104

Whipple KX, Meade BJ (2006) Orogen response to changes in climatic and tectonic forcing. Earth Planet Sci Lett 243(1–2):218–228

Wilder DT (2003) Relative motion history of the Pacific-Nazca (Farallon) plates since 30 million years ago [Master of Science]. University of South Florida, 105 p

Wilkinson BH (2005) Humans as geologic agents: a deep-time perspective. Geology 33(3):161–164

Wilkinson MT, Humphreys GS (2005) Exploring pedogenesis via nuclide-based soil production rates and OSL-based bioturbation rates. Soil Res 43(6):767–779

Wilkinson BH, McElroy BJ (2005) The impact of humans on continental erosion and sedimentation. GSA Bull 119(1–2):140–156

Wilkinson BH, McElroy BJ, Kesler SE, Peters SE, Rothman ED (2009) Global geologic maps are tectonic speedometers – rates of rock cycling from area-age frequencies. GSA Bull 121:760–779

Willett SD (1999) Orogeny and orography: the effects of erosion on the structure of mountain belts. J Geophys Res Solid Earth 104(B12):28957–28981

Willett SD, Slingerland R, Hovius N (2001) Uplift, shortening, and steady-state topography in active mountain belts. Am J Sci 301(4–5):455–485

Willett SD, Fisher D, Fuller C, En-Chao Y, Chia-Yu L (2003) Erosion rates and orogenic-wedge kinematics in Taiwan inferred from fission-track thermochronometry. Geology 31(11):945–948

Willett SD, Hovius N, Brandon MT, Fisher DM (2006) Tectonics, climate, and landscape evolution, Penrose Conference Volume Special Paper 398, Boulder, CO, Geological Society of America 447

Willgoose G (2005) Mathematical modeling of whole landscape evolution. Ann Rev Earth Planet Sci 33:443–459

Wolf RA, Farley KA, Kass DM (1998) A sensitivity analysis of the apatite (U-Th)/He thermochronometer. Chem Geol 148:105–114

Woodburne MO (2010) The great American biotic interchange: dispersals, tectonics, climate, sea level and holding pens. J Mamm Evol 17(4):245–264

Wyllie PJ (1976) The way the earth works. Wiley, New York

Yarce J, Monsalve G, Becker TW, Cardona A, Poveda E, Alvira D, Ordoñez-Carmona O (2014) Seismological observations in Northwestern South America: evidence for two subduction segments, contrasting crustal thicknesses and upper mantle flow. Tectonophysics 637:57–67

Zaun PE, Wagner GA (1985) Fission-track stability in zircons under geological conditions. Nucl Tracks Radiat Meas 10:303–307

Zeil, W. (1979) The Andes-a geological review. Gebruder Borntraeger, p 260

Zeitler PK, Meltzer AS, Koons PO, Craw D, Hallet B, Chamberlain C, Kidd WSF, Park SK, Seeber L, Bishop M, Shroder J (2001) Erosion, Himalayan geodynamics, and the geomorphology of metamorphism. GSA Today:4–8

Zuluaga MD, Houze RA (2015) Extreme convection of the near-equatorial Americas, Africa, and adjoining oceans as seen by TRMM. Monthly Weather Rev 143:298–316

Zuluaga J, Mattsson L (1981) Glaciaciones de la Cordillera Occidental de Colombia, Páramo de Frontino. Departamento de Antioquia. Revista CIAF 6:639–654

Chapter 12
The Romeral Shear Zone

César Vinasco

Within the study area to the south and west of Medellín, the principle lithotectonic units of the Romeral shear zone include:

1. Low- to medium-grade pre-Triassic and Cretaceous metavolcanic and metasedimentary rocks of the Arquía Complex, including the Sinifaná schists, interpreted to represent part of the disjointed paleo-autochthon
2. Mafic and ultramafic mid-Triassic supra-subduction zone intrusive rocks
3. An Early Cretaceous dominantly sedimentary sequence (Abejorral Formation) and a second generation of Late Cretaceous sedimentary-mafic volcanic rocks (Quebradagrande Complex) which appear to be of oceanic, arc-marginal basin affinity
4. Coal-bearing epiclastic sequences of the Oligo-Miocene Amagá Formation, unconformably deposited, in transtensional basins, upon the disjointed basement units
5. The mid-Late Miocene-Pliocene Combia Formation and associated arc-related subvolcanic intrusions, deposited upon and cross-cutting both the Amagá Formation and the underlying Romeral basement units

Contacts between the Mesozoic components of the Romeral assemblage are complex and structural in nature.

Reworking and emplacement of tectonic elements within the Romeral shear zone along the paleo-Colombian margin took place in various phases prior to and during the Meso-Cenozoic Northern Andean orogenic cycle. Structural evolution between the Upper Cretaceous and Eocene reflects the collision of allochthonous oceanic rocks of Pacific provenance to the west, yielding an early generation of tight,

---

C. Vinasco (✉)
Departamento De Geociencias Y Medio Ambiente, Universidad Nacional De Colombia, Facultad De Minas, Medellin, Colombia
e-mail: cvinasco@unal.edu.co

upright, dextral-oblique, west-verging folds and thrust faults, including top-to-the-west transport of slivers of the western paleo-continental margin (Cajamarca Complex) over pericratonic oceanic basement (e.g., Quebradagrande assemblage). These dextral-oblique reverse faults are common along the eastern Romeral margin and define most geological contacts in the basement assemblage, in a thick-skinned deformational style. They provide key elements in the understanding of the kinematics of the Early Northern Andean orogeny.

Subsequent deformational episodes along the Romeral fault zone and within the Central Cordillera to the east are related to continued uplift of the Central Cordillera, reaccommodation of the regional plate tectonic setting, and collision of South America with the Panamá-Chocó block, during the Late Northern Andean orogeny, beginning in the Miocene and extending through the Early Pliocene. This final episode generated discrete, east-verging structures, highlighted by the reactivation of Late Cretaceous faults/folds in both the oceanic and continental margin assemblages. The general modern-day geometry of the basement of the Romeral shear system corresponds to a folded and structurally telescoped, west-verging megastructure, containing an anastomosing array of discrete Late Cretaceous mylonitic shears, developed mainly within the metasedimentary units and metavolcanic rocks and eventually in the ultramafic fragments. The mylonitic shears wrap the more competent tectonic blocks, containing granitoid and ultramafic rocks, yielding a mélange-like regional tectonic configuration. Mid- to Late Miocene uplift along the Western and Central cordilleras led to superposition of Amagá Formation and equivalent epiclastic rocks in diachronous depocenters along the Cauca-Patia intermontane valley. Continued compression led to localized inversion, thrusting, and folding of the basal Amagá units prior to emplacement of the essentially undeformed Combia Formation volcanics in the Late Miocene.

Many of the lithotectonic units encompassed by the Romeral shear zone have a demonstrable genetic relationship with the autochthonous Late Paleozoic through mid-Cretaceous paleo-margin of Central Colombia. The polylithic nature and structural complexity recorded along this paleo-margin suggests the Romeral shear zone has served as a locus for repeated tectonic cycles, involving rifting, detachment, oblique transport, and re-accretion of tectonic blocks during the Late Paleozoic through Meso-Cenozoic tectonic evolution of northwestern South America.

12.1 Introduction

Beginning in the Early Cretaceous, the northwestern margin of South America has experienced the interplay between the South American, proto-Caribbean, Farallon-Caribbean, and (after ca. 25 Ma) Nazca plates (Meschede and Barckhausen 2000; Lonsdale 2005). This interaction involves episodes of extension, subduction, collision, and accretion, including the separation of blocks from the paleo-continental margin, the accretion of exotic oceanic terranes, and, since the Early to mid-Cretaceous, the subsequent inversion of the detached margin-oceanic terrane

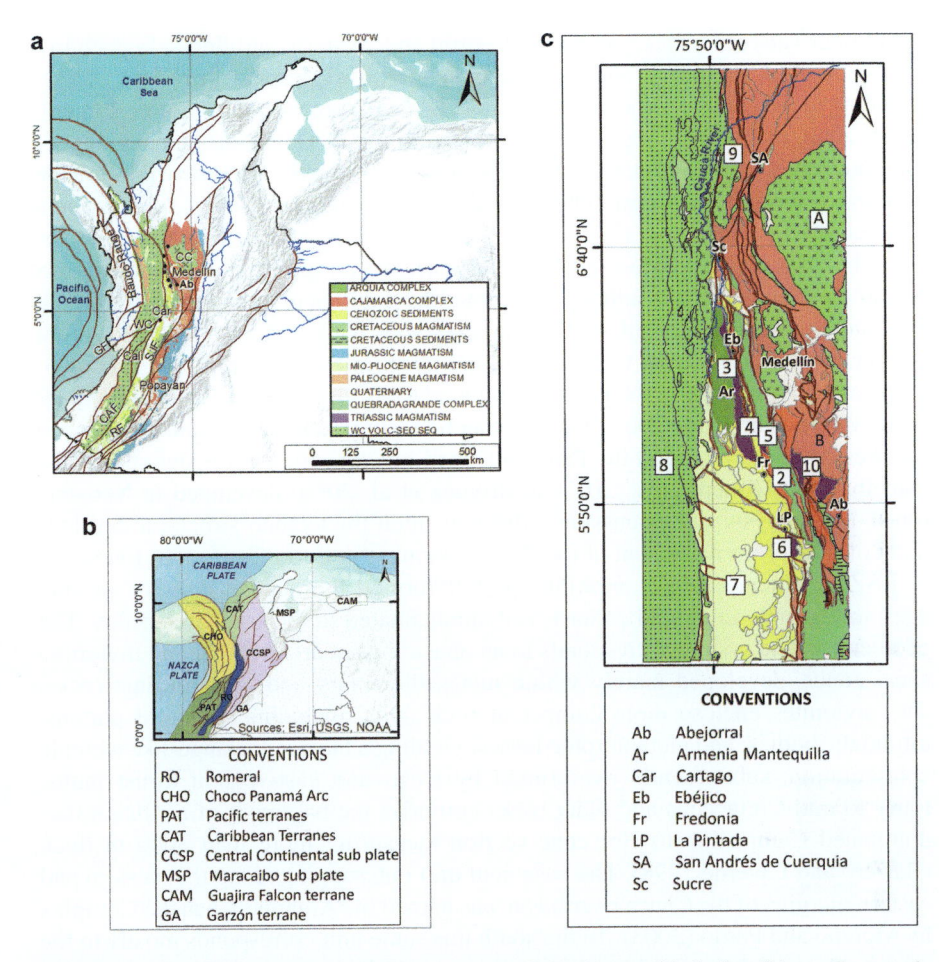

Fig. 12.1 (a) Simplified geological map of western Colombia modified after Gómez-Cruz et al. (1995). GF Garrapatas fault, CAF Cauca Almaguer fault, RF Romeral fault. (b) Major tectonic realms in the North Andean block as defined by Cediel et al. (2011). (c) Lithotectonic elements of the northern Romeral shear zone modified after Gómez et al. (2015). 1 = Quebradagrande Complex, 2 = Sinifana metasediments, 3 = Arquia Complex (including the Sabaletas schists), 4 = mafic and ultramafic rocks (Pueblito diorite), 5 = granitic rocks (Amaga stock), 6 = Cenozoic sedimentary rocks, 7 = Mio-Pliocene magmatism, 8 = Barroso Fm., 9 = Sabanalarga Batholith, 10 = La Honda and El Buey stocks, A = Cretaceous Antioquian Batholith, B = Cajamarca Complex

assemblage against the tectonized continental margin of the Colombian Central Cordillera. The allochthonous, oceanic western segment of the Colombian Andes includes the physiographic Western Cordillera (and Panamá-Chocó block) (Fig. 12.1a, b), the basement of which is of predominantly Farallon-Caribbean affinity.

Physiographically, the regional boundary between the continental autochthon to the east and exotic/para-autochthonous oceanic terranes to the west is broadly defined by the depression formed by the Cauca (and in southern Colombia, the

Patía) river valley, which separates the Central and Western cordilleras (Fig. 12.1a). The geological boundary between these major physiographic provinces, corresponding to the complex Romeral shear zone (RSZ) (Romeral fault system, Romeral suture zone, or Dolores megashear after Cline et al. 1981), and accompanying tectonic mélange, however, is less precisely defined. The RSZ is responsible for the accommodation of deformation between the oceanic and continental domains, in a dominantly transpressional regime since at least the Early Cretaceous (and perhaps since the Triassic, as a proto-fault/shear system), including subsequent reactivations during the Cenozoic. In southern Colombia, the geological boundary between the RSZ and the Western Cordilleran terranes coincides with the Cauca fault and suture system, while, in northern Colombia, to the north of the Garrapatas fault (Etayo-Serna and Barrero 1983), the geological boundary is given by the Cauca-Almaguer fault (Moreno and Pardo 2003). On the other hand, the geological boundary between the Western Cordillera and the Panamá-Chocó block to the west is defined by the Dabeiba fault (Botero et al. 2017; Rodriguez et al. 2016) developed in Neogene times. My analysis and discussion will focus upon the tectono-structural development of the northern segment of the RSZ, beginning in Early Cretaceous times.

RSZ basement encloses multiple, poly-deformed lithological units of varying ages and origins, in faulted contact, and amalgamated in the Late Cretaceous. The geometry of the system corresponds to an anastomosed array of discrete mylonitic shear zones, developed mainly within metasedimentary and metavolcanic rocks. The mylonites enclose more competent rock units, including granitoid plutons, ultramafic bodies, and metamorphic lenses, yielding a blocky, mélange-like tectonic configuration, subsequently overprinted by west- and east-vergent thrust faults. Early to mid-Cretaceous and older rocks comprise the basement of a siliciclastic-dominated Campanian to Holocene section measuring more than 3000 m thick (Barrero and Laverde 1998). The basement unit outcropping along the western and eastern margins of the Cauca river basin was named the Amaime-Chaucha Complex by Moreno and Pardo (2003). To the south this same unit corresponds mostly to the western flank of the Central Cordillera (Sierra and Marin-Cerón 2011). In the north, basement contains blocks of the pre-Cretaceous and Cretaceous Arquía Complex, the Triassic Pueblito diorite and Amagá stock, and volcanic rocks of the Quebradagrande Complex.

Classical studies with key field information pertaining to the RSZ come from Grosse (1926) and Botero (1963). Van der Hammen (1958) presented a time line for episodes of deformation and uplift along the Cauca-Patía river valley, which in many respects is still applicable. Regional-level tectonic studies pertaining to the RSZ have been presented by Restrepo and Toussaint (1988); Villagomez et al. (2011); Spikings et al. (2015), and others.

Elements critical to understanding the nature and evolution of the RSZ and associated tectonic mélange include understanding the provenance and deformational history and style and number of deformational episodes, recorded within the various basement blocks. Previous work has partially addressed these questions (Giraldo 2010; García 2010; Rodriguez 2010), whereas more recent work has been focused mainly in understanding evolution of the Quebradagrande Complex

(Cardona et al. 2012; Vinasco and Cordani 2012), the geochemistry of metabasites of the Arquía Complex (Ruiz-Jimenez et al. 2012), and the age of high-pressure rocks intimately associated with the RSZ (Bustamante et al. 2011). Sierra and Marin-Cerón (2011) presented a discussion based upon drill core and geophysical data along the entire Amagá-Cauca-Patía (ACP) basin, including a revision of the stratigraphy and structural geology of the region. State-of-the-art investigations include the study of morpho-tectonic response to deformation through low-temperature thermochronology, both in in situ vertical profiles in uplifted blocks and within modern detrital sediments (Vinasco et al. 2015; Botero et al. 2015; Noriega 2015).

Importantly, most recent studies (Vinasco et al. 2006; Martens et al. 2012) suggest that the lithotectonic units contained within the RSZ and associated mélange have a genetic relationship with the ancient Late Paleozoic continental paleomargin. This suggests that the margin has experienced repeated episodes of extension and compression, involving blocks detached and re-docked along the paleo-margin, challenging former ideas of a margin construction solely from accretion of exotic blocks (Toussaint and Restrepo 1989; Restrepo et al. 2009; Restrepo and Toussaint 1991).

In Colombia, the RSZ is well exposed in several E-W sections along the Cauca River, south and west of Medellín, where most of our work has been focused. As a major plate boundary, fault system, and tectonic mélange, the RSZ can be traced into southwestern Ecuador as the Peltetec assemblage (Aspden and Litherland 1992; Cediel et al. 2003). West-southwest-verging reverse faults are common along the RSZ, defining most geological contacts in a thick-skinned deformational style. Relationships between the RSZ and the thrust faults require additional definition, although the thrusts appear to represent key elements of a transpressional orogenic style, closely associated with the accretionary regime prevalent during the mid-Late Cretaceous and Cenozoic. Rotation of lithotectonic blocks within the RSZ is common, disjointing the along-strike continuity of rock units and early-formed structures.

To the south and west of Medellín, the RSZ corresponds to a kilometer-scale shear zone encompassing various lithotectonic units (Fig. 12.1c), including (1) the Cretaceous Quebradagrande Complex volcano-sedimentary sequence, (2) the low-grade Devonian (?) Sinifaná metasedimentary schist, (3) Triassic mafic and ultramafic supra-subduction zone intrusives, and (4) the Arquía Complex, Permian (?), and/or Cretaceous (?) low- to medium-grade, metavolcano-sedimentary rocks and N-type MORB. The Amagá Formation (Fm.), a coal-bearing, Oligo-Miocene epiclastic sequence, unconformably covers the older lithological units. Mio-Pliocene subaerial volcanic rocks and subvolcanic intrusions of the Combia Fm. cover and intrude the Amagá Fm. and older rocks. The Arquía Complex is interpreted to represent the core of the RSZ. It comprises an incipiently defined composite collection of rocks including pre-Mesozoic and Cretaceous fragments. Some of these fragments represent tectonic remobilization of the paleo-continental margin, while others appear to represent the exotic oceanic-continental margin assembly. I interpret the mélange-like relationships observed in the RSZ to reflect the poorly understood composite collection of rocks including pre-Mesozoic and Cretaceous fragments.

Some of these fragments would be remobilized pieces from both the Central and Western cordilleras, reorganized in the context of long-lasting shear zone dynamics.

^{39}Ar-^{40}Ar and K-Ar age data for Quebradagrande volcanic rocks and for neoformed micas in crosscutting mylonitic bands, within the easternmost branch of the RSZ (i.e., the San Jerónimo Fault, SJF), was presented by Vinasco and Cordani (2012). Samples of andesitic volcanic rocks collected in the area of influence of the SJF range from 91 to 102 Ma. These ages are interpreted to represent reset ages due to fault activity. Within the shear bands, plateau ages ranging from 90 to 87 Ma in biotite and 81–72 Ma in sericite suggest reactivation activity along the fault system reaching its peak during the Late Cretaceous. This age interval is interpreted to be related to the collision/accretion of the Caribbean-Colombian Oceanic Plateau/Caribbean Large Igneous Province (CCOP/CLIP; Nivia et al. 1996; Spikings et al. 2015) between the Late Cretaceous and mid-Eocene, during the early Northern Andean orogenic cycle, with ensuing uplift of the Central Cordillera. Subsequent collisional episodes are responsible for the Neogene deformational history of the orogen, involving accretion of the Panamá-Chocó block in Late Miocene-Pliocene. During the Late Miocene and Pliocene, voluminous arc-related volcanic rocks and associated intrusives (the Combia Fm.; Fig. 12.1c) invaded the fault system, generating important deposits of Ag, Zn, Pb, Cu, and Au (see Shaw et al. 2018). Eastward migration of this same arc generated the Nevado del Ruiz-Tolima volcanic complexes and other manifestations of the northern volcanic zone during the Pliocene to recent.

Over the last 10 years, new field studies, supported by radiometric age and lithochemical data, have enabled the reevaluation of geodynamic models for Northern Andean-Caribbean plate interaction. Long-standing and important questions regarding subduction and accretion models and deformational mechanisms during the Meso-Cenozoic can now be better addressed. Such questions relate to (1) the nature and provenance of para-autochthonous blocks vs. allochthonous terranes within/along the RSZ and accompanying tectonic mélange, (2) the emplacement of batholiths within continental and oceanic domains along the Colombian proto-Andean margin, and (3) the history, role, and evolving architecture of the RSZ and mélange, as a major, complex plate boundary within which the redistribution of lithotectonic elements sandwiched between the continental and oceanic domains took place, in consort with the evolution of the Meso-Cenozoic Northern Andean orogeny.

12.2 Geological Overview, Nomenclature, and Regional Fault Systems

The geological configuration of Colombia has long been envisaged as a mosaic of terranes and tectonic realms (Etayo-Serna and Barrero 1983). A recent interpretation based upon the work of Cediel et al. (2003) is presented in Fig. 12.1b. Several authors have contributed to the prevailing definition and nomenclature of the numerous tectonic domains, terranes, and terrane boundaries observed in the Colombian

Andes, based mainly upon contrasting geological evolution documented in the basement rocks comprising each terrane.

The RSZ has been included in the composite Calima terrane of Toussaint and Restrepo (1989) and Romeral terrane of Etayo-Serna and Barrero (1983). The Romeral terrane of Cediel et al. (2003) is contained within the Western Tectonic Realm, along with the Dagua, Gorgona, and Cañasgordas (western Chocó arc) terranes (Fig. 12.1). The allochthonous vs. autochthonous nature of the Romeral mélange with respect to the Northern Andean paleo-margin was called into question by Cediel et al. (2003, p. 829–830) versus the more clearly established exotic nature of the Dagua, Cañasgordas, and Gorgona terranes, which comprise basement to Colombia's physiographic Western Cordillera. The vision of Romeral as a strictly exotic terrane is challenged in the present work, at least for some components of the so-called exotic blocks. Within the present work, I demonstrate the correspondence and relationships between the paleo-continental margin and marginal basin assemblages and the collage of para-autochthonous blocks derived primarily from these essentially in situ crustal sources. In this context, it is difficult to demonstrate a well-defined, exotic basement component within the Romeral mélange. I will discuss developments along the Colombian continental margin since Permian times and tectonic assembly of the RSZ during the Late Mesozoic and Cenozoic. Eventually the Romeral terrane will be used for regional comparative purposes in the concluding remarks.

The Romeral terrane may be traced southward into Ecuador where sporadic exposures of the lithologically and temporally equivalent Peltetec, El Toro, and Raspas units indicate that it underlies the western margin of the Cordillera Real and probably much of the inter-Andean depression (CODIGEM 1993; Litherland et al. 1994; Leal-Mejía et al. 2011). Within the study area, the RSZ and associated mélange are bound to the east by metamorphic rocks, which formed the paleo-continental margin during pre- and Early Mesozoic times. These composite units have been given various names over the years, including the Cajamarca Group (Nelson 1957), the Cajamarca Complex (Maya and González 1995), the Tahamí terrane (Toussaint and Restrepo 1989), the Central Cordillera Polymetamorphic Complex (Toussaint and Restrepo 1989), and the Cajamarca-Valdivia terrane (Cediel et al. 2003). They form the basement to the physiographic Central Cordillera, which I herein refer to as the Central Cordillera block.

In general, Central Cordillera basement can be regarded as a Lower Paleozoic-Permian metamorphic assemblage, with localized in situ granitoids returning U-Pb (zircon) ages as old as 473 Ma (Leal-Mejía et al. 2018) and detrital U-Pb (zircon) ages ranging from 270 to 380 Ma (Vinasco et al. 2006; Villagómez et al. 2011). The complex is intruded by Permian crustal syntectonic and post-tectonic Triassic arc granitoids (Vinasco et al. 2006) and Late Triassic anatectic granitoids and amphibolites associated with the breakup of Pangea (Vinasco et al. 2006; Cochrane et al. 2014b). Important components of the Romeral terrane were derived from the Central Cordillera block, as tectonic rafts detached from the paleo-autochthon through extension, rifting, and marginal basin development along the Colombian Pacific paleo-margin during the Late Permian through Early Cretaceous.

To the west, the RSZ is limited by accreted terranes, which form the basement to the physiographic Western Cordillera, including the Cañasgordas terrane (Etayo-Serna and Barrero 1983 and Duque-Caro 1990) and Dagua terrane further south (Nivia 1996). These terranes constitute aseismic ridges, oceanic plateaus, and basaltic rocks of tholeiitic MORB-type affinity (Ortiz 1979). The Western Cordillera is comprised of at least four distinct terrane assemblages. In the south the Dagua terrane is comprised of E-MORB and N-MORB, with associated marine sediments, and is considered to represent an accreted fragment of the CCOP/CLIP plateau (Nivia et al. 1996; Cediel et al. 2003; Pindell and Kennan 2009). To the west of Dagua, the Gorgona terrane is cored by ultramafic complexes and contains komatiite with associated marine sediments. However, its relationship to the CCOP plateau has yet to be clearly established (Cediel et al. 2003; Kerr and Tarney 2005; Pindell and Kennan 2009). North of the Garrapatas paleo-transform fault (Etayo-Serna and Barrero 1983), accreted terranes include Cañasgordas and the Panamá-Chocó block. Cañasgordas contains mixed, primitive MORB and calc-alkaline volcanic rocks with associated marine sediments containing a wide range of faunal assemblages dating from the Barremian through Upper Cretaceous (González 2001). Recent investigations by Botero et al. (2017) reveal detrital zircon age populations in the upper part of the Penderisco Fm. which are interpreted to reflect a source region from the Central Cordillera block. In this context, the fully exotic character of the Cañasgordas terrane is uncertain, as zircon populations suggest that Cañasgordas was proximal to the continental block, at least by the Late Cretaceous. In reality, Cañasgordas is a mixed terrane constructed upon oceanic basement. Structural complications make identification of the nature and age of this basement difficult to ascertain, but it likely includes tectonic slices of the Farallon plate and possibly the CCOP/CLIP assemblage and potentially even intercalated slices of the para-autochthonous Romeral mélange, especially along the eastern margin, where it is in direct accretionary contact with the Quebradagrande and Arquía units.

Moreno and Pardo (2003) and Maya and González (1995) define structural complexes and lithostratigraphic units, respectively, for western Colombia, including the Quebradagrande and Arquía complexes, among others. The Quebradagrande and Arquía complexes are located within the RSZ and form the core components of the Romeral mélange, along with the low-grade Devonian (?) metasedimentary Sinifaná schists, Triassic mafic and ultramafic supra-subduction intrusive, and Late Cretaceous granitoids (Fig. 12.1c).

In southern Colombia, the main fault systems that comprise the RSZ have a regional NNE strike, while to the north, around Medellin, the strikes changes dominantly to N-NW. From east to west, the principle component faults comprising the RSZ include San Jeronimo, Silvia-Pijao, and Cauca-Almaguer (Maya and González 1995) (Fig. 12.1a), all interpreted to be of transpressional character. The San Jeronimo Fault is a fundamental structure of the RSZ and defines the eastern Romeral boundary, placing the Quebradagrande Complex directly against the Central Cordillera block. Along the middle Cauca Valley, Permian schists are overthrust, top-to-the-west, upon the Quebradagrande Complex. Regionally, Quebradagrande is in turn separated from the Arquía Complex along the Silvia-Pijao fault; however, at a local scale, this contact is complex and includes thrust slices of pre-Cretaceous mafic and ultramafic intrusives and Late Cretaceous granitoids. Finally, the Arquía Complex is limited on its western margin by the Cauca-Almaguer fault which places Arquía against Amaime volcanic rocks (Moreno and Pardo 2003) to the south or Barroso Fm. further north (Fig.12.1).

The main lithotectonic and lithostratigraphic components of the RSZ and the associated basement mélange are now described in detail.

12.2.1 Quebradagrande Complex

The Quebradagrande Complex (Botero 1963; Fig. 12.1c) includes (i) a para-autochthonous Late Cretaceous arc remnant, including slices of newly formed oceanic crust which have been interpreted to represent the opening of an ensialic marginal basin (Nivia et al. 2006) and (ii) an Early Cretaceous sedimentary sequence known as the Abejorral Fm. (Bürgl and Radelli 1962; Moreno and Pardo 2003; Nivia et al. 2006). In more recent work, Zapata (2015) distinguished two Cretaceous volcano-sedimentary sequences outcropping along the western flank of the Central Cordillera and differentiated the sedimentary-dominated Aptian-Albian Abejorral Fm. from the volcanic-dominant Late Cretaceous sequence of Quebradagrande Complex. He suggested that Albian-Aptian sequences, including an intercalated andesitic horizon with an age of 102 Ma, are related to the Abejorral Fm., while the dominantly volcanic sequences west of the San Jeronimo fault would be younger, having formed in Santonian-Campanian times. Zapata (2015) suggested that, following a period of Aptian-Albian extension derived from changes in regional plate convergence vectors, Late Cretaceous volcano-sedimentary sequences were built on a hyperextended continental basement. Pardo and Moreno (2001) implicitly separate these domains arguing that differences in lithofacies suggest a passive margin to the east and a volcanic arc to the west. The Quebradagrande Complex was constructed along the paleo-continental margin, presently exposed as a discontinuous fringe along the western flank of the Central Cordillera of Colombia. Regionally, the complex is structurally located between two pre-Cretaceous metamorphic complexes, the Arquía Complex to the west and the Cajamarca Complex to the east. It contains an ophiolitic assemblage of ultramafic rocks, gabbros, diorites, diabases, basalts, basaltic andesites, and andesites with mudrocks, cherts, feldspathic and quartzitic sandstones, lithic and quartzitic conglomerates, and pyroclastic rocks, usually affected by zeolite and prehnite-pumpellyite through greenschist facies dynamo-thermal metamorphism (Moreno and Pardo 2003; Nivia et al. 2006; Moreno-Sánchez et al. 2008). The depositional environments of sedimentary rocks associated with the Abejorral Fm. vary from fluvial, coastal, coarse-grained deltas, and platform (Rodríguez and Rojas 1985) to slope and submarine fan deposits containing a volcanic component (Gómez-Cruz et al. 1995). Locally, in the northern sector, close to the town of Olaya, Quebradagrande volcanic rocks are faulted toward the W-SW, over the Late Cretaceous Sucre amphibolite and syenogranites

(Vinasco et al. unpublished data) as a result of dextral transpression. Tectonic lenses of Triassic ultramafic rocks are embedded in deformed Quebradagrande volcanic-shale matrix. The Sucre amphibolite corresponds to a deformed and metamorphosed L-tectonite-type Gabbro, which could represent Quebradagrande late syntectonic arc magmatism (Vinasco et al. unpublished data). Further north, between Liborina and Sabanalarga, a well-defined top-to-the-SW thrust of Quebradagrande volcanic rocks over Oligo-Miocene sediments of the Amagá Fm. constrains the timing of late top-to-the-west thrusting to the post Late Miocene.

The regionally complex and structurally/stratigraphically disjointed nature of Quebradagrande has long been recognized. From a historic perspective, numerous lithostratigraphic names have been used in reference to this unit including its various components. From north to south, these names include:

The Formación Porfirítica (Grosse 1926)
The Quebradagrande (Botero 1963), Abejorral (Bürgl and Radelli 1962), and Valle Alto formations (González 1977; González 1980)
The Valle Alto, San Fèlix, and El Establo stratigraphic-tectonic intervals (Rodríguez and Rojas 1985)
The Grupo Quebradagrande and Complejo Ofiolítico de Pácora (Alvarez 1987)
The "eastern" and "western" intervals of Manizales (Gómez-Cruz et al. 1995)
The Aranzazu-Manizales Metasedimentary Complex (Lozano et al. 1975; Mosquera 1978)
The Quebradagrande Fm. (McCourt and Aspden 1984)

In southern Colombia this complex remains undifferentiated from allochthonous volcanic and sedimentary units of the Western Cordillera (Pardo and Moreno 2001).

The composition of clastic fragments contained within the Quebradagrande Complex suggests it developed in situ, along the continental margin to the east, with a strong volcanic influence sourced from the west (Rodríguez and Rojas 1985; Gómez-Cruz et al. 1995). The easternmost section of the complex, the Abejorral Fm. (Burgl and Radelli 1962), is comprised of continental sedimentary rocks, deposited upon metamorphic basement of the Central Cordillera block (González 2001) and intercalated to the west with immature sedimentary rocks (Moreno-Sánchez et al. 2008). The San Jeronimo fault is interpreted to have truncated the Quebradagrande Complex in the Late Cretaceous, leaving the continental Abejorral component of the complex out of the RSZ. Aptian-Albian ammonites are contained within the Abejorral Fm. (Burgl and Radelli 1962; González 1980; Botero and González 1983). Further to the south, Gómez-Cruz et al. (1995) reported Albian ammonites contained within Quebradagrande sedimentary rocks east of Manizales. Based upon bivalve, ammonite, gastropod, and lamellibranch assemblages in correlative marine sedimentary rocks located to the north and west of Medellín, a Berriasian-Albian (140–100 Ma) age has also been proposed for the western sector of Quebradagrande (González 1980; Botero and González 1983; Etayo-Serna 1985; Gómez-Cruz et al. 1995; Grosse 1926).

Various radiometric age dates have been derived for igneous rocks contained within the Quebradagrande Complex. Euhedral zircon crystals from a meta-tuff collected close to the San Jeronimo fault returned a U-Pb (zircon) age of 114.3 ± 3.8 Ma (Villagómez et al. 2011), which is in agreement with mid-Cretaceous fossil ages in the Abejorral Fm. Euhedral, zoned zircons from the Córdoba granodiorite, which intrudes Quebradagrande Complex along the westernmost flank of the Central Cordillera, yielded a mean age of 79.7 ± 2.5 Ma (Villagómez et al. 2011), providing a minimum age for Quebradagrande. In the same way, small bodies of nondeformed gabbroic, monzogabbroic, and syenogranitic rocks dated at ca. 74 Ma (Vinasco et al. unpublished data) intrude the volcanic and sedimentary rocks of the Quebragrande Complex.

U-Pb ages for detrital zircons collected from fine-grained sedimentary strata intercalated with andesitic volcanic rocks outcropping in the northern part of the Quebradagrande Complex (Barbudo creek, near Santafé de Antioquia) (Fig. 12.1c) yield ages of >145 Ma (Vinasco et al. 2011). The distribution of ages provided by 59 data points strongly suggests the sedimentary sequences of the westernmost Quebradagrande Complex were derived from the continental Northern Andean block, contrary to ideas of an allochthonous origin. Three percent of data ages were derived from Triassic sources from the Central Cordillera; 10% from Cambro-Ordovician sources, perhaps representing the Central or Eastern Cordillera basement rocks; 17% Neoproterozoic sources of the Central Cordillera; 25% from Grenvillian sources likely from the Eastern Cordillera basement; and finally 42% from cratonic sources, as direct or reworked zircons. Volcanic deposits in the western domain close to Manizales contain mainly a Cenomanian-Aptian (95–115 Ma) zircon population, probably defining the peak of the arc volcanism. U-Pb LA-ICP-MS analysis (Pardo 2015) of detrital zircons from a quartz-rich unit in the Manizales area yielded a similar age distribution to that described above, strengthening the regional character of the source region: (1) Carboniferous-Triassic sources (220–300 Ma), (2) Neoproterozoic-Ordovician sources (440–600 Ma), and (3) Mesoproterozoic sources (1000–1100 Ma). Geophysical studies carried out between 4° and 5.5° latitude north (Case et al. 1973; Meissner et al. 1976) show that these rocks overlie the metamorphic basement of the Central Cordillera.

Basalts associated with Quebradagrande mostly display calc-alkaline affinities typical of volcanic rocks generated in supra-subduction zone mantle wedges, in island arc, marginal basin, or active continental margin settings, independent of the Caribbean-Colombian Cretaceous igneous province of Nivia et al. (2006) and Caribbean-Colombian large igneous province of Kerr et al. (1997). Basaltic andesites, andesites, and a diorite studied by Villagomez et al. (2011) yield negative Nb-Ta and Ti anomalies on a primitive-mantle normalized multielement plot, high La/Yb ratios, and low Zr/Th values suggesting they are petrogenetically related to subduction with a calc-alkaline signature. Other basalts and gabbros of the Quebradagrande Complex, however, are characterized by flat to positive slopes on chondrite-normalized REE plots, suggesting a depleted mantle source, indicative of a mid-ocean ridge, or enriched OIB setting (Villagómez et al. 2011). These rocks may be interpreted as representing newly generated crust formed during rifting and marginal ocean basin formation in the Late Cretaceous, which detached the Quebradagrande Complex from the continental margin.

The Quebradagrande Complex is cut by an anastomosed system of north-south faults, sometimes yielding a discrete centimeter- to meter-scale mylonitic foliation, which is easily misinterpreted as regional schistosity. The volcano-sedimentary rocks of the Quebradagrande Complex acted as a regional matrix, enclosing units of higher competency, including meter- to kilometer-scale undeformed lenses of ultramafic rocks and granitoids. Much of the regional deformation recorded along the RSZ was focused within/absorbed by the fine-grained, clay-rich rock types within the Quebradagrande Complex.

The dominant structural style for the complex corresponds to steeply dipping, intercalated layers of dominantly andesitic rocks, black shales, and minor green chert (overturned in the Manizales area; Moreno and Pardo 2003), often defining tight isoclinal folds with axial planes dipping mainly to the east. Dextral kinematic indicators are observed in mylonitized domains that are in turn cut by west-verging reverse faults, suggesting a significant top-to-the-west transport direction. A subhorizontal stretching lineation associated with the strike-slip component is overprinted by a stretching S-SW-oriented lineation associated with the reverse component. Some undeformed felsic sills and dikes crosscut the sequence and postdate the thrusting.

In summary, Quebradagrande represents a highly varied assemblage of structurally dismembered lithotectonic units with continental, continent margin, and oceanic volcano-sedimentary affinities, which has yet to be completely defined along the entire length of the RSZ, especially in southern Colombia. Based upon paleontological studies, radiometric age dating, and crosscutting relationships, the age of Quebradagrande has at least locally been well constrained and ranges from Lower Cretaceous (Abejorral Fm.) to Upper Cretaceous (ca. 74 Ma). Much of the marine sedimentary component is ca. Aptian-Albian in age. Sedimentary rocks associated with Quebradagrande in at least two localities return detrital zircon age populations, which are characteristic of a source region along the paleo-Northern Andean continental margin. Due to widespread folding and faulting in at least two phases, and tectonic dismemberment and translation within a dextral-transpressive stress regime, it is difficult to establish lithostratigraphic continuity and the original spatial-temporal relationships between the various sedimentary, volcanic, and intrusive assemblages comprising the Quebradagrande Complex.

Numerous hypotheses have been proposed for the origin of the Quebradagrande Complex. Some of the more recent include:

1. An intracratonic marginal basin, eventually resulting in the generation of oceanic crust (Nivia et al. 2006)
2. A remnant of deposits originated between the NW South America and the pericratonic Chaucha-Amaime volcanic arc (Moreno and Pardo 2003)
3. A volcanic arc and some portions of the proto-Caribbean plate, accreted to the NW border of South America during the Late Cretaceous (Villagómez et al. 2011)
4. A continental volcanic arc (Toussaint 1996; Pindell and Kennan 2009; Cochrane et al. 2014a; Spikings et al. 2015)

12.2.2 Sinifaná Metamorphic Unit

These low-grade, predominantly metasedimentary rocks were first defined in outcrop along the western flank of the Central Cordillera block to the southwest of Medellín. The unit is elongated along a N-NW trend and occupies an area about 36 km^2 between the towns of Santa Barbara and Fredonia (Fig. 12.1c) (González 2001). The eastern boundary of the unit is defined by the Romeral fault (Grosse 1926), along which altered mafic volcanic rocks of the Quebradagrande Complex overthrust the Sinifaná unit (González 2001). The western limit is defined by the Amagá fault against siliciclastic sedimentary rocks of the Oligo-Miocene Amagá Fm. and the Triassic Pueblito Diorite (Calle and González 1980). Elsewhere, the Sinifaná unit has been documented in three additional localities, including a southern occurrence near Santa Barbara, a central occurrence near Ebejico, and a northern occurrence near Sucre (Fig. 12.1c). Neither the stratigraphic thickness nor a complete stratigraphic column for Sinifaná has yet to be established.

The protoliths of the Sinifaná metasediments were predominantly siliciclastic rocks including siltstones, slates, graywackes, psammites, and quartz arenites, mostly metamorphosed to greenschist grade (González 2001), although locally very-low-grade metasediments with well-preserved stratification and no penetrative foliation are recorded. The Sinifaná unit has been ascribed to the Devonian, based primarily on weak paleontological evidence (González 2001). It is intruded by the Triassic Amagá stock (Grosse 1926; González 2001) with an age of 227.6 ± 4.5 Ma (Vinasco 2004), evidenced by a 20-m-wide metamorphic aureole (Giraldo and Toro 1985), providing a minimum pre-Triassic age for the unit. Correlation with metamorphic rocks of the Central Cordillera block has been difficult due to the paucity of geochronological evidence and the complexity of regional tectonic patterns.

Martens et al. (2012) presented detrital U-Pb zircon geochronological data on a Sinifaná quartzite collected ca. 40 km southwest of Medellín, near Santa Barbara. Results revealed three age populations including (1) 640–500 Ma, (2) a minor population between 800–720 Ma, and (3) a population between 325–291 Ma. Additional, although minor, age populations at 1040 Ma and 2300 Ma were also obtained. Given the intrusion of the ca. 230 Ma Amagá Stock, Martens et al. (2012) conclude that the deposition of Sinifaná rocks and their metamorphism occurred between the Upper Carboniferous and Lower Triassic. The Sinifaná detrital zircon populations appear analogous to zircon populations revealed in Cajamarca migmatites and paragneisses (Vinasco et al. 2006) suggesting a link with Central Cordillera block. Martens et al. (2012) suggest the Sinifaná unit(s) represent blocks of Central Cordilleran metamorphic rock broken away from the paleo-continental margin.

Recent data obtained from U-Pb analysis of zircons recovered from a metapsammite collected near Sucre (Vinasco et al. unpublished data, Fig. 12.1c) provide additional age constraints for the Sinifaná unit. Thirty-five percent of analyses correspond to Permo-Triassic ages, 21% from Neoproterozoic sources, 24% from Grenvillian sources, and 13% from SA cratonic sources. We interpret the youngest age, ca. 239 Ma, to correspond to the maximum age of deposition at this locality and stratigraphic interval. Based upon the age of the Amagá pluton, this implies sedimentation of the some of the Sinifaná protoliths took place between ca. 239 and 227 Ma, that is, in the mid- to Late Triassic. Regardless, it is probable that the sample collected near Sucre belongs to a different Sinifaná horizon than that studied by previous authors. This is a plausible explanation for the age variations observed in Sinifaná, given that the presently mapped Sinifaná unit is likely comprised by several intervals, which have not yet been clearly differentiated.

With respect to source region, the zircon populations derived from the Sucre and Santa Barbara localities suggest Sinifaná was deposited in proximity to or upon the NW South American paleo-continental margin, including the Central Cordillera block, and the present para-autochthonous location of the Sinifaná blocks is due to postdepositional separation and tectonic reworking during the Late Triassic through Miocene.

Rodriguez (2010) presents structural cross sections for the Pueblito diorite, Sabaletas schist (Arquía Complex, see below), and Sinifaná metasediments near Ebejico, where tight NNW-verging, overturned, meter-scale, asymmetrical folds with subhorizontal axial planes are observed (Fig. 12.2). Rodriguez (2010) suggests the Sinifaná metasediments and Sabaletas schist formed part of a regional anticlinorium, which was intruded by the Pueblito diorite and later faulted. The age assigned to the entire metamorphic complex is pre-Triassic on the basis of a Triassic dike intruding the Sabaletas schist (Rodriguez 2010). In this interpretation, the Sinifaná unit would be part of the pre-Triassic Arquía Complex (see below).

In the northern sector near Sucre, Sinifaná meta-argillites reveal a well-defined fracture cleavage and development of asymmetrical meter-scale kink folds with a W-SW vergence with layers dipping less than 20° east. Subhorizontal mylonitic bands with a dextral component are well developed, indicative of a thrusting component toward the W. It is probable that these metasediments are thrust over clastic rocks of the Oligo-Miocene Amagá Fm., yielding a time constraint to the late, regional, west-verging overthrust event.

12.2.3 Arquía Complex

The Arquía Complex is a composite collection of sheared, mylonitized, and faulted rocks, including Late Paleozoic metasedimentary units, and fragments of Early and Late Cretaceous mafic volcanic and probably intrusive rocks, distributed along a discontinuous NNE to N trending belt along the western margin of the Central Cordillera and the easternmost margin of the physiographic Western Cordillera (Fig. 12.1a, c). The internal distribution of rock types within the Arquía Complex has yet to be mapped in detail, and structural disruption of the complex makes it difficult to establish lithostratigraphic continuity between the individual units. In this sense, the Arquía Complex corresponds to a heterogeneous association of metamorphic rocks bounded by regional faults, composed of several lithodems with

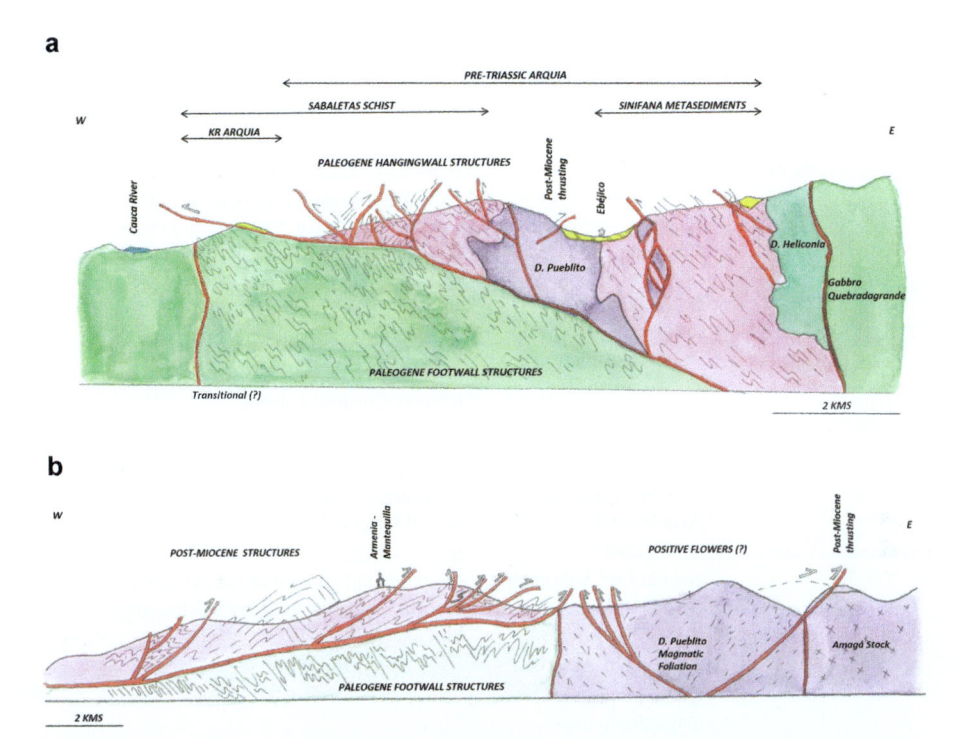

Fig. 12.2 Cross sections illustrating the main structural features of the Romeral shear zone. A. E-W cross section in the Ebejico area (Fig. 12.1) showing Late Cretaceous to Paleocene-Eocene west-verging structures and a complex block tectonic configuration. B. E-W cross section in the Armenia (Antioquia) area (Fig. 12.1) showing younger east-verging structures in hanging-wall position and tight isoclinal folds in footwall position. DH Diorita de Heliconia, QG Quebradagrande Complex

different origins and ages (Giraldo 2010; Moreno Sánchez et al. 2008), interpreted by González (1977) as a mélange zone. Arquía includes sericite schist, black schist, quartzite, green schist, amphibolite, garnet amphibolite, metagabbro, and ultramafic rocks, metamorphosed to actinolite-epidote facies (García 2010) in a Barrovian-type environment (Restrepo and Toussaint 1976).

Along the south-central RFZ, the eastern limit of Arquía is defined by the Silvia-Pijao fault, which places Arquía rocks against volcanic rocks of the Quebradagrande Complex (Maya and González 1995). The western boundary is defined by the Cauca-Almaguer fault, which separates the unit from Cretaceous rocks of the volcanic Barroso Fm. (Maya and González 1995).

The Cretaceous fragments contained within Arquía may be related to two different volcanic arcs of Early and Late Cretaceous age (Table 12.1). Fragments of volcanic arcs are described by Bustamante et al. (2011) and Bustamante et al. (2012) who interpreted high-pressure metamorphic assemblages as discrete Early and Late Cretaceous subduction channels (Table 12.1), developed in tectonic

Table 12.1 Main tectonic events affecting the RSZ in Colombia

Neogene	Holocene	Recent volcanism	
	Pliocene	Combia Fm. Siliciclastic molassic syntectonic sedimentation	Arc volcanism. Related hydrothermal deposits
	Miocene		Panamá-Chocó block arrival, regional east-verging structures. Closure of Cauca-Patía basin. Left-lateral kinematics to the south –right-lateral to the north of 5° N
		Andean orogeny Initiation of Andean orogeny	
Paleogene	Oligocene		Sediments of Central Cordillera provenance
	Eocene	Siliciclastic molassic syntectonic sedimentation	Sediments of Western Cordillera provenance with minor Central Cordillera contribution
	Paleocene	Pre-Andean climax	Central Cordillera uplift
Cretaceous	Late	Initiation of Pre-Andean orogeny CCOP/CLIP collision Arquía final amalgamation	Tight folding of Quebradagrande and Arquía complexes. Regional west-verging structures. Reactivation of main RSZ. Right lateral kinematics
		Arquía Fm. Subduction channels. High-pressure rocks	RSZ reactivations Opening and closure of the continental margin. Generation of para-autochthonous blocks
		Alkaline magmatism	
		Quebradagrande magmatism over hyperextended transitional crust	
	Early	Quebradagrande magmatism over CC basement (?)	Subduction and Rifting. Pacific arc collisions and reactivation of RSZ
		Abejorral sedimentation over Central Cordillera basement	
		Arquía Complex. Subduction channels. High-pressure rocks	
Jurassic		Continued Pangaea breakup	
Triassic		Post-tectonic magmatism	Proto-Romeral fault system? Left-lateral kinematics
		Beginning of Pangaea breakup	Rifting, subduction
Permian		Central Cordillera basement	
		Arquía Complex. Low-grade metasediments	Tectono-thermal event
		Pangaea Amalgamation. Alleghanian suture	Loose amalgamation of peri-Gondwanan terranes along paleo-Romeral margin

regimes intermediate between the Franciscan (rapid cooling) and Alpine (rapid decompression in compressional regimes) types of subduction. In Colombia, high-pressure blueschist and eclogite metamorphic assemblages are recorded in the Barragán and Jambaló complexes (McCourt and Feininger 1984; Bustamante 2008). Bustamante et al. (2011, 2012) present ^{39}Ar-^{40}Ar ages in retrograde dynamo-thermal paragonite and phengite associated with a mylonitic foliation cutting blueschist in the Jambalo area. They interpret the ca. 61–66 Ma plateau ages to represent exhumation ages for the blueschist.

Age constraints for the Early Cretaceous low- and medium-pressure metamorphic rocks are inferred from K-Ar data presented by Moreno and Pardo (2003), Restrepo and Toussaint (1991), and Toussaint and Restrepo (1989). A ^{40}Ar-^{39}Ar age of ca. 120 Ma for blueschist-greenschist facies transition metamorphic rocks in the Barragan area (Bustamante 2008; Bustamante et al. 2012) suggests high-pressure metamorphism peaked in the pre-Aptian. Bustamante et al. (2012) suggest the Early Cretaceous metavolcanic rocks were generated along the Pacific margin associated with a subduction complex of oceanic, N-MORB affinity which was accreted to the paleo-autochthon prior to the Late Cretaceous arrival and collision of the CCOP plateau. McCourt (1984) and McCourt and Aspden (1984) suggest that some high-pressure metavolcanic rocks of the Arquía Complex represent continental marginal arcs originally developed in the Carboniferous. They interpret ca. 125 Ma K-Ar ages obtained from samples of blueschist to represent reactivation along the paleo-continental margin during accretion of exotic blocks to the west (Table 12.1).

Late Cretaceous blocks of the Arquía Complex outcrop along the western margin of the RTZ. These fragments are represented by the westernmost part of the Sabaletas schist and are mainly composed of metabasite. A U-Pb zircon sample from this metabasite, sampled near Titiribí, yielded an age of 71.9 ± 1.1 Ma, considered to represent the age of the unit (Giraldo 2010).

An Arquía Late Cretaceous-Paleocene (?) transpressional event is characterized by the formation of a penetrative S1 cleavage (D1) affected by a later deformation (D2), generating upright to horizontal, inclined tight to isoclinals, west-verging folds, without formation of a second cleavage. The D2 event is responsible for intrafolial folding observed in intercalated graphitic schist, where transposition of D1 to D2 is evidenced. A third, gentle deformational event is evidenced by microcrenulation folding (D3) cut by discrete west-vergent reverse faults (D4). The sequence is crosscut by subvertical N-S striking shear zones (D5) (García 2010), which define a late, brittle deformational event.

García 2010 presents additional field observations, which further complicate the above scenario, as he interprets some Late Cretaceous fragments of the Arquía Complex to potentially represent remobilized blocks of the allochthonous Barroso Fm., which are observed in a transitional relationship along the Amagá River (García 2010). Texturally, the progression from undeformed volcanic rocks typical of the Barroso Fm. over tens of meters east of the Cauca River into a foliated metabasite typical of the Sabaletas schist is evident. In this interpretation, these Arquía fragments would be detached slivers of the Farallon or CCOP/CLIP oceanic basement, which collided against the paleo-margin in the Late Cretaceous, yielding the S1 and the subsequent tight to isoclinal folding episode. Subsequent deformational episodes generated crenulation, west-vergent thrusting and superimposed transcurrent N-S-striking vertical faults.

The pre-Cretaceous blocks of Arquía are dominated by low-grade metasedimentary rocks, including meta-psammite, metapelite, and carbonaceous schist, in which sedimentary structures are sporadically preserved, represented by the easternmost part of the Sabaletas schist (Fig. 12.2). This sequence is intruded by the Pueblito Pluton (Fig. 12.2) (Rodriguez 2010), a relationship well constrained by a Triassic dike intruding metasediments of the Sabaletas schist near Ebejico (Fig. 12.1c) (Rodriguez 2010) and supported by a small contact aureole and boulders of Triassic gabbros with hardened xenoliths of Sabaletas schist near Armenia Mantequilla. ^{39}Ar-^{39}Ar ages in the 203–229 Ma interval presented by Rodriguez and Arango (2013) are interpreted as metamorphic ages. These rocks are probably related to the Sinifaná schist outcropping further east (Fig. 12.2), which represents a lower grade of regional metamorphism, defining the regional anticlinorium mentioned for the Sinifaná metasediments. The pre-Cretaceous fragments are interpreted as detached fragments of Central Cordilleran provenance, separated during prolonged shear zone dynamics along the RSZ margin.

The structural architecture of the pre-Cretaceous components of Arquía is extremely complex and quite distinct, when compared to the Late Cretaceous components. The high strains experienced by these units are reflected in different styles of folding and faulting and in a discrete block tectonic configuration at the meter to tens of meters scale (Fig. 12.2). The lithostratigraphic and structural relationship(s) between the Cretaceous vs. pre-Cretaceous components of the Arquía Complex have yet to be fully established. Important observations are summarized in schematic cross sections derived from field transects and observations recorded near the towns of Armenia and Ebejico (Figs. 12.1c and 12.2). In the Armenia area (Fig. 12.2b), the systematic documentation of east- and minor west-verging structures thrusted over Oligo-Miocene rocks of the Amagá Fm., interpreted as Late Miocene structures, permits the definition of regional asymmetrical east-verging overturned folds, cut by west-dipping reverse faults.

As already mentioned, a very distinct structural style is recorded by the Late Cretaceous lithotectonic units contained within Arquía. In various localities this style is characterized by tight, upright, west-verging folds in metavolcanic rocks in a regional footwall position. It is possible to consider a gentle, east-verging detachment separating tight to isoclinal folding in the footwall from asymmetrical east-vergent folds in the hanging wall (Fig. 12.2b). This last deformation would be of Late Miocene age.

Similar contrasting structural domains are observed a few kilometers north, in the Ebejico area (Fig. 12.2a), where a complex structure is recorded in pre-Cretaceous metasedimentary fragments, which have been thrust over Late Cretaceous metavolcanic rocks. The metasediments in a hanging wall position yield meter-scale positive flowers, recumbent folds and west vergence faults while metavolcanic rocks in a foot wall position yield contrasting tight, upright west-verging folds. In the Ebejico area, the dominantly west-verging structures (Fig. 12.2a) are opposite to the mostly east-verging structures recorded near Armenia (Fig. 12.2b). Based upon the present level of understanding, the contrasting structural styles are difficult to explain. A thrust-back coupled fault system of Late Miocene age may explain this geometry. However, the west-verging structures are probably remnants of Late Cretaceous tectonism. The flower structures and final assembly of tectonic blocks are interpreted to have resulted from Late Miocene compression related to collision of the Panamá-Chocó arc and the reactivation of early-formed structures.

Further south, near Abejorral and Pantanillo, in a similar structural position along the RSZ corridor, a similar W-SW-vergent thrusting is observed (Zapata 2015), suggesting that this structural style is of regional importance along the eastern RSZ margin. Local north-vergent thrusting observed in the Sabaletas schist, close to Ebejico town, is interpreted to be related to shear-related block rotation.

During the last decade, various studies suggest that Cretaceous metabasites contained within Arquía were formed in oceanic island arc and oceanic ridge settings. To the south and west of Medellín, these units have been metamorphosed to greenschist and amphibolite grade (Giraldo 2010; García 2010). Farther south, along the eastern RSZ margin, complex greenschist, blueschist, and eclogite facies are recorded (Bustamante 2008). Geochemical and geochronological data suggest an origin for metabasites from tholeiitic basalts of MORB affinity (Rodriguez and Arango 2013), from low-K (sub-alkaline) tholeiitic basalt of mid-oceanic ridge affinity (Giraldo 2010), or potentially from primitive arcs build on oceanic crust. This interpretation is consistent with the formation of high-pressure rocks in subduction channels. Cretaceous quartz-muscovite and actinolite Sabaletas schists have been interpreted to represent metamorphosed oceanic crust, including interbedded layers of N-MORB basalt (Giraldo 2010; Ruiz-Jimenez et al. 2012), marine sediments, siliceous volcaniclastic material, carbonaceous sediments, and carbonates, reaching the greenschist facies, with temperature conditions exceeding 300 °C and depths in the range of 5–20 km (García 2010).

In summary, Arquía represents a highly varied assemblage of structurally dismembered lithotectonic units of pre-Cretaceous and Early and Late Cretaceous age, with a complex metamorphic history involving sub- and low-grade greenschist through blueschist- and eclogite-grade metamorphic events. The Cretaceous components are probably associated with subduction environments and island arcs close to the continent. The lithotectonic and structural complexity of Arquía is in many respects similar to that of the Quebradagrande Complex, and like Quebradagrande, it is difficult to establish lithostratigraphic continuity and the original spatial-temporal relationships between the various lithotectonic units comprising the Arquía Complex. Arquía is envisaged as a complex composite of blocks related to pre-Triassic fragments detached from the ancient Central Cordillera block and at least two arcs constructed upon oceanic crust, close to the continental margin. Some Late Cretaceous blocks may be associated with volcanics belonging to the Barroso Fm. Within Arquía, metamorphism and widespread folding and faulting, in at least five phases, are recorded. East- and west-verging structures, of differing structural styles and/or crustal levels, record deformation spanning the Late Cretaceous through Late Miocene. The latest deformations are largely responsible for the present-day litho-structural and tectonic architecture of the RSZ and associated mélange; however, they overprint and obscure the complex Permo-Triassic through Lower-mid-Cretaceous tectonic evolution of the NW South America paleo-margin.

12.2.4 Mafic and Ultramafic Igneous Rocks

Mafic and ultramafic igneous rocks contained within the RSZ were referred to as the Romeral ophiolitic belt by Alvarez (1983). They include peridotites, serpentinized dunites (González 2001), and gabbro-diorite, with locally gradational contacts. In general the mafic-ultramafic bodies are confined to fault-bound, discontinuous kilometer- to meter-scale lenses (Fig. 12.2) elongated along the general strike of the entire RFZ. Some occurrences are highly sheared and plastically deformed, while others have behaved as rigid bodies within ductile volcanic rocks of the Quebradagrande Complex, caught up within the kilometer-scale RSZ. Structural complications render the boundaries of many occurrences difficult to establish. The faulted and sheared contacts define wide mylonites and include the thrust emplacement of ultramafic rocks over Oligo-Miocene sediments of the Amagá Fm. to the SW of Medellín, constraining the age of a late episode of thrusting to Late or post-Miocene.

Two generations of ultramafic rocks exist in the study area. The oldest is associated with Triassic magmatism, which in turn may be related to Triassic ultramafic rocks of the Central Cordillera block, including the Aburra ophiolite (García 2010; Correa et al. 2005; Restrepo 2008). The second group is interpreted to be associated with the Cretaceous Quebradagrande Complex (González 1980). Near Medellín, the Triassic occurrences are represented by the Pueblito gabbro/diorite and numerous small and discontinuous lenses of ultramafic rocks. ^{40}Ar-^{39}Ar dating of amphibole from the Pueblito diorite and gabbro yielded plateau ages ranging from ca. 238 Ma to 224 Ma (Vinasco 2001). A U-Pb (zircon) age of 227.6 ± 4.5 Ma was obtained for the same intrusion (Vinasco 2004). The Pueblito diorite consists of an elongate body, the long axis of which is oriented N10 W. It can be traced over a length of ca. 56 km (Rodriguez 2010). It was interpreted to have been emplaced into metasedimentary rocks, intruding the pre-Cretaceous fragments of the Sabaletas schist (Arquía Complex) and the Sinifaná metasediments (Fig. 12.2), within a sinistral transpressional tectonic regime (Rodriguez 2010). Its characteristic elongate shape, subparallel to the RSZ, and internal structure as revealed by anisotropy in the magnetic susceptibility support this assertion (Rodriguez 2010).

Similar intrusive units to the south, including the Cambumbia stock, suggest the presence of a semi-discontinuous but presently fragmented mid-Late Triassic magmatic belt (Vinasco 2001). The Cambumbia stock returned a U-Pb (zircon) age of 233.41 ± 3.4 Ma (Rojas, 2012) and a ^{40}Ar-^{39}Ar (biotite) plateau age of 236.6 ± 0.7 Ma (Vinasco 2001) and was interpreted to have been generated in an island arc environment. Planar magmatic fabrics, compositional banding, and the presence of dikes characterize the I-type Pueblito diorite and gabbro unit. These mafic and ultramafic rocks are interpreted to have formed in an immature Triassic subduction environment, possibly related to the Aburrá ophiolite in the Central Cordillera (Rodriguez 2010). The presence of inherited zircons typical of Central Cordillera block basement suggests arc development close to or over the continental margin or possibly within a pericratonic fragment of the rifted continental margin. The southern part of the Pueblito diorite is uplifted against Oligo-Miocene Amagá Fm., defining an east-vergent thrust. However, internally, a westward vergence of brittle structures is systematically recorded. The northern termination thins out and disappears under Cenozoic sedimentary cover.

Emplacement models presented by Rodriguez (2010) suggest a general NW-SE transpressive subduction regime. Lithofabrics are mainly magmatic with some evidence of deformation, suggesting syntectonic emplacement during the final stages of regional deformation. Contact metamorphism and the presence of diorite dikes and roof pendants of Sabaletas schist characterize the intrusive contact between the Pueblito diorite and the Arquía Complex to the south and west of Medellín.

The second group of mafic and ultramafic rocks considered herein is thought to be Cretaceous in age. They are contained within a discontinuous belt, within which ultramafic rocks are intimately associated with gabbros and mafic volcanics. The belt was referred to as the Cauca ophiolitic complex by Restrepo and Toussaint (1974). It contains tectonized peridotites, dunites and gabbros of various textures and degrees of deformation, which have been metamorphosed to greenschist facies (Alvarez 1983). The Sucre amphibolite, located to the N and W of Medellín, corresponds to a heterogeneous LS-gabbro. It is interpreted to represent a syntectonic gabbro, the product of Late Cretaceous dynamic metamorphism. It is intruded by variably deformed intermediate dikes (Vinasco et al. unpublished data). Other gabbros, not directly associated with the Quebradagrande Complex, such as those spatially related to the Sabanalarga batholith, intrude the Cajamarca Complex (Nivia and Gomez-Tapias 2005), which forms part of the Central Cordillera block. The Sabanalarga gabbro exhibits well-developed and generally undeformed, holocrystalline magmatic texture.

12.2.5 Granitoid Rocks

Within the northern RSZ, the most voluminous granitoid is the Amagá Stock (Fig. 12.1c), a quartz monzonite to granodiorite pluton for which U-Pb (zircon) and ^{40}Ar-^{39}Ar suggest an Upper Triassic age of ca. 227 Ma (Vinasco 2004). The Amagá stock was named by Grosse (1926) in reference to a series of small plutons situated along the western slopes of the Central Cordillera block. The Amagá body itself is about 12 km in length with a maximum width of 3 km, contained within the regional shear fabric of the RSZ. The contact with the Sinifaná metasediments appears intrusive, while the eastern contact with the Quebradagrande Complex is faulted. Restrepo et al. (2009) propose that this local Sinifaná-Amagá assemblage, along with other independent blocks within the RSZ, corresponds to a discrete terrene; however, it is not possible to differentiate this assemblage from similar lithotectonic units within the RSZ and mélange. Indeed, to the east of the RSZ, over the continental margin itself, the El Buey and La Honda intrusive stocks purport similar age, compositional, lithochemical, and isotopic characteristics to those documented for the Amagá stock (Vinasco et al. 2006). From a tectonic perspective, the Amagá, El Buey, and La Honda granitoids are contained within distinct and contrasting geological blocks; however, it is possible to sustain a genetic relationship between all three bodies, supporting the hypothesis of a close relationship between many of the blocks within the RSZ and the ancient paleo-continental margin.

The emplacement of post-tectonic, Late Triassic granitoid stocks like La Honda, El Buey, Amagá, and others could be related to orogenic collapse in the Central Cordillera block, following the amalgamation of Pangaea, as recorded in Colombia and the northern Andes in general by a tectono-thermal event during the Late Permian to Early Triassic (Vinasco et al. 2006). Orogenic collapse was followed by the emplacement of anatectic granitoids and associated mafic melts marking rifting that gave rise to the beginning of breakup of Pangaea in the Late Triassic (Vinasco et al. 2006; Cochrane et al. 2014b). Additional granitoids contained within the RSZ to the west of Medellín include quartz diorites and monzonites of the Late Cretaceous Heliconia diorite suite (González 2001). These bodies also appear elongated within the regional trend of the fault, although careful inspection reveals they contain little internal deformation. Cardona et al. (2012) provided data reflecting an alkaline character to this suite, which led them to propose an origin related to a slab break during the Late Cretaceous collision of allochthonous rocks of the Farallon-CCOP assemblage of the Western Cordillera. To the north, in the Sucre area, a syenogranite, potentially of the same suite, yielded a U-Pb (zircon) age of ca. 71 Ma (Vinasco et al. unpublished data). This granitoid is characterized by undeformed, centimetric pink K-spar megacrysts. In this area, the contacts are defined by overthrusts along both the eastern and western margins of the intrusive. The unit is thus located as a tectonic slice between the volcanic rocks of Quebradagrande Complex to the east and ultramafic rocks to the west. The syenogranite body is disrupted along the strike as a result of post-ca. 74 Ma shearing, locally generating meter-scale muscovite-rich mylonites. Granitoid dykes measuring tens of meters thick intrude Quebradagrande volcanic rocks near Liborina, in the northern part of the shear zone, and are probably related to this alkaline magmatism. These granitoids are deformed along with the volcanic host, forming loose blocks within the volcanic matrix, transposed by mylonitization. Syenogranites have also been described in intrusive units cutting volcanic rocks of the Quebradagrande Complex further south (León et al. 2014).

12.2.6 Cenozoic Sedimentary Rocks

Cenozoic sedimentary rocks distributed along the RSZ were deposited within three sub-basins, physiographically constrained by the Cenozoic emergence of the proto-Central and Western cordilleras, to the east and west, respectively. These sub-basins include, from south to north, Patia, Cauca, and Amagá, collectively referred to herein as the ACP basins (Fig. 12.1a, c). The modern-day distribution of these sedimentary rocks coincides with an elongated intermontane depression (the Cauca-Patía river system) that extends over an approximately 590 km, NNE trend, broadly coincident with the trace of the RSZ and mélange. The ACP sedimentary deposits represent molassic sequences containing continental, continental margin and marine siliciclastic rocks and coals (Sierra and Marín-Cerón 2011).

Sedimentation along the ACP resulted from uplift associated with the dextral-oblique collision of oceanic terrains along the RSZ beginning in the Late Cretaceous (McCourt and Aspden 1984; Pindell and Barrett 1990; Cediel et al. 1998). This collision triggered thrust-faulting and dextral strike-slip displacement along the RSZ, inducing the formation of the ACP pull-apart basins during the Cenozoic (Sierra and Marín-Cerón 2011). A pull-apart origin (Kellogg et al. 1983; Alfonso et al. 1994) or a graben structure (Acosta 1978; MacDonald et al. 1996) has been proposed to explain the origin of the composite depression. Following Late Cretaceous terrane accretion, ensuing interaction between continental South America and the Chocó-Panamá indenter led to tectonic tightening and closure of the northern part of the Cauca Valley (Amagá Basin) in the Late Miocene (Suter et al. 2008). In this context, most of the ACP basin margins are shaped by late reverse faults associated with basin inversion and by the reactivation of branches of the Romeral and Cauca fault systems (Fig. 12.1a).

Lower Eocene deposition sedimentation along the southern RSZ is recorded by the Peña Morada Fm. and by the Upper Rio Guabas sequence in the Patía sub-basin and by the Chimborazo Fm. in the Cauca sub-basin (Sierra and Marín-Cerón 2011). The composition of this assemblage strongly suggests provenance of sediments from accreted oceanic rocks underlying the Western Cordillera with a minor contribution from the Central Cordillera block (Sierra and Marín-Cerón 2011). The Mosquera and Guachinte Fms. define the Late Eocene to Early Oligocene sequences in the Patía sub-basin (Sierra and Marín-Cerón 2011), with equivalent lithofacies recognized in the basal part of the "Lower" Cinta de Piedra Fm. of the Cauca sub-basin (Ecopetrol-ESRI 1988). An equivalent lithofacies was recognized in the lower member of the Amagá Fm. in the Amagá sub-basin further north (Sierra et al. 2005; Silva et al. 2008). The Lower Amagá is characterized by poorly selected and quartz-rich conglomerates, well-sorted sublitharenites, massive gray siltstones, and layers of coal deposited in a delta plain environment, with provenance of sediments from the Central Cordillera (Sierra and Marín-Cerón 2011).

The Late Oligocene-Miocene sequences in the ACP basin are composed of three major lithostratigraphic units: the Esmita in the Patía sub-basin, the Ferreira and

Upper Cinta de Piedra Fms. in the Cauca sub-basin, and the upper member of Amagá Fm. in the Amagá sub-basin (Sierra and Marín-Cerón 2011). The quartz-rich composition of the four lithofacies described above, together with the metamorphic origin for some of the components, strongly suggests a source of sediments from the Central Cordillera (Sierra and Marín-Cerón 2011).

Additional Oligo-Miocene components of the Cauca sub-basin are exposed within the Santa Barbara Range, near the town of Cartago (Fig. 12.1a), including three continental sedimentary formations: the Oligocene Cartago Fm. (Rios and Aranzazu 1989); the Miocene, syntectonic, La Paila Fm.; and the La Pobreza Fm. (Suter et al. 2008). According to Suter et al. (2008), the Cauca sub-basin in this area is located at the front of the Chocó-Panamá indenter, where the Romeral fault system changes kinematics from right-lateral in the south to left-lateral in the north. Additionally, at the point of indentation of the Panamá-Chocó block, a dominantly marine sedimentary sequence is observed to the south while inter-Andean continental sedimentation is observed to the north (e.g., Amagá Fm.) (Silva et al. 2008). Amagá is composed of siliciclastic rocks intercalated with coal, associated with a fluvial depositional environment within a tectonically controlled pull-apart basin (Sierra and Marín-Cerón 2011). Folding and deformation of the Amagá sequences are related to collision between continental South America and the Panamá-Chocó block. Folding and thrusting in the Santa Bárbara Range are still active (Suter et al. 2008). Folding within Santa Barbara began during the Upper Oligocene-Lower Miocene (Suter et al. 2008), possibly coinciding with a shift to W-E convergence between the oceanic plate and the continental margin, following rifting of the Farallon plate to form the Nazca-Cocos plate system (Taboada et al. 2000).

The Patía, Caucá, and Amagá sedimentary sequences were intruded by numerous porphyritic stocks of mid- through Late Miocene age (Fig. 12.1a, c) (Sierra and Marín-Cerón 2011). The Galeón and Popayán Fms. in the Patía sub-basin; the Zarzal, Jamundi, and Armenia Fms. in the Cauca sub-basin; and the upper part of the Combia Fm. in the Amagá sub-basin represent Late Pliocene to Holocene sedimentary successions (Sierra and Marín-Cerón 2011). These units form the youngest sequences in the basins. They overlie and unconformably coalesce with older sequences (Alfonso et al. 1994; Sierra and Marín-Cerón 2011). The Combia Fm. is a thick volcano-sedimentary unit, which is divided into two members (González 1980). The upper member comprises sedimentary rocks and volcanic ash. The lower member is characterized by pyroclastic rocks, which include tuffs, lapilli tuffs, agglomerates, and breccias with clasts of basalt and andesites. Subvolcanic andesite and dacite bodies emplaced within Combia yield K-Ar ages ranging from 9 Ma to 6 Ma (Restrepo et al. 1981). An ^{40}Ar-^{39}Ar (biotite) age of 6.7 ± 0.06 Ma from the Marmato stock was presented by Vinasco (2001). The Combia Fm. rocks are not internally deformed. Only local tilting of beds is observed (Piedrahita 2015), defining a reliable time constraint for deformation associated with accretion of the Panamá-Chocó block.

12.3 Tectonic Evolution and Timing

Tectonic models for development and evolution of the RSZ and associated tectonic mélange must explain the nature and age of the lithodemic and lithostratigraphic units, including their deformational age(s) and thermochronological record. It must also explain sedimentation associated with uplift along the physiographic Central and Western cordilleras and the geometry and tectonic significance of the regional structures. The model should be backed by the available geological mapping and structural, geophysical, and geochemical data, including lithochemical and isotopic constraints.

The RSZ and associated tectonic mélange encompasses the locus of tectonic development along the NW margin of continental South America and contains a long-standing record of the most important Meso-Cenozoic tectonic and deformational events along the Pacific margin, potentially including the assembly and breakup of Pangaea and the arrival and collision of the CCOP/CLIP assemblage and more recent events associated with accretion of the Panamá-Chocó block and even the Carnegie ridge to the south. Table 12.1 summarizes the principle geological tectonic events associated with the RSZ development, and Fig. 12.3 illustrates a series of schematic diagrams representing the most important geological/tectonic events since Permian times. This synopsis permits the local reconstruction of the paleo-continental and pericratonic margin, as depicted via a simplified geological map covering the city of Medellín and surrounding area (Fig. 12.3a).

Time constraints for the Meso-Cenozoic development of the RSZ are defined by:

1. Stratigraphic constraints provided by Villamil (1999)
2. ^{39}Ar-^{40}Ar results in neo-formed micas in mylonitic bands (Vinasco and Cordani 2012) and additional geochronological data
3. Thermochronological (fission track and He dating) analyses conducted on samples derived from vertical profiles through some of the massifs in the Colombian cordilleras (e.g., Vinasco et al. 2015; Noriega et al. 2015; Botero et al. 2015), from modern sediments within the Romeral fault system (Montoya-Betancur et al. 2015), and from regional thermochronological data presented by Villagomez and Spikings (2013) and Spikings et al. (2015)
4. The relative age of structures as derived from field observations

The early tectonic history of the region pertaining to the RSZ and associated tectonic mélange began in the Permo-Triassic, with the assembly and disruption of Pangaea (Fig. 12.3b). The detailed tectonic history of this time period remains unclear. Various authors depict the loose amalgamation of peri-Gondwanan terranes along the NW South American paleo-margin during the Late Permian, peripheral to distal to the principal Alleghanian (Ouachita-Marathon) suture (Weber et al. 2007; Cochrane et al. 2014b). In this context, the paleo-Romeral margin may have represented an unwelded contact between the numerous "Mexican" terranes and the autochthonous paleo-Northern Andean margin during this pre-Cretaceous event.

Fig. 12.3 (**a**) Simplified geological map showing position of lithological units in the region surrounding Medellín, modified after Gonzalez (2001). The western domain limited by dots is schematic and is shown for reconstruction purposes only. Red tones, Permian; purple, Triassic; blue, Jurassic; green, Cretaceous. SJF San Jeronimo fault. See Fig. 12.1 for location. (**b**) Global tectonic reconstruction after Blakey (2016). Permian rocks are highlighted. Pangaea assembly,

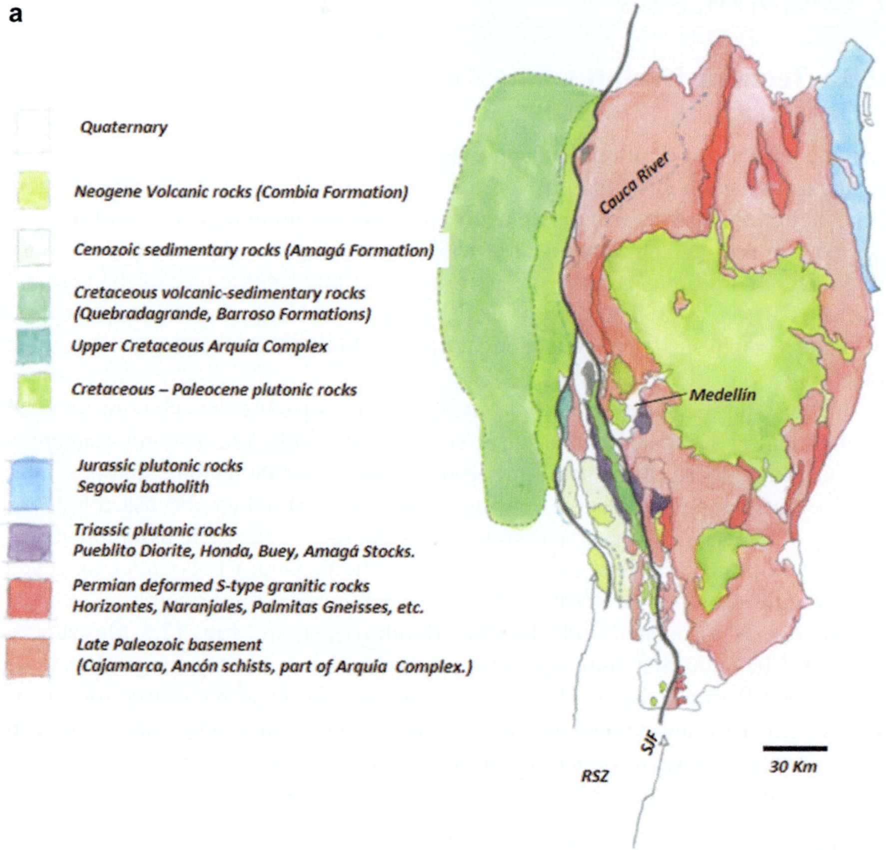

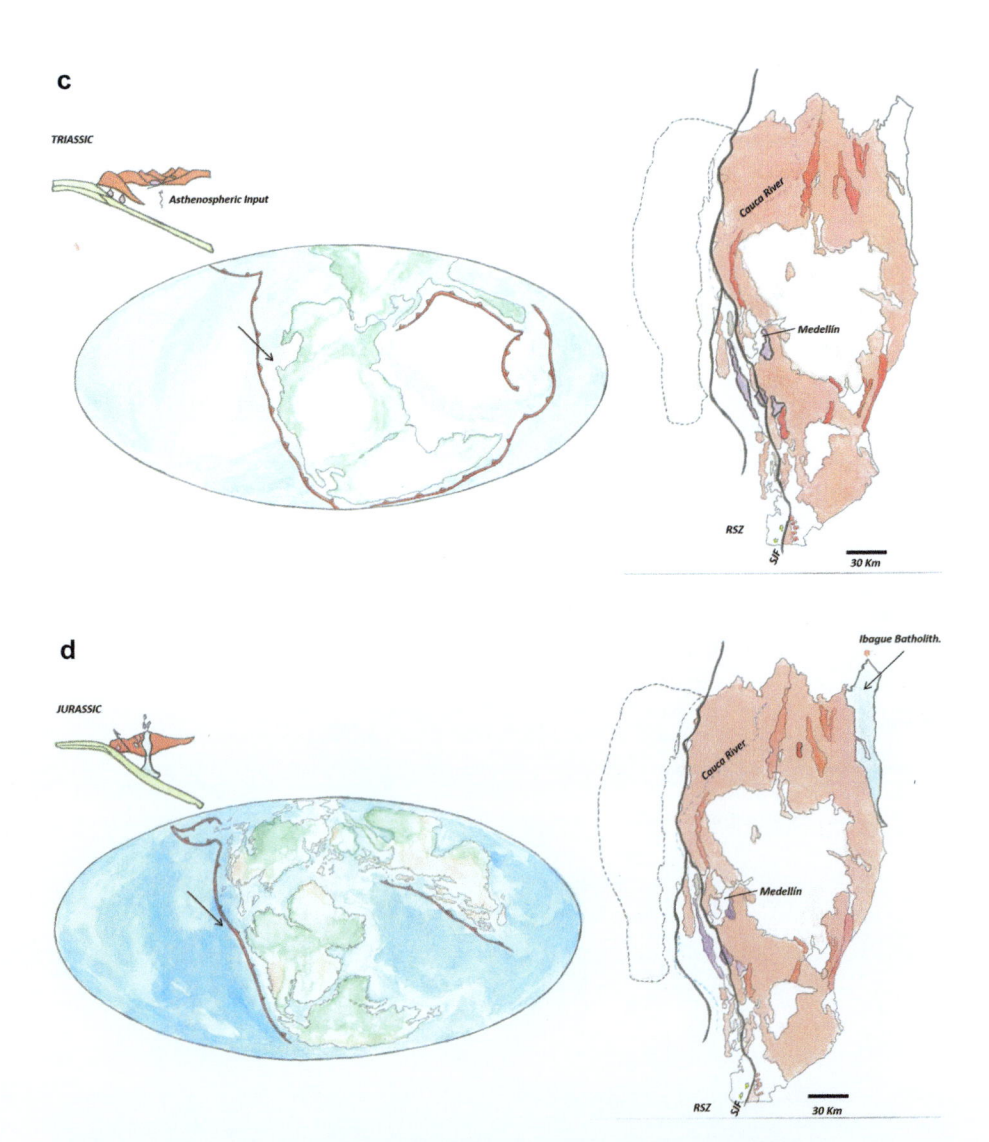

Fig. 12.3 (continued) compressional regime, crustal thickening, S-type magmatism. Note the presence of Permian components within the RSZ. (**c**) Global tectonic reconstruction after Scotese (2001). Permian and Triassic highlighted. Beginning of Pangaea breakup – orogenic collapse rifting – asthenospheric input – proto-Romeral fault system developed – post-syntectonic granites – RSZ ophiolites (Vinasco et al. 2006; Cochrane et al. 2014b; Spikings et al. 2015). (**d**) Global tectonic reconstruction after Blakey (2016). Permian, Triassic, and Jurassic highlighted. Continued subduction of Farallon plate, arc magmatism, and intracontinental rifting. (**e**) Global tectonic reconstruction after Blakey (2016). Permian, Triassic, Jurassic, and Early Cretaceous highlighted. Continued subduction of Farallon plate, arc magmatism, intracontinental rifting, crustal attenuation, and deposition of Abejorral Fm. SJF San Jeronimo fault. (**f**) Global tectonic reconstruction after Blakey (2016). Permian, Triassic, Jurassic, Early Cretaceous, and Late Cretaceous highlighted. Continued subduction of Farallon plate, arc magmatism, forearc magmatism giving rise to Quebradagrande Fm., and double subduction zone (?). SJF San Jeronimo fault. (**g**) Global reconstruction after Scotese (2001). Permian, Triassic, Jurassic, Early Cretaceous, and Late Cretaceous highlighted. Quebradagrande closure, final assembly of RSZ, and isoclinal folding. Note: RSZ high-pressure metamorphic rocks occurring to the south are not observed within the map area. (**h**) Global tectonic reconstruction after Scotese (2001) and Blakey (2016). Andean orogenic phase, uplift and syntectonic sedimentation, and collision between South America with the Panamá-Chocó Arc

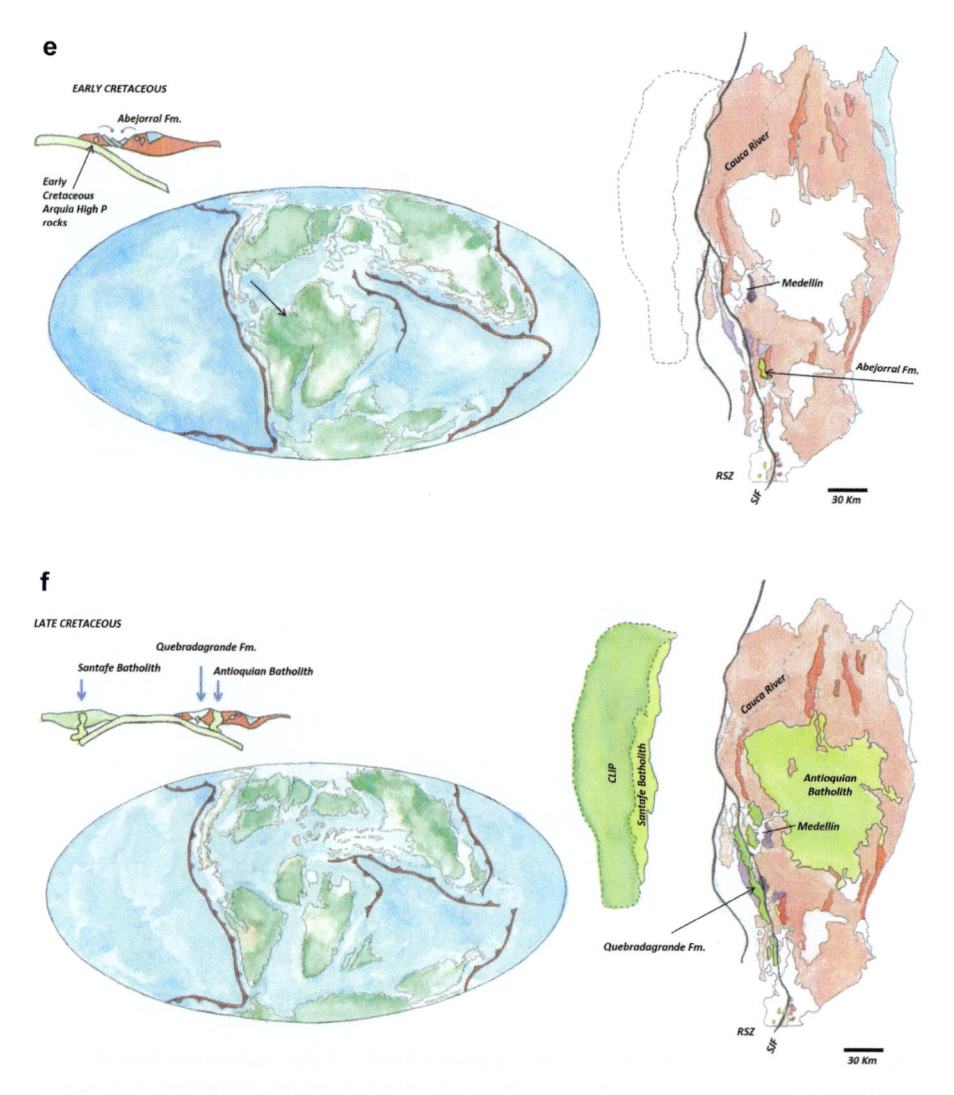

Fig. 12.3 (continued)

Based upon the geological record contained along the RSZ and elsewhere, the Permian continental (paleo Romeral) margin was characterized by syntectonic intrusion of S-type granitoids (dark red in Fig. 12.3b), strongly deformed in a dominant collisional environment (Vinasco et al. 2006; Cochrane et al. 2014b). The emplacement of synorogenic granitoids during the Late Permian-Early Triassic is interpreted to mark the culmination of Pangaea assembly, while the posterior emplacement of post-collisional granitoids (purple south of Medellín in Fig. 12.3c) and a bimodal, granitoid anatectite-metamafite suite appears to represent orogenic collapse followed by rifting and the breakup of Pangaea in the paleo-Northern

12 The Romeral Shear Zone

g

LATE CRETACEOUS-PALEOCENE

Late Cretaceous Arquia High P rocks

Upper Cretaceous Arquia rocks

Medellín

RSZ   SJF   30 Km

h

OLIGOCENE TO MIOCENE   MIOCENE TO PRESENT

Medellín

Syntectonic sedimentation and volcanism

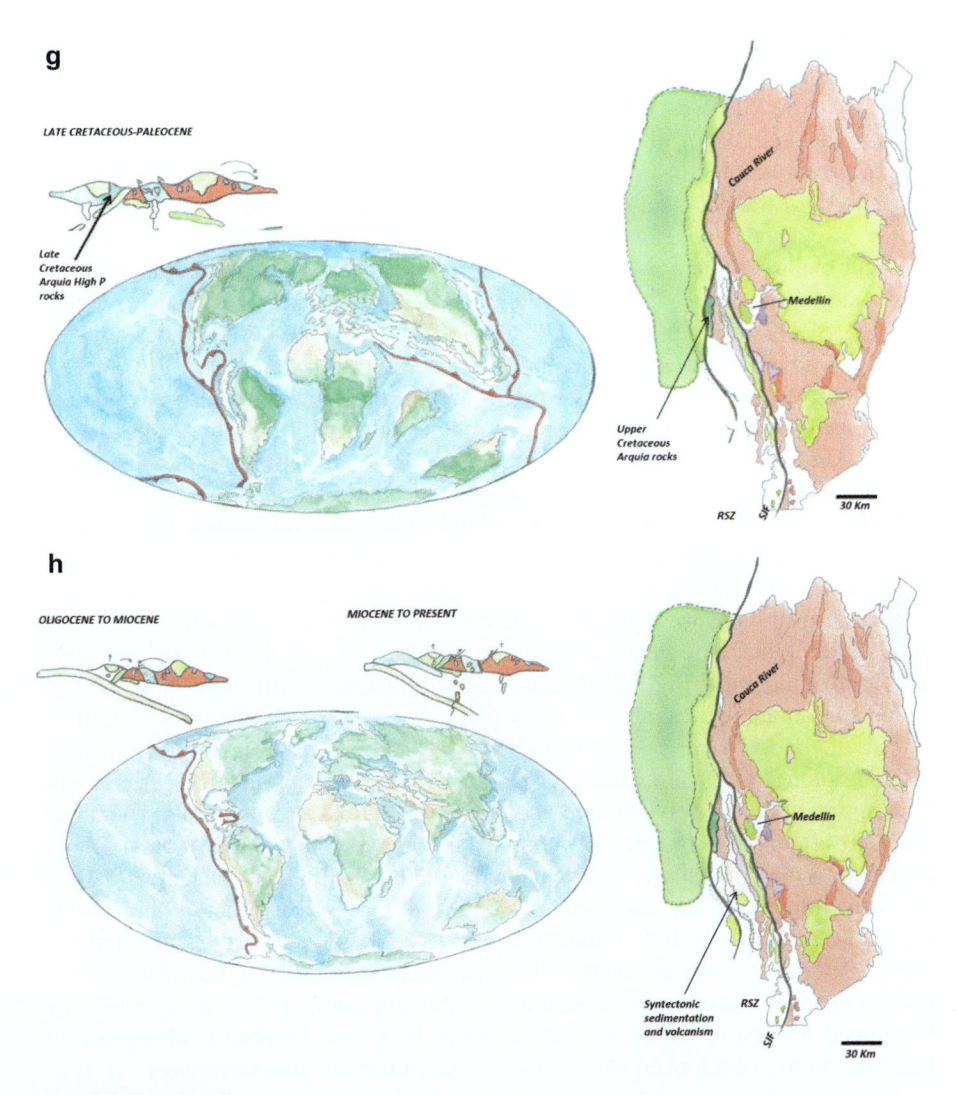

RSZ   SJF   30 Km

Fig. 12.3 (continued)

Andean region during the period from ca. 250–225 Ma with some subduction contribution (Fig. 12.3c) (Vinasco et al. 2006; Cochrane et al. 2014b; Spikings et al. 2015). The culmination of this breakup is marked by the formation of oceanic crust at ca. 216 Ma, as represented by the juvenile, MORB-associated Aburrá ophiolite (purple east of Medellín in Fig. 12.3c) (Martínez 2007; Cochrane et al. 2014b). It is presently unclear if the present tectonic position of the Aburrá ophiolite represents rifting directly within the continental margin or if the ophiolite has undergone significant tectonic transport to its present-day location over the continental margin as suggested by Toussaint (1996).

The Permian components of the metasedimentary rocks within the RSZ and associated mélange, including tectonic blocks within the Arquía Complex, are interpreted to represent slivers of the disrupted continental margin, possibly rifted off the paleo-autochthon during the breakup of Pangaea (red within RSZ in Figs. 12.3b, c). The full metamorphic history of these blocks has as yet to be deciphered, but it is suspected that they contain an early imprint of Permian tectono-thermal metamorphism associated with Pangaea assembly. The Arquía lithotecton, including the Sinifaná metasediments, was intruded by the post-tectonic Amagá stock and by mafic and ultramafic rocks belonging to the Pueblito intrusive suite (purple within the RSZ, SW of Medellín in Fig. 12.3c; Figs. 12.1 and 12.2); however, it is not possible to establish if intrusion took place prior to or after separation of the lithotectonic blocks from the Late Triassic paleo-margin. Based upon the primitive nature of the Pueblito suite and its tight spatial relationships with both the Amagá stock and the Arquía metasediments, these intrusives may represent a pericratonic assemblage built upon a rifted sliver of the paleo-continental margin.

Following the Late Triassic breakup of the paleo-Romeral margin, the development of an active margin was associated with subduction of the early Farallon plate, beginning at ca. 209 Ma. Leal-Mejía et al. (2011) and Spikings et al. (2015) interpret an east-dipping, west-migrating subduction zone associated with rollback of the Farallon plate. The major granitoid continental arcs of the Jurassic represent new continental crust (Fig. 12.3d) emplaced within a regionally extensional regime within the attenuated South American plate (Fig. 12.3e) (Spikings et al. 2015). During the period from ca. 210 to 115 Ma, the magmatic arc axis migrated westward. Regional extension after ca.152 Ma led to the opening of a north-south-trending marginal seaway (Pindell and Kennan 2009; Nivia et al. 2006; Cochrane et al. 2014a), which is interpreted herein to have included older (Late Jurassic-Early Cretaceous) components of the Quebradagrande Complex (Nivia et al. 2006) including blocks of the sedimentary Abejorral Fm. (green east of the SJF in Fig. 12.3e). Opening of this continental margin basin took place as westward trench retreat associated with the Farallon plate (Cochrane et al. 2014a) moved the axis of arc magmatism outside of the continental domain and into the marginal basin. Continued extension led to the development of proto-Caribbean lithosphere of MORB affinity, which overprinted the early margin basin arc assemblage.

South America migrated westward during the Early to mid-Cretaceous, closing the marginal seaway and leading to the tectonic reworking and accretion of previously rifted pre-Cretaceous fragments of the Arquía Complex, recently formed in a para-autochthonous, continental-oceanic crust margin setting, including possible mantle peridotites (Cochrane et al. 2014a; Villagomez and Spikings 2013). The first phase of Early Cretaceous high-pressure metamorphism of the Arquía Complex is considered to have taken place at this time (Fig. 12.3e) (Bustamante et al. 2011). Important lithotectonic components of the RSZ, including the pre-Cretaceous metasedimentary blocks of Arquía, were derived from the paleo-continent, as disrupted fragments incorporated during the Early Cretaceous evolution of the RSZ. These components contain a significant contribution of inherited zircons derived from the continental regime. Siliciclastic rocks of the Abejorral Fm. and other continental margin assemblages were deposited over the continent in a forearc position (Fig. 12.3e).

Reestablishment of subduction beneath the Colombian Pacific margin at ca. 96 Ma (Leal-Mejía et al. 2018) gave rise to the Antioquia batholith within the autochthonous, continental margin (green east of the SJF in Fig. 12.3f). Ages for this batholith and its satellite plutons range from ca. 96 Ma to 70 Ma (Ibañez-Mejía et al. 2007; Ordoñez et al. 2008; Villagómez et al. 2011; Leal-Mejía et al. 2018), overlapping with forearc magmatism related to the Late Cretaceous components of the Quebradagrande Complex (green west of the SJF in Fig. 12.3f). The occurrence of the ca. 100–92 Ma Santa Fe batholiths (Weber et al. 2015), intruding rocks of Barroso Fm. to the west of the RSZ, has led some authors to propose a model involving both east- and west-directed subduction (Fig. 12.3f) (e.g., León et al. 2014; Weber et al. 2015), to explain the penecontemporaneous development of the Antioquia batholith within the continental domain, and intra-oceanic granitoid arcs of the Farallon-CCOP/CLIP assemblage, developed to the west (see analysis and reconstructions presented by Leal-Mejía et al. 2018).

Based upon the foregoing, it has been established that by the mid-Cretaceous, the proto-Andean margin of the Central Cordillera block contained a complex and fragmented mosaic (i.e., tectonic mélange) containing metamorphic, magmatic, and sedimentary lithotectonic elements of autochthonous to para-autochthonous, continental to continental margin and oceanic affinity, representing a geologic history of extension and compression along the paleo-Romeral margin spanning ca. 190 m.y. (i.e., Permian through mid-Cretaceous). It is evidently very difficult to unravel and interpret each of these early events along the Early Romeral paleo-margin, because they have been overprinted by Andean-phase orogenic events during the Late Cretaceous through Miocene. Notwithstanding, it is important to recognize that the tectonic and structural architecture of the Andean-phase (i.e., ca. Late Cretaceous to Mio-Pliocene) RTZ and associated mélange is primarily overprinted upon pre- and proto-Andean lithostratigraphic components and tectonic events.

Beginning in the Late Cretaceous, coherent, mappable, correlatable tectono-structural elements (fault and fold patterns), lithostratigraphic sequences (e.g., sedimentation in the Cauca, Patía and Amagá basins), and thermochronological data, recorded within, upon, and adjacent to the RSZ, constrain and characterize tectonic development during the principle phases of Northern Andean orogenic development.

These features are associated with tectonic activity related to the approach and collision of oceanic assemblages belonging to the Farallon-Caribbean plate assemblages forming basement to the physiographic Western Cordillera, against the Romeral/continental paleo-margin. It is during this series of events that the modern-day structural architecture of the RSZ is established (Fig. 12.3g). Late Cretaceous fragments of the Arquía Complex from ca. 71 Ma were formed by this time (dark green within RSZ in Fig. 12.3g). Although high-pressure fragments of similar age are recorded further south, the metamafic volcanic rocks of the Arquía Complex in the north include greenschist assemblages which are not of high-pressure origin and appear to represent oceanic crust formed and accreted during this time (Fig. 12.3g).

Some fragments may be mylonitized fragments of the Barroso Fm. captured within the RSZ as suggested by García (2010) (dark green within RSZ in Fig. 12.3g).

The uplift recorded in the Central Cordillera block during collision of the Pacific terrane assemblage described above supplied sediments to the west; into the proto-Patía, Cauca, and Amagá basins; and, additionally, to the east, toward the Middle Magdalena Basin (Fig. 12.3g), whose tectonic development is temporally linked with Late Cretaceous tectonic development along the RSZ (Villamil 1999). In this context, uplift of the easternmost regions of the Eastern Cordillera and portions of the western foothills of the EC took place at this time, defining the beginning of the polyphase Andean orogeny (or pre-Andean orogeny as defined by Villamil 1999). The El Cobre, Cimarrona, and Monserrate formations in the Magdalena Valley are direct expressions of this uplift (Fig. 12.4) (Villamil 1999). Thermochronological data presented by Spikings et al. (2015) suggests uplift associated with the initial collision of the CCOP plateau along the Colombian margin at ca. 76 Ma. This is supported by shutdown of subduction-related magmatism in continental Colombia by ca. 72 Ma (Leal-Mejía et al. 2018). Helium degassing systematics on zircons from the western Central Cordillera block, including in El Buey, Belmira and San Andres de Cuerquia sectors (Vinasco et al. 2015; Noriega et al. 2015), reveal a well-defined thermal event from ca. 50 to 60 Ma, recording the climax of the pre-Andean orogeny, associated with molassic sediments of the El Hoyon Fm. (Fig. 12.4). The vertical distribution pattern revealed in the data suggest rapid exhumation rates for the western Central Cordillera block in the Paleogene; however, this uplift post-dates important Late Cretaceous activity within the RSZ, as recorded by ^{40}Ar-^{39}Ar data for neo-formed micas associated with dynamic metamorphism along the fault at ca. 70 Ma (Vinasco and Cordani 2012). The dextral transpressional regime characteristic of the RSZ was configured during this time. Older ages for the same mylonitic rocks within the San Jerónimo fault (SJF), along the easternmost limit of the RSZ, as derived from ^{40}Ar-^{39}Ar and K-Ar analyses of neo-formed micas in mylonitic bands developed in Quebradagrande volcanic rocks, shows plateau ages ranging from 87 to 90 Ma in biotite and 72–81 Ma in sericite, whereas K-Ar whole rock ages in samples collected in the area of influence of the SJF range between 91 and 102 Ma (Vinasco and Cordani 2012). These ages are considered as reactivation episodes within the prolonged shear zone dynamics of the Romeral system. By the Late Cretaceous (ca. 74 Ma), alkaline magmatism related to a slab break following terrane collision (Fig. 12.3g) (León et al. 2014) is contemporaneous with the last regional mylonitization event.

Fission track analyses of zircons collected from active detritus along the Sabaletas and Horcona creeks, near Armenia (Fig. 12.1c), yield different age peaks at 120, 50, and 40 Ma (Montoya-Betancur et al. 2015). The 120 Ma date is the oldest event recorded via low-temperature geochronometers along the northern RSZ and is perhaps a relict of the high-pressure metamorphic events recorded by blueschist rocks further south (Bustamante 2008). The 50 and 40 Ma events appear consistent with uplift along the Western and Central cordilleras during the Andean orogeny.

Inverse modeling of zircon and apatite fission track and (U-Th)/He data from throughout the physiographic Central and Western cordilleras was presented by

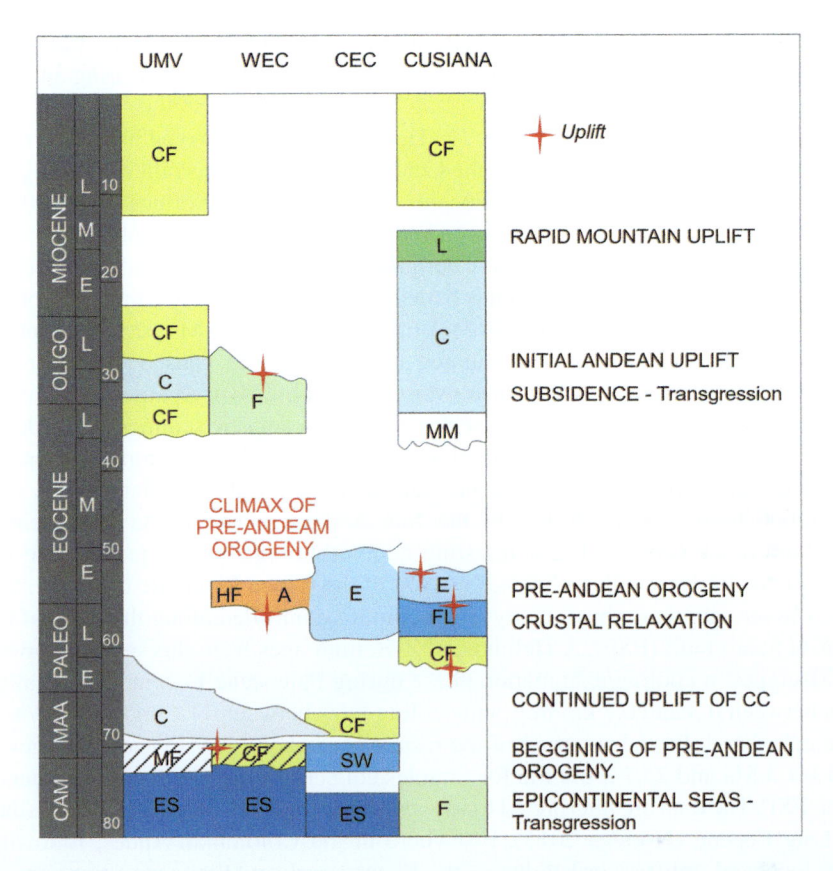

Fig. 12.4 Distribution and duration of Cretaceous unconformities, modified after Villamil (1999). Note the regional extent of Paleocene and middle Eocene unconformities. UMV Upper Magdalena Valley, WEC western foothills of Eastern Cordillera, CEC central part of Eastern Cordillera, F Fluvial, MM marginal marine, E estuarine, A alluvial fan-molassic, C Coastal, CF coastal-fluvial, ES epicontinental seas, SW shallow water, L lacustrine, Fl flooding coastal, HF Hoyón Fm., MF Monserrate Fm.; CF Cimarrona Fm

Villagomez and Spikings (2013). The data suggest three periods of rapid cooling associated with lithotectonic uplift and unroofing, since the Late Cretaceous. The first, from ca. 75–65 Ma, was driven by collision of the leading edge of the Farallon plate-CCOP/CLIP assemblage against the in situ Romeral margin. This resulted in important but unquantified tectonic shortening, yielding a generation of tight to isoclinal folds and west-verging thrusts, recorded primarily within the Quebradagrande Complex and the Late Cretaceous fragments of the Arquía Complex (Fig. 12.2). This event also resulted in the transposition of earlier structures as evidenced by intrafolial S1 folding in the Late Cretaceous Arquía units. Collision of the CCOP plateau telescoped the Quebradagrande Complex against the continental margin (Fig. 12.3g). Results suggest continuous deformation/uplift along the entire Late Cretaceous margin until at least 40 Ma, explaining the superposition of deformational fabrics due to prolonged or reactivated shearing and isoclinal folding.

Results for zircon and apatite (U-Th)/He dating suggest two additional periods of rapid cooling and exhumation in the Central and Western cordilleras, during the Pale-Eocene (ca. 51–56 Ma) and Eocene (ca. 40–42 Ma; Noriega et al. 2015), clearly recording the climax of the Early Andean orogeny. Helium analyses (apatite) from the RSZ in the San Andres de Cuerquia sector of the northern RSZ record a vague ca. 35–40 Ma episode, which is better reflected in the Mande batholith of the Western Cordillera (Botero et al. 2015). During the ca. 45–30 Ma period, the physiographic Central Cordillera was exhumed at moderate rates, caused by an increase in continent-ocean plate convergence (Martinod et al. 2010). Apatite fission-track low-temperature thermochronology performed on siliciclastic sedimentary rocks of the Amagá Fm. defines a clear exhumation pulse at ca. 43–45 Ma, similar to erosion-related cooling events at a similar latitude in the Central Cordillera to the east (Restrepo-Moreno et al. 2009). In the San Andres sector of the NW Central Cordillera block (Fig. 12.1c), contrasting geomorphologic styles, juxtaposing the Antioquian altiplano with the incised canyon of the Cauca River, are observed. The Cauca River canyon at this locality is structurally controlled along the trace of the Espíritu Santo fault (ESF). A-Helium and Z-Helium ages from this sector (Noriega 2015) suggest a cooling/exhumation phase during Paleocene-Eocene. Exhumation rates between 0.3 and 0.7 km/m.y. and cooling rates between 11.6 and 13.4 °C/m.y. were calculated. Partial resetting of the results is marked by AHe and ZHe ages at 45.3 ± 1.3 Ma and 25.1 ± 6.3 Ma for samples collected along the zone of influence of the ESF. The data are interpreted to suggest differential uplift along the ESF since the Late Eocene (Noriega 2015). Elsewhere in the Colombian Andes, following more localized uplift recorded during the Eocene-earliest Oligocene, strong inversion of the Eastern Cordillera and the Merida Andes (Venezuela) is recorded in the Miocene defining the beginning of Andean uplift (Fig. 12.4). Timing of this uplift is well supported by the development of foreland basins flanking both sides of these mountain ranges (Villamil 1999).

On the basis of He degassing analyses in apatite collected from samples of the Antioquian plateau, Restrepo-Moreno et al. (2009) reports a rapid exhumation pulse beginning at ca. 25 Ma. Beginning at ca. 23 Ma, the western Andean margin was affected by plate reorganization associated with rifting of the Farallon plate and the birth of the Nazca and Cocos plates (Lonsdale 2005). The Nazca plate further split along the Sandra rift, into the Cauca and Coiba segments in the mid-Miocene (Lonsdale 2005). Subduction of the Cauca segment beneath southern Colombia was underway by ca. 23 Ma, as recorded in the Piedrancha-Cuembí arc, while subduction of the Coiba segment is recorded somewhat later, by ca. 12 Ma (Leal-Mejía et al. 2018). Exhumation rates greatly increased in the middle through Late Miocene, with the greatest uplift occurring in southern Colombia as a consequence of the collision and subduction of the Carnegie Ridge beginning at ca. 15 Ma (Villagomez and Spikings 2013). The subduction vector of the Carnagie Ridge to in the

Ecuadorian Pacific suggests the Cauca segment at least is taking place in an oblique convergent setting (Cediel et al. 2003; Spikings et al. 2015).

Widespread regional unconformities along the eastern foothills of Central Cordillera (CC) define the intermediate phases of the Northern Andean orogeny beginning in the Eocene. This uplift event affected primarily the CC and produced an east-verging fold-belt that propagated to regions within the EC (Villamil 1999). Regional stages of uplift and erosion related to the Northern Andean orogeny are marked by five unconformities: (1) sub-lower Eocene, (2) sub-middle Eocene, (3) sub-Oligocene, (4) sub-Miocene, and (5) sub-Pliocene unconformities (Fig. 12.4) (Villamil 1999). Within the RSZ, the Cenozoic geological and tectonic history is well recorded by near continuous sedimentation within the Patía, Cauca, and Amagá basins. These basins have been interpreted to represent compartmentalized pull-apart depocenters, which developed as a result of the diachronous, oblique collision of terranes along the Pacific margin (Sierra and Marín-Cerón 2011). During the Early Eocene to Middle Eocene, sediments from a marginal arc and the proto-Central Cordillera were deposited as alluvial fans, fan deltas, and submarine fans upon a shallow platform and in a deltaic to transitional marine environment. Within the Cauca and Patía basins, deposition is related to periods of high tectonic activity and abundant sediment supply (Sierra and Marín-Cerón 2011). The basement terranes of the southern Western Cordillera were partially or completely immersed at this time (Sierra and Marín-Cerón 2011). The Guachinte Fm. was deposited in a marginal marine or transitional environment, with sediment provenance from the Central Cordillera. In the northern Cauca basin and the Amagá basin during the Oligocene, sediments of Cinta de Piedra Fm. and the upper member of the Amagá Fm. accumulated mainly in a continental setting, as deposits associated with meandering streams. Sediment was also sourced from the Central Cordillera (Fig. 12.3h). The syntectonic Amagá Fm. was deposited within a pull-apart basin developed within an Oligocene transpressional regime. Within the Patía basin to the south, the basal interval of the Esmita Fm. accumulated in a predominantly estuarine to shallow marine platform setting; however, the Arenáceo and Limolítico members were accumulated in purely continental settings.

In northwesternmost Colombia, collision and accretion of the Panamá-Chocó arc along the Pacific margin initiated in the Late Oligocene (Farris et al. 2011; Montes et al. 2012). The sedimentary record of this activity is recorded in Early to Middle Miocene sedimentation within the Amagá basin. The reactivation of Paleogene structures took place during this phase of renewed uplift and basin infilling. By the Late Miocene, mixed sedimentary lithofacies indicate multiple sources of sediments, including from the Central and Western cordilleras. During this time, Eocene deposits were uplifted and re-eroded, serving as sources of the Miocene sequences (Sierra and Marin-Cerón 2011).

Final deformational episodes along the RSZ and within the Central Cordillera are related to accretion of the Panamá-Chocó block in the Late Miocene-Early Pliocene (Fig. 12.3h). East-vergent thrust faults and west-vergent back thrusts define a transpressional deformation style associated with this collisional event. The best-exposed thrusts place Triassic ultramafic rocks and Cretaceous volcanic rock of Quebradagrande Complex over Oligo-Miocene sediments of Amagá Fm. This event is interpreted to be responsible for tectonic rotation of blocks within the RSZ and mélange. The siliciclastic Amagá sequences were moderately deformed into a series of anticlines and synclines bound by west- and east-verging thrust sheets. Accretion of the Panamá-Chocó block during the Miocene reactivated the RSZ with a left-lateral component (Ego and Sébrier 1995). Crustal thinning associated with pull-apart basin formation or back-arc extension east of the Farallon-El Cerro arc segment (Leal-Mejía et al. 2018) facilitated the emplacement of volcanic rocks of the Combia Fm. including numerous ca. 8.5–5 Ma subvolcanic bodies along the middle Cauca basin (Fig. 12.3h) (Leal-Mejía et al. 2018).

Panamá-Chocó block accretion inflicted significant deformation upon all pre-Middle Miocene lithotectonic and stratigraphic units along the northern RSZ. Within the western Chocó arc, tight folding along the Istmina deformation zone and deformation of Lower to Middle Miocene strata are observed (Duque-Caro 1990). Based upon Eocene detrital zircon populations interpreted to have been derived from the Panamá-Chocó block, Montes et al. (2015) suggested closure of the Central American Seaway by ca. 15 Ma. This interpretation is, however, challenged by the possible inaccuracy of Eocene zircons being exclusively derived from the Panamá-Chocó block (Restrepo et al. personal communication). At this point the precise age of collision remains elusive. New unpublished data presented by Botero (2018) suggest an initiation of collision about 20–25 Ma based upon detrital He analysis in sediments of Penderisco Fm. Notwithstanding, some constraints upon final collision between NW South America and the Panamá-Chocó arc suggest Late Miocene-Pliocene tectonic interaction. Constraints utilized in the present analysis include (1) thrusting of Triassic and Cretaceous rocks over the Miocene Amagá Fm.; (2) the initiation of Combia volcanism during the Late Miocene, given that the Combia Fm. is not strongly deformed; (3) low-temperature inverse modeling of Mande batholith (Vinasco et al. unpublished data); and (4) detailed provenance analyses in key Upper Miocene strata in the Western Cordillera (Botero et al. personal communication). With respect to neo-tectonic activity, preliminary forwarded thermal models performed using Pecube® code (Botero et al. personal communication) suggests that exhumation rates during the post-Miocene to present are slightly higher for the Mande batholith of the western Chocó arc when compared with those of the western Central Cordillera. Values somewhat under 1 mm/yr for the western Central Cordillera correspond to exhumation rates typical of Andean-type orogens.

12.4 Concluding Remarks

The RSZ and associated tectonic mélange records a prolonged history of autochthonous and para-autochthonous tectonic evolution along the Northern Andean paleomargin dating from at least the Permo-Triassic and extending through to modern times. Based upon the composite geologic, metamorphic, sedimentologic, magmatic, and structural record contained within the Romeral mélange, numerous pre-Andean tectonic settings and events can be interpreted to have evolved along the paleo-Romeral margin. Some of these events may have included:

1. Subduction-related magmatism and tectono-thermal metamorphism during the Late Permian assembly of Pangaea
2. Bimodal granitoid-mafic magmatism associated with orogenic collapse and rifting during the breakup of Pangaea in the mid- to Late Triassic
3. The deposition of trench-fill assemblages and the formation of an extended marginal basin associated with slab rollback of the Farallon plate during the Jurassic through Early Cretaceous
4. Development of a continental-margin magmatic arc assemblage in the Early Cretaceous
5. Marginal basin closure, obduction, and high-pressure metamorphism of the marginal arc assemblages during the Aptian (ca. 120 Ma)
6. Arc and forearc magmatism and ensuing closure along the continental margin, defining the beginning of the Northern Andean polyphase orogeny during the definitive juxtaposition of Pacific exotic domains against the Late Cretaceous Romeral assemblage and the Central Cordillera block to the east.

It is difficult to precisely define some of these events, including the spatial-temporal relationships between the individual lithotectonic elements, as the entire proto-Romeral assemblage has been strongly overprinted by Northern Andean-phase orogenic development, folding, faulting, metamorphism, and magmatism, beginning in the mid- to Late Cretaceous and extending through to modern times.

Despite the variable tectonic scenarios presented by different authors for the origin of the RSZ and associated tectonic mélange, it is concluded that a purely exotic character for the lithotectonic elements associated with the RSZ (e.g., Quebradagrande and Arquía complexes) cannot be supported. These lithotectonic units were derived primarily in situ as part of the marginal basin assemblage or as tectonic slivers separated from the Central Cordillera block during extensional events along the Early Romeral paleo-margin. It has been shown that, although strongly dismembered, lithotectonics within the Romeral mélange have a demonstrably intimate relationship with the pre-Cretaceous basement of the Central Cordillera block to the east. In this context, these units share a genetic relationship with the pre-Cretaceous paleo-autochthon. The Quebradagrande Complex corresponds to a volcano-sedimentary basin developed in the pericratonic realm. In itself, it has a prolonged and complex tectonic history and contains elements of a continental and marginal sedimentary basin, a primitive continental-margin arc, and an evolving fore-to-back-arc basin assemblage containing vestiges of oceanic crust dating from the Late Cretaceous and possibly the Early Cretaceous and even Jurassic. A divergent tectonic regime separated rocks of the Quebradagrande Complex from the continental margin prior to tectonic reorganization in the regional Pacific realm. Quebradagrande was located in a forearc position to the east (i.e., in the hanging wall) of the subduction trench along which the Farallon plate was subducting during emplacement of the Antioquia batholith between ca. 96 and 72 Ma. Final approach and accretion of the CCOP/CLIP plateau led to Late Cretaceous closure of the Quebradagrande forearc against the continental margin. Structural telescoping associated with this event is recorded as isoclinal folds both in Quebradagrande Complex rocks and in fragments of the Arquía Complex.

The dominant structural style imprinted upon the RSZ during Northern Andean (Late Cretaceous through Miocene to recent) tectonic development within the principle study area is highlighted by thick-skinned, W-SW-vergent, imbricated thrust sheets, which affect all of the pre-Cretaceous through Oligo-Miocene components of the RSZ. This style is exemplified in various areas where lithotectons of Central Cordilleran affinity (e.g., Paleozoic quartz sericite schist) are thrust over volcanic rocks of the Cretaceous Quebradagrande Complex, which in turn are thrust over Late Cretaceous lithodemic units (e.g., the ca. 74 Ma Sucre syenogranite) with structurally intercalated Late Cretaceous ultramafic rocks and meta-gabbros. In various localities, para-autochthonous or structurally dismembered pre-Cenozoic units are thrust over autochthonous Oligo-Miocene sedimentary rocks and basin-fill sequences marking uplift of the physiographic Central and Western cordilleras. These structures clearly constrain the late phases of accretion of allochthonous oceanic rocks comprising the basement to the physiographic Western Cordillera of Colombia.

References

Acosta C (1978) El graben interandino Colombo-Ecuatoriano (Fosa Tectónica del Cauca Patía y el corredor Andino-Ecuatoriano). Boletin de Geología, Universidad Industrial de Santander 12:63–199

Alfonso C, Sacks P, Secor D, Rine J, Pérez V (1994) A Tertiary fold and thrust belt in the Valle del Cauca Basin, Colombian Andes. J South Am Earth Sci 7:387–402

Alvarez J (1983) Mapa geologico generalizado y localizacion del muestreo geoquimico de la cordillera occidental, departamentos de Chocó y Antioquia. Ingeominas. Medellín

Alvarez J (1987) Tectonitas dunitas de Medellín, departamento de Antioquia, Colombia. Boletin Geologico Ingeominas 28(3):9–44

Aspden J, Litherland M (1992) The geology and Mesozoic collisional history of the Cordillera Real, Ecuador. In: Oliver RA, Vatin-PCrignon N, Laubacher G (eds) Andean Geodynamics. Tectonophysics, 205: 187–204

Barrero D, Laverde F (1998) Estudio Integral de evaluación geológica y potencial de hidrocarburos de la cuenca "intramontana" Cauca- Patía, ILEx- Ecopetrol report, Inf. No. 4977

Blakey R (2016) Global paleogeography and tectonics in deep time. https://www2.nau.edu/rcb7/index.html

Botero A (1963) Contribución al conocimiento de la geología de la zona central de Antíoquia: Anales Facultad de Minas (Medellín) 57:101

Botero M (2018) Proveniencia y estilo estructural de la Formación Penderisco y las sedimentitas de Beibaviejo en el corte Uramita - Dabeiba: relación con la evolución del bloque Panamá-Choco. MSc. Universidad Nacional de Colombia. Medellín.

Botero G, González H (1983) Algunas localidades fosiliferas cretacicas de la Cordillera Central, Antioquia y Caldas. Geologia Norandina 7:15–28

Botero M, Vinasco C, Restrepo S, Marín-Cerón M, Noriega S, Bermudez M, Bernet M, Min K, Foster D (2015) Modelamiento termal del batolito de Mandé a traves de la utilización de herramientas termocronológicas trazas de fision y (U-Th)/He en apatito y circón. XV Congreso Colombiano de Geología. Bucaramanga, Colombia

Botero M, Vinasco C, Foster D, Kyle M (2017) Panamá Choco block boundary defined by the Dabeiba Fault. Cosntraints from U-Pb analysis. Unpublished Data

Burgl H, Radelli L (1962) Nuevas localidades fosilíferas en la Cordillera Central de Colombia. Geologia Colombiana 3:133–138

Bustamante A (2008) Geotermobarometria, geoquímica, geocronologia e evolução tectônica das rochas da fácies xisto azul nas áreas de Jambaló (Cauca) e Barragán (Valle del Cauca), Colômbia. Instituto de Geociencias. Sao Paulo, Universidade de Sao Paulo. PhD: 179

Bustamante A, Juliani C, Hall C, Essene E (2011) $^{40}Ar/^{39}Ar$ ages from blueschists of the Jambaló region, Central Cordillera of Colombia: implications on the styles of accretion in the Northern Andes. Geol Acta 9(3–4):351–362

Bustamante A, Juliani C, Essene E, Hall C, Hyppolito T (2012) Geochemical constraints on blueschist-facies rocks of the Central Cordillera of Colombia: the Andean Barragán region. Int Geol Rev 54:1013–1030

Calle B, González H (1980) Geología y Geoquímica de la Plancha 166, Jericó. Informe No. 1822. Medellín, INGEOMINAS, 232p

Cardona A, Montes C, Ayala-Calvo C, Bustamante N, Hoyos O, Montenegro C, Ojeda H, Niño V, Ramirez D, Rincon J, Vervoort J, Zapata S (2012) From arc-continent collision to continuous convergence, clues from Paleogene conglomerates along the southern Caribbean-South America plate boundary. Tectonophysics 580:58–87

Case J, Barnes J, Paris G, González H, Vina A (1973) Trans-Andean geophysical profile, southern Colombia. Bull Geol Soc Am 84:2895–2904

Cediel F, Barrero D, Caceres, C (1998) Seismic expression of structural styles in the basins of Colombia in six volumes, prepared by Geotec for Ecopetrol: Ed. by Robertson Research, London

Cediel F, Shaw R, and Cáceres C (2003) Tectonic Assembly of the Northern Andean block. In Bartolini C, Buffler RT, Blickwede J (eds) The Circum-Gulf of Mexico and Caribbean: Hydrocarbon habitats, basin formation and plate tectonics: AAPG Memoir 79, 2003, p 815–848

Cediel F, Leal-Mejia H, Shaw R, Melgarejo J, Restrepo-Pace P (2011) Regional Geology of Colombia. Petroleum Geology of Colombia. ANH

Cline K, Page W, Gillam M, Cluff L, Arias L, Benalcázar L, López J (1981) Quaternary activity on the Romeral and Cauca faults, Northwest Colombia. Revista C.I.A.F. 6(1–3):115–116

Cochrane R, Spikings R, Gerdes A, Winkler W, Ulianov A, Mora A, Chiaradia M (2014a) Distinguishing between in-situ and accretionary growth of continents along active margins. Lithos 202–203(2014):382–394

Cochrane R, Spikings R, Gerdes A, Ulianov A, Mora A, Villagómez D, Putlitz B, Chiaradia M (2014b) Permo-Triassic anatexis, continental rifting and the disassembly of western Pangaea. Lithos 190(2014):383–402

CODIGEM (1993) Mapa Geológico de la República del Ecuador, Escala 1:1,000.000, compilado por CODIGEM-BGS, A. Zamora y M. Litherland, directores, Quito

Correa A, Martens U, Restrepo J, Ordoñez-Carmona O, Martins-Pimentel M (2005) Subdivisión de las metamorfitas básicas de los alrededores de Medellín-Cordillera Central de Colombia. Rev Acad Colomb Cienc 29(112):325–344

Duque-Caro H (1990) The Chocó Block in the northwestern corner of South America: structural, tectonostratigraphic and paleogeographic implications. J S Am Earth Sci 3(1):71–84

Ecopetrol/ESRI (1988) Valle del Cauca field report, v 1 (Text), Esri technical report 88-0012

Ego F, Sébrier M (1995) Is the Cauca-Patía and Romeral Fault System left or right-lateral? Geophys Res Lett 22(1):33–36

Etayo-Serna F (1985) Documentación paleontológica del Infracretácico de San Felíx y Valle Alto. Publicaciones Especiales, Ingeominas 16:XXV1–XXV7

Etayo-Serna F, Barrero D (1983) Mapa de terrenos geológicos de Colombia. Ingeominas Special Publication Colombia 14:48–56

Farris DW, Jaramillo C, Bayona G, Restrepo-Moreno S, Montes C, Cardona A, Mora A, Speakman RJ, Glasscock MD, Reiners P, Valencia V (2011) Fracturing of the Panamánian Isthmus during initial collision with South America. Geology 39:1007–1010

García D (2010) Caracterización de la deformación y metamorfismo de los Esquistos de Sabaletas, parte norte de la Cordillera Central de Colombia. Medellín, Universidad Nacional de Colombia Maestría en Ingeniería-Materiales y Procesos

Giraldo M (2010) Esquema geodinámico de la parte noroccidental de la Cordillera Central de Colombia. Facultad de Minas. Medellín, Universidad Nacional de Colombia Maestría en Ingeniería-Materiales y Procesos: 146

Giraldo B, Toro L (1985) Cartografía detallada del Stock de Amagá y sus rocas encajantes. Proyecto de grado. Universidad Nacional Sede Medellín, pp 1–200

Gómez J, Nivia A, Montes N, Almanza M, Alcárcel F, Madrid C (2015) Notas explicativas: Mapa Geológico de Colombia. En: Gómez J, Almanza MF (eds) Compilando la geología de Colombia: Una visión a 2015. Servicio Geológico Colombiano, Publicaciones Geológicas Especiales 33 Bogotá, pp 9–33

Gómez-Cruz AJ, Moreno M, Pardo A (1995) Edad y Origen del complejo metasedimentario Aranzazu-Manizales en los alrededores de Manizales (departamento de Caldas, Colombia). Geología Colombiana 19:83–93

González H (1977) Conceptos de metamorfismo dinamico. Su aplicacion a la zona de falla de Romeral. Boletin Ciencias de la Tierra. Universidad Nacional. 2:82–106

González H (1980) Geología de la plancha 167 (Sonsón), INGEOMINAS, p 174

González H (2001) Memoria explicativa del Mapa Geológico del Departamento de Antioquia. Ingeominas, Bogotá, pp 1–240 (CD)

Grosse E (1926) El Terciario carbonifero de Antioquia, en la parte occidental de la Cordillera Central de Colombia entre el río Arma y Sacaojal: Berlin, Dietrich Reimer (Ernst Vohsen), 361p

Ibañez Mejía M, Tassinari C, Jaramillo Mejía J (2007) U-Pb Circón Ages of the "Antioquian Batholith": Geochronological constraints of Late Cretaceous Magmatism in the Central Andes of Colombia. Memorias XI Congreso Colombiano de Geología, Bucaramanga, pp 1

Kellogg J, Godley V, Ropain C, Bermudez A (1983) Gravity anomalies and tectonic evolution of northwester South America. In: 10a Conferencia Geológica del Caribe, Memorias, Cartagena, pp 18–31

Kerr A, Tarney J (2005) Tectonic evolution of the Caribbean and northwestern South America: the case for accretion of two late cretaceous oceanic plateaus. Geology 33:269–272

Kerr A, Marriner G, Tarney J, Nivia A, Saunders A, Thirwall M, Sinton C (1997) Cretaceous basaltic terranes in Western Colombia: elemental, chronological and Sr-Nd isotopic constraints on petrogenesis. J Petrol 38(6):677–702

Leal-Mejia H, Shaw RP, Melgarejo J (2011) Phanerozoic granitoid magmatism in Colombia and the tectono-magmatic evolution of the Colombian Andes. In: Cediel F (ed) Petroleum Geology of Colombia. Department of Geology, University EAFIT. Medellín, Colombia

Leal-Mejia H, Shaw RP, Melgarejo JC (2018) Spatial/temporal migration of granitoid magmatism and the phanerozoic tectono-magmatic evolution of the Colombian Andes. In: Cediel F and Shaw RP (eds). Geology and Tectonics of Northwestern South America: The Pacific-Caribbean-Andean Junction, Springer, pp 253–397

León S, Jaramillo S, Cardona A, Valencia V, Vinasco C, Weber M (2014) Monzodioritic plutons in the Northwestern area of the Central Cordillera: alkaline magmatism as indicator of the cretaceous arc-continent collision. XI Semana Tecnica de Geología. UIS. Abstract

Litherland M, Aspden J, Jemielita R (1994) The metamorphic belts of Ecuador. Br Geol Surv Overseas Mem 11:147

Lonsdale P (2005) Creation of the Cocos and Nazca plates by fission of the Farallon plate. Tectonophysics 404:237–264

Lozano H, Pérez H, Mosquera D (1975) Prospección geoquímica en los Municipios de Salento. Quindio y Cajamarca, Tolima, Ingeominas, 103p

MacDonald W, Estrada J, Sierra G, González H (1996) Late Cenozoic tectonics and paleomagnetism of north Cauca Basin intrusions, Colombian Andes Dual rotation modes. Tectonophysics 261:277–289

Martens U, Restrepo J, Solari L (2012) Sinifaná Metasedimentites and relations with Cajamarca Paragneisses of the Central Cordillera of Colombia. Revista Boletín Ciencias de la Tierra, Nro. 32:99–10. Medellín

Martínez, A (2007) Petrogenesis and Evolution of Aburra Ophiolite, Colombian Andes, Central Range. Ph.D. thesis, University of Brasilia

Martinod L, Husson L, Roperch P, Guillaume B, Espurt N (2010) Horizontal subduction zones, convergence velocity and the building of the Andes. Earth Planet Sci Lett 299:299–309

Maya M, González H (1995) Unidades litodémicas en la Cordillera Central de Colombia. Boletín Geológico, Ingeominas 35(215–3):43–57

McCourt W (1984) The geology of Central Cordillera in the Departments of Valle del Cauca, Quindio and NW Tolima (sheets 243, 261, 262, 280 and 300). Ingeominas – Mision Britanica. Report N 8. Cali

McCourt W, Aspden J (1984) A plate tectonic model for the Phanerozoic evolution of Central and Southern Colombia. In: 10th Caribbean Geological Conference, Ingeominas, pp 38–47

McCourt WJ, Feininger T (1984) High pressure metamorphic rocks in the Central Cordillera of Colombia. Brit Geol Sur Repr Ser 84(1):28–35

Meissner R, Flueh E, Stibane F, Berg F (1976) Dynamics of the active plate boundary in southwest Colombia according to recent geophysical measurements. Tectonophysics 35:115–136

Meschede M, Barckausen U (2000) Plate tectonic evolution of the Cocos–Nazca spreading center. In: Silver EA, Kimura G, Shipley TH (eds) Proceedings of the ocean drilling program, scientific results, 170. A.M. University, pp 1–10

Montes C, Cardona A, McFadden RR, Moron S, Silva CA, Restrepo-Moreno S, Ramirez D, Hoyos N, Wilson J, Farris DW, Bayona G, Jaramillo C, Valencia V, Bryan J, Flores J-A (2012) Evidence for middle Eocene and younger emergence in Central Panamá: implications for Isthmus closure. Geol Soc Am Bull 124:780–799

Montes C, Cardona A, Jaramillo C, Pardo A, Silva J, Valencia V, Ayala C, Pérez-Angel L, Rodriguez-Parra L, Ramirez V, Niño H (2015) Middle Miocene closure of the Central American Seaway. Science 348:226

Montoya-Betancur E, Vinasco C, Restrepo S, Marín-Cerón M, Botero M, Noriega S, Bernet M, Bermudez M, Min K, Foster D (2015) Aplicación de la termocronología detrítica para el estudio de levantamiento de bloques tectónicos. Caso de estudio: cuenca de las quebradas Sabaletas y la Horcona. 2015. XV Congreso Colombiano de Geología. Bucaramanga, Colombia

Moreno M, Pardo A (2003) Stratigraphical and sedimentological constraints on Western Colombia: implications on the evolution of the Caribbean plate. In: Bartolini C, Buffler RT, Blickwede, J (eds) The Circum-Gulf of Mexico and the Caribbean: Hydrocarbon habitats, basin formation, and plate tectonics. American Association of Petroleum Geologists Memoir 79:891–924

Moreno-Sánchez M, Gómez-Cruz A, Toro L (2008) Proveniencia del material clástico del Complejo Quebradagrande y su relación con los complejos estructurales adyacentes. Boletín de ciencias de la tierra, Universidad Nacional de Colombia 22:27–38

Mosquera D (1978) Geología del cuadrángulo K8 Manizales (informe preliminar), Ingeominas, 63 p. Unpublished

Nelson H (1957) Contribution to the geology of the Central and Western Cordillera of Colombia in the section between Ibagué and Cali. Leidse Geologische Mededlingen 22:1–76

Nivia A (1996) El Complejo Estructural Dagua. Registro de deformación de la Provincia Litósferica Oceánica Cretácica Occidental: Un prisma acrecionario. VII Congreso Colombiano de Geología, Cali, Tomo III:54–67

Nivia A, Goméz-Tapias J (2005) El Gabro Santa Fe de Antioquia y la Cuarzodiorita Sabanalarga, una propuesta de nomenclatura litoestratigráfica para dos cuerpos plutónicos diferentes agrupados previamente como Batolito de Sabanalarga en el Departamento de Antioquia, Colombia

Nivia A, Marriner G, Kerr A (1996) El Complejo Quebradagrande; una posible cuenca marginal intracratónica del Cretáceo Inferior en la Cordillera Central de los Andes Colombianos. VII Congreso Colombiano de Geología, Cali, Tomo III:108–123

Nivia A, Marriner GF, Kerr AC, Tarney J (2006) The Quebradagrande Complex: a Lower Cretaceous ensialic marginal basin in the Central Cordillera of the Colombian Andes. J S Am Earth Sci 21:423–436

Noriega S (2015) Geomorfología tectónica del Noroccidente de la Cordilllera Central, Andes del Norte – Colombia. Tesis de maestría. Universidad Nacional de Colombia, Facultad de Minas. Departamento de Minerales y Materiales. Medellín, Colombia

Noriega S, Restrepo S, Vinasco C, Marín M, Bernet M, Foster D, Min K, Bermúdez M (2015) Termotectónica del borde noroccidental de la cordillera Central, Andes del Norte – Colombia. XV Congreso Colombiano de Geología. Bucaramanga, Colombia

Ordoñez O, Pimentel M, Laux J (2008) Edades U-Pb del Batolito Antioqueño. Bol. Ciencias de la Tierra, Universidad Nacional, Sede Medellín :129–130

Ortiz F (1979) Petroquímica del volcanismo básico de la Cordillera Occidental (Informe Preliminar). Boletín de Ciencias de la Tierra, Medellín 4:29–44

Pardo A (2015) Regional provenance from southwestern Colombia fore-arc and intra-arc basins: implications for Middle to Late Miocene orogeny in the Northern Andes. En: Colombia Terra 27:1–8

Pardo A, y Moreno M (2001) Estratigrafía del occidente Colombiano y su relación con la Evolución de la Provincia Ígnea Cretácica del Caribe Colombiano. VIII Congreso Colombiano de Geología. 19 p

Piedrahita V (2015) Evolución deformacional de rocas Cenozoicas en la Cuenca de Amagá. Trabajo de Grado. Eafit. Medellín, Colombia

Pindell J, Barrett S (1990) Geological evolution of the Caribbean region: A plate tectonic perspective. In Dengo G, Case, JE (eds) The Caribbean region: Geological Society of America, vol H. Geology of North America, pp 405–432

Pindell J, Kennan L (2009) Tectonic evolution of the Gulf of Mexico, Caribbean and northern South America in the mantle reference frame: an update. In: James K, Lorente MA, Pindell J (eds) The geology and evolution of the region between North and South America, vol 328. Geological Society of London, Special Publication, pp 1–55

Restrepo J (2008) Obducción y metamorfismo de ofiolitas triásicas en el flanco occidental del terreno Tahamí, cordillera Central de Colombia. Boletin de Ciencias de la Tierra 22. Edicion Especial

Restrepo J, Toussaint J (1974) Obducción cretácea en el Occidente Colombiano. Anales Fac. Minas, Univ. Nacional, Medellín 58:73–105

Restrepo J, Toussaint J (1976) Evolución cretácica del Occidente Colombiano. II Congr. Latín. Geol. México. Resúmen. 1p

Restrepo J, Toussaint J (1988) Terranes and continental accretions in the Colombian Andes. Episodes 11(3):189–193

Restrepo J, Toussaint J (1991) Terranes and continental Acretion in the Colombian Andes. Episodes 11(3):189–193

Restrepo J, Toussaint J, González H (1981) Edades Mio-Pliocenas del magmatismo asociado a la Formación Combia, Departamentos de Antioquia y Caldas, Colombia. Geología Norandina 3:22–26

Restrepo J, Ordóñez-Carmona O, Martens U, Correa A (2009) Terrenos, complejos y provincias en la cordillera Central de Colombia. Ingeniería Investigación y Desarrollo UPTC 9(2):49–56

Restrepo-Moreno S, Foster D, Stockli F, Parra-Sánchez L (2009) Long- term erosion and exhumation of the "Altiplano Antioqueño", Northern Andes (Colombia) from apatite (U–Th)/He thermochronology. Earth Planet Sci Lett 278:1–12

Rios P, Aranzazu J (1989) Analysis litofacial del intervalo Oligoceno-Mioceno en el sector noreste de la subcuenca del Valle del Cauca, Departamento del Valle, Colombia. Tesis de Grado, 257 pp, Universidad de Caldas, Manizales, Colombia (in Spanish)

Rodriguez V (2010) Fábrica y emplazamiento de la Diorita de Pueblito, NW Cordillera Central de Colombia: análisis de fábrica magnética y mineral. Medellín, Universidad Nacional de Colombia Maestría en Ingeniería-Materiales y Procesos

Rodríguez G, Arango M (2013) Formación Barroso: arco volcánico toleítico y Diabasas de San José de Urama: un prisma acrecionario T-MORB en el Segmento Norte de la Cordillera Occidental de Colombia

Rodriguez C, Rojas R (1985) Estratigrafía y tectónica de la serie inter Cretácica a los alrededores de San Felix. INGEOMINAS, p 121

Rodríguez G, Arango M, Zapata G, Bermúdez J (2016) Estratigrafía, petrografía y análisis multi-método de procedencia de la Formación Guineales, norte de la cordillera occidental de Colombia. Boletín de Geología 38(1)

Rojas, S (2012) Caracterización Petrográfica y Geoquímica del Stock de Cambumbia en el Departamento de Caldas, Colombia. Trabajo de Grado. Universidad nacional de Colombia sede Bogotá

Ruiz-Jiménez E, Blanco-Quintero I, Toro L, Moreno--Sánchez M, Vinasco C, García-Casco A, Morata D, Gómez-Cruz A (2012) Geoquímica y petrología de las metabasitas del complejo Arquía (municipio de Santafe de Antioquia y río Arquía, Colombia): implicaciones geodinámicas. Boletín Ciencias de la Tierra, No 32:65–80

Scotese CR (2001) Atlas of Earth History, Volume 1, Paleogeography, PALEOMAP Project, Arlington, Texas 1:52

Shaw RP, Leal-Mejía H, Melgarejo JC (2018) Phanerozoic metallogeny in the Colombian Andes: a tectono-magmatic analysis in space and time. In: Cediel F and Shaw RP (eds). Geology and Tectonics of Northwestern South America: The Pacific-Caribbean-Andean Junction, Springer pp 411–535

Sierra G, Marín-Cerón M (2011) Amagá, Cauca Patía Basin. In: Fabio Cediel (ed) Petroleum Geology of Colombia, vol 2. Fondo Editorial Universidad EAFIT, 104p

Sierra G, Ríos A, Sierra M (2005) Registro del volcanismo Neogeno y la sedimentacion fluvial en el suroeste antioqueño. En: Colombia, Boletin de Ciencias de la Tierra 17:135–152

Silva J, Sierra G, Correa L (2008) Tectonic and climate driven fluctuations in the stratigraphic base level of a Cenozoic continental coal basin, northwestern Andes

Spikings R, Cochrane R, Villagómez D, Van Der Lelij R, Vallejo C, Winklerf W, Beate B (2015) The geological history of northwestern South America: from Pangaea to the early collision of the Caribbean Large Igneous Province (290–75 Ma). Gondwana Res 27(1):95–139

Suter F, Sartori M, Neuwerth R, Gorin G (2008) Structural imprints at the front of the Chocó-Panamá indenter: field data from the North Cauca Valley Basin, Central Colombia. Tectonophysics 460(1–4):134–157

Taboada A, Rivera L, Fuenzalida A, Cisternas H, Philip H, BIjwaard J (2000) Geodynamics of the Norther Andes: Subduction and intracontinental deformation. Tectonics 19:787–813

Toussaint J (1996) Evolucion Geologica de Colomnbia – Cretacico. Ed. Univ. Nal Medellín: Medellín, Universidad Nacional De Colombia, 277p

Toussaint J, Restrepo J (1989) Acreciones sucesivas en Colombia: Un nuevo modelo de evolución geológica. V Congreso Colombiano de Geología, Bucaramanga. Tomo I:127–146

Van der Hammen T (1958) Estratigrafía del Terciario y Maastrichtiano continentales y tectogénesis de los Andes Colombianos. Servicio Geológico Nacional VI(1–3):67–128

Villagómez D, Spikings R (2013) Thermochronology and tectonics of the Central and Western Cordilleras of Colombia: early Cretaceous–Tertiary evolution of the Northern Andes. Lithos 160–161:228–249

Villagómez D, Spikings R, Magna T, Kammer A, Winkler W, Beltrán A (2011) Geochronology, geochemistry and tectonic evolution of the Western and Central Cordilleras of Colombia. Lithos 125:875–896

Villamil T (1999) Campanian-Miocene tectonostratigraphy, depocenter evolution and basin development of Colombia and western Venezuela. Palaeogeogr Palaeoclimatol Palaeoecol 153:239–275

Vinasco C (2001) A utilizacao da metodologia 40Ar 39Ar para o estudo de reativacoes tectonicas em zonas de cisalhamento. Tesis de maestría. Universidad de Sao Paulo, Instituto de Geociencias, pp 1–85

Vinasco C (2004) Evolucao crustal e historia tectonica dos granitoides Permo-Triassicos Dos Andes do norte. Ph.D. dissertation, Universidade de São Paulo, Brazil

Vinasco C, Cordani U (2012) Reactivation episodes of thermal fault system in northwestern part of central Andes, Colombia, through Ar-Ar and K-Ar results. Boletín de ciencias de la tierra 32:111–124

Vinasco C, Cordani U, González H, Weber M, Pelaez C (2006) Geochronological, isotopic, and geochemical data from Permo-Triassic granitic gneisses and granitoids of the Colombian Central Andes. J S Am Earth Sci 21:355–371

Vinasco C, Weber M, Cardona A, Areiza M, Restrepo S, Pindell J, Pardo A, Toro L, Lara M (2011) Geological transect through an accretionary margin, Western Colombia. Field trip September 3–5, 2011, IGCP Project 546 "Subduction zones of the Caribbean", Medellín

Vinasco C, Restrepo S, Marín-Cerón M, Botero M, Bernet M, Bermudez M, Noriega S, Montoya E, Jaramillo M, Londoño L, Min K, Foster D (2015) Evolución tectonotermal de la margen occidental de la cordillera Central y la zona de cizallamiento de Romeral: Respuesta Morfo-Tectónica a los principales pulsos deformacionales de la margen acrecionaria de los Andes del Norte. XV Congreso Colombiano de Geología. Bucaramanga, Colombia

Vinasco C, Restrepo-Moreno S, Botero M, Bermudez M, Marin-Ceron M, in prep. Low temperature inverse modeling of Central Cordillera and Romeral shear zone. Unpublished Data

Weber B, Iriondo A, Premo WR, Hecht L, Schaaf P (2007) New insights into the history and origin of the southern Maya block, SE Mexico: U–Pb–SHRIMP zircon geochronology from metamorphic rocks of the Chiapas massif. Int J Earth Sci 96:253–269

Weber M, Gomez-Tapias J, Cardona A, Duarte E, Pardo-Trujillo A, Valencia V (2015) Geochemistry of the Santa Fe Batholith and Buritica Tonalite in NWColombiaeEvidence of subduction initiation beneath the ColombianCaribbean Plateau. J S Am Earth Sci 62:257–274

Zapata S (2015) Mesozoic evolution of Colombia Central cordillera: from extensional tectonics to volcanic arc settings. Universidad Nacional de Colombia Maestría en Recursos Minerales. Medellín, Colombia

Part VI
Continental Uplift-Drift

Chapter 13
Exhumation-Denudation History of the Maracaibo Block, Northwestern South America: Insights from Thermochronology

Mauricio A. Bermúdez, Matthias Bernet, Barry P. Kohn, and Stephanie Brichau

Abbreviations

AFT	Apatite fission-track thermochronologic method
$^{40}Ar/^{39}Ar$	Argon-argon thermochronologic method
°C/km	Units for cooling rate (Celsius degree by kilometer)
GPS	Global Position System
km/Myr	Units for exhumation rate (kilometers by million years)
Ma	Million years ago
MCB	Maracaibo continental block
Myr	Million year
NW	Northwest
SE	Southeast
SM	Santander Massif
SNSM	Sierra Nevada de Santa Marta
SP	Serranía de Perijá or Perijá ranges

M. A. Bermúdez (✉)
Escuela de Ingeniería Geológica, Universidad Pedagógica y Tecnológica de Colombia, Sogamoso, Colombia
e-mail: maberce@gmail.com

M. Bernet
Institut des Sciences de la Terre, Université Grenoble Alpes, Grenoble, France

B. P. Kohn
School of Earth Sciences, University of Melbourne, Melbourne, VIC, Australia

S. Brichau
Geosciences Environment Toulouse, Université Paul Sabatier, Toulouse, France

© Springer Nature Switzerland AG 2019
F. Cediel, R. P. Shaw (eds.), *Geology and Tectonics of Northwestern South America*, Frontiers in Earth Sciences, https://doi.org/10.1007/978-3-319-76132-9_13

U-Pb	Uranium-lead geochronologic system
(U-Th)/He	Uranium-thorium-helium thermochronologic method
VA	Venezuelan or Mérida Andes
ZFT	Zircon fission-track thermochronologic method

13.1 Introduction

The Maracaibo continental block (MCB), northwestern South America, comprises a low lying area, which includes the Maracaibo basin, on its edges prominent mountain belts (the Sierra Nevada de Santa Marta (SNSM), the Santander Massif (SM), the Mérida or Venezuelan Andes (VA), the Trujillo Mountains (TM) part of the VA, and the inner part the Serranía de Perijá (SP)), and the Cesar-Ranchería and Maracaibo basins (Fig. 13.1). The MCB can be considered as the northwesternmost fragment of the Guiana Shield, overlain by extensive Phanerozoic supracrustal sequences. During the late Cretaceous, the MCB began to migrate northwestward along the Santa Marta-Bucaramanga and Oca fault systems (Fig. 13.1; Pindell 1993).

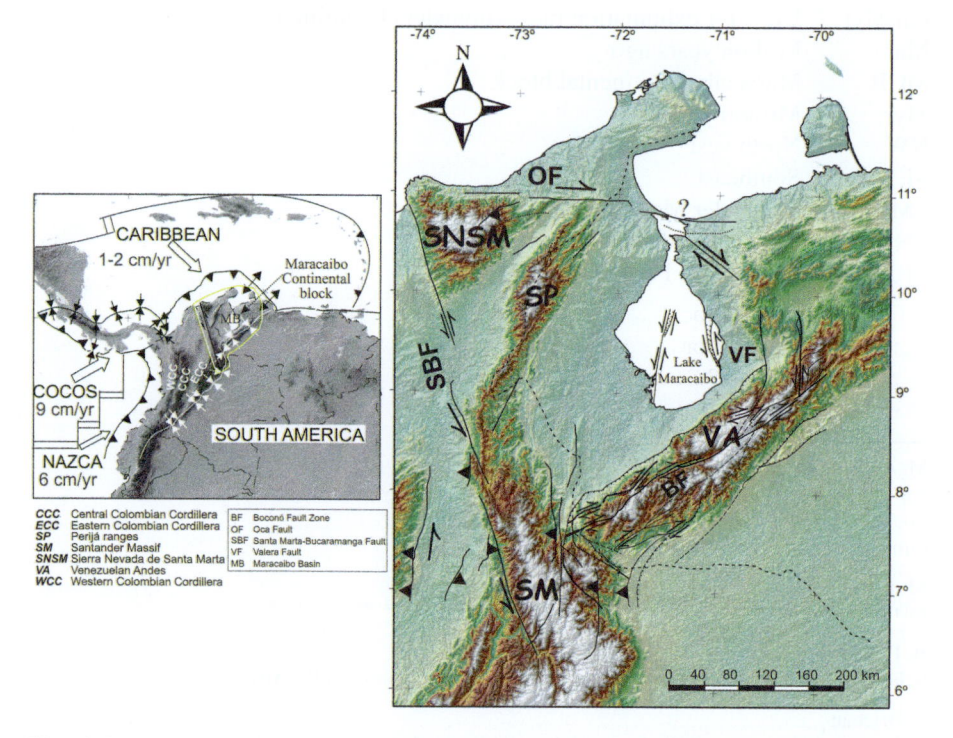

Fig. 13.1 Major tectonic features of northern South America. (Modified after Colmenares and Zoback 2003)

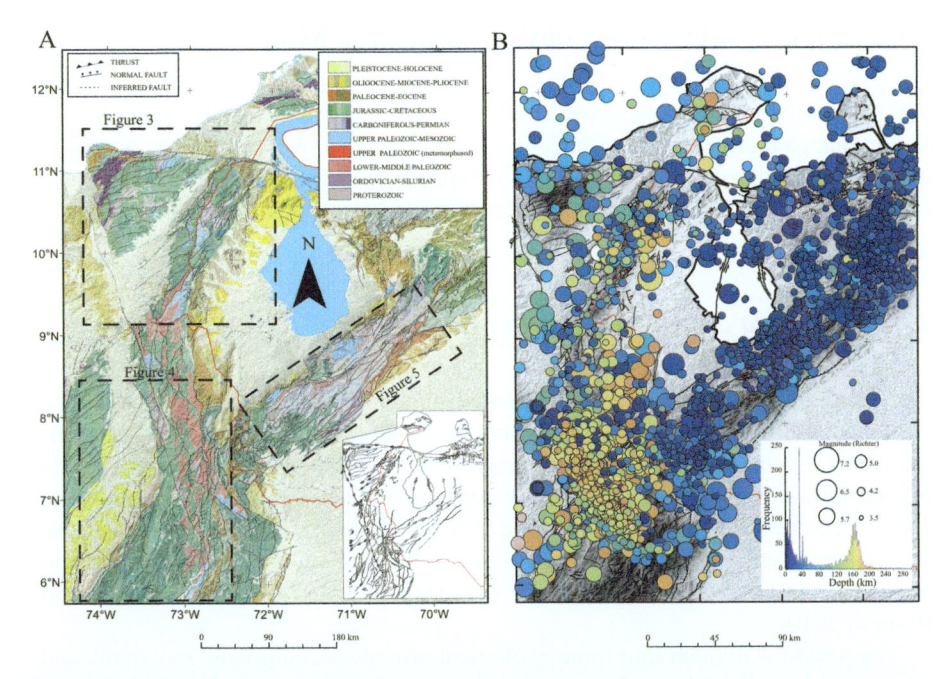

Fig. 13.2 (a) Simplified geological map of the MCB and surrounding belts. (Modified from Gómez et al. 2015; Hackley et al. 2005). (b) Seismicity between 1911 and 2013 using data available from the digital library of the Geophysics Laboratory of Universidad de Los Andes, Mérida (http://lgula.ciens.ula.ve/), data provided by the Venezuelan Foundation for Seismological Investigation (FUNVISIS; http://www.funvisis.gob.ve), the National Seismological Network of Colombia (http://seisan.sgc.gov.co/RSNC/), and the United States Geological Survey (USGS; https://earthquake.usgs.gov/earthquakes/feed/v1.0/csv.php). Faults and DEM were obtained from http://www.geomapapp.org; Ryan et al. (2009) and Bermúdez et al. (2013)

Transpressional forces generated during this process resulted in the development of different tectonic blocks such as the Mérida Andes, the Caparo block, and the Serranía de Trujillo or Trujillo block within the VA (inset in Fig. 13.2, Bermúdez et al. 2010; Figs. 13.2 and 13.5), SM, SP, and SNSM. The MCB is distinguished from other continental blocks to the northwest of the Guiana Shield by a unique and regionally constrained style of deformation, which originated during the Mesozoic-Cenozoic period as a result of interactions between the Pacific (Nazca), Caribbean, and continental South American plates (Fig. 13.1).

Detailed Global Positioning System (GPS) measurements and a recent compilation of displacement rate data (e.g., offsets of glacial moraines and pyroclastic flows) suggest that a large part of the Northern Andes, including the MCB, is currently escaping to the northeast relative to "stable" South America at rates close to 6 ± 2 mm/a (Egbue and Kellogg 2010).

The purpose of this chapter is to provide a general overview of the temporal and spatial morphotectonic evolution of the different mountain belts belonging to the MCB within the Caribbean and Northern Andes tectonic framework.

13.2 Structural Settings of the Maracaibo Mountain Belts

The Maracaibo mountain belts are crossed by different system faults: Santa Marta-Bucaramanga fault in the west, the Valera fault in the east, the Oca fault to the north, and Boconó fault to the south (Fig. 13.1). Figure 13.2 shows a simplified geological map of the SNSM, SP, SM, and VA (Gómez et al. 2015; Hackley et al. 2005).

13.2.1 Sierra Nevada de Santa Marta

The triangular SNSM massif is an isolated mountain range bounded by the Oca fault, Santa Marta-Bucaramanga fault, and the Cesar lineament (Fig. 13.3). It forms the world's highest coastal mountain range with ~5750 m elevation across ~40 km of coastline. During the Tertiary, dextral, and sinistral activity, movement of 65 and 110 km, respectively, occurred along the Oca and Santa Marta-Bucaramanga faults. Subsequently, erosional exhumation of several thousand meters led to development of the current geomorphological shape (Tschanz et al. 1974; Idárraga-García and Romero 2010).

The SNSM is divided into three geological provinces comprising magmatic and metamorphic rocks separated by the Sevilla and Cesar lineaments (see Colmenares et al.'s chapter in this book; Tschanz et al. 1974; Bustamante et al. 2009):

1. The Sierra Nevada province, an old granulite terrane (1.3 Ga) overlain by unmetamorphosed Paleozoic and Permian(?)-Triassic rocks (Fig. 13.3).
2. A metamorphic terrane consisting of Precambrian amphibolite-grade metamorphic rocks overlain by Silurian phyllites and unmetamorphosed Paleozoic and Mesozoic rocks, which are typical of the Colombian Western Cordillera (Doolan 1970; Tschanz et al. 1974; Restrepo-Pace et al. 1997; Ordoñez et al. 2002; Cordani et al. 2005; Cardona et al. 2006, 2010).
3. The youngest province consists of three northeast-trending regional metamorphic belts, which are intruded by Tertiary plutons. These belts comprise Permian-Triassic gneiss, Jurassic schist, and Cretaceous-Paleocene greenschist and were formed in the same subduction zone setting as the plutonic belts in province (2) above.

Cardona et al. (2006) proposed Triassic-Jurassic heating and a cooling pulse in the SNSM based on the oldest individual step-heating ages from discordant hornblende and biotite $^{40}Ar/^{39}Ar$ age spectra, for a paragneiss in the Paleozoic Sevilla Complex (minimum zircon U-Pb age of 529 ± 10 Ma; Cardona et al. 2006; Fig. 13.3a). Because the age spectra suggest the existence of thermal perturbations, the $^{40}Ar/^{39}Ar$ ages are considered to be inconclusive (Villagómez 2010). Cardona et al. (2010) reported zircon and apatite (U-Th)/He ages from crystalline rocks exposed along a NW-SE traverse within the northwestern region of the Santa Marta Province, which range from 20–27 Ma and 5–24 Ma, respectively. These authors

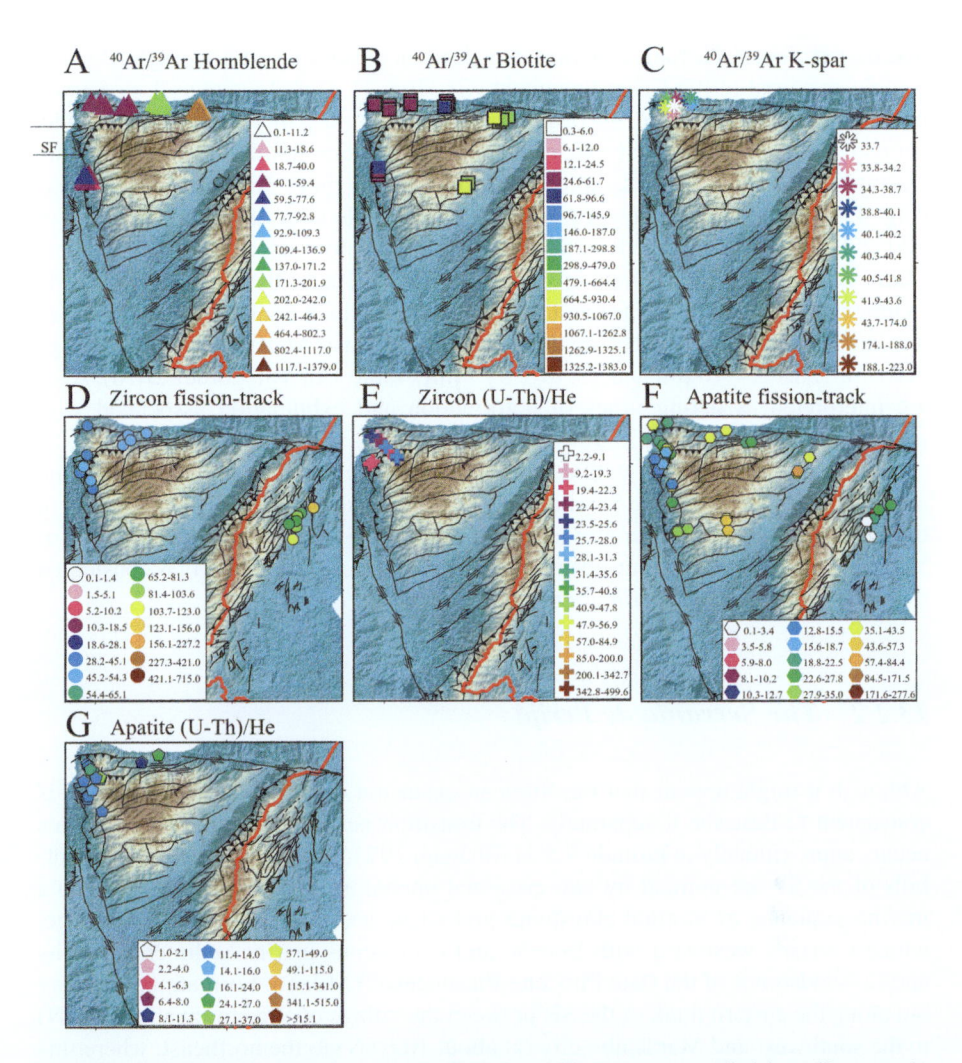

Fig. 13.3 Available radiometric age data for SNSM and SP (Dataset compiled from Hoorn et al. 2010; Herman et al. 2013; Gómez et al. 2015; Piraquive 2017 and references therein). Each panel shows the distribution of data derived by a different geo-thermochronological method or on a different mineral. Ages shown do not take into account any consideration of their uncertainties. In this figure SF corresponds to the Sevilla fault

assume a high geothermal gradient of 50 °C/km and propose moderate exhumation rates of 0.16 km/Myr during the late Oligocene and 0.33 km/Myr since the middle–late Miocene.

Villagómez (2010) reported zircon and apatite fission-track (AFT) ages across the SM and SNSM. From these, 14 samples were collected along a single traverse oriented approximately perpendicular to the northwestern corner of the SNSM (Fig. 13.3). The oldest AFT ages within the SNSM were obtained from the southern and eastern Sierra Nevada Province. Six Jurassic granitoids and a single rhyolite sample proximal to the Cesar lineament (CL, Fig. 13.3) yield apatite fission-track ages between ~60 Ma and ~40 Ma. These rocks form part of the Andean-wide, Jurassic continental arc that extends along the western South American margin. Despite the fact that these samples reside in three or more distinct faulted blocks (Fig. 13.3), almost all AFT ages are indistinguishable over an elevation range of 400–2700 m. Rocks located within 10 km of the deformation zone of the Santa Marta-Bucaramanga fault yield indistinguishable AFT ages between 23 and 30 Ma (Villagómez 2010; Villagómez and Spikings 2013), which are clearly younger than less deformed regions of the same province to the east. Mean AFT lengths range between 12.21 ± 1.70 µm and 13.96 ± 0.17 µm (Table 1 in Villagómez 2010). Ages are interpreted as dating times of deformation and exhumation associated with movement along the Santa Marta-Bucaramanga fault.

The youngest AFT ages within the SNSM, ranging between 16 Ma and 28 Ma, are found to the north of the Sevilla fault (see Colmenares's chapter in this book) in the Santa Marta and Sevilla provinces (SF, Figs. 13.3a and 13.4). Sample SN6 of Villagómez (2010) yielding an apatite fission-track age of 41.0 ± 9.6 Ma is considered to relate to the onset of exhumation.

13.2.2 The Serranía de Perijá

Although it might appear that the SP is an extension of the SM in Colombia, it is convenient to describe it separately. The transition between both mountain ranges occurs approximately at latitude 9.5°N (Shagam 1975; Fig. 13.3). The eastern foothills of the SP are defined by two erosional unconformities (Bucher 1952): (1) a marine sequence of mottled claystones and siltstones of Oligocene-Miocene age, which overlaps westward with Eocene and Cretaceous marine sedimentary units and (2) sandstones of the (late Pliocene-Pleistocene?) Milagro Formation, cropping out along the eastern flank of the SP, between the village of Riecito (at about 9.5°N) to the southwest and Maracaibo city (at about 10.6 °N) to the northeast, where this unit is in direct contact with pre-Cretaceous igneous rocks. AFT data in this area indicate cooling by the late Oligocene–early Miocene (22–27 Ma) for the Jurassic Palmar granite and Paleozoic metasedimentary rocks in the Palmar High and Pliocene (ca. 3 Ma) cooling in the Totumo-Inciarte arch that affected the Jurassic La Quinta Formation and Paleozoic Lajas granite (Shagam et al. 1984). (Plate 1 in Miller 1962; Shagam et al. 1984) (Fig. 13.3).

Fault systems and folds of the SP trend N35°E. The main structures of the SP, the Perijá and Tigre faults, play a significant mechanical role, dividing the chain into several tectonic blocks. Other faults present in the area, such as the Cesar fault on the northwestern side of the SP and the Cuiba fault on the southeastern side, form the boundary faults of the SP (Fig. 13.3; Cediel's chapter in this book). The Cuiba fault appears to be a high-angle reverse fault (Fig. 14 in Shagam 1975). The east-trending transverse Oca fault system defines the northern termination of the SP range and the SNSM block. In the northern part of this range, south-trending faults are connected with others in the Maracaibo Basin (Fig. 13.1). Crossing through the southern and south-central part of the SP, a system of transverse faults is recognized in the Cesar-Ranchería Basin (Arena Blanca trend in Miller 1962).

The Perijá-El Tigre fault system crosses diagonally through the SP and marks a syncline and structural depression that segments the range. Most of the northern SP lies to the west of the fault, including the Serranía de Valledupar and structural blocks between Cuiba and Cesar faults (Fig. 14 in Shagam 1975; Fig. 13.3). The entire southern segment of the range lies to the east of this last area forming the Rio Negro anticlinal feature and the structural feature known as the Totumo-Inciarte arch. The Perijá fault terminates or is sharply offset where it meets the Arena Blanca displacement closer to the town of El Tucuy (Plate 1 in Miller 1962).

Development of the Cretaceous Perijá trough and early Eocene movement on the margin of this trough are related to Oca fault activity. Several raised beaches formed across the Oca fault zone near Sinamaica show evidence of Pleistocene or recent movement on the fault.

On the east side of the SP, different structural features can be observed. For example, an en echelon pattern of anticlinal prongs controls a portion of the eastern front of the range, which reappears with poor development and reversed direction in the north. Other features include largely concealed anticlinal belts in the structurally high region west of Maracaibo, as well as the abrupt termination of this mountain range.

SP relief is more moderate in comparison with SNSM, SM, and VA, and the range was apparently not uplifted to the same degree as the VA (Shagam 1975). Perijá surface uplift was counterbalanced by subsidence in the Cesar Valley and the northwestern Maracaibo Basin. Miller (1962) proposed at least four distinct Tertiary surface uplift episodes in the SP: (1) Eocene growth of the Totumo arch; (2) Oligocene activity such as reactivation of the Oca fault; (3) post-Oligocene tectonism producing strong folding and faulting, which formed the current architecture of the chain; and (4) structural activity apparently extended into the late Pliocene but since that time has been somewhat subdued.

13.2.3 Santander Massif

The SM, located in the Eastern Cordillera, represents the western boundary of the MCB (Figs. 13.2 and 13.4; see Zuluaga et al. chapter in this book). The geological history of the SM is complex. Its basement is composed of igneous and metamorphic rocks of the Precambrian Bucaramanga gneiss and the pre-Devonian Silgará Formation, which are unconformably overlain by non-metamorphic Carboniferous and Permian clastic to calcareous sedimentary rocks of the Floresta Formation. The Silgará Formation was intruded by Mesozoic quartz monzonite, diorite, and other Cretaceous or younger porphyritic rocks. The Jurassic Jordan and Girón Formations are separated by unconformities from the Cretaceous Tambor, Rosa Blanca, Paja, Tablazo, Simití, La Luna, and Umir Formations. Igneous rocks with

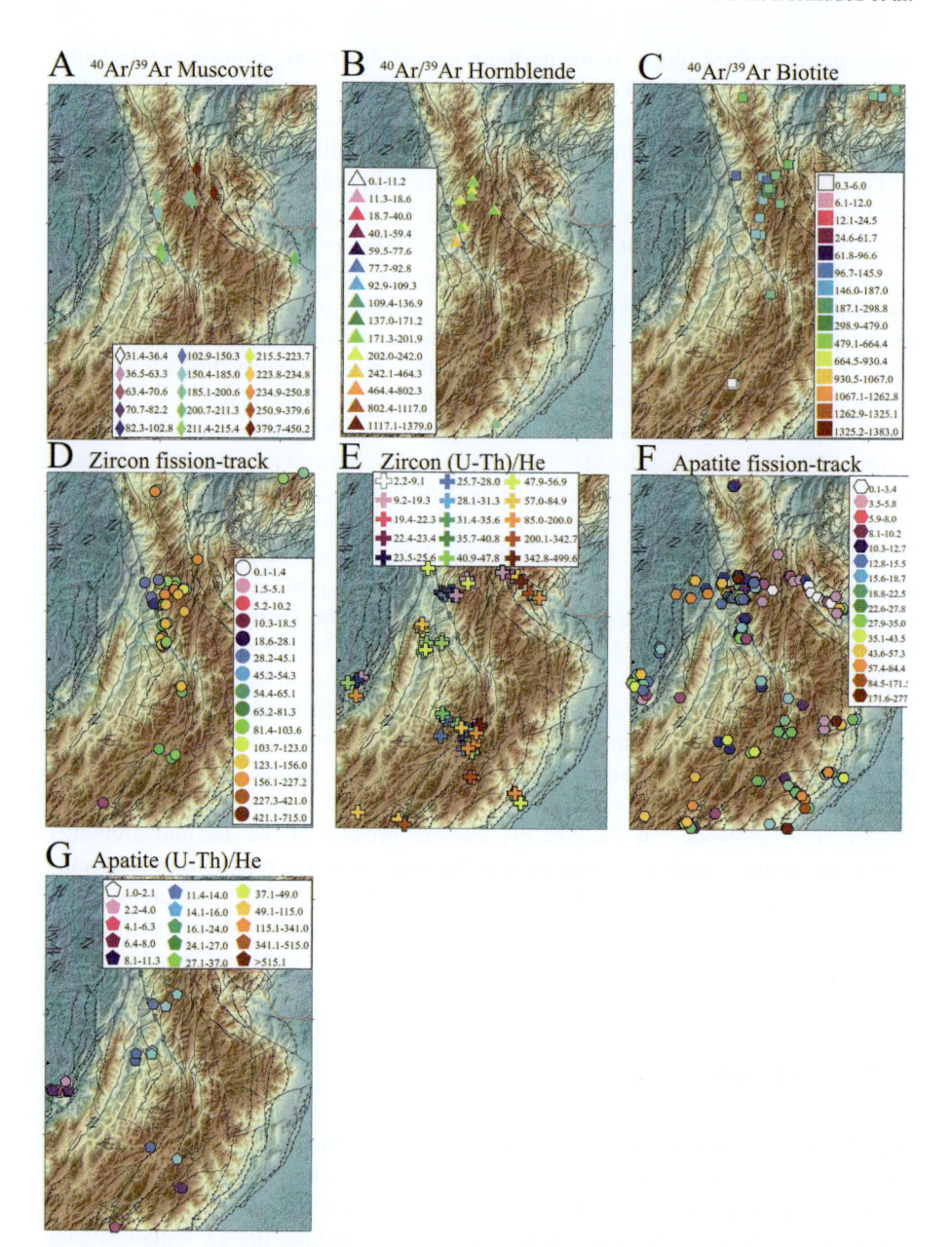

Fig. 13.4 Available radiometric age data for the SM and Bucaramanga fault (Dataset compiled from Hoorn et al. 2010; Herman et al. 2013; Mora et al. 2015; Gómez et al. 2015; Amaya 2016; Amaya et al. 2017 and references therein). Each panel shows the distribution of data derived by a different geo-thermochronological method or on a different mineral. Ages shown do not take into account any consideration of their uncertainties

porphyritic-aphanitic and porphyritic-phaneritic textures are related to the Miocene and most recent magmatic events across this massif (Mantilla et al. 2011). The SM is a structurally uplifted block, bordered to the west by the Santa Marta-Bucaramanga fault and to the east by the Pamplona fault system (see Zuluaga et al. in this book). This latter fault system defines the western boundary of the MCB.

Ward et al. (1974) suggested that foliation and fold orientations in the Bucaramanga Gneiss, Silgará Formation, and orthogneissic rocks of the SM are approximately similar. Regionally, these orientations are N-S, and tend to parallel the axis of the orogenic belt (Eastern Cordillera), showing local variations and discontinuities due to the presence of intrusive bodies of the Santander Plutonic Complex and a sedimentary cover. Restrepo-Pace (1995) proposed a shear-couple model for the SM fault system. The SM fault system requires a left-lateral sense of shear along faults such as the Bucaramanga fault and the Pamplona thrust (which can be considered a part of the Pamplona fault system). Displacements along Santa Marta-Bucaramanga faults generated a structural dissection of the SM, leading to the formation of rhomboidal blocks with Precambrian to Paleozoic rocks intruded by Mesozoic granites (van der Lelij et al. 2016). Along the western flank, an exhumed zircon fission-track partial annealing zone is exposed, and since the late Miocene based on AFT, exhumation rates in the SM are in the order of 0.3–0.4 km/Myr (Amaya 2016; Amaya et al. 2017).

13.2.4 Venezuelan or Mérida Andes

The VA form a complex orogen which extends over a length of ~500 km in SW-NE direction (Figs. 13.1 and 13.5). The SW-NE oriented Boconó strike-slip fault runs along this mountain belt and divides it almost symmetrically into two main tectonic blocks, the Sierra la Culata (to the north) and Sierra Nevada (to the south). The VA attains a maximum elevation of 4978 m at Pico Bolívar in the Sierra Nevada block. The orogen is bounded by northern and southern thrust faults (Fig. 13.5). The VA is formed due to convergence between the Caribbean, Nazca, and South American plates and the MCB (Colletta et al. 1997). Significant surface uplift of the central part of the VA had a major impact on the deviation of the Orinoco, Amazonas, and Magdalena paleo-drainage systems, which in turn had an important influence on the biodiversity of northern South America (Hoorn et al. 1995, 2010).

Bermúdez et al. (2010) demonstrated an important control of inherited faults and structures on exhumation patterns across the Mérida Andes and divided the orogen into at least seven individual active tectonic blocks (Fig. 13.5). Further, using fission-track dating and 3-D thermokinematic Pecube modeling (Braun 2003; Braun et al. 2012), Bermúdez et al. (2011) showed an asynchronous exhumation history of the two tectonic blocks located in the central part of this chain. The Sierra Nevada block located to the south of the Boconó fault was exhumed since 14–12 Ma at rates of 1.5 km/Myr until about 6–4 Ma. Later these rates decreased to 0.5 km/Myr. In contrast, the El Carmen intrusive body, a sliver of the Sierra La Culata block to

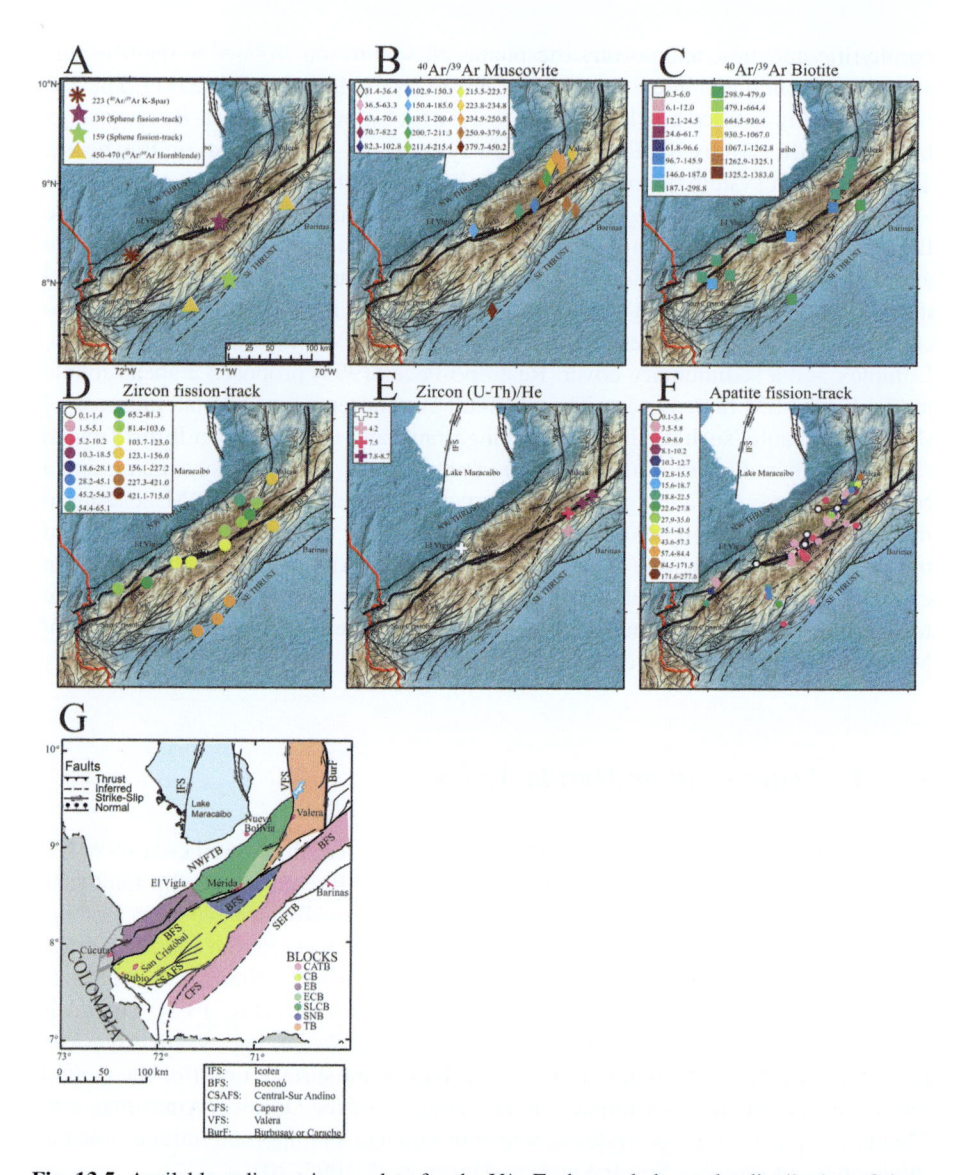

Fig. 13.5 Available radiometric age data for the VA. Each panel shows the distribution of data derived by a different geo-thermochronological method or on a different mineral. Ages shown do not take into account any consideration of their uncertainties. Note panel (G) shows main tectonic blocks (All compiled from Kohn et al. 1984; Bermúdez et al. 2010, 2011, 2014; van der Lelij et al. 2016). In this last panel, the abbreviations CATB, CB, EB, ECB, SLCB, SNB, and TB correspond to Cerro Azul Thrust, Caparo, Escalante, El Carmen, Sierra la Culata, Sierra Nevada, and Trujillo tectonic blocks, respectively

the north of the Boconó fault was exhumed from around 6 Ma to the present at rates of 1.5 km/Myr. This indicates that movement along the Boconó strike-slip fault includes an important oblique component, which allows for accommodation of deformation and exhumation. The exhumation rate however does not explain the high elevation of this chain. The surface uplift may be explained in part by compression of tectonic plates and in part by isostatic compensation. The Caparo and Trujillo blocks, located at the northeastern and southwestern terminations of this belt, exhibit lower mean elevations and slightly lower mean slopes than the central blocks (Sierra Nevada and Sierra La Culata blocks). Based on fission-track data, these two tectonic blocks cooled slowly from the Oligocene to late Miocene (Bermúdez et al. 2011). From a seismic point of view, the Caparo and Trujillo blocks are very different (Fig. 13.2b). Earthquakes to the south of the Caparo block are concentrated on the vertex formed by the Bramón, Central-Sur Andino, and Caparo fault systems. In this area earthquakes tend to be deep (>60 km) with magnitudes between 4.2 and 5.7. To the north of the Caparo block, seismicity is restricted to an active branch of the Boconó fault system (San Simón fault), and earthquakes with magnitudes between 3.5 and 4.5 are shallower (20–40 km). This seismic area coincides with the Escalante block, an active tectonic block with AFT ages and track length distributions similar to the Sierra La Culata block (2–6 Ma). Seismic activity in the Trujillo block is concentrated across the boundary of the blocks defined by the Valera, Boconó, and Burbusay-Carache faults (Bermúdez et al. 2010, 2013). Deeper earthquakes for this area (40–60 km) with magnitudes between 3.5 and 5.0 are focused on the Valera fault. Further, AFT ages >30 Ma are restricted to the Trujillo block (Figs. 13.6 and 13.7 in Bermúdez et al. 2010). The northern part of the Trujillo block shows an older AFT age of 145 ± 7 Ma (sample SVA-80-44; Bermúdez et al. 2010), which was obtained for the Valera Granite. This pattern reflects a pre-orogenic cooling history. Younger AFT ages (3.5–5.8 Ma) are observed in the southwest of the Valera block. The Trujillo block was not affected by the rapid exhumation observed in the central Mérida Andes. Seismicity across the Cerro Azul block, located in the southwestern foothills of the VA, is not uniform. AFT data from Proterozoic rocks located in this block suggest that the exhumation history is similar to that of the Sierra Nevada block (Figs. 13.2 and 13.5).

In summary, the VA experienced rapid exhumation starting in the middle Miocene, which was controlled by tectonic activity along the Boconó fault. Using a multidisciplinary approach, in which different types of data were used to reconstruct the timing of mountain building and the surface uplift history, it has been determined that the VA had reached present-day elevations by late Miocene to Pliocene time (Kohn et al. 1984; Bermúdez et al. 2010). The evolution of the VA is contemporaneous with the evolution of the Eastern Cordillera in Colombia (Bermúdez et al. 2017), for which climatic (Mora et al. 2008, 2010) and tectonic (Parra et al. 2010) controls have been documented, particularly over the past 35 Ma.

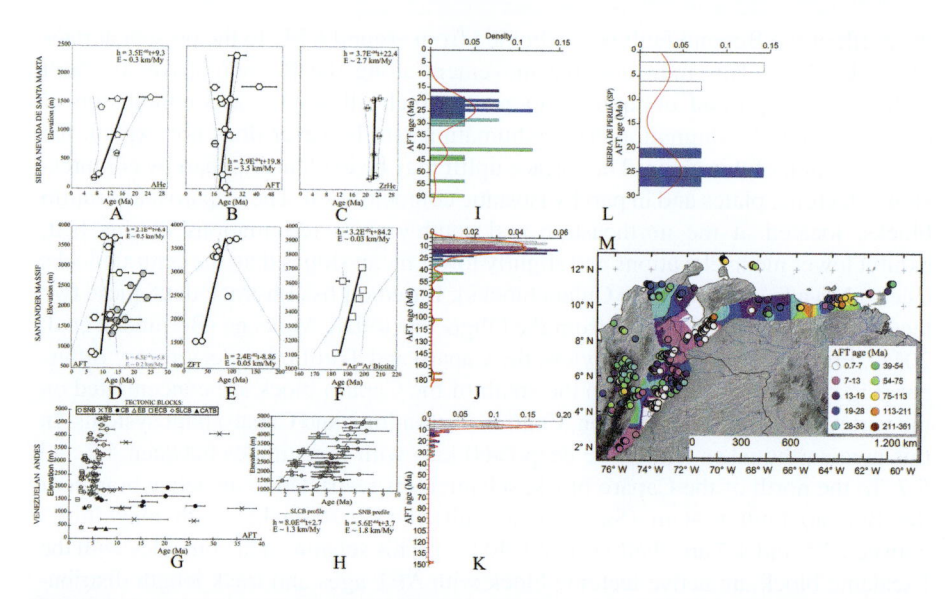

Fig. 13.6 Age-elevation relationships (AER) obtained using York's method for different thermochronometers from the SNSM, SM, and VA. For SNSM: (**a**) AHe-AER, (**b**) AFT-AER, (**c**) ZHe-AER. For SM: (**d**) AFT-AER (gray line and hexagons correspond to AER for Bucaramanga-Picacho profile, Mora et al. 2015; Amaya et al. 2017; black line and white hexagons correspond to a profile to the east of SMB fault, (**e**) ZFT-AER, (**f**) Biotite ^{40}Ar/^{39}Ar AER. For VA: (**g**) AFT-AER for different tectonic blocks, (**h**) AFT-AER ages only for Sierra Nevada and Sierra La Culata blocks. Probability density function of apatite fission-track ages in (**i**) SNSM, (**j**) SM, (**k**) VA, and (**l**) SP mountain belts. (**m**) Map of interpolated AFT ages for the circum-MCB mountain belts

13.3 Exhumation and Age-Elevation Relationships

For discriminating between differences in exhumation across the different mountain belts, we compiled a thermochronological dataset comprising 34 muscovite, 81 biotite, 55 hornblende, and 9 K-feldspar ^{40}Ar/^{39}Ar ages; 103 zircon and 46 apatite (U-Th)/He ages; and 2 sphene, 95 zircon, and 309 apatite fission-track ages (including ages obtained by both the LA-ICP-MS and external detector methods). The dataset used is based on previous compilations by the Colombian Geological Service (Gómez et al. 2015), as well as Herman et al. (2013), Hoorn et al. (2010), Mora et al. (2015), Amaya (2016), Piraquive (2017), and Amaya et al. (2017). The data used however are not spread evenly across the different mountain belts. For the SNSM the data are concentrated in the NW foothills. Hornblende, biotite, and K-feldspar ^{40}Ar/^{39}Ar ages are distributed across the western part of the Oca fault (Fig. 13.4). Zircon fission-track are also concentrated in the NW foothills and closer to the Oca fault; these vary from 36.1 to 56.8 Ma (Piraquive 2017). Apatite fission-track, zircon, and apatite (U-Th)/He data are available for the westernmost border of the SNSM (Villagómez 2010; Villagómez et al. 2011). These apatite ages are

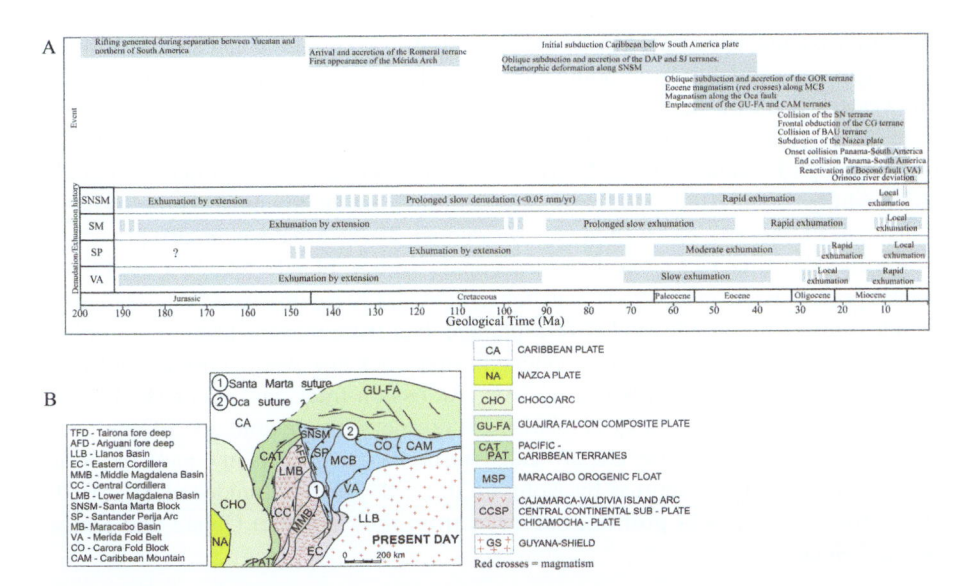

Fig. 13.7 (**a**) Summary of the cooling/denudation history of circum-MCB mountain belts. (**b**) Current tectonic configuration of accreted terranes and plates. In this figure DAP is Dagua-Piñon terrane, SJ is the San Jacinto terrane, GOR is the Gorgona terrane, CG is the Cañas Gordas terrane, BAU is the Baudó terrane, GU-FA is the Guajira-Falcón terrane, and CAM is the Caribbean Mountain terrane (see Cediel et al. 2003; and Cediel et al. chapter in this book)

based on samples located approximately perpendicular to the chain, along a unique elevation profile (24–2340 m) for this region. Some apatite fission-track ages have been reported from between the Tierra Nueva fault and the Cesar lineament (Figs. 13.2 and 13.3), but their distribution is not sufficient for estimating age-elevation relationships. No thermochronological data have been reported from the central part of the SNSM, where maximum elevations are close to 5800 m.

For the SP (Fig. 13.3), the only available data are zircon fission-track ages generated by Shagam et al. (1984) with ages ranging between 65.2 and 156 Ma. However, the sampling pattern does not permit an evaluation of any possible age-elevation relationships.

Muscovite, hornblende, and biotite $^{40}Ar/^{39}Ar$ ages are available for the SM. However, the data are not distributed in a systematic fashion across the different tectonic blocks. Further, the sampling strategy did not include collecting rocks from across a vertical transect orthogonal to the main structures, with exception of some $^{40}Ar/^{39}Ar$ biotite ages. For this last study, it is possible to estimate an exhumation rate for a tectonic block close to the Bucaramanga fault (Fig. 13.4). No zircon and apatite (U-Th)/He age data are available for the SM. So far such age data are only available for the Eastern Cordillera located between El Socorro and Cepitá, which is an area located in the western part of the Bucaramanga fault (Caballero et al. 2013) bounded by the Ocamonte and del Suárez faults, which themselves define a tectonic block (Diederix et al. 2009). Zircon and apatite fission-track ages are available from both sides of the Bucaramanga fault (Fig. 13.4). For the western flank of the SM, apatite and zircon fission-track and (U-Th)/He ages have been reported from along two different vertical profiles. These are a Bucaramanga-Picacho profile (Amaya 2016; Mora et al. 2015; van der Lelij et al. 2016; Amaya et al. 2017) and for an area delimited by the towns of Santa Bárbara, Zapatoga, El Socorro, and Cepitá (see Mora et al. 2015; Amaya et al. 2017).

For the VA (Fig. 13.5), one K-feldspar and two hornblende $^{40}Ar/^{39}Ar$ ages have been reported (Kohn et al. 1984). Van der Lelij et al. (2016) obtained $^{40}Ar/^{39}Ar$ hornblende, muscovite, biotite, and K-feldspar age data from igneous and metamorphic rocks. Twenty-one zircon and two sphene fission-track ages were reported by Kohn et al. (1984). The areal distribution of $^{40}Ar/^{39}Ar$, fission-track, and zircon (U-Th)/He ages reported by Bermúdez et al. (2014) do not permit an estimation of any age-elevation relationships in the Caparo and Trujillo blocks. This is in contrast to apatite fission-track ages reported by Kohn et al. (1984) and Bermúdez et al. (2010, 2011) from two vertical profiles, one each from the Sierra La Culata and Sierra Nevada blocks.

13.4 Discussion

13.4.1 Exhumation Rates

In order to estimate age-elevation relationships of the different mountain belts, we selected those areas where there was a greater density of ages and datasets were not cut by major fault systems. In Fig. 13.6, we summarize significant areas from the different mountain belts. For the SNSM, samples collected across a vertical profile by Villagómez et al. (2011) were used. For modeling the data, we distinguish three different exhumation rates as follows: 2.7 ± 0.4 km/Myr from 24 to 20 Ma, increasing to 3.5 ± 0.3 km/Myr between 20 and 16 Ma, and finally decreasing to 0.3 ± 0.1 km/Myr between 15 and 5 Ma. Villagómez et al. (2011) sampled along a vertical profile in the foothills of the SNSM massif, where steep slopes and high rainfall would be predicted to significantly increase erosive power. However, exhumation rates are relatively low. Long-term exhumation rates across the SM were calculated using $^{40}Ar/^{39}Ar$ biotite ages. These provide a rate of 0.03 ± 0.01 km/Myr, between 180 and 200 Ma, which seem to remain almost invariant between 80 and 50 Ma. More recently, between 10 Ma and 5 Ma, exhumation rates increased to 0.5 ± 0.2 km/Myr, which are very similar to those calculated for the SNSM.

For the SM, the cooling histories of tectonic blocks are similar to the VA (Mora et al. 2009; Parra et al. 2009), but the intense faulting pattern of the SM render rigorous exhumation rate estimates difficult. Two different profiles at the east of Bucaramanga fault (Fig. 13.6) provide different exhumation rates 0.5 ± 0.1 km/Ma and 0.2 ± 0.1 km/Ma, respectively, which support the notion of differential exhumation between independent tectonic blocks. Across the VA the pattern of exhumation is different than for the SM and the SNSM. The central part of the VA indicates exhumation rates ranging between 1.3 km/Myr and 1.8 km/Myr for the Sierra La Culata and Sierra Nevada blocks, respectively. However, because 3-D thermokinematic modeling has shown asymmetric behavior and changes in exhumation rates (Bermúdez et al. 2011), these rates should be considered as first-order approximations. Further, other effects such as climate, relief change, and fluid flow in areas closer to the faults need to be taken into account.

Finally, it is very important to take into account the effect of eroding topography on steady-state isotherms (Stüwe et al. 1994). In this chapter we were careful to consider such effects; further, exhumation rates across the MCB are not always high; thus the isotherms may not be unduly perturbed.

13.4.2 Periods of Cooling

The cooling history of the SNSM has been poorly constrained, mainly because several $^{40}Ar/^{39}Ar$ do not show plateau ages or because they exhibit disturbed spectra (Restrepo-Pace 1995; Restrepo-Pace et al. 1997; Cordani et al. 2005). These are interpreted as a consequence of reheating during the intrusion of several plutons located at shallow depths at ~200 Ma (Dörr et al. 1995; van der Lelij et al. 2016). Some of the plateau ages range between 184 ± 3 and 213 ± 3 Ma. This age range is similar to that observed in the Bucaramanga gneiss of the SM.

For the SNSM, Villagómez (2010) proposed three cooling pulses at (1) 65–58 Ma and 50–40 Ma (Jurassic granitoids), in the central Sierra Nevada Province; (2) 40–25 Ma (Precambrian gneisses) proximal to the Santa Marta-Bucaramanga Fault with higher exhumation rates at 26–29 Ma (Jurassic granitoid) close to the Sevilla lineament, in the western Sierra Nevada Province; and (3) 25–30 Ma (Paleogene quartz diorite and aplite) and 16–25 Ma (Paleogene aplite and quartz diorites and Upper Cretaceous schist) north of the Sevilla lineament. Further, this author also proposed that cooling events could be directly associated with exhumation because (i) they broadly correlate with steep gradients in the age-elevation relationships, (ii) they are not coeval with and do not immediately postdate proximal magmatic activity, and (iii) they are contemporaneous with sedimentation in the neighboring basins (Cesar and Lower Magdalena basins).

For the SM, $^{40}Ar/^{39}Ar$ thermochronometry on muscovite, hornblende, and biotite indicates two separate thermal events: (1) 150–200 Ma related to rifting processes and (2) 60–120 Ma. Zircon fission-track data collected from Jurassic and older basement rocks exposed in the SM define two age groups at 80–120 Ma and 50–70 Ma (Shagam et al. 1984), which support an exhumed partial annealing zone (Amaya et al. 2017). Apatite fission-track data from Jurassic and older basement rocks distinguish between two different cooling pulses, the first between 15 and 25 Ma and the second between 3.5 and 10 Ma, which suggest differences in timing and rates of cooling within discrete fault blocks. Shagam et al. (1984) also proposed an earliest Oligocene (~34 Ma) exhumation phase based on an interpretation that sedimentary rocks in the Eastern Piedmont were derived from erosion of the SM.

For the southern VA, $^{40}Ar/^{39}Ar$ biotite ages ranging between 185 and 480 Ma are older than those reported from SNSM and SM (van der Lelij et al. 2016). Zircon fission-track ages fall between 54 and 120 Ma, with older ages located closer to the main faults. For the apatite fission-track ages, it is possible to discriminate between four different age groups at (1) 1.8–3 Ma, (2) 4–9 Ma, (3) 15–30 Ma, and (4) 57–145 Ma (Miller 1962; Shagam 1975; Shagam et al. 1984; Kohn et al. 1984; Villagómez et al. 2011; van der Lelij et al. 2016).

13.4.3 Main Plate Tectonic Events

During the middle Jurassic (~170 Ma), Pangea breakup resulted in the formation of Gondwana and Laurasia, leaving a series of rifts and discontinuities in the crust. The reactivation of preexisting tectonic discontinuities of different age and origin, mostly developed along the former passive margin of the South American plate, has played an important role in the evolution of the circum-MCB mountain belts. Figure 13.7 summarizes the main interactions between the relevant plates and accretion of different tectonic blocks. Preexisting tectonic discontinuities correspond to rift structures generated in the Jurassic during separation between the North American and South American plates (Fig. 13.7a). From Aptian to Albian time, the Romeral terrane was pushed by the Farallon plate toward South America (Cediel et al. 2003), and at this time the first appearance of a system of arcs (Mérida Arch) occurs, which correspond with the first tectonic inversion of Jurassic rifts. For this period thermochronologic ages closer to ~150 Ma correspond with this inversion.

From late Cretaceous to Paleocene (70–65 Ma), exhumation of the SM commenced together with slow exhumation of the proto-Eastern Cordillera (Amaya et al. 2017). This is possibly related to oblique subduction and collision of the Dagua-Piñón (DA) and San Jacinto terranes (SJ; Cediel et al. 2003), which are remnants of the Caribbean plate. From the Paleocene to Eocene (65–56 Ma), fast exhumation occurred in the Colombian Central Cordillera as well as metamorphic deformation along the SNSM, and the first reactivation of the Oca fault is also recorded during this period.

From Eocene to early Oligocene (56–28 Ma), the following events are discerned: oblique subduction and accretion of the Gorgona terrane (GOR; Cediel et al. 2003), Eocene magmatism along the MCB, magmatism along the Oca Fault, and emplacement of the Guajira-Falcón (GU-FA) and Caribbean Mountain (CAM) terranes. These events are responsible for rapid exhumation throughout SNSM and SM, but at this time exhumation was moderate in the SP and slow in the VA (Fig. 13.7a).

From late Oligocene to early Miocene (28–20 Ma), thermochronological data reflect collision of the Sinú terrane (SN; Fig. 9e) and frontal obduction of the Cañas Gordas (CG) terrane, a remnant of the Farallon Plate. Collision of the Baudó (BAU) terrane and subduction of the Nazca plate occur during this period, resulting in the onset of collision between the Panamá arc and South America.

From middle Miocene to late Miocene (14–8 Ma), collision between the Panamá arc and South American plates terminates, the main lineation of the Boconó fault is reactivated, and deviation of the Orinoco River occurs (Hoorn et al. 1995). The main exhumation of the VA is also achieved during this period. Further, rapid exhumation on both sides of the Bucaramanga fault is reported from different areas.

From Pliocene to Pleistocene (5–1 Ma), local exhumation in several areas (SM, SNSM, SP, and VA) is triggered as a consequence of the northwest migration of the MCB.

13.5 Conclusions

Despite the caveats outlined above, some exhumation episodes have been significant enough to be recorded throughout the different mountain belts circumventing the MCB as follows:

1. Reactivation of normal faults during Mesozoic rifting between ~150 and 120 Ma across the SP, the Trujillo block in the VA, the Oca fault in the SNSM, and the northern SM.
2. Tectono-thermal pulses between 65 and 40 Ma are recorded in the SNSM, SP, and VA. However, with the available thermochronology dataset, these events are less evident within the SM. These pulses were triggered by oblique subduction and collision of the Dagua-Piñón (DA) and San Jacinto (SJ) terranes.
3. Relatively fast cooling from 35 to 16 Ma is recorded in the SNSM, SM, SP, and VA. This cooling was generated by accretion of Gorgona terrane and onset of subduction of the Caribbean plate below South America. During this period, tectonic inversion of the Guajira-Falcón terrane also took place.

Continuous cooling pulses between 15 and 2 Ma are commonly recorded across the different mountain belts, including some faults that were locally reactivated during last 4–2 Ma. This continuous pulse can be related to onset of collision between the Panamá arc and South America.

Acknowledgments We thank Fabio Cediel for motivating this work. To the Universidad de Ibagué Project 15-377-INT. MAB is grateful to the BEST project, a scientific agreement between IRD (Institut de Recherche pour le Développement, France) and the Universidad de Ibagué.

References

Amaya S (2016) Termocronología y geocronología del basamento metamórfico del Macizo de Santander, Departamento de Santander. Tesis Doctoral. Universidad Nacional de Colombia, p 174

Amaya S, Zuluaga C, Bernet M (2017) New fission-track age constraints on the exhumation of the central Santander Massif: implications for the tectonic evolution of the Northern Andes, Colombia. Lithos 282–283:388–402. https://doi.org/10.1016/j.lithos.2017.03.019

Bermúdez MA, Kohn BP, van der Beek PA, Bernet M, O'Sullivan PB, Shagam R (2010) Spatial and temporal patterns of exhumation across the Venezuelan Andes: implications for Cenozoic Caribbean geodynamics. Tectonics 29:TC5009. https://doi.org/10.1029/2009TC002635

Bermúdez MA, van der Beek P, Bernet M (2011) Asynchronous Miocene–Pliocene exhumation of the central Venezuelan Andes. Geology 39:139–142. https://doi.org/10.1130/G31582.1

Bermúdez MA, van der Beek PA, Bernet M (2013) Strong tectonic and weak climatic control on exhumation rates in the Venezuelan Andes. Lithosphere 5:3–16

Bermúdez MA, Kohn B, van der Beek P, Bernet M (2014) Patrones de exhumación de los Andes venezolanos: Un aporte de la termocronología y de la modelación numérica termocinemática 3D. Acta Científica 65(2): 17–27

Bermúdez MA, Hoorn C, Bernet M, Carrillo E, van der Beek PA, Garver JI, Mora JL, Mehrkian K (2017) The detrital record of late-Miocene to Pliocene surface uplift and exhumation of the Venezuelan Andes in the Maracaibo and Barinas foreland basins. Basin Res 29:370–395

Braun J (2003) Pecube: a new finite element code to solve the heat transport equation in three dimensions in the Earth's crust including the effects of a time-varying, finite amplitude surface topography. Comput Geosci 29:787–794

Braun J, van der Beek P, Valla P, Robert X, Herman F, Glotzbach C, Pedersen V, Perry C, Simon-Labric T, Prigent C (2012) Quantifying rates of landscape evolution and tectonic processes by thermochronology and numerical modeling of crustal heat transport using PECUBE. Tectonophysics 524-525:1–28

Bucher WH (1952) Geologic structure and orogenic history of Venezuela. Geol Soc Am Mem 49:1–113

Bustamante C, Cardona A, Saldarriaga M, García-Casco VV, Weber M (2009) Metamorfismo de los esquistos verdes y anfibolitas pertenecientes a los esquistos de Santa Marta, Sierra Nevada de Santa Marta (Colombia): ¿registro de la colisión entre el arco caribe y la margen suramericana? Revista Boletín Ciencias de la Tierra 25:7–26

Caballero V, Mora A, Quintero I, Blanco V, Parra M, Rojas LE, Lopez C, Sánchez N, Horton BK, Stockli D, Duddy I (2013) Tectonic controls on sedimentation in an intermontane hinterland basin adjacent to inversion structures: the Nuevo Mundo syncline, Middle Magdalena Valley, Colombia. Geol Soc Lond, Spec Publ 377(1):315–342

Cardona A, Cordani UG, MacDonald WD (2006) Tectonic correlations of pre-Mesozoic crust from the northern termination of the Colombian Andes, Caribbean region. J S Am Earth Sci 21:337–354

Cardona A, Valencia V, Garzón A, Montes C, Ojeda G, Ruiz J, Weber M (2010) Permian to Triassic I to S-type magmatic switch in the northeast Sierra Nevada de Santa Marta and adjacent regions, Colombian Caribbean: tectonic setting and implications within Pangea paleogeography. J S Am Earth Sci 29(4):772–783

Cediel F, Shaw RP, Cáceres C (2003) Tectonic assembly of the Northern Andean Block, in C. Bartolini, R.T. Buffler, and J. Blickwede, eds., The Circum-Gulf of Mexico and the Caribbean: Hydrocarbon habitats, basin formation, and plate tectonics, Am Assoc Petrol Geol, Memoir 79, 815–848

Colletta B, Roure F, De Toni B, Loureiro D, Passalacqua H, Gou Y (1997) Tectonic inheritance, crustal architecture, and contrasting structural styles in the Venezuelan Andes. Tectonics 16:777–794

Colmenares L, Zoback MD (2003) Stress field and seismotectonics of northern South America. Geology 31:721–724

Cordani UG, Cardona A, Jiménez DM, Liu D, Nutran AP (2005) Geochronology of Proterozoic basement inliers in the Colombian Andes: tectonic history of remnants of a fragmented Grenville belt. In: Vaughan APM, Leat PT, Pankhurst RJ (eds) Terrane processes at the margins of Gondwana. Geological society, London, special publications, vol 246, pp 329–346

Diederix H, Hernández C, Torres E, Osorio J, Botero P (2009) Preliminary results of the first paleoseismologic study along the Bucaramanga fault, Colombia. I+D (9) 2 p 18–23

Doolan BL (1970) The structure and metamorphism of the Santa Marta area Colombia, South America. PhD thesis, State University of New York (USA), Binghamton, N.Y. p 200

Dörr W, Grösser JR, Rodriguez GI, Kramm U (1995) Zircon U–Pb age of the Paramo Rico tonalite–granodiorite, Santander Massif (Cordillera Oriental, Colombia) and its geotectonic significance. J S Am Earth Sci 8(2):187–194

Egbue O, Kellogg J (2010) Pleistocene to present North Andean "escape". Tectonophysics 489:248–257. https://doi.org/10.1016/j.tecto.2010.04.021

Gómez J, Nivia Á, Montes NE, Almanza MF, Alcárcel FA, Madrid CA (2015) Notas explicativas: Mapa Geológico de Colombia. En: Gómez, J. & Almanza, M.F. (Editores), Compilando la geología de Colombia: Una visión a 2015. Servicio Geológico Colombiano, Publicaciones Geológicas Especiales 33 p 9–33 Bogotá

Hackley PC, Urbani F, Karlsen AW, Garrity CP (2005) Geologic shaded relief map of Venezuela, U.S. geological survey open file report, 2005–1038

Herman F, Seward D, Valla PG, Carter A, Kohn B, Willett SD, Ehlers TA (2013) Worldwide acceleration of mountain erosion under a cooling climate. Nature 504(7480):423–426

Hoorn C, Guerrero J, Sarmiento GA, Lorente MA (1995) Andean tectonics as a cause for changing drainage patterns in Miocene northern South America. Geology 23:237–240

Hoorn C, Wesselingh FP, ter Steege H, Bermúdez MA, Mora AJ, Sevink J, Sanmartín I, Sanchez-Meseguer A, Anderson CL, Figueiredo J, Jaramillo C, Riff D, Negri FR, Hooghiemstra H, Lundberg J, Stadler T, Sarkinen T, Antonelli A (2010) Amazonia through time: the far-reaching effect of Andean uplift on landscape evolution and biota. Science 330(6006):927–931. https://doi.org/10.1126/science.1194585

Idárraga-García J, Romero J (2010) Neotectonic study of the Santa Marta fault system, western foothills of the Sierra Nevada de Santa Marta, Colombia. J S Am Earth Sci 29:849–860

Kohn B, Shagam R, Banks P, Burkley L (1984) Mesozoic–Pleistocene fission track ages on rocks of the Venezuelan Andes and their tectonic implications. Geol Soc Am Mem 162:365–384

van der Lelij R, Spikings R, Mora A (2016) Thermochronology and tectonics of the Mérida Andes and the Santander massif, NW South America. Lithos 248–251:220–239. https://doi.org/10.1016/j.lithos.2016.01.006

Mantilla LC, Mendoza H, Bissig T, Craig H (2011) Nuevas evidencias sobre el magmatismo miocenico en el distrito minero de vetas-california (Macizo de Santander, Cordillera Oriental, Colombia). Boletín de Geología 33:43–58

Miller JB (1962) Tectonic trends in sierra de Perija and adjacent parts of Venezuela and Colombia. Am Assoc Pet Geol Bull 46(9):1565–1595

Mora A, Parra M, Strecker MR, Sobel ER, Hooghiemstra H, Torres V, Jaramillo JV (2008) Climatic forcing of asymmetric orogenic evolution in the Eastern Cordillera of Colombia. Geol Soc Am Bull 120:930–949

Mora A, Goana T, Kley J, Montoya D, Parra M, Quiroz LJ, Reyes G, Strecker MR (2009) The role of inherited extensional fault segmentation and linkage in contractional orogenesis: a reconstruction of lower Cretaceous inverted rift basins in the Eastern Cordillera of Colombia. Basin Res 21:111–137

Mora A, Parra M, Strecker MF, Sobel ER, Zeilinger G, Jaramillo C, Da Silva SF, Blanco M (2010) The eastern foothills of the Eastern Cordillera of Colombia: an example of multiple factors controlling structural styles and active tectonics. Geol Soc Am Bull 122:1846–1864

Mora A, Parra M, Forero G, Blanco V, Moreno N, Caballero V, Stockli D, Duddy I, Global B (2015) What drives orogenic asymmetry in the Northern Andes?: a case study from the apex of the Northern Andean Orocline, in C. Bartolini and P. Mann, eds., Petroleum geology and potential of the Colombian Caribbean Margin: AAPG Memoir 108 p 547–586, 13880_ch20_ptg01_547–586.indd 547 10/27/15 10:43 AM

Ordoñez O, Pimentel MM, De Moraes R (2002) Granulitas de Los Mangos: un fragmento grenviliano en la parte SE de la Sierra Nevada de Santa Marta: Revista Academia Colombiana de Ciencias, 26 p 169–179

Parra M, Mora A, Sobel ER, Strecker MR, Jaramillo C, González R (2009) Episodic orogenic-front migration in the northern Andes: constraints from low temperature thermochronology in the Eastern Cordillera, Colombia. Tectonics 28. https://doi.org/10.1029/2008TC002423

Parra M, Mora A, Jaramillo C, Torres V, Zeilinger G, Strecker M (2010) Tectonic controls on Cenozoic foreland basin development in the north-eastern Andes. Colombia Basin Res 22(p):874–903

Pindell JL (1993) Regional synopsis of Gulf of Mexico and Caribbean evolution. Mesozoic and Early Cenozoic development of the Gulf of Mexico and Caribbean region: a context for hydrocarbon exploration. Selected Papers Presented at the G.C.S.S.E.P.M Foundation Thirteenth Annual Research Conference: 251–274

Piraquive A (2017) Cadre structurel, déformations et exhumation des Schistes du Santa Marta: accumulation et histoire de déformation d'un terrain caraïbe au nord de la Sierra Nevada de Santa Marta. Sciences de la Terre. Université Grenoble Alpes, p. 395. https://tel.archives-ouvertes.fr/tel-01689912

Restrepo-Pace PA (1995) Late Precambrian to Early Mesozoic tectonic evolution of the Colombian Andes, based on new geochronological geochemical and isotopic data. [PhD Thesis, University of Arizona, USA], p 195

Restrepo-Pace PA, Ruiz J, Gehrels G, Cosca M (1997) Geochronology and Nd isotopic data of Grenvilleage rocks in the Colombian Andes: new constraints for late Proterozoic-early Paleozoic paleocontinental reconstructions of the Americas. Earth Planet Sci Lett 150:427–441

Ryan WBF, Carbotte SM, Coplan JO, O'Hara S, Melkonian A, Arko R, Weissel RA, Ferrini V, Goodwillie A, Nitsche F, Bonczkowski J, Zemsky R (2009) Global multi-resolution topography synthesis. Geochem Geophys Geosyst 10:Q03014. https://doi.org/10.1029/2008GC002332

Shagam R (1975) The northern termination of the Andes. In: Nairn AE, Stehli FG (eds) The ocean basins and margins, vol 3. Plenum Press, New York, pp 325–420

Shagam R, Kohn BP, Banks P, Dasch L, Vargas R, Rodríguez G, Pimentel N (1984) Tectonic implications of Cretaceous–Pliocene fission track ages from rocks of the circum-Maracaibo Basin region of western Venezuela and eastern Colombia. Geol Soc Am Mem 162:385–412

Stüwe K, White L, Brown R (1994) The influence of eroding topography on steady state isotherms. Application fission track analysis. Earth Planet Sci Lett 124:63–74

Tschanz CM, Marvin RF, Cruz J, Mehnert HH, Cebula GT (1974) Geologic evolution of the Sierra Nevada de Santa Marta, northeastern Colombia. Geol Soc Am Bull 85:273–284

Villagómez D (2010) Thermochronology, geochronology and geochemistry of the Western and Central Cordilleras and Sierra Nevada de Santa Marta, Colombia: the tectonic evolution of NW South America. PhD Thesis, Université de Géneve, p 143. ISBN 978-2940472-01-7

Villagómez D, Spikings RA (2013) Thermochronology and tectonics of the Central and Western cordilleras of Colombia: Early Cretaceous–tertiary evolution of the Northern Andes. Lithos 160–161:228–249

Villagómez D, Spikings R, Mora A, Guzmán G, Ojeda G, Cortés E, and van der Lelij R (2011) Vertical tectonics at a continental crust-oceanic plateau plate boundary zone: fission track thermochronology of the Sierra Nevada de Santa Marta, Colombia. Tectonics 30(4), TC4008:1–18

Ward DE, Goldsmith R, Jaime B, Restrepo HA (1974) Geology of quadrangles H-12, H-13, and parts of I-12 and I-13, (zone III) in northeastern Santander department. U.S. Geological Survey, Colombia

Part VII
Active Oceanic – Continental Collision

Part VII
Active Oceanic – Continental Collision

Chapter 14
The Geology of the Panama-Chocó Arc

Stewart D. Redwood

14.1 Introduction

This chapter describes the geology of the Panama-Chocó Arc (Fig. 14.1). Most publications treat Panama and the Chocó separately based on the border between Panama and Colombia. For the purposes of this chapter, the description extends from Panama City to Buenaventura, a distance of 630 km. Eastern Panama, also known as the Chocó Block, forms the eastern part of the Isthmus of Panama. The Chocó region of Colombia is part of the Western Cordillera of the Northern Andes. Although the Panama Canal Basin is not a terrane boundary, there is a major change in the geology to the west, which is defined as the Chorotega Block. It was also an important paleo-faunal limit before development of the isthmian land bridge.

The Panama-Chocó Arc is formed of oceanic crust, volcano-plutonic island arcs and sedimentary basins of Cretaceous to Quaternary age. The Chocó arc forms part of the Western Cordillera of Colombia and was accreted onto the continental margin which forms the Central Cordillera and is composed of metamorphic, sedimentary and plutonic rocks of Proterozoic to Mesozoic age, including Jurassic to Cretaceous arc-related magmatism. Post-accretion, subduction-related magmatic arcs of Miocene to Recent age occur in both the Western and Central Cordilleras.

14.1.1 Geography

The Panama-Chocó Arc has mountain ranges or massifs (*serranias*) on the northern to eastern side (from west to east these are Chagres, Mamoní, San Blas and Darien in Panama, Acandí and Mandé in Colombia) and the southern to western side

S. D. Redwood (✉)
Consulting Economic Geologist, Panama City, Panama

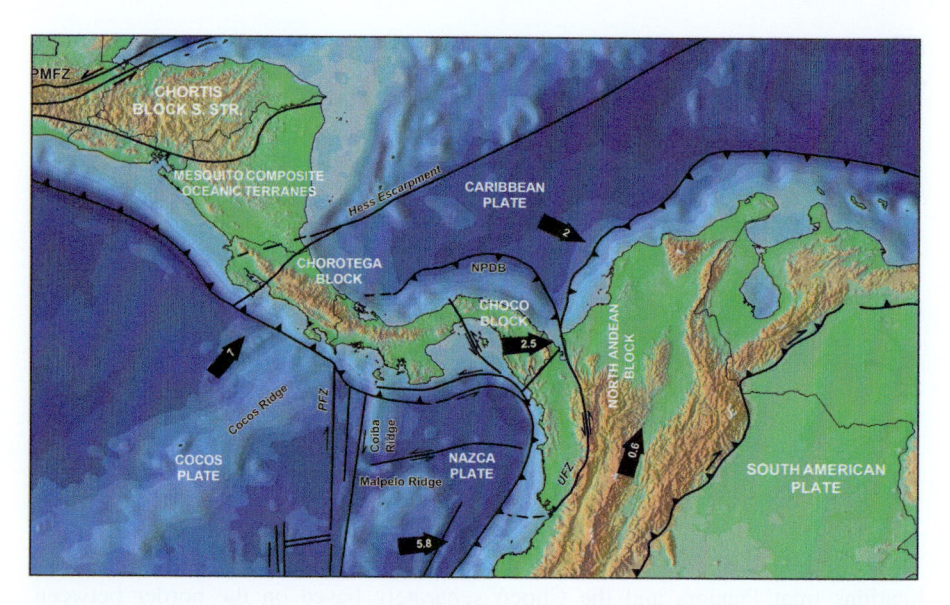

Fig. 14.1 Plate tectonic map of the Panama-Chocó Arc. Black arrows show present-day GPS velocities (cm/yr) relative to stable South America (Trenkamp et al. 2002). PFZ, Panama Farcture Zone; NPDB, North Panama Deformed Belt; UFZ, Uramita Fault Zone

(from west to east these are Majé, Sapo, Bagre, Jungurudo, Jurado, Pirre and Altos de Quia in Panama and Baudó in Colombia) which are formed by Cretaceous oceanic crust overlain by a Paleogene volcano-plutonic arc on the north and east sides and a Neogene volcano-plutonic arc on the south and west sides (Fig. 14.2). These mountain ranges are separated by broad central, longitudinal valleys with major rivers (the Bayano or Chepo Rivers, Chucunaque and Tuira Rivers in Panama, and Atrato and San Juan Rivers in Colombia) with deep sedimentary basins. The Tuira and Atrato Basins are now separated by the Altos de Quia mountain range that crosses the isthmus with a northeast trend and forms the frontier between the two countries. The Cenozoic Gulf of Panama Basin lies south of eastern Panama, with Neogene arc volcanism on a structural high in the Pearl Islands. The arc is curved and is a recumbent S shape from NE to E-W to NW trending in Panama, and N-S in Colombia, in an oroclinal bend formed by collision with north-western South America.

The region is covered by tropical rain forest and has poor access and is poorly known geologically. In addition, the public security situation in the Chocó Province for the past half century hindered access.

There is a single highway from Panama City along the central valley to Yaviza in eastern Panama, but there is no road beyond through the Darien Gap to Colombia. Access to most of eastern Panama is by boat on the coasts and rivers. Indeed, La Palma, the capital of Darien Province, is only accessible by boat. The only other roads are along the Caribbean coast from Colón to Portobello, Nombre de Dios and

14 The Geology of the Panama-Chocó Arc

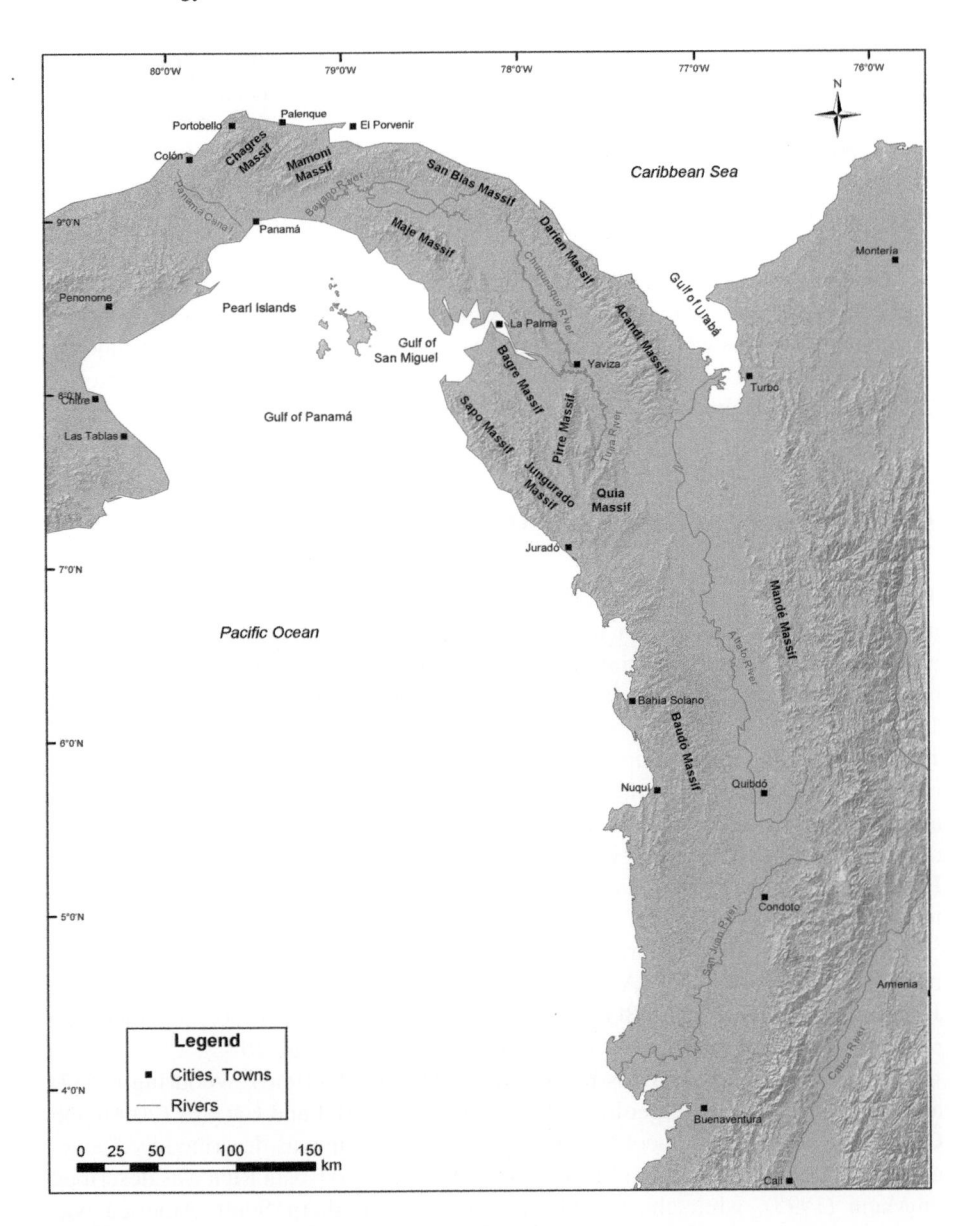

Fig. 14.2 Digital terrain model of the Panama-Chocó Arc

Palenque, known as Costa Arriba, and a new road across the San Blas Mountains to Cartí on the Caribbean coast.

In Colombia the highway from Medellin to Turbo on the Gulf of Uraba skirts the north-eastern side of the Chocó region. There is an isolated road in the coast of the Acandí range, north of the Atrato River. In the southern part of the Chocó, there are roads across the Western Cordillera from Medellin and Pereira to Quibdo, the departmental capital, and to Istmina, Novita and the port city of Buenaventura. Again, most transport in the Chocó is by boat on the rivers and coasts.

The area is sparsely populated. Chocó has a population of about 500,000 (2005), while eastern Panama is less than half of that (excluding the cities of Panama and Colón). The main inhabitants of eastern Panama are Guna (Kuna) Indians in the large indigenous territories (*comarcas*) of Guna Yala (San Blas), Madugandí and Wargandí, and Emberá and Wounaan Indians in the Emberá-Wounaan comarca in Darien. The Costa Arriba region of Colón Province (Portobello-Palenque) is inhabited mainly by Panamanians of Afro-Antillean descent, together with mestizo Panamanian farmers who also occupy the central valley. In the Chocó, the majority of the population is Afro-Colombian, with numerous small indigenous territories (*resguardos*) of Emberá and Wounaan Indians.

14.1.2 History of Geological Mapping and Research

The earliest geological studies of the Panama-Chocó region were carried out in the mid- to late nineteenth century to explore possible routes for a trans-isthmian canal in eastern Panama and the Atrato (Wyse 1877) and again in the mid-twentieth century to explore for sea-level canal routes (Governor of the Panama Canal 1947a, b, 1949; Schmidt et al. 1947; Interoceanic Canal Study Commission 1968; MacDonald 1969). The sedimentary basins were explored for oil by geological mapping along rivers, seismic surveys and limited drilling in the twentieth century and are generally better studied than the oceanic and arc rocks of the mountain belts. Mineral reconnaissance programmes were carried out in the mountain belts of eastern Panama by the United Nations Development Programme (UNDP) in the late 1960s to early 1970s, which added greatly to the geological knowledge of these areas. Similar programmes were carried out in the Mandé-Acandí Mountains of Colombia by UNDP and INGEOMINAS in the 1970s, with follow-up programmes by the Japanese JICA and German BGR missions in the late 1970s to 1990s, together with regional geological mapping by INGEOMINAS (Sillitoe et al. 1982).

The geology of Panama has been described by Terry (1956), Woodring (1957) and Weyl (1980), and the geology of Colombia by Cediel and Cáceres (2000) and Cediel et al. (2003). The geology of the Caribbean Basin was described by Dengo and Case (1990). The Pacific Plate margin of Panama and Costa Rica was described in Mann (1995), while the Caribbean margin of northern South America was described by Avé Lallemant and Sisson (2005). The most recent tectonic research on the Caribbean Plate and Northern Andes is James et al. (2009), with key papers by Pindell and Kennan (2009) on the Caribbean, Gulf of Mexico and northern South America and Kennan and Pindell (2009) about the Northern Andes.

Arc magmatism in Panama from the Late Cretaceous to present was described by Lissinna (2005), Wegner et al. (2011), Montes et al. (2012a) and Farris et al. (2011, 2017).

14.1.2.1 Plate Tectonic Setting

The Panama-Chocó Arc is located at the junction of the Cocos, Nazca, Caribbean and South American Plates (Fig. 14.1). The Panama-Chocó Arc formed as a volcanic arc on the southwestern, trailing edge of the Caribbean Plate from the late Campanian, related to the subduction of the Farallon Plate, and later the Cocos Plate. The southeastern or Chocó part of the arc collided with the South American Plate and was accreted to the Northern Andes to form part of the physiographic Western Cordillera of Colombia (Suter et al. 2008), while the western part forms the Isthmus of Panama, part of the Central America land bridge. Collision was oblique and propagated to the northwest and is ongoing (Barat 2013; Barat et al. 2014).

The western part of the arc is called the Panama Block or Microplate which includes Panama and Costa Rica (Kellogg and Vega 1995; Trenkamp et al. 2002), although there is considerable internal deformation and rotation along strike-slip faults and it is questioned whether it can really be considered as a stable rigid microplate (Rockwell et al. 2010a; Barat et al. 2014). The Panama Microplate formed after collision with South America and was detached from the Caribbean Plate to the north by the North Panama Deformed Belt, from the Nazca Plate to the south by the South Panama Deformed Belt, and from the Cocos Plate to the southwest by the Middle America subduction zone. The Cocos Plate is being subducted beneath Central America and the western part of the Panama Microplate at a rate of about 72 mm/year along the Middle America Trench, with its associated seismicity and volcanic arc (Trenkamp et al. 2002). The aseismic Cocos oceanic ridge is colliding with the subduction zone in western Panama – eastern Costa Rica. The boundary between the Cocos and Nazca Plates is the Panama Fracture Zone, which includes the Coiba and Balboa Fracture Zones, which are north-trending, seismically active, right-lateral oceanic transform faults (Westbrook et al. 1995).

The southern boundary of the Panama microplate changes from oblique ENE-directed subduction in the west (50 mm/y) to a sinistral strike-slip fault east of the Azuero Peninsula (80°W), marked by a lack of seismicity, due to a change in the strike of the margin (Kellogg and Vega 1995; Mann 1995; Westbrook et al. 1995). The Nazca Plate is bounded to the north by the South Panama Deformed Belt, an accretionary prism which has filled the trench in the eastward continuation of the Middle America Trench (Defant et al. 1991). Westbrook et al. (1995) estimated that the present tectonic regime started at about 3.5 Ma based on the 140 km displacement along the Southern Panama Fault Zone, while Lonsdale and Klitgord (1978) suggested that strike-slip motion started in the Late Miocene between 12 and 8 Ma and resulted in the cessation of subduction along the southern margin of eastern Panama. As a consequence of the lack of subduction, there is no active volcanic arc in eastern Panama, where the youngest volcanism is 15 Ma (Montes et al. 2011). This results in the markedly different topography of eastern and western Panama: eastern Panama has a longitudinal central valley in a synclinal sedimentary basin bounded by older coastal mountain ranges on the northern and southern sides, instead of a young volcanic arc forming a Central Cordillera as in western Panama.

The northern boundary of the Panama microplate is defined by the underthrusting at 10 mm/year (Reed and Silver 1995) or subduction of the Caribbean Plate along the north side of North Panama Deformed Belt, a wide accretionary wedge, which has been actively forming since the Middle Miocene. Camacho et al. (2010) showed the existence of subduction of the Caribbean Plate with a southward-dipping Wadati-Benioff Zone to a depth of 80 km beneath the isthmus.

The North Panama Deformed Belt passes through the Gulf of Uraba and joins the left-lateral Uramita Fault Zone which defines the eastern boundary of the Panama-Chocó Arc with the Northern Andean Block of the South American Plate (Mann and Kolarsky 1995; Barat et al. 2014).

The southern limit of the Panama-Chocó Arc is the NE-trending, right-lateral Garrapatas fault system which forms the southern margin of the San Juan Basin (Cediel et al. 2003).

The western limit of the Panama-Chocó Arc is placed at the Panama Canal Zone by most authors and is taken here for convenience in this paper. The postulated, left-lateral Panama Canal fault zone separates the Chocó Block of eastern Panama from the Chorotega Block to the west. The oceanic basement and Late Cretaceous to Paleogene arc continue to the west of the canal in the Azuero Peninsula, which is interpreted to be the left-lateral offset continuation of the San Blas-Mandé arc (Buchs et al. 2011; Corral et al. 2016).

The geology of eastern Panama and the Chocó is shown in Figs. 14.3, 14.4, and 14.5.

14.2 Terranes

A number of different terrane names have been given to the whole or parts of the Panama-Chocó Arc, resulting in some confusion. McCourt et al. (1984) and Aspden et al. (1987) first recognised the accretionary nature of the Colombian Andes. Restrepo and Toussaint (1988) simply named the arc the Panama-Baudó-Mandé Terrain. The *Preliminary Geological Terrane Map* of Colombia by INGEOMINAS (Etayo-Serna et al. 1985) defined two terranes, Cañas Gordas and Baudó, separated by the Atrato-San Juan Basin, and this model was followed by subsequent maps with the terranes extended northwest to the Panama Canal Fault Zone (Cediel and Cáceres 2000; Cediel et al. 2003; Bedoya et al. 2009; Cediel and Restrepo 2011). The Cañasgordas Terrane is bounded by the Garrapatas-Dabeiba suture of post-Maastrichtian age on the eastern side and the San Juan-Sebastián suture of Eocene age on the west. The Baudó Terrane is shown as being bordered by the Baudó collisional event of Pliocene age (4–6 Ma) on the eastern side.

These maps show the Mandé-Atrato magmatic arc to have formed on the western edge of the Cañasgordas Terrane. However, as described below, the Cañasgordas Terrane is formed of older oceanic and arc rocks and was accreted at an earlier time. The correct eastern limit of the Panama-Chocó Arc is the western side of the Cañasgordas Terrane at the contact with the eastern side of the Mandé Batholith and its contemporaneous volcanic rocks.

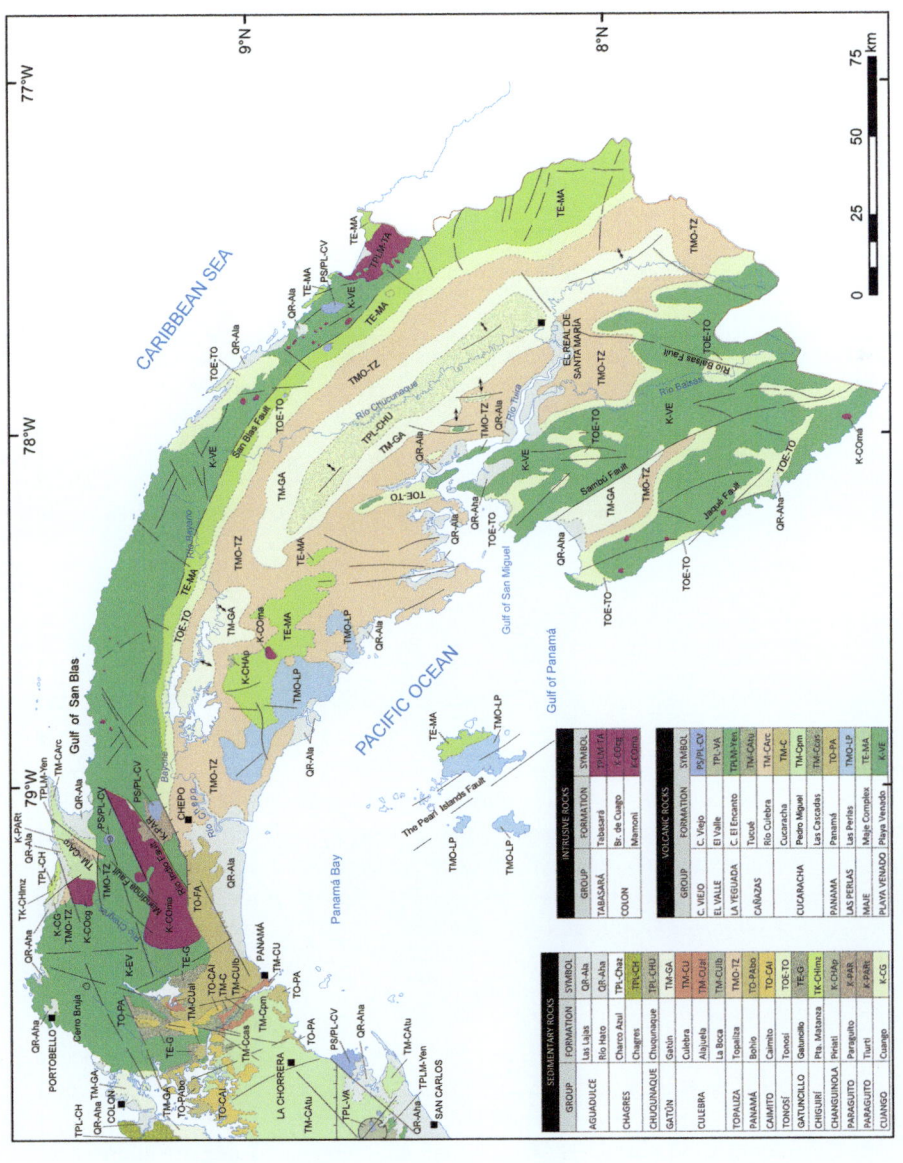

Fig. 14.3 Geological map of eastern Panama. (Based on Mapa Geológico República de Panamá, Dirección General de Minerales 1996 1:500,000 scale)

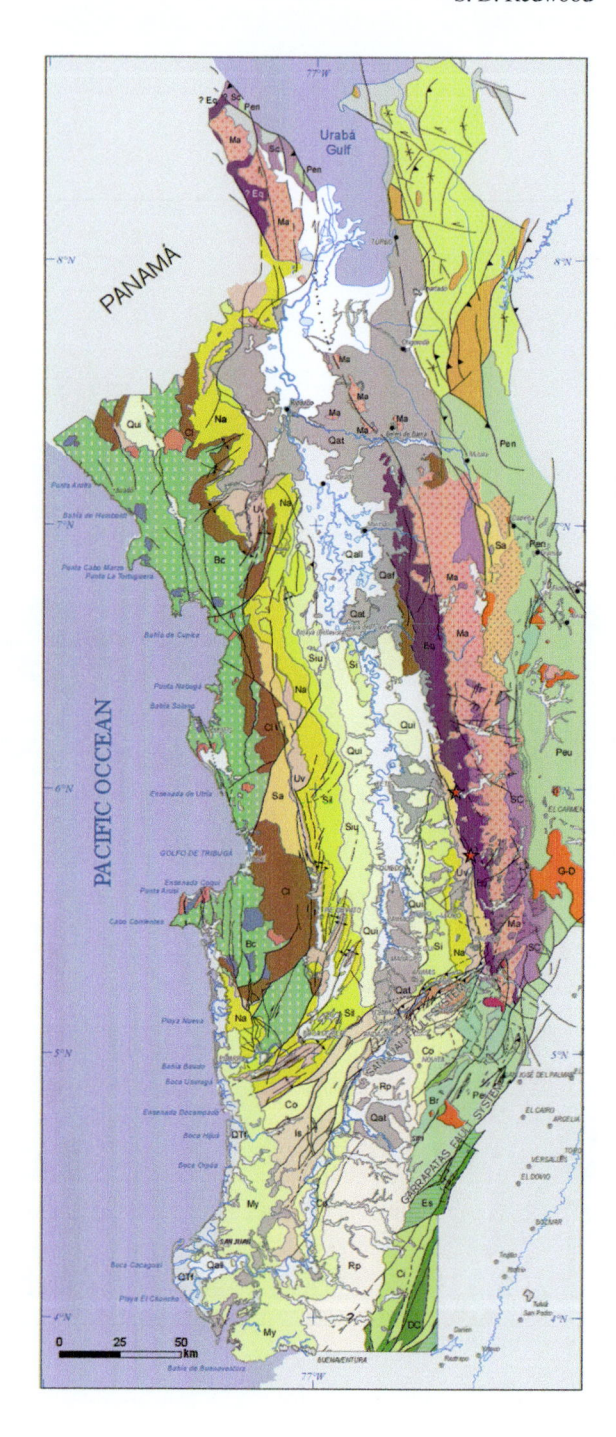

Fig. 14.4 Geological map of the Chocó showing the Atrato and San Juan Basins (after Cediel, Chapter 1 this volume)

14 The Geology of the Panama-Chocó Arc

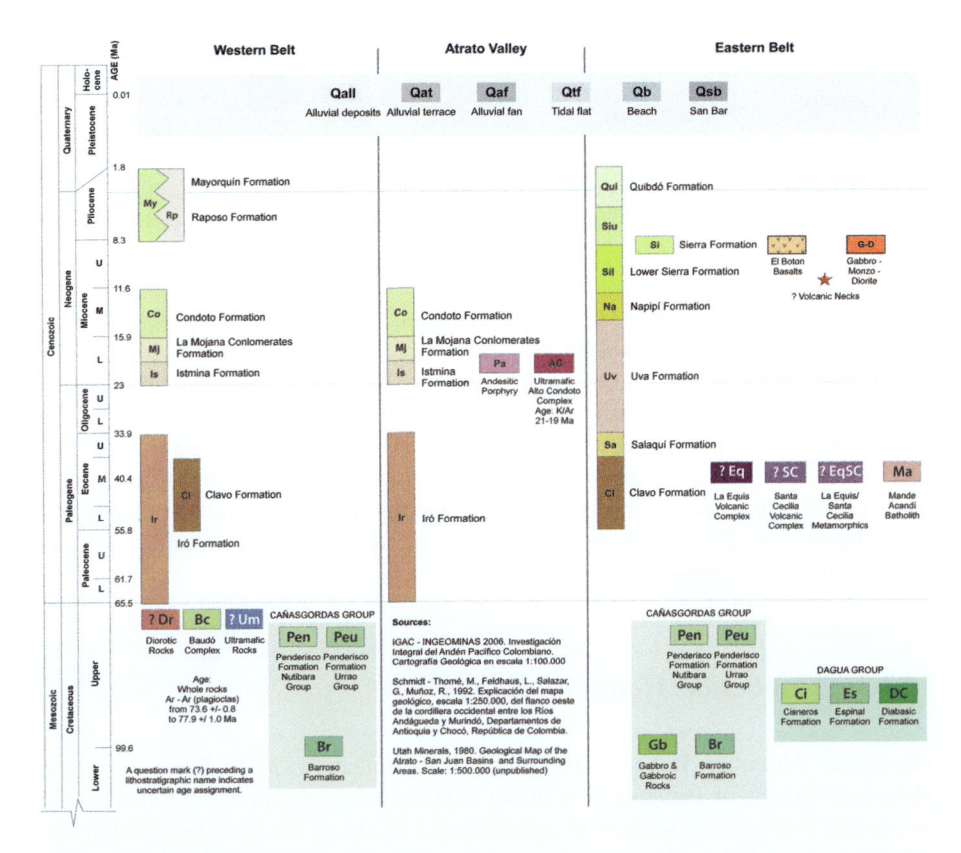

Fig. 14.5 Stratigraphy of the Chocó including the Atrato and San Juan Basins (after Cediel, Chapter 1 this volume)

The El Paso terrane, or El Paso-Baudó terrane, which is not shown on any map, was described as forming the basement to the Atrato Basin and to outcrop in the Istmina-Condoto High (Bedoya et al. 2009; Cediel and Restrepo 2011). It has a distinctive gravimetric-magnetic signature from the basement of the San Juan Basin which is interpreted to be the Cañasgordas Terrane.

Duque-Caro (1990) used the names Dabeiba Arch (San Blas-Mandé) and Baudó Arch, separated by the Chuqunaque-Atrato Basin, bounded on the south by the Isthmina fault zone. Kerr et al. (1997) called the Santa Cecilia-La Equis Complex volcanic rocks, east of the Mandé Batholith, the Dabeiba Volcanic Arc.

Kennan and Pindell (2009) included the Panama-Chocó Arc as part of the "Greater Panama" terrane 5 (T5) made up of the Baudó-Chocó Block and the Dabeiba Arc (Mandé-Atrato), bounded on the eastern side by the Atrato or Uramita suture, part of the Pallatanga fault zone. They correlated the island arc terranes and ocean crust basement of the Caribbean large igneous province (CLIP) of the Panama-Chocó Arc with terranes in southwestern Colombia and western Ecuador as part of the Greater Panama Terrane.

14.3 Cretaceous Oceanic Crust Basement

14.3.1 Eastern Panama

The stratigraphy of eastern Panama is shown in Fig. 14.6. The basement San Blas Complex of Coates et al. (2004) outcrops in the mountains of San Blas, Majé, Bagre, Sapo, Jaque, the Gulf of San Miguel and the Portobello Peninsula, but is poorly understood. It comprises basement of oceanic crust including pillow and massive basaltic lavas with red radiolarites and cherts (Case et al. 1971; Case 1974). The Sapo basalts were dated at 88 Ma (Turonian, Lissinna 2005), and the oceanic crust is dated at Coniacian to Lower Campanian. The oceanic crust is interpreted to have formed at the CLIP oceanic plateau (Kerr et al. 1997).

The initiation of the volcanic arc is marked at Portobello by the Late Campanian Ocú Formation of volcanic breccia with andesite basalt clasts overlain by limestone (with Late Campanian foraminifera) and siliceous tuffs cut by basaltic andesite dikes (Barat et al. 2014). This is overlain by a poorly known Paleocene to Eocene volcanic arc which forms the San Blas Mountains.

14.3.2 The Baudó Mountain Range

The Baudó Mountain Range is the southern continuation of the Sapo and Pirre Mountain Ranges of Panama. It is about 300 km long and up to 65 km wide. It trends northwest-southeast from the Panamanian border to the oroclinal bend in the coastline at 6°30′N, and then trends south to the mouth of the Baudó River at 4°56′N. Several fault-bounded mountain ranges in the northern part trend north-south to northeast-southwest transverse to the trend of the coast and are interpreted to be reverse faults in the continuation of the East Panama Deformed Belt.

The geology of the Baudó Mountain Range is very poorly known. A profile was mapped in 1949 along the River Atrato and the River Truandó across the Sierra de Los Saltos to Curiche in Humboldt Bay on the Pacific for investigations of the Atrato-Truandó isthmian canal route and incorporated oil company maps (Governor of the Panama Canal 1949). Mapping showed a basement complex of basalts, andesites, peridotites, tuffs, dacite and granite. This is overlain by the Cretaceous Sautata Formation of sandstones, shales and tuffs; the Eocene Barrial Formation of sandstone, limestone, shale and conglomerate; the Oligocene Truandó Formation of siltstone, tuff and sandstone; the late Oligocene Saltos Basalt which forms an extensive, thick laccolith in the high part of the mountain range; the Miocene Rio Salado Formation of mudstone, sandstone and shale; the Pliocene Barranca Colorada Formation of clays, gravels and sandstone; and by Quaternary alluvium.

More recent studies of coastal sections indicate that the Baudó Mountain Range is dominated by pillowed and massive basalts with basaltic breccias, dolerites and gabbros, and intercalations of fine-grained sandstone, chert and limestone. The basalts have tholeiitic geochemistry. The basalts are unconformably overlain by

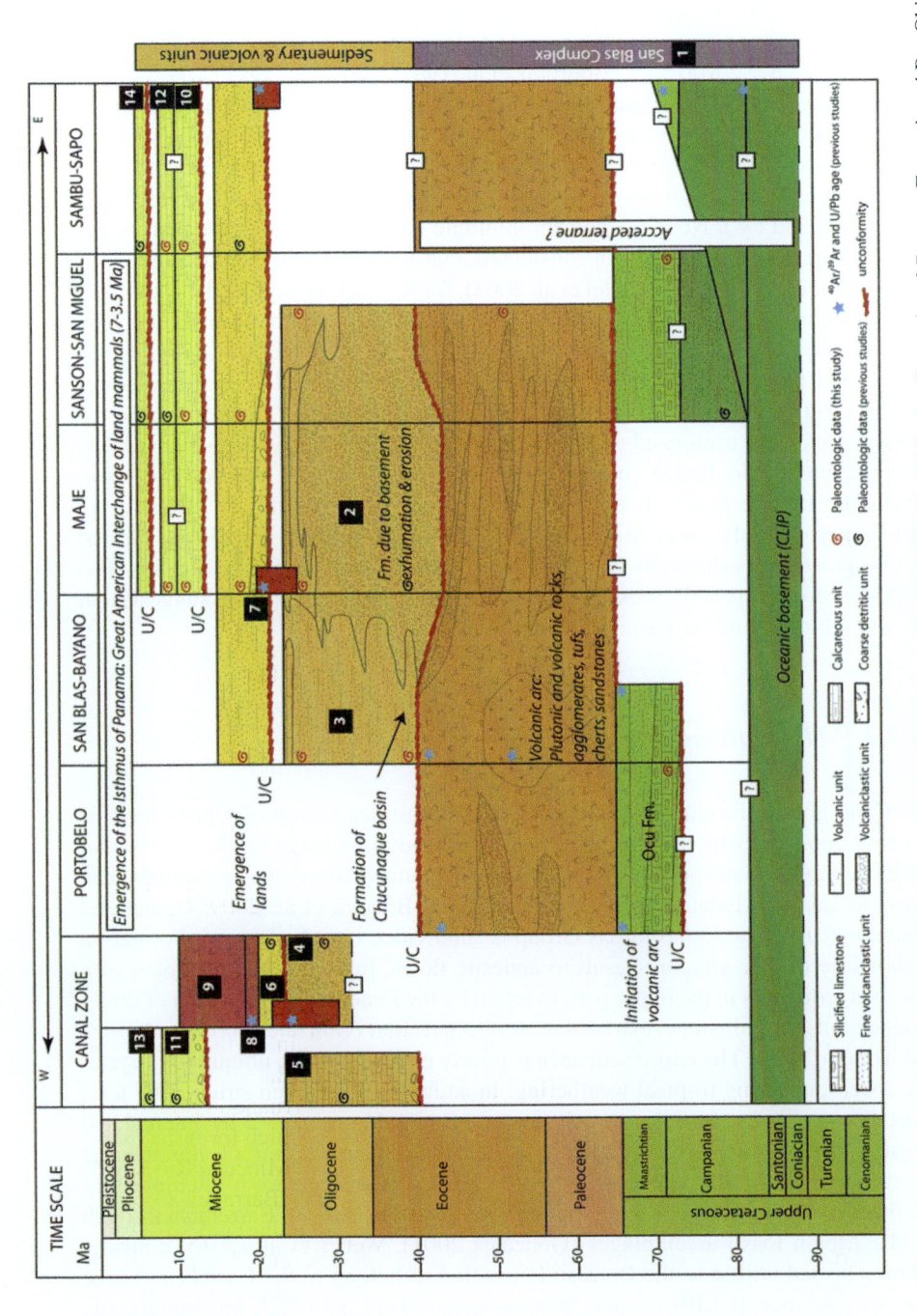

Fig. 14.6 Stratigraphy of the Chocó Block eastern Panama. (Barat et al. 2014). 1 San Blas Complex; 2 Darien Formation; 3 Porcona Formation; 4 Bas Obispo Formation; 5 Gatuncillo Formation; 6 Las Cascadas Formation, La Culebra Formation and Cucaracha Formation; 7 Clarita Formation; 8 Pedro Miguel Formation; 9 Late Basalts Formation; 10 Tapaliza Formation; 11 Gatun Formation; 12 Tuira Formation; 13 Chagres Formation; 14 Chuqunaque Formation

basalts intercalated with Eocene limestone (Duque-Caro 1990; Kerr et al. 1997; Gonzalez et al. 2014). Limited radiometric dating gives Late Cretaceous (76.2 ± 1.1 to 71.8 ± 1.4 Ma, Campanian-Maastrichtian), mid Eocene (41 ± 3 Ma) and late Oligocene (25.8 ± 2 Ma) ages (Bourgois et al. 1982; Kerr et al. 1997).

The origin of the Majé-Baudó Mountains or terrane is poorly understood, and it is one of the key areas of investigation required to improve tectonic models for the Panama-Chocó Arc. It is formed of oceanic tholeiitic basalts (Goosens et al. 1977) of Cretaceous age overlain by sediments of Late Cretaceous to Cenozoic age (Kerr et al. 1997) and by a Neogene volcano-plutonic arc. The oceanic basement of the Baudó Mountain Range has been considered a far-travelled exotic or allochthonous terrane (Duque-Caro 1990; Cediel et al. 2003), forming a fragment of the Caribbean Plate, which was accreted to the western margin of the Atrato Basin in the Late Miocene (Cediel et al. 2003; Cediel and Restrepo, 2011), coincident with the closure of the Chuqunaque Basin at 5.6 Ma (Coates et al. 2004). More recent interpretations consider the range as the western segment of the El Paso Terrane, interpreted as a segment of the trailing edge of the Caribbean Plate. Cediel and Restrepo (2011) interpret the Baudó Range to represent a positive flexure in the oceanic plate, uplifted during the Baudó Event between ca. 8 and 4 Ma. Similarly, Barat et al. (2014) interpreted the range to be an uplifted part of the fore-arc and to have formed the basin margin, as shown by Coates et al. (2004). The convergence of the San Blas and Majé Mountains at the west end of the Bayano-Chuqunaque Basin suggests that this was a hinge zone for basin extension in the fore-arc.

14.3.3 The Cañasgordas Terrane

The Cañasgordas Terrane is formed of the Cañasgordas Group of Upper Jurassic to Cretaceous age which was accreted to the Romeral melange on the continental margin along the Garrapatas-Dabeiba or Uramita Suture, sometime between the late Maastrichtian and Paleocene (Cediel et al. 2003; Bedoya et al. 2009; Cediel and Restrepo 2011). The Cañasgordas Group is subdivided into the Barroso Formation of tholeiitic to calc-alkaline basalt to andesite flows, tuffs and agglomerates with siliceous sediments in the upper part, overlain by the Penderisco Formation of thinly bedded mudstone, siltstone, marl, lithic arenite and chert (González 2001; Rodriguez and Arango 2013). The entire sequence is poorly exposed due to ubiquitous vegetative cover and strong tropical weathering. In addition, it has been structurally telescoped into a series of tight, mostly east-verging folds truncated by intercalated thrust faults. These factors preclude a detailed understanding of the internal structural-stratigraphic relationships within the Cañasgordas Group. Sedimentary interbeds within the Barroso and Penderisco Formations contain Barremian through middle Albian fossil assemblages (González 2001). Weber et al. (2015) consider gabbros hosted within to the Barroso Formation to belong to the Caribbean Large Igneous Province (CLIP) oceanic plateau assemblage, although biostratigraphic data suggests that the Barroso Formation is at least in part pre-CLIP.

New mapping and research on the Barroso Formation has divided it into two units – the San José de Urama Diabases and the Barroso Formation *sensu stricto* (Rodriguez and Arango 2013). The San José de Urama Diabases are diabase and pillow basalts of low-K tholeiite series and T-MORB-type mantle source. It was dated by whole rock Ar-Ar at 155.1 ± 11.2 Ma (Upper Jurassic) and is interpreted to be an accretionary prism of ocean crust.

The Barroso Formation *sensu strictu* is a volcano-sedimentary unit of basalts and andesites with agglomerates, tuffs and marine sedimentary rocks. The composition is tholeiitic to medium-K calc-alkaline, and it formed in a subduction-related arc, with plutonic rocks in the Sabanalarga Batholith and Buriticá Tonalite. The volcanic-plutonic arc was formed between 115 and 88 Ma (Aptian-Turonian) on basement of Upper Jurassic to Lower Cretaceous ocean crust of the San José de Urama Diabases (Rodriguez and Arango 2013).

The Cañasgordas Terrane is thus a diachronous unit, formed of older, latest Jurassic to early Cretaceous ocean crust and younger sedimentary, volcanic and plutonic arc-related rocks of mid Cretaceous age, possibly belonging to the Caribbean Plate.

14.3.4 The Istmina-Condoto High

The Istmina-Condoto High is comprised of ultramafic rock outcrops in the upper part of the San Juan Basin. It has been studied by the BGR-INGEOMINAS as the possible source of the alluvial platinum deposits of the Chocó (Muñoz et al. 1990; Tistl 1994; Tistl et al. 1994). It is a NE-SW trending composite structure about 130 km long and up to 30 km wide, which plunges to the SW beneath a thick cover of Cenozoic sediments. It has been exhumed in a transpressive shear zone with right-lateral displacement as shown by thrusts, high-angle reverse faults, normal faults and associated folds (Cediel and Restrepo 2011). It is interpreted to be oceanic crust and upper mantle of Late Cretaceous age and was exhumed in the Cenozoic. The most significant igneous rocks in the High are the El Paso Complex, the Viravira Complex, and the Alto Condoto Ultramafic Complex; the former is described here, while the other two are younger and are described below.

The El Paso Complex is the oldest lithological unit in the Chocó Arc. It is a strongly fractured sequence of high Mg tholeiitic basalts and ophitic diabases with interbeds of chert, claystone and mudstone. Cr/Y and Ti/Zr ratios suggest an oceanic (MORB) origin. No radiometric dates are available, but the regional geology indicates that the El Paso Complex is likely to be of Late Cretaceous to Paleocene age.

14.4 Paleogene Arcs

Volcano-plutonic arcs of Paleogene age form the Chagres, Mamoní, San Blas, Acandí and Mandé mountain ranges along the northern and eastern side of the Panama-Chocó Arc.

14.4.1 The Chagres and Mamoní Mountains

The Chagres and Mamoní Mountains occur in the central part of the Isthmus of Panama, east of the canal, and are the continuation of the San Blas Mountains. The area has very poor access and the geology has not been mapped in detail. The geology of the Lake Madden Dam (Lago Alajuela) on the Chagres River was described by Reeves and Ross (1930). The geology of the upper Chagres River was described by Wörner et al. (2005, 2006, 2009) and Wegner et al. (2011), who call the area the Chagres Igneous Complex or the Chagres-Bayano Arc. The Mamoní Mountains to the east were described by Montes et al. (2011, 2012b).

The Chagres Mountains are part of an undeformed, mafic ocean floor and island arc sequence of the Basement Complex of eastern Panama (Montes et al. 2011, 2012b). This is divided into the Volcaniclastic Basement Complex of basalt, diorite and basaltic andesite interbedded with pelagic sediments (chert and siliceous limestone), and intruded by the Plutonic Basement of plutonic bodies ranging from granodiorite to gabbro. Magmatic activity, east of the Panama Canal, was from about 70 to 38 Ma (Upper Cretaceous to Eocene), with a peak at 50 Ma (Lower Eocene), and stopped at 15 Ma (Montes et al. 2011, 2012b).

Further east, in the Mamoní Mountains, the volcaniclastic basement complex has been subdivided into three main units about 4000 m thick in an E-W trending, north-verging, anticline-syncline pair (Montes et al. 2011, 2012b). The tectonic environment shows development of a proto-arc and volcanic arc on oceanic plateau of the Caribbean Large Igneous Province (CLIP). The units are:

- Upper unit: Basaltic andesite lava flows and tuffs of a volcanic arc
- Middle unit: Massive basalt, pillow basalt, diorite and basaltic dikes, interlayered with sedimentary rocks interpreted as a proto-arc
- Lower unit: Massive and pillow basalt interlayered with chert and limestone interpreted as an oceanic plateau (CLIP)

Wörner et al. (2005, 2006, 2009) described the basement of the upper Chagres basin as highly deformed mafic basalts, basaltic andesites, gabbros, diorites with associated dyke swarms, as well as granodiorites, tonalities and granites. These are interpreted to represent a series of submarine volcanic centres and associated sheet flow/dyke complexes, magma chamber intrusions and marginal volcanoclastic aprons. Rare oxidized scoria is evidence for some subaerial eruption. Trace elements in Chagres igneous rocks show that they were derived from a Cretaceous to Paleocene (69–66 Ma) subduction system and are interpreted to mark the timing for the onset of arc magmatism at the Caribbean Plate margin.

Wegner et al. (2011) published $^{40}Ar/^{39}Ar$ dates for amphiboles from the Chagres-Bayano arc of 66–42 Ma. These are divided into earlier (66.4–61.5 Ma, Early Paleocene) and later (49.4–41.6 Ma, Eocene) magmatic suites. A quartz diorite pluton at Cerro Azul in the Mamoní Mountains yielded K-Ar dates of 61.58 ± 0.70 Ma on hornblende and 51.11 Ma ± 0.58 Ma on plagioclase (Kesler et al. 1977).

14.4.2 The San Blas and Acandí Mountain Ranges

The Acandí Mountains have been explored and studied by INGEOMINAS and others since the 1980s for porphyry copper deposits. However, the contiguous San Blas Mountains are very poorly known, with the only study by UNDP for copper exploration in the early 1970s and some recent research by STRI (Strong et al. 2011; Montes et al. 2015).

The Acandí Mountains, situated west of the Gulf of Uraba, are separated from the Mandé Mountains by the Darien Shear Zone, a NE-trending reverse fault with a topographic low on the east side occupied by the basin of the Atrato River which flows into the gulf. The Acandí Batholith is interpreted to be the continuation of the Mandé Batholith. The Acandí Mountains are continuous with the San Blas Mountains of the Caribbean coast of eastern Panama, which extend to the Chagres-Mamoní Mountains on the east side of the canal basin.

The Acandí and San Blas Mountains consist of volcanic rocks of basalt to rhyolite composition, the most common being hornblende basalts, with large plutons and batholiths of gabbro to granite composition. There are extensive pillow basalts between Puerto Obaldia and the border, associated with deep-sea sediments, dykes and gabbro which suggests oceanic crust (Strong et al. 2011; Barat et al. 2014). Limestones with caves have recently been recognised near Aligandi and are undated but believed to be of Eocene age (Strong et al. 2011; Cronin 2017). Different geological units are separated by major faults with dominant trends NW-SE with left-lateral movement and oblique reverse faults with a left-lateral SW-NE component in the extreme south of Guna Yala (Strong et al. 2011). These are the result of compression due to collision of the isthmus with South America.

The only published date for Acandí is by K-Ar on sericite from a tonalite which gave 48.1 ± 1.0 Ma (Sillitoe et al. 1982).

The Rio Pito porphyry in the eastern San Blas range gave K-Ar ages of 48.45 ± 0.55 Ma on hornblende and 49.23 ± 0.57 Ma for plagioclase from a quartz porphyry (Kesler et al. 1977), and K-Ar ages on hornblende from a tonalite of 43.4 ± 1.0 Ma and 41.9 ± 1.0 Ma (Sillitoe et al. 1982).

Montes et al. (2015) published six U-Pb magmatic zircon ages from the San Blas Mountains of plutons of gabbro to granite composition with ages between 59.0 and 49.7 Ma (late Paleocene to early Eocene). From west to east these are two diorites (58.3 ± 1.0 Ma and 59.0 ± 1.9 Ma), a medium-grained gabbro (58.6 ± 1.6 Ma), two pyroxene-hornblende gabbros (49.7 ± 1.6 Ma and 49.5 ± 1.1 Ma) and a mylonitized granodiorite with relict biotite and feldspars (49.5 ± 0.9 Ma). These ages are older than the two ages for the Mandé Batholith of 43.8 to 42.5 Ma (Montes et al. 2015) and suggest a younging of the batholith from NW to SE.

14.4.3 The Mandé Mountain Range

There is reasonable literature on the Mandé and Acandí Mountains by INGEOMINAS and others since the 1980s as a result of porphyry copper exploration (INGEOMINAS n.d.; Schmidt-Thomé et al. 1992).

The Mandé Batholith intrudes Upper Cretaceous to Paleocene island arc basement of submarine basalts, andesite and dacite lavas and minor sedimentary rocks called the Santa Cecilia-La Equis Complex, which outcrops on either side of the batholith. The composition of the volcanic rocks varies from tholeiitic to boninitic upwards to calc-alkaline. These volcanic rocks are co-magmatic with the Mandé Batholith.

The Mandé Batholith is a multiphase intrusion of Paleocene to Eocene age. Late-stage porphyry stocks host porphyry copper mineralization. Four main intrusive phases are described: tonalite-granodiorite, monzonite-monzodiorite, gabbro-diorite and porphyry tonalite to quartz diorite (INGEOMINAS n.d.). The latter are related to porphyry copper mineralization. The petrochemistry is calk-alkaline.

The intrusive rocks have been dated as Eocene age. The porphyries give K-Ar ages on hornblende of 54.7 ± 1.3 Ma (Murindo) and hydrothermal sericite of 42.7 ± 0.9 Ma (Pantanos-Pegadorcito) (Sillitoe et al. 1982; INGEOMINAS n.d.). More recently, magmatic zircon ages by U-Pb from the Murindo batholith gave middle Eocene ages of 43.8 ± 0.8 Ma for a hornblende-pyroxene diorite and 42.5 ± 1.3 Ma for a granodiorite (Montes et al. 2015). A tonalite at the Pantanos-Pegadorcito porphyry gave a U-Pb zircon age of ca. 45 Ma, with inherited zircon ages of ca. 59–67 Ma (Leal-Mejía 2011). Basalts yielded an Ar-Ar age of 43.1 ± 0.4 Ma (Kerr et al. 1997).

The Riosucio volcano-sedimentary complex overlies the Santa Cecilia-La Equis Complex east of the Mandé Batholith for up to 25 km to Dabeiba and in minor outcrops on the north and west sides. This comprises a deep to shallow water sedimentary sequence with abundant pyroclastic rocks and some lavas of Paleocene to Pliocene age.

The Mandé Batholith and Santa Cecilia-La Equis Complex are bounded to the east of Dabeiba by ocean floor basalts of the Barroso Formation which are overlain by turbidites, chert and limestone of the Penderisco Formation, both of the Cañasgordas Group.

14.5 Cenozoic Basins

14.5.1 *The Panama Canal Basin*

The Panama Canal Basin has been extensively studied for more than a century for engineering geology for canal construction (Bertrand and Zürcher 1899; Howe 1907; MacDonald 1915; Reeves and Ross 1930; Governor of the Panama Canal 1947b; Jones 1950) and palaeontology (Woodring 1957, 1982), with maps by Stewart et al. (1980) and Woodring (1982). Detailed recent studies of stratigraphy (Kirby and MacFadden 2005; Kirby et al. 2008; Montes et al. 2011, 2012b), palaeontology (MacFadden 2006), volcanic rocks (Rooney et al. 2011; Farris et al. 2011, 2017), neotectonics and palaeoseismicity (Rockwell et al. 2010a, b) were carried out in the past decade during the canal expansion.

Sedimentation in the canal basin started with Upper Eocene to Oligocene coarse detrital sediments with limestones and volcaniclastic rocks of the Gatuncillo Formation, followed by Oligocene subaerial volcanic and volcaniclastic rocks of the Bas Obispo, Bohio and Las Cascadas Formations. These are overlain by Early Miocene shallow marine to continental sediments of the Culebra and Cucaracha Formations (Montes et al. 2011, 2012b). These are capped and cross-cut by volcanic diatreme pipes of the Middle Miocene Pedro Miguel Formation, with basalt plugs and lavas of the Late Basalt Formation, and andesite-dacite plugs and lava flows of the Panama City Formation (Farris et al. 2017). Late Miocene, shallow marine sediments of the Gatun Formation occur in the NW part of the basin at Lake Gatun and on the Caribbean coast around Colón and are overlain by the Pliocene Chagres Sandstone, with the basal coquina Toro Limestone or Toro Point Member (Coates et al. 1992).

The volcanic rocks record a change from hydrous basaltic pyroclastic deposits, typical of mantle wedge-derived subduction zone magmas, to hot, dry, bimodal tholeiitic volcanism at the Oligocene-Miocene boundary (21–25 Ma; Farris et al. 2017). This transition is synchronous with formation of the Canal Basin and extensional faulting in a radial rift basin, which is the cause of the change in magma chemistry (Farris et al. 2011, 2017). The basin is a graben with the main normal faults parallel to the axis and normal to the arc. Extension ceased at about 3–6 Ma and changed to the modern strike-slip fault regime (Rockwell et al. 2010b).

14.5.2 The Chuqunaque-Tuira Basin

The stratigraphy of the Chuqunaque-Tuira and Sambu Basins was described by Coates et al. (2004) (Fig. 14.7) and Barat et al. (2014) based on field mapping along rivers, previous surveys for interoceanic canal routes and oil exploration and radar imagery mapping (MacDonald 1969). This sedimentary sequence passes into the Bayano Basin to the northwest and the Atrato Basin going south.

The Upper Cretaceous to Eocene basement crystalline rocks of the San Blas Complex of the mountains on both coasts are overlain by 4000 m of arc-related volcanic and sedimentary rocks of Eocene to Lower Miocene age. They consist of the Eocene-Oligocene Darien Formation of tuffs, agglomerate, radiolarian chert and basalt in the lower part, overlain by calcareous and siliceous mudstone, micritic calcarenite and volcaniclastic rocks in the upper part; the Oligocene Porcona Formation of calcareous, foraminiferal shale, limestone and tuff with radiolarians and resedimented blocks of glauconitic sandstone; and the Lower-Middle Miocene Clarita Formation of thick-bedded, crystalline limestone with sandy, shaly and tuffaceous mudstone units.

These are overlain unconformably by a 3000 m thick sequence of deposits which are coarse- to fine-grained siliciclastic sedimentary rocks and turbiditic sandstone of upper middle to latest Miocene age (Coates et al. 2004). This sequence comprises the upper middle Miocene Tapaliza Formation of foraminiferal mudstone and siltstone, volcanic sandstone, black shale and turbidites; the lower upper Miocene

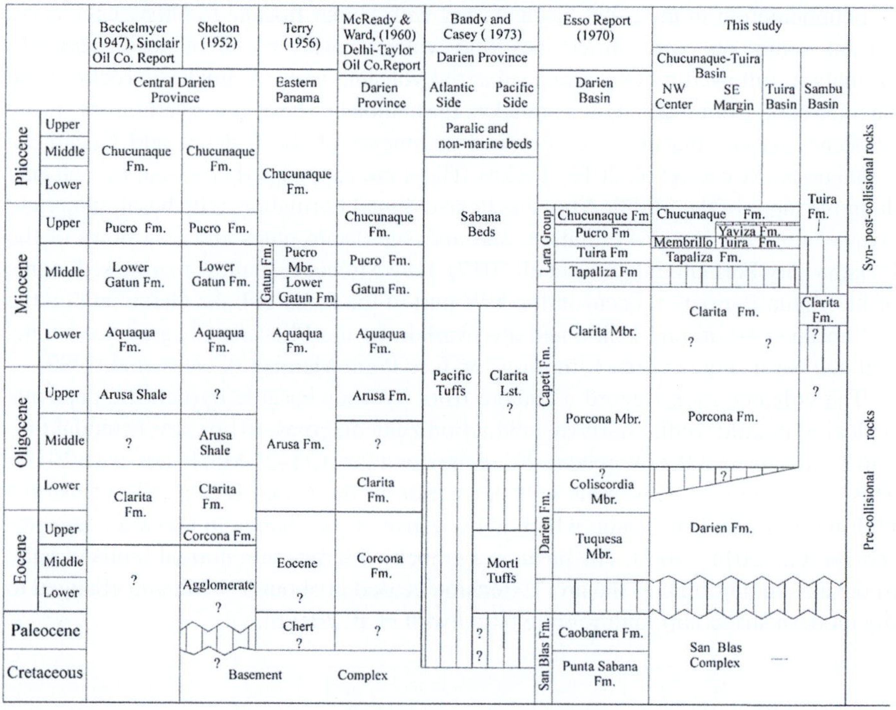

		Beckelmyer (1947), Sinclair Oil Co. Report	Shelton (1952)	Terry (1956)	McReady & Ward, (1960) Delhi-Taylor Oil Co.Report	Bandy and Casey (1973)		Esso Report (1970)	This study				
		Central Darien Province		Eastern Panama	Darien Province	Darien Province		Darien Basin	Chucunaque-Tuira Basin				
						Atlantic Side	Pacific Side		NW Center	SE Margin	Tuira Basin	Sambu Basin	
Pliocene	Upper	Chucunaque Fm.	Chucunaque Fm.	Chucunaque Fm.		Paralic and non-marine beds							Syn-post-collisional rocks
	Middle												
	Lower										Tuira Fm.		
Miocene	Upper	Pucro Fm.	Pucro Fm.	Pucro Fm. (Gatun Fm.)	Chucunaque Fm.	Sabana Beds		Chucunaque Fm / Pucro Fm / Tuira Fm / Tapaliza Fm	Chucunaque Fm. Membrillo/Tuira Fm. Yaviza Fm. Tapaliza Fm.				
	Middle	Lower Gatun Fm.	Lower Gatun Fm.	Pucro Mbr. Lower Gatun Fm	Pucro Fm. Gatun Fm.								
	Lower	Aquaqua Fm.	Aquaqua Fm.	Aquaqua Fm.	Aquaqua Fm.			Clarita Mbr.		?	?	Clarita Fm. ?	
Oligocene	Upper	Arusa Shale	?		Arusa Fm.	Pacific Tuffs	Clarita Lst. ?	Porcona Mbr.	Porcona Fm.			?	Pre-collisional rocks
	Middle	?	Arusa Shale	Arusa Fm.	?								
	Lower	Clarita Fm.	Clarita Fm.		Clarita Fm.			Coliscordia Mbr.	?				
Eocene	Upper		Corcona Fm.	Eocene	Corcona Fm.		Morti Tuffs	Tuquesa Mbr.	Darien Fm.				
	Middle	?	Agglomerate ?		?								
	Lower												
Paleocene			?	Chert ?	?		?	Caobanera Fm.	San Blas Complex				
Cretaceous			? Basement	Complex			?	Punta Sabana Fm.					

Fig. 14.7 Stratigraphy of the Chuqunaque Basin, eastern Panama. (Modified after Coates et al. 2004)

Tuira Formation of greywacke, sandstone, siltstone and claystone, and Membrillo Formation of shelly mudstone; the middle upper Miocene Yaviza Formation of shelly sandstone overlain by coquinoid limestone; and the middle to upper Miocene Chucunaque Formation of silty claystone and siltstone. The sequence was deformed into a synclinorium in the Chuqunaque-Tuira Basin. There is a shallowing of the sedimentary package in the Middle Miocene at 12.8–7.1 Ma which is interpreted as synchronous with the approach of the Panama-Chocó Arc with the Northern Andes.

The so-called pre-collisional open marine strata of Late Cretaceous to Middle Miocene age are separated from the overlying post-collisional sequence of Middle to Late Miocene age by a regional unconformity at 14.8–12.8 Ma. This unconformity was interpreted to be related to the initial collision between the Panama-Chocó Arc with South America (Coates et al. 2004).

The Bayano-Chuqunaque-Tuira syncline was formed after the Chuqunaque Formation (5.6 Ma) by shortening within the Panama Microplate, together with the NW-trending Jaque River Fault and the Sambu and Majé Faults. Continued deformation produced en echelon, doubly plunging and truncated folds along the southern side of the syncline. These were subsequently truncated by the left-lateral strike-slip Sanson Hills Fault. These are, in turn, cut by the NE-trending Pirre Fault, an east-dipping high-angle reverse fault. The left-lateral, strike-slip fault movement in the Darien indicates that internal deformation of the Panama Microplate was accompanied by northwestward "escape" of fault-bounded blocks (Coates et al. 2004). Undeformed late Pliocene-Pleistocene sediments bury deformed older Neogene sediments in the east Panama deformed belt and the western Gulf of Panama and Pearl Islands Basins (Mann and Kolarsky 1995).

14.5.3 The Gulf of Panama Basin

The geology of the eastern Gulf of Panama Basin was described by Mann and Kolarsky (1995) and Derksen et al. (2003) based on 2D seismic surveys and oil wells. Coates et al. (2004) correlated the former's stratigraphy of the Pearl Islands with that of the Chuqunaque-Tuira Basin, showing the San Blas Complex, Darien and Clarita Formations. The two basins are separated by the basement highs and Neogene magmatic arcs of the Majé and Bagre mountains. The Sambu Basin of the eastern gulf, which extends onshore, is a left-lateral pull-apart basin controlled by the NW-trending Sambu Fault and is separated from the Plaris Basin of the southern gulf by a deformed anticlinal belt exposed in the Pearl Islands (Derksen et al. 2003).

There is a thick sedimentary sequence of Eocene to Pleistocene age beneath the Gulf of Panama, with Neogene depocentres west of the Pearl Islands, and in the Sambu and Paris Basins (Mann and Kolarsky 1995; Derksen et al. 2003). In the eastern half of the gulf, these are deformed by east-dipping, west-verging reverse faults, which may have a left-slip component, of Middle Miocene to Plio-Pleistocene age which form a largely buried, 90-km-wide thrust belt which is part of the East Panama Deformed Belt (Mann and Kolarsky 1995).

14.5.4 The Atrato-San Juan Basins

The stratigraphy of the Atrato-San Juan Basins was described by Bedoya et al. (2009) and Cediel and Restrepo (2011). They form two distinct sedimentary basins separated by the Istmina-Condoto High, a basement high of oceanic crust. The Atrato sedimentary sequence is formed of six lithostratigraphic units of lower Eocene to Pliocene age and up to 10,000 m thick. They were deposited on a basement of Upper Cretaceous El Paso Complex in a marine environment with minor transitional and continental influences towards the Pliocene (Fig. 14.8). They comprise the lower to middle Eocene Clavo Formation of mudstones with thin calcareous siltstones and limestones (>300 m thick), the Upper Eocene Salaquí Formation of siliceous limestones (700–2700 m thick), the Oligocene to Lower Miocene Uva Formation of calcareous mudstones and sandstones (600–2300 m thick), the Middle Miocene Napipi Formation of calcareous mudstones (150–220 m thick), the upper Miocene Sierra Formation of mudstones and sandstones (300–1800 m thick), the lower Pliocene Munguidó Formation of mudstones and sandstones (sometimes included with the Sierra Formation), and the middle to upper Pliocene Quibdó Formation of mudstones with a basal conglomerate (200–750 m thick), overlain by

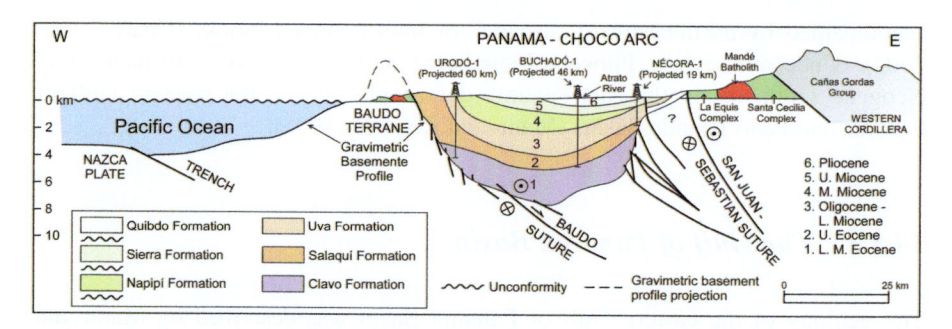

Fig. 14.8 Schematic cross section of the Chocó and Atrato Basin at 6°N. (Modified after Cediel and Restrepo 2011)

Pleistocene alluvial deposits. The sequence records a gradual shallowing from bathyal to abyssal depths (~2000–6000 m) in the Eocene, through outer neritic to bathyal depths (~100–1000 m) in the Miocene, and inner neritic depths (~0–30 m) in the Pliocene.

Surface structural studies based on air photos, radar images and a digital terrain model show that the Atrato Basin is limited to the west by faults which put the Baudó Complex in contact with several sedimentary units which appear to be growth faults of various ages (Cediel and Restrepo 2011). The eastern flank of the basin is a N-S fault system of normal or high-angle reverse faults which put sedimentary rocks in contact with igneous rocks of the Mandé Batholith. The fault system indicates right-lateral displacement throughout the Neogene. Regional Bouger gravity and aeromagnetic survey interpretation show that the west side of the basin has a gentle slope, suggesting onlap of the sedimentary units.

The San Juan sedimentary sequence is formed of five lithostratigraphic units up to 9000 m thick of Paleocene to Pliocene age (Cediel and Restrepo 2011). They were deposited on a basement of Upper Cretaceous (Cañasgordas Group?) in a marine environment with minor transitional and continental influences towards the Pliocene. They comprise the lower Paleocene and Upper Eocene Iró Formation of pelagic limestone, chert, siliceous mudstones and lesser fine grained sandstones (4252 m thick); the Lower Miocene Istmina Formation of siltstones, mudstones and lesser conglomerates (3990 m thick); the Lower Miocene Mojarra Conglomerate of thick conglomerates with sandstones and siltstones (1920 m thick); the Middle Miocene Condoto Formation of mudstones and siltstones (3170 m thick); and the Pliocene Mayorquín and Raposo Formations of conglomerates, sandstones and mudstones, with Pleistocene to recent alluvial and deltaic deposits. The environment of deposition was a carbonate shelf in the Paleocene-Eocene, followed by submarine fans at bathyal depths with continental alluvial fans in the Pliocene which record a major delta which prograded from NE to SW. The basin is controlled by two NE-trending, right-lateral transcurrent fault systems, the Garrapatas-Dabeiba Suture and the San Juan-Sebastián Suture. These are highlighted by regional aeromagnetic maps, which show a strong anomaly in the NE part of the basin which corresponds to the mafic-ultramafic Istmina-Condoto High.

14.6 Neogene Volcanic Arcs

Volcanic, subvolcanic and plutonic rocks form a very poorly known magmatic arc of Late Oligocene to Lower Miocene age in the Majé, Sapo, Bagre, Jungurudu and Pirre Mountain Ranges, which continues south into the Baudó Mountain Range. An outlier volcanic centre occurs in front of the main arc in the Pearl Islands in the Gulf of Panama.

14.6.1 The Majé, Sapo, Bagre, Jungurudu and Pirre Mountain Ranges

There are poorly studied Neogene volcanic and intrusive rocks in the Majé Mountains, Gulf of San Miguel and the Sapo-Bagre-Jungurudu-Pirre Mountains. Porphyry stocks were described in Majé by Kesler et al. (1977), and the Majé igneous complex was briefly described by Whattam et al. (2012) as porphyritic and hypabyssal intrusions of gabbro to granodiorite composition and adakitic-like geochemistry. U-Pb dating of zircon from a diorite gave an age of 18.9 ± 0.4 Ma (Lower Miocene, Whattam et al. 2012).

Neogene pyroclastic rocks, small felsic stocks and columnar-jointed basalts outcrop in the Gulf of San Miguel (Barat et al. 2014; author's observations).

Evidence for Neogene magmatism in the Sapo-Bagre-Jungurudu-Pirre Mountains is shown by andesite flows, fragmental rocks and porphyry stocks at the historic Cana occurrences in the Pirre Mountains (Nelson 2006). Lissinna (2005) obtained an Ar-Ar date of 21.7 ± 0.7 Ma (Lower Miocene) from a basaltic andesite at Bahia Piña in the Sapo range.

14.6.2 The Pearl Islands

Recent mapping shows that the large southern islands are formed of a basic submarine volcanic complex comprised of basaltic breccias, basalt lavas, sedimentary breccias, felsic tuffaceous and calcareous sediments, with a large gabbro-diorite intrusion and small dacite stocks and domes (S. Redwood and D. Buchs, unpublished data). These formed a seamount or possibly an island. The volcanic rocks have been dated at 21.9 ± 0.7 Ma to 18.4 ± 0.3 Ma (Lower Miocene, Lissinna 2005). The northern part of the archipelago is formed of glauconitic sandstones correlated with the Miocene Clarita Formation and cross-cut by dioritic intrusions. There is no evidence for outcrop of basement of the San Blas Complex (Mann and Kolarsky 1995; Coates et al. 2004).

14.6.3 The Baudó Mountains

Limited investigations show evidence for Late Oligocene volcanic and subvolcanic rocks in the Baudó Mountains. In the Sierra de Los Saltos, the Late Oligocene Saltos basalt laccolith was described by the Governor of the Panama Canal (1949), and elsewhere a radiometric date of 25.8 ± 2 Ma (late Oligocene) was reported (Bourgois et al. 1982).

14.6.4 Miocene Ultramafic Complexes, Condoto

The Viravira Complex and Alto Condoto Ultramafic Complex in the Istmina-Condoto High are of Neogene age (Muñoz et al. 1990; Tistl 1994; Tistl et al. 1994). The high is interpreted to be formed of oceanic crust and upper mantle of Late Cretaceous age that was exhumed in the Cenozoic.

The Viravira Complex comprises a sequence of komatiitic and high Mg basalt, basaltic breccia, peridotite, harzburgite and serpentinized dunite, together with mudstone, black claystone, fine-grained sandstone, some calcareous sediments and chert. The marine deposits are 500–1000 m thick. Radiolaria and globorotaloid fossils indicate an Upper Eocene to Lower Miocene age.

The Alto Condoto Ultramafic Complex is a zoned Alaskan-type (or Ural-Alaska-type) ultramafic pluton with dimensions of about 8 × 5 km. It cross-cuts the Viravira Complex with a 1.5 km wide thermal metamorphic aureole. The Alto Condoto Ultramafic Complex has a dunite core, grading outwards to wehrlite, olivine clinopyroxenite and hornblende-magnetite clinopyroxenite. On the outer rim there are small outcrops of hornblendite, diorite and dioritic dykes. It has been dated by K-Ar on hornblende at 21.5–17.8 Ma (Lower Miocene). This is the youngest known Alaskan-type zoned ultramafic pluton in the world. Several other such poorly known intrusions occur in a belt to the north of Alto Condoto (Tistl 1994). These intrusions are believed to be the source of the alluvial platinum deposits for which the Chocó is famous.

The Viravira and Alto Condoto complexes formed from a common parental magma of mafic tholeiitic composition. Tistl et al. (1994) attribute the occurrence of these and other ultramafic complexes further north along a major N-S strike-slip fault close to the terrane boundary to a deeply penetrating fault system which allowed a release of pressure and the formation of Si-undersaturated Mg-rich melts in the mantle. High heat flow and a vertical migration pathway allowed the intrusion of mantle-derived melts into shallow levels of the upper crust.

Alaskan-type intrusions are common in deeply eroded arc roots of Mesozoic age (Foley et al. 1997), and the lack of any other such plutons in the Northern Andes or Central America may be ascribed to the lack of deep erosion and deep fault sutures.

14.7 Closure of the Isthmian Seaway

The timing of the accretion of the Panama-Chocó Arc with the Northern Andes Block is the subject of debate. The collision closed the Central American Seaway, defined as the deep oceanic seaway between the Panama-Chocó Arc and South America, to form the Isthmus of Panama. This was responsible for the inter-American migration of terrestrial animals, known as the Great American Biotic Interchange (GABI) (Woodring 1966; Marshall et al. 1982; Keigwin 1982; Montes et al. 2015). It has been proposed that the emergence of the isthmus disrupted global oceanic circulation and climate and triggered the onset of the Northern Hemisphere glacial period about three million years ago, and the onset of the thermohaline circulation (Haug and Tiedemann 1998; Bartoli et al. 2005). However, Molnar (2017) argued that these changes are part of concurrent, worldwide phenomena that require a global, not a local, explanation.

Closure of the seaway has been placed at about 3.5 Ma (Pliocene) based on the evolutionary divergence of Pacific and Caribbean nearshore marine faunas in sedimentary sections of western Panama and Costa Rica (Coates et al. 1992). The vertebrate faunal interchange event started at 2.8–2.5 Ma, with the assumption this happened when the sea barrier closed and a land bridge was in place (Marshall et al. 1982; Webb 2006; Leigh et al. 2013). Alternatively, climatic changes may also have played a significant role in faunal interchanges. Molnar (2008), Groeneveld et al. (2014) and Bacon et al. (2016) suggested that the onset of the glacial period triggered the faunal interchange by the creation of savannas necessary for faunal migration, due to low sea levels and a drier climate, rather than the present-day rain forests. Retallack and Kirby (2007) suggested that the isthmus could have affected oceanographic conditions in the Caribbean and north Atlantic since about 16 Ma. In addition, Bacon et al. (2015), using molecular and fossil data, showed significant waves of dispersal of terrestrial organisms at approximately ca. 20 Ma and 6 Ma and corresponding events separating marine organisms in the Atlantic and Pacific oceans at ca. 23 Ma and 7 Ma. This, they argued, indicates that the biotic interchange was a long and complex process that began as early as about 23 Ma at the start of the Miocene.

Coates et al. (2004) associated a regional unconformity at 14.8–12.8 Ma in the Chuqunaque Basin with the initiation of the uplift of the isthmus. By the middle Late Miocene, they continued, neritic depths were widespread throughout the Darien region, and a regional unconformity suggests completion of collision by ca. 7.1 Ma. They argued that the absence of Pliocene deposits in the Darien and the Panama Canal Basin, and absence of sediments younger than 4.8 Ma in the Atrato Basin, suggest rapid uplift and emergence of the isthmus in the Late Miocene.

However, newer tectonic models by Farris et al. (2011, 2017) and Montes et al. (2012a, 2015) suggest that tectonic closure initiated in the late Eocene-Oligocene, while Barat et al. (2014) argue it started as far back as the middle Eocene, at about ca, 40 Ma. Montes et al. (2015) argued that detrital zircons of Eocene age in Middle Miocene fluvial (Amaga Formation) and shallow marine (Lower Magdalena Basin) deposits, in the Northern Andes, were derived from the Panama-Chocó Arc (the San Blas-Mandé Batholith). They interpret the data to indicate arc uplift and erosion and a fluvial connection between the Panama-Chocó Arc and South America in the Middle Miocene (ca. 15–13 Ma). This, they conclude, implies that the terrane had docked by this time and that the Central American Seaway had closed to the extent that no intervening deep seaway existed. Support for this hypothesis also comes from thermochronological studies by Ramirez al. (2016), who demonstrate that some areas of Central Panama and the Azuero Peninsula were exposed above sea level, as islands, by the middle Eocene. They interpret four cooling events, at 47–42 Ma, 32–28 Ma, 17–9 Ma and 7–0 Ma, which show episodic uplift from the Eocene to Pliocene.

The above hypotheses were challenged by Coates and Stallard (2013) and O'Dea et al. (2016), who argued for the formation of the isthmus at about 2.8 Ma, with rebuttals by Jaramillo et al. (2017) and Molnar (2017). Coates and Stallard (2013) pointed out that it is important to make a distinction between tectonic closure and seaway closure, as it is difficult to define when a seaway actually ceased to exist. As an example, they cite the Wallace Line, a major faunal barrier within a deep seaway between the islands of Bali and Lombok in the Indonesian archipelago, which is only 35 km wide (Coates and Stallard 2013).

The geological evidence for the timing of collision and formation of the isthmus is thus, at present, not definitive. Much of the evidence is based upon the effects, real or purported, of isthmus formation, rather than direct geological and palaeogeographical evidence. Detained mapping and research, at a regional level, of the arcs that make up the Panama-Chocó Arc, are required to better define and constrain the history of approach, collision and accretion between the Panama-Chocó Arc and northwestern South America.

14.8 Summary: Geological History and Tectonic Synthesis

The Panama-Chocó Arc is a composite volcano-plutonic island arc of Late Cretaceous to Miocene age, developed upon the western, trailing edge of the Caribbean Plate (CLIP), and accreted to the Northern Andean Block of the South American Plate along its southern and eastern margin. The arc probably extended to the western end of the Chorotega Block (Costa Rica), where it is presently overlain by the Neogene Central American volcanic arc, west of the Panama Canal Basin. A synthesis of the tectonic history of the arc is based on recent work by Buchs et al. (2010), Farris et al. (2011), Montes et al. (2012a, b, 2015), Whattam et al. (2012), Barat et al. (2014), and others (Fig. 14.9).

1. The basement of the Panama-Chocó Arc is the Caribbean Plate formed of Cretaceous ocean crust comprising basalts, cherts and limestone, which is interpreted to have formed in an oceanic plateau in the Pacific realm at the Galapagos hotspot (the Caribbean Large Igneous Province or CLIP).

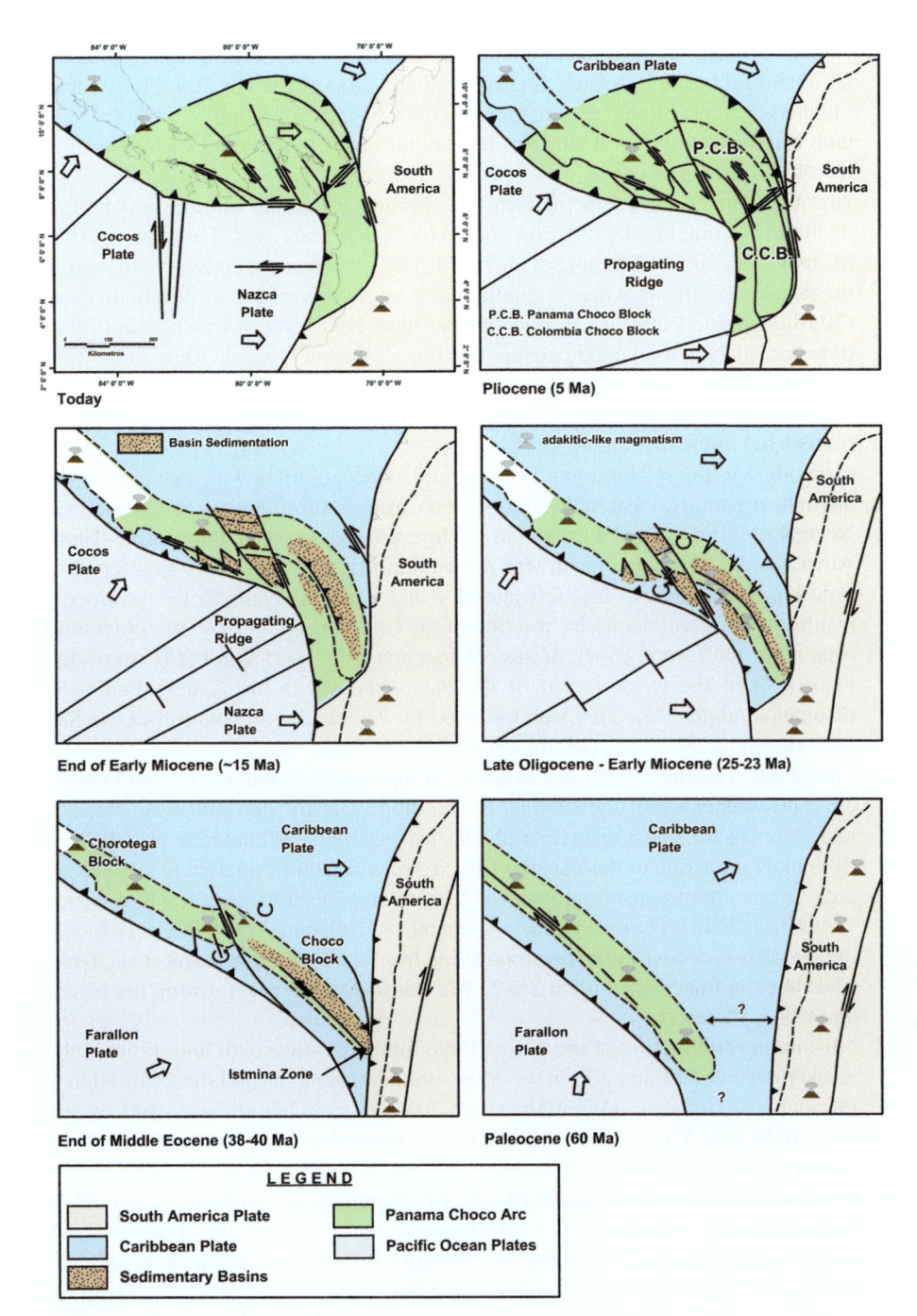

Fig. 14.9 Cartoon models showing the tectonic evolution of the Panama-Chocó Arc from the Paleocene to present. (Modified after Barat et al. 2014)

2. A Late Cretaceous to Eocene island arc developed on the trailing edge of the CLIP during northerly or north-easterly subduction of the Farallon Plate, forming the volcano-plutonic arcs mapped in the Chagres, Mamoní, San Blas, Atrato and Mandé Mountains. A 200 km left-lateral offset is recorded with respect to the Azuero Peninsula.
3. Arc magmatism ceased in the San Blas-Mandé arc in the middle Eocene (ca. 47–42 Ma), coincident with uplift shown by fission track ages (Farris et al. 2011; Montes et al. 2012a; Ramirez et al. 2016). The arc remained active to the west in the Azuero Peninsula where it continued to migrate northwards. Whattam et al. (2012) suggested that arc shutdown may have been caused by collision of an indentor, such as thickened oceanic crust or a plateau. Oceanic seamounts were accreted to the Azuero Peninsula at this time (Buchs et al. 2010).
4. There was a magmatic arc gap in eastern Panama-Chocó spanning the Late Eocene to Oligocene (ca. 40–20 Ma).
5. Coincident with arc shutdown in the Middle Eocene, there was a change in the tectonic regime of eastern Panama-Chocó from compressional to extensional. Normal faulting formed a graben in the fore-arc to form the Chuqunaque-Tuira-Atrato Basin. The canal basin was initiated at this time, probably as a result of transtensive tectonics. The left-lateral Panama Canal Fault Zone developed, resulting in counterclockwise rotation of the Chorotega Block by 20°, clockwise rotation of the Chocó Block of eastern Panama by 5°, and about 200 km of displacement of the western part of the Paleogene arc in the Azuero Peninsula (Montes et al. 2012a). This was followed by 25° clockwise rotation of the San Blas massif at 28–25 Ma (Montes et al. 2012a).
6. The Middle Eocene extension and arc break-up may be explained by two hypotheses: (a) steepening of the subducting Farallon Plate by slab roll-back and ultimately break-off, causing adakite-like intrusive activity (Whattam et al. 2012), or (b) initial interaction of the Panama-Chocó Arc with South America in the Istmina area, as a result of consumption of the Caribbean plateau beneath South America (Barat et al. 2014). The extensional regime during collision is explained by a model of oblique progressive collision propagating from south to north (Barat et al. 2014).
7. The Farallon Plate broke up at 25–23 Ma (latest Oligocene), forming the Nazca and Cocos Plates.
8. NE-dipping subduction of the Nazca Plate jumped to the south and initiated the weakly developed Later Arc in the Majé-Baudó Mountains and the Pearl Islands in the Late Oligocene (Whattam et al. 2012). Arc magmatism ceased by the Early to Middle Miocene in eastern Panama due to the change in the Nazca Plate convergence to strike-slip.
9. Oblique collision of the Chocó Block of Colombia was complete by about the Early Miocene. Transpressive deformation then formed the East Panama Deformed Belt in the Darien region and Gulf of Panama starting in the Middle Miocene as Panama collided against South America (Barat et al. 2014), resulting in the formation of left-lateral strike-slip faults, pull-apart basins, thrust-ramp anticlinal highs in Majé and the Pearl Islands (Mann and Kolarsky 1995), closure of the Chuqunaque Basin and uplift of the isthmus (Coates et al. 2004). Active seismicity on these faults indicates that collision is ongoing.

References

Aspden JA, McCourt WJ, Brook M (1987) Geometrical control of subduction-related magmatism: the Mesozoic and Cenozoic plutonic history of Western Colombia. J Geol Soc Lond 144:893–905

Avé Lallemant HG, Sisson VB (2005) Caribbean-South American plate interactions, Venezuela, Geological Society of America special paper no. 394. Geological Society of America, Boulder, p 331

Bacon CD, Silvestro D, Jaramillo C, Smith BT, Chakrabarty P, Antonelli A (2015) Biological evidence supports an early and complex emergence of the Isthmus of Panama. Proc Natl Acad Sci 112:6110–6115

Bacon CD, Molnar P, Antonelli A, Crawford AJ, Montes C, Vallejo-Pareja MC (2016) Quaternary glaciation and the Great American Biotic Interchange. Geology 44:357–378

Bandy OL, Casey RE (1973) Reflector horizons and paleobathymetric history, eastern Panama. Geological Society of AmericaBulletin 84:3081–3086

Barat F (2013) Nature et structure de l'isthme inter-américain, Panama: implication sur la reconstitution et l'évolution géodynamique de la plaque Caraïbe. Unpublished doctoral thesis, University of Nice-Sophia, p 273

Barat F, Mercier de Lépinay B, Sosson M, Müller C, Baumgartner PO, Baumgartner-Mora C (2014) Transition from the Farallon plate subduction to the collision between South and Central America: geological evolution of the Panama Isthmus. Tectonophysics 622:145–167

Bartoli G, Sarnthein M, Weinelt M, Erlenkeuser H, Garbe-Schönberg D, Lea DW (2005) Final closure of Panama and the onset of northern hemisphere glaciation. Earth Planet Sci Lett 237:33–44

Beckelmyer RL (1947) Columnar section of Rio Chico and tributaries Traverse, Darien province, Panama. Unpublished report, Sinclair Oil Company, p 1–7

Bedoya G, Cediel F, Restrepo-Correa I, Cuartas C, Montenegro G, Marin-Cerón MI, Mojica J, Cerón R (2009) Aportes al conocimiento de la evolución geológica de las cuencas Atrato y San Juan dentro del arco Panamá-Chocó. Boletin de Geología 31(2):69–81

Bertrand M, Zürcher P (1899) Etude géologique sur l'isthme de Panama. Paris, Compagnie Nouvelle de Canal de Panama, Rapport de la Commission, app 1

Bourgois J, Azéma J, Touron J, Bellon H, Calle B, Parra E, Toussaint J, Glaçon G, Feinberg H, De Weber P, Origlia I (1982) Ages et structures des complexes basiques et ultrabasiques de la façade pacifique entre 3° N et 12° N (Colombie, Panama et Costa-Rica). Bulletin de la Société géologique de France 24:545–554

Buchs DM, Arculus RJ, Baumgartner PO, Baumgartner-Mora C, Ulianov A (2010) Late Cretaceous arc development on the SW margin of the Caribbean Plate: Insights from the Golfito, Costa Rica, and Azuero, Panama, complexes. Geochem Geophys Geosyst 11:Q07S24

Buchs DM, Baumgartner PO, Baumgartner-Mora C, Flores D, Bandini AN (2011) Upper Cretaceous to Miocene tectonostratigraphy of the Azuero area (Panama) and the discontinuous accretion and subduction erosion along the Middle American margin. Tectonophysics 512:31–46

Camacho E, Hutton W, Pacheco JF (2010) A new look at evidence for a Wadati-Benioff zone and active convergence at the North Panama Deformed Belt. Bull Seismol Soc Am 100:343–348

Case JE (1974) Oceanic crust forms basement of eastern Panama. Geol Soc Am Bull 85:645–652

Case JE, Lopez RA, Moore WR (1971) Tectonic investigations in western Colombia and eastern Panama. Geol Soc Am Bull 82:2685–2712

Cediel F, Cáceres C (2000) Geological map of Colombia, 3rd edn. Geotec Ltda, Bogotá. 7 thematic maps at 1:1,000,000 scale

Cediel F, Restrepo I (2011) Geology and Hydrocarbon Potential, Atrato, San Juan and Urabá Basins. University EAFIT, Medellin, Colombia (Petroleum Geology of Colombia series, Vol. 3, F. Cediel (ed)), 104 p

Cediel F, Shaw RP, Cáceres C (2003) Tectonic assembly of the northern Andean block. In: Bartolini C, Buffler RT, Blickwede J (eds) The Circum-Gulf of Mexico and the Caribbean: hydrocarbon habitats, basin formation, and plate tectonics, AAPG memoir 79. American Association of Petroleum Geologists, Tulsa, pp 815–848

Coates AG, Stallard RF (2013) How old is the Isthmus of Panama? Bull Mar Sci 89:801–813

Coates AG, Jackson JBC, Collins LS, Cronin TS, Dowsett HJ, Bybell LM, Jung P, Obando JA (1992) Closure of the Isthmus of Panama: the near-shore marine record of Costa Rica and Panama. Geol Soc Am Bull 104:814–828

Coates AG, Collins LS, Aubry MP, Berggren WA (2004) The Geology of the Darien, Panama, and the late Miocene-Pliocene collision of the Panama arc with northwestern South America. Geol Soc Am Bull 116:1327–1344

Corral I, Cardellach E, Corbella M, Canals A, Gómez-Gras D, Griera A, Cosca MA (2016) Cerro Quema (Azuero Peninsula, Panama): geology, alteration, mineralization, and geochronology of a volcanic dome-hosted high-sulfidation Au-Cu deposit. Econ Geol 111:287–310

Cronin P (2017) Caving in Panama 2017. Report of the 2017 Anglo-Irish Expedition, 39 p (www.rgs.org)

Defant MJ, Richerson PM, De Boer JZ, Stewart RH, Maury RC, Bellon H, Drummond MS, Feigenson MD, Jackson TE (1991) Dacite genesis via both slab melting and differentiation: petrogenesis of La Yeguada volcanic complex, Panama. J Petrol 32:1101–1142

Dengo G, Case JE (eds) (1990) The Caribbean region, The geology of North America, Vol. H. Geological Society of America, Boulder, p 528

Derksen SJ, Coon HL, Shannon PJ (2003) Eastern gulf of Panama exploration potential. AAPG search and discovery article #90017, AAPG international conference, Barcelona, 21–24 Sept 2003, p 7

Duque-Caro H (1990) The Choco block in the northwestern corner of South America: structural, tectonostratigraphic, and paleogeographic implications. J S Am Earth Sci 3:71–84

Esso Exploration and Production Panama (1970) Annual Report Corresponding to the Second Year of Exploration Conforming to Contract 59, Darien Basin, October 19690September 1970. Unpublished report by Esso Exploration and Production Panama for Department of Mineral Resources, Government of Panama 44 p

Etayo Serna F, Barrero D, Lozano H, Espinosa A, González H, Orrego A, Ballesteros I, Forero H, Ramírez C, Zambrano Ortiz F, Duque Caro H, Vargas R, Núñez A, Álvarez J, Ropaín C, Cardozo E, Galvis N, Sarmiento L, Alberts JP, Case JE, Singer DA, Bowen RW, Berger BR, Cox DP, Hodges CA (1985) Mapa de terrenos geológicos de Colombia. INGEOMINAS, Bogotá. Publicaciones Geológicas Especiales del INGEOMINAS 14(1):135

Farris DW, Jaramillo C, Bayona G, Restrepo-Moreno SA, Montes C, Cardona A, Mora A, Speakman RJ, Glascock MD, Valencia V (2011) Fracturing of the Panamanian Isthmus during initial collision with South America. Geology 39:1007–1010

Farris DW, Cardona A, Montes C, Foster D, Jaramillo C (2017) Magmatic evolution of Panama Canal volcanic rocks: a record of arc processes and tectonic change. PLoS One 12(5):44 e0176010. https://doi.org/10/1371/journal.pone.0176010

Foley JY, Light TD, Nelson SW, Harris RA (1997) Mineral occurrences associated with mafic-ultramafic and related alkaline complexes in Alaska. In: Goldfarb RJ, Miller LD (eds) Mineral deposits of Alaska, Economic geology monograph. Economic Geology Publishing Company, El Paso, Texas, USA 9:396–449

González H (2001) Mapa Geológico del Departamento de Antioquia, Escala 1:400,000, Memoria Explicativa. INGEOMINAS, Bogotá, 240 p

González JL, Shen Z, Mauz B (2014) New constraints on Holocene uplift rates for the Baudo Mountain range, northwestern Colombia. J S Am Earth Sci 52:194–202

Goosens PJ, Rose WI, Flores D (1977) Geochemistry of Tholeiites of the basic igneous complex of north-western South America. Bull Geol Soc Am 88:1711–1720

Governor of the Panama Canal (1947a) Isthmian Canal Studies, Appendix 2, Routes Studied. Report of the Governor of the Panama Canal to 79th Congress of the United States, 1st Session, p 225

Governor of the Panama Canal (1947b) Isthmian Canal Studies, Appendix 8, Geology. Report of the Governor of the Panama Canal to 79th Congress of the United States, 1st Session, p 84

Governor of the Panama Canal (1949) Special Report of the Governor of the Panama Canal on the Atrato-Truando Canal Route. Report to 79th Congress of the United States, 1st Session, 20 p + 31 p Apendices

Groeneveld J, Hathorne EC, Steinke S, DeBey H, Mackensen A, Tiedemann R (2014) Glacial induced closure of the Panamanian gateway during marine isotope stages (MIS) 95–100 (~2.5 Ma). Earth Planet Sci Lett 404:296–306

Haug GH, Tiedemann R (1998) Effect of the formation of the Isthmus of Panama on Atlantic Ocean thermohaline circulation. Nature 393:673–676

Howe E (1907) Isthmian geology and the Panama Canal. Econ Geol 2:639–658

INGEOMINAS (n.d.) Batolito de Mandé. Catálogo de las Unidades Litoestratigráficas de Colombia. Report by INGEOMINAS, p 19

Interoceanic Canal Study Commission (1968) Panama Canal Investigation Route 10, p 32

James K, Lorente MA, Pindell J (eds) (2009) The origin and evolution of the Caribbean plate. Geol Soc Lond Spec Pub 328:686

Jaramillo C, Montes C, Cardona A, Silvestro D, Antonelli A, Bacon CD (2017) Comment (1) on "Formation of the Isthmus of Panama" by O'Dea et al. Sci Adv 3(2017):e1602321. p 8

Jones SM (1950) Geology of Gatun Lake and vicinity, Panama. Bull Geol Soc Am 61:893–922

Keigwin L (1982) Isotopic paleoceanography of the Caribbean and East Pacific: role of Panama uplift in late neogene time. Science 217:350–353

Kellogg JN, Vega V (1995) Tectonic development of Panama, Costa Rica, and the Colombian Andes: constraints from global positioning system geodetic studies and gravity. In: Mann P (ed) Geologic and tectonic development of the Caribbean plate boundary in southern Central America, Geological Society of America special paper 295. Geological Society of America, Boulder, pp 75–90

Kennan L, Pindell J (2009) Dextral shear, terrane accretion and basin formation in the Northern Andes: best explained by interaction with a Pacific-derived Caribbean plate? In: James K, Lorente MA, Pindell J (eds) The origin and evolution of the Caribbean plate, Geological Society of London, special publication, vol 328. Geological Society, London, pp 487–531

Kerr AC, Marriner GF, Tarney J, Nivia A, Saunders AD, Thirlwall MF, Sinton CW (1997) Cretaceous basaltic terranes in Western Colombia: elemental, chronological and Sr-Nd isotopic constraints on petrogenesis. J Petrol 38:677–702

Kesler SE, Sutter JF, Issigonis MJ, Jones LM, Walker RL (1977) Evolution of porphyry copper mineralization in an Oceanic Island Arc: Panama. Econ Geol 72:1142–1153

Kirby MX, MacFadden B (2005) Was southern Central America an archipelago or a peninsula in the middle Miocene? A test using land-mammal body size. Palaeogeogr Palaeoclimatol Palaeoecol 228:193–202

Kirby MX, Jones DS, MacFadden BJ (2008) Lower Miocene stratigraphy along the Panama Canal and its bearing on the Central American Peninsula. PLoS One 3(7):e2791. https://doi.org/10.1371/journal.pone.0002791

Leal-Mejía H (2011) Phanerozoic gold metallogeny in the Colombian Andes: a tectono-magmatic approach. Unpublished PhD thesis, Universitat de Barcelona, p 999

Leigh EG, O'Dea A, Vermeij GJ (2013) Historical biogeography of the Isthmus of Panama. Biol Rev 89:148–172

Lissinna B (2005) A profile through the Central American Landbridge in western Panama: 115 Ma Interplay between the Galápagos hotspot and the Central American subduction zone. PhD thesis, Christian-Albrechts-Univesität, Kiel, p 102

Lonsdale P, Klitgord KM (1978) Structure and tectonic history of the eastern Panama Basin. Geol Soc Am Bull 89:981–999

MacDonald DF (1915) Some engineering problems of the Panama Canal in their relation to geology and topography, U.S. Bureau of Mines bulletin no. 86. Govt. Print. Off, Washington, DC, p 88

MacDonald H (1969) Geological evaluation of radar imagery from Darien Province, Panama. Mod Geol 1:1–63

MacFadden BJ (2006) North American Miocene land mammals from Panama. J Vertebr Paleontol 26(3):720–734

Mann P (1995) Geologic and tectonic development of the Caribbean Plate boundary in Southern Central America, Geological Society of America special paper no. 295. Geological Society of America, Boulder, p 349

Mann P, Kolarsky RA (1995) East Panama deformed belt: structure, age, and neotectonic significance. In: Mann P (ed) Geologic and tectonic development of the Caribbean Plate boundary in Southern Central America, Geological Society of America special paper no. 295. Geological Society of America, Boulder, pp 111–130

Marshall LG, Webb SD, Sepkoski JJ Jr, Raup DM (1982) Mammalian evolution and the great American interchange. Science 215:1351–1357

McCourt WJ, Aspden JA, Brook M (1984) New geological and geochronological data from the Colombian Andes: continental growth by multiple accretion. J Geol Soc Lond 141:831–845

McReady W, Ward R (1960) Anticipated thickness and lithology of formations in Darien province, Panama. Unpublished report, Delhi-Taylor Oil Company, p 1–14

Molnar P (2008) Closing of the central American seaway and the ice age: a critical review. Paleoceanography 23:PA2201. https://doi.org/10.1029/2007/PA001574

Molnar P (2017) Comment (2) on "Formation of the Isthmus of Panama" by O'Dea et al. Sci Adv 3(2017):e1602320. p 4

Montes C, Cardona A, Bayona G, Farris D (2011) Field trip guide: evidence for middle Eocene and younger emergence in Central Panama. Smithsonian Tropical Research Institute, Panama, p 24

Montes C, Bayona G, Cardona A, Buchs DM, Silva CA, Morón S, Hoyos N, Ramírez DA, Jaramillo CA, Valencia V (2012a) Arc-continent collision and orocline formation: closing of the Central American seaway. J Geophys Res 117:B04105. https://doi.org/10.1029/2011JB008959

Montes C, Cardona A, McFadden R, Morón SE, Silva CA, Restrepo-Moreno S, Ramírez DA, Hoyos N, Wilson J, Farris D, Bayona GA, Jaramillo CA, Valencia V, Bryan J, Flores JA (2012b) Evidence for middle Eocene and younger land emergence in central Panama: implications for Isthmus closure. Geol Soc Am Bull 124:780–799

Montes C, Cardona A, Jaramillo C, Pardo A, Silva JC, Valencia V, Ayala C, Pérez-Angel LC, Rodriguez-Parra LA, Ramirez V, Niño H (2015) Middle Miocene closure of the Central American seaway. Science 348(6231):226–229

Muñoz R, Salinas R, James M, Bergmann H, Tistl M (1990) Mineralizaciones primarias de minerales del grupo del platino y oro en la cuenca de los ríos Condoto e Iro (Chocó-Colombia), Informe Tecnico Final. BGR & INGEOMINAS, Medellin, p 330

Nelson CE (2006) Metallic mineral resources. In: Bundschuh J, Alvarado G (eds) Central America: Geology, Resources and Hazards. Balkema, London. Chapter 32 p, pp 885–915

O'Dea A, Lessios HA, Coates AG, Eytan RI, Restrepo-Moreno SA, Cione AL, Collins LS, de Queiroz A, Farris DW, Norris RD, Stallard RF, Woodburne MO, Aguilera O, Aubry MP, Berggren WA, Budd AF, Cozzuol MA, Coppard SE, Duque-Caro H, Finnegan W, Gasparini GM, Grossman EL, Johnson KG, Keigwin LD, Knowlton N, Leigh RG, Leonard-Pingel JS, Marko PB, Pyenson ND, Rachello-Dolmen PG, Soibelzon E, Soibelzon L, Todd JA, Vermeij GJ, Jackson JBC (2016) Formation of the Isthmus of Panama. Sci Adv 2(8):e1600883. https://doi.org/10.1126/sciadv.1600883

Pindell J, Kennan L (2009) Tectonic evolution of the Gulf of Mexico, Caribbean and northern South America in the mantle reference frame: an update. In: James K, Lorente MA, Pindell J (eds) The origin and evolution of the Caribbean plate, Geological Society of London, special publication, vol 328. Geological Society, London, pp 1–55

Ramirez DA, Foster DA, Min K, Montes C, Cardona A, Sadove G (2016) Exhumation of the Panama basement complex and basins: implications for the closure of the Central American seaway. Geochem Geophys Geosyst 17:1758–1777. https://doi.org/10.1002/2016GC006289

Reed DL, Silver EA (1995) Sediment dispersal and accretionary growth of the North Panama deformed belt. In: Mann P (ed) Geologic and tectonic development of the Caribbean plate boundary in Southern Central America, Geological Society of America special paper 295. Geological Society of America, Boulder, pp 213–223

Reeves F, Ross CP (1930) A geologic study of the Madden Dam project, Alhajuela, Canal Zone, United States Geological Survey Bulletin 821-B. U.S. G.P.O, Washington, DC, pp 11–47

Restrepo JJ, Toussaint JF (1988) Terranes and continental accretion in the Colombian Andes. Episodes 11:189–193

Retallack GJ, Kirby MX (2007) Middle Miocene global change and paleogeography of Panama. PALAIOS 22:667–679

Rockwell TK, Bennett RA, Gath E, Franceschi P (2010a) Unhinging an indenter: a new tectonic model for the internal deformation of Panama. Tectonics 29. https://doi.org/10.1029/2009TC002571

Rockwell T, Gath E, González T, Madden C, Verdugo D, Lippincott C, Dawson T, Owen LA, Fuchs M, Cadena A, Williams P, Weldon E, Franceschi P (2010b) Neotectonics and paleoseismicity of the Limón and Pedro Miguel faults in Panamá: earthquake hazard to the Panamá Canal. Bull Seismol Soc Am 100:3097–3129

Rodriguez G, Arango MI (2013) Formación Barroso: arco volcanico toleitico y diabasas de San José de Urama: un prisma acrecionario T-MORB en el segmento norte de la Cordillera Occidental de Colombia. Boletín de Ciencias de la Tierra 33:p17–p38

Rooney TO, Franceschi P, Hall CM (2011) Water-saturated magmas in the Panama Canal region: a precursor to adakite-like magma generation? Contrib Mineral Petrol 161:373–388

Schmidt et al (1947) Panama Canal route 10. Isthmian Canal Commission report, p 10

Schmidt-Thomé M, Feldhaus L, Salazar G, Muñoz R (1992) Explicación del mapa geológico, escala 1:250,000, del flanco oeste de la Cordillera Occidental entre los ríos Andágueda y Murindó, Departamentos Antioquia y Choco, República de Colombia. Geologisches Jahrbuch 78:1–23. and 1 map at 1:250,000 scale

Shelton BJ (1952) Geology and petroleum prospects of Darien, southeastern Panama. Unpublished M.Sc. Thesis, Oregon State College, Corvallis, Oregon, USA, 62 p

Sillitoe RH, Jaramillo L, Damon PE, Shafiqullah M, Escovar R (1982) Setting, characteristics, and age of the Andean porphyry Copper Belt in Colombia. Econ Geol 77:1837–1850

Stewart RH, Stewart JL, Woodring WP (1980) Geologic map of the Panama Canal and vicinity, Republic of Panama. United States Geological Survey Miscellaneous Investigations Series Map I-1232, scale 1:100,000

Strong N, Hendrickson M, Farris D, Rodríguez F, Pérez A, Del Valle Y, De Gracia C, O'Dea A (2011) Geología de la Comarca de Kuna Yala. Poster, Smithsonian Tropical Research Institute, Panama

Suter F, Sartori M, Neuwerth R, Gorin G (2008) Structural imprints at the front of the Chocó-Panama indenter: field data from the North Cauca Valley basin, Central Colombia. Tectonophysics 460:134–157

Terry RA (1956) A geological reconnaissance of Panama, Occasional papers of the California Academy of Sciences, no. 23. California Academy of Sciences, San Francisco, p 91

Tistl M (1994) Geochemistry of platinum-group elements of the zoned ultramafic Alto Condoto Complex, Northwest Colombia. Econ Geol 89:158–167

Tistl M, Burgath KP, Höhndorf A, Kreuzer H, Muñoz R, Salinas R (1994) Origin and emplacement of Tertiary ultramafic complexes in northwest Colombia: evidence from geochemistry and K-Ar, Sm-ND and Rb-Sr isotopes. Earth Planet Sci Lett 126:41–59

Trenkamp R, Kellogg JN, Freymueller JT, Mora HP (2002) Wide plate margin deformation, southern Central America and northwestern South America, CASA GPS observations. J S Am Earth Sci 15:157–171

Webb SD (2006) The great American biotic interchange: patterns and processes. Ann Mo Bot Gard 93:245–257

Weber M, Gómez-Tapias J, Cardona A, Duarte E, Pardo-Trujillo A, Valencia VA (2015) Geochemistry of the Santa Fé Batholith and Buriticá Tonalite in NW Colombia and evidence of subduction initiation beneath the Colombian Caribbean Plateau. J S Am Earth Sci 62:257–274

Wegner W, Wörner G, Harmon RS, Jicha BR (2011) Magmatic history and evolution of the Central American Land Bridge in Panama since Cretaceous times. Geol Soc Am Bull 123:703–724

Westbrook GK, Hardy NC, Heath RP (1995) Structure and tectonics of the Panama-Nazca plate boundary. In: Mann P (ed) Geologic and tectonic development of the Caribbean Plate boundary in Southern Central America, Geological Society of America special paper 295. Geological Society of America, Boulder, pp 91–109

Weyl R (1980) Geology of Central America. Gebrüder Borntraeger, Berlin, p 371

Whattam SA, Montes C, McFadden RR, Cardona A, Ramirez D, Valencia V (2012) Age and origin of earliest adakitic-like magmatism in Panama: implications for the tectonic evolution of the Panamanian magmatic arc system. Lithos 142-143:226–244

Woodring WP (1957) Geology and paleontology of Canal Zone and adjoining parts of Panama: geology and descriptions of Tertiary molluscs (Gastropods: Trochidae to Turritellidae). US Geol Surv Prof Pap 306-A:145. + 2 plates

Woodring WP (1966) The Panama land bridge as a sea barrier. Proc Am Philos Soc 110(p):425–433

Woodring WP (1982) Geology and paleontology of Canal Zone and adjoining parts of Panama. Chapter 306-F: description of Tertiary molluscs (Pelecypods: Propeamussiidae to Cuspidariidae; Additions to Families Covered in P 306-E; Additions to Gastropods; Cephalopods). US Geol Surv Prof Pap 306-F:iv. p. + p. 541–759 + Plates 83–125. Plate 125: Geologic Map of the Panama Canal and Vicinity, Republic of Panamá, compiled by R. H. and J. L. Stewart with the collaboration of W. P. Woodring, 1:100,000 scale

Wörner G, Harmon RS, Hartmann G, Simon K (2005) Igneous geology and geochemistry of the upper Río Chagres Basin. In: Harmon RS (ed) The Río Chagres, Panama. A multidisciplinary profile of a tropical watershed, Water science and technology library, vol 52. Springer, Dordrecht, pp 65–81

Wörner G, Harmon RS, Wegner W, Singer B (2006) Linking America's backbone: geological development and basement rocks of central Panama. In: Abstracts with programs, Geological Society of America conference "Backbone of the Americas" Mendoza, Argentina, Geological Society of America, Boulder, 3–7 Apr 2006, p 60

Wörner G, Harmon RS, Wegner W (2009) Geochemical evolution of igneous rocks and changing magma sources during the formation and closure of the Central American land bridge of Panama. In: Kay SM, Ramos VA, Dickinson WR (eds) Backbone of the Americas: shallow subduction, plateau uplift, and ridge and terrane collision, Geological Society of America memoir 204. Geological Society of America, Boulder, pp 183–196

Wyse LNB (1877) L'exploration de L'Isthme du Darien en 1876–1877. Bulletin de la Societe de Geographie, Paris, Serie 6 14:561–580

Part VIII
Holocene – Anthropocene

Chapter 15
Sediment Transfers from the Andes of Colombia during the Anthropocene

Juan D. Restrepo

15.1 Global Sediment Transfers in the Anthropocene

The Earth has been pushed out of the Holocene epoch by human activities, with the mid-twentieth century as a strong candidate for the start date of the Anthropocene (Steffen et al. 2016), the proposed new geological epoch in Earth history (Crutzen and Stoermer 2000; Crutzen 2002; Zalasiewicz et al. 2008).

Determination of a start date for the stratigraphic Anthropocene requires an examination of how the magnitude and rate of contemporary Earth system change, driven largely by human impact, may be best represented by optimal selection of a stratigraphic marker or markers to allow tracing of a synchronous boundary globally (Steffen et al. 2016). Globally recognizable, geosynchronous change clearly began in the mid-twentieth century at the beginning of the Great Acceleration (Hibbard et al. 2006; Steffen et al. 2015; McNeill and Engelke 2016), which marks a rapid change in human activity. Thus, it is clear from chronostratigraphic and Earth system perspectives (Steffen et al. 2016) as well as from several key global biogeochemical cycles (Williams et al. 2016) that the Earth has entered the Anthropocene, and the mid-twentieth century is the most convincing start date (Waters et al. 2016).

Human impact on sediment production began 3000 years ago but accelerated more widely 1000 years ago. By the sixteenth century, societies were engineering their environment. Early twentieth century mechanization has led to global signals of increased sediment flux in most large rivers (Syvitski and Kettner 2011). Global data of sediments, nitrogen, phosphorous, and other elements down the river systems

---

J. D. Restrepo (✉)
Departamento de Ciencias de la Tierra, Universidad EAFIT, Medellín, Colombia
e-mail: jdrestre@eafit.edu.co

and into the coastal zone show how humans have changed the terrestrial hydrological system by engineering the landscape since the mid-twentieth century (Syvitski and Kettner 2011; Steffen et al. 2016). For most global rivers, the anthropogenic footprint increases sharply after World War II (Walling and Fang 2003), when humans increased the sediment transport through soil erosion by 2.3 ± 0.6 billion tons per year (Syvitski et al. 2005).

A significant proportion of the sediment transported by many of the world's rivers represents soil eroded from agricultural land, and the magnitude of this flux therefore also provides a measure of land degradation and the associated reduction in the global soil resource (Morgan 1986; Walling and Fang 2003). There is no doubt that human activity is an effective agent in altering the landscape, affecting erosion rates and consequently fluvial sediment transport. Some studies have documented the relevant role played by the so-called technological denudation, the human contribution to sediment generation (e.g., Cendrero et al. 2006; Bonachea et al. 2010). Human mobilization of sediments could be one to two orders of magnitude greater than natural denudation rates. In fact, global erosion rates from natural processes are between 0.1 and 0.01 mm y^{-1}, while soil denudation due to human activities accounts for 1 mm y^{-1} (Bonachea et al. 2010).

This chapter reviews data, models, and analyses on Anthropocene-impacted sediment fluxes in the Andes of Colombia (Fig. 15.1) and provides examples on how direct human alteration has increased sediment flux during the last decades. Firstly, it describes the context of the northern Andes in terms of sediment production within the whole Andes Cordillera. Secondly, it presents a summary of major land cover changes witnessed in the region from 8000 years ago to the beginning of large-scale land transformation that occurred in Colombia during the last three decades and analyzes major human-induced drivers of change. Also, trends in sediment load during the 1980–2010 period are documented. Finally, it compares modern and prehuman conditions of sediment flux by using some applied models in global and Colombian rivers.

15.2   The Sediment Production of the Andes Cordillera

The understanding of river basin sediment yield at a continental scale provides useful information for (i) developing quantitative models of landscape evolution, (ii) studying geochemical and sediment mass balance, (iii) estimating the intensity of continental and regional erosion, and (iv) assessing the volume of solids contributed from continents to the ocean and the trapping of sediments at the continental scale (Pinet and Souriau 1988; Summerfield and Hulton 1994; Harrison 2000; Hovius 2000; Latrubesse and Restrepo 2014).

Milliman and Syvitski (1992) have previously emphasized the importance of the large number of small- and medium-sized basins, many of which have high relief and, therefore, increased potential for human impact on their sediment fluxes, in accounting for a significant proportion of the total land-ocean sediment flux.

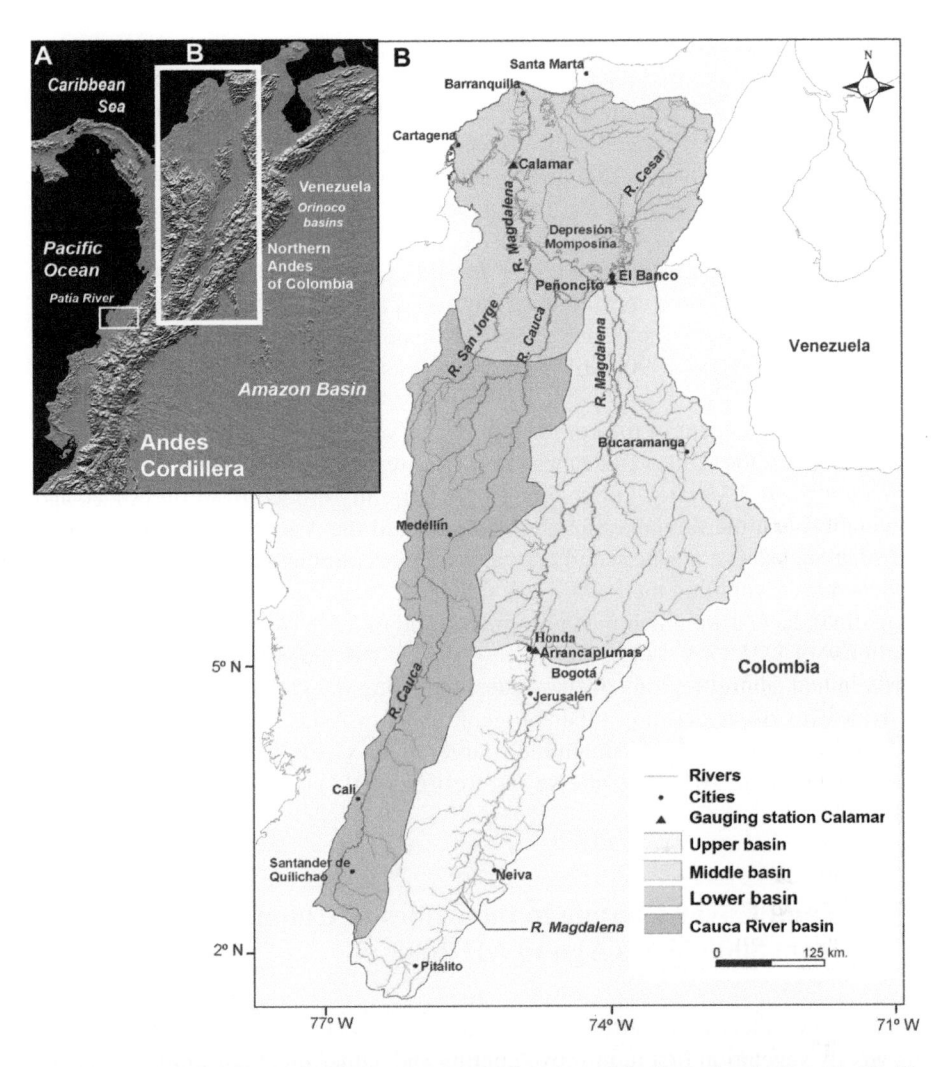

Fig. 15.1 (a) The Andes of Colombia, showing the location of the Patía river basin in the Pacific margin, the Orinoco catchments on the eastern Cordillera, and the largest fluvial system, the Magdalena (b); the upper, middle, and lower sections of the catchment and the most downstream hydrological gauging station, Calamar, are also shown

Catchments located in tropical mountain areas are highly susceptible to soil erosion due to their topography and erosive climate (Milliman and Syvitski 1992; Dadson et al. 2003; Molina et al. 2008) and deforestation (Wunder 1996; Restrepo and Syvitski 2006; Restrepo et al. 2015). In these mountainous basins, particularly sensitive to soil erosion by water and landsliding (Hess 1990; Wunder 1996; Hovius et al. 1997; Hovius 1998; Vanacker et al. 2003; Molina et al. 2008), the sediment budget is not only controlled by hydrological erosion processes on agricultural land but also by gully erosion over overgrazed steep lands as well as by erratic sediment fluxes from landslides and bank erosion, processes that integrate over larger spatial scales (Trimble 1975; Trimble and Crosson 2000; Molina et al. 2008).

The Andes is a tectonically active region characterized by active volcanism, ongoing uplift, earthquakes, and high-magnitude mass movements (Vanacker et al. 2003; Harden 2006; Molina et al. 2008). Uplift has caused rivers to incise and denudation rates to be high. In this region of steep slopes, mass movements are triggered by wet conditions and by earthquakes (Aalto et al. 2006; Harden 2006). Thus, catchments located in the cordilleras of the Andes are highly susceptible to soil erosion due to their topography and erosive climate and the occurrence of extreme geologic events (Hess 1990; Milliman and Syvitski 1992; Wunder 1996; Dadson et al. 2003).

A recent assessment of the continental budget of sediment yield and its regional variation along the Andes (Latrubesse and Restrepo 2014) indicates that between 2.57 Gt y^{-1} and 3.33 Gt y^{-1} of sediments are currently eroded from the Andes and flowing through the fluvial systems, mainly toward the Amazon, Paraná, Orinoco, and Magdalena. The northern and central Andes, accounting for ~46% of the entire Andean area (excluding the Pacific catchments of Peru, northern Chile, and central Argentina), have a weighted mean sediment yield of 2045 t km^{-2} y^{-1} and produce a minimum of 2.25 Gt y^{-1} of sediment (Fig. 15.2). The magnitude of yields for the northern and central regions of the Andes, including the Colombian cordilleras, is equivalent to rivers draining other orogenic belts in Asia, Insular Asia, and New Zealand and one order of magnitude larger than yields reported for the Alps (Vanmaercke et al. 2011; Latrubesse and Restrepo 2014) (Fig. 15.2).

15.3 Land Cover Change in the Andes of Colombia from 8000 Years Ago to AD 2000

Humans have altered the Earth's land surface since the Paleolithic mainly by clearing woody vegetation first to improve hunting and gathering opportunities and later to provide agricultural cropland. In the Holocene, agriculture was established on nearly all continents and led to widespread modification of terrestrial ecosystems (Kaplan et al. 2010).

For the Andes of Colombia, data on anthropogenic land cover change from 8000 years ago to the beginning of large-scale industrialization ($_{AD}$ 1850) was kindly shared by Jed Kaplan. This inventory is based on reconstruction of carbon emissions caused by anthropogenic land cover change (ALCC) over the Holocene based on contrasting scenarios of population and anthropogenic land use over time, including a new empirical model in which per capita ALCC declines over time and a conventional model that holds per capita ALCC roughly constant over time. These scenarios are used to drive a dynamic global vegetation model to estimate regional and global patterns of changes in terrestrial carbon storage through the last

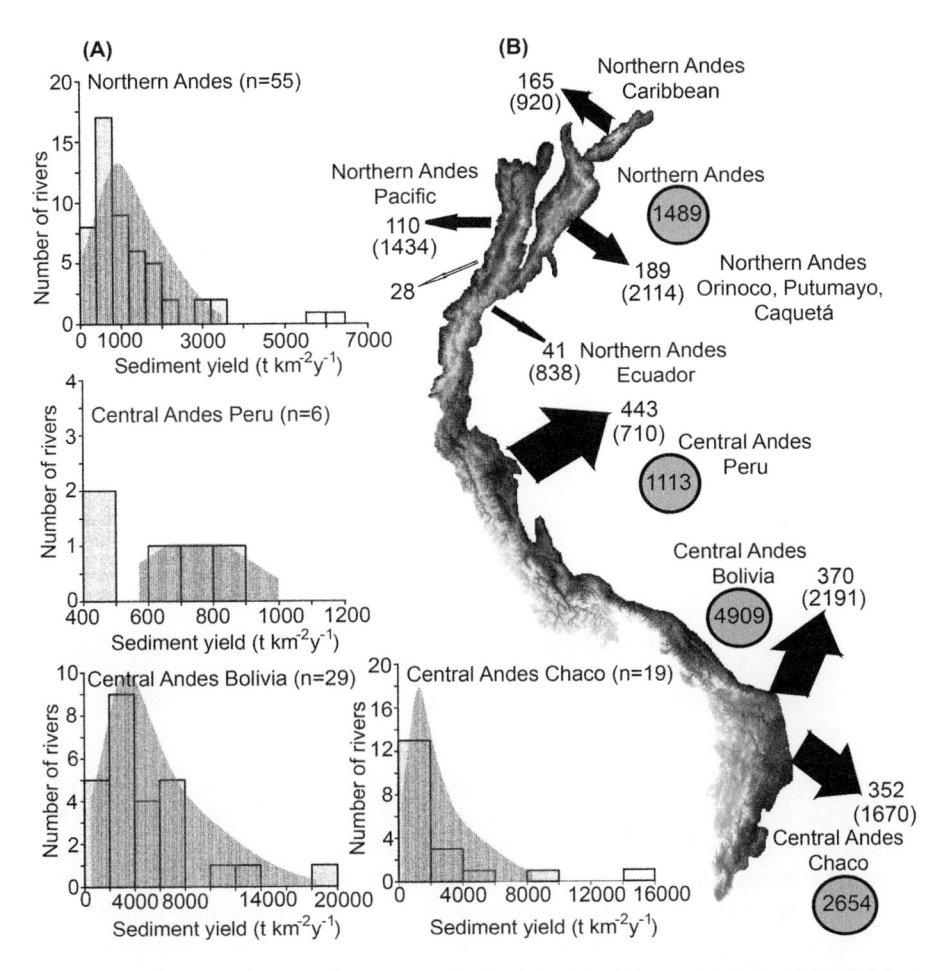

Fig. 15.2 Frequency distributions of sediment yield for the northern and central regions of the Andes (**a**) and geographic location of major sediment fluxes ($\times 10^6$ t y^{-1}) (**b**). Arrow width is proportional to values of sediment load. Average sediment yield (circles) for each Andean region. Values in parentheses indicate piedmont sediment yield after deposition processes in floodplains and aggradational fans. (From Latrubesse and Restrepo 2014)

8000 years (Kaplan et al. 2010). Also, land cover changes in the Colombian Andes during the last five centuries were analyzed from studies of land conversion (Etter et al. 2005, 2006a, b, c, 2008).

Figure 15.3 shows per capita anthropogenic land cover change (ALCC) from 8 ka to $_{AD}$ 2000 for the Andes of Colombia. Comparing ALCC of the Andes to the observed global and regional patterns, as noted by Kaplan et al. (2010), many similarities arrive. A nearly pristine world existed until 3 ka when clearing in Northeast China, the Middle East, Europe, and South America began to emerge. Two thousand years later, by $_{AD}$ 1, ALCC only slightly increases. Moreover, the ALCC scenarios

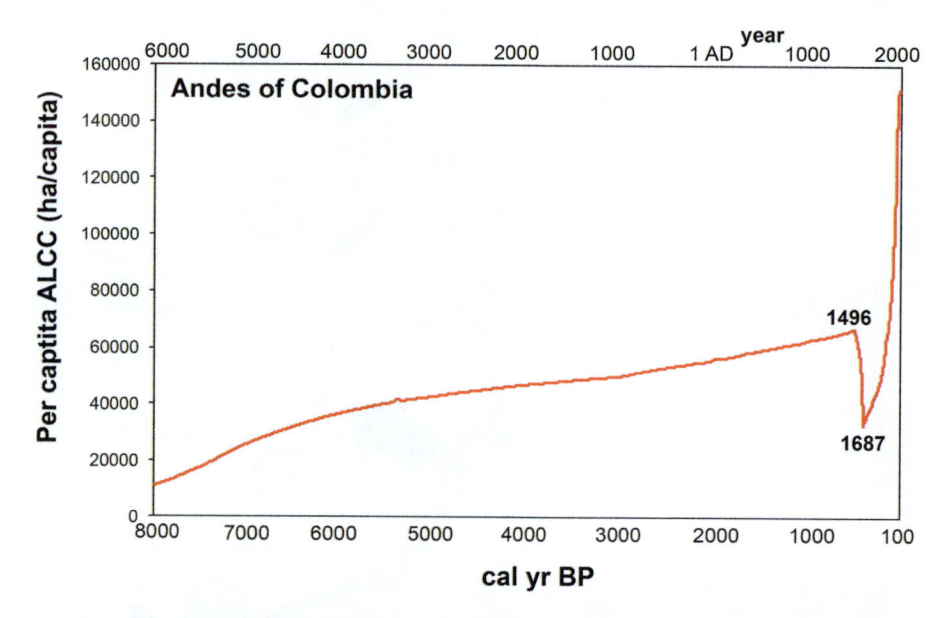

Fig. 15.3 Per capita ALCC from 8 ka to AD 2000 for the Andes of Colombia from HYDE 3.1 database. (Data kindly given by Kaplan et al. 2010)

show that parts of Mesoamerica, the Andes, Europe, and China were nearly 60% cleared at $_{AD}$ 1. From $_{AD}$ 1500 to $_{AD}$ 1600, a decrease in anthropogenic land use in the Western Hemisphere is visible, as the indigenous populations of the Americas succumbed to disease and war brought by European explorers and colonists. The collapse of large precontact populations with advanced agriculture, which were especially concentrated in Mesoamerica and the Andes, led to high amounts of land abandonment. The low levels of ALCC shown at $_{AD}$ 1500 are almost entirely abandoned 100 years after conquest. By $_{AD}$ 1800, anthropogenic land use in the Americas accelerated with the spread of colonies and nations founded by Europeans. The Americas only start to result in substantial amounts of ALCC emissions during the last centuries. In addition, ALCC for the Andes during the last four decades is of the same magnitude as the land cover changes documented by Etter et al. (2008) (Fig. 15.4) (see next section).

Contrary to the common understanding, which suggests that major land conversion took place since the 1970s, a study on historical patterns and drivers of landscape change in Colombia since 1500 (Etter et al. 2008) reveals that land conversion in the Andes started five centuries ago. The transformed area in the Andean region rose from 15 M ha in 1500 to 42 M ha in 2000.

During the last two centuries, the annual rate of forest-transformed area increased two orders of magnitude, from 4330 ha y^{-1} in 1800 to 171,190 ha y^{-1} in 2000 (Fig. 15.4). By the year 2000, 80% of the natural vegetation in the Andes was

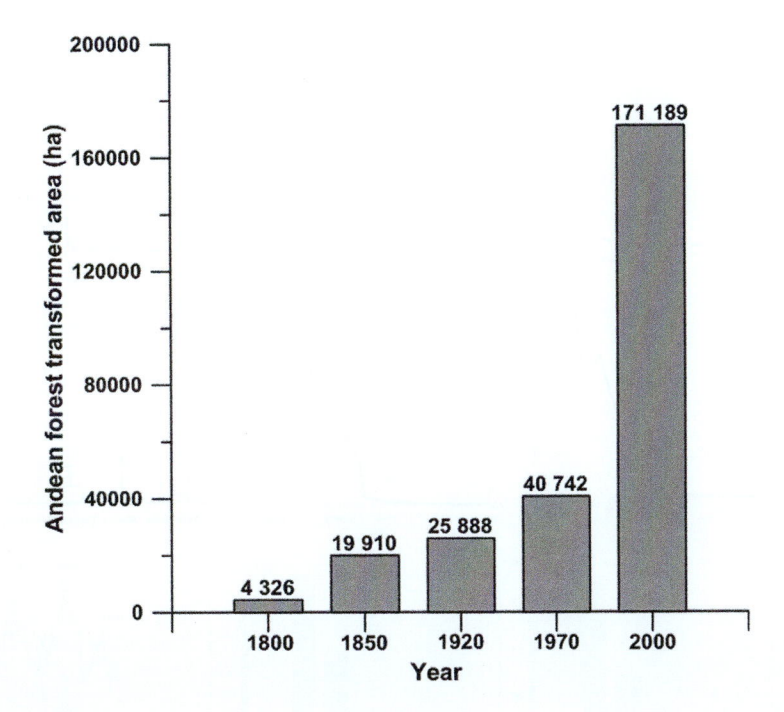

Fig. 15.4 Transformed forest area in Colombia between 1800 and 2000 (data from Etter et al. 2008)

cleared, with 20% remaining as scattered remnants. An assumed value of 30% was cleared in preconquest agricultural landscapes (before 1500), increasing to 80% in 2000. Demographic impacts of colonization and the introduction of cattle were the major drivers of this change (Etter et al. 2008).

The population of Colombia increased tenfold and surpassed 40 million in 2000 (Fig. 15.5). Historically, most of the Colombian population (<65%) has been concentrated in the Andean and Caribbean regions of the Magdalena basin (Etter et al. 2006b). In addition, beef cattle industry, the largest contributor to the spatial footprint of agricultural land uses in Colombia at both national and regional levels, has been a major driver of forest clearance. Biologically diverse tropical forests were transformed into ecologically simplified grasslands and cropping areas (Etter et al. 2006b), and cattle grazing is now the most widespread land use in the Andes of Colombia. The national herd size increased from 2 million cattle in 1920 to 11 million cattle in 2000. And this is still growing as estimates indicate to a size of 30 million cattle in 2005 (Etter et al. 2005) (Fig. 15.5). Currently, cattle dominate over 75% of the transformed landscapes in the Colombian Andes (Etter et al. 2008).

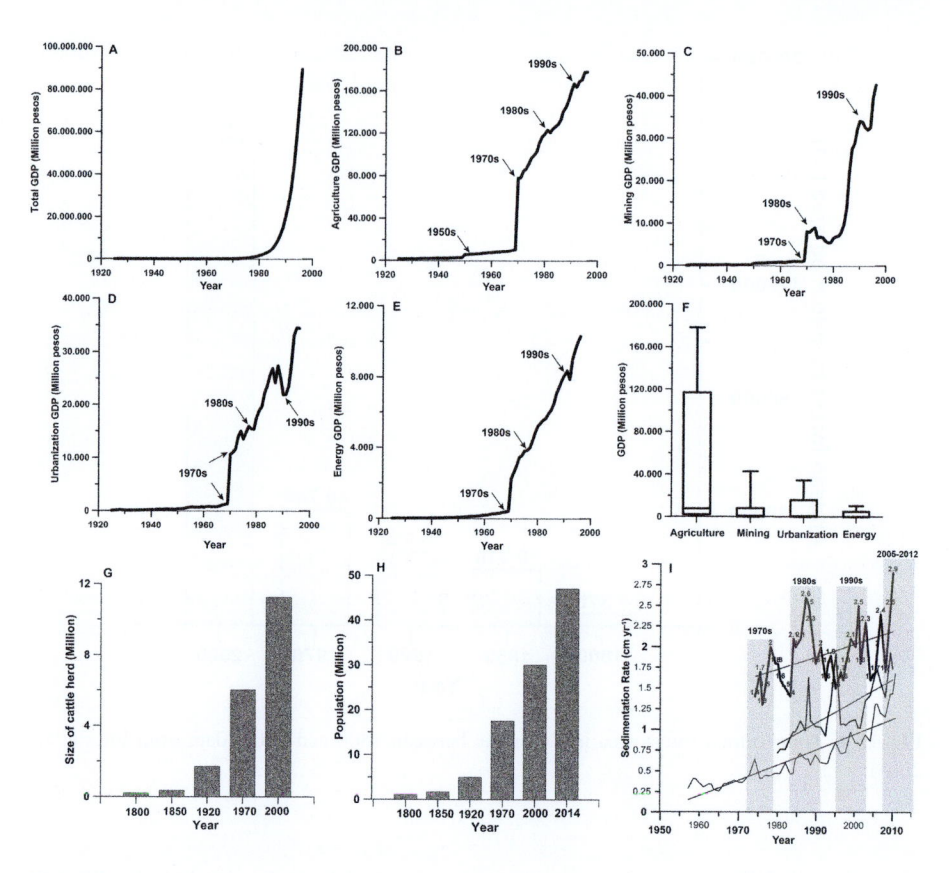

Fig. 15.5 (**a–e**) Major economic activities in the Magdalena drainage basin and their participation (million pesos) in the gross domestic product (GDP). (**f**) Box whisker plot of major economic activities altering soils within the Magdalena catchment (Data from Republic Bank of Colombia). (**g**) Size of cattle herd in the Andes of Colombia during the 1800–2000 year period (Data from Etter et al. 2008). (**h**) Population of Colombia between 1800 and 2014 (Data from Etter et al. 2008). (**i**) ^{210}Pb Sedimentation rates in the lower Magdalena and its distributary channel, the Canal del Dique, showing trends in two lagoon systems (low-center) and in the Pasacaballos delta at the El Dique entrance in the Cartagena Bay (up). (Data from Ecoral-Barú Project on the Environmental Base Line of Argos-Barú)

15.4 Human-Induced Drivers of Land Cover Change in the Andes of Colombia

When analyzing the participation (million pesos) in the gross domestic product of human activities that promote soil erosion in the Magdalena basin for the 1927–2000 period, including agriculture, mining, energy, and urbanization (Fig. 15.5a–e), a large part of the land conversion and further deforestation resulted from agriculture activities (Fig. 15.5f). Overall, all human drivers show very clear increases during the 1970s, 1980s, and 1990s. These trends match quite well with the observed increasing trends in sediment transport of the Magdalena tributaries and in the main Magdalena River at Calamar (see next section, Fig. 15.7). Also, cattle ranching and population in the Andes of Colombia increased exponentially during the last two centuries, with major increases between 1970s and 2000 (Fig. 15.5g–h).

While the clearing of forests began more than 10,000 years ago, the rate of clearing has accelerated since the 1900s when the area of cropland doubled (Houghton 1994). Deforestation accelerated again since the 1960s, coinciding with rapid global population growth, especially in the tropics (Etter et al. 2006a; Restrepo et al. 2015). The rate of net forest loss globally is presently 125,000 km^2 y^{-1} and increasing by 2000 km^2 y^{-1}. Of all the deforestation, 85% occurs in the tropics (Hansen et al. 2013) where forests are being converted to cropland and pasture for the production of soy, beef, palm oil, and timber (Ferretti-Gallon and Busch 2014).

In the tropical Andes of Colombia, 80% of the natural vegetation was cleared by 2000, with 20% remaining as remnants of forests (Etter et al. 2008). Some 180,600 km^2 (69%) of the Andean forests and 203,400 km^2 (30%) of the lowland forests were cut down by 2000 (Etter et al. 2006b), with the highest rates of forest clearing corresponding to the Andean region. The total national deforestation rates rose from an estimated 10,000 ha y^{-1} to more than 230,000 ha y^{-1} between 1500 and 2000. Thus the Andean forest belt has been constantly cleared over the last 500 years, with clearing accelerating to 1.4% y^{-1} during the second half of the twentieth century (Etter et al. 2008).

Clearance of natural vegetation for cattle ranching, land cultivation, and mining is known to have increased rates of soil erosion by several orders of magnitude (Walling and Fang 2003). According to Geist and Lambin (2001, 2002), agricultural expansion is by far the leading land use change associated with nearly all deforestation cases in tropical regions (96%). Also, this study on proximate causes and underlying drivers of tropical deforestation indicates that permanent agriculture displays low geographical variation in tropical areas; that is, regional values for permanent cultivation in Latin America, for example, are close to the global value (i.e., 50%). In the Colombian Andes, the agricultural spatial footprint in 1500 was approximately 7.5 M ha, dropping to 6.5 M ha in 1600, increasing to 12 M ha in 1850, and then exponentially increasing to 33 M ha in 2000 (Etter et al. 2008). There is no doubt that agriculture has been the major driver of land conversion and forest clearance in the Andes (Fig. 15.5f).

Recent assessments on global deforestation show that the tropics account 58% of the net global forest loss (e.g., Ferretti-Gallon and Busch 2014). A recent deforestation study spanning over 34 tropical countries that take into account the majority of the global land of humid tropical forests (Kim et al. 2015) indicates a 62% acceleration in net deforestation from the 1990s to the 2000s, contradicting a 25% reduction reported by FAO (2010). Analyzing the data presented by Restrepo et al. (2015), net forest loss in Colombia peaked from 1990 to 2010, a period of an exponential increase in forest clearance, from 170,000 ha y^{-1} between 1990 and 2000 to 499,000 ha y^{-1} during the 2005–2010 period. After Brazil, Colombia has the highest deforestation rate of all Latin American countries (Kim et al. 2015).

The above rates of forest loss for Colombia (Kim et al. 2015) do not even closely match the rates of forest clearance estimated by the Hydrological Institute of Colombia-IDEAM for different deforestation assessments. Where Kim et al. (2015) measure national estimates of deforestation of 170,000 ha y^{-1} during the 1990–2000 period and 490,000 ha y^{-1} during the 2005–2010 period, IDEAM-REDD (reducing emissions from deforestation and forest degradation) project database reports rates of 100,000 ha y^{-1} and 336,000 ha y^{-1}, respectively. Nevertheless, the scale of deforestation in the Andes of Colombia is dramatic as also noted by different assessments of forest loss (Fig. 15.6).

The area (in ha) occupied by each of the land cover types obtained in the classification of the 1980 and 2000 MSS and TM Landsat images shows that land cover in the Magdalena basin (Fig. 15.1) has undergone considerable change. The forest cover decreased by 40% over the period of study, while the area under agriculture and pasture cover (agricultural lands 1 and 2) increased by 65% during the same 20-year period (Restrepo and Syvitski 2006) (Fig. 15.6). In addition, many Magdalena sub-catchments, including the Cauca, Opón, Suarez, Negro, and Páez rivers, witnessed an order of magnitude higher deforestation rates than other tributaries during the 2005–2010 year period. The total forest clearance in the Magdalena basin of 510,565 ha between 2005 and 2010 represents 24% of the combined deforestation in Colombia (Restrepo et al. 2015).

Additional drivers of deforestation in the Andes of Colombia are illegal gold mining and agricultural cocaine crops. A regional assessment of deforestation due to gold mining in the tropical moist biome of South America shows that deforestation was significantly higher during the 2007–2013 period. More than 90% of the deforestation occurred in four major hotspots, including Guianan ecoregion (41%), Southwest Amazon region (28%), Tapajós-Xingú (11%), and Magdalena Valley mountain forests (9%) (Alvarez-Berríos and Aide 2015). In the Magdalena River, the Cauca basin shows a dramatic increase in sediment transport for the last 20 years. The most extensive and profitable gold mining is located in the lower course of the Cauca and its tributary, the Nechi. Extraction of only gold has increased from 6.6 t y^{-1} in 1990 to 25 t y^{-1} in 2008. High concentrations of suspended sediments, often greater than 1600 mg L^{-1}, are the result of rapid erosion of the lowlands, partly because of ongoing gold mining (Restrepo 2013). Gold mining has been an important economic activity in the Magdalena since 1990, and 86% of gold production in Colombia is estimated to be illegal (Alvarez-Berríos and Aide 2015).

The illicit drug trade of coca in Colombia, Peru, and Bolivia has also been identified as a contributor to deforestation in the tropical Andes (Harden 2006). The US State Department (2001) estimated that a minimum of 2.4 M ha of forest was cleared for coca production in the Andean region over the previous 20 years. The environmental report of Colombia, a study provided by the World Bank (Sánchez-Triana et al. 2007), estimated that at least 850,000 ha of forest were cleared for cocoa production in the Colombian Andes between 1978 and 1998. Overall, one third of the deforested area in Colombia is due to cocaine crops (Dávalos et al. 2011).

Fig. 15.6 (**a**) Land cover maps of the Magdalena drainage basin for 1980 and 2000 prepared from the classification of MSS and TM Landsat images (Restrepo and Syvitski, 2006). (**b**) Deforestation rates for the 21 tributary basins of interest for the 2005–2010 period, based on MODIS image classification by IDEAM – Colombia. (**c**) Deforestation rates in Colombia during the 2001–2012 period (data from IDEAM, 2014). (**d**) Deforestation rates in Colombia during the 1990–2000, 2000–2005, and 2005–2010 periods. (Data from Kim et al. 2015) (From Restrepo and Escobar, 2016)

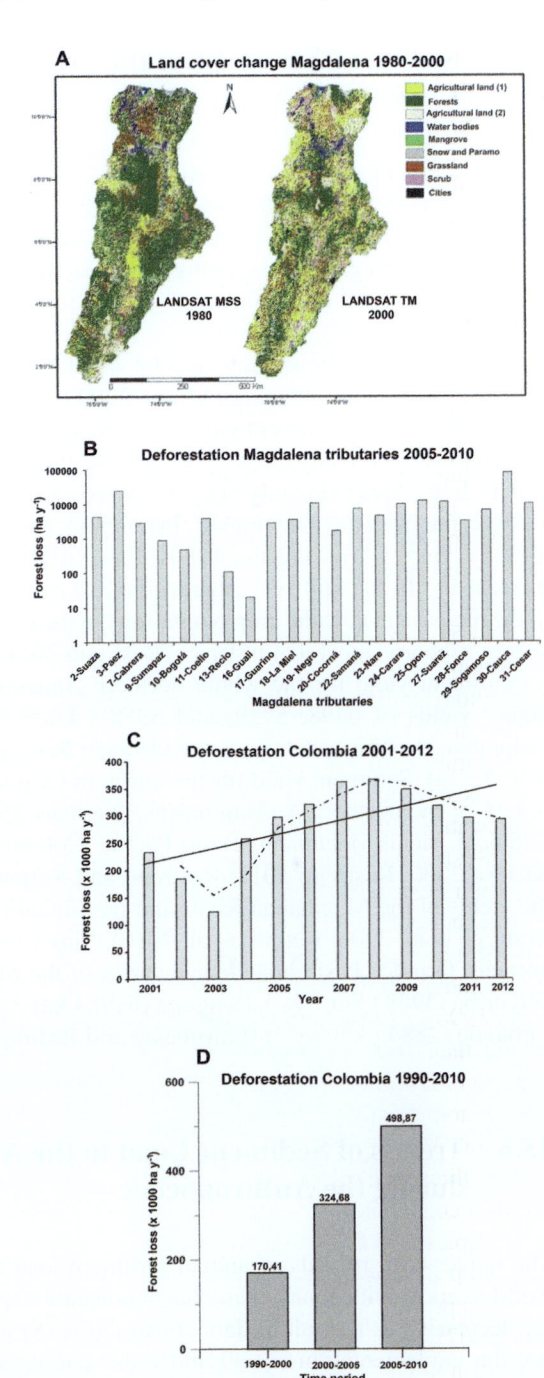

15.5 Sediment Fluxes from the Western and Eastern Cordilleras

The Patía River catchment with an area of 23,700 km² has the largest drainage basin of the Colombian Rivers draining into the Pacific (Fig. 15.1a). Based on daily sediment load data from 1972 to 2001, the best estimate of sediment load into the Pacific Ocean from both gauged and non-gauged Patía tributaries is 27 Mt. y^{-1}. This results in a sediment yield of 1500 t km^{-2} y^{-1} for the entire basin (Restrepo and Kettner 2012).

The Patía River shows high decadal variability in sediment yield. Between 1972 and 1980, average sediment yield was 2200 t km^{-2} y^{-1}. One of the human-induced drivers of this high sediment yield is deforestation. Since the 1960s, the Colombian government gave many licenses to remove the forests in the Patía drainage basin and its delta. Approximately 60% of the wood production in the country during the 1970s came from the Patía region. In contrast, the average sediment yield decreased to 825 t km^{-2} y^{-1} during the 1980s. Since 1991, the average sediment yield has increased to 1470 t km^{-2} y^{-1} as a result of the expansion in area of deforestation due to the increase in cocaine crop plantations in the Patía catchment and its main downstream tributary, the Telembí River (Restrepo 2012).

Sediment yield for the whole northern Andes averages 1485 t km^{-2} y^{-1}, with higher yields of 6498, 5739, and 3958 t km^{-2} y^{-1} in the Meta River basin in Colombia, a major tributary of the Orinoco River (Latrubesse and Restrepo 2014) (Fig. 15.1a). Sediment yield for the northern Orinoco tributary basin, which covers ~ 32% of the northern Andean basins, averages 2572 t km^{-2} y^{-1}, almost twice the sediment yields documented for Pacific Colombian watersheds (Restrepo and Kjerfve 2000; Restrepo 2012; Restrepo and Kettner 2012) and almost three times the yields of the Magdalena River and its tributaries (Restrepo and Syvitski 2006; Restrepo et al. 2006; Kettner et al. 2010). Major sediment yield contributors in the northern Orinoco basin include tributaries of the Meta and Guaviare rivers such as Guyuriba (3958 t km^{-2} y^{-1}), Lengupa (6498 t km^{-2} y^{-1}), Upía (3492 t km^{-2} y^{-1}), and Cobugón (2884 t km^{-2} y^{-1}) (Latrubesse and Restrepo 2014).

15.6 Trends of Sediment Load in the Andes during the Anthropocene

The longer-term records of annual sediment load and annual runoff available for world rivers provide many examples of nonstationary behavior. In some cases, loads are decreasing as a result of dam construction (Syvitski 2003; Syvitski et al. 2005) and the implementation of soil and water conservation and sediment control programs (Mou 1996; Xu 2003), while in others they are increasing as a result of land use change and deforestation, other forms of catchment disturbance, such as mining

(Restrepo and Syvitski 2006; Restrepo et al. 2015), and increased runoff caused by increased precipitation (Walling and Fang 2003).

In attempting to assess the relative importance of increasing, decreasing, and stable trends in the recent sediment loads, Walling and Fang (2003) assembled longer-term sediment records for 145 world rivers, mostly with catchment areas >10,000 km². Many of these rivers were tributaries of larger rivers, and the results are, therefore, not wholly representative of land-ocean fluxes. In most cases these records extended to >25 years, and simple trend analysis was applied to the records, to discriminate those with increasing, decreasing, and stable trends. The results of this analysis indicated that almost half (70) of the rivers showed no evidence of a statistically significant trend in the sediment loads, but of the remainder, most (68 rivers) show evidence of a significant decrease; only 7 rivers were characterized by a significant increase. The dataset for this analysis was inevitably highly constrained by the availability of longer-term sediment records, and all the data came from the Northern Hemisphere, with much of this coming from river basins of the developed world. If other catchments of the developing world had been represented, it is likely that more rivers would have shown an increasing trend (Walling 2006).

Few examples of strong human impacts have been recently documented on tropical Andean drainage basins. In highly degraded Andean catchments in southern Ecuador, land cover conversions are often followed by a phase of intense soil degradation that further exacerbates the anthropogenic impact on surface hydrology. For example, high erosion rates of 24 and 150 $t\ ha^{-1}\ y^{-1}$ are observed in pastures and croplands, respectively (Molina et al. 2008; Vanacker et al. 2014). In addition, observed changes in streamflow during the last four decades are not the result of long-term climate change. Despite increased precipitation in some Ecuadorian river basins, there is a remarkable decrease in streamflow that very likely results from direct anthropogenic disturbances after land cover change (Molina et al. 2015). Further analysis of landslide frequency and area distribution after forest conversion in the tropical Andes demonstrates that the majority of landslide-induced sediment is coming from anthropogenic environments. Thus, land cover change plays an important role in enhancing the overall soil denudation rates in tropical mountain regions (Guns and Vanacker 2013, 2014).

For the Andes of Colombia (Fig. 15.1), human activities appear to have played a more prominent role compared to rainfall (climate change) to mobilize sediment (Restrepo and Syvitski 2006; Restrepo et al. 2015). Deniers of land-use change and its impact on floods argue that climate change is the main trigger of erosion and floods experienced during the last decade. Nevertheless, recent studies on precipitation trends in Colombia during the last four decades (Carmona and Poveda 2011, 2014) show no regional signs of increasing trends in rainfall in the central Andes. In addition, Andean rivers of sedimentation rates (and, therefore, denudation) show signs of acceleration in the basin and some significant sub-basins within it (Restrepo et al. 2015). That augmentation can hardly be attributed to rainfall change alone, which in the northern part of South America shows an increase of around 5–7% during the last two decades (Bonachea et al. 2010).

Previous linear trend analysis of sediment load revealed that 17 watersheds on the Magdalena River basin (68% of the drainage basin area), the largest watershed draining the Andes of Colombia, showed increasing trends during the 1980–2000 period, whereas 12 locations or 31% of the land basin area displayed decreasing trends. Only three stations, representing 1% of the drainage basin area, showed no significant trend in sediment load (Restrepo and Syvitski 2006).

A study on sediment load trends for 21 main tributary systems in the central Andes of Colombia during the 1980–2010 period (Restrepo et al. 2015; Restrepo and Escobar, 2016) shows that the extent of erosion within the Andes of Colombia has severely increased over the last 30 years. For example, the last decade has been a period of increased pulses in sediment transport as seen by the statistical significant trends in load (Fig. 15.5). Overall, six sub-catchments, representing 55% of the analyzed Magdalena basin area, have witnessed increasing trends in sediment load during the last decade. Also, some major tributaries have experienced changes in their interannual mean sediment flux during the mid-1990s and 2005. Further findings show that increasing trends in sediment load match quite well with the marked increase in forest clearance during 1980–2000, 1990–2000, and 2005–2010 periods.

Monthly series of sediment load in 1972–2011, obtained at the most downstream station of the Magdalena River at Calamar (Fig. 15.1), which represents all erosion and depositional upstream processes, reveals significant trends in annual sediment load during the mid-1980s, 1990s, and post-2000 (Fig. 15.7). Between 2000 and 2010, the annual sediment load increased 33% with respect to the pre-2000 period. As a whole, the Magdalena drainage basin has witnessed an increase in erosion rates of 33%, from 550 t km^{-2} y^{-1} before 2000 to 710 t km^{-2} y^{-1} for the 2000–2010 period, and the average sediment load for the whole basin increased in 44 Mt. y^{-1} for the same period (Fig. 15.7). As highlighted by many studies focusing on global trends of sediment load, most large rivers show decreases in sediment load due to the large amount of sediment captured by reservoirs. In contrast, the Magdalena may be one of the few large world rivers experiencing such dramatic increases in sediment load during the last decade.

Trends in upstream erosion from the Andes (Fig. 15.7) are also in agreement with sedimentation rates in the lower floodplains of the Magdalena River. The analysis of the ^{210}Pb data indicates that sedimentation in the lower reach of the Magdalena River increased markedly during the period analyzed, particularly in the last 40 years. Clear sedimentation pulses are observed at the end of the 1970s and 1980s, beginning of 2000, and between 2005 and 2010 (Fig. 15.5i).

15.7 Modern and Pre-Anthropocene Conditions of Sediment Transfers from the Andes

Our inability to accurately model the human impact on sediment transport and erosion in fluvial systems remains one of the bottlenecks of the study of human-landscape interactions (Etter et al. 2006c; Syvitski and Milliman 2007). Many

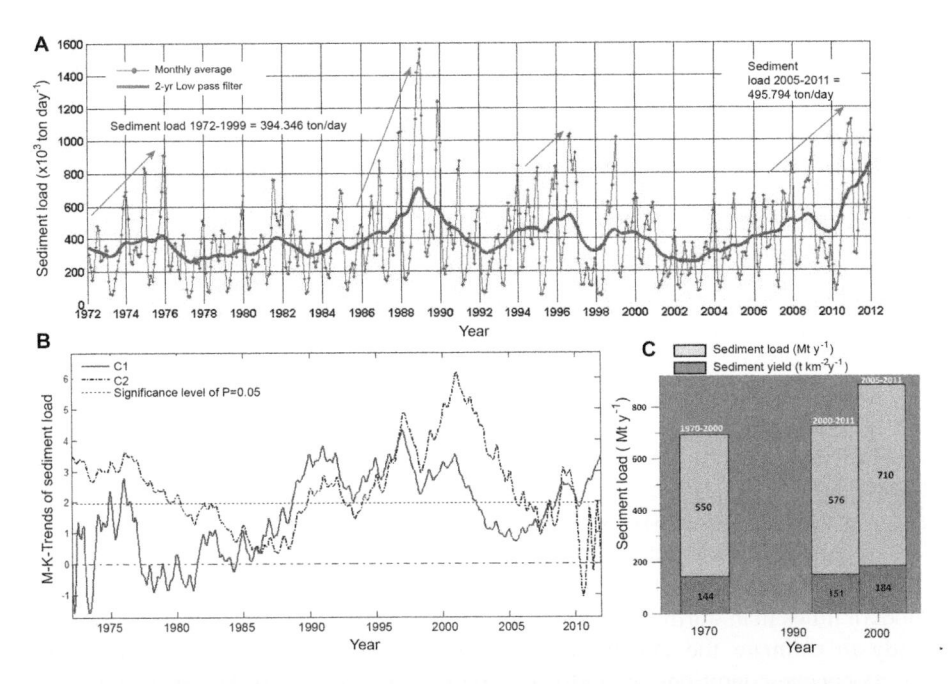

Fig. 15.7 Sediment load (**a–b**) and yield (**c**) for the whole Magdalena River basin 1972–2011, as measured at Calamar (Fig. 15.1). (**b**) The M-K trends of sediment load for the most downstream station of the main Magdalena at Calamar; progressive and retrograde series are shown as C1 and C2, respectively

algorithms to model the influence of human on sediment flux (Syvitski and Milliman 2007), including the Soil Conservation Service curve number method (Mishra et al. 2006), the revised universal soil loss equation (Erskine et al. 2002), and the water erosion prediction project model (Croke and Nethery 2006), are all designed to plot scale or, at best, small catchments and are not easily adapted to simulate human impacts on erosion for medium–large river basins (Syvitski and Milliman 2007). Also, determining the magnitude of the composite human disturbance is like trying to hit a moving target as each decade brings a new environmental situation (e.g., Restrepo and Syvitski 2006; Wang et al. 2006; Syvitski and Milliman 2007).

A recent study presenting a proxy for estimating the amount of sediment produced by human activities in the Andes of Colombia (e.g., deforestation) applied a robust model (BQART; Kettner et al. 2010; Restrepo et al. 2015), which combines natural and human forces, like basin area, relief, temperature, runoff, lithology, ice cover, and sediment trapping and soil erosion induced by humans (Syvitski and Milliman 2007). Overall, comparisons between observed and simulated sediment load of the Andean basins indicate that the BQART model (1) replicates successfully spatial distribution of sediment load with only a $10% bias over two orders of magnitude, (2) captures for 86% the between-tributary spatial variation in the 25–30-year period of observations of the 21 tributaries, (3) attributes 9%

between-river variability in sediment flux to deforestation, and (4) offers a useful method to estimate the amount of sediments produced by deforestation in tropical drainage basins. Overall, simulation results from the BQART model show that the anthropogenic-deforestation factor accounts for 10% of the between-tributary loads. The estimated 482 Mt. of sediments produced by forest clearance in the central Andes over the last three decades must be taken as a conservative value. A possible reason for this may be the underestimated rate of deforestation in 2005–2010 used in this study (Restrepo et al. 2015), which is on average 145% lower than the more accurate value obtained by Kim et al. (2015). For instance, the highest peak of forest loss on record in Colombia occurred during the 2005–2010 period (Fig. 15.4) (Kim et al. 2015).

Denudation processes are influenced by paleo conditions within the drainage basins. For example, landscape-scale erosion rates, estimated by the concentration of ^{10}Be in southeastern US river catchments, revealed that soil erosion and sediment transport during the late 1800s, when most of the region was cleared of native forest and was used most intensively for agriculture, exceeded background erosion rates by more than 100-fold (Reusser et al. 2014).

Similarly, levels of sediment load in the Andes, one order of magnitude higher in modern times than during prehuman conditions, were previously documented by a study to estimate the amount of sediments produced under modern and pre-Anthropocene conditions on a global scale (Syvitski et al. 2005). The differences between prehuman and modern sediment load in South American Rivers were more pronounced for the Magdalena River, with a difference ranging between −100 and −150 Mt. y^{-1}. Thus, during pristine conditions and according to the observed total load of the Magdalena estimated in this study, 184 Mt. y^{-1}, the Magdalena could have had an annual sediment load between 34 and 84 Mt. y^{-1} during prehuman times (Fig. 15.8).

15.8 Final Remarks

To the best of our knowledge, studies on human influence on denudation processes in the Magdalena River basin during pre-Columbian times and the nineteenth century are not available. Nevertheless, a study to estimate the amount of sediments produced under modern and pre-Anthropocene conditions on a global scale (Syvitski et al. 2005) reveals that levels of sediment transport in the Andes of Colombia are one order of magnitude higher in modern times than during prehuman conditions (Fig. 15.8).

The Andean rivers of Colombia have experienced high increases in erosion rates during the Anthropocene (last four decades) in agreement with the observed trends in land cover change. Much of the river catchments (79%) are under severe erosional conditions due in part to the clearance of more than 70% natural forest between 1970 and 2010.

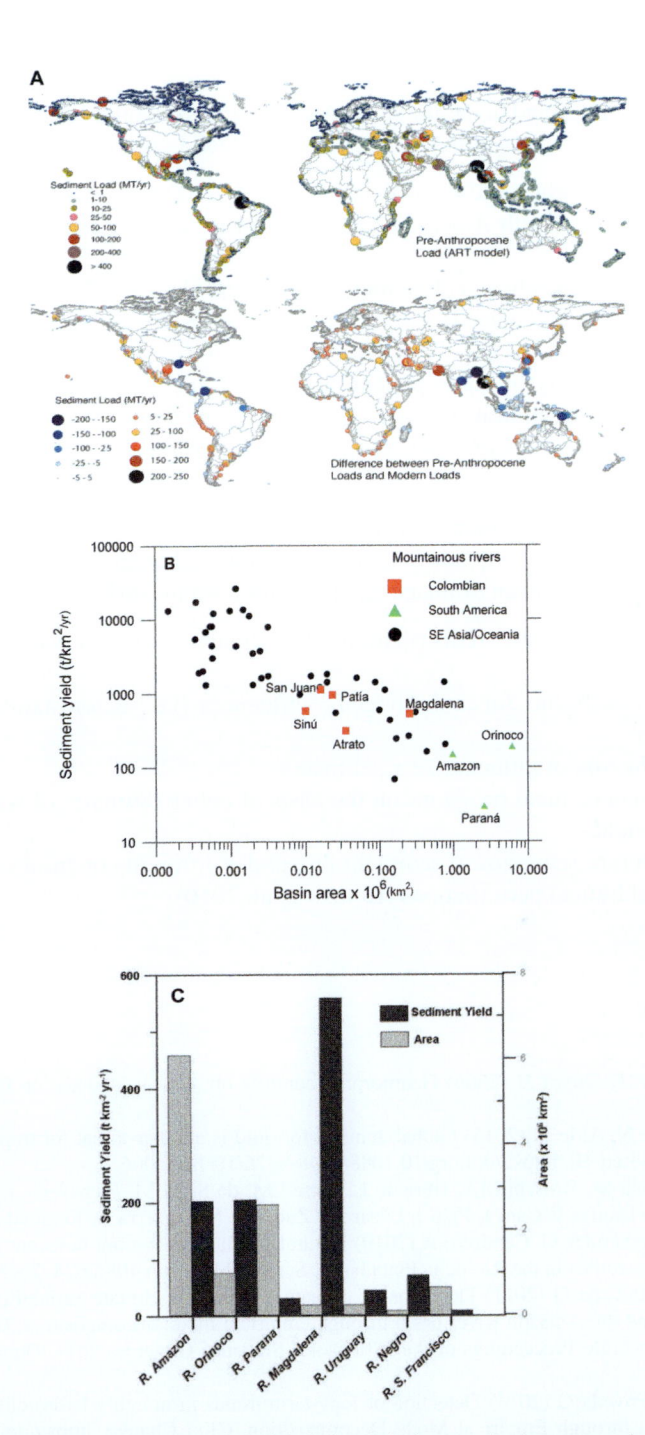

Fig. 15.8 (**a**) Sediment load of major global rivers during pre-Anthropocene (up) and modern conditions (down). Note the high deviation of sediment load from the Magdalena River between prehuman and modern loads (from Syvitski et al. 2005). (**b**) Sediment yield of major Andean rivers of Colombia compared to yields of large South American Rivers and catchments draining South East Asia and Oceania. (**c**) Sediment yield of large South American Rivers

In addition, the Magdalena, the main Andean river of Colombia, is a major and unique fluvial system in South America for the following reasons:

1. It ranks as the highest sediment-yielding river ($\sim$710 t km^2 y^{-1}) among the large rivers that drain South America (Latrubesse et al. 2005; Restrepo et al. 2015) (Fig. 15.8).
2. It is a morphologically and climatologically diverse basin with large floodplains, high mountains, and episodic local climate events (Kettner et al. 2010; Carmona and Poveda 2014).
3. There is a diversity of anthropogenic influences, including deforestation, poor soil conservation, mining practices, and increasing rates of urbanization, that have accounted for overall increasing trends in erosion and sedimentation on a regional scale.

This review about sediment transfers from the Andes of Colombia during the Anthropocene allows us to quantify spatially variable sediment fluxes and their temporal trends within Andean drainage and provides a framework to:

1. Better understand the redistribution of sediments through weathering and erosion.
2. Analyze river basins for anthropogenic influences (i.e., deforestation, mining, reservoirs).
3. Quantify factors of influence (e.g., climate).
4. Improve management decisions on the basis of better estimates of within-basin sediment yield.
5. Estimate future scenarios of sediment flux under different conditions of climate change and human perturbations (Kettner et al. 2010).

References

Aalto R, Dunne T, Guyot JJ (2006) Geomorphic controls on Andean denudation rates. J Geol 114:85–99

Alvarez-Berríos N, Aide M (2015) Global demand for gold is another threat for tropical forests. Environ Res Lett 10. https://doi.org/10.1088/1748-9326/10/1/014006

Bonachea J, Viola M, Bruschi MA, Hurtado L, Forte LM, da Silva M, Etcheverry R, Cavallotto J, Marcilene Dantas F, Osni J, Pejo J, Lázaro V, Zuquette M, Bezerra O, Remondo J, Rivas V, Gómez J, Fernández G, Cendrero A (2010) Natural and human forcing in recent geomorphic change; case studies in the Rio de la Plata basin. Sci Total Environ 408:2674–2695

Carmona AM, Poveda G (2011) Detection of climate change and climate variability signals in Colombia and the Amazon River basin through empirical mode decomposition. International Statistical Institute Proceedings of the 58th World Statistics Congress 2011, Dublin (Session IPS081)

Carmona AM, Poveda G (2014) Detection of long-term trends in monthly hydro-climatic series of Colombia through Empirical Mode Decomposition. Clim Change. https://doi.org/10.1007/s/10584-013-1046-3

Cendrero A, Remondo J, Bonachea J, Rivas V, Soto J (2006) Sensitivity of landscape evolution and geomorphic processes to direct and indirect human influence. Geogr Fis Din Quat 29:125–137

Croke J, Nethery M (2006) Modeling runoff and soil erosion in logged forests: scope and application of some existing models. Catena 67:35–49

Crutzen PJ (2002) Geology of mankind-the Anthropocene. Nature 415:23. https://doi.org/10.1038/415023a

Crutzen PJ, Stoermer EF (2000) The Anthropocene. Global change. Newsletter 41:17–18

Dadson SJ, Hovius N, Chen HG, Dade WB, Hsieh ML, Willet SD, Hu JC, Horng MJ, Chen MC, Stark CP, Lague D, Lin JC (2003) Links between erosion, runoff, variability and seismicity in the Taiwan orogen. Nature 426:648–651

Dávalos LM, Bejarando AC, Hall MA, Correa LH, Corthals A, Espejo OJ (2011) Forests and drugs: coca-driven deforestation in tropical biodiversity hotspots. Environ Sci Technol 45:1219–1227

Erskine WD, Mahmoudzadeh A, Myers C (2002) Land use effects on sediment yields and soil loss rates in small basins of Triassic sandstone near Sydney, NSW, Australia. CATENA 49(4):271–287

Etter A, McAlpine C, Pullar D, Possingham H (2005) Modeling the age of tropical moist forest fragments in heavily-cleared lowland landscapes of Colombia. For Ecol Manag 2018:249–260

Etter A, McAlpine C, Phinn S, Pullar D, Possingham H (2006a) Unplanned land clearing of Colombian rainforests: spreading like disease ? Landsc Urban Plan 77:240–254

Etter A, McAlpine C, Wilson K, Phinn S, Possingham H (2006b) Regional patterns of agricultural land and deforestation in Colombia. Agric Ecosyst Environ 114:369–386

Etter A, McAlpine C, Pullar D, Possingham H (2006c) Modelling the conversion of Colombia lowland ecosystems since 1940: drivers, patterns and rates. J Environ Manag 79:74–87

Etter A, McAlpine C, Possingham H (2008) Historical patterns and drivers of landscape change in Colombia since 1500: a regionalized spatial approach. Ann Assoc Am Geogr 98:2–23

FAO (2010) State of the World's forests 2009. Food Agr Organ United Nations Report 117.

Ferretti-Gallon K, Busch J (2014) What drives deforestation and what stops it? A meta analysis of spatially explicit econometric studies. Center for Global Development (GD), CGD climate and Forest paper series, 361

Geist HJ, Lambin EF (2001) What drives tropical deforestation? Land-Use and Land-Cover Change project (LUCC) report series no.4. Ciaco Printshop. Louvain-la- Neuve, Belgium, p 116

Geist HJ, Lambin EF (2002) Proximate causes and underlying driving forces of tropical deforestation. Bioscience 52:143–150

Guns M, Vanacker V (2013) Forest cove change trajectories and their impact on landslide occurrence in the tropical Andes. Environ Earth Sci 70:2941–2952

Guns M, Vanacker V (2014) Shifts in landslide frequency-area distribution after forest conversion in the tropical Andes. Anthropocene 6:75–78

Hansen MC, Potapov PV, Moore R, Hancher M, Turubanova SA, Tyukavina A, Loveland TR (2013) High-resolution global maps of 21st-century forest cover change. Science 342(6160):850–853

Harden CP (2006) Human impacts on headwater fluvial systems in the northern and central Andes. Geomorphology 79:249–263

Harrison CGA (2000) What factors control mechanical erosion rates? Int J Earth Sci 531:1–11

Hess CG (1990) Moving up moving down: agro-pastoral land- use patterns in the Ecuadorian paramos. Mt Res Dev 10:333–342

Hibbard KA, Crutzen PJ, Lambin EF, Liverman DM, Mantua NJ, McNeill JR, Messerli B, Steffen W (2006) Decadal interactions of humans and the environment. In: Costanza R, Graumlich L, Steffen W (eds) Integrated history and future of people on earth, Dahlem workshop report 96. The MIT Press, Cambridge, pp 341–375

Houghton RA (1994) The worldwide extent of land-use change: in the last few centuries, and particularly in the last several decades, effects of land-use change have become global. Bioscience 44:305–313

Hovius N (1998) Controls on sediment supply by large rivers. In: Relative role of eustacy, climate, and tectonism in continental rocks, vol 59. Soc. Sediment. Geol. Spec. Publ, pp 3–16

Hovius N (2000) Macroscale process systems of mountain belt erosion. In: Summerfield MA (ed) Geomorphology and global tectonics. Wiley, Hoboken, pp 77–105

Hovius N, Stark CP, Allen PA (1997) Sediment flux from a mountain belt derived by landslide mapping. Geology 25:231–234

IDEAM (2014) Deforestation assessment in Colombia 2005-2010. IDEAM-REED Project. MODIS Land Cover Data Base, Bogotá

Kaplan JO, Krumhardt KM, Ellis EC, Ruddiman WF, Lemmen C, Goldewijk KK (2010) The Holocene 21(5):775–791

Kettner A, Restrepo JD, Syvitski JPM (2010) Simulating spatial variability of sediment fluxes in an Andean drainage basin, the Magdalena River. J Geol 118:363–379

Kim DH, Sexton JO, Townshend JR (2015) Accelerated deforestation in the humid tropics from the 1990s to the 2000s. American Geophysical Union. https://doi.org/10.1002/2014GL062777

Latrubesse E, Restrepo J (2014) The role of Andean rivers on global sediment yield. Geomorphology 216:225–233

Latrubesse EM, Stevauxs JC, Sinha R (2005) Tropical rivers. Geomorphology 70:187–206

McNeill JR, Engelke P (2016) The great acceleration. Harvard Univ, Press, Cambridge Mass

Milliman JD, Syvitski JPM (1992) Geomorphic/tectonic control of sediment transport to the ocean: the importance of small mountainous rivers. J Geol 100:525–544

Mishra SK, Tyagi JV, Singh VP, Singh R (2006) SCS-CN-based modeling of sediment yield. J Hydrol 324(1–4):301–322

Molina A, Govers G, Poesen J, Van Hemelryck H, De Bievre B, Vanacker V (2008) Environmental factors controlling spatial variation in sediment yield in a central Andean mountain area. Geomorphology 98:176–186

Molina A, Vanacker V, Brisson E, Mora D, Balthazar V (2015) Long-term effects of climate and land cover change on freshwater provision in the tropical Andes. Hydrol Earth Syst Sci 19:1–32

Morgan RPC (1986) Soil erosion and conservation. Longman, Harlow

Mou J (1996) Recent studies of the role of soil conservation in reducing erosion and sediment yield in the loess plateau are of the Yellow River basin. In: Walling DE, Webb BW (eds) Erosion and sediment yield: global and regional perspectives. Proceedings of the Exeter symposium, July 1996, IAHS publication, vol 236. IAHS Press, Wallingford, pp 541–548

Pinet P, Souriau M (1988) Continental erosion and large-scale relief. Tectonics 7:563–584

Restrepo JD (2012) Assessing the effect of sea-level change and human activities on a major delta on the Pacific coast of northern South America: the Patía River. Geomorphology 151–152:207–223

Restrepo JD (2013) The perils of human activity on South American deltas: lessons from Colombia's experience with soil erosion. Deltas: landforms, ecosystems and human activities, 358. IAHS Publ:143–152

Restrepo JD, Escobar HA (2016) Sediment load trends in the Magdalena River basin (1980–2010): anthropogenic and climate-induced causes. Geomorphology. https://doi.org/10.1016/j.geomorph.2016.12.013

Restrepo JD, Kettner A (2012) Human induced discharge diversion in a tropical delta and its environmental implications: the Patía River, Colombia. J Hydrol 424:124–142

Restrepo JD, Kjerfve B (2000) Water discharge and sediment load from the western slopes of the Colombian Andes with focus on Rio San Juan. J Geol 108:17–33

Restrepo JD, Syvitski JPM (2006) Assessing the effect of natural controls and land use change on sediment yield in a major Andean river: the Magdalena Drainage Basin, Colombia. AMBIO J Hum Environ 35:44–53

Restrepo JD, Kjerfve B, Hermelin M, Restrepo JC (2006) Factors controlling sediment yield in a major South American drainage basin: the Magdalena River, Colombia. J Hydrol 316:213–232

Restrepo JD, Kettner A, Syvitski J (2015) Recent deforestation causes rapid increase in river sediment load in the Colombian Andes. Anthropocene 10:13–28

Reusser L, Bierman P, Rood D (2014) Quantifying human impacts on rates of erosion and sediment transport at a landscape scale. Geology. https://doi.org/10.1130/G36272.1

Sánchez-Triana E, Ahmed K, Awe Y (2007) Prioridades ambientales para la reducción de la pobreza en Colombia: un análisis ambiental del país para Colombia. Informe del Banco

Mundial, Direcciones para el Desarrollo, medio ambiente y desarrollo sustentable. Report No. 38610. Washington, DC, 522

Steffen W, Broadgate W, Deutsch L, Gaffney O, Ludwig C (2015) The trajectory of the Anthropocene: the great acceleration. Anthropocene Rev 2(1):81–98. https://doi.org/10.1177/2053019614564785

Steffen W et al (2016) Stratigraphic and earth system approaches to defining the Anthropocene. AGU Publ Earth's Fut 4

Summerfield MA, Hulton NJ (1994) Natural controls of fluvial denudation in major world drainage basins. J Geophys Res 99:13871–13884

Syvitski JPM (2003) Supply and flux of sediment along hydrological pathways: research for the 21st century. Glob Planet Change 39:1–11

Syvitski JPM, Kettner AJ (2011) Sediment flux and the Anthropocene. Phil Trans R Soc 369:957–975

Syvitski JPM, Milliman JD (2007) Geology, geography, and humans battle for dominance over the delivery of fluvial sediment to the Coastal Ocean. J Geol 115:1–19

Syvitski JPM, Vörösmartry CJ, Kettner AJ, Green P (2005) Impact of humans on the flux of terrestrial sediment to the Global Ocean. Science 308:376–380

Trimble SW (1975) Denudation studies: can we assume steady state? Science 188:1207–1208

Trimble SW, Crosson P (2000) Land use, US soil erosion rates myth and reality. Science 289:248–250

US Department of State (2001) The Andes under Siege; Environmental consequences of the drug trade. http://usinfo.state.gov/products/pubs/archive/

Vanacker V, Vanderschaeghe M, Govers G, Willems E, Poesen J, Deckers J, De Biévre B (2003) Linking hydrological, infinite slope stability and land use change models through GIS for assessing the impact of deforestation on landslide susceptibility in high Andean watersheds. Geomorphology 52:299–315

Vanacker V, Bellin N, Molina A, Kubik PW (2014) Erosion regulation as a function of human disturbances to vegetation cover: a conceptual model. Landsc Ecol 29:293–309

Vanmaercke M, Poesen J, Verstraeten G, de Vente J, Ocakoglu F (2011) Sediment yield in Europe: spatial patterns and scale dependency. Geomorphology 130:142–161

Walling DE (2006) Human impact on land–ocean sediment transfer by the world's rivers. Geomorphology 79:192–216

Walling DE, Fang D (2003) Recent trends in the suspended sediment loads of the world's rivers. Glob Planet Chang 39:111–126

Wang H, Yang Z, Saito Y, Liu JP, Sun X (2006) Interannual and seasonal variation of the Huanghe (Yellow River) water discharge over the past 50 years: connection to impacts from ENSO events and dams. Glob Planet Change 50:212–225

Waters CN et al (2016) The Anthropocene is functionally and stratigraphically distinct from the Holocene. Science 351(6269):137. https://doi.org/10.1126/science.aad2622

Williams M et al (2016) The Anthropocene: a conspicuous stratigraphical signal of anthropogenic changes in production and consumption across the biosphere. AGU Publ Earth's Fut 4:34–53

Wunder S (1996) Deforestation and the uses of wood in the Ecuadorian Andes. Mt Res Dev 16:367–382

Xu JX (2003) Sediment flux to the sea as influenced by changing human activities and precipitation: example of the Yellow River, China. Environ Manag 31:328–341

Zalasiewicz J et al (2008) Are we now living in the Anthropocene. GSA Today 18:4–8. https://doi.org/10.1130/GSAT01802A.1

Chapter 16
The Historical, Geomorphological Evolution of the Colombian Littoral Zones (Eighteenth Century to Present)

Iván D. Correa and Cristina I. Pereira

16.1 Introduction

It has long been recognized that human occupation of coastal areas during the last centuries has had profound, cumulative effects on the geomorphology and the environmental quality of populated littoral zones around the world (Morton 2002; Eurosion 2004; Pranzini and Williams 2013). In fact, all the factors considered by the International Geological Program (IGP) to formally define the Anthropocene as a new stratigraphic unit – land use practices, river diversions, sand and fluid extraction, and infrastructure building – had greatly modified the natural processes and caused major impacts in many places on the original, pristine nature of the littoral ecosystems (Subcommission on Quaternary Stratigraphy 2016). Thus, human activities are now considered as one of the main factors in all conceptual frame-works dealing with the evolution of coastal environments (Fig. 16.1).

Sedimentary balances result from multiple feedbacks at all time scales between climate, coastal processes, relative sea-level changes, and human activities. Positive sediment budgets set up local or regional regressive conditions and are commonly reflected by coastal accretion and the formation of depositional features, typically sedimentary prisms, deltas, tidal flats, beaches, offshore bars, and dunes. Negative budgets originate transgressive erosional conditions resulting in beach retreat, land losses, flooding and salinization of coastal deposits, among other undesirable impacts (Fig. 16.1). Transgressive conditions do not necessarily result in beach destruction, since sandy coastlines like barrier islands are able to migrate onshore and avoid destruction in the case of free evolution and enough accommodation space (Pilkey 1983; Kaufman and Pilkey Jr. 1983). Obviously, future accelerated sea-level rise resulting from climate change and/or coastal subsidence adds a new dimension to the areal extension and rates of expected changes along much of the

---

I. D. Correa (✉) · C. I. Pereira
Area de Ciencias del Mar, Universidad EAFIT, Medellín, Colombia

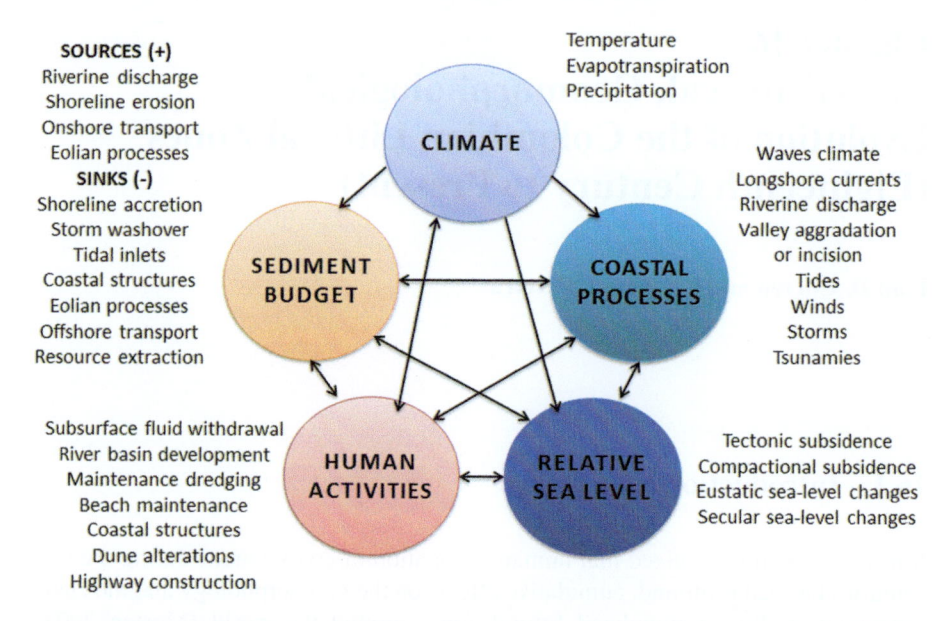

Fig. 16.1 Factors affecting coastal environments. After Morton and Pieper (1977) and Williams et al. (1995)

SOURCES (+)
Riverine discharge
Shoreline erosion
Onshore transport
Eolian processes
SINKS (−)
Shoreline accretion
Storm washover
Tidal inlets
Coastal structures
Eolian processes
Offshore transport
Resource extraction

Subsurface fluid withdrawal
River basin development
Maintenance dredging
Beach maintenance
Coastal structures
Dune alterations
Highway construction

CLIMATE — Temperature, Evapotranspiration, Precipitation

COASTAL PROCESSES — Waves climate, Longshore currents, Riverine discharge, Valley aggradation or incision, Tides, Winds, Storms, Tsunamis

SEDIMENT BUDGET

HUMAN ACTIVITIES

RELATIVE SEA LEVEL — Tectonic subsidence, Compactional subsidence, Eustatic sea-level changes, Secular sea-level changes

low coastal zones around the world, at times where human occupation of these areas is growing at unprecedented rates (Bird 2010; World Economic Forum 2015).

Although its occupation by European conquerors and settlers began not before the sixteenth century, the magnitudes and consequences of much of the historical morphological changes along the Colombian seaboards are good enough to represent both scientific and practical interests. The dynamic character of all natural factors of northwestern South America and the inappropriate development of the Colombian coastal watersheds have resulted in numerous critical cases of physical instability and strong environmental deterioration. Geological hazards and vulnerability related to flooding and shore erosion are growing rapidly in the Colombian seaboards, especially along the densely populated urban areas of the Caribbean coast main cities (Santa Marta, Cartagena, Turbo) presently subjected to extensive flooding during high tides and/or heavy weather (Ortiz 2012; Rangel-Buitrago and Anfuso 2015); on the Pacific coast, main cities (Buenaventura, Bahia Solano, Guapi and Tumaco) and all the littoral villages are highly vulnerable to earthquake and tsunamis (Correa 1996; Herd et al. 1981; Correa and Morton 2003; Posada et al. 2009; Corporación OSSO 2008; AIS 2009). Strong environmental deterioration caused by shore erosion, waste disposal, agriculture, mining, and pollution of coastal waters and soils is currently of primary interest in the environmental agenda of the country (Olivero and Johnson 2002; INVEMAR 2003; Restrepo et al. 2014, 2015). These facts urgently call for intensive research about the ecological evolution of the Colombian littoral zones from an anthropogenic point of view. The information about past trends and factors involved in the physical evolution of the different coastal types of Colombia is then necessary for predicting future trends and implementing adequate coastal management strategies.

The purpose of this chapter is to illustrate the magnitudes and causes of some of the major historical morphological changes along the Colombian Caribbean and Pacific seaboards. Some of these changes have been of kilometric magnitudes, and most of them can be easily identified by comparing shore contours and morphological features depicted in modern maps and remote sensing materials with those shown in some historical reliable charts dating back to the end of the eighteenth century. Among the first reliable cartographic charts available are those produced by the Spanish Brigadier Francisco Fidalgo (The Fidalgo Expedition) along the southern Caribbean Sea, including the coasts of Panama, Colombia, and Venezuela. These charts were based on field work during the period 1792–1812 and have been fully recognized for their quality and accuracy (Domínguez et al. 2012; Fuentes and Jaramillo 2015). They can be also complemented with other ancient charts available for the main Colombian Caribbean deltaic zones and urban areas elaborated during the nineteenth century.

16.2 Main Historical Morphological Changes along the Central Caribbean Coast of Colombia

Although noticeable geomorphological instability has been reported for numerous sectors between Santa Marta and Castilletes (Correa and Morton 2003; Posada and Henao 2008; Dimar-CIOH 2013; Rangel-Buitrago et al. 2015), the main historical morphological changes along the Caribbean are found along its central sector, here defined between the Magdalena River delta-Salamanca bar and the Sinú-Tinajones River delta, on the south westernmost part of the Morrosquillo Gulf (Fig. 16.2). Coastal evolution of this zone has been extremely dynamic in historical times and has included both erosional and depositional events that have greatly modified the littoral landscapes.

16.2.1 General Context

The tropical, mixed micro-tidal (40 cm tidal amplitude), wave-dominated Caribbean littoral of Colombia extends for about 1700 km (scale 1:100,000) between Punta Castilletes (Venezuelan border) and Cabo Tiburones, at the Panamá border, northern tip of The Urabá Gulf (Fig. 16.2). The Caribbean Coast of Colombia is a physiographic region with medium diurnal temperatures of 28°–30 ° C° and annual rainfalls varying between 300 and 600 mm in the northeastern sector and 1200 and 2000 mm in the southwestern part (IDEAM 2010). Semi-desert conditions are found at the northern part of La Guajira peninsula, and a dry climate is found between the littorals of Santa Marta and Cartagena. Humid conditions are dominant

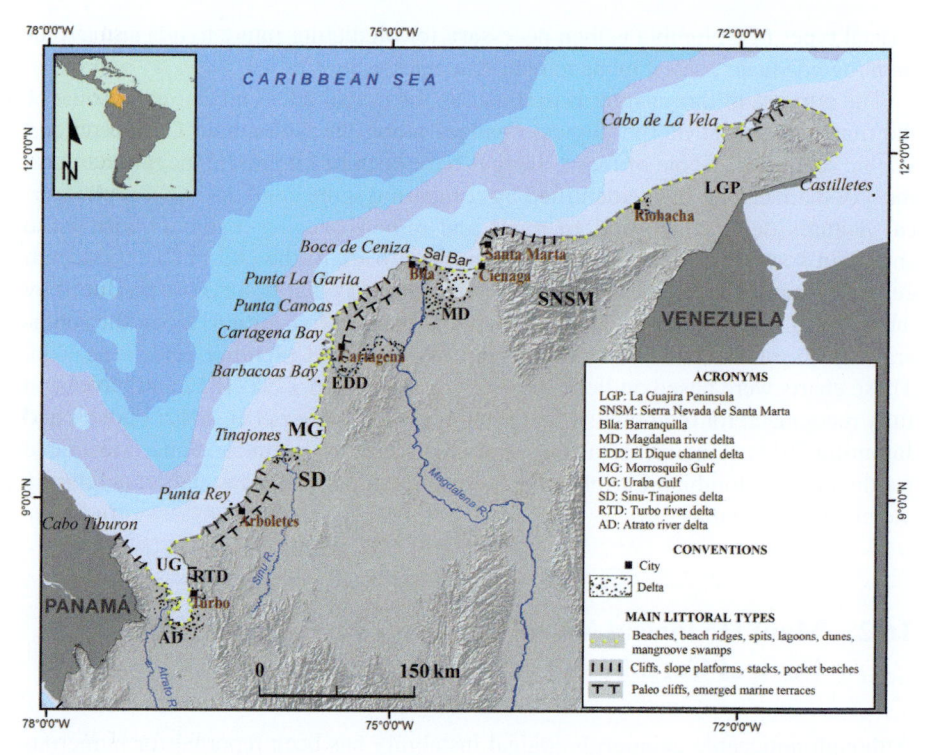

Fig. 16.2 Location map and main morphological types of the Colombia Caribbean littoral

along the Gulf of Morrosquillo area and the Sierra Nevada de Santa Marta Massif SNSM, the highest mountain in the world plunging directly to the sea and reaching its maximum height at Pico Bolivar (5755 masl).

Oblique, up to 7 Kt, NE-NW incidence of the trade winds (Alisios) on the Caribbean's shores generates high waves during summertime (November–April) which induces strong longshore currents and net sand transport in a SW direction along most of its length (Invemar 2006). This condition is morphologically reflected by the dominant, S to SW orientation of spits and by the plain-view configuration of the z-bays form orientation. During the rest of the year, longshore currents are reversed by EW and SW winds and waves when sand transport has a northern component (Correa 1990; Invemar 2006; Dimar-Cioh 2013).

Except for the internal shores of the Morrosquillo Gulf, the littoral fringe of the southern Caribbean coast of Colombia between Ciénaga and the Urabá Gulf is entirely located on the Sinú belt, a tectonically active, fragmented tertiary accretionary prism whose evolution has been strongly influenced by mud diapirism since Miocene times (Geotec 2003; Vernette 1985). Mud diapirism has been interpreted as one of the main controls of the sedimentation patterns, shelf topography, and the occurrences and distribution of reefs along the southern Caribbean coast of Colombia (Shepard et al. 1968; Vernette 1989; Restrepo et al. 2007; Ojeda et al. 2007; Carvajal and Mendivelso 2011; Vernette et al. 2011; Carvajal 2016).

Of direct interest in the context, the historical activity of mud diapirism in the area includes major landscape modifications through, in many cases, violent extrusions of mud and gases both in the continental shelf and inland (Ramírez 1959; Raasveldt and Tomic 1958; Correa 1990; Carvajal 2016). Late Holocene relative changes in sea level associated to mud diapirism, fault activity, and neotectonism are evidenced along the Galerazamba-Cartagena area by emerged coralline terraces and marsh deposits located between 1 and 15 masl. These deposits have been radiometrically dated between 2700 and 3600 BP and interpreted upheaving varying between app 3.8 mm/year and 1.5 mm/year (Richard and Broecker 1963; Burel and Vernette 1981; Vernette 1985, 1989; de Porta et al. 2008; Page 1983; Martínez et al. 2010; Carvajal 2016).

Main littoral types along the Caribbean seaboard reflect the inherent geology of each coastal strip and its Late Holocene history and climate (Correa and Morton 2003; Correa et al. 2007a, b; Posada and Henao 2008; Dimar-Cioh 2013). They range between steep-plunging cliffs cut on metamorphic and igneous rocks found at the SNSM, some short sectors of the Guajira peninsula and the northwestern part of the Uraba Gulf, and the low-relief, late Holocene sandy and muddy deposits forming offshore bars, spits, beach ridge lagoon complexes, and large areas of mangrove swamps located on the Magdalena, El Dique channel, the Sinú-Tinajones, and Atrato River deltas (Fig. 16.2). Minor areas of beach ridge lagoons and mangrove swamps are found in the embouchures of minor rivers of the Caribbean (Riohacha, Dibulla, Moñitos, Córdoba, Hobo, Mulatos), some of them regularizing the coastal indentations (plain view) in presently non-deltaic zones and small river mouths along the low-relief shores of La Guajira Peninsula and the Gulf of Morrosquillo. Longitudinal mobile dunes up to 10 meters high are found at the Caribbean coast of Colombia only to the north of Cartagena city where dry conditions, low relief, sand availability, and the incidence of the Alisios facilitate its formation. Major dune zones are found at the Galerazamba area, the Salamanca bar (in front of the Ciénaga Grande de Santa Marta (CGSM) lagoon, east of the Magdalena delta), and the northern part of La Guajira Peninsula (Raasveldt and Tomic 1958; Khobzi 1981; Posada and Henao 2008; DIMAR-CIOH 2009a, b; Gómez et al. 2016).

16.2.2 Historical Morphological Changes along the Magdalena River Delta Shores and Prodelta

The historical morphological changes along the shores and prodelta of the Magdalena River delta are the best available examples of dramatic, short-term littoral changes induced by human actions. Current active and severe erosional trends were triggered along the coastline of the Magdalena River delta and surrounding shores as a direct result of the construction of the Bocas de Ceniza jetties (Fig. 16.3). These structures were built to channel the Magdalena River's mouth and allow the entrance of big ships from the Caribbean Sea into the fluvial port at Barranquilla city. The first step in the construction of these 800-meter-long, parallel and rocky/

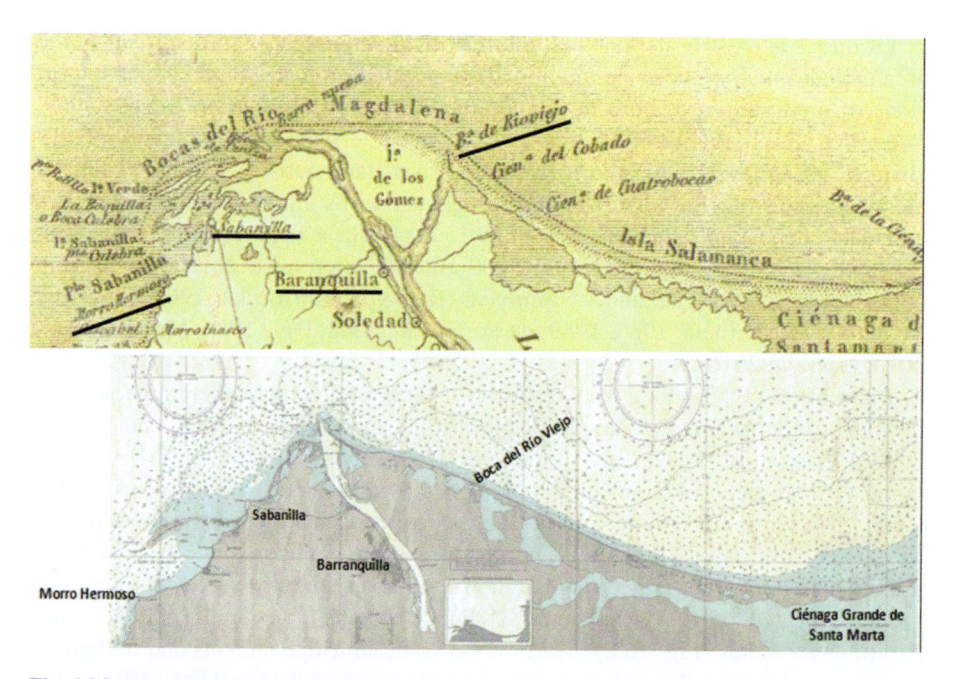

Fig. 16.3 Magdalena River delta shores, 1852 and 1934. Upper: Magdalena River delta coastline and offshore bars to Puerto Colombia, including Isla Verde; chart published in 1852, taken from *Comisión Corográfica y de orden del Gobierno General, por Manuel Ponce de León y Manuel María Carta corográfica del Estado de Bolívar, la Paz, Bogotá, 1864*. Down: USS Nokomis Chart 5688, after hydrographical surveys data obtained during 1935–1936

concrete structures was finished in 1934, and the short-term results show strong modifications to the sediment balances of the zone (Koopmans 1971; Alvarado 2007). By intercepting the E-W net longshore drift along the delta's shore, they triggered an acute deficit of sand in the down-current direction, resulting in the erosion of all emerged and submerged sandy shoals west of the river. This included the Sabanilla barrier island that in 1945 (7 years after the longshore disruption) was delineated as an erosional submarine sandy shoal in a photo interpretation made by Raasveldt and Tomic (1958) (Fig. 16.3). Recent studies for the last 50 years of evolution of the coast southwest of Bocas de Ceniza have shown the progressive formation of newer and smaller depositional sand bodies to the south of the initial position of river shoals and bars (Martínez et al. 1990; Anfuso et al. 2015). The importance of mud diapirism and possible structural control on the orientation of these features has been argued by Anfuso et al. (2015) and Hermelin (2015) among others.

Coastal instability along the Magdalena delta front and nearby shores has not been restricted to sand imbalances along its shore areas and shallow platform but also to deeper waters. Besides interfering with the east-west directed sand drift along the delta's shores, the jetties also concentrated the sand deposition (estimated in 30×10^6 m^3/year) just offshore of its ending points and induced strong conditions of instability on the delta submerged front (Laboratorio Central de Hidráulica de

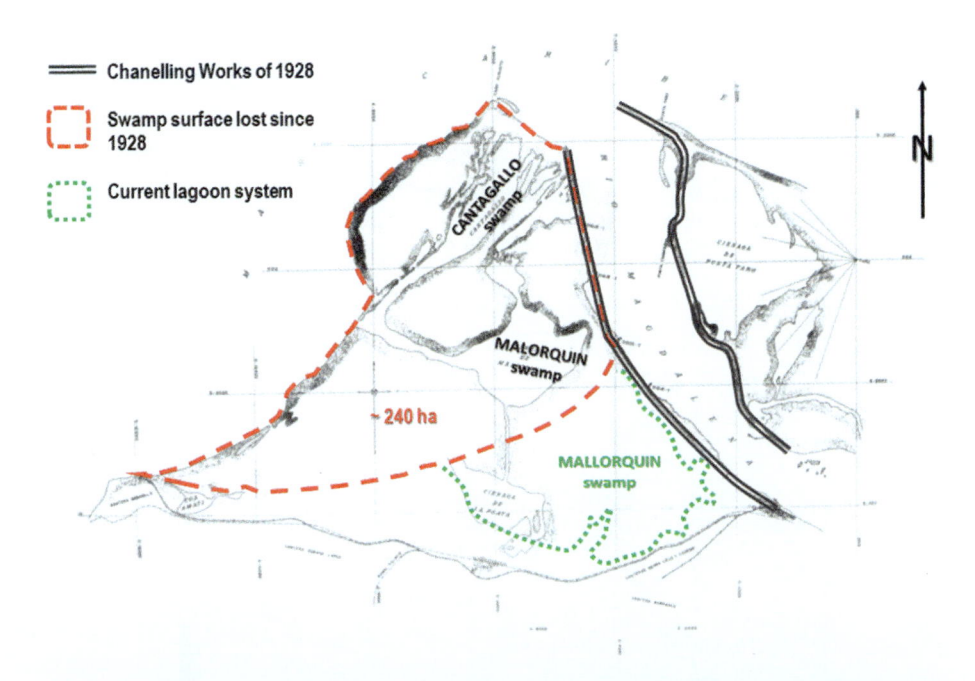

Fig. 16.4 Comparison of shoreline configuration west of the Magdalena River mouth between 1928 and present, indicating the swamp surface lost since the construction of the channeling works of Bocas de Ceniza in red and the remnant of the lagoon system of the current delta in green. The historical map represents the progress of the channeling works at Bocas de Ceniza by 1928. Sources: Rico Pulido (1969)

Francia 1958). As a result, and coinciding with times of high discharge picks of the river (August and November–December), at least five submarine slides and turbidity currents occurred in the area between 1935 and 1963. Two of these slides (in 1935 and 1945) started near the shore and eroded 248 m and 500 m of the outer ends of the eastern and western groins, respectively (Heezen 1956; Koopmans 1971; Alvarado 2007). Turbidity currents followed the ancient Magdalena River channels and involved bathymetrical changes up to 200 m. They caused the rupture of several submarine cables located up to 24 km distant from the river mouth and at 1400 m of water depth on at least five different occasions (Figs. 16.4 and 16.5).

16.2.3 Historical Morphological Changes between Galerazamba (La Garita Point) and the Sinú-Tinajones Delta

South of the Magdalena delta area, the eighteenth century configuration of the Caribbean littoral has been profoundly modified both by erosional and accretional trends involving kilometric modifications of the coastal contours. Most of the length

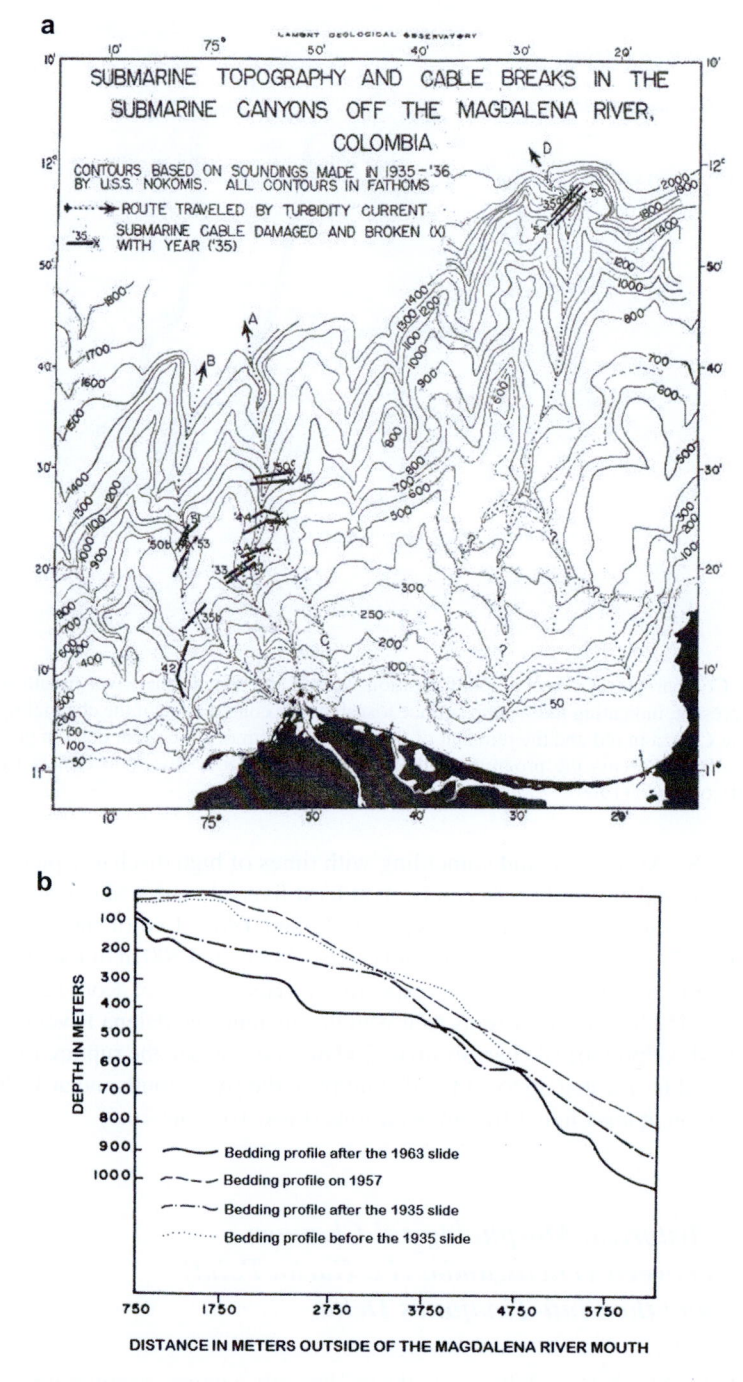

Fig. 16.5 Submarine slides at the Magdalena delta front. a: paths of submarine slides and location of cable's breaks. After Heezen (1956). b: submarine profiles changes along the Magdalena delta front after the 1935, 1953, 1957, and 1963 submarine slides; after Rico Pulido in Koopmans (1971)

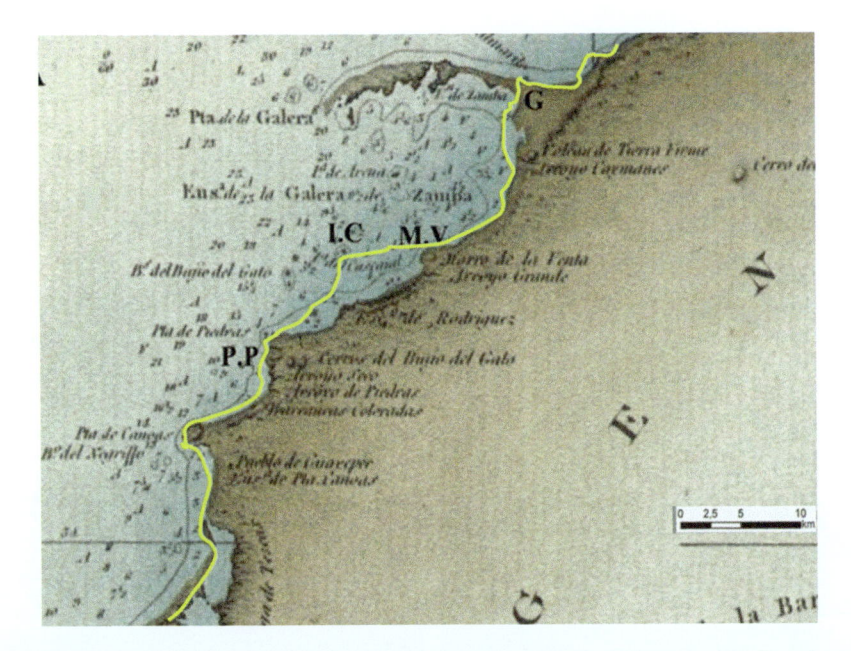

Fig. 16.6 Historical coastline changes at Galerazamba (La Garita Point) and Isla Cascajo area, Barranquilla-Cartagena littoral from Brigadier Fidalgo chart made with field data from the end of the eighteenth century. Current configuration of shoreline is depicted in yellow along with the conventions *G* La Garita Point, *IC* Cascajo Island, *MV* Morro de La Venta Point, northeastern extreme of Isla Cascajo tombolo, *PP* Punta de Piedra point, southwestern extreme of Isla Cascajo tombolo

of the central and southern Caribbean of Colombia is currently dominated by strong net erosional trends up to 1 m/year (40 m/year in some exceptional cases) due in part to the interruptions of the sand longshore drift by more than 600 hard coastal defenses. Negative littoral deficits of sediments are also produced by intensive sand and gravel mining dating back to the beginning of the last century (Correa 1990; Correa and Morton 2003; Posada and Henao 2008; Rangel et al. 2015).

Most notorious, entirely naturally-induced morphological changes along the central Caribbean coincide with zones of active mud diapirism and are well illustrated between Barranquilla and Cartagena by the disappearance of the 10 km-length Galerazamba spit (Figs 16.6 and 16.7) and by the formation of the Isla Cascajo tombolo (Fig. 16.8). This latter feature, with an area of 25 km^2 (measured on a Google Earth image of 2016) conforms a beach ridge-dune-lagoon complex capped in some places by fluvial deposits. Today it represents the surface of a sandy accretionary prism whose deposition was promoted by the strong wave diffraction of N-NE incident swells around the Isla Cascajo island, in conditions of high sediment supplies. The maximum water depths at the shallow platform were of approx. 5.5 m according to bathymetrical surveys of the USS Nokomis made in 1934. Its formation initiated at an unknown date between 1792 and 1934 (Raasveldt and Tomic 1958; Correa 1990; Anfuso et al. 2015).

Fig. 16.7 Galerazamba (Punta Garita) area coastline changes and geomorphological features. (**A**) net coastline changes between 1794 and 1996, including the erosion of the Galerazamba spit and the noticeable cliff retreat to the south, in absence of the previous wave protection offered by the spit (from Correa 1990); (**B**) El Totumo mud volcano, view to the east (Photo by Ivan Correa); (**C**) El Totumo mud volcano – 15 m height, located on the northern flank of a 70 m-height diapiric dome (Photo by Ivan Correa); (**D**) touristic use of the originally 1-m-wide El Totumo volcano crater (Photo by Ivan Correa); (**E**) High-energy dissipative beaches to the east of La Garita Point; (**F**) Active, 7 m-height retreating cliffs south of Galerazamba (Photo by Ivan Correa). The Galerazamba area is famous for several violent explosive events of offshore and onshore volcanoes (at least three in the past century), the last being the explosion of the Pueblo Nuevo mud volcano in 2008 that represented a serious risk for the inhabitants of Pueblo Nuevo village

On the western tip of the Morrosquillo Gulf, the deposition of the Tinajones delta is the most important event of historical littoral accretion on the entire Caribbean coast (Fig. 16.9). The digging of drainage channels between the river's course and the sea at the Tinajones area facilitated the deviation of the main course of the Sinú River toward the sea in 1942–1943. This intervention induced the formation of the new delta, a lobular deposit presently with an overall area of 29 km^2 whose evolution has been described and interpreted in detail by several authors including

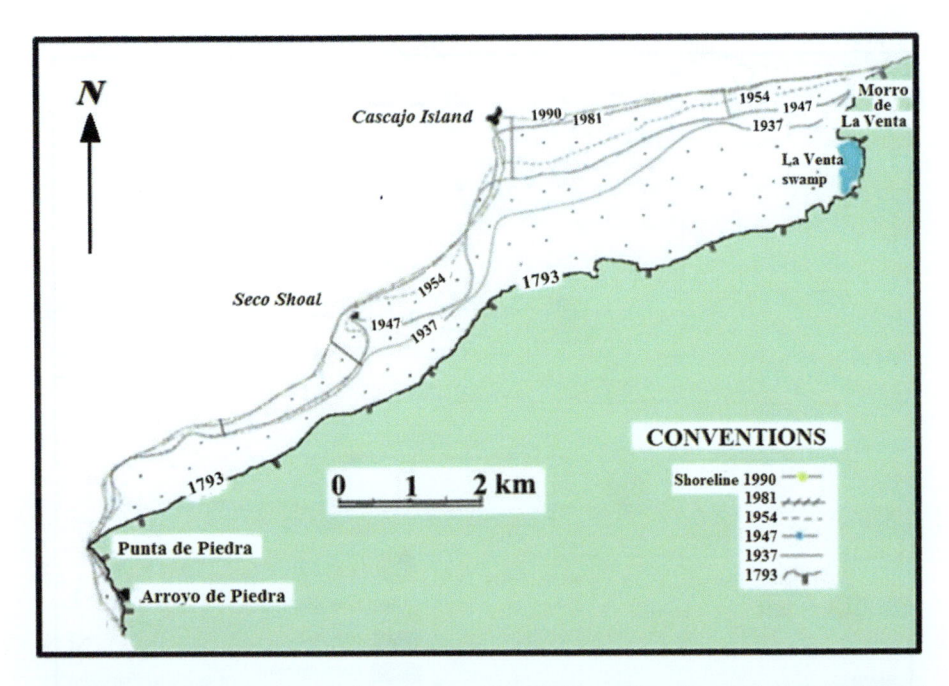

Fig. 16.8 Isla Cascajo area coastline changes. Historical coastline changes between 1793 and 1990 showing successive coastline positions in 1793, 1937, 1947, 1954, 1981, and 1990 (After Correa 1990). Tombolo formation initiated at some time between 1793 (Fidalgo's cliffed coastline) and 1937 (coastline depicted in the USS Nokomis navigation charts)

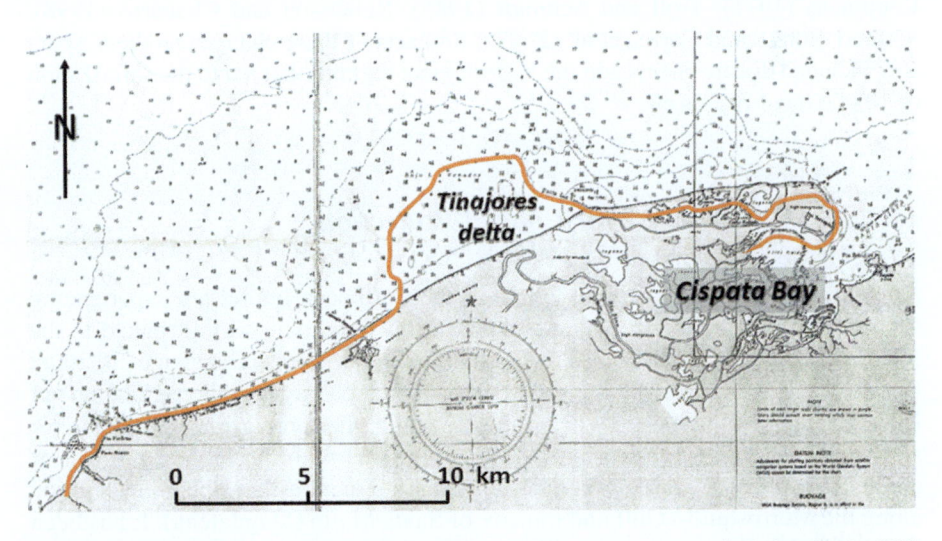

Fig. 16.9 Tinajones delta and Cispatá bay. 1938 coastline shown on a nautical chart by the USS Nokomis based on surveys made in 1934

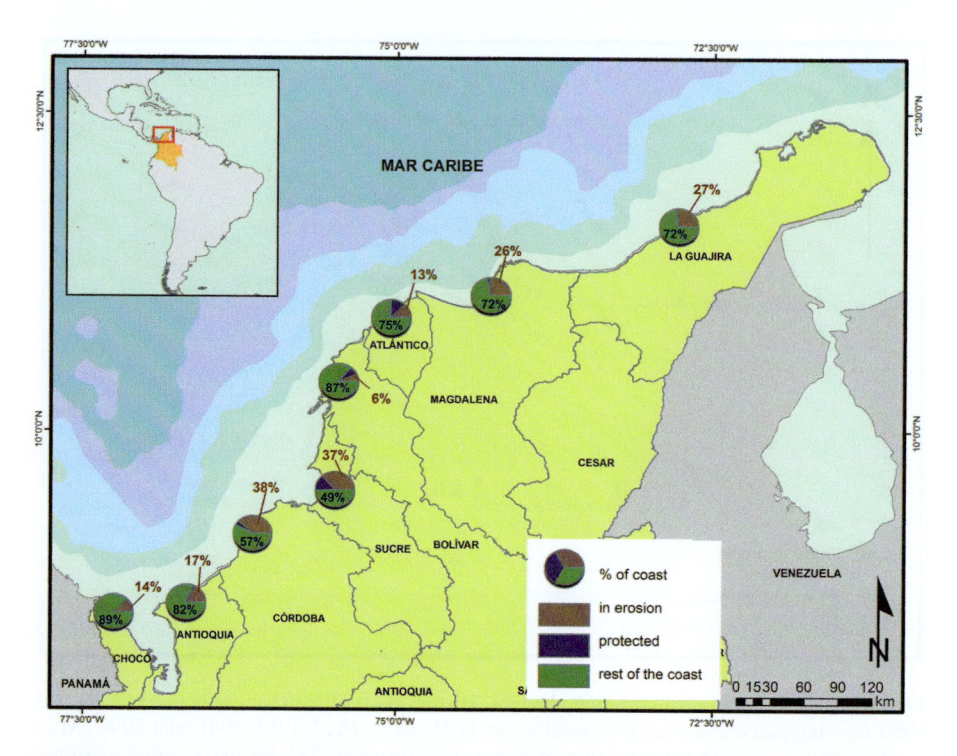

Fig. 16.10 Proportion of shoreline in erosion and with protection by department of the Caribbean coast. Data from Posada and Henao (2008)

Koopmans (1971), Troll and Schmidt (1985), Robertson and Chaparro (1998), Serrano (2004), and Correa et al. (2007a). Impacts of these changes on the Cispatá Bay included drastic changes in the hydrological regimes that led to the salinization of 10,000 ha of rice crops.

16.2.4 Infrastructure for Coastal Protection

Approximately 22% of the total length of the Caribbean littoral are experiencing severe erosion (see Fig. 16.10), and much of this trend is in part attributed to the interruption of sand drifts caused by coastal protection works (Guzman et al. 2008; Correa et al. 2007b; Rangel-Buitrago et al. 2015). According to the estimations of Posada and Henao (2008), the departments with higher proportion of coastal erosion are Sucre and Cordoba, but Bolivar is the one with the highest amount of coastal structures. A total of 496 coastal defense structures have been inventoried along the Morrosquillo Gulf and the city of Santa Marta (Magdalena). It has been concluded that these structures allowed the preservation of some local areas but caused important imbalances in the sediment budgets of other sectors (Correa et al. 2005; Rangel-Buitrago et al. 2012).

Most rigid defenses (90% of the coastal protection structures according to Posada and Henao 2008) along the Caribbean coast have been constructed to face shoreline erosion without a long-term framework for proper territorial planning and relocation of endangered goods. Soft methods and defenses that replicate natural dynamics to recover sedimentary balance (artificial reefs, beach nourishment, dune regeneration, conservation and planting of mangrove or cliff drainage) have been implemented in only a few, localized areas. Examples of hard coastal defenses and types of materials used in Colombia are shown on Fig. 16.11, where a crafted shore protection is depicted as a groin built up with "bolsacreto" or a resistant bag filled with a mixture of cement and sand.

16.3 Main Historical Geomorphological Changes along the Pacific Littoral of Colombia

Because of its low historical occupation and difficult accessibility, the morphological changes along the Pacific coast are much less documented and measured than those of the Caribbean coast. Available information indicates, however, a strong littoral instability evidenced by the rapid retreat of the Pacific deltaic coastlines as a short to long term effect of coastal subsidence. The influence of the human activities on the littoral changes and sediment budgets of the poorly engineered but strongly deforested and impacted by mining activities Pacific coast has not been studied at all.

16.3.1 General Context

The mixed meso-macro tidal (tidal amplitudes between 2.5 and 4.5 m), Pacific littoral of Colombia extends for about 1300 km (scale 1: 100,000) between Punta Ardita at the Panama border and Bahia Ancon, the southern tip of the Mira delta at the Ecuadorian border (Fig. 16.12). The Pacific Coast of Colombia is one of the rainiest regions in the world, with medium diurnal temperatures of 28°–30 ° C° and overall annual precipitation varying between 2000 and 12,000 mm (Restrepo and López 2008; Ideam 2016).

The Pacific coast has an approximated catchment area of 83,000 km^2 located on the coastal plain and on the western slopes of the Western Cordillera of Colombia. It is drained by more than 150 rivers, the most important of them being the San Juan River, the Patia River, and the Mira River (Fig. 16.12). The San Juan River at the central Pacific coast has a drainage area of 16,470 km^2 (352 km length) and carries a multiannual mean flow of 7200 m^3/s, with a sediment load of ~16.4 Mt./year (Correa and Restrepo 2002). Further south, the Patia River (mean flow of 400 m^3/s, 415 km length) collects sediments from a watershed of ~ 23,700 km^2 and delivers an annual sediment load of ~ 21.1 Mt./year. At the southernmost part of the Pacific coast, the Mira River (catchment area of 9530 km^2; mean flow 871 m^3/s) delivers a total sediment load of ~ 9.77 Mt./year (Restrepo et al. 2006) to the sea.

Fig. 16.11 Shore protection structures in the Caribbean littoral. (**A**) Sector of Marbella in Cartagena city (Bolivar), where a battery of groins is trapping sediments from the longshore drift toward the southwest (source: Esri Imagery); (**B**) southern part of the Gulf of Morrosquillo (Sucre), where a sequence of breakwaters and T-shape groins do not succeed in stabilizing the highly intervened sandy barrier in front of extensive mangrove swamps (Courtesy of Eafit University); (**C**) Impermeable breakwater forming a tombolo in Arboletes (Antioquia); (**D**) Groin of tetrapods in Zapata (Antioquia), where a localized effect of sand accumulation up drift contrast the severe deficit downdrift; (**E**) Groin of bolsacreto in Damaquiel (Antioquia). Photos C, D and E by Cristina Pereira, May 2016

The Pacific coast of Colombia is a high seismic-risk area characterized by the common occurrence of high-magnitude (M > 5) earthquakes, the best known being the 1836, 1868, 1906, 1979, and 1991 events (West 1957; Ramírez 1970, 2004; Herd et al. 1981; Meyer et al. 1992; Correa and Morton 2003; Corporación Osso 2008; AIS 2009; Martínez and López 2010). The earthquakes of 1906 and 1979 are proverbial in the zone because they generated at least two tsunami waves up to 2.5 m

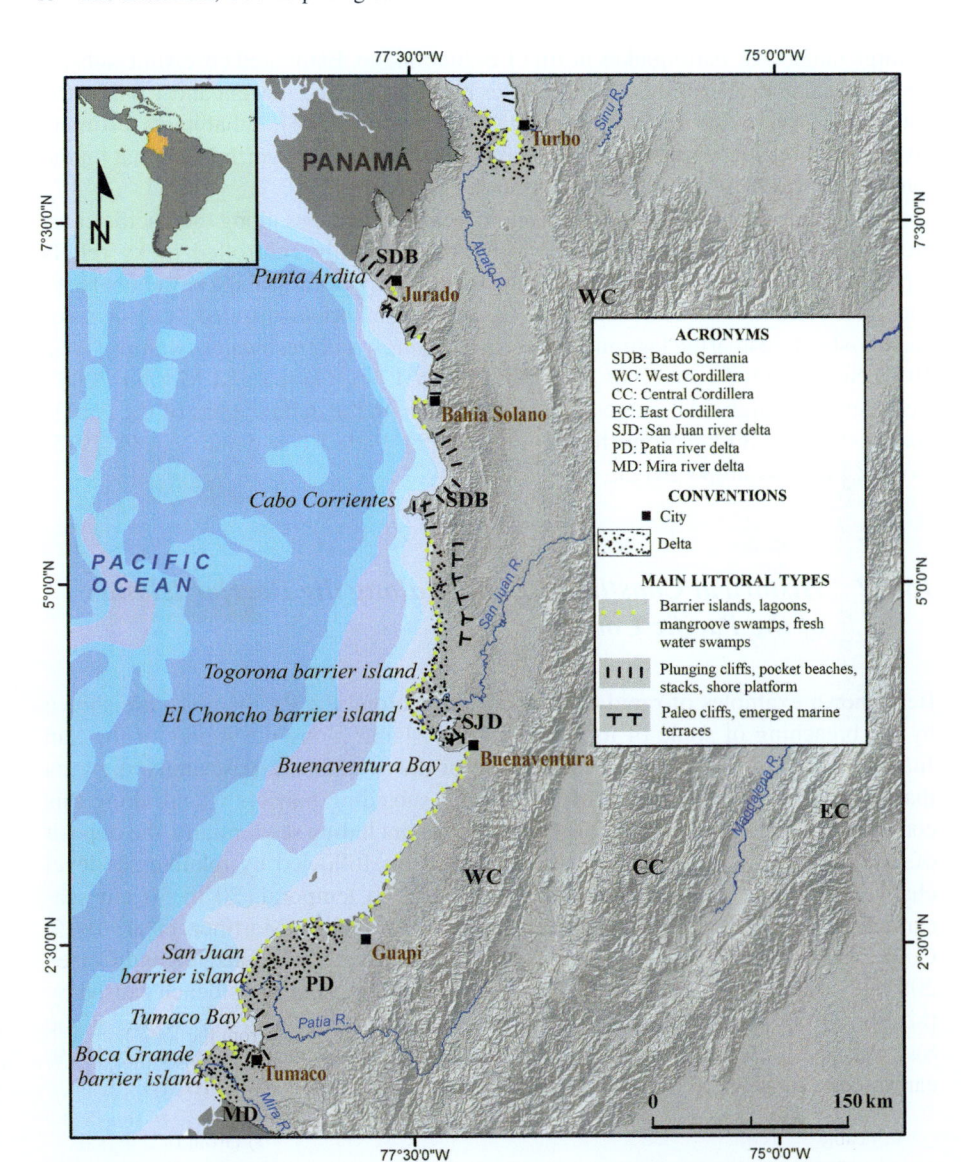

Fig. 16.12 Location map and main morphological littoral types along the Pacific Coast of Colombia

high that flooded the low deltaic plains of the Patia and Mira deltas and caused general destruction along the coastline fringe and up to 30 km inland on terrains located well above the maximum tidal penetration, including the city of Guapi. For the northern Pacific coast, Ramírez (1970) reports the destruction of the Bahia Solano village by an earthquake occurred in 1970, and Page and James (1981) reports the occurrence of several events of tectonic subsidence associated with the occurrence of large magnitude earthquakes north of Bahia Solano. Estimated coseismic subsidence values reported for these earthquakes range between a few cm and 1.6 m at the southern coast of the Patia River delta (Herd et al. 1981). Inhabitants estimate coseismic subsidence values up to 2 m associated to the 1991 earthquake that hit the San Juan River delta and surrounding northern areas.

From south to north, the main morphological littoral types along the Pacific coast are highly contrasting, varying between the structurally controlled rocky reliefs typical of the Serranía de Baudó range and the Buenaventura-Malaga bay (Figs 16.12 and 16.13) and the low Holocene depositional coastal prisms fronted by systems of barrier islands, estuarine lagoons, mangrove swamps, and freshwater swamps (West 1957, Smit 1972, Martínez et al. 1995, Correa 1996, Correa and Morton 2003). Because of the high tidal ranges, tidal penetration on the deltaic areas of the Pacific coast gets up to 30 km from the shoreline on the Patia River delta (Van Es 1975; Gómez 1986; Restrepo 2012).

16.3.2 Historical Coastline Changes along the Barrier Islands of the Pacific Coast

Best known examples of rapid coastal evolution along the Pacific coast are shown by the breaching of some of its major barrier islands along the shores of the San Juan, Patia, and Mira deltas. Interpretation of the available data strongly suggests that the erosion and breaching of the already subsiding barriers island along this coast result from a combination of natural events, including sequentially, the deposit of extensive sandy tidal flats at the river's mouths followed by relative sea-level changes associated to coseismic subsidence and to temporal, 20 to 30 cm sea-surface positive anomalies associated to El Niño events (Martínez et al. 1995; Correa 1996; Morton et al. 2000; Correa and Gonzalez 2000; González and Correa 2001; Restrepo et al. 2002). At the El Choncho barrier island, subsidence caused by the November 19, 1991, earthquake was estimated at 20–30 cm, while at the San Juan de La Costa barrier island (Patía delta) hit by the November 12, 1979, earthquake, subsidence was estimated up to 1.6 m (Figs. 16.13 and 16.14).

16.4 Final Remarks

The examples shown here of natural and man-induced changes in the budgets of sediments and the morphological responses along the coasts of Colombia illustrate only a part of its historical evolution. Strong erosional trends could also be reported all along the Caribbean littoral, where more recent evaluations classified 48.3% of coast (1182 km) as suffering serious erosion during the period of 1980–2014 (Rangel et al. 2015). As a matter of fact, practically all the urban beaches of

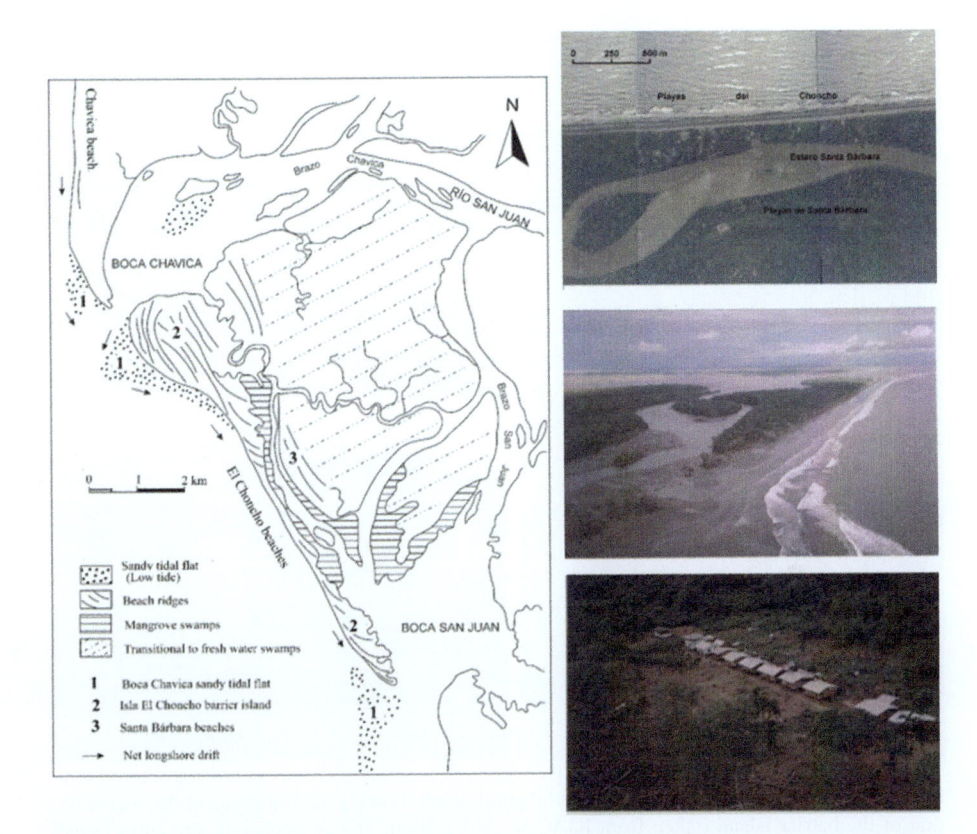

Fig. 16.13 El Choncho barrier island. Left: geomorphological units of the southern lobule of the San Juan river delta; radiometric data taken from beach ridges besides Boca Chavica mouth and comparisons with ancient charts suggest that El Choncho barrier island initiated its formation by the end of the seventeenth century (From Morton and Correa 2003). Upper right: El Choncho barrier island central part (photo by Ahmed Restrepo in 1996). Center right: aerial photograph illustrating the initial breaching at the central part of the barrier island, the same area illustrated above (photo by Iván Correa in 1997). Down right: The new Choncho after relocation to the ancient beach ridges – Santa Bárbara beaches. (Photo by Iván Correa, November 1998)

Caribbean cities are subject to erosional trends and are sustained with different success by engineering structures and/or beach replenishment projects often at exorbitant costs. The geological complexity of Colombian littorals points out a challenge for risk management at coastal zones and adds great uncertainties to the integrated assessment of coastal risks associated to natural hazards (Martínez et al. 1994; Rangel and Anfuso 2015; Restrepo and Cantera 2013; Freitas de et al. 2013). However, probably much more important than the physical changes and their direct impacts on land losses and infrastructure, the environmental conditions of Colombian coastal ecosystems are rapidly deteriorating due to anthropogenic actions in the Andes´ catchments and adjacent coastal plains.

Fig. 16.14 San Juan de La Costa. Left: aerial photograph showing the beginning of the segmentation of the barrier island (taken from IGAC in 1979). Right: the two concrete structures (small church) and school were the unique remnants of the two 2.5-m-high tsunami waves that hit the island in December 12, 1979. Two hundred fifty inhabitants were drowned by this event. (photo by Iván Correa in 1989)

Besides the challenges imposed by natural drivers of morphological changes, coastal management in Colombia also faces the pressure of an accelerated population growth that comes along with poorly planned territorial development, especially in the Caribbean domain (Anfuso et al. 2011; Barragán and de Andrés 2015). Human interventions linked to these developments, such as the jetties at Bocas de Ceniza in the Magdalena River mouth or the diversion of natural currents within Tinajones area, have triggered negative effects on the stability of coastal terrain due to changes in the patterns of coastal dynamics and in the sub-oceanic geological processes that modify the coastal reliefs. These man-made-induced perturbations have been responsible for the instability of coastal areas and the consequent deterioration of environmental conditions (Bernal 1996; Correa et al. 2005; González et al. 2010; Rangel et al. 2015). According to Vilardy (2009), there were approximately 60,000 Ha. of mangrove when the high road Ciénaga-Barranquilla was built; a few years later, during the construction of the Palermo-Sitio Nuevo road, there was already a reduction of 5000 ha. It wasn't until 1995, after the big expansion of agricultural frontiers into the Lagoon Complex, when the situation reached its most critical point because the mangrove was less than 30,000 ha.

The pressure imposed by uncontrolled human uses and activities causes coastal ecosystems to exceed their capacity for self-regulation, thereby increasing the vulnerability of coastal areas to natural threats of marine or terrestrial origin, such as storms, mud volcanism, river floods, mass movements, and the sea-level rise, among others (Anfuso et al. 2011; Montes and Sala 2007; Botero et al. 2016). The combined effects of linear, punctual, and scattered human interventions over coastal ecosystems have induced serious problems in the Caribbean of Colombia. They include salinization of swamps and soils, mangrove death within the lagoons, and habitat deterioration for aquatic and terrestrial species. Land colonization for agricultural purposes within swamps and lagoon territories involves the leaching of pesticides traces, heavy metals, and fertilizers, which alter the physiochemical composition in the natural system and translate into pollution (Ibarra et al. 2014).

Therefore, unplanned territorial development represents another challenge for risk management and entails excessive costs of social and environmental protection for coastal populations and settlements (Invemar 2003; Restrepo 2008; Cooper et al. 2009). For example, local and national territorial authorities have been seen in need of managing more than 15 million dollars to counteract the coastal erosion triggered by the Magdalena River mouth channeling works (Heraldo 2014). This intervention has been responsible for the loss of important ecosystem services related with beaches and lagoon systems affected, including resources for the economic support of local settlements, the discharge and recharge of aquifers, communications routes, or flood mitigation (Anfuso et al. 2015).

The examples of coastal interventions cited in this chapter show that negative effects derived from diverse types of coastal projects and activities have lacked adequate environmental evaluation, monitoring, and control. Such insufficiency is due either to an absence of a regulatory framework or the reduced scope of Colombian legislation concerning all the possible coastal interventions that currently take place in the country (control and protection structures, buildings, docks, ports, marinas and navigation infrastructure, roads and bridges, thermoelectric and desalination plants, water pipes and drains, agricultural farms, dredging and mining or beach nourishment). An example of a lack of regulation corresponds to the described case of Bocas de Ceniza, whose channeling works initiated by 1922 before the existence of the first environmental law of the country (Code of Natural Resources of 1974).

Four decades later, environmental licensing processes in Colombia still don't regulate the wide range of activities taking place in coastal areas. A review of the terms of reference for environmental impact assessments of projects or activities, published by the National Authority of Environmental Licensing in Colombia, comprises only two types of coastal interventions: maritime and fluvial harbors and structures for shore control and protection (PU-TER-1-01 2006; PU-TER-1-03 2006; EIA-TER-PC-1-01 2011; M-M-INA-05 2013). Although several highways in the country have been built near the coastline, especially on the Colombian Caribbean coast, terms of reference for road construction projects developed in 2013 do not include specifications regarding coastal conditions. This context reveals that there are still no specific criteria for projecting diverse types of interventions in the coastal environment and assessing their associated impacts on coastal stability.

Given the complexity of the physical elements and the biological fabrics that intervene in geomorphologic evolution of coastal zones, the evaluation, monitoring, and control of human interventions should consider how prone biotic and abiotic factors are to experience changes due to the perturbation induced by the construction, operation, or dismantling of projects, built structures, and activities performed by man. This characterization can be defined as the physical-biotic susceptibility of a littoral territory regarding the morphological changes induced by the emplacement of coastal interventions. Such susceptibility comprises intrinsic and extrinsic factors that may give a partial representation of the resilience of ecosystems and the character of natural stressors exposed to human perturbations (Toro et al. 2012).

Extrinsic factors comprise the forces inducing dynamic instability of littoral areas, such as the hydrodynamic, subaerial, geodynamic, and human elements considered by Morton and Pieper (1977). This approach conceives the property of physical-biotic susceptibility as a state of natural or artificially acquired exposition to morphological changes, in which previous human interventions play a significant role. Intrinsic factor refers to the ability of the natural system for recovering and toleration, which can be defined by the inherited geology of the littoral, along with indicators of health and functional integrity in coastal ecosystems (Unesco 2006; Rangel and Anfuso 2015). Sandy, rocky, marine, and wetland ecosystems play a key role both as indicators of morphological evolutions and predisposition to unnatural perturbations.

Several studies at local and regional scales have been done regarding the natural conditions of the Colombian coasts, their evolution and their vulnerability to specific hazards. At regional scale, the Maritime General Directions, throughout the Research Center of Oceanography and Hydrography, have performed a physical-biotic characterization of the Colombian Caribbean coast (Dimar-Cioh 2009a, b) and the Geomorphological Atlas of the Colombian Caribbean coast (Dimar-Cioh 2013). In addition, there are also two separate assessments of coastal vulnerability to the effects of the sea-level rise for both Pacific and Caribbean coasts, one developed by the Institute of Hydrology, Meteorology and Environmental Studies of Colombia and the other one by the José Benito Vives de Andreis Marine and Coastal Research Institute (Ideam 2001; Invemar 2003).

Despite these studies of coastal characterization and vulnerability, there is no tool to recognize the susceptibility of coastal areas against coastal morphological changes that are further enhanced by the installation of civil works or infrastructure. These studies have focused mainly on natural hazards, leaving aside the adverse effect generated by human interventions or considering it only as one more element in the general vulnerability assessment. Environmental licensing of coastal interventions in Colombia can be improved by institutionalizing adequate criteria in the assessment of the environmental factor that truly describe the intricate processes governing coastal dynamics. Therefore, it is pertinent to focus the analysis of physical-biotic susceptibility of littoral areas against the effects of coastal interventions, so that environmental licensing processes have a conceptual and methodological reference to reduce subjectivity in the environmental assessments regulated in Colombia (Toro et al. 2010).

References

AIS Asociación Colombiana de Ingeniería Sísmica (2009) Estudio general de Amenaza Sísmica en Colombia

Alvarado M (2007) Barranquilla. In: Hermelin M (ed) Entorno natural de 17 ciudades de Colombia. Sociedad Colombiana de Geología-Academia Colombiana de Ciencias Exactas Físicas y Naturales. Fondo Editorial Universidad Eafit, Medellín, pp 66–88

Anfuso G, Pranzini E, Vitale G (2011) An integrated approach to coastal erosion problems in northern Tuscany (Italy): littoral morphological evolution and cell distribution. Geomorphology 129:204–214

Anfuso G, Rangel N, Correa I (2015) Evolution of sand spits along the Caribbean coast of Colombia. In: Randazzo G, Jackson D, Cooper J (eds) Sand and gravel spits. Coastal research library. Springer, Switzerland, pp 1–19

Barragán JM, de Andrés M (2015) Analysis and trends of the world's coastal cities and agglomerations. Ocean Coast Manage 114:11–20

Bernal G (1996) Caracterización geomorfológica de la llanura deltaica del río Magdalena con énfasis en el sistema lagunar de la Ciénaga Grande de Santa Marta, Colombia. Bol. Invest. Mar. Cost. 28: 19-48

Bird ECF (ed) (2010) Encyclopedia of the World's coastal landforms. Springer Science+Businnes Media

Botero CM, Fanning LM, Milanes C, Planas JA (2016) An indicator framework for assessing progress in land and marine planning in Colombia and Cuba. Ecol Indic 64:181–193

Burel T, Vernette G (1981) Evidencias de cambios del nivel del mar en el Cuaternario de la región de Cartagena. Rev CIAF 6(1–3):77–92

Carvajal JH, Mendivelso D (2011) Catálogo de "volcanes de lodo" Caribe Central Colombiano. Ingeominas Bogotá 84

Carvajal JH (2016) Mud Diapirism in the central Colombian Caribbean coastal zone. In: Hermelin M (ed) Landscapes and landforms of Colombia. Springer World Geomorphological Landscapes, pp 35–53

Cooper JAG, Anfuso G, Rios LD (2009) Bad beach management: European perspectives. Sp Paper of the Geol Soc of Am 460:167–179

Correa ID (1990) Inventario de erosión y acreción litoral (1793-1990) entre Los Morros y Galerazamba, Departamento de Bolívar, Colombia. In: Hermelin M (ed) Environmental geology and natural hazards of the andes region. AGID, Medellin, pp 129–142

Correa ID (1996) Le Littoral Pacifique Colombien: Interdependendance des Agents Morphostructuraux et Hidrodinamiques. Ph. D Thesis Université Bordeaux I, No. Ordre 1538. 2 tomes

Correa ID, González JL (2000) Coastal erosion and village relocation: a Colombian case study. Ocean Coast Manage 43:51–64

Correa ID, Restrepo JD (eds) (2002) Geología y Oceanografía del Delta del Ríos San Juan, Litoral Pacífico colombiano. Colciencias-Fondo Editorial Universidad Eafit, Medellín, pp 13–168

Correa ID, Morton RA (2003) Coasts of Colombia. In: https://coastal.er.usgs.gov/coasts-colombia/. Accessed 21 Oct 2016

Correa ID, Alcántara-Carrió J, González D (2005) Historical and recent shore erosion along the Colombian Caribbean coast. J Coastal Res SI 49:52–57

Correa ID, Ríos A, González D, Toro MI, Ojeda G, Restrepo IC (2007a) Erosión Litoral entre Arboletes y Punta San Bernardo, Costa Caribe colombiana. Bol Geol UIS 29(2):115–128

Correa ID, Acosta S, Bedoya G (2007b) Análisis de las causas y monitoreo de la erosión litoral en el Departamento de Córdoba. Convenio de Transferencia Horizontal de Ciencia y Tecnología No. 30. Corporación Autónoma de los valles del Sinú y San Jorge (CVS)-Universidad Eafit, Departamento de Geología (Área de Ciencias del Mar). Fondo Editorial Universidad Eafit, Medellín, pp 33–128

Corporación OSSO-CVC (2008) Evaluación básica en investigación geológica, sismológica y red acelerográfica como insumo para la macrozonificación sísmica del área urbana y de expansión de Buenaventura. Tsunamitas y horizonte cosísmico Informe final, Convenio, pp 148–106

De Porta J, Barrera R, Julia R (2008) Raised marsh deposits near Cartagena de Indias, Colombia: evidence of eustatic and climatic instability during the late Holocene. Bol Geol UIS 30(1):1–14

Dimar-Cioh (2009a) Caracterización físico-biótica del litoral Caribe colombiano. Tomo I. Dirección General Marítima-Centro de Investigaciones Oceanográficas e Hidrográficas. Ed. DIMAR, Serie Publicaciones Especiales CIOH, vol 1. Cartagena de Indias, Colombia, p 154

Dimar-Cioh (2009b) Caracterización físico-biótica del litoral Caribe colombiano. Tomo II. Dirección General Marítima-Centro de Investigaciones Oceanográficas e Hidrográficas. Ed. DIMAR, Serie Publicaciones Especiales CIOH, vol 2. Cartagena de Indias, Colombia, p 100

Dimar-Cioh (2013) Atlas Geomorfológico del Litoral Caribe Colombiano. Dirección General Marítima-Centro de Investigaciones Oceanográficas e Hidrográficas del Caribe. Ed Dimar. Serie Publicaciones Especiales CIOH, vol 8. Cartagena de Indias, Colombia, p 225

Domínguez C, Salcedo H, Martin L (eds) (2012) Joaquín Francisco Fidalgo: Derrotero y cartografía de la Expedición Fidalgo por el Caribe Neogranadino (1792–1810), 2nd edn. El Ancora Editores, Bogotá, pp 5–430

EIA-TER-PC-1-01 (2011) Terms of reference for the environmental impact statement of shore protection and control works, developed in 2011. http://www.anla.gov.co/. Accessed 21 Oct 2016

Eurosion (2004) A guide to coastal erosion management practices in Europe: lessons learned:1–50

de Freitas DM, Smith T, Stokes A (2013) Planning for uncertainty: local scale coastal governance. Ocean & Coast Manage 86:72–74

Fuentes N, Jaramillo N (2015) Atlas histórico marítimo de Colombia Siglos XVI – XVIII. Comisión Colombiana del Océano, Bogotá, pp 2–153

Geotec Ltda (2003) Geología de los Cinturones Sinú-San Jacinto. 50 Puerto Escondido, 51 Lorica, 59 Mulatos, 60 canalete, 61 Montería, 69 Necoclí, 70 San Pedro de Urabá, 761 Planeta Rica, 79 Turbo, 80 Tierra alta. Mapas geológicos y Memoria explicativa. Ingeominas Bogotá

Gómez H (1986) Algunos aspectos neotectónicos hacia el suroeste del litoral Pacífico colombiano. Rev CIAF 11(1–3):281–289

Gómez, JF, Byrne ML, Hamilton J, Isla F (2016) Historical coastal evolution and dune vegetation in Isla Salamanca National Park, Colombia. doi:10.2012/J COASTRES-D-15-00189.1

González JL, Correa ID (2001) Late Holocene evidence of Coseismic subsidence on the san Juan Delta, Pacific coast of Colombia. J Coast Res 17(2):459–467

González C, Urrego LE, Martínez JI, Polanía L, Yokoyama Y (2010) Mangrove dynamics in the southern Caribbean since the "little ice age": a history of human and natural disturbances. The Holocene 20(6):849–861

Guzman W, Posada B, Guzman G, Morales D (2008) Programa nacional de investigación para la prevención, mitigación y control de la erosión costera en Colombia-PNIEC: Plan de Acción 2009–2019: Invemar

Heezen BC (1956) Corrientes de Turbidez del rio Magdalena. Bol Soc Geograf Colombia, Nos 51 y 52, XIV, Colombia, pp 135–140

Heraldo (2014) Anuncian 36 mil millones de pesos para prevenir erosión en playas de Puerto Colombia. Periódico El Heraldo: Barranquilla. http://www.elheraldo.co/local/36300-millones-para-contener-erosion-costera-en-puerto-165991/. Accessed May 2016

Herd DG, Leslie T, Meyer H, Arango JL, Person WJ, Mendoza C (1981) The great Tumaco, Colombia earthquake of December 12, 1979. Science 211(4481):441–445

Hermelin M (ed) (2015) Landscapes and landforms of Colombia. Springer, World Geomorphological Landscapes, pp 1–21

Ibarra K, Gómez M, Viloria E, Arteaga E, Cuadrado I, Martínez M, Nieto Y, Rodríguez A, Licero L, Perdomo L, Chávez S, Romero J, Rueda M (2014) Monitoreo de las condiciones ambientales y los cambios estructurales y funcionales de las comunidades vegetales y de los recursos pesqueros durante la rehabilitación de la Ciénaga Grande de Santa Marta. INVEMAR Informe Técnico Final Santa Marta 140

Ideam (2001) Vulnerabilidad y adaptación de la zona costera colombiana al ascenso acelerado del nivel del mar. Instituto de Hidrografía y Meteorología y Estudios Ambientales – IDEAM, Bogotá

Ideam (2010) Sistemas Morfogénicos del Territorio Colombiano. Instituto de Hidrología, Meteorología y Estudios Ambientales. Bogotá, D.C.., 252 p., 2 anexos, 26 planchas en DVD

Ideam (2016) Atlas Climatológico de Colombia. http://atlas.ideam.gov.co/cclimatologicas. Accesed 21 Oct 2016

Invemar (2003) Programa Holandés de asistencia para estudios en cambio climático, Colombia. Definición de la vulnerabilidad de los sistemas bio-geofísicos y socioeconómicos debido al cambio del nivel del mar en la zona costera colombiana (Caribe y Pacífico) y medidas de adaptación. Instituto de Investigaciones Marinas y Costeras José Venito de Andrei

Invemar (2006) Climatologie de la vitesse et la direction des vents pour la mer territoriale sous jurisdiction colombienne 8° a 19° N et 69 ° a 84° W. Atlas ERS 1 et 2 et Quicksat, Colombie. Laboratoire de Geographie Physique et Programa Geociencias. Instituto de Investsigaciones Marinas y Costeras, Santa Marta

Kaufman W, Pilkey H Jr (1983). The beaches are moving: the drowning of America's shoreline (living with the shore) Duke University Press: 1–329

Khobzi J (1981) Los campos de dunas del Norte de Colombia y de los llanos del Orinoco (Colombia y Venezuela). Rev CIAF 6(1):257–292

Koopmans BN (1971) Interpretación de Fotografías Aéreas en Morfología Costera relacionada con proyectos de Ingeniería. Ministerio de Obras Públicas-Centro Interamericano de Fotointerpretación, Bogotá, septiembre 1971: 2–23

Laboratorio Central de Hidráulica de Francia (1958) Bocas de Ceniza: Informes 1957-1958. In: Koopmans BN (ed) Interpretación de Fotografías Aéreas en Morfología Costera relacionada con proyectos de Ingeniería, vol 1971. Ministerio de Obras Públicas-Centro Interamericano de Fotointerpretación, Bogotá, pp 2–23

Martínez JO, Pilkey OH, Neal W (1990) Rapid formation of large coastal sand bodies after the emplacement of Magdalena river jetties, northern Colombia. Environ Geol Water Sci Vol 14(3):187–194

Martínez JM, Parra E, París G, Forero CA, Bustamante M, Cardona OD, Jaramillo JD (1994) Los sismos del Atrato medio 17 y 18 de octubre de 1992 Noroccidente de Colombia. Rev Ingeominas 4:35–66

Martínez JO, González JL, Pikey OH, Neal WJ (1995) Barrier Island on the subsiding Central Pacific coast, Colombia. J Coast Res 16(3):663–674

Martínez JO, López E (2010) High-resolution seismic stratigraphy of the late Neogene of the central sector of the Colombian Pacific continental shelf: a seismic expression of an active continental margin. J South Am Earth Scie 31:28–44

Martínez JI, Yokoyama Y, Gómez A, Delgado A, Matsuzaki H, Rendón E (2010) Late Holocene marine terraces of the Cartagena region, southern Caribbean: the product of neotectonism or a former high stand in sea-level? J S Am Earth Sci 29:214–224

Meyer H, Mejía JA, Velásquez A (1992) Informe preliminar sobre el terremoto del 19 de noviembre de 1991 en el Departamento del Chocó. Publicaciones ocasionales OSSO, Universidad del Valle, Cali, Colombia 13

M-M-INA-05. Terms of reference for the environmental impact statement of construction, enlargement and dredging of maritime harbors of great draft, developed in 2013. http://www.anla.gov.co/. Accessed 21 Oct 2016

Montes C, Sala O (2007) La Evaluación de los Ecosistemas del Milenio. Las relaciones entre el funcionamiento de los ecosistemas y el bienestar humano. Ecosistemas 16(3):137–147

Morton RA, Correa ID (2003) Introducción al uso de los Geoindicadores de cambios Ambientales en Costas Húmedas tropicales. Geología Norandina 12(1):1–56

Morton RA, Pieper M (1977) Shoreline changes in Mustang Island and north Padre Island (Aransas Pass to yarborough). Geol Circ 45:77–77. The Univ of Texas at Austin

Morton RA, González JL, López GI, Correa ID (2000) Frequent non-storm Washover of Barrier Islands, Pacific coast of Colombia. J Coast Res 16(1):82–87

Morton RA (2002) Coastal geoindicators of environmental changes in the humid tropics. Environ Geol 42:711–724

Ojeda GY, Restrepo IC, Correa ID, Ríos A (2007) Morfología y Arquitectura interna de una Plataforma continental cambiante: Golfo de Morrosquillo. Bol Geol 29:105–114

Olivero J, Johnson B (2002) El lado gris de la minería del oro -la contaminación por mercurio en el Norte de Colombia. Univ de Cartagena, Fac de Ciencias químicas y farmacéuticas. Cartagena Ed. Universitaria. p 94

Ortiz JC (2012) Exposure of the Colombian Caribbean coast, including San Andrés Island, to tropical storms and hurricanes, 1900-2010. Nat Hazards 61:815–827

Page W (1983) Holocene deformation of the Caribbean coast, Northwestern Colombia. Woodward &Clyde Consultants report, p 25

Page W, James M (1981) Tectonic subsidence and evidence for the recurrence of large magnitude earthquakes near Bahía Solano Colombia. Mem III Congr Col Geol, Ingeominas Bogotá, pp 14–20

Pilkey OH (1983) The beaches are moving. North Carolina Public TV Documental

Posada BO, Henao W (2008) Diagnóstico de la erosión en la zona costera del Caribe colombiano. INVEMAR, Serie Publicaciones Especiales No.13, Santa Marta, p 124

Posada BA, Henao W, Guzmán G (2009) Diagnóstico de la erosión y sedimentación en la zona costera del Pacífico Colombiano. INVEMAR, Serie Publicaciones Especiales No. 17, Santa Marta, p 178

Pranzini E, Williams A (eds) (2013) Coastal Erosión and Protección in Europe. Routledge Taylor and Francis Group: 457

PU-TER-1-01. Terms of reference for the environmental impact statement of deep dredging for access channels to large seaports, developed in 2006. http://www.anla.gov.co/. Accessed 21 Oct 2016

PU-TER-1-03. Terms of reference for the environmental impact statement of deep dredging for navigation channels in deltaic areas, developed in 2006. http://www.anla.gov.co/. Accessed 21 Oct 2016

Raasveldt H, Tomic A (1958) Lagunas colombianas: contribución a la geomorfología de la Costa del Mar Caribe con algunas observaciones sobre las Bocas de Ceniza. Rev Acad Col Cien Ex Fis y Nat 10:16–198

Ramírez JE (1959) El volcán submarino de Galerazamba. Rev Acad Com Cien Ex Fis Nat 10:301–314

Ramírez JE (1970) El terremoto de Bahía Solano. Rev Javeriana 370(70):573–585

Ramírez JE (2004) Actualización de la Historia de los terremotos en Colombia. Instituto Geofísico de la Universidad Javeriana. Editorial Pontificia Universidad Javeriana, Bogotá, p 186

Rangel-Buitrago N, Anfuso G, Correa I (2012) Obras de defensa costeras en el Caribe Colombiano ¿solución o problema?, in Proceedings I congreso Iberoamericano de Gestión Integrada de Áreas Litorales, Cádiz (España), p 6

Rangel-Buitrago NG, Anfuso G (2015) Risk assessment of storms in coastal zones: case studies form Cartagena and Cadiz. Springer briefs in earth science. Springer International Publishing, Switzerland, p 63

Rangel-Buitrago NG, Anfuso G, Williams AT (2015) Coastal erosion along the Caribbean coast of Colombia: magnitudes, causes and management. Ocean & Coast Manage 114:129–144

Restrepo JD, Kjerfve B, Correa ID, González JL (2002) Morphodynamics of a high discharge tropical delta, San Juan River, Pacific coast of Colombia. Mar Geol 192:355–381

Restrepo JD, Kjerfve B, Hermelin M, Restrepo JC (2006) Factors controlling sediment yield in a major south American drainage basin: the Magdalena River, Colombia. J Hydrol 316:213–332

Restrepo IC, Ojeda GY, Correa ID (2007) Geomorfología de la Plataforma somera del Departamento de Córdoba, Costa Caribe colombiana. Bol Ciencias Tierra Univ Nal 20:39–52

Restrepo JD (ed) (2008) Deltas de Colombia: morfodinámica y vulnerabilidad ante el cambio Global. Fondo Editorial Universidad Eafit, Medellín, p 280

Restrepo JD, López SA (2008) Morphodynamics of the Pacific and Caribbean deltas of Colombia, South America. J south am earth. Sciences 25(1):1–21

Restrepo JD (2012) Assessing the effect of sea-level change and human activities on a major delta on the Pacific coast of northern South America: the Patía River. Geomorphology 151-152:207–223

Restrepo JD, Cantera JR (2013) Discharge diversion in the Patía River Delta, the Colombian Pacific: geomorphic and ecological consequences for mangrove ecosystems. J S Am Earth Sci 46:183–198

Restrepo JC, Ortiz JC, Pierini J, Schrottke K, Maza M, Otero L, Aguirre J (2014) Freshwater discharge into the Caribbean Sea from rivers of northwestern South America (Colombia): magnitude, variability and recent changes. J Hydrol 509(1):266.281

Restrepo JD, Kettner AJ, Syvitsky JPM (2015) Recent deforestation causes rapid increase in river sediment load in the Colombia Andes. Anthropocene 10:13–25

Rico Pulido E (1969) Las Obras de Bocas de Ceniza. Colpuertos, p 100

Richard H, Broecker W (1963) Emerged Holocene south American shorelines. Science 141:1044–1045

Robertson K, Chaparro J (1998) Evolución histórica del Río Sinú. Cuadernos de Geografía Vol VII 1(2):70–84

Serrano BE (2004) The Sinú river delta on the northwestern Caribbean coast of Colombia: bay infilling associated with delta development. J S Am Earth Sci 16:639–647

Shepard FP, Dill RF, Heezen BC (1968) Diapiric intrusions in Foreset slope sediments off Magdalena Delta, Colombia. Am Assoc Petr Geol Bull 52(11):2197–2207

Smit GS (1972) Aplicación de las imágenes de radar en la fotointerpretación de los bosques húmedos tropicales: Región de Tumaco-Barbacoas-Guapi. Rev CIAF 18:59–70

Subcommission on Quaternary Stratigraphy-ICS Working Group (2016) What is the "Anthropocene"? current definitions and status. In: http://quaternary.stratigraphy.org/workinggroups/anthropocene/. Accesed 15 Oct 2016

Toro J, Requena I, Zamorano M (2010) Environmental impact assessment in Colombia: critical analysis and proposals for improvement. Environ Impact Assess Rev 30(4):247–261

Toro J, Duarte O, Requena I, Zamorano M (2012) Determining vulnerability importance in environmental impact assessment: the case of Colombia. Environ Impact Assess Rev 32(1):107–117

Troll C, Schmidt E (1985) Das Neue Delta des Rio Sinu an der karibishen Kúste Kolombiens. Geographische Interpretation und kartographische Auswwertung von Luftbildern. Mit 1 Abbildung, 3 Bildern und 3: 14–23

UNESCO (2006) Manual para la medición del progreso y de los efectos directos del manejo integrado de costas y océanos, Paris

Van Es E (1975) Análisis geológico-geomorfológico de las imágenes de radar de la llanura Pacífica de Nariño, Colombia, América del Sur. Rev CIAF 17:59–70

Vernette G (1985) La Plataforme continentale Caraibe de Colombie (du debouche du Magdalena au Golfe de Morrosquillo) Importance du diapirisme argileux sur la morphologie et la sedimentation. University of Bordeaux I

Vernette G (1989) Examples of diapiric control on shelf topography and sedimentation patterns on the Colombian Caribbean continental shelf. J S Am Earth Sci 2(4):391–400

Vernette G, Mauffret A, Bobier C, Briceño L, Gayet J (2011) Mud diapirism, fan sedimentation and strike-slip faulting, Caribbean Colombian margin. Tectonophysics 2002:335–349

Vilardy S (2009) Estructura y dinámica de la ecorregión Ciénaga Grande de Santa Marta: una aproximación desde el marco conceptual de los sistemas socioecológicos complejos y la teoría de resiliencia. Tesis de doctorado en Ecología y Medio Ambiente, Universidad Autónoma de Madrid, España, p 295

West RC (1957) The Pacific lowlands of Colombia: a Negroid area of the American tropics. Louisiana State Univ. Press, Baton Rouge, p 278

Williams SJ, Dodd K, Krafft K (1995) Coasts in crisis. US Geol. Circular 1075:1–32

World Economic Forum (2015) Insight Report Global

Index

A
Accreted oceanic terranes, Pacific domain, 733
ACIGEMI project, 415
Active *vs.* passive margin, 734
AFTSolve® (computer code), 769
Age-elevation relationship (AER), 769, 773
　See also Exhumation
Agua Blanca Granite (Agua Blanca Batholith), 225
Aguas Blancas Fm, 723
Alkali-basaltic volcanism, 626
Alkalinity index (AI), 345
Allochthonous terranes, 8, 11, 257, 261, 264, 310, 313, 321, 322, 325, 373, 393, 835
Altered oceanic crust (AOC), 606, 627
Amagá-Cauca-Patía (ACP), 837
Amazonian and Orinoquian basement, 115–119
　geoeconomic potential, 158
　Guiana Shield (*see* Guiana Shield)
Amazonian Craton, 151–152
Amphibolites, 130
Andean and subandean precambrian basement
　Central Cordillera, 175, 177, 178
　Colombian Andes, 158–161
　Garzón Massif, 161–168
　geological evolution, 179–182
　Guajira Peninsula, 175
　Santander Massif, 170–171
　Serranía de Macarena, 170
　Sierra Nevada de Santa Marta, 171–174
　Subandean basement, 168–170
Andean orogeny, 730
Andes
　cordilleras, 938
　described, 938
　northern and central, sediment yield, 938, 939
Andes Cordillera
　catchments, 937, 938
　sediment production of, 936–938
Andes of Colombia
　ALCC, 938–942
　anthropocene-impacted sediment fluxes, 936, 937
　BQART model, 949, 950
　denudation processes, 950
　human-induced drivers, 942–944
　sediment fluxes, 946
　sediment load trends, 946–948
　Soil Conservation Service, 949
Andesite formation, 626, 632, 635, 636
Anthropocene, 957
　global sediment transfers, 935
　sediment fluxes, in Andes of Colombia, 936, 937
Anthropogenic land cover change (ALCC), 938–942
Antioquian Batholith arc segment, 309, 313, 375
Apatite partial retention zone (A-PRZ), 773
Aptian sedimentation, paleo-UVM, 718
Aptian-Albian Cushabatay Fm, 729
Arc segment A (Cali) volcanic zone, 630
Arc segment D (Quito) volcanic zone, 631–632
Arc Segments B (Popayán) and C (Nariño) volcanic zone, 630–631

Arquía complex, 840
  age constraints, 849
  Cretaceous fragments, 847, 849
  Ebejico, 850, 851
  Late Cretaceous-Paleocene, 849
  lithotectonic units, 851
  metamorphic rocks, 846
  pre-Cretaceous blocks, 850
  Silvia-Pijao fault, 847
  structural style, 850
Atabapo granite, 133
Atrato-San Juan basins, 919–920
Au-Cu mineralization, 496

B

Back Arc Basin Basalt (BABB), 367
Bagre mountain ranges, 921
Bakhuis granulite belt, 118
Barinas-Apure basin, 675
Basement rocks, 149
Baudó mountain range, 910, 912, 922
Berlín orthogneiss, 200
Berriasian to Aptian sedimentation
  Cocuy sub-basin, 717
  in Cundinamarca sub-basin, 717, 718
Blueschist, 849, 851, 864
Bolivar Aulacogen, 363, 364, 369, 372, 373
Borde Llanero Fault System, 98
Brazilian Shields, 115, 116, 726
Bucaramanga Gneiss unit, 198–200
Bucaramanga-Santa Marta-Garzón fault, 421, 428, 489
Buenaventura fault, 431
Buga Batholith, 476
Buritaca Pluton (BP), 556, 563, 565, 570

C

Caguán-Putumayo Basin, 675
Cajamarca-Salento porphyry Au
  E-dipping Belgica fault, 520
  La Colosa deposit, 517, 520
  porphyritic magmatism, 521
  potassic alteration, 519
  pyrite, 519
  Romeral and Palestina, 518
  sulphide assemblage, 519
Cajamarca-Valdivia terrane (CA-VA), 26, 28, 262, 302, 325, 369, 423, 429, 430
Campanian sedimentation, late Aptian, 728–730
Cañasgordas Terrane, 840, 912, 913

Capa 2, 710
Caracanoa/Raudal Alto ridge, 138
Carbonate-rich sediments (CS), 627
Carboniferous granitoids, 273, 277
Caribbean-Colombian oceanic plateau
  (CCOP), 10, 65, 426, 469, 681, 682, 733, 734
Caribbean Colombian Realm (CCR), 52
Caribbean large igneous province (CLIP), 426, 733, 734, 909, 910, 912, 914, 924
Caribbean Plate
  basement, southern, 551, 552
  Great Caribbean Arc collision, 581
  plate interaction model, 592–594
  SCDB, 552
  "tectonic escape" of, 585
Caribbean Terrane Assemblage (CAT), 266
Caribbean terranes
  San Jacinto, 9
  Sinú, 9
Catatumbo sub-basin, 722, 723
Cauarane-Coeroeni belt, 116
Cauca-Almaguer fault system, 682
Cauca fault and suture system, 430, 836
Cauca-Patia basin, 43
Cauca-Romeral Fault System (CRFS), 788
CA-VA assemblage, 262
Cenomanian-Coniacian fine-grained pelagic deposits, 685
Cenomanian-Turonian boundary, 688, 708, 720, 723, 735
Cenozoic basins
  Atrato-San Juan basins, 919–920
  Chuqunaque-Tuira and Sambu basins, 917, 919
  Gulf of Panama basin, 919
  Panama canal basin, 916, 917
Cenozoic orogenesis, 776
Cenozoic paleogeographic and paleotopographic evolution, 807–810
Cenozoic reactivation of Mesozoic structure, 734
Cenozoic sedimentary rocks, 855, 856
Central Amazonia Province (CAP), 151
Central Cordillera and Lower Magdalena Valley (Plato-San Jorge area), 724
Central Cordillera (CC), 175–178, 334, 665–667, 781–789, 867
Central Guiana Granulite Belt, 116
Central Tectonic Realm (CTR), 5, 7, 261, 262, 421–424
  Cajamarca-Valdivia Terrane, 262

Index   985

CA-VA, 261, 262
granitoid assemblages, 263
late Mesozoic history, 263
Precambrian Chicamocha Terrane, 261
Proterozoic Chicamocha Terrane, 261
Cesar-Ranchería Basin (CR), 658, 659, 723, 724
Chagres and Mamoní mountains, 914
Chocó Arc assemblage (CHO), 266, 427, 428
Chocó indenter, 669, 670
Chocó-Panamá Arc (CHO), 10
Chocó-Panamá Indenter, 69
 Cañasgordas terrane, 72
 CCOP, 65, 67
 Paleomagnetic data, 70
 San Juan Basin, 72
 structural architecture, 68, 70
Chondrite-normalized rare earth elements (REE) plots, 617, 620
Chuqunaque-Tuira basin, 917, 919
Coastal Cordillera, 727
Coastal environments, 957, 958
Coastal Trough/intra-arc basin, 726
Cocuy sub-basin, Berriasian to Aptian sedimentation, 717
Colombia, 10–12, 715–719
 Au + co-metal metallogenesis, 417
 Catatumbo sub-basin, 722, 723
 Central Cordillera and Lower Magdalena Valley (Plato-San Jorge area), 724
 Chicamocha, 7
 CHO, 10 (see also Colombian Andes)
 Cretaceous post-rift sedimentation, 719–721
 Cretaceous-Eocene granitoids, 307
 early Cretaceous syn-rift Sedimentation
  aptian sedimentation in paleo-Upper Magdalena Valley (UVM), 718–719
  Berriasian to Aptian sedimentation in Cundinamarca sub-basin, 717, 718
  Berriasian to Aptian sedimentation on Cocuy sub-basin, 717
  Jurassic red beds, 715
  Santander-Floresta paleo-Massif, 717
  Tablazo sub-basin, 716–717
 Eastern Cordillera, 415
 and Ecuador, 28 (see also Andes of Colombia)
 emerald, 414
 gold mining, 412, 416
 mining districts, 413
 Perijá Range and Cesar-Ranchería Basin, 723, 724

phanerozoic basins
 Cenozoic, 12
 Cretaceous, 11
 geological mapping, 10
 pre-Cretaceous, 11
 radiometric dating techniques, 416 (see Santander Massif, Northern Andes (Colombia))
 Sinú-San Jacinto Basin, 722
 social and ethnocultural history, 413
 southern Cauca valley, 721, 722
Colombian Amazonian geology, 119
Colombian Andean system, 196
Colombian Andes, 54, 59, 256
 Andean orogeny, 40, 43, 46
 Bucaramanga-Santa Marta-Garzón fault, 428
 Buenaventura fault, 431
 Cauca fault, 430
 CA-VA, 26, 28
 Chocó-Panamá Indenter, 65, 67, 69, 70, 72
 distribution of Precambrian basement, 158–161
 Eastern Cordillera, 22, 23, 79, 80
 Farallon-Caribbean plateau, 61
 Garrapatas fault, 431
 geological interpretation, 12, 13
 geotectonic models, 52
 graben-rift-aulacogen, 34, 35, 38, 40
 GU-FA, 62
 litho-tectonic and structural evolution, 665
 Llanos basin, 26, 29
 Maracaibo orogenic float, 72, 74, 76, 79
 meso-neoproterozoic, 17, 19, 20
 morpho-structural expression, 16
 Otú fault, 429
 Palestina fault, 429
 Phanerozoic (see Phanerozoic)
 Phanerozoic orogenic systems, 21, 22
 plate collision, 29, 30, 34
 Quetame and Silgará, 28, 29
 Romeral fault, 430
 Roraima tectono-sedimentary, 85–87
 San juan-sebastian fault, 431
 southern Caribbean, 56
  fault systems, 59
  tectonic contact, 54
 structural framework, 267, 268
 taphrogenic tectonics, 29, 30, 34
 transpressional regime, 49–52
 transtensional regime, 49–52
 western Caribbean, 52, 57
 WETSA, 88, 89

Colombian Caribbean littoral
  beach ridge lagoons and mangrove
    swamps, 961
  coastal evolution, 959
  coastal protection, infrastructure, 968–970
  described, 959
  Galerazamba (La Garita point) and
    Sinú-Tinajones delta, 963, 965, 968
  morphological changes, 959, 960
  mud diapirism, 960, 961
  semi-desert conditions, 959
  shores and prodelta, Magdalena River
    Delta, 961–963
  trade winds, 960
Colombian Cordilleras, 271
Colombian Guajira Peninsula, 686
Colombian littoral zones, 959
  Caribbean littoral (*see* Colombian
    Caribbean littoral)
  cartographic charts, 959
  environmental deterioration, 958
  human interventions, 974–976
  seaboards, 958
Colombian Pacific littoral
  barrier islands, 972
  catchment area, 969
  coastal evolution, 972
  described, 969
  earthquakes, 970
  map and morphological littoral types, 969,
    971
  morphological changes, 969
  northern Pacific coast, 971
Colombian portion, Maracaibo Basin, 722,
  723
Colombian tectono-magmatic, 359–360
Colombian volcanic arc, 330, 333
Colombia of Guiana Shield
  Colombia Paleoproterozoic metamorphic
    basement, 119
  Ediacaran San José del Guaviare
    Nepheline Syenite, 148
  K'Mudku-Nickerie Tectonometamorphic
    Episode, 155
  late Paleoproterozoic older granites, 131
  late Proterozoic-Phanerozoic events, 155
  Meso-Neoproterozoic Mafic intrusives, 148
  Mesoproterozoic anorogenic granitoid
    magmatism, 154
  Mesoproterozoic mylonitization, 145
  Mesoproterozoic Parguaza Rapakivi
    granite, 133–135
  Mesoproterozoic sedimentation of Tunuí
    sandstone, 155
  Mesoproterozoic Tunuí Folded
    Metasandstone Formations, 135–145
  Mesoproterozoic younger granites, 131–133
  metasomatic conception, 119
  Mid-Paleoproterozoic Caicara
    Metavolcanics, 120–122
  Mitú complex, 120, 154
  Piraparaná Formation, 145
  PRORADAM project, 119
  proterozoic metamorphic basement,
    124–131
  Trans-Amazonian Orogeny, 152
Colombia Paleoproterozoic metamorphic
  basement, 119
Colombia-Venezuela
  Andean domain, 100
  basement exposures, 99
  Borde Llanero Fault System, 98, 100
  cratonic domain, 98
  Grenvillian-Orinoquiense rocks, 100
  Metapelitic rocks, 100
  Silurian rocks, 100
Combia Fm. volcanism (12–6 Ma), 617,
  624, 625
Continental arcs potassic (CAP), 242
Continental collision, 119, 181, 182
Continental margin domain, Northern
    Andes, 687
  Cenomanian-Turonian boundary, 688
  Colombia, 715–724
  Jurassic and Paleocene deposits, 688
  MFS (*see* Maximum flooding surfaces
    (MFS))
  sedimentary basins, 686
  sequence boundaries (SB), 687
  Venezuela, 689–715
Cooling period, 893–894
Copper (Cu)
  Au mineralization, 426
  Chilean Andes, 414
  Colombia, 442
  El Roble and El Dovio, 481
  La Quinta-Girón Fms, 442
  pluton and porphyry-related, 478, 479,
    483, 484
  Quetame Massif, 438
  Serranía de Perijá and Santander Massif, 442
  Supatá Zn (Cu), 461–464
CR, *see* Cesar-Ranchería Basin (CR)
Cratonization, 154
Cretaceous, 686–733
  continental margin domain (*see*
    Continental margin domain,
    Northern Andes)

post-rift sedimentation, 719–724
sedimentation in Northern Peru and Northwestern Brazil, 727
Venezuela, 678
Western Colombia, 678–682
Western Ecuador, 682–686
Cretaceous-Eocene granitoid, 309–315, 317, 320–322
    Antioquian and Sonsón batholiths, 306–307
    Colombia, 307
    distribution, 307, 309
    ICP-based analytical techniques, 313
    lithogeochemistry
        Eastern Group, 313–315, 317, 321, 322
        Western Group, 317, 320–322
    Pb isotope, 324–326
    Sr-Nd isotope, 322, 324, 326
    U-Pb ages
        Antioquian Batholith, 309
        Eastern Group, 312
        Eastern Group Manizales Stock, 311
        K-Ar-based database, 309
        Mandé-Acandí Batholith, 312
        Mistrató Batholith, 311
        Santa Marta Batholith, 311
        Sonsón Batholith, 310
        Western Group, 312
Cretaceous-Eocene metallogeny, 470, 472–480
    Berlin-Rosario Au (Ag), 468, 469
    CCOP, 469
    granitoids
        Antioquian Batholith, 470
        Buga Batholith, 476
        Jejénes stock, 477
        Mandé-Acandí arc, 477–479
        pre-collisional phase, 470, 472–475
    gypsum mine, 481
    laterite deposits, 483
    oceanic basement terranes
        Dagua Terrane, 480
        Guapí ophiolite, 480
        VMS, 480
    proto-Northern Andean orogeny, 468
Cretaceous oceanic crust
    Baudó mountain range, 910, 912
    Cañasgordas Terrane, 912, 913
    eastern Panama, 910, 911
    Istmina-Condoto high rock outcrops, 913
Cretaceous-Paleocene boundary, 684
Cretaceous sedimentary rocks, 735
CTR, *see* Central Tectonic Realm (CTR)

Cu-Au mineralization, 499
Cundinamarca sub-basin, 717, 718, 735

D

Dabeiba fault, 836
Dagua terrane, 8
Darién and Baudó serranías, 802, 803
Deforestation, in Andes of Colombia, 942–944
Digital elevation model (DEM), 269, 458, 493, 518, 526
Digital terrain model, 903
Durania granite, 207

E

Early Paleozoic granitoids, 272, 276
Early Paleozoic pulse, 196, 244
East Peruvian trough, 726, 728–731
Eastern Cordillera (EC), 22, 23, 79, 80, 261, 323, 335, 661, 662, 664, 790–799
Eastern Cordillera mineralization, 461–463
    emerald deposits, 456
    emerald mineralization, 456
    oolitic Fe formation, 460
    sulphide mineralization, 462
    Supatá
        Cu, 462
        emerald, 463
        SEDEX deposits, 463
        stratigraphy, 461
        sulphide occurrences, 463
    Valle Alto rift, 455
    Zn-Pb-Cu-Fe (Ba) sulphide, 460, 461
Eastern Flank of Central Cordillera
    Icarcó Complex, 177
    Las Minas Massif and La Plata Massif, 177
    Río Téllez-La Cocha Migmatitic Complex, 175, 176
    San Lucas Metamorphic Complex, 178
    Tierradentro gneisses and amphibolites, 177
Eastern Group post-collisional arc segments, 377–380
Eastern Maturin sub-basin, 675
Eastern Panama, 910, 911
Economic Geology and *Mineralium Deposita*, 411
Ecuador (Oriente Basin), 724–726
Ediacaran San José del Guaviare Nepheline Syenite, 148
E-dipping Belgica fault zone, 520
El Carmen-El Bagre Au district, 439, 440
El Carmen-El Cordero suite, 368

El Cerro igneous complex, 504, 505
El Choncho barrier island, 972, 973
Emerald mineralization, 456, 464
Epithermal, 450, 451, 508, 512, 513
  See also Middle Cauca epithermal Ag-Au-Zn (Pb-Cu)
Erosion
  denudation processes, 950
  and depositional upstream processes, 948
  in fluvial systems, 948
  in Magdalena basin, 942
Erosion rates in Northern Andes
  erosional exhumation trends, 804
  global erosion rates, 804, 805
  "hypererosive" regime, 805
  industrialized agriculture, 805
  local relief and erosion rate, 806
  surficial and deep-seated orogenetic process, 806
Exhumation
  apatite fission-track ages, 891
  40Ar/39Ar, fission-track and zircon (U-Th)/He ages, 892
  Bucaramanga fault, 891
  mountain belts, 890
  rates, 892–893
  thermochronological dataset, 890
External detector method (EDM), 772

F

Falcon Basin (FB), 552, 553
Farallones Batholith, 502
Farallon Plate, 397, 677, 678
Faulting and folding, 654, 655
Fe formation, 460
Feldspathoid silica-saturation index (FSSI), 345
Field mapping, 668
Fission-track analysis, 772
Fission-track dating method, 772, 773, 799
Fluvial sediment transport, 936, 938, 948, 952

G

Galeras Volcano, 529
Galerazamba (Punta Garita) area, 965, 966
Garrapatas-Dabeiba Fault System (GDFS), 60, 431
Garzón Massif, 161–168
  EC, 794, 795
  geochemistry, 164
  geochronology, 164, 168
  lithology, 161
  metamorphism, 164
  subdivision, 161, 163
Geochronological provinces, Amazonian Craton, 151–152
Geodynamic model, 838
Geoeconomic potential, 158
Global positioning system (GPS), 881
Gold, 414, 445, 449–451, 492
  Au-Cu mineralization, 484
  Colombian Andes, 490–491
  epithermal deposits, 445
  Farallones Batholith, 502
  Galeras Volcano, 529
  mineral districts
    Cerro San Carlos, 450
    Guamoco, 449
    Juana-El Piñal, 450
    Piedrancha-La Llanada-Cuembí, 492
    Pueblito Mejía, 451
  mining, 416
  occurrences (*see* San Lucas)
  pluton-related, 445
  production comparison, 412
Gold-silver mineralization
  Bosconia, 454
  Ibagué Batholith, 454, 455
  La Concepción, 497–499
  Pacarní, 452
  San Luis, 453
Gondwana-Laurentia suture, 368
Gorgona terrane, 8, 382
Graben-rift-aulacogen-type deposits, 36
Granitoid arc segments, 260, 267
Granitoid magmatism, 286, 289–300, 306, 326
  Cretaceous to Eocene (*see* Cretaceous to Eocene granitoid)
  distribution, 254
  geological analysis, 254
  isotope geochemistry, 254
  K-Ar and Rb-Sr radiometric age, 349
  latest Oligocene to Pliocene (*see* Latest Oligocene-Pliocene)
  late Triassic-Jurassic
    age constrains, 289–291
    distribution, 286, 289
    lithogeochemical plots, 295, 296
    lithogeochemistry, 294, 296–299
    Sr-Nd and Pb isotope, 300
    temporal-spatial analysis, 292, 293
    U-Pb ages, 292
  late Triassic to Pliocene, 267
  lithogeochemical analysis, 254

Index 989

Meso-Cenozoic granitoids, 255
Permo-Triassic, 256
Phanerozoic, 255
temporal development, 254
U-Pb age, 256
U-Pb (zircon) dating, 255
Granitoid metallogeny, *see* Cretaceous-Eocene metallogeny
Granitoid, RSZ, 853, 854
Granitoids, 273, 275–278
 age constraints
  Carboniferous, 277
  early Paleozoic, 276
  Permo-Triassic, 277, 278
 distribution
  carboniferous, 273
  Permo-Triassic, 275
 early Paleozoic, 272
Grenvillian Orogeny
 active continental margin sedimentation and early magmatic activity, 181
 continental collision, 181, 182
 early stages, 181
 Eastern boundary, 180, 181
GSR, *see* Guiana Shield Realm (GSR)
Guaca River Diorite, 207
Guadalupe Gp. sands, 720
Guajira allochthon
 litho-tectonic units, 651, 653
 Oca Fault, 654
 Puralapo and Cosinas Fault, 653
Guajira Peninsula, 175
Guajira-falcón terranes (GU-FA), 9, 62
Guiana Shield, 119–148
 Bakhuis granulite belt, 118
 basement rocks, 149
 and Brazilian Shield, 115, 116
 Cauarane-Coeroeni belt, 116
 Colombia
  late paleoproterozoic older granites, 131
  mesoproterozoic Tunuí Folded Metasandstone Formations, 145
  mesoproterozoic younger granites, 133
 Colombian Precambrian basement, 115
 crystalline basement, 118
 lineaments, 149
 shearing and low-grade thermal metamorphism, 118
 Trans-Amazonian Orogeny, 115, 116, 118
Guiana Shield Realm (GSR), 4, 5, 260, 420, 421
Gulf of Panama basin, 919
Gypsum mineralization, 481, 482

H

Heavy rare earth elements (HREE), 617
HeFTy® (computer code), 769
Helium dating, 750
Hemipelagic sediments (HS), 627
High-field-strength (HFS) element, 617
Holocrystalline plutonism, 291, 293
Human intervention, 974–976

I

Ibagué Batholith, 452, 454, 455
Içana medium-grained bi-mica granites, 131, 133
ICP-based analytical techniques, 313
INGEOMINAS, 415, 904
Inter-Andean and marginal basins, 767
International Geological Program (IGP), 957
Irra Fm. volcanism (6–3 Ma), 625
Irra stock, 314
Isthmian seaway, 923–924
Istmina-Condoto high rock outcrops, 913

J

Japan International Cooperation Agency (JICA), 415
Jejénes Pluton, 320
Jejénes stock, 477
Jungurudu mountain ranges, 921
Jurassic-Cretaceous boundary, 678
Jurassic magmatic arc, 196, 245
Jurassic metallogeny, 444, 445
 Au-Ag, 444
 Cu and Mo, 445
 deposit types
  intrusion-related gold, 444
  pluton, 444
 epithermal mineralization, 444
 gold region (*see* San Lucas)
 magmatic episodes/arc, 443
 sedimentary Cu, 442
Jurassic plate tectonic reconstructions, 677, 679
Jurassic red beds, 689, 715

K

Kinematic models, 349
K'Mudku-Nickerie Tectonometamorphic Episode, 155
Kübler crystallinity indexes, 203

L

La Campana fine-grained (subvolcanic) granites, 133
La Concepción
  Cu-Au mineralization, 499
  footwall zone, 497
  hydrothermal, 498
  metamorphic and sedimentary rocks, 498
La Corcova Quartz-monzonite, 225
LA-ICP-MS approach, 772
Land cover change, in Andes of Colombia
  capita anthropogenic, 939
  from 8000 years ago, 938
  human-induced drivers, 942–944
Landscape evolution
  and orogenesis, 768
  tectonic plates paradigm, 776
La Pastorera
  gypsum mine, 481, 482
  sulphide mineralization, 482
Large-ion lithophile (LIL) elements, 617
Latal Pluton (LP), 555, 557, 563–566
Late Albian maximum flooding surface, 706
Late Aptian, 728–730
Late Campanian, 730–731
Late Cenozoic magmatism, 606
Late Mesozoic basins, 734
Late Oligocene-Pleistocene metallogeny
  Buenos Aires-Suárez porphyry Au (Cu), 499, 500
  CCOP/CLIP, 485
  El Cerro igneous complex, 504, 505
  Farallones Batholith, 502
  Gorgona terrane, 485
  middle Cauca porphyry belt, 506
  Páramo de Frontino, 503, 504
  Piedrancha-La Llanada-Cuembí AU, 492, 494, 495
  Piedrancha-La Vega-Berruecos granitoids, 495–497
  tectonic framework, 485, 486, 489, 490
Late Paleoproterozoic Older Granites, Tiquié granite, 131
Late Precambrian-Paleozoic forensics, 101, 103
Late Proterozoic-early Paleozoic, 98, 109
Late Proterozoic-Phanerozoic Events, 155
Late Triassic-Early Jurassic pulse, 196
Late Triassic-Jurassic
  Bolívar Aulacogen, 370
  holocrystalline, 371
  slab rollback, 372
  time-space analysis, 370
  Valle Alto Rift, 372, 373
  WNW migration, 370, 371

Laterite deposits, 483
Latest Jurassic to early aptian sedimentation West Peruvian Trough, 727–728
Latest Oligocene-Pliocene, 331–335, 341–346
  categories, 327
  distribution, 327, 329, 330
  lithogeochemistry
    bimodal distribution, 346
    Cajamarca-Salento hypabyssal porphyry, 343
    Farallones-Páramo de Frontino-El Cerro, 341
    feldspathic igneous rocks, 345
    Middle Cauca hypabyssal porphyry, 342
    Paipa-Iza and Quetame, 344
    Patía-Upper Cauca hypabyssal porphyry, 342
    Piedrancha-La Llanada-Cuembi, 341
    Río Dulce hypabyssal porphyry, 343
    Santander Massif hypabyssal porphyry, 343
    whole-rock, 345
  Pb isotope, 347–349
  spatial *vs.* temporal relationships, 336, 337
  Sr-Nd isotope, 346–348
  U-Pb ages
    Combia volcanism, 333
    Eastern Cordilleran, 335
    Espíritu Santo-Santa Bárbara, 334
    gold mineralization, 335
    holocrystalline, 331
    hypabyssal porphyry, 332–334
    K-Ar, 331, 333
    porphyritic granitoid, 331
    Río Dulce, 334
Lead isotope geochemistry, 324
Light rare earth elements (LREE), 617
Lima province, 728
Lithogeochemistry, 254, 255, 271, 313, 314, 316, 317, 319, 320, 341–346
  Carboniferous granitoids, 279
  Cretaceous-Eocene granitoids, 321, 322
  early Paleozoic granitoids, 279
  Eastern Group
    Antioquian Batholith, 313
    El Bosque Batholith, 317
    Irra Stock, 314
    Mariquita Stock, 316
    Sonsón Batholith, 314, 316
  felsic granitoid, 282
  latest Oligocene-Pliocene
    bimodal distribution, 346
    Cajamarca-Salento hypabyssal porphyry, 343

Index   991

Farallones-Páramo de Frontino-El
   Cerro, 341
feldspathic igneous rocks, 345
middle Cauca hypabyssal porphyry, 342
Paipa-Iza and Quetame, 344
Patía-Upper Cauca hypabyssal
   porphyry, 342
Piedrancha-La Llanada-Cuembi, 341
REE, 345
Río Dulce hypabyssal porphyry, 343
Santander Massif hypabyssal porphyry,
   343
late Triassic-Jurassic granitoids, 298, 299
Mocoa-Garzón trend, 297
Norosí and San Martín batholiths, 296
northern Ibagué and Segovia batholiths, 297
northern Ibagué hypabyssal porphyry, 297
Permian-mid-Triassic granitoids, 281
Permo-Triassic granitoids, 281
Santander Plutonic Group, 294
SiO_2, 281
southern Ibagué Batholith, 297
Western Group
   Jejénes Pluton, 320
   Jejénes Stock, 319
   low-K behaviour, 317
   Mandé Batholith, 320
   Mistrató Batholith, 320
Lithologies, 203
Lithosphere-atmosphere-hydrosphere coupled
   systems, 750
Litho-tectonic elements, 353
Llanos Orientales Basin, 675
Lower crust interaction, 632–640
Lower Middle Albian surface, 705
Low-temperature thermochronology (LTTC),
   768
   crystalline basement rocks, 767
   description, 754 (*see also* Morphotectonic
      reconstructions, Colombian Andes)
   normal fault, crustal thinning, 754
   paleo-relief, 754
   post-magmatic cooling, 754
   tectonic geomorphology (*see* Tectonic
      geomorphology)
   topographic configuration, 754
   uplift-driven erosion, 754
Lu-Hf isotope, 284–286

M
Maastrichtian, 712
Mafic and ultramafic igneous rocks, 852, 853
Magdalena River

Cauca basin, 944
deviation, sediment load, 951
downstream station, 948
sediment load and yield, 946, 949
sediment transport, trends, 943
watersheds, 948
Magdalena River delta, 959, 961–964
Magmatic belts
   Agua Blanca Granite (Agua Blanca
      Batholith), 225
   Durania granite, 207
   granitic rocks, 204–206
   Guaca River Diorite, 207
   La Corcova Quartz-monzonite, 225
   mineralogical content, 205
   Mogotes Quartz-monzonite, 225
   Ocaña Alkaline Granite (Ocaña Batholith),
      225
   Onzaga Granodiorite, 225
   Paleozoic and Cenozoic intrusives, 204
   Páramo Rico Tonalite (Granodiorite), 225
   Pescadero Monzogranite, 226
   published radiometric ages, crystalline
      units, 204, 208–224
   Rionegro Batholith, 226
   Sanín-Villa Diorite, 207
   Santa Bárbara Quartz-monzonite, 225
   Santa Rosita Quartz-monzonite, 225
   Suratá River Pluton, 225
   types, 204
Majé mountain ranges, 921
Mandé-Acandí arc
   assemblage, 380, 381
   La Equis Zn-Pb-Cu, 479
   porphyry-related Cu, 478, 479
Mandé-Acandí Batholith, 312
Mandé Batholith, 320
Mandé mountain range, 915–916
Maracaibo Basin (MB), 552, 553, 586,
   591–593, 722, 723
Maracaibo block (MB), 555
   CR, 659
   geological history, 553
   northern boundary, 552
   Paleogene magmatism (*see* Paleogene
      magmatism, in MB)
   SNSM, 656, 657
   southern boundary, 553
   tectonic transposition, 656, 658
Maracaibo continental block (MCB)
   description, 880
   geological map and seismicity, 881
   GPS, 881 (*see also* Maracaibo mountain
      belts)

Maracaibo continental block (MCB) (*cont.*)
   periods of cooling, 893–894
   plate tectonic events, 894–895
   tectonic feature, 880
   transpressional forces, 881
Maracaibo mountain belts
   Santander Massif (SM), 885–887
   Serranía de Perijá (SP), 884–885
   SNSM, 882–884
   Venezuelan/Mérida Andes (VA), 887–889
Maracaibo orogenic float (MOF)
   geological setting and kinematic model, 73
   geophysical and structural features, 74
   Santander Massif, 74
   Serranía de Perijá, 74
   SJFS and OPTFS, 60
   SNStM, 73
   structural and tectonic evolution, 72
   tectonic realms, 7
Maracaibo Sub-plate Realm (MSP), 261, 421
Marañón High, 726
Marine *vs.* continental sedimentation, 734
Mariquita Stock, 316
Markov Chain Monte Carlo algorithm, 771
Marmato-Aguas Claras Suite (MACS), 514
Maximum flooding surfaces (MFS)
   Albian Esperanza Fm., 729
   Campanian, 722
   Cenomanian-Turonian boundary, 688, 708, 720, 735
   and condensed section development, 704, 705
   Late Albian, 706
   lower Turonian, 708
   Middle Albian, 719
   proposed stratigraphic sequences, 690–700
   Santonian-Early Campanian, 721
   and SB, 688, 689, 735, 738
   and shelf shale facies, 725
   in southern Ecuador, 732
   in stratigraphic sections, 687
   and transgressive surface, 706, 715
Mean track length (MTL), 773
Mérida Andes mountain belts, 887–889
Meso-Cenozoic, 258, 360, 369, 750, 838
Meso-Cenozoic granitoid magmatism, 292
Meso-Cenozoic Northern Andean orogeny, 418
Meso-Neoproterozoic Mafic Intrusives, 148
Mesoproterozoic anorogenic granitoid magmatism, 154
Mesoproterozoic mylonitization, 145
Mesoproterozoic Parguaza Rapakivi Granite, 133–135

Mesoproterozoic sedimentation of Tunuí Sandstone, 155
Mesoproterozoic Tunuí folded metasandstone formations
   Cerro El Carajo, 141
   geochronology, 141–143, 145
   higher-grade migmatitic gneisses, 135
   La Libertad range, 138, 139
   Machado ridge, 139, 141
   metasediments, 135
   sedimentology and stratigraphy, 137, 138
Mesoproterozoic younger granites
   Atabapo granite, 133
   Içana medium-grained bi-mica granites, 131, 133
   La Campana fine-grained (subvolcanic) granites, 133
   Mitú granite, 131
   Tijereto granophyre, 131
Mesozoic plate tectonic interpretations
   Cretaceous, 678–686
   Jurassic plate tectonic reconstructions, 677–679
Metallogeny, 438–443, 466, 484
   Cretaceous-Eocene (*see* Cretaceous-Eocene metallogeny)
   Eastern Cordillera, 415
   gold, 414
   information sources, 415
   jurassic (*see* Jurassic metallogeny) sedimentary Cu, 442
   late Oligocene-Pleistocene (*see* Late Oligocene-Pleistocene metallogeny)
   mineral deposits, 432
   MVT, 432
   pre-jurassic
      Bailadores, 438
      Caño Negro-Quetame-Cerro de Cobre, 438
      El Carmen-El Cordero, 439, 440
      Santa Elena chromitite, 441
      tectonic models, 438
   tectonic framework, 413
   temporal-spatial evolution, 412, 437
   time-space analysis, 417
   time-space charts, 432
Metamorphism
   amphibolite facies, 200
   Bucaramanga Gneiss, 200
   Famatinian orogeny, 244
   Kübler crystallinity indexes, 203
   lithologies, 203
   and migmatization, 196

Metasedimentitas de Guaca, La Virgen (MGV), 203–204
Middle Cauca epithermal Ag-Au-Zn (Pb-Cu)
  Buriticá, 516
  Caramanta-Valparaiso, 514
  epithermal deposits, 511
  Marmato, 513, 514
  Quinchia, 512, 513
  Supía-Riosucio, 513
  Titiribí, 515
  wallrock alteration, 512
Middle Cauca porphyry belt, 506
Mid-Paleoproterozoic Caicara Metavolcanics, 120–122
Migmatitic biotite-(muscovite) gneisses, 126, 127
Milankovitch cycles, 719
Mineral districts, 437, 449–451, 470, 472, 474–476
  Bosconia, 454
  gold
    Antioquian Batholith, 472
    Cerro Gramalote, 474
    Cerro San Carlos, 450
    Guamoco, 449
    Juana-El Piñal, 450
    Pueblito Mejía, 451
  granitoids
    Antioquian Batholith, 475, 476
    Segovia-Remedios, 470, 472
  Ibagué Batholith, 454, 455
  Pacarní, 452
  San Luis, 453
  San Pablo Fm., 464, 466
Miocene granitoid magmatism, 390
Miocene hypabyssal porphyry, 496
Miocene ultramafic complexes, Condoto, 922
Mio-Pliocene granitoids, 328, 489
Mio-Pliocene porphyry arc segments, 535
Mississippi Valley-type (MVT), 432
Mitú complex, 120
Mitú complex supracrustals, 154
Mitú granite, 131
MOF, *see* Maracaibo orogenic float (MOF)
Mogotes Quartz-monzonite, 225
Morphotectonic elements, 552
Morphotectonic reconstructions, Colombian Andes
  cenozoic orogenesis, 776
  Central Cordillera (CC), 781–789
  chronological framework, 776
  climatic and biotic patterns, 775
  detrital materials, 775
  Eastern Cordillera (EC), 790–799
  floral assemblages and paleoelevation, 778
  late Miocene–Pliocene, 778
  orogenetic events, 777
  palynology, 776
  SNC and Santander Massif, 777
  uplift and erosional exhumation, 777
  Western Cordillera (WC), 778–780
MSP, *see* Maracaibo Sub-plate Realm (MSP)
Mud diapirism, 960–962, 965
Mylonitic shears, 834, 836

N

Nazca Plate segments, 398, 611, 629, 638, 640, 641
Neogene volcanic arcs
  Baudó mountains, 922
  Majé, Sapo, Bagre, Jungurudu and Pirre mountain ranges, 921
  miocene ultramafic complexes, Condoto, 922
  the Pearl Islands, 921
Norosí and San Martin Batholiths, 445
Northern Andean Block (NAB), 414
  AFM, TAS diagrams and REE plots, 617
  alkali-basaltic volcanism, 626
  allochthonous and parautochthonous, 257
  amagmatic zones/volcanic gaps, 629
  arc segment A (Cali), 630
  arc segment D (Quito), 631–632
  arc segments B (Popayán) and C (Nariño), 630–631
  Colombian Andes localities, 614
  continental margin assemblages, 609
  dextral transpression-transtention, 610
  disequilibrium features, 627
  fluid-mobile elements, 627
  geodynamic setting, 606
  HFS, LREE, LIL and HREE elements, 617
  hydrothermal alteration, 617
  hypabyssal porphyry bodies, 614
  intra-oceanic and continental margin granitoids, 609
  isotopic signatures, 627
  late Paleogene to Neogene magmatism, 617, 621–623
  magmatic episodes, Colombian Andes, 612
  magmatism, 606, 608
  microplate, 609
  mid-Miocene to Pliocene magmatic rocks, 612, 613
  Nazca-Cocos Plate configuration, 612

Northern Andean Block (NAB) (*cont.*)
    nothwestern South America, time periods, 609
    oceanic affinity terranes, 606
    paleo-structure, 629
    Phanerozoic granitoids, 612
    plate reorientation, 609
    Plio-Pleistocene location, 606, 607, 610, 611
    pyroclastic and ignimbrite flows, 626
    subducted components and sediments, 627
    subduction zones, 606
    tectonic evolution of, 610
    tectonic models, 257
    tectonic reorganization, 606
    volcanic rocks, 614
Northern Andean orogeny, 7, 23, 40, 41, 43, 45, 46, 53, 258, 259, 327, 609
    Antioquian Batholith arc, 375
    Caribbean Plateau, 384, 385, 397
    Colombian volcanic arc, 389, 398
    Dagua terrane, 382
    earthquake activity, 386
    Eastern Cordillera, 391
    Eastern Group, 374, 377–380
    Farallon-Nazca-Cocos Plate, 386, 387
    Farallon Plate, 386, 397
    genesis and spatial evolution, 382
    Gorgona Terrane, 382
    Jurassic metallogeny, 467
    lithogeochemical and isotopic, 390
    Mandé-Acandí arc, 380, 381
    Mio-Pliocene granitoids, 390
    N and W migration, 397
    Nazca Plate, 387, 388, 398
    Neogene granitoid, 392
    Neogene reinitiation, 386, 387
    Paipa-Iza-Quetame, 391
    Rio Dulce, 390
    Santander Massif, 387, 390
    Tectonic Realm, 385
    time-space analysis, 374, 382
    Vetas-California, 391, 392
    Vetas-Paipa-Quetame, 389
    Western Group, 374–377
Northern Andes of Colombia, 754
    atmospheric circulation, 750
    convoluted mosaic, tectonic blocks, 755
    "cordilleran"/"collisional", 755
    Cretaceous oceanic terranes, 677
    deformation and uplift, 739
    features, 755
    geologic/geographic importance, 750
    geoscientists, 751
    inter-Andean and marginal basins, 767
    lithosphere-atmosphere-hydrosphere coupled systems, 750
    litho-structural domains/crustal blocks, 754
    lithotectonic and morphostructural features, 753, 757
    LTTC (*see* Low-temperature thermochronology (LTTC))
    mantle *vs.* crustal processes, 753
    mountainous relief and intermontane basins, 675–677
    orogenic belts, 751
    orogenic models, 750, 755
    physiographic configuration, 761–767
    sub-Andean foreland basins, 675
    subduction and accretionary dynamics, 753
    surface uplift and erosional exhumation, 753
    tectonic (internal) and geomorphic (external) processes, 751
    tectono- and litho-structural framework, 756–761
Northern Colombia Guajira peninsula, 686
Northern Peru, 726–733
Northern South America
    Appalachian region, 104
    Caparonensis-Quetame event, 103
    Grenvillian age, 103
    late Precambrian-Paleozoic forensics, 101, 103
    Mérida Andes, 103
    paleogeographic models, 102, 106, 108, 109
    Paleozoic faunal assemblages, 104, 106
    Pangaea, 108
Northern Venezuela rift basin, 735
Northwestern Brazil, 726–733
Northwestern Peru, 731–733
Northwestern South America, 677–686
    accreted oceanic terranes, Pacific domain, 733–734
    Aptian to Cenomanian, 736, 737
    Berriasian to Aptian, 736
    Berriasian to Early Aptian, 736
    Campanian to early Paleocene, 736
    Campanian to Paleocene, 737–738
    Cenomanian to Santonian, 736, 737
    Cretaceous oceanic terranes in Northern Andes, 677
    Cretaceous sedimentary rocks, 735
    Ecuador (Oriente Basin), 724–726

Index

eustacy and tectonics, 736
Jurassic rifting, 735
Late Mesozoic, 734
litho-tectonic and morpho-structural features, 675
map, 673, 674 (*see also* Maracaibo continental block (MCB))
mesozoic plate tectonic interpretations (*see* Mesozoic plate tectonic interpretations)
methodology, 675
mountainous relief and intermontane basins, 675
Northern Peru, 726–733
Northwestern Brazil, 726–733
paleogeographic reconstructions, 673
pre-Cambrian shields, 675
scale of, 736
sequence stratigraphy, 735
southern Colombia (Putumayo Basin), 724–726
southernmost Ecuador, 726–733
sub-Andean foreland basins, 675
syn-tectonic structures, 736
terranes containing pre-Cretaceous rocks, 677
unconformities, 738

O

Oblique subduction, 245
Oca-El Pilar Transform Fault System (OPTFS), 60
Oca fault, 654, 655
Ocaña Alkaline Granite (Ocaña Batholith), 225
Onzaga Granodiorite, 225
Orocaima event, 118
Orogenic belts, 751
Orogenic events, 98, 100, 103, 106
Orogenic float model, 593
Orogenic phases, 774, 778
"Orthogneiss" unit, 200, 201
Otú fault, 429

P

Pacific Coast of Colombia, 969–971
Pacific domain, accreted oceanic terranes, 733
Pacific/Farallon, 734
Pacific model, 592, 594
Pacific terrane assemblage (PAT), 264, 265, 425, 426

Pacific terranes
Dagua terrane, 8
Gorgona terrane, 8
Romeral terrane, 8
Paipa-Izá volcanism (5.9–1.8 Ma), 626
Paleocene granitoid magmatism, 310
Paleocene Sedimentation, Late Campanian, 730–731
Paleocene-Eocene Mandé-Acandí arc, 477
Paleocontinental re-constructions, 98, 105
Paleo-facies maps
ages of Cretaceous period, 673
color for sedimentary environment, 724
color legend, stratigraphic sections, 687
northwestern South America, 701–703, 705–707, 709–714
and stratigraphic sections, 675
Paleogene arcs
Chagres and Mamoní mountains, 914
Mandé mountain range, 915–916
San Blas and Acandí mountain ranges, 915
Paleogene magmatism, in MB, 555–557, 570–574
basin filling, 591
BP, 563
geochronological data
crystallization ages, 570, 571
eocene crystallization ages, 571
temperature geochronology, 571–574
thermochronology, 571
leucocratic granitoids, 565, 566
location, 555, 556
LP, 563, 564
magmatic belt
Atanques laccolith, 557
Latal and Toribio Plutons, 555
Santa Marta Intrusive Complex, 556
Socorro stock, 557
major elements, 575
SMB, 557–559, 562
TP, 564, 565
trace elements, 577, 578, 580
Paleogeographic maps, 673
Paleogeographic reconstructions, 673
Paleomagnetic analyses, 684
Paleomagnetic declination data, 684
Paleo-margin, 834, 837, 839, 850, 857, 862, 863, 868
Paleo-Upper Magdalena Valley (UVM), 718
Palestina fault system, 262, 429
Panama canal basin, 916, 917

Panama-Chocó Arc, 386
  basement of, 924
  Chorotega Block, 901
  composite volcano-plutonic island arc, 924
  Farallon Plate, 926
  geography, 901–904
  geology of, 904
  late Cretaceous to Eocene island arc, 926
  magmatism, 904, 926
  middle Eocene extension and arc break-up, 926
  NE-dipping subduction, 926
  plate tectonic map, 901, 902
  plate tectonic setting, 905–906
  sedimentary basins, 904
  UNDP and INGEOMINAS mapping, 904
  Western and Central Cordilleras, 901
Paramo de Frontino volcanic complex, 503, 504
Páramo Rico Tonalite (Granodiorite), 225
Pb isotope, 303–306
  analyses, 472
  Cretaceous-Eocene granitoid, 324
  latest Oligocene-Pliocene, 347–349
  latest Triassic, 300
  pre-Jurassic granitoids, 283
The Pearl Islands, 921
Penecontemporaneous, 734
Perijá Range, 723, 724
Permo-Triassic granitoids, 256, 273, 277, 278
Pescadero Monzogranite, 226
Petrogenesis, 285, 286, 325, 361, 367, 369, 374, 379, 390, 393, 394
Phanerozoic, 272–278, 280, 281
  Colombian Andes, 268
  data filtering, 271
  distribution, 259
  granitoid magmatism, 255
  lithogeochemical database, 271
  mid-Triassic granitoid
    age constraints, 275–278
    distribution, 272–275
    lithogeochemistry, 278, 281
    U-Pb, 272
  pre-Triassic granitoid
    lithogeochemical plots, 280
    trace element and REE, 281
  tectonic models, 258
  tectono-magmatic, lithogeochemical and isotopic, 361
  U-Pb ages, 270
Physiographic configuration, 761–767
Piedrancha-Cuembí arc segment, 336

Piedrancha-Cuembí suite, 346
Piedrancha-La Llanada-Cuembí trend, 492, 494, 495
  mineralization
    Au, 492, 495
    economic, 492
    sulphides, 494
    veining, 495
    wallrocks, 494, 495
Piraparaná Formation, 145–147
Pirre mountain ranges, 921
Plate tectonic events, 894–895
Plato-San Jorge area, 724
Plato-San Jorge Basin (PSJB), 592
Playa Salguero leucogranite, 558, 565, 568–570
Pleistocene-age granitoid arc segments, 490
Plutonism, 103, 288–289
Pluton-related, 444, 447, 448, 450, 484, 492
Poly-deformed, 836
Porphyry, 452, 473, 478, 484, 490, 511
Porphyry intrusions magmatism (17–6 Ma), 615, 618, 624
Porphyry-style Au-Cu mineralization
  Chuscalita-Mina Alemana, 510, 511
  K-feldspar, 507
  La Mina, 509
  La Quebradona, 509
  pyrite and chalcopyrite, 507
  Quinchía, 508
  sodic alteration, 507
  South Támesis, 508
  Titiribí, 510
Post-Berriasian time, 718
Pre-Andean orogeny, 609
Pre-Cambrian Guiana, 726
Pre-Cambrian shields, 675
Pre-Cretaceous rocks, 677
Pre-jurassic metallogeny
  Bailadores, 438
  Caño Negro-Quetame-Cerro de Cobre, 438
  El Carmen-El Cordero, 439, 440
  Santa Elena chromitite, 441
  tectonic models, 438
Pre-mesozoic crystalline units
  Bucaramanga Gneiss, 198–200
  MGV, 203–204
  "orthogneiss" unit, 200, 201
  Silgará Schist sequence, 200, 202 (see also Triassic-Jurassic plutons)
Proterozoic metamorphic basement, 124–131
  amphibolites, 130
  granulites, 130

migmatitic biotite-(muscovite) gneisses, 126, 127
quartzofeldspathic gneisses, 124, 125
Proto-Andean orogen, 98
Proto-Caribbean, 734
Provincia Litosférica Oceánica Cretácica del Occidente de Colombia (PLOCO), 264, 681
Pujilí-Pallatanga suture zone, 684

Q

Quartzofeldspathic gneisses, 124, 125
Quaternary, 957
Quebradagrande Complex, 681, 836
    Abejorral Fm., 841, 842
    basalts, 843
    folding and faulting, 844
    hypotheses, 844
    lithostratigraphic names, 842
    paleo-continental margin, 841
    para-autochthonous Late Cretaceous, 841
    radiometric age, 842, 844
    San Jeronimo fault, 842
    structural style, 844
    Triassic ultramafic rocks, 842
    U-Pb ages, 843
    volcano-sedimentary rocks, 844
Quebradagrande marginal basin, 681
Querarí Orogeny, 115, 154

R

Radiogenic isotopes, 299, 301–306
    late Triassic-Jurassic
        Pb, 303–306
        Sr-Nd, 299, 301–303
    Lu-Hf, 284–286
    Sr-Nd-Pb, 282–284
Radiometric dating techniques, 416
Radiometric systems, 772
Rear-arc volcanoes (RA), 630
Regional kinematic models, 419
Regional stratigraphy, 673
Ridge tholeiitic granitoid (RTG), 440
Río Dulce cluster
    Arboledas, 521
    Espiritu Santo, 522
    Santa Rita Sector, 522, 523
Rionegro Batholith, 226
Rio Negro Fm, 591
Río Negro-Juruena (RNJ), 151
Rivers

Ecuadorian river basins, 947
Magdalena River, 948
Meta River basin, Colombia, 946
Patía River catchment, 946
Romeral fault system, 430, 468
Romeral-Peltetec suture zone, 681
Romeral shear zone (RSZ), 857
    Arquía Complex, 837
    Calima terrane, 839
    Cañasgordas, 840
    CC, 834
    Cenozoic sedimentary rocks, 855, 856
    Central Cordillera, 839
    Colombia, 837
    Dabeiba fault, 836
    early Cretaceous, 834
    Ecuador, 839
    geological configuration, 838
    granitoid rocks, 853, 854
    late Cretaceous mylonitic, 834
    late Miocene and Pliocene, 838
    lithotectonic units, 833, 834, 837
    Medellín, 837
    paleo-Colombian margin, 833
    physiographically, 835
    Quebradagrande and Arquía complexes, 840
    Quebradagrande volcanic rocks, 838
    regional-level tectonic, 836
    tectonic models (*see* Tectonic mélange)
    Western Cordilleran terranes, 836, 840
Romeral terrane, 8
Rondinia paleocontinental reconstruction, 98
Roraima Formation, 118
Roraima tectono-sedimentary, 85–87

S

Sambu basins, 917, 919
San Blas and Acandí mountain ranges, 915
San Jacinto fault system (SJFS), 60
San Jacinto terrane, 9
San Jerónimo fault (SJF), 864
San Juan de La Costa barrier island, 972, 974
San Juan river delta, 969, 972, 973
San Juan-Sebastian fault system, 431
San Lucas
    basement-hosted mineralization, 449
    epithermal volcano-sedimentary-hosted, 450, 452
    pluton-related, 447, 448
San Pablo Fm., 464, 466
Sanín-Villa Diorite, 207

Santa Bárbara Quartz-monzonite, 225
Santa Elena chromitite, 441
Santa Marta Batholith (SMB)
    amphibole-biotite granodiorite to tonalite, 558, 560
    component, rock, 557
    cumulitic facies, 561
    fine-grained facies, 558
    mafic magmatic enclaves, 562
    magma mixing, 561
    magmatic facies map, 557–559
    magnetic foliation, 589
    plastic-state cataclastic deformation, 558
    poikilitic facies, 559, 561
    QAP Streckeisen classification diagram, 558
Santa Marta lagoon, 961
Santa Rosita Quartz-monzonite, 225
Santander Massif (SM), 170–171, 790–794, 885–887
Santander-Floresta paleo-Massif, 717
Santonian-Campanian age, 685
Sapo mountain ranges, 921
SEDEX base metal sulphide occurrences, 463
Sedimentary basins, 901, 902, 904, 905, 919
Sediment load
    in Andes, during anthropocene, 946–948
    Patía River catchment, 946
Sediment yield
    at continental scale, 936
    northern and central Andes, 938, 939
    variability, 946
    for whole northern Andes, 946
Sequence boundaries (SB), 687–700, 704, 708, 711, 714, 720, 725, 727, 729–732, 735, 738
Sequence stratigraphy, 673, 674, 704, 719, 735
Serranía de Macarena, 170
Serranía de Perijá (SP), 884–885
Sierra Nevada de Santa Marta (SNSM), 171–174, 585–590, 657, 800–802, 882–884
    alumina saturation index, 575, 577
    early Paleocene magmatism, 580, 581
    Falcon Basin (FB), 553
    later Paleocene-early Eocene magmatism, 581, 583, 584
    leucocratic granitoids, 565
    major elements, 575, 576
    MB, 552
    mountain ranges, 553
    SMB, 555
    tectonic implications
        exhumation rates, 588
        faulting pattern, 589
        mantle-referenced palinspastic reconstructions, 585
        pre-deformational shape, SMB, 590
        regional magmatic gap, 587
        sedimentary package, 586, 587
        thermobarometric calculations, 587
        TTG characteristics, 585
    trace elements, 577, 578, 580
Silgará Schist sequence, 200, 202
Sinemurian deposits, 38
Sinifaná metamorphic unit, 845, 846
Sinú-San Jacinto Basin, 722
Sinú terrane, 9, 60
Slab decarbonation, 606, 612, 627
SM, Northern Andes (Colombia), 204–226
    basement geology map, 196, 197
    Colombian Andean system, 196
    early Paleozoic pulse, 196
    late Triassic-Early Jurassic pulse, 196
    magmatic belts (*see* Magmatic belts)
    magmatic episode, 196
    Mesozoic sedimentary basins development, 196 (*see also* Pre-mesozoic crystalline units)
    structure, 198
    Triassic-Jurassic magmatic arc, 196
SNSM, *see* Sierra Nevada de Santa Marta (SNSM)
Sonsón Batholith, 310, 314
South America, 734
South Caribbean deformed belt (SCDB), 551, 552
South central Colombia, 738
Southern Cauca Valley, 721, 722
Southern coastal Ecuador, 685
Southern Colombia (Putumayo Basin), 724–726
Southern Ecuador, 731–733
Southern Ecuador *vs.* northernmost Peru, 677
Southernmost Ecuador, 726–733
Sr-Nd isotopes, 322–324
    Colombian cordilleras, 301
    Cretaceous-Eocene granitoid
        Antioquian Batholith, 323, 324
        Eastern and Western groups, 322
        Mandé Batholith, 324
    late Triassic-Jurassic, 300, 302
    latest Oligocene-Pliocene, 346–348
    Mocoa-Garzón, 302
    pre-Jurassic granitoids, 283
    Santander Plutonic Group, 299–301
    Segovia Batholith, 301
    southern Ibagué Batholith, 300

Sr-Nd-Pb isotope geochemistry, 282–284
Strike-slip faults, 49–52
Subandean basement, 168–170
Subducted sediments (SS), 606
Subduction
    Andean-type subduction zone, 584
    Caribbean flat slab, 593
    Caribbean Plateau, 588, 592
    component, 627, 633, 635, 638
    TTG characteristics, 585
    zones, 606, 632, 635
Sulphide mineralization, 462, 482
Sulphide mineralogy, 473
Suratá River Pluton, 225
Syn-collisional granites (syn-COLG), 240

T
Tablazo sub-basin, 716
Taphrogenic tectonics, 31, 34
Technological denudation, 936
Tectonic configuration, 734
Tectonic cycles, 834
Tectonic elements, 264–267
    CTR, 261–263
    GSR, 260
    MSP, 261
    WTR, 264
        CAT, 266
        CHO, 266, 267
        PAT, 264, 265
Tectonic geomorphology
    AER, 773
    A-PRZ, 773
    computational power and software/code development, 768
    cordilleran massifs and sedimentary basins, 768
    definition, 768
    fission-track analysis, 772
    in situ LA-ICP-MS techniques, 768
    internal and external processes, 768
    Oligo-Miocene transition, 774
    paleo-PRZ/PAZ, 774
    radiometric systems, 772
    sediment-stratigraphic analyses, 774
    surface and near-surface processes, 768
    upper crust process, 769–771
    (U-Th)/He analysis, 772
Tectonic mélange
    Aburrá ophiolite, 861
    apatite (U-Th)/He dating, 866
    Arquía Complex, 862
    CC, 867

CCOP/CLIP assemblage, 857
    Central Cordillera block, 864
    Colombian Pacific margin, 863
    early Eocene to middle Eocene, 867
    EC, 867
    Farallon-Caribbean plate assemblages, 863
    Farallon plate, 862
    Farallon plate-CCOP/CLIP, 865
    Guachinte Fm., 867
    helium analyses, 866
    lithodemic and lithostratigraphic units, 857
    lithotectonic components, 862
    Meso-Cenozoic development, 857
    mid-Cretaceous, 863
    Panamá-Chocó block, 868
    Quebradagrande Complex, 863
    SJF, 864
    S-type granitoids, 860
    thermochronological data, 864
    zircons, 864
Tectonic realms
    CTR, 5, 7
    geological setting, 14
    GSR, 4–5
    MOF, 7
    WTR, 7
Tectonic reconstructions, 366–369, 476, 478, 484, 485
    Bolivar Aulacogen, 363, 364
    Cajamarca-Valdivia, 363
    Carboniferous, 364, 365
    early Paleozoic, 362
    late Triassic-Jurassic, 369, 371–373 (*see also* Northern Andean Orogeny)
    Northern Andean-Caribbean Plate, 258
    Ordovician-Silurian Quetame Orogeny, 363
    Permian to mid-late Triassic
        age and nature, 367
        amphibolites, 367
        genetic model, 366
        lithogeochemical and isotopic, 367, 368
        Northern Andean, 367
        Pangaea, 366, 368, 369
        zircons, 366
    Quetame orogenic cycle, 362
Tectonic-related unconformity, 739
Tectonics
    Cajamarca-Valdivia terranes, 101
    Oaxaquia, 101
    and paleogeography, 102
Tectono-and litho-structural framework, 756–761

Tectono-magmatic framework, 424
  Colombian tectonic evolution, 418
  CTR, 421–424
  GS, 420, 421
  Meso-Cenozoic Northern Andean orogeny, 418
  MSP, 421
  Northern Andes, 418
  regional kinematic models, 419
  WTR (see Western Tectonic Realm (WTR))
Tectono-stratigraphic evolution
  Aptian-Albian, 53
  Eocene-Early Miocene, 53
  Miocene oblique collision, 53
  Paleocene-early Eocene, 53
Terranes
  Cañasgordas, 906
  El Paso/El Paso-Baudó terrane, 909
  Greater Panama, 909
  Panama-Baudó-Mandé Terrain, 906
Thermochronology, 571
Tijereto granophyre, 131
Time-space analysis, 370, 417, 432, 434, 436, 465, 467, 486, 488
  early Cretaceous, 357
  early Paleozoic, 350, 352
  latest Oligocene, 383
  latest Triassic, 354
Tiquié granite, 131
Tonalite, trondhjemite and granodiorite (TTG), 116
Toribio Pluton (TP), 555, 556, 564, 565, 567, 575
Total alkalis vs. silica (TAS) diagram, 616, 617
Trace element spider diagrams, 617–619
Trace element spider plot, 244
Trans-Amazonian Orogeny, 115, 116, 118, 152
Tres Esquinas Member, 724
Triassic and Jurassic sedimentary deposits, 35
Triassic-Jurassic plutons
  Jurassic igneous rocks, 226–238
  tectonic setting and magma affinity, 226–242
  trace elements and isotopic relations, 242–244
Trondhjemites, 575, 581

U

Ultra-high temperature (UHT), 118
United Nations Development Programme (UNDP), 904
United Nations (UN), 415
United States Geological Survey (USGS), 415
U-Pb ages, 256, 270, 292, 293
U-Pb dating techniques, 255, 275
U-Pb radiometric ages, 125
Upper Aptian-Lower Albian Tablazo Fm, 716
Upper crust process, 769–771
Upper Magdalena Valley, 720

V

Venezuela, 689, 701–715
  continental margin domain
    early Cretaceous sedimentation, 701–705
    Jurassic red beds, 689
    middle Albian and late Cretaceous sedimentation, 705–715
  Cretaceous, 678
Venezuelan/Mérida Andes (VA), 887–889
Vetas-California
  Au-Ag, 524
  Au district, 526
  Cu, 525
  district-scale field observation, 527
  epithermal mineralization, 528
  granitoids, 489
  La Baja, 525–527
  La Baja-La Alta, 524
  La Mascota and La Bodega deposits, 528
  magmatic crystallization, 527
  sulphide concentrations, 525
Volcanic arc granites (VAG), 240
Volcanic front volcanoes (VF), 630
Volcanism, 288–289, 529
Volcanogenic massive sulphide (VMS), 480

W

West-East Tectono-Sedimentary Anomaly (WETSA), 87–89
Western Caribbean Orogen, 57
Western Colombia, 678–682
Western Cordillera (WC), 668, 669, 778–780, 840
Western Ecuador, Cretaceous, 682–686

Western Group arc segments, 375, 376
Western Guárico sub-basin, 675
Western Tectonic Realm (WTR), 7, 264
    Caribbean Terrane assemblage, 426
    CAT, 266
    CHO, 266, 267
    Chocó Arc assemblage, 427, 428
    PAT, 264, 265, 425, 426

West Peruvian Trough, 726–730
W-E stretching belt, 118
Within plate granites (WPG), 240
WTR, *see* Western Tectonic Realm (WTR)

Z

Zn-Pb-Cu-Fe (Ba) sulphide, 460, 461